GRAPHS

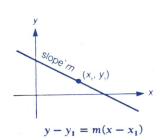

$$y - y_1 = m(x - x_1)$$

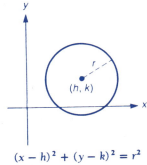

$$(x - h)^2 + (y - k)^2 = r^2$$

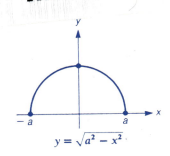

$$y = \sqrt{a^2 - x^2}$$

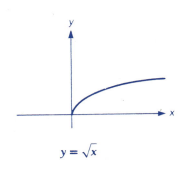

$$y = |x|$$

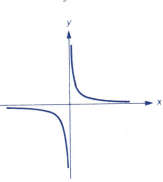

$$y = x^2$$

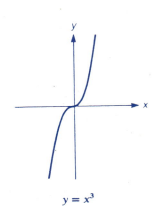

$$y = x^3$$

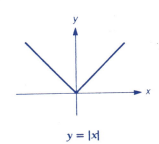

$$y = \sqrt{x}$$

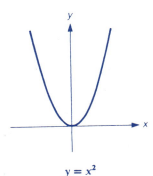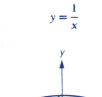

$$y = \frac{1}{x}$$

$$y = [x]$$

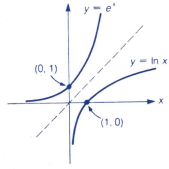

$$y = e^x \quad \text{and} \quad y = \ln x$$

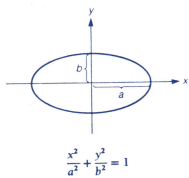

$$\frac{x^2}{a^2} + \frac{y^2}{b^2} = 1$$

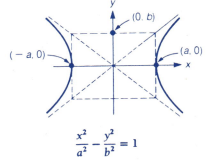

$$\frac{x^2}{a^2} - \frac{y^2}{b^2} = 1$$

ALGEBRA AND TRIGONOMETRY

Third Edition

ALGEBRA AND TRIGONOMETRY

Third Edition

DAVID COHEN
Department of Mathematics
University of California
Los Angeles

WEST PUBLISHING COMPANY
Minneapolis/St. Paul   New York   Los Angeles   San Francisco

FOR MY DAUGHTER EMILY

Copyediting: Susan Gerstein
Text Design: Janet Bollow
Technical Illustrations: TECHarts, Boulder, Colorado
Composition: The Clarinda Company
Cover Design: Diane Beasley
Cover Image: Photomicrograph by Michael Davidson© 1992. "A thin
    film of the geodesic buckminister-fullerene 60-carbon soccer
    ball-shaped molecule. This new form of carbon was discovered in
    1986."
Production, Prepress, Printing, and Binding by West Publishing Company.

WEST'S COMMITMENT TO THE ENVIRONMENT

In 1906, West Publishing Company began recycling materials left over from the production of books. This began a tradition of efficient and responsible use of resources. Today, up to 95 percent of our legal books and 70 percent of our college texts are printed on recycled, acid-free stock. West also recycles nearly 22 million pounds of scrap paper annually—the equivalent of 181,717 trees. Since the 1960s, West has devised ways to capture and recycle waste inks, solvents, oils, and vapors created in the printing process. We also recycle plastics of all kinds, wood, glass, corrugated cardboard, and batteries, and have eliminated the use of styrofoam book packaging. We at West are proud of the longevity and the scope of our commitment to our environment.

LIBRARY OF CONGRESS CATALOGING-IN-PUBLICATION DATA

Cohen, David, 1942 Sept. 28-
    Algebra and trigonometry / David Cohen. — 3rd ed.
        p.   cm.
    Includes index.
    ISBN 0-314-93363-8
        1. Algebra.   2. Trigonometry.   I. Title.
QA154.2.C58   1993
512'.13—dc20

                                        92-11099
                                      ∞ CIP

CONTENTS

NINE — ADDITIONAL TOPICS IN TRIGONOMETRY — 515

TEN — SYSTEMS OF EQUATIONS — 578

ELEVEN — ROOTS OF POLYNOMIAL EQUATIONS — 659

TWELVE — THE CONIC SECTIONS — 699

THIRTEEN — ADDITIONAL TOPICS IN ALGEBRA — 746

APPENDIX

TABLES

ANSWERS TO SELECTED EXERCISES

INDEX

PREFACE

This text develops the elements of college algebra and trigonometry in a straightforward manner. As in the earlier editions, my goal has been to create a book that is *accessible* to the student. The presentation is student-oriented in three specific ways. First, I've tried to talk to, rather than lecture at, the student. Second, examples are consistently used to introduce, to explain, and to motivate concepts. And third, all of the initial exercises for each section are carefully correlated with the worked examples in that section.

AUDIENCE
In writing *Algebra and Trigonometry,* I have assumed that the students have been exposed to intermediate algebra, but that they have not necessarily mastered that subject. Also, for many algebra students, there may be a gap of several years between their last mathematics course and the present one. For these reasons, the review material in Chapter 1 and in parts of Chapter 2 is unusually thorough.

FEATURES
1. *Word problems and applications.* Beginning with the review material in Chapter 1, there is a constant emphasis on learning to use the *language* of algebra as a tool in problem-solving. Word problems and strategies for solving them are explained and developed throughout the book. For examples of this, see any of the following sections.

> Sec. 2.1 (setting up equations)
>
> Sec. 2.4 (word problems leading to linear and quadratic equations)
>
> Sec. 5.3 (more on setting up equations)
>
> Sec. 5.4 (maximum–minimum problems relating to quadratic functions)
>
> Sec. 6.5 (applications of the exponential function)

2. *Emphasis on graphing.* Graphs and techniques for graphing are developed throughout the text, and graphs are used to explain and reinforce algebraic concepts. (See, for example, Sections 4.3 and 6.3.)

3. *Calculator Exercises.* There are two broad categories of calculator exercises in this text:

> (i) OPTIONAL GRAPHING CALCULATOR EXERCISES
> Over the past several years, all of us in the mathematics teaching community have become increasingly aware of the graphing calculator

and its potential for making a positive impact in our teaching. While it is clear that vast reforms lie ahead, many of the specific details are still evolving in workshops and in classrooms across the country. And indeed, at present, even within a given school, some instructors are teaching the course with the graphing calculator, while others are not. Thus, for this 1992 edition of *Algebra and Trigonometry,* I have labeled the graphing calculator exercises "optional," and I have placed them at the ends of the chapters. There are graphing calculator exercises for many of the major topics in Chapters 3 through 6 and 8 through 12. These exercises have been class-tested. Although the exercises can be adapted for use with any graphing calculator, they were written specifically for use with the Texas Instruments TI-81 Graphics Calculator. Since many of these exercises contain carefully detailed instructions for using this calculator, a minimum of class time is required for discussing its operation. Additionally, Section A.2 of the appendix (written by Professor Mickey Settle) contains a complete introduction to the basic features of the calculator.

(ii) (ORDINARY) CALCULATOR EXERCISES

As in the previous edition, there are many calculator exercises integrated throughout the regular exercise sets in the text. Some of these exercises reinforce or supplement the core material; some of these exercises contain surprising results that motivate subsequent theoretical questions; and a few of these exercises demonstrate that the use of a calculator cannot replace thinking or the need for mathematical proofs.

4. *Trigonometry.*   The development of most topics in this text proceeds from the specific to the general, and the trigonometry follows this pattern, too. Thus, Chapter 7 introduces the right-triangle definitions (in Section 7.1) before the unit-circle definitions (in Section 7.4).

The unit circle definitions of the trigonometric functions involve simple geometric ideas that should be presented as such, and not cluttered with excessive notation or terminology (that will never appear in subsequent science or calculus courses). In presenting these definitions in Sections 7.4 and 8.2, the text sets a pace that is deliberately measured. Rather than rushing on, we take time to help the student understand the definitions. Experience shows that this pays dividends later on when the student studies the more analytical portions of trigonometry in Chapter 8.

5. *Graded Exercise Sets.*   As a convenience for the instructor, most of the exercise sets (except for the Chapter Review sets) are divided into three categories: A, B, and C. The group A exercises are based directly on the examples and definitions in that section of the text. Moreover, these problems treat topics in roughly the same order in which they appear in the text. The group B exercises serve several key functions. Some of them require students to use several different techniques or topics in a single solution. Some group B problems, while not conceptually more difficult than their group A counterparts, require lengthier calculations or algebraic manipulations. Finally, some group B exercises are included simply because they are interesting or because they allow the students to see the material from a different perspective. The group C exercises contain the more challenging problems. (In many cases, though, detailed hints are provided.)

6. *End-of-Chapter Material.* Each chapter concludes with a detailed Chapter Summary, a *Writing Mathematics* section, an extensive Chapter Review exercise set, and a Chapter Test. Also, as described before, for Chapters 3–6 and 8–12, there are graphing calculator exercises.

The *Writing Mathematics* questions are new to this edition. In these exercises, the student is asked to organize his or her thoughts and respond in complete sentences. Some of the questions are simply *true-or-false* questions that can be explained in just a sentence or two. In other cases, a more elaborate response is required. For example, in Chapter 3 (page 166), the Writing Mathematics section describes a geometric technique for solving quadratic equations, and the student is asked to explain why the method is valid. And in Chapter 9, The Writing Mathematics section describes a new proof for the law of cosines from the *College Mathematics Journal,* and the student is asked to fill in the details. (See page 570.)

CHANGES IN THIS EDITION

Comments and suggestions from students, instructors, and reviewers have helped me to revise this text in a number of ways that I believe will make the book more useful to the instructor and more accessible to the student.

In many sections I've included boxed summaries entitled "Errors to Avoid." These summaries warn the student about some of the "popular" errors in algebra. For instance, the three messages

$$\frac{1}{x} + \frac{1}{y} \neq \frac{1}{x+y} \qquad \sqrt{a+b} \neq \sqrt{a} + \sqrt{b} \qquad f(a+b) \neq f(a) + f(b)$$

are delivered (with examples and comments) on pages 10, 34, and 177, respectively. To help the student internalize these messages, there are often follow-up exercises that ask for examples showing that a purported rule is, in general, false.

Section 1.9 on factoring has been revised. There are more examples and exercises, and many of these emphasize *patterns*. (See, for instance, Example 3 on page 51.)

At the beginning of Chapter 2, there is now a preparatory section on setting up equations. I have used George Polya's simple two-column format for translating English into algebraic notation. The pace and the subject matter in this warm-up section are deliberately limited, and all of the exercises are tied directly to the worked examples. Class testing has convinced me that this is a very effective way to begin the chapter.

In Chapter 3, the material on slopes and lines, formerly two sections, is now in one section (3.3). (The applications of slope, such as marginal cost and velocity, appear in Chapter 5 in the context of linear functions.)

In Chapter 4, Section 4.2, *The Graph of a Function*, has been expanded slightly to include the geometric ideas of increasing and decreasing functions, turning points, and maximum and minimum values. The material on techniques in graphing in Section 4.3 has been streamlined and the presentation is now, I believe, easier to teach and easier to learn. (Some of the material on translation and coordinates is presented again later, but from a different viewpoint, in Chapter 12.)

The treatment of linear functions at the beginning of Chapter 5 has been expanded in this edition to a full section. The idea of slope as a rate of change is emphasized through examples involving marginal cost and velocity. The concepts of a scatter diagram and the least-squares line are introduced in the text rather than the exercises, and the treatment is expanded.

In Section 5.5 on graphs of polynomial functions, the behavior of $f(x)$ when $|x|$ is very large is given greater emphasis both in the text and in the exercises. (See, for example, Figure 4 and Example 1 on pages 276–277.) The material on graphing polynomials in factored form has been revised. We now make use of the earlier work in Chapter 2 on solving inequalities. Also, more details are given about how to analyze a graph in the vicinity of an x-intercept.

Chapter 6 introduces exponential functions from a more practical or contemporary viewpoint. We indicate how exponential functions are used to model real-life situations. (See Figures 1 and 2 on page 306 and Example 9 on page 349.)

The section "Algebra and the Trigonometric Functions" from the second edition has been expanded and reorganized. Some of this material now appears much earlier (in Section 7.2). The portion of this material dealing with proofs of trigonometric identities appears in a new section, Section 7.5. The advantage to this format is that the unit-circle definitions now precede the work on identities, and so each identity needn't carry the restriction that θ be an acute angle.

In Chapter 8, the material on graphs of the trigonometric functions has been expanded and it is placed earlier in the chapter. Also in Chapter 8, the inverse trigonometric functions are now introduced earlier, in the context of solving trigonometric equations. The more formal properties of these functions are then developed in Section 8.9.

In Chapter 9, there are additional exercises on the law of sines and the law of cosines. (An extended example in the text shows how these laws are used in navigational problems.) Chapter 9 also contains a new section on parametric equations.

In Chapter 12, conics sections are discussed in greater detail than is found in most other algebra texts today. In this edition, there is a new section presenting the focus–directrix properties of these curves.

SUPPLEMENTARY MATERIALS

1. The *Student's Solutions Manual,* by Ross Rueger, contains complete solutions for the odd-numbered exercises and for the test questions at the end of each chapter.

2. The *Instructor's Solutions Manual,* by Ross Rueger, contains answers or solutions for every even-numbered exercise in *Algebra and Trigonometry*.

3. The *Computer-Generated Testing Programs* Brownstone's DIPLOMA IV© are available to schools adopting *Algebra and Trigonometry*. (There are versions for both the Macintosh and the IBM PCs or compatibles.)

4. The *Test Bank,* by Charles Heuer, is a package that contains two versions of a chapter test, as well as two multiple-choice versions, for each chapter in *Algebra and Trigonometry*.

5. *GraphToolz,* by Tom Saxton, is a software program for graphing and evaluating functions. This ingenious easy-to-use software, along with *Drill and Enrichment Exercises* by David Cohen, can make many of the topics in *Algebra and Trigonometry* really seem to come alive. *GraphToolz* is free to qualified adopters. (Available for the Macintosh family of computers.)

6. *Transparency masters* for many of the key figures or tables appearing in the text are available to schools adopting *Algebra and Trigonometry.*

7. *College Algebra: In Simplest Terms* (a series of video tapes produced by Annenberg/CPB Collection) is available to qualified adopters. This series covers the major topics in college algebra, and it includes some excellent applications.

8. *Natural Language Mathematics Tutorial Software,* by Mathens, Inc., contains algorithms for generating multiple examples of the basic problem types in *Algebra and Trigonometry.* For qualified adopters, the software also includes customized class-tested tutorials for each section in *Algebra and Trigonometry.* (There are versions for both IBM PC-compatible and Macintosh computers.)

ACKNOWLEDGMENTS   Many students and teachers from both colleges and high schools have made useful constructive suggestions about the text and exercises, and I thank them for that. Special thanks go to Professor Charles Heuer for his careful work in checking the text and the exercise solutions for accuracy. I am grateful to Professor Mickey Settle for allowing us to reproduce his copyrighted work, *Using the TI-81 Graphics Calculator* in the appendix. In preparing and revising the manuscript, I received valuable suggestions and comments from the following reviewers.

Lynn H. Brown
Illinois State University

Nathaniel Silver
The City College, City University of New York

Paula Gnepp
Cleveland State University

William Wallace
Phoenix College

Donald Gnirk
South Dakota State University

Paul Weichsel
University of Illinois-Urbana

Richard Nadel
Florida International University

James Wood
Virginia Commonwealth University

I would like to thank Professor Ross Rueger, who wrote the supplementary manuals and created the answer section for the text; and Susan Gerstein, who copyedited the manuscript. Finally, to Stacy Lenzen, Peter Marshall, and Jane Bass at West Publishing Company, thank you for all your work and help in bringing this manuscript into print.

David Cohen
Lunada Bay, California, 1992

CHAPTER ONE

FUNDAMENTAL CONCEPTS

INTRODUCTION

This first chapter sets the foundations for our work in algebra. The central topics here are the real numbers, polynomials, and operations with polynomials. In the last section of the chapter we describe the complex number system. The notation and the results in this chapter will be used repeatedly throughout the rest of the book.

1.1 NOTATION AND LANGUAGE

Perhaps Pythagoras was a kind of magician to his followers because he taught them that nature is commanded by numbers. There is a harmony in nature, he said, a unity in her variety, and it has a language: numbers are the language of nature.

Jacob Bronowski

Ever since the time of Descartes, the terms "real" and "imaginary" numbers have been used in mathematics, despite their misleading associations.

Alfred Hooper

Here, as in your previous mathematics courses, most of the numbers we deal with are *real numbers*. These are the numbers used in everyday life, in the sciences, in industry, and in business. Perhaps the simplest way to define a real number is this: A **real number** is any number that can be expressed in decimal form. Some examples of real numbers are

$$7 \ (= 7.000 \cdots) \qquad -\tfrac{2}{3} \ (= -0.\overline{6}) \qquad \sqrt{2} \ (= 1.4142 \cdots)$$

In algebra we often use letters to stand for numbers. This simple idea is what gives algebra much of its power and applicability in problem solving. By using letters, we can often see patterns and reach conclusions that might not be so apparent otherwise.

Arithmetic, and therefore algebra too, begins with the *operations* of addition, subtraction, multiplication, and division. Table 1 will serve to remind you

TABLE I

Operation	Arithmetic Example	Algebraic Example
Addition (+)	$5 + 7.01$	$p + q$
Subtraction (−)	$13 - \sqrt{5}$	$r - s$
Multiplication (×)	3×4	$a \times b$ or $a \cdot b$ or ab
Division (÷)	$5 \div 6$	$x \div y$ or $\dfrac{x}{y}$ or x/y

that, for the most part, the notation for these operations in both arithmetic and algebra is the same. One exception: as the table indicates, the product $a \times b$ is often abbreviated simply by ab.

There are several other notational conventions that will seem familiar to you from previous courses. A sum such as $a + a + a$ is denoted by $3a$, while a product such as $a \cdot a \cdot a$ is written a^3. In general, if n is a **positive integer** (i.e., one of the numbers 1, 2, 3, 4, ...), then

$$na \quad \text{means} \quad \underbrace{a + a + a + \cdots + a}_{n \text{ terms}}$$

and

$$a^n \quad \text{means} \quad \underbrace{a \cdot a \cdot a \cdots a}_{n \text{ factors}}$$

In the expression na, the number n is called the **coefficient** of a. For example, in the expression $3a$, the coefficient is 3. When we write a^n, we refer to a as the **base** and n as the **exponent** or **power** to which the base is raised. The process of raising a base to a power is called **exponentiation**. The notation a^2 is read "a squared," because the area of a square with sides each of length a is $a \times a$ or a^2. Similarly, a^3 is read "a cubed," because the volume of a cube with sides each of length a is $a \times a \times a$ or a^3.

There are several important conventions regarding the order in which the operations of arithmetic and algebra are to be performed. Again, these will seem familiar to you from previous courses, and we summarize them in the box on the next page.

Examples 1 through 3, which follow, serve to summarize the ideas discussed so far. In Example 4, we translate some expressions from words into algebraic notation. This is an important skill that will be developed in more detail in Chapters 2 and 5.

EXAMPLE I Rewrite each expression using exponential notation.

(a) $x \cdot x \cdot x \cdot y \cdot y \cdot y \cdot y$

(b) $(a + 3)(a + 3)(b - 1)(b - 1)(b - 1)$

Solution (a) $x \cdot x \cdot x \cdot y \cdot y \cdot y \cdot y = x^3 y^4$

(b) $(a + 3)(a + 3)(b - 1)(b - 1)(b - 1) = (a + 3)^2 (b - 1)^3$

PROPERTY SUMMARY   THE ORDER OF OPERATIONS

CONVENTION	EXAMPLE
1. If no grouping symbols (such as parentheses) are present:	
(a) First do the exponentiations, working from right to left;	$3 + 4^2 = 3 + 16 = 19; \qquad 5 \times 2^3 = 5 \times 8 = 40;$ $2^{3^2} = 2^9 = 512$
(b) Next do the multiplications and divisions, working from left to right;	$9 - 2 \times 3 = 9 - 6 = 3; \qquad 12 + 8 \div 2 = 12 + 4 = 16;$ $4 + 2 \times 3^2 = 4 + 2 \times 9 = 4 + 18 = 22$
(c) Then do the additions and subtractions, working from left to right.	$1 + 2 \times 3 - 4 = 1 + 6 - 4 = 7 - 4 = 3$ $15 + 6^2 \div 4 - 2 \times 3^2 = 15 + 36 \div 4 - 2 \times 9$ $= 15 + 9 - 18$ $= 24 - 18 = 6$
2. If grouping symbols are present:	
(a) First do the operations within the grouping symbols;	$(1 + 2) \times 5 = 3 \times 5 = 15; \qquad (3 + 4)^2 = 7^2 = 49$
(b) If grouping symbols appear within grouping symbols, the operations within the innermost set are to be done first.	$3[8 - 2(6 - 4)] = 3(8 - 2 \times 2)$ $= 3(8 - 4)$ $= 3 \times 4 = 12$

EXAMPLE 2   **(a)** Evaluate $x - 4y^2$ when $x = 19$ and $y = 2$.
(b) Evaluate $(x - 4)y^2$ using the values for x and y given in part (a).

Solution   **(a)** $x - 4y^2 = 19 - 4 \times 2^2$
$= 19 - 4 \times 4 = 19 - 16 = 3$
(b) $(x - 4)y^2 = (19 - 4) \times 2^2 = 15 \times 4 = 60$

EXAMPLE 3   **(a)** Evaluate each of the following expressions using $x = 5$ and $y = 2$.

$$(x + y)^2 \qquad x^2 + y^2 \qquad x^2 + 2xy + y^2$$

(b) Compare the results in part (a). What conclusions can be drawn?

Solution   **(a)** Using $x = 5$ and $y = 2$ in each of the expressions, we have

$$(x + y)^2 = (5 + 2)^2 = 7^2 = 49$$
$$x^2 + y^2 = 5^2 + 2^2 = 25 + 4 = 29$$
$$x^2 + 2xy + y^2 = 5^2 + 2(5)(2) + 2^2 = 49$$

(b) On the one hand, using $x = 5$ and $y = 2$, we found that the values of $(x + y)^2$ and $x^2 + y^2$ are not the same. This shows that, in general $(x + y)^2 \neq x^2 + y^2$. (It takes only one counterexample to show that something is not, in general, true.) On the other hand, again using $x = 5$ and $y = 2$, we did find that the values of $(x + y)^2$ and $x^2 + 2xy + y^2$ are the same. But this does not *prove*, in general, that $(x + y)^2 = x^2 + 2xy + y^2$. It only shows that the equation is valid in one particular case. [Actually, as you may recall from a previous course (and as you'll see in Section 1.8), this last equation is, in fact, true in general. But, the point is, what we just did does not constitute a proof of this fact.]

EXAMPLE 4   Translate each of the following into algebraic notation:

(a) five more than twice the number x;
(b) three times the sum of x and y^3;
(c) the sum of three times x and y^3.

Solution   (a) $2x + 5$

twice x —┘   └— five more

(b) $3 \times (x + y^3) = 3(x + y^3)$

three times ——┘
the sum of x and y^3 ——┘

(c) $3x + y^3$

└— the sum of $3x$ and y^3

We conclude this section with some comments about exponents and calculators. Suppose, for example, that we wish to evaluate the expression

$$3^{2^4}$$

According to Property 1(a) in the box on page 3, we have

$$3^{2^4} = 3^{16} \qquad \text{because} \qquad 2^4 = 16$$

Now we need to evaluate 3^{16}. Although this can certainly be computed "by hand," a more sensible approach here is to use a calculator. All scientific calculators have keys for computations involving exponents. The details vary, however, from brand to brand, and even between different models of the same brand. The examples that follow indicate how 3^{16} is calculated on four common types of calculators. If you have questions about how your calculator operates in this respect, you should consult the owner's manual.

CALCULATOR	SEQUENCE OF KEYSTROKES	OUTPUT
Sharp EL-506H	3 $\boxed{y^x}$ 16 $\boxed{=}$	43,046,721
Casio fx-7000G	3 $\boxed{x^y}$ 16 $\boxed{\text{EXE}}$	43,046,721
Hewlett Packard 11C	3 $\boxed{\text{ENTER}}$ 16 $\boxed{y^x}$	43,046,721
Texas Instruments TI-81	3 $\boxed{\wedge}$ 16 $\boxed{\text{ENTER}}$	43,046,721

EXERCISE SET 1.1

A

In Exercises 1–10, rewrite the given expressions using coefficients and exponents.

1. $x \cdot x \cdot x \cdot x \cdot x \cdot y \cdot y$ 2. $x \cdot x \cdot y \cdot y \cdot z$

3. $(x + 1)(x + 1)(x + 1)$

4. $(a^2 + b^2)(a^2 + b^2)(a^2 + b^2)$

5. (a) $x + x + x + x$
   (b) $(x^2 + 1) + (x^2 + 1) + (x^2 + 1)$

6. (a) $a^5 + a^5$
   (b) $(a^4 + 1) + (a^4 + 1)$

7. (a) $a + a + a + b + b$
   (b) $a^2 + a^2 + a^2 + b^2 + b^2$

8. (a) $t^4 + t^4 + t^4 + z^2 + z^2$
   (b) $(t^4 + 1) + (t^4 + 1) + (t^4 + 1)$

9. (a) $(2a + 1)(2a + 1)(2a + 1)(2b + 1)(2b + 1)$
   (b) $(1 + 2a)(2a + 1)(1 + 2a)(2b + 1)(1 + 2b)$

10. (a) $(x + y^2)(x + y^2)(x^2 + y)(x^2 + y)$
    (b) $(x + y^2)(y^2 + x)(x^2 + y)(y + x^2)$

For Exercises 11–26, evaluate the expressions using the given values of the variables.

11. $x + 3y$; $x = 4, y = 6$

12. $a - 4b^2$; $a = 20, b = 2$

13. $a^2 + b^2$; $a = 3, b = 4$

14. $(a + b)^2$; $a = 3, b = 4$

15. $x^2 - 4y^2$; $x = 10, y = 4$

16. $x^2 - xy + y^2$; $x = 1, y = 1$

17. $x^2 - x + 1$; $x = 5$

18. $(x + 1)(x^2 - x + 1)$; $x = 5$

19. $2x^3 - 3y^2$; $x = 3, y = 2$

20. $2(x^3 - 3y^2)$; $x = 3, y = 2$

21. $1 \div a^2 + b^2$; $a = 1, b = 1$

22. $1 \div (a^2 + b^2)$; $a = 1, b = 1$

23. $\dfrac{1}{a^2} + \dfrac{1}{b^2} - \dfrac{1}{a^2 + b^2}$; $a = \dfrac{1}{2}, b = \dfrac{1}{2}$

24. $\dfrac{1}{1 - 1/x}$; $x = 2$

25. A^{B^C} for (a) $A = B = 2, C = 3$; (b) $A = C = 2, B = 3$; (c) $B = C = 2, A = 3$

26. A^{B^C} for (a) $A = 2, B = 1, C = 3$; (b) $A = 2, B = 3, C = 1$; (c) $A = 1, B = 2, C = 3$

27. Which is larger: $(2^3)^4$ or 2^{3^4}?

28. Compute $2^{3^2} - (2^3)^2$.

In Exercises 29 and 30, use a calculator to evaluate A^{B^C}, using the given values.

29. (a) $A = 5, B = 2, C = 3$
    (b) $A = 5, B = 3, C = 2$

30. (a) $A = 4, B = 3, C = 2$
    (b) $A = 3, B = 4, C = 2$

31. (a) Without using a calculator, decide which of the quantities, 2^{2^5} or 5^{2^2}, is larger. (You should be able to do this without completely evaluating both expressions.)
    (b) Use a calculator to evaluate the expressions and check your answer in part (a).

32. Follow Exercise 31 using the expressions 2^{3^3} and 3^{3^2}.

In Exercises 33–46, translate each phrase into algebraic notation.

33. Two more than x

34. Two less than x

35. Four times the sum of a and b^2

36. The sum of four times a and b^2

37. The sum of x^2 and y^2

38. The square of the sum of x and y

39. The sum of x and twice the square of x

40. Twice the sum of x and the square of x

41. The average of x, y, and z

42. The square of the average of x, y, and z

43. The average of the squares of x, y, and z

44. The square of the average of the squares of x, y, and z

45. One less than twice the product of x and y

46. The cube of three more than the product of x^2 and y^3

B

For Exercises 47 and 48, evaluate the expressions using the given values of the variables.

47. (a) $\dfrac{1000}{19 - \dfrac{12}{1 - (1/x)}}$; $x = 3$

    (b) $\dfrac{1}{1 + \dfrac{1}{1 + \dfrac{1}{1 + (1/x)}}}$; $x = 2$

48. (a) $[p(p^2 - 2^p)]^{p+1}$; $p = 3$

    (b) $\dfrac{\dfrac{a+b}{a-b} + \dfrac{a-b}{a+b}}{\dfrac{a-b}{a+b} - \dfrac{a+b}{a-b}} \cdot \dfrac{ab^3 - a^3b}{a^2 + b^2}$; $a = \dfrac{5}{4}, b = \dfrac{1}{4}$

49. (a) In each case, use a calculator to determine which of the two quantities is larger.
    (i) 6^7; 7^6 (ii) 7^8; 8^7 (iii) 9^8; 8^9
    (b) Find a positive whole number a such that $(a + 1)^a > a^{a+1}$. (There are only two such numbers; some experimenting is called for.)

1.2 PROPERTIES OF THE REAL NUMBERS

Today's familiar plus and minus signs were first used in 15th-century Germany as warehouse marks. They indicated when a container held something that weighed over or under a certain standard weight.

Martin Gardner in "Mathematical Games" (*Scientific American*, June 1977)

I do not like × as a symbol for multiplication, as it is easily confounded with x; . . . often I simply relate two quantities by an interposed dot . . .

G. W. Leibniz in a letter dated July 29, 1698

In this section we first will list the basic properties for the real number system. After that we'll summarize some of the more down-to-earth implications of those properties, such as the procedures for working with signed numbers and fractions. (Section A.3 of the appendix, at the back of this book, presents a more theoretical treatment of some of these ideas.)

The set of real numbers is **closed** with respect to the operations of addition and multiplication. This just means that when we add or multiply two real numbers, the result (that is, the **sum** or the **product**) is again a real number. Some of the other most basic properties and definitions for the real-number system are listed in the following box. In the box, the lowercase letters a, b, and c denote arbitrary real numbers.

PROPERTY SUMMARY   SOME FUNDAMENTAL PROPERTIES OF THE REAL NUMBERS

Commutative properties	$a + b = b + a$ $ab = ba$
Associative properties	$a + (b + c) = (a + b) + c$ $a(bc) = (ab)c$
Identity properties	**1.** There is a unique real number 0 (called **zero** or the **additive identity**) such that $$a + 0 = a \quad \text{and} \quad 0 + a = a$$ **2.** There is a unique real number 1 (called **one** or the **multiplicative identity**) such that $$a \cdot 1 = a \quad \text{and} \quad 1 \cdot a = a$$
Inverse properties	**1.** For each real number a there is a real number $-a$ (called the **additive inverse** of a or the **negative** of a) such that $$a + (-a) = 0 \quad \text{and} \quad (-a) + a = 0$$ **2.** For each real number $a \neq 0$, there is a real number denoted by $\frac{1}{a}$ (or $1/a$ or a^{-1}), and called the **multiplicative inverse** or **reciprocal** of a, such that $$a \cdot \frac{1}{a} = 1 \quad \text{and} \quad \frac{1}{a} \cdot a = 1$$
Distributive properties	$a(b + c) = ab + ac$ $(b + c)a = ba + ca$

On reading this list of properties for the first time, many students ask the natural question, "Why do we even bother to list such obvious properties?" One reason is that all the other laws of arithmetic and algebra (including the "not-so-obvious" ones) can be derived from our rather short list. For example, the rule $0 \cdot a = 0$ can be proved using the distributive property, as can the rule that the product of two negative numbers is a positive number. (See Section A.3

of the appendix for these proofs.) The distributive properties in a sense indicate the way in which multiplication and addition interact with each other. More specifically, each of the following examples relies on the distributive properties.

EXAMPLE	REMARK
$12 \times 103 = 12(100 + 3)$ $= 1200 + 36 = 1236$	The distributive property can be used as an aid in mental calculations.
$3a^2(a + 2b) = 3a^2 \cdot a + 3a^2 \cdot 2b$ $= 3a^3 + 6a^2b$	An application of the distributive property, followed by a simplification.
$A(c + d) = Ac + Ad$	A direct application of the distributive property. In the next example we are going to replace A with the quantity $(a + b)$.
$(a + b)(c + d) = (a + b)c + (a + b)d$ $= ac + bc + ad + bd$	The distributive properties are used three times here.
$(x + 3)(x + 6) = (x + 3)x + (x + 3)6$ $= x^2 + 3x + 6x + 18$ $= x^2 + 9x + 18$	This is similar to the previous example. Examples of this type will be discussed in greater detail in Section 1.8.

The distributive and associative properties can be generalized. The **generalized distributive property** asserts that

$$a(b_1 + b_2 + \cdots + b_n) = ab_1 + ab_2 + \cdots + ab_n$$

The generalized associative property tells us that we can group the terms as we please in computing a sum or a product. For instance, we have

$$17 + 47 + 53 + 80 = 17 + [(47 + 53) + 80]$$
$$= 17 + [100 + 80]$$
$$= 17 + 180 = 197$$

EXAMPLE 1 Identify the property of the real numbers that justifies each statement.

(a) $(17 \times 50) \times 2 = 17 \times (50 \times 2) = 17 \times 100 = 1700$
(b) Since π and $\sqrt{2}$ are real numbers, so is $\pi + \sqrt{2}$.
(c) $3 + y^2 = y^2 + 3$
(d) $[y + (z + 1)]x = yx + (z + 1)x$
(e) $(z + 1) + [-(z + 1)] = 0$
(f) $(a^2 + 1) \cdot \left(\dfrac{1}{a^2 + 1}\right) = 1$

Solution
(a) Associative property of multiplication
(b) Closure property with respect to addition
(c) Commutative property of addition
(d) Distributive property
(e) Additive inverse property
(f) Multiplicative inverse property

If you look back at the box on page 6, you'll see that the operations of subtraction and division are not mentioned. This is because addition and multiplication are actually the more fundamental operations. In fact, we can define subtraction and division in terms of addition and multiplication as follows.

DEFINITIONS Subtraction and Division

Subtraction and division are defined in terms of addition and multiplication, respectively, as follows.

$$a - b = a + (-b)$$

$$a \div b = a \cdot \left(\frac{1}{b}\right) = ab^{-1} \qquad \text{provided that } b \neq 0$$

Division by zero is not defined. To see why this must be the case, suppose just for the moment that division by zero were defined. Then we would have $\frac{1}{0} = k$ for some real number k. But then, multiplying both sides of the equation by 0 would yield $1 = k \cdot 0$; that is, $1 = 0$, a contradiction. In summary, if we wish the rules of arithmetic and algebra to be consistent, then division by zero must remain undefined.

In basic algebra you were introduced to a number of procedures that were useful in working with algebraic expressions. In the three boxes that follow, we summarize the properties of the real numbers that are behind those procedures you learned. In the statements of these properties, the lowercase letters a, b, c, and d are used to denote arbitrary real numbers. In the properties of fractions, of course, we are assuming that none of the denominators is zero.

PROPERTY SUMMARY   PROPERTIES OF ZERO

PROPERTY	EXAMPLE
1. $a \cdot 0 = 0$	$(x^3 - \pi) \cdot 0 = 0$
2. If $\dfrac{a}{b} = 0$, then $a = 0$.	If $\dfrac{x^2 + x - 1}{2x} = 0$, then $x^2 + x - 1 = 0$.
3. If $ab = 0$, then $a = 0$ or $b = 0$.	If $5x = 0$, then $x = 0$.

PROPERTY SUMMARY   PROPERTIES OF NEGATIVES

PROPERTY	EXAMPLE
1. $-(-a) = a$	$-[-(x + 1)] = x + 1$
2. $(-a)b = -(ab) = a(-b)$	$(-3)4 = -12 = 3(-4)$
3. $(-a)(-b) = ab$	$[-(x + y)][-(x + y)] = (x + y)^2$
4. $-a = (-1)a$	$-(x + y - z) = -1(x + y - z) = -x - y - z$

PROPERTY SUMMARY   PROPERTIES OF FRACTIONS

(The denominators in each case are assumed to be nonzero.)

PROPERTY	EXAMPLE
1. $\dfrac{a}{b} = \dfrac{c}{d}$ if and only if $ad = bc$	$\dfrac{8}{12} = \dfrac{2}{3}$ since $8 \cdot 3 = 12 \cdot 2$
2. $\dfrac{a}{b} = \dfrac{ac}{bc}$	$\dfrac{4}{7} = \dfrac{4 \cdot 5}{7 \cdot 5} = \dfrac{(4 \cdot 5)\pi}{(7 \cdot 5)\pi}$
3. $\dfrac{a}{b} \pm \dfrac{c}{b} = \dfrac{a \pm c}{b}$	$\dfrac{10}{3} + \dfrac{3}{3} = \dfrac{13}{3}; \quad \dfrac{10}{3} - \dfrac{3}{3} = \dfrac{7}{3}$
4. $\dfrac{a}{b} \pm \dfrac{c}{d} = \dfrac{ad \pm bc}{bd}$	$\dfrac{2}{3} + \dfrac{3}{4} = \dfrac{2 \cdot 4 + 3 \cdot 3}{3 \cdot 4} = \dfrac{17}{12}; \quad \dfrac{2}{3} - \dfrac{3}{4} = \dfrac{2 \cdot 4 - 3 \cdot 3}{3 \cdot 4} = \dfrac{-1}{12}$
5. $\dfrac{a}{b} \cdot \dfrac{c}{d} = \dfrac{ac}{bd}$	$\dfrac{5}{4} \cdot \dfrac{7}{9} = \dfrac{35}{36}$
6. $\dfrac{a}{b} \div \dfrac{c}{d} = \dfrac{a}{b} \cdot \dfrac{d}{c} = \dfrac{ad}{bc}$	$\dfrac{3}{5} \div \dfrac{2}{7} = \dfrac{3}{5} \cdot \dfrac{7}{2} = \dfrac{21}{10}; \quad 1 \div \dfrac{1}{10} = \dfrac{1}{1} \cdot \dfrac{10}{1} = 10$
7. $-\dfrac{a}{b} = \dfrac{-a}{b} = \dfrac{a}{-b}$	$-\dfrac{3}{4} = \dfrac{-3}{4} = \dfrac{3}{-4}; \quad -\dfrac{x-1}{x+1} = \dfrac{-(x-1)}{x+1} = \dfrac{x-1}{-(x+1)}$
8. $\dfrac{a-b}{b-a} = -1$	$\dfrac{7-x}{x-7} = -1; \quad \dfrac{x^2 - y^2}{y^2 - x^2} = -1$

For the next example, you'll need to recall the definition of least common multiple. The **least common multiple** (l.c.m.) is the smallest positive integer that is a multiple of each of the given integers. *Examples:* The l.c.m. of 2 and 3 is 6; the l.c.m. of 3, 5, and 6 is 30.

  EXAMPLE 2   Simplify the following expression (that is, write it as a single fraction).

$$\frac{\frac{5}{4} + \frac{7}{3}}{\frac{1}{6} + \frac{3}{2}}$$

Solution   The individual denominators are 4, 3, 6, and 2. Since the least common multiple of these numbers is 12, we multiply the entire original fraction by $\frac{12}{12}$ (which is equal to 1) to obtain

$$\frac{12}{12} \cdot \frac{\frac{5}{4} + \frac{7}{3}}{\frac{1}{6} + \frac{3}{2}} = \frac{12\left(\frac{5}{4} + \frac{7}{3}\right)}{12\left(\frac{1}{6} + \frac{3}{2}\right)}$$

$$= \frac{15 + 28}{2 + 18} = \frac{43}{20}$$

The properties of fractions that we have listed in the box above will be used in Section 1.10 (and then throughout the rest of the book) when we work with fractional expressions. To help you gain some additional perspective here, we list several common types of errors to avoid when working with fractions.

Errors to Avoid

ERROR	CORRECTION	COMMENT
$\dfrac{1}{x} + \dfrac{1}{y} \neq \dfrac{1}{x+y}$	$\dfrac{1}{x} + \dfrac{1}{y} = \dfrac{1(y) + x(1)}{xy} = \dfrac{x+y}{xy}$	Add fractions with unlike denominators using Property 4 on page 9.
$\dfrac{a}{2} + \dfrac{3}{7} \neq \dfrac{a+3}{2+7}$	$\dfrac{a}{2} + \dfrac{3}{7} = \dfrac{a(7) + 2(3)}{(2)(7)} = \dfrac{7a+6}{14}$	Again, use Property 4.
$(1/a) + 3 \neq \dfrac{1}{a+3}$	$\dfrac{1}{a} + \dfrac{3}{1} = \dfrac{1(1) + a(3)}{a(1)} = \dfrac{1 + 3a}{a}$	Writing $1/a$ and 3 as $\dfrac{1}{a}$ and $\dfrac{3}{1}$, respectively, makes it clear that Property 4 applies.
$6(x + y) \neq 6x + y$	$6(x + y) = 6x + 6y$	According to the distributive property, the factor 6 must be distributed over *both* terms of the sum.
$6(xy) \neq (6x)(6y)$	$6(xy) = (6x)y = 6xy$	The distributive law does not apply.

EXERCISE SET 1.2

A

For Exercises 1–10, use the distributive properties to carry out the indicated multiplication.

1. $8 \times 104 = 8(100 + 4) = ?$

2. $17 \times 102 = 17(100 + 2) = ?$

3. $45 \times 98 = 45[100 + (-2)] = ?$

4. (a) $2a[3a + (-1)]$
    (b) $2a(3a - 4)$
    (c) $2a(3a + b - 4)$

5. (a) $(A + B)(A + B)$
    (b) $(A + B)(A - B)$
    (c) $(A - B)(A - B)$

6. (a) $(A + B)(C + D)$
    (b) $(A + B)(2A + 3B)$
    (c) $(A + B)(2A - 3B)$

7. (a) $(x + 2y)(x + 2y)$
    (b) $(x + 2y)(x - 2y)$
    (c) $(x - 2y)(x - 2y)$

8. (a) $(x + 3)(x + 3)$
    (b) $(x + 3)(x - 3)$
    (c) $(x^2 + 3)(x^2 - 3)$

9. (a) $A(x^2 - x + 1)$ (b) $(x + 1)(x^2 - x + 1)$

10. (a) $(2x + 3)(x - 1)$ (b) $(3x + 2)(x - 1)$

In Exercises 11–24, identify the property or properties of the real numbers that justify each statement.

11. $x + 3y = 3y + x$

12. $x(3y) = (3y)x$

13. $\dfrac{a^2}{b^2} + \left(-\dfrac{a^2}{b^2}\right) = 0$

14. $6(x + 8) = 6x + 48$

15. $(x + 1) + y = x + (1 + y)$

16. $(x + y) \cdot \left(\dfrac{1}{x + y}\right) = 1$

17. $(x + y)[(x + y)(x - y)] = (x + y)^2 (x - y)$

18. $(\sqrt{2})(1) = \sqrt{2}$ 19. $(5 + \pi) + 0 = 5 + \pi$

20. $[(x + y) + (x^2 + y^2)]x = (x + y)x + (x^2 + y^2)x$

21. Since $\sqrt{7}$ and $\sqrt{\pi}$ are real numbers, so is $\sqrt{7} + \sqrt{\pi}$.

22. $y(y + y^2 + y^3) = y^2 + y^3 + y^4$

23. $(x + 2)(x^2 + 4) = (x + 2)x^2 + (x + 2)4$

24. Since $\dfrac{1}{\sqrt{7}}$ and $\sqrt{5}$ are real numbers, so is $\left(\dfrac{1}{\sqrt{7}}\right)(\sqrt{5})$.

For Exercises 25–34, give numerical examples to show that the purported "rules" are not, in general, valid.

25. $\dfrac{1}{b} + \dfrac{1}{c} = \dfrac{1}{b + c}$

26. $\dfrac{1}{a} - \dfrac{1}{b} = \dfrac{1}{a - b}$

27. $3(x + y) = 3x + y$

28. $4(a - b) = 4a - b$

29. $x^2(x + y) = x^3 + y$

30. $a(b^2 - c^2) = ab^2 - c^2$

31. $3(xy) = (3x)(3y)$

32. $5(bc) = (5b)(5c)$

33. $a - (b - c) = a - b - c$

34. $a - (b + c) = a - b + c$

In Exercises 35–60, simplify the given expressions.

35. $1 + 2(-3)(4)$

36. $(97)(-98)(99)(0)$

37. $4 - (-3 + 5)$

38. $(-1)^4$

39. $-[(-1)^4]$

40. -1^4

41. $1 - \{1 - [-(-1 - 1)]\}$

42. $4^2 - (-4)^3$

43. $\frac{7}{5} + \frac{2}{3}$

44. $\frac{3}{4} - \frac{4}{3}$

45. $\frac{7}{x} - \frac{12}{x}$

46. $\frac{a + 1}{a^2 + 1} + \frac{a - 1}{a^2 + 1}$

47. $\frac{x^2 + y^2}{x + y} - \frac{x^2 - y^2}{x + y}$

48. $\frac{7}{12} \cdot \frac{5}{4}$

49. $\frac{7}{12} \cdot \left(\frac{-7}{3}\right)$

50. $1 \div \frac{1}{3}$

51. $\frac{2}{5} \div \frac{7}{3}$

52. $1 \div \frac{2}{3} + 1$

53. $1 \div \left(\frac{2}{3} + 1\right)$

54. $\dfrac{\frac{4}{5} + \frac{1}{2}}{\frac{5}{2}}$

55. (a) $\dfrac{\frac{4}{3} + \frac{1}{2}}{\frac{11}{6} + \frac{5}{4}}$ (b) $\dfrac{\frac{4}{3} - \frac{1}{2}}{\frac{1}{2} - \frac{4}{3}}$

56. (a) $\dfrac{b - a}{a - b}$ (b) $\dfrac{x - 7}{7 - x}$ (c) $\dfrac{3 - x^4}{x^4 - 3}$

57. $\dfrac{1 - (x + y)}{x + y - 1}$

58. $\dfrac{\frac{1}{5} - \frac{1}{4}}{\frac{1}{5} + \frac{1}{4}}$

59. $\dfrac{\frac{1}{2} - \frac{1}{3} + \frac{1}{4}}{\frac{1}{2} + \frac{1}{3} - \frac{1}{4}}$

60. $1 + \dfrac{1}{1 - \dfrac{1}{1 - \frac{1}{5}}}$

In Exercises 61–65, evaluate each expression using the given value for x.

61. $x^2 - 8x - 4$; $x = -2$

62. $2x^3 + 4x^2 - 1$; $x = -3$

63. $8x^3 - 4x^2 - 4x$; $x = -\frac{1}{2}$

64. $\dfrac{x^2 - 2x + 1}{x^3 - 3x^2 + 3x - 1}$; $x = -4$

65. $1 - x - x^2 - x^3 - x^4$; $x = -1$

66. If $m = \dfrac{y_2 - y_1}{x_2 - x_1}$, find m when $y_2 = 3$, $y_1 = -1$, $x_2 = 1$, and $x_1 = 4$.

67. If $d^2 = (x_2 - x_1)^2 + (y_2 - y_1)^2$, compute the value of d^2 when $x_2 = -1$, $x_1 = -2$, $y_2 = 6$, and $y_1 = -6$.

68. Let $A = \dfrac{x - y}{y - x}$. What is the value of A when

(a) $x = 3$, $y = 7$;

(b) $x = 25$, $y = 35$;

(c) $x = \frac{1}{2}$, $y = \frac{2}{3}$?

B

Exercises 69–72 ask you to show that certain fractions are equal. To do this, make use of Property 1 in the summary box for fractions on page 9. (In each of the exercises assume that the denominators are nonzero.)

69. If $\dfrac{a}{b} = \dfrac{c}{d}$, show that $\dfrac{a}{c} = \dfrac{b}{d}$.

70. If $\dfrac{a}{b} = \dfrac{c}{d}$, show that $\dfrac{a + b}{b} = \dfrac{c + d}{d}$.

71. If $\dfrac{a}{b} = \dfrac{c}{d}$, show that $\dfrac{a - b}{b} = \dfrac{c - d}{d}$.

72. If $\dfrac{a}{b} = \dfrac{c}{d}$, show that $\dfrac{a + b}{a - b} = \dfrac{c + d}{c - d}$.

In Exercises 73–76, use a calculator.

73. If $y = x^3 - 3x^2 - 6x - 4$, find y when $x = -0.75$.

74. If $y = \dfrac{x^2 - 4x + 1}{x^3 - x^2 - 3x - 1}$, compute y when $x = -1.01$.

75. Let $y = (x^2 - 1)/x^3$.

(a) Complete the following table.

x	1	1.250	1.500	1.750	2.000	2.250
y						

(b) Which x-value in the table corresponds to the largest y-value?

(c) It can be shown using calculus that the (positive) x-value that produces the largest possible y-value for the equation $y = (x^2 - 1)/x^3$ is $x = \sqrt{3}$. Use a calculator to compute this largest y-value and compare your answer with the y-values obtained in part (a).

76. Let $y = x^4 - 2x^3 - 12x^2 + 13$.

(a) Complete the following table.

x	3.000	3.100	3.200	3.300	3.400	3.500
y						

(b) Which x-value in the table yields the smallest y-value?

(c) It can be shown using calculus that the x-value that produces the smallest possible y-value for the equation $y = x^4 - 2x^3 - 12x^2 + 13$ is $x = (3 + \sqrt{105})/4$. Which x-value in the table in part (a) is closest to this? Compute the y-value corresponding to $x = (3 + \sqrt{105})/4$.

1.3   SETS OF REAL NUMBERS

Natural numbers have been used since time immemorial; fractions were employed by the ancient Egyptians as early as 1700 B.C.; and the Pythagoreans, in ancient Greece, about 400 B.C., discovered numbers, like $\sqrt{2}$, which cannot be fractions.

Stefan Drobot in *Real Numbers* (Englewood Cliffs, NJ: Prentice-Hall, Inc., 1964)

What secrets lie hidden in decimals?

Stephen P. Richards in *A Number for Your Thoughts* (New Providence, NJ: S. P. Richards, 1982)

Certain collections of real numbers are referred to often enough to be given special names. These are summarized in the box that follows.

PROPERTY SUMMARY   SETS OF REAL NUMBERS

NAME	DEFINITION AND COMMENTS	EXAMPLES
Natural numbers	These are the ordinary counting numbers, 1, 2, 3, and so on.	1, 4, 29, 1066
Integers	These are the natural numbers along with their negatives and zero.	$-26, 0, 1, 1992$
Rational numbers	As the name suggests, these are the real numbers that are *ratios* of two integers (with nonzero denominators, of course). It can be proved that a real number is rational if and only if its decimal expansion terminates (e.g., 3.15) or repeats (e.g., $2.\overline{43}$).	$4\left(=\frac{4}{1}\right), -\frac{2}{3},$ $1.7\left(=\frac{17}{10}\right),$ $4.\overline{3}, 4.1\overline{73}$
Irrational numbers	These are the real numbers that are not rational. It can be shown that any number of the form $\sqrt{n}$, where n is a natural number that is not a perfect square, is irrational.* Also, any number that is the sum or the (nonzero) product of a rational number and an irrational number is irrational.	$\sqrt{2}, 3+\sqrt{2}, 3\sqrt{2},$ $\pi, 4+\pi, 4\pi$

According to the comments in the box, the repeating decimal number $2.\overline{4}$ is rational. In other words, there is some fraction p/q such that $2.\overline{4} = p/q$. How can we find this fraction? Example 1 shows one way to do this.

EXAMPLE 1   Express the repeating decimal $2.\overline{4}$ in the form p/q, where p and q are integers and $q \neq 0$.

Solution   Let $x = 2.444\ldots$ Then $10x = 24.44\ldots$, and we have

$$10x - x = 24.\overline{4} - 2.\overline{4} = 22$$

*Section A.4 of the Appendix contains a proof of the fact that the number $\sqrt{2}$ is irrational. The proof that π is irrational is more difficult. The first person to prove that π is irrational was the Swiss mathematician J. H. Lambert (1728–1777).

or

$$9x = 22 \qquad \text{and therefore} \qquad x = \frac{22}{9}$$

We now have $2.\overline{4} = \frac{22}{9}$, as required.

Actually, a rigorous justification of each step used in Example 1 would require certain techniques from calculus. Nevertheless, you can easily verify for yourself that the answer in Example 1 is correct; just divide 22 by 9 and see what you find.

In previous courses you learned that the real numbers could be viewed as points on a *number line*. Because this is such an important idea, let's take a moment to review the essential features. Beginning with a straight line, we choose two points, O and P, as shown in Figure 1(a). We call O the **origin**. With the points O and P we identify the numbers 0 and 1, as shown in Figure 1(b).

FIGURE 1

(a)

(b)

Now, using the distance from O to P as our unit of length, we mark off points at one-unit intervals on both sides of the origin. As indicated in Figure 2, the points that we've marked to the right of the origin are identified with the numbers 1, 2, 3, and so on; the points to the left of the origin are identified with $-1, -2, -3$, and so on.

FIGURE 2

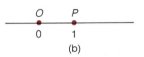

A fundamental fact is that there is a **one-to-one correspondence** between the set of real numbers and the set of points on the line. This means that each real number is identified with exactly one point on the line; conversely, with each point on the line we identify exactly one real number. The real number associated with a given point is called the **coordinate** of the point. As a practical matter, we're usually more interested in relative locations than precise locations on a number line. For instance, since π is approximately 3.1, we show π slightly to the right of 3 in Figure 3. Similarly, since $\sqrt{2}$ is approximately 1.4, we show $\sqrt{2}$ slightly less than halfway from 1 to 2 in Figure 3.

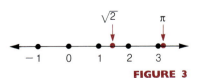

FIGURE 3

There are times when it is convenient to use number lines that show reference points other than the integers used in Figure 2. For instance, Figure 4(a) displays a number line with reference points that are multiples of π. In this case, of course, it is the integers that we then locate approximately. For example, in Figure 4(b) we show the approximate location of the number 1 on such a line.

FIGURE 4

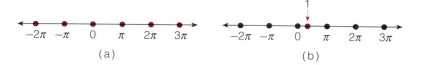

(a)   (b)

Two of the most basic relations for real numbers are **less than** and **greater than**, symbolized by $<$ and $>$, respectively. For ease of reference, we define these and two related symbols in the box that follows. [Incidentally, the symbols $<$ and $>$ were invented by the English mathematician and astronomer Thomas Harriot (1560–1621).]

■▶ **PROPERTY SUMMARY**	**NOTATION FOR LESS THAN AND GREATER THAN**	
NOTATION	**DEFINITION**	**EXAMPLES**
$a < b$	a is less than b. On a number line, oriented as in Figure 2, 3, or 4, the point a lies to the left of b.	$2 < 3$; $-4 < 1$
$a \leq b$	a is less than or equal to b.	$2 \leq 3$; $3 \leq 3$
$b > a$	b is greater than a. On a number line oriented as in Figure 2, 3, or 4, the point b lies to the right of a. [$b > a$ is equivalent to $a < b$.]	$3 > 2$; $0 > -1$
$b \geq a$	b is greater than or equal to a.	$3 \geq 2$; $3 \geq 3$

In general, relationships involving real numbers and any of the four symbols $<, \leq, >$, and $\geq$ are called **inequalities**. One of the simplest uses of inequalities is in defining certain sets of real numbers called *intervals*. Roughly speaking, any uninterrupted portion of the number line is referred to as an **interval**. In the definitions that follow, you'll see notation such as $a < x < b$. This means that *both* of the inequalities $a < x$ and $x < b$ hold; in other words, the number x is between a and b.

(a) The open interval (a, b) contains all real numbers from a to b, excluding a and b.

(b) The closed interval $[a, b]$ contains all real numbers from a to b, including a and b.

FIGURE 5

> **DEFINITION   Open Intervals and Closed Intervals**
>
> The **open interval** (a, b) consists of all real numbers x such that $a < x < b$. See Figure 5(a).
>
> The **closed interval** $[a, b]$ consists of all real numbers x such that $a \leq x \leq b$. See Figure 5(b).

Notice that the brackets in Figure 5(b) are used to indicate that the numbers a and b are included in the interval $[a, b]$, whereas the parentheses in Figure 5(a) indicate that a and b are excluded from the interval (a, b). At times you'll see notation such as $[a, b)$. This stands for the set of all real numbers x such that $a \leq x < b$. Similarly, $(a, b]$ denotes the set of all numbers x such that $a < x \leq b$.

EXAMPLE 2 Show the following intervals on a number line, and specify inequalities describing the numbers x in each interval.

$$[-1, 2] \qquad (-1, 2) \qquad (-1, 2] \qquad [-1, 2)$$

Solution   See Figure 6.

FIGURE 6

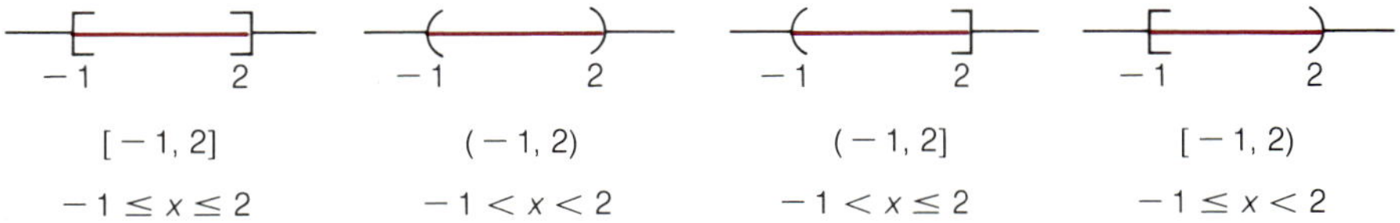

| $[-1, 2]$ | $(-1, 2)$ | $(-1, 2]$ | $[-1, 2)$ |
| $-1 \leq x \leq 2$ | $-1 < x < 2$ | $-1 < x \leq 2$ | $-1 \leq x < 2$ |

In addition to the four types of intervals shown in Figure 6, we can also consider **unbounded intervals.** These are intervals that extend indefinitely in one direction or the other, as shown, for example, in Figure 7. We also have a convenient notation for unbounded intervals. As an example, we indicate the unbounded interval in Figure 7 with the notation $(2, \infty)$.

FIGURE 7

The set of all real numbers x such that $x > 2$.

Comment and Caution   The symbol ∞ is read *infinity*.* It is not a real number, and its use in the context $(2, \infty)$ is only to indicate that the interval has no right-hand boundary. In the box that follows we define the five types of unbounded intervals. Notice that the last interval, $(-\infty, \infty)$, is actually the entire real number line.

PROPERTY SUMMARY   UNBOUNDED INTERVALS

NOTATION	DEFINING INEQUALITY	EXAMPLE
(a, ∞)	$x > a$	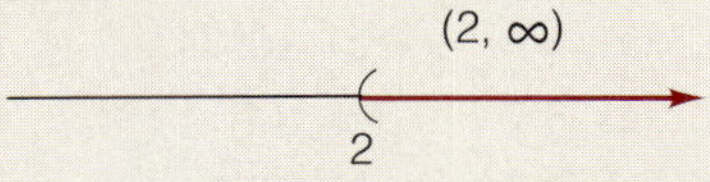 $(2, \infty)$
$[a, \infty)$	$x \geq a$	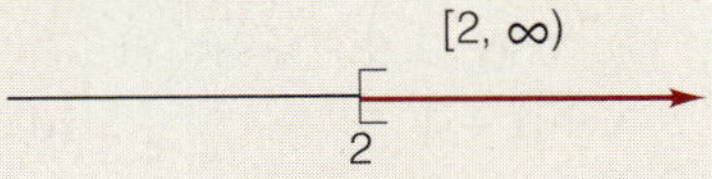 $[2, \infty)$
$(-\infty, a)$	$x < a$	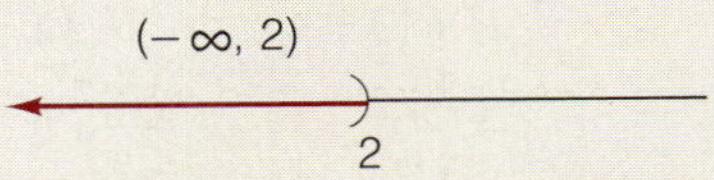 $(-\infty, 2)$
$(-\infty, a]$	$x \leq a$	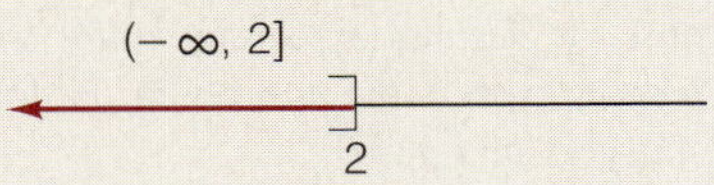 $(-\infty, 2]$
$(-\infty, \infty)$		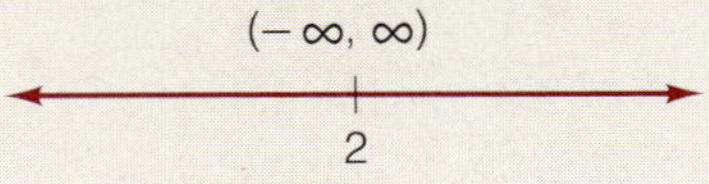 $(-\infty, \infty)$

---

*The mathematician John Wallis (1616–1703) introduced the symbol ∞ in 1655. Some historians, knowing that Wallis was also a classical scholar, have suggested that he derived the infinity symbol from CID, which was one of the Roman notations for 1000.

EXAMPLE 3 Indicate each set of real numbers on a number line.

(a) $(-\infty, 4]$ **(b)** $(-3, \infty)$

Solution **(a)** The interval $(-\infty, 4]$ consists of all real numbers that are less than or equal to 4. See Figure 8.

FIGURE 8
$(-\infty, 4]$

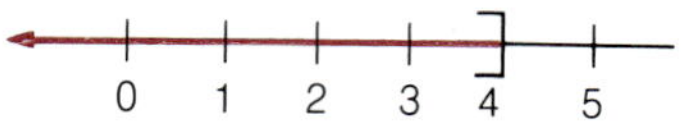

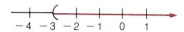

(b) The interval $(-3, \infty)$ consists of all real numbers that are greater than -3. See Figure 9.

FIGURE 9
$(-3, \infty)$

EXERCISE SET 1.3

A

For each of Exercises 1–12, determine if the number is a natural number, an integer, a rational number, or an irrational number. (Some numbers belong to more than one category.)

1. (a) 7   (b) -7
2. (a) -203 (b) $\frac{204}{2}$
3. (a) $\frac{27}{4}$ (b) $\sqrt{\frac{27}{4}}$
4. (a) $\sqrt{7}$ (b) $4\sqrt{7}$
5. (a) 10^6 (b) $10^6/10^7$
6. (a) 8.7   (b) $8.\overline{7}$
7. (a) 8.74   (b) $8.\overline{74}$
8. (a) $\sqrt{99}$ (b) $\sqrt{99} + 1$
9. $3\sqrt{101} + 1$ 10. $(\sqrt{5} + 1)/4$
11. $(3 - \sqrt{2}) + (3 + \sqrt{2})$ 12. $\dfrac{0.1234}{0.5677}$

In Exercises 13–18, each of the given numbers is rational because the decimal expansion either terminates or is periodic. In each case, express the number in the form p/q, where p and q are integers and $q \neq 0$.

13. (a) 5.4   (b) $5.\overline{4}$
14. (a) 2.8   (b) $2.\overline{8}$
15. (a) 0.99   (b) $0.\overline{9}$
16. (a) 0.66   (b) $0.\overline{6}$
17. (a) 1.7   (b) $1.\overline{7}$
18. (a) 9.001   (b) $9.\overline{1}$

In Exercises 19–28, show the approximate location of each number on a number line. For Exercises 19–24, use a number line similar to the one shown in Figure 2. For Exercises 25–28, use a number line similar to the one in Figure 4(a). In Exercises 21 and 22, use a calculator.

19. (a) $\sqrt{3} + 2$ *Suggestion:* You don't need a calculator for this one if you use the approximation $\sqrt{3} \approx$ 1.732. (This fact is easy to remember because 1732 is the year that George Washington was born!)
    (b) $-\sqrt{3}/2$
20. (a) $5 + \sqrt{2}$ (Use the approximation $\sqrt{2} \approx 1.4$.)
    (b) $5 - \sqrt{2}$
21. $\sqrt{3} + \sqrt{7}$ 22. $\sqrt{3} - \sqrt{7}$
23. $-\pi/2$ 24. $\pi/4$
25. 6   26. -2
27. -4 28. 5

For Exercises 29–34, say whether each statement is true or false.

29. $-5 < -100$ 30. $0 \leq -1$
31. $-1 \leq -1$ 32. $\sqrt{2} + \sqrt{5} \geq 0$
33. $\frac{15}{16} > \frac{17}{18}$ 34. $0.\overline{8} > 0.8$

In each of Exercises 35–52, show the intervals on a number line. For Exercises 35–47, specify inequalities describing the numbers x in each interval.

35. $(2, 5)$ 36. $(-2, 2)$ 37. $[1, 4]$
38. $\left[-\frac{3}{2}, \frac{1}{2}\right]$ 39. $[0, 3)$ 40. $(-4, 0]$

41. $(-3, \infty)$ 42. $\left(\sqrt{2}, \infty\right)$ 43. $[-1, \infty)$

44. $[0, \infty)$ 45. $(-\infty, 1)$ 46. $(-\infty, -2)$

47. $(-\infty, \pi]$ 48. $(-\infty, \infty)$

49. All real numbers x such that $x > -2$

50. All real numbers x such that $x \leq 5$

51. All real numbers x such that $-1 < x < 1$

52. All real numbers x such that $x \geq \sqrt{2}$

B

In Exercises 53–56, each of the given numbers is rational because the decimal expansion is periodic. In each case, express the number in the form p/q, where p and q are integers and $q \neq 0$.

53. $0.\overline{19}$ *Hint:* Let $x = 0.\overline{19}$ and consider $100x - x$.

54. $61.\overline{26}$ (See the hint given for Exercise 53.)

55. $0.3121212\ldots$ *Hint:* Let $x = 0.3\overline{12}$ and consider $1000x - 10x$.

56. $0.\overline{142857}$ (Adapt the hint given in Exercise 53.)

57. From grade school on, we all acquire a good deal of experience in working with rational numbers. Certainly we can tell when two rational numbers are equal, although it may require some calculation, as, for example, in the case of $\frac{129}{31} = \frac{2193}{527}$. The point of this exercise is to show you that, in working with irrational numbers, sometimes it's hard to tell (or even to show) that two numbers are equal. In each of the following cases, use a calculator to verify that the quantities on both sides of the equation agree, to six decimal places. (In each case it can be shown that the two numbers are indeed equal.)

(a) $\sqrt{6} + \sqrt{2} = 2\sqrt{2 + \sqrt{3}}$

(b) $\sqrt{3 + \sqrt{5}} + \sqrt{3 - \sqrt{5}} = \sqrt{10}$

(c) $\sqrt{\sqrt{6 + 4\sqrt{2}}} = \sqrt{2 + \sqrt{2}}$

(d) $\sqrt{\frac{3}{2} + \frac{2}{\sqrt{2}}} + \sqrt{\frac{3}{2} - \frac{2}{\sqrt{2}}} = 2$

58. The value of the irrational number π, correct to ten decimal places (without rounding off), is 3.1415926535. By using a calculator, determine to how many decimal places each of the following quantities agrees with π.

(a) $\frac{22}{7}$ (b) $\frac{355}{113}$ (c) $\frac{63}{25}\left(\frac{17 + 15\sqrt{5}}{7 + 15\sqrt{5}}\right)$

Remarks: A simple approximation that agrees with π through the first 14 decimal places is $\frac{355}{113}\left(1 - \frac{0.0003}{3533}\right)$. This approximation was discovered by the Indian mathematician Srinivasa Ramanujan (1887–1920). For a fascinating account of the history of π, see the book by Petr Beckmann, *A History of π*, 3rd ed. (New York: St. Martin's Press, 1974).

1.4 ABSOLUTE VALUE

There has been a real need in analysis for a convenient symbolism for "absolute value" … and the two vertical bars introduced in 1841 by Weierstrass, as in $|z|$, have met with wide adoption; …

Florian Cajori in *A History of Mathematical Notations*, vol. 1 (La Salle, Illinois: The Open Court Publishing Company, 1928)

As an aid in measuring distances on the number line, we introduce the concept of *absolute value*. We will begin with a definition of absolute value that is geometric in nature. Then, after you have developed some familiarity with the concept, we will explain a more algebraic approach that is often useful in analytical work.

DEFINITION Absolute Value (geometric version)

The **absolute value** of a real number x, denoted by $|x|$, is the distance from x to the origin.

For instance, because the numbers 5 and -5 are both five units from the origin, we have $|5| = 5$ and $|-5| = 5$. Here are three more examples:

$$|17| = 17 \qquad \left|-\frac{2}{3}\right| = \frac{2}{3} \qquad |0| = 0$$

In dealing with an expression such as $|-5 + 3|$, the convention is to compute the quantity $-5 + 3$ first and then to take the absolute value. We therefore have, in this case.

$$|-5 + 3| = |-2| = 2$$

EXAMPLE 1   Evaluate each expression.

(a) $5 - |6 - 7|$ (b) $||-2| - |-3||$

Solution   (a) $5 - |6 - 7| = 5 - |-1|$ (b) $||-2| - |-3|| = |2 - 3|$
$$= 5 - 1 = 4 \qquad\qquad\qquad = |-1| = 1$$

As we said at the beginning of this section, there is an equivalent, more algebraic way to define absolute value. According to this equivalent definition, the value of $|x|$ is x itself when $x \geq 0$, and the value of $|x|$ is $-x$ when $x < 0$. We can write this symbolically as follows:

DEFINITION   Absolute Value (algebraic version)

$$|x| = \begin{cases} x & \text{when } x \geq 0 \\ -x & \text{when } x < 0 \end{cases}$$

EXAMPLE

$$|-7| = -(-7) = 7$$

By looking at examples with specific numbers, you should be able to convince yourself that both definitions yield the same results. We use the algebraic definition of absolute value in Examples 2 and 3.

EXAMPLE 2   Rewrite each expression in a form that does not contain absolute values.

(a) $|\pi - 4| + 1$
(b) $|x - 5|$ given that $x \geq 5$
(c) $|t - 5|$ given that $t < 5$

Solution   (a) The quantity $\pi - 4$ is negative (since $\pi \approx 3.14$), and therefore its absolute value is equal to $-(\pi - 4)$. In view of this, we have

$$|\pi - 4| + 1 = -(\pi - 4) + 1 = -\pi + 5$$

(b) Since $x \geq 5$, the quantity $x - 5$ is nonnegative, and therefore its absolute value is equal to $x - 5$ itself. Thus, we have

$$|x - 5| = x - 5 \quad \text{when } x \geq 5$$

(c) Since $t < 5$, the quantity $t - 5$ is negative. Therefore its absolute value is equal to $-(t - 5)$, which in turn is equal to $5 - t$. In view of this, we have

$$|t - 5| = 5 - t \quad \text{when } t < 5$$ ▮▮▮

EXAMPLE 3 Simplify the expression $|x - 1| + |x - 2|$, given that x is in the open interval $(1, 2)$.

Solution Since x is greater than 1, the quantity $x - 1$ is positive and, consequently,

$$|x - 1| = x - 1$$

On the other hand, we are also given that x is less than 2. Therefore the quantity $x - 2$ is negative, and we have

$$|x - 2| = -(x - 2) = -x + 2$$

Putting things together now, we can write

$$|x - 1| + |x - 2| = (x - 1) + (-x + 2)$$
$$= -1 + 2 = 1$$ ▮▮▮

The solutions in Examples 2 and 3 used the algebraic version of the definition of absolute value. In the error box that follows, the first two items indicate possible misuses of this definition. *Suggestion:* Before reading further, cover up the columns headed "Correction" and "Comment" in the box and try to decide for yourself where the error lies.

Errors to Avoid

ERROR	CORRECTION	COMMENT
$\lvert 3 - \sqrt{2} \rvert \neq -(3 - \sqrt{2}) = -3 + \sqrt{2}$	$\lvert 3 - \sqrt{2} \rvert = 3 - \sqrt{2}$	Because the quantity $3 - \sqrt{2}$ is positive, the appropriate formula is $\lvert x \rvert = x$, not $\lvert x \rvert = -x$.
$\lvert x - y \rvert \neq -(x - y) = -x + y$	$\lvert x - y \rvert = \begin{cases} x - y & \text{if } x - y \geq 0 \\ -(x - y) & \text{if } x - y < 0 \end{cases}$	The error arises in assuming that the quantity $x - y$ is negative. Unless additional information is available, $\lvert x - y \rvert$ cannot be simplified.
$\lvert a + b \rvert \neq \lvert a \rvert + \lvert b \rvert$	$\lvert a + b \rvert = \begin{cases} a + b & \text{if } a + b \geq 0 \\ -(a + b) & \text{if } a + b < 0 \end{cases}$	The absolute value of a sum is, in general, not equal to the sum of the individual absolute values. For example, if $a = -1$ and $b = 2$, then $\lvert a + b \rvert = \lvert -1 + 2 \rvert = \lvert 1 \rvert = 1$, but $\lvert a \rvert + \lvert b \rvert = \lvert -1 \rvert + \lvert 2 \rvert = 3$.

In the box that follows, we list several basic properties of the absolute value. Each of these properties can be derived from the definitions. (With the exception of the *triangle inequality,* we shall omit the derivations. For a proof of the triangle inequality, see Exercise 71.)

PROPERTY SUMMARY   PROPERTIES OF ABSOLUTE VALUE

1. For all real numbers x, we have
(a) $|x| \geq 0$;
(b) $x \leq |x|$ and $-x \leq |x|$.

2. For all real numbers a and b, we have
(a) $|ab| = |a||b|$ and $|a/b| = |a|/|b|$ $(b \neq 0)$;
(b) $|a + b| \leq |a| + |b|$ (the triangle inequality).

EXAMPLE 4 Write the expression $|-2 - x^2|$ in an equivalent form that does not contain absolute values.

Solution We have

$$|-2 - x^2| = |-1(2 + x^2)|$$
$$= |-1||2 + x^2| \qquad \text{using Property 2(a)}$$
$$= 2 + x^2$$

The last equality follows from the fact that x^2 is nonnegative for any real number x and, consequently, the quantity $2 + x^2$ is positive. ∎

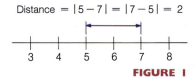

Distance $= |5 - 7| = |7 - 5| = 2$

FIGURE 1

If we think of the real numbers as points on a number line, the distance between two numbers a and b is given by the absolute value of their difference. For instance, as indicated in Figure 1, the distance between 5 and 7, namely 2, is given by either $|5 - 7|$ or $|7 - 5|$. For reference, we summarize this simple but important fact as follows.

PROPERTY SUMMARY   DISTANCE ON A NUMBER LINE

The **distance** between a and b is $|a - b| = |b - a|$.

EXAMPLE 5 Rewrite each of the following statements using absolute values.

(a) The distance between 12 and -5 is 17.
(b) The distance between x and 2 is 4.
(c) The distance between x and 2 is less than 4.
(d) The point t is more than five units from the origin.

Solution (a) $|12 - (-5)| = 17$ or $|-5 - 12| = 17$
(b) $|x - 2| = 4$ or $|2 - x| = 4$
(c) $|x - 2| < 4$ or $|2 - x| < 4$
(d) $|t| > 5$ ∎

EXAMPLE 6 In each case, the set of real numbers satisfying the given inequality is one or more intervals on the number line. Show the intervals on a number line.

(a) $|x| < 2$ (b) $|x| > 2$ (c) $|x - 3| < 1$ (d) $|x - 3| \geq 1$

Solution   (a) The given inequality tells us that the distance from x to the origin is less then two units. So, as indicated in Figure 2(a), the point x must lie in the open interval $(-2, 2)$.

(b) The condition $|x| > 2$ means that x is more than two units from the origin. Thus, as indicated in Figure 2(b), the point x lies either to the right of 2 or to the left of -2.

(c) The given inequality tells us that x must be less than one unit away from 3 on the number line. Looking one unit to either side of 3, then, we see that x must lie between 2 and 4 and x cannot equal 2 or 4. See Figure 2(c).

(d) The inequality $|x - 3| \geq 1$ says that x is at least one unit away from 3 on the number line. This means that either $x \geq 4$ or $x \leq 2$, as shown in Figure 2(d). [Here's an alternate way of thinking about this. The numbers satisfying the given inequality are precisely those numbers that *do not* satisfy the inequality in part (c). So for part (d), we need to shade that portion of the number line that was not shaded in part (c).]

FIGURE 2

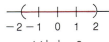
(a) $|x| < 2$

(b) $|x| > 2$

(c) $|x - 3| < 1$ (d) $|x - 3| \geq 1$

EXERCISE SET 1.4

A

In Exercises 1–16, evaluate each expression.

1. $|3|$
2. $3 + |-3|$
3. $|-6|$
4. $-6 - |-6|$
5. $|-1 + 3|$
6. $|-6 + 3|$
7. $\left|-\frac{4}{5}\right| - \frac{4}{5}$
8. $\left|\frac{4}{5}\right| - \frac{4}{5}$
9. $|-6 + 2| - |4|$
10. $|-3 - 4| - |-4|$
11. $||-8| + |-9||$
12. $||-8| - |-9||$
13. $\left|\dfrac{27 - 5}{5 - 27}\right|$
14. $\dfrac{|27 - 5|}{|5 - 27|}$
15. $|7(-8)| - |7||-8|$
16. $|(-7)^2| + |-7|^2 - (-|-3|)^3$

In Exercises 17–24, evaluate each expression, given that $a = -2$, $b = 3$, and $c = -4$.

17. $|a - b|^2$
18. $a^2 - |bc|$
19. $|c| - |b| - |a|$
20. $|b + c| - |b| - |c|$
21. $|a + b|^2 - |b + c|^2$
22. $\dfrac{|a| + |b| + |c|}{|a + b + c|}$
23. $\dfrac{a + b + |a - b|}{2}$
24. $\dfrac{a + b - |a - b|}{2}$

In Exercises 25–38, rewrite each expression in a form that does not contain absolute values.

25. $\left|\sqrt{2} - 1\right| - 1$
26. $\left|1 - \sqrt{2}\right| + 1$
27. $|x - 3|$ given that $x \geq 3$
28. $|x - 3|$ given that $x < 3$
29. $|t^2 + 1|$
30. $|x^4 + 1|$
31. $\left|-\sqrt{3} - 4\right|$
32. $\left|-\sqrt{3} - \sqrt{5}\right|$
33. $|x - 3| + |x - 4|$ given that $x < 3$
34. $|x - 3| + |x - 4|$ given that $x > 4$
35. $|x - 3| + |x - 4|$ given that $3 < x < 4$
36. $|x - 3| + |x - 4|$ given that $x = 4$
37. $|x + 1| + 4|x + 3|$ given that $-\frac{5}{2} < x < -\frac{3}{2}$
38. $|x + 1| + 4|x + 3|$ given that $x < -3$

In Exercises 39–50, rewrite each statement using absolute values, as in Example 5.

39. The distance between x and 4 is 8.
40. The distance between x and 4 is at least 8.
41. The distance between x and 1 is $\frac{1}{2}$.
42. The distance between x and 1 is less than $\frac{1}{2}$.
43. The distance between x and 1 is at least $\frac{1}{2}$.

44. The distance between x and 1 exceeds $\frac{1}{2}$.

45. The distance between y and -4 is less than 1.

46. The distance between x^3 and -1 is at most 0.001.

47. The number y is less than three units from the origin.

48. The number y is less than one unit from the number t.

49. The distance between x^2 and a^2 is less than M.

50. The sum of the distances of a and b from the origin is greater than or equal to the distance of $a + b$ from the origin.

In Exercises 51–62, the set of real numbers satisfying the given inequality is one or more intervals on the number line. Show the intervals on a number line.

51. $|x| < 4$
52. $|x| < 2$
53. $|x| > 1$
54. $|x| > 0$
55. $|x - 5| < 3$
56. $|x - 4| < 4$
57. $|x - 3| \leq 4$
58. $|x - 1| \leq \frac{1}{2}$
59. $\left|x + \frac{1}{3}\right| < \frac{3}{2}$
60. $\left|x + \frac{\pi}{2}\right| > 1$
61. $|x - 5| \geq 2$
62. $|x + 5| \geq 2$

B

63. In parts (a) and (b), sketch the interval or intervals corresponding to the given inequality.
    (a) $|x - 2| < 1$
    (b) $0 < |x - 2| < 1$
    (c) In what way do your answers in (a) and (b) differ? (The distinction is important in the study of *limits* in calculus.)

64. Given two real numbers a and b, the notation **max(a, b)** denotes the larger of the two numbers. For instance, if $a = 1$ and $b = 2$, then max(1, 2) = 2. In cases where $a = b$, then max(a, b) denotes the common value of a and b. It can be shown (see Exercise 69) that max(a, b) can be expressed in terms of absolute value as follows:

$$\max(a, b) = \frac{a + b + |a - b|}{2}$$

Verify this equation in each of the following cases.
    (a) $a = 1$ and $b = 2$
    (b) $a = -1$ and $b = -5$
    (c) $a = b = 10$

65. Given two real numbers a and b, the notation **min(a, b)** denotes the smaller of the two numbers. In cases where $a = b$, then min(a, b) denotes the common value of a and b. It can be shown (see Exercise 70)

that min (a, b) can be expressed in terms of absolute value as follows:

$$\min(a, b) = \frac{a + b - |a - b|}{2}$$

Verify this equation in each of the following cases.
    (a) $a = 6$ and $b = 1$
    (b) $a = 1$ and $b = -6$
    (c) $a = b = -6$

66. Show that for all real numbers a and b, we have

$$|a| - |b| \leq |a - b|$$

Hint: Beginning with the identity $a = (a - b) + b$, take the absolute value of each side and then use the triangle inequality.

67. Show that

$$|a + b + c| \leq |a| + |b| + |c|$$

for all real numbers a, b, and c.   *Hint:*   The left-hand side can be written $|a + (b + c)|$. Now use the triangle inequality.

68. Explain why there are no real numbers that satisfy the equation $|x^2 + 4x| = -12$.

C

69. (As background for this exercise, you'll need to have worked Exercise 64.) Prove that

$$\max(a, b) = \frac{a + b + |a - b|}{2}$$

Hint: Consider three separate cases: $a = b$; $a > b$; $b > a$.

70. (As background for this exercise, you'll need to have worked Exercise 65.) Prove that

$$\min(a, b) = \frac{a + b - |a - b|}{2}$$

71. Complete the following steps to prove the triangle inequality.
    (a) Let a and b be real numbers. Which property in the box on page 20 tells us that $a \leq |a|$ and $b \leq |b|$?
    (b) Add the two inequalities in part (a) to obtain $a + b \leq |a| + |b|$.
    (c) In a similar fashion, add the two inequalities $-a \leq |a|$ and $-b \leq |b|$ and deduce that $-(a + b) \leq |a| + |b|$.
    (d) Why do the results in parts (b) and (c) imply that $|a + b| \leq |a| + |b|$?

1.5 INTEGER EXPONENTS. SCIENTIFIC NOTATION

I write a^{-1}, a^{-2}, a^{-3}, etc., for $\dfrac{1}{a}$, $\dfrac{1}{aa}$, $\dfrac{1}{aaa}$, etc.

Isaac Newton (June 13, 1676)

In Section 1.1 we defined the exponential notation a^n, where a is a real number and n is a natural number. Since that definition is basic to all that follows in this and the next two sections, we repeat it here.

DEFINITION 1 Base and Exponent

Given a real number a and a natural number n, we define a^n by

$$a^n = \underbrace{a \cdot a \cdot a \cdots a}_{n \text{ factors}}$$

In the expression a^n, the number a is the **base** and n is the **exponent** or **power** to which the base is raised.

EXAMPLE 1 Rewrite each expression using algebraic notation.

(a) x to the fourth power, plus five
(b) the fourth power of the quantity x plus five
(c) x plus y, to the fourth power
(d) x plus y to the fourth power

Solution (a) $x^4 + 5$ (b) $(x + 5)^4$ (c) $(x + y)^4$ (d) $x + y^4$ ▮▮▮

CAUTION Note that (c) and (d) differ only in the use of the comma. So, for spoken purposes, the idea in (c) would be more clearly conveyed by saying "the fourth power of the quantity x plus y." In the case of (d), if you read it aloud to another student or your instructor, chances are you'll be asked, "Do you mean $x + y^4$ or do you mean $(x + y)^4$?"

MORAL Algebra is a precise language; use it carefully. Learn to ask yourself if what you've written will be interpreted in the manner you intended.

In basic algebra the four properties in the box at the top of page 24 are developed for working with exponents that are natural numbers. Each of these properties is a direct consequence of the definition of a^n. For instance, according to the first property we have $a^2 a^3 = a^5$. To verify that this is indeed correct, we note that

$$a^2 a^3 = (aa)(aaa) = a^5$$

PROPERTY SUMMARY   **PROPERTIES OF EXPONENTS**

PROPERTY	EXAMPLES
1. $a^m a^n = a^{m+n}$	$a^5 a^6 = a^{11};$ $(x+1)(x+1)^2 = (x+1)^3$
2. $(a^m)^n = a^{mn}$	$(2^3)^4 = 2^{12};$ $[(x+1)^2]^3 = (x+1)^6$
3. $\dfrac{a^m}{a^n} = \begin{cases} a^{m-n} & \text{if } m > n \\[4pt] \dfrac{1}{a^{n-m}} & \text{if } m < n \\[4pt] 1 & \text{if } m = n \end{cases}$	$\dfrac{a^6}{a^2} = a^4;$ $\dfrac{a^2}{a^6} = \dfrac{1}{a^4};$ $\dfrac{a^5}{a^5} = 1$
4. $(ab)^m = a^m b^m;$ $\left(\dfrac{a}{b}\right)^m = \dfrac{a^m}{b^m}$	$(2x^2)^3 = 2^3 \cdot (x^2)^3 = 8x^6;$ $\left(\dfrac{x^2}{y^3}\right)^4 = \dfrac{x^8}{y^{12}}$

Now we want to extend our definition of a^n to allow for exponents that are integers but not necessarily natural numbers. We begin by defining a^0.

DEFINITION 2   Zero Exponent

For any nonzero real number a,   EXAMPLES

$\qquad a^0 = 1$

$(0^0$ is not defined$)$

(a) $2^0 = 1$
(b) $(-\pi)^0 = 1$
(c) $\left(\dfrac{3}{1 + a^2 + b^2}\right)^0 = 1$

It's easy to see the motivation for defining a^0 to be 1. Assuming that the exponent zero is to have the same properties as do exponents that are natural numbers, we can write

$$a^0 a^n = a^{0+n}$$

That is,

$$a^0 a^n = a^n$$

Now divide both sides of this last equation by a^n to obtain $a^0 = 1$, which agrees with our definition.

Our next definition (see the box on the top of page 25) assigns a meaning to the expression a^{-n} when n is a natural number. Again, it's easy to see the motivation for this definition. We have

$$a^n a^{-n} = a^{n+(-n)} = a^0 = 1$$

That is,

$$a^n a^{-n} = 1$$

Now divide both sides of this last equation by a^n to obtain $a^{-n} = 1/a^n$, in agreement with Definition 3.

DEFINITION 3 Negative Exponent

EXAMPLES

$$a^{-n} = \frac{1}{a^n}$$

where $a \neq 0$ and n is a natural number

(a) $2^{-1} = \frac{1}{2^1} = \frac{1}{2}$

(b) $\left(\frac{1}{10}\right)^{-1} = \frac{1}{\left(\frac{1}{10}\right)^1} = 10$

(c) $x^{-2} = \frac{1}{x^2}$

(d) $(a^2 b)^{-3} = \frac{1}{(a^2 b)^3} = \frac{1}{a^6 b^3}$

(e) $\frac{1}{2^{-3}} = \frac{1}{\frac{1}{2^3}} = 2^3 = 8$

It can be shown that the four properties of exponents that we listed earlier continue to hold now for all integer exponents. We make use of this fact in the next three examples.

EXAMPLE 2 Simplify the following expression. Write the answer in such a way that only positive exponents appear.

$$(a^2 b^3)^2 (a^5 b)^{-1}$$

Solution FIRST METHOD

$$(a^2 b^3)^2 (a^5 b)^{-1} = (a^2 b^3)^2 \cdot \frac{1}{a^5 b}$$

$$= \frac{a^4 b^6}{a^5 b} = \frac{b^{6-1}}{a^{5-4}} = \frac{b^5}{a}$$

ALTERNATE METHOD

$$(a^2 b^3)^2 (a^5 b)^{-1} = (a^4 b^6)(a^{-5} b^{-1})$$

$$= a^{4-5} b^{6-1} = a^{-1} b^5$$

$$= \frac{b^5}{a} \qquad \text{as obtained previously}$$

EXAMPLE 3 Simplify the following expression, writing the answer so that negative exponents are not used.

$$\left(\frac{a^{-5} b^2 c^0}{a^3 b^{-1}}\right)^3$$

Solution We show two solutions. The first makes immediate use of the property $(a^m)^n = a^{mn}$. In the second solution we begin by working within the parentheses.

FIRST SOLUTION

$$\left(\frac{a^{-5} b^2 c^0}{a^3 b^{-1}}\right)^3 = \frac{a^{-15} b^6}{a^9 b^{-3}}$$

$$= \frac{b^{6-(-3)}}{a^{9-(-15)}} = \frac{b^9}{a^{24}}$$

ALTERNATE SOLUTION

$$\left(\frac{a^{-5} b^2 c^0}{a^3 b^{-1}}\right)^3 = \left(\frac{b^3}{a^8}\right)^3$$

$$= \frac{b^9}{a^{24}}$$

EXAMPLE 4 Simplify the following expressions, writing the answers so that negative exponents are not used. (Assume that p and q are natural numbers.)

(a) $\dfrac{a^{4p+q}}{a^{p-q}}$ (b) $(a^p b^q)^2 (a^{3p} b^{-q})^{-1}$

Solution (a) $\dfrac{a^{4p+q}}{a^{p-q}} = a^{4p+q-(p-q)}$

$$= a^{4p+q-p+q} = a^{3p+2q}$$

(b) $(a^p b^q)^2 (a^{3p} b^{-q})^{-1} = (a^{2p} b^{2q})(a^{-3p} b^q)$

$$= a^{-p} b^{3q}$$

$$= \dfrac{b^{3q}}{a^p}$$

EXAMPLE 5 Use the properties of exponents to compute the quantity $\dfrac{2^{10} \cdot 3^{13}}{27 \cdot 6^{12}}$.

Solution $\dfrac{2^{10} \cdot 3^{13}}{27 \cdot 6^{12}} = \dfrac{2^{10} \cdot 3^{13}}{(3^3)(3 \cdot 2)^{12}}$

$$= \dfrac{2^{10} \cdot 3^{13}}{3^3 \cdot 3^{12} \cdot 2^{12}} = \dfrac{2^{10} \cdot 3^{13}}{3^{15} \cdot 2^{12}}$$

$$= \dfrac{1}{(3^{15-13})\,(2^{12-10})} = \dfrac{1}{3^2 \cdot 2^2} = \dfrac{1}{36}$$

In Examples 2 through 5 we used the definitions and properties of exponents to simplify expressions. The box that follows shows some common errors to avoid in working these types of problems. *Suggestion:* Before reading further, cover up the columns headed "Correction" and "Comment" in the box and try to decide for yourself where the error lies.

Errors to Avoid

ERROR	CORRECTION	COMMENT
$x^2 x^3 \neq x^6$	$x^2 x^3 = x^5$	Add the exponents. (This is Property 1 in the summary box on page 24.)
$(2^x)(2^x) \neq 4^x$	$(2^x)(2^x) = 2^{2x}$	Add the exponents, not the bases.
$(y^2)^3 \neq y^5$	$(y^2)^3 = y^6$	Multiply the exponents. (This is Property 2 in the summary box on page 24.)
$\dfrac{2^{12}}{2^4} \neq 2^3$	$\dfrac{2^{12}}{2^4} = 2^8$	Subtract, don't divide the exponents. (This is Property 3 in the summary box.)
$(3a)^2 \neq 3a^2$	$(3a)^2 = 9a^2$	Because of the parentheses, the exponent applies to both factors. (This is Property 4 in the summary box.)
$x^{-3} \neq -\dfrac{1}{x^3}$	$x^{-3} = \dfrac{1}{x^3}$	The error involves a misuse of the definition (on page 25) of negative exponents.
$(a+b)^{-1} \neq \dfrac{1}{a} + \dfrac{1}{b}$	$(a+b)^{-1} = \dfrac{1}{a+b}$	Because of the parentheses, the exponent applies to the *quantity* $a + b$.

As an application of some of the ideas in this section, we briefly discuss **scientific notation**, which is a convenient form for writing very large or very small numbers. Such numbers occur often in the sciences. For instance, the speed of light in a vacuum is

29,979,000,000 cm/sec

As written, this number would be awkward to work with in calculating. In fact, the number as written cannot even be displayed on a hand-held calculator, since there are too many digits. To write the number 29,979,000,000 (or any positive number) in scientific notation, we express it as a number between 1 and 10, multiplied by an appropriate power of 10. That is, we write it in the form

$$a \times 10^n \qquad \text{where } 1 \le a < 10 \quad \text{and} \quad n \text{ is an integer}$$

According to this convention, the number 4.03×10^6 is in scientific notation, but the same quantity written as 40.3×10^5 is not in scientific notation. In order to convert a given number into scientific notation, we'll rely on the following two-step procedure.

To Express a Number Using Scientific Notation

1. First move the decimal point until it is to the immediate right of the first non-zero digit.

2. Then multiply by 10^n or 10^{-n}, depending on whether the decimal point was moved n places to the left or to the right, respectively.

For example, to express the number 29,979,000,000 in scientific notation, first move the decimal point 10 places to the left so that it's located between the 2 and the 9. Then multiply by 10^{10}. The result is

$$29{,}979{,}000{,}000 = 2.9979000000 \times 10^{10}$$

or, more simply,

$$29{,}979{,}000{,}000 = 2.9979 \times 10^{10}$$

As additional examples we list the following numbers expressed in both ordinary and scientific notation. For practice you should verify each conversion for yourself using our two-step procedure.

$$55708 = 5.5708 \times 10^4$$
$$0.000099 = 9.9 \times 10^{-5}$$
$$0.0000002 = 2 \times 10^{-7}$$

All scientific calculators have keys for working with numbers in scientific notation. We'll show three representative examples here using the number 3.519×10^{20}. (This gargantuan number is the diameter, in miles, of a galaxy at the center of a distant star cluster known as Abell 2029. As of 1991, this is the largest galaxy ever discovered.) If you have questions about how your calculator operates with respect to scientific notation, you should consult the owner's manual. (Also in the owner's manual, you will find that the calculator can convert numbers expressed in ordinary decimal notation into scientific notation.)

CALCULATOR	SEQUENCE OF KEYSTROKES
Casio fx-300V	3.519 (EXP) 20
Hewlett Packard 11C	3.519 (EEX) 20 (ENTER)
Texas Instruments TI-81	3.519 (EE) 20

EXERCISE SET 1.5

A

In Exercises 1–6, evaluate each expression using the given value of x.

1. $2x^3 - x + 4$; $x = -2$

2. $1 - x + 2x^2 - 3x^3$; $x = -1$

3. $\dfrac{1 - 2x^2}{1 + 2x^3}$; $x = -\dfrac{1}{2}$

4. $\dfrac{1 - (x - 1)^2}{1 + (x - 1)^2}$; $x = -1$

5. $\dfrac{x^2 + x^3 - x^x}{2^x + 3^x - (x + 1)^2}$; $x = 2$

6. $\dfrac{2^{x-1} - \dfrac{1}{2^{x-1}}}{2^{x-1} + \dfrac{1}{2^{x-1}}}$; $x = 3$

For Exercises 7–14, rewrite each expression using algebraic notation.

7. The square of the quantity x plus y

8. The square of the product of x and y

9. Three more than the square of the quantity x plus y

10. The square of x, minus twice the cube of y

11. The square of half of the quantity $x^2 - 2y^3$

12. The cube of the average of x^2 and $-2y^2$

13. The cube of the absolute value of the quantity $x - 1$

14. The absolute value of the fifth power of the average of $-x^2$ and y^2

In Exercises 15–26, use the properties of exponents to simplify each expression.

15. (a) $a^3 a^{12}$ (b) $(a + 1)^3 (a + 1)^{12}$
     (c) $(a + 1)^{12} (a + 1)^3$

16. (a) $(3^2)^3 - (2^3)^2$ (b) $(x^2)^3 - (x^3)^2$ (c) $(x^2)^a - (x^a)^2$

17. (a) $yy^2 y^8$ (b) $(y + 1)(y + 1)^2 (y + 1)^8$
     (c) $[(y + 1)(y + 1)^8]^2$

18. (a) $\dfrac{x^{12}}{x^{10}}$ (b) $\dfrac{x^{10}}{x^{12}}$ (c) $\dfrac{2^{10}}{2^{12}}$

19. (a) $\dfrac{(x^2 + 3)^{10}}{(x^2 + 3)^9}$ (b) $\dfrac{(x^2 + 3)^9}{(x^2 + 3)^{10}}$ (c) $\dfrac{12^{10}}{12^9}$

20. (a) $(y^2 y^3)^2$ (b) $y^2 (y^3)^2$ (c) $2^m (2^n)^2$

21. (a) $\dfrac{t^{15}}{t^9}$ (b) $\dfrac{t^9}{t^{15}}$ (c) $\dfrac{(t^2 + 3)^{15}}{(t^2 + 3)^9}$

22. (a) $\dfrac{x^6}{x}$ (b) $\dfrac{x}{x^6}$ (c) $\dfrac{(x^2 + 3x - 2)^6}{x^2 + 3x - 2}$

23. (a) $\dfrac{x^6 y^{15}}{x^2 y^{20}}$ (b) $\dfrac{x^2 y^{20}}{x^6 y^{15}}$ (c) $\left(\dfrac{x^2 y^{20}}{x^6 y^{15}}\right)^2$

24. (a) $(x^2 y^3 z)^4$ (b) $2(x^2 y^3 z)^4$ (c) $(2x^2 y^3 z)^4$

25. (a) $4(x^3)^2$ (b) $(4x^3)^2$ (c) $\dfrac{(4x^2)^3}{(4x^3)^2}$

26. (a) $2(x - 1)^7 - (x - 1)^7$
     (b) $[2(x - 1)]^7 - (x - 1)^7$
     (c) $2(x - 1)^7 - [2(x - 1)]^7$

For Exercises 27–54, simplify each expression. Write the answers in such a way that negative exponents do not appear. In Exercises 49–54, assume that the letters p and q represent natural numbers.

27. (a) 64^0 (b) $(64^3)^0$ (c) $(64^0)^3$

28. (a) $2^0 + 3^0$ (b) $(2^0 + 3^0)^3$ (c) $(2^3 + 3^3)^0$

29. (a) $10^{-1} + 10^{-2}$ (b) $(10^{-1} + 10^{-2})^{-1}$
     (c) $[(10^{-1})(10^{-2})]^{-1}$

30. (a) $4^{-2} + 4^{-1}$ (b) $\left[\left(\tfrac{1}{4}\right)^{-1} + \left(\tfrac{1}{4}\right)^{-2}\right]^{-1}$
     (c) $\left[\left(\tfrac{1}{4}\right)^{-1}\left(\tfrac{1}{4}\right)^{-2}\right]^{-1}$

31. $\left(\tfrac{1}{3}\right)^{-1} + \left(\tfrac{1}{4}\right)^{-1}$ 32. $\left(\tfrac{1}{3} + \tfrac{1}{4}\right)^{-1}$

33. $\left(\tfrac{2}{3} + \tfrac{3}{2}\right)^{-1}$ 34. $\left(\tfrac{2}{3}\right)^{-1} + \left(\tfrac{3}{2}\right)^{-1}$

35. (a) $5^{-2} + 10^{-2}$
(b) $(5 + 10)^{-2}$

36. (a) $\left(\frac{1}{4}\right)^{-2} + \left(\frac{3}{4}\right)^{-2}$
(b) $\left(\frac{1}{4} + \frac{3}{4}\right)^{-2}$

37. $(a^2 b c^0)^{-3}$

38. $(a^3 b)^3 \, (a^2 b^4)^{-1}$

39. $(a^{-2} b^{-1} c^3)^{-2}$

40. $(2^{-2} + 2^{-1} + 2^0)^{-2}$

41. $\left(\dfrac{x^3 y^{-2} z}{x y^2 z^{-3}}\right)^{-3}$

42. $\left(\dfrac{x^4 y^{-8} z^2}{x y^2 z^{-6}}\right)^2$

43. $\left(\dfrac{x^4 y^{-8} z^2}{x y^2 z^{-6}}\right)^{-2}$

44. $\left(\dfrac{a^3 b^{-9} c^2}{a^5 b^2 c^{-4}}\right)^0$

45. $\left(\dfrac{a^{-2} b^{-3} c^{-4}}{a^2 b^3 c^4}\right)^2$

46. $(-2x)^{-3}(-2x^{-2})^{-1}$

47. $\dfrac{x^2}{y^{-3}} \div \dfrac{x^2}{y^3}$

48. $(2x^2)^{-3} - 2(x^2)^{-3}$

49. $\dfrac{b^{p+1}}{b^p}$

50. $\dfrac{b^{p+2q}}{(b^2)^q}$

51. $(x^p)(x^p)$

52. $(2^{p-q})(2^{q-p+1})$

53. $(a^{2q} b^{-p})^3$

54. $1/(a^{4q} b^{-p})^{-4}$

In Exercises 55–58, use the properties of exponents in computing each quantity, as in Example 5. (The point here is to do as little arithmetic as possible.)

55. $\dfrac{2^8 \cdot 3^{15}}{9 \cdot 3^{10} \cdot 12}$

56. $\dfrac{2^{12} \cdot 5^{13}}{10^{12}}$

57. $\dfrac{24^5}{32 \cdot 12^4}$

58. $\left(\dfrac{144 \cdot 125}{2^3 \cdot 3^2}\right)^{-1}$

For Exercises 59–68, express each number in scientific notation.

59. The average distance (in miles) from Earth to the Sun:

92,900,000

60. The average distance (in miles) from the planet Pluto to the Sun:

3,666,000,000

61. The average orbital speed (in miles per hour) of Earth:

66,800

62. The average orbital speed (in miles per hour) of the planet Mercury:

107,300

63. The average distance (in miles) from the Sun to the nearest star:

25,000,000,000,000,000,000

64. The equatorial diameter (in miles) of
(a) Mercury: 3031
(b) Earth: 7927
(c) Jupiter: 88,733
(d) the Sun: 865,000

65. The length of the "year," that is, the time to orbit the Sun once (in terms of Earth days), for the planet
(a) Mercury: 86.688
(b) Saturn: 10604.772
(c) Pluto: 89424

66. The mass (in grams) of
(a) Earth:

6000000000000000000000000000

(b) Jupiter:

1900000000000000000000000000000

67. The time (in seconds) for light to travel
(a) one foot:

0.000000001

(b) across an atom:

0.000000000000000001

(c) across the nucleus of an atom:

0.0000000000000000000000001

68. The mass (in grams) of
(a) a proton:

0.00000000000000000000000167

(b) an electron:

0.000000000000000000000000000911

B

For Exercises 69–72, use the properties of exponents to simplify each expression.

69. $\dfrac{a^{3x+y}}{a^{2x} \cdot a^{x+y}}$

70. $\dfrac{a^{4p+2q}}{a^{3p} a^p (a^q)^2}$

71. $\dfrac{(x^{5n+1})^n}{(x^n)^{5n}} \cdot \dfrac{1}{x^{n-2}}$

72. $\dfrac{x^{3a+2b-c}}{(x^{2a})\,(x^b)} \cdot x^{3c-a-b}$

73. Use the following data to compute the time required, in minutes, for light to travel from the Sun to Earth. Round off your answer to the nearest minute.

speed of light: 2.9979×10^{10} cm/sec

distance from the Sun to Earth: 92.9×10^6 miles

1 mile = 160930 cm

74. Each planet in our solar system travels around the Sun in an elliptical orbit. Let T denote the time for a planet to complete one orbit around the Sun. For example, for Earth, $T = 365$ days. (Actually, more precisely, $T = 365.26$ days.) Let a denote the length of the *semimajor axis* of the ellipse. (See the figure.)

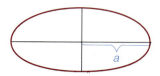

Planet	a (km)	T (Earth Days)	$\dfrac{a^3}{T^2}$
Mercury	5.791×10^7	87.95	
Venus	1.082×10^8	224.71	
Earth	1.496×10^8	365.26	
Mars	2.279×10^8	687.02	
Jupiter	7.783×10^8	4332.79	
Saturn	1.427×10^9	10759.72	
Uranus	2.869×10^9	30686.59	
Neptune	4.497×10^9	60191.2	
Pluto	5.913×10^9	90803.6	

In 1618 the German astronomer Johannes Kepler announced his Third Law of Planetary Motion, which stated that the ratio a^3/T^2 is the same for all planets in the solar system. Verify Kepler's Third Law empirically by completing the following table. Express the required values of a^3/T^2 in scientific notation, rounding off to three significant digits. (See Appendix A1 for the definition of significant digits.)

C

75. Suppose that $p = b^x$, $q = b^y$, and $b^2 = (p^y q^x)^z$, where all the letters denote natural numbers and $b \neq 1$. Show that $xyz = 1$.

76. Which pair of numbers is closer together: 10^{27} and 10^{98} or 10^{105} and 10^{106}?

1.6 nTH ROOTS

Hindu mathematicians first recognized negative roots, and the two square roots of a positive number, ... though they were suspicious also.

David Wells in *The Penguin Dictionary of Curious and Interesting Numbers* (Harmondsworth, Middlesex, England; Penguin Books Ltd., 1986)

In this section we generalize the notion of square root that you studied in elementary algebra. The new idea is that of an nth root; the definition and some basic examples are given in the box that follows.

DEFINITION 1 nth Roots

	EXAMPLES
Let n be a natural number. If a and b are real numbers and $$a^n = b$$ then we say that a is an **nth root** of b. When $n = 2$ and when $n = 3$, we refer to the roots as **square roots** and **cube roots**, respectively.	Both 3 and -3 are square roots of 9 because $3^2 = 9$ and $(-3)^2 = 9$. Both 2 and -2 are fourth roots of 16 because $2^4 = 16$ and $(-2)^4 = 16$. 2 is a cube root of 8 because $2^3 = 8$. -3 is a fifth root of -243 because $(-3)^5 = -243$.

As the examples in the box suggest, square roots, fourth roots, and all *even* roots of positive numbers occur in pairs, one positive and one negative. In these cases, we use the notation $\sqrt[n]{b}$ to denote the positive, or **principal**, *n*th root of *b*. As examples of this notation, we can write

$$\sqrt[4]{81} = 3 \qquad \text{(The principal fourth root of 81 is 3.)}$$
$$\sqrt[4]{81} \neq -3 \qquad \text{(The principal fourth root of 81 is not } -3.)$$
$$-\sqrt[4]{81} = -3 \qquad \text{(The negative of the principal fourth root of 81 is } -3.)$$

The symbol $\sqrt{}$ is called a **radical sign**, and the number within the radical sign is the **radicand**.* The natural number *n* used in the notation $\sqrt[n]{}$ is called the **index** of the radical. For square roots, as you know from basic algebra, we suppress the index and simply write $\sqrt{}$ rather than $\sqrt[2]{}$. So, for example, $\sqrt{25} = 5$.

On the other hand, as we saw in the examples, cube roots, fifth roots, and, in fact, all *odd* roots occur singly, not in pairs. In these cases, we again use the notation $\sqrt[n]{b}$ for the *n*th root. The definition and examples in the following box summarize our discussion up to this point.

DEFINITION 2   Principal *n*th Root

EXAMPLES

1. Let *n* be a natural number, If *a* and *b* are nonnegative real numbers, then

$$\sqrt[n]{b} = a \quad \text{if and only if} \quad b = a^n.$$

The number *a* is the **principal *n*th root** of *b*.

$\sqrt{25} = 5;\quad \sqrt{25} \neq -5$
$-\sqrt{25} = -5$

$\sqrt[4]{\dfrac{1}{16}} = \dfrac{1}{2};\quad \sqrt[3]{8} = 2$

2. If *a* and *b* are negative and *n* is an odd natural number, then

$$\sqrt[n]{b} = a \quad \text{if and only if} \quad b = a^n.$$

$\sqrt[5]{-\dfrac{1}{32}} = -\dfrac{1}{2}$

There are five properties of *n*th roots that are frequently used in simplifying certain expressions. The first four are similar to the properties of square roots that are developed in elementary algebra. For reference, we list these properties side by side in the following box (on page 32). Property 5 is listed here only for the sake of completeness; we'll postpone discussing it until Section 1.7.

Our immediate use for these properties will be in simplifying expressions involving *n*th roots. In general, we try to factor the expression under the radical so that one factor is the largest perfect *n*th power that we can find. Then we apply Property 2 or 3. For instance, the expression $\sqrt{72}$ is simplified as follows:

$$\sqrt{72} = \sqrt{(36)(2)} = \sqrt{36}\sqrt{2} = 6\sqrt{2}$$

---

*The radical sign $\sqrt{}$ was introduced in 1525 by the German mathematician Christoff Rudolff. Historian Howard Eves suggests (as did Leonhard Euler) that Rudolff may have obtained this symbol by modifying the letter *r*, the first letter of the word *radix* (root). However, research into the 15th- and 16th-century German manuscripts suggests that the radical sign may have actually evolved from a dot. For details see *A History of Mathematical Notations*, by Florian Cajori, pp. 366–368 (La Salle, Illinois: The Open Court Publishing Company, 1928.)

PROPERTY SUMMARY	PROPERTIES OF nTH ROOTS	CORRESPONDING PROPERTIES FOR SQUARE ROOTS
	Suppose that x and y are real numbers and that m and n are natural numbers. Then each of the following properties holds, provided only that the expressions on both sides of the equation are defined (and so represent real numbers).	
	1. $\left(\sqrt[n]{x}\right)^n = x$	$\left(\sqrt{x}\right)^2 = x$
	2. $\sqrt[n]{xy} = \sqrt[n]{x}\sqrt[n]{y}$	$\sqrt{xy} = \sqrt{x}\sqrt{y}$
	3. $\sqrt[n]{\dfrac{x}{y}} = \dfrac{\sqrt[n]{x}}{\sqrt[n]{y}}$	$\sqrt{\dfrac{x}{y}} = \dfrac{\sqrt{x}}{\sqrt{y}}$
	4. n even: $\sqrt[n]{x^n} = \|x\|$ n odd: $\sqrt[n]{x^n} = x$	$\sqrt{x^2} = \|x\|$
	5. $\sqrt[m]{\sqrt[n]{x}} = \sqrt[mn]{x}$	

In this procedure, we began by factoring 72 as $(36)(2)$. Note that 36 is the largest factor of 72 that is a perfect square. If we had begun instead with a different factorization, say $72 = (9)(8)$, we could still arrive at the same answer, but it would take longer. (Check this for yourself.) As another example, let us simplify $\sqrt[3]{40}$. First, what (if any) is the largest perfect cube factor of 40? Since the first few perfect cubes are

$$1^3 = 1 \qquad 2^3 = 8 \qquad 3^3 = 27 \qquad 4^3 = 64$$

we see that 8 is a perfect cube factor of 40, and we write

$$\sqrt[3]{40} = \sqrt[3]{(8)(5)} = \sqrt[3]{8}\sqrt[3]{5} = 2\sqrt[3]{5}$$

EXAMPLE 1   Simplify: **(a)** $\sqrt{12} + \sqrt{75}$ **(b)** $\sqrt{\dfrac{162}{49}}$

Solution   **(a)** $\begin{aligned}\sqrt{12} + \sqrt{75} &= \sqrt{(4)(3)} + \sqrt{(25)(3)} \\ &= \sqrt{4}\sqrt{3} + \sqrt{25}\sqrt{3} \\ &= 2\sqrt{3} + 5\sqrt{3} = (2+5)\sqrt{3} \\ &= 7\sqrt{3}\end{aligned}$

(b) $\sqrt{\dfrac{162}{49}} = \dfrac{\sqrt{162}}{\sqrt{49}} = \dfrac{\sqrt{81}\sqrt{2}}{7} = \dfrac{9\sqrt{2}}{7}$

EXAMPLE 2   Simplify: $\sqrt[3]{16} + \sqrt[3]{250} - \sqrt[3]{128}$.

Solution $\begin{aligned}\sqrt[3]{16} + \sqrt[3]{250} - \sqrt[3]{128} &= \sqrt[3]{8}\sqrt[3]{2} + \sqrt[3]{125}\sqrt[3]{2} - \sqrt[3]{64}\sqrt[3]{2} \\ &= 2\sqrt[3]{2} + 5\sqrt[3]{2} - 4\sqrt[3]{2} \\ &= 3\sqrt[3]{2}\end{aligned}$

EXAMPLE 3 Simplify each of the following expressions by removing the largest possible perfect square or perfect cube factor from within the radical.

(a) $\sqrt{8x^2}$

(b) $\sqrt{8x^2}$ assuming $x \geq 0$

(c) $\sqrt{18a^7}$ assuming $a \geq 0$

(d) $\sqrt[3]{16y^5}$

Solution (a) $\sqrt{8x^2} = \sqrt{(4)(2)(x^2)} = \sqrt{4}\sqrt{2}\sqrt{x^2} = 2\sqrt{2}\,|x|$

(b) $\sqrt{8x^2} = \sqrt{4}\sqrt{2}\sqrt{x^2}$
$$= 2\sqrt{2}x \qquad \sqrt{x^2} = x \text{ because } x \geq 0$$

(c) $\sqrt{18a^7} = \sqrt{(9a^6)(2a)} = \sqrt{9a^6}\,\sqrt{2a}$
$$= 3a^3\sqrt{2a}$$

(d) $\sqrt[3]{16y^5} = \sqrt[3]{8y^3}\,\sqrt[3]{2y^2} = 2y\sqrt[3]{2y^2}$

EXAMPLE 4 Simplify each of the following, assuming that a, b, and c are positive.

(a) $\sqrt{8ab^2c^5}$ (b) $\sqrt[4]{\dfrac{32a^6b^5}{c^8}}$

Solution (a) $\sqrt{8ab^2c^5} = \sqrt{(4b^2c^4)(2ac)}$
$$= \sqrt{4b^2c^4}\,\sqrt{2ac}$$
$$= 2bc^2\sqrt{2ac}$$

(b) $\sqrt[4]{\dfrac{32a^6b^5}{c^8}} = \dfrac{\sqrt[4]{32a^6b^5}}{\sqrt[4]{c^8}}$
$$= \dfrac{\sqrt[4]{16a^4b^4}\,\sqrt[4]{2a^2b}}{c^2} = \dfrac{2ab\sqrt[4]{2a^2b}}{c^2}$$

In Examples 1 through 4, we used the definitions and properties of *n*th roots to simplify certain expressions. The box that follows (on page 34) shows some common errors to avoid in working with roots. As suggested in the previous sections, use the error box to test yourself: cover up the columns labeled "Correction" and "Comment" and try to decide for yourself where the error lies.

There are times when it is convenient to rewrite fractions involving radicals in alternate forms. Suppose, for example, that we want to rewrite the fraction $5/\sqrt{3}$ in an equivalent form not involving a radical in the denominator. This is called **rationalizing the denominator.** The procedure here is to multiply by 1 in this way:

$$\frac{5}{\sqrt{3}} = \frac{5}{\sqrt{3}} \cdot 1 = \frac{5}{\sqrt{3}} \cdot \frac{\sqrt{3}}{\sqrt{3}} = \frac{5\sqrt{3}}{3}$$

That is, $\dfrac{5}{\sqrt{3}} = \dfrac{5\sqrt{3}}{3}$, as required.

Errors to Avoid

ERROR	CORRECTION	COMMENT
$\sqrt[4]{16} \neq \pm 2$	$\sqrt[4]{16} = 2$	Although -2 is one of the fourth roots of 16, it is not the *principal* fourth root. The notation $\sqrt[4]{16}$ is reserved for the principal fourth root.
$\sqrt{25} \neq -5$	$\sqrt{25} = 5$	Although -5 is one of the square roots of 25, it is not the *principal* square root.
$\sqrt{a + b} \neq \sqrt{a} + \sqrt{b}$	The expression $\sqrt{a + b}$ cannot be simplified.	The properties of roots differ with respect to addition versus multiplication. For $\sqrt{ab}$, we do have the simplification $\sqrt{ab} = \sqrt{a}\sqrt{b}$ (assuming a and b are nonnegative).
$\sqrt[3]{a + b} \neq \sqrt[3]{a} + \sqrt[3]{b}$	The expression $\sqrt[3]{a + b}$ cannot be simplified.	For $\sqrt[3]{ab}$, we do have $\sqrt[3]{ab} = \sqrt[3]{a}\sqrt[3]{b}$

To rationalize a denominator of the form $a + \sqrt{b}$, we need to multiply the fraction not by $\dfrac{\sqrt{b}}{\sqrt{b}}$, but rather by $\dfrac{a - \sqrt{b}}{a - \sqrt{b}}$ ($= 1$). To see why this is necessary, notice that

$$\left(a + \sqrt{b}\right)\sqrt{b} = a\sqrt{b} + b$$

which still contains a radical, whereas

$$\left(a + \sqrt{b}\right)\left(a - \sqrt{b}\right) = a^2 - a\sqrt{b} + a\sqrt{b} - b$$
$$= a^2 - b$$

which is free of radicals. Similarly, to rationalize a denominator of the form $a - \sqrt{b}$, we multiply the fraction by $\dfrac{a + \sqrt{b}}{a + \sqrt{b}}$. (The quantities $a + \sqrt{b}$ and $a - \sqrt{b}$ are said to be **conjugates** of each other.) The next two examples make use of these ideas.

EXAMPLE 5 Simplify: $\dfrac{1}{\sqrt{2}} - 3\sqrt{50}$.

Solution First, we rationalize the denominator in the fraction $1/\sqrt{2}$:

$$\frac{1}{\sqrt{2}} = \frac{1}{\sqrt{2}} \cdot 1 = \frac{1}{\sqrt{2}} \cdot \frac{\sqrt{2}}{\sqrt{2}} = \frac{\sqrt{2}}{2}$$

Next, we simplify the expression $3\sqrt{50}$:

$$3\sqrt{50} = 3\sqrt{(25)(2)} = 3\sqrt{25}\sqrt{2} = (3)(5)\sqrt{2} = 15\sqrt{2}$$

Now, putting things together, we have

$$\frac{1}{\sqrt{2}} - 3\sqrt{50} = \frac{\sqrt{2}}{2} - 15\sqrt{2} = \frac{\sqrt{2}}{2} - \frac{30\sqrt{2}}{2}$$

$$= \frac{\sqrt{2} - 30\sqrt{2}}{2} = \frac{(1 - 30)\sqrt{2}}{2}$$

$$= \frac{-29\sqrt{2}}{2}$$

EXAMPLE 6   Rationalize the denominator in the expression $\dfrac{4}{2 + \sqrt{3}}$.

Solution   We multiply by 1, writing it as $\dfrac{2 - \sqrt{3}}{2 - \sqrt{3}}$:

$$\frac{4}{2 + \sqrt{3}} \cdot 1 = \frac{4}{2 + \sqrt{3}} \cdot \frac{2 - \sqrt{3}}{2 - \sqrt{3}}$$

$$= \frac{4(2 - \sqrt{3})}{4 - (\sqrt{3})^2} = \frac{4(2 - \sqrt{3})}{4 - 3} = \frac{8 - 4\sqrt{3}}{1}$$

$$= 8 - 4\sqrt{3}$$

Note   Check for yourself that multiplying the original fraction by $\dfrac{2 + \sqrt{3}}{2 + \sqrt{3}}$ does *not* eliminate radicals in the denominator.

In the next example, we are asked to rationalize the numerator rather than the denominator. This is useful at times in calculus.

EXAMPLE 7   Rationalize the *numerator*: $\dfrac{\sqrt{x} - \sqrt{3}}{x - 3}$ $(x \geq 0,\ x \neq 3)$.

Solution
$$\frac{\sqrt{x} - \sqrt{3}}{x - 3} \cdot 1 = \frac{\sqrt{x} - \sqrt{3}}{x - 3} \cdot \frac{\sqrt{x} + \sqrt{3}}{\sqrt{x} + \sqrt{3}}$$

$$= \frac{(\sqrt{x})^2 - (\sqrt{3})^2}{(x - 3)(\sqrt{x} + \sqrt{3})} = \frac{x - 3}{(x - 3)(\sqrt{x} + \sqrt{3})}$$

$$= \frac{1}{\sqrt{x} + \sqrt{3}}$$

The strategy for rationalizing numerators or denominators involving nth roots is similar to that used for square roots. To rationalize a numerator or a denominator involving an nth root, we multiply that numerator or denominator by a factor that yields a product that itself is a perfect nth power. The next example displays two instances of this.

EXAMPLE 8 (a) Rationalize the denominator: $\dfrac{6}{\sqrt[3]{7}}$.

(b) Rationalize the denominator: $\dfrac{ab}{\sqrt[4]{a^2b^3}}$, where $a>0$ and $b>0$.

Solution (a) $\dfrac{6}{\sqrt[3]{7}} \cdot 1 = \dfrac{6}{\sqrt[3]{7}} \cdot \dfrac{\sqrt[3]{7^2}}{\sqrt[3]{7^2}}$

$= \dfrac{6\sqrt[3]{7^2}}{\sqrt[3]{7^3}}$

$= \dfrac{6\sqrt[3]{49}}{7}$ as required

(b) $\dfrac{ab}{\sqrt[4]{a^2b^3}} \cdot 1 = \dfrac{ab}{\sqrt[4]{a^2b^3}} \cdot \dfrac{\sqrt[4]{a^2b}}{\sqrt[4]{a^2b}}$

$= \dfrac{ab\sqrt[4]{a^2b}}{\sqrt[4]{a^4b^4}} = \dfrac{ab\sqrt[4]{a^2b}}{ab} = \sqrt[4]{a^2b}$

EXERCISE SET 1.6

A

In Exercises 1–10, determine whether each statement is true or false.

1. $\sqrt{81} = -9$
2. $\sqrt{256} = 16$
3. $-\sqrt{100} = -10$
4. $\sqrt{49} = -7$
5. $\sqrt{4/3} = 2/\sqrt{3}$
6. $\sqrt{10+6} = \sqrt{10} + \sqrt{6}$
7. $(\sqrt[3]{10})^3 = 10$
8. $\sqrt{(-5)^2} = -5$
9. $\sqrt[5]{32a^5} = 2a$
10. $\sqrt[4]{x^4} = x$ $(x<0)$

For Exercises 11–20, evaluate each expression. If the expression is undefined (i.e., does not represent a real number), say so.

11. (a) $\sqrt[3]{-64}$ (b) $\sqrt[4]{-64}$
12. (a) $\sqrt[5]{32}$ (b) $\sqrt[5]{-32}$
13. (a) $\sqrt[3]{8/125}$ (b) $\sqrt[3]{-8/125}$
14. (a) $\sqrt[3]{-1/1000}$ (b) $\sqrt[6]{-1/1000}$
15. (a) $\sqrt{-16}$ (b) $\sqrt[4]{-16}$
16. (a) $-\sqrt[4]{16}$ (b) $-\sqrt[4]{-16}$
17. (a) $\sqrt[4]{256/81}$ (b) $\sqrt[3]{-27/125}$
18. (a) $\sqrt[6]{64}$ (b) $\sqrt[6]{-64}$
19. (a) $\sqrt[5]{-32}$ (b) $-\sqrt[5]{-32}$
20. (a) $\sqrt[4]{(-10)^4}$ (b) $\sqrt[3]{(-10)^3}$

In Exercises 21–50, simplify each expression. (Unless otherwise specified, assume that all letters in Exercises 37–50 represent positive numbers.)

21. (a) $\sqrt{18}$ (b) $\sqrt[3]{54}$
22. (a) $\sqrt{150}$ (b) $\sqrt[3]{375}$
23. (a) $\sqrt{98}$ (b) $\sqrt[5]{-64}$
24. (a) $\sqrt{27}$ (b) $\sqrt[5]{-108}$
25. (a) $\sqrt{25/4}$ (b) $\sqrt[4]{16/625}$
26. (a) $\sqrt{225/49}$ (b) $\sqrt[5]{-256/243}$
27. (a) $\sqrt{2} + \sqrt{8}$ (b) $\sqrt[3]{2} + \sqrt[3]{16}$
28. (a) $4\sqrt{3} - 2\sqrt{27}$ (b) $2\sqrt[3]{81} + 3\sqrt[3]{24}$
29. (a) $4\sqrt{50} - 3\sqrt{128}$ (b) $\sqrt[4]{32} + \sqrt[4]{162}$
30. (a) $\sqrt{3} - \sqrt{12} + \sqrt{48}$
    (b) $\sqrt[5]{-2} + \sqrt[5]{-64} - \sqrt[5]{486}$
31. (a) $\sqrt{0.09}$ (b) $\sqrt[3]{0.008}$
32. (a) $\sqrt[3]{-2} + \sqrt[3]{2}$ (b) $\sqrt[3]{81/121} - \sqrt[3]{-8/1331}$
33. $4\sqrt{24} - 8\sqrt{54} + 2\sqrt{6}$
34. $\sqrt[3]{192} + \sqrt[3]{-81} + \sqrt[3]{\sqrt[3]{9}}$
35. $\sqrt{\sqrt[3]{64}}$
36. $\sqrt[3]{\sqrt{4096}}$
37. (a) $\sqrt{36x^2}$ (b) $\sqrt{36y^2}$, $y<0$
38. (a) $\sqrt{225x^4y^3}$ (b) $\sqrt[4]{16a^4}$, $a<0$

39. (a) $\sqrt{ab^2}\,\sqrt{a^2b}$ (b) $\sqrt{ab^3}\,\sqrt{a^3b}$

40. (a) $\sqrt[3]{125x^6}$ (b) $\sqrt[4]{64y^4}, \quad y<0$

41. $\sqrt{72a^3b^4c^5}$

42. $\sqrt{\dfrac{(a+b)^5}{16a^2b^2}}$

43. $\sqrt[4]{16a^4b^5}$

44. $\sqrt[3]{8a^4b^6}$

45. $\sqrt{18a^3b^2}$

46. $\sqrt[5]{64a^6b^{12}}$

47. $\sqrt[3]{\dfrac{16a^{12}b^2}{c^9}}$

48. $\sqrt[4]{ab^3}\,\sqrt[4]{a^3b}$

49. $\sqrt[6]{\dfrac{5a^7}{a^{-5}b^6}}$

50. $\sqrt[3]{a^2b}\,\sqrt[3]{ab}\,\sqrt[3]{b^4}$

In Exercises 51–72, rationalize the denominators and simplify where possible. (Assume that all letters represent positive quantities.)

51. $\dfrac{4}{\sqrt{7}}$

52. $\dfrac{3}{\sqrt{3}}$

53. $\dfrac{1}{\sqrt{8}}$

54. $\dfrac{\sqrt{2}}{\sqrt{3}}$

55. $\dfrac{1}{1+\sqrt{5}}$

56. $\dfrac{\sqrt{2}}{1-\sqrt{2}}$

57. $\dfrac{1+\sqrt{3}}{1-\sqrt{3}}$

58. $\dfrac{\sqrt{a}+\sqrt{b}}{\sqrt{a}-\sqrt{b}}$

59. $\dfrac{1}{\sqrt{5}}+4\sqrt{45}$

60. $\dfrac{3}{\sqrt{8}}-\sqrt{450}$

61. $\dfrac{1}{\sqrt[3]{25}}$

62. $\dfrac{4}{\sqrt[3]{16}}$

63. $\dfrac{3}{\sqrt[4]{3}}$

64. $\dfrac{\sqrt[3]{5}}{\sqrt[3]{6}}$

65. $\dfrac{1}{\sqrt[4]{2ab^5}}$

66. $\dfrac{1}{\sqrt[3]{4a^2b^8}}$

67. $\dfrac{3}{\sqrt[5]{16a^4b^9}}$

68. $\dfrac{3}{\sqrt[4]{27a^5b^{11}}}$

69. $\dfrac{x}{\sqrt{x}-2}$

70. $\dfrac{\sqrt{x}}{5-\sqrt{x}}$

71. $\dfrac{\sqrt{x}-\sqrt{a}}{\sqrt{x}+\sqrt{a}}$

72. $\dfrac{1}{\sqrt{x+h}-\sqrt{x}}$

In Exercises 73–76, rationalize the numerator.

73. $\dfrac{\sqrt{x}-\sqrt{5}}{x-5}$

74. $\dfrac{\sqrt{a}-\sqrt{b}}{a-b}$

75. $\dfrac{\sqrt{2+h}-\sqrt{2}}{h}$

76. $\dfrac{\sqrt{x+h}-\sqrt{x}}{h}$

In Exercises 77–80, refer to the following table. The left-hand column of the table lists four errors to avoid in working with expressions containing radicals. In each case, give a numerical example showing that the expressions on each side of the equation are, in general, not equal. Use a calculator as necessary.

Errors to Avoid	Numerical example showing that the formula is not, in general, valid
77. $\sqrt{a+b} \neq \sqrt{a}+\sqrt{b}$	
78. $\sqrt{x^2+y^2} \neq x+y$	
79. $\sqrt[3]{u+v} \neq \sqrt[3]{u}+\sqrt[3]{v}$	
80. $\sqrt[3]{p^3+q^3} \neq p+q$	

B

81. (a) Use a calculator to evaluate $\sqrt{8-2\sqrt{7}}$ and $\sqrt{7}-1$. What do you observe?

    (b) Prove that $\sqrt{8-2\sqrt{7}}=\sqrt{7}-1$.   *Hint:* In view of the definition of a principal square root, you need to check that $(\sqrt{7}-1)^2=8-2\sqrt{7}$.

82. Verify that the principal square root of $16-2\sqrt{55}$ is $\sqrt{11}-\sqrt{5}$.   *Hint:* In other words, check that $(\sqrt{11}-\sqrt{5})^2=16-2\sqrt{55}$.

83. Simplify: $\dfrac{\sqrt{a}}{\sqrt{a}+\sqrt{b}}+\dfrac{\sqrt{b}}{\sqrt{a}-\sqrt{b}}$.   *Hint:* First rationalize the denominators.

84. Use a calculator to provide empirical evidence indicating that both of the following equations may be correct:

$$\dfrac{2-\sqrt{3}}{\sqrt{2}-\sqrt{2-\sqrt{3}}}+\dfrac{2+\sqrt{3}}{\sqrt{2}+\sqrt{2+\sqrt{3}}}=\sqrt{2}$$

$$\sqrt{\sqrt{97.5-\dfrac{1}{11}}}=\pi$$

The point of this exercise is to remind you that as useful as calculators may be, there is still the need for proofs in mathematics. In fact, it can be shown that the first equation is indeed correct, but the second is not.

85. Rationalize the denominator: $\dfrac{1}{1+\sqrt{2}+\sqrt{3}}$.

    Hint: First multiply by $\dfrac{1+\sqrt{2}-\sqrt{3}}{1+\sqrt{2}-\sqrt{3}}$.

1.7   RATIONAL EXPONENTS

... Nicole Oresme, a bishop in Normandy (about 1323–1382), first conceived the notion of fractional powers, afterwards rediscovered by [Simon] Stevin [1548–1620]...

Florian Cajori in *A History of Elementary Mathematics* (New York: Macmillan Company, 1917)

If Descartes ... had discarded the radical sign altogether and had introduced the notation for fractional as well as integral exponents,... it is conceivable that generations upon generations of pupils would have been saved the necessity of mastering the operations with two ... notations when one alone (the exponential) would have answered all purposes.

Florian Cajori, *A History of Mathematical Notations,* vol. 1 (La Salle, Illinois: The Open Court Publishing Company, 1928)

We can use the concept of an *n*th root to give a meaning to fractional exponents that is useful and, at the same time, consistent with our earlier work. First, by way of motivation, suppose that we want to assign a value to $5^{1/3}$. Assuming that the usual properties of exponents continue to apply here, we can write

$$(5^{1/3})^3 = 5^1 = 5$$

That is,

$$(5^{1/3})^3 = 5$$

or

$$5^{1/3} = \sqrt[3]{5}$$

By replacing 5 and 3 with *b* and *n*, respectively, we can see that we want to define $b^{1/n}$ to mean $\sqrt[n]{b}$. Also, by thinking of $b^{m/n}$ as $(b^{1/n})^m$, we see that the definition for $b^{m/n}$ ought to be $(\sqrt[n]{b})^m$. These definitions are formalized in the box that follows.

DEFINITION   **Rational Exponents**

                                              EXAMPLES

1. Let *b* denote a real number and *n* a natural number. We define $b^{1/n}$ by

$$b^{1/n} = \sqrt[n]{b}$$

(If *n* is even, we require that $b \geq 0$.)

$$4^{1/2} = \sqrt{4} = 2$$

$$(-8)^{1/3} = \sqrt[3]{-8} = -2$$

2. Let *m/n* be a rational number reduced to lowest terms. Assume that *n* is positive and that $\sqrt[n]{b}$ exists. Then,

$$b^{m/n} = (\sqrt[n]{b})^m$$

or, equivalently,

$$b^{m/n} = \sqrt[n]{b^m}$$

$$8^{2/3} = (\sqrt[3]{8})^2 = 2^2 = 4$$

or, equivalently,

$$8^{2/3} = \sqrt[3]{8^2} = \sqrt[3]{64} = 4$$

It can be shown that the four properties of exponents that we listed in Section 1.5 (on page 24) continue to hold for rational exponents in general. In fact, we'll take this for granted rather than follow the lengthy argument needed for its verification. We will also assume that these properties apply to irrational exponents. So, for instance, we have

$$\left(2^{\sqrt{5}}\right)^{\sqrt{5}} = 2^5 = 32$$

(The definition of irrational exponents is discussed in Section 6.1.) In the next three examples, we display the basic techniques for working with rational exponents.

EXAMPLE 1 Simplify each of the following quantities. Express the answers using positive exponents. If an expression does not represent a real number, say so.

(a) $49^{1/2}$ (b) $-49^{1/2}$ (c) $(-49)^{1/2}$ (d) $49^{-1/2}$

Solution (a) $49^{1/2} = \sqrt{49} = 7$
(b) $-49^{1/2} = -(49^{1/2}) = -\sqrt{49} = -7$
(c) The quantity $(-49)^{1/2}$ does not represent a real number because there is no real number x such that $x^2 = -49$.

(d) $49^{-1/2} = \sqrt{49^{-1}} = \sqrt{\dfrac{1}{49}} = \dfrac{\sqrt{1}}{\sqrt{49}} = \dfrac{1}{7}$

Alternatively, we have

$$49^{-1/2} = (49^{1/2})^{-1} = 7^{-1} = \frac{1}{7}$$

EXAMPLE 2 Simplify each of the following. Write the answers using positive exponents. (Assume that $a > 0$.)

(a) $(5a^{2/3})(4a^{3/4})$

(b) $\sqrt[5]{\dfrac{16a^{1/3}}{a^{1/4}}}$

(c) $(x^2 + 1)^{1/5}(x^2 + 1)^{4/5}$

Solution (a) $(5a^{2/3})(4a^{3/4}) = 20a^{(2/3)+(3/4)}$

$$= 20a^{17/12} \qquad \text{because } \frac{2}{3} + \frac{3}{4} = \frac{17}{12}$$

(b) $\sqrt[5]{\dfrac{16a^{1/3}}{a^{1/4}}} = \left(\dfrac{16a^{1/3}}{a^{1/4}}\right)^{1/5}$

$$= (16a^{1/12})^{1/5} \qquad \text{because } \frac{1}{3} - \frac{1}{4} = \frac{1}{12}$$

$$= 16^{1/5}a^{1/60}$$

(c) $(x^2 + 1)^{1/5} (x^2 + 1)^{4/5} = (x^2 + 1)^1 = x^2 + 1$

EXAMPLE 3   Simplify: **(a)** $32^{-2/5}$ **(b)** $(-8)^{4/3}$

Solution   **(a)** $32^{-2/5} = \left(\sqrt[5]{32}\right)^{-2} = 2^{-2} = \dfrac{1}{2^2} = \dfrac{1}{4}$

Alternatively, we have

$$32^{-2/5} = (2^5)^{-2/5} = 2^{-2} = \dfrac{1}{2^2} = \dfrac{1}{4}$$

(b) $(-8)^{4/3} = \left(\sqrt[3]{-8}\right)^4 = (-2)^4 = 16$

Alternatively, we can write

$$(-8)^{4/3} = [(-2)^3]^{4/3} = (-2)^4 = 16$$

Rational exponents can be used to simplify certain expressions containing radicals. For example, one of the properties of nth roots, which was listed but not discussed in the previous section, is $\sqrt[m]{\sqrt[n]{x}} = \sqrt[mn]{x}$. Using exponents, it is easy to verify this property. We have

$$\sqrt[m]{\sqrt[n]{x}} = (x^{1/n})^{1/m}$$
$$= x^{1/mn} = \sqrt[mn]{x}, \qquad \text{as we wished to show.}$$

EXAMPLE 4   Consider the expression $\sqrt{x}\,\sqrt[3]{y^2}$, where x and y are positive.

(a) Rewrite the expression using rational exponents.
(b) Rewrite the expression using only one radical sign.

Solution   **(a)** $\sqrt{x}\,\sqrt[3]{y^2} = x^{1/2}y^{2/3}$

(b) $\sqrt{x}\,\sqrt[3]{y^2} = x^{1/2}y^{2/3}$
$$= x^{3/6}y^{4/6} \qquad \text{rewriting the fractions using a common denominator}$$
$$= \sqrt[6]{x^3}\,\sqrt[6]{y^4} = \sqrt[6]{x^3 y^4}$$

EXAMPLE 5   Rewrite the following expression using rational exponents. (Assume that x, y, and z are positive.)

$$\sqrt{\dfrac{\sqrt[3]{x}\,\sqrt[4]{y^3}}{\sqrt[5]{z^4}}}$$

Solution
$$\sqrt{\dfrac{\sqrt[3]{x}\,\sqrt[4]{y^3}}{\sqrt[5]{z^4}}} = \left(\dfrac{\sqrt[3]{x}\,\sqrt[4]{y^3}}{\sqrt[5]{z^4}}\right)^{1/2} = \left(\dfrac{x^{1/3}y^{3/4}}{z^{4/5}}\right)^{1/2}$$
$$= \dfrac{x^{1/6}y^{3/8}}{z^{2/5}} = x^{1/6}y^{3/8}z^{-2/5}$$

EXERCISE SET 1.7

A

Exercises 1–14 are warmup exercises to help you become familiar with the definition $b^{m/n} = \left(\sqrt[n]{b}\right)^m = \sqrt[n]{b^m}$. In Exercises 1–8, write the expression in the two equivalent forms

$\left(\sqrt[n]{b}\right)^m$ and $\sqrt[n]{b^m}$. In Exercises 9–14, write the expression in the form $b^{m/n}$.

1. $a^{3/5}$ **2.** $x^{3/7}$ **3.** $5^{2/3}$

4. $10^{4/5}$ **5.** $(x^2 + 1)^{3/4}$ **6.** $(a + b)^{7/10}$

7. $2^{xy/3}$ **8.** $2^{(a+1)/b}$ **9.** $\sqrt[3]{p^2}$

10. $\sqrt[5]{R^3}$ **11.** $\sqrt[7]{(1+u)^4}$ **12.** $\sqrt[6]{(1+u)^5}$

13. $\sqrt[p]{(a^2+b^2)^3}$ **14.** $\sqrt[3p]{(a^2+b^2)^{2p}}$

In Exercises 15–50, evaluate or simplify each expression. Express the answers using positive exponents. If an expression is undefined (i.e., does not represent a real number), say so. (Assume that all letters represent positive numbers.)

15. $16^{1/2}$ **16.** $100^{1/2}$ **17.** $\left(\frac{1}{36}\right)^{1/2}$

18. $0.09^{1/2}$ **19.** $(-16)^{1/2}$ **20.** $(-1)^{1/2}$

21. $625^{1/4}$ **22.** $\left(\frac{1}{81}\right)^{1/4}$ **23.** $8^{1/3}$

24. $0.001^{1/3}$ **25.** $8^{2/3}$ **26.** $64^{2/3}$

27. $(-32)^{1/5}$ **28.** $\left(-\frac{1}{125}\right)^{1/3}$ **29.** $(-1000)^{1/3}$

30. $(-243)^{1/5}$ **31.** $49^{-1/2}$ **32.** $121^{-1/2}$

33. $(-49)^{-1/2}$ **34.** $(-64)^{-1/3}$ **35.** $36^{-3/2}$

36. $(-0.001)^{-2/3}$ **37.** $125^{2/3}$

38. $125^{-2/3}$ **39.** $(-1)^{3/5}$

40. $27^{4/3} + 27^{-4/3} + 27^0$ **41.** $32^{4/5} - 32^{-4/5}$

42. $64^{1/2} + 64^{-1/2} - 64^{4/3}$ **43.** $\left(\frac{9}{16}\right)^{-5/2} - \left(\frac{1000}{27}\right)^{4/3}$

44. $(256^{-3/4})^{4/3}$ **45.** $(2a^{1/3})(3a^{1/4})$

46. $\sqrt[5]{3a^4}$ **47.** $\sqrt[4]{\dfrac{64a^{2/3}}{a^{1/3}}}$

48. $(x^2+1)^{2/3}(x^2+1)^{4/3}$ **49.** $\dfrac{(x^2+1)^{3/4}}{(x^2+1)^{-1/4}}$

50. $\dfrac{(2x^2+1)^{-6/5}(2x^2+1)^{6/5}(x^2+1)^{-1/5}}{(x^2+1)^{9/5}}$

For Exercises 51–58, follow Example 4 in the text to rewrite each expression in two ways: (a) using rational exponents; (b) using only one radical sign. (Assume x, y, and z are positive.)

51. $\sqrt{3}\sqrt[3]{6}$ **52.** $\sqrt{5}\sqrt[3]{7}$

53. $\sqrt[3]{6}\sqrt[4]{2}$ **54.** $\sqrt[3]{2}\sqrt[5]{2}$

55. $\sqrt[3]{x^2}\sqrt[5]{y^4}$ **56.** $\sqrt{x}\sqrt[3]{y}\sqrt[4]{z}$

57. $\sqrt[4]{x^a}\sqrt[3]{x^b}\sqrt{x^{a/6}}$ **58.** $\sqrt[3]{27\sqrt{64x}}$

In Exercises 59–66, rewrite each expression using rational exponents rather than radicals. (Assume that x, y, and z are positive.)

59. $\sqrt[3]{(x+1)^2}$ **60.** $\dfrac{1}{\sqrt{x}} + \sqrt{x}$

61. $\left(\sqrt[5]{x+y}\right)^2$ **62.** $\sqrt{\sqrt{x}}$

63. $\sqrt[3]{\sqrt{x}} + \sqrt{\sqrt[3]{x}}$ **64.** $\sqrt[3]{\sqrt{2}}$

65. $\sqrt{\sqrt[3]{x}\sqrt[4]{y}}$ **66.** $\sqrt[5]{\dfrac{\sqrt{x}\sqrt[3]{y}}{\sqrt[4]{z^2}}}$

67. Use a calculator to determine which is larger, $9^{10/9}$ or $10^{9/10}$.

68. Use a calculator to determine which number is closer to 3: $13^{3/7}$ or $6560^{1/8}$.

In Exercises 69–74, refer to the following table. The left-hand column of the table lists six errors to avoid in working with expressions containing fractional exponents. In each case, give a numerical example showing that the expressions on each side of the equation are not equal. Use a calculator as necessary.

Errors to Avoid	Numerical example showing that the formula is not, in general, valid
69. $(a+b)^{1/2} \ne a^{1/2} + b^{1/2}$	
70. $(x^2+y^2)^{1/2} \ne x + y$	
71. $(u+v)^{1/3} \ne u^{1/3} + v^{1/3}$	
72. $(p^3+q^3)^{1/3} \ne p + q$	
73. $x^{1/m} \ne \dfrac{1}{x^m}$	
74. $x^{-1/2} \ne \dfrac{1}{x^2}$	

75. Consider the expression $n^{1/n}$, where n is a natural number. In calculus, it is shown that as n takes on larger and larger values, the resulting value of the expression approaches 1. Confirm this empirically by completing the table. (Round off your results to four decimal places.)

n	2	5	10	100	10^3	10^4	10^5	10^6
$n^{1/n}$								

B

76. In working with roots and radicals, it is not always apparent from the outset when two numbers are equal. (We've already seen examples of this in Exercise Set 1.3, Exercise 57.) In each of the following cases, use a calculator to verify that the quantity on the left side of the equation agrees with the quantity on the right side through the first six decimal places. (In each case, it can be shown that the numbers are indeed equal.)

(a) $\left(176 + 80\sqrt{5}\right)^{1/5} = 1 + \sqrt{5}$

(b) $\left(7\sqrt[3]{20} - 19\right)^{1/6} = \sqrt[3]{\frac{5}{3}} - \sqrt[3]{\frac{2}{3}}$

(c) $\left(2 + \sqrt{5}\right)^{1/3} + \left(2 - \sqrt{5}\right)^{1/3} = 1$

77. Without using a calculator, decide which number is larger in each case.
   (a) $2^{2/3}$ or $2^{3/2}$ (b) $5^{1/2}$ or 5^{-2}
   (c) $2^{1/2}$ or $2^{1/3}$ (d) $\left(\frac{1}{2}\right)^{1/2}$ or $\left(\frac{1}{2}\right)^{1/3}$
   (e) $10^{1/10}$ or $\left(\frac{1}{10}\right)^{10}$

78. (a) Give an example in which a rational number raised to a rational power is irrational.
   (b) Give an example in which an irrational number raised to a rational power is rational.

79. Without using a calculator, decide which number is closer to zero: $(0.5)^{1/3}$ or $(0.5)^{1/4}$. Then use a calculator to check your answer.

80. Without using a calculator, decide which number is smaller: $(-0.5)^{1/3}$ or $(-0.4)^{1/3}$. Then use a calculator to check your answer.

C

81. Show that $\dfrac{a-b}{a+b}\sqrt{\dfrac{a+b}{a-b}} = \left(\dfrac{a-b}{a+b}\right)^{1/2}$. (Assume that $a > b > 0$.)

82. Given $a + b = 2c$, evaluate the following expression:
$$\left[\frac{(2^{a-b})^b(2^{b-c})^{c-a}}{(2^{c+b})^{c-b}}\right]^{1/c}$$

1.8   POLYNOMIALS

The Cartesian use of letters near the beginning of the alphabet for parameters and those near the end as unknown quantities, the adaptation of exponential notation to these, and the use of the Germanic symbols + and − all combined to make Descartes' algebraic notation look like ours, for, of course, we took ours from him.

Carl B. Boyer, *A History of Mathematics*, 2nd ed., revised by Uta C. Merzbach (New York: John Wiley & Sons, Inc., 1991)

As background for our work on polynomials, we first review the terms *constant* and *variable*. By way of example, consider the familiar expression for the area of a circle of radius r, namely, πr^2. Here π is a constant; its value never changes throughout the discussion. On the other hand, r is a variable; we can substitute any positive number for r to obtain the area of a particular circle. More generally, by a **constant** we mean either a particular number (such as π, or -17, or $\sqrt{2}$) or a letter whose value remains fixed (although perhaps unspecified) throughout a given discussion. In contrast, a **variable** is a letter for which we can substitute any number selected from a given set of numbers. The given set of numbers is called the **domain** of the variable.

Some expressions will make sense only for certain values of the variable. For instance, $1/(x - 3)$ will be undefined when x is 3 (for then the denominator is zero). So in this case we would agree that the domain of the variable x consists of all real numbers except $x = 3$. Similarly, throughout this chapter (with the exception of Section 1.11 on complex numbers), we adopt the following convention:

> **The Domain Convention**
>
> The domain of a variable in a given expression is the set of all real-number values of the variable for which the expression is defined.

In algebra it's customary (but certainly not mandatory) to use letters near the end of the alphabet for variables; letters from the beginning of the alphabet are generally used for constants. So, for example, in the expression $ax + b$, the letter x is the variable and a and b are constants.

EXAMPLE 1 Specify the variable, the constants, and the domain of the variable for each expression.

(a) $3x + 4$ (b) $\dfrac{1}{(t-1)(t+3)}$ (c) $ay^2 + by + c$ (d) $4x + 3x^{-1}$

Solution

	VARIABLE	CONSTANTS	DOMAIN
(a) $3x + 4$	x	$3, 4$	The set of all real numbers.
(b) $\dfrac{1}{(t-1)(t+3)}$	t	$1, -1, 3$	The set of all real numbers except $t = 1$ and $t = -3$.
(c) $ay^2 + by + c$	y	a, b, c	The set of all real numbers.
(d) $4x + 3x^{-1}$	x	$4, 3$	The set of all real numbers except $x = 0$. (Remember, $x^{-1} = 1/x$.)

The expressions in parts (a) and (c) of Example 1 are *polynomials*. By a **polynomial in x**, we mean an expression of the form

$$a_n x^n + a_{n-1} x^{n-1} + \cdots + a_1 x + a_0$$

where n is a nonnegative integer and $a_n \neq 0$. The individual expressions $a_k x^k$ making up the polynomial are callled **terms**. In this chapter, the **coefficients** a_k will always be real numbers. For example, the terms of the polynomial $x^2 - 7x + 3$ are x^2, $-7x$, and 3; the coefficients are 1, -7, and 3. In writing a polynomial, it's customary (but not mandatory) to write the terms in order of decreasing powers of x. For instance, we would usually write $x^2 - 7x + 3$ rather than $x^2 + 3 - 7x$. The highest power of x in a polynomial is called the **degree** of the polynomial. For example, the degree of the polynomial $x^2 - 7x + 3$ is 2. In the case of a polynomial consisting only of a nonzero constant a_0, we say that the degree is zero (because $a_0 = a_0 x^0$). No degree is defined for the polynomial whose only term is zero.

Some additional terminology is useful in describing polynomials that have only a few terms. A polynomial with only one term (such as $3x^2$) is a **monomial**; a polynomial with only two terms (such as $2x + 3$) is a **binomial**; and a polynomial with only three terms (such as $5x^2 - 6x + 3$) is a **trinomial**. The table that follows provides examples of the terminology.

		If Polynomial:		
Expression	Polynomial?	Degree	Terms	Coefficients
$2x^3 - 3x^2 + 4x - 1$	yes	3	$2x^3, -3x^2, 4x, -1$	$2, -3, 4, -1$
$t^2 + 1$	yes	2	$t^2, 1$	$1, 1$
-12	yes	0	-12	-12
$2x^{-4} + 5$	no			
$\dfrac{1}{2x - 3}$	no			
$\sqrt{4x^2 + 1}$	no			
$\sqrt{2}\, x + 1$	yes	1	$\sqrt{2}\, x, 1$	$\sqrt{2}, 1$

Just as we can combine two numbers through addition, subtraction, multiplication, or division, so can we combine polynomials. To add or subtract two polynomials, we follow the familiar process of combining like terms. For example, to add the polynomials $3x - 1$ and $x^2 + 8x - 4$, we have

$$(3x - 1) + (x^2 + 8x - 4) = \underbrace{x^2}_{\uparrow \atop 0 + x^2} \quad + \quad \underbrace{11x}_{\uparrow \atop 3x + 8x} \quad \underbrace{-5}_{\uparrow \atop -1 + (-4)}$$

Similarly, to subtract we write

$$(3x - 1) - (x^2 + 8x - 4) = \underbrace{-x^2}_{\uparrow \atop 0 - x^2} \quad \underbrace{-5x}_{\uparrow \atop 3x - 8x} \quad + \quad \underbrace{3}_{\uparrow \atop -1 - (-4)}$$

Since all terms in a polynomial ultimately stand for real numbers, we can rely on the properties of real numbers in simplifying polynomial expressions. For instance, the following rules are useful in dealing with sums and differences of polynomials.

Rules for Sums and Differences of Polynomials

RULE	EXAMPLE
(a) If the parentheses are preceded by a plus sign, or no sign appears, then the parentheses can be omitted.	$(3x^2 + 4x) + (2x^2 - 5x)$ $= 3x^2 + 4x + 2x^2 - 5x$ $= 5x^2 - x$
(b) If the parentheses are preceded by a minus sign, then the parentheses can be omitted if the sign of every term within is reversed.	$(x^2 - 3x + 1) - (3x^2 - 8x + 4)$ $= x^2 - 3x + 1 - 3x^2 + 8x - 4$ $= -2x^2 + 5x - 3$

As we saw in Section 1.2, multiplication of polynomials can be carried out by means of the distributive properties. Example 2 displays three instances of this.

EXAMPLE 2   Multiply:

(a) $3x(4x^2 - 8x + 1)$
(b) $(x^2 - 4x + 2)(x + 3)$
(c) $(2x - 3)(4x + 5)$

Solution
(a) $3x(4x^2 - 8x + 1) = 3x(4x^2) + 3x(-8x) + 3x(1)$
$\qquad\qquad\qquad\quad = 12x^3 - 24x^2 + 3x$
(b) $(x^2 - 4x + 2)(x + 3) = (x^2 - 4x + 2)x + (x^2 - 4x + 2)3$
$\qquad\qquad\qquad\qquad\quad = x^3 - 4x^2 + 2x + 3x^2 - 12x + 6$
$\qquad\qquad\qquad\qquad\quad = x^3 - x^2 - 10x + 6$
(c) $(2x - 3)(4x + 5) = (2x - 3)4x + (2x - 3)5$
$\qquad\qquad\qquad\quad = 8x^2 - 12x + 10x - 15$
$\qquad\qquad\qquad\quad = 8x^2 - 2x - 15$

In part (c) of the example we just concluded, we showed the multiplication of two binomials. Because this type of product occurs so frequently, it's very useful to be able to carry out the work mentally. The so-called FOIL method allows us to accomplish this. "FOIL" is an acronym standing for First, Outer, Inner, Last. Here's how the procedure works in computing the product $(2x - 3)(4x + 5)$.

STEP 1 Multiply the first two terms: $(2x)(4x) = 8x^2$.

$$(2x - 3)(4x + 5)$$
first

STEP 2 Add the products of the outer and inner terms: $(2x)(5) + (-3)(4x) = -2x$.

outer
$$(2x - 3)(4x + 5)$$
inner

STEP 3 Multiply the last two terms: $(-3)(5) = -15$.

$$(2x - 3)(4x + 5)$$
last

STEP 4 Add the results: $8x^2 - 2x - 15$ (as obtained previously).

EXAMPLE 3 Compute each product:

(a) $(2x + 3y)(5x - y)$ (b) $(A - B)(A + B)$

Solution (a) $(2x + 3y)(5x - y) = 10x^2 + 13xy - 3y^2$ (FOIL method)
(b) $(A - B)(A + B) = A^2 - B^2$ (The middle term, $AB - AB$, is zero.)

There are several products that occur so frequently that it's convenient to memorize the results. The product $(A - B)(A + B) = A^2 - B^2$ in the example just concluded is one of these. The list that follows displays these special products. (Exercise 84 asks you to verify these results by carrying out the multiplication.) In memorizing these formulas, remember it's the *form* or *pattern* that's important, not the specific choice of letters.

PROPERTY SUMMARY   SPECIAL PRODUCTS

1. $(A - B)(A + B) = A^2 - B^2$

2. (a) $(A + B)^2 = A^2 + 2AB + B^2$
(b) $(A - B)^2 = A^2 - 2AB + B^2$

3. (a) $(A + B)^3 = A^3 + 3A^2B + 3AB^2 + B^3$
(b) $(A - B)^3 = A^3 - 3A^2B + 3AB^2 - B^3$

4. (a) $(A + B)(A^2 - AB + B^2) = A^3 + B^3$
(b) $(A - B)(A^2 + AB + B^2) = A^3 - B^3$

EXAMPLE 4 Use the Special Product formulas to compute each product:

(a) $\left(3\sqrt{xy} - z\right)\left(3\sqrt{xy} + z\right)$ **(b)** $(4x^3 - 3)^2$ **(c)** $(2x + 5)^3$

Solution **(a)** Using Special Product 1, we have

$$\left(3\sqrt{xy} - z\right)\left(3\sqrt{xy} + z\right) = \left(3\sqrt{xy}\right)^2 - z^2$$
$$= 9xy - z^2$$

(b) Using Special Product 2(b) we have

$$(4x^3 - 3)^2 = (4x^3)^2 - 2(4x^3)(3) + 3^2$$
$$= 16x^6 - 24x^3 + 9$$

(c) Using Special Product 3(a) we have

$$(2x + 5)^3 = (2x)^3 + 3(2x)^2(5) + 3(2x)(5)^2 + (5)^3$$
$$= 8x^3 + 60x^2 + 150x + 125$$

We conclude this section by describing the process of long division for polynomials. The terms *quotient, remainder, divisor,* and *dividend* are used here in the same way they're used in ordinary division of numbers. For instance, when 7 is divided by 2, the quotient is 3 and the remainder is 1. We write this

$$\frac{7}{2} = 3 + \frac{1}{2}$$

or, equivalently,

$$7 = 2 \times 3 + 1$$

dividend ⌐ ↑ ↑ ⌐ remainder
divisor ——⌐ |
quotient ——⌐

The process of long division for polynomials follows the same four-step cycle used in ordinary long division of numbers: divide, multiply, subtract, bring down. As a first example, we divide $2x^2 - 7x + 8$ by $x - 2$. Notice that in setting up the division, we write both the dividend and the divisor in decreasing powers of x.

$$\begin{array}{r}
2x - 3 \\
x - 2 \overline{\smash{\big)}\ 2x^2 - 7x + 8} \\
\underline{2x^2 - 4x} \\
-3x + 8 \\
\underline{-3x + 6} \\
2
\end{array}$$

1. Divide the first term of the dividend by the first term of the divisor: $\dfrac{2x^2}{x} = 2x$. The result becomes the first term of the quotient, as shown.

2. Multiply the divisor $x - 2$ by the term $2x$ obtained in the previous step. This yields the quantity $2x^2 - 4x$, which is written below the dividend, as shown.

3. From the quantity $2x^2 - 7x$ in the dividend, subtract the quantity $2x^2 - 4x$. This yields $-3x$.

4. Bring down the $+8$ in the dividend, as shown. The resulting quantity, $-3x + 8$, is now treated as the dividend and the entire process is repeated.

We've now found that when $2x^2 - 7x + 8$ is divided by $x - 2$, the quotient is $2x - 3$ and the remainder is 2. This is summarized by writing either

$$\frac{2x^2 - 7x + 8}{x - 2} = 2x - 3 + \frac{2}{x - 2}$$

or, after multiplying through by $x - 2$,

$$\underbrace{2x^2 - 7x + 8}_{\text{dividend}} = \underbrace{(x - 2)}_{\text{divisor}}\underbrace{(2x - 3)}_{\text{quotient}} + \underbrace{2}_{\text{remainder}}$$

As a second example, we divide $3x^4 - 2x^3 + 2$ by $x^2 - 1$. Notice in what follows that we have inserted the terms whose coefficients are zero in the divisor and in the dividend. These terms serve as place holders.

$$
\begin{array}{r}
3x^2 - 2x + 3 \\
x^2 + 0x - 1 \overline{\smash{\big)}\, 3x^4 - 2x^3 + 0x^2 + 0x + 2} \\
\underline{3x^4 + 0x^3 - 3x^2} \\
-2x^3 + 3x^2 + 0x \\
\underline{-2x^3 + 0x^2 + 2x} \\
3x^2 - 2x + 2 \\
\underline{3x^2 + 0x - 3} \\
-2x + 5
\end{array}
$$

We can write this result as

$$\frac{3x^4 - 2x^3 + 2}{x^2 - 1} = 3x^2 - 2x + 3 + \frac{-2x + 5}{x^2 - 1}$$

or, multiplying through by $x^2 - 1$,

$$\underbrace{3x^4 - 2x^3 + 2}_{\text{dividend}} = \underbrace{(x^2 - 1)}_{\text{divisor}}\underbrace{(3x^2 - 2x + 3)}_{\text{quotient}} + \underbrace{(-2x + 5)}_{\text{remainder}} \qquad (1)$$

In the example we just concluded, note that the degree of the remainder is 1, whereas the degree of the divisor is 2. So in this case we have

(degree of remainder) $<$ (degree of divisor)

In fact, it can be shown that this inequality remains valid whenever we obtain a nonzero remainder in dividing two nonconstant polynomials. (We'll return to this idea in a later chapter.)

EXERCISE SET 1.8

A

In Exercises 1–6, specify the domain of each variable.

1. (a) $2x^2 - 3x + 5$ (b) $2x^{1/2} - 3x + 5$

2. (a) $y - 1$ (b) $\dfrac{2y}{y - 1}$

3. (a) $ax + b$ (b) $ax^{1/3} + b$

4. (a) $x + x^{-1}$ (b) $t^{-1} + 2t^{-2}$

5. (a) $4\sqrt{t} + t^2$ (b) $\dfrac{4}{(t - 1)\sqrt{t}}$

6. (a) $\dfrac{1}{x}$ (b) $\dfrac{1}{(x - 1)(x + 2)(x - 3)}$

For Exercises 7–10, specify the degree and the (nonzero) coefficients of each polynomial.

7. (a) $4x^3 - 2x^2 - 6x - 1$
    (b) $ax^2 + bx + c$ $(a, b,$ and c nonzero)

8. (a) $-4t^3 + 3t^4 + \sqrt{2}$ (b) $3\sqrt{2}x^2$

9. (a) $6x$ (b) $6x + x^2 - x^3$

10. (a) 18 (b) $18x^{10}$

In Exercises 11–40, carry out the indicated operations.

11. $(12x^2 - 4x + 2) + (8x^2 + 6x - 1)$

12. $(4x^2 - 1) + (4x^2 + 1)$

13. $(x^2 - x - 1) - (x^2 + x + 1)$

14. $(x^3 + 2x^2 + 1) - (x^2 - 1)$

15. $(2x^2 - 4x - 4) - (6x^2 + 5) - (8x - 1)$

16. $(x^3 - 1) - (x^3 - 1)$

17. $(2x^2 - 6x - 1) + (5x^2 - 5x - 1) - (3x^2 + 8x + 12)$

18. $(ax + b) - (2ax - b)$

19. $(ax^2 + bx + c) - (2ax^2 - 3bx + c)$

20. $x^3 - 4x^2 + x - 2(x^3 + 2x^2 + 1)$

21. $2x(x^2 - 4x - 5)$ 22. $4ab(a^2 + 2ab + b^2)$

23. $(x - 1)(x - 2)$ 24. $(y^2 + 1)(y^2 - 3)$

25. $(2x + 4)(x + 1)$ 26. $(a + 2b)(a - 3b)$

27. $(x^2 + 3x)(x^2 + x)$ 28. $(\sqrt{x} + 1)(\sqrt{x} + 2)$

29. $(2xy - 3)(2xy - 1)$ 30. $(5xy^2 - 2)(xy^2 - 3)$

31. $\left(x + \dfrac{1}{x}\right)\left(x + \dfrac{1}{x}\right)$ 32. $(2pq + r)(pq - 2r)$

33. $(\sqrt{a+b} + 1)(\sqrt{a+b} + 3)$

34. $(x^{1/2} + y^{1/2})(x^{1/2} + 2y^{1/2})$

35. $(x^{1/2} - 1)(x^{1/2} - 2)$

36. $(x - 1)(x^2 + 4x + 1)$

37. $(y + 2)(y^2 - 3y - 5)$

38. $(a + b + c)(a + b - c)$

39. $(x - 2y + z)(x + 2y - z)$

40. $a(b - c) + b(c - a) + c(a - b)$

For Exercises 41–52, use the Special Product formulas to compute the products.

41. (a) $(x - y)(x + y)$ (b) $(x^2 - 5)(x^2 + 5)$

42. (a) $(3y - a^2)(3y + a^2)$
    (b) $(3\sqrt{y} - a^2)(3\sqrt{y} + a^2)$

43. (a) $(A - 4)(A + 4)$
    (b) $[(a + b) - 4][(a + b) + 4]$

44. (a) $(\sqrt{ab} + \sqrt{c})(\sqrt{ab} - \sqrt{c})$
    (b) $(a^{1/2} + b^{1/2})(a^{1/2} - b^{1/2})$

45. (a) $(x - 8)^2$ (b) $(2x^2 - 5)^2$

46. (a) $(2^m + 1)^2$ (b) $(a^m - a^{-m})^2$

47. (a) $(\sqrt{x} + \sqrt{y})^2$ (b) $(\sqrt{x+y} - \sqrt{x})^2$

48. (a) $(2x + y)^3$ (b) $(x - 2y)^3$

49. (a) $(a + 1)^3$ (b) $(3x^2 - 2a^2)^3$

50. (a) $(x - y)(x^2 + xy + y^2)$
    (b) $(x^2 - y^2)(x^4 + x^2y^2 + y^4)$

51. (a) $(x + 1)(x^2 - x + 1)$
    (b) $(x^2 + 1)(x^4 - x^2 + 1)$

52. (a) $(5 - y)(25 + 5y + y^2)$
    (b) $(3 + x)(9 - 3x + x^2)$

In Exercises 53–68, use long division to find the quotient and remainder. Also, express the result in an equation [as in equation (1) on page 47].

53. $\dfrac{x^2 - 8x + 4}{x - 3}$ 54. $\dfrac{x^3 - 4x^2 + x - 2}{x - 5}$

55. $\dfrac{x^2 - 6x - 2}{x + 5}$ 56. $\dfrac{3x^2 + 4x - 1}{x - 1}$

57. $\dfrac{6x^3 - 2x + 3}{2x + 1}$ 58. $\dfrac{x^4 - 4x^3 + 6x^2 - 4x + 1}{x - 1}$

59. $\dfrac{t^5 + 2}{t + 3}$ 60. $\dfrac{4t^3 - t^2 + 8t - 1}{t^2 - t + 1}$

61. $\dfrac{u^6 - 64}{u - 2}$ 62. $\dfrac{u^6 + 64}{u - 2}$

63. $\dfrac{5x^4 - 3x^2 + 2}{x^2 - 3x + 5}$ 64. $\dfrac{8x^6 - 36x^4 + 54x^2 - 27}{2x^2 - 3}$

65. $\dfrac{t^4 - 4t^3 + 4t^2 - 16}{t^2 - 2t + 4}$ 66. $\dfrac{2t^5 - 6t^4 - t^2 + 2t + 3}{t^3 - 2}$

67. $\dfrac{z^5 - 1}{z - 1}$ 68. $\dfrac{1 + z + z^2 + z^3}{1 + z + z^2}$

B

In Exercises 69–76, use the Special Product formulas to compute the products.

69. $(a + b - 3)(a + b + 3)$ 70. $(u + v + 4)(u + v - 4)$

71. $(a + b + x + y)(a + b - x - y)$

72. $(2x - y + xy)(2x - y - xy)$

73. $(\sqrt{a} + \sqrt{b})^2$ 74. $(\sqrt{a} - \sqrt{b})^2$

75. $(x^{1/3} - y^{1/3})(x^{2/3} + x^{1/3}y^{1/3} + y^{2/3})$

76. $(a^{1/3} + 5^{1/3})(a^{2/3} - a^{1/3} \cdot 5^{1/3} + 5^{2/3})$

77. Notice that the expression $(x^2 - 16)/(x - 4)$ is undefined when $x = 4$. In this exercise we investigate the values of the expression when x is very close to 4.

(a) Complete the tables.

x	$\dfrac{x^2 - 16}{x - 4}$
3.9	
3.99	
3.999	
3.9999	
3.99999	

x	$\dfrac{x^2 - 16}{x - 4}$
4.1	
4.01	
4.001	
4.0001	
4.00001	

(b) What value does the expression $\dfrac{x^2 - 16}{x - 4}$ seem to be approaching as x gets closer and closer to 4? This "target value" is referred to as the *limit* of $\dfrac{x^2 - 16}{x - 4}$ as x approaches 4. (The notion of a limit is made more precise in calculus.)

(c) How could Special Product 1 have been used to obtain the limit without the work in part (a)?

78. Notice that the expression $\dfrac{x^3 - 8}{x - 2}$ is undefined when $x = 2$. In this exercise we investigate the values of the expression when x is very close to 2.

(a) Complete the tables.

x	$\dfrac{x^3 - 8}{x - 2}$
1.9	
1.99	
1.999	
1.9999	
1.99999	

x	$\dfrac{x^3 - 8}{x - 2}$
2.1	
2.01	
2.001	
2.0001	
2.00001	

(b) On the basis of the tables, what "target value" or limit does the expression $\dfrac{x^3 - 8}{x - 2}$ seem to be approaching as x gets closer and closer to 2?

(c) How could Special Product 4(b) have been used to obtain this limit without the work in part (a)?

For Exercises 79–82, verify each statement by carrying out the operations on the left-hand side of the equation.

79. $(b - c)(b + c - a) + (c - a)(c + a - b)$
$$+ (a - b)(a + b - c) = 0$$

80. $(b^{-1} + c^{-1})(b + c - a) + (c^{-1} + a^{-1})(c + a - b)$
$$+ (a^{-1} + b^{-1})(a + b - c) = 6$$

81. $(x^2 + 8 - 4x)(x^2 + 8 + 4x) = x^4 + 64$

82. $\left(\sqrt{a + x} + \sqrt{a - x}\right)\left(\sqrt{a + x} - \sqrt{a - x}\right) = 2x$

83. (a) Use long division to determine the remainder when $ax^2 + bx + c$ is divided by $x - r$.
(b) Use long division to determine the remainder when $ax^3 + bx^2 + cx + d$ is divided by $x - r$.
(c) On the basis of your answers in parts (a) and (b), guess the remainder when $ax^4 + bx^3 + cx^2 + dx + e$ is divided by $x - r$.

84. Carry out the multiplication to verify each of the Special Products listed on page 45.

85. (a) Complete the table.

n	1	2	3	4	5	6
$\dfrac{n^2}{n + 1}$						

(b) Are there any natural numbers n for which $n^2/(n + 1)$ is itself a natural number? *Hint:* Use long division.

1.9 FACTORING

fac·tor (*fak′tər*), n. ... **2.** Math. *one of two or more numbers, algebraic expressions, or the like, that when multiplied together produce a given product; ... v.t.* **10.** Math. *to express (a mathematical quantity) as a product of two or more quantities of like kind, as* $30 = 2 \cdot 3 \cdot 5$, *or* $x^2 - y^2 = (x + y)(x - y)$.

The Random House Dictionary of the English Language, 2nd ed. (New York: Random House, 1987)

There are many cases in algebra in which the process of *factoring* simplifies the work at hand. To **factor** a polynomial means to write it as a product of two or more nonconstant polynomials. For instance, a factorization of $x^2 - 9$ is given by

$$x^2 - 9 = (x - 3)(x + 3)$$

In this case, $x - 3$ and $x + 3$ are the **factors** of $x^2 - 9$.

There is one convention that we need to agree on at the outset. If the polynomial or expression that we wish to factor contains only integer coefficients, then the factors (if any) should involve only integer coefficients. For example, according to this convention, we will not consider the following type of factorization in this section:

$$x^2 - 2 = \left(x - \sqrt{2}\right)\left(x + \sqrt{2}\right)$$

because it involves coefficients that are irrational numbers. (We should point out, however, that factorizations such as this are useful at times, particularly in calculus.) As it happens, $x^2 - 2$ is an example of a polynomial that cannot be factored using integer coefficients. We say in such a case that the polynomial is **irreducible over the integers.**

If the coefficients of a polynomial are rational numbers, then we do allow factors with rational coefficients. For instance, the factorization of $y^2 - \frac{1}{4}$ over the rational numbers is given by

$$y^2 - \tfrac{1}{4} = \left(y - \tfrac{1}{2}\right)\left(y + \tfrac{1}{2}\right)$$

We'll consider five techniques for factoring in this section. These techniques will be applied in Section 1.10 (Fractional Expressions), in Chapter 2 (Equations and Inequalities), and throughout the text. In Table 1 (on the next page) we summarize these techniques. Notice that three of the formulas in the table are just restatements of Special Product formulas. Remember, it is the form, or pattern, in the formula that is important—not the specific choice of letters.

The idea in factoring is to use one or more of these techniques until each of the factors obtained is irreducible. The examples that follow show how this works in practice.

EXAMPLE 1   Factor: **(a)** $x^2 - 49$;   **(b)** $(2a - 3b)^2 - 49$.

Solution   **(a)** $x^2 - 49 = x^2 - 7^2$
$$= (x - 7)(x + 7) \qquad \text{difference of squares}$$

(b) Notice that the pattern or form is the same as in part (a); it's a difference of squares. We have

$$(2a - 3b)^2 - 49 = (2a - 3b)^2 - 7^2$$
$$= [(2a - 3b) - 7][(2a - 3b) + 7] \qquad \text{difference of squares}$$
$$= (2a - 3b - 7)(2a - 3b + 7)$$

EXAMPLE 2   Factor: **(a)** $2x^3 - 50x$;   **(b)** $3x^5 - 3x$.

Solution   **(a)** $2x^3 - 50x = 2x(x^2 - 25) \qquad \text{common factor}$
$$= 2x(x - 5)(x + 5) \qquad \text{difference of squares}$$

TABLE I

Basic Factoring Techniques

Technique	Example or Formula	Remark
Common factor	$3x^4 + 6x^3 - 12x^2 = 3x^2(x^2 + 2x - 4)$ $4(x^2 + 1) - x(x^2 + 1) = (x^2 + 1)(4 - x)$	In any factoring problem, the first step always is to look for the common factor of highest degree.
Difference of squares	$x^2 - a^2 = (x - a)(x + a)$	There is no corresponding formula for a sum of squares; $x^2 + a^2$ is irreducible over the integers.
Trial and error	$x^2 + 2x - 3 = (x + 3)(x - 1)$	In this example, the only possibilities, or trials, are: **(a)** $(x - 3)(x - 1)$ **(b)** $(x - 3)(x + 1)$ **(c)** $(x + 3)(x - 1)$ **(d)** $(x + 3)(x + 1)$ By inspection or by carrying out the indicated multiplications, we find that only case (c) checks.
Difference of cubes Sum of cubes	$x^3 - a^3 = (x - a)(x^2 + ax + a^2)$ $x^3 + a^3 = (x + a)(x^2 - ax + a^2)$	Verify these formulas for yourself by carrying out the multiplications. Then memorize the formulas.
Grouping	$x^3 - x^2 + x - 1 = (x^3 - x^2) + (x - 1)$ $\qquad = x^2(x - 1) + (x - 1) \cdot 1$ $\qquad = (x - 1)(x^2 + 1)$	This is actually an application of the common factor technique.

$$\textbf{(b)}\ 3x^5 - 3x = 3x(x^4 - 1) \qquad\qquad\qquad \text{common factor}$$
$$= 3x[(x^2)^2 - 1^2]$$
$$= 3x(x^2 - 1)(x^2 + 1) \qquad\qquad \text{difference of squares}$$
$$= 3x(x - 1)(x + 1)(x^2 + 1) \qquad \text{difference of squares, again}$$

EXAMPLE 3 Factor: **(a)** $x^2 - 4x - 5$; **(b)** $(a + b)^2 - 4(a + b) - 5$.

Solution **(a)** $x^2 - 4x - 5 = (x - 5)(x + 1)$ trial and error

(b) Note that the form of this expression is

$$(\ \ )^2 - 4(\ \ ) - 5$$

This is the same *form* as the expression in part (a). So, we need only replace x with the quantity $a + b$ in the solution for part (a). This yields

$$(a + b)^2 - 4(a + b) - 5 = [(a + b) - 5][(a + b) + 1]$$
$$= (a + b - 5)(a + b + 1)$$

EXAMPLE 4 Factor: $2z^4 + 9z^2 + 4$.

Solution We use trial and error to look for a factorization of the form $(2z^2 + ?)(z^2 + ?)$. There are three possibilities:

(i) $(2z^2 + 2)(z^2 + 2)$ (ii) $(2z^2 + 1)(z^2 + 4)$ (iii) $(2z^2 + 4)(z^2 + 1)$

Each of these yields the appropriate first term and last term, but (after checking) only possibility (ii) yields $9z^2$ for the middle term. The required factorization is then $2z^4 + 9z^2 + 4 = (2z^2 + 1)(z^2 + 4)$.

QUESTION Why didn't we consider any possibilities with subtraction signs in place of addition signs?

EXAMPLE 5   Factor: **(a)** $x^2 + 9$;   **(b)** $x^2 + 2x + 3$.

Solution   **(a)** The expression $x^2 + 9$ is irreducible over the integers. (This can be discovered by trial and error.) If the given expression had instead been $x^2 - 9$, then it could have been factored as a difference of squares. Sums of squares, however, cannot in general be factored over the integers.

(b) The expression $x^2 + 2x + 3$ is irreducible over the integers. (Check this for yourself by trial and error.) ▪▪▪

EXAMPLE 6   Factor: $x^2 - y^2 + 10x + 25$.

Solution   Familiarity with the Special Products in Section 1.8 suggests that we try grouping the terms this way:

$$(x^2 + 10x + 25) - y^2$$

Then we have

$$
\begin{aligned}
x^2 - y^2 + 10x + 25 &= (x^2 + 10x + 25) - y^2 \\
&= (x + 5)^2 - y^2 \\
&= [(x + 5) - y][(x + 5) + y] \qquad \text{difference of squares} \\
&= (x - y + 5)(x + y + 5)
\end{aligned}
$$

▪▪▪

EXAMPLE 7   Factor: $ax + ay^2 + bx + by^2$.

Solution   We factor a from the first two terms and b from the second two, to obtain

$$ax + ay^2 + bx + by^2 = a(x + y^2) + b(x + y^2)$$

We now recognize the quantity $(x + y^2)$ as a common expression that can be factored out. We have then

$$a(x + y^2) + b(x + y^2) = (x + y^2)(a + b)$$

The required factorization is therefore

$$ax + ay^2 + bx + by^2 = (x + y^2)(a + b)$$

As you may wish to check for yourself, this factorization can also be obtained by trial and error. ▪▪▪

EXAMPLE 8   Factor: **(a)** $t^3 - 125$;   **(b)** $8 + (a - 2)^3$.

Solution   **(a)** $t^3 - 125 = t^3 - 5^3$
$= (t - 5)(t^2 + 5t + 25) \qquad \text{difference of cubes}$

(b) $8 + (a - 2)^3 = 2^3 + (a - 2)^3$
$= [2 + (a - 2)][2^2 - 2(a - 2) + (a - 2)^2] \qquad \text{sum of cubes}$
$= a(4 - 2a + 4 + a^2 - 4a + 4)$
$= a(a^2 - 6a + 12)$

As you can check now, the expression $a^2 - 6a + 12$ is irreducible over the integers. Therefore the required factorization is $a(a^2 - 6a + 12)$. ▪▪▪

For some calculations (particularly in calculus), it's helpful to be able to factor an expression involving fractional exponents. For instance, suppose (as in the next example), we want to factor the expression

$$x(2x - 1)^{-1/2} + (2x - 1)^{3/2}$$

The common expression to factor out here is $(2x - 1)^{-1/2}$; the technique is to *choose the expression with the smaller exponent.*

EXAMPLE 9   Factor: $x(2x - 1)^{-1/2} + (2x - 1)^{3/2}$.

Solution $x(2x - 1)^{-1/2} + (2x - 1)^{3/2} = (2x - 1)^{-1/2}[x + (2x - 1)^{3/2 + 1/2}]$
$$= (2x - 1)^{-1/2}[x + (2x - 1)^2]$$
$$= (2x - 1)^{-1/2}(4x^2 - 3x + 1)$$

We conclude this section with examples of four common errors to avoid in factoring. The first two errors (in the box that follows) are easy to detect; simply multiplying out the supposed factorizations shows that they do not check. The third error may result from a lack of familiarity with the basic factoring techniques listed on page 51. The fourth error indicates a misunderstanding of what is required in factoring. In factoring, the final quantity must be expressed as a *product,* not a sum, of terms or expressions.

Errors to Avoid

ERROR	CORRECTION	COMMENT
$x^2 + 6x + 9 \neq (x - 3)^2$	$x^2 + 6x + 9 = (x + 3)^2$	Check the middle term.
$x^2 + 64 \neq (x + 8)(x + 8)$	$x^2 + 64$ is irreducible over the integers.	A sum of squares is, in general, irreducible over the integers. (A difference of squares, however, can always be factored.)
$x^3 + 64$ is irreducible over the integers.	$x^3 + 64 = (x + 4)(x^2 - 4x + 16)$	Although a sum of squares is irreducible, a sum of cubes can be factored.
$x^2 - 2x + 3 \neq x(x - 2) + 3$	$x^2 - 2x + 3$ is irreducible over the integers.	The polynomial $x^2 - 2x + 3$ is the *sum* of the expression $x(x - 2)$ and the constant 3. But, by definition, the factored form of a polynomial must be a *product* (of two or more nonconstant polynomials).

EXERCISE SET 1.9

A

In Exercises 1–66, factor each polynomial or expression. If a polynomial is irreducible, state this. (In Exercises 1–6, the factoring techniques are specified.)

1. (Common factor and difference of squares)
    (a) $x^2 - 64$ (b) $7x^4 + 14x^2$
    (c) $121z - z^3$ (d) $a^2b^2 - c^2$

2. (Common factor and difference of squares)
    (a) $1 - t^4$ (b) $x^6 + x^5 + x^4$
    (c) $u^2v^2 - 225$ (d) $81x^4 - x^2$

3. (Trial and error)
    (a) $x^2 + 2x - 3$ (b) $x^2 - 2x - 3$
    (c) $x^2 - 2x + 3$ (d) $-x^2 + 2x + 3$

4. (Trial and error)
   (a) $2x^2 - 7x - 4$ (b) $2x^2 + 7x - 4$
   (c) $2x^2 + 7x + 4$ (d) $-2x^2 - 7x + 4$

5. (Sum and difference of cubes)
   (a) $x^3 + 1$ (b) $x^3 + 216$
   (c) $1000 - 8x^6$ (d) $64a^3x^3 - 125$

6. (Grouping)
   (a) $x^4 - 2x^3 + 3x - 6$
   (b) $a^2x + bx - a^2z - bz$

7. (a) $144 - x^2$
   (b) $144 + x^2$
   (c) $144 - (y - 3)^2$

8. (a) $4a^2b^2 + 9c^2$
   (b) $4a^2b^2 - 9c^2$
   (c) $4a^2b^2 - 9(ab + c)^2$

9. (a) $h^3 - h^5$
   (b) $100h^3 - h^5$
   (c) $100(h + 1)^3 - (h + 1)^5$

10. (a) $x^4 - x^2$
    (b) $3x^4 - 48x^2$
    (c) $3(x + h)^4 - 48(x + h)^2$

11. (a) $x^2 - 13x + 40$
    (b) $x^2 - 13x - 40$

12. (a) $x^2 + 10x + 16$
    (b) $x^2 - 10x + 16$

13. (a) $x^2 + 5x - 36$
    (b) $x^2 - 13x + 36$

14. (a) $x^2 - x + 6$
    (b) $x^2 + x - 6$

15. (a) $3x^2 - 22x - 16$
    (b) $3x^2 - x - 16$

16. (a) $x^2 - 4x + 1$
    (b) $x^2 - 4x - 3$

17. (a) $6x^2 + 13x - 5$
    (b) $6x^2 - x - 5$

18. (a) $16x^2 + 18x - 9$
    (b) $16x^2 - 143x - 9$

19. (a) $t^4 + 2t^2 + 1$
    (b) $t^4 - 2t^2 + 1$
    (c) $t^4 - 2t^2 - 1$

20. (a) $t^4 - 9t^2 + 20$
    (b) $t^4 - 19t^2 + 20$
    (c) $t^4 - 19t^2 - 20$

21. (a) $4x^3 - 20x^2 - 25x$
    (b) $4x^3 - 20x^2 + 25x$

22. (a) $x^3 + x^2 + x$
    (b) $x^3 + 2x^2 + x$

23. (a) $ab - bc + a^2 - ac$
    (b) $(u + v)x - xy + (u + v)^2 - (u + v)y$

24. (a) $3(x + 5)^3 + 2(x + 5)^2$
    (b) $a(x + 5)^3 + b(x + 5)^2$

25. $x^2z^2 + xzt + xyz + yt$

26. $a^2t^2 + b^2t^2 - cb^2 - ca^2$

27. $a^4 - 4a^2b^2c^2 + 4b^4c^4$

28. $A^2 - B^2 + 16A + 64$

29. $A^2 + B^2$

30. $x^2 + 64$

31. $x^3 + 64$

32. $27 - (a - b)^3$

33. $(x + y)^3 - y^3$

34. $(a + b)^3 - 8c^3$

35. $x^3 - y^3 + x - y$

36. $8a^3 + 27b^3 + 2a + 3b$

37. (a) $p^4 - 1$
    (b) $p^8 - 1$

38. (a) $p^4 + 4$
    (b) $p^4 - 4$

39. $x^3 + 3x^2 + 3x + 1$

40. $-1 + 6x - 12x^2 + 8x^3$

41. $x^2 + 16y^2$

42. $4u^2 + 25v^2$

43. $\frac{25}{16} - c^2$

44. $\frac{81}{4} - y^2$

45. $z^4 - \frac{81}{16}$

46. $\frac{(a + b)^2}{4} - \frac{a^2b^2}{9}$

47. $\frac{125}{m^3n^3} - 1$

48. $\frac{x^3}{8} - \frac{512}{x^3}$

49. $\frac{1}{4}x^2 + xy + y^2$

50. $x^2 + x + 1$

51. $64(x - a)^3 - x + a$

52. $64(x - a)^4 - x + a$

53. $x^2 - a^2 + y^2 - 2xy$

54. $a^4 - (b + c)^4$

55. $21x^3 + 82x^2 - 39x$

56. $x^3a^2 - 8y^3a^2 - 4x^3b^2 + 32y^3b^2$

57. $12xy + 25 - 4x^2 - 9y^2$

58. $ax^2 + (1 + ab)xy + by^2$

59. $ax^2 + (a + b)x + b$

60. $(5a^2 - 11a + 10)^2 - (4a^2 - 15a + 6)^2$

61. $(x + 1)^{1/2} - (x + 1)^{3/2}$

62. $(x^2 + 1)^{3/2} + (x^2 + 1)^{7/2}$

63. $(x + 1)^{-1/2} - (x + 1)^{-3/2}$

64. $(x^2 + 1)^{-2/3} + (x^2 + 1)^{-5/3}$

65. $(2x + 3)^{1/2} - \frac{1}{3}(2x + 3)^{3/2}$

66. $(ax + b)^{-1/2} - \sqrt{ax + b}/b$

In Exercises 67 and 68, evaluate the expressions using factoring techniques. (The point here is to do as little actual arithmetic as possible.)

67. (a) $100^2 - 99^2$ (b) $8^3 - 6^3$ (c) $1000^2 - 999^2$

68. (a) $10^3 - 9^3$ (b) $50^2 - 49^2$ (c) $\frac{15^3 - 10^3}{15^2 - 10^2}$

B

In Exercises 69–73, factor each expression.

69. $A^3 + B^3 + 3AB(A + B)$

70. $p^3 - q^3 - p(p^2 - q^2) + q(p - q)^2$

71. $2x(a^2 + x^2)^{-1/2} - x^3(a^2 + x^2)^{-3/2}$

72. $\frac{1}{2}(x - a)^{-1/2}(x + a)^{-1/2} - \frac{1}{2}(x + a)^{1/2}(x - a)^{-3/2}$

73. $y^4 - (p + q)y^3 + (p^2q + pq^2)y - p^2q^2$

74. (a) Factor: $x^4 + 2x^2y^2 + y^4$.
    (b) Factor: $x^4 + x^2y^2 + y^4$.   *Hint:*   Add and subtract a term. [Keep part (a) in mind.]
    (c) Factor: $x^6 - y^6$ as a difference of squares.
    (d) Factor: $x^6 - y^6$ as a difference of cubes. [Use the result in part (b) to obtain the same answer as in part (c).]

C

In Exercises 75–82, factor the expressions.

75. $(1 + t)^2(1 + u^2) - (1 + u)^2(1 + t^2)$

76. $4p^2q^2 - (p^2 + q^2 - r^2)^2$

77. $x^4 + 64$ *Hint:* Add and subtract a term, as in Exercise 74(b).

78. $x^4 - 15x^2 + 9$

79. $(x + y)^2 + (x + z)^2 - (z + t)^2 - (y + t)^2$

80. $(a - a^2)^3 + (a^2 - 1)^3 + (1 - a)^3$

81. $(b - c)^3 + (c - a)^3 + (a - b)^3$

82. $(a + b + c)^3 - a^3 - b^3 - c^3$

1.10 FRACTIONAL EXPRESSIONS

But if you do use a rule involving mechanical calculation, be patient, accurate, and systematically neat in the working. It is well known to mathematical teachers that quite half the failures in algebraic exercises arise from arithmetical inaccuracy and slovenly arrangement.

George Chrystal (1893)

The rules of arithmetic for fractions were listed in Section 1.2. These same rules are used for fractions involving algebraic expressions. In algebra, as in arithmetic, we say that a fraction is *reduced to lowest terms* or *simplified* when the numerator and denominator contain no common factors (other than 1 and -1). The factoring techniques developed in the previous section are used to reduce fractions. For example, to reduce the fraction $\dfrac{x^2 - 9}{x^2 + 3x}$ we write

$$\frac{x^2 - 9}{x^2 + 3x} = \frac{(x - 3)(x + 3)}{x(x + 3)} = \frac{x - 3}{x}$$

In the box that follows we review two properties of negatives and fractions (from Section 1.2) that will be useful in this section. Notice that Property 2 follows from Property 1 because

$$\frac{a - b}{b - a} = \frac{a - b}{-(a - b)} = -1$$

PROPERTY SUMMARY   **NEGATIVES AND FRACTIONS**

PROPERTY	EXAMPLE
1. $b - a = -(a - b)$	$2 - x = -(x - 2)$
2. $\dfrac{a - b}{b - a} = -1$	$\dfrac{2x - 5}{5 - 2x} = -1$

EXAMPLE 1   Simplify: $\dfrac{4 - 3x}{15x - 20}$.

Solution $\dfrac{4 - 3x}{15x - 20} = \dfrac{4 - 3x}{5(3x - 4)}$

$= \dfrac{1}{5}\left(\dfrac{4 - 3x}{3x - 4}\right) = -\dfrac{1}{5}$ using Property 2

In Example 2 we display two more instances in which factoring is used to reduce a fraction. After that, Example 3 indicates how these skills are used to multiply and divide fractional expressions.

EXAMPLE 2   Simplify: **(a)** $\dfrac{x^3 - 8}{x^2 - 2x}$;   **(b)** $\dfrac{x^2 - 6x + 8}{a(x - 2) + b(x - 2)}$.

Solution   **(a)** $\dfrac{x^3 - 8}{x^2 - 2x} = \dfrac{(x - 2)(x^2 + 2x + 4)}{x(x - 2)} = \dfrac{x^2 + 2x + 4}{x}$

(b) $\dfrac{x^2 - 6x + 8}{a(x - 2) + b(x - 2)} = \dfrac{(x - 2)(x - 4)}{(x - 2)(a + b)} = \dfrac{x - 4}{a + b}$

EXAMPLE 3   Carry out the indicated operations and simplify:

(a) $\dfrac{x^3}{2x^2 + 3x} \cdot \dfrac{12x + 18}{4x^2 - 6x}$ **(c)** $\dfrac{x^2 - 144}{x^2 - 4} \div \dfrac{x + 12}{x + 2}$

(b) $\dfrac{2x^2 - x - 6}{x^2 + x + 1} \cdot \dfrac{x^3 - 1}{4 - x^2}$

Solution   **(a)** $\dfrac{x^3}{2x^2 + 3x} \cdot \dfrac{12x + 18}{4x^2 - 6x} = \dfrac{x^3}{x(2x + 3)} \cdot \dfrac{6(2x + 3)}{2x(2x - 3)}$

$= \dfrac{6x^3}{2x^2(2x - 3)} = \dfrac{3x}{2x - 3}$

(b) $\dfrac{2x^2 - x - 6}{x^2 + x + 1} \cdot \dfrac{x^3 - 1}{4 - x^2} = \dfrac{(2x + 3)(x - 2)}{(x^2 + x + 1)} \cdot \dfrac{(x - 1)(x^2 + x + 1)}{(2 - x)(2 + x)}$

$= \dfrac{-(2x + 3)(x - 1)}{2 + x}$ using the fact that $\dfrac{x - 2}{2 - x} = -1$

$= \dfrac{-2x^2 - x + 3}{2 + x}$

(c) $\dfrac{x^2 - 144}{x^2 - 4} \div \dfrac{x + 12}{x + 2} = \dfrac{x^2 - 144}{x^2 - 4} \cdot \dfrac{x + 2}{x + 12}$

$= \dfrac{(x - 12)(x + 12)}{(x - 2)(x + 2)} \cdot \dfrac{x + 2}{x + 12} = \dfrac{x - 12}{x - 2}$

As in arithmetic, to add or subtract two fractions, the denominators must be the same. The rules in this case are

$$\frac{a}{b} + \frac{c}{b} = \frac{a + c}{b} \qquad \text{and} \qquad \frac{a}{b} - \frac{c}{b} = \frac{a - c}{b}$$

For example, we have

$$\frac{4x - 1}{x + 1} - \frac{2x - 1}{x + 1} = \frac{4x - 1 - (2x - 1)}{x + 1} = \frac{4x - 1 - 2x + 1}{x + 1} = \frac{2x}{x + 1}$$

Fractions with unlike denominators are added or subtracted by first converting to a common denominator. For instance, to add $9/a$ and $10/a^2$ we write

$$\frac{9}{a} + \frac{10}{a^2} = \frac{9}{a} \cdot \frac{a}{a} + \frac{10}{a^2}$$

$$= \frac{9a}{a^2} + \frac{10}{a^2} = \frac{9a + 10}{a^2}$$

Notice that the common denominator used was a^2. This is the **least common denominator.** In fact, other common denominators (such as a^3 or a^4) could be used here, but that would be less efficient. In general, the least common denominator for a given group of fractions is chosen as follows. Write down a product involving the irreducible factors from each denominator. The power of each factor should be equal to (but not greater than) the highest power of that factor appearing in any of the individual denominators. For example, the least common denominator for the two fractions $\dfrac{1}{(x+1)^2}$ and $\dfrac{1}{(x+1)(x+2)}$ is $(x+1)^2(x+2)$. In the example that follows, notice that the denominators must be in factored form before the least common denominator can be chosen.

EXAMPLE 4   Combine into a single fraction and simplify:

(a) $\dfrac{3}{4x} + \dfrac{7x}{10y^2} - 2;$ **(b)** $\dfrac{x}{x^2-9} - \dfrac{1}{x+3};$ **(c)** $\dfrac{15}{x^2+x-6} + \dfrac{x+1}{2-x}.$

Solution   **(a)** Denominators: $2^2 \cdot x;$ $2 \cdot 5 \cdot y^2;$ 1
Least common denominator; $2^2 \cdot 5xy^2 = 20xy^2$

$$\frac{3}{4x} + \frac{7x}{10y^2} - \frac{2}{1} = \frac{3}{4x} \cdot \frac{5y^2}{5y^2} + \frac{7x}{10y^2} \cdot \frac{2x}{2x} - \frac{2}{1} \cdot \frac{20xy^2}{20xy^2}$$

$$= \frac{15y^2 + 14x^2 - 40xy^2}{20xy^2}$$

(b) Denominators: $x^2 - 9 = (x-3)(x+3);$ $x + 3$
Least common denominator: $(x-3)(x+3)$

$$\frac{x}{x^2-9} - \frac{1}{x+3} = \frac{x}{(x-3)(x+3)} - \frac{1}{x+3}$$

$$= \frac{x}{(x-3)(x+3)} - \frac{1}{x+3} \cdot \frac{x-3}{x-3}$$

$$= \frac{x - (x-3)}{(x-3)(x+3)} = \frac{3}{(x-3)(x+3)}$$

(c) $\dfrac{15}{x^2+x-6} + \dfrac{x+1}{2-x} = \dfrac{15}{(x-2)(x+3)} - \dfrac{x+1}{x-2}$ using the fact that $2 - x = -(x-2)$

$$= \frac{15}{(x-2)(x+3)} - \frac{x+1}{x-2} \cdot \frac{x+3}{x+3}$$

$$= \frac{15 - (x^2 + 4x + 3)}{(x-2)(x+3)} = \frac{-x^2 - 4x + 12}{(x-2)(x+3)}$$

$$= \frac{(x-2)(-x-6)}{(x-2)(x+3)} = \frac{-x-6}{x+3}$$

Notice in the last line that we were able to simplify the answer by factoring the numerator and reducing the fraction. (This is why we prefer to leave the least common denominator in factored form, rather than multiply it out, in this type of problem.) ■■■

EXAMPLE 5 Simplify: $\dfrac{\dfrac{1}{3a} - \dfrac{1}{4b}}{\dfrac{5}{6a^2} + \dfrac{1}{b}}$.

Solution The least common denominator for the four individual fractions is $12a^2b$. Multiplying the given expression by $\dfrac{12a^2b}{12a^2b}$, which equals 1, yields

$$\frac{12a^2b}{12a^2b} \cdot \frac{\dfrac{1}{3a} - \dfrac{1}{4b}}{\dfrac{5}{6a^2} + \dfrac{1}{b}} = \frac{4ab - 3a^2}{10b + 12a^2}$$

■■■

EXAMPLE 6 Simplify: $(x^{-1} + y^{-1})^{-1}$. (The answer is *not* $x + y$.)

Solution After applying the definition of negative exponents to rewrite the given expression, we'll use the method shown in Example 5.

$$(x^{-1} + y^{-1})^{-1} = \left(\frac{1}{x} + \frac{1}{y}\right)^{-1} = \frac{1}{\dfrac{1}{x} + \dfrac{1}{y}}$$

$$= \frac{xy}{xy} \cdot \frac{1}{\dfrac{1}{x} + \dfrac{1}{y}} = \frac{xy}{y + x}$$

■■■

EXAMPLE 7 Simplify: $\dfrac{\dfrac{1}{x + h} - \dfrac{1}{x}}{h}$. (This type of expression occurs in calculus.)

Solution $\dfrac{\dfrac{1}{x + h} - \dfrac{1}{x}}{h} = \dfrac{(x + h)x}{(x + h)x} \cdot \dfrac{\dfrac{1}{x + h} - \dfrac{1}{x}}{h}$

$$= \frac{x - (x + h)}{(x + h)xh} = \frac{-h}{(x + h)xh} = -\frac{1}{(x + h)x}$$

■■■

EXAMPLE 8 Simplify: $\dfrac{x - \dfrac{1}{x^2}}{\dfrac{1}{x^2} - 1}$.

$$\text{Solution} \quad \frac{x - \dfrac{1}{x^2}}{\dfrac{1}{x^2} - 1} = \frac{x^2}{x^2} \cdot \frac{x - \dfrac{1}{x^2}}{\dfrac{1}{x^2} - 1}$$

$$= \frac{x^3 - 1}{1 - x^2} = -\frac{x^3 - 1}{x^2 - 1}$$

$$= -\frac{(x - 1)(x^2 + x + 1)}{(x - 1)(x + 1)} = -\frac{x^2 + x + 1}{x + 1}$$

EXERCISE SET 1.10

A

In Exercises 1–12, reduce the fractions to lowest terms.

1. $\dfrac{x^2 - 9}{x + 3}$

2. $\dfrac{25 - x^2}{x - 5}$

3. $\dfrac{x + 2}{x^4 - 16}$

4. $\dfrac{x^2 - x - 20}{2x^2 + 7x - 4}$

5. $\dfrac{x^2 + 2x + 4}{x^3 - 8}$

6. $\dfrac{a + b}{ax^2 + bx^2}$

7. $\dfrac{9ab - 12b^2}{6a^2 - 8ab}$

8. $\dfrac{a^3b^2 - 27b^5}{(ab - 3b^2)^2}$

9. $\dfrac{a^3 + a^2 + a + 1}{a^2 - 1}$

10. $\dfrac{(x - y)^2(a + b)}{(x^2 - y^2)(a^2 + 2ab + b^2)}$

11. $\dfrac{x^3 - y^3}{(x - y)^3}$

12. $\dfrac{x^4 - y^4}{(x^4y + x^2y^3 + x^3y^2 + xy^4)(x - y)^2}$

In Exercises 13–78, carry out the indicated operations and simplify where possible.

13. $\dfrac{2}{x - 2} \cdot \dfrac{x^2 - 4}{x + 2}$

14. $\dfrac{ax + 3}{2a + 1} \div \dfrac{a^2x^2 + 3ax}{4a^2 - 1}$

15. $\dfrac{x^2 - x - 2}{x^2 + x - 12} \cdot \dfrac{x^2 - 3x}{x^2 - 4x + 4}$

16. $(3t^2 + 4tx + x^2) \div \dfrac{3t^2 - 2tx - x^2}{t^2 - x^2}$

17. $\dfrac{x^3 + y^3}{x^2 - 4xy + 3y^2} \div \dfrac{(x + y)^3}{x^2 - 2xy - 3y^2}$

18. $\dfrac{a^2 - a - 42}{a^4 + 216a} \div \dfrac{a^2 - 49}{a^3 - 6a^2 + 36a}$

19. $\dfrac{x^2 + xy - 2y^2}{x^2 - 5xy + 4y^2} \cdot \dfrac{x^2 - 7xy + 12y^2}{x^2 + 5xy + 6y^2}$

20. $\dfrac{x^4y^2 - xy^5}{x^4 - 2x^2y^2 + y^4} \cdot \dfrac{x^2 + 2xy + y^2}{x^3y^2 + x^2y^3 + xy^4}$

21. $\dfrac{4}{x} - \dfrac{2}{x^2}$

22. $\dfrac{1}{3x} + \dfrac{1}{5x^2} - \dfrac{1}{30x^3}$

23. $\dfrac{6}{a} - \dfrac{a}{6}$

24. $\dfrac{1}{a} + \dfrac{1}{b} + \dfrac{1}{c}$

25. $\dfrac{1}{x + 3} + \dfrac{3}{x + 2}$

26. $\dfrac{4}{x - 4} - \dfrac{4}{x + 1}$

27. $\dfrac{3x}{x - 2} - \dfrac{6}{x^2 - 4}$

28. $1 + \dfrac{1}{x} - \dfrac{1}{x^2}$

29. $\dfrac{a}{x - 1} + \dfrac{2ax}{(x - 1)^2} + \dfrac{3ax^2}{(x - 1)^3}$

30. $\dfrac{a^2 + 5a - 4}{a^2 - 16} - \dfrac{2a}{2a^2 + 8a}$

31. $\dfrac{x}{x^2 - 9} + \dfrac{x - 1}{x^2 - 5x + 6}$

32. $\dfrac{1}{x - 1} + \dfrac{1}{1 - x}$

33. $\dfrac{4}{x - 5} - \dfrac{4}{5 - x}$

34. $\dfrac{x}{x + a} + \dfrac{a}{a - x}$

35. $\dfrac{a^2 + b^2}{a^2 - b^2} + \dfrac{a}{a + b} + \dfrac{b}{b - a}$

36. $\dfrac{3}{2x + 2} - \dfrac{5}{x^2 - 1} + \dfrac{1}{x + 1}$

37. $\dfrac{1}{x^2 + x - 20} - \dfrac{1}{x^2 - 8x + 16}$

38. $\dfrac{4}{6x^2 + 5x - 4} + \dfrac{1}{3x^2 + 4x} - \dfrac{1}{2x - 1}$

39. $\dfrac{2q + p}{2p^2 - 9pq - 5q^2} - \dfrac{p + q}{p^2 - 5pq}$

40. $\dfrac{1}{x - 1} + \dfrac{1}{x^2 - 1} + \dfrac{1}{x^3 - 1}$

41. $\dfrac{x}{(x-y)(x-z)} + \dfrac{y}{(y-z)(y-x)} + \dfrac{z}{(z-x)(z-y)}$

42. $\dfrac{3x}{x-2} + \dfrac{4x^2}{x^2+2x+1} - \dfrac{x^2}{x^2-x-2}$

43. $\dfrac{y+z}{x^2-xy-xz+yz} - \dfrac{x+z}{xy-xz-y^2+yz} + \dfrac{x+y}{xy-yz-xz+z^2}$

44. $\left(x+\dfrac{1}{x}\right) \cdot \left(1-\dfrac{1}{x}\right)^2 \div \left(x-\dfrac{1}{x}\right)$

45. $\dfrac{x^2+x-a-1}{a^2-x^2} + \dfrac{x+1}{x+a} \div \dfrac{x-a}{x-1}$

46. $\dfrac{1}{2ax-2a^2} - \dfrac{1}{2ax+2a^2}$

47. $\dfrac{\dfrac{1}{x}+1}{\dfrac{1}{x}-1}$

48. $\dfrac{\dfrac{4}{a}-a}{\dfrac{2}{a}+1}$

49. $\dfrac{\dfrac{1}{x}-\dfrac{1}{a}}{x-a}$

50. $\dfrac{\dfrac{1}{a}+\dfrac{1}{b}}{\dfrac{1}{a}-\dfrac{1}{b}}$

51. $\dfrac{1+\dfrac{4}{x}}{\dfrac{3}{x}-2}$

52. $\dfrac{\dfrac{1}{a}}{1+\dfrac{b}{c}}$

53. $\dfrac{a-\dfrac{1}{a}}{1+\dfrac{1}{a}}$

54. $\dfrac{\dfrac{1}{x^2}-\dfrac{1}{y^2}}{\dfrac{1}{x}+\dfrac{1}{y}}$

55. $\dfrac{\dfrac{1}{2+h}-\dfrac{1}{2}}{h}$

56. $\dfrac{\dfrac{3}{x^2+h}-\dfrac{3}{x^2}}{h}$

57. $\dfrac{\dfrac{a}{x^2}+\dfrac{x}{a^2}}{a^2-ax+x^2}$

58. $\dfrac{x+\dfrac{xy}{y-x}}{\dfrac{y^2}{x^2-y^2}+1}$

59. $(x^{-1}+2)^{-1}$

60. $\dfrac{(x^{-2}+2x)^{-1}}{x^2}$

61. $\left(\dfrac{1}{a^{-1}}+\dfrac{1}{a^{-2}}\right)^{-1}$

62. $(a^{-1}+a^{-2})^{-1}$

63. $x(x+y)^{-1}+y(x-y)^{-1}$

64. $x(2x-2y)^{-1}+y(2y-2x)^{-1}$

65. $(t^2+t^{-2}+2)(t+t^{-1})^{-1}$

66. $(a+a^{-1}x^2)(a-x^4a^{-3})^{-1}$

B

67. $\dfrac{x+y}{2(x^2+y^2)} - \dfrac{1}{2(x+y)} + \dfrac{x-y}{x^2-y^2} - \dfrac{x^3-y^3}{x^4-y^4}$

68. $\dfrac{4y}{x^2-y^2} + \dfrac{1}{x+y} + \dfrac{1}{y-x} - \dfrac{2y}{x^2+y^2}$

69. $\dfrac{\dfrac{a+b}{a-b}+\dfrac{a-b}{a+b}}{\dfrac{a-b}{a+b}-\dfrac{a+b}{a-b}} \cdot \dfrac{ab^3-a^3b}{a^2+b^2}$

70. $\dfrac{\sqrt{x+a}}{\sqrt{x-a}} - \dfrac{\sqrt{x-a}}{\sqrt{x+a}}$

71. $[x-(x+x^{-1})^{-1}]^{-1}-[x+(x-x^{-1})^{-1}]^{-1}$

72. $\dfrac{x^{-1}y^{-1}(xy^{-1}+yx^{-1})^{-1}}{x+y}$

C

73. $\dfrac{ap+q}{ax-bx-a^2+ab} + \dfrac{bp+q}{bx-ax-b^2+ab}$

74. $\dfrac{b-c}{a^2-ab-ac+bc} + \dfrac{a-c}{b^2-bc-ab+ac} + \dfrac{a-b}{c^2-ac-bc+ab}$

75. $\dfrac{\dfrac{1}{a}-\dfrac{a-x}{a^2+x^2}}{\dfrac{1}{x}-\dfrac{x-a}{x^2+a^2}} + \dfrac{\dfrac{1}{a}-\dfrac{a+x}{a^2+x^2}}{\dfrac{1}{x}-\dfrac{x+a}{x^2+a^2}}$

76. $\dfrac{\left(a+\dfrac{1}{b}\right)^a\left(a-\dfrac{1}{b}\right)^b}{\left(b+\dfrac{1}{a}\right)^a\left(b-\dfrac{1}{a}\right)^b}$

77. $\dfrac{x^2-qr}{(p-q)(p-r)} + \dfrac{x^2-rp}{(q-r)(q-p)} + \dfrac{x^2-pq}{(r-p)(r-q)}$

78. $\left(\dfrac{x^{-2}+y^{-2}}{x^{-2}-y^{-2}} - \dfrac{x^{-2}-y^{-2}}{x^{-2}+y^{-2}}\right)$
$\div \left[\left(\dfrac{x+y}{x-y}+\dfrac{x-y}{x+y}\right)\left(\dfrac{x^2}{y^2}+\dfrac{y^2}{x^2}-2\right)\right]^{-1}$

79. Consider the three fractions

$$\dfrac{b-c}{1+bc}, \quad \dfrac{c-a}{1+ca}, \quad \text{and} \quad \dfrac{a-b}{1+ab}$$

(a) If $a=1$, $b=2$, and $c=3$, find the sum of the three fractions. Also compute their product. What do you observe?

(b) Show that the sum and the product of the three given fractions are, in fact, always equal.

80. Evaluate the expression

$$\dfrac{y-a}{x-a} - \dfrac{y+a}{x+a} + \dfrac{y-b}{x-b} - \dfrac{y+b}{x+b}$$

when $x=\sqrt{ab}$.   *Hint:* Keep the first and second terms separate from the third and fourth as long as possible.

81. Show that $(1+a^{x-y})^{-1}+(1+a^{y-x})^{-1}=1$.

82. Simplify:

$$\dfrac{x^3y^{-3}-y^3x^{-3}}{(xy^{-1}-yx^{-1})(xy^{-1}+yx^{-1}-1)} \cdot \dfrac{y^{-1}-x^{-1}}{x^{-2}+y^{-2}+x^{-1}y^{-1}}$$

1.11 THE COMPLEX NUMBER SYSTEM

In the following I shall denote the expression $\sqrt{-1}$ by the letter i so that ii = −1.

Leonhard Euler in a paper presented to the Saint Petersburg Academy in 1777.

In elementary algebra, complex numbers appear as expressions a + bi, where a and b are ordinary real numbers and $i^2 = -1$…. Complex numbers are manipulated by the usual rules of algebra, with the convention that i^2 is to be replaced by −1 whenever it occurs.

Ralph Boas in *Invitation to Complex Analysis* (New York: Random House, 1987)

The preceding quotation from Professor Boas summarizes the basic approach we will follow in this section. Near the end of the section, after you've become accustomed to working with complex numbers, we'll present a formal list of some of the basic definitions and properties that can be used to develop the subject more rigorously, as is required in more advanced courses.

When we solve equations in Chapter 2, you'll see instances in which the real-number system proves to be inadequate. In particular, since the square of a real number is never negative, there is no real number x such that $x^2 = -1$. To overcome this inconvenience, mathematicians define the symbol i by the equation

$$i^2 = -1$$

For reasons that are more historical than mathematical, i is referred to as the **imaginary unit.** This name is unfortunate in a sense, because to an engineer or a mathematician, i is neither less "real" nor less tangible than any real number. Having said this, however, we do have to admit that i does not belong to the real-number system.

Algebraically, we operate with the symbol i as if it were any letter in a polynomial expression. However, when we see i^2, we must remember to replace it by −1. Here are four sample calculations involving i.

1. $3i + 2i = 5i$
2. $-2i^2 + 6i = -2(-1) + 6i = 2 + 6i$
3. $(-i)^2 = i^2 = -1$
4. $0i = 0$

An expression of the form $a + bi$, where a and b are real numbers, is called a **complex number.*** Thus, four examples of complex numbers are:

$2 + 3i$

$4 + (-5)i$ (usually written $4 - 5i$)

$1 - \sqrt{2}\,i$ (also written $1 - i\sqrt{2}$)

$\dfrac{1}{2} + \dfrac{3}{2}i$ $\left(\text{also written } \dfrac{1 + 3i}{2}\right)$

---

*The phrase *complex number* is attributed to Carl Friedrich Gauss (1777–1855). The phrase *imaginary number* originated with René Descartes (1596–1650).

Given a complex number $a + bi$, we say that a is the **real part** of $a + bi$, and b is the **imaginary part** of $a + bi$. For example, the real part of $3 - 4i$ is 3, and the imaginary part is -4. Observe that both the real part and the imaginary part of a complex number are themselves real numbers.

We define the notion of *equality* for complex numbers in terms of their real and imaginary parts. Two complex numbers are said to be **equal** if their corresponding real and imaginary parts are equal. We can write this definition symbolically as follows:

$$a + bi = c + di \quad \text{if and only if} \quad a = c \quad \text{and} \quad b = d$$

EXAMPLE 1   Determine real numbers c and d such that $10 + 4i = 2c + di$.

Solution   Equating the real parts of the two complex numbers gives us $2c = 10$, and therefore $c = 5$. Similarly, equating the imaginary parts yields $d = 4$. These are the required values for c and d.   ∎

As the next example indicates, addition, subtraction, and multiplication of complex numbers are carried out using the usual rules of algebra, with the understanding (as mentioned before) that i^2 is always to be replaced with -1. (We'll discuss division of complex numbers subsequently.)

EXAMPLE 2   Let $z = 2 + 5i$ and $w = 3 - 4i$. Compute each of the following.

(a) $w + z$ **(b)** $3z$ **(c)** $w - 3z$ **(d)** zw **(e)** wz

Solution
(a) $w + z = (3 - 4i) + (2 + 5i) = (3 + 2) + (-4i + 5i) = 5 + i$
(b) $3z = 3(2 + 5i) = 6 + 15i$
(c) $w - 3z = (3 - 4i) - (6 + 15i) = (3 - 6) + (-4i - 15i) = -3 - 19i$
(d) $zw = (2 + 5i)(3 - 4i) = 6 - 8i + 15i - 20i^2$
$\qquad\qquad = 6 + 7i - 20(-1) = 26 + 7i$
(e) $wz = (3 - 4i)(2 + 5i) = 6 + 15i - 8i - 20i^2$
$\qquad\qquad = 6 + 7i - 20(-1) = 26 + 7i$ ∎

If you look over the result of part (a) in Example 2, you can see that the sum is obtained simply by adding the corresponding real and imaginary parts of the given numbers. Likewise in part (c), the difference is obtained by subtracting the corresponding real and imaginary parts. In part (d), however, notice that the product is *not* obtained in a similar fashion. (Exercise 54 provides some perspective on this.) Finally, notice that the results in parts (d) and (e) are identical. In fact, it can be shown that for any two complex numbers z and w, we always have $zw = wz$. In other words, just as with real numbers, multiplication of complex numbers is commutative. Furthermore, along the same lines, it can be shown that all of the properties of real numbers listed in Section 1.2 continue to hold for complex numbers. We'll return to this point at the end of this section and in the exercises.

As background for the discussion of division of complex numbers, we introduce the notion of a *complex conjugate,* or simply a *conjugate.**

*The term "conjugates" (*conjuguées*) was introduced by the nineteenth century French mathematician A. L. Cauchy in his text *Cours d'Analyse algébrique* (Paris: 1821)

DEFINITION   Complex Conjugate

	EXAMPLES
Let $z = a + bi$. The **complex conjugate** of z, denoted by $\bar{z}$, is defined by $$\bar{z} = a - bi$$	If $z = 3 + 4i$, then $\bar{z} = 3 - 4i$. If $w = 9 - 2i$, then $\bar{w} = 9 + 2i$.

EXAMPLE 3   **(a)** If $z = 6 - 3i$, compute $z\bar{z}$.   **(b)** If $w = a + bi$, compute $w\bar{w}$.

Solution   **(a)** $z\bar{z} = (6 - 3i)(6 + 3i)$ **(b)** $w\bar{w} = (a + bi)(a - bi)$
$\qquad\qquad = 36 + 18i - 18i - 9i^2$ $\qquad\qquad\qquad = a^2 - abi + abi - b^2i^2$
$\qquad\qquad = 36 + 9 = 45$ $\qquad\qquad\qquad\qquad = a^2 + b^2$

NOTE   The result in part (b) shows that the product of a complex number and its conjugate is always a real number. ■■■

Quotients of complex numbers are easy to compute using conjugates. Suppose, for example, that we wish to compute the quotient.

$$\frac{5 - 2i}{3 + 4i}$$

To do this, take the conjugate of the denominator, namely, $3 - 4i$, and then multiply the given fraction by $\dfrac{3 - 4i}{3 - 4i}$, which equals 1 (assuming that the usual rules of algebra are in force). This yields

$$\frac{5 - 2i}{3 + 4i} = \frac{5 - 2i}{3 + 4i} \cdot \frac{3 - 4i}{3 - 4i}$$

$$= \frac{15 - 26i + 8i^2}{9 - 16i^2}$$

$$= \frac{7 - 26i}{25} \qquad \text{since } 8i^2 = -8 \text{ and } -16i^2 = 16$$

$$= \frac{7}{25} - \frac{26}{25}i \qquad \text{as required}$$

In the box that follows, we summarize our procedure for computing quotients. The condition $w \neq 0$ means that w is any complex number other than $0 + 0i$.

Procedure for Computing Quotients

Let z and w be two complex numbers, and $w \neq 0$. Then z/w is computed as follows:

$$\frac{z}{w} = \frac{z}{w} \cdot \frac{\bar{w}}{\bar{w}}$$

EXAMPLE 4 Let $z = 3 + 4i$ and $w = 1 - 2i$. Compute each of the following:

(a) $\dfrac{1}{z}$; (b) $\dfrac{z}{w}$.

Solution (a) $\dfrac{1}{z} = \dfrac{1}{z} \cdot \dfrac{\bar{z}}{\bar{z}} = \dfrac{1}{3 + 4i} \cdot \dfrac{3 - 4i}{3 - 4i}$

$= \dfrac{3 - 4i}{9 - 16i^2} = \dfrac{3 - 4i}{25}$

$= \dfrac{3}{25} - \dfrac{4}{25}i$

(b) $\dfrac{z}{w} = \dfrac{z}{w} \cdot \dfrac{\bar{w}}{\bar{w}} = \dfrac{3 + 4i}{1 - 2i} \cdot \dfrac{1 + 2i}{1 + 2i}$

$= \dfrac{3 + 10i + 8i^2}{1 - 4i^2} = \dfrac{-5 + 10i}{5}$

$= -1 + 2i$

We began this section by defining i by the equation $i^2 = -1$. This can be rewritten

$$i = \sqrt{-1}$$

provided that we agree to certain conventions regarding principal square roots and negative numbers. In dealing with the principal square root of a negative real number, say $\sqrt{-5}$, we shall write

$$\sqrt{-5} = \sqrt{(-1)(5)} = \sqrt{-1}\sqrt{5} = i\sqrt{5}$$

In other words, we are allowing the use of the rule $\sqrt{ab} = \sqrt{a}\sqrt{b}$ when a is -1 and b is a positive real number. However, the rule $\sqrt{ab} = \sqrt{a}\sqrt{b}$ *cannot* be used when both a and b are negative. If that were allowed, we could write

$$1 = (-1)(-1)$$

and then

$$\sqrt{1} = \sqrt{(-1)(-1)} = \sqrt{-1}\sqrt{-1} = (i)(i)$$

Consequently,

$$1 = \sqrt{1} = i^2 = -1$$

or

$$1 = -1$$

which is a contradiction. Again, the point here is that the rule $\sqrt{ab} = \sqrt{a}\sqrt{b}$ cannot be applied when both a and b are negative.

EXAMPLE 5 Simplify:

(a) i^4; (b) i^{101}; (c) $\sqrt{-12} + \sqrt{-27}$; (d) $\sqrt{-9}\sqrt{-4}$.

Solution   **(a)** We make use of the defining equation for i, which is $i^2 = -1$. Thus, we have

$$i^4 = (i^2)^2 = (-1)^2 = 1$$

(The result, $i^4 = 1$, is worth remembering.)

(b) $i^{101} = i^{100}i = (i^4)^{25}i = 1^{25}i = i$

(c) $\sqrt{-12} + \sqrt{-27} = \sqrt{12}\sqrt{-1} + \sqrt{27}\sqrt{-1}$
$$= \sqrt{4}\sqrt{3}\,i + \sqrt{9}\sqrt{3}\,i$$
$$= 2\sqrt{3}\,i + 3\sqrt{3}\,i = 5\sqrt{3}\,i$$

(d) $\sqrt{-9}\sqrt{-4} = (3i)(2i) = 6i^2 = -6$

Note: $\sqrt{-9}\sqrt{-4} \neq \sqrt{36}$; why?

Near the beginning of this section, we mentioned that we'd eventually present a formal list of some of the basic definitions and properties that can be used to develop the subject more rigorously. These are given in the two boxes that follow. As you'll see in some of the exercises, these definitions and properties are indeed consistent with our work in this section and with the properties of real numbers that were listed in Section 1.2.

DEFINITION   Addition, Subtraction, Multiplication, and Division for Complex Numbers

Let $z = a + bi$ and $w = c + di$. Then $z + w$, $z - w$, and zw are defined as follows.

1. $z + w = (a + bi) + (c + di) = (a + c) + (b + d)i$

2. $z - w = (a + bi) - (c + di) = (a - c) + (b - d)i$

3. $zw = (a + bi)(c + di) = (ac - bd) + (ad + bc)i$

Furthermore, if $w \neq 0$, then z/w is defined as follows.

4. $\dfrac{z}{w} = \dfrac{a + bi}{c + di} \cdot \dfrac{c - di}{c - di} = \left(\dfrac{ac + bd}{c^2 + d^2}\right) + \left(\dfrac{bc - ad}{c^2 + d^2}\right)i$

PROPERTY SUMMARY   PROPERTIES OF COMPLEX CONJUGATES

If $z = a + bi$, the complex conjugate, $\bar{z} = a - bi$, has the following properties.

1. $\bar{\bar{z}} = z$

2. $z = \bar{z}$ if and only if z is a real number

3. $\overline{z + w} = \bar{z} + \bar{w}$; $\overline{z - w} = \bar{z} - \bar{w}$

4. $\overline{zw} = \bar{z}\,\bar{w}$; $\overline{\left(\dfrac{z}{w}\right)} = \left(\dfrac{\bar{z}}{\bar{w}}\right)$

5. $(\bar{z})^n = \overline{(z^n)}$ for each natural number n

EXERCISE SET 1.11

A

1. Complete the table.

i^2	i^3	i^4	i^5	i^6	i^7	i^8
-1						

2. Simplify the following expression, and write the answer in the form $a + bi$.

$$1 + 3i - 5i^2 + 4 - 2i - i^3$$

For Exercises 3 and 4, specify the real and imaginary parts of each complex number.

3. (a) $4 + 5i$ (b) $4 - 5i$
   (c) $\frac{1}{2} - i$ (d) $16i$

4. (a) $-2 + \sqrt{7}i$ (b) $1 + 5^{1/3}i$
   (c) $-3i$ (d) 0

5. Determine the real numbers c and d such that

$$8 - 3i = 2c + di$$

6. Determine the real numbers a and b such that

$$27 - 64i = a^3 - b^3i$$

7. Simplify each of the following.
   (a) $(5 - 6i) + (9 + 2i)$ (b) $(5 - 6i) - (9 + 2i)$

8. If $z = 1 + 4i$, compute $z - 10i$.

9. Compute each of the following.
   (a) $(3 - 4i)(5 + i)$ (b) $(5 + i)(3 - 4i)$
   (c) $\dfrac{3 - 4i}{5 + i}$ (d) $\dfrac{5 + i}{3 - 4i}$

10. Compute each of the following.

   (a) $(2 + 7i)(2 - 7i)$ (b) $\dfrac{-1 + 3i}{2 + 7i}$

   (c) $\dfrac{1}{2 + 7i}$ (d) $\dfrac{1}{2 + 7i} \cdot (-1 + 3i)$

In Exercises 11–36, evaluate each expression using the values $z = 2 + 3i$, $w = 9 - 4i$, *and* $w_1 = -7 - i$.

11. (a) $z + w$ (b) $\bar{z} + w$ (c) $z + \bar{z}$
12. (a) $\bar{z} + \bar{w}$ (b) $\overline{(z + w)}$ (c) $w - \bar{w}$
13. $(z + w) + w_1$
14. $z + (w + w_1)$
15. zw
16. wz
17. $z\bar{z}$
18. $w\bar{w}$
19. $z(ww_1)$
20. $(zw)w_1$
21. $z(w + w_1)$
22. $zw + zw_1$

23. $z^2 - w^2$
24. $(z - w)(z + w)$
25. $(zw)^2$
26. z^2w^2
27. z^3
28. z^4
29. $\dfrac{z}{w}$
30. $\dfrac{w}{z}$
31. $\dfrac{\bar{z}}{\bar{w}}$
32. $\overline{\left(\dfrac{z}{w}\right)}$
33. $\dfrac{z}{\bar{z}}$
34. $\dfrac{\bar{z}}{z}$
35. $\dfrac{w - \bar{w}}{2i}$
36. $\dfrac{w + \bar{w}}{2}$

For Exercises 37–40, compute each quotient.

37. $\dfrac{i}{5 + i}$
38. $\dfrac{1 - i\sqrt{3}}{1 + i\sqrt{3}}$
39. $\dfrac{1}{i}$
40. $\dfrac{i + i^2}{i^3 + i^4}$

In Exercises 41–48, simplify each expression.

41. $\sqrt{-49} + \sqrt{-9} + \sqrt{-4}$
42. $\sqrt{-25} + i$
43. $\sqrt{-20} - 3\sqrt{-45} + \sqrt{-80}$
44. $\sqrt{-4}\sqrt{-4}$
45. $1 + \sqrt{-36}\sqrt{-36}$ **46.** $i - \sqrt{-100}$
47. $3\sqrt{-128} - 4\sqrt{-18}$ **48.** $64 + \sqrt{-64}\sqrt{-64}$

49. Let $z = a + bi$ and $w = c + di$. Compute each of the following, and then check that your results agree with the definitions in the first box on page 65.

   (a) $z + w$ (b) $z - w$ (c) zw (d) $\dfrac{z}{w}$

B

50. Show that $\left(\dfrac{-1 + i\sqrt{3}}{2}\right)^2 + \left(\dfrac{-1 - i\sqrt{3}}{2}\right)^2 = -1$.

51. Let $z = \dfrac{-1 + i\sqrt{3}}{2}$ and $w = \dfrac{-1 - i\sqrt{3}}{2}$. Verify the following statements.
   (a) $z^3 = 1$ and $w^3 = 1$ (b) $zw = 1$
   (c) $z = w^2$ and $w = z^2$
   (d) $(1 - z + z^2)(1 + z - z^2) = 4$

52. Let $z = a + bi$ and $w = c + di$.
   (a) Show that $\bar{\bar{z}} = z$.
   (b) Show that $\overline{(z + w)} = \bar{z} + \bar{w}$.

53. Show that the complex number $0(= 0 + 0i)$ has the following properties.
    (a) $0 + z = z$ and $z + 0 = z$, for all complex numbers z.
       Hint: Let $z = a + bi$.
    (b) $0 \cdot z = 0$ and $z \cdot 0 = 0$, for all complex numbers z.

54. This exercise indicates one of the reasons why multiplication of complex numbers is not carried out by simply multiplying the corresponding real and imaginary parts of the numbers. (Recall that addition and subtraction *are* carried out in this manner.) Suppose for the moment that we were to define multiplication in this seemingly less complicated way:

    $$(a + bi)(c + di) = ac + (bd)i \qquad (*)$$

    (a) Compute $(2 + 3i)(5 + 4i)$, assuming that multiplication is defined by $(*)$.
    (b) Still assuming that multiplication is defined by $(*)$, find two complex numbers z and w such that $z \neq 0$, $w \neq 0$, but $zw = 0$ (where 0 denotes the complex number $0 + 0i$).

    Now notice that the result in part (b) is contrary to our expectation or desire that the product of two nonzero numbers should be nonzero, as is the case for real numbers. On the other hand, it can be shown that when multiplication is carried out as described in the text, then the product of two complex numbers is nonzero if and only if both factors are nonzero.

55. (a) Show that addition of complex numbers is commutative. That is, show that $z + w = w + z$ for all complex numbers z and w.   *Hint:* Let $z = a + bi$ and $w = c + di$.
    (b) Show that multiplication of complex numbers is commutative. That is, show that $zw = wz$ for all complex numbers z and w.

56. Let $z = a + bi$.
    (a) Show that $(\overline{z})^2 = \overline{(z^2)}$.
    (b) Show that $(\overline{z})^3 = \overline{(z^3)}$.

C

57. Let a and b be real numbers. Find the real and imaginary parts of the quantity

    $$\frac{a + bi}{a - bi} + \frac{a - bi}{a + bi}$$

58. Find the real and imaginary parts of the quantity

    $$\left(\frac{a + bi}{a - bi}\right)^2 - \left(\frac{a - bi}{a + bi}\right)^2.$$

59. Find the real part of $\dfrac{(a + bi)^2}{a - bi} - \dfrac{(a - bi)^2}{a + bi}$.

60. (a) Let $\alpha = \left(\dfrac{\sqrt{a^2 + b^2} + a}{2}\right)^{1/2}$ and $\beta = \left(\dfrac{\sqrt{a^2 + b^2} - a}{2}\right)^{1/2}$. Show that the square of the complex number $\alpha + \beta i$ is $a + bi$.
    (b) Use the result in part (a) to find a complex number z such that $z^2 = i$.
    (c) Use the result in part (a) to find a complex number z such that $z^2 = -7 + 24i$.

One of the basic properties of real numbers is that if a product is equal to zero, then at least one of the factors is zero. Exercises 61 and 62 show that this property also holds for complex numbers. For Exercises 61 and 62, assume that $zw = 0$, where $z = a + bi$ and $w = c + di$.

61. If $a \neq 0$, prove that $w = 0$. (That is, prove that $c = d = 0$.)

62. If $b \neq 0$, prove that $w = 0$. (That is, prove that $c = d = 0$.)

CHAPTER ONE   SUMMARY OF PRINCIPAL TERMS AND NOTATION

TERMS OR NOTATIONS	PAGE REFERENCE	COMMENTS
1. Coefficient, base, exponent	2	In the expression ax^n, the number a is the coefficient, x is the base, and n is the exponent.
2. Commutative, associative, identity, inverse, and distributive properties	6	These are some of the most basic properties of the real number system. All of these properties are also valid for the complex number system.

TERMS OR NOTATIONS	PAGE REFERENCE	COMMENTS
3. Natural numbers, integers, rational numbers, and irrational numbers	12	The box on page 12 provides both definitions and examples. Also note the theorem in the box that explains how to distinguish between rationals and irrationals in terms of their decimal representations.
4. $a < b$	14	a is less than b.
$b > a$		b is greater than a.
5. (a, b)	14	The open interval (a, b) consists of all real numbers between a and b, excluding a and b.
$[a, b]$		The closed interval $[a, b]$ consists of all real numbers between a and b, including a and b.
6. (a, ∞)	15	The unbounded interval (a, ∞) consists of all real numbers x such that $x > a$. The infinity symbol, ∞, does not denote a real number. It is used in the context (a, ∞) to indicate that the interval has no right-hand boundary. For the definitions of the other types of unbounded intervals, see the box on page 15.
7. $|x|$	17, 18	The absolute value of a real number x is the distance of x from the origin. This is equivalent to the following algebraic definition: $$|x| = \begin{cases} x & \text{if } x \geq 0 \\ -x & \text{if } x < 0 \end{cases}$$
8. $a^0 = 1$ $a^{-n} = 1/a^n$ $a^{m/n} = \sqrt[n]{a^m} = (\sqrt[n]{a})^m$	24 25 38	These are the defining equations for zero as an exponent, for negative exponents, and for rational exponents. The definition $a^{m/n}$ presupposes that m/n is in lowest terms and that $\sqrt[n]{a}$ exists.
9. Scientific notation	27	This is a convenient and standard form that is used to express very large and very small numbers. The box on page 27 gives directions for converting to scientific notation. See also the comments in Appendix A1.
10. Constant, variable, and the domain convention	42	By a constant we mean either a particular number (such as -8 or π) or a letter whose value remains fixed (although perhaps unspecified) throughout a given discussion. In contrast, a variable is a letter for which we can substitute any number from a given set of numbers. The given set is called the domain of the variable. According to the domain convention, the domain of a variable in a given expression consists of all real numbers for which the expression is defined.
11. Polynomial	43	A polynomial is an expression of the form $a_n x^n + a_{n-1} x^{n-1} + \cdots + a_1 x + a_0$. The individual expressions $a_k x^k$ making up the polynomial are called *terms*. The numbers a_k are called *coefficients*.
12. Degree of a polynomial	43	For a polynomial that is not simply a constant, the degree is the highest power (that is, the exponent) of x that appears. The degree of a nonzero constant polynomial is zero. Degree is undefined for the zero polynomial.
13. Special products	45	The box on page 45 lists four basic types of products that you should memorize because they occur so frequently.

TERMS OR NOTATIONS	PAGE REFERENCE	COMMENTS
14. Factoring techniques	51	The techniques for factoring listed in the box on page 51 are fundamental for simplifying expressions (in Section 1.10) and for solving equations (in Chapter 2).
15. $i^2 = -1$	61	This is the defining equation for the symbol i. The quantity i is not a real number.
16. Complex numbers	61	An expression of the form $a + bi$, where a and b are real numbers, is a complex number. With the exception of division, complex numbers are combined using the usual rules of algebra, along with the convention $i^2 = -1$. Division is carried out using complex conjugates (defined below), as indicated in the box on page 63. All of the properties of real numbers listed on page 6 also apply to complex numbers.
17. If $z = c + di$, then $\bar{z} = c - di$.	63	$\bar{z}$ is called the *conjugate* of z.

WRITING MATHEMATICS

Many common errors in algebra can be traced to a confusion in the roles of addition and multiplication. For instance, the error $a^2 a^3 = a^6$ occurs from mistakenly multiplying rather than adding, the exponents. As another example, the error

$$\frac{2a + b}{2} = a + b$$

may have to do with the (visual) similarity of the expressions $\dfrac{2a + b}{2}$ and $\dfrac{2a \times b}{2}$. For the latter expression, we do have

$$\frac{2a \times b}{2} = a \times b$$

In Exercises 1–5, (a) give a numerical example to show that the equation is not, in general, true; (b) explain (in complete sentences) how the error may be due to a confusion in the roles of multiplication and addition. Also, if you feel that there is a different or additional reason behind the error, describe it.

1. $[(2a - 3)^2]^3 = (2a - 3)^5$
2. $\sqrt{b + x} = b + \sqrt{x}$
3. $(a + b^2)^3 = a^3 + b^6$
4. $a(b - 2) = ab - 2$
5. $(a + b)^{-2} = a^{-2} + b^{-2}$

CHAPTER ONE REVIEW EXERCISES

Factor each expression in Exercises 1–20.

1. $a^2 - 16b^2$
2. $x^2 + ax + yx + ay$
3. $8 - (a + 1)^3$
4. $x^2 - 18x + 81$
5. $a^2x^3 + 2ax^2b + b^2x$
6. $8a^2x^2 + 16a^3$
7. $8x^2 + 6x + 1$
8. $12x^2 - 2x - 4$
9. $a^4x^4 - x^8a^8$
10. $2x^2 - 2bx + ax - ab$
11. $8 + 12a + 6a^2 + a^3$
12. $(x^2 + 2x - 8)^2 - (2x + 1)^2$
13. $4x^2y^2z^3 - 3xyz^3 - z^3$
14. $(x + y - 1)^2 - (x - y + 1)^2$
15. $1 - x^6$
16. $12x^3 + 44x^2 - 16x$
17. $a^2x^2 + 2abx + b^2 - 4a^2b^2x^2$
18. $a^2 + 2ab + b^2 + a + b$
19. $a^2 - b^2 + ac - bc + a^2b - b^2a$
20. $5(a + 1)^2 + 29(a + 1) - 144$

Simplify each expression in Exercises 21–36. (Assume that the letters represent positive quantities.)

21. $4^{3/2}$
22. $49^{-1/2}$
23. $[(3025)^{1/2}]^0$
24. $(-125)^{2/3}$
25. $8^{-4/3}$
26. $10^{-3} + 10^3$
27. $(-243)^{-2/5}$
28. $\left(\frac{625}{16}\right)^{3/4}$
29. $(a^2b^6c^8)^{1/2}$
30. $\sqrt{25a^4b^{10}}$
31. $\sqrt{a^3b^5}\sqrt{4ab^3}$
32. $\sqrt{28a^3} + a\sqrt{63a}$
33. $\sqrt[3]{16} - \sqrt[3]{-54}$
34. $\sqrt[5]{\dfrac{-32a^{15}b^{10}}{c^5}}$
35. $\sqrt{24a^2b^3} + ba\sqrt{54b}$
36. $\sqrt{\sqrt{256x^8}}$

Simplify each expression in Exercises 37–40. (Do not assume that the letters represent positive quantities.)

37. $\sqrt{t^2}$
38. $\sqrt[3]{t^3}$
39. $\sqrt[4]{16x^4}$
40. $\sqrt[3]{8x^4}$

In Exercises 41–44, rewrite each expression in two ways: (a) using rational exponents; (b) using a single radical. Assume that all of the letters represent nonnegative quantities.

41. $\sqrt[3]{x}\sqrt[4]{x^3}$
42. $\sqrt[4]{\sqrt[3]{x}}$
43. $\sqrt{\sqrt[3]{t}\sqrt[5]{t^4}}$
44. $\sqrt[3]{\dfrac{\sqrt{a}\sqrt[4]{b^2}}{\sqrt[5]{c^3}}}$

Rationalize the denominators in Exercises 45–51.

45. $6/\sqrt{3}$
46. $\sqrt{3}/\sqrt{2}$
47. $\dfrac{1}{\sqrt{6} - \sqrt{3}}$
48. $\dfrac{1}{\sqrt{x + 2} - \sqrt{x}}$
49. $\dfrac{\sqrt{a^2 + x^2} + \sqrt{a^2 - x^2}}{\sqrt{a^2 + x^2} - \sqrt{a^2 - x^2}}$
50. $\dfrac{1}{\sqrt{a} + \sqrt{b} + \sqrt{c}}$ *Hint:* First multiply numerator and denominator by $\sqrt{a} + \sqrt{b} - \sqrt{c}$. Then multiply numerator and denominator by $(a + b - c) - 2\sqrt{ab}$.
51. $\dfrac{1}{\sqrt{2} + \sqrt{3} + \sqrt{5}}$ *Hint:* See Exercise 50.

In Exercises 52–55, rationalize the numerators.

52. $\dfrac{\sqrt{x} - 5}{x - 25}$
53. $\dfrac{\sqrt{t} - a}{t - a^2}$
54. $\dfrac{\sqrt{a^2 + x^2} - \sqrt{a^2 - x^2}}{\sqrt{a^2 + x^2} + \sqrt{a^2 - x^2}}$
55. $\dfrac{\sqrt[3]{x} - 2}{x - 8}$ *Hint:* The denominator can be written $(x^{1/3})^3 - 2^3$.

Use the Special Products formulas in Section 1.8 to compute the products in Exercises 56–70.

56. $(9x - y)(9x + y)$
57. $(3x + 1)^2$
58. $(2ab - 3)^2$
59. $(3x^2 + y^2)^2$
60. $(ax + b)^2(ax - b)^2$
61. $(2a + 3)^3$
62. $(x^2 - 2)^3$
63. $(1 - 3a)(1 + 3a + 9a^2)$
64. $(x^n + y^n)(x^{2n} - x^ny^n + y^{2n})$
65. $(x^{1/2} - y^{1/2})(x^{1/2} + y^{1/2})$
66. $(\sqrt{x} - \sqrt{3})(\sqrt{x} + \sqrt{3})$
67. $(x^{1/3} + y^{1/3})^3$
68. $(x^{1/3} + y^{1/3})(x^{2/3} - x^{1/3}y^{1/3} + y^{2/3})$
69. $(x^2 - 3x + 1)^2$ *Hint:* Rewrite the expression as $[x^2 - (3x - 1)]^2$.
70. $(x^2 + x - 1)(x^2 + x + 1)$

For Exercises 71–78, carry out the indicated operations and express your answer in the standard form $a + bi$.

71. $(3 - 2i)(3 + 2i) + (1 + 3i)^2$
72. $2i(1 + i)^2$
73. $(1 + i\sqrt{2})(1 - i\sqrt{2}) + (\sqrt{2} + i)(\sqrt{2} - i)$

74. $\dfrac{2 + 3i}{1 + i}$

75. $\dfrac{3 - i\sqrt{3}}{3 + i\sqrt{3}}$

76. $\dfrac{1 + i}{1 - i} + \dfrac{1 - i}{1 + i}$

77. $-\sqrt{-2}\sqrt{-9} + \sqrt{-8} - \sqrt{-72}$

78. $\dfrac{\sqrt{-4} - \sqrt{-3}\sqrt{-3}}{\sqrt{-100}}$

79. The **real part** of a complex number z is denoted by Re(z). For instance, Re($2 + 5i$) = 2. Show that for any complex number z, we have Re(z) = $\frac{1}{2}(z + \bar{z})$.
Hint: Let $z = a + bi$.

80. The **imaginary part** of a complex number z is denoted by Im(z). For instance, Im($2 + 5i$) = 5. Show that for any complex number z, we have Im(z) = $\dfrac{1}{2i}(z - \bar{z})$.

81. The **absolute value** of the complex number $a + bi$ is defined by $|a + bi| = \sqrt{a^2 + b^2}$.
(a) Compute $|6 + 2i|$ and $|6 - 2i|$.
(b) As you know, the absolute value of the real number -3 is 3. Now write -3 in the form $-3 + 0i$ and compute its absolute value using the new definition. (The point here is to observe that the two results agree.)
(c) Let $z = a + bi$. Show that $z\bar{z} = |z|^2$.

In Exercises 82–86, identify the properties of the real numbers that justify each statement. (The properties are listed in Section 1.2.)

82. $(x + y)^2 + z^2 = z^2 + (x + y)^2$

83. $(x + 1)(x + 2) = (x + 1)x + (x + 1)2$
$= (x^2 + x) + (2x + 2)$

84. $(ab)c = a(bc) = a(cb) = (ac)b$

85. $(x + a)y + (x + a)z = (x + a)(y + z)$

86. Since 3 and $\sqrt{\pi}$ are real numbers, $3\sqrt{\pi}$ is also a real number.

For Exercises 87–91, rewrite the statements using absolute values and inequalities or equalities.

87. The distance between x and 6 is 2.
88. The distance between x and a is less than $\frac{1}{2}$.
89. The distance between a and b is 3.
90. The distance between x and -1 is 5.
91. The distance between x and 0 exceeds 10.
92. What can you say about x if $|x - 5| = 0$?

Rewrite each of the expressions in Exercises 93–98 in a form not containing absolute values.

93. $\left|\sqrt{6} - 2\right|$

94. $\left|2 - \sqrt{6}\right|$

95. $\left|x^4 + x^2 + 1\right|$

96. (a) $|x - 3|$ if $x < 3$
(b) $|x - 3|$ if $x > 3$

97. $|x - 2| + |x - 3|$ if:
(a) $x < 2$ (b) $2 < x < 3$ (c) $x > 3$

98. $|x + 2| + |x - 1|$ if:
(a) $x < -2$ (b) $-2 < x < 1$ (c) $x > 1$

99. Determine whether each of the following is true or false.
(a) $-1 \leq 0$ (b) $\sqrt{\frac{3}{2}} \geq 0$
(c) $\frac{11}{10} \geq \frac{12}{11}$ (d) $0.9 \geq 1$

In Exercises 100–104, sketch each interval on a number line.

100. (3, 5)

101. (3, 5]

102. The set of all negative real numbers that are in the interval $[-5, 2]$.

103. $(-\infty, 4)$

104. $[-1, \infty)$

In Exercises 105–108, write the numbers using scientific notation.

105. 0.0014

106. 186,000

107. 12.001

108. 81×10^{12}

In Exercises 109–116, carry out the indicated operations.

109. $1 - 2[3 - 4(1 - 5)]$

110. $x^2 - 4x - (1 - 2x^2)$

111. $1 + 3(x^2 - 5x - 4) - [1 - (15x - 3x^2)]$

112. $(ax^2 - 1)(ax^2 + 1)$

113. $(x - 1)(x + 1)(x + 2)$

114. $(x + a)(x^2 + ax + 1)$

115. $x^2 + 4 - (x - 1)(x - 2)$

116. $(x^2 - x + 1)(x^2 + x - 1) + [(x - 1)^2 - x^4]$

For Exercises 117–120, use long division to determine the quotient and remainder.

117. $(x^3 - 2x^2 + x + 4) \div (x - 1)$

118. $(x^3 + 4x^2 - 2x - 5) \div (x + 1)$

119. $(x^4 - x^2 + 1) \div (x^2 + 1)$

120. $(2x^4 + x^3 - 4x^2 - 8x - 1) \div (x + 2)$

In Exercises 121–125, sketch the intervals described by the given inequalities.

121. $|x - 6| < 3$

122. $\left|x - \frac{1}{2}\right| < 1$

123. $|x + 1| \geq 1$

124. $|x| \geq 5$

125. (a) $0 < |x - 4| < 5$ (b) $|x - 4| < 5$

In Exercises 126–128, express the number in the form p/q, where p and q are integers.

126. $12.\overline{7}$ **127.** $12.\overline{71}$ **128.** $12.3\overline{91}$

129. Polynomials can be used to approximate more complicated expressions. For example, in calculus it is shown that when x is close to zero, the expression $\sqrt{1+x}$ can be approximated by the polynomial $1 + \frac{1}{2}x$. Complete the table to see evidence of this. (Use a calculator to compute $\sqrt{1+x}$; report the values to six decimal places without rounding off.)

x	$1 + \dfrac{x}{2}$	$\sqrt{1+x}$
0.1		
0.01		
0.001		

130. Using calculus, it can be shown that when x is close to zero, the expression $1/\sqrt{1-x^2}$ can be approximated by the polynomial $1 + \frac{1}{2}x^2 + \frac{3}{8}x^4$. Complete the table to see evidence of this. (Use a calculator and round off your answers to six significant digits.)

x	$1 + \dfrac{x^2}{2} + \dfrac{3x^4}{8}$	$1/\sqrt{1-x^2}$
0.1		
0.01		

131. (a) Use a calculator to evaluate the expression
$$\frac{\sqrt{3}-\sqrt{5}}{\sqrt{2}+\sqrt{7-3\sqrt{5}}}$$
(b) Use a calculator to evaluate the expression $\sqrt{5}/5$. If you do the calculator work carefully, your results in parts (a) and (b) will suggest (but not prove!) that the two expressions are equal. Follow

steps (c), (d), and (e) to prove that the two expressions are indeed equal.

(c) Multiply the expression in part (a) by
$$\frac{\sqrt{2}-\sqrt{7-3\sqrt{5}}}{\sqrt{2}-\sqrt{7-3\sqrt{5}}},$$
which equals 1. After carrying out the indicated operations, you should obtain
$$\frac{\sqrt{6-2\sqrt{5}}-2\sqrt{9-4\sqrt{5}}}{-5+3\sqrt{5}}$$

(d) Verify that $\sqrt{6-2\sqrt{5}} = \sqrt{5}-1$. Also verify that $\sqrt{9-4\sqrt{5}} = \sqrt{5}-2$.

(e) Use the results in part (d) to simplify the expression in part (c). The final result, after rationalizing the denominator, should be $\sqrt{5}/5$.

In Exercises 132–136, verify that the formulas are correct by carrying out the operations indicated on the left-hand side of the equations. (This list of formulas appears in A Treatise on Algebra, by George Peacock, published in 1845.)

132. $\dfrac{1}{a+bi} + \dfrac{1}{a-bi} = \dfrac{2a}{a^2+b^2}$

133. $\dfrac{1}{a-bi} - \dfrac{1}{a+bi} = \dfrac{2bi}{a^2+b^2}$

134. $\dfrac{a+bi}{a-bi} + \dfrac{a-bi}{a+bi} = \dfrac{2(a^2-b^2)}{a^2+b^2}$

135. $\dfrac{a+bi}{a-bi} - \dfrac{a-bi}{a+bi} = \dfrac{4abi}{a^2+b^2}$

136. $\dfrac{a+bi}{c+di} + \dfrac{a-bi}{c-di} = \dfrac{2(ac+bd)}{c^2+d^2}$

Exercises 137 and 138 are taken (verbatim) from Algebra, by Isaac Todhunter, first published in 1870.

137. Simplify:
$$\frac{bc(x-a)^2}{(a-b)(a-c)} + \frac{ca(x-b)^2}{(b-c)(b-a)} + \frac{ab(x-c)^2}{(c-a)(c-b)}$$

138. If $x = \sqrt[3]{a+\sqrt{a^2+b^3}} + \sqrt[3]{a-\sqrt{a^2+b^3}}$, shew that $x^3 + 3bx - 2a = 0$.

CHAPTER ONE TEST

In Exercises 1–4, factor each expression. (If an expression is irreducible over the integers, state this.)

1. (a) $64 - (x-2)^3$ (b) $w^2y^2 + 16$

2. (a) $2x^2 + 3x - 8$ (b) $6x^2y^2z + xyz - z$

3. $6x(1-x^2)^{-1/2} - 3x^2(1-x^2)^{-3/2}$

4. $9ax + 6bx - 6ay - 4by$

5. Simplify: $\sqrt[3]{54} - \sqrt[3]{-2000}$.

6. Rewrite the expression $(\sqrt[5]{x^2})(\sqrt[7]{x^3})$ in two ways:
   (a) using rational exponents; (b) using a single radical.

7. Rationalize the denominator: $\dfrac{\sqrt{x^2 + 1} + \sqrt{x^2 - 1}}{\sqrt{x^2 + 1} - \sqrt{x^2 - 1}}$.

8. Use one of the Special Products to compute $(2a + 3b^2)^3$.

9. Simplify the following expression. Write the answer in the form $a + bi$.

$$(4 + 2i)(4 - 2i) + \frac{3 + i}{1 + 2i}$$

10. Identify the property (or properties) of the real numbers that justify each statement.
    (a) $(a + b)c + (a + b)d = (a + b)(c + d)$
    (b) $[x^2 y^2](c + 1) = [y^2 x^2](c + 1) = y^2[x^2(c + 1)]$

11. Simplify the expression $|x - 6| + |x - 7|$, given that $6 < x < 7$.

12. Write each number using scientific notation.
    (a) the distance (in light-years) to the Hydra Galaxy: 3,960,000,000
    (b) the wavelength (in meters) of red light: 0.0000006750

13. Use long division to determine the quotient and remainder:
    $(2x^3 - 4x - 9) \div (2x + 3)$.

14. Express the number $31.\overline{2}$ in the form p/q, where p and q are integers.

15. Specify the intervals that are described by the inequalities.
    (a) $|x - 4| < \frac{1}{10}$ (b) $x \geq -2$

CHAPTER TWO

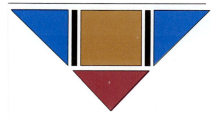

EQUATIONS AND INEQUALITIES

... mathematical studies in the late medieval Latin Europe were stimulated by Latin translations from the Arabic. An important source of information was the treatise on equations written by [the Persian mathematician and astronomer] Mohammed Al-Khwārizmī (c. A.D. 825). The Latin translation of its Arabic title, Liber algebrae et almucabala, *gave the name* algebra *to the theory of equations ...*

From *A Source Book in Mathematics, 1200–1800,* D.J. Struik, ed. (Princeton: Princeton University Press, 1986)

INTRODUCTION

In this chapter we solve equations and inequalities. The first section of the chapter is a warmup; we practice setting up equations that describe some simple situations. The techniques for solving these equations are then developed in Sections 2.2 (Linear Equations) and 2.3 (Quadratic Equations). These techniques are then used in Section 2.4 to solve a variety of applied problems. In Section 2.5 several methods are developed for solving equations that are neither linear nor quadratic. The remaining two sections of the chapter, Sections 2.6 and 2.7, deal with the solution of inequalities rather than equalities.

2.1 SETTING UP EQUATIONS (A WARMUP)

Algebra is a language which does not consist of words but of symbols. If we are familiar with it we can translate into it appropriate sentences of everyday language.

George Polya in *Mathematical Discovery* (New York: John Wiley & Sons, 1981)

In this first section we practice setting up equations that describe some simple situations. In later sections, we'll study techniques for solving (or sometimes analyzing) these equations.

EXAMPLE I   Consider the following problem:

> The sum of two numbers is 4, and the sum of their squares is 106. What are the two numbers?

(a) What is the problem asking for? Assign variables to denote the required quantities.

(b) Restate the problem using algebraic notation.

(c) Find an equation that could be solved for one of the variables (or "unknowns") introduced in part (a).

Solution (a) We first read and reread the problem until we can say just what the problem requires us to find. (By focusing on this, you can initially, at least, skim over many of the details that sometimes seem so overwhelming in a word problem. Often these details are best handled *after* the variables have been assigned.)

The problem asks us to find two numbers. Let x and y denote the two numbers.

(b) See Table 1. [The format in Table 1 is due to the mathematician and master teacher George Polya (1887–1985).]

TABLE 1

Restate the Problem

In English	In Algebraic Notation
Find two numbers	x and y
Their sum is 4	$x + y = 4$
The sum of their squares is 106	$x^2 + y^2 = 106$

(c) From part (b) we have the equations

$$x + y = 4 \tag{1}$$
$$x^2 + y^2 = 106 \tag{2}$$

Each of these equations involves both x and y. To obtain an equation involving only one variable, we first use basic algebra to solve equation (1), the simpler of the two equations, for y in terms of x. [Alternatively, we could solve equation (1) for x in terms of y.] From equation (1), we have $y = 4 - x$, and therefore

$$x^2 + (4 - x)^2 = 106 \qquad \text{replacing } y \text{ with } 4 - x \text{ in equation (2)}$$

This is an equation involving only one of the variables, as required. However, rather than leave the equation in this form, we carry out the multiplication on the left-hand side and collect like terms, as in Chapter 1. As you can readily check, this yields

$$2x^2 - 8x + 16 = 106 \tag{3}$$
$$2x^2 - 8x - 90 = 0 \qquad \text{subtracting 106 from both sides} \tag{4}$$
$$x^2 - 4x - 45 = 0 \qquad \text{dividing both sides by 2} \tag{5}$$

This last equation is our final form for the answer. *Note:* In obtaining equations (4) and (5), we relied on two procedures that should seem familiar to you from basic algebra. To obtain equation (4), we subtracted the same number from both sides of equation (3). To obtain equation (5), we divided both sides of equation (4) by the same nonzero number. Although we won't discuss these types of procedures in detail until the next section, we will make use of them in the present section, assuming, as we have, that they are familiar to you from basic algebra.

EXAMPLE 2   Consider the following problem:

The difference of the cubes of two consecutive integers is 217. What are the integers?

(a) What is the problem asking for? Assign variables or expressions to denote the required quantities.
(b) Restate the problem using algebraic notation.
(c) Find an equation that could be solved for one of the variables introduced in part (a).

Solution   (a) The problem asks us to find two *consecutive* integers. Let x and $x + 1$ denote these consecutive integers.
(b) In Table 2, we've restated the problem using algebraic notation. *Question:* In the table, why is the difference of the cubes denoted by $(x + 1)^3 - x^3$ rather than $x^3 - (x + 1)^3$?

TABLE 2

Restate the Problem

In English	In Algebraic Notation
Find two consecutive integers	x and $x + 1$
The difference of their cubes is 217	$(x + 1)^3 - x^3 = 217$

(c) Working from the equation $(x + 1)^3 - x^3 = 217$ obtained in part (b), we have

$$x^3 + 3x^2 + 3x + 1 - x^3 = 217 \qquad \text{multiplying out } (x + 1)^3$$
$$3x^2 + 3x + 1 = 217 \qquad \text{combining like terms}$$
$$3x^2 + 3x - 216 = 0 \qquad \text{subtracting 217 from both sides}$$
$$x^2 + x - 72 = 0 \qquad \text{dividing both sides by 3}$$

This is the required equation. In Section 2.3 we will develop several methods that can be used to solve it.

The next example that we consider involves a geometric figure. In this case, it's important to draw a sketch of the situation. This helps us to summarize the information and to see relationships that may exist between the variables. In Example 3 we use two familiar facts about a rectangle with width w and length l: the area is lw and the perimeter is $2w + 2l$. (For a summary of other basic formulas from geometry, see the inside front cover of this book.)

EXAMPLE 3   Consider the following problem:

The perimeter of a rectangle is 26 cm and the area is 40 cm². What are the dimensions of the rectangle?

(a) What is the problem asking for? Draw a sketch of the situation, and assign variables to denote the required quantities.

(b) Restate the problem using algebraic notation.
(c) Find an equation that could be solved for one of the variables introduced in part (a).

Solution   **(a)** The problem is asking for the dimensions (i.e., the length and the width) of a rectangle. In Figure 1 we've sketched a rectangle and labeled the length l and the width w.

FIGURE 1

area = 40 cm²
perimeter = 26 cm

(b) See Table 3.
Before going on to part (c), let us note that the equation $2w + 2l = 26$ in Table 3 will be simpler if we divide both sides by 2 to obtain $w + l = 13$.

TABLE 3

Restate the Problem

In English	In Algebraic Notation
Find the width and length of a rectangle	w and l
The perimeter is 26 cm	$2w + 2l = 26$
The area is 40 cm²	$lw = 40$

(c) From part (b) we have two equations:

$$w + l = 13 \qquad\qquad (6)$$
$$lw = 40 \qquad\qquad (7)$$

Each of these equations involves both w and l. We want an equation involving only one of the variables. From equation (6), we have $l = 13 - w$. Using this last equation to substitute for l in equation (7) yields

$$(13 - w)w = 40$$

or

$$13w - w^2 = 40$$

This is an equation involving only one variable, as required. (*Question*: Suppose instead that we had solved equation (7) for l and then substituted in equation (6). Would this lead to the same final equation?) ▮▮▮

The next example involves *simple interest* earned on money in a savings account. As an illustration of simple interest, suppose that you place $100 in a savings account for one year at 10% simple interest. This means that at the end of the year, the bank contributes 10% of $100, or $10, to your account. Furthermore, assuming that you make no subsequent withdrawals or deposits, the bank continues to contribute $10 to your account at the end of each year. The following general formula can be derived for computing simple interest.

> **$I = Prt$**
>
> where I is the simple interest earned on an initial deposit P (the *principal*) after t years at an annual rate r.

To use this formula, we need to convert the annual rate r from a percent to a decimal. For instance, to compute the simple interest earned on a principal of $1000 invested for six years at 10%, we write

$$I = Prt = (\$1000)(0.1)(6) = \$600$$

EXAMPLE 4   Consider the following problem:

Suppose that you open two savings accounts. In the first account you deposit $5000 at an annual simple interest rate of 8%. In the second you deposit $10,000 at 12% (also simple interest). After how many years will the total interest amount to $16,000?

(a) What is the problem asking for? Assign a variable to denote the required quantity.
(b) Find an equation that could be solved for the variable introduced in part (a). (Don't simplify or solve the equation.)

Solution   (a) The problem asks for the number of years required until the interest earned on both accounts adds up to $16,000. Let t denote the required number of years.
(b) We apply the formula $I = Prt$ to each account. After t years, the interest on the first account is $(5000)(0.08)t$, while that on the second is $(10,000)(0.12)t$. Since we want the sum of these two amounts to be $16,000, the required equation is

$$(5000)(0.08)t + (10,000)(0.12)t = 16000$$

Examples 1 through 4 are all similar in that we were given a word problem and asked to obtain an equation containing only one of the variables or unknowns. In the next two examples, our goal again is an equation, but this time one involving two variables. In each case, we want to find an equation or formula that expresses one variable in terms of another.

EXAMPLE 5   The formula for the volume of a right circular cylinder is $V = \pi r^2 h$. (See Figure 2 on page 79.)

(a) Suppose that in a right circular cylinder, the height is twice the radius. Write an equation relating r and h.
(b) Express the volume of the cylinder described in part (a) in terms of the radius r.
(c) Express the volume of the cylinder [in part (a)] in terms of the height h.

Solution   (a) We translate the English into algebraic notation as follows:

$$\underbrace{\text{The height}}_{h} \quad \underbrace{\text{is}}_{=} \quad \underbrace{\text{twice the radius}}_{2r}$$

That is, we have

$$h = 2r$$

(b) The general formula for the volume is $V = \pi r^2 h$. Replacing h with $2r$, we obtain

$$V = \pi r^2 (2r)$$

or

$$V = 2\pi r^3$$

(c) We can isolate r in the equation $h = 2r$ by dividing both sides by 2. This yields $r = h/2$. Now we use this to substitute for r in the equation obtained in part (b):

$$V = 2\pi \left(\frac{h}{2}\right)^3$$

$$= 2\pi \left(\frac{h^3}{8}\right) = \frac{\pi h^3}{4}$$

The required equation is therefore $V = \pi h^3/4$.

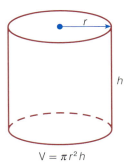

$V = \pi r^2 h$

FIGURE 2

EXAMPLE 6 A piece of wire 2 m long is cut into two pieces. Let x denote the length of the first piece and $2 - x$ the length of the second. The first piece is bent into a square and the second piece into a rectangle in which the length is three times the width. Express the combined area A of the square and the rectangle in terms of x.

Solution Figure 3 shows a sketch of the situation.

FIGURE 3

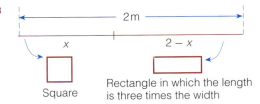

The variable A denotes the total area. In other words

$$A = (\text{area of square}) + (\text{area of rectangle})$$

We want to express each
of these in terms of x.

For the square, the perimeter is x and so the length of one side is $x/4$. Therefore, we have

$$\text{area of square} = \left(\frac{x}{4}\right)^2 = \frac{x^2}{16}$$

FIGURE 4

Next consider the rectangle. Since the length is three times the width, we can label the sides as indicated in Figure 4. The area of this rectangle is given by

$$\text{area of rectangle} = \text{length} \times \text{width}$$

$$= (3w) \times w = 3w^2 \qquad (8)$$

Equation (8) expresses the area of the rectangle in terms of w, but we want it in terms of x. From the statement of the problem we know that the perimeter of the rectangle must be $2 - x$. Therefore

$$2w + 2(3w) = 2 - x$$

$$8w = 2 - x$$

$$w = \frac{2 - x}{8}$$

dividing both sides of the equation by 8

Using this last equation to substitute for w in equation (8) yields

$$\text{area of rectangle} = 3w^2 = 3\left(\frac{2 - x}{8}\right)^2$$

$$= 3\left(\frac{4 - 4x + x^2}{64}\right) = \frac{12 - 12x + 3x^2}{64}$$

The combined area A of the square and the rectangle is therefore

$$A = \frac{x^2}{16} + \frac{12 - 12x + 3x^2}{64}$$

$$= \frac{7x^2 - 12x + 12}{64}$$

adding the fractions and combining like terms

EXERCISE SET 2.1

A

1. Consider the following problem.

> The sum of two numbers is -9, and the sum of their squares is 153. What are the two numbers?

(a) What is the problem asking for? Assign variables to denote the required quantities.

(b) Complete the following table, as in Example 1.

In English	In Algebraic Notation
Find two numbers	_____
Their sum is -9	_____
The sum of their squares is 153	_____

(c) Find and simplify an equation that could be solved for one of the variables or "unknowns" introduced in part (a).

2. Consider the following problem.

> Twice the sum of two numbers is 4, and the sum of their squares is 52. What are the two numbers?

(a) What is the problem asking for? Assign variables to denote the required quantities.

(b) Complete the following table, as in Example 1.

In English	In Algebraic Notation
Find two numbers	_____
Twice their sum is 4	_____
The sum of their squares is 52	_____

(c) Find and simplify an equation that could be solved for one of the variables or "unknowns" introduced in part (a).

3. Consider the following problem.

Find two numbers with sum 17 and product 52.

(a) What is the problem asking for? Assign variables to denote the required quantities.

(b) Complete the following table, as in Example 1.

In English	In Algebraic Notation
Find two numbers	_____
Their sum is 17	_____
Their product is 52	_____

(c) Find and simplify an equation that could be solved for one of the variables or "unknowns" introduced in part (a).

4. Consider the following problem.

The difference of two numbers is 20 and their product is −75. Find the two numbers.

(a) What is the problem asking for? Assign variables to denote the required quantities.

(b) Complete the following table, as in Example 1.

In English	In Algebraic Notation
Find two numbers	_____
Their difference is 20	_____
Their product is −75	_____

(c) Find and simplify an equation that could be solved for the larger of the two numbers. *Hint:* If x and y are two numbers and $x - y$ is positive, then x is the larger of the two numbers.

5. Consider the following problem.

The sum of the cubes of two consecutive integers is 341. Find the two integers.

(a) What is the problem asking for? Assign variables or expressions to denote the required quantities.

(b) Complete the following table, as in Example 2.

In English	In Algebraic Notation
Find two consecutive integers	_____
The sum of their cubes is 341	_____

(c) Find and simplify an equation that could be solved for the smaller of the two integers.

6. Consider the following problem.

Find two consecutive integers such that the sum of the square of the smaller and the cube of the larger is 141.

(a) What is the problem asking for? Assign variables or expressions to denote the required quantities.

(b) Complete the following table, as in Example 2.

In English	In Algebraic Notation
Find two consecutive integers	_____
The square of the smaller one	_____
The cube of the larger one	_____
The sum of the square of the smaller and the cube of the larger is 141	_____

(c) Find and simplify an equation that could be solved for one of the required integers.

7. Consider the following problem.

The average of two numbers is 4 and the average of their squares is 80. Find the numbers.

(a) What is the problem asking for? Assign variables to denote the required quantities.

(b) Complete the following table, as in Example 1.

In English	In Algebraic Notation
Find two numbers	_____
Their average is 4	_____
The average of their squares is 80	_____

(c) Find and simplify an equation that could be solved for one of the variables introduced in part (a).

8. Consider the following problem.

Find two numbers that have an average of 11 and a product of 72

(a) What is the problem asking for? Assign variables to denote the required quantities.

(b) Complete the following table, as in Example 1.

In English	In Algebraic Notation
Find two numbers	_____
The average of the two numbers is 11	_____
The product of the two numbers is 72	_____

(c) Find and simplify an equation that could be solved for one of the variables introduced in part (a).

9. Consider the following problem:

The perimeter of a rectangle is 40 cm and the area is 64 cm^2. What are the dimensions of the rectangle?

(a) What is the problem asking for? Draw a sketch of the situation, and assign variables to denote the required quantities.

(b) Complete the following table as in Example 3.

In English	In Algebraic Notation
Find the width and length of a rectangle	_____
The perimeter is 40 cm	_____
The area is 64 cm^2	_____

(c) Find and simplify an equation that could be solved for the width of the rectangle.

10. Consider the following problem:

Determine the base and the height of a triangle given that the area is 96 m^2 and that the base is three-fourths of the height.

(a) What is the problem asking for? Draw a sketch of the situation, and assign variables to denote the required quantities.

(b) Complete the following table as in Example 3.

In English	In Algebraic Notation
Find the base and height of a triangle	_____
The area is 96 m^2	_____
The base is three-fourths of the height	_____

(c) Find and simplify an equation that could be solved for the height of the triangle.

11. Consider the following problem:

Suppose that you open three savings accounts (each earning simple interest). In the first account you deposit $3000 at an annual simple interest rate of 7%. In the second you deposit $5000 at 8%, and in the third you deposit $10,000 at 8.5%. After how many years will the total interest amount to $8,760?

(a) What is the problem asking for? Assign a variable to denote the required quantity.

(b) Find an equation that could be solved for the variable introduced in part (a). (Don't simplify or solve the equation.)

12. Consider the following problem:

Suppose that you open three savings accounts (each earning simple interest). In the first account you deposit $2500 at an annual simple interest rate of 6%. In the second you deposit $3500 at 6.5%, and in the third you deposit $10,000 at 8.5%. After how many years will the total interest amount to $9,820?

(a) What is the problem asking for? Assign a variable to denote the required quantity.

(b) Find an equation that could be solved for the variable introduced in part (a). (Don't simplify or solve the equation.)

13. Consider the following problem:

Two savings accounts are opened with initial deposits of $7400 and $12,600. Both accounts earn simple interest, and the rate for the larger account is 1% higher than for the smaller. After ten years the total value of the two accounts is $31,260. Find the interest rate (expressed as a percent) for the account in which the initial deposit was $7400.

Find an equation that can be solved for the required interest rate. (Don't simplify or solve the equation.)

14. Consider the following problem:

Two savings accounts are opened with initial deposits of $14,000 and $25,000. Both accounts earn simple interest, and the rate for the smaller account is 2% less than for the larger. After five years the total value of the two accounts is $54,165. Find the interest rate (expressed as a percent) for the account in which the initial deposit was $14,000.

Find an equation that can be solved for the required interest rate. (Don't simplify or solve the equation.)

In Exercises 15–20, refer to the formulas on the inside front cover of this book.

15. Suppose that in a right circular cylinder, the height is four times the radius.
    (a) Write an equation relating the radius r and the height h.
    (b) Express the volume of the cylinder [described in part (a)] in terms of the radius r.
    (c) Express the volume of the cylinder in terms of the height h.

16. Suppose that in a right circular cylinder, the radius is two-thirds the height.
    (a) Write an equation relating the radius r and the height h.
    (b) Express the volume of the cylinder [described in part (a)] in terms of the height h.
    (c) Express the volume of the cylinder in terms of the radius r.

17. Suppose that in a right circular cylinder, the sum of the radius and the height is 4 cm.
    (a) Write an equation relating the radius r and the height h.
    (b) Express the volume V of the cylinder in terms of the radius r.
    (c) Complete the following table using the formula you obtained in part (b). (Use a calculator and round off the answers to two decimal places.)

r	1.0	1.5	2.0	2.5	3.0	3.5
V						

    (d) Which r-value in the table yields the largest volume?
    (e) Experiment with your calculator: Find a value for r (between 0 and 4) that yields a volume greater than 29.7 cm^3.
    (f) Experiment with your calculator: Can you find a value for r (between 0 and 4) that yields a volume greater than 29.8 cm^3?

18. Suppose that in a right circular cylinder, the sum of the radius and the height is 1 m.
    (a) Write an equation relating the radius r and the height h.
    (b) Express the volume V of the cylinder in terms of the height h.
    (c) Complete the following table using the formula you obtained in part (b). (Use a calculator and round off the answers to two decimal places.)

h	0.2	0.4	0.6	0.8
V				

    (d) Which h-value in the table yields the largest volume?
    (e) Experiment with your calculator: Find a value for h (between 0 and 1) that yields a volume greater than 0.46 m^3.
    (f) Experiment with your calculator: Can you find a value for h (between 0 and 1) that yields a volume greater than 0.465 m^3?

In Exercises 19 and 20, you will need to use the formulas for the volume and the lateral surface area of a right circular cone. These are given on the inside front cover of this book.

19. Suppose that the volume of a right circular cone is π cm^3.
    (a) Find an equation relating the radius r and the height h.
    (b) Express the lateral surface area S of the cone in terms of the radius r. Show that the result can be written

$$S = \frac{\pi}{r}\sqrt{r^6 + 9}$$

    (c) Complete the following table using the formula you obtained in part (b). (Use a calculator and round off the answers to three decimal places.)

r	0.25	0.50	0.75	1.00	1.25	1.50	1.75	2.00
S								

    (d) Which r-value in the table yields the least surface area?
    (e) Experiment with your calculator: Find a value for r (between 0 and 2) that yields a surface area less than 8.99 cm^2.
    (f) Experiment with your calculator: Can you find a (positive) value for r that yields a surface area less than 8.985 cm^2?

20. Suppose that the volume of a right circular cone is 10 m^3.
    (a) Find an equation relating the radius r and the height h.
    (b) Express the lateral surface area S of the cone in terms of the radius r. Show that the result can be written

$$S = \frac{\sqrt{\pi^2 r^6 + 900}}{r}$$

(c) Complete the following table using the formula you obtained in part (b). (Use a calculator and round off the answers to three decimal places.)

r	0.10	0.50	1.00	1.50	2.00	2.50	10.00
S							

(d) Which r-value in the table yields the least surface area?

(e) Experiment with your calculator: Can you find a (positive) value for r that yields a surface area less than 19.4 m²?

21. A piece of wire 1 m long is cut into two pieces. Let x denote the length of the first piece and $1 - x$ the length of the second. The first piece is bent into a square and the second piece into a rectangle in which the length is twice the width. Express the combined area A of the square and the rectangle in terms of x.

22. A piece of wire 4 m long is cut into two pieces. Let x denote the length of the first piece and $4 - x$ the length of the second. The first piece is bent into a square and the second piece into a rectangle in which the length is four times the width. Express the combined area A of the square and the rectangle in terms of x.

2.2   LINEAR EQUATIONS

A considerable part of the work in solving equations is mechanical. ... However, there is a reason for doing each step that is taken, and one should be able to give this reason.

C. I. Palmer and S. F. Bibb in *Practical Mathematics*, first published by McGraw-Hill in 1912.

In this section and the next we review the terminology and skills used in solving certain basic kinds of equations. Perhaps the simplest type of equation is the *linear* or *first-degree* equation in one variable.

DEFINITION   Linear Equation in One Variable

A **linear** or **first-degree equation in one variable** is an equation that can be written in the form

$$ax + b = 0 \qquad \text{with } a \text{ and } b \text{ real numbers and } a \neq 0$$

As examples of linear equations in one variable, we cite the following:

$$2x - 10 = 0 \qquad 3m + 1 = 2 \qquad \frac{y}{2} = \frac{y}{3} + 1$$

As with any equation involving a variable, each of these equations is neither true nor false *until* we replace the variable with a number. By a **solution** or a **root** of an equation in one variable, we mean a value for x (or m or y in the foregoing examples) that makes the equation a true statement. For example, the value $x = 5$ is a solution of the equation $2x = 10$ since, with $x = 5$, the equation becomes $2(5) = 10$, which is certainly true. We also say in this case that the value $x = 5$ **satisfies** the equation.

Equations that become true statements for all values in the domain of the variable are called **identities.** Two examples of identities are

$$x^2 - 9 = (x - 3)(x + 3) \qquad \text{and} \qquad \frac{4x^2}{x} = 4x$$

In contrast to this, a **conditional equation** is true only for some (or perhaps even none) of the values of the variable. Two examples of conditional equations are $2x = 10$ and $x = x + 1$. The first of these is true only when $x = 5$. The second equation has no solution (because, intuitively at least, no number can be one more than itself).

We say that two equations are **equivalent** when they have exactly the same solutions. In this section, and throughout the text, the basic method for solving an equation in one variable involves writing a sequence of equivalent equations until we finally reach an equation of the form

variable = a number

In generating equivalent equations, we rely on the following three principles. (These can be justified using the properties of real numbers discussed in Section 1.2.)

Procedures That Yield Equivalent Equations

1. Adding or subtracting the same quantity on both sides of an equation produces an equivalent equation.

2. Multiplying or dividing both sides of an equation by the same nonzero quantity produces an equivalent equation.

3. Simplifying an expression on either side of an equation (using the techniques from Chapter 1) produces an equivalent equation.

The examples that follow show how these principles are applied in solving various equations.

EXAMPLE 1 (a) Solve: $4x + 5 = 33$.
(b) Solve: $ax + b = c$ (a, b, and c are constants and $a \neq 0$).

Solution (a) $4x + 5 = 33$
$4x = 28$ subtracting 5 from both sides
$x = 7$ dividing both sides by 4

Check Replacing x with 7 in the original equation yields

$4(7) + 5 = 33$
$28 + 5 = 33$ True

(b) $ax + b = c$
$ax = c - b$ subtracting b from both sides
$x = \dfrac{c - b}{a}$ dividing both sides by a

Check   Replacing x with $\dfrac{c-b}{a}$ in the original equation yields

$$a\left(\frac{c-b}{a}\right) + b = c$$
$$c - b + b = c$$
$$c = c \qquad \text{True}$$

EXAMPLE 2   Solve $3[1 - 2(x + 1)] = 2 - x$.

Solution
$$3[1 - 2(x + 1)] = 2 - x$$
$$3[1 - 2x - 2] = 2 - x \qquad \text{simplifying the left-hand side}$$
$$3(-1 - 2x) = 2 - x$$
$$-3 - 6x = 2 - x$$
$$-3 - 5x = 2 \qquad \text{adding } x \text{ to both sides}$$
$$-5x = 5 \qquad \text{adding 3 to both sides}$$
$$x = -1 \qquad \text{dividing both sides by } -5$$

Check   Replacing x with -1 in the original equation yields
$$3[1 - 2(0)] = 2 - (-1)$$
$$3(1) = 2 + 1 \qquad \text{True}$$

In the next two examples the equations involve fractions. The strategy in such cases is to multiply through by the least common denominator. That eliminates the need to work with fractions.

EXAMPLE 3   Solve: **(a)** $\dfrac{x}{2} - \dfrac{3x}{4} = 1$;   **(b)** $\dfrac{1}{x-2} - \dfrac{4}{x+2} = \dfrac{1}{x^2 - 4}$.

Solution   **(a)**
$$\frac{x}{2} - \frac{3x}{4} = 1$$
$$4\left(\frac{x}{2}\right) - 4\left(\frac{3x}{4}\right) = 4(1) \qquad \begin{array}{l}\text{multiplying both sides by the}\\\text{least common denominator}\end{array}$$
$$2x - 3x = 4 \qquad \text{simplifying}$$
$$-x = 4 \qquad \text{simplifying}$$
$$x = -4 \qquad \text{multiplying both sides by } -1$$

Check   Replacing x with -4 in the original equation yields
$$\frac{-4}{2} - \frac{3(-4)}{4} = 1$$
$$-2 + 3 = 1 \qquad \text{True}$$

(b) By factoring the denominator on the right-hand side, we obtain
$$\frac{1}{x-2} - \frac{4}{x+2} = \frac{1}{(x-2)(x+2)}$$

From this we see that the least common denominator for the three fractions is $(x - 2)(x + 2)$. Now, multiplying both sides by this least common denominator, we have

$$\frac{(x - 2)(x + 2)}{x - 2} - \frac{4(x - 2)(x + 2)}{x + 2} = \frac{(x - 2)(x + 2)}{(x - 2)(x + 2)}$$

$$x + 2 - 4(x - 2) = 1 \qquad \text{simplifying}$$

$$-3x + 10 = 1 \qquad \text{simplifying}$$

$$-3x = -9 \qquad \text{subtracting 10 from both sides}$$

$$x = 3 \qquad \text{dividing both sides by } -3$$

Check   Replacing x with 3 in the original equation yields

$$\frac{1}{3 - 2} - \frac{4}{3 + 2} = \frac{1}{(3 - 2)(3 + 2)}$$

$$1 - \frac{4}{5} = \frac{1}{5} \qquad \text{True}$$

In part (a) of Example 3, we multiplied both sides of the original equation by 4 to produce an equivalent equation. In part (b), however, we multiplied both sides of the equation by $(x - 2)(x + 2)$. Since we didn't know at that point whether the quantity $(x - 2)(x + 2)$ was nonzero, we could not be certain that the resulting equation was actually an equivalent equation. For this reason, it is always necessary to check any solutions you obtain as a result of multiplying or dividing both sides of an equation by an expression involving the variable. Part (a) of the next example serves to emphasize this point.

EXAMPLE 4   Solve: **(a)** $\dfrac{1}{x + 5} = \dfrac{2}{x - 3} + \dfrac{2x + 2}{(x + 5)(x - 3)}$;

(b) $\dfrac{1}{x} = \dfrac{1}{a} + \dfrac{1}{b}$ (a and b are nonzero constants and $a + b \neq 0$).

Solution   **(a)** The least common denominator for the fractions is $(x + 5)(x - 3)$. After multiplying both sides of the given equation by this least common denominator, we have

$$x - 3 = 2(x + 5) + 2x + 2$$

$$x - 3 = 2x + 10 + 2x + 2 \qquad \text{simplifying}$$

$$x - 3 = 4x + 12 \qquad \text{simplifying}$$

$$-3x - 3 = 12 \qquad \text{subtracting } 4x \text{ from both sides}$$

$$-3x = 15 \qquad \text{adding 3 to both sides}$$

$$x = -5 \qquad \text{dividing both sides by } -3$$

Check   The preceding steps show that *if* the equation has a solution, then the solutin is $x = -5$. With $x = -5$, however, the left-hand side of the original equation becomes $\dfrac{1}{-5 + 5}$, or $\dfrac{1}{0}$, which is undefined. We conclude therefore that the given equation has no solution.

(b)
$$\frac{1}{x} = \frac{1}{a} + \frac{1}{b}$$

$$(xab) \cdot \frac{1}{x} = (xab) \cdot \frac{1}{a} + (xab) \cdot \frac{1}{b} \qquad \text{multiplying both sides by the least common denominator, } xab$$

$$ab = xb + xa \qquad \text{simplifying}$$

$$ab = x(b + a) \qquad \text{factoring}$$

$$\frac{ab}{b + a} = x \qquad \text{dividing both sides by } b + a$$

Check As Exercise 77 asks you to verify, the value $x = ab/(a + b)$ indeed satisfies the original equation.

In Example 4(a), the value $x = -5$, which did not check in the original equation, is called an **extraneous solution** or an **extraneous root**. The box that follows reminds you about the need to check for extraneous solutions. (We will update this box in later sections when extraneous solutions arise with other types of equations.)

PROPERTY SUMMARY   EXTRANEOUS SOLUTIONS

Multiplying or dividing both sides of an equation by an expression involving the variable may introduce extraneous solutions that do not check in the original equation. Therefore, it is always necessary to check any candidates for solutions that you obtain in this manner.

EXAMPLE 5 Solve for x in terms of the other letters: $y = \dfrac{ax + b}{cx + d}$. (Assume that $cx + d \neq 0$ and $yc - a \neq 0$.)

Solution Multiplying both sides of the given equation by the quantity $cx + d$ yields

$$y(cx + d) = ax + b$$
$$ycx + yd = ax + b$$
$$ycx - ax = b - yd$$
$$x(yc - a) = b - yd$$
$$x = \frac{b - yd}{yc - a}$$

You should check for yourself now that the expression for x on the right-hand side of this last equation indeed satisfies the original equation.

The techniques used in this section can also be applied in solving equations in which absolute values appear. The next example shows how this is done.

EXAMPLE 6 Solve: $|2x - 1| = 7$.

Solution There are two cases to consider.

If $2x - 1 \geq 0$, the equation becomes

$$2x - 1 = 7$$
$$2x = 8$$
$$x = 4$$

If $2x - 1 < 0$, the equation becomes

$$-(2x - 1) = 7$$
$$-2x + 1 = 7$$
$$-2x = 6$$
$$x = -3$$

We've now obtained the values $x = 4$ and $x = -3$. You should check for yourself that both of these numbers indeed satisfy the original equation.

Because our work in this section has focused solely on solving equations, we conclude with a reminder about the difference between simplifying expressions (as in Section 1.10) and solving equations. As indicated in the box that follows, you need to be aware of the context in which you are working. If you are solving an equation for a variable, you can indeed multiply both sides by the same nonzero quantity. However, if you are given only an expression, you can not unilaterally multiply it by any quantity (other than 1).

Error to Avoid

ERROR

To simplify $\dfrac{2}{3x} + \dfrac{4}{5x^2}$, multiply through by the least common denominator, $15x^2$, to obtain

$$\dfrac{2}{3x} + \dfrac{4}{5x^2} \times 2(5x) + 4(3)$$

$$= 10x + 12$$

CORRECTION

The least common denominator is indeed $15x^2$. Adding the two fractions, we have

$$\dfrac{2}{3x} + \dfrac{4}{5x^2} = \dfrac{2(5x) + 4(3)}{15x^2}$$

$$= \dfrac{10x + 12}{15x^2}$$

COMMENT

$\dfrac{2}{3x} + \dfrac{4}{5x^2}$ is an expression, not an equation. If instead we had been given an equation, such as

$$\dfrac{2}{3x} + \dfrac{4}{5x^2} = 7$$

then we could multiply both sides by the same quantity to eliminate the denominators.

EXERCISE SET 2.2

A

Solve each equation in Exercises 1–46. For Exercises 43–46, use a calculator and round off the final answers to two decimal places.

1. $2x - 3 = -5$
2. $x + 4 = 0$
3. $8x - 4 = 20$
4. $-6x + 1 = 49$
5. $1 - y = 12$
6. $3 - 4y = 8y + 3$
7. $2m - 1 + 3m + 5 = 6m - 8$
8. $1 - (2m + 5) = -3m$
9. $(x + 2)(x + 1) = x^2 + 11$
10. $t - \{4 - [t - (4 + t)]\} = 6$
11. $1 - (x - 2)^2 = -x^2 + 5x + 3$
12. $(x + 2)^2 = x^2 + 4$

13. $(x - 1)^2 - (x^2 - 1) = 16$

14. $(x - 1)^2 - (x^2 - 1) = 2 - 2x$

15. $\dfrac{x}{3} + \dfrac{2x}{5} = \dfrac{-11}{5}$

16. $\dfrac{x}{2} + 1 = 0$

17. $1 - \dfrac{y}{3} = 6$

18. $y - \dfrac{y}{2} - \dfrac{y}{3} = 2$

19. $3 - \dfrac{x - 1}{4} = \dfrac{x}{9}$

20. $1 + \dfrac{x}{3} - \dfrac{x}{4} = x - \dfrac{5x}{6}$

21. $\dfrac{x - 1}{4} + \dfrac{2x + 3}{-1} = 0$

22. $\dfrac{x - 1}{2} - \dfrac{x - 2}{4} = 1 + \dfrac{x}{3}$

23. $\dfrac{1}{x - 3} - \dfrac{2}{x + 3} = \dfrac{1}{x^2 - 9}$

24. $\dfrac{1}{x - 5} + \dfrac{1}{x + 5} = \dfrac{6}{x^2 - 25}$

25. $\dfrac{1}{x - 5} + \dfrac{1}{x + 5} = \dfrac{2x}{x^2 - 25}$

26. $\dfrac{1}{x - 5} + \dfrac{1}{x + 5} = \dfrac{2x + 1}{x^2 - 25}$

27. $\dfrac{4}{x + 2} + \dfrac{1}{x - 2} = \dfrac{4}{x^2 - 4}$

28. $\dfrac{1}{2x + 1} + \dfrac{1}{x + 1} = \dfrac{5}{2x^2 + 3x + 1}$

29. $\dfrac{2(x + 1)}{x - 1} - 3 = \dfrac{5x - 1}{x - 1}$

30. $\dfrac{x - 1}{x + 1} = \dfrac{x + 1}{x - 1}$

31. $\dfrac{1}{x} = \dfrac{4}{x} - 1$

32. $\dfrac{1}{y} + 1 = \dfrac{3}{y} - \dfrac{1}{2y}$

33. (a) $\dfrac{2}{3x} = \dfrac{3}{x}$

   (b) $\dfrac{2}{3x} = \dfrac{3}{x + 1}$

   (c) $\dfrac{2}{3x} = \dfrac{3}{x} + 1$

34. (a) $\dfrac{3}{x - 2} = \dfrac{5}{9x}$

   (b) $\dfrac{3}{x - 2} = \dfrac{5}{9x - 2}$

   (c) $\dfrac{3}{x - 2} = \dfrac{5}{\frac{5}{3}x - 2}$

35. (a) $\dfrac{1}{3x - 2} = 1$

   (b) $\dfrac{1}{3x - 2} = \dfrac{1}{x}$

   (c) $\dfrac{1}{3x - 2} = \dfrac{1}{3x + 2}$

36. (a) $\dfrac{4}{1 - x} = -8$

   (b) $\dfrac{4}{1 - x} = \dfrac{-8}{x}$

   (c) $\dfrac{4}{1 - x} = \dfrac{-8}{1 - 2x}$

37. $|x - 5| = 1$

38. $|x - 4| - 5 = 2$

39. $|x + 6| = \frac{1}{2}$

40. $|x + 6| + \frac{1}{2} = 0$

41. $|6x - 5| = 25$

42. $|5 - 6x| = 0$

43. $4.50x - 9.11 = x$

44. $\dfrac{1.28}{3.21x - 0.95} = 1.42$

45. $\dfrac{x - 0.18}{x + 0.23} = \dfrac{x + 1.16}{x - 2.24}$

46. $\dfrac{951}{644 - 105x} = \dfrac{105}{951 - 644x}$

For Exercises 47–58, solve each equation for x in terms of the other letters.

47. $3ax - 2b = b + 3$

48. $ax + b = bx - a$

49. $ax + b = bx + a$

50. $\dfrac{x}{a} + \dfrac{x}{b} = 1$

51. $\dfrac{1}{x} = a + b$

52. $\dfrac{1}{ax} = \dfrac{1}{bx} - \dfrac{1}{c}$

53. $(x + b)(x + c) - a(a + b) = (x - a)(x + a) + bc$

54. $\dfrac{1}{a} - \dfrac{1}{x} = \dfrac{1}{x} - \dfrac{1}{b}$

55. $(a - x)^2 = x^2 + b^2$

56. (a) $y = mx + b$ $(m \neq 0)$

   (b) $y - y_1 = m(x - x_1)$ $(m \neq 0)$

   (c) $\dfrac{x}{a} + \dfrac{y}{b} = 1$

   (d) $Ax + By + C = 0$ $(A \neq 0)$

57. (a) $\dfrac{a}{bx} = d$

   (b) $\dfrac{a}{bx + c} = d$

58. (a) $\dfrac{a}{bx + c} = \dfrac{d}{x}$

   (b) $\dfrac{a}{bx + c} = \dfrac{d}{ex + f}$

As background for Exercises 59–62, review the Error to Avoid *box on page 89. In part (a) of each exercise, carry out the indicated operations and simplify where possible. In part (b), solve the equation.*

59. (a) $\dfrac{1}{3x} - \dfrac{2}{5x}$

   (b) $\dfrac{1}{3x} - \dfrac{2}{5x} = -2$

60. (a) $\dfrac{3}{4x} - \dfrac{8}{x}$

   (b) $\dfrac{3}{4x} - \dfrac{8}{x} = -\dfrac{29}{2}$

61. (a) $\dfrac{8}{5x} + \dfrac{1}{x^2}$

   (b) $\dfrac{8}{5x} + \dfrac{1}{x^2} = -\dfrac{3}{2x}$

62. (a) $\dfrac{7}{x - 2} + 1$

   (b) $\dfrac{7}{x - 2} + 1 = 8$

B

63. Solve for x: $\dfrac{|x - 3|}{2} + \dfrac{|x - 3|}{3} = \dfrac{-5x}{3}$.

   Hint: Consider separately the cases $x \geq 3$ and $x < 3$.

64. Solve for x: $\dfrac{|x + 4|}{4} - |x + 1| = -\dfrac{9}{2}$.

In Exercises 65–70, solve each equation for x in terms of the other letters.

65. $a^2(a - x) = b^2(b + x) - 2abx$ (Assume that $a \neq b$.)

66. $\dfrac{b}{ax - 1} - \dfrac{a}{bx - 1} = 0$ (Assume that $a \neq b$.)

67. $\dfrac{a - x}{a - b} - 2 = \dfrac{c - x}{b - c}$

68. $\dfrac{x + 2p}{2q - x} + \dfrac{x - 2p}{2q + x} - \dfrac{4pq}{4q^2 - x^2} = 0$

69. $\dfrac{x - a}{x - b} = \dfrac{b - x}{a - x}$

70. $1 - \dfrac{a}{b}\left(1 - \dfrac{a}{x}\right) - \dfrac{b}{a}\left(1 - \dfrac{b}{x}\right) = 0$

In Exercises 71–76, solve each equation for the indicated variable.

71. $d = \dfrac{r}{1 + rt}$; for r

72. $\dfrac{x_1 x}{a^2} + \dfrac{y_1 y}{b^2} = 1$; for y

73. $S = 2\pi r^2 + 2\pi rh$; for h

74. $s = \dfrac{n}{2}(a + l)$; for a

75. $S = \dfrac{rl - a}{r - 1}$; for r

76. $3y^2 y' + 5x^2 yy' - x^2 = 0$; for y'

77. Verify that $x = ab/(a + b)$ is a solution of the equation $\dfrac{1}{x} = \dfrac{1}{a} + \dfrac{1}{b}$. [This is to check the answer obtained in Example 4(b).]

C

For Exercises 78 and 79, solve for x in terms of the other letters.

78. $\dfrac{q}{x + q} - \dfrac{p}{x + p} = \dfrac{q - p}{x - 2pq}$

79. $\dfrac{1}{x - p} - \dfrac{1}{x - q} = \dfrac{p - q}{x^2 - pq}$

80. Solve for x and simplify the answer:
$$\frac{x - a}{b} + \frac{x - b}{a} + \frac{x - 1}{ab} = \frac{2}{a} + \frac{2}{b} + 2$$
(Assume both a and b are nonzero and $a + b \neq -1$.)

2.3 QUADRATIC EQUATIONS

The formula for the quadratic equation seems first to have been discovered by the Moslems around 900 A.D., although quadratic equations had been solved by the Babylonians 3000 years earlier.

Charles Robert Hadlock in *Field Theory and Its Classical Problems* (Washington, D.C.: Mathematical Association of America, 1978)

An equation that can be written in the form

$$ax^2 + bx + c = 0 \qquad \text{with } a, b, \text{ and } c \text{ real numbers and } a \neq 0$$

is called a **quadratic equation**. Examples of quadratic equations are

$$x^2 - 8x - 9 = 0 \qquad 2y^2 - 5 = 0 \qquad 3m^2 = 1 - m$$

Recall that a **solution**, or **root**, of an equation is a value for the unknown that makes the equation a true statement. For example, the value $x = -1$ is a solution of the quadratic equation $x^2 - 8x - 9 = 0$, because when x is replaced by -1 in this equation, we have

$$(-1)^2 - 8(-1) - 9 = 0$$
$$1 + 8 - 9 = 0$$
$$0 = 0 \qquad \text{True}$$

When applicable, one of the simplest techniques for solving quadratic equations involves factoring. This method relies on the following familiar and important property of the real number system.

PROPERTY SUMMARY	**ZERO-PRODUCT PROPERTY OF REAL NUMBERS**

$$pq = 0 \quad \text{if and only if} \quad p = 0 \text{ or } q = 0 \quad \text{(or both)}$$

For example, to solve the equation $x^2 - 2x - 3 = 0$ by factoring, we have

$$x^2 - 2x - 3 = 0$$
$$(x - 3)(x + 1) = 0$$
$$x - 3 = 0 \quad \big| \quad x + 1 = 0$$
$$x = 3 \quad \big| \quad x = -1$$

As you can easily check, the values $x = 3$ and $x = -1$ both satisfy the given equation.

Unfortunately, not all quadratic expressions can be factored so easily. Consider, for example, the equation $x^2 - 2x - 4 = 0$. There are only three possible factorizations with integer coefficients, but none yields the appropriate middle term, $-2x$, when multiplied out:

$$(x - 4)(x + 1) \qquad (x + 4)(x - 1) \qquad (x - 2)(x + 2)$$

Moreover, it is not always obvious whether a particular equation, such as $x^2 + 156x + 5963 = 0$, can be solved by factoring. (We will come back to this equation in Exercise 50.) Clearly, we need a systematic approach to solving those quadratic equations that are not readily solvable through factoring. The technique of **completing the square** provides this approach. We'll demonstrate this technique by solving the equation

$$x^2 - 2x - 4 = 0$$

First, we rewrite the equation in the form

$$x^2 - 2x = 4 \tag{1}$$

with the x-terms isolated on the left-hand side of the equation. To complete the square, we follow these two steps:

STEP 1 Take half of the coefficient of x and square it.
STEP 2 Add the number obtained in Step 1 to both sides of the equation.

For equation (1), the coefficient of x is -2. Taking half of -2 and then squaring it gives us $(-1)^2$, or 1. Now, as directed in Step 2, we add 1 to both sides of equation (1). This yields

$$x^2 - 2x + 1 = 4 + 1$$
$$(x - 1)^2 = 5$$
$$x - 1 = \pm\sqrt{5}$$
$$x = 1 \pm \sqrt{5}$$

We have now obtained the two solutions, $1 + \sqrt{5}$ and $1 - \sqrt{5}$. (Exercise 6 will ask you to verify that these values indeed satisfy the original equation.) In the box that follows, we summarize the technique of completing the square, and we indicate how the process got its name.

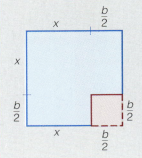

ALGEBRAIC PROCEDURE FOR COMPLETING THE SQUARE IN THE EXPRESSION $x^2 + bx$

Add the square of half of the x-coefficient:

$$(x^2 + bx) + \left(\frac{b}{2}\right)^2 = \left(x + \frac{b}{2}\right)^2$$

GEOMETRIC INTERPRETATION OF COMPLETING THE SQUARE FOR $x^2 + bx$

The blue region in the figure represents the quantity $x^2 + bx$ $\left(\text{since the area is } x^2 + \frac{b}{2}x + \frac{b}{2}x\right)$. By adding the red square to the blue region, we fill out or "complete" the larger square. The area of the red region that completes the square is $\frac{b}{2} \cdot \frac{b}{2} = \frac{b^2}{4}$.

The description in the box explains how to complete the square for expressions of the form $x^2 + bx$. To complete the square for a quadratic equation in which the coefficient of x^2 is not 1, we can first divide both sides of the equation by the coefficient of x^2. This is done in the next example.

EXAMPLE I   Solve by completing the square: $9x^2 + 36x - 1 = 0$.

Solution

$$9x^2 + 36x - 1 = 0$$

$$x^2 + 4x - \frac{1}{9} = 0 \qquad \text{dividing both sides by 9}$$

$$x^2 + 4x = \frac{1}{9}$$

$$x^2 + 4x + 4 = \frac{1}{9} + 4 \qquad \begin{array}{l}\text{completing the square by adding} \\ \left(\frac{1}{2} \cdot 4\right)^2 = 4 \text{ to both sides}\end{array}$$

$$(x + 2)^2 = \frac{37}{9}$$

$$x + 2 = \pm\sqrt{\frac{37}{9}} = \pm\frac{\sqrt{37}}{3}$$

$$x = -2 \pm \frac{\sqrt{37}}{3} = \frac{-6 \pm \sqrt{37}}{3}$$

The two solutions are therefore

$$\frac{-6 + \sqrt{37}}{3} \quad \text{and} \quad \frac{-6 - \sqrt{37}}{3}$$

The technique of completing the square can be used to derive the **quadratic formula**, a useful formula that gives the solutions to any quadratic equation. The derivation of this formula runs as follows. We start with the general quadratic equation $ax^2 + bx + c = 0$ $(a \neq 0)$ and divide both sides by a to obtain

$$x^2 + \frac{b}{a}x + \frac{c}{a} = 0$$

Subtracting c/a from both sides yields

$$x^2 + \frac{b}{a}x = -\frac{c}{a}$$

Now, to complete the square, we add $\left[\frac{1}{2}(b/a)\right]^2$, or $b^2/4a^2$, to both sides. That gives us

$$x^2 + \frac{b}{a}x + \frac{b^2}{4a^2} = \frac{b^2}{4a^2} - \frac{c}{a}$$

$$\left(x + \frac{b}{2a}\right)^2 = \frac{b^2 - 4ac}{4a^2}$$

$$x + \frac{b}{2a} = \pm\sqrt{\frac{b^2 - 4ac}{4a^2}} = \pm\frac{\sqrt{b^2 - 4ac}}{2|a|}$$

$$= \pm\frac{\sqrt{b^2 - 4ac}}{2a}$$

This last equality follows from the fact that for any real number $a \neq 0$, the expressions $\pm 2|a|$ and $\pm 2a$ both represent the same two numbers. We now conclude that the solutions are

$$x = -\frac{b}{2a} + \frac{\sqrt{b^2 - 4ac}}{2a} \quad \text{and} \quad x = -\frac{b}{2a} - \frac{\sqrt{b^2 - 4ac}}{2a}$$

These solutions are usually written in the more compact form displayed in the following box.

The Quadratic Formula

The solutions of the equation $ax^2 + bx + c = 0$ $(a \neq 0)$ are given by

$$x = \frac{-b \pm \sqrt{b^2 - 4ac}}{2a}$$

EXAMPLE 2   Use the quadratic formula to solve $2x^2 = 4 - x$.

Solution   We first rewrite the equation $2x^2 + x - 4 = 0$, so that it has the form $ax^2 + bx + c = 0$. By comparing these last two equations, we see that $a = 2$, $b = 1$, and $c = -4$. Therefore,

$$x = \frac{-b \pm \sqrt{b^2 - 4ac}}{2a} = \frac{-1 \pm \sqrt{1^2 - 4(2)(-4)}}{2(2)} = \frac{-1 \pm \sqrt{33}}{4}$$

Thus, the two solutions are $\dfrac{-1 + \sqrt{33}}{4}$ and $\dfrac{-1 - \sqrt{33}}{4}$.

EXAMPLE 3   Figure 1 shows a right circular cylinder with radius r and height h. The formula for the total surface area S of the cylinder is

$$S = 2\pi r^2 + 2\pi rh \tag{2}$$

Solve equation (2) for r in terms of h and S.

Solution   By rewriting equation (2) in the following form, we can see that the equation is quadratic in r, and we can determine what values to use for a, b, and c in the quadratic formula:

$$(2\pi)r^2 + (2\pi h)r - S = 0$$

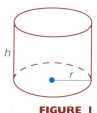

FIGURE 1

We have $a = 2\pi$, $b = 2\pi h$, and $c = -S$ and, therefore,

$$r = \frac{-2\pi h \pm \sqrt{4\pi^2 h^2 - 4(2\pi)(-S)}}{4\pi} = \frac{-2\pi h \pm 2\sqrt{\pi^2 h^2 + 2\pi S}}{4\pi}$$

$$= \frac{-\pi h \pm \sqrt{\pi^2 h^2 + 2\pi S}}{2\pi}$$

This last expression gives us two solutions to equation (2). In the present context, however, the radius r cannot be negative, so we discard the "minus" option within the plus-or-minus sign, since that would lead to a negative value for r. Thus we have

$$r = \frac{-\pi h + \sqrt{\pi^2 h^2 + 2\pi S}}{2\pi}$$

QUESTION   How do we know that this last expression is indeed positive?

In Examples 1 through 3 we found that each quadratic equation had two real roots (although in Example 3 we were interested only in the positive real root). As the next example indicates, however, it is also possible for quadratic equations to have only one real root, or even no real roots.

EXAMPLE 4   Use the quadratic formula to solve each equation:

(a) $4x^2 - 12x + 9 = 0$;   **(b)** $x^2 + x + 2 = 0$.

Solution   (a) Here $a = 4$, $b = -12$, and $c = 9$, so

$$x = \frac{12 \pm \sqrt{(-12)^2 - 4(4)(9)}}{2(4)} = \frac{12 \pm \sqrt{144 - 144}}{8}$$

$$= \frac{12 \pm 0}{8} = \frac{12}{8} = \frac{3}{2}$$

In this case our only solution is $\frac{3}{2}$. We refer to $\frac{3}{2}$ as a **double root**. (*Note:* We've used the quadratic formula here only for an illustration; in this case we can solve the original equation more efficiently by factoring.)

(b) Since $a = 1$, $b = 1$, and $c = 2$, we have

$$x = \frac{-1 \pm \sqrt{(1)^2 - 4(1)(2)}}{2(1)} = \frac{-1 \pm \sqrt{-7}}{2}$$

$$= \frac{-1 \pm i\sqrt{7}}{2}$$

In this case our roots are the two complex numbers

$$\frac{-1 + i\sqrt{7}}{2} \quad \text{and} \quad \frac{-1 - i\sqrt{7}}{2}$$

The quantity $b^2 - 4ac$ that appears under the radical sign in the quadratic formula is called the **discriminant**. In Examples 2 and 3, the discriminant was positive and consequently each equation had two real solutions. In Example 4(a), in which the discriminant was zero, we obtained but one (real) solution. In Example 4(b) the discriminant was negative, and consequently we obtained two solutions involving the imaginary unit i. These observations are generalized in the box that follows.

The Discriminant $b^2 - 4ac$

Consider the quadratic equation $ax^2 + bx + c = 0$, where a, b, and c are real numbers and $a \neq 0$. The expression $b^2 - 4ac$ is called the **discriminant**.

1. If $b^2 - 4ac > 0$, then the equation has two distinct real roots.

2. If $b^2 - 4ac = 0$, then the equation has exactly one real root.

3. If $b^2 - 4ac < 0$, then the equation has no real root. There are two (nonreal) complex-number roots. These two roots are complex conjugates.

EXAMPLE 5   (a) Compute the discriminant to determine how many real solutions there are for the equation $x^2 + x - 1 = 0$.

(b) Find a value for k such that the quadratic equation $x^2 + \sqrt{2}\,x + k = 0$ has exactly one real solution.

Solution   (a) Here $a = 1$, $b = 1$, $c = -1$ and, therefore,

$$b^2 - 4ac = 1^2 - 4(1)(-1) = 5$$

Since the discriminant is positive, the equation has two distinct real solutions.

(b) For the equation to have exactly one real solution, the discriminant must be zero, that is,

$$b^2 - 4ac = 0$$
$$(\sqrt{2})^2 - 4(1)(k) = 0$$
$$2 - 4k = 0$$
$$k = \tfrac{1}{2}$$

The required value for k is $\tfrac{1}{2}$.

EXERCISE SET 2.3

A

In Exercises 1–5, determine if the given value is a solution of the equation.

1. $2x^2 - 6x - 36 = 0$; $x = -3$
2. $(y - 4)(y - 5) = 0$; $y = -1$
3. $4x^2 - 1 = 0$; $x = -\tfrac{1}{4}$
4. $m^2 + m - \tfrac{5}{16} = 0$; $m = \tfrac{1}{4}$
5. $x^2 - 2x - 6 = 0$; $x = -1 - \sqrt{7}$
6. Verify that the numbers $1 + \sqrt{5}$ and $1 - \sqrt{5}$ both satisfy the equation $x^2 - 2x - 4 = 0$.

In Exercises 7–24, solve each equation by factoring.

7. $x^2 - 5x - 6 = 0$
8. $x^2 - 5x = -6$
9. $x^2 - 100 = 0$
10. $4y^2 + 4y + 1 = 0$
11. $25x^2 - 60x + 36 = 0$
12. $144 - t^2 = 0$
13. $10z^2 - 13z - 3 = 0$
14. $3t^2 - t - 4 = 0$
15. $(x + 1)^2 - 4 = 0$
16. $x^2 + 3x - 40 = 0$
17. $x(2x - 13) = -6$
18. $x(3x - 23) = 8$
19. $x(x + 1) = 156$
20. $x^2 + (2\sqrt{5})x + 5 = 0$
21. $x^2 - 49 = 0$
22. $x^2 - \tfrac{16}{9} = 0$
23. $\tfrac{1}{4} - x^2 = 0$
24. $2x^2 - 50 = 0$

In Exercises 25–36, solve each equation in two ways: (a) by completing the square; (b) by using the quadratic formula.

25. $x^2 + 10x + 9 = 0$
26. $x^2 - 10x + 25 = 0$
27. $x^2 - x - 5 = 0$
28. $x^2 - x + 5 = 0$
29. $2x^2 + 3x - 4 = 0$
30. $4x^2 - 3x - 9 = 0$
31. $12x^2 + 32x + 5 = 0$
32. $10x^2 - x - 1 = 0$
33. $2x^2 = x + 5$
34. $3 - 2x = -4x^2$
35. $-6x^2 + 12x = -1$
36. $-\sqrt{2}x^2 + x = -\sqrt{2}$

In Exercises 37–52, solve the equations using any method you choose.

37. $x(x + 18) = 81$
38. $x(x - 12) + 36 = 0$
39. $t^2 + 6 = 2t$
40. $t^2 + 10 = 6t$
41. $10 = 11u + 8u^2$
42. $11u = 10 + 8u^2$
43. $12 + 12y^2 = 10y$
44. $12 - 10y = 12y^2$
45. $x^2 = 24$
46. $x^2 + 16x = 0$
47. $x(x - 1) = 1$
48. $x(x - 1) = -4$
49. $\tfrac{1}{2}x^2 - x - \tfrac{1}{3} = 0$
50. $x^2 + 156x + 5963 = 0$
51. $2\sqrt{5}x^2 - x - 2\sqrt{5} = 0$
52. $\sqrt{2}x^2 + x = 10\sqrt{2}$

In Exercises 53–58, solve for x in terms of the other letters.

53. $2y^2x^2 - 3yx + 1 = 0$ $(y > 0)$
54. $(ax + b)^2 - (bx + a)^2 = 0$ $(a \neq \pm b)$
55. $(x - p)^2 + (x - q)^2 = p^2 + q^2$
56. $21x^2 - 2kx - 3k^2 = 0$ $(k > 0)$
57. $12x^2 = ax + 20a^2$ $(a > 0)$
58. $3Ax^2 - 2Ax - 3Bx + 2B = 0$ $(A \neq 0)$

In Exercises 59–64, solve for the indicated letter.

59. $2\pi r^2 + 2\pi rh = 20\pi$; for r
60. $2\pi y^2 + \pi yx = 12$; for y
61. $-16t^2 + v_0 t = 0$; for t
62. $-\tfrac{1}{2}gt^2 + v_0 t + h_0 = 0$; for t
63. $x^3 + bx^2 - 2b^2x = 0$; for x
64. $\dfrac{1}{y^3} + \dfrac{b}{y^2} - \dfrac{2b^2}{y} = 0$; for y

In Exercises 65–72, use the discriminant to determine how many real roots the equation has.

65. $x^2 - 12x + 16 = 0$
66. $2x^2 - 6x + 5 = 0$

67. $4x^2 - 5x - \frac{1}{2} = 0$

68. $4x^2 - 28x + 49 = 0$

69. $x^2 + \sqrt{3}x + \frac{3}{4} = 0$

70. $\sqrt{2}x^2 + \sqrt{3}x + 1 = 0$

71. $y^2 - \sqrt{5}y = -1$

72. $\dfrac{m^2}{4} - \dfrac{4m}{3} + \dfrac{16}{9} = 0$

In Exercises 73–76, find values for k such that the equation has exactly one real root.

73. $x^2 + 12x + k = 0$

74. $3x^2 + (\sqrt{2k})x + 6 = 0$

75. $x^2 + kx + 5 = 0$

76. $kx^2 + kx + 1 = 0$

In Exercises 77 and 78, use the quadratic formula and a calculator to solve for x. Round off each answer to two decimal places. (*Note:* *After completing Exercises 77 and 78, compare the two equations and their solutions. The point here is that a slight change in one of the coefficients can sometimes radically alter the nature of the solutions.*)

77. $x^2 + 3x + 2.249 = 0$

78. $x^2 + 3x + 2.251 = 0$

B

Solve the equations in Exercises 79–86. [In these exercises, you'll need to multiply both sides of the equations by expressions involving the variable. Remember (as indicated in Section 2.1) to check your answers in these cases.]

79. $\dfrac{3}{x + 5} + \dfrac{4}{x} = 2$

80. $\dfrac{5}{x + 2} - \dfrac{2x - 1}{5} = 0$

81. $1 - x - \dfrac{2}{6x + 1} = 0$

82. $\dfrac{x^2 - 3x}{x + 1} = \dfrac{4}{x + 1}$

83. $\dfrac{3x^2 - 6x - 3}{(x + 1)(x - 2)(x - 3)} + \dfrac{5 - 2x}{x^2 - 5x + 6} = 0$

84. $\dfrac{6}{x^2 - 1} + \dfrac{x}{x + 1} = \dfrac{3}{2}$

85. $\dfrac{2x}{x^2 - 1} - \dfrac{1}{x + 3} = 0$

86. $\dfrac{x}{x - 2} + \dfrac{x}{x + 2} = \dfrac{8}{x^2 - 4}$

87. In this section we solved quadratic equations by factoring, by completing the square, and by using the quadratic formula. This exercise shows how to solve a quadratic equation by the *method of substitution*. As an example, we use the equation

$$x^2 + x - 1 = 0 \qquad (1)$$

(a) In equation (1), make the substitution $x = y + k$. Show that the resulting equation can be written

$$y^2 + (2k + 1)y = 1 - k - k^2 \qquad (2)$$

(b) Find a value for k so that the coefficient of y in equation (2) is 0. Then, using this value of k, show that equation (2) becomes $y^2 = \frac{5}{4}$.

(c) Solve the equation $y^2 = \frac{5}{4}$. Then use the equation $x = y + k$ to obtain the two solutions of equation (1).

88. Use the substitution method (explained in Exercise 87) to solve the quadratic equation $2x^2 - 3x + 1 = 0$.

89. For which values of A and B will the roots of the equation $x^2 + Ax + B = 0$ be A and B?

90. For which values of k will the roots of the equation $x^2 = 2x(3k + 1) - 7(2k + 3)$ be equal?

91. Let α and β be the roots of the quadratic equation $ax^2 + bx + c = 0$. Verify each of the following statements.

(a) $\alpha + \beta = -b/a$ (b) $\alpha\beta = c/a$

(c) $\alpha^2 + \beta^2 = \dfrac{b^2 - 2ac}{a^2}$ *Suggestion:* Use (a) and (b) along with the fact that $\alpha^2 + \beta^2 = (\alpha + \beta)^2 - 2\alpha\beta$.

(d) $\dfrac{1}{\alpha^2} + \dfrac{1}{\beta^2} = \dfrac{b^2 - 2ac}{c^2}$ *Suggestion:* Add the fractions $1/\alpha^2$ and $1/\beta^2$.

92. Here is an outline for a slightly different derivation of the quadratic formula. The advantage to this method is that fractions are avoided until the very last step. Fill in the details.

(a) Beginning with $ax^2 + bx = -c$, multiply both sides by $4a$. Then add b^2 to both sides.

(b) Now factor the resulting left-hand side and take square roots.

C

93. Consider the quadratic equation $x^2 + px + q = 0$, where both p and q are nonzero.

(a) If one root is three times the other, show that $p^2/q = 16/3$.

(b) If one root is n times the other, show that $p^2/q = (n + 1)^2/n$. (Assume $n \neq \pm 1$.)

In Exercises 94–96, solve each equation for x in terms of the other letters.

94. $(a + b + c)x^2 - 2(a + b)x + (a + b - c) = 0$
$(c > 0, a + b + c \neq 0)$ *Hint:* Let $d = a + b$.

95. $ab(a + b)x^2 - (a^2 + b^2)x - 2(a + b) = 0$
(Assume that a and b are nonzero and $a \neq -b$.)

96. $x^2 - \left(\sqrt{\dfrac{a}{b}} + \sqrt{\dfrac{b}{a}}\right)x + 1 = 0$

2.4 APPLICATIONS

Solving problems is a practical art, like swimming, or skiing, or playing the piano: you can learn it only by imitation and practice. This book cannot offer you a magic key that opens all doors and solves all the problems, but it offers you good examples for imitation and many opportunities for practice....

George Polya in *Mathematical Discovery* (New York: John Wiley & Sons, 1981)

The techniques developed in the previous two sections for solving linear and quadratic equations can be used in solving a wide variety of applied problems. As you approach each example and exercise in this section for the first time, you may find it useful to refer to the following suggestions for organizing your thoughts and getting started. Eventually, of course, you'll probably want to modify this list to fit your own style.

Steps for Problem Solving

1. Read the problem carefully. When appropriate, draw a picture or a chart to summarize the situation.

2. State in your own words, as specifically as you can, what the problem is asking for. (This usually requires rereading the problem.) Assign variables or expressions to denote the required quantities. Then, as necessary, assign auxiliary variables to any other quantities that appear relevant. Are there equations relating these auxiliary variables?

3. Set up an equation (or equations) involving the required variables. Where necessary, use the results from Step 2 to obtain an equation involving but one required unknown.

4. Solve the equation that you obtained in Step 3.

5. Check your answer. Relate it to the original problem. Does it satisfy the conditions of the *original* problem?

In Example 1 we solve two problems that appear to be very similar. As you'll see, however, the first problem involves a linear equation, the second a quadratic.

EXAMPLE 1 The sum of two numbers is 20.

(a) Find the two numbers, assuming that their difference is 96.
(b) Find the two numbers, assuming that their product is 96.

Solution **(a)** The problem asks us to find two numbers that have a sum of 20 and a difference of 96. Let x and y denote the two numbers. In Table 1 (on the next page) we restate the problem using algebraic notation (as we did in Section 2.1).

From Table 1 we have the equations

$$x + y = 20 \tag{1}$$
$$x - y = 96 \tag{2}$$

TABLE I

Restate the Problem

In English	In Algebraic Notation
Find two numbers	x and y
Their sum is 20	$x + y = 20$
Their difference is 96	$x - y = 96$

Each of these equations involves both x and y. To obtain an equation involving only one variable, we first solve equation (1) for y in terms of x. This yields

$$y = 20 - x \tag{3}$$

and therefore

$$x - (20 - x) = 96 \qquad \text{using equation (3) to substitute for } y \text{ in equation (2)}$$

$$2x = 116 \tag{4}$$

$$x = 58$$

With $x = 58$, the second number is $y = 20 - x = 20 - 58 = -38$. As you check now for yourself, 58 and -38 are indeed the required numbers: their sum is 20 and their difference is 96.

TWO COMMENTS ON THIS SOLUTION

1. After we have equations (1) and (2), there are alternate methods that can be used. For instance, adding equations (1) and (2) takes us directly to equation (4), without the need for any intermediate steps. We will develop this technique in a later chapter when we study systems of equations.

2. We were given that the two numbers have a sum of 20, and initially we denoted the numbers by x and y. However, in view of equation (3), we could have begun by calling the two numbers x and $20 - x$, thereby eliminating the need to introduce a second variable. (Note that the sum of x and $20 - x$ is indeed 20; it would be incorrect to denote the numbers by x and $x - 20$ for then the sum is no longer 20.)

(b) As pointed out at the end of part (a), two numbers with a sum of 20 can be denoted by x and $20 - x$. This time we are assuming that their product (rather than their difference) is 96. Therefore we have

$$x(20 - x) = 96$$
$$20x - x^2 = 96$$
$$x^2 - 20x + 96 = 0$$
$$(x - 8)(x - 12) = 0$$

$$x - 8 = 0 \quad \bigg| \quad x - 12 = 0$$
$$x = 8 \quad \bigg| \quad x = 12$$

We've now found two possibilities for the first number, 8 or 12. If $x = 8$, then the second number is $20 - x = 20 - 8 = 12$; if $x = 12$, then the second number is $20 - 12 = 8$. So in either case we obtain the two numbers 8 and 12. As you can see, these are the required numbers: their sum is 20 and their product is 96. ▮▮▮

EXAMPLE 2 If you have three exam scores of 91, 76, and 84, what score do you need on the fourth exam to raise your average to 85?

Solution Let

$$x = \text{the score on the fourth exam}$$

The average of the four scores 91, 76, 84, and x is $(91 + 76 + 84 + x) \div 4$. Since this average is to be 85, we write

$$\frac{91 + 76 + 84 + x}{4} = 85$$

$$91 + 76 + 84 + x = 340 \qquad \text{multiplying by 4}$$

$$251 + x = 340$$

$$x = 340 - 251 = 89$$

We've now found that $x = 89$. This is the required score because the average of the four numbers 91, 76, 84, and 89 is, in fact, 85, as required. ▮▮▮

If you look back over Examples 1 and 2, you'll see that the equations we solved were obtained by directly translating the words in the problems into algebraic notation. Example 3 is one step removed from this in that we obtain the necessary equation by means of a basic formula from geometry, the Pythagorean theorem. For reference, we state this theorem in the box that follows. (For a proof of this theorem, see Exercise Set 3.1, Exercise 45.)

Pythagorean Theorem

In any right triangle, the lengths of the three sides are related by the equation

$$a^2 + b^2 = c^2$$

where a and b are the lengths of the sides forming the right angle and c is the length of the hypotenuse (the side opposite the right angle).

EXAMPLE 3 Are there right triangles other than the 3-4-5 right triangle in which the lengths of the sides are three consecutive integers?

Solution Any three consecutive integers can be denoted by x, $x + 1$, and $x + 2$ (see Figure 1). If these three quantities are to be the lengths of the sides of a right triangle, then according to the Pythagorean theorem we must have

$$x^2 + (x + 1)^2 = (x + 2)^2$$

$$x^2 + x^2 + 2x + 1 = x^2 + 4x + 4$$

$$x^2 - 2x - 3 = 0 \qquad \text{simplifying}$$

$$(x - 3)(x + 1) = 0$$

$$x - 3 = 0 \quad | \quad x + 1 = 0$$

$$x = 3 \quad | \quad x = -1$$

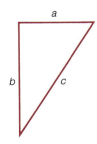

FIGURE 1

Thus, $x = 3$ or $x = -1$. We can immediately discard the value $x = -1$, for x is supposed to represent a length. On the other hand, with $x = 3$ we find that

$$x + 1 = 3 + 1 = 4 \qquad \text{and} \qquad x + 2 = 3 + 2 = 5$$

From this we conclude that if the lengths of the three sides in the right triangle are consecutive integers, then those integers must be 3, 4, and 5. Or, to answer the original question, there are no other right triangles in which the lengths of the sides are three consecutive integers. ▮▮▮

The next example uses the formula for simple interest, $I = Prt$, which was discussed in Section 2.1.

EXAMPLE 4 Suppose that you open two savings accounts. In the first you deposit $5000 at an annual simple interest rate of 8%. In the second you deposit $10,000 at 12% (also simple interest). After how many years will the total interest amount to $16,000?

Solution Let

$$t = \text{the required number of years}$$

Then in t years the interest on the first account is $(5000)(0.08)t$, while that on the second is $(10,000)(0.12)t$. Since the total interest is to be $16,000, we have

$$
\begin{aligned}
(5000)(0.08)t + (10,000)(0.12)t &= 16{,}000 \\
400t + 1200t &= 16{,}000 \\
1600t &= 16{,}000 \\
t &= 10
\end{aligned}
$$

Thus, after ten years the total interest will amount to $16,000.

Check After 10 years the interest from the first account will be ($5000)(0.08)(10), which is $4000. The interest from the second account after 10 years will be ($10,000)(0.12)(10), which is $12,000. So the total interest will be the sum of $4000 and $12,000, that is, $16,000, as required. ▮▮▮

In some problems it's helpful to set up a chart or table to see the given data at a glance. This is done in the next example.

EXAMPLE 5 Suppose that a chemistry student can obtain two acid solutions from the stockroom. The first solution is 20% acid and the second is 45% acid. (The percents are by volume.) How many cubic centimeters (cm^3) of each solution should the student mix together to obtain 100 cm^3 of a 30% acid solution?

Solution We begin by assigning letters or expressions to denote the required quantities. Let x denote the number of cm^3 of the 20% solution to be used. Then

$$100 - x$$

denotes the number of cm^3 of the 45% solution to be used. All the data can now be summarized in Table 2.

TABLE 2

Type of Solution	Number of cm^3	Percent Acid	Total Acid (cm^3)
First solution (20% acid)	x	20	$(0.20)x$
Second solution (45% acid)	$100 - x$	45	$(0.45)(100 - x)$
Mixture	100	30	$(0.30)(100)$

Looking at the data in the right-hand column of the table, we can write

$$\underbrace{0.20x}_{\substack{\text{Amount of acid} \\ \text{in } x \text{ cm}^3 \text{ of the} \\ \text{20\% solution}}} + \underbrace{0.45(100 - x)}_{\substack{\text{Amount of acid in} \\ (100 - x) \text{ cm}^3 \text{ of} \\ \text{the 45\% solution}}} = \underbrace{(0.30)(100)}_{\substack{\text{Amount of acid in} \\ \text{the final mixture}}}$$

So we have

$$0.20x + 0.45(100 - x) = 30$$
$$20x + 4500 - 45x = 3000 \qquad \text{multiplying by 100}$$
$$-25x = -1500$$
$$x = 60$$

Thus $x = 60$, and therefore $100 - x = 100 - 60 = 40$. Consequently, the student should use 60 cm^3 of the first solution and 40 cm^3 of the second solution.

Check The amount of acid in 60 cm^3 of the first solution is $(60 \text{ cm}^3)(0.20)$, which is 12 cm^3; and the amount of acid in 40 cm^3 of the second solution is $(40 \text{ cm}^3)(0.45)$, which is 18 cm^3. So the total amount of acid in the mixture is the sum of 12 cm^3 and 18 cm^3, which is 30 cm^3. Since the volume of the final mixture is 100 cm^3, the amount of acid there is $\frac{30}{100}(100)$; this is 30%, as required. ∎

EXAMPLE 6 The radius of a circle is 10 cm. By how many centimeters should the radius be increased so that the area increases by 5π cm^2? Round off the answer to two decimal places.

Solution Let

$$x = \text{the number of cm by which the radius should be increased}$$

Then, as the calculations accompanying Figure 2 (on the next page) indicate, the area of the original circle is 100π cm^2, while the area of the new circle is $\pi(10 + x)^2$. Now, according to the statement of the problem, we want the two areas to differ by 5π cm^2. Consequently, we have

$$\pi(10 + x)^2 - 100\pi = 5\pi$$
$$(10 + x)^2 - 100 = 5 \qquad \text{dividing by } \pi$$
$$(10 + x)^2 = 105$$
$$10 + x = \sqrt{105} \qquad \text{disregarding the root } -\sqrt{105}. \text{ (Why?)}$$
$$x = \sqrt{105} - 10$$

FIGURE 2

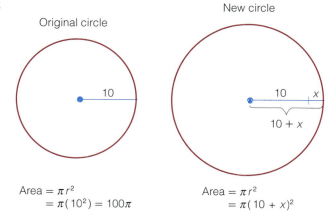

Area = πr^2
   = $\pi(10^2) = 100\pi$

Area = πr^2
   = $\pi(10 + x)^2$

Thus, the radius of the original circle should be increased by $\left(\sqrt{105} - 10\right)$ cm. Using a calculator and rounding off to two decimal places, we obtain $x \approx 0.25$ cm. As with all word problems, you should check for yourself here that the answer obtained satisfies the conditions of the original problem. (In checking, of course, use the exact value, $\sqrt{105} - 10$, not the calculator approximation, 0.25.)   ▮▮▮

Our final example is a motion problem. For travel at a constant speed or rate, the basic relationship is

distance = rate × time

or, more concisely,

$d = rt$

For example, if you drive 55 mph for three hours, then the distance you cover is $d = (55)(3) = 165$ miles. If a trip involves several parts, each at a different constant speed, then the *average speed* for the trip is defined by

$$\text{average speed} = \frac{\text{total distance}}{\text{total time}}$$

EXAMPLE 7   Andy and Jerry are going to run on a quarter-mile track. Andy can run at a pace of 8 minutes per mile $\left(= \frac{1}{8} \text{ mile/min}\right)$, while Jerry can run at a pace of 7 minutes per mile $\left(= \frac{1}{7} \text{ mile/min}\right)$. If they start together at the same time (running in the same direction), how long will it take for Jerry to lap Andy?

Solution   As in Example 5, we set up a table to organize the data. Table 3(a) indicates the format. (Notice that the top row is just taken from the formula $d = rt$.) In the r column of Table 3(a), we've filled in the given rates for Andy and Jerry.

TABLE 3(a)

	d (miles)	r (miles/min)	t (min)
Andy		$\frac{1}{8}$	
Jerry		$\frac{1}{7}$	

TABLE 3(b)

	d (miles)	r (miles/min)	t (min)
Andy	$\frac{1}{8}T$	$\frac{1}{8}$	T
Jerry	$\frac{1}{7}T$	$\frac{1}{7}$	T

Now we complete the table as shown in Table 3(b). In the t column, we've used T to represent the (unknown) number of minutes for Jerry to lap Andy; the d column was filled in using the fact that $d = rt$. Thus, Andy's and Jerry's distances are represented by $\frac{1}{8}T$ and $\frac{1}{7}T$, respectively. Consequently, the condition that Jerry be exactly $\frac{1}{4}$ mile ahead of Andy is

$$\frac{1}{7}T = \frac{1}{8}T + \frac{1}{4}$$

$$8T = 7T + 14 \qquad \text{multiplying by 56}$$

$$T = 14$$

Thus, after 14 minutes Jerry will have lapped Andy. (Check for yourself that this answer indeed satisfies the conditions of the problem.) ▌▌▌

EXERCISE SET 2.4

A

1. Find two numbers with sum 7 and difference 17.

2. The larger of two numbers is one less than twice the smaller. The sum of the two numbers is 29. Find the two numbers.

3. Find two numbers with sum 17 and product 52.

4. The sum of two numbers is 6 and the sum of their squares is 90. What are the two numbers?

5. The sum of a number and its double is 63. What is the number?

6. Find three consecutive integers such that twice the first plus half of the second is nine more than twice the third.

7. The product of a certain positive number and four more than that number is 96. What is the number?

8. The sum of the squares of three consecutive positive integers is 365. Find the integers.

9. What number must be added to both the numerator and the denominator of the fraction $\frac{3}{7}$ to yield a fraction that is equal to $\frac{2}{3}$?

10. Find two consecutive even integers such that when they are divided by 8 and 9, respectively, the sum of the quotients is 4.

11. The difference of two positive numbers is one-third of their sum. The sum of the squares of the two numbers is 180. Find the two numbers.

12. (a) Is there a positive real number such that the sum of the number and its square is 1? If so, find the number.
    (b) Is there a positive real number such that the difference between the number and its square is 1? If so, find the number.

13. If you have exam scores of 70, 77, and 75, what score do you need on the fourth exam to raise your average to 75?

14. The average of three numbers is 42. The first two numbers differ by 1, and the average of the second and third numbers is 47. What are the three numbers?

15. The average of four numbers is 52. Find the fourth number, given that the first three are 80, 22, and 62.

16. The average of the reciprocals of two consecutive integers is $\frac{9}{40}$. Find the smaller of the two integers.

17. In a certain right triangle, the hypotenuse is 8 cm longer than the first leg and 1 cm longer than the second. Find the lengths of the sides of the triangle.

18. In a certain right triangle, the lengths of the two shorter sides are consecutive integers (measured in cm). Find the area of the triangle given that the sum of the lengths of the hypotenuse and longer leg is 50 cm.

19. When the sides of a square are each increased by 2 cm, the area increases by 14 cm^2. Find the length of a side in the original square.

20. In a certain rectangle, the length exceeds the width by 2 m. If the length is increased by 4 m and the width is decreased by 2 m, the area of the new rectangle is the same as that of the original rectangle. Find the dimensions of the original rectangle.

21. The perimeter and area of a rectangle are 104 cm and 640 cm^2, respectively. Find the length of the shorter side of the rectangle.

22. A rectangular picture with dimensions 5 in. by 8 in. is surrounded by a frame of uniform width. If the area of the frame (not including the space for the picture) is 114 in^2 greater than the area of the picture, find the width of the frame.

23. A piece of wire 12 in. long is bent to form a right triangle in which the shortest side is 2 in. Find the lengths of the other two sides.

24. A piece of wire 24 cm long is cut into two pieces. Each piece is then formed into a square. If the sum of the areas of the two squares is 20 cm^2, find the lengths of the two pieces of wire.

25. In a certain isosceles triangle, each of the two equal legs is three times as long as the base. If the perimeter of the triangle is 70 cm, find the lengths of the sides.

26. In a certain triangle, the second angle is four times the first, while the third angle is 60° more than the first angle. Find the three angles. (Use the fact that the sum of the angles in any triangle is 180°.)

27. Suppose that you open two savings accounts. In the first you deposit $4000 at an annual simple interest rate of 7%. In the second you deposit $5000 at 9% (also simple interest). After how many years will the total interest amount to $5840?

28. You invest $600 in stock paying a 6% annual dividend and $800 in a stock paying an 8% annual dividend. How many years will it take for the total earnings to equal the total initial investment? (Use the formula $I = Prt$.)

29. If you deposit $10,000 at 11% (simple interest), how long will it be until the total amount in the account is $21,000?

30. Mr. X has $5000 to deposit. He puts part of this in an account earning 6%. The remainder he puts in an account earning 8.5%. (Both rates are simple interest.) If the total interest earned each year is $387.50, how much was initially deposited in each account?

31. You deposit $7500 at 7.5% and $4000 at 5%, simple interest. What is the **effective** interest rate? (In other words, what interest rate on $11,500 would yield the same total interest at year's end?) Express your answer as a percent, rounded off to the nearest tenth of one percent.

32. You deposit $2000 at 8% and $3000 at 9%, simple interest. What additional amount would you need to invest at 12% to yield an overall return of 10% after one year?

33. Ms. X deposits $8000 in one account and $9000 in another. The interest rate on the $8000 account is one percent less than that on the $9000 account. If the $9000 account earns $165 more per year than the $8000 account, find the interest rates on each account.

34. Jenny inherits $42,900 and invests it for a year in two stocks. Stock A earns 12.5% annually and stock B earns 8.5% annually. How much was invested in each stock if stock B earned three-quarters as much as stock A?

35. A student in a chemistry laboratory has two acid solutions available to her. The first solution is 10% acid and the second is 35% acid. (The percentages are by volume.) How many cm^3 of each solution should she mix to obtain 200 cm^3 of a 25% acid solution?

36. One salt solution is 15% salt and another is 20% salt. How many cm^3 of each solution must be mixed to obtain 50 cm^3 of a 16% salt solution?

37. A shopkeeper has two types of coffee beans. The first type sells for $5.20/lb, the second for $5.80/lb. How many pounds of the first type must be mixed with 5 lb of the second to produce a blend selling for $5.35/lb?

38. How much pure copper must be melted with 300 lb of an alloy that is 20% copper to yield an alloy that is 37.5% copper? (The percentages are by weight.)

39. Two sources of iron ore are available to a steel company. The first source of ore contains 20% iron (by weight). The second source contains 35% iron. How many tons of the second ore must be added to 16 tons of the first ore to produce a mixture that is 25% iron?

40. How much water must be evaporated from 400 lb of a 10% salt solution to yield a solution that is 16% salt?

41. The two shorter sides of a right triangle differ by 7 cm. If the area is 60 cm^2, find the hypotenuse.

42. If the sides of a square are doubled, the area increases by 147 cm^2. Find the sides of the original square.

43. The area of a square with sides of length $1 - x$ is equal to the area of a rectangle with dimensions 1 by x. Find x.

44. In the accompanying figure, $\overline{PT}$ is tangent to the circle at T and $PT = AB = 1$. Find PB. (Use the theorem from geometry stating that $PT^2 = PA \cdot PB$.)

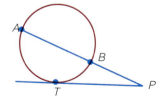

45. In the following figure, $AB = 1$ and B divides $\overline{AC}$ in such a way that the ratio of the whole segment to its larger part equals the ratio of the larger part to the smaller part. Find this ratio.

FIGURE FOR EXERCISE 45

46. For a certain right circular cylinder, the height is 3 m and the total surface area is $8\pi\text{m}^2$. Find the radius of the cylinder. (The formula for the total surface area S is $S = 2\pi r^2 + 2\pi rh$, where r is the radius and h is the height.)

47. A ball is thrown straight up. Suppose that the height of the ball at time t is given by $h = -16t^2 + 96t$, where h is in feet and t is in seconds, with $t = 0$ corresponding to the instant that the ball is first tossed.
    (a) How long does it take before the ball lands?
    (b) At what time is the height 80 ft? Why are there two answers here?

48. During a flu epidemic in a small town, a public health official finds that the total number P of people who have caught the flu after t days is closely approximated by the formula
$$P = -t^2 + 26t + 106 \qquad (1 \le t \le 13)$$
    (a) How many have caught the flu after 10 days?
    (b) After approximately how many days will 250 people have caught the flu?

49. Emily and Ernie are going to jog on a quarter-mile track. Emily can jog at a pace of 9 minutes per mile $\left(= \frac{1}{9}\text{ mile/min}\right)$, while Ernie can jog at a pace of 11 minutes per mile $\left(= \frac{1}{11}\text{ mile/min}\right)$. If they start together at the same time (running in the same direction), how long will it take for Emily to lap Ernie?

50. Consider the following motion problem.

    Suppose that you drive from town A to town B at 40 mph. At what (constant) speed should you make the return trip so as to average 50 mph for the entire round trip? (The answer is not 60 mph.)

    Solve this problem by completing the steps that follow.
    (a) Let D denote the distance (in miles) from A to B, and let R denote the speed (in mph) on the return trip. Complete the table.

	d (miles)	r (mph)	t (h)
Going	D		
Returning		R	

    (b) Explain why the following equation is valid.
$$\frac{2D}{(D/40) + (D/R)} = 50$$

(c) Solve the equation in part (b) for R, and round off your result to the nearest mph.

51. The speed of an eastbound train is 5 mph greater than the speed of a westbound train. If the trains leave at the same time from stations 630 miles apart and pass each other after 6 hours, find the speed of the eastbound train.

52. Jim and Juan are going to drive their motorcycles from San Francisco to Los Angeles. Jim leaves at 8:00 A.M. but Juan can't leave until 8:30 A.M. If Jim drives at 50 mph and Juan at 55 mph, at what time will Juan catch up with Jim?

53. Two airplanes take off from the airport at the same time, heading in opposite directions. After 2 hours the planes are 1500 miles apart. If the speeds of the planes differ by 50 mph, find the speed of the slower plane.

54. Annie drives from town A to town B in 3 hours (at a fixed speed). The traffic is heavier on the return trip. As a result, Annie's speed is 16 mph less, and the return trip takes 4 hours. What were the speeds on both parts of the trip?

55. A charter boat captain has 2 hours in which to leave the dock, head out to sea as far as possible, and then return to the dock. The boat's speeds on the outward and return journeys are 20 mph and 12 mph, respectively. How far can the boat travel out to sea?

56. You invest $5200 in a stock paying a 6.75% annual dividend and $8400 in a stock paying a 7.85% annual dividend. How many years will it take for the total earnings to be $2500. (Use a calculator and round off your answer to the nearest one-half year.)

57. You deposit $2650 at 8.1% and $3300 at 9.2% simple interest. What additional amount would you need to deposit at 11.5% to yield an overall annual return of 10%? (Use a calculator and round off your answer to the nearest dollar.)

58. In a certain right triangle, the hypotenuse is 205 cm and the two legs differ by 23 cm. Find the length of the shorter leg.

B

59. At noon Larry and Linda start walking in the same direction. Larry's speed is $\frac{1}{15}$ mile/minute; Linda's is less than that. However, when Larry has walked 2 miles, Linda doubles her pace and catches up with Larry at 1:15 P.M. How fast was Linda walking initially?

60. The length of a rectangular sheet of metal is 4 in. more than the width. As indicated in the figure, 3-in. squares are cut out from each corner of the rectangular sheet,

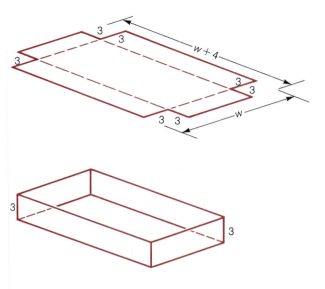

and the sides are then folded up to form a box with no top. If the volume of the box is 180 in.^{3}, determine the dimensions of the original rectangular sheet.

61. Suppose that you travel from town A to town B at a speed of v_1 mph and then return at a speed of v_2 mph. What is your average speed for the round trip (in terms of v_1 and v_2)?

62. If you travel for a hours at b mph and then c hours at d mph, what is your average speed for the trip?

63. In one savings account you deposit d dollars at an annual simple interest rate of $r\%$. At the same time you also deposit D dollars at $R\%$ in a second account. After how many years will the total interest amount to $d + D$ dollars?

64. A chemistry student has two acid solutions available to him. The first is $a\%$ acid and the second is $b\%$ acid. (The percentages are by volume.) How many cm^3 of each solution should be mixed to obtain d cm^3 of a solution that is $c\%$ acid? (Assume that a, b, c, and d are positive and that $a < c < b$.)

65. A piece of wire a inches long is bent to form a right triangle in which the shortest side is b inches. Find the length of the hypotenuse.

66. How much water must be evaporated from b pounds of an $a\%$ salt solution to obtain a solution that is $c\%$ salt? (Assume that the percentages are by weight and that $c > a$.)

67. If the length of each side of a square is multiplied by a, the area increases by b^2 square units. Find the sides of the original square.

68. A piece of wire L inches long is cut into two pieces. Each piece is then bent to form a square. If the sum of the areas of the two squares is $5L^2/128$, how long are the two pieces of wire?

69. A piece of wire 16 cm long is cut into two pieces, and each piece is then bent to form a circle. If the sum of the areas of the two circles is 12 cm^2, how long is the shorter piece of wire? (Round off your answer to two decimal places.)

70. The height of a right circular cyclinder is 1 ft. Find the radius, given that the total surface area is 660 ft^2. (Round off your answer to two decimal places.)

Exercises 71 and 72 are restatements of problems that were posed and solved in the ancient Chinese text "Nine Chapters on the Mathematical Art" (Chiu Chang Suan Shu), written sometime during the period 206 B.C.–221 A.D.

71. Find the side of a square inscribed in a right triangle with legs a and b. (See the accompanying figure.)

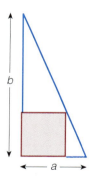

72. Find the diameter of a circle inscribed in a right triangle with legs a and b.

<hr/>

$\dfrac{2.5}{}$ **OTHER TYPES OF EQUATIONS**

If error is corrected wherever it is recognized as such, the path of error is the path of truth.

Hans Reichenbach (1891–1953)

The previous sections of this chapter dealt with linear and quadratic equations. In this section we consider several techniques that are useful in solving other types of equations. Throughout this section we'll concern ourselves only with solutions that are real numbers.

When n is a natural number, equations of the form $x^n = a$ are solved simply by rewriting the equation in terms of the appropriate nth root or roots. For example, if $x^5 = 34$, then $x = \sqrt[5]{34}$. If the exponent n is even, however, we have to remember that there are two real nth roots for each positive number. For instance, if

$$x^4 = 81$$

then

$$x = \pm\sqrt[4]{81} = \pm 3$$

EXAMPLE 1 Solve: **(a)** $(x - 1)^4 = 15$; **(b)** $3x^4 = -48$.

Solution **(a)** $(x - 1)^4 = 15$
$$x - 1 = \pm\sqrt[4]{15}$$
$$x = 1 \pm \sqrt[4]{15}$$

As you can check for yourself now, both of the values, $1 + \sqrt[4]{15}$ and $1 - \sqrt[4]{15}$, satisfy the given equation.

NOTE   You can estimate these values without a calculator. For instance, for $x = 1 + \sqrt[4]{15}$ we have

$$1 + \sqrt[4]{15} \approx 1 + \sqrt[4]{16} = 1 + 2 = 3$$

Therefore, $1 + \sqrt[4]{15}$ is a little less than 3. (In fact, a calculator shows that $1 + \sqrt[4]{15} \approx 2.97$.)

(b) For any real number x, the quantity x^4 is nonnegative. Therefore, the left-hand side of the equation is nonnegative, whereas the right-hand side is negative. Consequently, there are no real numbers satisfying the given equation.

In Example 2 we solve two equations by factoring. As with quadratic equations, the factoring method is justified by the zero-product property of real numbers discussed in Section 2.3.

EXAMPLE 2 Solve: **(a)** $3x^3 - 12x^2 - 15x = 0$; **(b)** $x^4 + x^2 - 6 = 0$.

Solution **(a)** $3x^3 - 12x^2 - 15x = 0$
$$3x(x^2 - 4x - 5) = 0$$
$$3x(x - 5)(x + 1) = 0$$

Now, by setting each factor equal to zero, we obtain the three values $x = 0$, $x = 5$, and $x = -1$. As you can check, each of these numbers indeed satisfies the original equation. So we have three solutions: 0, 5, and -1.

NOTE   If initially we'd divided both sides of the given equation by $3x$ to obtain $x^2 - 4x - 5 = 0$, then the solution $x = 0$ would have been overlooked.

(b)
$$x^4 + x^2 - 6 = 0$$
$$(x^2 - 2)(x^2 + 3) = 0$$

$x^2 - 2 = 0$	$x^2 + 3 = 0$
$x^2 = 2$	$x^2 = -3$
$x = \pm\sqrt{2}$	

By setting the first factor equal to zero, we've obtained the two values $x = \pm\sqrt{2}$. As you can check, these values do satisfy the given equation. When we set the second factor equal to zero, however, we obtained $x^2 = -3$, which has no real solutions (because the square of a real number must be nonnegative). We conclude therefore that the given equation in this case has just two real solutions, $\sqrt{2}$ and $-\sqrt{2}$.

The equation $x^4 + x^2 - 6 = 0$ that we considered in Example 2(b) is called **an equation of quadratic type.** This means that with a suitable substitution, the resulting equation becomes quadratic. In particular here, suppose that we let

$$x^2 = t \qquad \text{and therefore} \qquad x^4 = t^2$$

Then the equation $x^4 + x^2 - 6 = 0$ becomes $t^2 + t - 6 = 0$, which is quadratic. This idea is exploited in the next three examples.

EXAMPLE 3   Solve: $x^4 - 6x^2 + 4 = 0$.

Solution   This equation cannot readily be solved by factoring. However, it is of quadratic type. We let $x^2 = t$. Then $x^4 = t^2$, and the equation becomes

$$t^2 - 6t + 4 = 0$$

The quadratic formula then yields

$$t = \frac{6 \pm \sqrt{36 - 4(4)}}{2} = \frac{6 \pm \sqrt{20}}{2}$$
$$= \frac{6 \pm 2\sqrt{5}}{2} = 3 \pm \sqrt{5}$$

Consequently, we have (in view of the definition of t)

$$x^2 = 3 \pm \sqrt{5}$$

and therefore

$$x = \pm\sqrt{3 \pm \sqrt{5}}$$

We've now found four values for x:

$$\sqrt{3 + \sqrt{5}} \qquad -\sqrt{3 + \sqrt{5}} \qquad \sqrt{3 - \sqrt{5}} \qquad -\sqrt{3 - \sqrt{5}}$$

As Exercise 83(a) asks you to check, each of these values indeed satisfies the original equation.

EXAMPLE 4   Solve: $6x^{-2} - x^{-1} - 2 = 0$.

Solution   We'll show two methods. The first method depends on recognizing the equation as one of quadratic type.

FIRST METHOD   Let $x^{-1} = t$. Then $x^{-2} = t^2$ and the equation becomes

$$6t^2 - t - 2 = 0$$
$$(3t - 2)(2t + 1) = 0$$

$$3t - 2 = 0 \quad | \quad 2t + 1 = 0$$
$$t = \frac{2}{3} \quad | \quad t = -\frac{1}{2}$$

If $t = \frac{2}{3}$, then $x^{-1} = \frac{2}{3}$. Therefore, $\frac{1}{x} = \frac{2}{3}$ and, consequently, $x = \frac{3}{2}$. On the other hand, if $t = -\frac{1}{2}$, then $\frac{1}{x} = -\frac{1}{2}$ and, consequently, $x = -2$. In summary, the two solutions are $\frac{3}{2}$ and -2.

ALTERNATE METHOD   Multiplying both sides of the given equation by x^2 yields

$$6 - x - 2x^2 = 0$$
$$2x^2 + x - 6 = 0$$
$$(2x - 3)(x + 2) = 0$$

$$2x - 3 = 0 \quad | \quad x + 2 = 0$$
$$x = \frac{3}{2} \quad | \quad x = -2$$

Thus we have $x = \frac{3}{2}$ or $x = -2$, as obtained using the first method.

▌▌▌

EXAMPLE 5   Solve: $4x^{4/3} + 15x^{2/3} - 4 = 0$.

Solution   Let $x^{2/3} = t$. Then $x^{4/3} = t^2$, and the equation becomes

$$4t^2 + 15t - 4 = 0$$
$$(4t - 1)(t + 4) = 0$$

$$4t - 1 = 0 \quad | \quad t + 4 = 0$$
$$t = \tfrac{1}{4} \quad | \quad t = -4$$

We now have two cases to consider: $t = \frac{1}{4}$ and $t = -4$. With $t = \frac{1}{4}$ we have (in view of the definition of t)

$$x^{2/3} = \frac{1}{4}$$

$$x^2 = \frac{1}{64} \qquad \text{Two real numbers are equal if and only if their cubes are equal.}$$

$$x = \pm\frac{1}{8}$$

As Exercise 83(b) asks you to verify, the values $\frac{1}{8}$ and $-\frac{1}{8}$ satisfy the original equation. On the other hand, with $t = -4$, we have

$$x^{2/3} = -4$$

or

$$x^2 = -64 \qquad \text{cubing both sides}$$

This last equation has no real solutions, since the square of a real number is never negative. In summary, then, the only (real-number) solutions of the original equation are $\pm\frac{1}{8}$.

Some equations can be solved by raising both sides to the same power. In fact, if you review the solution given for Example 5, you'll see that we've already made use of this idea in cubing both sides of one of the equations there. As another example, consider the equation

$$\sqrt{x - 3} = 5$$

By squaring both sides of this equation we obtain $x - 3 = 25$ and, consequently, $x = 28$. As you can easily check, the value $x = 28$ does satisfy the original equation.

There is a complication, however, that may arise in raising both sides of an equation to the same power. Consider, for example, the equation

$$x - 2 = \sqrt{x}$$

By squaring both sides, we obtain

$$x^2 - 4x + 4 = x$$
$$x^2 - 5x + 4 = 0$$
$$(x - 4)(x - 1) = 0$$

Therefore

$$x = 4 \qquad \text{or} \qquad x = 1$$

The value $x = 4$ checks in the original equation, but the value $x = 1$ does not. (Verify this.) Therefore, $x = 1$ is an extraneous solution, and the only solution of the given equation is $x = 4$.

The extraneous solution in this case arises from the fact that if $a^2 = b^2$, then it need not be true that $a = b$. For instance, $(-3)^2 = 3^2$, but certainly $-3 \neq 3$. In the box that follows, we summarize this remark about extraneous solutions. For reference, we also include the results from Section 2.2.

PROPERTY SUMMARY   EXTRANEOUS SOLUTIONS

1. Squaring both sides of an equation (or raising both sides to an even integral power) may introduce extraneous solutions that do not check in the original equation.

2. Multiplying or dividing both sides of an equation by an expression involving the variable may introduce extraneous solutions.

Therefore, it is always necessary to check any candidates for solutions that you obtain in either of these ways.

Our final example for this section involves an equation containing several square root expressions. Before squaring both sides in such cases, it's usually a good idea to see if you can first isolate one of the radical expressions on one side of the equation.

EXAMPLE 6   Solve: $\sqrt{9 + x} + \sqrt{1 + x} - \sqrt{x + 16} = 0$.

Solution
$$\sqrt{9 + x} + \sqrt{1 + x} = \sqrt{x + 16} \qquad \text{adding } \sqrt{x + 16} \text{ to both sides in preparation for squaring}$$

$$\left(\sqrt{9 + x} + \sqrt{1 + x}\right)^2 = \left(\sqrt{x + 16}\right)^2$$

$$9 + x + 2\sqrt{9 + x}\,\sqrt{1 + x} + 1 + x = x + 16$$

$$2\sqrt{9 + x}\,\sqrt{1 + x} = 6 - x$$

$$\left(2\sqrt{9 + x}\,\sqrt{1 + x}\right)^2 = (6 - x)^2$$

$$4(9 + 10x + x^2) = 36 - 12x + x^2 \qquad \text{(Why?)}$$

$$3x^2 + 52x = 0$$

$$x(3x + 52) = 0$$

$$x = 0 \quad \bigg| \quad 3x + 52 = 0$$

$$x = -\tfrac{52}{3}$$

We've now found two possible solutions: 0 and $-\frac{52}{3}$. [Since these were obtained by squaring both sides of an equation, we need to check to see if either (or both) satisfy the original equation.] With $x = 0$, the original equation becomes $\sqrt{9} + \sqrt{1} - \sqrt{16} = 0$, or $3 + 1 - 4 = 0$, which is certainly true. On the other hand, with $x = -\frac{52}{3}$, the quantities underneath the radicals in the original equation are negative, and consequently the (real) square roots are undefined. So $-\frac{52}{3}$ is an extraneous solution, and $x = 0$ is the only solution of the given equation.

EXERCISE SET 2.5

A

In Exercises 1–54, find all the real solutions of each equation. For Exercises 21–26, give two forms for each answer: an exact answer (involving a radical) and a calculator approximation rounded off to two decimal places.

1. $3x^2 - 48x = 0$
2. $3x^3 - 48x = 0$
3. $t^3 - 125 = 0$
4. $t - t^3 = 0$
5. $7x^4 - 28x^2 = 0$
6. $y^4 - 81 = 0$
7. $225(x - 1) - x^2(x - 1) = 0$
8. $x^2(x + 4)^2 + 6x(x + 4)^2 + 9(x + 4)^2 = 0$
9. $4y^3 - 20y^2 + 25y = 0$
10. $y^4 + 27y = 0$
11. $t^4 + 2t^3 - 3t^2 = 0$
12. $2t^5 + 5t^4 - 12t^3 = 0$
13. $6x = 23x^2 + 4x^3$
14. $x^5 = 36x$
15. $x^4 - x^2 = 6$
16. $x^4 - 5x^2 = -6$
17. $4y^2 = 5 - y^4$
18. $6y^2 = -5 - y^4$
19. $3t^2 + 2 = 9t^4$
20. $5t^2 - 1 = 4t^4$
21. $(x - 2)^3 - 5 = 0$
22. $(x + 4)^3 + 2 = 0$

23. $(x + 4)^5 + 16 = 0$
24. $(1 - x)^5 - 40 = 0$
25. (a) $(x - 3)^4 - 30 = 0$
    (b) $(x - 3)^4 + 30 = 0$
26. (a) $(x^2 - 1)^4 - 81 = 0$
    (b) $(x^2 - 1)^4 + 81 = 0$
27. $y^4 + 4y^2 - 5 = 0$
28. $y^4 + 6y^2 + 5 = 0$
29. $9t^4 - 3t^2 - 2 = 0$
30. $4t^4 - 5t^2 + 1 = 0$
31. $x^6 - 10x^4 + 24x^2 = 0$
32. $2x^5 - 15x^3 - 27x = 0$
33. $x^4 + x^2 - 1 = 0$
34. $x^4 - x^2 + 1 = 0$
35. $x^4 + 3x^2 - 2 = 0$
36. $2x^4 + x^2 + 2 = 0$
37. $x^6 + 7x^3 = 8$
38. $x^6 + 27 = -28x^3$
39. $t^{-2} - 7t^{-1} + 12 = 0$
40. $8t^{-2} - 17t^{-1} + 2 = 0$
41. $12y^{-2} - 23y^{-1} = -5$
42. $y^{-2} - y^{-1} = 0$
43. $4x^{-4} - 33x^{-2} - 27 = 0$
44. $x^{-4} + x^{-2} + 1 = 0$
45. $t^{2/3} = 9$
46. $t^{3/2} = 8$
47. $x^4 = 81$
48. $x^4 = -81$
49. $(y - 1)^3 = 7$
50. $(2y + 3)^4 = 5$
51. $(t + 1)^5 = -243$
52. $(t + 3)^4 = 625$
53. $9x^{4/3} - 10x^{2/3} + 1 = 0$
54. $x^{4/3} + 3x^{2/3} - 28 = 0$

In Exercises 55–70, determine all the real-number solutions for each equation. (Remember to check for extraneous solutions.)

55. $\sqrt{1 - 3x} = 2$

56. $\sqrt{x^2 + 5x - 2} = 2$

57. $\sqrt{x + 6} = x$

58. $x - \sqrt{x} = 20$

59. $x - \sqrt{3 - x} = -3$

60. $\sqrt{2 - x} - 10 = x$

61. $4x + \sqrt{2x + 5} = 0$

62. $\sqrt{7 - 3x} - 6(x + 1) = 1$

63. $\sqrt{x^4 - 13x^2 + 37} = 1$

64. $\sqrt{y + 2} = y - 4$

65. $\sqrt{1 - 2x} + \sqrt{x + 5} = 4$

66. $\sqrt{x - 5} - \sqrt{x + 4} + 1 = 0$

67. $\sqrt{3 + 2t} + \sqrt{-1 + 4t} = 1$

68. $\sqrt{2t + 5} - \sqrt{8t + 25} + \sqrt{2t + 8} = 0$

69. $\sqrt{2y - 3} - \sqrt{3y + 3} + \sqrt{3y - 2} = 0$

70. $\sqrt{a - x} + \sqrt{b - x} = \sqrt{a + b - 2x}$ $(b > a > 0)$

In Exercises 71–73, solve the equations. Use a calculator and round off the answers to two decimal places.

71. $x^4 - 2x^2 - 4 = 0$

72. $(3x - 5)^7 = 405$

73. $2x\sqrt{x^2 + 4} + 2x\sqrt{x^2 + 1} = 3$

74. The radii of three lead spheres are 10.01 cm, 15.43 cm, and 20.14 cm. Suppose that these three spheres are melted down and recast to form one large sphere. Find the radius of that sphere. Use a calculator and round off the final answer to two decimal places. *Hint:* Use the formula $V = \frac{4}{3}\pi r^3$ for the volume of a sphere.

B

In Exercises 75–82, find all the real solutions of each equation.

75. $\sqrt{\sqrt{x} + \sqrt{a}} + \sqrt{\sqrt{x} - \sqrt{a}} = \sqrt{2\sqrt{x} + 2\sqrt{b}}$

76. $x = \sqrt{3x + x^2} - 3\sqrt{3x + x^2}$

77. $\dfrac{\sqrt{x} - a}{\sqrt{x}} - \dfrac{\sqrt{x} + a}{\sqrt{x} - b} = 0$ $(a > 0, b > 0)$
    Suggestion: Let $\sqrt{x} = t$.

78. $x - \sqrt{x^2 - x} = \sqrt{x}$ $(x > 0)$
    Suggestion: If $x \neq 0$, then you can divide through by $\sqrt{x}$ to obtain a simpler equation.

79. $\sqrt{x^2 - x - 1} - \dfrac{2}{\sqrt{x^2 - x - 1}} = 1$
    Hint: Let $t = x^2 - x - 1$.

80. $\sqrt{x^2 + 3x - 4} - \sqrt{x^2 - 5x + 4} = x - 1$ $(x > 4)$
    Hint: Factor the expressions beneath the radicals.

Then note that $\sqrt{x - 1}$ is a factor of both sides of the equation.

81. $\sqrt{\dfrac{x - a}{x}} + 4\sqrt{\dfrac{x}{x - a}} = 5$ $(a \neq 0)$

    Hint: Let $t = \dfrac{x - a}{x}$. Then $\dfrac{1}{t} = \dfrac{x}{x - a}$.

82. $\sqrt{p + 4q - 5t} + \sqrt{4p + q - 5t} = 3\sqrt{p + q - 2t}$
    (Assume that p and q are constants and $q > p$.)

83. (a) Verify that the four solutions obtained in Example 3 indeed satisfy the original equation.
    (b) Verify that the values $x = \frac{1}{8}$ and $x = -\frac{1}{8}$ satisfy the equation $4x^{4/3} + 15x^{2/3} - 4 = 0$.

Exercise 84 (including the clever solution that is outlined) appears in the book The USSR Olympiad Problem Book *by D. O. Shklarsky et al. (San Francisco: W. H. Freeman and Co., 1962)*

84. Follow steps (a)–(d) to solve the equation

    $$\sqrt{a - \sqrt{a + x}} = x \qquad (a \geq 1)$$

    (a) By squaring as usual, obtain the equation

    $$x^4 - 2ax^2 - x + a^2 - a = 0.$$

    (b) Although the equation obtained in part (a) is a fourth-degree equation in x, it is only a quadratic in a. Solve for a in terms of x to obtain $a = x^2 + x + 1$ or $a = x^2 - x$.

    (c) Solve each of the equations in part (b) for x in terms of a.

    (d) Check your results in part (c) to eliminate any extraneous roots.

C

For Exercises 85–89, find all real solutions of each equation. Exercises 88 and 89 are taken from Higher Algebra by H. S. Hall and S. R. Knight, first published in 1887.

85. $\sqrt{8 + 2t} + \sqrt{5 + t} = \sqrt{15 + 3t}$

86. $-x = \sqrt{1 - \sqrt{1 + x}}$ *Hint:* After squaring once and rearranging, look for a common factor appearing on both sides of the equation.

87. $\sqrt{x^2 + x} + \dfrac{1}{\sqrt{x^2 + x}} = \dfrac{5}{2}$ (Round off the answers to two decimal places.)

88. $\dfrac{\sqrt{x + 1} + \sqrt{x - 1}}{\sqrt{x + 1} - \sqrt{x - 1}} = \dfrac{4x - 1}{2}$

89. $\sqrt[m]{(a + x)^2} + 2\sqrt[m]{(a - x)^2} = 3\sqrt[m]{a^2 - x^2}$

2.6 INEQUALITIES

The fundamental results of mathematics are often inequalities rather than equalities.

E. Beckenbach and R. Bellman in *An Introduction to Inequalities* (New York: Random House, 1961)

If we replace the equal sign in an equation with any one of the four symbols $<$, $\leq$, $>$, or $\geq$, we obtain an **inequality**. As with equations in one variable, a real number is a **solution** of an inequality if we obtain a true statement when the variable is replaced by the real number. For example, the value $x = 5$ is a solution of the inequality $2x - 3 < 8$, because when $x = 5$ we have

$$2(5) - 3 < 8$$
$$7 < 8 \qquad \text{which is true}$$

We also say in this case that the value $x = 5$ **satisfies** the inequality. To **solve** an inequality means to find all of the solutions. The set of all solutions of an inequality is called (naturally enough) the **solution set.**

Recall that two equations are said to be equivalent if they have exactly the same solutions. Similarly, two inequalities are **equivalent** if they have the same solution set. Most of the procedures used for solving inequalities are similar to those for equalities. For example, adding or subtracting the same number on both sides of an inequality produces an equivalent inequality. We need to be careful, however, in multiplying or dividing both sides of an inequality by the same nonzero number. For instance, suppose that we start with the inequality $2 < 3$ and multiply both sides by 5. That yields $10 < 15$, which is certainly true. But if we multiply both sides of the inequality $2 < 3$ by -5, we obtain $-10 < -15$, which is false. Multiplying both sides of an inequality by the same *positive* number preserves the inequality, whereas multiplying by a *negative* number reverses the inequality. In the following box, we list some of the principal properties of inequalities. In general, whenever we use Property 1 or 2 in solving an inequality, we obtain an equivalent inequality. Also, note that each property can be rewritten to reflect the fact that $a < b$ is equivalent to $b > a$. For example, Property 3 can just as well be written this way: If $b > a$ and $c > b$, then $c > a$.

PROPERTY SUMMARY   **PROPERTIES OF INEQUALITIES**

PROPERTY	EXAMPLE
1. If $a < b$, then $a + c < b + c$ and $a - c < b - c$.	If $x - 3 < 0$, then $(x - 3) + 3 < 0 + 3$ and, consequently, $x < 3$.
2. (a) If $a < b$ and c is positive, then $ac < bc$ and $a/c < b/c$.	If $\frac{1}{2}x < 4$, then $2\left(\frac{1}{2}x\right) < 2(4)$ and, consequently, $x < 8$.
(b) If $a < b$ and c is negative, then $ac > bc$ and $a/c > b/c$.	If $-\dfrac{x}{5} < 6$, then $(-5)\left(-\dfrac{x}{5}\right) > (-5)(6)$ and, consequently, $x > -30$.
3. The transitive property: If $a < b$ and $b < c$, then $a < c$.	If $a < x$ and $x < 2$, then $a < 2$.

EXAMPLE 1   Solve: $4t + 8 \leq 7(1 + t)$.

Solution   Our work follows the pattern that we would use to solve the equation $4t + 8 = 7(1 + t)$. We have

$$4t + 8 \leq 7(1 + t)$$
$$4t + 8 \leq 7 + 7t$$
$$-3t \leq -1 \qquad \text{subtracting } 7t \text{ and } 8 \text{ from both sides}$$
$$t \geq \tfrac{1}{3} \qquad \text{dividing by } -3 \text{ (this reverses the inequality)}$$

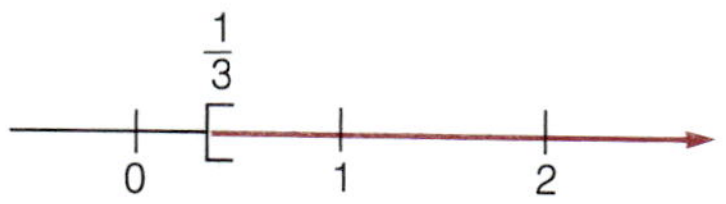

FIGURE 1
$t \geq \frac{1}{3}$

The solution set is therefore $\left[\frac{1}{3}, \infty\right)$. See Figure 1.

In the next example, we solve the inequality

$$-\frac{1}{2} < \frac{3 - x}{-4} < \frac{1}{2}$$

By definition, this is equivalent to the pair of inequalities

$$-\frac{1}{2} < \frac{3 - x}{-4} \qquad \text{and} \qquad \frac{3 - x}{-4} < \frac{1}{2}$$

One way to proceed here would be first to determine the solution set for each inequality. Then the set of real numbers common to both solution sets would be the solution set for the original inequality. However, the method shown in Example 2 is more efficient.

EXAMPLE 2   Solve: $-\frac{1}{2} < \frac{3 - x}{-4} < \frac{1}{2}$.

Solution   We begin by multiplying through by -4. Remember, this will reverse the inequalities:

$$2 > 3 - x > -2$$

Next, with a view toward isolating x, we first subtract 3 to obtain

$$-1 > -x > -5$$

Finally, multiplying through by -1, we have

$$1 < x < 5$$

The solution set is therefore the interval $(1, 5)$.

EXAMPLE 3   Suppose that you inherit $5000 under the stipulation that it be invested in two stocks, A and B. Stock A pays an 8% annual dividend, and stock B pays a 9% annual dividend. What is the largest amount that you can invest in stock A if the total dividends each year are to be at least $435? (Dividends are computed using the formula for simple interest: $I = Prt$.)

Solution   First, just to get a feeling for the problem, let's see what would happen if the entire $5000 were invested in one stock. If you invested all $5000 in stock A, the annual dividend would be ($5000)(0.08)(1) = $400. That falls short of the required $435. On the other hand, if the $5000 were invested in stock B, the annual dividend would be ($5000)(0.09)(1), which is $450. That exceeds the required minimum of $435. The question the problem asks is, What is the most that can be invested in stock A if the total annual dividends are to be at least $435? Let x denote the amount that can be invested in stock A. Then $5000 - x$ represents the amount (in dollars) to be invested in stock B. Since the combined dividends each year are to be at least $435, we have

$$x(0.08) + (5000 - x)(0.09) \geq 435$$
$$8x + (5000 - x)9 \geq 43500$$
$$8x + 45000 - 9x \geq 43500$$
$$-x \geq -1500$$
$$x \leq 1500$$

Thus, at most $1500 can be invested in stock A at 8%.

In the next example, reference is made to the Celsius and Fahrenheit scales for measuring temperature.* The formula relating the temperature readings on the two scales is

$$F = \frac{9}{5}C + 32$$

EXAMPLE 4 Over the temperature range $32° \leq F \leq 39.2°$ on the Fahrenheit scale, water contracts (rather than expands) with increasing temperature. What is the corresponding temperature range on the Celsius scale?

Solution

$$32 \leq F \leq 39.2 \qquad \text{given}$$

$$32 \leq \frac{9}{5}C + 32 \leq 39.2 \qquad \text{substituting } \frac{9}{5}C + 32 \text{ for } F$$

$$0 \leq \frac{9}{5}C \leq 7.2 \qquad \text{subtracting 32}$$

$$0 \leq C \leq \frac{5}{9}(7.2) \qquad \text{multiplying by } \frac{5}{9}$$

$$0 \leq C \leq 4$$

Thus, a range of $32°F - 39.2°F$ on the Fahrenheit scale corresponds to $0°C - 4°C$ on the Celsius scale.

In the next three examples, we solve inequalities that involve absolute values. The following theorem is very useful in this context.

Theorem: Absolute Value and Inequalities

If $a > 0$, then

$$|u| < a \qquad \text{if and only if} \qquad -a < u < a$$

and

$$|u| > a \qquad \text{if and only if} \qquad u < -a \quad \text{or} \quad u > a$$

---

*The Celsius scale was devised in 1742 by the Swedish astronomer Anders Celsius. The Fahrenheit scale was first used by the German physicist Gabriel Fahrenheit in 1724.

$-a < u < a$

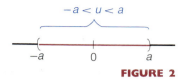

FIGURE 2

You can see why this theorem is valid if you think in terms of distance and position on a number line. The condition $-a < u < a$ means that u lies between $-a$ and a, as indicated in Figure 2. But this is the same as saying that the distance from u to zero is less than a, which in turn can be written $|u| < a$. (The second part of the theorem can be justified in a similar manner.)

EXAMPLE 5 Solve: **(a)** $|x| < 1$; **(b)** $|x| \geq 1$.

Solution **(a)** By the theorem we just discussed, the condition $|x| < 1$ is equivalent to

$$-1 < x < 1$$

The solution set is therefore the open interval $(-1, 1)$.

(b) In view of the second part of the theorem, the inequality $|x| \geq 1$ is satisfied when x satisfies either of the inequalities

$$x \leq -1 \qquad \text{or} \qquad x \geq 1$$

The solution sets for these last two inequalities are $(-\infty, -1]$ and $[1, \infty)$, respectively. Consequently, the solution set for the given inequality $|x| \geq 1$ consists of the two intervals $(-\infty, -1]$ and $[1, \infty)$. ∎

In Example 5(b), we found that the solution set consisted of two intervals on the number line. We have a convenient notation for describing such sets. Given any two sets A and B, we define the set $A \cup B$ (read **A union B**) to be the set of all elements that are in A or in B (or in both). For example, if $A = \{1, 2, 3\}$ and $B = \{4, 5\}$, then $A \cup B = \{1, 2, 3, 4, 5\}$. As another example, the union of the two closed intervals $[3, 5]$ and $[4, 7]$ is given by

$$[3, 5] \cup [4, 7] = [3, 7]$$

because the numbers in the interval $[3, 7]$ are precisely those numbers that are in $[3, 5]$ or $[4, 7]$ (or in both). Using this notation, we can write the solution set for Example 5(b) as

$$(-\infty, -1] \cup [1, \infty)$$

EXAMPLE 6 Solve: $|x - 3| < 1$.

Solution We'll show two methods.

FIRST METHOD We use the theorem preceding Example 5. With $u = x - 3$ and $a = 1$, the theorem tells us that the given inequality is equivalent to

$$-1 < x - 3 < 1$$

or (by adding 3)

$$2 < x < 4$$

The solution set is therefore the open interval $(2, 4)$.

ALTERNATE METHOD The given inequality tells us that x must be less than one unit away from 3 on the number line. Looking one unit to either side of 3, then, we see that x must lie strictly between 2 and 4. The solution set is therefore $(2, 4)$, as obtained using the first method. ∎

EXAMPLE 7 Solve: $\left|1 - \dfrac{t}{2}\right| > 5$.

Solution Referring again to the theorem, we use the fact that $|u| > a$ means that either $u < -a$ or $u > a$. So, in the present example the given inequality means that either

$$1 - \frac{t}{2} < -5 \qquad \text{or} \qquad 1 - \frac{t}{2} > 5$$

$$-\frac{t}{2} < -6 \qquad\qquad\qquad -\frac{t}{2} > 4$$

$$t > 12 \qquad\qquad\qquad\qquad t < -8$$

This tells us that the given inequality is satisfied when t is in either of the intervals $(12, \infty)$ or $(-\infty, -8)$. Thus the solution set is $(-\infty, -8) \cup (12, \infty)$.

EXERCISE SET 2.6

A

In Exercises 1–36, solve the inequality and specify the answer using interval notation.

1. $x + 5 < 4$
2. $2x - 7 < 11$
3. $1 - 3x \leq 0$
4. $6 - 4x \leq 22$
5. $4x + 6 < 3(x - 1) - x$
6. $2(t - 1) - 3(t + 1) \leq -5$
7. $1 - 2(t + 3) - t \leq 1 - 2t$
8. $t - 4[1 - (t - 1)] > 7 + 10t$
9. $\dfrac{3x}{5} - \dfrac{x - 1}{3} < 1$
10. $\dfrac{2x + 1}{2} + \dfrac{x - 1}{3} < x + \dfrac{1}{2}$
11. $\dfrac{x - 1}{4} - \dfrac{2x + 3}{5} \leq x$
12. $\dfrac{x}{2} - \dfrac{8x}{3} + \dfrac{x}{4} > \dfrac{23}{6}$
13. $-2 \leq x - 6 \leq 0$
14. $-3 \leq 2x + 1 \leq 5$
15. $-1 \leq \dfrac{1 - 4t}{3} \leq 1$
16. $\dfrac{2}{3} \leq \dfrac{5 - 3t}{-2} \leq \dfrac{3}{4}$
17. $0.99 < \dfrac{x}{2} - 1 < 0.999$
18. $\dfrac{9}{10} < \dfrac{3x - 1}{-2} < \dfrac{91}{100}$
19. (a) $|x| \leq \frac{1}{2}$ (b) $|x| > \frac{1}{2}$
20. (a) $|x| > 2$ (b) $|x| \leq 2$
21. (a) $|x| > 0$ (b) $|x| < 0$
22. (a) $|t| \geq 0$ (b) $|t| \leq 0$
23. (a) $x - 2 < 1$ (b) $|x - 2| < 1$
    (c) $|x - 2| > 1$
24. (a) $x - 4 \geq 4$ (b) $|x - 4| \geq 4$
    (c) $|x - 4| \leq 4$

25. (a) $1 - x \leq 5$ (b) $|1 - x| \leq 5$
    (c) $|1 - x| > 5$
26. (a) $3x + 5 < 17$ (b) $|3x + 5| < 17$
    (c) $|3x + 5| > 17$
27. (a) $a - x < c$ (b) $|a - x| < c$
    (c) $|a - x| \geq c$
28. (a) $|x - a| + b < c$ (Assume $b < c$ throughout this exercise.)
    (b) $|x + a| + b > c$ (c) $|x + a| + b < c$
29. $\left|\dfrac{x - 2}{3}\right| < 4$
30. $\left|\dfrac{4 - 5x}{2}\right| > 1$
31. $\left|\dfrac{x + 1}{2} - \dfrac{x - 1}{3}\right| < 1$
32. $\left|\dfrac{3(x - 2)}{4} + \dfrac{4(x - 1)}{3}\right| \leq 2$
33. (a) $|(x + h)^2 - x^2| < 3h^2$ $(h > 0)$
    (b) $|(x + h)^2 - x^2| < 3h^2$ $(h < 0)$
34. (a) $|3(x + 2)^2 - 3x^2| < \frac{1}{10}$
    (b) $|3(x + 2)^2 - 3x^2| < \epsilon$ $(\epsilon > 0)$
35. $1.86 < 4.11x - 0.15 < 1.95$ (Round off your answer to two decimal places.)
36. $2.713 < 1 - 3.332x < 3.001$ (Round off your answer to three decimal places.)
37. Suppose that you inherit $6000 under the stipulation that it be invested in two particular stocks, one paying a 7% dividend annually and the other paying 9%.

What is the largest amount that you can invest in the stock paying the 7% dividend if you want the total yearly dividends to be at least $500?

38. Data from the *Apollo 11* moon mission in July 1969 showed that temperature readings on the lunar surface vary over the interval $-183° \leq C \leq 112°$ on the Celsius scale. What is the corresponding interval on the Fahrenheit scale? (Round off the numbers you obtain to the nearest integers.)

39. From the cloud tops of Venus to the planet's surface, temperature readings range over the interval $-25° \leq C \leq 475°$ on the Celsius scale. What is the corresponding range on the Fahrenheit scale?

40. Measurements sent back to Earth from the *Viking I* spacecraft in June 1976 indicated temperature readings on the surface of Mars in the range $-139° \leq C \leq -28°$. What is the corresponding range on the Fahrenheit scale? (Round off the numbers you obtain to the nearest integers.)

41. The temperature of the variable star Delta Cephei varies over the interval $5100° \leq C \leq 6500°$ on the Celsius scale. What is the corresponding range on the Fahrenheit scale? (Round off the numbers in your answer to the nearest 100°F.)

42. Data from the *Mariner 10* spacecraft (launched November 3, 1973) indicate that the surface temperature on the planet Mercury varies over the interval $-170° \leq C \leq 430°$ on the Celsius scale. What is the corresponding interval on the Fahrenheit scale? (Round off the values that you obtain to the nearest 10°F.)

B

43. Given two positive numbers a and b, we define the **geometric mean** and the **arithmetic mean** as follows:

$$\text{G.M.} = \sqrt{ab} \qquad \text{A.M.} = \frac{a + b}{2}$$

(a) Complete the table (at the top of the next column). Use a calculator, as necessary, so that the entries in the third and fourth columns are in decimal form.

(b) Prove that the following inequality is valid for all positive numbers a and b.

$$\sqrt{ab} \leq \frac{a + b}{2}$$

(The geometric mean is less than or equal to the arithmetic mean.)

Hint: Use the following property of inequalities. If x and y are positive, then the inequality $x \leq y$ is equivalent to $x^2 \leq y^2$.

a	b	$\sqrt{ab}$ (G.M.)	$(a+b)/2$ (A.M.)	Which is larger, G.M. or A.M.?
1	2			
1	3			
1	4			
2	3			
3	4			
5	10			
9	10			
99	100			
999	1000			

44. Given two positive numbers a and b, we define the **root mean square** as follows:

$$\text{R.M.S.} = \sqrt{\frac{a^2 + b^2}{2}}$$

(a) Complete the following table. (Use a calculator, as necessary, so that the entries in the third and fourth columns are in decimal form.)

a	b	$(a+b)/2$ (A.M.)	$\sqrt{(a^2+b^2)/2}$ (R.M.S.)	Which is larger, A.M. or R.M.S.?
1	2			
1	3			
1	4			
2	3			
3	4			
5	10			
9	10			
99	100			
999	1000			

(b) Prove that the following inequality is valid for all positive numbers a and b.

$$(a + b)/2 \leq \sqrt{(a^2 + b^2)/2}$$

(The arithmetic mean is less than or equal to the root mean square.)

[Use the hint in Exercise 43(b).]

45. (a) Complete the table; use a calculator and round off to four decimal places.

x	y	$\dfrac{x}{y}+\dfrac{y}{x}$	True or False $\dfrac{x}{y}+\dfrac{y}{x}\geq 2$
1	1		
2	3		
3	5		
4	7		
5	9		
9	10		
49	50		
99	100		

(b) Prove that for all positive numbers x and y, we have $\dfrac{x}{y}+\dfrac{y}{x}\geq 2$. *Hint:* Use the inequality in Exercise 43(b). Take $a=x/y$ and $b=y/x$.

46. Solve the inequality $|x-1|+|x-2|<3$. *Hint:* Begin by considering three cases: $x<1$; $1\leq x<2$; $x\geq 2$.

47. Suppose that $|x-2|<0.1$ and $|y-3|<0.01$. Show that $|xy-6|\leq 0.321$. *Hint:* First check that $xy-6=x(y-3)+3(x-2)$. Then use the triangle inequality (page 20).

C

48. Let a and b be positive numbers. In Exercises 43 and 44, you were asked to show that
$$\sqrt{ab}\leq\frac{a+b}{2}\leq\sqrt{\frac{a^2+b^2}{2}}$$

This exercise shows how to establish these important inequalities using geometric, rather than algebraic, methods. In the following figure, $\overline{CF}$ is the diameter of a semicircle with center E. Let $CD=a$ and $DF=b$.
(a) Explain why $CE=EH=(a+b)/2$.
(b) Show that $DG=\sqrt{ab}$ and $DH=\sqrt{(a^2+b^2)/2}$.
(c) Explain why $DG\leq EH\leq DH$, and conclude from this that $\sqrt{ab}\leq (a+b)/2\leq\sqrt{(a^2+b^2)/2}$.

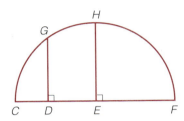

49. Let a, b, c, and d be positive real numbers. Use the inequality in Exercise 43(b) to prove each of the following inequalities.
(a) $\sqrt{abcd}\leq (ab+cd)/2$
(b) $\sqrt{ab}+\sqrt{cd}\leq\sqrt{(a+c)(b+d)}$

50. Let x denote the width of a rectangle with perimeter 30 ft.
(a) Show that the area A (in square feet) of the rectangle is given by $A=x(15-x)$.
(b) Use the inequality in Exercise 43(b) to show that $A\leq 225/4$.
(c) For which value of x is $A=225/4$? What are the dimensions of the rectangle in this case?

51. Let x denote the width of a rectangle with area 25 ft^2.
(a) Show that the perimeter P (in feet) of the rectangle is given by $P=2x+(50/x)$.
(b) Use the inequality in Exercise 43(b) to show that $P\geq 20$.
(c) For which value of x is $P=20$? What are the dimensions of the rectangle in this case?

2.7 MORE ON INEQUALITIES

In this section we are going to solve inequalities involving polynomials and quotients of polynomials. Some observations about linear polynomials and inequalities will be helpful in introducing the main ideas. Consider, for example, the linear inequality $2x-5<0$, along with its solution set $\left(-\infty,\frac{5}{2}\right)$. Let us call a solution of the corresponding equality $2x-5=0$ a **key number** for the given inequality. So, in this case, the only key number is $x=\frac{5}{2}$. As indicated in Figure 1, the key number $\frac{5}{2}$ divides the number line into two intervals, namely, $\left(-\infty,\frac{5}{2}\right)$ and $\left(\frac{5}{2},\infty\right)$. Notice that on each of the intervals determined by the key number,

FIGURE I

the algebraic sign of $2x - 5$ is constant. (That is, for $x < \frac{5}{2}$ the value of $2x - 5$ is always negative; for $x > \frac{5}{2}$ the value of $2x - 5$ is always positive.) More generally, it can be shown that this same type of behavior regarding *persistence of sign* occurs with all polynomials, and indeed with quotients of polynomials as well. This important fact, along with the definition of a key number, is presented in the box that follows.

Key Numbers and Persistence of Sign

Let P and Q be polynomials with no common factor (other than constants), and consider the following four inequalities:

$$\frac{P}{Q} < 0 \qquad \frac{P}{Q} \leq 0 \qquad \frac{P}{Q} > 0 \qquad \frac{P}{Q} \geq 0$$

The **key numbers** for each of these inequalities are the real numbers for which $P = 0$ or $Q = 0$. It can be proved that the algebraic sign of P/Q is constant on each of the intervals determined by these key numbers.

We'll show how this result is applied by solving the polynomial inequality

$$x^3 - 2x^2 - 3x > 0$$

First, using factoring techniques from Section 1.9, we rewrite the inequality in the equivalent form

$$x(x + 1)(x - 3) > 0$$

The key numbers, then, are the solutions of the equation $x(x + 1)(x - 3) = 0$. That is, the key numbers are $x = 0$, $x = -1$, and $x = 3$. Next, we locate these numbers on a coordinate line. As indicated in Figure 2, this divides the number line into four distinct intervals.

FIGURE 2

Now, according to the result stated in the box just prior to this example, no matter what x-value we choose in the interval $(-\infty, -1)$, the resulting sign of $x^3 - 2x^2 - 3x \ [= x(x + 1)(x - 3)]$ will always be the same. Thus, to see what that sign is, we first choose any convenient *test number* in the interval $(-\infty, -1)$, say, $x = -2$. Then, using $x = -2$, we determine the sign of $x(x + 1)(x - 3)$ simply by considering the sign of each factor, as indicated in Table 1.

From this table we conclude that the values of $x^3 - 2x^2 - 3x$ are negative

TABLE 1

On the interval $(-\infty, -1)$, the sign of $x^3 - 2x^2 - 3x \ [= x(x + 1)(x - 3)]$ is negative because it is the product of three negative factors.

Interval	Test Number	x	$x + 1$	$x - 3$	$x(x + 1)(x - 3)$
$(-\infty, -1)$	-2	neg.	neg.	neg.	neg.

throughout the interval $(-\infty, -1)$, and, consequently, no number in this interval satisfies the given inequality. Next, we carry out similar analyses for the remaining three intervals, as shown in the following table. (You should verify for yourself that the entries in the table are correct.)

TABLE 2

On the interval $(-1, 0)$, the product $x(x + 1)(x - 3)$ is positive because it has two negative factors and one positive factor. On $(0, 3)$, the product is negative because it has two positive factors and one negative factor. And on $(3, \infty)$, the product is positive because all the factors are positive.

Interval	Test Number	x	$x + 1$	$x - 3$	$x(x + 1)(x - 3)$
$(-1, 0)$	$-\frac{1}{2}$	neg.	pos.	neg.	pos.
$(0, 3)$	1	pos.	pos.	neg.	neg.
$(3, \infty)$	4	pos.	pos.	pos.	pos.

Looking at the two tables now, we conclude that the sign of $x(x + 1)(x - 3)$ is positive throughout both of the intervals $(-1, 0)$ and $(3, \infty)$ and, consequently, all the numbers in these intervals satisfy the given inequality. Moreover, our work also shows that the other two intervals that we considered are not part of the solution set. As you can readily check, the key numbers themselves do not satisfy the given inequality for this example. In summary, then, the solution set for the inequality $x^3 - 2x^2 - 3x > 0$ is $(-1, 0) \cup (3, \infty)$. Furthermore, it is important to notice that the work we just carried out also provides us with three additional pieces of information.

The solution set for $x^3 - 2x^2 - 3x \geq 0$ is $[-1, 0] \cup [3, \infty)$.

The solution set for $x^3 - 2x^2 - 3x < 0$ is $(-\infty, -1) \cup (0, 3)$.

The solution set for $x^3 - 2x^2 - 3x \leq 0$ is $(-\infty, -1] \cup [0, 3]$.

In the box that follows, we summarize the steps for solving polynomial inequalities.

Steps for Solving Polynomial Inequalities

1. If necessary, rewrite the inequality so that the polynomial is on the left-hand side and zero is on the right-hand side.

2. Find the key numbers for the inequality and locate them on a number line.

3. List the intervals determined by the key numbers.

4. From each interval, choose a convenient test number. Then use the test number to determine the sign of the polynomial throughout the interval.

5. Use the information obtained in the previous step to specify the required solution set. [Don't forget to take into account whether the original inequality is strict ($<$ or $>$) or nonstrict ($\leq$ or $\geq$).]

EXAMPLE 1 Solve: $x^4 \leq 14x^3 - 48x^2$.

Solution First, we rewrite the inequality so that zero is on the right-hand side. Then we factor the left-hand side as follows:

$$x^4 - 14x^3 + 48x^2 \leq 0$$
$$x^2(x^2 - 14x + 48) \leq 0$$
$$x^2(x - 6)(x - 8) \leq 0$$

From this last line, we see that the key numbers are $x = 0$, 6, and 8. As indicated in Figure 3, these numbers divide the number line into four distinct intervals. We need to choose a test number from each interval and see whether the polynomial is positive or negative on the interval. This work is carried out in the following table.

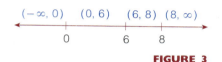

$(-\infty, 0)$ $(0, 6)$ $(6, 8)$ $(8, \infty)$

FIGURE 3

Interval	Test Number	x^2	$x - 6$	$x - 8$	$x^2(x - 6)(x - 8)$
$(-\infty, 0)$	-1	pos.	neg.	neg.	pos.
$(0, 6)$	1	pos.	neg.	neg.	pos.
$(6, 8)$	7	pos.	pos.	neg.	neg.
$(8, \infty)$	9	pos.	pos.	pos.	pos.

The table shows that the quantity $x^2(x - 6)(x - 8)$ is negative only for x-values in the interval $(6, 8)$. Also, as noted at the start, the quantity is equal to zero when $x = 0$, 6, or 8. Thus, the solution set consists of the numbers in the closed interval $[6, 8]$, along with the number 0. We can write this set

$$[6, 8] \cup \{0\}$$

where $\{0\}$ denotes the set that has zero as its only member. ■■■

EXAMPLE 2 Solve: **(a)** $x^2 - x + 15 > 0$;   **(b)** $x^2 - x + 15 < 0$.

Solution **(a)** The equation $x^2 - x + 15 = 0$ has no real solution (because the discriminant of the quadratic is -59, which is negative). So there is no key number, and, consequently, the polynomial $x^2 - x + 15$ never changes sign. To see what that sign is, choose the most convenient test number, namely, $x = 0$, and evaluate the polynomial: $0^2 - 0 + 15 > 0$. So the polynomial is positive for every value of x, and the solution set is $(-\infty, \infty)$, the set of all real numbers.

(b) Our work in part (a) shows that no real number satisfies the inequality $x^2 - x + 15 < 0$. ■■■

The technique used in the previous examples can also be used to solve inequalities involving quotients of polynomials. For these cases recall that the definition of a key number also includes the x-values for which the denominator is zero. For example, the key numbers for the inequality $\dfrac{x + 3}{x - 4} \geq 0$ are -3 and 4.

EXAMPLE 3 Solve: $\dfrac{x+3}{x-4} \geq 0$.

Solution The key numbers are -3 and 4. As indicated in Figure 4, these numbers divide the number line into three intervals. In the table that follows, we've chosen a test number from each interval and determined the sign of the quotient $\dfrac{x+3}{x-4}$ for each interval.

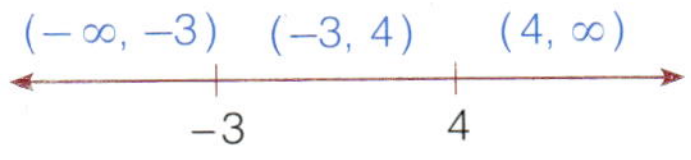

$(-\infty, -3)$ $(-3, 4)$ $(4, \infty)$

-3 4

FIGURE 4

Interval	Test Number	$x + 3$	$x - 4$	$\dfrac{x+3}{x-4}$
$(-\infty, -3)$	-4	neg.	neg.	pos.
$(-3, 4)$	0	pos.	neg.	neg.
$(4, \infty)$	5	pos.	pos.	pos.

From these results, we conclude that the solution set for $\dfrac{x+3}{x-4} \geq 0$ contains the two intervals $(-\infty, -3)$ and $(4, \infty)$. However, we still need to consider the two endpoints -3 and 4. As you can easily check, the value $x = -3$ does satisfy the given inequality, but $x = 4$ does not. In summary, then, the solution set is $(-\infty, -3] \cup (4, \infty)$. ∎

EXAMPLE 4 Solve: $\dfrac{2x+1}{x-1} - \dfrac{2}{x-3} < 1$.

Solution Our first inclination here might be to multiply through by $(x-1)(x-3)$ to eliminate fractions. This strategy is faulty, however, since we don't know whether the quantity $(x-1)(x-3)$ is positive or negative. Thus, we begin by rewriting the inequality in an equivalent form, with zero on the right-hand side and a single fraction on the left-hand side.

$$\frac{2x+1}{x-1} - \frac{2}{x-3} - 1 < 0$$

$$\frac{(2x+1)(x-3) - 2(x-1) - 1(x-1)(x-3)}{(x-1)(x-3)} < 0$$

$$\frac{x^2 - 3x - 4}{(x-1)(x-3)} < 0 \qquad \text{Check the algebra!}$$

$$\frac{(x+1)(x-4)}{(x-1)(x-3)} < 0 \qquad\qquad (1)$$

The key numbers are those x-values for which the denominator or the numerator is zero. By inspection then, we see that these numbers are -1, 4, 1, and 3. As Figure 5 (on the next page) indicates, these numbers divide the number line into five distinct intervals. Now, just as in the previous examples, we choose a test number from each interval and determine the sign of the quotient for that interval. (You should check each entry in the following table for yourself.)

FIGURE 5

Interval	Test Number	$(x+1)(x-4)$	$(x-1)(x-3)$	$\dfrac{(x+1)(x-4)}{(x-1)(x-3)}$
$(-\infty, -1)$	-2	pos.	pos.	pos.
$(-1, 1)$	0	neg.	pos.	neg.
$(1, 3)$	2	neg.	neg.	pos.
$(3, 4)$	$\frac{7}{2}$	neg.	pos.	neg.
$(4, \infty)$	5	pos.	pos.	pos.

From these results, we can see that the quotient on the left-hand side of inequality (1) is negative (as required) on the two intervals $(-1, 1)$ and $(3, 4)$. Now we need to check the endpoints of these intervals. When $x = -1$ or $x = 4$, the quotient is zero, and so, in view of the original inequality, we exclude these two x-values from the solution set. Furthermore, the quotient is undefined when $x = 1$ or $x = 3$, so we must also exclude those two values from the solution set. In summary, then, the solution set is $(-1, 1) \cup (3, 4)$. ∎

EXERCISE SET 2.7

A

Solve the inequalities in Exercises 1–62. *Suggestion: A calculator is useful for approximating the key numbers in Exercises 19, 20, 37, 46, and 48.*

1. $x^2 + x - 6 < 0$
2. $x^2 + 4x - 32 < 0$
3. $x^2 - 11x + 18 > 0$
4. $2x^2 + 7x + 5 > 0$
5. $9x - x^2 \le 20$
6. $3x^2 + x \le 4$
7. $x^2 - 16 \ge 0$
8. $24 - x^2 \ge 0$
9. $16x^2 + 24x < -9$
10. $x^4 - 16 < 0$
11. $x^3 + 13x^2 + 42x > 0$
12. $2x^3 - 9x^2 + 4x \ge 0$
13. $225x \le x^3$
14. $81x^2 \le x^6$
15. $2x^2 + 1 \ge 0$
16. $1 + x^2 < 0$
17. $12x^3 + 17x^2 + 6x < 0$
18. $8x^4 < x^2 - 2x^3$
19. $x^2 + x - 1 > 0$
20. $2x^2 + 9x - 1 > 0$
21. $x^2 - 8x + 2 \le 0$
22. $3x^2 - x + 5 \le 0$
23. $(x-1)(x+3)(x+4) \ge 0$
24. $x^4(x-2)(x-16) \ge 0$
25. $(x+4)(x+5)(x+6) < 0$
26. $\left(x - \frac{1}{2}\right)\left(x + \frac{1}{2}\right)\left(x + \frac{3}{2}\right) < 0$
27. $(x-2)^2(3x+1)^3(3x-1) > 0$
28. $(2x-1)^3(2x-3)^5(2x-5) > 0$
29. $(x-3)^2(x+1)^4(2x+1)^4(3x+2) \le 0$
30. $x^4 - 25x^2 + 144 \le 0$
31. $20 \ge x^2(9 - x^2)$
32. $x^2(3x^2 + 11) \ge 4$
33. $9(x-4) - x^2(x-4) < 0$
34. $(x+1)^2 - 5(x+1) > 14$
35. $4(x^2 - 9) - (x^2 - 9)^2 > -5$
36. $x(1 - x^2)^4 + (x+3)(1-x^2)^4 \ge 0$
37. $(x-4)(2x^2 - 6x - 1) < 0$
38. $(x+2)^3(x^2 - 4x - 2) < 0$
39. $x^3 + 2x^2 - x - 2 > 0$
40. $2x^4 + x^3 - 16x - 8 > 0$
41. $\dfrac{x-1}{x+1} \le 0$
42. $\dfrac{x+4}{2x-5} \le 0$
43. $\dfrac{2-x}{3-2x} \ge 0$
44. $\dfrac{x^2 - 1}{x^2 + 8x + 15} \ge 0$
45. $\dfrac{x^2 - 8x - 9}{x} < 0$
46. $\dfrac{x^2 - 3x + 1}{1 - x} < 0$
47. $\dfrac{2x^3 + 5x^2 - 7x}{3x^2 + 7x + 4} > 0$
48. $\dfrac{x^2 - x - 1}{x^2 + x - 1} > 0$

49. $\dfrac{x}{x+1} > 1$

50. $\dfrac{2x}{x-2} < 3$

51. $\dfrac{1}{x} \leq \dfrac{1}{x+1}$

52. $\dfrac{2}{x} < \dfrac{x}{2}$

53. $\dfrac{x+2}{x+5} \leq 1$

54. $\dfrac{2}{x-4} - \dfrac{1}{x-3} \geq \dfrac{2}{3}$

55. $\dfrac{1}{x-2} - \dfrac{1}{x-1} \geq \dfrac{1}{6}$

56. $\dfrac{2x}{x+5} + \dfrac{x-1}{x-5} < \dfrac{1}{5}$

57. $\dfrac{1+x}{1-x} - \dfrac{1-x}{1+x} < -1$

58. $\dfrac{x+1}{x+2} > \dfrac{x-3}{x+4}$

59. $\dfrac{3-2x}{3+2x} > \dfrac{1}{x}$

60. $\dfrac{x}{x-2} - \dfrac{3}{x+1} \geq 2$

61. $1 + \dfrac{1}{x} \geq \dfrac{1}{1+x}$

62. $x - \dfrac{10}{x-1} \geq 4$

For Exercises 63 and 64, determine the domain of the variable in each expression. (Recall the domain convention from Section 1.8: The domain is the set of all real-number values for the variable for which the expression is defined.)

63. (a) $\sqrt{x^2 - 4x - 5}$　　　(b) $\sqrt{1/(x^2 - 4x - 5)}$

64. (a) $\sqrt{\dfrac{x+2}{x-4}}$　　　(b) $\sqrt[3]{\dfrac{x+2}{x-4}}$

B

65. For which values of b will the equation $x^2 + bx + 1 = 0$ have real solutions?

66. The sum of the first n natural numbers is given by

$$1 + 2 + 3 + \cdots + n = \frac{n(n+1)}{2}$$

For which values of n will the sum be less than 1225?

67. For which values of a is $x = 1$ a solution of the inequality $\dfrac{2a + x}{x - 2a} < 1$?

68. Solve: $\dfrac{ax + b}{\sqrt{x}} > 2\sqrt{ab}$　$(a > 0, b > 0)$.

69. The two shorter sides in a right triangle have lengths x and $1 - x$　$(x > 0)$. For which values of x will the hypotenuse be less than $\sqrt{17}/5$?

70. A piece of wire 12 cm long is cut into two pieces. Denote the lengths of the two pieces by x and $12 - x$. Both pieces are then bent into squares. For which

values of x will the combined areas of the squares exceed 5 cm²?

71. Let V and S denote the volume and total surface area, respectively, for a right circular cylinder of radius r and height 1 unit. For which r-values will the ratio V/S be less than $\frac{1}{3}$?

72. Let V and S denote the volume and total surface area, respectively, for a right circular cone of radius r and height 1 unit. For which r-values will the ratio V/S be less than $\frac{4}{27}$?

73. (a) Complete the table. (Round off the entries in the right-hand column to two decimal places.)

x	$x^2 + 1000$	$2x^2 + x$	$\dfrac{x^2 + 1000}{2x^2 + x}$
1			
2			
5			
10			
100			
200			
10,000			

(b) For which values of x will we have
$$\frac{x^2 + 1000}{2x^2 + x} < 0.5?$$

(c) What is the smallest *natural number* n for which
$$\frac{n^2 + 1000}{2n^2 + n} < 0.5?$$

C

74. Find a nonzero value for c so that the solution set for the inequality
$$x^2 + 2cx - 6c < 0$$
is the open interval $(-3c, c)$.

75. Solve: $(x - a)^2 - (x - b)^2 > (a - b)^2/4$. Assume that $a > b$.

76. Solve: $x^3 + (1/x^3) \geq 3$. (Use a calculator to approximate the key numbers.)

CHAPTER TWO SUMMARY OF PRINCIPAL TERMS AND NOTATION

TERMS OR NOTATIONS	PAGE REFERENCE	COMMENTS
1. Linear (or first-degree) equation in one variable	84	These are equations that can be written in the form $ax + b = 0$, with $a \neq 0$.
2. Solution (or root) of an equation	84, 91	A solution or root is a number that, when substituted for the variable in an equation, yields a true statement.
3. Equivalent equations	85	Two equations are equivalent if they have the same set of solutions.
4. Extraneous solution (or extraneous root)	88, 112	In solving equations, certain processes (such as squaring both sides) can lead to answers that do not check in the original equation. These numbers are called extraneous solutions (or extraneous roots).
5. Quadratic equation	91	A quadratic equation is an equation that can be written in the form $ax^2 + bx + c = 0$ where $a \neq 0$.
6. Zero-product property	92	This property of real numbers (and of complex numbers in general) can be stated as follows. If $pq = 0$ then $p = 0$ or $q = 0$; conversely, if $p = 0$ or $q = 0$, then $pq = 0$. (We used this property in solving quadratic equations by factoring.)
7. Quadratic formula $$x = \frac{-b \pm \sqrt{b^2 - 4ac}}{2a}$$	94	This is the quadratic formula; it provides the solutions of the quadratic equation $ax^2 + bx + c = 0$. The formula is derived on page 94 by means of the useful technique of completing the square.
8. Discriminant	96	The discriminant of the quadratic equation $ax^2 + bx + c = 0$ is the number $b^2 - 4ac$. As indicated in the box on page 96, the discriminant provides information about the roots of the equation.
9. Solution set of an inequality	115	This is the set of numbers that satisfy the inequality.
10. $A \cup B$	118	The set $A \cup B$ consists of all elements that are in at least one of the two sets A and B.
11. Key numbers of an inequality	122	Let P and Q denote polynomials and consider the following four inequalities: $$\frac{P}{Q} < 0 \qquad \frac{P}{Q} \leq 0 \qquad \frac{P}{Q} > 0 \qquad \frac{P}{Q} \geq 0$$ The key numbers for each of these inequalities are the real numbers for which $P = 0$ or $Q = 0$. It can be proved that the algebraic sign of P/Q is constant on each of the intervals determined by these key numbers.

WRITING MATHEMATICS

1. After working the following problem for yourself, write out the solution in complete sentences, as if you were explaining it to a classmate. Be sure to let the

classmate know where you are headed and why each of the main steps is necessary.

A piece of wire one meter long is cut into two pieces. The first piece is bent into a square and the second piece is bent into a rectangle in which the length is twice the width. Express the combined area A of the square and rectangle in terms of x, where x denotes the length of the piece used for the square.

2. What is an extraneous solution of an equation? What are two of the ways in which extraneous solutions can arise? Use the following two equations as examples in your writing: $\sqrt{3x + 7} = x - 1$ and $\dfrac{2x}{x^2 - 9} - \dfrac{1}{x - 3} = \dfrac{2}{x + 3}$.

CHAPTER TWO REVIEW EXERCISES

In Exercises 1–12, answer T *if the statement is true without exception. Otherwise, answer* F.

1. If $x < 3$, then $x + 7 < 10$.
2. If $x < y$, then $x^2 < y^2$.
3. If $x \geq -4$, then $x \leq 4$.
4. If $-x > y$, then $x < -y$.
5. If $x < 2$, then $\dfrac{1}{x} < \dfrac{1}{2}$.
6. $x \leq x^2$.
7. If $\sqrt{x + 1} = 3$, then $x + 1 = 9$.
8. If $0 < x < 1$, then $\dfrac{1}{x} > x$.
9. If $x < 2$ and $y < 3$, then $x < y$.
10. If $x < 2$ and $y > 3$, then $x < y$.
11. If $a - b \leq a^2 - b^2$, then $1 \leq a + b$.
12. If $0 < a - b \leq a^2 - b^2$, then $1 \leq a + b$.

In Exercises 13–50, find all the real solutions of each equation.

13. $5 - 9x = 2$

14. $\dfrac{x}{3} - \dfrac{3x}{5} = \dfrac{x}{6} - 13$

15. $(t - 4)(t + 3) = (t + 5)^2$

16. $\dfrac{1 - x}{1 + x} = 1$

17. $\dfrac{2t - 1}{t + 2} = 5$

18. $\dfrac{1}{1 + \dfrac{1}{x + 1}} = \dfrac{5}{6}$

19. $\dfrac{2y - 5}{4y + 1} = \dfrac{y - 1}{2y + 5}$

20. $\dfrac{2x - 3}{x - 2} = \dfrac{1}{x - 2}$

21. $|y + 4| = 2$

22. $|x + 3| - |3x - 5| = -16$

23. $12x^2 + 2x - 2 = 0$

24. $4y^2 - 21y = 18$

25. $\frac{1}{2}x^2 + x - 12 = 0$

26. $x^2 + \frac{13}{2}x + 10 = 0$

27. $\dfrac{x}{5 - x} = \dfrac{-2}{11 - x}$

28. $\dfrac{x^2}{(x - 1)(x + 1)} = \dfrac{4}{x + 1} + \dfrac{4}{(x - 1)(x + 1)}$

29. $\dfrac{1}{3x - 7} - \dfrac{2}{5x - 5} - \dfrac{3}{3x + 1} = 0$

30. $4x^2 + x - 2 = 0$

31. $t^2 + t - \frac{1}{2} = 0$

32. $\dfrac{1}{x - 1} = x\sqrt{2}$

33. $y^4 = 9$

34. $(y - 1)^4 = 81$

35. $x^5 - 2x^3 - 2x = 0$

36. $x^3 - 6x^2 + 7x = 0$

37. $1 + 14x^{-1} + 48x^{-2} = 0$

38. $x^{-2} - x^{-1} - 1 = 0$

39. $x^{1/2} - 13x^{1/4} + 36 = 0$

40. $y^{2/3} - 21y^{1/3} + 80 = 0$

41. $\sqrt{4 - 3x} = 5$

42. $\sqrt{x} = \sqrt{x + 27} - 1$

43. $\sqrt{4x + 3} = \sqrt{11 - 8x} - 1$

44. $2 - \sqrt{3}\sqrt{2x - 1} + x = 0$

45. $\sqrt{x + 48} - \sqrt{x} = 4$

46. $1 + \sqrt{2x^2 + 5x - 9} = \sqrt{2x^2 + 5x - 2}$

47. $\dfrac{2}{\sqrt{x^2 - 36}} + \dfrac{1}{\sqrt{x + 6}} - \dfrac{1}{\sqrt{x - 6}} = 0$

48. $\sqrt{x} - \sqrt{1 - x} + \sqrt{x - 1} = 0$

49. $\sqrt{x + 7} - \sqrt{x + 2} = \sqrt{x - 1} - \sqrt{x - 2}$

50. $\dfrac{\sqrt{x} - 4}{\sqrt{x} - 18} = 3$

In Exercises 51–62, solve each equation for x in terms of the other letters.

51. $(a^2 + b^2)x = a^3 + b^3 + abx$

52. $\dfrac{2}{x} = \dfrac{1}{a} + \dfrac{1}{b}$ $(a + b \neq 0)$

53. $\dfrac{x}{a} + \dfrac{a}{x} = 2$

54. $\dfrac{1}{a + x} + \dfrac{1}{b + x} = \dfrac{a + b}{ab}$ $(a + b \neq 0)$

55. $4x^2y^2 - 4xy = -1$ $(y \neq 0)$

56. $4x^4y^4 + 2x^2y^2 + \frac{1}{4} = 0$ $(y \neq 0)$

57. $x + \dfrac{1}{a} - \dfrac{1}{b} = \dfrac{2}{a^2x} + \dfrac{2}{abx}$

58. $\dfrac{a}{x - b} + \dfrac{b}{x - a} = \dfrac{x - a}{b} + \dfrac{x - b}{a}$ $(a \neq -b)$

Suggestion: Before clearing fractions, carry out the indicated additions on each side of the equation.

59. $\dfrac{1}{x + a + b} = \dfrac{1}{x} + \dfrac{1}{a} + \dfrac{1}{b}$ $(a + b \neq 0)$

60. $\dfrac{1}{x} + \dfrac{1}{a - x} + \dfrac{1}{x + 3a} = 0$ $(a \neq 0)$

61. $\dfrac{a^2 - b^2}{x} - \dfrac{2(a^2 + b^2)}{\sqrt{x}} = b^2 - a^2$ $(a > b > 0)$

Hint: Let $t = 1/\sqrt{x}$ and $t^2 = 1/x$.

62. $x^2 + \dfrac{1}{x^2} = a^2 + \dfrac{1}{a^2}$

Solve the inequalities in Exercises 63–83.

63. $-1 < \dfrac{1 - 2(1 + x)}{3} < 1$

64. $3 < \dfrac{x - 1}{-2} < 4$

65. $|x| \leq \frac{1}{2}$

66. $|t - 2| < 1$

67. $|x + 4| < \frac{1}{10}$

68. $|3 - 5x| < 2$

69. $|2x - 1| \geq 5$

70. $x^2 - 21x + 108 \leq 0$

71. $x^2 + 3x - 40 < 0$

72. $x^2 \geq 15x$

73. $x^2 - 6x - 1 < 0$

74. $(x + 12)(x - 1)(x - 8) < 0$

75. $x^4 - 34x^2 + 225 < 0$

76. $\dfrac{x + 12}{x - 5} > 0$

77. $\dfrac{(x - 7)^2}{(x + 2)^3} \geq 0$

78. $\dfrac{(x - 6)^2(x - 8)(x + 3)}{(x - 3)^2} \leq 0$

79. $\dfrac{x^2 - 10x + 9}{x^3 + 1} \leq 0$

80. $\dfrac{3x + 1}{x - 4} < 1$

81. $\dfrac{1 - 2x}{1 + 2x} \leq \dfrac{1}{2}$

82. $x^2 + \dfrac{1}{x^2} > 3$ *Suggestion:* Use a calculator to evaluate the key numbers.

83. $\sqrt{x} - \dfrac{5}{\sqrt{x}} \leq 4$

For Exercises 84–87, find the values of k for which the roots of the equations are real numbers.

84. $kx^2 - 6x + 5 = 0$

85. $x^2 + x + k^2 = 0$

86. $kx(x + 2) = -1$

87. $x^2 + (k + 1)x + 2k = 0$

88. How many pounds of an alloy that is 12% copper must be added to 1000 lb of an alloy that is 9% copper to yield a new alloy that is 11.2% copper?

89. The sum of the digits in a certain two-digit number is 11. If the order of the digits is reversed, the number is increased by 27. Find the original number.

90. Robin deposits $6000, part at 10% and the remainder at 12%. If the net interest rate for the entire $6000 is $11\frac{1}{3}$%, how much was deposited at each rate?

91. The four corners of a square *ABCD* have been cut off to form a regular octagon, as shown in the figure. If each side of the square is 1 cm long, how long is each side of the octagon?

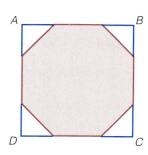

92. Determine *p* if the larger root of the equation $x^2 + px - 2 = 0$ is *p*.

93. A piece of wire *x* cm long is bent into a square. For which values of *x* will the area be (numerically) greater than the perimeter?

94. A piece of wire 6*x* cm long is bent into an equilateral triangle. For which values of *x* will the area be (numerically) less than the perimeter?

95. Find three consecutive positive integers such that the sum of their squares is 1454.

96. The sum of two numbers is 1 and the sum of their cubes is 91. What are the numbers?

97. In a 10-mile race, Joan covers the first 4 miles at a constant rate. Then she speeds up a bit and runs the last 6 miles at a rate that is $\frac{1}{2}$ mph faster. If the entire race had been run at this faster pace, Joan's overall time would have been 2 minutes better. What was Joan's speed for the first 4 miles, and what was her time for the race?

98. The length of a rectangular piece of tin exceeds the width by 8 cm. A 1-cm square is cut from each corner of the piece of tin, and then the resulting flaps are turned up to form a box with no top. What are the dimensions of the box if its volume is 48 cm³?

99. The length and width of a rectangular flower garden are a and b, respectively. The garden is bordered on all four sides by a gravel path of uniform width. Find the width of the path, given that the area of the garden equals the area of the path.

If an object is thrown vertically upward from a height of h_0 feet with an initial speed of v_0 ft/sec, then its height h (in feet) after t seconds is given by

$$h = -16t^2 + v_0t + h_0$$

Make use of this formula in working Exercises 100–102.

100. A ball is thrown vertically upward from ground level with an initial speed of 64 ft/sec.
   (a) At what time will the height of the ball be 15 ft? (Two answers.)
   (b) For how long an interval of time will the height exceed 63 ft?

101. One ball is thrown vertically upward from a height of 50 ft with an initial speed of 40 ft/sec. At the same instant, another ball is thrown vertically upward from a height of 100 ft with an initial speed of 5 ft/sec. Which ball hits the ground first?

102. An object is projected vertically upward. Suppose that its height is H ft at t_1 sec and again at t_2 sec. Express the initial speed in terms of t_1 and t_2.
   Answer: $16(t_1 + t_2)$

103. A rectangle is inscribed in a semicircle of radius 1 cm, as shown. For which value of x is the area of the rectangle 1 cm²?   *Note:* x is defined in the figure.

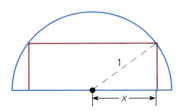

104. The height of an isosceles triangle is $a + b$ units $(a > b > 0)$ and the area does not exceed $a^2 - b^2$ square units. What is the range of possible values for the base of the triangle?

Exercises 105 and 106 appear in the text Introduction to Algebra by George Chrystal, first published in 1898.

105. Show that it is impossible to find a positive integer such that the sum of its square and its cube is an integral multiple of the square of the next highest integer.   [*Hint:*   Use the result in Exercise 85 of Exercise Set 1.8.]

106. A circle is inscribed in a quadrant of a larger circle of radius r (as shown in the figure). Find the radius of the inscribed circle.

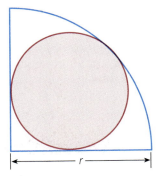

CHAPTER TWO TEST

In Problems 1–7, find all the real solutions of each equation.

1. $\dfrac{2}{x + 4} - \dfrac{1}{x - 4} = \dfrac{-7}{x^2 - 16}$

2. $\dfrac{1}{1 - x} + \dfrac{4}{2 - x} = \dfrac{11}{6}$

3.   (a) $x^2 + 4x = 5$
       (b) $x^2 + 4x = 1$

4. $x^2(x^2 - 7) + 12 = 0$

5. $4x^{4/3} - 13x^{2/3} + 9 = 0$

6. $\sqrt{5 - 2x} - \sqrt{2 - x} - \sqrt{3 - x} = 0$

7. $|3x - 1| = 2$

In Problems 8 and 9, solve the equations for x in terms of the other letters.

8. $\dfrac{ax + b}{cx + d} = e \quad (a \neq ce)$

9. $ax = bx + a^2 - b^2 \quad (a \neq b)$

In Problems 10–14, solve the inequalities. Write the answers using interval notation.

10. $4(1 + x) - 3(2x - 1) \geq 1$

11. $\dfrac{3}{5} < \dfrac{3 - 2x}{-4} < \dfrac{4}{5}$

12. $|3x - 8| \leq 1$

13. $(x - 4)^2(x + 8)^3 \geq 0$

14. $\dfrac{1}{x} + \dfrac{1}{x + 1} + \dfrac{1}{x + 2} \geq 0$

15. Kona coffee beans are sold for \$6/lb and Colombian beans for \$5/lb. How many pounds of each bean must be mixed to yield 280 lb of a blend selling for \$5.60/lb?

16. At 6 A.M. a train traveling at 50 mph leaves Los Angeles for San Francisco. Three hours later a train traveling 40 mph leaves San Francisco for Los Angeles. What time will the trains pass each other? Assume that the distance between the two cities is 450 miles.

17. Find the values of k for which the roots of the equation $x^2 + 3x + k^2$ are real numbers.

18. The height of a right circular cylinder is twice the radius.
     (a) Express the total surface area S of the cylinder in terms of the height. (Use the formula $S = 2\pi r^2 + 2\pi rh$.)
     (b) Express the height of the cylinder in terms of the surface area S.

CHAPTER THREE

COORDINATES AND GRAPHS

... [Descartes and Fermat] were moved by the same spirit: they wanted to show how the Renaissance algebra of Cardan and his successors could be applied to the geometry of the Greeks, ...

From *A Source Book in Mathematics, 1200–1800*, D. J. Struik, ed. (Princeton: Princeton University Press, 1986)

The mathematics I was taught in my freshman year was algebra (old stuff), trigonometry (too much of it), and analytical geometry (a revelation).

Paul R. Halmos in *I Want To Be a Mathematician* (N.Y.: Springer-Verlag, 1985)

INTRODUCTION

In Chapter 1 we saw an important connection between algebra and geometry: the points on a line can be labeled using the real numbers. In Chapter 2 we used this idea in solving inequalities and in graphing their solution sets. Now, just as each point on a line can be labeled with a unique real number, so each point in a *plane* can be labeled with a unique *pair* of real numbers. The details of this are explained in Section 3.1, Rectangular Coordinates. These ideas are then used in the next two sections to graph equations. The basic theme running through the work on graphing equations is this: We start with a given equation relating two variables, say, x and y. By choosing values for x, corresponding y-values are then determined. The resulting (x, y) number pairs (*ordered pairs*, they're called) are then interpreted as points in the plane. This provides another basic connection between algebra and geometry: with an equation in two variables we associate its geometric picture, its graph. Sometimes it is easier to work directly from a graph, other times from the equation. But in any event, having the latitude to think either algebraically or geometrically (or both) will clarify many problems later on.

3.1 RECTANGULAR COORDINATES

The name coordinate *does not appear in the work of Descartes. This term is due to Leibniz, and so are* abscissa *and* ordinate *(1692).*

David M. Burton in *The History of Mathematics, an Introduction,* 2nd ed. (Dubuque, Iowa: Wm. C. Brown Publishers, 1991)

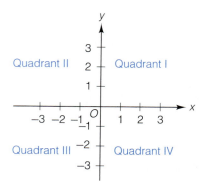

FIGURE 1

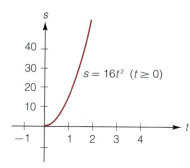

FIGURE 2

A graph of the formula $s = 16t^2$ in a t-s coordinate system. [The formula relates the distance s (in feet) and the time t (in seconds) for an object falling in a vacuum.]

In previous courses you learned to work with a rectangular coordinate system such as that shown in Figure 1. In this section we review some of the most basic formulas and techniques that are useful here.

The point of intersection of the two perpendicular number lines, or **axes**, is called the **origin** and is denoted by the letter O. The horizontal and vertical axes are often labeled the **x-axis** and the **y-axis**, respectively; but any other variables will do just as well for labeling the axes. For instance, Figure 2 shows a t-s coordinate system. (We'll discuss curves, or graphs, like the one in the figure in later sections.)

Notice that in Figures 1 and 2 the axes divide the plane into four regions, or **quadrants**, labeled I through IV, as shown in Figure 1. Unless indicated otherwise, we assume that the same unit of length is used on both axes. In Figure 1, the same scales are used on both axes; in Figure 2, the scale used on the s-axis is different than the scale on the t-axis.

Now look at the point P in Figure 3(a). Starting from the origin O, one way to reach P is to move three units in the positive x-direction and then two units in the positive y-direction. That is, the location of P relative to the origin and the axes is "right 3, up 2." We say that the **coordinates** of P are $(3, 2)$. The first number within the parentheses conveys the information "right 3," and the second number conveys the information "up 2." We say that the **x-coordinate** of P is 3 and the **y-coordinate** of P is 2. Likewise, the coordinates of point Q in Figure 3(a) are $(-2, 4)$. With this coordinate notation in mind, observe in Figure 3(b) that $(3, 2)$ and $(2, 3)$ represent different points; that is, the order in which the two numbers appear within the parentheses affects the location of the point. Figure 3(c) displays various points with given coordinates; you should check for yourself that the coordinates correspond correctly to the location of each point.

Some terminology and notation: The x-y coordinate system that we have described is often called a **Cartesian coordinate system**. The term *Cartesian* is used in honor of René Descartes, the seventeenth-century French philosopher and mathematician. The coordinates (x, y) of a point P are referred to as an **ordered pair**. Recall, for example, that $(3, 2)$ and $(2, 3)$ represent different points; that is, the order of the numbers matters. The x-coordinate of a point is sometimes referred to as the **abscissa** of the point; the y-coordinate is the **ordinate**. The notation $P(x, y)$ means that P is a point whose coordinates are (x, y).

FIGURE 3

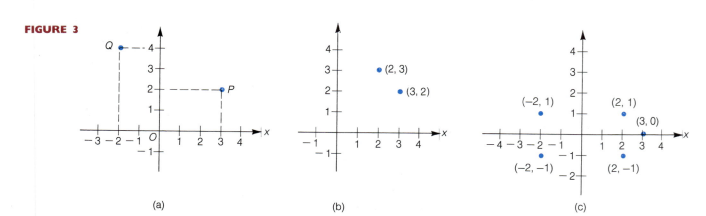

(a)   (b)   (c)

At times, we abbreviate the phrase *the point whose coordinates are* (x, y) to simply *the point* (x, y).

In Example 1 we are going to use the Pythagorean theorem to calculate the distance between two given points. Although we've already worked with this theorem (in Section 2.4), for emphasis and reference we restate it here, along with its converse, in the box that follows. (For a proof of the theorem, see Exercise 45.)

The Pythagorean Theorem and Its Converse

1. Pythagorean Theorem
(See Figure 4) In a right triangle, the lengths of the sides are related by the equation

$$a^2 + b^2 = c^2$$

where a and b are the lengths of the sides forming the right angle and c is the length of the hypotenuse (the side opposite the right angle).

2. Converse
If the lengths of the sides of a triangle are related by an equation of the form $a^2 + b^2 = c^2$, then the triangle is a right triangle, and c is the length of the hypotenuse.

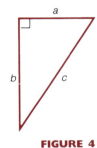

FIGURE 4

EXAMPLE 1   Use the Pythagorean theorem to calculate the distance d between points $(2, 1)$ and $(6, 3)$.

Solution   We plot the two given points and draw a line connecting them, as shown in Figure 5. Then we draw the broken lines as shown, parallel to the axes, and apply the Pythagorean theorem to the right triangle that is formed. The base of the triangle is four units long. You can see this by simply counting spaces or by using absolute value, as discussed in Section 1.4: $|6 - 2| = 4$. The height of the triangle is found to be two units, either by counting spaces or by computing the absolute value: $|3 - 1| = 2$. Thus, we have

$$d^2 = 4^2 + 2^2 = 20$$
$$d = \sqrt{20} = \sqrt{4}\sqrt{5} = 2\sqrt{5}$$

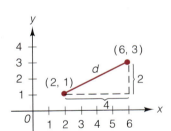

FIGURE 5

The method used in Example 1 can be applied to derive a general formula for the distance d between any two points (x_1, y_1) and (x_2, y_2) (see Figure 6). Just as before, we draw in the right triangle and apply the Pythagorean theorem. We have

$$d^2 = |x_2 - x_1|^2 + |y_2 - y_1|^2$$
$$= (x_2 - x_1)^2 + (y_2 - y_1)^2 \qquad \text{(Why?)}$$

and therefore

$$d = \sqrt{(x_2 - x_1)^2 + (y_2 - y_1)^2}$$

This last equation is referred to as the **distance formula**. For reference, we restate it in the box that follows.

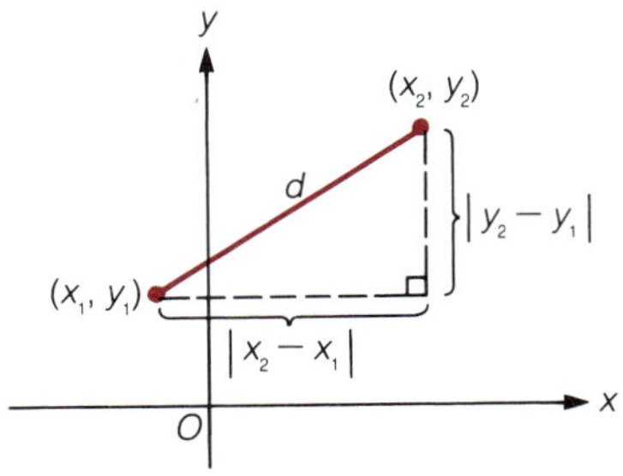

FIGURE 6

> **The Distance Formula**
>
> The distance d between the points (x_1, y_1) and (x_2, y_2) is given by
> $$d = \sqrt{(x_2 - x_1)^2 + (y_2 - y_1)^2}$$

Examples 2 through 4 demonstrate some simple calculations involving the distance formula.   *Note:*   In computing the distance between two given points, it does not matter which one you treat as (x_1, y_1) and which as (x_2, y_2). This is because quantities such as $(x_2 - x_1)$ and $(x_1 - x_2)$ are negatives of each other, and so their squares are equal.

EXAMPLE 2   Calculate the distance between the points $(2, -6)$ and $(5, 3)$.

Solution   Substituting $(2, -6)$ for (x_1, y_1) and $(5, 3)$ for (x_2, y_2) in the distance formula, we have

$$d = \sqrt{(5 - 2)^2 - [3 - (-6)]^2}$$
$$= \sqrt{3^2 + 9^2} = \sqrt{90}$$
$$= \sqrt{9}\sqrt{10} = 3\sqrt{10}$$

You should check for yourself that the same answer is obtained using $(2, -6)$ as (x_2, y_2) and $(5, 3)$ as (x_1, y_1).

EXAMPLE 3   Is the triangle with vertices $D(-2, -1)$, $E(4, 1)$, $F(3, 4)$ a right triangle?

Solution   First, we sketch the triangle in question; see Figure 7. From the sketch it "appears" that angle E could be a right angle, but certainly this is not a proof. The idea then is to use the distance formula to calculate the lengths of the three sides, then check whether any relation of the form $a^2 + b^2 = c^2$ holds. The calculations are as follows.

$$DE = \sqrt{[4 - (-2)]^2 + [1 - (-1)]^2} = \sqrt{36 + 4} = \sqrt{40}$$
$$EF = \sqrt{(4 - 3)^2 + (1 - 4)^2} = \sqrt{1 + 9} = \sqrt{10}$$
$$DF = \sqrt{[3 - (-2)]^2 + [4 - (-1)]^2} = \sqrt{25 + 25} = \sqrt{50}$$

FIGURE 7

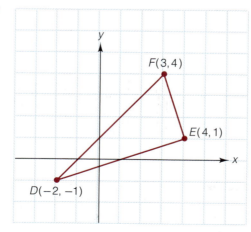

Because $\left(\sqrt{40}\right)^2 + \left(\sqrt{10}\right)^2 = \left(\sqrt{50}\right)^2$, we are indeed guaranteed that $\triangle DEF$ is a right triangle. (In Section 3.3, you'll see that this result can be obtained more efficiently using the concept of "slope.") ▌▌▌

EXAMPLE 4 Refer to Figure 8. A point $P(x, y)$ is equidistant from the points $A(2, 1)$ and $B(5, 3)$. Find and simplify an equation relating x and y.

Solution We are given that $PA = PB$. Using the distance formula, we have

$$PA = \sqrt{(x - 2)^2 + (y - 1)^2} \qquad \text{and} \qquad PB = \sqrt{(x - 5)^2 + (y - 3)^2}$$

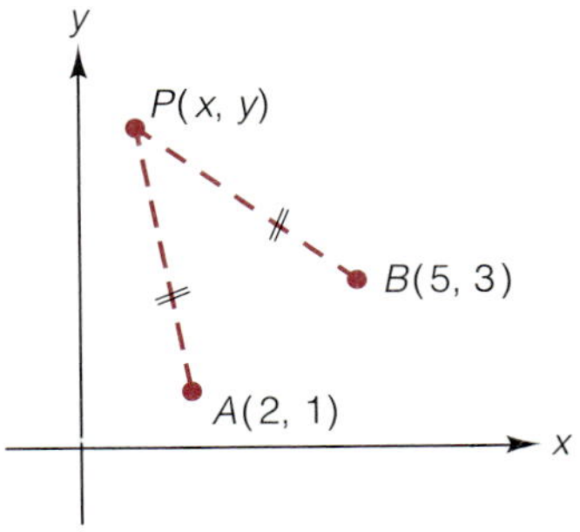

Therefore

$$\sqrt{(x - 2)^2 + (y - 1)^2} = \sqrt{(x - 5)^2 + (y - 3)^2}$$

or

$$\sqrt{x^2 - 4x + y^2 - 2y + 5} = \sqrt{x^2 - 10x + y^2 - 6y + 34}$$

We can simplify this last equation a great deal by squaring both sides. (We'll obtain an equivalent equation because two nonnegative quantities are equal if and only if their squares are equal.) Squaring yields

$$x^2 - 4x + y^2 - 2y + 5 = x^2 - 10x + y^2 - 6y + 34$$

and consequently

$$-4x - 2y + 5 = -10x - 6y + 34 \qquad \text{subtracting } x^2 \text{ and } y^2 \text{ from both sides}$$

or

$$6x + 4y - 29 = 0$$

This is the required equation. (In Section 3.3 we'll see that this is the equation of a line. The line is the perpendicular bisector of the segment joining the given points A and B.) ▌▌▌

As we've seen in this section, the basic idea behind coordinate geometry is to associate a unique ordered pair of numbers (x, y) with each point in the plane. This allows us to deal with geometric relationships in a purely numerical or algebraic manner. At times this can be a real advantage. For instance, by using the methods of coordinate geometry, proofs of theorems from elementary geometry can be constructed almost at will, without the need for arguments involving congruent triangles, angles, or the like. Example 5 illustrates this idea. *Note:* You needn't memorize the result in Example 5. Of primary importance for our purposes are the *methods* used to obtain that result.

In Example 5 we make use of a simple result that you may recall from previous courses: the midpoint formula. This result is summarized in the box on the next page. (For a proof of the formula, see Exercise 44.)

EXAMPLE 5 Use Figure 9 (on the next page) to prove the following theorem from geometry: The midpoint of the hypotenuse in a right triangle is equidistant from the vertices. (In Figure 9, $\triangle OBA$ is the right triangle and M is the midpoint of the hypotenuse.)

FIGURE 8

The Midpoint Formula

EXAMPLE

The midpoint of the line segment joining the points $P(x_1, y_1)$ and $Q(x_2, y_2)$ is

$$\left(\frac{x_1 + x_2}{2}, \frac{y_1 + y_2}{2}\right)$$

The midpoint of the line segment joining $(2, -15)$ and $(4, 5)$ is

$$\left(\frac{2 + 4}{2}, \frac{-15 + 5}{2}\right) = (3, -5)$$

Solution With reference to Figure 9, we have to show that $OM = MA = MB$. Now, by hypothesis, we already have that $MA = MB$, so we need only establish that $OM = MA$. The calculations are as follows.

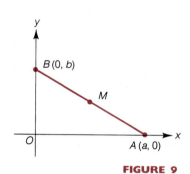

FIGURE 9

Coordinates of M: $\left(\frac{a}{2}, \frac{b}{2}\right)$ using the midpoint formula

$$OM = \sqrt{\left(\frac{a}{2} - 0\right)^2 + \left(\frac{b}{2} - 0\right)^2} = \sqrt{\frac{a^2 + b^2}{4}} = \frac{\sqrt{a^2 + b^2}}{2}$$

$$MA = \sqrt{\left(a - \frac{a}{2}\right)^2 + \left(0 - \frac{b}{2}\right)^2} = \sqrt{\frac{a^2}{4} + \frac{b^2}{4}} = \sqrt{\frac{a^2 + b^2}{4}} = \frac{\sqrt{a^2 + b^2}}{2}$$

From these calculations we conclude that $OM = MA$ (since both quantities are equal to $\frac{1}{2}\sqrt{a^2 + b^2}$). Thus $OM = MA = MB$, as required.

NOTE   We could have avoided working with fractions had we initially made a more clever choice for the coordinates of A and B. Since a and b were arbitrary, we could have used $(2a, 0)$ for A and $(0, 2b)$ for B.

EXERCISE SET 3.1

A

1. Plot the points $(5, 2)$, $(-4, 5)$, $(-4, 0)$, $(-1, -1)$, and $(5, -2)$.

2. Draw the square $ABCD$ with vertices (corners) at $A(1, 0)$, $B(0, 1)$, $C(-1, 0)$, and $D(0, -1)$.

3. (a) Draw the right triangle PQR with vertices $P(1, 0)$, $Q(5, 0)$, and $R(5, 3)$.
   (b) Use the formula for the area of a triangle, $A = \frac{1}{2}bh$, to find the area of triangle PQR in part (a).

4. (a) Draw the trapezoid $ABCD$ with vertices $A(0, 0)$, $B(7, 0)$, $C(6, 4)$, and $D(4, 4)$.
   (b) Compute the area of the trapezoid. (See the inside front cover of this book for the appropriate formula.)

In Exercises 5–10, calculate the distance between the given points.

5. (a) $(0, 0)$ and $(-3, 4)$ (b) $(2, 1)$ and $(7, 13)$

6. (a) $(-1, -3)$ and $(-5, 4)$ (b) $(6, -2)$ and $(-1, 1)$

7. (a) $(-5, 0)$ and $(5, 0)$ (b) $(0, -8)$ and $(0, 1)$

8. (a) $(-5, -3)$ and $(-9, -6)$
   (b) $\left(\frac{9}{2}, 3\right)$ and $\left(-2\frac{1}{2}, -1\right)$

9. $\left(1, \sqrt{3}\right)$ and $\left(-1, -\sqrt{3}\right)$

10. $(-3, 1)$ and $(374, -335)$ (Use a calculator.)

11. Which point is farther from the origin?
    (a) $(3, -2)$ or $\left(4, \frac{1}{2}\right)$?   (b) $(-6, 7)$ or $(9, 0)$?

12. Use the distance formula to show that the triangles with given vertices are isosceles triangles.
(a) $(0, 2)$, $(7, 4)$, $(2, -5)$
(b) $(-1, -8)$, $(0, -1)$, $(-4, -4)$
(c) $(-7, 4)$, $(-3, 10)$, $(1, 3)$

13. Which of the triangles with the given vertices are right triangles? *Hint:* Find the lengths of the sides and then use the *converse* of the Pythagorean theorem.
(a) $(7, -1)$, $(-3, 5)$, $(-12, -10)$
(b) $(4, 5)$, $(-3, 9)$, $(1, 3)$
(c) $(-8, -2)$, $(1, -1)$, $(10, 19)$

14. (a) Two of the three triangles specified in Exercise 13 are right triangles. Find their areas.
(b) Calculate the area of the remaining triangle in Exercise 13 by using the following formula for the area A of a triangle with vertices (x_1, y_1), (x_2, y_2), and (x_3, y_3):

$$A = \tfrac{1}{2}|x_1y_2 - x_2y_1 + x_2y_3 - x_3y_2 + x_3y_1 - x_1y_3|$$

(This formula is derived in Exercise 46.)
(c) Use the formula given in part (b) to check your answers to part (a).

15. Use the formula given in Exercise 14(b) to calculate the area of the triangle with vertices $(1, -4)$, $(5, 3)$, and $(13, 17)$. Conclusion?

16. The coordinates of points A, B, and C are $A(-4, 6)$, $B(-1, 2)$, and $C(2, -2)$.
(a) Show that $AB = BC$, using the distance formula.
(b) Show that $AB + BC = AC$, using the distance formula.
(c) What can you conclude from parts (a) and (b)?

In Exercises 17 and 18, find the midpoint of the line segment joining the given points P and Q.

17. (a) $P(3, 2)$ and $Q(9, 8)$
(b) $P(-4, 0)$ and $Q(5, -3)$
(c) $P(3, -6)$ and $Q(-1, -2)$

18. (a) $P(12, 0)$ and $Q(12, 8)$
(b) $P\left(\tfrac{3}{5}, -\tfrac{2}{3}\right)$ and $Q(0, 0)$
(c) $P(1, \pi)$ and $Q(3, 3\pi)$

19. The coordinates of A and B are $(-1, 2)$ and $(5, -3)$, respectively. If B is the midpoint of line segment AC, what are the coordinates of C?

20. The coordinates of the points S and T are $S(4, 6)$ and $T(10, 2)$. If M is the midpoint of $\overline{ST}$, find the midpoint of $\overline{SM}$.

In Exercises 21–28, a point P(x, y) satisfies a given condition. Find and simplify an equation relating x and y.

21. The point $P(x, y)$ is equidistant from the points $(-2, 4)$ and $(3, 0)$.

22. The point $P(x, y)$ is equidistant from the points $(-5, 1)$ and $(-6, 2)$.

23. The distance from $P(x, y)$ to the origin is 5 units.

24. The distance from $P(x, y)$ to the origin is $\sqrt{3}$ units.

25. The point $P(x, y)$ is twice as far from $(1, 0)$ as it is from $(-1, 0)$.

26. The point $P(x, y)$ is half as far from $(2, 1)$ as it is from $(-3, -4)$.

27. The distance from $P(x, y)$ to the point $(-6, 5)$ is one unit.

28. The distance from $P(x, y)$ to the point $(-8, 4)$ is 10 units.

29. (a) Sketch the parallelogram with vertices $A(-7, -1)$, $B(4, 3)$, $C(7, 8)$ and $D(-4, 4)$.
(b) Compute the midpoints of the diagonals $\overline{AC}$ and $\overline{BD}$.
(c) What conclusion can you draw from part (b)?

30. The vertices of $\triangle ABC$ are $A(1, 1)$, $B(9, 3)$, and $C(3, 5)$.
(a) Find the perimeter of $\triangle ABC$.
(b) Find the perimeter of the triangle that is formed by joining the midpoints of the three sides of $\triangle ABC$.
(c) Compute the ratio of the perimeter in part (a) to the perimeter in part (b).
(d) What theorem from geometry provides the answer for part (c) without the need for the results in (a) and (b)?

31. This exercise refers to $\triangle ABC$ with vertices specified in Exercise 30.
(a) Compute the sum of the squares of the lengths of the sides of $\triangle ABC$.
(b) Compute the sum of the squares of the lengths of the three medians. (A **median** is a line segment drawn from a vertex to the midpoint of the opposite side.)
(c) Check that the ratio of the answer in part (a) to that in part (b) is 4/3. *Remark:* As Exercise 37 asks you to show, this ratio is 4/3 for any triangle.

32. The diagonals of a parallelogram bisect each other. Steps (a), (b), and (c) outline a proof of this theorem.
(a) In the parallelogram $OABC$ in the accompanying figure, check that the coordinates of B must be $(a + b, c)$.

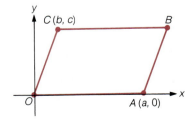

(b) Use the midpoint formula to calculate the midpoints of diagonals $\overline{OB}$ and $\overline{AC}$.

(c) Note that the two answers in part (b) are identical. This shows that the two diagonals do indeed bisect each other, as was to be proved.

33. Use the Pythagorean theorem to find the length a in the accompanying figure. Then find b, c, d, e, f, and g.

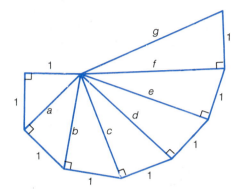

B

34. (A numerologist's delight)   Using the Pythagorean theorem and your calculator, compute the area of a right triangle in which the lengths of the hypotenuse and one leg are 2045 and 693, respectively.

35. Prove that in a parallelogram, the sum of the squares of the lengths of the diagonals equals the sum of the squares of the lengths of the four sides. (Use the figure in Exercise 32.)

36. Use the figure in Exercise 32 to prove the following theorem: If the lengths of the diagonals of a parallelogram are equal, then the parallelogram is a rectangle.   *Hint:*   Equate the two expressions for the lengths of the diagonals and conclude that $ab = 0$. The case in which $a = 0$ can be discarded.

37. The sum of the squares of the sides of any triangle equals 4/3 of the sum of the squares of the medians. Prove this, taking the vertices to be $(0, 0)$, $(2, 0)$, and $(2a, 2b)$.   *Note:*   The expression *squares of the sides* means *squares of the lengths of the sides*.

38. In $\triangle ABC$, let M be the midpoint of side $\overline{BC}$. Prove that $AB^2 + AC^2 = 2(BM^2 + AM^2)$.   *Hint:*   Let the coordinates be $A(0, 0)$, $B(2, 0)$, and $C(2a, 2b)$.

In Exercises 39 and 40, a point $P(x, y)$ satisfies a given condition. Find and simplify an equation relating x and y.

39. (a) The sum of the distances from $P(x, y)$ to the points $(-2, 0)$ and $(2, 0)$ is 6 units.
    (b) The sum of the distances from $P(x, y)$ to the points $(-2, 1)$ and $(2, -1)$ is 6 units.

40. (a) The sum of the distances from $P(x, y)$ to the points $(0, -4)$ and $(0, 4)$ is 10 units.
    (b) The sum of the distances from $P(x, y)$ to the points $(3, -4)$ and $(-3, 4)$ is 10 units.

41. Find the values for t such that points $(0, 2)$ and $(12, t)$ are 13 units apart.   *Hint:*   By the distance formula, $13 = \sqrt{(12 - 0)^2 + (t - 2)^2}$. Now square both sides and solve for t.

42. Find values for t such that the points $(-2, 3)$ and $(t, 1)$ are six units apart.

43. Can you find a real number t such that the points $(-2, 3)$ and $(t, 1)$ are one unit apart? Why or why not?

44. Suppose that the coordinates of points P, Q, and M are as follows:

$$P(x_1, y_1) \qquad Q(x_2, y_2) \qquad M\left(\frac{x_1 + x_2}{2}, \frac{y_1 + y_2}{2}\right)$$

Follow steps (a) and (b) to prove that M is the midpoint of the line segment from P to Q.

(a) By computing both of the distances PM and MQ, show that $PM = MQ$. (This shows that M lies on the perpendicular bisector of line segment $\overline{PQ}$, but it does not show that M actually lies on $\overline{PQ}$.)

(b) Show that $PM + MQ = PQ$. (This shows that M does lie on $\overline{PQ}$.)

C

45. This problem outlines one of the shortest proofs of the Pythagorean theorem. In the accompanying figure, we are given a right triangle ACB, with the right angle at C, and we want to prove that $a^2 + b^2 = c^2$. In the figure, CD is drawn perpendicular to AB.

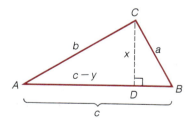

(a) Check that $\angle CAD = \angle DCB$ and that $\triangle BCD$ and $\triangle BAC$ are similar.

(b) Use the result in part (a) to obtain the equation $a/y = c/a$ and conclude that $a^2 = cy$.

(c) Show that $\triangle ACD$ is similar to $\triangle ABC$ and use this to deduce that $b^2 = c^2 - cy$.

(d) Combine the two equations deduced in parts (b) and (c) to arrive at $a^2 + b^2 = c^2$.

46. This problem indicates a method for calculating the area of a triangle when the coordinates of the three vertices are given.

(a) Calculate the area of $\triangle ABC$ in the figure.
   Hint: First calculate the area of the rectangle enclosing $\triangle ABC$ and then subtract the areas of the three right triangles.

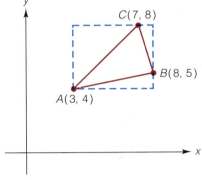

(b) Calculate the area of the triangle with vertices (1, 3), (4, 1), and (10, 4). *Hint:* Work with an enclosing rectangle and three right triangles, as in part (a).

(c) Using the same technic (a) and (b), show that the following figure is g

$$A = \tfrac{1}{2}(x_1 y_2 - x_2 y_1 + .$$

Remark: If we use abso the parentheses, then the regardless of the relative ... quadrants of the three vertices. Thus, the area of a triangle with vertices (x_1, y_1), (x_2, y_2), (x_3, y_3) is given by

$$A = \tfrac{1}{2}|x_1 y_2 - x_2 y_1 + x_2 y_3 - x_3 y_2 + x_3 y_1 - x_1 y_3|$$

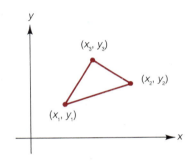

3.2 GRAPHS AND SYMMETRY

Symmetry is a working concept. If all the object is symmetrical, then the parts must be halves (or some other rational fraction) and the amount of information necessary to describe the object is halved (etc.).

Alan L. Macay, Department of Crystallography, Univesity of London

One of the most useful ways to obtain information about an equation relating two variables is through a graph. The key definition is as follows.

DEFINITION   The Graph of an Equation

The **graph** of an equation in two variables is the set of all points whose coordinates satisfy the equation.

Suppose, for example, that we want to graph the equation $y = 3x - 2$. We begin by noting that the domain of the variable x in the expression $3x - 2$ is the set of all real numbers. Now we choose (at will) various values for x and in each case compute the corresponding y-value from the equation $y = 3x - 2$. For example, if x is zero, then $y = 3(0) - 2 = -2$. Table 1 summarizes the results of

TABLE 1

$y = 3x - 2$

x	y
0	−2
1	1
2	4
3	7
−1	−5
−2	−8

some of these calculations. The first line in Table 1 tells us that the point with coordinates $(0, -2)$ is on the graph of $y = 3x - 2$. Reading down the table, we see that some other points on the graph are $(1, 1)$, $(2, 4)$, $(3, 7)$, $(-1, -5)$, and $(-2, -8)$.

We now plot (locate) the points we have determined and note, in this example, that they all appear to lie on a straight line. We draw the line indicated; this is the graph of $y = 3x - 2$ (see Figure 1). In Example 1, we ask a simple question that will test your understanding of the process we've just described and of graphing in general.

EXAMPLE 1   Does the point $\left(1\frac{2}{3}, 2\frac{2}{3}\right)$ lie on the graph of $y = 3x - 2$?

Solution   Looking at the graph in Figure 1, we certainly see that the point $\left(1\frac{2}{3}, 2\frac{2}{3}\right)$ is very close to the line. But on this visual inspection alone, we cannot be certain that this point actually lies on the line. To settle the question, then, we check to see if the values $x = 1\frac{2}{3} = \frac{5}{3}$ and $y = 2\frac{2}{3} = \frac{8}{3}$ together satisfy the equation $y = 3x - 2$. We have

$$\frac{8}{3} \overset{?}{=} 3\left(\frac{5}{3}\right) - 2$$

$$\frac{8}{3} \overset{?}{=} 3 \qquad \text{No!}$$

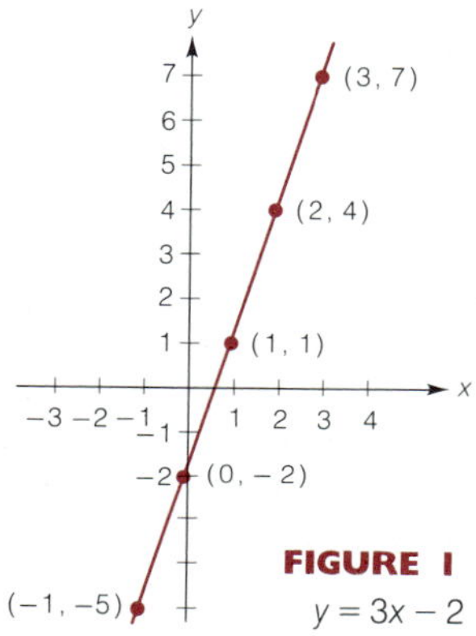

Since the coordinates $\left(1\frac{2}{3}, 2\frac{2}{3}\right)$ evidently do not satisfy the equation $y = 3x - 2$, we conclude that the point $\left(1\frac{2}{3}, 2\frac{2}{3}\right)$ is not on the graph. [The calculation further tells us that the point $\left(1\frac{2}{3}, 3\right)$ *is* on the graph.]

FIGURE 1

$y = 3x - 2$

EXAMPLE 2   As indicated in Figure 2 (on the next page), the graph of the equation $y = -x - 1$ is a line. Find the points on this line that are 5 units from the origin.

Solution   The coordinates of an arbitrary point on the given graph can be denoted by

$$(x, -x - 1)$$

[The reason for this is that, by definition, a point (x, y) is on the graph if and only if x and y satisfy the equation $y = -x - 1$.] Now, since the distance from the point $(x, -x - 1)$ to the origin $(0, 0)$ is to be 5 units, we have

$$\sqrt{(x - 0)^2 + [(-x - 1) - 0]^2} = 5$$

or

$$\sqrt{2x^2 + 2x + 1} = 5$$

We have solved this type of equation in Section 2.5. (Recall that the first step is to square both sides.) As you can check, the solutions are $x = -4$ and $x = 3$. When $x = -4$, the corresponding y-coordinate of a point on the given graph is

$$y = -x - 1 = -(-4) - 1 = 3$$

Similarly, using $x = 3$, we obtain $y = -x - 1 = -3 - 1 = -4$. We have now determined two points on the graph of the given equation that are 5 units from the origin: the points are $(-4, 3)$ and $(3, -4)$. See Figure 3.

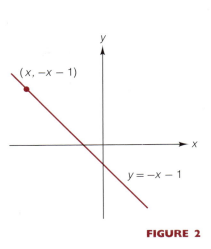

FIGURE 2

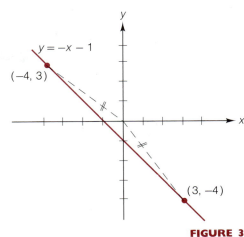

FIGURE 3

The points $(-4, 3)$ and $(3, -4)$ lie on the line $y = -x - 1$ and are 5 units from the origin.

When we graph an equation, it's helpful to know where the curve or line crosses the x- or y-axis. By a **y-intercept** of a graph, we mean the y-coordinate of a point where the graph crosses the y-axis. For instance, returning to Figure 1 (on page 142), the y-intercept is -2. Notice that this y-intercept can be obtained algebraically just by setting x equal to zero in the given equation $y = 3x - 2$. In a similar fashion, an **x-intercept** of a graph is the x-coordinate of a point where the graph crosses the x-axis. In Figure 1 we can see that the x-intercept is a number between 0 and 1. To determine this x-intercept, set y equal to zero in the equation $y = 3x - 2$. That yields $0 = 3x - 2$ and, consequently, $x = \frac{2}{3}$. So, the x-intercept of the graph in Figure 1 is $\frac{2}{3}$.

EXAMPLE 3 Figure 4 shows a graph of the equation $y = 2x^3 - 2x^2 - x$. Determine the x-intercepts of the graph.

Solution At the points where the graph crosses the x-axis, the y-coordinates are zero. Setting y equal to zero in the given equation yields

$$2x^3 - 2x^2 - x = 0$$
$$x(2x^2 - 2x - 1) = 0$$

Thus, $x = 0$ or $2x^2 - 2x - 1 = 0$. This last equation can be solved by using the quadratic formula. As you should check for yourself, the roots are $(1 \pm \sqrt{3})/2$. We've now found the three x-intercepts of the graph in Figure 4: they are 0, $(1 + \sqrt{3})/2$, and $(1 - \sqrt{3})/2$. *Suggestion:* Use your calculator to check that these last two values are consistent with Figure 4. ▪▪▪

In elementary graphing there is always the question of how many points must be plotted before the essential features of a graph are clear. As you'll see throughout this text, there are a number of techniques and concepts that make it unnecessary to plot a large number of points. In this section we introduce three types of *symmetry* that are useful in graphing equations.

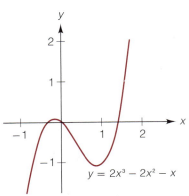

FIGURE 4

$y = 2x^3 - 2x^2 - x$

Three Types of Symmetry

TYPE OF SYMMETRY	EXAMPLE	DEFINITION
1. Symmetry about the x-axis	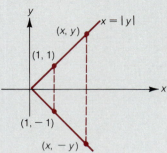 **FIGURE 5** Symmetry about the x-axis	For each point (x, y) on the graph, the point $(x, -y)$ is also on the graph. The points (x, y) and $(x, -y)$ are *reflections of one another about* (or *in*) *the x-axis.* In Figure 5, the portions of the graph above and below the x-axis also are said to be reflections of one another about the x-axis.
2. Symmetry about the y-axis	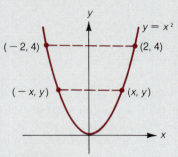 **FIGURE 6** Symmetry about the y-axis	For each point (x, y) on the graph, the point $(-x, y)$ is also on the graph. The points (x, y) and $(-x, y)$ are *reflections of one another about* (or *in*) *the y-axis.* In Figure 6, the portions of the graph to the right and left of the y-axis also are said to be reflections of one another about the y-axis.
3. Symmetry about the origin	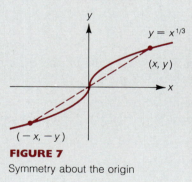 **FIGURE 7** Symmetry about the origin	For each point (x, y) on the graph, the point $(-x, -y)$ is also on the graph. The points (x, y) and $(-x, -y)$ are *reflections of one another about the origin.* In Figure 7, the first- and third-quadrant portions of the curve also are said to be reflections of one another about the origin. [In terms of the two previous symmetries, the point $(-x, -y)$ can be obtained from (x, y) as follows: first reflect (x, y) about the y-axis to obtain $(-x, y)$, then reflect $(-x, y)$ about the x-axis to obtain $(-x, -y)$.]

EXAMPLE 4   A line segment $\mathcal{L}$ has endpoints $(1, 2)$ and $(5, 3)$. Sketch the reflection of $\mathcal{L}$ about

(a) the x-axis;   **(b)** the y-axis;   **(c)** the origin.

Solution   **(a)** First reflect the endpoints $(1, 2)$ and $(5, 3)$ about the x-axis to obtain the new endpoints $(1, -2)$ and $(5, -3)$, respectively. Then join these two points as indicated in Figure 8(a), in quadrant IV.

(b) Reflect the given endpoints about the y-axis to obtain the new endpoints $(-1, 2)$ and $(-5, 3)$; then join these two points as shown in Figure 8(a), in quadrant II.

(c) As described in the box preceding this example, reflection about the origin can be carried out in two steps: first reflect about the y-axis, then reflect about the x-axis. In part (b) we obtained the reflection of $\mathscr{L}$ in the y-axis. So now we need only reflect the line segment obtained in part (b) about the x-axis. See Figure 8(b). ▮▮▮

FIGURE 8

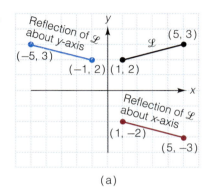

(a)

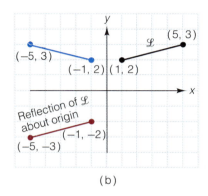

(b)

In the box that follows we list three rules for testing whether the graph of an equation possesses any of the types of symmetry we've been discussing. (The validity of each rule follows directly from the definitions of symmetry.)

Three Tests for Symmetry

1. The graph of an equation is symmetric about the y-axis if replacing x with $-x$ yields an equivalent equation.

2. The graph of an equation is symmetric about the x-axis if replacing y with $-y$ yields an equivalent equation.

3. The graph of an equation is symmetric about the origin if replacing x and y with $-x$ and $-y$, respectively, yields an equivalent equation.

EXAMPLE 5 **(a)** Test for symmetry about the y-axis: $y = x^4 - 3x^2 + 1$.

(b) Test for symmetry about the x-axis: $x = \dfrac{y^2}{y - 1}$.

Solution **(a)** Replacing x with $-x$ in the original equation gives us

$$y = (-x)^4 - 3(-x)^2 + 1$$

or

$$y = x^4 - 3x^2 + 1$$

Since this last equation is the same as the original equation, we conclude that the graph is symmetric about the y-axis.

(b) Replacing y with $-y$ in the original equation gives us

$$x = \frac{(-y)^2}{(-y) - 1} \quad \text{or} \quad x = \frac{y^2}{-y - 1}$$

Since this last equation is not equivalent to the original equation, we conclude that the graph is not symmetric about the x-axis. ■■■

EXAMPLE 6   Is the graph of $y = 2x^3 - x$ symmetric about the origin?

Solution   Replacing x and y with $-x$ and $-y$, respectively, gives us

$$-y = 2(-x)^3 - (-x) = -2x^3 + x$$
$$y = 2x^3 - x \qquad \text{multiplying through by } -1$$

This last equation is identical to the given equation. Therefore the graph is symmetric about the origin. ■■■

In the next two examples we use the notions of symmetry and intercepts as guides for drawing the required graphs.

EXAMPLE 7   Graph the equation: $y = -x^2 + 5$.

Solution   The domain of the variable x in the expression $-x^2 + 5$ is the set of all real numbers. However, since the graph must be symmetric about the y-axis (why?), we need only to be able to sketch the graph to the right of the y-axis; the portion to the left will then be the mirror image of this (with the y-axis as the mirror). The x- and y-intercepts (if any) are computed as follows:

x-INTERCEPTS	y-INTERCEPTS
$-x^2 + 5 = 0$	$y = -(0)^2 + 5$
$x^2 = 5$	$y = 5$
$x = \pm\sqrt{5} \ (\approx \pm 2.2)$	

In Figure 9(a) we've set up a short table of values and sketched the graph for $x \geq 0$. The complete graph is then obtained by reflection about the y-axis, as shown in Figure 9(b). [The curve in Figure 9(b) is a *parabola*. This type of curve will be considered in detail in later chapters.] ■■■

FIGURE 9

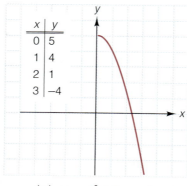

(a) $y = -x^2 + 5$, $x \geq 0$

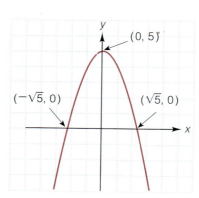

(b) The graph of $y = -x^2 + 5$ is symmetric about the y-axis.

EXAMPLE 8 Graph: $y = 1/x$.

Solution Before doing any calculations, we note that the domain of the variable x consists of all real numbers other than zero. Thus, however the resulting graph may look, it cannot contain a point with x-coordinate zero. In other words, the graph cannot cross the y-axis. (Check for yourself that the graph does not cross the x-axis.) Now, since the graph is symmetric about the origin (why?), we need only to be able to sketch the graph for $x > 0$; the portion corresponding to $x < 0$ can then be obtained by reflection about the origin. In Figure 10(a) we've set up a table of values and sketched the graph for $x > 0$. In Figure 10(b) we have reflected the first-quadrant portion of the graph, first about the y-axis, then about the x-axis, to obtain the required reflection about the origin. Figure 10(c) shows our final graph of $y = 1/x$.

FIGURE 10

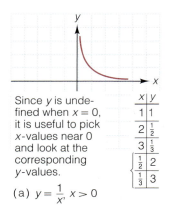

Since y is undefined when $x = 0$, it is useful to pick x-values near 0 and look at the corresponding y-values.

x	y
1	1
2	$\frac{1}{2}$
3	$\frac{1}{3}$
$\frac{1}{2}$	2
$\frac{1}{3}$	3

(a) $y = \frac{1}{x},\ x > 0$

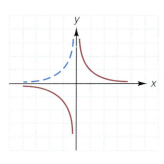

(b) Reflections about the y-axis and then the x-axis yield a reflection about the origin.

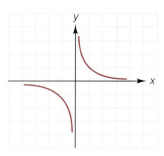

(c) The graph of $y = 1/x$ is symmetric about the origin.

We can use the graph in Figure 10(a) to introduce the idea of an *asymptote*. A line is an **asymptote** for a curve if the distance between the line and the curve approaches zero as we move out farther and farther along the line. So, for the curve in Figure 10(a), both the x-axis and the y-axis are asymptotes. (We'll return to this idea several times later in the text. For other pictures of asymptotes, see Figure 2 in Section 5.6.)

Since we've been discussing symmetry in this section, it seems appropriate to conclude with a discussion of the circle, because in some sense, this is the most symmetric curve. We can use the distance formula to obtain the equation of a circle. Figure 11 shows a circle with center (h, k) and radius r. By definition, a point (x, y) is on this circle if and only if the distance from (x, y) to (h, k) is r. Thus we have

$$\sqrt{(x - h)^2 + (y - k)^2} = r$$

or, equivalently,

$$(x - h)^2 + (y - k)^2 = r^2 \tag{1}$$

(These last two equations are equivalent because two nonnegative quantities are equal if and only if their squares are equal.)

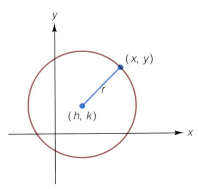

FIGURE 11

The work in the previous paragraph shows us two things. First, if a point (x, y) lies on the circle in Figure 11, then x and y together satisfy equation (1). Second, if a pair of numbers x and y satisfies equation (1), then the point (x, y) lies on the circle in Figure 11. (Does it sound to you as if the previous two sentences say the same thing? They don't! Think about it.) Equation (1) is called the **standard form for the equation of a circle.** For reference, we record the result in the box that follows.

The Equation of a Circle in Standard Form

The equation of a circle with center (h, k) and radius r is
$$(x - h)^2 + (y - k)^2 = r^2$$

EXAMPLE 9   Write the equation of the circle with center $(-2, 5)$ and radius 3.

Solution   In the equation $(x - h)^2 + (y - k)^2 = r^2$, we substitute the given values $h = -2$, $k = 5$, and $r = 3$. This yields

$$[x - (-2)]^2 + (y - 5)^2 = 3^2$$
$$(x + 2)^2 + (y - 5)^2 = 9$$

This is the standard form for the equation of the given circle. An alternative form of this answer is found by carrying out the indicated algebra. We have

$$(x + 2)^2 + (y - 5)^2 = 9$$

Therefore,

$$x^2 + 4x + 4 + y^2 - 10y + 25 = 9$$

and, consequently,

$$x^2 + 4x + y^2 - 10y + 20 = 0$$

The disadvantage to this alternative form is that information regarding the center and radius is no longer readily visible.

When the equation of a circle is not in standard form, the technique of completing the square (discussed in Section 2.3) can be used to convert the equation to standard form. Example 10 shows how this is done.

EXAMPLE 10   Determine the center and radius of the circle given by
$$4x^2 - 24x + 4y^2 + 16y + 51 = 0$$

Solution   First, divide through by 4 so that the coefficients of x^2 and y^2 are both 1:

$$x^2 - 6x + y^2 + 4y + \frac{51}{4} = 0$$

Now, in preparation for completing the squares, we write

$$(x^2 - 6x + \underline{\quad}) + (y^2 + 4y + \underline{\quad}) = -\frac{51}{4}$$

To complete the square in x, we need to add $\left(-\frac{6}{2}\right)^2$, or 9. To complete the square in y, we need to add $\left(\frac{4}{2}\right)^2$, or 4. Thus, we have

$$(x^2 - 6x + 9) + (y^2 + 4y + 4) = -\frac{51}{4} + 9 + 4$$

or

$$(x - 3)^2 + (y + 2)^2 = -\frac{51}{4} + \frac{52}{4} = \frac{1}{4}$$

$$(x - 3)^2 + (y + 2)^2 = \left(\frac{1}{2}\right)^2$$

This is the equation in standard form. The center of the circle is $(3, -2)$; the radius is $\frac{1}{2}$. ∎

EXERCISE SET 3.2

A

In Exercises 1–6, determine whether the point lies on the graph of the equation, as in Example 1. (Note: You are not asked to draw the graph.)

1. $(8, 6)$; $y = \frac{1}{2}x + 3$
2. $\left(\frac{3}{5}, -\frac{17}{5}\right)$; $y = -\frac{2}{3}x - 3$
3. $(4, 3)$; $3x^2 + y^2 = 52$
4. $(4, -2)$; $3x^2 + y^2 = 52$
5. $(a, 4a)$; $y = 4x$
6. $(a - 1, a + 1)$; $y = x + 2$
7. (a) Graph the equation $y = 2x + 2$.
   (b) Find the points on the graph that are 5 units from the origin.
8. (a) Graph the equation $y = -\frac{1}{2}x + 11$.
   (b) Find the points on the graph that are 10 units from the origin.
9. (a) Graph the equation $y = -3x - 6$.
   (b) Find a point on the graph where the x-coordinate is half of the y-coordinate.
10. (a) Graph the equation $y = \frac{1}{3}x + 1$.
    (b) Find a point on the graph where the x- and y-coordinates are the same.

In Exercises 11–16, the graph of the equation is a straight line. Graph the equation after finding the x- and y-intercepts.

11. $3x + 4y = 12$
12. $3x - 4y = 12$
13. $y = 2x - 4$
14. $x = 2y - 4$
15. $x + y = 1$
16. $2x - 3y = 6$

In Exercises 17–28, determine any x- or y-intercepts for the graph of the equation. Note: You're not asked to draw the graph.

17. (a) $y = x^2 + 3x + 2$ (b) $y = x^2 + 2x + 3$
18. (a) $y = x^2 - 4x - 12$ (b) $y = x^2 - 4x + 12$
19. (a) $y = x^2 + x - 1$ (b) $y = x^2 + x + 1$
20. (a) $y = 6x^3 + 9x^2 + x$ (b) $y = 9x^3 + 6x^2 + x$
21. $\frac{x}{2} + \frac{y}{3} = 1$
22. $y = |x - 1|$
23. $3x - 5y = 10$
24. $xy = 1$
25. $y = x^3 - 8$
26. $y = (x - 8)^3 + 1$
27. $y = \sqrt{x - 1} - x + 3$
28. $y = 7 + x - \sqrt{5 - x}$

In Exercises 29–32, each figure shows a graph of an equation. Find the x- and y-intercepts of the graph. If an intercept involves a radical, give both the radical form of the answer and a calculator approximation rounded to two decimal places. (Check to see that your answer is consistent with the given figure.)

29.

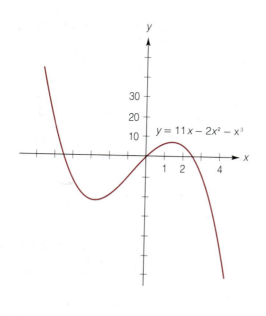

$y = 11x - 2x^2 - x^3$

30.

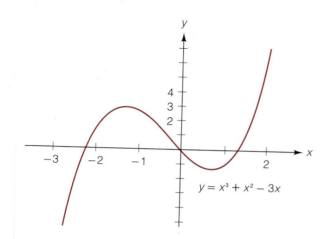

$y = x^3 + x^2 - 3x$

31.

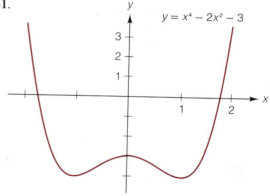

$y = x^4 - 2x^2 - 3$

32.

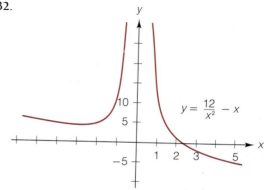

$y = \dfrac{12}{x^2} - x$

*In Exercises 33–36, the endpoints of a line segment $\overline{AB}$ are given. Sketch the reflection of $\overline{AB}$ about **(a)** the x-axis; **(b)** the y-axis; **(c)** the origin.*

33. $A(1, 4)$ and $B(3, 1)$

34. $A(-1, -2)$ and $B(-5, -2)$

35. $A(-2, -3)$ and $B(2, -1)$

36. $A(-3, -3)$ and $B(-3, -1)$

In Exercises 37–46, test each equation for symmetry about the x-axis, the y-axis, and the origin.

37.   (a) $3x^2 + y = 16$ (b) $3x + y^2 = 16$

38.   (a) $x^2y + xy^2 = 0$ (b) $x^2y + xy^2 = 1$

39.   (a) $y = x^4 - 3$ (b) $y = x^3 - 3$
     (c) $y^2 = x^3 - 3$ (d) $y^3 = x$

40.   (a) $y = x^5$ (b) $y = x^5 - 4x^3$
     (c) $y = x^5 - 4x^3 - 1$ (d) $y = x^4 - 4x^2 - 1$

41. $x^2 + y^2 = 16$ **42.** $y = |x - 2|$

43. $y = x^2 + x^3$ **44.** $x = y^2 - 4y^6$

45. $x + y = 1$ **46.** $|x| + |y| = 1$

In Exercises 47–64, graph the equation after determining the x- and y-intercepts and whether the graph possesses any of the three types of symmetry discussed in this section.

47. $y = x^2$

48. $y = -x^3$

49. $y = 1/x$

50. $x = y^2 - 1$

51. $y = -x^2$

52. $y = 1/x^2$

53. $y = -1/x^3$

54. $y = |x|$

55. $y = \sqrt{x^2}$

56. $y = x + 1$

57. $y = x^2 - 2x + 1$

58. $x = y^3 - 1$

59. $y^2 = 2x - 4$

60. $|y| = 2x - 4$

61. $y = 2x^2 + x - 4$

62. (a) $y = x^2 - 4$ (b) $y = |x^2 - 4|$

63. (a) $y = 3x - 6$ (b) $y = |3x - 6|$

64. (a) $x + y = 2$ (b) $|x| + y = 2$
    (c) $x + |y| = 2$ (c) $|x + y| = 2$

In Exercises 65–72, determine the center, the radius, and the x- and y-intercepts (if any) for each circle.

65. $(x - 3)^2 + (y - 1)^2 = 25$

66. $x^2 + (y + 1)^2 = 20$

67. $x^2 + y^2 = \sqrt{2}$

68. $x^2 + y^2 - 10x + 2y + 17 = 0$

69. $x^2 + y^2 + 8x - 6y = -24$

70. $4x^2 - 4x + 4y^2 - 63 = 0$

71. $9x^2 + 54x + 9y^2 - 6y + 64 = 0$

72. $3x^2 + 3y^2 + 5x - 4y = 1$

B

73. Find the area and the circumference of the circle $x^2 - 2ax + y^2 - 2by = c^2 - a^2 - b^2$. (Assume that a, b, and c are constants and $c > 0$.)

74. (a) Find the equation of the circle tangent to the x-axis and with center $(3, 5)$.
    (b) Find the equation of the circle tangent to the y-axis and with center $(3, 5)$.
    (c) Find the equation of the circle passing through the origin and with center $(3, 5)$.

In Exercises 75–78, determine the equation of the circle satisfying the given conditions. Write the equation in standard form.

75. The line segment $\overline{PQ}$ is a diameter and the coordinates of P and Q are $P(-3, -4)$ and $Q(2, 8)$.

76. The circle passes through the origin and it is concentric with the circle $x^2 - 6x + y^2 - 4y + 4 = 0$.

77. The center of the circle is the point $(-3, 2)$, and the radius is equal to twice the distance between the two y-intercepts of the circle $x^2 - 8x + y^2 - 4 = 0$.

78. The circle passes through $(-4, 1)$, and its center is the midpoint of the line segment joining the centers of the two circles $x^2 + y^2 - 6x - 4y + 12 = 0$ and $x^2 + y^2 - 14x + 47 = 0$.

79. (a) Graph the equation $s = 16t^2$, which gives the distance s (in feet) traveled by a freely falling object in t seconds. (The time $t = 0$ corresponds to the instant the object begins its fall; thus, your graph should lie only in the first quadrant, where $t \geq 0$.)
    (b) Indicate the portion of the s-axis that shows the distance covered during the first second, from $t = 0$ to $t = 1$.
    (c) Indicate the portion of the s-axis that shows the distance covered during the next second, from $t = 1$ to $t = 2$.

80. Suppose that the formula $C = 5 + x$ gives the manufacturer's cost in dollars to produce x units of a certain commodity.
    (a) Graph this equation.
    (b) Indicate on the C-axis the cost corresponding to a production level of $x = 2$ units.
    (c) Indicate on the x-axis the number of units corresponding to a cost of $9.

81. Draw the following graphs on the same set of axes and in each case include only the portion of the graph between $x = 0$ and $x = 1$. Draw the graphs carefully enough to ensure that you can make accurate comparisons among them.
    (a) $y = x$ (b) $y = x^2$ (c) $y = x^3$ (d) $y = x^4$
    What is the pattern here? What would you say that $y = x^{100}$ must look like in this interval?

82. The accompanying figure shows the graph of $y = \sqrt{x}$. Use this graph to estimate the following quantities (to one decimal place).
    (a) $\sqrt{2}$ (b) $\sqrt{3}$
    (c) $\sqrt{6}$ *Hint:* $\sqrt{ab} = \sqrt{a}\sqrt{b}$

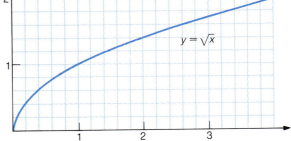

83. In a certain biology experiment, the number N of bacteria increases with time t as indicated in the following figure.
   (a) How many bacteria are initially present when $t = 0$?
   (b) Approximately how long does it take for the original colony to double in size?
   (c) For which value of t is the population approximately 2500?
   (d) During which time interval does the population increase more rapidly, between $t = 0$ and $t = 1$ or between $t = 3$ and $t = 4$?

84. Determine the intercepts and symmetry for the equation $y = x^{2/3}$, and then graph the equation. (Use a calculator.)

85. Find the intercepts and symmetry for $y = x^3 - 27x$. Then use a calculator (as necessary) to set up a table of x- and y-values, with x running from 0 to 6 in increments of $\frac{1}{2}$. Finally, graph the equation.

86. Consider the equation $y = 4x^2/(1 + x^2)$.
   (a) Use a calculator to set up a table with x-values running from 0 to 5 in increments of $\frac{1}{2}$.
   (b) Graph the equation using your table of values and symmetry.

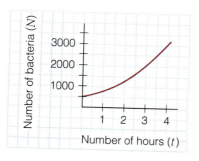

3.3 EQUATIONS OF LINES

He [Pierre de Fermat (1601–1665)] introduced perpendicular axes and found the general equations of straight lines and circles and the simplest equations of parabolas, ellipses, and hyperbolas; and he further showed in a fairly complete and systematic way that every first- or second-degree equation can be reduced to one of these types.

George F. Simmons in *Calculus with Analytic Geometry* (New York: McGraw-Hill Book Company, 1985)

In this section we take a systematic look at equations of lines and their graphs. We begin by recalling the concept of *slope*, which you've seen in previous courses. The slope of a nonvertical line is a number that measures the slant or direction of the line; it is defined as follows.

DEFINITION   Slope

The **slope** of a nonvertical line passing through the two points (x_1, y_1) and (x_2, y_2) is the number m defined by

$$m = \frac{y_2 - y_1}{x_2 - x_1}$$

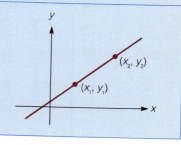

Note that the quantity $x_2 - x_1$ appearing in the definition of slope is the amount by which x changes as we move from (x_1, y_1) to (x_2, y_2) along the line. We denote this change in x by the symbol Δx (read *delta x*). Thus $\Delta x = x_2 - x_1$ (see Figure 1). Similarly, the symbol Δy is defined to mean the change in y: $\Delta y = y_2 - y_1$. Using these ideas, we can rewrite our definition of slope as $m = \Delta y / \Delta x$.

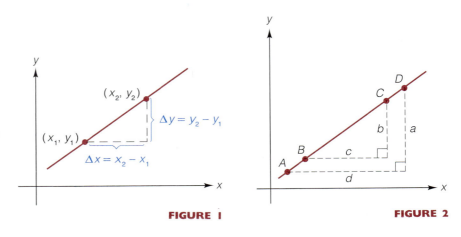

FIGURE 1   **FIGURE 2**

The slope of a line does not depend on which two particular points on the line are used in the calculation. To see why this is so, consider Figure 2. The two right triangles are similar (because the corresponding angles are equal). This implies that the corresponding sides of the two triangles are proportional, and so we have

$$\frac{a}{d} = \frac{b}{c}$$

Now notice that the left-hand side of this equation represents the slope $\Delta y / \Delta x$ calculated using the points A and D, and the right-hand side represents the slope calculated using the points B and C. Thus the values we obtain for the slope are indeed equal.

EXAMPLE 1   Compute and compare the slopes of the three lines shown in Figure 3.

FIGURE 3

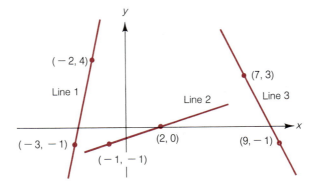

Solution First, we'll calculate the slope of line 1, using the formula $m = (y_2 - y_1)/(x_2 - x_1)$. Which point should we use as (x_1, y_1) and which as (x_2, y_2)? In fact, it does not matter how we label our points. Using $(-3, -1)$ for (x_1, y_1) and $(-2, 4)$ for (x_2, y_2), we have

$$m = \frac{y_2 - y_1}{x_2 - x_1} = \frac{4 - (-1)}{-2 - (-3)} = 5$$

So the slope of line 1 is 5. If, instead, we had used $(-2, 4)$ for (x_1, y_1) and $(-3, -1)$ for (x_2, y_2), then we'd have $m = (-1 - 4)/(-3 + 2) = 5$, the same result. This is not accidental because, as we know from Chapter 1 (page 9)

$$\frac{y_2 - y_1}{x_2 - x_1} = \frac{y_1 - y_2}{x_1 - x_2}$$

Next, we calculate the slopes of lines 2 and 3.

Slope of line 2: $\dfrac{0 - (-1)}{2 - (-1)} = \dfrac{1}{3}$ Slope of line 3: $\dfrac{3 - (-1)}{7 - 9} = -2$

Lines 1 and 2 both have positive slopes and slant upward to the right. Note that line 1 is steeper than line 2 and correspondingly has the larger slope, 5. Line 3 has a negative slope, -2, and slants downward to the right. ■■■

The observations made in Example 1 are true in general. Lines with a positive slope slant upward to the right, the steeper line having the larger slope. Likewise, lines with a negative slope slant downward to the right. See Figure 4. We have yet to mention slopes for horizontal or vertical lines. In Figure 5, line 1 is horizontal; it passes through the points (a, b) and (c, b). Note that the two y-coordinates must be the same in order for the line to be horizontal. Line 2 in Figure 5 is vertical; it passes through (d, e) and (d, f). Note that the two x-coordinates must be the same in order for the line to be vertical. For the slope of line 1 we have $m = \dfrac{b - b}{c - a} = \dfrac{0}{c - a} = 0$ (provided $a \neq c$). Thus the slope of line 1, a horizontal line, is zero. For the vertical line in Figure 5, the calculation of slope begins with writing $\dfrac{e - f}{d - d}$. But then the denominator is zero, and since

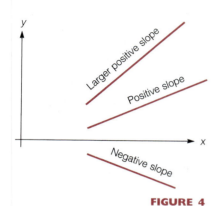

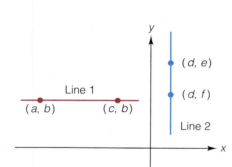

FIGURE 4

FIGURE 5

FIGURE 6

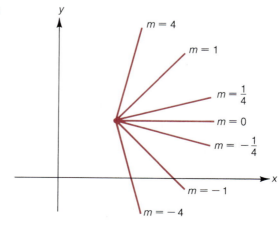

division by zero is undefined, we conclude that slope is undefined for vertical lines. We summarize these results in the box that follows. Figure 6 shows for comparison some values of m for various lines.

Horizontal and Vertical Lines

1. The slope of a horizontal line is zero.

2. Slope is not defined for vertical lines.

We can use the concept of slope to find the equation of a line. Suppose we have a line with slope m and passing through the point (x_1, y_1), as shown in Figure 7(a). Let (x, y) be any other point on the line, as in Figure 7(b). Then the slope of the line is given by $m = (y - y_1)/(x - x_1)$ and, therefore, $y - y_1 = m(x - x_1)$. Note that the given point (x_1, y_1) also satisfies this last equation, because in that case we have $y_1 - y_1 = m(x_1 - x_1)$, or $0 = 0$. The equation $y - y_1 = m(x - x_1)$ is called the **point–slope formula**. We have shown that any point on the line satisfies this equation. (Conversely, it can be shown that if a point satisfies this equation, then the point does lie on the given line.)

FIGURE 7

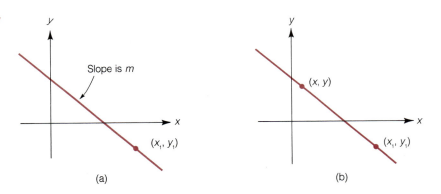

(a)   (b)

The Point–Slope Formula

The equation of a line with slope m passing through the point (x_1, y_1) is

$$y - y_1 = m(x - x_1)$$

EXAMPLE 2 Write the equation of the line passing through $(-3, 1)$ with a slope of -2. Sketch a graph of the line.

Solution Since the slope and a point are given, we use the point–slope formula:

$$y - y_1 = m(x - x_1)$$
$$y - 1 = -2[x - (-3)]$$
$$y - 1 = -2x - 6$$
$$y = -2x - 5$$

This is the required equation. One way to graph this line is to pick values for x and then compute the corresponding y-values, as done in Section 3.2. The table and graph are displayed in Figure 8. Notice that if we know ahead of time that the graph is a line, a table as brief as that in Figure 8 is sufficient. ▪▪▪

There is another way to go about graphing the line $y = -2x - 5$ in Example 2. Since slope is (change in y)/(change in x), a slope of -2 (or $-2/1$) can be interpreted as telling us that if we start at $(-3, 1)$ and let x increase by one unit, then y must decrease by two units to bring us back to the line. Following this path in Figure 9 takes us from $(-3, 1)$ to $(-2, -1)$. We now draw the line through these two points, as shown in Figure 9.

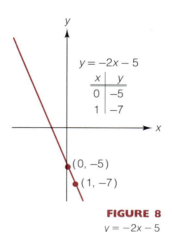

FIGURE 8
$y = -2x - 5$

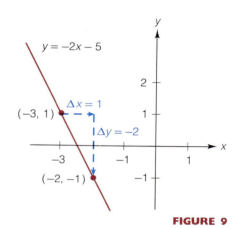

FIGURE 9

EXAMPLE 3   Find the equation of the line passing through the points $(-2, -3)$ and $(2, 5)$.

Solution   The slope of the line is

$$m = \frac{y_2 - y_1}{x_2 - x_1} = \frac{5 - (-3)}{2 - (-2)} = \frac{8}{4} = 2$$

Knowing the slope, we can apply the point–slope formula, making use of either $(-2, -3)$ or $(2, 5)$. Using the point $(2, 5)$ as (x_1, y_1), we have

$$y - y_1 = m(x - x_1)$$
$$y - 5 = 2(x - 2)$$
$$y - 5 = 2x - 4$$
$$y = 2x + 1$$

Thus the required equation is $y = 2x + 1$. You should check for yourself that the same answer is obtained using the point $(-2, -3)$ instead of $(2, 5)$ in the last set of calculations. ▪▪▪

EXAMPLE 4   Find the equation of the horizontal line passing through the point $(4, -2)$. See Figure 10 (on the next page).

Solution Since the slope of a horizontal line is zero, we have

$$y - y_1 = m(x - x_1)$$
$$y - (-2) = 0(x - 4)$$
$$y = -2$$

Thus the equation of the horizontal line passing through $(4, -2)$ is $y = -2$.

By using the point–slope formula exactly as we did in Example 4, we can show more generally that the equation of the horizontal line in Figure 11 is $y = b$. What about the equation of the vertical line in Figure 12 passing through the point (a, b)? Because slope is not defined for vertical lines, the point–slope formula is not applicable this time. However, in Figure 12, note that as you move along the vertical line, the x-coordinate is always a; it is only the y-coordinate that varies. The equation $x = a$ expresses exactly these two facts; it says that x must always be a and it places no restrictions on y.

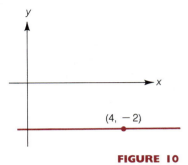

FIGURE 10

FIGURE 11   FIGURE 12

In the box that follows, we summarize our results concerning horizontal and vertical lines.

Equations of Horizontal and Vertical Lines

1. The equation of a horizontal line through the point (a, b) is $y = b$. (See Figure 11.)

2. The equation of a vertical line through the point (a, b) is $x = a$. (See Figure 12.)

Another basic form for the equation of a line is the *slope–intercept form*. We are given the slope m and y-intercept b, as shown in Figure 13, and we want to find the equation of the line. To say that the line has a y-intercept of b is the same as saying that the line passes through $(0, b)$. The point–slope formula is applicable now, using the slope m and the point $(0, b)$. We have

$$y - y_1 = m(x - x_1)$$
$$y - b = m(x - 0)$$
$$y = mx + b$$

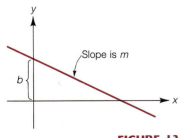

FIGURE 13   This last equation is called the **slope–intercept formula**.

The Slope–Intercept Formula

The equation of a line with slope m and y-intercept b is

$$y = mx + b$$

EXAMPLE 5 Write the equation of a line with slope 4 and y-intercept 1. Graph the line.

Solution Substituting $m = 4$ and $b = 1$ in the equation $y = mx + b$ yields

$$y = 4x + 1$$

This is the required equation. We could draw the graph by first setting up a simple table, but for purposes of emphasis and review we proceed as we did at the end of Example 2. Starting from the point $(0, 1)$, we interpret the slope of 4 as saying that if x increases by 1, then y increases by 4. This takes us from $(0, 1)$ to the point $(1, 5)$ and the line can now be sketched as in Figure 14. ▪▪▪

FIGURE 14
$y = 4x + 1$

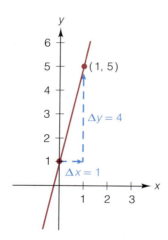

EXAMPLE 6 Find the slope and y-intercept of the line $3x - 5y = 15$.

Solution First we solve for y so that the equation is in the form $y = mx + b$:

$$3x - 5y = 15$$

$$-5y = -3x + 15 \qquad \text{and therefore} \qquad y = \frac{3}{5}x - 3$$

The slope m and the y-intercept b can now be read directly from the equation: $m = \frac{3}{5}$ and $b = -3$. ▪▪▪

As a consequence of our work up to this point, we can say that the graph of any **linear equation** $Ax + By + C = 0$, where A and B are not both zero, is a line, since an equation of this type can always be rewritten in one of the following three forms: $y = mx + b$, or $x = a$, or $y = b$. In the box that follows, we summarize our basic results on equations of lines.

PROPERTY SUMMARY   EQUATIONS OF LINES

EQUATION	COMMENT
$y - y_1 = m(x - x_1)$ (the point–slope formula)	This is the equation of a line with slope m, passing through the point (x_1, y_1).
$y = mx + b$ (the slope–intercept formula)	This is the equation of a line with slope m and y-intercept b.
$Ax + By + C = 0$	The graph of any equation of this form (where A and B are not both zero) is a line. Special cases: If $A = 0$, then the equation can be written $y = -C/B$, which is the equation of a horizontal line with y-intercept $-C/B$. If $B = 0$, then the equation can be written $x = -C/A$, which represents a vertical line with x-intercept $-C/A$.

We conclude this section by discussing two useful relationships regarding the slopes of parallel lines and the slopes of perpendicular lines. First, nonvertical parallel lines have the same slope. This should seem reasonable if you recall that slope is a number indicating the direction or slant of a line. The relationship concerning slopes of perpendicular lines is not so obvious. The slopes of two nonvertical perpendicular lines are negative reciprocals of each other. That is, if m_1 and m_2 denote the slopes of the two perpendicular lines, then $m_1 = -1/m_2$, or, equivalently, $m_1 m_2 = -1$. For example, if a line has a slope of $\frac{2}{3}$, then any line perpendicular to it must have a slope of $-\frac{3}{2}$. For reference we summarize these facts in the box that follows. (Proofs of these facts are outlined in detail in Exercises 61 and 62.)

Parallel and Perpendicular Lines

Let m_1 and m_2 denote the slopes of two nonvertical lines. Then:

1. The lines are parallel if and only if $m_1 = m_2$;

2. The lines are perpendicular if and only if $m_1 = -1/m_2$.

EXAMPLE 7   Are the two lines $3x - 6y - 8 = 0$ and $2y = x + 1$ parallel?

Solution   By solving each equation for y, we can see what the slopes are:

$$3x - 6y - 8 = 0 \qquad\qquad 2y = x + 1$$
$$-6y = -3x + 8 \qquad\qquad y = \frac{1}{2}x + \frac{1}{2}$$
$$y = \frac{-3}{-6}x + \frac{8}{-6}$$
$$y = \frac{1}{2}x - \frac{4}{3}$$

From this we see that both lines have the same slope, namely, $m = \frac{1}{2}$. It follows therefore that the lines are parallel.

EXAMPLE 8   Find the equation of the line parallel to $5x + 6y = 30$ and passing through the origin.

Solution   First, find the slope of the line $5x + 6y = 30$:

$$5x + 6y = 30$$

$$6y = -5x + 30 \qquad \text{and therefore} \qquad y = -\frac{5}{6}x + 5$$

The slope is the x-coefficient in the last equation: $m = -\frac{5}{6}$. A parallel line will have the same slope. Since the required line is to pass through $(0,0)$, we have

$$y - y_1 = m(x - x_1)$$

$$y - 0 = -\frac{5}{6}(x - 0)$$

$$y = -\frac{5}{6}x$$

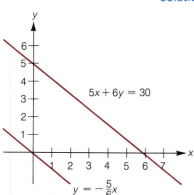

FIGURE 15

This is the equation of a line parallel to $5x + 6y = 30$ and passing through the origin, as required. See Figure 15.

EXAMPLE 9   Find the equation of the line tangent to the circle $x^2 + y^2 = 25$ at the point $(3, -4)$ (see Figure 16). Write the answer in the form $y = mx + b$.

Solution   We have one point on the tangent line, namely, $(3, -4)$. If we knew the slope, then we could apply the point–slope formula. From elementary geometry, we know that the tangent line is perpendicular to the radius. Since the radius passes through $(0,0)$ and $(3, -4)$, its slope is

$$\frac{-4 - 0}{3 - 0} = -\frac{4}{3}$$

and, consequently, the slope of the tangent is $\frac{3}{4}$. The point–slope formula is now applicable:

$$y - y_1 = m(x - x_1)$$

$$y - (-4) = \frac{3}{4}(x - 3)$$

FIGURE 16

Check now for yourself that this can be simplified to $y = \frac{3}{4}x - \frac{25}{4}$, which is the required equation of the tangent line.

EXERCISE SET 3.3

A

In Exercises 1–3, compute the slope of the line passing through the two given points. In Exercise 3, include a sketch with your answers.

1. (a) $(-3, 2), (1, -6)$
   (b) $(2, -5), (4, 1)$
   (c) $(-2, 7), (1, 0)$
   (d) $(4, 5), (5, 8)$

2. (a) $(-3, 0), (4, 9)$
   (b) $(-1, 2), (3, 0)$
   (c) $\left(\frac{1}{2}, -\frac{3}{5}\right), \left(\frac{3}{2}, \frac{3}{4}\right)$
   (d) $\left(\frac{17}{3}, -\frac{1}{2}\right), \left(-\frac{1}{2}, \frac{17}{3}\right)$

3. (a) $(1, 1), (-1, -1)$
   (b) $(0, 5), (-8, 5)$
   (c) $(-1, 1), (1, -1)$
   (d) $(a, b), (b, a)$
   (assume $a \neq b$)

4. Compute the slope of the line in the following figure using the two points indicated.
   (a) A and B (b) B and C (c) A and C
   The principle involved here is that no matter which pair of points you choose, the slope is the same.

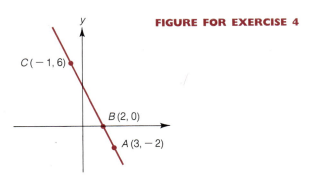

FIGURE FOR EXERCISE 4

$C(-1, 6)$

$B(2, 0)$

$A(3, -2)$

5. The slopes of four lines are indicated in the figure. List the slopes m_1, m_2, m_3, m_4 in order of increasing value.

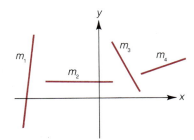

6. Refer to the accompanying figure.

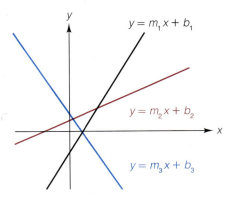

(a) List the slopes m_1, m_2, and m_3 in order of increasing size.

(b) List the numbers b_1, b_2, and b_3 in order of increasing size.

In Exercises 7–9, three points A, B, and C are specified. Determine if A, B, and C are collinear (lie on the same line) by checking to see whether the slope of $\overline{AB}$ equals the slope of $\overline{BC}$.

7. $A(-8, -2)$, $B(2, \frac{1}{2})$, $C(11, -1)$

8. $A(4, -3)$, $B(-1, 0)$, $C(-4, 2)$

9. $A(0, -5)$, $B(3, 4)$, $C(-1, -8)$

10. If the area of the "triangle" formed by three points is zero, then the points must in fact be collinear. Use this observation, along with the formula in Exercise 14(b) of Exercise Set 3.1, to rework Exercise 9.

In Exercises 11 and 12, find the equation for the line having the given slope and passing through the given point. Write your answers in the form $y = mx + b$.

11. (a) $m = -5$; through $(-2, 1)$
(b) $m = 4$; through $(4, -4)$
(c) $m = \frac{1}{3}$; through $\left(-6, -\frac{2}{3}\right)$
(d) $m = -1$; through $(0, 1)$

12. (a) $m = 22$; through $(0, 0)$
(b) $m = -222$; through $(0, 0)$
(c) $m = \sqrt{2}$; through $(0, 0)$

In Exercises 13 and 14, find the equation for the line passing through the two given points. Write your answer in the form $y = mx + b$.

13. (a) $(4, 8)$ and $(-3, -6)$
(b) $(-2, 0)$ and $(3, -10)$
(c) $(-3, -2)$ and $(4, -1)$

14. (a) $(7, 9)$ and $(-11, 9)$
(b) $\left(\frac{5}{4}, 2\right)$ and $\left(\frac{3}{4}, 3\right)$
(c) $(12, 13)$ and $(13, 12)$

In Exercises 15 and 16, write the equation of a vertical line passing through the given point. In Exercises 17 and 18, write the equation of a horizontal line passing through the given point.

15. $(-3, 4)$ **16.** $(8, 5)$

17. $(-3, 4)$ **18.** $(8, 5)$

19. Is the graph of the line $x = 0$ the x-axis or the y-axis?

20. Is the graph of the line $y = 0$ the x-axis or the y-axis?

In Exercises 21 and 22, find the equation of the line with the given slope and y-intercept.

21. (a) slope -4; y-intercept 7
(b) slope 2; y-intercept $\frac{3}{2}$
(c) slope $-\frac{4}{3}$; y-intercept 14

22. (a) slope 0; y-intercept 14
(b) slope 14; y-intercept 0

In Exercises 23–28, find an equation for the line that is described, and sketch the graph. For Exercises 23–25, write the final answer in the form $y = mx + b$; for Exercises 26–28, write the answer in the form $Ax + By + C = 0$.

23. (a) Passing through $(-3, -1)$, with slope 4
(b) Passing through $\left(\frac{5}{2}, 0\right)$, with slope $\frac{1}{2}$
(c) With x-intercept 6, y-intercept 5
(d) With x-intercept -2, slope $\frac{3}{4}$
(e) Passing through $(1, 2)$ and $(2, 6)$

24. (a) Passing through $(-7, -2)$ and $(0, 0)$
    (b) Passing through $(6, -3)$, with y-intercept 8
    (c) Passing through $(0, -1)$ and with the same slope as the line $3x + 4y = 12$
    (d) Passing through $(6, 2)$ and with the same x-intercept as the line $-2x + y = 1$
    (e) With x-intercept -6, y-intercept $\sqrt{2}$

25. With the same x- and y-intercepts as the circle $x^2 + y^2 + 4x - 4y + 4 = 0$

26. Passing through $(3, -5)$ and through the center of the circle $4x^2 + 8x + 4y^2 - 24y + 15 = 0$

27. Passing through $(-3, 4)$ and parallel to the x-axis

28. Passing through $(-3, 4)$ and parallel to the y-axis

In Exercises 29 and 30, find the x- and y-intercepts of the line, and find the area and perimeter of the triangle formed by the line and the axes.

29. (a) $3x + 5y = 15$ (b) $3x - 5y = 15$

30. (a) $5x + 4y = 40$ (b) $2x + 4y = \sqrt{2}$

31. Determine whether the two lines are parallel, perpendicular, or neither.
    (a) $3x - 4y = 12$; $4x - 3y = 12$
    (b) $y = 5x - 16$; $y = 5x + 2$
    (c) $5x - 6y = 25$; $6x + 5y = 0$
    (d) $y = -\frac{2}{3}x - 1$; $y = \frac{3}{2}x - 1$
    (e) $-2x - 5y = 1$; $y - \frac{2}{5}x - 4 = 0$
    (f) $x = 8y + 3$; $4y - \frac{1}{2}x = 32$

32. Are the lines $y = x + 1$ and $y = 1 - x$ parallel, perpendicular, or neither?

In Exercises 33–38, find an equation for the line that is described. Write the answer in the two forms $y = mx + b$ and $Ax + By + C = 0$.

33. Parallel to $2x - 5y = 10$ and passing through $(-1, 2)$

34. Parallel to $4x + 5y = 20$ and passing through $(0, 0)$

35. Perpendicular to $4y - 3x = 1$ and passing through $(4, 0)$

36. Perpendicular to $x - y + 2 = 0$ and passing through $(3, 1)$

37. Parallel to $3x - 5y = 25$ and with the same y-intercept as $6x - y + 11 = 0$

38. Perpendicular to $9y - 2x = 3$ and with the same x-intercept as the circle $x^2 + y^2 - 12x - 2y = -36$

39. (a) Sketch the circle $x^2 + y^2 = 25$.
    (b) Find the equation of the line tangent to this circle at the point $(-4, -3)$. Include a sketch.

40. (a) Tangents are drawn to the circle $x^2 + y^2 = 169$ at the points $(5, 12)$ and $(5, -12)$. Find the equations of the tangents.

(b) Find the coordinates of the point where these two tangents intersect.

B

41. Show that the slope of the line passing through the two points $(3, 9)$ and $(3 + h, (3 + h)^2)$ is $6 + h$.

42. Show that the slope of the line passing through the two points (x, x^2) and $(x + h, (x + h)^2)$ is $2x + h$.

43. Show that the slope of the line passing through the two points (x, x^3) and $(x + h, (x + h)^3)$ is $3x^2 + 3xh + h^2$.

44. Show that the slope of the line passing through the points $(x, \sqrt{x})$ and $(x + h, \sqrt{x + h})$ can be written
$$\frac{1}{\sqrt{x + h} + \sqrt{x}}$$

45. Show that the slope of the line passing through the two points $\left(x, \dfrac{1}{x}\right)$ and $\left(x + h, \dfrac{1}{x + h}\right)$ is $\dfrac{-1}{x(x + h)}$.

46. Why is the letter m used to denote slope? Look up the verb *monter* in a French-to-English dictionary.

47. Sketch the curve $y = x^2$ and indicate the point $T(3, 9)$ on the graph. In each case that follows, you're given the x-coordinate of a point P on the curve. Find the corresponding y-coordinate, and then calculate the slope of the line through P and T. Display your results in a table.
    (a) $x = 2.5$ (b) $x = 2.9$ (c) $x = 2.99$
    (d) $x = 2.999$ (e) $x = 2.9999$
    As P gets closer to T, what numerical value does the slope of $\overline{PT}$ seem to approach? What would you estimate to be the slope of the line that is tangent to the curve $y = x^2$ at point T?

48. Follow the instructions for Exercise 47 using the curve $y = \sqrt{x}$, the point $T(1, 1)$, and the following sequence of x-values.
    (a) $x = 1.1$ (b) $x = 1.01$ (c) $x = 1.001$
    (d) $x = 1.0001$ (e) $x = 1.00001$
    (Round your answers to six decimal places.)

49. Find a point P on the curve $y = x^3$ such that the slope of the line passing through P and $(1, 1)$ is $\frac{3}{4}$.

50. Find a point P on the curve $y = \sqrt{x}$ such that the slope of the line through P and $(1, 1)$ is $\frac{1}{4}$.

51. Find a point P on the curve $y = 1/x$ such that the slope of the line through P and $\left(2, \frac{1}{2}\right)$ is $-\frac{1}{16}$.

52. Find a point P on the circle $x^2 + y^2 = 20$ such that the slope of the radius drawn from $(0, 0)$ to P is 2. (You will get two answers.)

53. A line with a slope of -5 passes through the point $(3, 6)$. Find the area of the triangle in the first quadrant formed by this line and the coordinate axes.

54. The y-intercept of the line in the figure is 6. Find the slope of the line if the area of the shaded triangle is 72 square units.

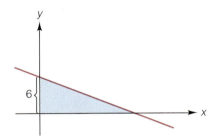

55. (a) Sketch the line $y = \frac{1}{2}x - 5$ and the point $P(1, 3)$. Follow parts (b)–(d) to calculate the perpendicular distance from point $P(1, 3)$ to the line.

(b) Find the equation of the line that passes through $P(1, 3)$ and is perpendicular to the line $y = \frac{1}{2}x - 5$.

(c) Find the coordinates of the point where these two lines intersect.

(d) Use the distance formula to find the perpendicular distance from $P(1, 3)$ to the line $y = \frac{1}{2}x - 5$.

56. (a) Show that the equation of the line tangent to the circle $x^2 + y^2 = a^2$ at the point (x_1, y_1) on the circle is

$$x_1 x + y_1 y = a^2$$

(b) Use the result in part (a) to rework Exercises 39(b) and 40(a).

57. Follow the instructions in parts (a)–(e) to determine the slope m of the line in the figure.

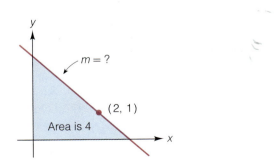

(a) Use the point–slope formula to show that the equation of the line is

$$y - 1 = m(x - 2)$$

(b) Show that the y-intercept for the line is $-2m + 1$. Show that the x-intercept is $(2m - 1)/m$.

(c) Show that the area of the triangle, being one-half base times height, is $\frac{1}{2}\left(\frac{2m - 1}{m}\right)(-2m + 1)$.

(d) Set the expression for area in part (c) equal to 4, as given in the figure, and simplify the resulting equation to obtain $4m^2 + 4m + 1 = 0$.

(e) Solve the equation for m. You should obtain $m = -\frac{1}{2}$.

58. A line with a negative slope passes through the point $(3, 1)$. The area of the triangle bounded by this line and the axes is 6 square units. Find the possible slopes of the line.   *Hint:*   See Exercise 57.

59. By analyzing sales figures, the accountant for College Stereo Company knows that 280 units of a compact disc player can be sold each month when the price is $P = \$195$ per unit. The figures also show that for each \$15 hike in the price, 10 fewer units are sold monthly.

(a) Let x denote the number of units sold per month and P the price per unit. Find an equation that expresses x in terms of P, assuming that this relationship is linear.
Hint: $\Delta x/\Delta P = -10/15 = -2/3$.

(b) Use the equation that you found in part (a) to determine how many units can be sold in a month when the price is \$270 per unit.

(c) What should the price be to sell 205 units per month?

60. Imagine that you own a grove of orange trees, and suppose that from past experience you know that when 100 trees are planted, each tree will yield approximately 240 oranges per year. Furthermore, you've noticed that when additional trees are planted in the grove, the yield per tree decreases. Specifically, you have noted that the yield per tree decreases by about 20 oranges for each additional tree planted.

(a) Let y denote the yield per tree when x trees are planted. Find a linear equation relating x and y.

(b) Use the equation in part (a) to determine how many trees should be planted to obtain a yield of 400 oranges per tree.

(c) If the grove contains 95 trees, what yield can you expect from each tree?

61. This exercise outlines a proof of the fact that two nonvertical lines are parallel if and only if their slopes are equal. The proof relies on the following observation for the given figure: The lines $y = m_1 x + b_1$ and $y = m_2 x + b_2$ will be parallel if and only if the two vertical distances AB and CD are equal. (In the figure, the points C and D both have x-coordinate 1.)

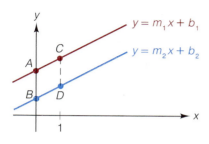

(a) Verify that the coordinates of A, B, C, and D are

$$A(0, b_1) \quad B(0, b_2) \quad C(1, m_1 + b_1) \quad D(1, m_2 + b_2)$$

(b) Using the coordinates in part (a), check that $AB = b_1 - b_2$ and $CD = (m_1 + b_1) - (m_2 + b_2)$.

(c) Use part (b) to show that the equation $AB = CD$ is equivalent to $m_1 = m_2$.

62. This exercise outlines a proof of the fact that two nonvertical lines with slopes m_1 and m_2 are perpendicular if and only if $m_1 m_2 = -1$. In the figure below, we've assumed that our two nonvertical lines $y = m_1 x$ and $y = m_2 x$ intersect at the origin. [If they did not intersect there, we could just as well work with lines parallel to these that do intersect at $(0, 0)$, recalling that parallel lines have the same slope.] The proof relies on the following geometric fact:

$$\overline{OA} \perp \overline{OB} \quad \text{if and only if} \quad OA^2 + OB^2 = AB^2$$

(a) Verify that the coordinates of A and B are $A(1, m_1)$ and $B(1, m_2)$.

(b) Show that

$$OA^2 = 1 + m_1^2$$
$$OB^2 = 1 + m_2^2$$
$$AB^2 = m_1^2 - 2m_1 m_2 + m_2^2$$

(c) Use part (b) to show that the equation $OA^2 + OB^2 = AB^2$ is equivalent to $m_1 m_2 = -1$.

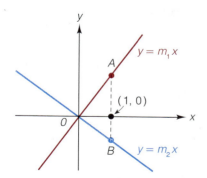

63. Find the equation of the line that is obtained when $y = mx + b$ is reflected about **(a)** the x-axis; **(b)** the y-axis; **(c)** the origin.

64. The following figure shows two tangent lines drawn from the point (a, b) to the circle $x^2 + y^2 = R^2$. Follow steps (a)–(d) to show that the equation of the line passing through the two points of tangency is

$$ax + by = R^2$$

(This problem, along with the following clever solution, appears in the classic text by Isaac Todhunter, *A Treatise on Plane Co-ordinate Geometry*, first published in 1855.)

(a) Let (x_1, y_1) be one of the points of tangency. Show that the equation of the tangent line through this point is $x_1 x + y_1 y = R^2$.

(b) Using the result in part (a), explain why $x_1 a + y_1 b = R^2$.

(c) In a similar fashion, explain why $x_2 a + y_2 b = R^2$, where (x_2, y_2) is the other point of tangency.

(d) The equation $ax + by = R^2$ represents a line. Explain why this line must pass through (x_1, y_1) and (x_2, y_2).

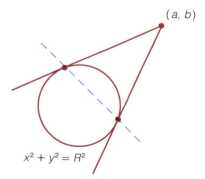

C

65. The vertices of a triangle are $A(-4, 0)$, $B(2, 0)$, and $C(0, 6)$. Let M_1, M_2, and M_3 be the midpoints of $\overline{AB}$, $\overline{BC}$, and $\overline{AC}$, respectively. Let H_1, H_2, and H_3 be the feet of the altitudes on sides $\overline{AB}$, $\overline{BC}$, and $\overline{AC}$, respectively.

(a) Find the equation of the circle passing through M_1, M_2, and M_3. *Answer:* $3x^2 + 3y^2 + 3x - 11y = 0$

(b) Find the equation of the circle passing through H_1, H_2, and H_3. *Answer:* $3x^2 + 3y^2 + 3x - 11y = 0$

(c) Find the point P at which the three altitudes intersect. *Answer:* $\left(0, \frac{4}{3}\right)$

(d) Let N_1, N_2, and N_3 be the midpoints of $\overline{AP}$, $\overline{BP}$, and $\overline{CP}$, respectively. Show that the circle obtained in parts (a) and (b) passes through N_1, N_2, and N_3. *Note:* This circle is called the *nine-point circle* of $\triangle ABC$.

(e) The *circumcircle* of $\triangle ABC$ is the circle passing through the three points A, B, and C. Show that the coordinates of the center Q of the circumcircle are $Q\left(-1,\frac{7}{3}\right)$. *Hint:* Use the result from geometry that the perpendicular bisectors of the sides of the triangle intersect at the center of the circumcircle.

(f) For $\triangle ABC$, show that the radius of the nine-point circle is one-half the radius of the circumcircle.

(g) Show that the midpoint of line segment $\overline{QP}$ is the center of the nine-point circle.

(h) Find the point G where the three medians of $\triangle ABC$ intersect.

(i) Show that G lies on line segment $\overline{QP}$ and that $PG = 2(GQ)$.

CHAPTER THREE   SUMMARY OF PRINCIPAL TERMS AND FORMULAS

TERMS OR FORMULAS	PAGE REFERENCE	COMMENTS
1. Pythagorean theorem $a^2 + b^2 = c^2$	135	In a right triangle, the lengths of the sides are related by this equation, where c is the length of the hypotenuse. Conversely, if the lengths of the sides of a triangle are related by an equation of the form $a^2 + b^2 = c^2$, then the triangle is a right triangle, and c is the length of the hypotenuse.
2. $d = \sqrt{(x_2 - x_1)^2 + (y_2 - y_1)^2}$	136	d is the distance between (x_1, y_1) and (x_2, y_2).
3. $\left(\dfrac{x_1 + x_2}{2}, \dfrac{y_1 + y_2}{2}\right)$	138	This is the midpoint of the line segment joining (x_1, y_1) and (x_2, y_2).
4. Graph of an equation	141	The graph of an equation in the variables x and y is the set of all points (x, y) with coordinates that satisfy the equation.
5. x-intercept and y-intercept	143	An x-intercept of a graph is the x-coordinate of a point where the graph intersects the x-axis. Similarly, a y-intercept is the y-coordinate of a point where the graph intersects the y-axis.
6. Symmetry about the x-axis	144	A graph is symmetric about the x-axis if, for each point (x, y) on the graph, the point $(x, -y)$ is also on the graph. The points (x, y) and $(x, -y)$ are *reflections* of each other about (or in) the x-axis.
7. Symmetry about the y-axis	144	A graph is symmetric about the y-axis if, for each point (x, y) on the graph, the point $(-x, y)$ is also on the graph. The points (x, y) and $(-x, y)$ are *reflections* of each other about (or in) the y-axis.
8. Symmetry about the origin	144	A graph is symmetric about the origin if, for each point (x, y) on the graph, the point $(-x, -y)$ is also on the graph. The points (x, y) and $(-x, -y)$ are *reflections* of each other about the origin.
9. Symmetry tests	145	(i) The graph of an equation is symmetric about the y-axis if replacing x with $-x$ yields an equivalent equation. (ii) The graph of an equation is symmetric about the x-axis if replacing y with $-y$ yields an equivalent equation. (iii) The graph of an equation is symmetric about the origin if replacing x and y with $-x$ and $-y$, respectively, yields an equivalent equation.
10. $(x - h)^2 + (y - k)^2 = r^2$	148	This is the equation of a circle with center at (h, k) and radius r.
11. $m = \dfrac{y_2 - y_1}{x_2 - x_1}$	152	m is the slope of a nonvertical line passing through the points (x_1, y_1) and (x_2, y_2).

TERMS OR FORMULAS	PAGE REFERENCE	COMMENTS
12. $y - y_1 = m(x - x_1)$	155	This is the point–slope form for the equation of a line passing through the point (x_1, y_1) with slope m.
13. $y = mx + b$	158	This is the slope–intercept form for the equation of a line with slope m and y-intercept b.
14. $m_1 = m_2$	159	Nonvertical parallel lines have the same slope.
15. $m_1 = -1/m_2$	159	Two nonvertical lines are perpendicular if and only if the slopes are negative reciprocals of one another.

WRITING MATHEMATICS

1. The following geometric method for solving quadratic equations of the form $x^2 - ax + b = 0$ is due to the British writer Thomas Carlyle (1795–1881).

   To solve the equation $x^2 - ax + b = 0$, plot the points $A(0, 1)$ and $B(a, b)$. Then draw the circle with $\overline{AB}$ as diameter. The x-intercepts of this circle are the roots of the equation.

   Use the techniques of this chapter to verify for yourself that this method indeed yields the roots for the equation $x^2 - 6x + 5 = 0$. (You need to find the equation of the circle, determine the x-intercepts, and then check that the numbers you obtain are the roots of the given equation.) After you have done this, carefully write out your verification in detail (using complete sentences), as if you were explaining the method and why it works to a classmate. Be sure to make explicit reference to any formulas or equations that you use from the chapter.

2. In Section 3.3 we developed the point-slope formula, $y - y_1 = m(x - x_1)$, and the slope-intercept formula, $y = mx + b$. Although these two formulas look quite different from one another, both are merely restatements of the definition of slope. Write a paragraph (or two, at the most) to explain this last sentence. Make use of the following two figures.

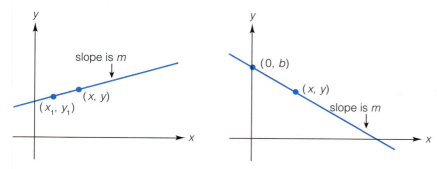

CHAPTER THREE REVIEW EXERCISES

In Exercises 1–18, find an equation for the line satisfying the given conditions. For Exercises 1–9, write the answer in the form $y = mx + b$; in Exercises 10–18, write the answer in the form $Ax + By + C = 0$.

1. Passing through $(-4, 2)$ and $(-6, 6)$
2. $m = -2$; y-intercept 5
3. $m = \frac{1}{4}$; passing through $(-2, -3)$

4. $m = \frac{1}{3}$; x-intercept -1

5.   x-intercept -4; y-intercept 8

6. $m = -10$; y-intercept 0

7.   y-intercept -2; parallel to the x-axis

8.   Passing through $(0, 0)$ and parallel to $6x - 3y = 5$

9.   Passing through $(1, 2)$ and perpendicular to the line $x + y + 1 = 0$

10.   Passing through $(1, 1)$ and through the center of the circle $x^2 - 4x + y^2 - 8y + 16 = 0$

11.   Passing through the centers of the circles $x^2 + 4x + y^2 + 2y = 0$ and $x^2 - 4x + y^2 - 16y = 0$

12. $m = 3$; the same x-intercept as the line $3x - 8y = 12$

13.   Passing through the origin and the midpoint of the line segment joining the points $(-2, -3)$ and $(6, -5)$

14.   Tangent to the circle $x^2 + y^2 = 20$ at the point $(-2, 4)$.

15.   Tangent to the circle $x^2 - 6x + y^2 + 8y = 0$ at the point $(0, 0)$

16.   Passing through $(2, 4)$; the y-intercept is twice the x-intercept

17.   Passing through $(2, -1)$; the sum of the x- and y-intercepts is 2 (There are two answers.)

18.   Passing through $(4, 5)$; no x-intercept

19.   Find the distance between the points $(-1, 2)$ and $(4, -10)$.

20.   (a) Find the perimeter of the triangle with vertices $A(3, 1)$, $B(7, 4)$, and $C(-2, 13)$.
      (b) Find the perimeter of the triangle formed by joining the midpoints of the sides of the triangle in part (a).

21.   Which point is farther from the origin, $(15, 6)$ or $(16, 2)$?

22.   Find the midpoint of the line segment joining
      (a) $(4, 9)$ and $(10, -2)$;   (b) $\left(\frac{2}{3}, -1\right)$ and $\left(-\frac{3}{4}, -\frac{1}{3}\right)$.

In Exercises 23–32, test each equation for symmetry about the x-axis, the y-axis, and the origin.

23. $y = x^4 - 2x^2$

24. $y = x^4 - 2x^2 + 1$

25. $y = x^3 + 5x$

26. $y = x^3 + 5x + 1$

27. $y^2 = (x + y)^4$

28. $y = 2/(1 - x^2)$

29. $y = 3x - \dfrac{1}{x}$

30. $y = 2^x + 2^{-x}$

31. $y = 2^x - 2^{-x}$

32. $x^2 + y^2 + 2x = 2\sqrt{x^2 + y^2}$

For Exercises 33–40, tell whether each graph is symmetric about the x-axis, the y-axis, or the origin.

33.

34.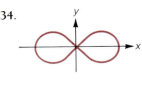

35.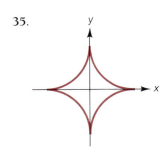

36.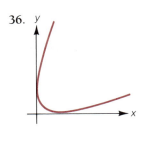

37.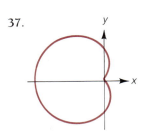

38.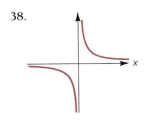

39.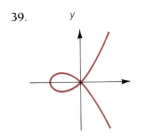

40.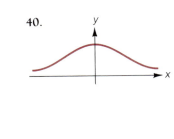

In Exercises 41–44, determine the coordinates of the point (or points) P satisfying the given conditions.

41.   The point P lies on the line $y = -\frac{2}{3}x + 1$ and the distance from P to the origin is $\sqrt{10}$ units.

42.   The point P lies on the line $y = 17 - x$ and the distance from P to the origin is 13 units.

43.   The point P lies on the curve $y = x^2$ and the sum of the x- and y-coordinates of P is 2.

44.   The point P lies on the curve $y = \sqrt{x}$ and the product of the x- and y-coordinates is $\frac{27}{64}$.

Graph the equations in Exercises 45–62.

45. $x = 9 - y^2$

46. $y = -2x + 6$

47. $x^2 + y^2 = 1$

48. $(x + 2)^2 + y^2 = 1$

49. $y = |x| + 1$

50. $y = 1 - |x|$

51. $y = x^2 - 4x + 4$

52. $y = -x$

53. $y = |x - 2| + 2$

54. $y = |x - 2| - 2$

55. $3x + y = 0$

56. $|x| + |y| = 1$

57. $x^2 - 4x + y^2 + 6y = 0$

58. $x^2 + 2x + y^2 + 8y = 8$

59. $y = \dfrac{x^2 - 16}{x - 4}$

60. $y = 1 + \dfrac{1}{x}$

61. $(4x - y + 4)(4x + y - 4) = 0$

62. $(y - x)(y - x + 2)(y - x - 2) = 0$

63. Find a value for t such that the slope of the line passing through $(2, 1)$ and $(5, t)$ is 6.

64. Express in terms of x the slope of the line passing through (x, x^2) and $(-3, 9)$.

65. Express in terms of x the slope of the line passing through $(-2, -8)$ and (x, x^3).

66. The vertices of a parallelogram are $(0, 0)$, $(5, 2)$, $(8, 7)$, and $(3, 5)$. Find the midpoint of each diagonal. Observation?

67. The vertices of a right triangle are $A(0, 0)$, $B(0, 2b)$, and $C(2c, 0)$. Let M be the midpoint of the hypotenuse. Compute the three distances MA, MB, and MC. What do you observe?

68. The vertices of parallelogram $ABCD$ are $A(-4, -1)$, $B(2, 1)$, $C(3, 3)$, and $D(-3, 1)$.

    (a) Compute the sum of the squares of the two diagonals.

    (b) Compute the sum of the squares of the four sides. What do you observe?

In Exercises 69–72, two points are given. In each case compute **(a)** *the distance between the two points,* **(b)** *the slope of the line segment joining the two points, and* **(c)** *the midpoint of the line segment joining the two points.*

69. $(2, 5)$ and $(5, -6)$

70. $\left(\frac{3}{2}, 1\right)$ and $\left(\frac{7}{2}, 9\right)$

71. $\left(\sqrt{3}/2, 1/2\right)$ and $\left(-\sqrt{3}/2, -1/2\right)$

72. $\left(\sqrt{3}, \sqrt{2}\right)$ and $\left(\sqrt{2}, \sqrt{3}\right)$

73. A line passes through the points $(1, 2)$ and $(4, 1)$. Find the area of the triangle bounded by this line and the coordinate axes.

74. Let $P_1(x_1, y_1)$ and $P_2(x_2, y_2)$ be two given points. Let Q be the point

    $$\left(\tfrac{1}{3}x_1 + \tfrac{2}{3}x_2, \tfrac{1}{3}y_1 + \tfrac{2}{3}y_2\right)$$

    (a) Show that the points P_1, Q, and P_2 are collinear.   *Hint:*   Compute the slope of $\overline{P_1Q}$ and the slope of $\overline{QP_2}$.

    (b) Show that $P_1Q = \frac{2}{3}P_1P_2$. (In other words, Q is on the line segment $\overline{P_1P_2}$ and two-thirds of the way from P_1 to P_2.)

75. (a) Let the vertices of $\triangle ABC$ be $A(-5, 3)$, $B(7, 7)$, and $C(3, 1)$. Find the point on each median that is two-thirds of the way from the vertex to the midpoint of the opposite side. (Recall that a median of a triangle is a line segment drawn from a vertex to the midpoint of the opposite side.) What do you observe?   *Hint:*   Use the result in Exercise 74.

    (b) Follow part (a) but take the vertices to be $A(0, 0)$, $B(2a, 0)$, and $C(2b, 2c)$. What do you observe? What does this prove?

76. Suppose that the circle $(x - h)^2 + (y - k)^2 = r^2$ has two x-intercepts, x_1 and x_2, and two y-intercepts, y_1 and y_2. Compute the quantity $x_1x_2 - y_1y_2$.

77. The point $(1, -2)$ is the midpoint of a chord of the circle $x^2 - 4x + y^2 + 2y = 15$. Find the length of the chord.

78. In the accompanying figure, points P and Q trisect the hypotenuse in $\triangle ABC$. Prove that the square of the hypotenuse is equal to $\frac{9}{5}$ the sum of the squares of the distances from the trisection points to the vertex A of the right angle.   *Hint:*   Let the coordinates of B and C be $(0, 3b)$ and $(3c, 0)$, respectively. Then the coordinates of P and Q are $(c, 2b)$ and $(2c, b)$, respectively.

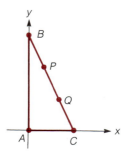

79. Figure (a) shows a triangle with sides of lengths s, t, and u and a median of length m. Prove that

    $$m^2 = \frac{1}{2}(s^2 + t^2) - \frac{1}{4}u^2$$

    Hint: Set up a coordinate system as indicated in Figure (b). Then each of the quantities m^2, s^2, t^2, and u^2 can be computed in terms of a, b, and c.

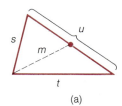

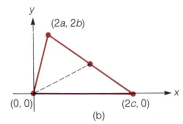

(a)

(b)

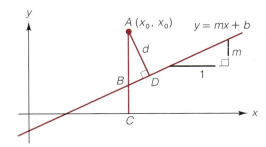

(c) Conclude from parts (a) and (b) that

$$d = \frac{y_0 - mx_0 - b}{\sqrt{1 + m^2}}.$$

For the general case (in which the point and line may not be situated as in our figure), we need to use the absolute value of the quantity in the numerator to assure that AB and d are nonnegative.

80. In this exercise you'll derive a useful formula for the (perpendicular) distance d from the point (x_0, y_0) to the line $y = mx + b$. The formula is

$$d = \frac{|y_0 - mx_0 - b|}{\sqrt{1 + m^2}}$$

(a) Refer to the accompanying figure (in the next column). Use similar triangles to show that

$$\frac{d}{AB} = \frac{1}{\sqrt{1 + m^2}}$$

Therefore, $d = AB/\sqrt{1 + m^2}$.

(b) Check that $AB = AC - BC = y_0 - mx_0 - b$.

81. Use the formula given in Exercise 80 to find the distance from the point $(1, 2)$ to the line $y = \frac{1}{2}x - 5$.

82. Use the formula given in Exercise 80 to demonstrate that the distance from the point (x_0, y_0) to the line $Ax + By + C = 0$ is

$$d = \frac{|Ax_0 + By_0 + C|}{\sqrt{A^2 + B^2}}$$

83. Use the formula given in Exercise 82 to find the distance from the point $(-1, -3)$ to the line $2x + 3y - 6 = 0$.

84. Find the equation of the circle with center $(2, 3)$ that is tangent to the line $x + y - 1 = 0$.   *Hint:*   Use the formula given in Exercise 82 to find the radius.

 OPTIONAL TI-81 GRAPHING CALCULATOR EXERCISES

EXERCISES FOR CHAPTER 3

In Exercises 1–10, graph the equation. Does the graph appear to possess any of the three types of symmetry discussed in Section 3.2? In cases where your answer is "yes," use the symmetry tests on page 144 to verify your answer. Note: Exercises 1 through 4, 9, and 10 contain detailed suggestions for the settings in the RANGE menu. In Exercises 5–8, you are on your own; try to use settings that show the essential features of the graph.

1. $y = x^2 - 3x$ *Suggestions:*   After entering the given function in the Y= menu, press the ZOOM key and then the "6" key to see the graph in the *standard viewing rectangle*. (The standard viewing rectangle

extends from -10 to $+10$ in both the x- and the y-directions.) Next, for a better view of the graph, press the RANGE key and make the following adjustments: $x_{min} = -2$, $x_{max} = 5$, and $y_{min} = -4$. (Retain the initial setting $y_{max} = 10$.) Now, press the GRAPH key.

2. $y = x^3 - 3x$ *Suggestions:*   Press the Y= key. Rather than typing in the entire expression $x^3 - 3x$, it's easier just to edit the expression that you entered in Exercise 1. Using the arrow keys, move the cursor to the location of the exponent "2" that you entered in Exercise 1, and then type "3". Now press the ZOOM

key and then the "6" key to see the graph in the standard viewing rectangle. Finally, for a better view, press the RANGE key and set $x_{min} = -3$ and $x_{max} = 3$. Then press the GRAPH key. (Additional question: As the graph shows, one of the x-intercepts is located between $x = 1$ and $x = 2$. What is the *exact* value for that intercept?)

3. $y = 2^x$ *Suggestions:* After entering the given function in the Y= menu, press the ZOOM key and then the "6" key to see the graph in the standard viewing rectangle. Next, for a better view, press the RANGE key and make the following adjustments: $x_{min} = -3$, $x_{max} = 3$, $y_{min} = 0$, $y_{max} = 8$. Now press the GRAPH key.

4. (a) $y = 2^x + 2^{-x}$ *Suggestions:* First look at the graph in the standard viewing rectangle. Then make the following adjustments in the RANGE menu: $x_{min} = -3$, $x_{max} = 3$, $y_{min} = 0$, $y_{max} = 9$.
    (b) $y = 2^x - 2^{-x}$ *Suggestion:* Use the standard viewing rectangle.

5. $y = \sqrt{|x|}$

6. $y = \sqrt{|x|^3}$

7. $y = 2x - x^3 - x^5 + x^7$

8. $y = |2x - x^3 - x^5 + x^7|$

9. $y = 1/(x^2 - x)$ *Hint:* In the standard viewing rectangle (obtained by pressing the ZOOM key and then the "6" key, the graph is difficult to analyze because of the scale that is involved. For a better view, press the RANGE key and set $x_{min} = -2$ and $x_{max} = 2$. (Then press the GRAPH key.)

10. $y = 1/(x^3 - x)$ *Hint:* Use the RANGE settings given in Exercise 9.

11. (a) Using the Y= key, enter the function $Y_1 = x^2 - 2x^3$. Next, press the ZOOM key and then the "6" key to obtain the graph of this equation in the standard viewing rectangle. Which type of symmetry does the graph appear to possess?
    (b) The graph in part (a) is deceiving because of the scale that is involved. For a different view, press the RANGE key and set $x_{min} = -2$ and $x_{max} = 2$. (Then press the GRAPH key.) Note that the graph no longer appears to be symmetric about the origin.
    (c) Use a symmetry test from page 144 to confirm that the graph is not symmetric about the origin.

12. This exercise shows how to graph circles. As an example, we will graph the circle $x^2 + y^2 = 4$. As preparation, press the Y= key and delete any formulas that appear. Next, press the ZOOM key and then the "6" key to obtain the standard viewing rectangle.

(a) Solve the equation $x^2 + y^2 = 4$ for y. This yields $y = \pm\sqrt{4 - x^2}$. Using the Y= key, enter the equations $Y_1 = \sqrt{4 - x^2}$ and $Y_2 = -\sqrt{4 - x^2}$. Now press the GRAPH key. As you can see, some adjustment needs to be made. Go on to part (b).

(b) Press the RANGE key and set $x_{min} = -2$, $x_{max} = 2$, $y_{min} = -2$, and $y_{max} = 2$. Now press the GRAPH key. The "circle" is larger, but it appears distorted. This is because different scales have been used on the x- and the y-axes. Go on to part (c).

(c) Press the ZOOM key and then the "5" key. This ensures that the same scale is used on both axes. The resulting graph is our completed view of the circle $x^2 + y^2 = 4$. As the graph shows, the center of the circle is $(0, 0)$ and the radius is 2.

(d) Here is a shortcut that could have been used in part (a) for entering the second equation $Y_2 = -\sqrt{4 - x^2}$. Press the Y= key and, for the moment, delete the existing expression for Y_2. Now, rather than typing in the expression $-\sqrt{4 - x^2}$, we will enter the simpler (but equivalent) expression $-Y_1$. This is done as follows. With the cursor on the line corresponding to Y_2, press the "negative sign" key $(-)$ on the bottom row of the calculator. Next, press the "2nd" key and then the VARS key. In the resulting menu, note that the first item is Y_1. Also note that the cursor is on this item. Thus, to select Y_1, just press ENTER. This takes you back to the Y= menu, and your screen display should read

$$Y_1 = \sqrt{(4 - x^2)}$$
$$Y_2 = -Y_1$$

Now press the GRAPH key to see the circle.

In Exercises 13–16, graph the circles using the technique explained in Exercise 12.

13. $x^2 + y^2 = 9$

14. $x^2 + y^2 = 81$

15. $(x - 2)^2 + (y - 1)^2 = 16$ *Hint:* Solve for y to obtain the two equations $y = 1 + \sqrt{16 - (x - 2)^2}$ and $y = 1 - \sqrt{16 - (x - 2)^2}$. For the RANGE settings, use $x_{min} = -2$, $x_{max} = 6$, $y_{min} = -3$, and $y_{max} = 5$.

16. $(x + 3)^2 + (y + 2)^2 = 1$

You know from the text that that graph of the equation $x^2 + y^2 = 9$ is a circle with center $(0, 0)$ and radius 3. In Exercises 17–20, use the calculator to investigate what happens when the equation is changed slightly.

17. (a) Graph the equation $x^2 + 2y^2 = 9$. The resulting curve is an *ellipse*. (We'll study this curve in a later chapter.)

(b) On the same set of axes, graph the ellipse $x^2 + 2y^2 = 9$ and the circle $x^2 + y^2 = 9$. At which points do the curves appear to intersect?

(c) Carry out the calculations to show that the points you have indicated in part (b) indeed lie on both graphs.

18. (a) Graph the equation $x^2 - y^2 = 9$. The resulting curve is a *hyperbola*. (We'll study this curve in a later chapter.)

(b) On the same set of axes, graph the hyperbola $x^2 - y^2 = 9$ and the circle $x^2 + y^2 = 9$. At which points do the curves appear to intersect?

(c) Carry out the calculations to show that the points you have indicated in part (b) indeed lie on both graphs.

19. (a) Graph the equation $x^2 - y^2 = 0$.

(b) On the same set of axes, graph the equations $x^2 - y^2 = 9$ and $x^2 - y^2 = 0$.

20. Graph each equation: (a) $x^2 + y^3 = 9$; (b) $x^3 + y^2 = 9$.

21. In this exercise, you will use the DRAW key to draw the triangle in Figure 7 on page 136. The vertices of this triangle are $D(-2, -1)$, $E(4, 1)$, and $F(3, 4)$.

(a) Press the Y= key and delete or deselect any functions that appear. Then press ZOOM 6.

(b) We will start with a 10-by-10 viewing rectangle. To arrange this, press the RANGE key and enter the following values: $x_{min} = -5$, $x_{max} = 5$, $y_{min} = -5$, and $y_{max} = 5$. Now press the GRAPH key. The coordinate system shown on the screen will accommodate all of the given points, but it will not show the true proportions of $\triangle DEF$. (Why?) Press ZOOM 5. Finally, press the "2nd" key followed by the CLEAR key to return to the home screen.

(c) To draw the segment connecting the points D and E, press the "2nd" key and then the DRAW key. The DRAW menu (with seven choices) will appear on the screen. Use the "down" arrow to move the cursor to choice 2, then press ENTER. The screen display will show

Line (

Now, enter the coordinates of the points D and E (separated by commas) so that the screen display looks as follows:

Line $(-2, -1, 4, 1)$

(To access the comma, press the ALPHA key and then the key used for the decimal point.) Next press ENTER. The line segment $\overline{DE}$ will appear on the screen.

(d) Now we will draw $\overline{EF}$. Press CLEAR, then "2nd," then DRAW. Move the cursor to choice 2 and press ENTER. The screen display should read as follows:

Line $(-2, -1, 4, 1)$

Line (

In the second line of the display, enter the coordinates of the points E and F so that the screen display now reads

Line $(-2, -1, 4, 1)$

Line $(4, 1, 3, 4)$

Press ENTER. The line segments $\overline{DE}$ and $\overline{EF}$ will both appear on the screen.

(e) Use the procedure in part (d) to add the line segment $\overline{DF}$ to the picture. Check to see that your picture agrees with Figure 7 on page 136.

22. What are the equations of the four lines passing through the origin with slopes $m = 1, 2, 3$, and 4? On the same set of axes, graph these four lines.

23. What are the equations of the three lines passing through the origin with slopes $m = -1, -4$, and -8? On the same set of axes, graph these three lines.

24. (a) The following four lines are parallel: $y = 2x - 1$, $y = 2x$, $y = 2x + 1$, $y = 2x + 2$. Without drawing the graphs, how do you know this to be true?

(b) On the same set of axes, graph the four lines in part (a).

25. Graph the following set of parallel lines on the same set of axes:

$$y - 1 = -0.5(x - 2) \qquad y - 1 = -0.5(x + 2)$$
$$y - 1 = -0.5(x - 4)$$

26. In each case, graph the pair of perpendicular lines on the same set of axes: (a) $y = 2x$ and $y = -\frac{1}{2}x$; (b) $y = \frac{8}{3}x$ and $y = -\frac{3}{8}x$; (c) $y = 3x - 4$ and $y = -\frac{1}{3}x - 4$.

27. In parts (a) through (d), solve each equation for y and then graph the line. In each case, make a note of the x- and y-intercepts.

(a) $\dfrac{x}{2} + \dfrac{y}{3} = 1$ (b) $\dfrac{x}{-2} + \dfrac{y}{-3} = 1$

(c) $\dfrac{x}{6} + \dfrac{y}{5} = 1$ (d) $\dfrac{x}{-6} + \dfrac{y}{-5} = 1$

(e) Based on your results in parts (a) through (d), describe, in general, the graph of $\dfrac{x}{a} + \dfrac{y}{b} = 1$.

28. In this exercise, you'll draw Figure 16 in the text on page 160. (As background for this, you should first reread Example 9 on page 160 in the text.) Using the $Y=$ key, enter the following three functions: $Y_1 = \sqrt{25 - x^2}$; $Y_2 = -\sqrt{25 - x^2}$; $Y_3 = \frac{3}{4}x - \frac{25}{4}$. Press the GRAPH key. Check that your result agrees with Figure 16 on page 160.

CHAPTER THREE TEST

In Problems 1–4, find an equation for the line satisfying the given conditions. Write the answer in the form $y = mx + b$.

1. Passing through $(1, -2)$ and $(3, 8)$

2. Passing through $(2, -1)$ and perpendicular to the line $5x + 6y = 30$

3. Passing through the origin and the midpoint of the line segment joining the points $(-5, -2)$ and $(3, 8)$

4. Tangent to the circle $x^2 - 8x + y^2 + 10y = 0$ at the origin

5. Which point is farther from the origin, $(3, 9)$ or $(5, 8)$?

6. Test each equation for symmetry about the x-axis, the y-axis, and the origin:
   (a) $y = x^3 + 5x$; (b) $y = 3^x + 3^{-x}$; (c) $y^2 = 5x^2 + x$.

In Problems 7–9, graph the equations and specify all x- and y-intercepts.

7. $3x - 5y = 15$

8. $2x^2 - 10x + 2y^2 + 6y = 17$

9. $y = |x + 1| + 2$

10. Find a value for t such that the slope of the line passing through $(-4, 3)$ and $(t, 13)$ is 2.

11. Find the point (or points) on the graph of the equation $y = x^2$ where the sum of the x- and y-coordinates is 12.

12. Does the point $\left(-\frac{1}{2}, 5\right)$ lie on the graph of the equation $y = 4x^2 - 8x$?

13. A point $P(x, y)$ is equidistant from the points $(-3, -4)$ and $(7, 10)$. Find (and simplify) an equation relating x and y.

14. A line passes through the points $(6, 2)$ and $(1, 4)$. Find the area of the triangle bounded by this line and the x- and y-axes.

CHAPTER FOUR

FUNCTIONS

From the beginning of modern mathematics in the 17th century the concept of function has been at the very center of mathematical thought.

Richard Courant and Fritz John in *Introduction to Calculus and Analysis* (New York: Wiley-Interscience, 1965)

The word "function" was introduced into mathematics by Leibniz, who used the term primarily to refer to certain kinds of mathematical formulas. It was later realized that Leibniz's idea of function was much too limited in scope, and the meaning of the word has since undergone many steps of generalization.

Tom M. Apostol in *Calculus*, second edition (New York: John Wiley & Sons, 1967)

INTRODUCTION

We begin the study of functions in this chapter. The first section deals with matters of definition and notation. In Sections 4.2 and 4.3 we discuss graphs of functions; several techniques are developed that allow us to graph functions with a minimum of calculation. In Sections 4.4 and 4.5 we find that two functions can be combined in various ways to produce new functions. One of these ways of combining two functions is known as *composition of functions*. As you'll see, this is the unifying theme between Sections 4.4 and 4.5. Finally, in Section 4.6 we study certain types of functions and formulas that occur so frequently in the sciences that a special terminology exists to describe them.

4.1 THE DEFINITION OF A FUNCTION

There are numerous instances in mathematics and its applications in which one quantity corresponds to or depends on another according to some definite rule. Consider, for example, the equation $y = 3x - 2$. Each time that we select an x-value, a corresponding y-value is determined, in this case according to the rule *multiply by 3, then subtract 2*. In this sense, the equation $y = 3x - 2$ is an example of a *function*. It is a *rule* specifying a y-value corresponding to each x-value. It is useful to think of the x-values as inputs and the corresponding y-values as outputs. The function, or rule, then tells us what output results from a given input. This is indicated schematically in Figure 1. As another example,

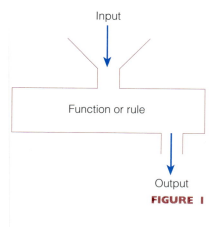

Input

Function or rule

Output

FIGURE I

TABLE I

Mercury	0
Venus	0
Earth	1
Mars	2
Jupiter	16
Saturn	18
Uranus	15
Neptune	8
Pluto	1

the area A of a circle depends on the radius r according to the rule or function $A = \pi r^2$. For each value of r there is a corresponding value for A obtained by the rule *square r and multiply the result by π*. In this case the inputs are the values of r and the outputs are the corresponding values of A.

For most of the functions studied in this text (and in beginning calculus), the inputs and outputs are real numbers and the function or rule is specified by means of an equation. This was the case with the two examples we have given; however, this need not always be the case. Consider, for example, the correspondences set up in Table 1, which indicate the number of moons each planet in the solar system was known to have as of 1992. If we think of the inputs as the planets listed in the left-hand column of Table 1 and the outputs as the numbers in the right-hand column, then these correspondences constitute a function. The rule for this function may be stated, *Assign to each planet in the solar system the number of moons it was known to have in 1992.*

The following is one definition of the term *function*. As you will see, the definition is broad enough to encompass all the examples we have just looked at.

DEFINITION   Function

Let A and B be two nonempty sets. A **function** from A to B is a rule of correspondence that assigns to each element in A exactly one element in B.

The set A in the definition just given is called the **domain** of the function. Think of the domain as the set of all possible inputs. The set of all outputs, on the other hand, is called the **range** of the function. When a function is defined by means of an equation, the letter representing elements from the domain (that is, the inputs) is called the **independent variable**. For example, in the equation $y = 3x - 2$, the independent variable is x. The letter representing elements from the range (that is, the outputs) is called the **dependent variable**. In the equation $y = 3x - 2$, the dependent variable is y; its value *depends* on x. This is also expressed by saying that y is a function of x.

For functions defined by equations, we'll agree to the following convention regarding the domain: Unless otherwise indicated, the domain is assumed to be the set of all real numbers that lead to unique real-number outputs. (This is essentially the domain convention described in Section 1.8.) Thus, the domain of the function defined by $y = 3x - 2$ is the set of all real numbers, whereas the domain of the function defined by $y = 1/(x - 5)$ is the set of all real numbers except 5. [The expression $1/(x - 5)$ is undefined when $x = 5$ because the denominator is then zero.]

EXAMPLE I   Find the domain of the function defined by each equation.

(a) $y = \sqrt{2x + 6}$ **(b)** $s = \dfrac{1}{t^2 - 6t - 7}$

Solution   (a) The quantity under the radical sign must be nonnegative, so we have

$$2x + 6 \geq 0$$
$$2x \geq -6$$
$$x \geq -3$$

The domain is therefore the interval $[-3, \infty)$.

(b) Since division by zero is undefined, the domain of this function consists of all real numbers t except those for which the denominator is zero. Thus to find out which values of t to exclude, we solve the equation $t^2 - 6t - 7 = 0$. We have

$$t^2 - 6t - 7 = 0$$
$$(t - 7)(t + 1) = 0$$
$$t - 7 = 0 \quad \mid \quad t + 1 = 0$$
$$t = 7 \quad \mid \quad t = -1$$

It follows now that the domain of the function defined by $s = \dfrac{1}{t^2 - 6t - 7}$ is the set of all real numbers except $t = 7$ and $t = -1$. ∎

EXAMPLE 2   Find the range of the function defined by $y = \dfrac{x + 2}{x - 3}$.

Solution   The range of this function is the set of all outputs y. One way to see what restrictions the given equation imposes on y is to solve the equation for x as follows:

$$y(x - 3) = x + 2 \qquad \text{multiplying by } (x - 3)$$
$$xy - 3y = x + 2$$
$$xy - x = 3y + 2$$
$$x(y - 1) = 3y + 2$$
$$x = \frac{3y + 2}{y - 1}$$

From this last equation we see that the value of y cannot be 1. (The denominator is zero when $y = 1$.) The range therefore consists of all real numbers except $y = 1$. ∎

We often use single letters in order to name functions. If f is a function and x is an input for the function, then the resulting output is denoted by $f(x)$. This is read f of x or *the value of f at x*. As an example of this notation, suppose that f is the function defined by

$$f(x) = x^2 - 3x + 1 \tag{1}$$

$$f(-2) = 11$$

Output

Input

Name of function

FIGURE 2

Then $f(-2)$ denotes the output that results when the input is -2. To calculate this output, just replace x with -2 throughout equation (1). This yields

$$f(-2) = (-2)^2 - 3(-2) + 1$$
$$= 4 + 6 + 1 = 11$$

That is, $f(-2) = 11$. Figure 2 summarizes this result and the notation.

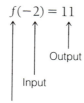

EXAMPLE 3   Let $f(x) = \dfrac{1}{x-1}$. Compute: **(a)** $f(0)$;   **(b)** $f(t)$;   **(c)** $f(x-1)$.

Solution   **(a)** $f(0) = \dfrac{1}{0-1} = -1$

(b) $f(t) = \dfrac{1}{t-1}$

(c) Replace x with the quantity $x-1$ throughout the given equation. This yields

$$f(x-1) = \frac{1}{(x-1)-1} = \frac{1}{x-2}$$

EXAMPLE 4   Let $g(x) = 1 - x^2$. Compute $g(x-1)$.

Solution   In the equation $g(x) = 1 - x^2$, we substitute the quantity $x - 1$ in place of each occurrence of x. This gives us

$$g(x-1) = 1 - (x-1)^2$$
$$= 1 - (x^2 - 2x + 1)$$
$$= -x^2 + 2x$$

Thus, $g(x-1) = -x^2 + 2x$.

Here is a slightly different perspective on function notation that we can apply in Example 4 and in similar problems. Instead of writing $g(x) = 1 - x^2$, we can write $g(\ ) = 1 - (\ )^2$, with the understanding that whatever quantity goes in the parentheses on the left-hand side of the equation must also be placed in the parentheses on the right-hand side of the equation. In particular, if we want $g(x-1)$, we simply write $x - 1$ inside each set of parentheses:

$$g(\ ) = 1 - (\ )^2 \qquad \text{and therefore} \qquad g(x-1) = 1 - (x-1)^2$$

From here on, the algebra is the same as in Example 4. We again obtain $g(x-1) = -x^2 + 2x$.

In the box that follows, we indicate several errors to avoid when using function notation. The first error indicated in the box involves a confusion between a function and a number. For any real numbers r, a, and b, it is always true that $r(a+b) = ra + rb$; this is the familiar *distributive law* for real numbers. If f is a function, however, it is not in general true that $f(a+b) = f(a) + f(b)$.

Errors to Avoid

ERROR	EXAMPLE USING $f(x) = x + 1$, $a = 2$, $b = 3$
$f(a + b) \neq f(a) + f(b)$	$f(2 + 3) = f(5) = 6$; $\quad f(2) + f(3) = 3 + 4 = 7$; therefore $f(2 + 3) \neq f(2) + f(3)$.
$f(ab) \neq f(a) \cdot f(b)$	$f(2 \cdot 3) = f(6) = 7$; $\quad f(2) \cdot f(3) = 3 \cdot 4 = 12$; therefore $f(2 \cdot 3) \neq f(2) \cdot f(3)$.
$f\left(\dfrac{1}{a}\right) \neq \dfrac{1}{f(a)}$	$f\left(\dfrac{1}{2}\right) = \dfrac{1}{2} + 1 = \dfrac{3}{2}$; $\quad \dfrac{1}{f(2)} = \dfrac{1}{2 + 1} = \dfrac{1}{3}$; therefore $f\left(\dfrac{1}{2}\right) \neq \dfrac{1}{f(2)}$.
$\dfrac{f(a)}{f(b)} \neq \dfrac{a}{b}$	$\dfrac{f(2)}{f(3)} = \dfrac{2 + 1}{3 + 1} = \dfrac{3}{4} \neq \dfrac{2}{3}$.

The next two examples using function notation involve calculations with a **difference quotient.** This is an expression of the form

$$\frac{f(x + h) - f(x)}{h} \qquad \text{or} \qquad \frac{f(x) - f(a)}{x - a}$$

For now we'll concentrate on the algebraic techniques used in calculating such quantities. (Later you'll see some applications, in Exercise Set 4.2, Exercises 44 and 45.)

EXAMPLE 5 Let $f(x) = x^2 + 3x$. Compute $\dfrac{f(x) - f(2)}{x - 2}$.

Solution

$$\frac{f(x) - f(2)}{x - 2} = \frac{(x^2 + 3x) - [2^2 + 3(2)]}{x - 2}$$

$$= \frac{x^2 + 3x - 10}{x - 2}$$

$$= \frac{(x - 2)(x + 5)}{x - 2} = x + 5$$

The difference quotient is $x + 5$.

EXAMPLE 6 Let $G(x) = 2/x$. Find $\dfrac{G(x + h) - G(x)}{h}$.

Solution

$$\frac{G(x + h) - G(x)}{h} = \frac{2/(x + h) - 2/x}{h}$$

An easy way to simplify this last expression is to multiply it by $\dfrac{(x + h)x}{(x + h)x}$, which equals 1. This yields

$$\frac{G(x + h) - G(x)}{h} = \frac{(x + h)x}{(x + h)x} \cdot \frac{\dfrac{2}{x + h} - \dfrac{2}{x}}{h}$$

$$= \frac{2x - 2(x + h)}{h(x + h)x}$$

$$= \frac{2x - 2x - 2h}{h(x + h)x} = \frac{-2h}{h(x + h)x}$$

$$= \frac{-2}{(x + h)x}$$

In Examples 1 through 6 we've considered functions that are defined by means of equations. It is important to understand, however, that not all equations and rules define functions. For example, consider the equation $y^2 = x$ and the input $x = 4$. Then we have $y^2 = 4$ and, consequently, $y = \pm 2$. So we have *two* outputs in this case, whereas the definition of a function requires that there be *exactly one* output. Example 7 provides some additional perspective on this situation.

EXAMPLE 7   Let $A = \{b, g\}$ and $B = \{s, t, u, z\}$. Which of the four correspondences in Figure 3 represent functions from A to B? For those correspondences that do represent functions, specify the range in each case.

FIGURE 3

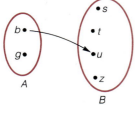
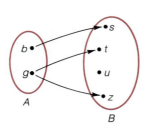
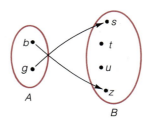
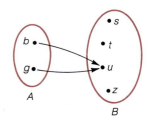

(a)   (b)   (c)   (d)

Solution   FIGURE 3(a)   Not a function. The definition requires that *each* element in A be assigned an element in B. The element g in this case has no assignment.

FIGURE 3(b)   Not a function. The definition requires *exactly one* output for a given input. In this case there are two outputs for the input g.

FIGURES 3(c) AND (d)   Both of these rules qualify as functions from A to B. For each input there is exactly one output. (Regarding the function in Figure 3(d) in particular, notice that nothing in the definition of a function prohibits two different inputs from producing the same output.) For the function in Figure 3(c), the outputs are s and z, and so the range is the set $\{s, z\}$. For the function in Figure 3(d), the only output is u, and consequently the range is the set $\{u\}$.

Because the function concept is so central to the rest of the material in this text, for reference we conclude here with a summary of the terminology introduced in this section.

PROPERTY SUMMARY   TERMINOLOGY FOR FUNCTIONS

TERM	DEFINITION AND COMMENTS	EXAMPLE
Function	Given two nonempty sets A and B, a function f from A to B is a rule that assigns to each element x in A exactly one element $f(x)$ in B. We think of each element x in A as an input and the corresponding element $f(x)$ in B as an output.	The equation $f(x) = x^2$ defines a function. The rule in this case is: *For each real number* x, *compute its square.* Given the input $x = 3$, for example, the output is $f(3) = 9$.
Domain and range	The domain of a function is the set of all inputs; this is the set A in the definition. The range is the set of all outputs; see Figure 4.	The domain and the range of the function defined by $f(x) = x^2$ are the sets $(-\infty, \infty)$ and $[0, \infty)$, respectively.
Independent and dependent variables	The letter used to represent the elements in the domain of a function (the inputs) is the independent variable. The letter used for elements in the range (the outputs) is the dependent variable.	For the function $y = x^2$, the independent variable is x and the dependent variable is y. The value of y *depends* on x.

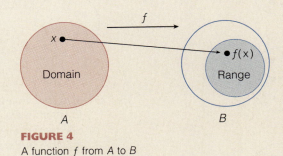

FIGURE 4
A function f from A to B

EXERCISE SET 4.1

A

In Exercises 1–6, find the domain of each function.

1. (a) $y = -5x + 1$ (b) $y = \sqrt{-5x + 1}$

(c) $y = \dfrac{x}{\sqrt{|-5x + 1|}}$

2. (a) $y = x^2 - 3x + 4$
(b) $y = \sqrt{x^2 - 3x + 4}$ *Hint:* Section 2.6 explains a method for solving inequalities such as $x^2 - 3x + 4 \geq 0$.
(c) $y = \sqrt{x^2 - 3x - 4}$

3. (a) $y = \dfrac{x + 4}{x - 4}$ (b) $y = \dfrac{x^2 + 4}{x^2 - 4}$ (c) $y = \sqrt{\dfrac{x^2 + 4}{x^2 - 4}}$

4. (a) $y = \dfrac{1}{x^3 - 16x}$ (b) $y = \dfrac{1}{x^3 - 16x^2}$

(c) $y = \sqrt[3]{x^3 - 16x}$

5. (a) $y = \sqrt{x^2 - 4x - 5}$ (b) $y = \dfrac{2x}{\sqrt{x^2 - 4x - 5}}$
(c) $y = \sqrt[3]{x^2 - 4x - 5}$

6. (a) $y = x^3 - 14x^2 - 6x - 1$
(b) $y = \dfrac{1}{x^3 + x^2 - 2x - 2}$ (c) $y = \dfrac{1}{x^3 + x^2 - 2x}$

In Exercises 7–10, find the range of each function.

7. (a) $y = \dfrac{x + 3}{x - 5}$ (b) $y = \dfrac{x - 5}{x + 3}$

8. (a) $y = \dfrac{3x - 2}{x + 4}$ (b) $y = \dfrac{1}{x + 4}$

9. (a) $y = x^2 + 4$ (b) $y = x^3 + 4$

10. (a) $y = \dfrac{ax + b}{x}$ (a and b are real numbers and $b \neq 0$)

     (b) $y = cx^3 + d$ ($c \neq 0$)

11. Let $A = \{x, y, z\}$ and $B = \{1, 2, 3\}$. Which of the rules displayed in the figure represent functions from A to B?

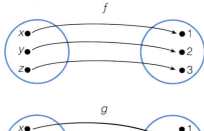

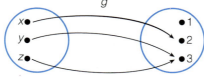

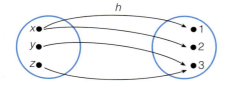

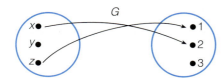

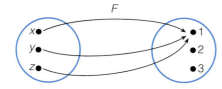

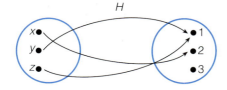

12. Let $D = \{a, b\}$ and $C = \{i, j, k\}$. Which of the rules displayed in the figure (in the next column) represents a function from D to C?

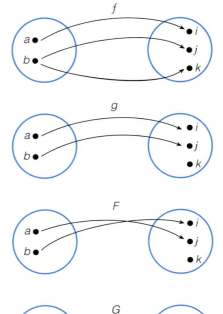

13. (a) Specify the range for each rule that represents a function in Exercise 11.
     (b) Specify the range for each rule that represents a function in Exercise 12.

14. (a) Suppose that in Exercise 11 all the arrows were reversed. Which, if any, of the new rules would be functions from B to A?
     (b) Suppose that in Exercise 12 all the arrows were reversed. Which, if any, of the new rules would be functions from C to D?

15. Each of the following rules defines a function whose domain is the set of all real numbers. Express each rule by means of an equation.

EXAMPLE

The rule *For each real number, compute its square* can be written $y = x^2$.

(a) For each real number, subtract 3 and then square the result.
(b) For each real number, compute its square and then subtract 3 from the result.
(c) For each real number, multiply it by 3 and then square the result.
(d) For each real number, compute its square and then multiply the result by 3.

16. Each of the following rules defines a function with domain equal to the set of all real numbers. Express each rule in words.

(a) $y = 2x^3 + 1$ (b) $y = 2(x + 1)^3$
(c) $y = (2x + 1)^3$ (d) $y = (2x)^3 + 1$

17. Let $f(x) = x^2 - 3x + 1$. Compute the following.

(a) $f(1)$ (b) $f(0)$ (c) $f(-1)$
(d) $f\left(\frac{3}{2}\right)$ (e) $f(z)$ (f) $f(x + 1)$
(g) $f(a + 1)$ (h) $f(-x)$ (i) $|f(1)|$
(j) $f\left(\sqrt{3}\right)$ (k) $f\left(1 + \sqrt{2}\right)$ (l) $|1 - f(2)|$

18. Let $H(x) = 1 - x + x^2 - x^3$.
(a) Which number is larger, $H(0)$ or $H(1)$?
(b) Find $H\left(\frac{1}{2}\right)$. Does $H\left(\frac{1}{2}\right) + H\left(\frac{1}{2}\right) = H(1)$?

19. Let $f(x) = 3x^2$. Find the following.

(a) $f(2x)$ (b) $2f(x)$ (c) $f(x^2)$
(d) $[f(x)]^2$ (e) $f(x/2)$ (f) $f(x)/2$
For checking: No two answers are the same.

20. Let $f(x) = 4 - 3x$. Find the following.

(a) $f(2)$ (b) $f(-3)$ (c) $f(2) + f(-3)$
(d) $f(2 + 3)$ (e) $f(2x)$ (f) $2f(x)$
(g) $f(x^2)$ (h) $f(1/x)$ (i) $f[f(x)]$
(j) $x^2 f(x)$ (k) $1/f(x)$ (l) $f(-x)$
(m) $-f(x)$ (n) $-f(-x)$

21. Let $H(x) = 1 - 2x^2$. Find the following.

(a) $H(0)$ (b) $H(2)$
(c) $H\left(\sqrt{2}\right)$ (d) $H\left(\frac{5}{6}\right)$
(e) $H(x + 1)$ (f) $H(x + h)$
(g) $H(x + h) - H(x)$ (h) $\dfrac{H(x + h) - H(x)}{h}$

22. (a) If $f(x) = 2x + 1$, does $f(3 + 1) = f(3) + f(1)$?
(b) If $f(x) = 2x$, does $f(3 + 1) = f(3) + f(1)$?
(c) If $f(x) = \sqrt{x}$, does $f(3 + 1) = f(3) + f(1)$?

23. Let $R(x) = \dfrac{2x - 1}{x - 2}$. Find the following.

(a) The domain and range of R (b) $R(0)$
(c) $R\left(\frac{1}{2}\right)$ (d) $R(-1)$
(e) $R(x^2)$ (f) $R(1/x)$
(g) $R(a)$ (h) $R(x - 1)$

24. Let $g(x) = 2$, for all x. Find each output.

(a) $g(0)$ (b) $g(5)$ (c) $g(x + h)$

25. Let $d(t) = -16t^2 + 96t$.
(a) Compute $d(1)$, $d\left(\frac{3}{2}\right)$, $d(2)$, $d(t_0)$.
(b) For which values of t is $d(t) = 0$?
(c) For which values of t is $d(t) = 1$?

26. Let $A(x) = |x^2 - 1|$. Compute $A(2)$, $A(1)$, and $A(0)$.

27. Let $g(t) = |t - 4|$. Find $g(3)$. Find $g(x + 4)$.

28. Let $f(x) = x^2/|x|$.
(a) What is the domain of f?
(b) Find $f(2)$, $f(-2)$, $f(20)$, and $f(-20)$.
(c) What is the range of f? *Hint:* Look over your results in part (b).

29. Compute $\dfrac{f(x + h) - f(x)}{h}$ for the function f specified in each case.

(a) $f(x) = x^2$ (b) $f(x) = 2x^2 - 3x + 1$
(c) $f(x) = x^3$

30. Compute $\dfrac{f(x) - f(a)}{x - a}$ for the three functions given in Exercise 29.

31. Let $f(x) = \dfrac{x}{x - 1}$. Compute each difference quotient.

(a) $\dfrac{f(x) - f(a)}{x - a}$ (b) $\dfrac{f(x) - f(3)}{x - 3}$
(c) $\dfrac{f(x + h) - f(x)}{h}$ (d) $\dfrac{f(3 + h) - f(3)}{h}$

32. Let $g(t) = \dfrac{1}{3t - 5}$. Compute each difference quotient.

(a) $\dfrac{g(t) - g(a)}{t - a}$ (b) $\dfrac{g(t) - g(2)}{t - 2}$
(c) $\dfrac{g(t + h) - g(t)}{h}$ (d) $\dfrac{g(2 + h) - g(2)}{h}$

33. In each case a pair of functions f and g are given. Find all real numbers x_0 for which $f(x_0) = g(x_0)$.
(a) $f(x) = 4x - 3$; $g(x) = 8 - x$
(b) $f(x) = x^2 - 4$; $g(x) = 4 - x^2$
(c) $f(x) = x^2$; $g(x) = x^3$
(d) $f(x) = 2x^2 - x$; $g(x) = 3$

In Exercises 34–38, use a calculator for the computations.

34. Let $h(x) = \sqrt{x}$.
(a) Compute $\dfrac{h(5) - h(1)}{5 - 1}$ and $\dfrac{1}{h(5) + 1}$. Round both answers to three decimal places.
(b) Why are the two answers in part (a) the same?

35. When $1000 is deposited in a savings account at an annual rate of 12%, compounded quarterly, the amount in the account after t years is given by

$$A(t) = 1000\left(1 + \frac{0.12}{4}\right)^{4t}$$

(a) Compute $A(1) - A(0)$. This is the amount by which the account will grow in the first year.
(b) Compute $A(10) - A(9)$. This is the amount by which the account will grow in the tenth year.

36. Let $f(n) = [1 + (1/n)]^n$.

    (a) Complete the table. (Round results to three decimal places.)

n	1	2	5	10	15	20
$f(n)$						

    (b) By trial and error, find the smallest natural number n such that $f(n) > 2.7$.

    (c) Using your calculator, can you find a number n such that $f(n) \geq 2.8$?

37. Let $g(n) = n^{1/n}$.

    (a) Complete the table. (Round results to four decimal places.)

n	2	3	4	5	6	7	8
$g(n)$							

    (b) By trial and error, find the smallest natural number n such that $g(n) < 1.2$.

38. Consider the function f defined by

$$f(x) = x^2 + \frac{2}{x^2}, \qquad x > 0$$

    (a) Complete the table. (Round the results to four decimal places.)

x	1	1.05	1.10	1.15	1.20	1.25
$f(x)$						

    (b) Which x-value in the table yields the smallest value for $f(x)$?

    (c) It can be shown using calculus that the input x yielding the smallest possible output $f(x)$ for this function is $x = 2^{1/4}$. Which x-value in the table is closest to this?

    (d) Compute $f(2^{1/4})$. Which value of $f(x)$ in the table is closest to $f(2^{1/4})$?

B

39. Let $f(x) = \dfrac{x - a}{x + a}$.

    (a) Find $f(a)$, $f(2a)$, and $f(3a)$. Is it true that $f(3a) = f(a) + f(2a)$?

    (b) Show that $f(5a) = 2f(2a)$.

40. Let $M(x) = \dfrac{x - a}{x + a}$. Compute $M\left(\dfrac{1}{x}\right)$.

41. Let $\phi(y) = 2y - 3$. Show that $\phi(y^2) \neq [\phi(y)]^2$.

42. Let $k(x) = 5x^3 + \dfrac{5}{x^3} - x - \dfrac{1}{x}$. Show that $k(x) = k(1/x)$.

43. Let $f(x) = 2x + 3$. Find values for a and b such that the equation $f(ax + b) = x$ is true for all values of x.

44. If $p(x) = 2^x$, verify each identity.

    (a) $2p(x) = p(x + 1)$ (b) $p(a + b) = p(a) \cdot p(b)$

    (c) $p(x) \cdot p(-x) - 1 = 0$

45. Let $f(t) = \dfrac{t - x}{t + y}$. Show that

$$f(x + y) + f(x - y) = \frac{-2y^2}{x^2 + 2xy}$$

46. Let $f(z) = \dfrac{3z - 4}{5z - 3}$. Find $f\left(\dfrac{3z - 4}{5z - 3}\right)$.

47. Let $F(x) = \dfrac{ax + b}{cx - a}$. Show that $F\left(\dfrac{ax + b}{cx - a}\right) = x$.

    (Assume that $a^2 + bc \neq 0$.)

48. If $f(x) = -2x^2 + 6x + k$ and $f(0) = -1$, find k.

49. If $g(x) = x^2 - 3xk - 4$ and $g(1) = -2$, find k.

50. If $f(x) = 1/x^2$, show that

$$\frac{f(x + h) - f(x)}{h} = \frac{-2x - h}{(x + h)^2 x^2}$$

51. A function doesn't always have to be given by an algebraic formula. For example, let the function L be defined by the following rule: *$L(x)$ is the exponent to which 2 must be raised to yield x.* (For the moment, we won't concern ourselves with the domain and range.) Then $L(8) = 3$, for example, since the exponent to which 2 must be raised to yield 8 is 3 ($8 = 2^3$). Find the following outputs.

    (a) $L(1)$ (b) $L(2)$ (c) $L(4)$ (d) $L(64)$

    (e) $L\left(\frac{1}{2}\right)$ (f) $L\left(\frac{1}{4}\right)$ (g) $L\left(\frac{1}{64}\right)$ (h) $L(\sqrt{2})$

The function L is called a *logarithm function*. The usual notation for $L(x)$ in this example is $\log_2 x$. Logarithm functions will be studied in Chapter 6.

52. Let $f(x) = ax^2 + bx + c$. Show that

$$\frac{f(x + h) - f(x)}{h} = 2ax + ah + b$$

53. Let $q(x) = ax^2 + bx + c$. Evaluate

$$q\left(\frac{-b + \sqrt{b^2 - 4ac}}{2a}\right)$$

54. By definition, a **fixed point** for the function f is a number x_0 such that $f(x_0) = x_0$. For instance, to find any fixed points for the function $f(x) = 3x - 2$, we write $3x_0 - 2 = x_0$. On solving this last equation, we find that $x_0 = 1$. Thus, 1 is a fixed point for f. Calculate the fixed points (if any) for each function.

(a) $f(x) = 6x + 10$ (b) $g(x) = x^2 - 2x - 4$

(c) $S(t) = t^2$ (d) $R(z) = \dfrac{z + 1}{z - 1}$

55. Let $f(x) = \dfrac{3x - 4}{x - 3}$.

(a) Find $f[f(x)]$.

(b) Find $f\left[f\left(\frac{22}{7}\right)\right]$. Try not to do it the hard way!

56. Consider the following two rules, F and G, where F is the rule that assigns to each person his or her mother, and G is the rule that assigns to each person his or her aunt. Explain why F is a function but G is not.

*In Exercises 57–59, use this definition: A **prime number** is a positive whole number with no factors other than itself and 1. For example, 2, 13, and 37 are primes, but 24 and 39 are not. By convention, 1 is not considered prime, so the list of the first few primes is as follows:*

2, 3, 5, 7, 11, 13, 17, 19, 23, 29, . . .

57. Let G be the rule that assigns to each positive integer the nearest prime. For example, $G(8) = 7$, since 7 is the prime nearest 8. Explain why G is not a function. How could you alter the definition of G to make it a function?

58. Let f be the function that assigns to each natural number x the number of primes that are less than or equal to x. For example, $f(12) = 5$, because, as you can easily check, five primes are less than or equal to 12. Similarly, $f(3) = 2$, because two primes are less than or equal to 3. Find $f(8)$, $f(10)$, and $f(50)$.

59. If $P(x) = x^2 - x + 17$, find $P(1)$, $P(2)$, $P(3)$, and $P(4)$. Can you find a natural number x for which $P(x)$ is not prime?

4.2   THE GRAPH OF A FUNCTION

In my own case, I got along fine without knowing the name of the distributive law until my sophomore year in college; meanwhile I had drawn lots of graphs.

Professor Donald E. Knuth (judged by many to be the world's preeminent computer scientist) in *Mathematical People* (Boston: Birkhäuser, 1985)

When the domain and range of a function are sets of real numbers, the function can be graphed in the same way in which equations were graphed in Chapter 3. In graphing functions, the usual practice is to reserve the horizontal axis for the independent variable and the vertical axis for the dependent variable. The function or rule then tells you how you must pick your y-coordinate, once you have selected an x-coordinate.

> **DEFINITION   Graph of a Function**
>
> The **graph** of a function f in the x-y plane consists of those points (x, y) such that x is in the domain of f and $y = f(x)$. See Figure 1.

FIGURE 1

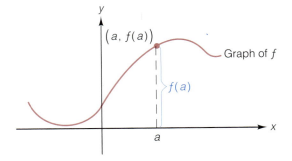

EXAMPLE 1   Graph the functions f and g defined as follows:

(a) $f(x) = x^2 - 1$;   (b) $g(x) = |x^2 - 1|$.

Solution   (a) The graph of this function is by definition the graph of the equation $y = x^2 - 1$. (Whether we label the vertical axis y or $f(x)$ is immaterial.) As preparation for drawing the graph, we first determine the domain of f, the x- and y-intercepts of the graph, and any symmetries the graph may have. As you can readily check, the results are as follows.

$$\text{domain:} \quad (-\infty, \infty)$$
$$x\text{-intercepts:} \quad 1, -1$$
$$y\text{-intercept:} \quad -1$$
$$\text{symmetry:} \quad \text{about the } y\text{-axis}$$

The required graph is then sketched in Figure 2(a).

(b) The graph of g is shown in Figure 2(b). It is obtained from the graph of f in Figure 2(a) as follows. First, for $x \geq 1$ or $x \leq -1$, the graph of g is identical to that of f because in this case, we have

$$g(x) = |x^2 - 1| = x^2 - 1 = f(x)$$

Second, on the interval $(-1, 1)$, the graph of g is obtained by reflecting the graph of f in the x-axis. This is because on the interval $(-1, 1)$, the quantity $x^2 - 1$ is negative and therefore

$$g(x) = |x^2 - 1| = -(x^2 - 1) = -f(x)$$ ∎

FIGURE 2

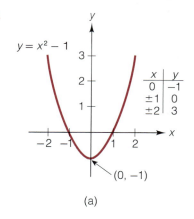

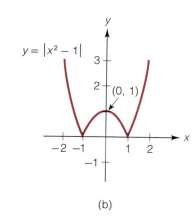

(a)   (b)

EXAMPLE 2   Specify the domain and the range of the function g whose graph appears in Figure 3(a) (on the next page).

Solution   The domain of g is just that portion of the x-axis (the inputs) utilized in graphing g. As Figure 3(b) indicates, this amounts to all real numbers x from 1 to 5, inclusive: $1 \leq x \leq 5$. Recall from Section 1.3 that this set of numbers is denoted by [1, 5]. To find the range of g, we need to check which part of the y-axis is utilized in graphing g. As Figure 3(b) indicates, this is the set of all real numbers y between 2 and 4, inclusive: $2 \leq y \leq 4$. Our shorthand notation for this interval of numbers is [2, 4]. ∎

FIGURE 3

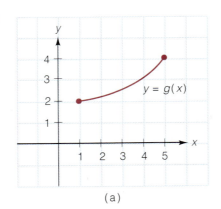

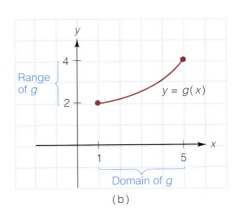

(a)   (b)

EXAMPLE 3   The graph of a function h is shown in Figure 4. The open circle in the figure is used to indicate that the point $(3, 3)$ does not belong to the graph of h.

(a) Specify the domain and the range of the function h.
(b) Determine each value: **(i)** $h(-2)$;   **(ii)** $h(3)$;   **(iii)** $|h(-4)|$

FIGURE 4

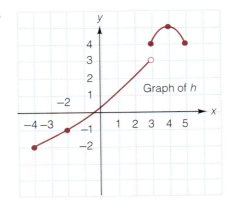

Graph of h

Solution   **(a)** The domain of h is the interval $[-4, 5]$. The range is the set $[-2, 3) \cup [4, 5]$.

(b) (i) The function notation $h(-2)$ stands for the y-coordinate of that point on the graph of h whose x-coordinate is -2. Since the point $(-2, -1)$ is on the graph of h, we conclude $h(-2) = -1$.

(ii) We conclude that $h(3) = 4$ because the point $(3, 4)$ lies on the graph of h. Note that h would not be considered a function if the point $(3, 3)$ were also part of the graph. (Why?)

(iii) Since the point $(-4, -2)$ lies on the graph of h, we write $h(-4) = -2$. Thus $|h(-4)| = |-2| = 2$.   ∎

Most of the graphs that we looked at in Chapter 3 are graphs of functions. However, it's important to understand that not every graph represents a function. Consider, for example, the graph in Figure 5 (on the next page).

FIGURE 5

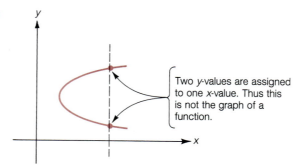

Two y-values are assigned to one x-value. Thus this is not the graph of a function.

Figure 5 also shows a vertical line intersecting the graph in two distinct points. The specific coordinates of the two points are unimportant. What the vertical line helps us to see is that two different y-values have been assigned to the same x-value, and therefore the graph cannot be the graph of a function $y = f(x)$.

The preceding remarks can be summarized as follows.

Vertical Line Test

A graph in the x-y plane represents a function $y = f(x)$, provided that any vertical line intersects the graph in at most one point.

EXAMPLE 4   The vertical line test implies that the graph in Figure 6(a) represents a function and that the graph in Figure 6(b) does not.

FIGURE 6

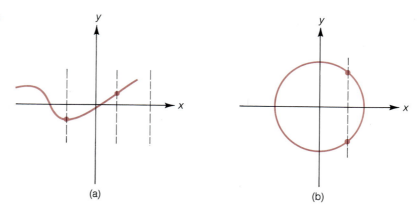

(a)   (b)

Six basic functions arise frequently enough to make it worth your while to memorize the basic shapes and features of their graphs. Figure 7 displays these graphs. Exercise 16 at the end of this section asks you to set up tables and verify for yourself that the graphs in Figure 7 are indeed correct. From now on (with the exception of Exercise 16), if you need to sketch a graph of one of these basic functions, you should do so from memory.

FIGURE 7

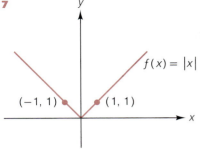

The absolute value
function $f(x) = |x|$
Domain: $(-\infty, \infty)$
Range: $[0, \infty)$

(a)

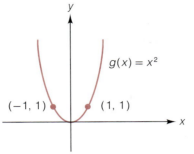

The squaring function
$g(x) = x^2$
Domain: $(-\infty, \infty)$
Range: $[0, \infty)$

(b)

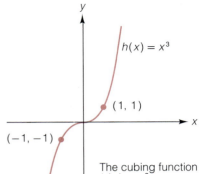

The cubing function
$h(x) = x^3$
Domain: $(-\infty, \infty)$
Range: $(-\infty, \infty)$

(c)

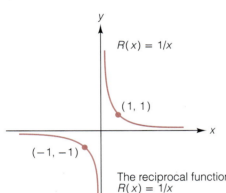

The reciprocal function
$R(x) = 1/x$
Domain: $(-\infty, 0) \cup (0, \infty)$
Range: $(-\infty, 0) \cup (0, \infty)$

(d)

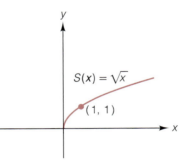

The square root function
$S(x) = \sqrt{x}$
Domain: $[0, \infty)$
Range: $[0, \infty)$

(e)

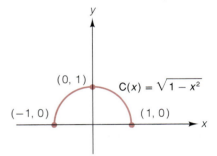

The function $C(x) = \sqrt{1 - x^2}$
(The graph is the upper half
of the circle $x^2 + y^2 = 1$.)
Domain: $[-1, 1]$
Range: $[0, 1]$

(f)

All the graphs in Figure 7 arise from functions defined by rather simple equations. In some instances, however, functions may be defined by combinations of equations. The next two examples display instances of this.

EXAMPLE 5 A function g is defined by

$$g(x) = \begin{cases} x^2 & \text{if } x < 2 \\ 1/x & \text{if } x \geq 2 \end{cases}$$

(a) Find $g(1)$, $g(2)$, and $g(3)$.   **(b)** Sketch the graph of g.

Solution **(a)** To find $g(1)$, do we substitute the value $x = 1$ in the expression x^2 or in the expression $1/x$? According to the instructions contained in the given definition of g, we should use the expression x^2 whenever the inputs are less than 2. Thus we have $g(1) = 1^2 = 1$, that is, $g(1) = 1$. On the other hand, the definition of g tells us to use the expression $1/x$ whenever the inputs are greater than or equal to 2. So in this case we have $g(2) = \frac{1}{2}$ and $g(3) = \frac{1}{3}$.

(b) For the graph of g, we look back at Figures 7(b) and (d) and choose the appropriate portion of each. The result is displayed in Figure 8. The open circle in the figure is used to indicate that the point (2, 4) does not belong to the graph of g. The filled-in circle, on the other hand, is used to indicate that the point $(2, \frac{1}{2})$ does belong to the graph of g.   ▮▮▮

FIGURE 8

A graph of the function

$$g(x) = \begin{cases} x^2, & x < 2 \\ \dfrac{1}{x}, & x \geq 2 \end{cases}$$

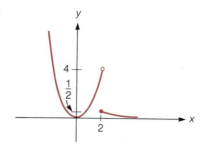

EXAMPLE 6   Graph the function f defined by

$$f(x) = \begin{cases} x^2 - 1 & \text{if } x \leq 2 \\ 5 - x & \text{if } x > 2 \end{cases}$$

Solution   Given this definition of f, we'll set up two tables. For the first table, the inputs x will all be less than or equal to 2, and the outputs will be computed using the expression $x^2 - 1$. In the second table, the inputs x will be greater than 2, and the ouputs will be computed from the expression $5 - x$. These tables and the graph derived from them are shown in Figure 9. [Actually, we could save a few steps by referring to the graph of $y = x^2 - 1$ shown in Figure 2(a).]

We make two observations here. First, notice that, in setting up a table for $y = 5 - x$ with $x > 2$, only two points are needed, since the graph is a portion of a straight line. Second, notice that in this example, as opposed to the function in Example 5, the two portions of the graph together form a graph with no break or gap in it. We say that the function f (whose graph appears in Figure 9) is *continuous* at $x = 2$ but that the function g (whose graph appears in Figure 8) is *discontinuous* at $x = 2$. A rigorous definition of continuity is properly a subject for calculus. However, even at the intuitive level at which we've presented the idea here, you'll find that the concept is useful in helping you to organize your thoughts about the graph of a function.   ▮▮▮

FIGURE 9

$$f(x) = \begin{cases} x^2 - 1, & x \leq 2 \\ 5 - x, & x > 2 \end{cases}$$

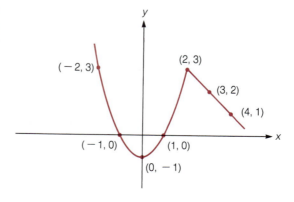

$y = x^2 - 1$, $x \leq 2$

x	± 2	± 1	0
y	3	0	-1

$y = 5 - x$, $x > 2$

x	3	4
y	2	1

We conclude this section with some terminology that is helpful in analyzing functions. To introduce this terminology we'll use the graphs in Figures 10 and 11.

The points P and Q in Figures 10 and 11 are called *turning points*. At a **turning point**, the graph changes from rising to falling, or vice versa. In Figure 10, the highest point on the graph of the function G is $P(2, 102)$. We say that the *maximum value* of the function G is 102 and that this maximum value occurs at $t = 2$. More generally, we say that $f(x_0)$ is the **maximum value** of a function f if the inequality $f(x_0) \geq f(x)$ holds for every x in the domain of f. Minimum values are defined similarly: $f(x_0)$ is the **minimum value** for a function f if the inequality $f(x_0) \leq f(x)$ holds for every x in the domain of f. Assuming that the domain of the function in Figure 10 is $[0, 6]$, the minimum value of the function G is 99, and it occurs when $t = 6$. In Figure 11, the minimum value of the function is -1, and this minimum occurs when $x = 3$. Not every function has a maximum or minimum value. For example, the function $y = 1/x$ possesses neither a maximum nor a minimum value; you can see this by looking back at the graph of this function in Figure 7(d).

The function G in Figure 10 is said to be *increasing* on the open interval $(0, 2)$, and *decreasing* on the interval $(2, 6)$. In terms of the temperature interpretation for Figure 10, the patient's temperature is rising between noon and two o'clock and falling between two and six o'clock. In Figure 11, the function is decreasing on the interval $(-\infty, 3)$ and increasing on the interval $(3, \infty)$. For theoretical work, the terms increasing and decreasing can be defined in terms of inequalities. A function f is **increasing** on an interval if the following condition holds: If x_1 and x_2 are in the interval and $x_1 < x_2$, then $f(x_1) < f(x_2)$. Similarly, a function f is said to be **decreasing** on an interval if the following condition holds: If x_1 and x_2 are in the interval and $x_1 < x_2$, then $f(x_1) > f(x_2)$.

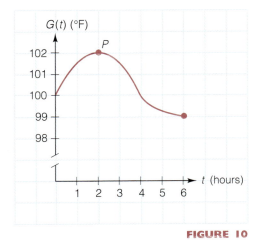

FIGURE 10

The graph of the function G is a fever graph, where $G(t)$ is a patient's temperature t hours after 12 noon, $0 \leq t \leq 6$.

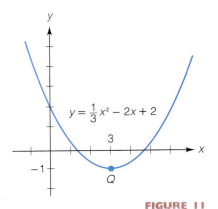

FIGURE 11

The point Q is a turning point. The minimum value of the function is -1. The function is decreasing on $(-\infty, 3)$ and increasing on $(3, \infty)$.

EXERCISE SET 4.2

A

In Exercises 1–8, the graph of a function is given. In each case, specify the domain and the range of the function. (The axes are marked off in one-unit intervals.)

1.

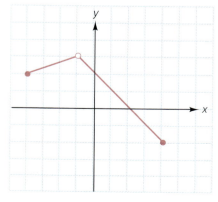

2.

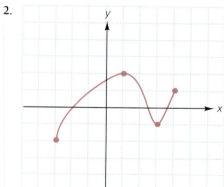

3.

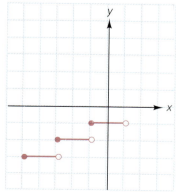

4.

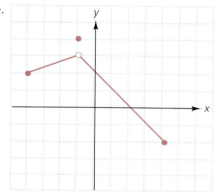

5.

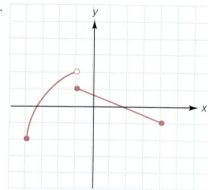

6.

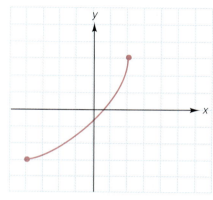

7.

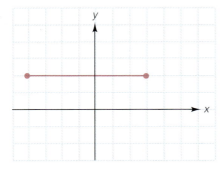

8.

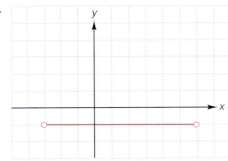

In Exercises 9 and 10, refer to the graph of the function F in the figure. (Assume that the axes are marked off in one-unit intervals.)

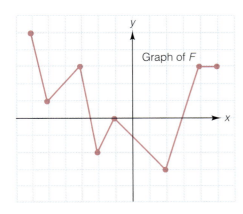

9. (a) Find $F(-5)$.
   (b) Find $F(2)$.
   (c) Is $F(1)$ positive?
   (d) For which value of x is $F(x) = -3$?
   (e) Find $F(2) - F(-2)$.

10. (a) Find $F(4)$.
   (b) Find $F(-1)$.
   (c) Is $F(-4)$ positive?
   (d) For which value of x is $F(x) = 5$?
   (e) Find $F(5) - F(-3)$.

11. The following figure displays the graph of a function f.

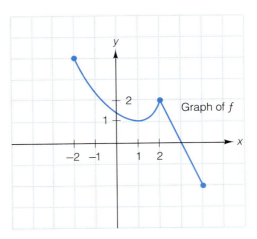

(a) Is $f(0)$ positive or negative?
(b) Find $f(-2)$, $f(1)$, $f(2)$, and $f(3)$.
(c) Which is larger, $f(2)$ or $f(4)$?
(d) Find $f(4) - f(1)$.
(e) Find $|f(4) - f(1)|$.
(f) Write the domain and range of f using the interval notation $[a, b]$.

12. The following figure shows the graph of a function h.

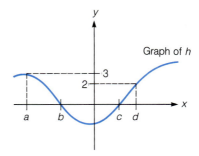

(a) Find $h(a)$, $h(b)$, $h(c)$, and $h(d)$.
(b) Is $h(0)$ positive or negative?
(c) For which values of x does $h(x) = 0$?
(d) Which is larger, $h(b)$ or $h(0)$?
(e) As x increases from c to d, do the corresponding values of $h(x)$ increase or decrease?
(f) As x increases from a to b, do the corresponding values of $h(x)$ increase or decrease?

In Exercises 13–15, refer to the graphs of the functions f and g in the figure. Assume that the domain of each function is [−3, 3] and that the axes are marked off in one-unit intervals.

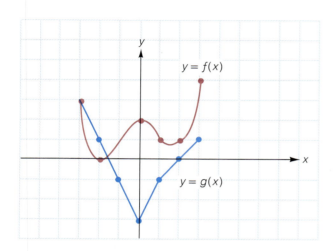

13. (a) Which is larger, $f(-2)$ or $g(-2)$?
   (b) Compute $f(0) - g(0)$.
   (c) Which among the following three quantities is the smallest?

$$f(1) - g(1) \qquad f(2) - g(2) \qquad f(3) - g(3)$$

   (d) For which value(s) of x does $g(x) = f(1)$?
   (e) Is the number 4 in the range of f or in the range of g?

14. (a) For the interval $[0, 3]$, is the quantity $g(x) - f(x)$ positive or negative?
   (b) For the interval $(-3, -2)$, is the quantity $g(x) - f(x)$ positive or negative?
   (c) Compute $\dfrac{f(x) - f(2)}{x - 2}$ when $x = 3$.
   (d) Compute $\dfrac{g(x) - g(-2)}{x + 2}$ when $x = -3$.

15. Specify the range of f and the range of g.

16. Set up a table and graph each function. (The symmetry tests from Section 3.2 are helpful here.)
   (a) $f(x) = |x|$ (b) $g(x) = x^2$
   (c) $h(x) = x^3$ (d) $R(x) = 1/x$
   (e) $S(x) = \sqrt{x}$ (f) $C(x) = \sqrt{1 - x^2}$

17. Complete the following table.

Function	$|x|$	x^2	x^3
Turning Point			
Maximum Value			
Minimum Value			
Interval(s) Where Increasing			
Interval(s) Where Decreasing			

18. Set up and complete a table as in Exercise 17 for the three functions $1/x$, $\sqrt{x}$, and $\sqrt{1 - x^2}$.

In Exercises 19–22, you are given a function with domain [0, 4]. Specify (a) the range of the function; (b) the maximum value of the function; (c) the minimum value; (d) interval(s) where the function is increasing; and (e) interval(s) where the function is decreasing.

19.

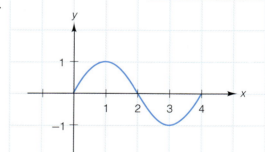

20.

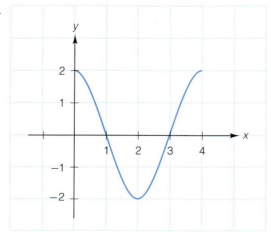

21.

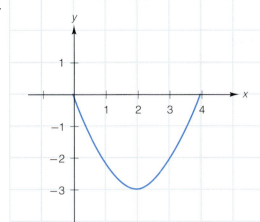

22.

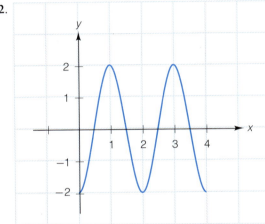

In Exercises 23 and 24, use the vertical line test to determine if each graph represents a function y = f(x).

23.

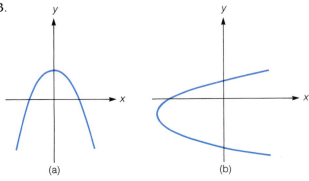

(a)   (b)

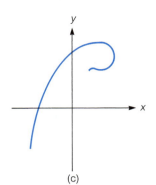

(c)

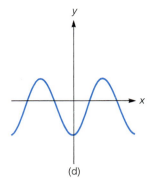

(d)

24.

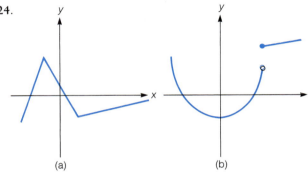

(a)   (b)

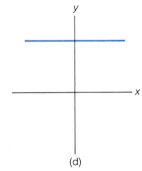

(c)   (d)

25. Let $f(x) = x^2$. Find the slope of the straight line that passes through the points $(3, f(3))$ and $(4, f(4))$. Include a sketch.

26. Let $G(x) = x^3$. Find the slope of the straight line that passes through the points $(0, G(0))$ and $(-2, G(-2))$. Include a sketch.

27. Let $T(x) = \sqrt{x}$. Which line has the larger slope, the line that passes through $(1, T(1))$ and $(4, T(4))$ or the line that passes through $(4, T(4))$ and $(9, T(9))$?

28. What are the domain and range of the function whose graph is the horizontal line $y = 3$?

In Exercises 29–38, graph the function defined by the given rules.

29. $f(x) = \begin{cases} |x| & \text{if } x \le 0 \\ x^2 & \text{if } x > 0 \end{cases}$

30. (a) $g(x) = \begin{cases} |x| & \text{if } x \le 0 \\ x + 1 & \text{if } x > 0 \end{cases}$

   (b) $F(x) = \begin{cases} |x| & \text{if } x < 0 \\ x + 1 & \text{if } x \ge 0 \end{cases}$

31. $A(x) = \begin{cases} x^3 & \text{if } -2 \le x \le -1 \\ x^2 & \text{if } x > -1 \end{cases}$

32. $B(x) = \begin{cases} \sqrt{1 - x^2} & \text{if } -1 \le x < 1 \\ 1/x & \text{if } x \ge 1 \end{cases}$

33. $C(x) = \begin{cases} x^3 & \text{if } x < 1 \\ \sqrt{x} & \text{if } x > 1 \end{cases}$

34. (a) $f(x) = \begin{cases} x^2/|x| & \text{if } x \ne 0 \\ 0 & \text{if } x = 0 \end{cases}$

   (b) $F(x) = \begin{cases} x^2/|x| & \text{if } x \ne 0 \\ 1 & \text{if } x = 0 \end{cases}$

35. (a) $g(x) = \begin{cases} x/|x| & \text{if } x \ne 0 \\ 0 & \text{if } x = 0 \end{cases}$

   (b) $G(x) = \begin{cases} x/|x| & \text{if } x \ne 0 \\ 1 & \text{if } x = 0 \end{cases}$

36. $U(x) = \begin{cases} 1 & \text{if } x \le -2 \\ -1 & \text{if } x > -2 \end{cases}$

37. $f(x) = \begin{cases} 1/x & \text{if } x < -1 \\ x & \text{if } -1 \le x \le 1 \\ 1/x & \text{if } x > 1 \end{cases}$

38. $g(x) = \begin{cases} 1/x & \text{if } x < -\frac{1}{2} \\ 1 & \text{if } -\frac{1}{2} \le x \le 1 \\ x^3 & \text{if } x > 1 \end{cases}$

B

39. Let $f(x) = 1/x$. Find a number t such that the slope of the straight line passing through the two points $(1, 1)$ and $\big(t, f(t)\big)$ on the graph of f is $-\frac{1}{5}$. Include a sketch.

40. Find the coordinates of a point P on the graph of $f(x) = \sqrt{x}$ if the slope of the straight line passing through P and the point $(1, 1)$ is $\frac{1}{7}$. Include a sketch.

41. The graph of a function f is a straight line. As the figure shows, A and B are two points on the line and the x-coordinates of A and B are h units apart. The coordinates of A are $\big(x, f(x)\big)$. What are the coordinates of B? Show that the slope of the line is

$$\frac{f(x + h) - f(x)}{h}$$

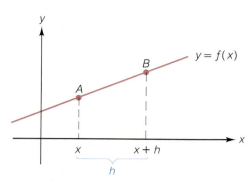

42. Let $F(x) = 2x + 3$. Using the values $x = 1$ and $h = \frac{1}{2}$, compute the value of $\dfrac{F(x + h) - F(x)}{h}$. Include a sketch of the type shown in Exercise 41.

43. The **average rate of change** of a function f between $x = a$ and $x = b$ is defined to be the quantity

$$\frac{f(b) - f(a)}{b - a}.$$

   (a) By means of a sketch, indicate why the average rate of change represents the slope of the line joining the two points on the graph of f with x-coordinates a and b.

   (b) Which is larger, the average rate of change of $f(x) = x^2$ between 0 and 1 or between 10 and 11?

   (c) Which of the following functions has the largest average rate of change between 3 and 4?

$$f(x) = x^2 \qquad g(x) = x^3 \qquad h(x) = 1/x$$

   (d) Find the average rate of change for the function $f(x) = x^2$ over the following intervals.
      (i) $a = 1$ to $b = 1.1$
      (ii) $a = 1$ to $b = 1.01$
      (iii) $a = 1$ to $b = 1.001$
      As b gets closer and closer to 1, what value does your average rate of change seem to be approaching? This "target value" is sometimes called the **instantaneous rate of change**.

44. Suppose that during the first 4 hours of a laboratory experiment, the temperature of a certain substance is given by the formula $f(t) = t^3 - 6t^2 + 9t$, where t is measured in hours with $t = 0$ corresponding to the time the experiment begins, and where $f(t)$ is the temperature (degrees Fahrenheit) of the substance at time t. Calculate the average rate of change of temperature between the following times.
   (a) $t = 0$ and $t = 1$ (b) $t = 1$ and $t = 2$
   (c) $t = 0$ and $t = 3$ (d) $t = 0$ and $t = 4$
   What are the units associated with your answers?

45. Assume that the distance covered by a falling object is given by the formula $f(t) = 16t^2$, where t is measured

in seconds, $f(t)$ is in feet, and $t = 0$ corresponds to the instant the object first begins to fall. Thus, after 1 sec, the object has fallen $16(1)^2$, or 16 ft; after 2 sec, the object has fallen a total of $16(2)^2$, or 64 ft; and so on.

(a) Calculate $f(2) - f(1)$, the distance the object has fallen during the time interval from $t = 1$ to $t = 2$ (i.e., during the second second). Include a sketch of the graph of f. On the t-axis, indicate the time interval from $t = 1$ to $t = 2$; on the y- or $f(t)$-axis, indicate the interval corresponding to the distance $f(2) - f(1)$.

(b) Calculate $f(3) - f(2)$, the distance the object has fallen during the time interval from $t = 2$ to $t = 3$ (i.e., during the third second). Include a sketch as you did in part (a).

(c) The **average velocity** during the time interval from $t = a$ to $t = b$ is defined to be the quantity $\dfrac{f(b) - f(a)}{b - a}$. Note that this is just distance divided by elapsed time and that the units are feet per second. Calculate the average velocity from $t = 1$ to $t = 3$. On a graph of f, locate the two points $\left(1, f(1)\right)$ and $\left(3, f(3)\right)$. What is the slope of the line joining these two points?

(d) Calculate the average velocity over each of the following intervals.
(i) $a = 1$ to $b = 1.1$
(ii) $a = 1$ to $b = 1.001$
(iii) $a = 1$ to $b = 1.00001$
As b comes closer and closer to 1, what value does your average velocity seem to be approaching? This "target value" is called the **instantaneous velocity** (or just **velocity**) at $t = 1$.

46. The notation $[x]$ denotes the greatest integer that is less than or equal to x. For instance, $[3] = 3, \left[4\frac{1}{2}\right] = 4$, and $\left[-\frac{4}{3}\right] = -2$.
(a) Complete the following table.

x	0	0.1	0.5	0.9	1.0
$[x]$					

(b) Graph the **greatest integer function** $y = [x]$ on the interval $-2 \le x < 3$.
(c) The domain of the greatest integer function is $(-\infty, \infty)$. What is the range?

4.3 TECHNIQUES IN GRAPHING

...geometrical figures are graphic formulas.

David Hilbert (1862–1943)

The simple geometric concepts of reflection and translation can be used to great advantage in graphing. We discussed the idea of reflection in the x-axis and in the y-axis in Section 3.2. By a **translation** of a graph, we mean a shift or movement in its location that does not alter the size or the shape of the graph.

EXAMPLE 1 The graph of a function G is the quarter-circle shown in Figure 1 (on the next page). In each case, sketch the resulting graph after the following operations are carried out on the graph of G:

(a) a translation of four units to the right followed by a translation of one unit vertically downward;

(b) a translation of three units to the right followed by a reflection in the y-axis;

(c) a reflection in the y-axis followed by a translation of three units to the right. [Note that these are the same two operations used in part (b), but here the order is reversed.]

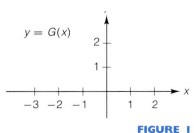

$y = G(x)$

FIGURE 1

Solution   **(a)** Figure 2 shows the result of translating the graph of G four units to the right and then one unit down. (As you can check by drawing a sketch, the same end result is obtained if we first translate one unit down and then four units to the right.   *Fact:*   If several translations are to be carried out in succession, the end result will be the same, no matter in what order the individual translations are carried out.)

FIGURE 2
The red curve is the end result when the graph of G is translated four units to the right and then one unit down.

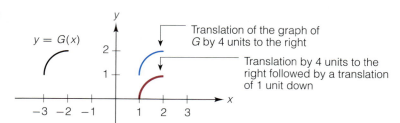

FIGURE 3
Figure (a) shows the result of translating the graph of G three units to the right. When this is followed by a reflection in the y-axis, we obtain the red curve in Figure (b).

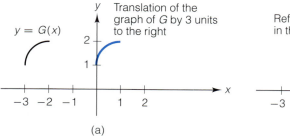
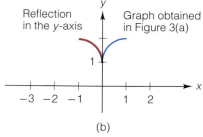

(a)   (b)

FIGURE 4
The reflection of the graph of G in the y-axis followed by a translation of three units to the right.

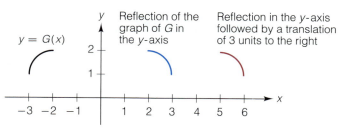

(b) Figure 3(a) shows the graph of G after a translation of three units to the right. When this is followed by a reflection in the y-axis, we obtain the result shown in Figure 3(b).

(c) Figure 4 shows what happens to the graph of G if we use the same two operations that were used in part (b), but we reserve the order in which they are carried out. Note that the end result is quite different than that obtained in part (b). ■

Given the graph of a function f, it is useful to know how to sketch efficiently the graphs of the following closely related functions. (In what follows, c denotes a positive constant.)

$$y = f(x) + c \qquad y = f(x + c) \qquad y = -f(x)$$
$$y = f(x) - c \qquad y = f(x - c) \qquad y = f(-x)$$

Each of these can be graphed by translating or reflecting the graph of $y = f(x)$. In the box that follows, we summarize the techniques that are involved. (Exercises 45–48 will help you to see why these techniques are valid.)

PROPERTY SUMMARY   TRANSLATIONS AND REFLECTIONS

(In items 1–4, the letter c denotes a positive constant.)

EQUATION	HOW TO OBTAIN THE GRAPH FROM THAT OF $y = f(x)$
1. $y = f(x) + c$	Translate c units vertically upward
2. $y = f(x) - c$	Translate c units vertically downward
3. $y = f(x + c)$	Translate c units to the left
4. $y = f(x - c)$	Translate c units to the right
5. $y = -f(x)$	Reflect in the x-axis
6. $y = f(-x)$	Reflect in the y-axis

In Figure 5 we show examples of the first four techniques in the box, the techniques involving translation. Figure 6 displays examples of the remaining two techniques, the ones involving reflection. You should look over these figures carefully before going on to Examples 2 through 5, in which the various graphing techniques are combined.

FIGURE 5

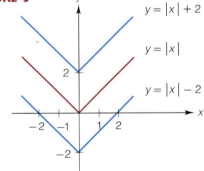

(a) To graph $y = |x| + 2$, translate $y = |x|$ up 2 units. To graph $y = |x| - 2$, translate $y = |x|$ down 2 units.

(b) To graph $y = |x + 5|$, translate $y = |x|$ to the left 5 units. To graph $y = |x - 5|$, translate $y = |x|$ to the right 5 units.

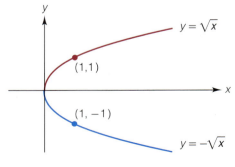

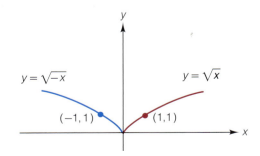

FIGURE 6

(a) To graph $y = -\sqrt{x}$, reflect $y = \sqrt{x}$ in the x-axis. More generally, to graph $y = -f(x)$, reflect the graph of $y = f(x)$ in the x-axis.

(b) To graph $y = \sqrt{-x}$, reflect $y = \sqrt{x}$ in the y-axis. More generally, to graph $y = f(-x)$, reflect the graph of $y = f(x)$ in the y-axis.

EXAMPLE 2 Graph $y = (x - 2)^2 + 1$.

Solution Begin with the graph of $y = x^2$ in Figure 7(a). Move the curve two units to the right to obtain the graph of $y = (x - 2)^2$; see Figure 7(b). Then move the curve in Figure 7(b) up one unit to obtain the graph of $y = (x - 2)^2 + 1$; see Figure 7(c).

FIGURE 7

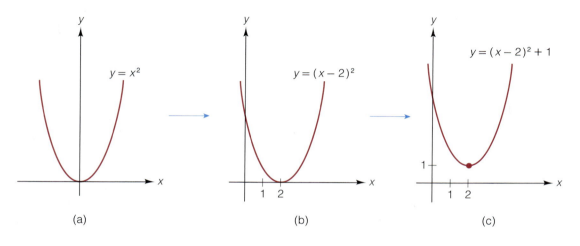

(a)   (b)   (c)

EXAMPLE 3 Graph the functions defined by $y = \dfrac{1}{x - 1}$ and $y = \dfrac{1}{x - 1} + 1$.

Solution Begin with the graph of $y = 1/x$ in Figure 8(a). As we explained in Section 3.2, the x- and y-axes are asymptotes for this graph. Moving this graph to the right one unit yields the graph of $y = \dfrac{1}{x - 1}$, shown in Figure 8(b). (Note that the vertical asymptote moves one unit to the right also, but the horizontal asymptote is unchanged.) Next, we move the graph in Figure 8(b) up one unit (why?) to obtain the graph of $y = \dfrac{1}{x - 1} + 1$, as shown in Figure 8(c). (You should verify for yourself that the y-intercepts in Figures 8(b) and 8(c) are -1 and 0 respectively.)

FIGURE 8

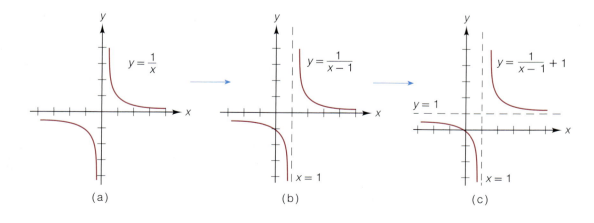

(a)   (b)   (c)

EXAMPLE 4 Graph each equation: **(a)** $y = -|x|$; **(b)** $y = -|x - 2| + 3$.

Solution See Figure 9.

FIGURE 9

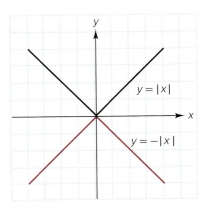

(a) Reflecting the graph of $y = |x|$ about the x-axis yields the graph of $y = -|x|$.

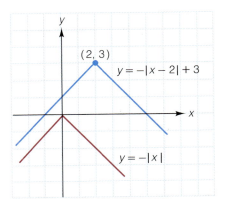

(b) Translating the graph of $y = -|x|$ two units to the right and three units up yields the graph of $y = -|x - 2| + 3$.

EXAMPLE 5 The graph of a function f is a line segment joining the points $(-3, 1)$ and $(2, 4)$. Graph each of the following functions.

(a) $y = f(-x)$ **(b)** $y = -f(x)$ **(c)** $y = -f(-x)$
(d) $y = f(x + 1)$ **(e)** $y = -f(x + 1)$ **(f)** $y = f(1 - x)$

Solution See Figure 10.

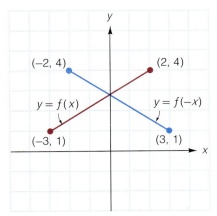

(a) Reflect the graph of $y = f(x)$ in the y-axis to obtain the graph of $y = f(-x)$.

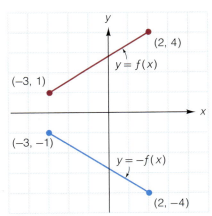

(b) Reflect the graph of $y = f(x)$ in the x-axis to obtain the graph of $y = -f(x)$.

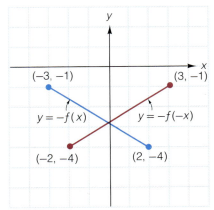

(c) Reflect the graph of $y = -f(x)$ in the y-axis to obtain the graph of $y = -f(-x)$. [Or, reflect the graph of $y = f(-x)$ in the x-axis.]

FIGURE 10

(Continued.)

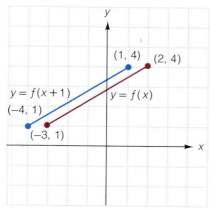

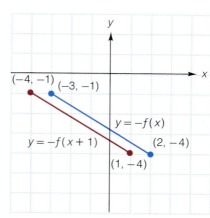

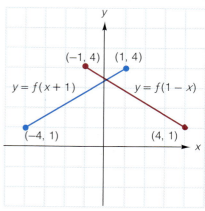

(d) Translate the graph of $y = f(x)$ one unit to the left to obtain the graph of $y = f(x + 1)$.

(e) Translate the graph of $y = -f(x)$ one unit to the left to obtain the graph of $y = -f(x + 1)$. [Or, reflect the graph of $y = f(x + 1)$ in the x-axis.]

(f) Reflect the graph of $y = f(x + 1)$ in Figure 10(d) in the y-axis to obtain the graph of $y = f(1 - x)$. [Reason: Replacing x with $-x$ in $y = f(x + 1)$ yields $y = f(1 - x)$.]

FIGURE 10 cont'd

EXERCISE SET 4.3

A

In Exercises 1 and 2, the right-hand column contains instructions for translating and/or reflecting the graph of $y = f(x)$. With each equation in the left-hand column, match an appropriate set of instructions in the right-hand column.

1. (a) $y = f(x - 1)$
    (b) $y = f(x) - 1$

    (c) $y = f(x) + 1$
    (d) $y = f(x + 1)$

    (e) $y = f(-x) + 1$

    (f) $y = f(-x) - 1$
    (g) $y = -f(x) + 1$

    (h) $y = -f(x + 1)$

    (i) $y = -f(x) - 1$
    (j) $y = f(1 - x) + 1$
       Hint: See Example 5, part (f)
    (k) $y = -f(-x) + 1$

    (A) Translate left 1 unit.
    (B) Reflect in the x-axis, then translate left 1 unit.
    (C) Translate right 1 unit.
    (D) Reflect in the x-axis, then translate up 1 unit.
    (E) Reflect in the x-axis, then translate down 1 unit.
    (F) Translate down 1 unit.
    (G) Reflect in the x-axis, reflect in the y-axis, then translate up 1 unit.
    (H) Translate left 1 unit, then reflect in the y-axis, then translate up 1 unit.
    (I) Translate up 1 unit.
    (J) Reflect in the y-axis, then translate up 1 unit.
    (K) Reflect in the y-axis, then translate down 1 unit.

2. (a) $y = f(x + 2) + 3$

    (b) $y = f(x + 3) + 2$

    (c) $y = f(x - 2) + 3$

    (d) $y = f(x - 2) - 3$

    (e) $y = f(x + 2) - 3$

    (f) $y = f(x - 3) + 2$

    (g) $y = f(x - 3) - 2$

    (h) $y = f(x + 3) - 2$

    (i) $y = -f(x + 2)$

    (j) $y = -f(x - 2)$

    (k) $y = f(2 - x)$

    (l) $y = f(-x) + 2$

    (A) Translate left 2 units, then translate down 3 units.
    (B) Translate left 3 units, then translate up 2 units.
    (C) Translate right 3 units, then translate up 2 units.
    (D) Translate left 3 units, then translate down 2 units.
    (E) Translate right 3 units and down 2 units.
    (F) Reflect in the y-axis, then translate up 2 units.
    (G) Reflect in the x-axis, then translate right 2 units.
    (H) Reflect in the x-axis, then translate left 2 units.
    (I) Translate left 2 units, then reflect in the y-axis.
    (J) Translate right 2 units, then translate up 3 units.
    (K) Translate left 2 units, then translate up 3 units.
    (L) Translate right 2 units and down 3 units.

In Exercises 3–24, sketch the graph of the function.

3. $y = x^3 - 3$

4. $y = x^2 + 3$

5. $y = (x + 4)^2$

6. $y = (x + 4)^2 - 3$

7. $y = (x - 4)^2$

8. $y = (x - 4)^2 + 1$

9. $y = -x^2$

10. $y = -x^2 - 3$

11. $y = -(x - 3)^2$

12. $y = -(x - 3)^2 - 3$

13. $y = \sqrt{x - 3}$

14. $y = \sqrt{x - 3} + 1$

15. $y = -\sqrt{x + 1}$

16. $y = -\sqrt{x + 1} + 1$

17. $y = \dfrac{1}{x + 2} + 2$

18. $y = \dfrac{1}{x - 3} - 1$

19. $y = (x - 2)^3$

20. $y = (x - 2)^3 + 1$

21. $y = -x^3 + 4$

22. $y = -(x - 1)^3 + 4$

23. (a) $y = |x + 4|$ (b) $y = |4 - x|$
     (c) $y = -|4 - x| + 1$

24. (a) $y = \sqrt{x + 2}$ (b) $y = \sqrt{2 - x}$
     (c) $y = -\sqrt{2 - x}$

In Exercises 25–40, sketch the graph of the function, given that f, F, and g are defined as follows:

$$f(x) = |x| \qquad F(x) = 1/x \qquad g(x) = \sqrt{1 - x^2}$$

25. $y = f(x - 5)$

26. $y = -f(x - 5)$

27. $y = f(5 - x)$

28. $y = -f(5 - x)$

29. $y = 1 - f(x - 5)$

30. $y = f(-x)$

31. $y = F(x + 3)$

32. $y = F(x) + 3$

33. $y = -F(x + 3)$

34. $y = F(-x) + 3$

35. $y = g(x - 2)$

36. $y = -g(x - 2)$

37. $y = 1 - g(x - 2)$

38. $y = g(-x)$

39. $y = g(2 - x)$

40. $y = -g(2 - x)$

In Exercises 41 and 42, refer to the graph of the function f shown in the figure to graph each function.

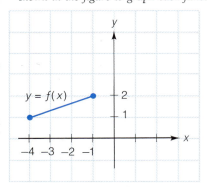

41. (a) $y = -f(x)$ (b) $y = f(-x)$
     (c) $y = -f(-x)$

42. (a) $y = f(x) - 2$ (b) $y = f(x - 2)$
     (c) $y = f(x - 2) - 2$

In Exercises 43 and 44, refer to the graph of the function g shown in the figure in order to graph each function.

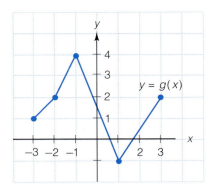

43. (a) $y = g(-x)$ (b) $y = -g(x)$
     (c) $y = -g(-x)$

44. $y = -g(x - 3) - 1$

45. (a) Complete the table.

x	x^2	$x^2 - 1$	$x^2 + 1$
0			
±1			
±2			
±3			

(b) Using the results in the table, graph the functions $y = x^2$, $y = x^2 - 1$, and $y = x^2 + 1$ on the same set of axes. How are the graphs related?

46. (a) Complete the table.

x	x^2	$(x - 1)^2$	$(x + 1)^2$
0			
1			
2			
3			
−1			
−2			
−3			

(b) Using the results in the table, graph the functions $y = x^2$, $y = (x - 1)^2$, and $y = (x + 1)^2$ on the same set of axes. How are the graphs related?

47. (a) Complete the table. (Use a calculator where necessary.)

x	$\sqrt{x}$	$-\sqrt{x}$
0		
1		
2		
3		
4		
5		

(b) Using the results in the table, graph the functions $y = \sqrt{x}$ and $y = -\sqrt{x}$ on the same set of axes. How are the two graphs related?

48. (a) Complete the tables. (Use a calculator where necessary.)

x	0	1	2	3	4	5
$\sqrt{x}$						

x	0	−1	−2	−3	−4	−5
$\sqrt{-x}$						

(b) Using the tables, graph the functions $y = \sqrt{x}$ and $y = -\sqrt{x}$ on the same set of axes. How are the graphs related?

B

For Exercises 49 and 50, assume that (a, b) is a point on the graph of $y = f(x)$, and specify the corresponding point on the graph of each equation. [For example, the point that corresponds to (a, b) on the graph of $y = f(x - 1)$ is $(a + 1, b)$.]

49. (a) $y = f(x - 3)$ (b) $y = f(x) - 3$
    (c) $y = f(x - 3) - 3$ (d) $y = -f(x)$
    (e) $y = f(-x)$ (f) $y = -f(-x)$
    (g) $y = f(3 - x)$ (h) $y = -f(3 - x) + 1$

50. (a) $y = f(-x) + 2$ (b) $y = -f(-x) + 2$
    (c) $y = -f(x - 3)$ (d) $y = 1 - f(x + 1)$
    (e) $y = f(1 - x)$ (f) $y = -f(1 - x) + 1$

51. Show that $\dfrac{x}{x - 1} = \dfrac{1}{x - 1} + 1$, provided that $x \neq 1$. Then use this fact to graph the function g defined by $g(x) = x/(x - 1)$.

52. Verify that $-x^2 + 2x + 2 = -(x - 1)^2 + 3$. Then use this fact to graph the function h defined by $h(x) = -x^2 + 2x + 2$.

53. What is the range of the function f defined by
$$f(x) = \frac{x^2}{1 - \sqrt{1 - x^2}}?$$
Hint: The function is easy to graph after rationalizing the denominator.

54. (a) A function f is said to be **even** if the equation $f(-x) = f(x)$ is satisfied by all values of x in the domain of f. Explain why the graph of an even function must be symmetric about the y-axis.
    (b) Show that each function is even by computing $f(-x)$ and then noting that $f(x)$ and $f(-x)$ are equal.
    (i) $f(x) = x^2$ (ii) $f(x) = 2x^4 - 6$
    (iii) $f(x) = 3x^6 - \dfrac{4}{x^2} + 1$

55. (a) A function f is said to be **odd** if the equation $f(-x) = -f(x)$ is satisfied by all values of x in the domain of f. Show that if (x, y) is a point on the graph of an odd function f, then the point $(-x, -y)$ is also on the graph. (This implies that the graph of an odd function must be symmetric about the origin.)
    (b) Show that each function is odd by computing $f(-x)$ as well as $-f(x)$ and then noting that the two expressions obtained are equal.
    (i) $f(x) = x^3$ (ii) $f(x) = -2x^5 + 4x^3 - x$
    (iii) $f(x) = \dfrac{|x|}{x + x^7}$

56. Is each function odd, even, or neither? (See Exercises 54 and 55 for definitions.)
    (a) $f(x) = \dfrac{1 - x^2}{2 + x^2}$ (b) $g(x) = \dfrac{x - x^3}{2x + x^3}$
    (c) $h(x) = x^2 + x$ (d) $F(x) = (x^2 + x)^2$
    (e) $G(x) = \begin{cases} 1 & \text{if } x > 0 \\ 0 & \text{if } x = 0 \\ -1 & \text{if } x < 0 \end{cases}$
    Suggestion for part (e): Look at the graph.

57. Suppose that the function f is increasing on the interval $(2, 4)$ and decreasing on the intervals $(-\infty, 2)$ and $(4, \infty)$. On what interval(s) is each function increasing?
    (a) $y = f(-x)$ (b) $y = -f(x)$

58. Suppose that the function g is decreasing on $(-\infty, 1)$ and increasing on $(1, \infty)$. On what interval(s) is the function $y = -f(-x)$ increasing?

4.4 METHODS OF COMBINING FUNCTIONS

The notation ϕx to indicate a function of x was introduced by . . . [John Bernoulli] in 1718, . . . but the general adoption of symbols like f, F, ϕ, and ψ, . . . to represent functions, seems to be mainly due to Euler and Lagrange.

W. W. Rouse Ball in *A Short Account of the History of Mathematics* (New York: Dover Publications, 1960)

Two given numbers a and b can be combined in various ways to produce a third number. For instance, we can form the sum $a + b$ or the difference $a - b$ or the product ab. Also, if $b \neq 0$, we can form the quotient a/b. Similarly, two functions can be combined in various ways to produce a third function. Suppose, for example, that we start with the two functions $y = x^2$ and $y = x^3$. It would seem natural to define their sum, difference, product, and quotient as follows:

$$\text{sum:} \quad y = x^2 + x^3$$
$$\text{difference:} \quad y = x^2 - x^3$$
$$\text{product:} \quad y = x^2 x^3 = x^5$$
$$\text{quotient:} \quad y = \frac{x^2}{x^3} = \frac{1}{x}$$

Indeed, this is the idea behind the formal definitions we now give.

DEFINITIONS **Arithmetical Operations with Functions**

Let f and g be two functions. Then the **sum** $f + g$, the **difference** $f - g$, the **product** fg, and the **quotient** f/g are functions defined by the following equations:

$$(f + g)(x) = f(x) + g(x) \tag{1}$$
$$(f - g)(x) = f(x) - g(x) \tag{2}$$
$$(fg)(x) = f(x) \cdot g(x) \tag{3}$$
$$(f/g)(x) = f(x)/g(x) \quad \text{provided } g(x) \neq 0 \tag{4}$$

For the functions defined by equations (1), (2), and (3), the domain is the set of all inputs x belonging to both the domain of f and the domain of g. For the quotient function in equation (4), we impose the additional restriction that the domain exclude all inputs x for which $g(x) = 0$.

EXAMPLE 1 Let $f(x) = 3x + 1$ and $g(x) = x - 1$. Compute each of the following functions.

(a) $(f + g)(x)$ **(b)** $(f - g)(x)$ **(c)** $(fg)(x)$ **(d)** $(f/g)(x)$

Solution
(a) $(f + g)(x) = f(x) + g(x) = (3x + 1) + (x - 1) = 4x$
(b) $(f - g)(x) = f(x) - g(x) = (3x + 1) - (x - 1) = 2x + 2$
(c) $(fg)(x) = [f(x)][g(x)] = (3x + 1)(x - 1) = 3x^2 - 2x - 1$
(d) $(f/g)(x) = \dfrac{f(x)}{g(x)} = \dfrac{3x + 1}{x - 1} \quad (x \neq 1)$

For the remainder of this section, we are going to discuss a method for combining functions known as **composition of functions**. As you will see, this method is based on the familiar algebraic process of substitution. Suppose, for example, that f and g are two functions defined by

$$f(x) = x^2 \qquad g(x) = 3x + 1$$

Choose any number in the domain of g, say $x = -2$. We can compute $g(-2)$:

$$g(-2) = 3(-2) + 1 = -5$$

Now let's use the output -5 that g has produced as an *input* for f. We obtain

$$f(-5) = (-5)^2 = 25$$

and, consequently,

$$f[g(-2)] = 25$$

So, beginning with the input -2, we've successively applied g and then f to obtain the output 25. Similarly, we could carry out this same procedure for any other number in the domain of g. Here is a summary of the procedure:

1. Start with an input x and calculate $g(x)$.
2. Use $g(x)$ as an input for f; that is, calculate $f[g(x)]$.

We use the notation $f \circ g$ to denote the function, or rule, that tells us to assign the output $f[g(x)]$ to the initial input x. In other words, $f \circ g$ denotes the rule consisting of two steps: *First apply g; then apply f.* We read the notation $f \circ g$ as f *circle g* or f *composed with g.* In Figure 1, we summarize these ideas.

FIGURE I
Diagram for the function $f \circ g$

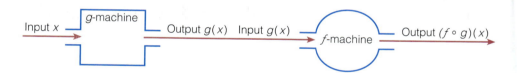

When we write $g(x)$, we assume that x is in the domain of the function g. Likewise, for the notation $f[g(x)]$ to make sense, the outputs $g(x)$ must themselves be acceptable inputs for the function f. Our formal definition, then, for the composite function $f \circ g$ is as follows.

DEFINITION Composition of Functions: $f \circ g$

Given two functions f and g, the function $f \circ g$ is defined by

$$(f \circ g)(x) = f[g(x)]$$

The domain of $f \circ g$ consists of those inputs x (in the domain of g) for which $g(x)$ is in the domain of f.

EXAMPLE 2 Let $f(x) = x^2$ and $g(x) = 3x + 1$. Compute $(f \circ g)(x)$ and $(g \circ f)(x)$.

Solution
$$
\begin{aligned}
(f \circ g)(x) &= f[g(x)] && \text{definition of } f \circ g \\
&= f(3x + 1) && \text{definition of } g \\
&= (3x + 1)^2 && \text{definition of } f \\
&= 9x^2 + 6x + 1 \\
(g \circ f)(x) &= g[f(x)] && \text{definition of } g \circ f \\
&= g(x^2) && \text{definition of } f \\
&= 3(x^2) + 1 && \text{definition of } g \\
&= 3x^2 + 1
\end{aligned}
$$

Notice that the two results obtained in Example 2 are not the same. This shows that, in general, $f \circ g$ and $g \circ f$ represent different functions.

EXAMPLE 3 Let f and g be defined as in Example 2: $f(x) = x^2$ and $g(x) = 3x + 1$. Compute $(f \circ g)(-2)$.

Solution We will show two methods.

FIRST METHOD Using the formula for $(f \circ g)(x)$ developed in Example 2, we have

$$(f \circ g)(x) = 9x^2 + 6x + 1$$

and therefore,

$$
\begin{aligned}
(f \circ g)(-2) &= 9(-2)^2 + 6(-2) + 1 \\
&= 36 - 12 + 1 \\
&= 25
\end{aligned}
$$

ALTERNATIVE METHOD Working directly from the definition of $f \circ g$, we have

$$
\begin{aligned}
(f \circ g)(-2) &= f[g(-2)] \\
&= f[3(-2) + 1] \\
&= f(-5) \\
&= (-5)^2 = 25
\end{aligned}
$$

In Examples 2 and 3, the domain of both f and g is the set of all real numbers. And as you can easily check, the domain of both $f \circ g$ and $g \circ f$ is also the set of all real numbers. In Example 4, however, some care needs to be taken in describing the domain of the composite function.

EXAMPLE 4 Let f and g be defined as follows:

$$f(x) = x^2 + 1 \qquad g(x) = \sqrt{x}$$

Compute $(f \circ g)(x)$. Find the domain of $f \circ g$ and sketch its graph.

Solution
$$
\begin{aligned}
(f \circ g)(x) &= f[g(x)] \\
&= f(\sqrt{x}) = (\sqrt{x})^2 + 1 = x + 1
\end{aligned}
$$

So we have $(f \circ g)(x) = x + 1$. Now what about the domain of $f \circ g$? Our first inclination may be to say (incorrectly!) that the domain is the set of all real numbers, since any real number can be used as an input in the expression $x + 1$. However, the definition of $f \circ g$ on page 204 tells us that the inputs for $f \circ g$ must first of all be acceptable inputs for g. Given the definition of g, then, we must require that x be nonnegative. On the other hand, for any nonnegative input x, the number $g(x)$ will be an acceptable input for f. (Why?) In summary,

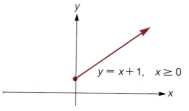

FIGURE 2
Graph of the function $f \circ g$ in Example 4

then, the domain of $f \circ g$ is the interval $[0, \infty)$. The graph of $f \circ g$ is shown in Figure 2.

One reason for studying the composition of functions is that it lets us express a given function in terms of simpler functions. This is often useful in calculus. Suppose, for example, that we wish to express the function C defined by

$$C(x) = (2x^3 - 5)^2$$

as a composition of simpler functions. That is, we want to come up with two functions f and g so that the equation

$$C(x) = (f \circ g)(x)$$

holds for every x in the domain of C.

We begin by thinking what we would do to compute $(2x^3 - 5)^2$ for a given value of x. First, we would compute the quantity $2x^3 - 5$, then we would square the result. Therefore, recalling that the rule $f \circ g$ tells us to do g first, we let $g(x) = 2x^3 - 5$. And then, since the next step is squaring, we let $f(x) = x^2$. Now let us see if these choices for f and g are correct; that is, let us calculate $(f \circ g)(x)$ and see if it really is the same as $C(x)$.

Using $f(x) = x^2$ and $g(x) = 2x^3 - 5$, we have

$$(f \circ g)(x) = f[g(x)] = f(2x^3 - 5)$$
$$= (2x^3 - 5)^2 = C(x)$$

This shows that our choices for f and g were indeed correct, and we have expressed C as a composition of two simpler functions.*

Note that in expressing C as $f \circ g$, we chose g to be the "inner" function, that is, the quantity inside parentheses: $g(x) = 2x^3 - 5$. This observation is used in Example 5.

EXAMPLE 5   Let $s(x) = \sqrt{1 + x^4}$. Express the function s as a composition of two simpler functions f and g.

Solution   Let g be the "inner" function; that is, let $g(x)$ be the quantity inside the radical:

$$g(x) = 1 + x^4$$

And let us take f to be the square root function:

$$f(x) = \sqrt{x}$$

Now we need to verify that these are appropriate choices for f and g; that is, we need to check that the equation $(f \circ g)(x) = s(x)$ is true for every x in the domain of s. We have

$$(f \circ g)(x) = f[g(x)] = f(1 + x^4)$$
$$= \sqrt{1 + x^4} = s(x)$$

Thus, $(f \circ g)(x) = s(x)$, as required.

---

*Other answers are possible, too. For instance, if $F(x) = (x - 5)^2$ and $G(x) = 2x^3$, then (as you should verify for yourself) $C(x) = F[G(x)]$.

EXERCISE SET 4.4

A

In Exercises 1–10, compute each expression, given that the functions f, g, h, k, and m are defined as follows:

$$f(x) = 2x - 1 \qquad k(x) = 2, \quad \text{for all } x$$
$$g(x) = x^2 - 3x - 6 \qquad m(x) = x^2 - 9$$
$$h(x) = x^3$$

1. (a) $(f + g)(x)$ (b) $(f - g)(x)$ (c) $(f - g)(0)$
2. (a) $(fh)(x)$ (b) $(h/f)(x)$ (c) $(f/h)(1)$
3. (a) $(m - f)(x)$ (b) $(f - m)(x)$
4. (a) $(fg)(x)$ (b) $(fg)\left(\frac{1}{2}\right)$
5. (a) $(fk)(x)$ (b) $(kf)(x)$ (c) $(fk)(1) - (kf)(2)$
6. (a) $(g + m)(x)$ (b) $(g + m)(x) - (g - m)(x)$
7. (a) $(f/m)(x) - (m/f)(x)$ (b) $(f/m)(0) - (m/f)(0)$
8. (a) $[h \cdot (f + m)](x)$ *Note:*   h and $(f + m)$ are two functions; the notation $h \cdot (f + m)$ denotes the product function.
   (b) $(hf)(x) + (hm)(x)$
9. (a) $[m \cdot (k - h)](x)$ (b) $(mk)(x) - (mh)(x)$
   (c) $(mk)(-1) - (mh)(-1)$
10. (a) $(g + g)(x)$ (b) $(g - g)(x)$
    (c) $(kg)(x)$ (d) $(g + g)(-3) - (kg)(-3)$
11. Let $f(x) = 3x + 1$ and $g(x) = -2x - 5$. Compute the following.
    (a) $(f \circ g)(x)$ (b) $(f \circ g)(10)$
    (c) $(g \circ f)(x)$ (d) $(g \circ f)(10)$
12. Let $f(x) = 1 - 2x^2$ and $g(x) = x + 1$. Compute the following.
    (a) $(f \circ g)(x)$ (b) $(f \circ g)(-1)$
    (c) $(g \circ f)(x)$ (d) $(g \circ f)(-1)$
    (e) $(f \circ f)(x)$ (f) $(g \circ g)(-1)$
13. Compute $(f \circ g)(x)$, $(f \circ g)(-2)$, $(g \circ f)(x)$, and $(g \circ f)(-2)$ for each pair of functions.
    (a) $f(x) = 1 - x$; $g(x) = 1 + x$
    (b) $f(x) = x^2 - 3x - 4$; $g(x) = 2 - 3x$
    (c) $f(x) = x/3$; $g(x) = 1 - x^4$
    (d) $f(x) = 2^x$; $g(x) = x^2 + 1$
    (e) $f(x) = x$; $g(x) = 3x^5 - 4x^2$
    (f) $f(x) = 3x - 4$; $g(x) = \frac{1}{3}(x + 4)$
14. Let $h(x) = 4x^2 - 5x + 1$, $k(x) = x$, and $m(x) = 7$ for all x. Compute the following.
    (a) $h[k(x)]$ (b) $k[h(x)]$ (c) $h[m(x)]$
    (d) $m[h(x)]$ (e) $k[m(x)]$ (f) $m[k(x)]$
15. Let $F(x) = \frac{3x - 4}{3x + 3}$ and $G(x) = \frac{x + 1}{x - 1}$. Compute the following.
    (a) $(F \circ G)(x)$ (b) $F[G(t)]$ (c) $(F \circ G)(2)$
    (d) $(G \circ F)(x)$ (e) $G[F(y)]$ (f) $(G \circ F)(2)$

16. Let $f(x) = \frac{1}{x^2} + 1$ and $g(x) = \frac{1}{x - 1}$.
    (a) Compute $(f \circ g)(x)$.
    (b) What is the domain of $f \circ g$?
    (c) Graph the function $f \circ g$.
17. Let $M(x) = \frac{2x - 1}{x - 2}$.
    (a) Compute $M(7)$ and then $M[M(7)]$.
    (b) Compute $(M \circ M)(x)$.
    (c) Compute $(M \circ M)(7)$, using the formula you obtained in part (b). Check that your answer agrees with that obtained in part (a).
18. Let $F(x) = (x + 1)^5$, $f(x) = x^5$, and $g(x) = x + 1$. Which of the following is true for all x?
    $$(f \circ g)(x) = F(x) \qquad \text{or} \qquad (g \circ f)(x) = F(x)$$
19. Refer to the graphs of the functions f, g, and h to compute the required quantities. Assume that all the axes are marked off in one-unit intervals.
    (a) $f[g(3)]$ (b) $g[f(3)]$ (c) $f[h(3)]$
    (d) $(h \circ g)(2)$ (e) $h\{f[g(3)]\}$
    (f) $(g \circ f \circ h \circ f)(2)$ *Note:*   This means first do f, then h, then f, then g.

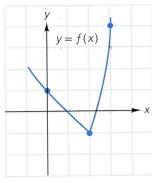

(a)

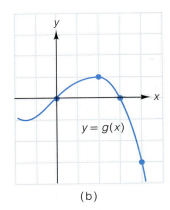

(b)

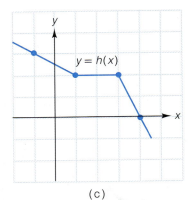

(c)

20. Let $f(x) = 2x - 1$. Find $f[f(x)]$. If $z = f[f(x)]$, find $f(z)$.

21. (a) Let $T(x) = 4x^3 - 3x^2 + 6x - 1$ and $I(x) = x$. Find $(T \circ I)(x)$ and $(I \circ T)(x)$.
    (b) Let $G(x) = ax^2 + bx + c$ and $I(x) = x$. Find $(G \circ I)(x)$ and $(I \circ G)(x)$.
    (c) What general conclusion do you arrive at from the results of parts (a) and (b)?

22. The figure shows the graphs of two functions, F and G. Use the graphs to compute each quantity.
    (a) $(G \circ F)(1)$
    (b) $(F \circ G)(1)$
    (c) $(F \circ F)(1)$
    (d) $(G \circ G)(-3)$
    (e) $(G \circ F)(5)$
    (f) $(F \circ F)(5)$
    (g) $(F \circ G)(-1)$
    (h) $(G \circ F)(2)$

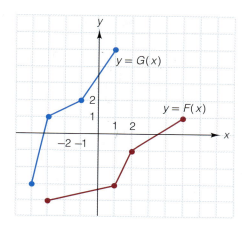

23. The domain of a function f consists of the numbers $-1, 0, 1, 2,$ and 3. The following table shows the output f assigns to each input.

x	-1	0	1	2	3
$f(x)$	2	2	0	3	1

The domain of a function g consists of the numbers $0, 1, 2, 3,$ and 4. The following table shows the output g assigns to each input.

x	0	1	2	3	4
$g(x)$	3	2	0	4	-1

Use this information to complete the tables for $f \circ g$ and $g \circ f$. *Note:* Two of the entries will be undefined.

x	0	1	2	3	4
$(f \circ g)(x)$					

x	-1	0	1	2	3	4
$(g \circ f)(x)$						

24. The following two tables show certain pairs of inputs and outputs for functions f and g.

x	0	$\pi/6$	$\pi/4$	$\pi/3$	$\pi/2$
$f(x)$	0	$1/2$	$\sqrt{2}/2$	$\sqrt{3}/2$	1

y	0	$1/4$	$\sqrt{2}/4$	$1/2$	$\sqrt{2}/2$	$3/4$	$\sqrt{3}/2$	1
$g(y)$	$\pi/2$	π	0	$\pi/3$	$\pi/4$	0	$\pi/6$	0

Use this information to complete the following table of values for $(g \circ f)(x)$.

x	0	$\pi/6$	$\pi/4$	$\pi/3$	$\pi/2$
$(g \circ f)(x)$					

25. Let $f(x) = 2x + 1$ and $g(x) = 3x - 4$.
    (a) Find $(f \circ g)(x)$ and graph the function $f \circ g$.
    (b) Find $(g \circ f)(x)$ and graph the function $g \circ f$.

26. Let $F(x) = x^2$ and $G(x) = x - 4$.
    (a) Find $F[G(x)]$ and graph the function $F \circ G$.
    (b) Find $G[F(x)]$ and graph the function $G \circ F$.

27. Let $g(x) = \sqrt{x} - 3$ and $f(x) = x - 1$.
    (a) Sketch a graph of g. Specify the domain and range.
    (b) Sketch a graph of f. Specify the domain and range.
    (c) Compute $(f \circ g)(x)$. Graph the function $f \circ g$ and specify its domain and range.
    (d) Find a formula for $g[f(x)]$. Which values of x are acceptable inputs here? That is, what is the domain of $g \circ f$?
    (e) Use the results of part (d) to sketch a graph of the function $g \circ f$.

28. Let $F(x) = -x^2$ and $G(x) = \sqrt{x}$. Determine the domains of $F \circ G$ and $G \circ F$.

29. Let $C(x) = (3x - 1)^4$. Express C as a composition of two simpler functions.

30. Let $C(x) = (1 + x^2)^3$. Find functions f and g so that $C(x) = (f \circ g)(x)$ is true for all values of x.

31. Express each function as a composition of two functions.
    (a) $F(x) = \sqrt[3]{3x + 4}$
    (b) $G(x) = |2x - 3|$
    (c) $H(x) = (ax + b)^5$
    (d) $T(x) = 1/\sqrt{x}$

32. Let $a(x) = x^2$, $b(x) = |x|$, $c(x) = 3x - 1$. Express each function as a composition of two of the given functions.
    (a) $f(x) = (3x - 1)^2$
    (b) $g(x) = |3x - 1|$
    (c) $h(x) = 3x^2 - 1$

33. Let $a(x) = 1/x$, $b(x) = \sqrt[3]{x}$, $c(x) = 2x + 1$, and $d(x) = x^2$. Express each function as a composition of two of the given functions.
    (a) $f(x) = \sqrt[3]{2x + 1}$
    (b) $g(x) = 1/x^2$
    (c) $h(x) = 2x^2 + 1$
    (d) $K(x) = 2\sqrt[3]{x} + 1$
    (e) $l(x) = \dfrac{2}{x} + 1$
    (f) $m(x) = \dfrac{1}{2x + 1}$
    (g) $n(x) = x^{2/3}$

34. Express $G(x) = 1/(1 + x^4)$ as a composition of two simpler functions, one of which is $f(x) = 1/x$.

B

35. The circumference of a circle with radius r is given by $C(r) = 2\pi r$. Suppose that the circle is shrinking in size and that the radius at time t is given by

$$r = f(t) = \frac{1}{t^2 + 1}$$

    Assume that t is measured in minutes and that r, or $f(t)$, is in feet. Find $(C \circ f)(t)$ and use this to find the circumference when $t = 3$ min.

36. A spherical weather balloon is being inflated in such a way that the radius is given by

$$r = g(t) = \tfrac{1}{2}t + 2$$

    Assume that r is in meters and t is in seconds, with $t = 0$ corresponding to the time that inflation begins. If the volume of a sphere of radius r is given by

$$V(r) = \tfrac{4}{3}\pi r^3$$

    compute $V[g(t)]$ and use this to find the time at which the volume of the balloon is 36π m^3.

37. Suppose that a manufacturer knows that the daily production cost to build x bicycles is given by the function C, where

$$C(x) = 100 + 90x - x^2 \qquad (0 \le x \le 40)$$

That is, $C(x)$ represents the cost in dollars of building x bicycles. Furthermore, suppose that the number of bicycles that can be built in t hr is given by the function f, where

$$x = f(t) = 5t \qquad (0 \le t \le 8)$$

(a) Compute $(C \circ f)(t)$.
(b) Compute the production cost on a day that the factory operates for $t = 3$ hr.
(c) If the factory runs for 6 hr instead of 3 hr, is the cost twice as much?

38. Suppose that in a certain biology lab experiment, the number of bacteria is related to the temperature of the environment by the function

$$N(T) = -2T^2 + 240T - 5400 \qquad (40 \le T \le 90)$$

Here, $N(T)$ represents the number of bacteria present when the temperature is T degrees Fahrenheit. Also, suppose that t hr after the experiment begins, the temperature is given by

$$T(t) = 10t + 40 \qquad (0 \le t \le 5)$$

(a) Compute $N[T(t)]$.
(b) How many bacteria are present when $t = 0$? When $t = 2$ hr? When $t = 5$ hr?

39. Let $g(x) = 4x - 1$. Find $f(x)$, given that the equation $(g \circ f)(x) = x + 5$ is true for all values of x.

40. Let $g(x) = 2x + 1$. Find $f(x)$, given that $(g \circ f)(x) = 10x - 7$.

41. Let $f(x) = -2x + 1$ and $g(x) = ax + b$. Find a and b so the equation $f[g(x)] = x$ holds for all values of x.

42. Let $f(x) = \dfrac{3x - 4}{x - 3}$.
    (a) Compute $(f \circ f)(x)$.
    (b) Find $f\left[f\left(\frac{113}{355}\right)\right]$. (Try not to do it the hard way.)

43. Let $f(x) = x^2$ and $g(x) = 2x - 1$.
    (a) Compute $\dfrac{f[g(x)] - f[g(a)]}{g(x) - g(a)}$.
    (b) Compute $\dfrac{f[g(x)] - f[g(a)]}{x - a}$.

44. Suppose that $y = f(x)$ and $g(y) = x$. Show that $(g \circ f)(x) - (f \circ g)(y) = x - y$.

45. Given three functions f, g, and h, we define the function $f \circ g \circ h$ by the formula

$$(f \circ g \circ h)(x) = f\{g[h(x)]\}$$

For example, if $f(x) = x^2$, $g(x) = x + 1$, and $h(x) = x/2$, we evaluate $(f \circ g \circ h)(x)$ as

$$(f \circ g \circ h)(x) = f\{g[h(x)]\}$$

$$= f\left[g\left(\frac{x}{2}\right)\right] = f\left(\frac{x}{2} + 1\right)$$

$$= \left(\frac{x}{2} + 1\right)^2 = \frac{x^2}{4} + x + 1$$

Using $f(x) = x^2$, $g(x) = x + 1$, and $h(x) = x/2$, compute each composition.

(a) $(g \circ h \circ f)(x)$ (b) $(h \circ f \circ g)(x)$
(c) $(g \circ f \circ h)(x)$ (d) $(f \circ h \circ g)(x)$
(e) $(h \circ g \circ f)(x)$

     Hint for checking: No two answers are the same.

46. Let $F(x) = 2x - 1$, $G(x) = 4x$, and $H(x) = 1 + x^2$. Compute each composition.
(a) $(F \circ G \circ H)(x)$ (b) $(G \circ F \circ H)(x)$
(c) $(F \circ H \circ G)(x)$ (d) $(H \circ F \circ G)(x)$
(e) $(H \circ G \circ F)(x)$ (f) $(G \circ H \circ F)(x)$

47. Let $f(x) = x^2$, $g(x) = 1 - x$, and $h(x) = 3x$. Express each function as a composition of f, g, and h.
(a) $p(x) = 1 - 9x^2$ (b) $q(x) = 3 - 3x^2$
(c) $r(x) = 1 - 6x + 9x^2$ (d) $s(x) = 3 - 6x + 3x^2$

48. Suppose that three functions f', g, and g' are defined as follows (f' is read f *prime*):

$$f'(x) = 3x^2 \qquad g(x) = 2x^2 + 5 \qquad g'(x) = 4x$$

(a) Find $f'[g(x)] \cdot g'(x)$. (The dot denotes multiplication, not composition.)
(b) Evaluate $f'[g(-1)] \cdot g'(-1)$.
(c) Find $f'[g(t)] \cdot g'(t)$.

49. Suppose that three functions F', G, and G' are defined by

$$F'(x) = \frac{1}{2\sqrt{x}} \qquad G(x) = x^2 + 2x + 2 \qquad G'(x) = 2x + 2$$

(a) Find $F'[G(x)] \cdot G'(x)$. (The dot denotes multiplication.)
(b) Find $F'[G(9)] \cdot G'(9)$.

C

50. Let p and q be two numbers whose sum is 1. Define the function F by

$$F(x) = p - \frac{1}{x + q}$$

Show that $F\{F[F(x)]\} = x$.

51. Let $i(x) = x$, $a(x) = -x$, $b(x) = 1/x$, and $c(x) = -1/x$. Assume that the domain of each function is $(0, \infty)$.
(a) Complete the following table, where the operation is composition of functions. For example, c is the proper entry in the second row, third column, since if you compute $(a \circ b)(x)$, you'll get $-1/x$, which is $c(x)$.

$\circ$	i	a	b	c
i				
a			c	
b				
c				

(b) According to your table, does $a \circ b = b \circ a$?
(c) Use your table to find a^2 (meaning $a \circ a$), b^2, c^2, and c^3.
(d) Use your table to find out if $(a \circ b) \circ c = a \circ (b \circ c)$.
(e) Let G denote the set consisting of the four functions you've worked with in this exercise. Now define a function f, with domain G, as follows: $f(t) = t^2$ for each element t in the set G. [For example, $f(a) = a^2 = a \circ a = i$.] What is the range of f?
(f) Verify that $f(a \circ b) = [f(a)] \circ [f(b)]$
(g) Find out if $f(a \circ b \circ c) = [f(a)] \circ [f(b)] \circ [f(c)]$.

ZOLUTION
PROBLEM

PROBLEM
SOLUTION

4.5   INVERSE FUNCTIONS

If anybody ever told me why the graph of $y = x^{1/2}$ is the reflection of the graph of $y = x^2$ in a 45° line, it didn't sink in. To this day, there are textbooks that expect students to think that it is so obvious as to need no explanation. This is a pity, if only because [in calculus] it is such a common practice to define the natural logarithm first and then define the exponential function as its inverse.

Professor Ralph P. Boas, recalling his student days in the article "Inverse Functions" in *The College Mathematics Journal* 16(1985): 42.

We shall introduce the idea of inverse functions through an easy example. Consider the function f defined by

$$f(x) = 2x$$

A short table of values for f and the graph of f are displayed in Figure 1.

Now we ask this question: What happens if we take the table for f and interchange the entries in the x and y columns to obtain a new table that looks like the following?

x	y
−4	−2
−2	−1
0	0
2	1
4	2

← New table formed by interchanging the inputs and outputs for f

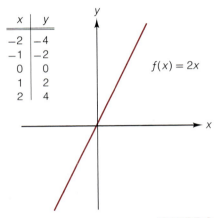

x	y
−2	−4
−1	−2
0	0
1	2
2	4

$f(x) = 2x$

FIGURE I

Several observations are in order. Our new table does itself represent a function (in this case); for each input x there is exactly one output y. Let us call this new function g. Note that each output is half the corresponding input. So the function g can be described by the formula $g(x) = x/2$. In summary, we began with the "doubling function," $f(x) = 2x$; then, by interchanging the inputs with the outputs, we obtained the "halving function," $g(x) = x/2$.

The doubling function f and the halving function g are in a certain sense opposites; each reverses, or undoes, the effect of the other. That is, if you begin with x and then calculate $g[f(x)]$, you get x again; similarly, $f[g(x)]$ is also equal to x. (Verify these last two statements for yourself.)

The two functions f and g that we've been discussing are an example of a pair of *inverse functions*. The general definition for inverse functions is given in the box that follows.

DEFINITION   Inverse Functions

Two functions f and g are **inverses** of one another provided that

$$f[g(x)] = x \qquad \text{for each } x \text{ in the domain of } g$$

and

$$g[f(x)] = x \qquad \text{for each } x \text{ in the domain of } f$$

EXAMPLE I   Verify that the functions f and g are inverses, where

$$f(x) = \frac{1}{3}x + 2 \qquad \text{and} \qquad g(x) = 3x - 6$$

Solution   In view of the definition, we must check that $f[g(x)] = x$ and $g[f(x)] = x$. We have

$$f[g(x)] = f(3x - 6)$$
$$= \tfrac{1}{3}(3x - 6) + 2 = x \qquad \text{(Check the algebra.)}$$

Thus, $f[g(x)] = x$. Now we still need to check that $g[f(x)] = x$. We have

$$g[f(x)] = g\left(\tfrac{1}{3}x + 2\right)$$
$$= 3\left(\tfrac{1}{3}x + 2\right) - 6 = x \qquad \text{(Again, check the algebra.)}$$

Having shown that $f[g(x)] = x$ and $g[f(x)] = x$, we conclude that f and g are indeed inverse functions.   ▪▪▪

EXAMPLE 2   Suppose that f and g are a pair of inverse functions. If $f(2) = 3$, what is $g(3)$?

Solution   We can use either of two methods.

FIRST METHOD   The quantities 3 and $f(2)$ are declared to be equal. Thus, whether we use 3 or $f(2)$ as an input for g, the result must be the same. We then have

$$3 = f(2)$$
$$g(3) = g[f(2)]$$
$$g(3) = 2 \qquad \text{using the fact that } g[f(x)] = x \text{ for any } x \text{ in the domain of } f$$

Thus $g(3) = 2$.

ALTERNATIVE METHOD   Think of a table of values for the function f. Since $f(2) = 3$, one entry in the table must look like this:

x		2	
y		3	

Now, g reverses the roles of x and y, so in the table for g, one entry must look like this:

x		3	
y		2	

Thus, $g(3) = 2$, as obtained previously. In Figure 2 (on the next page), we show a graphic interpretation of this method.   ▪▪▪

It is customary to use the notation f^{-1} (read f *inverse*) for the function that is the inverse of f. So in Example 1, for instance, we have

$$f(x) = \frac{1}{3}x + 2 \qquad \text{and} \qquad f^{-1}(x) = 3x - 6$$

FIGURE 2

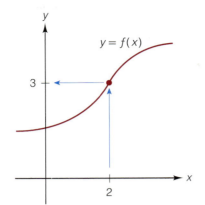

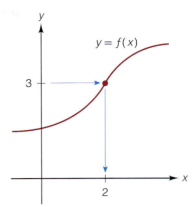

(a) The function f assigns the output 3 to the input 2. (Note the direction of the arrows.)

(b) The inverse of the function f assigns the output 2 to the input 3. (Note the direction of the arrows.)

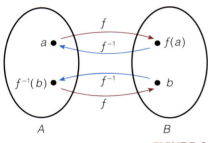

FIGURE 3

The action of inverse functions: The set A in the figure is both the domain of f and the range of f^{-1}. The set B is both the domain of f^{-1} and the range of f.

Note: In the context of functions, f^{-1} does *not* in general mean $1/f$. (For an example of this, see Exercise 4 at the end of this section.) In the following box, we rewrite the defining equations for inverse functions, using the f^{-1} notation rather than g. Figure 3 provides a graphic summary of these equations.

$$f[f^{-1}(x)] = x \qquad \text{for each } x \text{ in the domain of } f^{-1}$$

and

$$f^{-1}[f(x)] = x \qquad \text{for each } x \text{ in the domain of } f$$

EXAMPLE 3 Suppose $f(x) = 4x^3 + 7$. Find $f[f^{-1}(5)]$. (Assume that f^{-1} exists and that 5 is in its domain.)

Solution By definition, $f[f^{-1}(x)] = x$. Substituting $x = 5$ on both sides of this identity directly gives us

$$f[f^{-1}(5)] = 5 \qquad \text{as required}$$

Note that we didn't need to make use of the formula $f(x) = 4x^3 + 7$, nor did we need to find a formula for $f^{-1}(x)$. ∎

EXAMPLE 4 Solve the following equation for x, given that the domain of both f and f^{-1} is $(-\infty, \infty)$ and that $f(1) = -2$.

$$3 + f^{-1}(x - 1) = 4$$

Solution
$$3 + f^{-1}(x - 1) = 4$$
$$f^{-1}(x - 1) = 1$$
$$f[f^{-1}(x - 1)] = f(1) \qquad \text{applying } f \text{ to both sides}$$
$$x - 1 = f(1) \qquad \text{definition of inverse function}$$
$$x = 1 + f(1) = 1 + (-2) = -1 \qquad$$ ∎

We'll postpone until the end of this section a discussion of which functions have inverses and which do not. For functions that do have inverses, however, there's a simple method that can often be used to determine those inverses. This method is illustrated in Examples 5 and 6.

EXAMPLE 5   Let $f(x) = 3x - 4$. Find $f^{-1}(x)$.

Solution   We begin by rewriting the given equation as

$$y = 3x - 4$$

We know that f^{-1} interchanges the inputs and outputs of f. So to determine f^{-1}, we only need to switch the x's and y's in the equation $y = 3x - 4$. That gives us $x = 3y - 4$. Now we solve for y as follows:

$$x = 3y - 4$$
$$x + 4 = 3y$$
$$\frac{x + 4}{3} = y$$

Thus, the inverse function is $y = \dfrac{x + 4}{3}$. We can also write this as

$$f^{-1}(x) = \frac{x + 4}{3}$$

Actually, we should call this result just a *candidate* for f^{-1}, since certain technical matters remain to be discussed. However, if you compute $f[f^{-1}(x)]$ and $f^{-1}[f(x)]$ and find that they both do equal x, then the matter is settled, by definition. Exercise 13 at the end of this section asks you to carry out those calculations.

Here is a summary of our procedure for calculating $f^{-1}(x)$.

To Find $f^{-1}(x)$ for the Function $y = f(x)$

1. Interchange x and y in the equation $y = f(x)$.

2. Solve the resulting equation for y.

EXAMPLE 6   Let $f(x) = \dfrac{2x - 1}{3x + 5}$. Find $f^{-1}(x)$.

Solution   STEP 1   Write the given function as $y = \dfrac{2x - 1}{3x + 5}$. Interchange x and y to get

$$x = \frac{2y - 1}{3y + 5}$$

STEP 2 Solve the resulting equation for y.

$$x = \frac{2y - 1}{3y + 5}$$

$3xy + 5x = 2y - 1 \qquad$ multiplying by $3y + 5$

$3xy - 2y = -5x - 1 \qquad$ rearranging

$y(3x - 2) = -5x - 1$

$$y = \frac{-5x - 1}{3x - 2}$$

Thus, the inverse function is given by

$$f^{-1}(x) = \frac{-5x - 1}{3x - 2}$$

A certain symmetry always exists between the graph of a function and the graph of its inverse. As background for this discussion, we need the following definition of *symmetry about a line*.

DEFINITION   Symmetry About a Line

Refer to Figure 4. Two points P and Q are **symmetric about the line** $\mathscr{L}$ provided that

$\overline{PQ}$ is perpendicular to $\mathscr{L}$

and

points P and Q are equidistant from $\mathscr{L}$

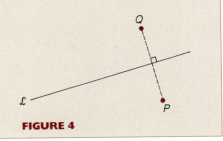

FIGURE 4

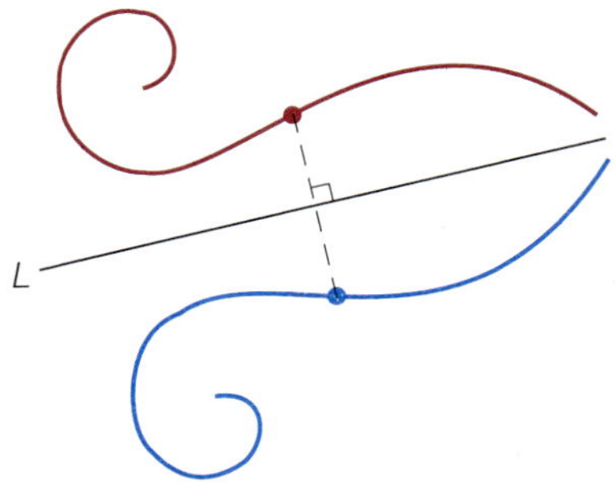

In other words, P and Q are symmetric about the line $\mathscr{L}$ if $\mathscr{L}$ is the perpendicular bisector of line segment $\overline{PQ}$. We say that $\mathscr{L}$ is the **axis of symmetry** and that P and Q are mirror images, or **reflections**, of each other through $\mathscr{L}$. (Actually, you already encountered two instances of this type of symmetry in Section 3.2, where we discussed symmetry about the x- and y-axes.) In addition, we say that two curves are symmetric about a line if each point on one curve is the reflection of a corresponding point on the other curve, and vice versa (see Figure 5).

FIGURE 5

EXAMPLE 7   Verify that the points $P(4, 1)$ and $Q(1, 4)$ are symmetric about the line $y = x$. See Figure 6 (on the next page).

Solution   We have to show that the line $y = x$ is the perpendicular bisector of the line segment $\overline{PQ}$. Now, the slope of $\overline{PQ}$ is

$$m = \frac{4 - 1}{1 - 4} = -1$$

We know that the slope of the line $y = x$ is 1. Since these two slopes are negative reciprocals, we conclude that the line $y = x$ is perpendicular to $\overline{PQ}$. Next,

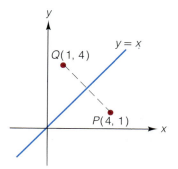

FIGURE 6

we must show that the line $y = x$ passes through the midpoint of $\overline{PQ}$. But (as you should check for yourself), the midpoint of $\overline{PQ}$ is $\left(\frac{5}{2}, \frac{5}{2}\right)$, which does lie on the line $y = x$. In summary, then, we've shown that the line $y = x$ passes through the midpoint of $\overline{PQ}$ and is perpendicular to $\overline{PQ}$. Thus P and Q are symmetric about the line $y = x$, as we wished to show. ◼◼◼

Just as we've shown that $(4, 1)$ and $(1, 4)$ are symmetric about the line $y = x$, we can also show in general that (a, b) and (b, a) are always symmetric about the line $y = x$. Figure 7 displays two examples of this, and Exercise 41 at the end of this section asks you to supply a proof for the general situation.

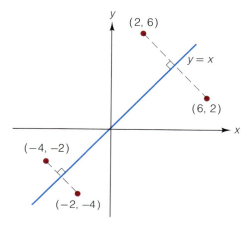

FIGURE 7

Now let us see how these ideas about symmetry relate to inverse functions. By way of example, consider the functions f and f^{-1} from Example 5:

$$f(x) = 3x - 4 \qquad f^{-1}(x) = \tfrac{1}{3}x + \tfrac{4}{3}$$

Figure 8 shows the graphs of f, f^{-1}, and the line $y = x$. Note that the graphs of f and f^{-1} are symmetric about the line $y = x$. This kind of symmetry, in fact, always occurs for the graphs of any pair of inverse functions.

FIGURE 8

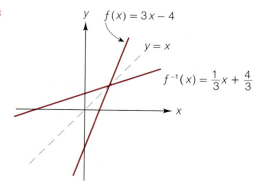

Why are the graphs of f and f^{-1} always mirror images of each other about the line $y = x$? First, recall that the function f^{-1} switches the inputs and outputs of f. Thus, (a, b) is on the graph of f if and only if (b, a) is on the graph of f^{-1}.

But as we have stated, the points (a, b) and (b, a) are mirror images in the line $y = x$. It follows that the graphs of f and f^{-1} are reflections of each other about the line $y = x$. For reference, we restate this useful fact in the box that follows.

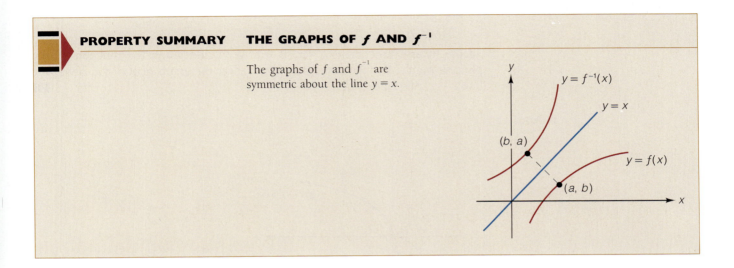

PROPERTY SUMMARY   THE GRAPHS OF f AND f^{-1}

The graphs of f and f^{-1} are symmetric about the line $y = x$.

EXAMPLE 8 The graph of a function f consists of the line segment joining the points $(-2, -3)$ and $(-1, 4)$, as shown in Figure 9. Sketch a graph of f^{-1}.

Solution The graph of f^{-1} is obtained by reflecting the graph of f about the line $y = x$. The reflection of $(-2, -3)$ is $(-3, -2)$, and the reflection of $(-1, 4)$ is $(4, -1)$. To graph f^{-1}, then, we plot the reflected points $(-3, -2)$ and $(4, -1)$ and connect them with a line segment, as shown in Figure 10. For reference, Figure 10 also shows the graphs of f and $y = x$.

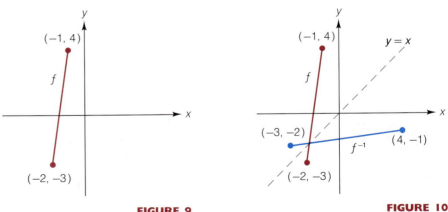

FIGURE 9   **FIGURE 10**

EXAMPLE 9 Let $g(x) = x^3$. Find $g^{-1}(x)$ and then, on the same set of axes, sketch the graphs of g, g^{-1}, and $y = x$.

Solution   We begin by writing $y = x^3$, then switching x and y and solving for y. We first have

$$x = y^3$$

To solve this equation for y, we take the cube root of both sides to obtain

$$\sqrt[3]{x} = y$$

So the inverse function is $g^{-1}(x) = \sqrt[3]{x}$. We could graph the function g^{-1} by plotting points. But the easier way is to reflect the graph of $g(x) = x^3$ about the line $y = x$. See Figure 11.   ■■■

FIGURE 11

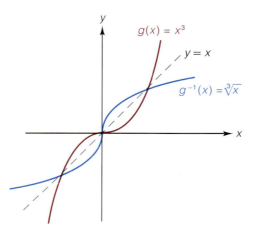

Earlier in this section, we mentioned that not every function has an inverse function. For instance, Tables 1(a) and (b) indicate what happens if we begin with the function defined by $F(x) = x^2$ and interchange the inputs and outputs. The resulting table, Table 1(b), does not define a function. For instance, the input 4 in Table 1(b) produces two distinct outputs, 2 and -2, but the definition of a function requires exactly one output for a given input.

TABLE 1

(a) $F(x) = x^2$

x	y
1	1
2	4
3	9
-1	1
-2	4
-3	9

Reverse inputs
and outputs

(b) The table formed by reversing the inputs and outputs for $F(x) = x^2$

x	y
1	1
4	2
9	3
1	-1
4	-2
9	-3

Why didn't this difficulty arise when we looked for the inverse of $f(x) = 2x$ at the beginning of this section? The answer is that there is an essential difference between the functions defined by $f(x) = 2x$ and $F(x) = x^2$. With $f(x) = 2x$, distinct inputs never yield the same output. That is, if $x_1 \neq x_2$, then $f(x_1) \neq f(x_2)$. This condition guarantees that interchanging the inputs and outputs of f will yield a function. With $F(x) = x^2$, however, distinct inputs can yield the same output. For instance

$$2 \neq -2$$

but

$$F(2) = F(-2) \quad \text{because } 2^2 = 4 = (-2)^2$$

It's useful to have a name for the type of function in which distinct inputs always yield distinct outputs. We call such a function a **one-to-one function**. So in view of the discussion in the previous paragraph, the function defined by $f(x) = 2x$ is one-to-one, but $F(x) = x^2$ is not. Using graphs, there is an easy way to tell which functions are one-to-one:

Horizontal Line Test

A function f is one-to-one if and only if each horizontal line intersects the graph of $y = f(x)$ in at most one point.

EXAMPLE 10   Determine if each function is one-to-one.

    (a) $F(x) = 2x$ **(b)** $H(x) = x^2$ **(c)** $h(x) = \left(\sqrt{x}\right)^4$

Solution   We sketch the graphs and apply the horizontal line test, as in Figure 12. The horizontal line test tells us that F and h are one-to-one, but H is not.   ∎∎∎

FIGURE 12

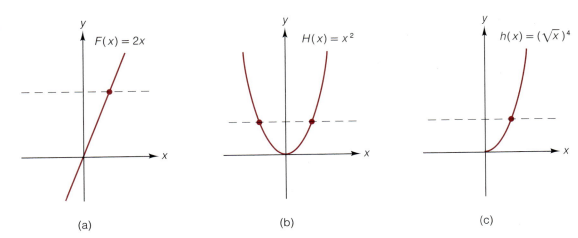

(a)   (b)   (c)

The following theorem tells us that the functions with inverses are precisely the one-to-one functions. (The proof of the theorem isn't difficult, but we'll omit giving it here.)

Theorem

A function f has an inverse function if and only if f is one-to-one.

EXAMPLE 11 Which of the three functions in Example 10 have inverse functions?

Solution According to our theorem, the functions with inverses are those that are one-to-one. In Example 10, we found that F and h were one-to-one, but H was not. So both F and h each have inverses, but H does not. ∎

EXERCISE SET 4.5

 A

1. Verify that the given pairs of functions are inverse functions.
   (a) $f(x) = 3x$; $g(x) = x/3$
   (b) $f(x) = 4x - 1$; $g(x) = \frac{1}{4}x + \frac{1}{4}$
   (c) $g(x) = \sqrt{x}$; $h(x) = x^2$ [Assume that the domain of both g and h is $[0, \infty)$.]

2. Which pairs of functions are inverses?
   (a) $f(x) = -3x + 2$; $g(x) = \frac{2}{3} - \frac{1}{3}x$
   (b) $F(x) = 2x + 1$ $G(x) = \frac{1}{2}x - 1$
   (c) $G(x) = x^3$; $H(x) = 1 - x^3$
   (d) $f(t) = t^3$; $g(t) = \sqrt[3]{t}$

3. Let $f(x) = x^3 + 2x + 1$ and assume that the domain of f^{-1} is $(-\infty, \infty)$. Simplify each expression.
   (a) $f[f^{-1}(4)]$ (b) $f^{-1}[f(-1)]$
   (c) $(f \circ f^{-1})(\sqrt{2})$ (d) $f[f^{-1}(t + 1)]$
   (e) $f(0)$; use your answer to find $f^{-1}(1)$.
   (f) $f(-1)$; use your answer to find $f^{-1}(-2)$.

4. Let $f(x) = 2x + 1$.
   (a) Find $f^{-1}(x)$.
   (b) Calculate $f^{-1}(5)$ and $1/f(5)$. Are your answers the same?

5. Let $f(x) = 3x - 1$.
   (a) Compute $f^{-1}(x)$.
   (b) Verify that $f[f^{-1}(x)] = x$ and that $f^{-1}[f(x)] = x$.
   (c) On the same set of axes, sketch the graphs of f, f^{-1}, and the line $y = x$. Note that the graphs of f and f^{-1} are symmetric about the line $y = x$.

6. Follow Exercise 5, but use $f(x) = \frac{1}{3}x - 2$.

7. Follow Exercise 5, but use $f(x) = \sqrt{x - 1}$. [The domain of f^{-1} will be $[0, \infty)$.]

8. Follow Exercise 5, but use $f(x) = 1/x$.

9. Let $f(x) = \dfrac{x + 2}{x - 3}$.
   (a) Find the domain and range of the function f.
   (b) Find $f^{-1}(x)$.
   (c) Find the domain and range of the function f^{-1}. What do you observe?

10. Let $f(x) = \dfrac{2x - 3}{x + 4}$. Find $f^{-1}(x)$. Find the domain and range for f and f^{-1}. What do you observe?

11. Let $f(x) = 2x^3 + 1$. Find $f^{-1}(x)$.

12. In our preliminary discussion at the beginning of this section, we considered the functions $f(x) = 2x$ and $g(x) = x/2$. Verify that $f[g(x)] = x$ and that $g[f(x)] = x$. On the same set of axes, sketch the graphs of f, g, and $y = x$.

13. This exercise refers to the comments made at the end of Example 5. Compute $f[f^{-1}(x)]$ and $f^{-1}[f(x)]$ using the functions $f(x) = 3x - 4$ and $f^{-1}(x) = \frac{1}{3}x + \frac{4}{3}$. By actually carrying out the calculations, you can see in each case that the answer is indeed x. Then, on the same set of axes, sketch the graphs of f, f^{-1}, and the line $y = x$.

14. Let $f(x) = (x - 3)^2$ and take the domain to be $[3, \infty)$.
    (a) Find $f^{-1}(x)$. Give the domain of f^{-1}.
    (b) On the same set of axes, graph both f and f^{-1}.

15. Let $f(x) = (x - 3)^3 - 1$.
    (a) Compute $f^{-1}(x)$.
    (b) On the same set of axes, sketch the graphs of f, f^{-1}, and the line $y = x$. *Hint:* Sketch f using the ideas of Section 4.3. Then reflect f to get f^{-1}.

16. The following figure shows the graph of a function f. (The axes are marked off in one-unit intervals.) Graph each of the following functions.
    (a) $y = f^{-1}(x)$ (b) $y = f(x - 2)$
    (c) $y = f(x) - 2$ (d) $y = f(-x)$
    (e) $y = -f(x)$ (f) $y = f^{-1}(x - 2)$
    (g) $y = f^{-1}(x) - 2$ (h) $y = f^{-1}(x - 2) - 2$
    (i) $y = f^{-1}(-x)$ (j) $y = -f^{-1}(x)$

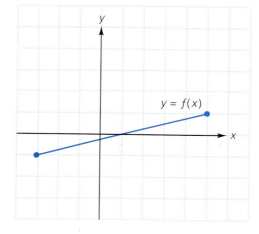

17. The following figure shows the graph of a function g. (The axes are marked off in one-unit intervals.) Graph each of the following functions.
    (a) $y = g^{-1}(x)$
    (b) $y = g^{-1}(x) - 1$
    (c) $y = g^{-1}(x - 1)$
    (d) $y = g^{-1}(-x)$
    (e) $y = -g^{-1}(x)$
    (f) $y = -g^{-1}(-x)$

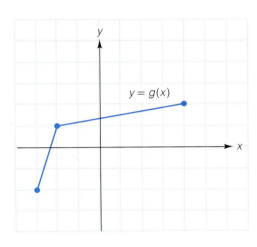

18. The following figure shows the graph of a function $y = h(x)$.
    (a) Sketch a graph that is the reflection of h about the y-axis. What is the equation of the graph you obtain?
    (b) On the same set of axes, sketch graphs of h, h^{-1}, and the line $y = x$.
    (c) Let k be the function whose graph is the reflection of h in the x-axis. Graph the function k^{-1}.

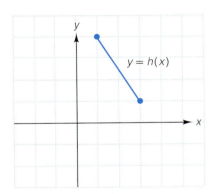

In Exercises 19–30, use the horizontal line test to determine if the function is one-to-one (and therefore has an inverse).

19. $y = x^2 + 1$
20. $y = \sqrt{x}$
21. $f(x) = 1/x$
22. $g(x) = |x|$
23. $y = x^3$
24. $y = 1 - x^3$
25. $y = \sqrt{1 - x^2}$
26. $y = 3x$

27. $g(x) = 5$ (for all x)
28. $y = mx + b$ $(m \neq 0)$
29. $f(x) = \begin{cases} x^2 & \text{if } -1 \leq x \leq 0 \\ x^2 + 1 & \text{if } x > 0 \end{cases}$
30. $g(x) = \begin{cases} x^2 & \text{if } -1 \leq x < 0 \\ x^2 + 1 & \text{if } x \geq 0 \end{cases}$
31. Let $f(x) = \dfrac{3x - 2}{5x - 3}$. Show that $f[f(x)] = x$. This says that f is its own inverse.
32. Let $f(x) = \dfrac{ax + b}{cx - a}$. Show that f is its own inverse. (Assume that $a^2 + bc \neq 0$.)

In Exercises 33–36, assume that the domain of f and f^{-1} is $(-\infty, \infty)$. Solve the equation for x or for t (whichever is appropriate) using the given information.

33. (a) $7 + f^{-1}(x - 1) = 9$; $f(2) = 6$
    (b) $4 + f(x + 3) = -3$; $f^{-1}(-7) = 0$
34. (a) $f^{-1}(2x + 3) = 5$; $f(5) = 13$
    (b) $f(1 - 2x) = -4$; $f^{-1}(-4) = -5$
35. $f^{-1}\left(\dfrac{t + 1}{t - 2}\right) = 12$; $f(12) = 13$
36. $f\left(\dfrac{1 - 2t}{1 + 2t}\right) = 7$; $f^{-1}(7) = -3$

B

37. Let $f(x) = \sqrt{x}$.
    (a) Find $f^{-1}(x)$. What is the domain of f^{-1}? [The domain is not $(-\infty, \infty)$.]
    (b) In each case, determine whether the given point lies on the graph of f or f^{-1}.
    (i) $(4, 2)$
    (ii) $(2, 4)$
    (iii) $(5, \sqrt{5})$
    (iv) $(\sqrt{5}, 5)$
    (v) $(a, f(a))$ (Assume $a \geq 0$.)
    (vi) $(f(a), a)$
    (vii) $(b, f^{-1}(b))$ (Assume $b \geq 0$.)
    (viii) $(f^{-1}(b), b)$

38. Let $F(x) = x^3 + 7x - 5$. Find $F^{-1}(-5)$. [Assume that the domain of F^{-1} is $(-\infty, \infty)$.]

39. Reread the definition of symmetry about a line given on page 215. Then use the method shown in Example 7 to show that the points $(7, -1)$ and $(-1, 7)$ are symmetric about the line $y = x$.

40. Use the method shown in Example 7 to show that $(5, 2)$ and $(2, 5)$ are symmetric about $y = x$.

41. Use the method of Example 7 to show that the points (a, b) and (b, a) are symmetric about the line $y = x$.

42. Points P and Q are both reflected in the line $y = x$ to obtain points P' and Q', respectively. Does the distance from P to Q equal the distance from P' to Q'?

43. Pick any three points on the line $y = \frac{1}{2}x + 3$. Reflect each point in the line $y = x$. Show that the three reflected points all lie on one line. What is the equation of that line?

4.6   VARIATION

Perhaps the most dramatic moment in the development of physics during the 19th century occurred to J. C. Maxwell one day in the 1860's.... Maxwell noted that the equations or the laws [for electricity, magnetism, and light] ... were mutually inconsistent ..., and in order for the whole system to be consistent, he had to add another term to his equations. With this new term there came an amazing prediction, which was that a part of the electric and magnetic fields would fall off much more slowly with the distance than the inverse square, namely, inversely as the first power of the distance! And so he realized that electric currents in one place can affect other charges far away, and he predicted the basic effects with which we are familiar today—radio transmission, radar, and so on.

Richard Feynman in *The Feynman Lectures on Physics* (Reading, Mass.: Addison-Wesley, 1963)

Certain formulas or relationships between variables occur so frequently in the applications of mathematics that there is a special terminology used to describe them. For example, if $y = 9x$, we say that y **varies directly as** x or that y is **directly proportional** to x. The number 9 in this case is called the **constant of proportionality**. As another example of this special terminology, suppose that the variables A and B are related by the equation $A = 3/B$. Then we say that A **varies inversely as** B or that A is **inversely proportional** to B. The constant of proportionality in this case is the number 3. Before going on to list other forms of this terminology, let's look at examples of what we have just described.

EXAMPLE I   Suppose that F varies directly as a. If $F = 60$ when $a = 5$, find F when $a = 6$.

Solution   Since F varies directly as a, we write

$$F = ka$$

To find k, we substitute the given pair of values $F = 60$ and $a = 5$ in this equation. This yields

$$60 = k \cdot 5 \qquad \text{and therefore} \qquad k = 12$$

Now that we know $k = 12$, the equation $F = ka$ becomes

$$F = 12a$$

From this equation we can calculate F whenever a is given. In particular, then, if $a = 6$, we obtain $F = 12(6)$. Thus $F = 72$, as required. The formula $F = 12a$ is called the **law of variation** for this problem.

EXAMPLE 2   At a constant speed, the distance traveled varies directly as the time. Suppose a runner covers 1 mile in 7 min. At this rate, what distance would be covered in 30 min?

Solution   Let d and t denote distance and time, respectively. The statement that distance varies directly as time is written algebraically as

$$d = kt$$

As in Example 1, we first use the given data to calculate k. Since we are given that $d = 1$ when $t = 7$, we obtain

$$1 = k \cdot 7 \qquad \text{and therefore} \qquad k = \frac{1}{7}$$

Now that we know $k = \frac{1}{7}$, the equation $d = kt$ becomes

$$d = \frac{1}{7} t$$

Using this law of variation, we can calculate d whenever t is given. In particular, then, if $t = 30$, we obtain $d = \frac{1}{7}(30)$. Thus in 30 min, the runner covers $\frac{30}{7} \approx 4.3$ miles. ▪

EXAMPLE 3 Given that y varies inversely as x and that y is 6 when x is -2, determine y when x is 4.

Solution The statement that y varies inversely as x translates algebraically to

$$y = \frac{k}{x}$$

To first find k, we substitute the given pair of values $x = -2$ and $y = 6$ in this equation. This yields

$$6 = \frac{k}{-2} \qquad \text{and therefore} \qquad k = -12$$

Now that we know $k = -12$, the equation $y = k/x$ becomes

$$y = -12/x$$

Using this law of variation, we can calculate y whenever x is given. In particular, then, if $x = 4$, we obtain

$$y = -12/4 = -3$$

Thus when x is 4, we have y is -3; this is the required answer. ▪

The list that follows shows some common formulas that can be expressed in the language of variation. In this list, k denotes a constant in each case, the so-called **constant of proportionality**.

Formula or Law of Variation	Terminology
1. $y = kx^n$	y varies directly as x^n; or y is directly proportional to x^n.
2. $y = \dfrac{k}{x^n}$	y varies inversely as x^n; or y is inversely proportional to x^n.
3. $y = kxz$	y varies jointly as x and z; or y varies directly as the product of x and z.
4. $y = \dfrac{kx}{z}$	y varies directly as x and inversely as z.

The rest of this section and the exercises that follow deal with this terminology. Notice that we have already discussed formulas 1 and 2 for the case $n = 1$.

EXAMPLE 4   The volume of a sphere varies directly as the cube of the radius. When the radius is 2 cm, the volume is $32\pi/3$ cm^3. Find the constant of proportionality k. Write the law of variation.

Solution   Let V denote the volume and r the radius. The statement that the volume varies directly as the cube of the radius translates algebraically to

$$V = kr^3$$

To find the constant of proportionality k, we substitute the given values for r and V in this equation to obtain

$$\frac{32\pi}{3} = k \cdot 2^3 = 8k$$

$$k = \frac{32\pi}{3 \cdot 8} = \frac{4\pi}{3}$$

Now that we have found the constant k, we can write the law of variation simply by substituting this value of k in the equation $V = kr^3$. This yields $V = (4\pi/3)r^3$.

EXAMPLE 5   The quantity B varies inversely as the square of C. What happens to the value of B if C is tripled?

Solution   The statement "B varies inversely as the square of C" is written algebraically as

$$B = \frac{k}{C^2}$$

To see what happens when C is tripled, replace C by the quantity $(3C)$ in our equation.

$$B = \frac{k}{(3C)^2} = \frac{k}{9C^2}$$

Now compare the original expression for B with the new expression.

original expression for B: $\dfrac{k}{C^2}$

new expression for B after C is tripled: $\dfrac{1}{9} \cdot \dfrac{k}{C^2}$

From this we can see that if C is tripled, B is one-ninth of its original value. This is the required solution.

Example 5 differs from those preceding it in this section in that we did not need to evaluate the constant of proportionality k specifically. Indeed, the problem did not supply enough information to calculate a value for k. Example 6 is similar in this respect.

EXAMPLE 6   The force of attraction between two objects varies directly as the product of their masses and inversely as the square of the distance between them. (This is Newton's Law of Gravitation.) What happens to the force if both masses are tripled and the distance is doubled?

Solution   We begin by assigning letters to denote the relevant quantities and then writing the law of variation. Let F, m, M, and d denote, respectively, the force, the two masses, and the distance. Then the given statement translates algebraically to

$$F = \frac{kmM}{d^2}$$

We want to find out what happens to F when both masses are tripled and the distance is doubled. To do this, we make the following substitutions in our law of variation:

replace m by $(3m)$

replace M by $(3M)$

replace d by $(2d)$

We then have

$$F = \frac{k(3m)(3M)}{(2d)^2} = \frac{9kmM}{4d^2}$$

Now compare the original expression for F with the expression obtained after tripling the masses and doubling the distance.

original expression for F: $\dfrac{kmM}{d^2}$

new expression for F: $\dfrac{9}{4}\left(\dfrac{kmM}{d^2}\right)$

From this we can see that the new force is nine-fourths of the original force. In other words, the new force is $2\frac{1}{4}$ times as strong as the original. This is the required solution.

EXAMPLE 7   Suppose that A varies jointly as X and Y. Also, suppose that X varies inversely as the square of Y. Show that A varies inversely as Y.

Solution   Translating the two given statements into algebraic equations gives us

$$A = kXY \tag{1}$$

and

$$X = \frac{c}{Y^2} \tag{2}$$

Both k and c are constants here. We do need the two different letters since there is no reason to suppose that $k = c$. Now we are asked to show that A varies inversely as Y. To do this, we just substitute for X in equation (1) using the value given in equation (2). We have

$$A = kXY = k\left(\frac{c}{Y^2}\right)Y \quad \text{and therefore} \quad A = \frac{kc}{Y}$$

This last equation says that A equals a constant divided by Y; that is, A varies inversely as Y. If we want, we can denote kc by the single letter K. Then we have $A = K/Y$, which again says that A varies inversely as Y. ∎

EXERCISE SET 4.6

A

In Exercises 1–5, translate the given statement into an equation.

1. (a) y varies directly as x.
    (b) A varies inversely as B.

2. (a) y varies directly as the cube of z.
    (b) s varies inversely as t^2.

3. (a) x varies jointly as u and v^2.
    (b) z varies jointly as A^2 and B^3.

4. (a) y varies directly as u and inversely as v.
    (b) y varies directly as the square root of x and inversely as the product of w and z.

5. (a) F is inversely proportional to r^2.
    (b) V^2 is directly proportional to the sum of U^2 and T^2.

6. If y varies directly as x, and y is 20 when $x = 4$, find y when $x = -6$.

7. Suppose A varies inversely as B, and $A = -1$ when $B = 2$.
    (a) Find the constant of proportionality.
    (b) Write the law of variation.
    (c) Find A when $B = \frac{5}{4}$.

8. Suppose T varies directly as V^2, and $T = 8$ when $V = 4$.
    (a) Find the constant of proportionality and write the law of variation.
    (b) Find T when $V = 10$.
    (c) Find T when $V = \sqrt{10}$.

9. Assume x varies jointly as y and z. Also $x = 9$ when $y = 2$ and $z = -3$. Find x when both y and z are 4.

10. Suppose F varies directly as the product of q and Q and inversely as the square of d. When $q = 2$, $Q = 4$, and $d = \frac{1}{2}$, the value of F is 8.
    (a) Find the constant of proportionality and write the law of variation.
    (b) Find F when $q = 3$, $Q = 6$, and $d = 2$.

11. A varies jointly as B and C. If B is tripled and C is doubled, what happens to A?

12. If S varies inversely as the square of T, what happens to S if T is cut in half?

13. If x varies jointly as B and C and inversely as the square root of A, what happens to x if A, B, and C are all quadrupled?

14. Suppose that $xy = \frac{3}{4}$. Does y vary directly or inversely with x? What is the constant of proportionality in this case?

15. The surface area of any sphere varies directly as the square of the radius. When the radius is 2 cm, the surface area is 16π cm^2.
    (a) Find the constant of proportionality and write the law of variation.
    (b) Find the surface area when the radius is $\sqrt{3}$ cm.

16. For motion at a constant speed, the distance varies directly as the time. Suppose a car covers 10 miles in 15 min. At this rate, what distance would be covered in 50 min?

17. The force of attraction between two objects varies jointly as their masses and inversely as the square of the distance between them. What happens to the force if one mass is tripled, the other mass is quadrupled, and the distance is cut in half?

18. The magnitude of the force between two electrical charges Q_1 and Q_2 varies directly as the product of Q_1 and Q_2 and inversely as the square of the distance between the charges. (This is Coulomb's law, named after the French physicist Charles Coulomb, who verified the law experimentally in 1785.) What happens to the magnitude of the force if Q_1 is doubled, Q_2 is quadrupled, and the distance is tripled?

19. Neglecting air resistance, the distance any object falls starting from rest is directly proportional to the square of the time. (This law was discovered experimentally by Galileo, sometime before the year 1590. In his first experiments, Galileo used his own pulse to measure

time.) Given that an object starting from rest will fall 490 m in 10 sec, how far will an object fall in 5 sec, starting from rest?

20. The volume of a right circular cone varies jointly as the height and the square of the radius of the base. When the radius and height are both 3 cm, the volume is 9π cm^3. Find the volume of a cone having a height of 4 cm and a radius of $\frac{3}{2}$ cm.

21. (Boyle's law) In 1662, Robert Boyle observed that, for a sample of gas at a constant temperature, the volume is inversely proportional to the pressure.
   (a) Suppose that at a certain fixed temperature, the pressure of a 2-liter sample of gas is 1.025 atm. (atm is the abbreviation for atmospheres, a unit of pressure, where 1 atm = 14.7 lb/in^2, the air pressure at sea level.) Calculate the constant of proportionality and write the law of variation.
   (b) Suppose that, with the temperature still fixed, the pressure on our sample of gas is reduced to 1 atm. Calculate the new volume of the gas.

22. (Charles's law) In 1787, Jacques Charles stated that, for a sample of gas held at a constant pressure, the volume is directly proportional to the Kelvin temperature. Suppose that initially you have 300 ml of a gas at 300°K. Then holding the pressure constant, you heat the gas to 310°K. What is the new volume?

23. (a) For a satellite in a circular orbit about the earth, the kinetic energy varies directly as the mass of the satellite and inversely as the radius of the orbit. What happens to the kinetic energy if we cut the radius in half and triple the mass?
   (b) The velocity of an earth satellite in a circular orbit varies inversely as the square root of the radius of the orbit. What happens to the velocity if we cut the radius in half?

24. (Hooke's law) In 1676, the English physicist Robert Hooke found that the distance a spring is stretched varies directly as the force pulling on the spring. Suppose that, for a certain spring, a pull of 5 lb results in a stretch of 3 in.
   (a) Find the constant of proportionality and write the law of variation.
   (b) How far will a pull of 12 lb stretch the spring?

25. The weight of an object above the earth's surface varies inversely as the square of its distance from the center of the earth. How much would an astronaut weigh 500 miles above the surface of the earth, assuming her weight at the surface is 140 lb? (Assume that the radius of the earth is 4000 miles.)

26. One of the ways in which the human body adjusts blood flow is through changes in the radii of small arteries called *arterioles*. According to Poiseuille's law [named after the French physiologist J. L. Poiseuille (1779–1869)], the blood flow in a given arteriole (measured, for example, in cm^3/sec) varies directly as the fourth power of the radius. By what fraction is the blood flow through an arteriole multiplied if the radius is decreased by 20%?

B

27. The graphs of six functions are displayed here. Which graph best represents the given statement?
   (a) y varies directly as x, where the constant of proportionality is negative.
   (b) y varies inversely as x.
   (c) y varies directly as x^2, with a positive constant of proportionality.
   (d) y varies directly as x, with a positive constant of proportionality.
   (e) y varies directly as x^2, with a negative constant of proportionality.

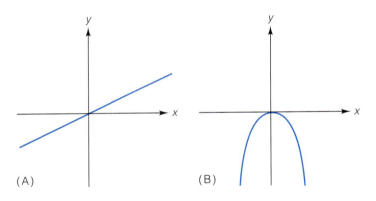

(A) (B)

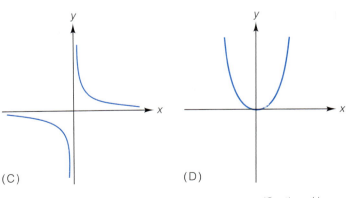

(C) (D)

(Continued.)

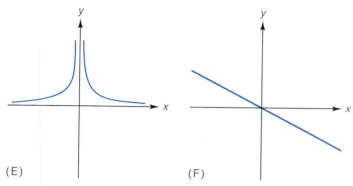

(E)

(F)

28. Three solid lead spheres of radii a, b, and c are melted down and recast into one large sphere of radius R. Show that $R = \sqrt[3]{a^3 + b^3 + c^3}$. Use the fact that the volume of a sphere varies directly as the cube of the radius.

29. The *period* of a pendulum is the time required for one complete swing, forward and back. If the period of a pendulum varies directly as the square root of the length of the pendulum, by how much should a pendulum of length L be shortened in order to cut the period in half?

30. At highway speeds, the wind resistance against a car varies directly as the square of the speed.
   (a) If the speed is increased from 40 mph to 60 mph, by what factor is the wind resistance multiplied?
   (b) At what speed is the wind resistance twice as much as that at 50 mph?

31. The intensity of illumination due to a small light source varies inversely as the square of the distance from the light. Suppose a book is 6 ft from a lamp. At what distance from the lamp will the illumination be twice as great?

32. The volume of a coin varies jointly as the thickness and the square of the radius. Suppose that two silver dollars are melted down (illegal!) and cast into one coin of the same thickness as a silver dollar. Show that the diameter of the new coin is $\sqrt{2}$ times the diameter of a silver dollar.

33. Show that the area of the shaded triangle in the figure varies jointly as m and the square of x.

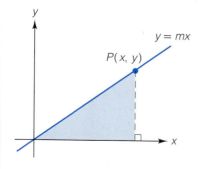

34. The weight of an object on or above the earth's surface varies inversely as the square of the object's distance from the center of the earth. At what height x above the earth's surface will an object weigh half as much as it does on the surface? Assume the radius of the earth is 4000 miles.

   Answer: $4000(\sqrt{2} - 1) \approx 1656.9$ miles.

35. In the year 1619, Johannes Kepler announced his Third Law of Planetary Motion. This law states that the square of the time it takes for a planet to complete one revolution about the sun varies directly as the cube of the planet's average distance from the sun. A table giving the average distance of each planet from the sun follows. Use a calculator to find the time it takes for each planet to complete one revolution about the sun. For Earth, assume the time is 365.2564 days.

Planet	Average Distance from Sun Measured in Astronomical Units
Mercury	0.387099
Venus	0.723332
Earth	1.000000
Mars	1.523691
Jupiter	5.204377
Saturn	9.577971
Uranus	19.26020
Neptune	30.09421
Pluto	39.82984

36. Most astronomers accept the theory that the universe is expanding and that the galaxies are receding from one another at velocities that vary directly as their distances. Given that a galaxy in Virgo, which is 78,000,000 light years from us, is receding at 745 miles/sec, estimate the distance from us of the Hydra galaxy, which is receding from us at 37,881 miles/sec.

C

37. Suppose that x varies directly as z and that y also varies directly as z.
   (a) Does $x + y$ vary directly as z?
   (b) Does xy vary directly as z?
   (c) Does $\sqrt{xy}$ vary directly as z? (Assume x, y, and z are positive.)

38. Suppose that A varies directly as B. Show that AB varies directly as $A^2 + B^2$.

39. Suppose that $x + y$ varies directly as $x - y$. Also suppose that the constant of proportionality is not ± 1.
   (a) Show that y varies directly as x.
   (b) Show that $x^2 + y^2$ varies jointly as x and y.
   (c) Show that $x^3 + y^3$ varies jointly as x, y, and $x + y$.

40. Suppose that $A^3 + \dfrac{1}{B^3}$ varies directly as $A^3 - \dfrac{1}{B^3}$. Also suppose that the constant of proportionality is not ± 1. Show that B varies inversely as A.

41. Show that the area of the shaded triangle in the figure varies directly as a^2.

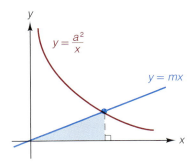

42. For this exercise, refer to the following figure.
(a) Show that the area of $\triangle OAB$ varies inversely as b.
(b) Show that the area of triangle I varies directly as a and inversely as the product of b and $a - b$.
(c) Show that the area of triangle II varies inversely as $a - b$.

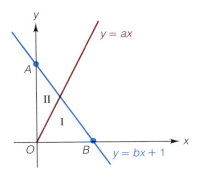

CHAPTER FOUR   SUMMARY OF PRINCIPAL TERMS AND FORMULAS

TERMS OR FORMULAS	PAGE REFERENCE	COMMENTS
1. Function	174	Given two nonempty sets A and B, a function from A to B is a rule of correspondence that assigns to each element of A exactly one element in B.
2. Domain	174	The domain of a function is the set of all inputs for that function. If f is a function from A to B, then the domain is the set A.
3. Range	174	The range of a function from A to B is the set of all elements in B that are actually used as outputs.
4. $f(x)$	175	Given a function f, the notation $f(x)$ denotes the output that results from the input x.
5. Graph of a function	183	The graph of a function f consists of those points (x, y) such that x is in the domain of f and $y = f(x)$.
6. Vertical line test	186	A graph in the x-y plane represents a function $y = f(x)$ provided that any vertical line intersects the graph in at most one point.
7. Turning point	189	A turning point on a graph is a point where the graph changes from rising to falling, or vice versa. See, for example, Figures 10 and 11 on page 189.
8. Maximum value and minimum value	189	An output $f(a)$ is a maximum value of the function f if $f(a) \geq f(x)$ for every x in the domain of f.
		An output $f(b)$ is a minimum value if $f(b) \leq f(x)$ for every x in the domain of f.
9. Increasing function	189	A function f is increasing on an interval if the following condition holds: If a and b are in the interval and $a < b$, then $f(a) < f(b)$. Geometrically, this means that the graph is rising as we move in the positive x-direction.

TERMS OR FORMULAS	PAGE REFERENCE	COMMENTS
10. Decreasing function	189	A function f is decreasing on an interval if the following condition holds: If a and b are in the interval and $a<b$ then $f(a)>f(b)$. Geometrically, this means that the graph is falling as we move in the positive x- direction.
11. Techniques for graphing functions	197	The box on page 197 lists six techniques involving translation and reflection. For a comprehensive example, see Example 5 on pages 199-200.
12. $f \circ g$	204	A composition of functions f and g: $(f \circ g)(x) = f[g(x)]$
13. Inverse functions	211	Two functions f and g are inverses of one another provided that the following two conditions are met. First, $f[g(x)] = x$ for each x in the domain of g. Second, $g[f(x)] = x$ for each x in the domain of f.
14. f^{-1}	212	f^{-1} denotes the inverse function for f. *Note:* In this context, f^{-1} does not mean $1/f$. For an example, see Exercise 4 on page 220.
15. Symmetry about a line	215	Two points P and Q are symmetric about a line if that line is the perpendicular bisector of $\overline{PQ}$.
16. One-to-one function	219	A function f is said to be one-to-one provided that distinct inputs always yield distinct outputs. That is, if x_1 and x_2 are in the domain of f and $x_1 \neq x_2$, then $f(x_1) \neq f(x_2)$. The relationship between inverse functions and the notion of one-to-one is given by the following theorem: A function f has an inverse if and only if f is one-to-one.
17. Horizontal line test	219	A function f is one-to-one if and only if each horizontal line intersects the graph of $y = f(x)$ in at most one point.
18. y varies directly as x	222	$y = kx$, where k is a constant
19. y varies inversely as x	222	$y = k/x$, where k is a constant
20. y varies jointly as x and z	223	$y = kxz$, where k is a constant

WRITING MATHEMATICS

1. Section 4.5 on inverse functions begins with the following quotation from Professor R. P. Boas:

   > If anybody ever told me why the graph of $y = x^{1/2}$ is the reflection of the graph of $y = x^2$ $[x \geq 0]$ in a 45° line, it didn't sink in. To this day, there are textbooks that expect students to think that it is so obvious as to need no explanation.

   Show Prof. Boas that you are not in the dark about inverse functions. Write out an explanation of why the graphs of $y = x^{1/2}$ and $y = x^2$ $[x \geq 0]$ are indeed reflections of one another in a 45° line. (As preparation for your writing, you'll probably need to look over Section 4.5 and make a few notes.)

2. Decide which of the following rules are functions. Write out your reasons in complete sentences. (In each case, assume that the domain is the set of students in your school.)

(a) F is the rule that assigns to each person his or her brother.
(b) G is the rule that assigns to each person his or her aunt.
(c) H is the rule that assigns to each person his or her mother.
(d) K is the rule that assigns to each person his or her mother or father.

3. A student evaluated the difference quotient $\dfrac{f(x + h) - f(x)}{h}$ in the following two

    cases: (a) $f(x) = 4x - 6$; (b) $f(x) = 3x - 5$. In each case, she found that the answer was the coefficient of x in the given function. Explain why this pattern holds in general. You can make use of algebra or diagrams, but tie your explanation together with complete English sentences, as if you were presenting the ideas to a classmate or to your instructor.

 CHAPTER FOUR REVIEW EXERCISES

1. **(a)** Find the domain of the function defined by
    $f(x) = \sqrt{15 - 5x}$.
    (b) Find the range of the function defined by
    $$g(x) = \frac{3 + x}{2x - 5}.$$

2. Let $f(x) = 3x^2 - 4x$ and $g(x) = 2x + 1$. Compute each of the following: **(a)** $(f - g)(x)$; **(b)** $(f \circ g)(x)$; **(c)** $f[g(-1)]$.

3. A **linear function** is a function defined by an equation of the form $F(x) = ax + b$, where a and b are real numbers. Suppose f and g are linear functions. Is the function $f \circ g$ a linear function?

4. A **quadratic function** is a function defined by an equation of the form $F(x) = ax^2 + bx + c$, where a, b, and c are real numbers and $a \neq 0$.
    (a) Give an example of quadratic functions f and g for which $f + g$ is also a quadratic function.
    (b) Give an example of quadratic functions f and g for which $f + g$ is not a quadratic function.
    (c) Suppose f and g are quadratic functions. Is the function $f \circ g$ a quadratic function?

5. **(a)** Compute $\dfrac{F(x) - F(a)}{x - a}$ given that $F(x) = 1/x$.
    (b) Compute $\dfrac{g(x + h) - g(x)}{h}$ for the function defined by $g(x) = x - 2x^2$.

6. The y-intercept for the graph of $y = f(x)$ is 4. What is the y-intercept for the graph of $y = -f(x) + 1$?

7. **(a)** Find $g^{-1}(x)$ given that $g(x) = \dfrac{1 - 5x}{3x}$.
    (b) The figure displays the graph of a function f. Sketch the graph of f^{-1}.

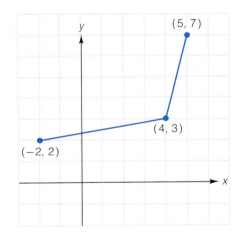

8. Let d denote the distance from the point $(1, 0)$ to a point (x, y) on the circle $x^2 + y^2 = 1$.
    (a) Show that $d = \sqrt{2 - 2x}$
    (b) What is the domain of the distance function in part (a). *Hint:* The domain is *not* the interval $(-\infty, 1)$.

9. Graph each function and specify the intercepts:

    (a) $y = |x + 2| - 3$;   **(b)** $y = \dfrac{1}{x + 2} - 1$.

10. Refer to the graph of $y = g(x)$ in the figure (on the next page). The domain of g is $[-5, 2]$.
    (a) What are the coordinates of the turning point(s)?
    (b) What is the maximum value of g?
    (c) Which input yields a minimum value for g?
    (d) On which interval(s) is g increasing?

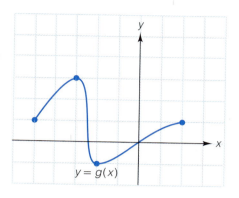

$y = g(x)$

11. Let $f(x) = 3x^2 - 2x$.
    (a) Find $f(-1)$.   (b) Find $f(1 - \sqrt{2})$.

12. Graph the function G defined by

$$G(x) = \begin{cases} \sqrt{1 - x^2} & \text{if } -1 \le x < 0 \\ \sqrt{x} & \text{if } x \ge 0 \end{cases}$$

13. Express the slope of a line passing through the points $(5, 25)$ and $\left(5 + h, (5 + h)^2\right)$ as a function of h.

14. Graph the function $y = |9 - x^2|$. Specify symmetry and intercepts.

15. The figure shows the graph of a function $y = f(x)$. Sketch the graph of $y = f(-x)$.

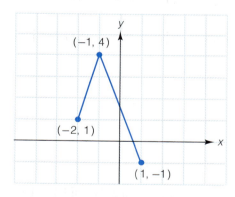

16. Given that the domains of f and f^{-1} are both $(-\infty, \infty)$ and that $f^{-1}(1) = -4$, solve the following equation for t:

$$2 + f(3t + 5) = 3$$

17. Suppose that (a, b) is a point on the graph of $y = f(x)$. Match the functions defined in the left-hand column with the points in the right-hand column. For example, the appropriate match for (a) in the left-hand column is determined as follows. The graph of $y = f(x) + 1$ is obtained by translating the graph of $y = f(x)$ up one unit. Thus, the point (a, b) moves up

to $(a, b + 1)$ and, consequently, (E) is the appropriate match for (a).

(a) $y = f(x) + 1$ (A) $(-a, b)$
(b) $y = f(x + 1)$ (B) (b, a)
(c) $y = f(x - 1) + 1$ (C) $(a - 1, b)$
(d) $y = f(-x)$ (D) $(-b, a + 1)$
(e) $y = -f(x)$ (E) $(a, b + 1)$
(f) $y = -f(-x)$ (F) $(1 - a, b)$
(g) $y = f^{-1}(x)$ (G) $(-a, -b)$
(h) $y = f^{-1}(x) + 1$ (H) $(-b, 1 - a)$
(i) $y = f^{-1}(x - 1)$ (I) $(b, -a)$
(j) $y = f^{-1}(-x) + 1$ (J) $(a, -b)$
(k) $y = -f^{-1}(x)$ (K) $(b + 1, a)$
(l) $y = -f^{-1}(-x) + 1$ (L) $(a + 1, b + 1)$
(m) $y = 1 - f^{-1}(x)$ (M) $(b, a + 1)$
(n) $y = f(1 - x)$ (N) $(b, 1 - a)$

In Exercises 18–40, sketch the graph and specify any x- or y-intercepts.

18. $y = 4 - x^2$ 19. $y = -(x - 1)^2 + 2$

20. $y = \dfrac{1}{x} + 1$ 21. $f(x) = \dfrac{1}{x + 1}$

22. $y = \dfrac{1}{x + 1} + 1$ 23. $y = |x + 3|$

24. $g(x) = -\sqrt{x - 4}$ 25. $h(x) = \sqrt{1 - x^2}$

26. $f(x) = \sqrt{1 - x^2} + 1$ 27. $y = 1 - (x + 1)^3$

28. $y = 4 - \sqrt{-x}$ 29. $y = \left(\sqrt{x}\right)^2$

30. $f \circ g$, where $g(x) = x + 3$ and $f(x) = x^2$

31. $f \circ g$, where $g(x) = \sqrt{x - 1}$ and $f(x) = -x^2$

32. $f(x) = \begin{cases} \sqrt{1 - x^2} & \text{if } -1 \le x \le 0 \\ \sqrt{x} + 1 & \text{if } x > 0 \end{cases}$

33. $F(x) = \begin{cases} -\sqrt{1 - x^2} & \text{if } -1 \le x < 0 \\ \sqrt{x} & \text{if } x \ge 0 \end{cases}$

34. $y = \begin{cases} |x - 1| & \text{if } 0 \le x \le 2 \\ |x - 3| & \text{if } 2 < x \le 4 \end{cases}$

35. $y = \begin{cases} 1/x & \text{if } 0 < x \le 1 \\ 1/(x - 1) & \text{if } 1 < x \le 2 \end{cases}$

36. $(y + |x| - 1)(y - |x| + 1) = 0$

37. f^{-1}, where $f(x) = \frac{1}{2}(x + 1)$

38. g^{-1}, where $g(x) = \sqrt[3]{x + 2}$

39. $f \circ f^{-1}$, where $f(x) = \sqrt{x - 2}$

40. $f^{-1} \circ f$, where $f(x) = \sqrt{x - 2}$

In Exercises 41–50, find the domain of the function.

41. $y = 1/(x^2 - 9)$ 42. $y = x^3 - x^2$

43. $y = \sqrt{8 - 2x}$ 44. $y = \dfrac{x}{6x^2 + 7x - 3}$

45. $y = \sqrt{|2 - 5x|}$

46. $y = (25 - x^2)/\sqrt{x^2 + 1}$

47. $y = \sqrt{x^2 - 2x - 3}$

48. $y = \sqrt{5 - x^2}$

49. $y = x + \dfrac{1}{x}$

50. $z = \dfrac{1}{t - \sqrt{t + 2}}$

In Exercises 51–56, determine the range of the function.

51. $y = \dfrac{x + 4}{3x - 1}$

52. $y = \dfrac{2x - 3}{x - 2}$

53. $f \circ g$, where $f(x) = 1/x$ and $g(x) = 3x + 4$

54. $g \circ f$, where $f(x) = \dfrac{x + 2}{x - 1}$ and $g(x) = \dfrac{x + 1}{x + 4}$

55. f^{-1}, where $f(x) = \dfrac{x}{3x - 6}$

56. f^{-1}, where $f(x) = \dfrac{5 - x}{1 + x}$

In Exercises 57–64, express the function as a composition of two or more of the following functions:

$$f(x) = 1/x \qquad g(x) = x - 1 \qquad F(x) = |x| \qquad G(x) = \sqrt{x}$$

57. $a(x) = \dfrac{1}{x - 1}$

58. $b(x) = \dfrac{1}{x} - 1$

59. $c(x) = \sqrt{x - 1}$

60. $d(x) = \sqrt{x} - 1$

61. $A(x) = 1/\sqrt{x} - 1$

62. $B(x) = |x - 2|$

63. $C(x) = \sqrt[4]{x} - 1$

64. $D(x) = 1/\sqrt{x - 3}$

For Exercises 65–98, compute the indicated quantity using the functions f, g, and F defined as follows:

$$f(x) = x^2 - x \qquad g(x) = 1 - 2x \qquad F(x) = \dfrac{x - 3}{x + 4}$$

65. $f(-3)$

66. $f(1 + \sqrt{2})$

67. $F\left(\frac{3}{4}\right)$

68. $f(t)$

69. $f(-t)$

70. $g(2x)$

71. $f(x - 2)$

72. $g(x + h)$

73. $g(2) - g(0)$

74. $f(x) - g(x)$

75. $|f(1) - f(3)|$

76. $f(x + h) - f(x)$

77. $f(x^2)$

78. $f(x)/x \quad (x \neq 0)$

79. $[f(x)][g(x)]$

80. $f[f(x)]$

81. $f[g(x)]$

82. $g[f(3)]$

83. $(g \circ f)(x)$

84. $(g \circ f)(x) - (f \circ g)(x)$

85. $(F \circ g)(x)$

86. $\dfrac{g(x + h) - g(x)}{h}$

87. $\dfrac{f(x + h) - f(x)}{h}$

88. $\dfrac{F(x) - F(a)}{x - a}$

89. $F^{-1}(x)$

90. $F[F^{-1}(x)]$

91. $F^{-1}[F(x)]$

92. $F^{-1}(0) - \dfrac{1}{F(0)}$

93. $(g \circ g^{-1})(x)$

94. $g^{-1}(x)$

95. $g^{-1}(-x)$

96. $\dfrac{g^{-1}(x + h) - g^{-1}(x)}{h}$

97. $F^{-1}\left[F\left(\frac{22}{7}\right)\right]$

98. $T^{-1}(x)$, where $T(x) = f(x)/x \quad (x \neq 0)$

In Exercises 99–112, refer to the graph of the function f in the figure. (The axes are marked off in one-unit intevals.)

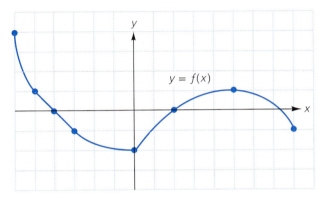

99. Is $f(0)$ positive or negative?

100. Specify the domain and range of f.

101. Find $f(-3)$.

102. Which is larger, $f\left(-\frac{5}{2}\right)$ or $f\left(-\frac{1}{2}\right)$?

103. Compute $f(0) - f(8)$.

104. Compute $|f(0) - f(8)|$.

105. Specify the coordinates of the turning points.

106. What are the minimum and the maximum values of f?

107. On which interval(s) is f decreasing?

108. For which x-values is it true that $1 \leq f(x) \leq 4$?

109. What is the largest value of $f(x)$ when $|x| \leq 2$?

110. Is f a one-to-one function?

111. Does f possess an inverse function?

112. Compute $f[f(-4)]$.

For Exercises 113–128, refer to the graphs of the functions f and g in the figure at the top of the next page. Assume that the domain of each function is $[0, 10]$.

113. For which x-value is $f(x) = g(x)$?

114. For which x-values is it true that $g(x) \leq f(x)$?

115. (a) For which x-value is $f(x) = 0$?
(b) For which x-value is $g(x) = 0$?

116. Compute $f(0) + g(0)$.

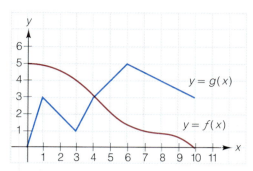

117. Compute each of the following:
(a) $(f + g)(8)$;
(b) $(f - g)(8)$;
(c) $(fg)(8)$;
(d) $(f/g)(8)$.

118. Compute each of the following:
(a) $g[f(5)]$;
(b) $f[g(5)]$;
(c) $(g \circ f)(5)$;
(d) $(f \circ g)(5)$.

119. Which is larger, $(f \circ f)(10)$ or $(g \circ g)(10)$?

120. Compute $g[f(10)] - f[g(10)]$.

121. For which x-values is it true that $f(x) \geq 3$?

122. For which x-values is it true that $|f(x) - 3| \leq 1$?

123. What is the largest number in the range of g?

124. Specify the coordinates of the highest point on the graph of each of the following equations:
(a) $y = g(-x)$;
(b) $y = -g(x)$;
(c) $y = g(x - 1)$;
(d) $y = f(-x)$;
(e) $y = -f(x)$
(f) $y = -f(-x)$.

125. On which intervals is the function g decreasing?

126. What are the coordinates of the turning points of g?

127. For which values of x in the interval $(4, 7)$ is the quantity $\dfrac{f(x) - f(5)}{x - 5}$ negative?

128. For which values of x in the interval $(0, 5)$ is the quantity $\dfrac{f(x) - f(2)}{x - 2}$ positive?

129. Suppose A varies jointly as x and y^2. What happens to the value of A when x is tripled and y is doubled?

130. Suppose B varies directly as x and inversely as y^3. When $x = 6$ and $y = 2$, the value of B is 9. Find B when $x = 3$ and $y = 4$.

131. The intensity of illumination from a small light source varies inversely as the square of the distance from the light. Suppose that a book is 10 ft from a lamp. At what distance from the lamp will the illumination be five times as great?

132. The velocity of an earth satellite in a circular orbit varies inversely as the square root of the radius of the orbit. What happens to the velocity if the radius is decreased by 10 percent?

133. The length of a side in an equilateral triangle varies directly as the square root of the area of the triangle. When the area is $16\sqrt{3}$ cm^2, the length of a side is 8 cm. Find the length of a side when the area is 20 cm^2.

 ## OPTIONAL TI-81 GRAPHING CALCULATOR EXERCISES
EXERCISES FOR CHAPTER FOUR

Note: As preparation for this set of exercises, read the instructions in Appendix A-2 (at the back of this book) for using the ZOOM BOX feature of the calculator.

1. In this exercise we graph the following three functions.

$$Y_1 = |x| \qquad Y_2 = x^2 \qquad Y_3 = x^3$$

(These are three of the basic six functions that are graphed in the text on page 187.) Press the $Y=$ key and type in these three functions.
(a) In the $Y=$ menu, select (only) the function $Y_1 = |x|$. Then press ZOOM 6 to see the graph in the standard viewing rectangle. Check that the graph you obtain is similar to the one shown in Figure 7 in the text on page 187.
(b) In the $Y=$ menu, select (only) the function $Y_2 = x^2$. Then press ZOOM 6 to see the graph in the

standard viewing rectangle. Notice how flat the graph appears to be near the origin. For a closer look, use the ZOOM BOX. For the upper left corner of the box, use a point as close as possible to $(-2, 4)$; for the lower right corner use a point as close as possible to $(2, -1)$.
(c) In the $Y=$ menu, select (only) the function $Y_3 = x^3$. Then press ZOOM 6 to see the graph in the standard viewing rectangle. Notice how flat the graph appears to be near the origin. For a closer look, use the ZOOM BOX. For the lower left corner of the box, use a point as close as possible to $(-2, -8)$; for the upper right corner use a point as close as possible to $(2, 8)$.
(d) Compare the graphs of $Y_2 = x^2$ and $Y_3 = x^3$ on the interval $[0, 1]$. In the $Y=$ menu, select both of

these functions and then press ZOOM 6. Next, press the RANGE key and enter the settings $x_{min} = 0$, $x_{max} = 1$, $y_{min} = 0$, and $y_{max} = 1$. Now press the GRAPH key. Where do the graphs intersect? For x-values in the interval $(0, 1)$, which quantity is larger, x^2 or x^3?

(e) Carry out an analysis similar to the one in part (d) to compare $Y_2 = x^2$ and $Y_3 = x^3$ when $x \geq 1$.

2. In this exercise we graph the following three functions.

$$Y_1 = 1/x \qquad Y_2 = \sqrt{x} \qquad Y_3 = \sqrt{1 - x^2}$$

(These are three of the basic six functions that are graphed in the text on page 187.) Press the $Y=$ key and type in these three functions.

(a) In the $Y=$ menu, select (only) the function $Y_1 = 1/x$. Then press ZOOM 6 to see the graph in the standard viewing rectangle. Next, for a better view, use the ZOOM BOX. For the upper left corner of the box, use a point as close as possible to $(-4, 4)$; for the lower right corner use a point as close as possible to $(4, -4)$. Check that the graph you obtain is similar to the one shown in Figure 7 in the text on page 187.

(b) In the $Y=$ menu, select (only) the function $Y_2 = \sqrt{x}$. Then press ZOOM 6 to see the graph in the standard viewing rectangle. Next, for a better view, use the ZOOM BOX. For the lower left corner of the box, use a point as close as possible to $(-1, -1)$; for the upper right corner use a point as close as possible to $(9, 3)$. Check that the graph you obtain is similar to the one shown in Figure 7 in the text on page 187.

(c) Press ZOOM 6 [with the function from part (b) still selected] to return to the view of the square root function in the standard viewing rectangle. Now press the TRACE key. You will see a blinking cursor located on the curve near the origin. The numbers on the bottom of the screen are the coordinates of the cursor. By repeatedly pressing the right-arrow key, move the cursor along the curve until the x-coordinate is 2. As you can see on the screen, the corresponding y-coordinate of the point on the curve is approximately 1.4142136. What is the *exact* value of the y-coordinate?

(d) In the $Y=$ menu, select (only) the function $Y_3 = \sqrt{1 - x^2}$. Then press ZOOM 6 to see the graph in the standard viewing rectangle. Next, for a better view, use the ZOOM BOX. For the lower left corner of the box, use a point as close as possible to $(-1, 0)$; for the upper right corner use a point as close as possible to $(1, 1)$. This yields a larger picture, but the semicircle is still distorted

because different scales are used on the axes. To correct for this, press ZOOM 5. (What defects still remain in the view shown on your calculator screen?)

3. This exercise shows how to graph functions that are defined piecewise. As an example, we will use

$$f(x) = \begin{cases} x^2 - 1 & \text{if } x \leq 2 \\ 5 - x & \text{if } x > 2 \end{cases}$$

As preparation, clear all functions that appear in the $Y=$ menu, and press ZOOM 6 for the standard viewing rectangle. Now press the $Y=$ key.

(a) In the $Y=$ menu, we are going to define Y_1 to be the function $x^2 - 1$ with the restriction $x \leq 2$. Begin by typing in the following sequence of keystrokes.

$$\boxed{(} \quad \boxed{\text{X|T}} \quad \boxed{\wedge} \quad \boxed{2} \quad \boxed{-} \quad \boxed{1} \quad \boxed{)}$$

Do not press ENTER now. Rather, continue with the following keystrokes:

$$\boxed{(} \quad \boxed{\text{X|T}} \quad \boxed{\text{2nd}} \quad \boxed{\text{MATH}} \quad \boxed{5} \quad \boxed{2} \quad \boxed{)}$$

Your screen display in the $Y=$ menu should now read: $Y_1 = (x\hat{\ }2 - 1)(x \leq 2)$. Press the GRAPH key to see the portion of $x^2 - 1$ corresponding to $x \leq 2$. In order to remove the extraneous points that appear in the graph, press the MODE key and select "Dot" rather than "Connected"; then press the ENTER key, then the GRAPH key.

(b) Next, in the $Y=$ menu, we are going to define Y_2 to be the function $5 - x$ with the restriction $x > 2$. The sequence of keystrokes is similar to that in part (a). In the Y_2 line, begin by typing

$$\boxed{(} \quad \boxed{5} \quad \boxed{-} \quad \boxed{\text{X|T}} \quad \boxed{)}. \text{ Do not press ENTER.}$$

Rather, continue with the following sequence of keystrokes.

$$\boxed{(} \quad \boxed{\text{X|T}} \quad \boxed{\text{2nd}} \quad \boxed{\text{MATH}} \quad \boxed{3} \quad \boxed{2} \quad \boxed{)}$$

Your screen display for Y_2 should read: $Y_2 = (5 - x)(x > 2)$. By pressing the GRAPH key now, you will obtain the required graph. Check that your result is consistent with Figure 9 in the text on page 188.

4. Graph the function g defined by

$$g(x) = \begin{cases} -x^2 & \text{if } -2 \leq x \leq 0 \\ \frac{1}{2}x & \text{if } 0 < x \leq 4 \end{cases}$$

Hint: You need to create the following screen in the $Y=$ menu (as indicated in the previous exercise, the symbols for inequality are accessed by pressing the "2nd" key and then the MATH key):

$$:Y_1 = (-x\hat{\ }2)(x \geq -2)(x \leq 0)$$
$$:Y_2 = (x/2)(x > 0)(x \leq 4)$$

As in Exercise 3, you will want to view the graph in the "Dot" mode rather than the "Connected" mode.

In Exercises 5–10, graph the functions using the techniques indicated in Exercises 3 and 4.

5. $y = \begin{cases} x^2 & \text{if } 0 \le x < 1 \\ 1/x & \text{if } 1 \le x \le 4 \end{cases}$

6. $y = \begin{cases} |x| & \text{if } -4 \le x < 0 \\ x^2 & \text{if } 0 \le x < 1 \\ 1/x & \text{if } 1 \le x \le 4 \end{cases}$

7. $y = \begin{cases} \sqrt{1 - x^2} & \text{if } -1 \le x < 0 \\ \sqrt{x} & \text{if } 0 \le x \le 4 \end{cases}$

8. $y = \begin{cases} x^3 & \text{if } -1 \le x < 0 \\ \sqrt{x} & \text{if } 0 \le x \le 4 \end{cases}$

9. $y = \begin{cases} -1/x & \text{if } x < -1 \\ x^2 & \text{if } -1 \le x < 1 \\ 1/x & \text{if } x \ge 1 \end{cases}$

10. $y = \begin{cases} |x| & \text{if } x < -1 \\ x^2 & \text{if } -1 \le x < 1 \\ |x| & \text{if } x \ge 1 \end{cases}$

For Exercises 11–16, tell how the graph in part (a) should be translated and/or reflected to obtain the remaining graphs. Then draw the graphs and check your statements.

11. (a) $y = x^2$ (b) $y = (x + 3)^2$
    (c) $y = -(x + 3)^2$ (d) $y = (-x + 3)^2$

12. (a) $y = x^3$ (b) $y = (x + 2)^3$
    (c) $y = -(x + 2)^3$ (d) $y = (-x + 2)^3$

13. (a) $y = |x|$ (b) $y = |x + 4|$
    (c) $y = |-x + 4|$ (d) $y = |-x + 4| - 2$

14. (a) $F(x) = \frac{1}{2}x^3 - \frac{3}{2}x^2$ (b) $F(-x)$
    (c) $-F(x)$ (d) $-F(-x)$

15. (a) $G(x) = 2/(1 + x^2)$ (b) $G(x + 3)$
    (c) $G(x - 3)$ (d) $-G(x) + 2$

16. (a) $y = \sqrt{x + 2}$ (b) $y = \sqrt{2 - x}$
    (c) $y = 3 - \sqrt{2 - x}$

In Exercises 17–22, the two given functions are inverse functions. In each case, graph the pair of inverse functions on the same set of axes along with the line $y = x$. Note the symmetry. Make use of the ZOOM 5 setting. (Why?)

17. $y = 4x + 8$ and
    $y = \frac{1}{4}x - 2$

18. $y = 3x - 5$ and
    $y = (x + 5)/3$

19. $y = \sqrt{x - 2}, \quad x \ge 2$
    $y = x^2 + 2, \quad x \ge 0$

20. $y = \sqrt{2x + 3}, \quad x \ge -1.5$
    $y = (x^2 - 3)/2, \quad x \ge 0$

21. $y = \dfrac{1}{\sqrt{x}}$ and
    $y = \dfrac{1}{x^2}, \quad x > 0$

22. $y = \dfrac{1}{2x - 3}$ and
    $y = \dfrac{3x + 1}{2x}$

In Exercises 23–28, graph each function and then use the horizontal line test to say whether or not the function is one-to-one.

23. $f(x) = x^2 + 2x$

24. $f(x) = x^3 + 2x$

25. $f(x) = 2x^3 + x^2$ *Hint:* First looks may be deceiving. After looking at the graph in the standard viewing rectangle (ZOOM 6), make use of the ZOOM BOX to look more closely at the graph.

26. (a) $g(x) = 0.01x^4 - 1$
    (b) $g(x) = 2x^5 + x - 1$

27. (a) $y = x^3 + x^2 + x$
    (b) $y = x^3 - x^2 + x$
    (c) $y = x^3 - x^2 - x$

28. (a) $y = x^x, \quad x > 1$
    (b) $y = x^x, \quad x > 0$

29. Let $F(x) = \sqrt{1 - x^2}$, with domain $-1 \le x \le 0$. Find $F^{-1}(x)$ and then graph both functions on the same set of axes. (See Exercises 3 and 4 for graphing functions with domain restrictions.) Check your results: are the two graphs symmetric about the line $y = x$? (Why will the ZOOM 5 setting be necessary here?)

30. Let $f(x) = 2x + 1$ and $g(x) = \frac{1}{4}x - 3$.
    (a) Compute each of the following: (i) $f(g(x))$; (ii) $g(f(x))$; (iii) $f^{-1}(x)$; (iv) $g^{-1}(x)$; (v) $f^{-1}(g^{-1}(x))$; (vi) $g^{-1}(f^{-1}(x))$
    (b) On the same set of axes, graph the two answers that you obtained in (i) and (v) in part (a). Note that the graphs are *not* symmetric about $y = x$. The conclusion here is that the inverse function for $f(g(x))$ is not $f^{-1}(g^{-1}(x))$.
    (c) On the same set of axes, graph the two answers that you obtained in (i) and (vi) in part (a); also put the line $y = x$ into the picture. Note that the two graphs *are* symmetric about the line $y = x$. (If this symmetry does not appear in your graph, press ZOOM 5 so that the same scales are used on both axes.) The conclusion here is that the inverse function for $f(g(x))$ is $g^{-1}(f^{-1}(x))$. For reference, we summarize this fact about the inverse of a composite function as follows.

$$\boxed{(f \circ g)^{-1} = g^{-1} \circ f^{-1}}$$

31. Let $f(x) = \dfrac{3x}{1 + |x|}$.
    (a) Graph this function. Note that, according to the horizontal line test, the function is one-to-one.

(b) Since the function is one-to-one, it has an inverse. Compute this inverse. *Hint:* Consider two cases. First, if $x \geq 0$, then the given equation is $y = 3x/(1 + x)$. Second, if $x < 0$, the given equation is equivalent to $y = 3x/(1 - x)$. (Why?) When you think you have found f^{-1}, graph it and f on the same set of axes and check that the graphs are symmetric about $y = x$.

    Answer: For the inverse, you should obtain

$$y = \begin{cases} x/(3 - x) & \text{if } 0 \leq x < 3; \\ x/(3 + x) & \text{if } -3 < x < 0. \end{cases}$$

32. (a) Graph the following two functions on the same set of axes. (For Y_1 use the positive square root; for Y_2 use the negative square root. Use ZOOM 6 for the standard viewing rectangle.)

$$y = \pm \sqrt{\frac{6x^2 - x^3}{2 + x}}$$

(The resulting curve is called the *Trisectrix of Maclaurin*. The curve, studied by Colin Maclaurin in 1742, can be used to trisect an angle.) If you've entered the equations correctly, your graph should be similar to the following one.

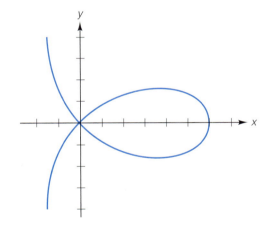

(b) For the rest of this problem, we focus on the first-quadrant portion of the trisectrix. To get a better look at it, press the Y= key and deselect Y_2. Now move the cursor to the end of the expression for Y_1 and add the restriction ($x \geq 0$). (Recall that inequalities are accessed by pressing the "2nd" key followed by the MATH key.) Now press the GRAPH key. Finally, use the ZOOM BOX. For the lower left corner of the box, choose a point as close as possible to $(-1, 1)$; for the upper right corner, choose a point as close as possible to $(7, 3)$.

(c) Let's let $y = g(x)$ denote the function whose graph is the arch in the first quadrant, which you just drew in part (b). By looking carefully at the graph, estimate the largest value of c for which $y = g(x)$ is one-to-one on the closed interval $[0, c]$. (This is also the input for which g is maximum.) For example, from the picture, it's clear that the function is one-to-one on the interval $[0, 2]$, and it's also clear that the function is not one-to-one on the interval $[0, 5]$. So c is something between 2 and 5. Now use the TRACE key and cursor arrows to obtain a more accurate value for c. Make a note of this estimate for c. Next, use the ZOOM BOX to take a closer look at the graph near its turning point. Then use the TRACE key to obtain a better estimate for c. Repeat this process several times, then go on to part (d).

(d) Using calculus, it can be shown that the exact value of c is $2\sqrt{3}$. Use your calculator to evaluate this expression. Did you come close to this value in part (c)?

CHAPTER FOUR TEST

1. Find the domain of the function defined by $f(x) = \sqrt{x^2 - 5x - 6}$.

2. Find the range of the function defined by $g(x) = \dfrac{2x - 8}{3x + 5}$.

3. Let $f(x) = 2x^2 - 3x$ and $g(x) = 2 - x$. Compute each of the following:
   (a) $(f - g)(x)$;   (b) $(f \circ g)(x)$;   (c) $f[g(-4)]$.

4. Let $f(t) = 2/t$. Compute $\dfrac{f(t) - f(a)}{t - a}$.

5. Let $g(x) = 2x^2 - 5x$. Compute $\dfrac{g(x + h) - g(x)}{h}$.

6. Find $g^{-1}(x)$ given that $g(x) = -4x/(6x + 1)$.

7. The graph of a function f is a line segment joining the two points $(-3, 1)$ and $(5, 6)$. Determine the slope of the line segment that results from graphing the equation $y = -f^{-1}(x)$.

8. Graph each function and specify the intercepts:

    (a) $y = -|x - 3| + 1$;   (b) $y = \dfrac{1}{x + 3} - 2$.

9. The figure shows the graph of a function g with domain $[-4, 2]$.

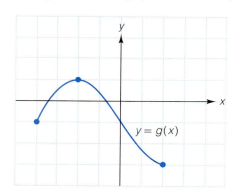

$y = g(x)$

    (a) Specify the range of g.
    (b) What are the coordinates of the turning point?
    (c) What is the minimum value of g?
    (d) Which input yields a maximum value for g? What is that maximum value?
    (e) On which interval is g decreasing?

10. Let $f(x) = x^2 - 3x - 1$. Compute each of the following:

    (a) $f\left(-\frac{3}{2}\right)$;   (b) $f(\sqrt{3} - 2)$.

11. Graph the function F defined by

$$F(x) = \begin{cases} |x + 1| & \text{if } x < -1 \\ -x^2 + 1 & \text{if } x > -1 \end{cases}$$

    What is the domain of this function?

12. The graph of f is a line segment joining the points $(1, 3)$ and $(5, -2)$. Sketch the graph of $y = f(-x)$.

13. Given that the domains of f and f^{-1} are both $(-\infty, \infty)$ and that $f^{-1}(-3) = 1$, solve the equation $5 + f(4t - 3) = 2$ for t.

14. Suppose that z varies directly as x^2 and inversely as the square root of t. When $x = 6$ and $t = 9$, the value of z is 12. Find z when $x = 3$ and $t = 16$.

CHAPTER FIVE

POLYNOMIAL AND RATIONAL FUNCTIONS: APPLICATIONS TO OPTIMIZATION

INTRODUCTION

In the previous chapter we studied some rather general rules for working with functions and their graphs. Now we focus our attention on several particular types of functions. In Section 5.1 we discuss linear functions and their applications. As you'll see, these applications are quite diverse, ranging from business and economics to physics and statistics. Section 5.2 contains a discussion of quadratic functions. Section 5.3 continues the work we began in Chapter 2 on setting up equations; the objective in Section 5.3 is to prepare you to solve the maximum and minimum problems that are discussed in Section 5.4. (If a friend of yours happens to be taking calculus, you will find that you can solve some of your friend's maximum and minimum problems without using calculus.) The last two sections of this chapter, 5.5 and 5.6, present a number of techniques for graphing polynomial and rational functions. Although these graphs can be obtained by using a graphing calculator, our goal in Sections 5.5 and 5.6 is to *understand* why the graphs look as they do.

5.1 LINEAR FUNCTIONS

All decent functions are practically linear.

Professor Andrew Gleason

By a **linear function**, we mean a function defined by an equation of the form

$$f(x) = Ax + B$$

where A and B are constants. In this chapter, the constants A and B will always be real numbers. From our work in Chapter 3, we know that the graph of $y = Ax + B$ is a straight line.

EXAMPLE 1   Suppose f is a linear function. If $f(1) = 0$ and $f(2) = 3$, find an equation defining f.

Solution   From the statement of the problem, we know that the graph of f is a straight line passing through the points $(1, 0)$ and $(2, 3)$. Thus the slope of the line is

$$m = \frac{y_2 - y_1}{x_2 - x_1} = \frac{3 - 0}{2 - 1} = 3$$

Now we can use the point–slope formula to find the required equation. We have

$$y - y_1 = m(x - x_1)$$
$$y - 0 = 3(x - 1)$$
$$y = 3x - 3$$

This is the equation defining f. If we wish, we can rewrite it using function notation: $f(x) = 3x - 3$.   ■

One basic application of linear functions in business and economics is *linear* or *straight-line depreciation*. In this situation we assume that the value V of an asset (such as a machine or an apartment building) decreases linearly over time t; that is, $V = mt + b$, where the slope m is negative.

EXAMPLE 2   A factory owner buys a new machine for $8000. After ten years, the machine has a salvage value of $500.

(a) Assuming linear depreciation (as indicated in Figure 1), find a formula for the value V of the machine after t years, where $0 \le t \le 10$.
(b) Use the formula derived in part (a) to find the value of the machine after five years.

Solution   **(a)** We need to determine m and b in the equation $V = mt + b$. From Figure 1 we see that the V-intercept of the line segment is $b = 8000$. Furthermore, since the line segment passes through the two points $(10, 500)$ and $(0, 8000)$, the slope m is $\dfrac{8000 - 500}{0 - 10}$, or -750. So we have $m = -750$ and $b = 8000$, and the required equation is

$$V = -750t + 8000 \qquad (0 \le t \le 10)$$

(b) Substituting $t = 5$ in the equation $V = -750t + 8000$ yields

$$V = -750(5) + 8000$$
$$= -3750 + 8000 = 4250$$

Thus the value of the machine after five years is $4250.   ■

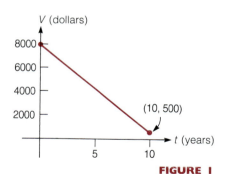

FIGURE 1

It is instructive to keep track of the units associated with the slope in Example 2. Repeating the slope calculation and keeping track of the units, we have

$$m = \frac{\$8000 - \$500}{0 \text{ years} - 10 \text{ years}} = -\$750/\text{year}$$

Thus, slope is a *rate of change*. In this example, the slope represents the rate of change of the value of the machine.

As a second example of slope as a rate of change, let us suppose that a small manufacturer of handmade running shoes knows that her total cost, $C(x)$,

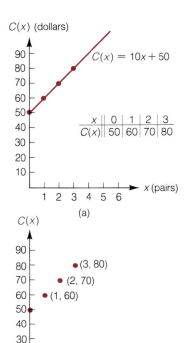

$C(x)$ (dollars)

$C(x) = 10x + 50$

x	0	1	2	3
$C(x)$	50	60	70	80

x (pairs)

(a)

$C(x)$

•(3, 80)
• (2, 70)
•(1, 60)

x

(b)

FIGURE 2

in dollars, for producing x pairs of shoes each business day is given by the linear function

$$C(x) = 10x + 50$$

Figure 2(a) displays a table and graph for this function. Actually, since x represents the daily number of pairs of shoes, x can assume only whole number values. So, technically, the graph that should be given is the one in Figure 2(b). However, the graph in Figure 2(a) turns out to be useful in practice, and we shall follow this convention.

The slope of the line $C(x) = 10x + 50$ in Figure 2(a) is 10, the coefficient of x. But to understand the units involved, let's calculate the slope using two of the points in Figure 2(a). Using the points $(0, 50)$ and $(1, 60)$ and keeping track of the units, we have

$$m = \frac{\$60 - \$50}{1 \text{ pair} - 0 \text{ pair}} = \frac{\$10}{1 \text{ pair}} = \$10/\text{pair}$$

Again the slope is a rate. In this case, the slope represents the rate of increase of cost; each additional pair of shoes produced costs the manufacturer \$10.

In the study of economics, an equation that gives the cost $C(x)$ for producing x units of a commodity is called a **cost equation** or **cost function**. When the graph of the cost equation is a line, we define the **marginal cost** as the additional cost to produce one more unit. Thus, in the preceding example, the marginal cost is \$10 per pair, and we see that the slope of the line in Figure 2(a) represents this marginal cost.

EXAMPLE 3   Suppose that the cost $C(x)$ in dollars of producing x bicycles is given by

$$C(x) = 625 + 45x$$

(a) Find the cost of producing 10 bicycles.
(b) What is the marginal cost?
(c) Use the answers in parts (a) and (b) to find the cost of producing 11 bicycles. Then check the answer by evaluating $C(11)$.

Solution   (a) Using $x = 10$ in the cost equation, we have

$$C(10) = 625 + 45(10) = 1075$$

Thus, the cost of producing ten bicycles is \$1075.

(b) Since C is a linear function, the marginal cost is the slope and we have

$$\text{marginal cost} = \$45 \text{ per bicycle}$$

(c) According to the result in part (b), each additional bicycle costs \$45. Therefore, we can compute the cost of 11 bicycles by adding \$45 to the cost for 10 bicycles:

$$\text{cost of 11 bicycles} = \text{cost of 10 bicycles} + \text{marginal cost}$$
$$= \$1075 + \$45$$
$$= \$1120$$

So, the cost of producing 11 bicycles is $1120. We can check this result by using the cost function $C(x) = 625 + 45x$ to compute $C(11)$ directly:

$$C(11) = 625 + 45(11) = 625 + 495$$

$$= \$1120 \qquad \text{as obtained previously}$$

Interpreting slope as a rate of change is not restricted to applications in business or economics. Suppose, for example, that you are driving a car at a steady rate of 50 mph. Using the distance formula from elementary mathematics,

$$\text{distance} = \text{rate} \times \text{time}$$

or

$$d = rt$$

we have, in this case, $d = 50t$, where d represents the distance traveled (in miles) in t hours. In Figure 3, we show a graph of the linear function $d = 50t$. The slope of this line is 50, the coefficient of t. But 50 is also the given rate of speed, in miles per hour. So again, slope is a rate of change. In this case, slope is the **velocity**, or *rate of change of distance with respect to time*.

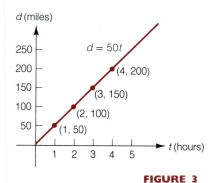

FIGURE 3

The slope of the line is the rate of change of distance with respect to time.

TABLE 1

Evolution of the Record for the Mile Run, 1911–1980

x (Year)	y (Time)	
1911	4:15.4	(John Paul Jones, U.S.)
1913	4:14.6	(John Paul Jones, U.S.)
1915	4:12.6	(Norman Taber, U.S.)
1923	4:10.4	(Paavo Nurmi, Finland)
1931	4:09.2	(Jules Ladoumegue, France)
1933	4:07.6	(Jack Lovelock, New Zealand)
1934	4:06.8	(Glen Cunningham, U.S.)
1937	4:06.4	(Sidney Wooderson, Great Britain)
1942	4:06.2	(Gunder Haegg, Sweden)
1942	4:06.2	(Arne Andersson, Sweden)
1942	4:04.6	(Gunder Haegg, Sweden)
1943	4:02.6	(Arne Andersson, Sweden)
1944	4:01.6	(Arne Andersson, Sweden)
1945	4:01.4	(Gunder Haegg, Sweden)
1954	3:59.4	(Roger Bannister, Great Britain)
1954	3:58.0	(John Landy, Australia)
1957	3:57.2	(Derek Ibbotson, Great Britain)
1958	3:54.5	(Herb Elliott, Australia)
1962	3:54.4	(Peter Snell, New Zealand)
1964	3:54.1	(Peter Snell, New Zealand)
1965	3:53.6	(Michel Jazy, France)
1966	3:51.3	(Jim Ryun, U.S.)
1967	3:51.1	(Jim Ryun, U.S.)
1975	3:51.0	(Filbert Bayi, Tanzania)
1975	3:49.4	(John Walker, New Zealand)
1979	3:49.0	(Sebastian Coe, Great Britain)
1980	3:48.8	(Steve Ovett, Great Britain)

We conclude this section with an example that shows a statistical application of linear functions. Table 1 shows the evolution of the record for the one-mile run during the years 1911–1980. (In case you're wondering why the table begins with 1911, John Paul Jones was the first twentieth-century runner to break the previous century's record of 4:15.6, set in 1895.) In Figure 4(a) we've plotted the (x, y) pairs given in the table. The resulting plot is called a **scatter diagram.**

A striking feature of the scatter diagram in Figure 4(a) is that the records do not appear to be leveling off; rather, they seem to be decreasing in an approximately linear fashion. Using the *least-squares technique* from the field of statistics, it can be shown that the linear function that best fits this particular set of data is

$$f(x) = -0.41x + 1040.94 \qquad (1)$$

The graph of this line is shown in Figure 4(b). The line itself is referred to as the **regression line,** or the **least-squares line.** (The formulas for determining this line are given in Exercise 35.)

As an intuitive check on the least-squares line, let's use it to estimate what the record for the mile run might have been in 1985 (then we'll check our prediction against the actual record). With $x = 1985$, we have

$$f(1985) = -0.41 (1985) + 1040.94$$
$$= 227.09 \text{ seconds}$$

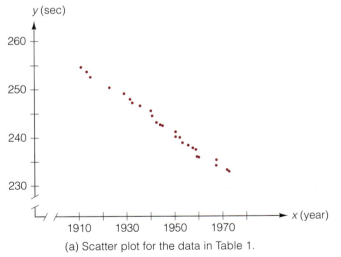

(a) Scatter plot for the data in Table 1.

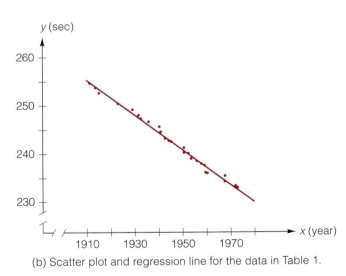

(b) Scatter plot and regression line for the data in Table 1.

FIGURE 4

Thus (after converting to minutes and rounding off), our least-squares estimate for the 1985 record is 3:47.1. This agrees favorably with the actual record of 3:46.31 set by Steve Cram of Great Britain.

EXAMPLE 4 The June 1976 issue of *Scientific American* contained an article entitled "Future Performance in Footracing" by H. W. Ryder, H. J. Carr, and P. Herget. According to the authors of the article, "It appears likely that within 50 years the record [for the mile] will be down to 3:30." Use the regression line defined in equation

(1) to make a projection for the mile time in the year 2026 (which will be 50 years after the article appeared). Is your projection close to the one given in the *Scientific American* article?

Solution   The given equation for the regression line is

$$f(x) = -0.41x + 1040.94$$

Setting $x = 2026$ and using a calculator, we have

$$f(2026) = -0.41\,(2026) + 1040.94$$
$$= 210.28 \text{ seconds}$$

Converting this result to minutes and then rounding off, our projection for the year 2026 is 3:30.3, which is essentially the same as the projection in *Scientific American*. ∎

EXERCISE SET 5.1

A

In Exercises 1–10, find the linear functions satisfying the given conditions.

1. $f(-1) = 0$ and $f(5) = 4$

2. $f(3) = 2$ and $f(-3) = -4$

3. $g(0) = 0$ and $g(1) = \sqrt{2}$

4. The graph passes through the points $(2, 4)$ and $(3, 9)$.

5. $f(\frac{1}{2}) = -3$ and the graph of f is a line parallel to the line $x - y = 1$.

6. $g(2) = 1$ and the graph of g is perpendicular to the line $6x - 3y = 2$.

7. The graph of f is a horizontal line that passes through the larger of the two y-intercepts of the circle $x^2 - 2x + y^2 - 3 = 0$.

8. The x- and y-intercepts of the graph of g are 1 and 4, respectively.

9. The graph of the inverse function passes through the points $(-1, 2)$ and $(0, 4)$.

10. The x- and y-intercepts of the inverse function are 5 and -1, respectively.

11. Let $f(x) = 3x - 4$ and $g(x) = 1 - 2x$. Determine if the function $f \circ g$ is linear.

12. Explain why there is no linear function with a graph that passes through all three of the points $(-3, 2)$, $(1, 1)$, and $(5, 2)$.

13. A factory owner buys a new machine for $20,000. After eight years, the machine has a salvage value of $1000. Find a formula for the value of the machine after t years, where $0 \le t \le 8$.

14. A manufacturer buys a new machine costing $120,000. It is estimated that the machine has a useful lifetime of ten years and a salvage value of $4000 at that time.
    (a) Find a formula for the value of the machine after t years, where $0 \le t \le 10$.
    (b) Find the value of the machine after eight years.

15. A factory owner installs a new machine costing $60,000. Its expected lifetime is five years, and at the end of that time the machine has no salvage value.
    (a) Find a formula for the value of the machine after t years, where $0 \le t \le 5$.
    (b) Complete the following depreciation schedule.

End of Year	Yearly Depreciation	Accumulated Depreciation	Value V
0	0	0	60,000
1			
2			
3			
4			
5		60,000	0

16. Let x denote a temperature on the Celsius scale, and let y denote the corresponding temperature on the Fahrenheit scale.
    (a) Find a linear function relating x and y; use the facts that 32° F corresponds to 0° C, and 212° F

corresponds to 100° C. Write the function in the form $y = Ax + B$.

(b) What Celsius temperature corresponds to 98.6° F?

(c) Find a number z for which $z°$ F $= z°$ C.

17. Suppose that the cost $C(x)$, in dollars, of producing x electric fans is given by $C(x) = 450 + 8x$.
    (a) Find the cost to produce 10 fans.
    (b) Find the cost to produce 11 fans.
    (c) Use your answers in parts (a) and (b) to find the marginal cost. (Then check that your answer is the slope of the line.)

18. Suppose that the cost to a manufacturer of producing x units of a certain motorcycle is given by $C(x) = 220x + 4000$, where $C(x)$ is in dollars.
    (a) Find the marginal cost.
    (b) Find the cost of producing 500 motorcycles.
    (c) Use your answers in parts (a) and (b) to find the cost of producing 501 motorcycles.

19. Suppose that the cost $C(x)$, in dollars, of producing x units of a certain cassette tape player is given by $C(x) = 400 + 50x$.
    (a) Compute $C(n + 1) - C(n)$.
    (b) What is the marginal cost?
    (c) What is the relationship between the answers in parts (a) and (b)?

20. Suppose that the cost $C(x)$, in dollars, of producing x cassette tapes is given by $C(x) = 0.5x + 500$.
    (a) Graph the given equation.
    (b) Compute $C(150)$.
    (c) If you add the marginal cost to the answer in part (b), you obtain a certain dollar amount. What does this amount represent?

21. The following graphs each relate distance and time for a moving object. Determine the velocity in each case.

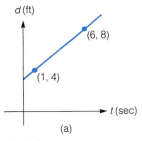

(a)

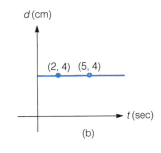

(b)

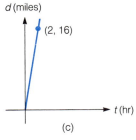

(c)

22. The distance d, in feet, covered by a particular object in t sec is given by the equation $d = 5t$.
    (a) Find the velocity of the object.
    (b) Find the distance covered in 15 sec.
    (c) Use your answers in parts (a) and (b) to find the distance covered in 16 sec. Check your answer by substituting $t = 16$ in the given distance formula.

23. Two points A and B move along the x-axis. After t sec, their positions are given by the equations

$$A: \quad x = 3t + 100$$
$$B: \quad x = 20t - 36$$

(a) Which point is traveling faster, A or B?

(b) Which point is farther to the right when $t = 0$?

(c) At what time t do A and B have the same x-coordinate?

24. A point moves along the x-axis, and its x-coordinate after t seconds is $x = 4t + 10$. (Assume that x is in centimeters.)
    (a) What is the velocity?
    (b) What is the x-coordinate when $t = 2$ sec?
    (c) Use your answers in parts (a) and (b) to find the x-coordinate when $t = 3$ sec. Check your answer by letting $t = 3$ in the given equation.

25. (a) On graph paper, plot the following points: $(1, 0)$, $(2, 3)$, $(3, 6)$, $(4, 7)$.
    (b) In your scatter diagram from part (a), sketch a line that, you feel, seems to best fit the data. Estimate the slope and y-intercept of the line.
    (c) The actual regression line in this case is $y = 2.4x - 2$. Add the graph of this line to your sketch from parts (a) and (b).

26. (a) On graph paper, plot the following points: $(1, 2)$, $(3, 2)$, $(5, 4)$, $(8, 5)$, $(9, 6)$.
    (b) In your scatter diagram from part (a), sketch a line that, you feel, seems to best fit the data.
    (c) Using your sketch from part (b), estimate the y-intercept and the slope of the regression line.
    (d) The actual regression line is $y = 0.518x + 1.107$. Check to see if your estimates in part (c) are consistent with the actual y-intercept and slope. Graph this line along with the points given in part (a). (In sketching the line, use the approximation $y = 0.5x + 1.1$.)

27. The following table shows the population of the city of Los Angeles over the years 1930–1988. The accompanying graph shows the corresponding scatter plot and regression line. The equation of the regression line (after some rounding off) is

$$y = 37{,}025x - 70{,}226{,}566$$

(a) Predict whether the population of Los Angeles will reach 4 million by the year 2000.
(b) Make a projection for the population of Los Angeles in the year 2010. Round off your answer to the nearest 5,000.

Year	Population of Los Angeles
1930	1,238,048
1940	1,504,277
1950	1,970,358
1960	2,479,015
1970	2,816,061
1980	2,966,850
1988	3,361,500

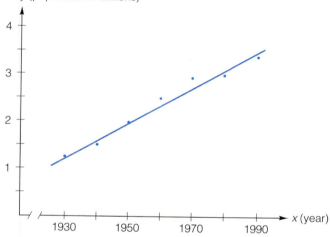

Year	Expenditure (millions of dollars)
1975	30,177
1977	37,566
1979	48,736
1980	53,538
1981	57,470
1982	57,680
1983	61,357
1985	73,824

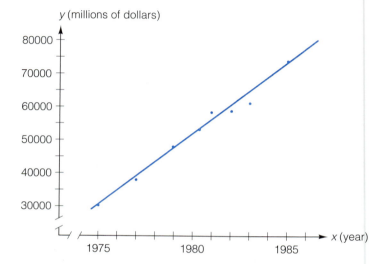

28. The following table shows expenditures on pollution control and abatement in the United States during the period 1975–1985. (The figures include both government and private-sector expenditures.) The accompanying graph displays the scatter plot and regression line for the data. The equation of the regression line (after rounding off) is

$$f(x) = (4198.8)x - 8262216$$

(a) Use the regression line (and a calculator, of course) to make a projection for the expenditures in the year 1995.
(b) Find $f^{-1}(x)$.
(c) Use your answer in part (b) to find the year when the expenditures might reach 125 billion dollars.

29. Many companies in the United States manufacture or assemble their products abroad to save on labor costs. For example, the toy industry in the United States imports approximately 30% of its products from China. (GI Joe and Teenage Mutant Ninja Turtles are assembled in China.) The following table shows the dollar value of these imports during the period 1986–1990. The graph (on the next page) displays the scatter plot and regression line for the data. The equation of the regression line is

$$f(x) = (0.3428)x - 680.6114$$

Year	Imports (billions of dollars)
1986	0.237
1987	0.509
1988	0.765
1989	1.317
1990	1.547

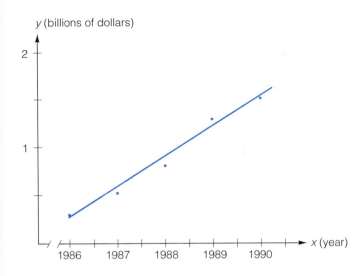

y (billions of dollars)

(a) Use the regression line to make a projection for the value of the toy imports from China in 1996.
(b) Predict whether the value of the toy imports will exceed 5 billion dollars by the year 2000.

30. This exercise illustrates one of the pitfalls that can arise in using a regression line: large discrepancies can occur when the regression line is used to make long-term projections. The following table shows the population of California during the years 1860–1900. The accompanying graph displays the scatter plot of the data and the regression line

$$f(x) = (28632.69)x - 52928780$$

As you can see from the scatter plot, the population growth over this period is very nearly linear.

(a) Use the regression line to estimate the population of California in 1990.
(b) According to the U.S. Bureau of the Census, the population of California in 1990 was 29,839,250. Is your estimate in part (a) close to this figure?

B

31. Find a linear function $f(x) = mx + b$ such that m is positive and $(f \circ f)(x) = 9x + 4$.

32. Let $f(x) = Ax + B$. Find $\dfrac{f(x + h) - f(x)}{h}$.

33. (a) Let f be a linear function. Show that

$$f\left(\frac{x_1 + x_2}{2}\right) = \frac{f(x_1) + f(x_2)}{2}$$

(In words: The output of the average is the average of the outputs.)
(b) Show that the equation in part (a) does not hold for the function $f(x) = x^2$.

34. Let f be a linear function such that

$$f(a + b) = f(a) + f(b)$$

for all real numbers a and b. Show that the graph of f passes through the origin. *Hint:* Let $f(x) = Ax + B$ and show that $B = 0$.

35. This exercise shows how to compute the slope and the y-intercept of the regression line. As an example, we'll work with the simple data set given in Exercise 25:

x	1	2	3	4
y	0	3	6	7

Year	Population
1860	379,994
1870	560,247
1880	864,694
1890	1,213,398
1900	1,485,053

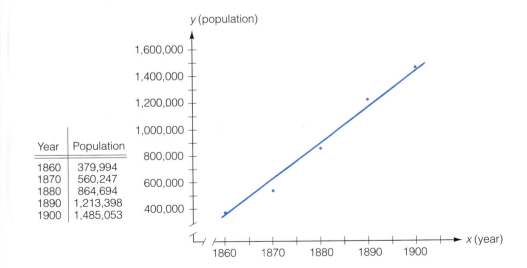

y (population)

(a) Let Σx denote the sum of the x-coordinates in the data set, and let Σy denote the sum of the y-coordinates. Check that $\Sigma x = 10$ and $\Sigma y = 16$.

(b) Let Σx^2 denote the sum of the squares of the x-coordinates, and let Σxy denote the sum of the products of the corresponding x- and y-coordinates. Check that $\Sigma x^2 = 30$ and $\Sigma xy = 52$.

(c) The slope m and the y-intercept b of the regression line satisfy the following pair of simultaneous equations [in the first equation, n denotes the number of points (x, y) in the data set]:

$$\begin{cases} nb + (\Sigma x)m = \Sigma y \\ (\Sigma x)b + (\Sigma x^2)m = \Sigma xy \end{cases}$$

So, in the present example, these equations become

$$\begin{cases} 4b + 10m = 16 \\ 10b + 30m = 52 \end{cases}$$

Solve this pair of equations for m and b, and check that your answers agree with the values in Exercise 25(c).

In Exercises 36–39, use the method described in Exercise 35 to find the equation of the regression line for the given data sets.

36.

x	2	4	8	10
y	-7	-5	-2	-1

37.

x	1	2	3	4	5
y	2	3	9	9	11

38.

x	1	2	3	4	5
y	16	13.1	10.5	7.5	2

39.

x	520	740	560	610	650
y	81	98	83	88	95

5.2 QUADRATIC FUNCTIONS

After the linear functions, the next simplest functions are the **quadratic functions**, which are defined by equations of the form

$$f(x) = ax^2 + bx + c \qquad (a \neq 0)$$

where a, b, and c are constants and a is not zero. In this chapter the constants a, b, and c will always be real numbers. We will see that the graph of any quadratic function is a curve called a **parabola**, which is similar or identical in shape to the graph of $y = x^2$. Figure 1 displays the graphs of two typical parabolas.

FIGURE 1

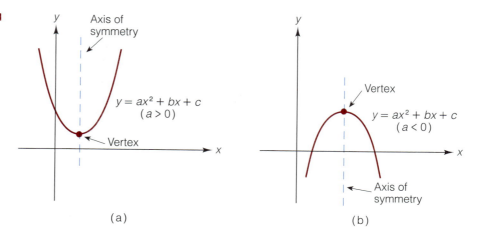

(a)

(b)

Subsequent examples will demonstrate that the parabola $y = ax^2 + bx + c$ opens upward when $a > 0$ and downward when $a < 0$. As Figure 1 indicates, the turning point on the parabola is called the **vertex**. The **axis of symmetry** of the parabola $y = ax^2 + bx + c$ is the vertical line passing through the vertex.

The methods that we will use for dealing with quadratic functions have already been developed in Chapters 2 and 4. In particular, the following two topics are prerequisites for understanding the examples in this section:

1. Completing the square; for a review, see Section 2.3.
2. Translations and reflections; for a review, see Section 4.3.

EXAMPLE 1 Graph the function $y = x^2 - 2x + 3$.

Solution The idea here is to use the technique of completing the square; this will enable us to obtain the required graph simply by shifting the basic $y = x^2$ graph. We begin by writing the given equation

$$y = x^2 - 2x \qquad + 3$$

To complete the square for the x-terms we want to add 1. (Check this.) Of course, to keep the equation in balance we have to account for this by writing

$$y = (x^2 - 2x + 1) + 3 - 1 \qquad \text{adding zero to the right side}$$

or

$$y = (x - 1)^2 + 2$$

Now, as we know from Section 4.3, the graph of this last equation is obtained by moving the parabola $y = x^2$ one unit in the positive x-direction and two units in the positive y-direction. This shifts the vertex from the origin to the point $(1, 2)$. See Figure 2.

Note As a guide to sketching the graph, you'll want to know the y-intercept. To find the y-intercept, substitute $x = 0$ in the given equation to obtain $y = 3$. Then, given the vertex $(1, 2)$ and the point $(0, 3)$, a reasonably accurate graph can be quickly sketched. [Actually, once you find that $(0, 3)$ is on the graph, you also know that the reflection of this point about the axis of symmetry is on the graph. This is why the point $(2, 3)$ is shown in Figure 2.] ∎

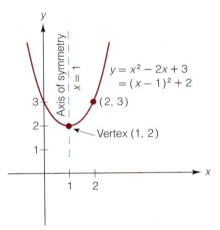

FIGURE 2

FIGURE 3

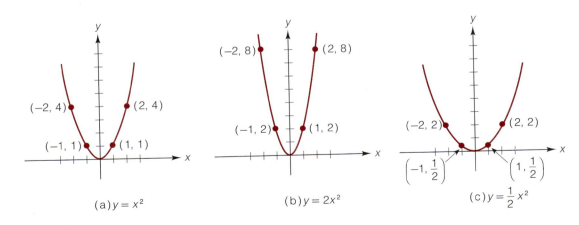

(a) $y = x^2$

(b) $y = 2x^2$

(c) $y = \frac{1}{2}x^2$

Now we want to compare the graphs of $y = x^2$, $y = 2x^2$, and $y = \frac{1}{2}x^2$. The last two equations were not discussed in Chapter 4, so for the moment you can think about graphing them by first setting up tables. Figure 3 (on the previous page) displays the graphs. All three graphs are parabolas that open upward; but notice that the shapes are not identical. The parabola $y = 2x^2$ is narrower than $y = x^2$, while $y = \frac{1}{2}x^2$ is wider than $y = x^2$. These observations about shape would also apply to $y = -x^2$, $y = -2x^2$, and $y = -\frac{1}{2}x^2$; in these cases, though, the parabolas open downward rather than upward. The observations that we have just made are generalized in the box that follows.

PROPERTY SUMMARY   THE GRAPH OF $y = ax^2$

1. The graph of $y = ax^2$ is a parabola with vertex at the origin. It is similar in shape to $y = x^2$.

2. The parabola $y = ax^2$ opens upward if $a > 0$, downward if $a < 0$.

3. The parabola $y = ax^2$ is narrower than $y = x^2$ if $|a| > 1$, wider than $y = x^2$ if $|a| < 1$.

EXAMPLE 2   Sketch the graph of $y = 3(x - 1)^2$.

Solution   Because of the $x - 1$, we shift the basic parabola $y = x^2$ one unit to the right. The factor of 3 in the given equation tells us that we want to draw a parabola that is narrower than $y = x^2$ but that has the same vertex, $(1, 0)$. To see exactly how narrow to draw $y = 3(x - 1)^2$, we would like to know another point on the graph other than the vertex $(1, 0)$. An easy point to obtain is the y-intercept. Setting $x = 0$ in the equation yields

$$y = 3(0 - 1)^2 = 3$$

Now that we know the vertex, $(1, 0)$, and the y-intercept, 3, a reasonably accurate graph can be sketched. See Figure 4.   ∎

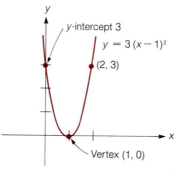

FIGURE 4   FIGURE 5

EXAMPLE 3   Sketch the graph of $y = -3(x - 1)^2$.

Solution   In Example 2 we sketched the graph of $y = 3(x - 1)^2$. By reflecting that graph in the x-axis, we obtain the graph of $y = -3(x - 1)^2$. See Figure 5.   ∎

EXAMPLE 4 Graph the function $f(x) = -2x^2 + 12x - 16$ and specify the vertex, axis of symmetry, maximum or minimum value of f, and x- and y-intercepts.

Solution The idea is to complete the square, as in Example 1. We have

$$
\begin{aligned}
y &= -2x^2 + 12x &&- 16 \\
&= -2(x^2 - 6x) - 16 \\
&= -2(x^2 - 6x + 9) - 16 + 18 &&\text{adding } (2)(9) = 18 \text{ to} \\
&&&\text{keep the equation in balance} \\
&= -2(x - 3)^2 + 2 &&(1)
\end{aligned}
$$

From equation (1), we see that the required graph is obtained simply by shifting the graph of $y = -2x^2$ "right 3, up 2," so that the vertex is $(3, 2)$. As a guide to sketching the graph, we want to compute the intercepts. The y-intercept is -16. (Why?) For the x-intercepts, we replace y with 0 in equation (1) to obtain

$$
\begin{aligned}
-2(x - 3)^2 &= -2 \\
(x - 3)^2 &= 1 \\
x - 3 &= \pm 1 \\
x = 4 \quad &\text{or} \quad 2
\end{aligned}
$$

Thus the x-intercepts are $x = 4$ and $x = 2$. Knowing these intercepts and the vertex, we can sketch the graph as in Figure 6. You should check for yourself that the information accompanying Figure 6 is correct.

FIGURE 6

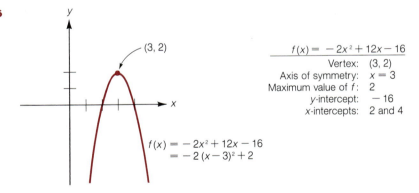

$$
\begin{aligned}
f(x) &= -2x^2 + 12x - 16 \\
\hline
\text{Vertex:} \quad &(3, 2) \\
\text{Axis of symmetry:} \quad &x = 3 \\
\text{Maximum value of } f: \quad &2 \\
y\text{-intercept:} \quad &-16 \\
x\text{-intercepts:} \quad &2 \text{ and } 4
\end{aligned}
$$

We conclude this section with several examples involving maximum and minimum values of functions. As background for this, we first summarize our basic technique for graphing parabolas.

PROPERTY SUMMARY   **THE GRAPH OF THE PARABOLA $y = ax^2 + bx + c$**

By completing the square, the equation of the parabola $y = ax^2 + bx + c$ can always be rewritten in the form

$$y = a(x - h)^2 + k$$

In this form, the vertex of the parabola is (h, k) and the axis of symmetry is the line $x = h$. The parabola opens upward if $a > 0$ and downward if $a < 0$.

EXAMPLE 5   Among all possible inputs for the function

$$g(x) = 3x^2 - 2x - 6$$

which yields the smallest output? What is this minimum output?

Solution   Think in terms of a graph. Since the graph of this function is a parabola opening upward, the input we want is the x-coordinate of the vertex. By completing the square (as shown in Example 4), we find that the given equation can be rewritten

$$g(x) = 3\left(x - \frac{1}{3}\right)^2 - \frac{19}{3}$$

(You should verify this for yourself, using Example 4 as a model.) Thus, the vertex of the parabola is $\left(\frac{1}{3}, -\frac{19}{3}\right)$. From this, and from the fact that the parabola opens upward, we conclude that the input $x = \frac{1}{3}$ produces the smallest output, and this minimum output is $g\left(\frac{1}{3}\right) = -\frac{19}{3}$. ∎∎∎

For the previous two examples, keep in mind that we were able to find the maximum or minimum easily only because the functions were quadratics. In contrast, you cannot expect to find the minimum of $y = x^4 - 8x$ using the method of Examples 4 and 5, because this is not a quadratic function. In general, the techniques of calculus are required to find maxima and minima for functions other than quadratics. There are some cases, however, in which our present method can be adapted to functions that are closely related to quadratics, as shown in the next two examples.

EXAMPLE 6   Let $D = \sqrt[3]{x^2 - x + 1}$. For which value of x is D a minimum?

Solution   The same x-value that makes D as small as it can be will also cause the quantity $D^3 = x^2 - x + 1$ to be its smallest. That is, we only need to find the x-value that minimizes the quantity $x^2 - x + 1$. Call this quantity y, and complete the square as follows,

$$y = x^2 - x \qquad + 1$$
$$= x^2 - x + \frac{1}{4} + 1 - \frac{1}{4}$$
$$= \left(x - \frac{1}{2}\right)^2 + \frac{3}{4}$$

This shows that the vertex of the parabola $y = x^2 - x + 1$ is $\left(\frac{1}{2}, \frac{3}{4}\right)$. Since this parabola opens upward, we conclude that the input $x = \frac{1}{2}$ will produce the minimum value for the quantity $x^2 - x + 1$. According to our initial remark, this same value, $x = \frac{1}{2}$, also minimizes the quantity $D = \sqrt[3]{x^2 - x + 1}$. (Our work also shows that the minimum value of D is $\sqrt[3]{\frac{3}{4}}$.) ∎∎∎

EXAMPLE 7 Find the maximum value of the function $y = 16t^2 - 4t^4$.

Solution Although this is not a quadratic function, we can complete the square as follows:

$$y = -4(t^4 - 4t^2 \quad)$$
$$= -4(t^4 - 4t^2 + 4) + 16$$
$$= -4(t^2 - 2)^2 + 16$$
$$= 16 - 4(t^2 - 2)^2$$

From this last equation we see that y never exceeds 16 (because the quantity being subtracted from 16 is nonnegative). Furthermore, y does attain the value 16 when $t^2 = 2$ or $t = \pm\sqrt{2}$. It now follows that the maximum value of the given function is 16.

EXERCISE SET 5.2

A

In Exercises 1–18, graph the quadratic function. Specify the vertex, axis of symmetry, maximum or minimum value, and intercepts.

1. $y = (x + 2)^2$
2. $y = -(x + 2)^2$
3. $y = 2(x + 2)^2$
4. $y = 2(x + 2)^2 + 4$
5. $y = -2(x + 2)^2 + 4$
6. $y = x^2 + 6x - 1$
7. $f(x) = x^2 - 4x$
8. $F(x) = x^2 - 3x + 4$
9. $g(x) = 1 - x^2$
10. $y = 2x^2 + \sqrt{2}x$
11. $y = x^2 - 2x - 3$
12. $y = 2x^2 + 3x - 2$
13. $y = -x^2 + 6x + 2$
14. $y = -3x^2 + 12x$
15. $s = 16t^2$
16. $s = -\frac{1}{2}t^2 - t$
17. $s = 2 + 3t - 9t^2$
18. $s = -\frac{1}{4}t^2 + t - 1$

For Exercises 19–24, determine the input that produces the largest or smallest output (whichever is appropriate). State whether the output is largest or smallest.

19. $y = 2x^2 - 4x + 11$
20. $f(x) = 8x^2 + x - 5$
21. $g(x) = -6x^2 + 18x$
22. $s = -16t^2 + 196t + 80$
23. $f(x) = x^2 - 10$
24. $h(x) = x^2 - 10x$

In Exercises 25–30, find the maximum or minimum value for each function (whichever is appropriate). State whether the value is a maximum or minimum.

25. $y = x^2 - 8x + 3$
26. $y = \frac{1}{2}x^2 + x + 1$
27. $y = -2x^2 - 3x + 2$
28. $y = -\frac{1}{3}x^2 - 2x$
29. $f(t) = -12t^2 + 1000$
30. $g(t) = 400t^2$
31. How far from the origin is the vertex of the parabola $y = x^2 - 6x + 13$?

32. Find the distance between the vertices of the parabolas $y = -\frac{1}{2}x^2 + 4x$ and $y = 2x^2 - 8x - 1$.

For Exercises 33–38, the functions f, g, and h are defined as follows.

$$f(x) = 2x - 3 \qquad g(x) = x^2 + 4x + 1 \qquad h(x) = 1 - 2x^2$$

In each exercise, classify the function as linear, quadratic, or neither.

33. $f \circ g$
34. $g \circ f$
35. $g \circ h$
36. $h \circ g$
37. $f \circ f$
38. $h \circ h$

In Exercises 39 and 40, determine the inputs that yield the minimum values for each function. Compute the minimum value in each case.

39. (a) $f(x) = \sqrt{x^2 - 6x + 73}$
    (b) $g(x) = \sqrt[3]{x^2 - 6x + 73}$
    (c) $h(x) = x^4 - 6x^2 + 73$
40. (a) $F(x) = (4x^2 - 4x + 109)^{1/2}$
    (b) $G(x) = (4x^2 - 4x + 109)^{1/3}$
    (c) $H(x) = 4x^4 - 4x^2 + 109$

In Exercises 41 and 42, determine the maximum value for each function.

41. (a) $f(x) = \sqrt{-x^2 + 4x + 12}$
    (b) $g(x) = \sqrt[3]{-x^2 + 4x + 12}$
    (c) $h(x) = -x^4 + 4x^2 + 12$
42. (a) $F(x) = (27x - x^2)^{1/2}$
    (b) $G(x) = (27x - x^2)^{1/3}$
    (c) $H(x) = 27x^2 - x^4$

B

In Exercises 43–46, find quadratic functions satisfying the given conditions.

43. The graph passes through the origin, and the vertex is $(2, 2)$.

44. The graph is obtained by translating $y = x^2$ four units in the negative x-direction and three units in the positive y-direction.

45. The vertex is $(3, -1)$, and one x-intercept is 1.

46. The axis of symmetry is the line $x = 1$. The y-intercept is 1. There is only one x-intercept.

47. By completing the square, show that the coordinates of the vertex of the parabola $y = ax^2 + bx + c$ are

$$\left(\frac{-b}{2a}, \frac{4ac - b^2}{4a}\right)$$

48. Find the equation of the circle that passes through the origin and is centered at the vertex of the parabola $y = 2x^2 + 12x + 14$.

49. For which value of c will the minimum value of the function $f(x) = x^2 + 2x + c$ be $\sqrt{2}$?

50. Explain why the graph of $y = -x^2 - 10x - \frac{51}{2}$ has no x-intercepts by finding the vertex and noting whether the parabola opens up or down.

51. Find the x-coordinate of the vertex of the parabola $y = (x - a)(x - b)$, where a and b are constants.

52. Compute the average of the two x-intercepts of the graph of $y = ax^2 + bx + c$. (Assume $b^2 - 4ac > 0$.) How does your answer relate to the result in Exercise 47?

C

53. Consider the quadratic function $y = px^2 + px + r$, where $p \neq 0$.
    (a) Show that if the vertex lies on the x-axis, then $p = 4r$.
    (b) Show that if $p = 4r$, then the vertex lies on the x-axis.

54. Let $g(x) = x^2 + 2(a + b)x + 2(a^2 + b^2)$, where a and b are constants.
    (a) Show that the coordinates of the vertex are
    $$(-(a + b), (a - b)^2)$$
    (b) Use the result in part (a) to explain why the graph of g has no x-intercepts unless $a = b$, in which case there is only one x-intercept.

55. As you have seen in this section, if the quadratic equation $f(x) = x^2 + Bx + C = 0$ has two real roots, there is a geometric interpretation: these roots are the x-intercepts for the graph of f. Less well known is the fact that there is a geometric interpretation when the quadratic equation $f(x) = x^2 + Bx + C = 0$ has nonreal complex roots. If these roots are $a + bi$ and $a - bi$, where a and b are real numbers, show that the coordinates of the vertex for the graph of f are (a, b^2). *Hint:* If r_1 and r_2 are the roots of the equation $x^2 + Bx + C = 0$, then $x^2 + Bx + C$ can be written $(x - r_1)(x - r_2)$.

$\dfrac{5.3}{}$ ## APPLIED FUNCTIONS: MORE ON SETTING UP EQUATIONS

I hope that I shall shock a few people in asserting that the most important single task of mathematical instruction in the secondary schools is to teach the setting up of equations to solve word problems.

George Polya (1887–1985)

Each problem that I solved became a rule which afterwards served to solve other problems.

René Descartes (1596–1650)

One of the first steps in problem solving often involves defining a function. The function then serves to describe or summarize a given situation in a way that is both concise and (one hopes) revealing. In this section we practice setting up equations that define such functions. (As background, you may wish to review our earlier work on setting up equations in Section 2.1.) In Section 5.4 we will use this skill in solving an important class of applied problems involving quadratic functions.

For many of the examples in this section, we'll rely on the following four-step procedure to set up the required equation. You may eventually want to modify this procedure to fit your own style. The important point, however, is that it is possible to approach these problems in a systematic manner. A word of advice: You're accustomed to working mathematics problems in which the answers are numbers. In this section, the answers are functions (or, more precisely, equations defining functions); you'll need to get used to this.

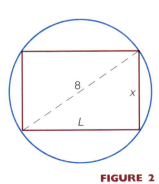

FIGURE 1

> ### Steps for Setting Up Equations
>
> STEP 1 After reading the problem carefully, draw a picture that conveys the given information.
>
> STEP 2 State in your own words, as specifically as you can, what the problem is asking for. (This usually requires rereading the problem.) Now, assuming that the problem asks you to find a particular quantity (or a formula for a particular quantity), assign a variable to denote that key quantity.
>
> STEP 3 Label any other quantities in your figure that appear relevant. Are there equations relating these quantities?
>
> STEP 4 Find an equation involving the key variable that you identified in step 2. (Some people prefer to do this right after step 2.) Now, as necessary, substitute in this equation using the auxiliary equations from step 3 to obtain an equation involving only the required variables.

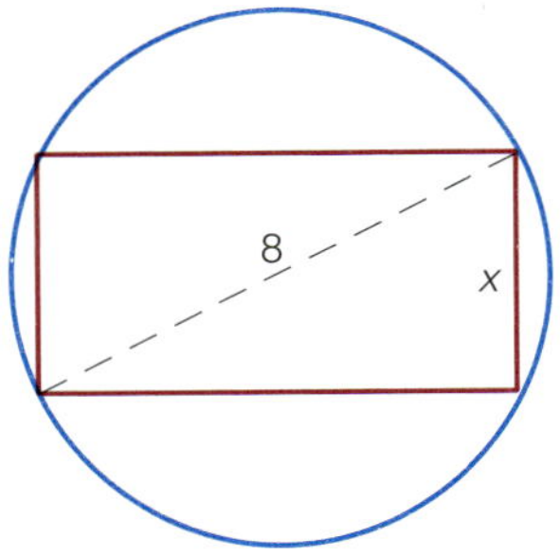

EXAMPLE 1   A rectangle is inscribed in a circle of diameter 8 cm. Express the perimeter of the rectangle as a function of its width x.

Solution   Let's follow our four-step procedure.

STEP 1 See Figure 1. Note that the diagonal of the rectangle is a diameter of the circle.

STEP 2 The problem asks us to come up with a formula or function that gives the perimeter of the rectangle in terms of x, the width. Let P denote the perimeter.

STEP 3 Let L denote the length of the rectangle, as shown in Figure 2. Then, by the Pythagorean theorem, we have

$$L^2 + x^2 = 8^2 = 64$$

This equation relates the length L and the width x. Rather than leaving the equation in this form, however, we'll solve for L in terms of x (because the instructions for the problem mention x, not L):

$$L^2 = 64 - x^2$$
$$L = \sqrt{64 - x^2} \tag{1}$$

FIGURE 2

STEP 4 The perimeter of the rectangle is the sum of the lengths of the four sides. Thus,

$$P = x + x + L + L = 2x + 2L \tag{2}$$

This equation expresses P in terms of x and L. However, the problem asks for P in terms of just x. Using equation (1) to substitute for L in equation (2), we have

$$P = 2x + 2\sqrt{64 - x^2} \tag{3}$$

This is the required equation. It expresses the perimeter P as a function of the width x, so if you know x, you can calculate P. To emphasize this dependence of P on x, we can employ function notation to rewrite equation (3):

$$P(x) = 2x + 2\sqrt{64 - x^2} \tag{4}$$

Before leaving this example, we need to specify the domain of the perimeter function in equation (4). An easy way to do this is to look back at Figure 1. Since x represents a width, we certainly want $x > 0$. Furthermore, Figure 1 tells us that $x < 8$, because in any right triangle, a leg is always shorter than the hypotenuse. Putting these observations together, we conclude that the domain of the perimeter function in equation (4) is the open interval $(0, 8)$.

Note Although it would make sense *algebraically* to use an input such as $x = -1$ in equation (4), it does not make sense in our *geometric* context, where x denotes the width. ∎

EXAMPLE 2 The perimeter of a rectangle is 100 cm. Express the area of the rectangle in terms of the width x.

Solution STEP 1 See Figure 3.
STEP 2 We want to express the area of the rectangle in terms of x, the width. Let A stand for the area of the rectangle.

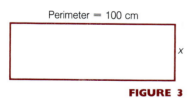

Perimeter = 100 cm

x

FIGURE 3

STEP 3 Call the length of the rectangle L, as indicated in Figure 4. Then, since the perimeter is given as 100 cm, we have

$$2x + 2L = 100$$
$$x + L = 50$$
$$L = 50 - x \tag{5}$$

STEP 4 The area of a rectangle equals width times length:

$$A = x \cdot L$$
$$= x(50 - x) \qquad \text{substituting for } L \text{ using equation (5)}$$
$$= 50x - x^2 \tag{6}$$

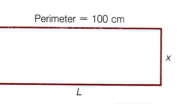

Perimeter = 100 cm

x

L

FIGURE 4

This is the required equation expressing the area of the rectangle in terms of the width x. To emphasize this dependence of A on x, we can use function notation to rewrite equation (6):

$$A(x) = 50x - x^2$$

The domain of this area function is the open interval $(0, 50)$. To see why this is so, first note that $x > 0$ because x denotes a width. Furthermore, in view of equation (5), we must have $x < 50$ (otherwise L, the length, would be zero or negative). ▪▪▪

EXAMPLE 3   Let $P(x, y)$ be a point on the curve $y = \sqrt{x}$. Express the distance from P to the point $(1, 0)$ as a function of x.

Solution   STEP 1   See Figure 5.

STEP 2   We want to express the length of the broken line in Figure 5 in terms of x. Call this length D.

STEP 3   There are no other quantities in Figure 5 that need labeling. But don't forget that we are given

$$y = \sqrt{x} \tag{7}$$

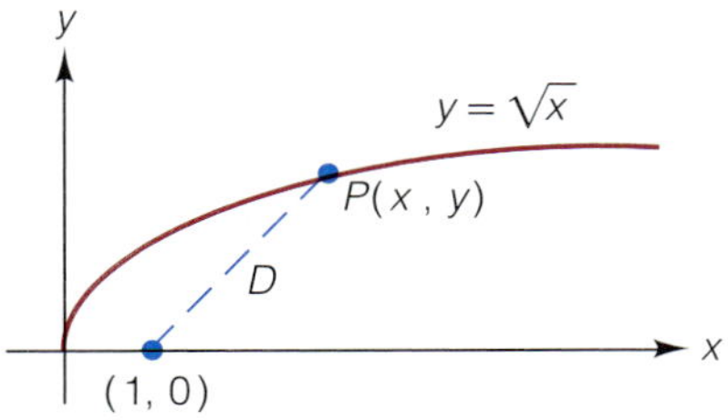

STEP 4   By the distance formula, we have

$$D = \sqrt{(x - 1)^2 + (y - 0)^2}$$
$$= \sqrt{x^2 - 2x + 1 + y^2} \tag{8}$$

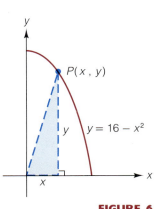

$y = \sqrt{x}$

$P(x, y)$

D

$(1, 0)$

FIGURE 5

Now we can use equation (7) to eliminate y in equation (8):

$$D(x) = \sqrt{x^2 - 2x + 1 + \left(\sqrt{x}\right)^2}$$
$$= \sqrt{x^2 - 2x + 1 + x}$$
$$= \sqrt{x^2 - x + 1} \tag{9}$$

Equation (9) expresses the distance as a function of x, as required. What about the domain of this distance function? Since the x-coordinate of a point on the curve $y = \sqrt{x}$ can be any nonnegative number, the domain of the distance function is $[0, \infty)$. ▪▪▪

EXAMPLE 4   A point $P(x, y)$ lies in the first quadrant on the parabola $y = 16 - x^2$, as indicated in Figure 6. Express the area of the triangular region in Figure 6 as a function of x.

Solution   STEP 1   See Figure 6.

STEP 2   We want to express the area of the shaded triangle in terms of x. Let A denote the area of this triangle.

STEP 3   Since the coordinates of P are (x, y), the base of our triangle is x and the height is y. Also, x and y are related by the given equation

$$y = 16 - x^2 \tag{10}$$

$P(x, y)$

y $y = 16 - x^2$

x

FIGURE 6

STEP 4   The area of a triangle equals $\frac{1}{2}$(base)(height):

$$A = \tfrac{1}{2}(x)(y)$$

so

$$A(x) = \tfrac{1}{2}(x)(16 - x^2) \qquad \text{substituting for } y \text{ using equation (10)}$$
$$= 8x - \tfrac{1}{2}x^3$$

This last equation expresses the area of the triangle as a function of x, as required. (Exercise 46 at the end of this section will ask you to specify the domain of this area function.) ▪▪▪

EXAMPLE 5   A piece of wire x inches long is bent into the shape of a circle. Express the area of the circle in terms of x.

Solution   STEP 1   See Figure 7.

STEP 2   We are supposed to express the area of the circle in terms of x, the circumference. Let A denote the area.

STEP 3   The general formula for circumference in terms of radius is $C = 2\pi r$. Since in our case the circumference is given as x, our equation becomes

$$x = 2\pi r$$

or

$$r = \frac{x}{2\pi} \tag{11}$$

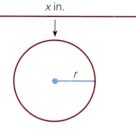

x in.

Circumference is x in.

FIGURE 7

STEP 4   The general formula for the area of a circle in terms of the radius is

$$A = \pi r^2 \tag{12}$$

This expresses A in terms of r, but we want A in terms of x. So we replace r in equation (12) by the quantity given in equation (11). This yields

$$A(x) = \pi \left(\frac{x}{2\pi}\right)^2 = \frac{\pi x^2}{4\pi^2} = \frac{x^2}{4\pi}$$

Thus, we have

$$A(x) = \frac{x^2}{4\pi}$$

This is the required equation. It expresses the area of the circle in terms of the circumference x. In other words, if we know the length of the piece of wire that is to be bent into a circle, we can use that length to calculate what the area of the circle will be.   ■■■

EXAMPLE 6   Figure 8 displays a right circular cylinder, along with the formulas for its volume V and total surface area S. Given that the volume is 10 cm^3, express the surface area S as a function of r, the radius of the base.

Solution   STEP 1   See Figure 8.

STEP 2   We are given a formula that expresses the surface area S in terms of both r and h. We want to express S in terms of just r.

STEP 3   We are given that $V = 10$ and also that $V = \pi r^2 h$. Thus,

$$\pi r^2 h = 10$$

and, consequently, expressing h in terms of r,

$$h = \frac{10}{\pi r^2} \tag{13}$$

STEP 4   We take the given formula for S, namely,

$$S = 2\pi r^2 + 2\pi r h$$

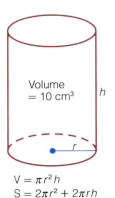

Volume = 10 cm³

$V = \pi r^2 h$
$S = 2\pi r^2 + 2\pi rh$

FIGURE 8

and replace h with the quantity given in equation (13). We get

$$S(r) = 2\pi r^2 + 2\pi r \left(\frac{10}{\pi r^2}\right)$$

$$= 2\pi r^2 + \frac{20}{r}$$

This is the required equation. It expresses the total surface area in terms of the radius r. Since the only restriction on r is that it be positive, the domain of the area function is $(0, \infty)$. ■■■

We have followed the same four-step procedure in Examples 1 through 6. Of course, no single method can cover all possible cases. As usual, common sense and experience are often necessary. Also, you should not feel compelled to follow this procedure at any cost. Keep this in mind as you study the last two examples in this section.

EXAMPLE 7   Two numbers add up to 8. Express the product P of these two numbers in terms of a single variable.

Solution   If we call the two numbers x and $8 - x$, then their product P is given by

$$P(x) = x(8 - x) = 8x - x^2$$

That's it. This last equation expresses the product as a function of the variable x. Since there are no restrictions on x (other than it's being a real number), the domain of this function is $(-\infty, \infty)$. ■■■

EXAMPLE 8   In economics, the revenue R generated by selling x units at a price of p dollars per unit is given by

$$R = x \cdot p$$

price per unit
number of units

In Figure 9, we are given a hypothetical function relating the selling price of a certain item to the number of units sold. Such a function is called a **demand function**. Express the revenue as a function of x.

Solution   More than anything else, this problem is an exercise in reading. After reading the problem several times, we find that it comes down to this:

given: $R = x \cdot p$ and $p = -\frac{1}{3}x + 4$ $(0 \le x \le 11)$

find: an equation expressing R in terms of x

In view of this, we write

$$R = x \cdot p$$

$$R(x) = x\left(-\frac{1}{3}x + 4\right)$$

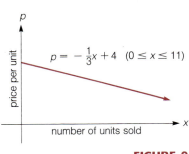

p

price per unit

$p = -\frac{1}{3}x + 4$ $(0 \le x \le 11)$

number of units sold

x

FIGURE 9

Thus, we have

$$R(x) = -\frac{1}{3}x^2 + 4x \qquad (0 \le x \le 11)$$

This is the required function. It allows us to calculate the revenue when we know the number of units sold. Note that this revenue function is a quadratic function.

Question for review and also preview How would you find the maximum revenue in this case?

EXERCISE SET 5.3

A

1. A rectangle is inscribed in a circle of diameter 12 in.
   (a) Express the perimeter of the rectangle as a function of its width x. *Suggestion:* First reread Example 1.
   (b) Express the area of the rectangle as a function of its width x.

2. (a) The perimeter of a rectangle is 16 cm. Express the area of the rectangle in terms of the width x. *Suggestion:* First reread Example 2.
   (b) The area of a rectangle is 85 cm². Express the perimeter as a function of the width x.

3. A point $P(x, y)$ is on the curve $y = x^2 + 1$.
   (a) Express the distance from P to the origin as a function of x. *Suggestion:* First reread Example 3.
   (b) In part (a), you expressed the length of a certain line segment as a function of x. Now express the slope of that line segment in terms of x.

4. A point $P(x, y)$ lies on the curve $y = \sqrt{x}$, as shown in the figure.
   (a) Express the area of the shaded triangle in terms of x. *Suggestion:* First reread Example 4.
   (b) Express the perimeter of the shaded triangle in terms of x.

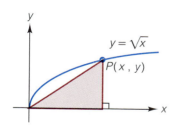

5. A piece of wire πy inches long is bent into a circle.
   (a) Express the area of the circle as a function of y. *Suggestion:* First reread Example 5.

   (b) If the original piece of wire were bent into a square instead of a circle, how would you express the area in terms of y?

6. The volume of a right circular cylinder is 20 in³. Express the total surface area of the cylinder as a function of r, the radius of the base. *Suggestion:* First reread Example 6.

7. Two numbers add to 16.
   (a) Express the product of the two numbers in terms of a single variable. *Suggestion:* First reread Example 7.
   (b) Express the sum of the squares of the two numbers in terms of a single variable.
   (c) Express the difference of the cubes of the two numbers in terms of a single variable. (There are two answers.)
   (d) What happens when you try to express the average of the two numbers in terms of one variable?

8. The product of two numbers is 16. Express the sum of the squares of the two numbers as a function of a single variable.

9. Given a demand function $p = -\frac{1}{4}x + 8$, express the revenue as a function of x. (See Example 8 for terminology and definitions.)

10. In Example 2, we considered a rectangle with perimeter 100 cm. We found that the area of such a rectangle is given by $A(x) = 50x - x^2$, where x is the width of the rectangle. Compute the numbers $A(1)$, $A(10)$, $A(20)$, $A(25)$, and $A(35)$. Which width x seems to yield the largest area $A(x)$?

11. In Example 1, we considered a rectangle inscribed in a circle of diameter 8 cm. We found that the perimeter of such a rectangle is given by

$$P(x) = 2x + 2\sqrt{64 - x^2}$$

where x is the width of the rectangle.

(a) Use a calculator to complete the following table. Round off each answer to two decimal places.

x	1	2	3	4	5	6	7
$P(x)$							

(b) In your table, what is the largest value for $P(x)$? What is the width x in this case?

(c) Using calculus, it can be shown that among all possible widths x, the width $x = 4\sqrt{2}$ cm yields the largest possible perimeter. Use a calculator to compute the perimeter in this case, and check to see that the value you obtain is indeed larger than all of the values obtained in part (a).

12. In economics, the demand function for a given commodity tells us how the unit price p is related to the number of units x that are sold. Suppose we are given a demand function

$$p = 5 - \frac{x}{4} \qquad (p \text{ in dollars})$$

(a) Graph this demand function.

(b) How many units can be sold when the unit price is \$3? Locate the point on the graph of the demand function that conveys this information.

(c) To sell 12 items, how should the unit price be set? Locate the point on the graph of the demand function that conveys this information.

(d) Find the revenue function corresponding to the given demand function. (Use the formula $R = x \cdot p$.) Graph the revenue function.

(e) Find the revenue when $x = 2$, when $x = 8$, and when $x = 14$.

(f) According to your graph in part (d), which x-value yields the greatest revenue? What is that revenue? What is the corresponding unit price?

13. Let $2s$ denote the length of the side of an equilateral triangle.

(a) Express the height of the triangle as a function of s.

(b) Express the area of the triangle as a function of s.

(c) Use the function you found in part (a) to determine the height of an equilateral triangle, each side of which is 8 cm long.

(d) Use the function you found in part (b) to determine the area of an equilateral triangle, each side of which is 5 in. long.

14. If x denotes the length of a side of an equilateral triangle, express the area of the triangle as a function of x.

15. The height of a right circular cylinder is twice the radius. Express the volume as a function of the radius.

16. Using the information given in Exercise 15, express the radius as a function of the volume.

17. The volume of a right circular cylinder is 12π in.³

(a) Express the height as a function of the radius.

(b) Express the total surface area as a function of the radius.

18. The total surface area of a right circular cylinder is 14 in.² Express the volume as a function of the radius.

19. The volume V and the surface area S of a sphere of radius r are given by the formulas $V = \frac{4}{3}\pi r^3$ and $S = 4\pi r^2$. Express V as a function of S.

20. The base of a rectangle lies on the x-axis, while the upper two vertices lie on the parabola $y = 10 - x^2$. Suppose that the coordinates of the upper right vertex of the rectangle are (x, y). Express the area of the rectangle as a function of x.

21. The hypotenuse of a right triangle is 20 cm. Express the area of the triangle as a function of the length x of one of the legs.

22. (a) Express the area of the shaded triangle in the accompanying figure as a function of x.

(b) Express the perimeter of the shaded triangle as a function of x.

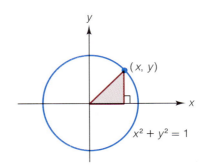

23. For the following figure, express the length of $\overline{AB}$ as a function of x. *Hint:* Note the similar triangles.

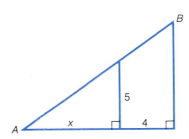

24. Five hundred feet of fencing are available to enclose a rectangular pasture alongside a river, which serves as one side of the rectangle (so only three sides require

fencing). Express the area of the rectangular pasture as a function of x. (See the figure.)

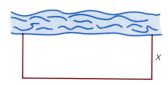

25. After rereading Example 2, complete the following table. Which x-value in the table yields the largest area A? What is the corresponding value of the length L in that case?

x	5	10	20	24	24.8	24.9	25	25.1	25.2	45
$A(x)$										

26. After rereading Example 3, complete the following three tables. For each table, specify the x-value that yields the smallest distance D. (For Table 2, round off the values of D to two significant digits. For Table 3, round off the values of D to seven significant digits.)

TABLE I

x	1	2	3	4	5
D					

TABLE 2

x	0.25	0.50	0.75
D			

TABLE 3

x	0.498	0.499	0.500	0.501	0.502
D					

27. (a) After rereading Example 4, complete the following three tables. For each table, specify the x-value that yields the largest area A. (For Tables 2 and 3, round off the values of A to six significant digits. In part (b), you'll see why we are asking for that many digits.)
    (b) Using calculus, it can be shown that the x-value yielding the largest area A is $x = 4\sqrt{3}/3$. Which x-value in the tables is closest to this x-value? To six significant digits, what is the area A when $x = 4\sqrt{3}/3$?

TABLE I

x	1	2	3	4
A				

TABLE 2

x	1.75	2.00	2.25	2.50	2.75
A					

TABLE 3

x	2.15	2.20	2.25	2.30	2.35
A					

28. A piece of wire 4 m long is cut into two pieces, and then each piece is bent into a square. Express the combined area of the two squares in terms of one variable.

29. A piece of wire 3 m long is cut into two pieces. Let x denote the length of the first piece and $3 - x$ the length of the second. The first piece is bent into a square and the second into a rectangle in which the width is half the length. Express the combined area of the square and the rectangle as a function of x. Is the resulting function a quadratic function?   *Hint:*   Review Example 6 in Section 2.1.

B

In Exercises 30–33, refer to the following figure, which displays a right circular cone along with the formulas for the volume V and the lateral surface area S.

$$V = \tfrac{1}{3}\pi r^2 h$$

$$S = \pi r \sqrt{r^2 + h^2}$$

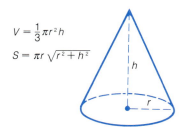

30. The volume of a right circular cone is 12π cm^3.
    (a) Express the height as a function of the radius.
    (b) Express the radius as a function of the height.

31. Suppose that the height and radius of a right circular cone are related by the equation $h = \sqrt{3}r$.
    (a) Express the volume as a function of r.
    (b) Express the lateral surface area as a function of r.

32. The volume of a right circular cone is 2 ft³. Show that the lateral surface area as a function of r is given by

$$S = \frac{\sqrt{\pi^2 r^6 + 36}}{r}$$

33. In a certain right circular cone, the volume is numerically equal to the lateral surface area.
    (a) Express the radius as a function of the height.
    (b) Express the height as a function of the radius.

34. A line is drawn from the origin O to a point $P(x, y)$ in the first quadrant on the graph of $y = 1/x$. From point P, a line is drawn perpendicular to the x-axis, meeting the x-axis at B.
    (a) Draw a figure of the situation described.
    (b) Express the perimeter of $\triangle OPB$ as a function of x.
    (c) Try to express the area of $\triangle OPB$ as a function of x. What happens?

35. A piece of wire 14 in. long is cut into two pieces. The first piece is bent into a circle, the second into a square. Express the combined total area of the circle and the square as a function of x, where x denotes the length of the wire that is used for the circle.

36. A wire of length L is cut into two pieces. The first piece is bent into a square, the second into an equilateral triangle. Express the combined total area of the square and the triangle as a function of x, where x denotes the length of wire used for the triangle. (L is to be considered a constant here, not another variable.)

37. An athletic field with a perimeter of $\frac{1}{4}$ mile consists of a rectangle with a semicircle at each end, as shown in the figure. Express the area of the field as a function of r, the radius of the semicircle.

38. A square of side x is inscribed in a circle. Express the area of the circle as a function of x.

39. An equilateral triangle of side x is inscribed in a circle. Express the area of the circle as a function of x.

40. Refer to the accompanying figure. An offshore oil rig is located at point A, which is 10 miles out to sea. An oil pipeline is to be constructed from A to a point C on the shore and then to an oil refinery at point D, farther up the coast. If it costs $8000 per mile to lay the pipeline in the sea and $2000 per mile on land, express the cost of laying the pipeline in terms of x, where x is the distance from B to C.

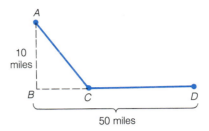

41. An open-top box is constructed from a 6-by-8-inch rectangular sheet of tin by cutting out equal squares at each corner and then folding up the flaps, as shown in the figure. Express the volume of the box as a function of x, the length of the side of each cutout square.

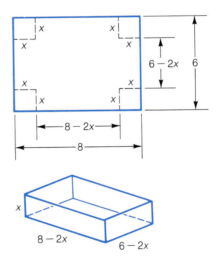

42. Follow Exercise 41, but assume that the original piece of tin is a square, 12 in. on each side.

43. A Norman window is in the shape of a rectangle surmounted by a semicircle, as shown in the figure (next page). Assume that the perimeter of the window is 32 ft.
    (a) Express the area of the window as a function of r, the radius of the semicircle.
    (b) The function you were asked to find in part (a) is a quadratic function, so its graph is a parabola. Does the parabola open upward or downward? Does it pass through the origin? Show that the vertex of the parabola is

$$\left(\frac{32}{\pi + 4}, \frac{512}{\pi + 4} \right)$$

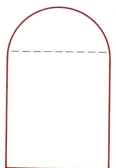

FIGURE FOR EXERCISE 43

44. Refer to the following figure. Express the lengths of $\overline{CB}$, $\overline{CD}$, $\overline{BD}$, and $\overline{AB}$ in terms of x.   *Hint:*   Recall the theorem from geometry stating that in a 30°–60°–90° right triangle, the side opposite the 30° angle is half the hypotenuse.

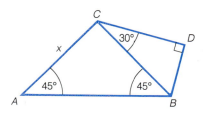

45. Refer to the following figure. Let s denote the ratio of y to z.
(a) Express y as a function of s.
(b) Express s as a function of y.
(c) Express z as a function of s.
(d) Express s as a function of z.

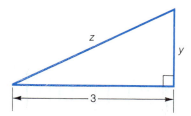

46. Refer to Example 4.
(a) What is the x-intercept of the curve $y = 16 - x^2$ in Figure 6?
(b) What is the domain of the area function in Example 4, assuming that the point P does not lie on the x- or y-axis?

47. The following figure shows the parabola $y = x^2$ and a line segment $\overline{AP}$ drawn from the point $A(0, -1)$ to the point $P(a, a^2)$ on the parabola.
(a) Express the slope of $\overline{AP}$ in terms of a.
(b) Show that the area of the shaded triangle in the figure is given by

$$\text{area} = \frac{a^5}{2(a^2 + 1)}$$

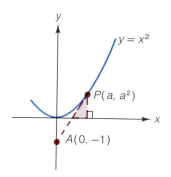

48. A rancher who wishes to fence off a rectangular area finds that the fencing in the east-west direction will require extra reinforcement owing to strong prevailing winds. Fencing in the east-west direction will therefore cost $12 per (linear) yard, as opposed to a cost of $8 per yard for fencing in the north-south direction. Given that the rancher wants to spend $4800 on fencing, express the area of the rectangle as a function of x, its width. The required function is in fact a quadratic, so its graph is a parabola. Does the parabola open upward or downward? By considering this graph, find which width x yields the rectangle of largest area. What is this maximum area?

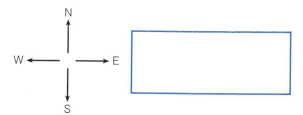

C

49. The following figure shows two concentric squares. Express the area of the shaded triangle as a function of x.

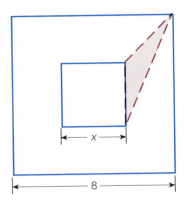

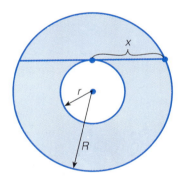

50. The following figure shows two concentric circles of radii r and R. Let A denote the area within the larger circle but outside the smaller one. Express A as a function of x (where x is defined in the figure).

51. A straight line with slope m ($m < 0$) passes through the point $(1, 2)$ and intersects the line $y = 4x$ at a point in the first quadrant. Let A denote the area of the triangle bounded by $y = 4x$, the x-axis, and the given line of slope m. Express A as a function of m.

52. A line with slope m ($m < 0$) passes through the point (a, b) in the first quadrant and intersects the line $y = Mx$ ($M > 0$) at another point in the first quadrant. Let A denote the area of the triangle bounded by $y = Mx$, the x-axis, and the given line with slope m. Express A in terms of m, M, a, and b.

Answer: $A = \dfrac{M(am - b)^2}{2m(m - M)}$

53. A line with slope m ($m < 0$) passes through the point (a, b) in the first quadrant. Express the area of the triangle bounded by this line and the axes in terms of m.

54. One corner of a page of width a is folded over and just reaches the opposite side, as indicated in the following figure. Express L, the length of the crease, in terms of x and a.

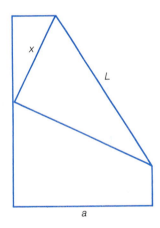

5.4 MAXIMUM AND MINIMUM PROBLEMS

. . . problems on maxima and minima, although new features in an English textbook, stand so little in need of apology with the scientific public that I offer none.

G. Chrystal in his preface to *Textbook of Algebra* (1886)

In daily life it is constantly necessary to choose the best possible (optimal) solution. A tremendous number of such problems arise in economics and in technology. In such cases it is frequently useful to resort to mathematics.

V. M. Tikhomirov in *Stories about Maxima and Minima* (translated from the Russian by Abe Shenitzer) (Providence, Rhode Island: The American Mathematical Society, 1990)

TABLE I

x	y	$x + y$	xy
-2	11	9	-22
-1	10	9	-10
0	9	9	0
1	8	9	8
2	7	9	14
3	6	9	18
4	5	9	⑳

TABLE 2

x	y	$x + y$	xy
1	8	9	8
1.5	7.5	9	11.25
2	7	9	14
2.5	6.5	9	16.25
3	6	9	18
3.5	5.5	9	19.25
4	5	9	20
4.5	4.5	9	20.25

You have already seen several examples of maximum and minimum problems in the second section of this chapter. Before reading further in the present section, you should first review Examples 5 through 7 on pages 252–253.

Actually, we begin this section's discussion of maximum and minimum problems at a more intuitive level. Consider, for example, the following question: If two numbers add to 9, what is the largest possible value of their product? To gain some insight here, we carry out a few preliminary calculations (see Table 1).

We have circled the 20 in the right-hand column of Table 1 because it *appears* to be the largest product. We say "appears" because our table is incomplete. For instance, what if we allowed x- and y-values that are not whole numbers? Might we get a product exceeding 20? Table 2 shows the results of some additional calculations along these lines.

As you can see from Table 2, there is a product exceeding 20, namely, 20.25. Now the question is, if we further expand our tables, can we find yet an even larger product, one exceeding 20.25? And here have come about as far as we want to go using this approach involving arithmetic and tables. For no matter what candidate we come up with for the largest product, there will always be the question of whether we might do still better using a larger table.

Nevertheless, this approach was useful, for it showed us what is really at the heart of a typical maximum or minimum problem. Essentially, we are trying to sort through an infinite number of possible cases and pick out the required extreme case. In the example at hand, there are infinitely many pairs of numbers x and y adding to 9. We are asked to look at the products of all of these pairs and see which (if any!) is the largest. Example 1 shows how to apply our knowledge of quadratic functions to solve this problem in a definitive manner.

EXAMPLE I   Two numbers add to 9. What is the largest possible value for their product?

Solution   Call the two numbers x and $9 - x$. Then their product P is given by

$$P = x(9 - x) = 9x - x^2$$

FIGURE I

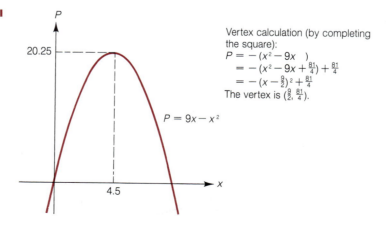

Vertex calculation (by completing the square):
$$P = -(x^2 - 9x \quad)$$
$$= -(x^2 - 9x + \tfrac{81}{4}) + \tfrac{81}{4}$$
$$= -(x - \tfrac{9}{2})^2 + \tfrac{81}{4}$$
The vertex is $(\tfrac{9}{2}, \tfrac{81}{4})$.

The graph of this quadratic function is the parabola in Figure 1. Note the accompanying calculation for the vertex. As Figure 1 and the accompanying calculations show, the largest value of the product P is 20.25. This is the required solution. (Note from the graph that there is no smallest value of P.) ▪▪▪

Example 1 illustrates the general strategy we will follow for solving the maximum and minimum problems in this section.

Strategy for Solving the Maximum and Minimum Problems in This Section

1. Express the quantity to be maximized or minimized in terms of a single variable. For instance, in Example 1, we found $P = 9x - x^2$. In setting up such functions, you'll want to keep in mind the four-step procedure used in Section 5.3.

2. Assuming that the function you have determined is a quadratic, note whether its graph, a parabola, opens upward or downward. Check whether this is consistent with the requirements of the problem. For instance, the parabola in Figure 1 opens downward; so it makes sense to look for a *largest*, not a smallest, value of P. Now complete the square to locate the vertex. (If the function is not a quadratic but is closely related to a quadratic, these ideas may still apply. See, for instance, Examples 6 and 7 in Section 5.2.)

3. After you have determined the vertex, you must relate that information to the original question. In Example 1, for instance, we were asked for the product P, not for x.

EXAMPLE 2 Among all rectangles having a perimeter of 10 ft, find the dimensions (length and width) of the one with the greatest area.

Solution First we want to set up a function that expresses the area of the rectangle in terms of a single variable. In doing this, we'll be guided by the four-step procedure that we used in Section 5.3. Figure 2(a) displays the given information. Our problem is to determine the dimensions of the rectangle that has the greatest area. As Figure 2(b) indicates, we can label the dimensions x and y. Since the perimeter is given as 10 ft, we have

$$2x + 2y = 10$$
$$x + y = 5$$
$$y = 5 - x \tag{1}$$

Letting A denote the area of the rectangle, we can write

$$A = xy$$
$$A(x) = x(5 - x) \qquad \text{substituting for } y \text{ using equation (1)}$$
$$= 5x - x^2$$

This expresses the area of the rectangle in terms of the width x. Since the graph of this quadratic function is a parabola that opens downward, it does make

Perimeter is 10 ft

(a)

x

y

Perimeter is 10 ft

(b)

sense to talk about the maximum. We can find the vertex of the parabola by completing the square:

$$A(x) = -(x^2 - 5x \qquad)$$

$$= -\left(x^2 - 5x + \frac{25}{4}\right) + \frac{25}{4} \qquad \text{adding } 0$$

$$= -\left(x - \frac{5}{2}\right)^2 + \frac{25}{4}$$

So the vertex is $\left(\frac{5}{2}, \frac{25}{4}\right)$, and, in particular, the width $x = \frac{5}{2}$ ft yields the maximum area. Now the problem asks us for the length and width of this rectangle, not its area. To compute the length y, we again use equation (1):

$$y = 5 - x = 5 - \frac{5}{2} = \frac{5}{2}$$

Thus, among all rectangles having a perimeter of 10 ft, the one with the greatest area is actually the square with dimensions of $2\frac{1}{2}$ ft by $2\frac{1}{2}$ ft. ■■■

EXAMPLE 3   Suppose that a baseball is tossed straight up and that its height as a function of time is given by the formula

$$h = -16t^2 + 64t + 6$$

In this formula, h is measured in feet and t in seconds, with $t = 0$ corresponding to the instant that the ball is released. What is the maximum height of the ball? When does the ball reach this height?

Solution   The given function tells us how the height h of the ball depends on the time t. We want to know the largest possible value for h. Since the graph of the given function is a parabola opening downward, we can determine the largest value of h just by finding the vertex of the parabola. After completing the square, we find that the original equation can be rewritten

$$h = -16(t - 2)^2 + 70 \tag{2}$$

[You should verify this for yourself. (If you get stuck, follow the model for completing the square in Example 4 of Section 5.2)] From equation (2) we see that the vertex of the parabola $h = -16t^2 + 64t + 6$ is (2, 70). Therefore the maximum height of the ball is 70 ft, and the ball reaches this height at $t = 2$ sec. ■■■

EXAMPLE 4   Which point on the curve $y = \sqrt{x}$ is closest to the point (1, 0)?

Solution   In Figure 3, we let D denote the distance from a point (x, y) on the curve to the point (1, 0). We are asked to find out exactly which point (x, y) will make the distance D as small as possible. Using the distance formula, we have

$$D = \sqrt{(x - 1)^2 + (y - 0)^2}$$
$$= \sqrt{x^2 - 2x + 1 + y^2} \tag{3}$$

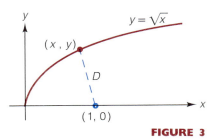

FIGURE 3

This expresses D in terms of both x and y. To express D in terms of x alone, we use the given equation $y = \sqrt{x}$ to substitute for y in equation (3). This yields

$$D(x) = \sqrt{x^2 - 2x + 1 + \left(\sqrt{x}\right)^2}$$
$$= \sqrt{x^2 - x + 1}$$

Since the same x-value that makes $D(x)$ its smallest will also cause the quantity $[D(x)]^2 = x^2 - x + 1$ to be its smallest, we need only find the x-value that minimizes the quantity $x^2 - x + 1$. Denoting this quantity by $f(x)$ for the moment (so that we have an equation whose graph is a parabola), we have

$$f(x) = x^2 - x + 1$$
$$= \left(x - \frac{1}{2}\right)^2 + \frac{3}{4} \qquad \text{completing the square}$$

From this last equation we conclude that the quantity $x^2 - x + 1$ will be smallest when $x = \frac{1}{2}$.

Now that we have x, we want to calculate y. (The problem did not ask for D; we needn't calculate it.) We know that $y = \sqrt{x}$ and $x = \frac{1}{2}$. Therefore,

$$y = \sqrt{\frac{1}{2}} = \frac{1}{\sqrt{2}} \cdot \frac{\sqrt{2}}{\sqrt{2}} = \frac{\sqrt{2}}{2}$$

Thus, among all points (x, y) on the curve $y = \sqrt{x}$, the point closest to $(1, 0)$ is $\left(\frac{1}{2}, \sqrt{2}/2\right)$. This is the required solution.

Question Why would it have made no sense to ask instead for the point farthest from $(1, 0)$?

EXAMPLE 5 Suppose that you have 600 m of fencing with which to build two adjacent rectangular corrals. The two corrals are to share a common fence on one side, as shown in Figure 4. Find the dimensions x and y so that the total enclosed area is as large as possible.

Solution You have 600 m of fencing to be set up as shown in Figure 4. The question is how you should choose x and y so that the total area is a maximum. Since the total length of fencing is 600 m, we can relate x and y by writing

$$x + x + x + y + y = 600$$
$$3x + 2y = 600$$
$$2y = 600 - 3x$$
$$y = 300 - \frac{3}{2}x \qquad (4)$$

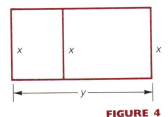

FIGURE 4

Letting A denote the total area, we have

$$A = xy$$
$$A(x) = x\left(300 - \frac{3}{2}x\right) \qquad \text{using equation (4) to substitute for } y$$
$$= 300x - \frac{3}{2}x^2$$

This last equation expresses the area as a function of x. Note that the graph of this function is a parabola opening downward, so it does make sense to talk about a maximum. We complete the square, as usual, to locate the vertex. As you can check, the result is

$$A(x) = -\frac{3}{2}(x - 100)^2 + 15000$$

Thus the x-coordinate of the vertex is 100, and this is the x-value that maximizes the area. The corresponding y-value can now be calculated using equation (4):

$$y = 300 - \frac{3}{2}x = 300 - \frac{3}{2}(100)$$

$$= 300 - 3(50) = 150$$

Thus, by choosing x to be 100 m and y to be 150 m, the total area in Figure 4 will be as large as possible. Incidentally, note that the exact location of the fence dividing the two corrals does not influence our work or final answer. ▪▪▪

EXAMPLE 6   Suppose that the following function relates the selling price, p, of an item to the quantity sold, x:

$$p = -\frac{1}{3}x + 40 \qquad (p \text{ in dollars})$$

For which value of x will the corresponding revenue be a maximum? What are the maximum revenue and the unit price in this case?

Solution   First, let us recall the formula for revenue given on page 259:

$$R = \text{number of units} \times \text{price per unit}$$

Using this, we have

$$R = x \cdot p$$

$$R(x) = x\left(-\frac{1}{3}x + 40\right) = -\frac{1}{3}x^2 + 40x \qquad (5)$$

We want to know which value of x yields the largest revenue R. Since the graph of the revenue function in equation (5) is a parabola opening downward, the required x-value is the x-coordinate of the vertex of the parabola. We complete the square to locate the vertex:

$$R(x) = -\frac{1}{3}(x^2 - 120x \qquad)$$

$$= -\frac{1}{3}(x^2 - 120x + 3600) + 1200 \qquad \text{adding } 0 \ (= -1200 + 1200)$$

$$= -\frac{1}{3}(x - 60)^2 + 1200$$

The vertex is therefore (60, 1200). Consequently, the revenue is a maximum when $x = 60$ items, and this maximum revenue is $1200. For the corresponding unit price, we have

$$p(x) = -\frac{1}{3}x + 40$$

$$p(60) = -\frac{1}{3}(60) + 40 = 20$$

Thus the unit price corresponding to the maximum revenue is $20. ▮▮▮

EXAMPLE 7 Figure 5 shows a graph of the function

$$f(x) = x^4 - 3x^2$$

From the graph, it is clear that the minimum of the function is slightly less than -2. Find the exact value of this minimum and the corresponding x-values at which it occurs.

Solution Although this is not a quadratic function, we can nevertheless use the technique of completing the square in this particular case. (You have already seen one instance of this in Example 7 of Section 5.2.) We have

$$f(x) = x^4 - 3x^2 + \frac{9}{4} - \frac{9}{4}$$

$$= \left(x^2 - \frac{3}{2}\right)^2 - \frac{9}{4}$$

$$= -\frac{9}{4} + \left(x^2 - \frac{3}{2}\right)^2$$

From this last equation we see that $f(x)$ is never less than $-\frac{9}{4}$ (because the quantity being added to $-\frac{9}{4}$ is nonnegative). Furthermore, $f(x)$ does attain the value $-\frac{9}{4}$ when $x^2 = \frac{3}{2}$; that is, when $x = \pm\sqrt{\frac{3}{2}}$. Thus the minimum value of f is $-\frac{9}{4}$ and the corresponding x-values are

$$x = \pm\frac{\sqrt{3}}{\sqrt{2}} = \pm\frac{\sqrt{6}}{2} \qquad \text{rationalizing the denominator}$$

Note Use your calculator to approximate these x-values, and then check that they are consistent with the graph in Figure 5. ▮▮▮

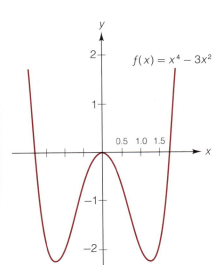

$f(x) = x^4 - 3x^2$

FIGURE 5

EXERCISE SET 5.4

A

1. Two numbers add to 5. What is the largest possible value of their product?

2. Find two numbers adding to 20 such that the sum of their squares is as small as possible.

3. The difference of two numbers is 1. What is the smallest possible value for the sum of their squares?

4. For each quadratic function, state whether it would make sense to look for a highest or a lowest point on

the graph. Then determine the coordinates of that point.

(a) $y = 2x^2 - 8x + 1$ (b) $y = -3x^2 - 4x - 9$
(c) $h = -16t^2 + 256t$ (d) $f(x) = 1 - (x + 1)^2$
(e) $g(t) = t^2 + 1$ (f) $f(x) = 1000x^2 - x + 100$

5. Among all rectangles having a perimeter of 25 m, find the dimensions of the one with the largest area.

6. What is the largest possible area for a rectangle whose perimeter is 80 cm?

7. What is the largest possible area for a right triangle in which the sum of the lengths of the two shorter sides is 100 in.?

8. The perimeter of a rectangle is 12 m. Find the dimensions for which the diagonal is as short as possible.

9. Two numbers add to 6.
(a) Let T denote the sum of the squares of the two numbers. What is the smallest possible value for T?
(b) Let S denote the sum of the first number and the square of the second. What is the smallest possible value for S?
(c) Let U denote the sum of the first number and twice the square of the second number. What is the smallest possible value for U?
(d) Let V denote the sum of the first number and the square of twice the second number. What is the smallest possible value for V?

10. Suppose that the height of an object shot straight up is given by $h = 512t - 16t^2$ (here h is in feet and t is in seconds). Find the maximum height and the time at which the object hits the ground.

11. A baseball is thrown straight up, and its height as a function of time is given by the formula $h = -16t^2 + 32t$ (where h is in feet and t is in seconds).
(a) Find the height of the ball when $t = 1$ and when $t = \frac{3}{2}$.
(b) Find the maximum height of the ball and the time at which that height is attained.
(c) At what times is the height 7 ft?

12. Find the point on the curve $y = \sqrt{x}$ that is nearest to the point $(3, 0)$.

13. Which point on the curve $y = \sqrt{x - 2} + 1$ is closest to the point $(4, 1)$? What is this minimum distance?

14. Find the coordinates of the point on the line $y = 3x + 1$ closest to $(4, 0)$.

15. (a) What number exceeds its square by the greatest amount?
(b) What number exceeds twice its square by the greatest amount?

16. Suppose that you have 1800 m of fencing with which to build three adjacent rectangular corrals, as shown in

the figure. Find the dimensions so that the total enclosed area is as large as possible.

17. Five hundred feet of fencing is available for a rectangular pasture alongside a river, the river serving as one side of the rectangle (so only three sides require fencing). Find the dimensions yielding the greatest area.

18. Let $A = 3x^2 + 4x - 5$ and $B = x^2 - 4x - 1$. Find the minimum value of $A - B$.

19. Let $R = 0.4x^2 + 10x + 5$ and $C = 0.5x^2 + 2x + 101$. For which value of x is $R - C$ a maximum?

20. Suppose that the revenue generated by selling x units of a certain commodity is given by $R = -\frac{1}{5}x^2 + 200x$. Assume that R is in dollars. What is the maximum revenue possible in this situation?

21. Suppose that the function $p = -\frac{1}{4}x + 30$ relates the selling price p of an item to the number of units x that are sold. Assume that p is in dollars. For which value of x will the corresponding revenue be a maximum? What is this maximum revenue and what is the unit price?

22. The action of sunlight on automobile exhaust produces air pollutants known as *photochemical oxidants*. In a study of cross-country runners in Los Angeles, it was shown that running performances can be adversely affected when the oxidant level reaches 0.03 part per million. Suppose that on a given day, the oxidant level L is approximated by the formula

$$L = 0.059t^2 - 0.354t + 0.557 \qquad (0 \le t \le 7)$$

where t is measured in hours, with $t = 0$ corresponding to 12 noon, and L is in parts per million. At what time is the oxidant level L a minimum? At this time, is the oxidant level high enough to affect a runner's performance?

23. (a) Find the smallest possible value of the quantity $x^2 + y^2$ under the restriction that $2x + 3y = 6$.
(b) Find the radius of the circle whose center is at the origin and that is tangent to the line $2x + 3y = 6$. How does this answer relate to your answer in part (a)?

B

24. Through a type of chemical reaction known as *autocatalysis*, the human body produces the enzyme trypsin from the enzyme trypsinogen. (Trypsin then breaks down proteins into amino acids, which the body needs for growth.) Let r denote the rate of the chemical reaction in which trypsin is formed from trypsinogen. It has been shown experimentally that $r = kx(a - x)$, where k is a positive constant, a is the initial amount of trypsinogen, and x is the amount of trypsin produced (so x increases as the reaction proceeds). Show that the reaction rate r is a maximum when $x = a/2$. In other words, the speed of the reaction is greatest when the amount of trypsin formed is half the original amount of trypsinogen.

25. (a) Let $x + y = 15$. Find the minimum value of the quantity $x^2 + y^2$.
    (b) Let C be a constant and $x + y = C$. Show that the minimum value of $x^2 + y^2$ is $C^2/2$. Then use this result to check your answer in part (a).

26. Suppose that A, B, and C are positive constants and that $x + y = C$. Show that the minimum value of $Ax^2 + By^2$ occurs when $x = BC/(A + B)$ and $y = AC/(A + B)$.

27. The following figure shows a square inscribed within a unit square. For which value of x is the area of the inscribed square a minimum? What is the minimum area? *Hint:* Denote the lengths of the two segments that make up the base of the unit square by t and $1 - t$. Now use the Pythagorean theorem and congruent triangles to express x in terms of t.

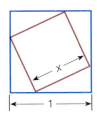

28. (a) Find the coordinates of the point on the line $y = mx + b$ that is closest to the origin.

    Answer: $\left(\dfrac{-mb}{1 + m^2}, \dfrac{b}{1 + m^2} \right)$

    (b) Show that the perpendicular distance from the origin to the line $y = mx + b$ is $|b|/\sqrt{1 + m^2}$. *Suggestion:* Use the result in part (a).
    (c) Use part (b) to show that the perpendicular distance from the origin to the line $Ax + By + C = 0$ is $|C|/\sqrt{A^2 + B^2}$.

29. The point P lies in the first quadrant on the graph of the line $y = 7 - 3x$. From the point P, perpendiculars are drawn to both the x-axis and the y-axis. What is the largest possible area for the rectangle thus formed?

30. Show that the largest possible area for the shaded rectangle shown in the figure is $-b^2/4m$. Then use this to check your answer to Exercise 29.

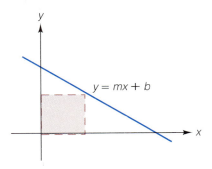

31. Show that the maximum possible area for a rectangle inscribed in a circle of radius R is $2R^2$. *Hint:* Maximize the square of the area.

32. An athletic field with a perimeter of $\frac{1}{4}$ mile consists of a rectangle with a semicircle at each end, as shown in the figure. Find the dimensions x and r that yield the greatest possible area for the rectangular region.

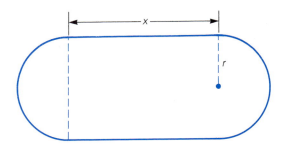

33. A rancher who wishes to fence off a rectangular area finds that the fencing in the east-west direction will require extra reinforcement owing to strong prevailing winds. Because of this, the cost of fencing in the east-west direction will be $12 per (linear) yard, as opposed to a cost of $8 per yard for fencing in the north-south direction. Find the dimensions of the largest possible rectangular area that can be fenced for $4800.

34. Let $f(x) = (x - a)^2 + (x - b)^2 + (x - c)^2$, where a, b, and c are constants. Show that $f(x)$ will be a minimum when x is the average of a, b, and c.

35. Let $y = a_1(x - x_1)^2 + a_2(x - x_2)^2$, where a_1, a_2, x_1, and x_2 are all constants. In addition, suppose that a_1 and a_2 are both positive. Show that the minimum of this function occurs when

$$x = \frac{a_1 x_1 + a_2 x_2}{a_1 + a_2}$$

36. Among all rectangles with a given perimeter P, find the dimensions of the one with the shortest diagonal.

37. By analyzing sales figures, the economist for a stereo manufacturer knows that 150 units of a tape player can be sold each month when the price is set at $p = \$200$ per unit. The figures also show that for each $10 hike in price, 5 fewer units are sold each month.
   (a) Let x denote the number of units sold per month and let p denote the price per unit. Find a linear function relating p and x.
      Hint: $\Delta p / \Delta x = 10/(-5) = -2$.
   (b) The revenue R is given by $R = xp$. What is the maximum revenue? At what level should the price be set to achieve this maximum revenue?

38. Let $f(x) = x^2 + px + q$, and suppose that the minimum value of this function is 0. Show that $q = p^2/4$.

39. For which numbers t will the value of the function $f(t) = t^2 - t^4$ be as large as possible?

40. Find the minimum value of the function $f(t) = t^4 - 8t^2$.

41. Among all possible inputs for the function $f(t) = -t^4 + 6t^2 - 6$, which ones yield the largest output?

42. Let $f(x) = x - 3$ and $g(x) = x^2 - 4x + 1$.
   (a) Find the minimum value of $g \circ f$.
   (b) Find the minimum value of $f \circ g$.
   (c) Are the results in parts (a) and (b) the same?

In Exercises 43–50, use the following result to lessen the amount of computation required: The x-coordinate of the vertex of the parabola $y = ax^2 + bx + c$ is given by the formula $x = -b/(2a)$. (As indicated in Exercise 47 in Section 5.2, this result can be derived by completing the square.)

43. A piece of wire 16 in. long is to be cut into two pieces. Let x denote the length of the first piece and $16 - x$ the length of the second. The first piece is to be bent into a circle and the second piece into a square.
   (a) Express the total combined area A of the circle and the square as a function of x.
   (b) For which value of x is the area A a minimum?
   (c) Using the x-value that you found in part (b), find the ratio of the lengths of the shorter to the longer piece of wire.   *Answer:* $\pi/4$

44. A 30-in. piece of string is to be cut into two pieces. The first piece will be formed into the shape of an equilateral triangle and the second piece into a square. Find the length of the first piece if the combined area of the triangle and the square is to be as small as possible.

C

45. Repeat Exercise 43, but assume that the length of the wire is L in.

46. Let $f(x) = x^2 + bx + 1$. Find a positive value for b such that the distance from the origin to the vertex of the parabola is as small as possible.

47. The next figure shows a rectangle inscribed in a given triangle of base b and height h. Find the ratio of the area of the triangle to the area of the rectangle when the area of the rectangle is a maximum.

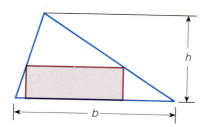

48. A Norman window is in the shape of a rectangle surmounted by a semicircle, as shown in the figure. Assume that the perimeter of the window is P, a constant. Show that the area of the window is a maximum when both x and r are equal to $P/(\pi + 4)$. Show that this maximum area is $\frac{1}{2}P^2/(\pi + 4)$.

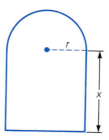

49. A triangle is inscribed in a semicircle of diameter $2R$, as shown in the figure (on the next page). Show that the smallest possible value for the area of the shaded region is $\frac{1}{2}(\pi - 2)R^2$.
   Hint: The area of the shaded region is a minimum when the area of the triangle is a maximum. Find the value of x that maximizes the *square* of the area of the triangle. This will be the same x that maximizes the area of the triangle.

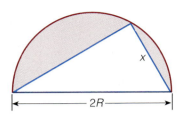

50. Let $f(x) = 1/(2x^2 - x + 1)$.
    (a) Experiment with your calculator: what is the largest output you can find for the function f?
    (b) Prove that for every value of x, we have $f(x) \leq 8/7$.

5.5 POLYNOMIAL FUNCTIONS

We can rephrase the definitions of linear and quadratic functions using the terminology for polynomials given in Section 1.8.

A function f is **linear** if $f(x)$ is a **polynomial of degree 1**:

$$f(x) = a_1 x + a_0 \tag{1}$$

A function f is **quadratic** if $f(x)$ is a **polynomial of degree 2**:

$$f(x) = a_2 x^2 + a_1 x + a_0 \tag{2}$$

In view of equation (1), linear functions are sometimes called **polynomial functions of degree 1**. Similarly, in view of equation (2), quadratic functions are **polynomial functions of degree 2**. More generally, by a **polynomial function of degree n**, we mean a function defined by an equation of the form

$$f(x) = a_n x^n + a_{n-1} x^{n-1} + \cdots + a_1 x + a_0 \tag{3}$$

where n is a nonnegative integer and $a_n \neq 0$. Throughout the remainder of this chapter, the coefficients a_k will always be real numbers.

In principle, we can obtain the graph of any polynomial function by setting up a table and plotting a sufficient number of points. Indeed, this is just the way a computer equipped with a curve plotter operates. However, to *understand* why the graphs look as they do, we want to discuss some additional methods for graphing polynomial functions.

FIGURE I

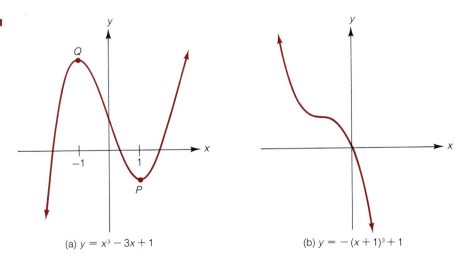

(a) $y = x^3 - 3x + 1$
(b) $y = -(x+1)^3 + 1$

There are three facts that we shall need. By way of example, look at the graphs of the polynomial functions in Figure 1 (on the previous page). First, notice that both graphs are unbroken, smooth curves, with no "corners." As is shown in calculus, this is true for the graph of every polynomial function. By way of contrast, the graphs in Figures 2 and 3 cannot represent polynomial functions. The graph in Figure 2 has a break in it, and the graph in Figure 3 has a **cusp**.

Now look back at the graph in Figure 1(a). Recall (from Section 4.2) that points such as P and Q are called **turning points**. These are points where the graph changes from rising to falling, or vice versa. It is a fact (proved in calculus) that the graph of a polynomial function of degreee n has *at most* $n - 1$ turning points. For instance, in Figure 1(a), there are two turning points, while the degree of the polynomial is 3. However, as Figure 1(b) indicates, we needn't have any turning points at all.

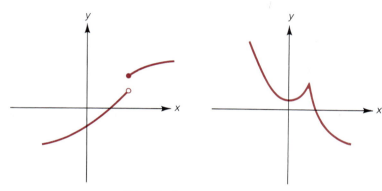

FIGURE 2
Since the graph has a break, it cannot represent a polynomial function.

FIGURE 3
Since the graph has a cusp, it cannot represent a polynomial function.

A third property of polynomial functions concerns their behavior when $|x|$ is very large. We'll illustrate the property using the function $y = x^3 - 3x + 1$, which was graphed in Figure 1(a). Now, in Figure 1(a), the x-values are relatively small; for instance, the x-coordinates of P and Q are 1 and -1, respectively. In Figure 4, however, we show the graph of this same function using units of 100 on the x-axis. On this scale, the graph appears indistinguishable from that of $y = x^3$. In particular, note that as $|x|$ gets very large, $|y|$ grows very large.

It's easy to see why the function $y = x^3 - 3x + 1$ resembles $y = x^3$ when $|x|$ is very large. First, let's rewrite the equation $y = x^3 - 3x + 1$ as

$$y = x^3 \left(1 - \frac{3}{x^2} + \frac{1}{x^3} \right)$$

Now, when $|x|$ is very large, both $3/x^2$ and $1/x^3$ are close to zero. So we have

$$y \approx x^3 \, (1 - 0 + 0)$$
$$\approx x^3 \quad \text{when } |x| \text{ is very large}$$

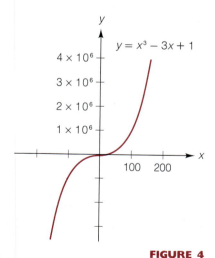

FIGURE 4
When $|x|$ is very large, the graph of $y = x^3 - 3x + 1$ appears indistinguishable from that of $y = x^3$.

The same technique that we've just used in analyzing $y = x^3 - 3x + 1$ can be applied to any (nonconstant) polynomial function. The result is summarized in item 3 in the following box.

PROPERTY SUMMARY   GRAPHS OF POLYNOMIAL FUNCTIONS

1. The graph of a polynomial function of degree 2 or greater is an unbroken smooth curve. (For degrees 1 and 0, the graph is a line.)

2. The graph of a polynomial function of degree n has at most $n - 1$ turning points.

3. For the graph of any polynomial function (other than a constant function), as $|x|$ gets very large, $|y|$ grows very large. If

$$f(x) = a_n x^n + a_{n-1} x^{n-1} + \cdots + a_1 x + a_0 \qquad (a_n \neq 0)$$

then

$$f(x) \approx a_n x^n \qquad \text{when } |x| \text{ is very large.}$$

EXAMPLE I   A function f is defined by

$$f(x) = -x^3 + x^2 + 9x + 9$$

Which of the graphs in Figure 5 might represent this function?

FIGURE 5

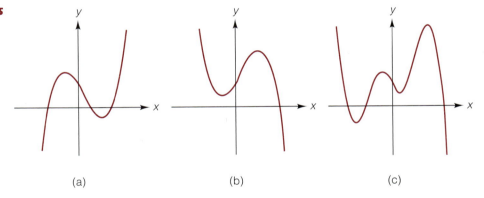

(a)   (b)   (c)

Solution   When $|x|$ is very large, $f(x) \approx -x^3$. This rules out the graph in Figure 5(a). The graph in Figure 5(c) can also be ruled out, but for a different reason. That graph has four turning points, whereas the graph of the cubic function f can have at most two turning points. The graph in Figure 5(b), on the other hand, does have two turning points; furthermore, that graph does behave like $y = -x^3$ when $|x|$ is very large. The graph in Figure 5(b) might be (in fact, it is) the graph of the given function f. ∎

The simplest polynomial functions to graph are those of the form $y = x^n$. From our work in Chapter 4 we are already familiar with the graphs of $y = x$, $y = x^2$, and $y = x^3$. For reference, these are shown in Figure 6.

FIGURE 6

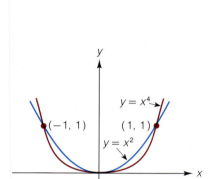

(a)   (b)   (c)

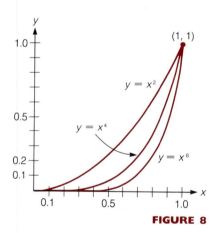

FIGURE 7

The graph of $y = x^n$, when n is greater than 3, resembles the graph of $y = x^2$ or $y = x^3$, depending on whether n is even or odd. Consider, for instance, the graph of $y = x^4$, shown in Figure 7 along with the graph of $y = x^2$. Just as with $y = x^2$, the graph of $y = x^4$ is a symmetric, U-shaped curve passing through the three points $(-1, 1)$, $(1, 1)$, and $(0, 0)$. However, in the interval $(-1, 1)$, the graph of $y = x^4$ is flatter than that of $y = x^2$. Similarly, the graph of $y = x^6$ in this interval would be flatter still. The data in Table 1 show why this is true. Figure 8 displays the graphs of $y = x^2$, $y = x^4$, and $y = x^6$ for the interval $0 \leq x \leq 1$.

TABLE 1

x	0.2	0.4	0.6	0.8	1.0
x^2	0.04	0.16	0.36	0.64	1.0
x^4	0.0016	0.0256	0.1296	0.4096	1.0
x^6	0.000064	0.004096	0.046656	0.262144	1.0

Incidentally, Figure 8 indicates one of the practical difficulties you may encounter in trying to draw an accurate graph of $y = x^n$. Suppose, for instance, that you want to graph $y = x^6$, and the lines you draw are 0.01 cm thick. Also suppose that you use the same scale on both axes, taking the common unit to be 1 cm. Then, in the first quadrant, your graph of $y = x^6$ will be indistinguishable from the x-axis when $x^6 < 0.01$, or $x < \sqrt[6]{0.01} \approx 0.46$ cm (using a calculator). This explains why sections of the graphs in Figure 8 appear horizontal.

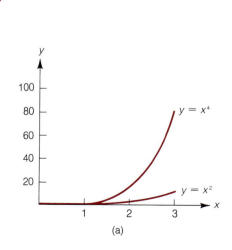

FIGURE 8

FIGURE 9

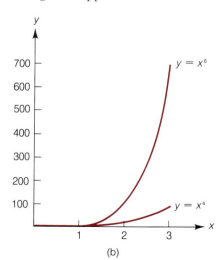

(a)   (b)

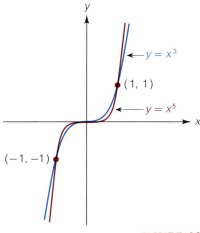

FIGURE 10

For $|x| > 1$, the graph of $y = x^4$ rises more rapidly than that of $y = x^2$. Similarly, the graph of $y = x^6$ rises still more rapidly. This is shown in Figures 9(a) and (b) (on the previous page). (Note the different scales used on the y-axes in the two figures.)

As mentioned earlier, when n is odd, the graph of $y = x^n$ resembles that of $y = x^3$. In Figure 10 we compare the graphs of $y = x^3$ and $y = x^5$. Notice that both curves pass through $(0, 0)$, $(1, 1)$, and $(-1, -1)$. For reasons similar to those explained for even n, the graph of $y = x^5$ is flatter than that of $y = x^3$ in the interval $-1 < x < 1$, and the graph of $y = x^7$ would be flatter still. For $|x| > 1$, the graph of $y = x^5$ is steeper than that of $y = x^3$, and $y = x^7$ would be steeper still.

In Section 5.2, we observed the effect of the constant a on the graph of $y = ax^2$. Those same comments apply to the graph of $y = ax^n$. For instance, the graph of $y = \frac{1}{2}x^4$ is wider than that of $y = x^4$, while the graph of $y = -\frac{1}{2}x^4$ is obtained by reflecting $y = \frac{1}{2}x^4$ in the x-axis.

EXAMPLE 2 Sketch the graph of $y = (x + 2)^5$ and specify the y-intercept.

Solution The graph of $y = (x + 2)^5$ is obtained by moving the graph of $y = x^5$ two units to the left. As a guide to drawing the curve, we recall that $y = x^5$ passes through the points $(1, 1)$ and $(-1, -1)$. Thus, $y = (x + 2)^5$ must pass through $(-1, 1)$ and $(-3, -1)$, as shown in Figure 11. Although the curve rises and falls very sharply, it is important to realize that it is never really vertical. For instance, the curve eventually crosses the y-axis. To find the y-intercept, we set $x = 0$ to obtain $y = 2^5 = 32$. Thus, the y-intercept is 32. ▪▪▪

FIGURE 11

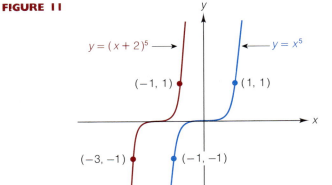

EXAMPLE 3 Graph the function $y = -2(x - 3)^4$.

Solution We begin with the graph of $y = -2x^4$ in Figure 12(a). The points $(1, -2)$ and $(-1, -2)$ are obtained by substituting $x = 1$ and $x = -1$, respectively, in the equation $y = -2x^4$. Now if we replace x with $x - 3$ in the equation $y = -2x^4$, we have $y = -2(x - 3)^4$, whose graph is obtained by translating the graph in Figure 12(a) three units to the right. See Figure 12(b). ▪▪▪

FIGURE 12

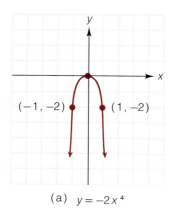

(a) $y = -2x^4$

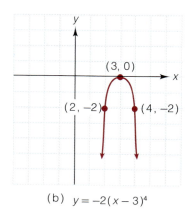

(b) $y = -2(x - 3)^4$

We can use our work on solving inequalities (in Section 2.7) to graph polynomial functions that are in factored form. Consider, for example, the function

$$f(x) = x(x + 1)(x - 3)$$

First of all, by inspection, we see that $f(x) = 0$ when $x = 0$, $x = -1$, or $x = 3$. These are the x-intercepts for the graph. Also, note that the y-intercept is 0. (Why?) Next, we want to know what the graph looks like in the intervals between the x-intercepts. To do this, we solve the two inequalities $f(x) > 0$ and $f(x) < 0$ using the technique in Section 2.7. Table 2 shows the results. (You should check these results for yourself; if you need a review, the details of this very example are worked out on pages 122–123.)

TABLE 2

Interval	$f(x) = x(x + 1)(x - 3)$
$(-\infty, -1)$	negative
$(-1, 0)$	positive
$(0, 3)$	negative
$(3, \infty)$	positive

Now we interpret the results in Table 2 graphically. When x is in either of the intervals $(-\infty, -1)$ or $(0, 3)$, the graph lies below the x-axis (because $f(x) < 0$); and when x is in either of the intervals $(-1, 0)$ or $(3, \infty)$, the graph lies above the x-axis (because $f(x) > 0$). This information is summarized in Figure 13(a). The three dots in the figure indicate the x-intercepts of the graph. The shaded regions are the **excluded regions** through which the graph cannot pass. The graph must pass only through the unshaded regions (and through the three x-intercepts). In Figure 13(b) we have drawn a rough sketch of a curve satisfying these conditions.

Notice that to draw a smooth curve satisfying the conditions of Figure 13(a), we need at least two turning points: one between $x = -1$ and $x = 0$, and another between $x = 0$ and $x = 3$. On the other hand, since the degree of $f(x)$ is 3, there can be no more than two turning points. Thus Figure 13(b) has exactly two turning points. As another check on our rough sketch, we note that for

FIGURE 13

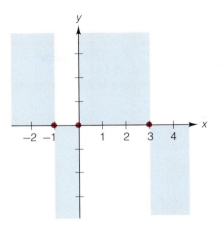

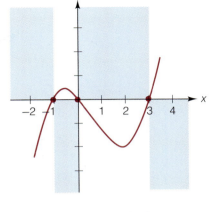

(a) Excluding regions and x-intercepts for the graph of $f(x) = x(x+1)(x-3)$

(b) A rough graph of $f(x) = x(x+1)(x-3)$

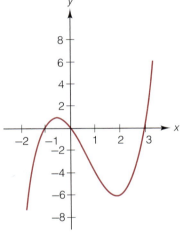

FIGURE 14

$f(x) = x(x+1)(x-3)$

large values of $|x|$, the graph indeed resembles that of $y = x^3$. (We are using the third property in the summary box on page 277.)

While the precise location of the turning points is a matter for calculus, we can nevertheless improve upon the sketch in Figure 13(b) by computing $f(x)$ for some specific values of x. Some reasonable choices in this case are the inputs -2, $-\frac{1}{2}$, 1, 2, and 4. As you can check, the resulting points on the graph are $(-2, -10)$, $\left(-\frac{1}{2}, \frac{7}{8}\right)$, $(1, -4)$, $(2, -6)$, and $(4, 20)$. We can now sketch the graph as shown in Figure 14.

In the example just concluded, the polynomial $f(x) = x(x+1)(x-3)$ has no *repeated factors*. That is, none of the factors is squared or cubed or raised to a higher power. Now let us look at two examples in which there are repeated factors. The observations we make will help us in determining the general shape of a graph without the need for plotting a large number of individual points. In Figure 15(a), we show the graph of $g(x) = x(x+1)(x-3)^2$. Notice that in the

FIGURE 15

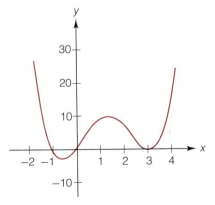

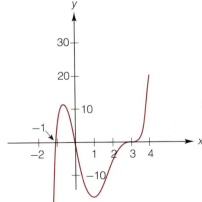

(a) $g(x) = x(x+1)(x-3)^2$
Near the x-intercept at 3, the graph has the same general shape as that of $y = A(x-3)^2$.

(b) $h(x) = x(x+1)(x-3)^3$
Near the x-intercept at 3, the graph has the same general shape as that of $y = A(x-3)^3$.

immediate vicinity of the intercept at $x = 3$, the graph has the same general shape as that of $y = A(x - 3)^2$. Similarly, in Figure 15(b), we show the graph of $h(x) = x(x + 1)(x - 3)^3$; notice that in the immediate vicinity of $x = 3$, the graph has the same general shape as that of $y = A(x - 3)^3$. In the box that follows, we state the general principle underlying these observations. (The principle can be justified using calculus.)

The Behavior of a Polynomial Function near an x-Intercept

Let $f(x)$ be a polynomial and suppose that $(x - a)^n$ is a factor of $f(x)$. [Furthermore, assume that none of the other factors of $f(x)$ contains $(x - a)$.] Then, in the immediate vicinity of the x-intercept at a, the graph of $y = f(x)$ closely resembles that of $y = A(x - a)^n$.

The principle that we have just stated is easy to apply because we already know how to graph functions of the form $y = A(x - a)^n$. The next example shows how this works.

EXAMPLE 4   Describe the behavior of each function in the immediate vicinity of the indicated x-intercept.

(a) $f(x) = \frac{1}{2}(x - 3)(x - 1)^3$; intercept: $x = 1$
(b) $g(x) = (x + 1)(x + 4)(x + 3)^2$; intercept: $x = -3$

Solution   (a) We make the following observation:

$$f(x) = \frac{1}{2}\underbrace{(x - 3)}_{\uparrow}(x - 1)^3$$

When x is close to 1, this factor is close to $1 - 3$, or -2.

So, if x is very close to 1, we have the approximation

$$f(x) \approx \tfrac{1}{2}(1 - 3)(x - 1)^3 = \tfrac{1}{2}(-2)(x - 1)^3 = -(x - 1)^3$$

Thus, in the immediate vicinity of $x = 1$, the graph of f closely resembles $y = -(x - 1)^3$. Notice the technique used to obtain this result. We retained the factor corresponding to the intercept $x = 1$, and we approximated the remaining factor using the value $x = 1$. See Figure 16 (on the next page).

(b) We use the approximation technique shown in part (a).

$$g(x) = \underbrace{(x + 1)}_{\uparrow}\underbrace{(x + 4)}_{\uparrow}(x + 3)^2$$

When x is close to -3, this factor is close to $-3 + 1$, or -2.   When x is close to -3, this factor is close to $-3 + 4$, or 1.

Thus, when x is close to -3, we have the approximation

$$g(x) \approx (-3 + 1)(-3 + 4)(x + 3)^2 = -2(x + 3)^2$$

This tells us that in the immediate vicinity of $x = -3$, the graph of g resembles $y = -2(x + 3)^2$. See Figure 17.

FIGURE 16

When x is close to 1, the graph of $f(x) = \frac{1}{2}(x-3)(x-1)^3$ closely resembles $y = -(x-1)^3$.

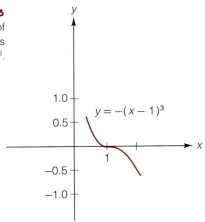

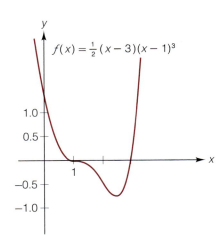

$f(x) = \frac{1}{2}(x-3)(x-1)^3$

FIGURE 17

When x is close to -3, the graph of $g(x) = (x+1)(x+4)(x+3)^2$ closely resembles $y = -2(x+3)^2$.

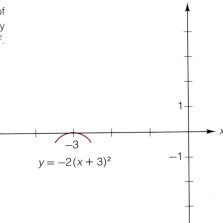

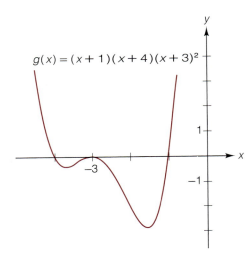

$g(x) = (x+1)(x+4)(x+3)^2$

EXERCISE SET 5.5

A

In Exercises 1–4, give a reason (as in Example 1) why each graph cannot represent a polynomial function of degree 3.

1.

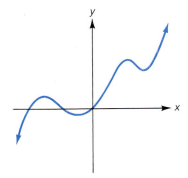

2.

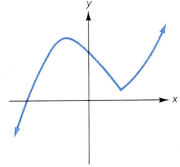

3.

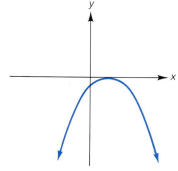

4.

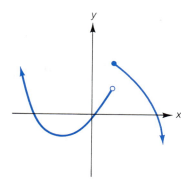

In Exercises 5–8, give a reason why each graph cannot represent a polynomial function that has highest-degree term $2x^5$.

5.

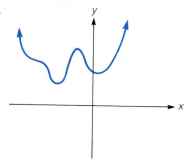

6.

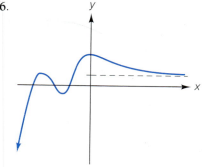

7.

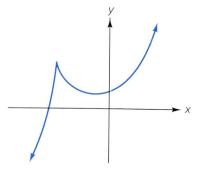

8.

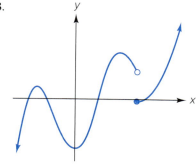

In Exercises 9–20, sketch the graph of each function and specify all x- and y-intercepts.

9. $y = (x - 2)^2 + 1$ 10. $y = -3x^4$

11. $y = -(x - 1)^4$ 12. $y = -(x + 2)^3$

13. $y = (x - 4)^3 - 2$ 14. $y = -(x - 4)^3 - 2$

15. $y = -2(x + 5)^4$ 16. $y = -2x^4 + 5$

17. $y = \frac{1}{2}(x + 1)^5$ 18. $y = \frac{1}{2}x^5 + 1$

19. $y = -(x - 1)^3 - 1$ 20. $y = x^8$

In Exercises 21–28: (a) Determine the x- and the y-intercepts and the excluded regions for the graph of the given function. Specify your results using a sketch similar to Figure 13(a). In Exercises 25–28, you will first need to factor the polynomial. (b) Graph the function.

21. $y = (x - 2)(x - 1)(x + 1)$

22. $y = (x - 3)(x + 2)(x + 1)$

23. $y = 2x(x - 2)(x - 1)$

24. $y = (x - 3)(x - 2)(x + 2)$

25. $y = x^3 - 4x^2 - 5x$ 26. $y = x^3 - 9x$

27. $y = x^3 + 3x^2 - 4x - 12$ 28. $y = x^3 - 5x^2 - x + 5$

In Exercises 29–38: (a) Determine the x- and the y-intercepts and the excluded regions for the graph of the given function. Specify your results using a sketch similar to Figure 13(a). (b) Describe the behavior of the function at each x-intercept that corresponds to a repeated factor. Specify your results using a sketch similar to the left-hand portion of Figure 17. (c) Graph the function.

29. $y = x^3(x + 2)$

30. $y = (x - 1)(x - 4)^2$

31. $y = 2(x - 1)(x - 4)^3$

32. $y = (x - 1)^2(x - 4)^2$

33. $y = (x + 1)^2(x - 1)(x - 3)$

34. $y = x^2(x - 4)(x + 2)$

35. $y = -x^3(x - 4)(x + 2)$

36. $y = 4(x - 2)^2(x + 2)^3$

37. $y = -4x(x - 2)^2(x + 2)^3$

38. $y = -3x^3(x + 1)^4$

B

In Exercises 39–44, six functions are defined as follows:

$$f(x) = x \qquad g(x) = x^2 \qquad h(x) = x^3$$
$$F(x) = x^4 \qquad G(x) = x^5 \qquad H(x) = x^6$$

Refer also to the following figure.

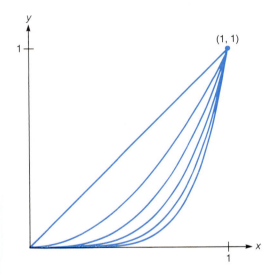

39. The six graphs in the figure are the graphs of the six given functions for the interval $[0, 1]$, but the graphs are not labeled. Which is which?

40. For which x-values in $[0, 1]$ will the graph of g lie strictly below the horizontal line $y = 0.1$? Use a calculator to evaluate your answer. Round off the result to two significant figures.

41. Follow Exercise 40, using the function H instead of g.

42. Find a number t in $[0, 1]$ such that the vertical distance between $f(t)$ and $g(t)$ is $\frac{1}{4}$.

43. Is there a number t in $[0, 1]$ such that the vertical distance between $g(t)$ and $F(t)$ is 0.26?

44. Find all numbers t in $[0, 1]$ such that $F(t) = G(t) + H(t)$.

45. Do the graphs of $y = x$ and $y = \frac{1}{100}x^2$ intersect anywhere other than at the origin? (See the following figure.)

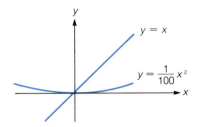

46. (a) Graph the function $y = 4x^2 - x^4$.
   (b) Find the coordinates of the turning points. *Hint:* See Example 7 in Section 5.2.

47. (a) Graph the function $D(x) = x^2 - x^4$.
   (b) Find the turning points of the graph. *Hint:* See Example 7 in Section 5.2.
   (c) On the same set of axes, sketch the graphs of $y = x^2$ and $y = x^4$ for $0 \le x \le 1$. What is the maximum vertical distance between the graphs?

48. (a) An open-top box is to be constructed from a 6-in.-by-8-in. rectangular sheet of tin by cutting out equal squares at each corner and then folding up the resulting flaps. Let x denote the length of the side of each cutout square. Show that the volume $V(x)$ is

$$V(x) = x(6 - 2x)(8 - 2x)$$

   (b) What is the domain of the volume function in part (a)? [The answer is *not* $(-\infty, \infty)$.]
   (c) Graph the volume function.
   (d) Use your graph to estimate the maximum possible volume for the box.

49. A cylinder is inscribed in a sphere of radius 6 cm. Let r and h denote the radius and height of the cylinder, as shown in the figure. We are going to estimate the maximum possible volume for the cylinder.

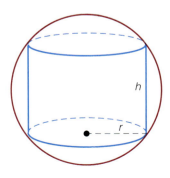

(a) Show that $h = 2\sqrt{36 - r^2}$.
(b) Let V denote the volume of the cylinder. Check that $V = 2\pi r^2\sqrt{36 - r^2}$.
(c) What is the domain of the volume function in part (b)?
(d) The expression for V in part (b) is not a polynomial, but the square of the expression is. So, we'll work with the function $f(r) = 4\pi^2 r^4(36 - r^2)$.

At the end, we can convert back to V by taking the square root of $f(r)$. Graph the function $f(r) = 4\pi^2 r^4(36 - r^2)$. (To fine-tune your graph, use a calculator to compute $f(r)$ for r running from 0 to 6 in increments of 0.5.) From the graph, estimate the maximum value of $f(r)$. Now estimate the maximum possible volume of the cylinder.

5.6   GRAPHS OF RATIONAL FUNCTIONS

We shall not attempt to explain the numerous situations in life sciences where the study of such graphs is important nor the chemical reactions which give rise to the rational functions $r(x)$ .... Let it suffice to mention that, to the experimental biochemist, theoretical results concerning the shapes of rational functions are of considerable interest.

W. G. Bardsley and R. M. W. Wood in "Critical Points and Sigmoidicity of Positive Rational Functions," *The American Mathematical Monthly,* **92** (1985) 37–42.

After the polynomial functions, the next simplest functions are the **rational functions**. These are functions defined by equations of the form

$$y = \frac{f(x)}{g(x)}$$

where $f(x)$ and $g(x)$ are polynomials. In general, throughout this section, when we write a function such as $y = f(x)/g(x)$, we assume that $f(x)$ and $g(x)$ contain no common factors (other than constants). [Exercises 36 and 37 ask you to consider several cases in which $f(x)$ and $g(x)$ do contain common factors.] Also, for each of the examples that we discuss, the degree of $f(x)$ is less than or equal to the degree of $g(x)$. [A case in which the degree of $f(x)$ exceeds the degree of $g(x)$ is developed in the exercises.]

EXAMPLE I   Specify the domain of the rational function defined by

$$y = \frac{3x - 2}{x^2 - 1}$$

Also, find the x- and y-intercepts (if any) for the graph of this function.

Solution   By factoring the denominator, we can rewrite the given equation

$$y = \frac{3x - 2}{(x - 1)(x + 1)}$$

Since the denominator is zero when $x = 1$ and when $x = -1$, it follows that the domain of the given function consists of all real numbers except 1 and -1. To determine the x-intercepts of the graph, we set $y = 0$ (as usual) in the given equation to obtain

$$\frac{3x - 2}{x^2 - 1} = 0 \qquad (x \neq 1, \ x \neq -1)$$

$$3x - 2 = 0 \qquad \begin{array}{l}\text{multiplying both sides of the equation by the} \\ \text{nonzero quantity } x^2 - 1, \text{ since } x \neq \pm 1\end{array}$$

$$x = \tfrac{2}{3}$$

Thus, the only x-intercept is $x = \tfrac{2}{3}$. For the y-intercept, we set $x = 0$ (as usual) in the given equation. As you can readily check, this yields $y = 2$. In summary then, the x- and y-intercepts are $\tfrac{2}{3}$ and 2, respectively. ■■■

In the box that follows, we summarize the ideas used in Example 1 for determining the domain and x-intercepts of a rational function.

> Let R be a rational function defined by
>
> $$R(x) = \frac{f(x)}{g(x)}$$
>
> where $f(x)$ and $g(x)$ are polynomials with no common factors (other than constants). Then the domain of R consists of all real numbers for which $g(x) \neq 0$; the x-intercepts (if any) are the real solutions of the equation $f(x) = 0$.

You are already familiar with the graph of one rational function, because in Chapter 4 we graphed $y = 1/x$. For convenience, this graph is shown again in Figure 1. The graph in Figure 1 differs from the graph of every polynomial function in two important aspects. First, the graph has a break in it; it is composed of two distinct pieces or **branches**. (Recall from Section 5.5 that the graph of a polynomial function never has a break in it.) In general, the graph of a rational function has one more branch than the number of distinct real values for which the denominator is zero. For instance, referring to the function in Example 1, we can expect the graph of $y = \dfrac{3x - 2}{x^2 - 1}$ to have three branches.

The second way that the graph in Figure 1 differs from that of a polynomial function is related to *asymptotes*. Recall that a line is an **asymptote** for a curve if the distance between the line and the curve approaches zero as we move out farther and farther along the line. Thus, the x-axis is a horizontal asymptote for the graph in Figure 1, while the y-axis is a vertical asymptote. It can be shown that the graph of a polynomial function never has an asymptote. Figure 2 (on the next page) displays additional examples of curves with asymptotes.

If k is a positive constant, the graph of $y = k/x$ resembles that of $y = 1/x$. Figure 3 (next page) shows the graphs of $y = 1/x$ and $y = 4/x$. Once we know about the graph of $y = k/x$, we can graph any rational function of the form

$$y = \frac{ax + b}{cx + d}$$

The next three examples show how this is done using the translation and reflection techniques developed in Chapter 4.

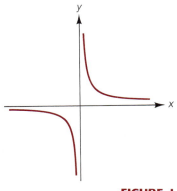

FIGURE 1

The graph of $y = \dfrac{1}{x}$

FIGURE 2
Curves with asymptotes

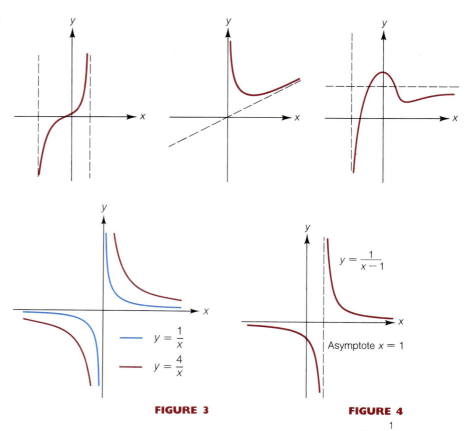

FIGURE 3

FIGURE 4
$$y = \frac{1}{x - 1}$$

EXAMPLE 2 Graph $y = \dfrac{1}{x - 1}$.

Solution By translating the graph of $y = 1/x$ one unit to the right, we obtain the graph of $y = 1/(x - 1)$ shown in Figure 4. Notice that the horizontal asymptote is unchanged by the translation. However, the vertical asymptote moves one unit to the right. (The y-intercept is -1; why?) ∎

EXAMPLE 3 Graph $y = \dfrac{2}{x - 1}$ and $y = \dfrac{-2}{x - 1}$.

Solution The graph of $y = 2/(x - 1)$ has the same basic shape and location as the graph of $y = 1/(x - 1)$ shown in Figure 4. As a further guide to sketching $y = 2/(x - 1)$, we can pick several convenient x-values on either side of the asymptote $x = 1$ and then compute the corresponding y-values. After doing this, we obtain the graph shown in Figure 5(a). By reflecting this graph about the x-axis, we obtain the graph of $y = -2/(x - 1)$, which is shown in Figure 5(b). ∎

FIGURE 5

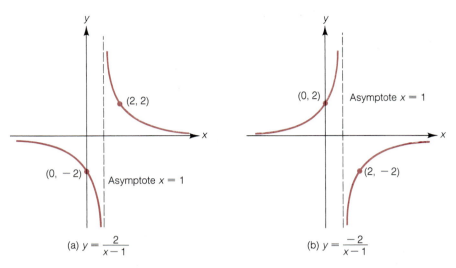

(a) $y = \dfrac{2}{x-1}$

(b) $y = \dfrac{-2}{x-1}$

EXAMPLE 4 Graph $y = \dfrac{4x-2}{x-1}$.

Solution First, as you can readily check, the x- and y-intercepts are $\frac{1}{2}$ and 2, respectively. Next, using long division, we find that

$$\frac{4x-2}{x-1} = 4 + \frac{2}{x-1}$$

We conclude that the required graph can be obtained by moving up the graph of $y = 2/(x-1)$ four units in the y-direction (see Figure 6). Notice that the vertical asymptote is still $x = 1$, but the horizontal asymptote is now $y = 4$ instead of the x-axis.

FIGURE 6
$$y = \frac{4x-2}{x-1}$$

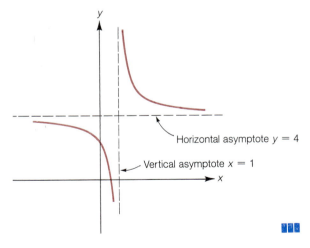

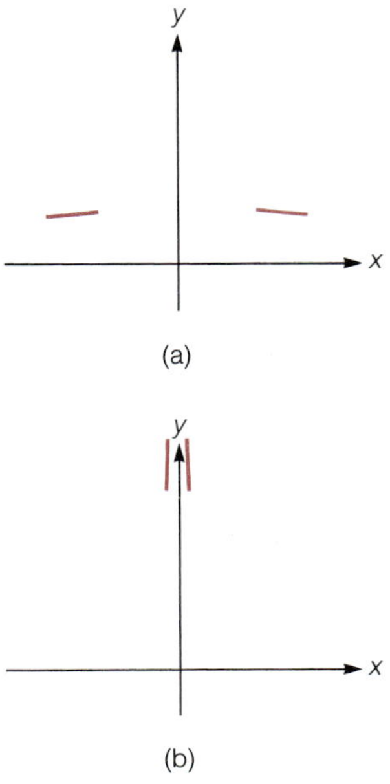

Horizontal asymptote $y = 4$

Vertical asymptote $x = 1$

Now let's look at rational functions of the form $y = 1/x^n$. First we'll consider $y = 1/x^2$. As with $y = 1/x$, the domain consists of all real numbers except $x = 0$. For $x \neq 0$, the quantity $1/x^2$ is always positive. This means that the graph will always lie above the x-axis. Furthermore, the graph will be symmetric about the y-axis. (Why?) As $|x|$ becomes very large, the quantity $1/x^2$ approaches zero; this is true whether x itself is positive or negative. So when $|x|$ is very large, we ex-

(a)

(b)

FIGURE 7

pect the graph of $y = 1/x^2$ to look as shown in Figure 7(a). On the other hand, when x is a very small fraction close to zero, either negative or positive, the quantity $1/x^2$ is very large. For instance, if $x = \frac{1}{10}$, we find that $y = 100$, and if $x = \frac{1}{100}$, we find that $y = 10,000$. Thus, as x approaches zero, from the right or from the left, the graph of $y = 1/x^2$ must look as shown in Figure 7(b). Now, by plotting several points and taking Figures 7(a) and (b) into account, we obtain the graph of $y = 1/x^2$, shown in Figure 8. Also, by following a similar line of reasoning, we find that the graph of $y = 1/x^3$ looks as shown in Figure 9. (Note that $y = 1/x^3$ is symmetric about the origin.

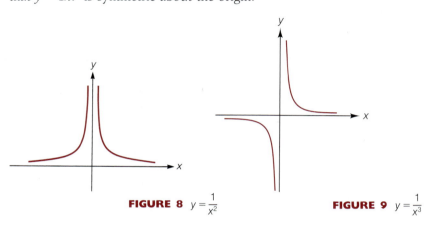

FIGURE 8 $y = \dfrac{1}{x^2}$ **FIGURE 9** $y = \dfrac{1}{x^3}$

In general, when n is an even integer greater than 2, the graph of $y = 1/x^n$ resembles that of $y = 1/x^2$. When n is an odd integer greater than 3, the graph of $y = 1/x^n$ resembles that of $y = 1/x^3$.

EXAMPLE 5 Graph $y = \dfrac{-1}{(x + 3)^2}$.

Solution Refer to Figure 10. Begin with the graph of $y = 1/x^2$. By moving the graph three units to the left, we obtain the graph of $y = 1/(x + 3)^2$. Then by reflecting the graph of $y = 1/(x + 3)^2$ about the x-axis, we get the graph of $y = -1/(x + 3)^2$.

Question   What are the y-intercepts in Figures 10(b) and (c)?

FIGURE 10

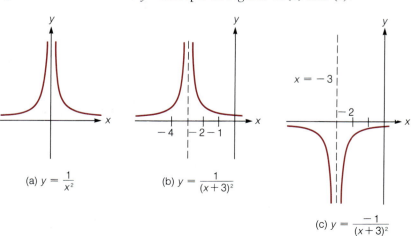

(a) $y = \dfrac{1}{x^2}$ (b) $y = \dfrac{1}{(x+3)^2}$

(c) $y = \dfrac{-1}{(x+3)^2}$

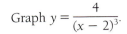

EXAMPLE 6 Graph $y = \dfrac{4}{(x-2)^3}$.

Solution Moving the graph of $y = 1/x^3$ two units to the right gives us the graph of the function $y = 1/(x-2)^3$. The graph of $y = 4/(x-2)^3$ will have the same basic shape and location. As a further guide to sketching the required graph, we can pick several convenient x-values near the asymptote $x = 2$ and compute the corresponding y-values. Using $x = 0$, $x = 1$, and $x = 3$, we find that the points $\left(0, -\frac{1}{2}\right)$, $(1, -4)$, and $(3, 4)$ are on the graph. With this information, the graph can be sketched as in Figure 11. ∎

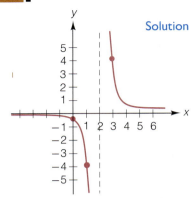

FIGURE 11
$$y = \frac{4}{(x-2)^3}$$

As we have seen in Examples 2 through 6, the vertical asymptotes of $y = f(x)/g(x)$ are found by solving the equation $g(x) = 0$. In Example 7 we demonstrate a method for determining the horizontal asymptote. The method involves dividing both the numerator and the denominator by the highest power of x that appears.

EXAMPLE 7 Determine the horizontal asymptote for the graph of $y = \dfrac{3x^2}{x^2 - 4x + 1}$.

Solution First we look for the highest power of x that appears in the given fractional expression; this is x^2. Now we divide both the numerator and the denominator by x^2 (you'll see the reason for doing this in a moment). We have

$$y = \frac{3x^2}{x^2 - 4x + 1} = \frac{\dfrac{3x^2}{x^2}}{\dfrac{x^2 - 4x + 1}{x^2}} = \frac{3}{1 - \dfrac{4}{x} + \dfrac{1}{x^2}}$$

When $|x|$ grows very large, the two fractions $4/x$ and $1/x^2$ both approach zero. We therefore have

$$y \approx \frac{3}{1 - 0 + 0} = 3 \qquad \text{when } |x| \text{ is very large}$$

This tells us that the line $y = 3$ is a horizontal asymptote for the graph of the given function. (You will see another example of this type of calculation in Example 8.) ∎

In the previous section you learned a technique for determining the general shape of a polynomial function near an x-intercept. That same technique is also very useful in graphing rational functions in which both the numerator and denominator can be expressed as products of linear factors. Example 8 shows how this works.

EXAMPLE 8 Graph the function $y = \dfrac{(x-3)(x+2)}{(x+1)(x-2)}$.

Solution   By inspection, we see that the x-intercepts are $x = 3$ and $x = -2$, the y-intercept is 3, and the vertical asymptotes are the lines $x = -1$ and $x = 2$. To find the horizontal asymptote, we write

$$y = \frac{x^2 - x - 6}{x^2 - x - 2}$$

$$= \frac{1 - \dfrac{1}{x} - \dfrac{6}{x^2}}{1 - \dfrac{1}{x} - \dfrac{2}{x^2}} \approx \frac{1 - 0 - 0}{1 - 0 - 0} \qquad \text{when } |x| \text{ is large}$$

The horizontal asymptote is therefore the line $y = 1$.

Now we want to see how the graph looks in the immediate vicinity of the x-intercepts and the vertical asymptotes. To do this, we use the approximation technique explained in the previous section for polynomial functions. Let's start with the x-intercept at $x = 3$. As in Section 5.5, we'll retain the factor $x - 3$ and approximate the remaining factors using $x = 3$. So, for x near 3, we have

$$y = \frac{(x - 3)(x + 2)}{(x + 1)(x - 2)} \approx \frac{(x - 3)(3 + 2)}{(3 + 1)(3 - 2)} = \frac{5}{4}(x - 3)$$

Thus, in the immediate vicinity of the x-intercept $x = 3$, the required graph will closely resemble the line $y = \frac{5}{4}(x - 3)$. The remaining calculations for approximating the graph near the other x-intercept and near the two vertical asymptotes are carried out in exactly the same manner. As Exercise 35 will ask you to verify, the results are:

x near 3: $y \approx \frac{5}{4}(x - 3)$	x near -1: $y \approx \dfrac{4/3}{x + 1}$
x near -2: $y \approx -\frac{5}{4}(x + 2)$	x near 2: $y \approx \dfrac{-4/3}{x - 2}$

We summarize these results as follows. As the graph passes through the points $(3, 0)$ and $(-2, 0)$, it resembles the lines $y = \frac{5}{4}(x - 3)$ and $y = -\frac{5}{4}(x + 2)$, respectively. Near the vertical asymptote $x = -1$, the graph has the same basic shape as $y = 1/(x + 1)$. And near the vertical asymptote $x = 2$, the graph resembles $y = -1/(x - 2)$. See Figure 12.

Finally, we want to find out how the graph approaches the horizontal asymptote $y = 1$. Perhaps the simplest way to decide this is to do some calculations. In Table 1 we have computed the outputs for some relatively large values of $|x|$. From the table, we see that when $|x|$ is large, y is less than 1. Graphically, this means that the curve approaches the asymptote $y = 1$ from below. In Figure 13 we summarize what we have discovered up to this point. Then, using Figure 13 as a guide, we can sketch the required graph, as shown in Figure 14.

Note   A general method for finding the coordinates of the lowest point on the middle branch of the graph is studied in calculus. Those coordinates can also be found, however, by applying algebraic techniques. (See Exercise 38.)   ∎

TABLE I

When $|x|$ is large, y is less than I.

x	y
10	0.95
100	0.99
-10	0.96
-100	0.9996

FIGURE 12

The graph of $y = \dfrac{(x-3)(x+2)}{(x+1)(x-2)}$ near its x-intercepts and vertical asymptotes.

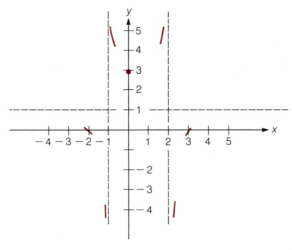

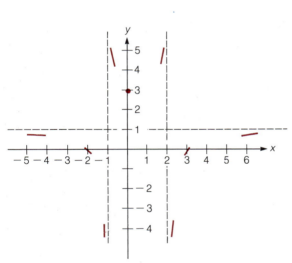

FIGURE 13

The graph of $y = \dfrac{(x-3)(x+2)}{(x+1)(x-2)}$ near its x-intercepts and horizontal and vertical asymptotes.

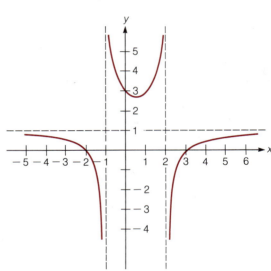

FIGURE 14

$y = \dfrac{(x-3)(x+2)}{(x+1)(x-2)}$

EXERCISE SET 5.6

A

In Exercises 1–6, find the domain and the x- and y-intercepts for each rational function.

1. $y = \dfrac{3x+15}{4x-12}$

2. $y = \dfrac{(x+6)(x+4)}{(x-1)^2}$

5. $y = \dfrac{(x^2-4)(x^3-1)}{x^6}$

3. $y = \dfrac{x^2-8x-9}{x^2-x-6}$

4. $y = \dfrac{2x^2+x-5}{x^2+1}$

6. $y = \dfrac{x^5-2x^4-9x+18}{8x^3+2x^2-3x}$

In Exercises 7–28, sketch the graph of each rational function. Specify the intercepts and the asymptotes.

7. $y = \dfrac{1}{x + 4}$

8. $y = \dfrac{-1}{x + 4}$

9. $y = \dfrac{3}{x + 2}$

10. $y = \dfrac{-3}{x + 2}$

11. $y = \dfrac{-2}{x - 3}$

12. $y = \dfrac{x - 1}{x + 1}$

13. $y = \dfrac{x - 3}{x - 1}$

14. $y = \dfrac{2x}{x + 3}$

15. $y = \dfrac{4x - 2}{2x + 1}$

16. $y = \dfrac{3x + 2}{x - 3}$

17. $y = \dfrac{1}{(x - 2)^2}$

18. $y = \dfrac{-1}{(x - 2)^2}$

19. $y = \dfrac{3}{(x + 1)^2}$

20. $y = \dfrac{-3}{(x + 1)^2}$

21. $y = \dfrac{1}{(x + 2)^3}$

22. $y = \dfrac{-1}{(x + 2)^3}$

23. $y = \dfrac{-4}{(x + 5)^3}$

24. $y = \dfrac{x}{(x + 1)(x - 3)}$

25. $y = \dfrac{-x}{(x + 2)(x - 2)}$

26. $y = \dfrac{2x}{(x + 1)^2}$

27. $y = \dfrac{x}{(x - 1)(x + 3)}$

28. $y = \dfrac{x^2 + 1}{x^2 - 1}$

29. (a) $f(x) = \dfrac{(x - 2)(x - 4)}{x(x - 1)}$

　　(b) $g(x) = \dfrac{(x - 2)(x - 4)}{x(x - 3)}$

[Compare the graphs you obtain in parts (a) and (b). Notice how a change in only one constant can radically alter the nature of the graph.]

30. (a) $f(x) = \dfrac{(x - 1)(x + 2.75)}{(x + 1)(x + 3)}$

　　(b) $g(x) = \dfrac{(x - 1)(x + 3.25)}{(x + 1)(x + 3)}$

[Compare the graphs you obtain in parts (a) and (b). Notice how a relatively small change in one of the constants can radically alter the graph.]

B

In Exercises 31–34, graph the functions.   Note: Each graph intersects its horizontal asymptote once. Find the intersection point before sketching your final graph.

31. $y = \dfrac{(x - 4)(x + 2)}{(x - 1)(x - 3)}$

32. $y = \dfrac{(x - 1)(x - 3)}{(x + 1)^2}$

33. $y = \dfrac{(x + 1)^2}{(x - 1)(x - 3)}$

34. $y = \dfrac{2x^2 - 3x - 2}{x^2 - 3x - 4}$

35. (This exercise refers to Example 8.) Let

$$y = \dfrac{(x - 3)(x + 2)}{(x + 1)(x - 2)}$$

Verify each of the following approximations.
(a) When x is close to -2, then $y \approx -\frac{5}{4}(x + 2)$.

(b) When x is close to -1, then $y \approx \dfrac{\frac{4}{3}}{x + 1}$.

(c) When x is close to 2, then $y \approx \dfrac{-\frac{4}{3}}{x - 2}$.

In Exercises 36 and 37, graph the functions. Notice in each case that the numerator and denominator contain at least one common factor. Thus, you can simplify each quotient; but don't lose track of the domain of the function as it was initially defined.

36. (a) $y = \dfrac{x + 2}{x + 2}$

　　(b) $y = \dfrac{x^2 - 4}{x - 2}$

　　(c) $y = \dfrac{x - 1}{(x - 1)(x - 2)}$

37. (a) $y = \dfrac{x^2 - 9}{x + 3}$

　　(b) $y = \dfrac{x^2 - 5x + 6}{x^2 - 2x - 3}$

　　(c) $y = \dfrac{(x - 1)(x - 2)(x - 3)}{(x - 1)(x - 2)(x - 3)(x - 4)}$

38. This exercise shows you how to determine the coordinates of the lowest point on the middle branch of the curve in Figure 14. The basic idea is as follows. Suppose that the required coordinates are (h, k). Then the horizontal line $y = k$ is the unique horizontal line intersecting the curve in one and only one point. (Any other horizontal line intersects the curve either in two points or not at all.) In steps (a) through (c) that follow, we use these observations to determine the point (h, k).
(a) Given any horizontal line $y = k$, its intersection with the curve in Figure 14 is determined by solving the following pair of simultaneous equations:

$$\begin{cases} y = k \\ y = \dfrac{x^2 - x - 6}{x^2 - x - 2} \end{cases}$$

In the second equation of the system, replace y with k and show that the resulting equation can be written

$$(k - 1)x^2 - (k - 1)x + (6 - 2k) = 0 \qquad (1)$$

(b) If k is indeed the required y-coordinate, then equation (1) must have exactly one real solution. Set the discriminant of the quadratic equation equal to zero to obtain

$$(k-1)^2 - 4(k-1)(6-2k) = 0$$

and deduce from this that $k=1$ or $k=\frac{25}{9}$. The solution $k=1$ can be discarded. (To see why, look at Figure 14.)

(c) Using the value $y = \frac{25}{9}$, show that the corresponding x-coordinate is $\frac{1}{2}$. Thus, the required point is $\left(\frac{1}{2}, \frac{25}{9}\right)$.

39. Graph the function $y = x/(x-3)^2$. Use the technique explained in Exercise 38 to find the coordinates of any turning points on the graph.

40. Graph the function $y = 2/(x-x^2)$. Use the technique explained in Exercise 38 to find the coordinates of any turning points on the graph.

An asymptote that is neither horizontal nor vertical is called a **slant, or oblique, asymptote.** *For example, as indicated in the following figure, the line $y = x$ is a slant asymptote for the graph of $y = (x^2 + 1)/x$. To understand why the line $y = x$ is an asymptote, we carry out the indicated division and write the function in the form*

$$y = x + \frac{1}{x} \tag{2}$$

From equation (2), we see that if $|x|$ is very large, then $y \approx x + 0$; that is, $y \approx x$, as we wished to show. Equation (2) actually tells us more than this. When $|x|$ is very close to zero, equation (2) yields $y \approx 0 + (1/x)$. In other words, as we approach the y-axis, the curve looks more and more like the graph of $y = 1/x$. In general, if we have a rational function

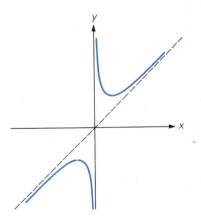

The curve $y = (x^2 + 1)/x$ with slant asymptote $y = x$

$f(x)/g(x)$ in which the degree of $f(x)$ is 1 greater than the degree of $g(x)$, then the graph has a slant asymptote that is obtained as follows. Divide $f(x)$ by $g(x)$ to obtain an equation of the form

$$\frac{f(x)}{g(x)} = (mx + b) + \frac{h(x)}{g(x)}$$

where the degree of $h(x)$ is less than the degree of $g(x)$. Then the equation of the slant asymptote is $y = mx + b$. (For instance, using the example in the previous paragraph, we have $mx + b = x$ and $h(x)/g(x) = 1/x$.) In Exercises 41–43, you are asked to graph functions that have slant asymptotes.

41. Let $y = F(x) = \dfrac{x^2 + x - 6}{x - 3}$.

(a) Use long division to show that

$$\frac{x^2 + x - 6}{x - 3} = (x + 4) + \frac{6}{x - 3}$$

(b) The result in part (a) shows that the line $y = x + 4$ is a slant asymptote for the graph of the function F. Verify this fact empirically by completing the following two tables (Use a calculator.)

x	$x + 4$	$\dfrac{x^2 + x - 6}{x - 3}$
10		
100		
1000		

x	$x + 4$	$\dfrac{x^2 + x - 6}{x - 3}$
-10		
-100		
-1000		

(c) Determine the vertical asymptote and the x- and y-intercepts for the graph of F.

(d) Graph the function F. (Use the techniques in this section along with the fact that $y = x + 4$ is a slant asymptote.)

(e) Use the technique explained in Exercise 38 to find the coordinates of the two turning points on the graph of F.

42. (a) Show that the line $y = x - 2$ is a slant asymptote for the graph of $F(x) = x^2/(x+2)$.

(b) Sketch the graph of F.

43. Show that the line $y = -x$ is a slant asymptote for the graph of $y = (1 - x^2)/x$. Then sketch the graph of this function.

CHAPTER FIVE SUMMARY OF PRINCIPAL TERMS

TERMS	PAGE REFERENCE	COMMENTS
1. Linear function	239	A linear function is a function of the form $f(x) = Ax + B$. The graph of a linear function is a straight line. An important idea that arose in several of the examples is that the slope of a line can be interpreted as a rate of change. Two instances of this are marginal cost and velocity.
2. Quadratic function	248	A quadratic function is a function of the form $f(x) = ax^2 + bx + c$. The graph of a quadratic function is a parabola. See the boxes on pages 250 and 251.
3. Polynomial function	275	A polynomial function of degree n is a function of the form $$f(x) = a_n x^n + a_{n-1} x^{n-1} + \cdots + a_1 x + a_0$$ where n is a nonnegative integer and $a_n \neq 0$. Three basic properties of polynomial functions are summarized in the box on page 277.
4. Rational function	286	A rational function is a function of the form $y = f(x)/g(x)$, where $f(x)$ and $g(x)$ are polynomials.
5. Asymptote	287	A line is said to be an asymptote for a curve if the distance between the line and the curve approaches zero as we move out farther and farther along the line. See, for example, Figure 2 on page 288, in which the dashed lines are asymptotes.

WRITING MATHEMATICS

Consider the following two problems.

(A) The perimeter of a rectangle is 2 m.
   (i)   Express the area of the rectangle as a function of the width w.
   (ii) Find the maximum possible area for such a rectangle.
(B) The area of a rectangle is 2 m².
   (i)   Express the perimeter of the rectangle as a function of the width w.
   (ii) Find the minimum possible perimeter for such a rectangle.

1. After working out problem A for yourself, write out the solution in complete sentences, as if you were explaining it to a classmate or to your instructor. Be sure to let the reader know where you are headed and why each of the main steps is necessary.

2. After working out part i of problem B for yourself, write out the solution in complete sentences.

3. Explain why the methods of Section 5.4 are not applicable for solving the second part of problem B.

4. The following result is known as the *arithmetic-geometric mean inequality:*

   For all nonnegative real numbers a and b, we have $a + b \geq 2\sqrt{ab}$, with equality holding when and only when $a = b$.

   By working with a classmate or your instructor, find out how to apply this result to solve the second part of problem B. Then write out your solution in complete sentences. How does your final answer here compare with that in problem A?

CHAPTER FIVE   REVIEW EXERCISES

1. Find $G(0)$ if G is a linear function such that $G(1) = -2$ and $G(-2) = -11$.

2. (a) Let $f(x) = 3x^2 + 6x - 10$. For which input x is the value of the function a minimum? What is that minimum value?
   (b) Let $g(t) = 6t^2 - t^4$. For which input t is the value of the function a maximum?

3. Suppose the function $p = -\frac{1}{8}x + 100 \ (0 \le x \le 12)$ relates the selling price p of an item to the quantity x that is sold. Assume that p is in dollars. What is the maximum revenue possible in this situation?

4. Graph the function $y = -1/(x + 1)^3$.

5. Graph the function $y = (x - 4)(x - 1)(x + 1)$.

6. Graph the function $f(x) = x^2 + 4x - 5$. Specify the vertex, the x- and y-intercepts, and the axis of symmetry.

7. A factory owner buys a new machine for $1000. After five years, the machine has a salvage value of $100. Assuming linear depreciation, find a formula for the value V of the machine after t years, where $0 \le t \le 5$.

8. Graph the function $y = 2(x - 3)^4$. Does the graph cross the y-axis? If so, where?

9. Graph the function $y = \dfrac{3x + 5}{x + 2}$. Specify all intercepts and asymptotes.

10. What is the largest area possible for a right triangle in which the sum of the lengths of the two shorter sides is 12 cm?

11. Graph the function $x/[(x + 2)(x - 4)]$.

12. Let $P(x, y)$ be a point [other than $(-1, -1)$] on the graph of $f(x) = x^3$. Express the slope of the line passing through the points P and $(-1, -1)$ as a function of x. Simplify your answer as much as possible.

13. A rectangle is inscribed in a circle. The circumference of the circle is 12 cm. Express the perimeter of the rectangle as a function of its width w.

14. Give a reason why each of the following two graphs cannot represent a a polynomial function with highest-degree term $-\frac{1}{3}x^3$.

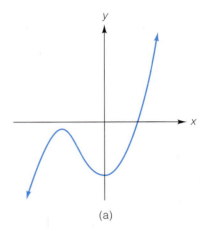

(a)

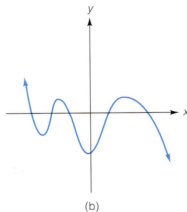

(b)

In Exercises 15–20, find equations for the linear functions satisfying the given conditions. Write each answer in the form $f(x) = mx + b$.

15. $f(3) = 5$ and $f(-2) = 0$

16. $f(1) = 5$ and the graph of f passes through the origin

17. $f(4) = -1$ and the graph of f is parallel to the line $3x - 8y = 16$

18. The graph passes through $(6, 1)$ and the x-intercept is twice the y-intercept

19. $f(-3) = 5$, and the graph of the inverse function passes through $(2, 1)$

20. The graph of f passes through the vertices of the two parabolas $y = x^2 + 4x + 1$ and $y = \frac{1}{2}x^2 + 9x + \frac{81}{2}$

In Exercises 21–26, graph the quadratic functions. In each case, specify the vertex and the x- and y-intercepts.

21. $y = x^2 + 2x - 3$

22. $f(x) = x^2 - 2x - 15$

23. $y = -x^2 + 2\sqrt{3}x + 3$

24. $f(x) = 2x^2 - 2x + 1$

25. $y = -3x^2 + 12x$

26. $f(x) = -4x^2 + 16x$

27. Find the distance between the vertices of the two parabolas $y = x^2 - 4x + 6$ and $y = -x^2 - 4x - 5$.

28. Find the value of a, given that the maximum value of the function $f(x) = ax^2 + 3x - 4$ is 5.

29. The sum of two numbers is $\sqrt{3}$. Find the largest possible value for their product.

30. The sum of two numbers is $\frac{2}{3}$. What is the smallest possible value for the sum of their squares?

31. Suppose that an object is thrown vertically upward (from ground level) with an initial velocity of v_0 ft/sec. It can be shown that the height h (in feet) after t sec is given by the formula $h = v_0 t - 16t^2$.
    (a) At what time does the object reach its maximum height? What is that maximum height?
    (b) At what time does the object strike the ground?

32. Let $f(x) = 4x^2 - x + 1$ and $g(x) = (x - 3)/2$.
    (a) For which input will the value of the function $f \circ g$ be a minimum?
    (b) For which input will the value of $g \circ f$ be a minimum?

33. (a) Let P be a point on the parabola $y = x^2$. Express the distance from P to the point $(0, 2)$ in terms of x.
    (b) Which point in the second quadrant on the parabola $y = x^2$ is closest to the point $(0, 2)$?

34. Find the coordinates of the point on the line $y = 2x - 1$ closest to $(-5, 0)$.

35. Find all the values of b such that the minimum distance from the point $(2, 0)$ to the line $y = \frac{4}{3}x + b$ is 5.

36. What number exceeds one-half its square by the greatest amount?

37. Suppose that $x + y = \sqrt{2}$. Find the minimum value of the quantity $x^2 + y^2$.

38. For which numbers t will the value of $9t^2 - t^4$ be as large as possible?

39. Find the maximum area possible for a right triangle with a hypotenuse of 15 cm.   *Hint:* Let x denote the length of one leg. Show that the area is $A = \frac{1}{2}x\sqrt{225 - x^2}$. Now work with A^2.

40. For which point (x, y) on the curve $y = 1 - x^2$ is the sum $x + y$ a maximum?

41. Let $f(x) = x^2 - (a^2 + 2a)x + 2a^3$, where $0 < a < 2$. For which value of a will the distance between the x-intercepts of f be a maximum?

42. Suppose that the revenue R (in dollars) generated by selling x units of a certain product is given by $R = 300x - \frac{1}{4}x^2$ ($0 \le x \le 1200$). How many units should be sold to maximize the revenue? What is that maximum revenue?

43. Suppose that the function $p = 160 - \frac{1}{5}x$ relates the selling price p of an item to the quantity x that is sold. Assume that p is in dollars. For which value of x will the revenue R ($= xp$) be a maximum? What is the selling price p in this case?

44. A piece of wire 16 cm long is cut into two pieces. Let x denote the length of the first piece and $16 - x$ the length of the second. The first piece is formed into a rectangle in which the length is twice the width. The second piece of wire is also formed into a rectangle, but with the length three times the width. For which value of x is the total area of the two rectangles a minimum?

In Exercises 45–60, graph each function and specify the x- and y-intercepts and asymptotes, if any.

45. $y = (x + 4)(x - 2)$

46. $y = (x + 4)(x - 2)^2$

47. $y = -(x + 5)^3$

48. $y = -x(x + 1)$

49. $y = -x^2(x + 1)$

50. $y = -x^3(x + 1)$

51. $y = x(x - 2)(x + 2)$

52. $y = (x - 3)(x + 1)(x + 5)$

53. $y = \dfrac{3x + 1}{x}$

54. $y = \dfrac{1 - 2x}{x}$

55. $y = \dfrac{-1}{(x - 1)^2}$

56. $y = \dfrac{x + 1}{x + 2}$

57. $y = \dfrac{x - 2}{x - 3}$

58. $y = \dfrac{x}{(x - 2)(x + 4)}$

59. $y = \dfrac{x^2 - 2x + 1}{x^2 - 4x + 4}$

60. $y = \dfrac{x(x - 2)}{(x - 4)(x + 4)}$

61. Let $f(x) = x^2 + 2bx + 1$.
    (a) If $b = 1$, find the distance from the vertex of the parabola to the origin.
    (b) If $b = 2$, find the distance from the vertex of the parabola to the origin.
    (c) For which real numbers b will the distance from the vertex to the origin be as small as possible?

62. Find the range of the function $f(x) = -2x^2 + 12x - 5$.
    Hint: Look at the graph.

63. The range of the function $y = x^2 - 2x + k$ is the interval $[5, \infty)$. Find the value of k.

64. Find a value for b such that the range of the function $f(x) = x^2 + bx + b$ is the interval $[-15, \infty)$.

65. Find the range of the function $y = \dfrac{(x - 1)(x - 3)}{x - 4}$.

    Hint: Solve the equation for x in terms of y using the quadratic formula. If you're careful with the algebra, you will find that the expression under the resulting radical sign is $y^2 - 8y + 4$. The range of the given function can then be found by solving the inequality $y^2 - 8y + 4 \ge 0$.

66. In the following figure, triangle OAB is equilateral and $\overline{AB}$ is parallel to the x-axis. Find the length of a side and the area of the triangle OAB.

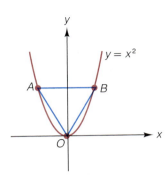

67. Let A denote the area of the right triangle in the first quadrant that is formed by the y-axis and the lines $y = mx$ and $y = m$. (Assume $m > 0$.) Express the area of the triangle as a function of m.

68. (a) Express the distance from the origin to the vertex of the parabola $y = x^2 - 2bx$ ($b > 0$) as a function of b.
   (b) Let V denote the vertex of the parabola in part (a), and let A and B denote the points where the curve meets the x-axis. Express the area of $\triangle VAB$ as a function of b.
   (c) A circle is drawn with center V and passing through A and B. Then, through the smaller y-intercept of the circle, a tangent line is drawn. Express the x-intercept of this tangent line as a function of b.

69. In the accompanying figure, the radius of the circle is $OC = 1$. Express the area of $\triangle ABC$ as a function of x.

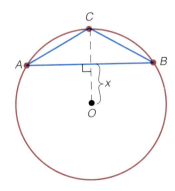

70. (a) Factor the expression $x^3 - 3x^2 + 4$. *Hint:* Add and subtract 1, then factor by grouping.
   (b) Use the factorization from part (a) to graph the function $y = x^3 - 3x^2 + 4$.

71. (a) Draw a scatter diagram (i.e., plot the points) for the data in the following table.

x (year)	y (Number of Refugees)
1983	12083
1984	10285
1985	9350
1986	8713
1987	8606
1988	7818

Refugees admitted into the United States from Eastern Europe during the (fiscal) years 1983–1988. *Source:* U.S. State Department

   (b) In your scatter diagram, draw the line that you feel seems to fit the data best. Estimate the slope and the y-intercept of the line.
   (c) Using the formulas in Exercise 35 of Section 5.1, it can be shown that the equation of the regression line is

$$f(x) = (-771.4)x + 1{,}541{,}090$$

   Use this equation (and your calculator, of course) to estimate (to the nearest 100) the number of refugees from Eastern Europe that will arrive in the United States in 1995, assuming that the current trend continues.
   (d) Find $f^{-1}(x)$.
   (e) Use your answer in part (d) to make a projection for the year when the number of refugees will dip below 1000.

72. (a) Let $f(x) = ax^2 + bx + c$ ($a \neq 0$). Find a constant x_0 such that the following equation is an identity.

$$f(x_0 + x) = f(x_0 - x)$$

   (b) What is the geometric significance of your answer in part (a)?

73. In Section 5.2, we saw how the sign of the constant a influences the graph of the parabola $f(x) = ax^2 + bx + c$. In this exercise you'll see how the sign of b affects the graph.
   (a) On the same set of axes, graph the parabolas $y = x^2 + 4x + 1$ and $y = x^2 - 4x + 1$. What type of symmetry do you observe?
   (b) Let $f(x) = ax^2 + bx + c$. Compute $f(-x)$.
   (c) Use the result in part (b) to describe what happens to the graph of $f(x) = ax^2 + bx + c$ when the sign of b is reversed.

OPTIONAL TI-81 GRAPHING CALCULATOR EXERCISES
EXERCISES FOR CHAPTER FIVE

1. Reread Example 2 in Section 5.1. Then graph the function $f(x) = -750x + 8000$. Use the following settings in the RANGE menu so that your picture is similar to Figure 1 on page 240.

    $$x_{min} = 0 \qquad y_{min} = 0$$
    $$x_{max} = 10 \qquad y_{max} = 8000$$
    $$x_{scl} = 5 \qquad y_{scl} = 2000$$

    Now press the TRACE key and use the arrow keys to move the cursor to a point where the x-coordinate is as close to 5 as possible. What is the corresponding y-coordinate? How close is your result to the answer $4250 obtained in the text?

2. This exercise shows how to use the calculator to determine a regression line. We will use the data set in the following table, which shows the world grain consumption for the years 1970–1980.

Year	World Grain Consumption (million metric tons)
1970	1130
1972	1192
1974	1213
1976	1303
1978	1418
1980	1475

    Source: Lester R. Brown, project director, *State of the World, 1989*, (New York: W. W. Norton and Company, 1989)

    (a) Clear any existing statistical data by first pressing the "2nd" key, then the STAT key, the left-hand arrow key, and the "2" key. After that, press ENTER.
    (b) To enter the data in the table, begin by pressing the "2nd" key, then the STAT key, and then the left-hand arrow key. But do not press ENTER yet. Instead press the number "1" to select the edit option. As you can see, the screen display shows

    DATA

    $$x_1 =$$
    $$y_1 = 1$$

    For the x_1 value, type 1970, and then press ENTER to move the cursor down to the line for y_1. For the y_1 value, type 1130 and then press ENTER. The remaining values are then entered in the same fashion.

    (c) After you have entered all of the data, press the "2nd" key and then the STAT key. Now use the arrow to move the cursor to item 2 (which is the choice for linear regression) and press ENTER. Then press ENTER again. You should see the following values displayed on your screen:

    $$a = -69049.71429$$
    $$b = 35.61428571$$

    (There is also a value for r, the *coefficient of correlation*, but we will not need that here.) These are the values for the constants a and b in the regression line $y = a + bx$.
    (d) Use the regression determined in part (c) to estimate the world grain consumption for the year 1985. (The actual value for 1985 was 1594 million metric tons. Is your estimate too high or too low?)

3. Linear functions can be used to approximate more complicated functions. This is one of the meanings or implications of the quotation on page 239. This exercise illustrates that idea. As preparation, clear any functions that are entered in the Y= menu. Press ZOOM 6 and then make the following adjustments in the RANGE window.

    $$x_{min} = -2 \qquad y_{min} = -3$$
    $$x_{max} = 3 \qquad y_{max} = 4$$

    (a) Using calculus, it can be shown that the equation of the line that is tangent to the curve $y = x^2$ at the point (1, 1) is $y = 2x - 1$. Verify this visually by entering the two functions $Y_1 = x^2$ and $Y_2 = 2x - 1$ and then pressing the GRAPH key. Note that the line is virtually indistinguishable from the curve in the immediate vicinity of (1, 1).
    (b) For numerical rather than visual evidence of how well the linear function $y = 2x - 1$ approximates the function $y = x^2$, use your calculator to complete the following tables.

x	0.9	0.99	0.999
$2x - 1$			
x^2			

x	1.1	1.01	1.001
$2x - 1$			
x^2			

    Exercises 4 and 5 show how the shape of the parabola $y = ax^2$ changes as a changes. As preparation for Exercise 4, clear any functions that are entered in the Y= menu. Press ZOOM 6 and then in the RANGE window, set $x_{min} = -3$ and $x_{max} = 3$.

4. (a) Press the Y= key and define the following four functions.

$$Y_1 = x^2 \qquad Y_3 = 3x^2$$
$$Y_2 = 2x^2 \qquad Y_4 = 8x^2$$

Graph all four functions on the same set of axes. As the coefficient of x^2 increases from 1 to 8, how does the resulting graph change?

(b) Where do you think the graph of $y = \frac{1}{2}x^2$ would fit in with the graphs in part (a)? Press the Y= key and redefine Y_4 to be $\frac{1}{2}x^2$. Now press the GRAPH key and see if you are right.

5. (a) Press the Y= key and define the following four functions.

$$Y_1 = x^2 \qquad Y_3 = \frac{1}{3}x^2$$
$$Y_2 = \frac{1}{2}x^2 \qquad Y_4 = \frac{1}{8}x^2$$

Press ZOOM 6 to graph all four functions in the standard viewing rectangle. As the coefficient of x^2 decreases from 1 to $\frac{1}{8}$, how does the resulting graph change?

(b) Where do you think the graph of $y = (0.01)x^2$ would fit in with the graphs in part (a)? Press the Y= key and redefine Y_4 to be $(0.01)x^2$. Now press the GRAPH key. Where is the graph of Y_4?

6. This exercise provides some additional perspective on Example 6 in Section 5.2.

(a) Press the Y= key and enter the functions $Y_1 = x^2 - x + 1$ and $Y_2 = (x^2 - x + 1)^{1/3}$. Now press ZOOM 6 to see the graphs. According to Example 6 in Section 5.2, the same input $\left(\text{namely, } x = \frac{1}{2}\right)$ minimizes both functions. To see this more clearly, press the RANGE key and set $x_{min} = -2$, $x_{max} = 2$, $x_{scl} = 0.5$, $y_{min} = 0$, and $y_{max} = 2$.

(b) Press the ZOOM key, then the "2" key, and then move the cursor so that it appears to be on the lowest point of the graph of $Y_2 = (x^2 - x + 1)^{1/3}$. Now press ENTER. Next press the TRACE key and use the arrow keys to move the cursor to a point on the graph of Y_2 where the x-coordinate is as close to 0.5 as possible. The corresponding y-coordinate on the screen readout is a calculator approximation for the minimum value of the function. Make a note of this number, and then go on to part (c).

(c) According to the text, the minimum value for $Y_2 = (x^2 - x + 1)^{1/3}$ is $\sqrt[3]{\frac{3}{4}}$. Use your calculator to evaluate this expression. Is your approximation in part (b) close to this?

Problems 7–13 are maximum–minimum problems, just as you practiced in Section 5.4, but now the function that you set up will turn out to be a rational function rather than a quadratic. To find the required maximum or minimum and/or input, first set up the appropriate function. (Your calculator won't help with this part of the work.) Then use the calculator to graph the function. The ZOOM key and the TRACE key will then allow you to approximate the required values.

7. The accompanying figure shows a line with slope m passing through the point (2, 1). Find the slope of this line so that the area of the shaded triangle is a minimum.   *Hint:* You should obtain the following function for the area A as a function of the slope.

$$A = \frac{-4m^2 + 4m - 1}{2m}$$

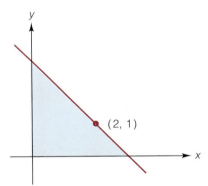

8. The product of two positive numbers is 6.
(a) Express the sum S of the two numbers as a function of one variable.   *Answer:* $S = (x^2 + 6)/x$
(υ) For which value of x will the sum be as small as possible? What is the other number in this case, and what is the corresponding minimum sum?   *Hint for checking:* Using calculus, it can be shown that the exact value for x is $\sqrt{6}$. Evaluate this using a calculator. How close were you to this value?

9. Find the point P on the first quadrant portion of the parabola $y = x^2$, with the property that the slope of the line through (0, −1) and P is a minimum. (After you've finished this problem, use the calculator to draw the parabola and the line through P and (0, −1). What do you observe?)

10. What is the smallest possible value for the sum of a positive number and its reciprocal?

11. The area of a rectangle is 6 m².
(a) Express the perimeter of the rectangle as a function of the width x.

(b) For which value of x is the perimeter a minimum? What is the length in this case?

12. A printer designing a small book decides to make the area of each page 60 in^2. The margins on the top and on the bottom of the page are to be 2 in. and 1 in., respectively. The margins on the sides are each to be 1.5 in.
    (a) Express the area of the printed portion of the page as a function of x, the width of the whole page.
    (b) For which width x will the printed area be a maximum? What is the corresponding length in this case?
    (c) *Hint for checking:* Using calculus, it can be shown that the exact value for x is $2\sqrt{15}$. Evaluate this using a calculator. How close were you to this value in part (b)?

13. The volume of a right circular cylinder is to be 54π in^3.
    (a) Express the height h of the cylinder in terms of r, the radius of the base.
    (b) Express the total (outer) surface area of the cylinder in terms of r.
    (c) Graph the function you found in part (b) and use the graph to estimate the value of r for which the surface area is a minimum.

(d) Using your formula from part (a) and the value of r that you obtained in part (c), find the corresponding value of h. How are the values of h and r related? *Hint for checking:* If you've worked out the correct formula in part (b), then the first-quadrant portion (the relevant portion) of the graph in (c) should be similar to the following figure.

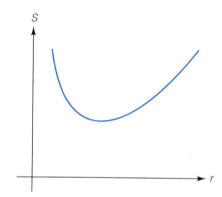

CHAPTER FIVE TEST

1. A linear function L satisfies the conditions $L(-2) = -4$ and $L(5) = 1$. Find $L(0)$.

2. (a) Let $F(x) = 4x - 2x^2$. On which interval is this function increasing? What is the maximum value for this function?
   (b) Let $G(t) = 9t^4 + 6t^2 + 2$. For which value of t is this function a minimum?

3. Let $f(x) = (x - 3)(x + 4)^2$.
   (a) Determine the intercepts and the excluded regions for the graph of f.
   (b) Determine the behavior of f when x is very close to -4.
   (c) Sketch the graph of f.

4. Graph each function: (a) $-1/(x - 3)^3$ (b) $-1/(x - 3)^2$

5. Graph the function $y = -x^2 + 7x + 6$. Specify the coordinates of the turning point, the intercepts, and the axis of symmetry.

6. Suppose that the function $p = -\frac{1}{6}x + 80$ $(0 \leq x \leq 400)$ relates the selling price of an item to the quantity x that is sold. Assume that p is in dollars. What is the maximum revenue possible in this situation? Which price p generates this maximum revenue?

7. The price of a new machine is \$14,000. After ten years, the machine has a salvage value of \$750. Assuming linear depreciation, find a formula for the value of the machine after t years, where $0 \leq t \leq 10$.

8. Graph the function $y = -\frac{1}{2}(3 - x)^2$. Specify the intercepts. *Hint:*   First graph $y = -\frac{1}{2}(x + 3)^2$.

9. Graph the function $y = \dfrac{2x - 3}{x + 1}$. Specify the intercepts and asymptotes.

10. (a) Suppose that $P(x, y)$ is a point on the line $y = 3x - 1$ and Q is the point $(-1, 3)$. Express the length PQ as a function of x.
     (b) For which value of x is the length PQ a minimum?

11. Let $f(x) = \dfrac{x(x - 2)}{x^2 - 9}$

     (a) Find the vertical and the horizontal asymptotes.
     (b) Use a sketch to show the behavior of f near the x-intercepts.
     (c) Use a sketch to show the behavior of f near the asymptotes.
     (d) Sketch the graph of f.

12. A rectangle is inscribed in a semicircle of diameter 8 cm. (See the accompanying figure.) Express the area of the rectangle as a function of the width w of the rectangle.

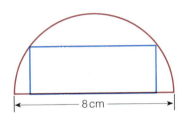

13. Explain why each of the following graphs cannot represent a polynomial function with the highest-degree term $-x^3$.

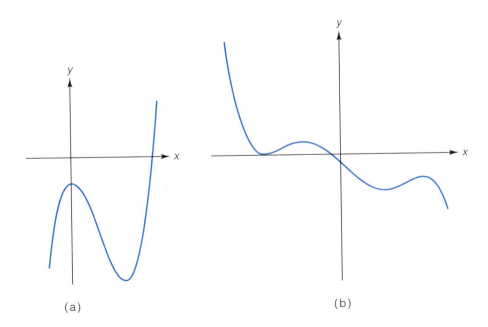

(a)   (b)

14. The following table shows the number of automobiles in use (worldwide) during the period from 1965 to 1980.

x (Year)	y (Automobiles in Use) (millions)
1965	140
1970	195
1975	260
1980	321

Source: Lester R. Brown, project director, *State of the World, 1989* (New York: W. W. Norton and Company, 1989)

(a) Draw a scatter diagram (i.e., plot the points) for the data in the table. Then, draw the line that you feel seems to best fit the data.

(b) Using formulas from statistics, it can be shown that the equation for the regression line is $y = -23757 + 12.16x$. Use this equation to estimate the number of automobiles in use worldwide in the year 1985. (The actual estimate from the publication indicated above was 375 million. Is your estimate too high or too low?)

CHAPTER SIX

EXPONENTIAL AND LOGARITHMIC FUNCTIONS

The exponential growth of population and its attendant assault on the environment is so recent that it is difficult for people to appreciate how much damage is being done.

Nathan Keyfitz in "The Growing Human Population," *Scientific American* **261** (September 1989) 119–126

... in 1958 ... C. D. Keeling of The Scripps Institution of Oceanography ... began a series of painstaking measurements of CO_2 concentration on a remote site at what is now the Mauna Loa Observatory in Hawaii. His observations, continued to the present, show an exponential growth of atmospheric carbon dioxide.... Keeling's observations have been duplicated at other stations in various parts of the world over shorter periods of time. In all sets of observations, the exponential increase is clear....

Gordon MacDonald in "Scientific Basis for the Greenhouse Effect," from *The Challenge of Global Warming*, Dean Edwin Abrahamson, ed. (Washington, D.C.: Island Press, 1989)

INTRODUCTION

In the previous chapter we saw that linear functions can be used to describe or *model* certain data sets. For instance, when we looked at the records for the one-mile run, we found that the times decreased in a linear fashion (see Figure 4 in Section 5.1). In this chapter we consider two other important types of functions that are often useful as models: the *exponential functions* and their inverses, the *logarithmic functions*. Figure 1 (on the next page) provides an example. Figure 1(a) shows the population of the United States over the years 1800–1900 along with a scatter plot of the data. The scatter plot makes it clear that the population did *not* increase in a linear fashion. Figure 1(b) shows the graph of a function that does fit this data very closely; this function is an exponential function. (The actual equation for this particular function is $y = ab^t$, where $a \approx 4.33 \times 10^{-15}$ and $b \approx 1.027$.)

As you will see throughout this chapter, the applications of exponential functions are quite diverse and not restricted to population growth. For instance, in studying the greenhouse effect, scientists obtain graphs like the one in Figure 2. This graph shows the concentration of carbon dioxide in the atmosphere over the years 1800–1988. Again, the growth can be described or modeled very closely by an exponential function.

Year	U.S. Population
1800	5,308,483
1810	7,239,881
1820	9,638,453
1830	12,866,020
1840	17,069,453
1850	23,191,876
1860	31,443,321
1870	38,558,371
1880	50,189,209
1890	62,947,714
1900	75,994,575

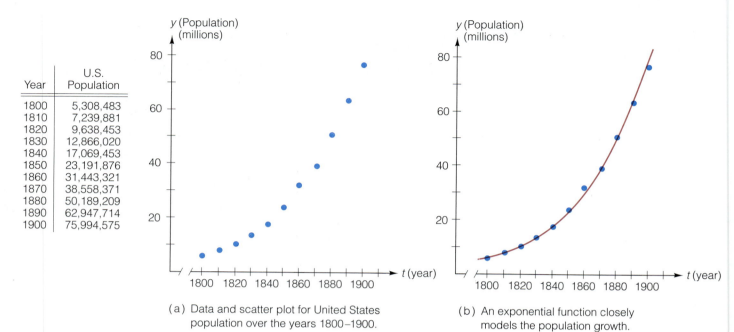

(a) Data and scatter plot for United States population over the years 1800–1900.

(b) An exponential function closely models the population growth.

FIGURE 1

FIGURE 2

The exponential growth of carbon dioxide levels in the atmosphere

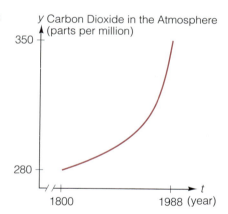

6.1 EXPONENTIAL FUNCTIONS

We begin with an example. Suppose that your mathematics instructor, in an effort to improve classroom attendance, offers to pay you each day for attending class! Suppose you are to receive 2¢ on the first day you attend class, 4¢ the second day, 8¢ the third day, and so on, as shown in Table 1. How much money will you receive for attending class on the 30th day?

As you can see by looking at Table 1, the amount y earned on day x is given by the rule, or *exponential function,*

$$y = 2^x$$

TABLE I

x (day number)	y (amount earned that day)
1	2¢ (= 2^1)
2	4¢ (= 2^2)
3	8¢ (= 2^3)
4	16¢ (= 2^4)
5	32¢ (= 2^5)
⋮	⋮
x	2^x

Thus on the 30th day (when $x = 30$), you will receive

$$y = 2^{30} \text{ cents}$$

If you use a calculator, you will find this amount to be well over 10 million dollars! The point here is simply this: Although we begin with a small amount, $y = 2$¢, repeated doubling quickly leads to a very large amount. In other words, the exponential function grows very rapidly.

Before leaving this example, we mention a simple method for quickly estimating numbers such as 2^{30} (or any power of two) in terms of the more familiar powers of ten. Begin by observing that

$$2^{10} \approx 10^3 \qquad \text{(a useful coincidence, worth remembering)}$$

Now just cube both sides to obtain

$$(2^{10})^3 \approx (10^3)^3 \qquad \text{or} \qquad 2^{30} \approx 10^9$$

Thus 2^{30} is about one billion. To convert this number of cents to dollars, we divide by 100 or 10^2 to obtain

$$\frac{10^9}{10^2} = 10^7 \text{ dollars}$$

which is 10 million dollars, as mentioned before.

EXAMPLE 1 Estimate 2^{40} in terms of a power of 10.

Solution Take the basic approximation $2^{10} \approx 10^3$ and raise both sides to the fourth power. This yields

$$(2^{10})^4 \approx (10^3)^4$$

or

$$2^{40} \approx 10^{12}, \qquad \text{as required}$$

EXAMPLE 2 Estimate the power to which 10 must be raised to yield 2.

Solution We begin with our approximation

$$10^3 \approx 2^{10}$$

Raising both sides to the power $\frac{1}{10}$ yields

$$(10^3)^{1/10} \approx (2^{10})^{1/10}$$

and, consequently,

$$10^{3/10} \approx 2$$

Thus the power to which 10 must be raised to yield 2 is approximately $\frac{3}{10}$.

In Section 1.7 we defined the expression b^x, where x is a rational number. We also stated that if x is irrational, then b^x can be defined so that the usual properties of exponents continue to hold. Although a rigorous definition of irrational exponents requires concepts from calculus, we can nevertheless convey the basic idea by means of an example. (We need to do this before we give the general definition for exponential functions.)

How shall we assign a meaning to $2^{\sqrt{2}}$, for example? The basic idea is to evaluate the expression 2^x successively by using rational numbers x that are closer and closer to $\sqrt{2}$. Table 2 displays the results of some calculations along these lines.

TABLE 2

Values of 2^x for Rational Numbers x Approaching $\sqrt{2}$ ($= 1.41421356\ldots$)

x	1.4	1.41	1.414	1.4142	1.41421	1.414213
2^x	2.6...	2.65...	2.664...	2.6651...	2.66514...	2.665143...

The data in the table suggest that as x approaches $\sqrt{2}$, the corresponding values of 2^x approach a unique real number, call it t, whose decimal expansion begins as 2.665. Furthermore, by continuing this process we could obtain (in theory, at least) as many places in the decimal expansion of t as we wished. The value of the expression $2^{\sqrt{2}}$ is then defined to be this number t. The following results (stated here without proof) summarize this discussion and also pave the way for the definition we will give for exponential functions.

PROPERTY SUMMARY   **REAL NUMBER EXPONENTS**

Let b denote an arbitrary positive real number. Then:

1. For each real number x, the quantity b^x is a unique real number.

2. When x is irrational, we can approximate b^x as closely as we wish by evaluating b^r, where r is a rational number sufficiently close to the number x.

3. The properties of rational exponents continue to hold for irrational exponents.

4. If $b^x = b^y$ and $b \neq 1$, then $x = y$.

EXAMPLE 3   Solve the equation $4^x = 8$.

Solution   First, let's estimate x, just to get a feeling for what kind of answer to expect. Since $4^1 = 4$, which is less than 8, and $4^2 = 16$, which is more than 8, we know that our final answer should be a number between 1 and 2. To obtain this answer, we take advantage of the fact that both 4 and 8 are powers of 2. Using this fact, we can write the given equation as

$$(2^2)^x = 2^3$$
$$2^{2x} = 2^3$$
$$2x = 3 \qquad \text{using Property 4 in the box}$$
$$x = \frac{3}{2}$$

Note that the answer $x = \frac{3}{2}$ is indeed between 1 and 2, as we had estimated.

For the remainder of this section, b denotes an arbitrary positive constant other than 1. In the box that follows, we define the exponential function with base b.

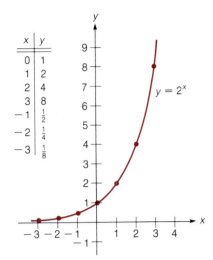

x	y
0	1
1	2
2	4
3	8
-1	$\frac{1}{2}$
-2	$\frac{1}{4}$
-3	$\frac{1}{8}$

FIGURE 1

DEFINITION   The Exponential Function with Base b

Let b denote an arbitrary positive constant other than 1. The **exponential function with base b** is defined by the equation

$$y = b^x$$

Note   In many scientific applications, functions of the form $y = ab^{kx}$ (where a, b, and k are constants) are also referred to as exponential functions.

EXAMPLES

1. The equations $y = 2^x$ and $y = 3^x$ define the exponential functions with bases 2 and 3, respectively.

2. The equation $y = (1/2)^x$ defines the exponential function with base 1/2.

3. The equations $y = x^2$ and $y = x^3$ do not define exponential functions.

To help with our analysis of exponential functions, let's set up a table and use it to graph the exponential function $y = 2^x$. This is done in Figure 1. In drawing a smooth and unbroken curve, we are actually relying on the results in the Property Summary on page 308. The key features of the exponential function $y = 2^x$ and its graph are:

1. The domain of $y = 2^x$ is the set of all real numbers. The range is the set of all *positive* real numbers.
2. The y-intercept of the graph is 1. The graph has no x-intercept.
3. For $x > 0$, the function increases or grows very rapidly. For $x < 0$, the graph rapidly approaches the x-axis; the x-axis is a horizontal asymptote for the graph. (Recall from Section 5.6 that a line is an **asymptote** for a curve if the separation distance between the curve and the line approaches zero as we move out farther and farther along the line.)

You should memorize the basic shape and features of the graph of $y = 2^x$ so that you can sketch it as needed without first setting up a table. The next example shows why this is useful.

EXAMPLE 4   Graph each of the following functions. In each case specify the domain, range, intercept(s), and asymptote.

(a) $y = -2^x$ (b) $y = 2^{-x}$ (c) $y = \left(\frac{1}{2}\right)^x$ (d) $y = 2^{-x} - 2$

Solution   (a) Recall that -2^x means $-(2^x)$, not $(-2)^x$. The graph of $y = -2^x$ is obtained by reflecting the graph of $y = 2^x$ in the x-axis. See Figure 2(a).

(b) Similarly, the graph of $y = 2^{-x}$ is obtained from the graph of $y = 2^x$ by reflection in the y-axis. See Figure 2(b).

(c) Next, regarding $y = \left(\frac{1}{2}\right)^x$, observe that

$$\left(\frac{1}{2}\right)^x = \frac{1^x}{2^x} = \frac{1}{2^x} = 2^{-x}$$

FIGURE 2

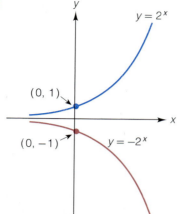

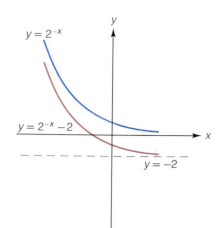

$y = -2^x$

Domain: $(-\infty, \infty)$
Range: $(-\infty, 0)$
y-intercept: -1
x-intercept: none
Asymptote: x-axis

(a)

$y = 2^{-x}$ (also $y = (\frac{1}{2})^x$)

Domain: $(-\infty, \infty)$
Range: $(0, \infty)$
y-intercept: 1
x-intercept: none
Asymptote: x-axis

(b)

$y = 2^{-x} - 2$

Domain: $(-\infty, \infty)$
Range: $(-2, \infty)$
y-intercept: -1
x-intercept: -1
Asymptote: $y = -2$

(c)

In other words, $y = (\frac{1}{2})^x$ is really the same function as $y = 2^{-x}$, which we already graphed in Figure 2(b).

(d) Finally, to graph $y = 2^{-x} - 2$, take the graph of $y = 2^{-x}$ in Figure 2(b) and move it two units in the negative y-direction, as shown in Figure 2(c). Note that the asymptote and y-intercept will also move down two units. To find the x-intercept, we set $y = 0$ in the given equation to obtain

$$2^{-x} - 2 = 0$$
$$2^{-x} = 2^1$$
$$-x = 1 \qquad \text{using Property 4 on page 308}$$
$$x = -1$$

Thus, the x-intercept is -1.

In the next example, we apply our knowledge about the graph of $y = 2^x$ to solve an equation. In particular, we use the fact that the graph of $y = 2^x$ always lies above the x-axis; for no value of x is 2^x ever zero.

EXAMPLE 5 Solve the equation $x^2 2^x - 2^x = 0$.

Solution First, we factor the left-hand side of the equation; the common term is 2^x. This gives us

$$2^x(x^2 - 1) = 0$$

Since 2^x is never zero, we may now divide both sides of this last equation by 2^x (without losing a solution) to obtain

$$x^2 - 1 = 0$$
$$(x - 1)(x + 1) = 0$$

and, consequently,

$$x = 1 \quad \text{or} \quad x = -1$$

Thus, the required solutions are $x = 1$, $x = -1$. (You should check for yourself that each of these values satisfies the original equation.) ■■■

Now what about exponential functions with bases other than 2? As Figure 3 indicates, the graphs are similar to $y = 2^x$. As you would expect, the graph of $y = 4^x$ rises more rapidly than $y = 2^x$ when x is positive. For negative x-values, the graph of $y = 4^x$ is below that of $y = 2^x$. You can see why this happens by taking $x = -1$, for example, and comparing the values of 4^x and 2^x. If $x = -1$, then

$$2^x = 2^{-1} = \frac{1}{2} \quad \text{but} \quad 4^x = 4^{-1} = \frac{1}{4}$$

Therefore $4^x < 2^x$ when $x = -1$. Notice also in Figure 3 that all three graphs have the same y-intercept of 1. This follows from the fact that $b^0 = 1$ for any positive number b.

FIGURE 3

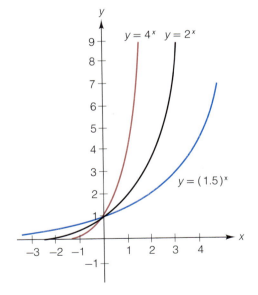

The exponential functions in Figure 3 all have bases larger than 1. To see examples in which the bases are in the interval $0 < b < 1$, we need only reflect the graphs in Figure 3 in the y-axis. For instance, the reflection of $y = 4^x$ in the

y-axis gives us the graph of $y = \left(\frac{1}{4}\right)^x$. (We discussed the idea behind this in Example 4; see Figure 2(b), for instance.)

In the box that follows, we summarize what we've learned up to this point regarding the exponential function $y = b^x$.

PROPERTY SUMMARY   THE EXPONENTIAL FUNCTION $y = b^x$

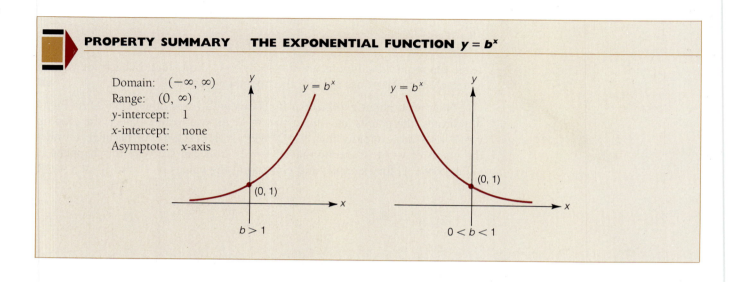

Domain: $(-\infty, \infty)$
Range: $(0, \infty)$
y-intercept: 1
x-intercept: none
Asymptote: x-axis

EXAMPLE 6   Graph the function $y = -3^{-x} + 1$.

Solution   The required graph is obtained by reflecting and translating the graph of $y = 3^x$, as shown in Figure 4.

FIGURE 4

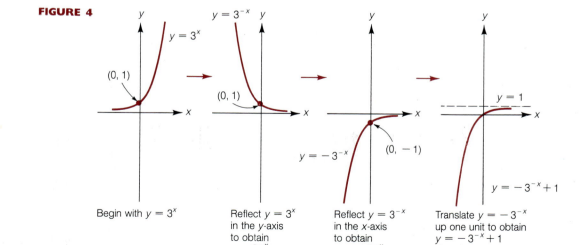

Begin with $y = 3^x$

Reflect $y = 3^x$
in the y-axis
to obtain
$y = 3^{-x}$

Reflect $y = 3^{-x}$
in the x-axis
to obtain
$y = -3^{-x}$

Translate $y = -3^{-x}$
up one unit to obtain
$y = -3^{-x} + 1$
Note that the y-intercept is 0
and the asymptote
is the line $y = 1$

EXERCISE SET 6.1

A

In Exercises 1 and 2, estimate each quantity in terms of powers of 10, as in Example 1.

1. (a) 2^{30} (b) 2^{50}

2. (a) 2^{90} (b) 4^{20}

In Exercises 3 and 4, solve each equation, as in Example 3.

3. (a) $3^x = 27$ (b) $9^t = 27$
    (c) $3^{1-2y} = \sqrt{3}$ (d) $3^z = 9\sqrt{3}$

4. (a) $2^x = 32$ (b) $2^t = \frac{1}{4}$
    (c) $2^{3y+1} = \sqrt{2}$ (d) $8^{z+1} = 32\sqrt{2}$

In Exercises 5–8, specify the domain of the function.

5. $y = 2^x$

6. $y = \dfrac{1}{2^x}$

7. $y = \dfrac{1}{2^{x-1}}$

8. $y = \dfrac{1}{2^x - 1}$

In Exercises 9–16, graph the pair of functions on the same set of axes.

9. $y = 2^x$, $y = 2^{-x}$ 10. $y = 3^x$, $y = 3^{-x}$

11. $y = 3^x$, $y = -3^x$ 12. $y = 4^x$, $y = -4^x$

13. $y = 2^x$, $y = 3^x$ 14. $y = \left(\frac{1}{3}\right)^x$, $y = 3^x$

15. $y = \left(\frac{1}{2}\right)^x$, $y = \left(\frac{1}{3}\right)^x$ 16. $y = \left(\frac{1}{2}\right)^{-x}$, $y = \left(\frac{1}{3}\right)^{-x}$

For Exercises 17–24, graph the function and specify the domain, range, intercept(s), and asymptote.

17. $y = -2^x + 1$ 18. $y = -3^x + 3$

19. $y = 3^{-x} + 1$ 20. $y = 3^{-x} - 3$

21. $y = 2^{x-1}$ 22. $y = 2^{x-1} - 1$

23. $y = 3^{x+1} + 1$ 24. $y = 1 - 3^{x-1}$

Exercises 25–28, solve the equation.

25. $3x(10^x) + 10^x = 0$ 26. $4x^2(2^x) - 9(2^x) = 0$

27. $3(3^x) - 5x(3^x) + 2x^2(3^x) = 0$

28. $\dfrac{(x+4)10^x}{x-3} = 2x(10^x)$

B

29. Let $f(x) = 2^x$. Show that $\dfrac{f(x+h) - f(x)}{h} = 2^x\left(\dfrac{2^h - 1}{h}\right)$

30. Let $\phi(t) = 1 + a^t$. Show that $\dfrac{1}{\phi(t)} + \dfrac{1}{\phi(-t)} = 1$

31. Let $f(x) = 2^x$ and let g denote the function that is the inverse of f.
    (a) On the same set of axes, sketch the graphs of f, g, and the line $y = x$.
    (b) Using the graph you obtained in part (a), specify the domain, range, intercept, and asymptote for the functon g.

32. Let $S(x) = \dfrac{2^x - 2^{-x}}{2}$ and $C(x) = \dfrac{2^x + 2^{-x}}{2}$. Compute $[C(x)]^2 - [S(x)]^2$.

For Exercises 33–40, refer to the following figure, which shows portions of the graphs of $y = 2^x$, $y = 3^x$, and $y = 5^x$. In each case, (a) use the figure to estimate the indicated quantity; (b) use a calculator to compute the indicated quantity, and round off the result to two decimal places.

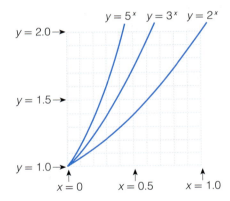

33. $\sqrt{2}$ *Hint:* $\sqrt{2} = 2^{\frac{1}{2}}$ 34. $\sqrt[5]{2}$

35. $2^{\frac{2}{5}}$ 36. $\sqrt[5]{8}$ *Hint:* $8^{\frac{1}{5}} = 2^?$

37. $\sqrt{3}$ 38. $\sqrt[3]{3}$

39. $5^{\frac{3}{10}}$ 40. $\sqrt[4]{5}$

In Exercises 41–44, refer to the following graph of the exponential function $y = 10^x$. Use the graph to estimate (to the nearest tenth) the solution of each equation. (After we've studied logarithmic functions later in the chapter, we will be able to obtain more precise solutions.)

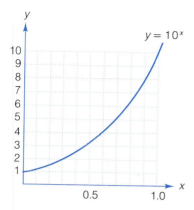

41. $10^x = 2$
42. $10^x = 4$
43. $10^x = 5$
44. $10^x = 8$

45. This exercise serves as a preview for the work on logarithms in Section 6.3. Follow steps (a)–(f) to complete the table. (Notice that one entry in the table is already filled in. Reread Example 2 in the text to see how that entry was obtained.)
   (a) Fill in the entries in the right-hand column corresponding to $x = 1$ and $x = 10$.
   (b) Note that 4 and 8 are powers of 2. Use this information along with the approximation $10^{0.3} \approx 2$ to find the entries in the table corresponding to $x = 4$ and $x = 8$.
   (c) Find the entry corresponding to $x = 5$.
      Hint: $5 = \dfrac{10}{2} \approx \dfrac{10}{10^{0.3}}$

(d) Find the entry corresponding to $x = 7$.
   Hint: $7^2 \approx 50 = 5 \times 10$; now make use of your answer in part (c).
(e) Find the entry corresponding to $x = 3$.
   Hint: $3^4 \approx 80 = 8 \times 10$
(f) Find the entries corresponding to $x = 6$ and $x = 9$.
   Hint: $6 = 3 \times 2$ and $9 = 3^2$
Remark   This table is called a *table of logarithms to the base 10*. We say, for example, that the logarithm of 2 to the base 10 is (about) 0.3. We write this symbolically as $\log_{10} 2 \approx 0.3$.

x	Exponent to Which 10 Must Be Raised to Yield x
1	
2	≈ 0.3
3	
4	
5	
6	
7	
8	
9	
10	

6.2 THE EXPONENTIAL FUNCTION $y = e^x$

This is undoubtedly the most important function in mathematics.

Walter Rudin in *Real and Complex Analysis* (New York: McGraw-Hill, 1966)

From the standpoint of calculus and scientific applications, one particular base for exponential functions is by far the most useful. This base is a certain irrational number that lies between 2 and 3 and is denoted by the letter e.* For purposes of approximation, you'll need to know that

$$e \approx 2.7$$

At the precalculus level, it's hard to escape the feeling that $y = 2^x$ or $y = 10^x$ is by far more simple and more natural than $y = e^x$. So as we work with

*The Swiss mathematician Leonhard Euler (1707–1783) introduced the letter e to denote this number. To six decimal places, the value of e is 2.718281....

FIGURE I

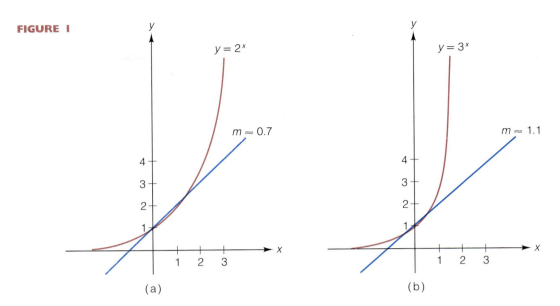

e and e^x in this chapter, you will need to take it on faith that, in the long run, the number e and the function $y = e^x$ make life simpler, not more complex.

Here is one way to define the number e. (For another approach, see Exercise 49; also see Table 3 in Section 6.5.) You know that the graph of each exponential function $y = b^x$ passes through the point $(0, 1)$. Figure 1(a) shows the exponential function $y = 2^x$ along with a line that is tangent to the curve at the point $(0, 1)$. By carefully measuring rise and run, it can be shown that the slope of this tangent line is about 0.7. Figure 1(b) shows a similar situation with the curve $y = 3^x$. Here the slope of the tangent line through $(0, 1)$ is approximately 1.1. Now, since the slope of the tangent to $y = 2^x$ is a bit less than 1, while that for $y = 3^x$ is a bit more than 1, it seems reasonable to suppose that there is a number between 2 and 3, call it e, with the property that the slope of the tangent to $y = e^x$ through $(0, 1)$ is exactly 1. See Figure 2 and the summary box that follows.

FIGURE 2

The slope of the tangent to the curve
$y = e^x$ at the point $(0, 1)$ is $m = 1$.

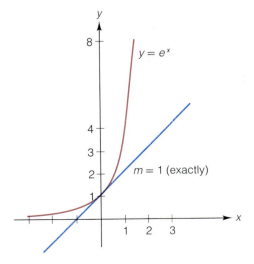

PROPERTY SUMMARY   **THE EXPONENTIAL FUNCTION** $y = e^x$

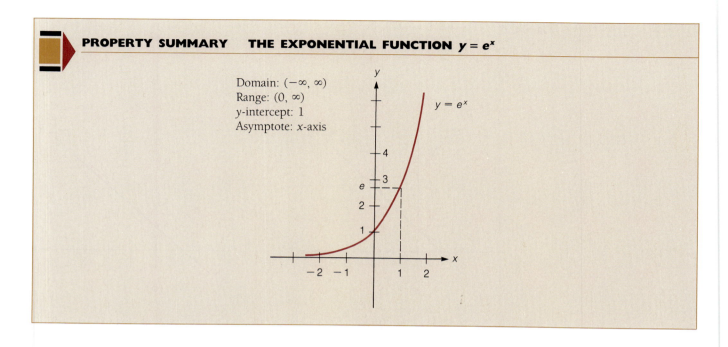

Domain: $(-\infty, \infty)$
Range: $(0, \infty)$
y-intercept: 1
Asymptote: x-axis

$y = e^x$

EXAMPLE 1   Graph each of the following functions. Specify the domain, range, intercept, and asymptote.

(a) $y = e^{x-1}$ (b) $y = -e^{x-1}$ (c) $y = -e^{x-1} + 1$

Solution   (a) We begin with the graph of $y = e^x$ in the Property Summary figure. Moving the graph to the right one unit yields the graph of $y = e^{x-1}$ shown in Figure 3(a). The x-axis is still an asymptote for this translated graph, but the y-

---

FIGURE 3

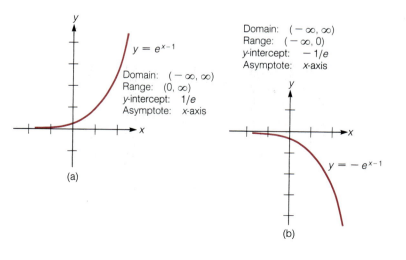

Domain: $(-\infty, \infty)$
Range: $(0, \infty)$
y-intercept: $1/e$
Asymptote: x-axis

$y = e^{x-1}$

(a)

Domain: $(-\infty, \infty)$
Range: $(-\infty, 0)$
y-intercept: $-1/e$
Asymptote: x-axis

$y = -e^{x-1}$

(b)

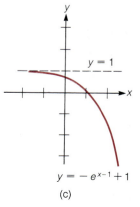

Domain: $(-\infty, \infty)$
Range: $(-\infty, 1)$
x-intercept: 1
y-intercept: $-1/e + 1$
Asymptote: $y = 1$

$y = 1$

$y = -e^{x-1} + 1$

(c)

intercept will no longer be 1. For the y-intercept, we replace x with 0 in the given equation to obtain

$$y = e^{-1} = \frac{1}{e} \approx 0.37 \qquad \text{using a calculator}$$

(b) Reflecting the graph from part (a) in the x-axis yields the graph of $y = -e^{x-1}$ [see Figure 3(b)]. Note that under this reflection, the y-intercept moves from $1/e$ to $-1/e$.

(c) Translating the graph in Figure 3(b) up one unit produces the graph of $y = -e^{x-1} + 1$, shown in Figure 3(c). Under this translation, the asymptote moves from $y = 0$ (the x-axis) to $y = 1$. Also, the y-intercept moves from $-1/e$ to $-1/e + 1$ (≈ 0.63). The x-intercept in Figure 3(c) is obtained by setting $y = 0$ in the equation $y = -e^{x-1} + 1$. This yields

$$0 = -e^{x-1} + 1$$
$$e^{x-1} = 1 = e^0$$
$$x - 1 = 0 \qquad \text{(Why?)}$$
$$x = 1$$

In some of the applications that we will discuss here and in Section 6.5, we will need to work with equations of the form

$$e^{ax} = b \tag{1}$$

where a and b are constants. In Section 6.3 we'll see how to solve this equation. (It will require logarithms.) For now, however, it will be sufficient to isolate the quantity e^x in equation (1). To do this, we just raise both sides of equation (1) to the power $1/a$:

$$(e^{ax})^{1/a} = b^{1/a}$$
$$e^x = b^{1/a}$$

EXAMPLE 2 Solve the equation $e^{4k} = 3$ for the quantity e^k.

Solution To isolate e^k, we raise both sides of the given equation to the power $\frac{1}{4}$. This yields

$$(e^{4k})^{1/4} = 3^{1/4}$$
$$e^k = 3^{1/4}$$

Thus, the value of e^k is $3^{1/4}$. As you can check using a calculator, this is approximately 1.3.

We now turn to some applications involving the number e. In Section 6.5 we'll look at these and other applications in greater detail. (Thus your immediate goals for the remainder of this section are to start getting used to seeing formulas involving e^x and to master the algebra that you will see in Examples 3 and 4.)

Under ideal conditions involving unlimited food and space, a colony of bacteria increases according to the *growth law*:

$$\mathcal{N} = \mathcal{N}_0 e^{kt}$$

In this formula, $\mathcal{N}$ is the population at time t and k is a positive constant related to the growth rate of the population. The number $\mathcal{N}_0$ is also a constant; it represents the size of the population at time $t = 0$. Sometimes, to emphasize the fact that $\mathcal{N}$ depends on (is a function of) t, we employ function notation to rewrite the growth law as

$$\mathcal{N}(t) = \mathcal{N}_0 e^{kt}$$

EXAMPLE 3 Suppose that at the start of an experiment in a biology laboratory, 1200 bacteria are present in a colony. Four hours later, the population is found to be 3600. How many bacteria were there three hours after the experiment began?

Solution We begin with the growth law $\mathcal{N} = \mathcal{N}_0 e^{kt}$. Our strategy will be to evaluate the quantity e^k; then we can determine the required value of $\mathcal{N}$. (*Note:* In Section 6.3 we'll see how to find k itself, rather than e^k.) We are given that $\mathcal{N}_0 = 1200$. Thus we can write

$$\mathcal{N} = 1200 e^{kt} \tag{2}$$

We are also given that $\mathcal{N} = 3600$ when $t = 4$. Substituting these values in equation (2) gives us

$$3600 = 1200 e^{4k}$$

Now we divide both sides of this last equation by 1200 to obtain

$$3 = e^{4k} \tag{3}$$

To isolate e^k, we raise both sides of the equation to the power $\frac{1}{4}$ (as in Example 2).

$$3^{1/4} = (e^{4k})^{1/4}$$

Therefore

$$3^{1/4} = e^k \tag{4}$$

Now we can use this value of e^k to determine the size of the colony after three hours, as required. We take equation (2) and replace t by 3.

$$\mathcal{N} = 1200 e^{3k} = 1200(e^k)^3$$
$$= 1200(3^{1/4})^3 \qquad \text{using equation (4) to substitute for } e^k$$
$$= 1200(3^{3/4})$$

Using a calculator, we find that $1200(3^{3/4}) \approx 2735$. Rather than claim that there are precisely 2735 bacteria after three hours, we follow common sense and round our answer to the nearest hundred. Thus we say that after three hours there are about 2700 bacteria in the colony.

In Example 3 we used the function $\mathcal{N} = \mathcal{N}_0 e^{kt}$ to describe population growth. It is a remarkable fact, shown in calculus, that the same general equation also governs radioactive decay, but in that case the constant k is negative, not positive. Thus we will assume here the following *decay law* for radioactive substances:

$$\mathcal{N} = \mathcal{N}_0 e^{kt}$$

where $\mathcal{N}_0$ is the original amount present at time $t = 0$, $\mathcal{N}$ is the amount present at time t, and k is a negative constant related to the rate of decay of the substance. In discussing radioactive decay, it is convenient to introduce the term *half-life*.

DEFINITION   Half-life

	EXAMPLE

The **half-life** of a radioactive substance is the time required for half of a given sample to disintegrate. The half-life is an intrinsic property of the substance; it does not depend on the given sample size.	Iodine-131 is a radioactive substance with a half-life of 8 days. Suppose that 2 g are present initially. Then:
	
	at $t = 0$, 2 g are present;
	at $t = 8$ days, 1 g is left;
	at $t = 16$ days, $\frac{1}{2}$ g is left;
	at $t = 24$ days, $\frac{1}{4}$ g is left;
	at $t = 32$ days, $\frac{1}{8}$ g is left.

EXAMPLE 4   Hospitals utilize the radioactive substance iodine-131 (with a half-life of 8 days) in the diagnosis of the thyroid gland. If a hospital acquires 2 g of iodine-131, how much of this sample will remain after 30 days?

Solution   First, let's estimate the answer to get a feeling for the situation. We noted just prior to this example that, after 24 days, $\frac{1}{4}$ g will remain, while after 32 days, $\frac{1}{8}$ g will remain. Since 30 is between 24 and 32, it follows that, after 30 days, the amount remaining will be something between $\frac{1}{4}$ g and $\frac{1}{8}$ g.

Our actual calculations for the answer begin with the decay law

$$\mathcal{N} = \mathcal{N}_0 e^{kt} \tag{5}$$

We are asked to find $\mathcal{N}$ when $t = 30$. As in Example 3, we first determine the quantity e^k. To do this, we use the half-life information. This says that when $t = 8$, the value of $\mathcal{N}$ is $\frac{1}{2}\mathcal{N}_0$. Using these values in equation (5), we find that

$$\frac{1}{2}\mathcal{N}_0 = \mathcal{N}_0 e^{8k}$$
$$\frac{1}{2} = e^{8k}$$
$$\left(\frac{1}{2}\right)^{1/8} = (e^{8k})^{1/8}$$
$$\left(\frac{1}{2}\right)^{1/8} = e^k \tag{6}$$

Returning to equation (5) now, we substitute $\mathcal{N}_0 = 2$ and $t = 30$ to obtain

$$\mathcal{N} = 2e^{30k}$$
$$= 2(e^k)^{30}$$
$$= 2\left[\left(\frac{1}{2}\right)^{1/8}\right]^{30} \qquad \text{using equation (6) to substitute for } e^k$$

Using the properties for exponents, this can be simplified to

$$\mathcal{N} = 2^{-11/4}$$

(Exercise 50 at the end of this section asks you to verify this.) Using a calculator now, we obtain

$$\mathcal{N} = 0.148\ldots$$

Thus, after 30 days, approximately 0.15 g of the iodine-131 remains. As you can check, this figure is indeed between $\frac{1}{4}$ g and $\frac{1}{8}$ g, as we first estimated.   ■■■

EXERCISE SET 6.2

A

In Exercises 1–12, graph the function and specify the domain, range, intercept(s), and asymptote.

1. $y = e^x$
2. $y = e^{-x}$
3. $y = -e^x$
4. $y = -e^{-x}$
5. $y = e^x + 1$
6. $y = e^{x+1}$
7. $y = e^{x+1} + 1$
8. $y = e^{x-1} - 1$
9. $y = -e^{x-2}$
10. $y = -e^{x-2} - 2$
11. $y = e - e^x$
12. $y = e^{-x} - e$

13. On the same set of axes, graph the functions $y = 2^x$, $y = e^x$, and $y = 3^x$.

14. On the same set of axes, graph the functions $y = 2^{-x}$, $y = e^{-x}$, and $y = 3^{-x}$.

In Exercises 15–18, solve the given equation for the quantity indicated. For Exercises 17 and 18, give two forms for the answer: an exact expression and a calculator approximation rounded off to three decimal places.

15. $e^{5k} = 32$; e^k
16. $e^{2k} = 225$; e^k
17. $e^{3t} = 4$; e^t
18. $e^{0.2t} = 10.3$; e^t

In Exercises 19–26, answer true or false. You do not need a calculator for these exercises. Rather, use the fact that e is approximately 2.7.

19. $e < \frac{1}{2}$
20. $e < \frac{5}{2}$
21. $\sqrt{e} < 1$
22. $e^2 < 4$
23. $e^2 < 9$
24. $e^3 < 27$
25. $e^{-1} < 0$
26. $e^0 = 1$

In Exercises 27–34, refer to the following graph of $y = e^x$ $(-2 \leq x \leq 1)$. In each case, use the graph to estimate (to the nearest tenth) the indicated quantity. For Exercises 27–32, use a calculator to obtain a second estimate. Round off the calculator value to two decimal places.

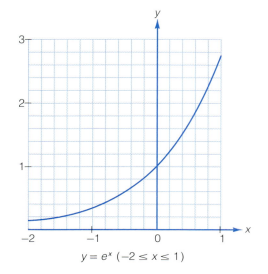

$y = e^x$ $(-2 \leq x \leq 1)$

27. $e^{0.2}$
28. $e^{0.6}$
29. $e^{4/5}$
30. e^{-1}
31. $\sqrt{e}$
32. $e^{7/10}$
33. The value of x for which $e^x = 0.8$
34. The value of x for which $e^x = 0.2$

35. At the start of an experiment, 2000 bacteria are present in a colony. In 2 hours, the population has tripled. Assume that the growth law $\mathcal{N} = \mathcal{N}_0 e^{kt}$ applies.
    (a) Determine e^k.
    (b) Determine the population 10 hours after the start of the experiment.

36. At the start of an experiment 2×10^4 bacteria are present in a colony. After 8 hours the population is 3×10^4. Assume that the growth law $\mathcal{N} = \mathcal{N}_0 e^{kt}$ applies.
    (a) Determine the population 1 hour after it is 3×10^4.
    (b) Determine the population 24 hours after the start of the experiment.

37. A colony of bacteria grows according to the law $\mathcal{N} = \mathcal{N}_0 e^{kt}$. It is known that the value of e^k is $\sqrt[5]{2}$. If the initial population is 3200, what will the population be 5 hours later?

38. Suppose that you are helping a friend with his homework on the growth law $N = N_0 e^{kt}$, and on his paper you see the equation $e^k = -0.75$. How do you know at this point that your friend must have made an error?

In Exercises 39 and 40, use the half-life information to complete each table. (The formula $N = N_0 e^{kt}$ is not required.)

39. (a) Uranium-228: half-life = 550 seconds

t (seconds)	0	550	1100	1650	2200
N (grams)	8				

(b) Uranium-238: half-life = 4.9×10^9 years

t (years)	0				
N (grams)	10	5	2.5	1.25	0.625

40. (a) Polonium-210: half-life = 138.4 days

t (days)	0	138.4	276.8	415.2	
N (grams)	0.4				0.025

(b) Polonium-214: half-life = 1.63×10^{-4} second

t (seconds)	0				6.52×10^{-4}
N (grams)	0.1	0.05	0.25	0.125	

41. The half-life of iodine-131 is 8 days. How much of a 1-g sample will remain after 7 days?

42. The half-life of strontium-90 is 28 years. How much of a 10-g sample will remain after
(a) 1 year?
(b) 10 years?

43. Plutonium-239 is a product of nuclear reactors. The half-life of plutonium-239 is about 24,000 years. What percentage of a given sample will remain after 1000 years?

44. Krypton-91 has a half-life of 10 sec. What percentage of a given sample will remain after 5 min?

B

45. If $E(x) = e^x$, show that
$$\frac{E(x + h) - E(x)}{h} = e^x \left(\frac{e^h - 1}{h} \right)$$

46. The **hyperbolic sine** function S is defined by
$$S(x) = \frac{e^x - e^{-x}}{2}$$

(a) What is the domain of S?
(b) Find the y-intercept for the graph of S.
(c) Show that $S(-x) = -S(x)$. What does this say about the graph of S?
(d) Sketch the graph of S for $x \geq 0$. (Use your calculator to set up a table of values.) Then use the result in part (c) to complete the graph.

47. The **hyperbolic cosine** function C is defined by
$$C(x) = \frac{e^x + e^{-x}}{2}$$

(a) What is the domain of C?
(b) Find the y-intercept for the graph of C.
(c) Show that $C(-x) = C(x)$. What does this say about the graph of C?
(d) Sketch the graph of C for $x \geq 0$. (Use your calculator to set up a table of values.) Then use the result in part (c) to complete the graph.

48. Refer to the graph of $y = e^x$ in Figure 2.
(a) Show that the equation of the tangent line in Figure 2 is $y = x + 1$.
(b) When x is close to zero, e^x can be approximated by the quantity $x + 1$. (This is because the tangent line and the curve are virtually indistinguishable in the immediate vicinity of $x = 0$.) Complete the following tables to see just how good this approximation is.

x	0.3	0.2	0.1	0.01	0.001	0.0001
$x + 1$						
e^x						

x	-0.3	-0.2	-0.1	-0.01	-0.001	-0.0001
$x + 1$						
e^x						

49. Use a calculator to complete the following table. It can be shown that if the table is continued indefinitely, the numbers in the right-hand column approach the value e. (In fact, this is one way of defining e in calculus.)

n	$\left(1 + \dfrac{1}{n}\right)$	$\left(1 + \dfrac{1}{n}\right)^n$
1		
10		
. 100		
1000		
10000		
100000		

50. Carry out the simplification referred to in Example 4 to show that $2\left[\left(\frac{1}{2}\right)^{1/8}\right]^{30} = 2^{-11/4}$.   *Hint:*   First replace $\frac{1}{2}$ with 2^{-1}.

51. Assume that the growth law $\mathcal{N} = \mathcal{N}_0 e^{kt}$, which describes bacterial growth, also approximates the Earth's human population.
   (a) Use the following data to estimate the population for the year 1990. In 1850 (call this $t = 0$), the population was estimated to be one billion. In 1930, it was 2 billion.
   (b) Statistics from the United Nations say that the actual 1990 population was about 5.3 billion. How does your estimate from part (a) compare with this? Does the formula $\mathcal{N} = \mathcal{N}_0 e^{kt}$ predict too high or too low a figure.

52. Let $f(x) = e^x$ and $g(x) = x - 1$. Graph the functions F and G defined as follows, and specify any intercepts or asymptotes.

$$F(x) = (f \circ g)(x)$$
$$G(x) = (g \circ f)(x)$$

53. Let $f(x) = e^x$. Let L denote the function that is the inverse of f.
   (a) On the same set of axes, sketch the graphs of f and L.   *Hint:*   You do not need the equation for $L(x)$.
   (b) Specify the domain, range, intercept, and asymptote for the function L and its graph.
   (c) Graph each of the following functions. Specify the intercept and asymptote in each case.
      (i) $y = -L(x)$
      (ii) $y = L(-x)$
      (iii) $y = L(x - 1)$

C

54. Let S and C denote the functions defined in Exercises 46 and 47, respectively. Prove each of the following identities. [The identities in parts (b) and (c) have already appeared in Exercises 46 and 47, but they are included here for the sake of completeness.]
   (a) $[C(x)]^2 - [S(x)]^2 = 1$
   (b) $S(-x) = -S(x)$
   (c) $C(-x) = C(x)$
   (d) $S(x + y) = S(x)C(y) + C(x)S(y)$
   (e) $C(x + y) = C(x)C(y) + S(x)S(y)$
   (f) $S(2x) = 2S(x)C(x)$
   (g) $C(2x) = [C(x)]^2 + [S(x)]^2$

6.3   LOGARITHMIC FUNCTIONS

[John Napier] *hath set my head and hands a work with his new and admirable logarithms. I hope to see him this summer, if it please God, for I never saw book that pleased me better, or made me more wonder.*

Henry Briggs, March 10, 1615

It has been thought that the earliest reference to the logarithmic curve was made by the Italian Evangelista Torricelli in a letter of the year 1644, but Paul Tannery made it practically certain that Descartes knew the curve in 1639.

Florian Cajori in *A History of Mathematics*, 4th ed. (New York: Chelsea Publishing Co., 1985)

For the number whose logarithm is unity, let e be written, . . .

Leonhard Euler in a letter written in 1727 or 1728

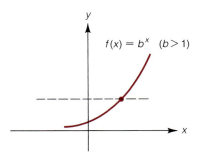

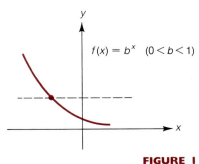

FIGURE 1
The exponential function $y = b^x$ is one-to-one.

In the previous two sections we studied exponential functions. Now we consider functions that are inverses of exponential functions. These inverse functions are called *logarithmic functions*.

Having said this, let's back up for a moment to review briefly some of the basic ideas behind inverse functions (as discussed in Section 4.5). We start with a given function F, say $F(x) = 3x$, for example, that is one-to-one. (That is, for each output there is but one input.) Then, by interchanging the inputs and outputs, we obtain a new function, the so-called inverse function. In the case of $F(x) = 3x$, it's easy to find an equation defining the inverse function. We just interchange x and y in the equation $y = 3x$ to obtain $x = 3y$. Solving for y in this last equation then gives us $y = \frac{1}{3}x$, which defines the inverse function. Using function notation, we can summarize the situation by writing $F(x) = 3x$ and $F^{-1}(x) = \frac{1}{3}x$.

In the preceding paragraph we saw that a particular linear function had an inverse, and we found a formula for that inverse. Now let's repeat that same reasoning beginning with an exponential function. First, we must make sure the exponential function f defined by $f(x) = b^x$ is one-to-one. We can see this by applying the horizontal line test, as indicated in Figure 1. Next, since $f(x) = b^x$ is one-to-one, it has an inverse function. Let's study this inverse.

We begin by writing the exponential function $f(x) = b^x$ in the form

$$y = b^x \tag{1}$$

Then, in order to obtain an equation for f^{-1}, we interchange x and y in equation (1). This gives us

$$x = b^y \tag{2}$$

The crucial step now is to express equation (2) in words:

$$\text{y is the exponent to which b must be raised to yield x} \tag{3}$$

Statement (3) defines the function that is the inverse of $y = b^x$. Now we introduce a notation that will allow us to write this statement in a more compact form.

DEFINITION $\log_b x$

EXAMPLES

We define the expression **$\log_b x$** to mean "the exponent to which b must be raised to yield x." ($\log_b x$ is read *log base b of x* or the *logarithm of x to the base b*.)

(a) $\log_2 8 = 3$, since 3 is the exponent to which 2 must be raised to yield 8.

(b) $\log_{10} \frac{1}{10} = -1$, since -1 is the exponent to which 10 must be raised to yield $\frac{1}{10}$.

(c) $\log_5 1 = 0$, since 0 is the exponent to which 5 must be raised to yield 1.

Using this notation, statement (3) becomes

$$y = \log_b x$$

TABLE 1

Exponential Form of Equation	Logarithmic Form of Equation
$8 = 2^3$	$\log_2 8 = 3$
$\dfrac{1}{9} = 3^{-2}$	$\log_3 \dfrac{1}{9} = -2$
$1 = e^0$	$\log_e 1 = 0$
$a = b^c$	$\log_b a = c$

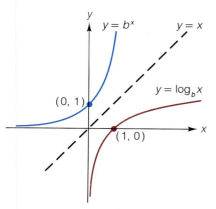

FIGURE 2

Since equations (2) and (3) are equivalent, we have the following important relationship.

$$y = \log_b x \qquad \text{is equivalent to} \qquad x = b^y.$$

We say that the equation $y = \log_b x$ is in **logarithmic form** and that the equivalent equation $x = b^y$ is in **exponential form**. Table 1 displays some examples.

Let us now summarize our discussion up to this point.

1. According to the horizontal line test, the function $f(x) = b^x$ is one-to-one and therefore possesses an inverse. This inverse function is written

$$f^{-1}(x) = \log_b x$$

2. $\log_b a = c$ means that $a = b^c$.

To graph the function $y = \log_b x$, we recall from Chapter 4 that the graph of a function and its inverse are reflections of one another about the line $y = x$. Thus, to graph $y = \log_b x$, we need only reflect the curve $y = b^x$ about the line $y = x$. This is shown in Figure 2.

Note For the rest of this chapter we assume that the base b is greater than 1 when we use the expression $\log_b x$.

With the aid of Figure 2, we can make the following observations about the function $y = \log_b x$.

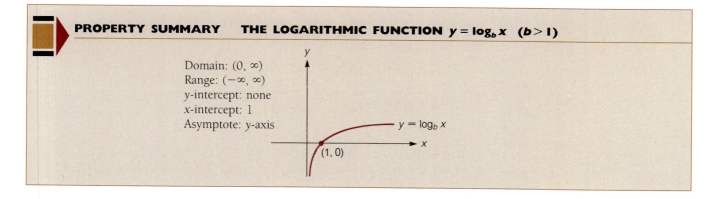

PROPERTY SUMMARY **THE LOGARITHMIC FUNCTION $y = \log_b x$ $(b > 1)$**

Domain: $(0, \infty)$
Range: $(-\infty, \infty)$
y-intercept: none
x-intercept: 1
Asymptote: y-axis

One aspect of the function $y = \log_b x$ may not be immediately apparent to you from Figure 2. The function grows or increases *very* slowly. Consider $y = \log_2 x$, for example. Let us ask how large x must be before the curve reaches the height $y = 10$ (see Figure 3).

To answer this question, we substitute $y = 10$ in the equation $y = \log_2 x$. This gives us

$$10 = \log_2 x$$

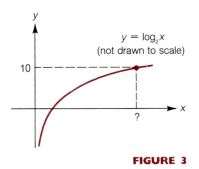

FIGURE 3

Writing this equation in exponential form yields

$$x = 2^{10} = 1024$$

In other words, we must go out beyond 1000 on the x-axis before the curve $y = \log_2 x$ reaches a height of 10 units. Exercise 49 at the end of this section asks you to show that the graph of $y = \log_2 x$ doesn't reach a height of 100 until x is greater than 10^{30}. (Numbers as large as 10^{30} rarely occur in any of the sciences. For instance, the distance in inches to the Andromeda galaxy is less than 10^{24}.) The point we are emphasizing here is this: The graph of $y = \log_2 x$ (or $\log_b x$) is always rising, but very slowly.

Before going on to consider some numerical examples, let us pause for a moment to think about the notation we have been using. In the expression

$$\log_b x$$

$\log_b$ is the name of a function and x is an input. To emphasize this, we might be better off writing $\log_b x$ as $\log_b(x)$, so that the similarity to the familiar $f(x)$ notation is clear. However, for historical reasons,* the convention is to suppress the parentheses, and we follow that convention here. In the box that follows, we indicate some errors that can occur when one forgets that $\log_b$ is the name of a function, not a number.

Errors to Avoid		
ERROR	**CORRECTION**	**COMMENT**
$\dfrac{\log_2 8}{\log_2 4} \ne \dfrac{8}{4}$	$\dfrac{\log_2 8}{\log_2 4} = \dfrac{3}{2}$	$\log_2$ is the name of a function. It is not a factor that that can be "cancelled" from the numerator and denominator.
$\dfrac{\log_2 16}{16} \ne \log_2$	$\dfrac{\log_2 16}{16} = \dfrac{4}{16} = \dfrac{1}{4}$	This "equation" is nonsense. On the left-hand side there is a quotient of two real numbers, which is a real number. On the right-hand side is the name of a function.

We conclude this section with a set of examples involving logarithms and logarithmic functions. In one way or another, every example makes use of the key fact that the equation $\log_b a = c$ is equivalent to $b^c = a$.

EXAMPLE 1 Which quantity is larger: $\log_3 10$ or $\log_7 40$?

Solution First we estimate $\log_3 10$. This quantity represents the exponent to which 3 must be raised to yield 10. Since $3^2 = 9$ (less than 10), but $3^3 = 27$ (more than 10), we conclude that the quantity $\log_3 10$ lies between 2 and 3. In a similar way we can estimate $\log_7 40$; this quantity represents the exponent to which 7 must be raised to yield 40. Since $7^1 = 7$ (less than 40), while $7^2 = 49$ (more than 40), we conclude that the quantity $\log_7 40$ lies between 1 and 2. It now follows from these two estimates that $\log_3 10$ is larger than $\log_7 40$.

EXAMPLE 2 Evaluate $\log_4 32$.

*The notation *Log* was introduced in 1624 by the astronomer Johannes Kepler (1571–1630). In Leonhard Euler's text *Introduction to Analysis of the Infinite*, first published in 1748, appears the statement, "It has been customary to designate the logarithm of y by the symbol log y."

Solution   Let $y = \log_4 32$. The exponential form of this equation is $4^y = 32$. Now, since both 4 and 32 are powers of 2, we can rewrite the equation $4^y = 32$ using the same base on both sides:

$$(2^2)^y = 2^5$$
$$2^{2y} = 2^5$$
$$2y = 5 \qquad \text{using Property 4 on page 308}$$
$$y = \frac{5}{2} \qquad \text{as required}$$

EXAMPLE 3   Graph the equations: **(a)** $y = \log_{10} x$ **(b)** $y = -\log_{10} x$

Solution   **(a)** The function $y = \log_{10} x$ is the inverse function for the exponential function $y = 10^x$. Thus we obtain the graph of $y = \log_{10} x$ by reflecting the graph of $y = 10^x$ in the line $y = x$. See Figure 4(a).

(b) To graph $y = -\log_{10} x$, we reflect $y = \log_{10} x$ in the x-axis. See Figure 4(b).

FIGURE 4

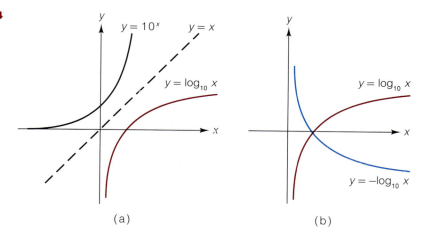

(a)   (b)

EXAMPLE 4   Find the domain of the function defined by $y = \log_2(12 - 4x)$.

Solution   As you can see by looking back at the figure in the box on page 324, the inputs for the logarithmic function must be positive. So, in the case at hand, we require that the quantity $12 - 4x$ be positive. Consequently, we have

$$12 - 4x > 0$$
$$-4x > -12$$
$$x < 3$$

Therefore the domain of the function defined by $y = \log_2(12 - 4x)$ is the interval $(-\infty, 3)$.

The next example concerns the exponential function $y = e^x$ and its inverse function, $y = \log_e x$. Many books, as well as calculators, abbreviate the expression $\log_e x$ by $\ln x$, read *natural log of x*.* For reference and emphasis, we repeat

*According to the historian Florian Cajori, the notation ln x was used by (and perhaps first introduced by) Irving Stringham in his text *Uniplanar Algebra* (San Francisco: University Press, 1893).

this fact in the following box. (Incidentally, on most calculators, *log* is an abbreviation for $\log_{10}$.)

DEFINITION

$$\ln x \qquad \text{means} \qquad \log_e x.$$

EXAMPLE 5   Graph the equations: **(a)** $y = \ln x$ **(b)** $y = \ln(x - 1)$

Solution   **(a)** The function $y = \ln x$ $(=\log_e x)$ is the inverse of $y = e^x$. Thus its graph is obtained by reflecting $y = e^x$ in the line $y = x$, as in Figure 5(a).

   (b) To graph $y = \ln(x - 1)$, we take the graph of $y = \ln x$ and move it one unit in the positive *x*-direction. See Figure 5(b). ■■■

FIGURE 5

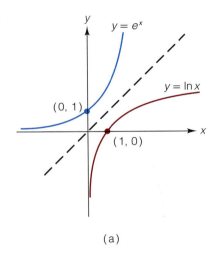
(a)

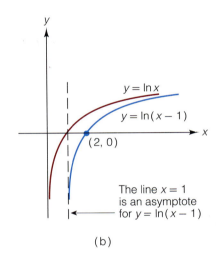
(b)

EXAMPLE 6   Simplify the expressions: **(a)** $\ln e$ **(b)** $\ln 1$

Solution   **(a)** The expression $\ln e$ denotes the exponent to which e must be raised to yield e. Clearly this exponent is 1. Thus $\ln e = 1$.

   (b) Similarly, $\ln 1$ denotes the exponent to which e must be raised to yield 1. Since $e^0 = 1$, zero is the required exponent. Thus $\ln 1 = 0$. ■■■

EXAMPLE 7   Solve the equations: **(a)** $10^{2x} = 200$ **(b)** $e^{3t-1} = 2$

Solution   **(a)** We write the equation $10^{2x} = 200$ in logarithmic form to obtain

$$2x = \log_{10} 200$$

$$x = \frac{\log_{10} 200}{2}$$

Without a calculator or tables, we leave the answer in this form. Alternatively, using a calculator we obtain

$$x \approx 1.15$$

(b) To solve $e^{3t-1} = 2$, we write the equation in logarithmic form:

$$3t - 1 = \log_e 2$$

that is,

$$3t - 1 = \ln 2$$
$$3t = 1 + \ln 2$$
$$t = \frac{1 + \ln 2}{3}$$

Without a calculator or tables, this is the final answer. On the other hand, using a calculator we obtain

$$t \approx 0.56$$

In the previous section, we used the equation $N = N_0 e^{kt}$ to describe radioactive decay as well as population growth. If you look back at the examples and problems there, you'll see that every question was of the same type: Given t, find N. In other words, we were always asked *how much* or *how many*, but never *when*. Now that we have defined logarithms, we can solve a much wider range of problems, as the next example indicates.

EXAMPLE 8   Suppose that the half-life of a certain radioactive substance is 4 days.

(a) Compute the *decay constant k* in the formula $N = N_0 e^{kt}$.
(b) If you begin with a 2-g sample, how long will it be until only 0.01 g remains?

Solution   **(a)** The half-life is 4 days. This means that when t is 4, the value of N is $\frac{1}{2}N_0$. Using these values for t and N in the formula $N = N_0 e^{kt}$, we obtain

$$\tfrac{1}{2}N_0 = N_0 e^{k4} \qquad \text{and therefore} \qquad \tfrac{1}{2} = e^{4k}$$

We solve for k by writing this last equation in logarithmic form:

$$4k = \ln \tfrac{1}{2} \qquad \text{and therefore} \qquad k = \frac{\ln \frac{1}{2}}{4} = \frac{\ln 0.5}{4}$$

If a numerical value for k is required, we can use tables or a calculator to obtain

$$k \approx -0.17$$

(b) We are given that $N_0 = 2$ g and we want to find the time t at which N is 0.01 g. Substituting the values $N_0 = 2$ and $N = 0.01$ into the decay law $N = N_0 e^{kt}$ yields

$$0.01 = 2e^{kt} \qquad \text{where } k \text{ has the value determined in part (a)}$$

or, dividing by 2,

$$0.005 = e^{kt}$$

Converting this last equation into its equivalent logarithmic form gives us

$$kt = \ln(0.005) \qquad \text{and therefore} \qquad t = \frac{\ln(0.005)}{k}$$

Finally, upon replacing k by the expression $\frac{1}{4}\ln 0.5$, we obtain

$$t = \frac{\ln(0.005)}{\frac{1}{4}\ln 0.5} = \frac{4\ln(0.005)}{\ln 0.5}$$

$$\approx 30.58 \quad \text{(using a calculator)}$$

Thus after about $30\frac{1}{2}$ days, only 0.01 g remains of the original 2-g sample.

EXERCISE SET 6.3

A

Exercises 1–6 are review exercises dealing with inverse functions.

1. Which of the following functions is one-to-one and therefore has an inverse?
    (a) $y = x^2 + 1$ (b) $y = 3x$ (c) $y = (x+1)^3$

2. Does each of the following functions have an inverse?
    (a) $f(x) = \begin{cases} x^2 & \text{if } -1 \le x \le 0 \\ x^2 + 1 & \text{if } x > 0 \end{cases}$

    (b) $g(x) = \begin{cases} x^2 & \text{if } -1 \le x < 0 \\ x^2 + 1 & \text{if } x \ge 0 \end{cases}$

3. Let $f(x) = \dfrac{2x-1}{3x+4}$. Find each quantity.
    (a) $f^{-1}(x)$ (b) $1/f(x)$
    (c) $f^{-1}(0)$ (d) $1/f(0)$

4. Let $f(x) = x^3 + 2x + 1$. Evaluate $f[f^{-1}(5)]$. [Assume that the domain of f^{-1} is $(-\infty, \infty)$.]

5. The graph of $y = f(x)$ is a line segment joining the two points $(3, -2)$ and $(-1, 5)$. What are the corresponding endpoints for the graph of $y = f^{-1}(x - 1)$?

6. Which (if either) of the following two conditions tells us that a function is one-to-one?
    (a) For each input there is exactly one output.
    (b) For each output there is exactly one input.

In Exercises 7 and 8, write each equation in logarithmic form.

7. (a) $9 = 3^2$ (b) $1000 = 10^3$
    (c) $7^3 = 343$ (d) $\sqrt{2} = 2^{1/2}$

8. (a) $\frac{1}{125} = 5^{-3}$ (b) $e^0 = 1$
    (c) $5^x = 6$ (d) $e^{3t} = 8$

9. Write each equation in exponential form.
    (a) $\log_2 32 = 5$ (b) $\log_{10} 1 = 0$
    (c) $\log_e \sqrt{e} = \frac{1}{2}$ (d) $\log_3 \frac{1}{81} = -4$
    (e) $\log_t u = v$

10. Complete the tables.
    (a)

x	1	10	10^2	10^3	10^4	10^{-1}	10^{-2}	10^{-3}
$\log_{10} x$								

(b)

x	1	e	e^2	e^3	e^4	e^{-1}	e^{-2}	e^{-3}
$\ln x$								

11. Which quantity is larger, $\log_5 30$ or $\log_8 60$?

12. Which quantity is larger?
    (a) $\log_{10} 90$ or $\log_e e^5$ (b) $\log_2 3$ or $\log_3 2$

In Exercises 13 and 14, evaluate each expression.

13. (a) $\log_9 27$ (b) $\log_4 \frac{1}{32}$ (c) $\log_5 5\sqrt{5}$

14. (a) $\log_{25} \frac{1}{625}$ (b) $\log_{16} \frac{1}{64}$
    (c) $\log_{10} 10$ (d) $\log_2 8\sqrt{2}$

In Exercises 15 and 16, solve each equation for x by converting to exponential form. In Exercises 15(b) and 16, give two forms for each answer: one involving e and the other a calculator approximation rounded off to two decimal places.

15. (a) $\log_4 x = -2$ (b) $\ln x = -2$

16. (a) $\log_5 x = e$ (b) $\ln x = -e$

In Exercises 17 and 18, find the domain of each function.

17. (a) $y = \log_4 5x$ (b) $y = \log_{10}(3 - 4x)$
    (c) $y = \ln(x^2)$ (d) $y = (\ln x)^2$
    (e) $y = \ln(x^2 - 25)$

18. (a) $y = \ln(2 - x - x^2)$ (b) $y = \log_{10}\dfrac{2x+3}{x-5}$

19. In the accompanying figure, what are the coordinates of the four points A, B, C, and D?

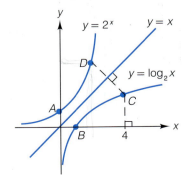

20. In the accompanying figure, what are the coordinates of the points A, B, C, and D?

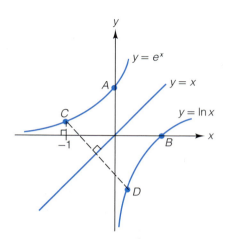

In Exercises 21–26, graph each function and specify the domain, range, intercept(s), and asymptote.

21. (a) $y = \log_2 x$ (b) $y = -\log_2 x$
    (c) $y = \log_2(-x)$ (d) $y = -\log_2(-x)$

22. (a) $y = \ln x$ (b) $y = -\ln x$
    (c) $y = \ln(-x)$ (d) $y = -\ln(-x)$

23. $y = -\log_3(x - 2) + 1$ 24. $y = -\log_{10}(x + 1)$

25. $y = \ln(x + e)$ 26. $y = \ln(-x) + e$

In Exercises 27 and 28, simplify each expression.

27. (a) $\ln e^4$ (b) $\ln(1/e)$ (c) $\ln \sqrt{e}$

28. (a) $\ln e$ (b) $\ln e^{-2}$ (c) $(\ln e)^{-2}$

In Exercises 29–36, find all the real-number solutions for each equation. In each case, give both the exact value of the answer and a calculator approximation rounded off to two decimal places.

29. $10^x = 25$ 30. $10^x = 145$

31. $10^{x^2} = 40$ 32. $(10^x)^2 = 40$

33. $e^{2t+3} = 10$ 34. $e^{t-1} = 16$

35. $e^{1-4t} = 12.405$ 36. $e^{3x^2} = 112$

37. The half-life of uranium-238 is 4.5×10^9 years.
    (a) Find the decay constant k.
    (b) What percentage of a given amount N_0 remains after 1000 years?

38. Compute the half-life of carbon-14 if the value of the decay constant is $k = -0.00012$. (Assume that t is in years.)

39. The half-life of a certain radioactive substance is 1 year.
    (a) Find the decay constant k.

    (b) Find the time required for 90% of a given 4-g sample to decay.

40. The initial population of a bacteria culture is 1.5×10^5. Three hours later, the population is found to be 2×10^5. Assume that the growth law $N = N_0 e^{kt}$ applies.
    (a) Find the growth constant k.
    (b) How long does it take for the initial population to double?

41. Initially a bacteria culture has a population of 2×10^7. Two hours later, the population is 3×10^8.
    (a) How long does it take for the culture to double its original size?
    (b) When does the population reach one billion?

42. The half-life of radium-226 is 1620 years.
    (a) Find the decay constant k.
    (b) How much of a 0.1-g sample remains after 100 years?

B

43. Let $f(x) = e^{x+1}$. Find $f^{-1}(x)$ and sketch its graph. Specify any intercept or asymptote.

44. Let $g(t) = \ln(t - 1)$. Find $g^{-1}(t)$ and draw its graph. Specify any intercept or asymptote.

45. Sketch the region bounded by $y = e^x$, $y = e^{-x}$, the x-axis, and the vertical lines $x = \pm 1$. Why must the area of this region be less than two square units?

46. Solve $2^{2x} - 2^{x+1} - 15 = 0$ for x. *Hint:* First show that the equation is equivalent to $(2^x)^2 - 2(2^x) - 15 = 0$.

47. Solve $e^{2x} - 5e^x - 6 = 0$ for x.

48. Solve $e^x - 3e^{-x} = 2$ for x. *Hint:* Multiply both sides by e^x.

49. Estimate a value for x such that $\log_2 x = 100$. Use the approximation $10^3 \approx 2^{10}$ to express your answer as a power of 10.   *Answer:* 10^{30}

50. (a) How large must x be before the graph of $y = \ln x$ reaches a height of $y = 100$?
    (b) How large must x be before the graph of $y = e^x$ reaches a height of (i) $y = 100$?   (ii) $y = 10^6$?

51. Suppose that $A = A_0 e^{k_1 t}$ and $B = B_0 e^{k_2 t}$. Find the value of t for which $A = B$.   *Answer:* $t = \dfrac{\ln(A_0/B_0)}{k_2 - k_1}$
How would you interpret this question and answer in terms of radioactive decay?

52. Chemists define pH by the formula $pH = -\log_{10}[H^+]$, where $[H^+]$ is the hydrogen ion concentration, measured in moles per liter. For example, if $[H^+] = 10^{-5}$, then $pH = 5$. Solutions with a pH of 7 are said to be *neutral*; a pH below 7 indicates an *acid* and a pH above 7 indicates a *base*.

(a) For some fruit juices, $[H^+] = 3 \times 10^{-4}$. Determine the pH and classify as acid or base.

(b) For sulfuric acid, $[H^+] = 1$. Find the pH.

(c) An unknown substance has a hydrogen ion concentration of 3.5×10^{-9}. Classify the substance as acid or base.

53. This exercise indicates one of the ways the natural logarithm function is used in the study of prime numbers. Recall that a prime number is a natural number greater than 1 with no factors other than itself and 1. For example, the first ten prime numbers are 2, 3, 5, 7, 11, 13, 17, 19, 23, and 29.

(a) Let $P(x)$ denote the number of prime numbers that do not exceed x. For instance, $P(6) = 3$, since there are three prime numbers (2, 3, and 5) that do not exceed 6. Compute $P(10)$, $P(18)$, and $P(19)$.

(b) According to the **prime number theorem**, $P(x)$ can be approximated by $x/\ln x$ when x is large, and in fact, the ratio $\dfrac{P(x)}{x/\ln x}$ approaches 1 as x grows larger and larger. Verify this empirically by completing the following table. Round your results to three decimal places. (These facts were discovered by Carl Friedrich Gauss in 1792, when he was 15 years old. It was more than 100 years later, however, in 1896, before the prime number theorem was formally proved by the French mathematician J. Hadamard and also by the Belgian mathematician C. J. de la Vallée-Poussin).

x	$P(x)$	$\dfrac{x}{\ln x}$	$\dfrac{P(x)}{x/\ln x}$
10^2	25		
10^4	1229		
10^6	78498		
10^8	5761455		
10^9	50847534		
10^{10}	455052512		

(c) In 1808, A. M. Legendre found that he could improve upon Gauss's approximation for $P(x)$ by using the expression $\dfrac{x}{\ln x - 1.08366}$ rather than $\dfrac{x}{\ln x}$. Complete the following table to see how well Legendre's expression approximates $P(x)$. Round off your results to four decimal places.

x	$P(x)$	$\dfrac{x}{\ln x - 1.08366}$	$\dfrac{P(x)}{x/(\ln x - 1.08366)}$
10^2	25		
10^4	1229		
10^6	78498		
10^8	5761455		
10^9	50847534		
10^{10}	455052512		

54. The figure shows a portion of the graph of $y = e^x$ for $1.5 \le x \le 1.7$. Using the figure, estimate the quantity $\ln 5$ to two decimal places. Then use a calculator to determine the percent error in your approximation. *Hint:* Write the equation $x = \ln 5$ in exponential form.

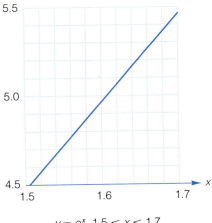

$y = e^x$, $1.5 \le x \le 1.7$

55. Using, as usual, x for the inputs and y for the outputs, graph the function defined by the equation $\ln y = x$.

C

56. (a) Find the domain of the function f defined by $f(x) = \ln(\ln x)$.

(b) Find $f^{-1}(x)$ for the function f in part (a).

(c) Find the domain of the function g defined by $g(x) = \ln[\ln(\ln x)]$.

6.4 PROPERTIES OF LOGARITHMS

... he [John Napier] invented the word "logarithms," using two Greek words, ... arithmos, "number" and logos, "ratio." It is impossible to say exactly what he had in mind when making up this word.

Alfred Hooper in *Makers of Mathematics* (New York: Random House, Inc., 1948)

The teaching of logarithms as a computing device is vanishing from the schools....
The logarithmic function, however, will never die, for the simple reason that logarithmic and exponential variations are a vital part of nature and of analysis.

Howard Eves in *An Introduction to the History of Mathematics*, 6th ed. (Philadelphia: Saunders College Publishing, 1990)

A few basic properties of logarithms are used repeatedly. Our procedure in this section will be to state these properties, discuss their proofs, and then look at examples.

Properties of Logarithms

1. (a) $\log_b b = 1$ (b) $\log_b 1 = 0$

2. $\log_b PQ = \log_b P + \log_b Q$
The log of a product is the sum of the logs of the factors.

3. $\log_b(P/Q) = \log_b P - \log_b Q$
The log of a quotient is the log of the numerator minus the log of the denominator.
As a useful particular case, we have $\log_b(1/Q) = -\log_b Q$.

4. $\log_b P^n = n \log_b P$

5. $b^{\log_b P} = P$

Note P and Q are assumed to be positive in Properties 2–5.

In essence, each of these properties follows from the equivalence of the two equations $y = \log_b x$ and $b^y = x$. For instance, the equivalent exponential forms for properties 1(a) and 1(b) are $b^1 = b$ and $b^0 = 1$, respectively, both of which are certainly valid.

To prove Property 2, we begin by letting $x = \log_b P$. The equivalent exponential form of this equation is

$$P = b^x \tag{1}$$

Similarly, we let $y = \log_b Q$. The exponential form of this equation is

$$Q = b^y \tag{2}$$

If we multiply equation (1) by equation (2), we get

$$PQ = b^x b^y$$

and, therefore,

$$PQ = b^{x+y} \tag{3}$$

Next we write equation (3) in its equivalent logarithmic form. This yields

$$\log_b PQ = x + y$$

But, using the definitions of x and y, this last equation is equivalent to

$$\log_b PQ = \log_b P + \log_b Q$$

That completes the proof of Property 2.

The proof of Property 3 is quite similar to the proof given for Property 2. Exercise 70(a) asks you to carry out the proof of Property 3.

We turn now to the proof of Property 4. We begin by letting $x = \log_b P$. In exponential form, this last equation becomes

$$b^x = P \tag{4}$$

Now we raise both sides of equation (4) to the power n. This yields

$$(b^x)^n = b^{nx} = P^n$$

The logarithmic form of this last equation is

$$\log_b P^n = nx$$

or (from the definition of x)

$$\log_b P^n = n \log_b P$$

The proof of Property 4 is now complete.

Property 5 is again just a restatement of the meaning of $\log_b P$. To derive this property, let $x = \log_b P$. Therefore,

$$b^x = P$$

Now, in this last equation, we simply replace x with $\log_b P$. The result is

$$b^{\log_b P} = P \qquad \text{as required.}$$

Now let's see how these properties are used. To begin with, we display some simple numerical examples in the box that follows (on the next page).

The preceding examples showed how we can simplify or shorten certain expressions involving logarithms. The next example is also of this type.

EXAMPLE 1   Express as a single logarithm with a coefficient of 1:

$$\frac{1}{2}\log_b x - \log_b(1 + x^2)$$

Solution $\frac{1}{2}\log_b x - \log_b(1 + x^2) = \log_b x^{1/2} - \log_b(1 + x^2)$ using Property 4 on page 332

$$= \log_b \frac{x^{1/2}}{1 + x^2}$$ using Property 3

This last expression is the required answer.

PROPERTY SUMMARY   **PROPERTIES OF LOGARITHMS**

PROPERTY	EXAMPLE
$\log_b P + \log_b Q = \log_b PQ$	Simplify: $\log_{10} 50 + \log_{10} 2$ Solution: $\log_{10} 50 + \log_{10} 2 = \log_{10}(50 \cdot 2)$ $= \log_{10} 100$ $= 2$
$\log_b P - \log_b Q = \log_b \dfrac{P}{Q}$	Simplify: $\log_8 56 - \log_8 7$ Solution: $\log_8 56 - \log_8 7 = \log_8 \dfrac{56}{7}$ $= \log_8 8 = 1$
$\log_b P^n = n \log_b P$	Simplify: $\log_2 \sqrt[5]{16}$ Solution: $\log_2 \sqrt[5]{16} = \log_2(16^{1/5})$ $= \dfrac{1}{5}\log_2 16$ $= \dfrac{1}{5} \cdot 4 = \dfrac{4}{5}$
$b^{\log_b P} = P$	Simplify: $3^{\log_3 7}$ Solution: $3^{\log_3 7} = 7$

Property 2 says that the logarithm of a product of two factors is equal to the sum of the logarithms of the two factors. This can be generalized to any number of factors. For instance, with three factors, we have

$$\log_b(ABC) = \log_b[A(BC)]$$
$$= \log_b A + \log_b BC \qquad \text{using Property 2}$$
$$= \log_b A + \log_b B + \log_b C \qquad \text{using Property 2 again}$$

The next example makes use of this idea.

EXAMPLE 2   Express as a single logarithm with a coefficient of 1:

$$\ln(x^2 - 9) + 2\ln\frac{1}{x+3} + 4\ln x \qquad (x > 3)$$

Solution $\ln(x^2 - 9) + 2\ln\dfrac{1}{x+3} + 4\ln x = \ln(x^2 - 9) + \ln\left[\left(\dfrac{1}{x+3}\right)^2\right] + \ln x^4$

$$= \ln(x^2 - 9) + \ln\frac{1}{(x+3)^2} + \ln x^4$$

$$= \ln\left[(x^2 - 9) \cdot \frac{1}{(x+3)^2} \cdot x^4\right]$$

$$= \ln\frac{(x^2 - 9)x^4}{(x+3)^2}$$

This last expression can be simplified still further by writing $x^2 - 9$ as $(x - 3)(x + 3)$. Then a factor of $x + 3$ can be divided out of the numerator and denominator of the fraction. The result is

$$\ln(x^2 - 9) + 2 \ln \frac{1}{x + 3} + 4 \ln x = \ln \frac{(x - 3)x^4}{x + 3}$$

$$= \ln \frac{x^5 - 3x^4}{x + 3} \qquad \blacksquare\blacksquare\blacksquare$$

In the examples considered so far, we've used the properties of logarithms to shorten given expressions. We can also use these properties to expand an expression. (This is useful in calculus.)

EXAMPLE 3 Write each of the following quantities as sums and differences of simpler logarithmic expressions. Express each answer in such a way that no logarithm of products, quotients, or powers appears.

(a) $\log_{10} \sqrt{3x}$ **(b)** $\log_{10} \sqrt[3]{\dfrac{2x}{3x^2 + 1}}$ **(c)** $\ln \dfrac{x^2 \sqrt{2x - 1}}{(2x + 1)^{3/2}}$

Solution **(a)** $\log_{10} \sqrt{3x} = \log_{10}(3x)^{1/2} = \frac{1}{2} \log_{10}(3x)$
$$= \tfrac{1}{2}(\log_{10} 3 + \log_{10} x)$$

(b) $\log_{10} \sqrt[3]{\dfrac{2x}{3x^2 + 1}} = \log_{10} \left[\left(\dfrac{2x}{3x^2 + 1} \right)^{1/3} \right]$

$$= \tfrac{1}{3} \log_{10} \frac{2x}{3x^2 + 1}$$

$$= \tfrac{1}{3} [\log_{10} 2x - \log_{10}(3x^2 + 1)]$$

$$= \tfrac{1}{3} [\log_{10} 2 + \log_{10} x - \log_{10}(3x^2 + 1)]$$

(c) $\ln \dfrac{x^2 \sqrt{2x - 1}}{(2x + 1)^{3/2}} = \ln \dfrac{x^2 (2x - 1)^{1/2}}{(2x + 1)^{3/2}}$

$$= \ln x^2 + \ln(2x - 1)^{1/2} - \ln(2x + 1)^{3/2}$$

$$= 2 \ln x + \tfrac{1}{2} \ln(2x - 1) - \tfrac{3}{2} \ln(2x + 1) \qquad \blacksquare\blacksquare\blacksquare$$

EXAMPLE 4 Given that $\log_{10} A = a$, $\log_{10} B = b$, and $\log_{10} C = c$, express $\log_{10} \dfrac{A^3}{B^4 \sqrt{C}}$ in terms of a, b, and c.

Solution $\log_{10} \dfrac{A^3}{B^4 \sqrt{C}} = \log_{10} \dfrac{A^3}{B^4 C^{1/2}}$

$$= \log_{10} A^3 - \log_{10} B^4 C^{1/2}$$

$$= \log_{10} A^3 - (\log_{10} B^4 + \log_{10} C^{1/2})$$

$$= 3 \log_{10} A - (4 \log_{10} B + \tfrac{1}{2} \log_{10} C)$$

$$= 3 \log_{10} A - 4 \log_{10} B - \tfrac{1}{2} \log_{10} C$$

$$= 3a - 4b - \tfrac{1}{2}c \qquad \blacksquare\blacksquare\blacksquare$$

The properties of logarithms are also useful in solving equations, as the next two examples demonstrate.

EXAMPLE 5 Use logarithms to the base e to solve the equation $10 = 3e^{1-2x}$ for x.

Solution We take the natural (base e) logarithm of both sides of the given equation to obtain

$$\ln 10 = \ln(3e^{1-2x})$$
$$= \ln 3 + \ln e^{1-2x} \qquad \text{using Property 2}$$
$$= \ln 3 + (1 - 2x) \qquad \text{using the definition of ln}$$
$$2x = \ln 3 - \ln 10 + 1$$
$$x = \frac{\ln 3 - \ln 10 + 1}{2} \approx -0.102$$

This is the required solution. If we wish, we can use Property 3 to rewrite this solution as

$$x = \frac{\ln \frac{3}{10} + 1}{2}$$

(There are yet other ways to write this answer. See Exercise 69.)

EXAMPLE 6 Solve for x: $\log_3 x + \log_3(x + 2) = 1$

Solution Using Property 2, we can write the given equation as

$$\log_3[x(x + 2)] = 1 \qquad \text{or} \qquad \log_3(x^2 + 2x) = 1$$

Writing this last equation in exponential form yields

$$x^2 + 2x = 3^1$$
$$x^2 + 2x - 3 = 0$$
$$(x + 3)(x - 1) = 0$$

Thus, we have

$$x + 3 = 0 \qquad \text{or} \qquad x - 1 = 0$$

and, consequently,

$$x = -3 \qquad \text{or} \qquad x = 1$$

Now let us check these values in the original equation to see if they are indeed solutions.

If $x = -3$, the equation becomes

$$\log_3(-3) + \log_3(-1) \stackrel{?}{=} 1$$

Neither expression is defined, since the domain of the logarithm function does not contain negative numbers.

If $x = 1$, the equation becomes

$$\log_3 1 + \log_3 3 \stackrel{?}{=} 1$$
$$0 + 1 \stackrel{?}{=} 1 \qquad \text{true}$$

Thus, the value $x = 1$ is a solution of the original equation, but $x = -3$ is not.

In the example we just concluded, an extraneous solution ($x = -3$) was generated along with the correct solution, $x = 1$. How did this happen? It happened because we used the property $\log_b PQ = \log_b P + \log_b Q$ in the second line of the solution. But this property is valid only when both P and Q are positive. In the box that follows, we generalize this remark about extraneous solutions.

PROPERTY SUMMARY   EXTRANEOUS SOLUTIONS FOR LOGARITHMIC EQUATIONS

Using the properties of logarithms (on page 332) to solve logarithmic equations may introduce extraneous solutions that do not check in the original equation. (This occurs because the logarithm function requires positive inputs, but in solving an equation, we may not know ahead of time the sign of an input involving a variable.) Therefore, it is always necessary to check any candidates for solutions that are obtained in this manner.

It is sometimes necessary to convert logarithms in one base to logarithms in another base. After the next example, we will state a formula for this. However, as the next example indicates, it is easy to work this type of problem from the basics, without relying on a formula.

EXAMPLE 7   Express the quantity $\log_2 5$ in terms of base 10 logarithms.

Solution   Let $z = \log_2 5$. The exponential form of this equation is

$$2^z = 5$$

We now take the base 10 logarithm of each side of this equation to obtain

$$\log_{10} 2^z = \log_{10} 5$$
$$z \log_{10} 2 = \log_{10} 5 \qquad \text{using Property 4}$$
$$z = \frac{\log_{10} 5}{\log_{10} 2}$$

Given our definition of z, this last equation can be written

$$\log_2 5 = \frac{\log_{10} 5}{\log_{10} 2}$$

This is the required answer.

The method shown in Example 7 can be used to convert between any two bases. Exercise 70(b) at the end of this section asks you to follow this method, using letters rather than numbers, to arrive at the following general formula.

Change of Base Formula

$$\log_a x = \frac{\log_b x}{\log_b a}$$

Examples 1–7 have dealt with applications of the five properties of logarithms. However, you also need to understand what the properties *don't* say. For instance, Property 3 does not apply to an expression such as $\dfrac{\log_{10} 5}{\log_{10} 2}$. (Property 3

would apply if the expression were $\log_{10} \frac{5}{2}$.) In the box that follows, we list some errors to avoid in working with logarithms.

Errors to Avoid

ERROR	CORRECTION	COMMENT
$\log_b(x + y) \neq \log_b x + \log_b y$	$\log_b(xy) = \log_b x + \log_b y$	In general, there is no simple identity involving $\log_b(x + y)$. This is similar to the situation for $\sqrt{x + y}$ (which is not equal to $\sqrt{x} + \sqrt{y}$).
$\dfrac{\log_b x}{\log_b y} \neq \log_b x - \log_b y$	$\log_b\left(\dfrac{x}{y}\right) = \log_b x - \log_b y$	In general, there is no simple identity involving the quotient $(\log_b x)/(\log_b y)$.
$(\ln x)^3 \neq 3 \ln x$	$(\ln x)^3 = (\ln x)(\ln x)(\ln x)$	$(\ln x)^3$ is not the same as $\ln(x^3)$. Regarding the latter, we do have $\ln(x^3) = 3 \ln x$, for $x > 0$.
$\ln \dfrac{x}{2} \neq \dfrac{\ln x}{2}$	$\ln \dfrac{x}{2} = \ln x - \ln 2$, for $x > 0$ and $\dfrac{\ln x}{2} = \dfrac{1}{2} \ln x = \ln \sqrt{x}$, for $x > 0$	The confusion between $\ln \dfrac{x}{2}$ and $\dfrac{\ln x}{2}$ is sometimes due simply to careless handwriting.

EXERCISE SET 6.4

A

In Exercises 1–10, simplify the expression by using the definition and properties for logarithms.

1. $\log_{10} 70 - \log_{10} 7$
2. $\log_{10} 40 + \log_{10} \frac{5}{2}$
3. $\log_7 \sqrt{7}$
4. $\log_9 25 - \log_9 75$
5. $\log_3 108 + \log_3 \frac{3}{4}$
6. $\ln e^3 - \ln e$
7. $-\frac{1}{2} + \ln \sqrt{e}$
8. $e^{\ln 3} + e^{\ln 2} - e^{\ln e}$
9. $2^{\log_2 5} - 3 \log_5 \sqrt[3]{5}$
10. $\log_b b^b$

In Exercises 11–19, write the expression as a single logarithm with a coefficient of 1.

11. $\log_{10} 30 + \log_{10} 2$
12. $2 \log_{10} x - 3 \log_{10} y$
13. $\log_5 6 + \log_5 \frac{1}{3} + \log_5 10$
14. $p \log_b A - q \log_b B + r \log_b C$
15. (a) $\ln 3 - 2 \ln 4 + \ln 32$
    (b) $\ln 3 - 2(\ln 4 + \ln 32)$
16. (a) $\log_{10}(x^2 - 16) - 3 \log_{10}(x + 4) + 2 \log_{10} x$
    (b) $\log_{10}(x^2 - 16) - 3[\log_{10}(x + 4) + 2 \log_{10} x]$
17. $\log_b 4 + 3[\log_b(1 + x) - \frac{1}{2} \log_b(1 - x)]$
18. $\ln(x^3 - 1) - \ln(x^2 + x + 1)$
19. $4 \log_{10} 3 - 6 \log_{10}(x^2 + 1) + \frac{1}{2}[\log_{10}(x + 1) - 2 \log_{10} 3]$

In Exercises 20–26, write the quantity using sums and differences of simpler logarithmic expressions. Express the answer so that logarithms of products, quotients, and powers do not appear.

20. (a) $\log_{10} \sqrt{(x + 1)(x + 2)}$
    (b) $\ln \sqrt{\dfrac{(x + 1)(x + 2)}{(x - 1)(x - 2)}}$

21. (a) $\log_{10} \dfrac{x^2}{1 + x^2}$
    (b) $\ln \dfrac{x^2}{\sqrt{1 + x^2}}$

22. (a) $\log_b \dfrac{\sqrt{1 - x^2}}{x}$
    (b) $\ln \dfrac{x\sqrt[3]{4x + 1}}{\sqrt{2x - 1}}$

23. (a) $\log_{10} \sqrt{9 - x^2}$
    (b) $\ln \dfrac{\sqrt{4 - x^2}}{(x - 1)(x + 1)^{3/2}}$

24. (a) $\log_b \sqrt[3]{\dfrac{x + 3}{x}}$
    (b) $\ln \dfrac{1}{\sqrt{x^2 + x + 1}}$

25. (a) $\log_b \sqrt{x/b}$
    (b) $2 \ln \sqrt{(1 + x^2)(1 + x^4)(1 + x^6)}$

26. (a) $\log_b \sqrt[3]{\dfrac{(x - 1)^2(x - 2)}{(x + 2)^2(x + 1)}}$
    (b) $\ln \left(\dfrac{e - 1}{e + 1}\right)^{3/2}$

In Exercises 27 and 28, suppose that $\log_{10} A = a$, $\log_{10} B = b$, and $\log_{10} C = c$. Express the following logarithms in terms of a, b, and c.

27. (a) $\log_{10} AB^2 C^3$ (b) $\log_{10} 10\sqrt{A}$
     (c) $\log_{10} \sqrt{10ABC}$ (d) $\log_{10}\left(10A/\sqrt{BC}\right)$

28. (a) $\log_{10} A + 2\log_{10}(1/A)$ (b) $\log_{10}(A/10)$
     (c) $\log_{10}\dfrac{100A^2}{B^4\sqrt[3]{C}}$ (d) $\log_{10}\dfrac{(AB)^5}{C}$

In Exercises 29 and 30, suppose that $\ln x = t$ and $\ln y = u$. Write each expression in terms of t and u.

29. (a) $\ln(ex)$ (b) $\ln xy - \ln(x^2)$
     (c) $\ln \sqrt{xy} + \ln(x/e)$ (d) $\ln\left(e^2 x\sqrt{y}\right)$

30. (a) $\ln(e^{\ln x})$ (b) $e^{\ln(\ln xy)}$
     (c) $\ln\left(\dfrac{ex}{y}\right) - \ln\left(\dfrac{y}{ex}\right)$ (d) $\dfrac{(\ln x)^3 - \ln(x^4)}{\left(\ln\dfrac{x}{e^2}\right)\ln(xe^2)}$

In Exercises 31–37, solve the equation. Express the answer in terms of base e logarithms.

31. $5 = 2e^{2x-1}$ **32.** $100 = 3e^x$ **33.** $3e^{1+t} = 2$

34. $4e^{1-2t} = 7$ **35.** $2^x = 9$ **36.** $5^{3x-1} = 27$

37. $10 \cdot 2^x = 5^x$

In Exercises 38–47, solve the equation for x; check the answers to remove any extraneous roots.

38. $\log_6 x + \log_6(x+1) = 1$

39. $\log_9(x+1) = \frac{1}{2} + \log_9 x$

40. $\log_2(x+4) = 2 - \log_2(x+1)$

41. $\log_{10}(2x+4) + \log_{10}(x-2) = 1$

42. $\ln x + \ln(x+1) = \ln 12$

43. $\log_{10}(x+3) - \log_{10}(x-2) = 2$

44. $\ln(x+1) = 2 + \ln(x-1)$

45. $\log_b(x+1) = 2\log_b(x-1)$

46. $\log_2(2x^2+4) = 5$

47. $\log_{10}(x-6) + \log_{10}(x+3) = 1$

48. Solve for x in terms of a:
$$\log_2(x+a) - \log_2(x-a) = 1$$

49. Solve for x in terms of y.
     (a) $\log_{10} x - y = \log_{10}(3x-1)$
     (b) $\log_{10}(x-y) = \log_{10}(3x-1)$

50. Solve for x in terms of b: $\log_b(1-3x) = 3 + \log_b x$.

In Exercises 51–56, express the quantity in terms of base 10 logarithms.

51. $\log_2 5$ **52.** $\log_5 10$ **53.** $\ln 3$

54. $\ln 10$ **55.** $\log_b 2$ **56.** $\log_2 b$

In Exercises 57–61, express the quantity in terms of natural logarithms.

57. $\log_{10} 6$ **58.** $\log_2 10$ **59.** $\log_{10} e$

60. $\log_b 2$, where $b = e^2$ **61.** $\log_{10}(\log_{10} x)$

62. Give specific examples showing that each statement is false.
     (a) $\log(x+y) = \log x + \log y$
     (b) $(\log x)/(\log y) = \log x - \log y$
     (c) $(\log x)(\log y) = \log x + \log y$
     (d) $(\log x)^k = k\log x$

63. True or false?
     (a) $\log_{10} A + \log_{10} B - \frac{1}{2}\log_{10} C = \log_{10}\left(AB/\sqrt{C}\right)$
     (b) $\log_e \sqrt{e} = \frac{1}{2}$ (c) $\ln \sqrt{e} = \frac{1}{2}$
     (d) $\ln x^3 = \ln 3x$ (e) $\ln x^3 = 3\ln x$
     (f) $\ln 2x^3 = 3\ln 2x$ (g) $\log_a c = b$ means $a^b = c$.
     (h) $\log_5 24$ is between 5^1 and 5^2.
     (i) $\log_5 24$ is between 1 and 2.
     (j) $\log_5 24$ is closer to 1 than to 2.
     (k) The domain of $g(x) = \ln x$ is the set of all real numbers.
     (l) The range of $g(x) = \ln x$ is the set of all real numbers.
     (m) The function $g(x) = \ln x$ is one-to-one.

Use a calculator for Exercises 64 and 65.

64. (a) Check Property 2 using the values $b = 10$, $P = \pi$, and $Q = \sqrt{2}$.
     (b) Let $P = 3$ and $Q = 4$. Show that $\ln(P+Q) \neq \ln P + \ln Q$.
     (c) Check Property 3 using the values $b = 10$, $P = 2$, and $Q = 3$.
     (d) If $P = 10$ and $Q = 20$, show that $\ln(PQ) \neq (\ln P)(\ln Q)$.
     (e) Check Property 3 using natural logarithms and the values $P = 19$ and $Q = 89$.
     (f) Show that $(\log_{10} 19)/(\log_{10} 89) \neq \log_{10} 19 - \log_{10} 89$.
     (g) Show that $(\ln 19)/(\ln 89) \neq \ln 19 - \ln 89$.

65. (a) Check Property 4 using the values $b = 10$, $P = \pi$, and $n = 7$.
     (b) Using the values given for b, P, and n in part (a), show that $\log_b P^n \neq (\log_b P)^n$.
     (c) Verify Property 5 using the values $b = 10$ and $P = 1776$.
     (d) Verify that $\ln 2 + \ln 3 + \ln 4 = \ln 24$.
     (e) Verify that $\log_{10} A + \log_{10} B + \log_{10} C = \log_{10}(ABC)$, using the values $A = 11$, $B = 12$, and $C = 13$.

(f) Let $f(x) = e^x$ and $g(x) = \ln x$. Compute $f[g(2345.6)]$.

(g) Let $f(x) = 10^x$ and $g(x) = \log_{10} x$. Compute $g[f(0.123456)]$.

B

66. As indicated in the figure, the graph of $N = N_0 e^{kt}$ passes through the two points $(2, 10)$ and $(8, 80)$. Follow steps (a) through (c) to determine the values of the constants N_0 and k.

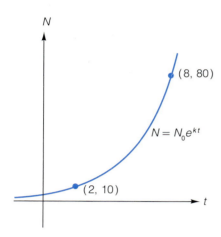

(a) Since the point $(8, 80)$ lies on the graph, the pair of values $t = 8$ and $N = 80$ satisfies the equation $N = N_0 e^{kt}$. That is,

$$80 = N_0 e^{8k} \qquad (1)$$

Similarly, since the point $(2, 10)$ lies on the graph, we have

$$10 = N_0 e^{2k} \qquad (2)$$

Now use equations (1) and (2) to show that $k = (\ln 8)/6$.

(b) In equation (2), substitute for k using the expression obtained in part (a). Show that the resulting equation can be written $10 = N_0 (e^{\ln 8})^{1/3}$.

(c) Use one of the properties of logarithms to simplify the right-hand side of the equation obtained in part (b). Then solve for N_0. (You should obtain $N_0 = 5$.)

In Exercises 67 and 68, you are given the coordinates of two points on the graph of the curve $N = N_0 e^{kt}$. In each case, determine the values of k and N_0. Hint: Use the method explained in Exercise 66.

67. $(-2, 324)$ and $\left(\frac{1}{2}, \frac{4}{3}\right)$ 68. $(1, 2)$ and $(4, 8)$

69. In Example 5 we solved the equation $10 = 3e^{1-2x}$ and found that $x = (\ln 0.3 + 1)/2$.
(a) Show that this x-value indeed satisfies the given equation.
(b) Show that an equivalent form for x is $\frac{1}{2} + \ln \sqrt{0.3}$.
(c) Show that another equivalent form for x is $\ln \sqrt{3e/10}$.

70. (a) Prove that $\log_b(P/Q) = \log_b P - \log_b Q$.
    Hint: Study the proof of Property 2 in the text.
(b) Prove the **change of base formula.**

$$\log_a x = \frac{\log_b x}{\log_b a}$$

    Hint: Use the method of Example 7 in the text.

71. Show that $\log_b \dfrac{\sqrt{3} + \sqrt{2}}{\sqrt{3} - \sqrt{2}} = 2 \log_b(\sqrt{3} + \sqrt{2})$.

72. (a) Show that $\log_b(P/Q) + \log_b(Q/P) = 0$.
(b) Simplify: $\log_a x + \log_{1/a} x$.

73. Simplify: $b^{3 \log_b x}$.

74. Let $\log_{10} 2 = a$ and $\log_{10} 3 = b$. Express each logarithm in terms of a and/or b.
    (a) $\log_{10} 4$ (b) $\log_{10} 8$ (c) $\log_{10} 5$
    (d) $\log_{10} 6$ (e) $\log_{10} \frac{5}{64}$ (f) $\log_{10} 108$
    (g) $\log_{10} \sqrt[3]{12}$ (h) $\log_{10} 0.0027$

75. Simplify: $\log_2 \sqrt[5]{4\sqrt{2}}$.

In Exercises 76 and 77, solve for x in terms of α and β.

76. $\alpha \ln x + \ln \beta = 0$ 77. $3 \ln x = \alpha + 3 \ln \beta$

78. Is there a constant k such that the equation $e^x = 2^{kx}$ holds for all values of x?

79. Prove that $\log_b a = 1/(\log_a b)$.

80. Simplify: $(\log_2 3)(\log_3 4)(\log_4 5)$.

C

81. Prove that $(\log_a x)/(\log_{ab} x) = 1 + \log_a b$.

82. Simplify a^x when $x = [\log_b(\log_b a)]/\log_b a$.

83. If $a^2 + b^2 = 7ab$, and a and b are positive, show that

$$\log[\tfrac{1}{3}(a + b)] = \tfrac{1}{2}(\log a + \log b)$$

no matter which base is used for the logarithms. (But it is understood that the same base is used throughout.)

84. Solve for x (assuming $a > b > 0$):

$$(a^4 - 2a^2 b^2 + b^4)^{x-1} = (a - b)^{2x}(a + b)^{-2}$$

    Answer: $x = \dfrac{\ln(a - b)}{\ln(a + b)}$

85. Let $f(x) = \ln(x + \sqrt{x^2 + 1})$. Find $f^{-1}(x)$.

86. Suppose that $\log_{10} 2 = a$ and $\log_{10} 3 = b$. Solve for x in terms of a and b:

$$6^x = \frac{10}{3} - 6^{-x}$$

87. Let $y = \ln[\ln(\ln x)]$. First, complete the table using a calculator. After doing that, disregard the evidence in your table and prove (without a calculator, of course) that the range of the given function is actually the set of all real numbers.

x	100	1000	10^6	10^{20}	10^{30}	10^{99}
y						

6.5 APPLICATIONS

$S = Pe^{rt}$. This result is remarkable both because of its simplicity and the occurrence of e. (Who would expect that number to pop up in finance theory?)

Philip Gillett in *Calculus and Analytic Geometry,* 3rd ed. (Lexington, Mass.: D.C. Heath & Company, 1988)

Drawing by Professor Ann Jones, University of Colorado, Boulder. From the cover of *The Physics Teacher,* Vol. 14, No. 7 (October 1976).

We begin this section by considering how money accumulates in a savings account. Eventually, this will lead us back to the number e and the growth law $N = N_0 e^{kt}$, both of which we used in different contexts in Section 6.2.

The following idea from arithmetic is a prerequisite for our discussion. To increase a given quantity by, say, 15%, we multiply the quantity by 1.15. For instance, suppose that we want to increase \$100 by 15%. The calculations can be written

$$100 + 0.15(100) = 100(1 + 0.15) = 100(1.15)$$

Similarly, to increase a quantity by 30%, we would multiply by 1.30; and so on. The next example displays some calculations involving percentage increase. The results may surprise you unless you're already familiar with this topic.

EXAMPLE 1 An amount of $100 is increased by 15% and then the new amount is increased by 15%. Is this the same as an overall increase of 30%?

Solution To increase $100 by 15%, we multiply by 1.15 to obtain $100(1.15). Now to increase this new amount by 15%, we multiply it by 1.15 to obtain

$$[(\$100)(1.15)](1.15) = \$100(1.15)^2 = \$132.25$$

Alternatively, if we increase the original $100 by 30%, we obtain

$$\$100(1.30) = \$130$$

Comparing, we see that the result of two successive 15% increases is greater than the result of a single 30% increase. ∎

Now let's look at another example and use it to introduce some terminology. Suppose that you place $1000 in a savings account at 10% interest *compounded annually*. This means that at the end of each year, the bank contributes to your account 10% of the amount that is in the account at that time. Interest compounded in this manner is called **compound interest**. The original deposit of $1000 is called the **principal**, denoted P. The interest rate, expressed as a decimal, is denoted by r. Thus, $r = 0.10$ in this example. The variable A is used to denote the **amount** in the account at any given time. The calculations displayed in Table 1 show how the account grows.

TABLE 1

Time Period	Algebra	Arithmetic
After 1 year	$A = P(1 + r)$	$A = 1000(1.10)$ $= \$1100$
After 2 years	$A = [P(1 + r)](1 + r)$ $= P(1 + r)^2$	$A = 1000(1.10)^2$ $= \$1210$
After 3 years	$A = [P(1 + r)^2](1 + r)$ $= P(1 + r)^3$	$A = 1000(1.10)^3$ $= \$1331$

We can learn several things from Table 1. First, consider how much interest is paid each year.

Interest paid for first year: $\$1100 - \$1000 = \$100$

Interest paid for second year: $\$1210 - \$1100 = \$110$

Interest paid for third year: $\$1331 - \$1210 = \$121$

Thus the interest earned each year is not a constant; it increases each year. However, notice that the *ratio* of interest earned in successive years is constant:

$$\frac{110}{100} = 1.1$$

$$\frac{121}{110} = 1.1$$

This is because the given growth rate of 10% per year is constant.

If you look at the algebra in Table 1, you can see what the general formula should be for the amount after t years.

Compound Interest Formula (Interest compounded annually)

Suppose that a principal of P dollars is invested at an annual rate r that is compounded annually. Then the amount A after t years is given by

$$A = P(1 + r)^t$$

EXAMPLE 2   Suppose that $2000 is invested at $7\frac{1}{2}$% interest compounded annually. How many years will it take for the money to double?

Solution   In the formula $A = P(1 + r)^t$, we use the given values $P = \$2000$ and $r = 0.075$. We want to find how long it will take for the money to double; that is, we want to find t when $A = \$4000$. Making these substitutions in the formula, we obtain

$$4000 = 2000(1 + 0.075)^t$$

and, therefore,

$$2 = 1.075^t \qquad \text{dividing by 2000}$$

We can solve this exponential equation by taking the logarithm of both sides. We will use base e logarithms. (Base 10 would be just as convenient here.) This yields

$$\ln 2 = \ln 1.075^t$$

and, consequently,

$$\ln 2 = t \ln 1.075 \qquad \text{(Why?)}$$

To isolate t, we divide both sides of this last equation by $\ln 1.075$. This yields

$$t = \frac{\ln 2}{\ln 1.075} \approx 9.6 \text{ years}$$

TABLE 2

$A = 2000(1.075)^t$

t (years)	A (dollars)
9	3834.48
10	4122.06

Now, assuming that the bank computes the compound interest only at the end of the year, we must round the preliminary answer of 9.6 years and say that when $t = 10$ years, the initial $2000 will have *more than* doubled. Table 2 adds some perspective to this. The table shows that after 9 years, something less than $4000 is in the account; whereas after 10 years, the amount exceeds $4000.

In Example 2, the interest was compounded annually. In practice, though, interest is usually computed more often. For instance, a bank may advertise a rate of 10% per year compounded semiannually. This means that after half a year, the interest is compounded at 5%, and then after another half year, the interest is again compounded at 5%. If you review Example 1, you'll see that one compounding at a rate of r is not the same as two compoundings, each at a rate of $r/2$. The formula $A = P(1 + r)^t$ can be generalized to cover such cases where interest is compounded more than once each year.

> **Compound Interest Formula**
> **(Interest compounded n times per year)**
>
> Suppose that a principal of P dollars is invested at an annual rate r that is compounded n times per year. Then the amount A after t years is given by
>
> $$A = P\left(1 + \frac{r}{n}\right)^{nt}$$

EXAMPLE 3   Suppose that $1000 is placed in a savings account at 10% per annum. How much is in the account at the end of one year if the interest is **(a)** compounded once each year ($n = 1$)? **(b)** compounded quarterly ($n = 4$)?

Solution   We use the formula $A = P[1 + (r/n)]^{nt}$.

(a) For $n = 1$, we obtain

$$A = 1000\left(1 + \frac{0.10}{1}\right)^{(1)(1)}$$

$$= 1000(1.1)$$

$$= \$1100$$

(b) For $n = 4$, we obtain

$$A = 1000\left(1 + \frac{0.10}{4}\right)^{4(1)}$$

$$= 1000(1.025)^4$$

$$= \$1103.81 \quad \text{using a calculator}$$

Notice here that compounding the interest quarterly rather than annually yields the greater amount. This is in agreement with our observations in Example 1. ∎

   The results in Example 3 will serve to illustrate some additional terminology used by financial institutions. In that example, the interest for the year under quarterly compounding was

$$\$1103.81 - \$1000 = \$103.81$$

Now, $103.81 is 10.381% of $1000. We say in this case that the **effective rate** of interest is 10.381%. The given rate of 10% per annum compounded once a year is called the **nominal rate.*** The next example further illustrates these ideas.

EXAMPLE 4   A bank offers a nominal interest rate of 12% per annum for certain accounts. Compute the effective rate if interest is compounded monthly.

Solution   Let P denote the principal earning 12% ($r = 0.12$) compounded monthly ($n = 12$). Then with $t = 1$, our formula yields

$$A = P\left(1 + \frac{0.12}{12}\right)^{12(1)} = P(1.01)^{12}$$

Using a calculator to approximate the quantity $(1.01)^{12}$, we obtain

$$A \approx P(1.12683)$$

This shows that the effective interest rate is about 12.68%. ∎

*The nominal rate and the effective rate are also referred to as the *annual rate* and the *annual yield*, respectively.

There are two rather natural questions to ask when one first encounters compound interest calculations:

QUESTION 1 For a fixed period of time (say one year), does more and more frequent compounding of interest continue to yield greater and greater amounts?

QUESTION 2 Is there a limit on how much money can accumulate in a year when interest is compounded more and more frequently?

The answer to both of these questions is yes. If you look back over Example 1, you'll see evidence for the affirmative answer to question 1. For additional evidence, and for the answer to question 2, let's do some calculations. To keep things as simple as possible, suppose a principal of $1 is invested for 1 year at the nominal rate of 100% per annum. (More realistic figures could be used here, but the algebra becomes more cluttered.) With these data, our formula becomes

$$A = 1\left(1 + \frac{1}{n}\right)^{n(1)}$$

or

$$A = \left(1 + \frac{1}{n}\right)^{n}$$

Table 3 shows the results of compounding the interest more and more frequently.

TABLE 3

Number of Compoundings, n	Amount, $\left(1 + \dfrac{1}{n}\right)^{n}$
$n = 1$ (annually)	$\left(1 + \dfrac{1}{1}\right)^{1} = 2$
$n = 2$ (semiannually)	$\left(1 + \dfrac{1}{2}\right)^{2} = 2.25$
$n = 4$ (quarterly)	$\left(1 + \dfrac{1}{4}\right)^{4} \approx 2.44$
$n = 12$ (monthly)	$\left(1 + \dfrac{1}{12}\right)^{12} \approx 2.61$
$n = 365$ (daily)	$\left(1 + \dfrac{1}{365}\right)^{365} \approx 2.7146$
$n = 8760$ (hourly)	$\left(1 + \dfrac{1}{8760}\right)^{8760} \approx 2.7181$
$n = 525{,}600$ (each minute)	$\left(1 + \dfrac{1}{525{,}600}\right)^{525{,}600} \approx 2.71827$
$n = 31{,}536{,}000$ (each second)	$\left(1 + \dfrac{1}{31{,}536{,}000}\right)^{31{,}536{,}000} \approx 2.71828$

Table 3 shows that the amount does increase with the number of compoundings. But assuming that the bank rounds off to the nearest penny, Table 3 also shows that there is no difference between compounding hourly, compounding each minute, and compounding each second. In each case, the amount, when rounded off, is $2.72.

The data in Table 3 suggest that the quantity $[1 + (1/n)]^n$ gets closer and closer to the number e as n becomes larger and larger. In symbols,

$$\left(1 + \frac{1}{n}\right)^n \approx e \qquad \text{when } n \text{ is large}$$

Indeed, in many calculus books, the number e is defined as the *limiting value* or *limit* of the quantity $[1 + (1/n)]^n$ as n grows ever larger. Admittedly, we have not defined here the meaning of limiting value or limit. That is a topic for calculus. Nevertheless, Table 3 should give you a reasonable, if intuitive, appreciation of the idea.

Some banks advertise interest compounded not monthly, daily, or even hourly, but *continuously*—that is, at each instant. The formula for the amount earned under continuous compounding of interest is as follows.

**Compound Interest Formula
(Interest compounded continuously)**

Suppose that a principal of P dollars is invested at an annual rate r that is compounded continuously. Then the amount A after t years is given by

$A = Pe^{rt}$

EXAMPLE 5 A sum of $100 is placed in a savings account at 5% per annum compounded continuously. Assuming no subsequent withdrawals or deposits, when will the balance reach $150?

Solution   Substitute the values $A = 150$, $P = 100$, and $r = 0.05$ in the formula $A = Pe^{rt}$ to obtain

$$150 = 100e^{0.05t}$$

and, therefore,

$$1.5 = e^{0.05t}$$

To solve this last equation for t, we rewrite it in its equivalent logarithmic form:

$$0.05t = \ln 1.5$$
$$t = \frac{\ln 1.5}{0.05} \text{ years}$$

Using a calculator, we find that $t \approx 8.1$ years. In other words, it will take slightly more than 8 years 1 month for the balance to reach $150.

In the next example, we compare the nominal rate with the effective rate under continuous compounding of interest.

EXAMPLE 6 (a) Given a nominal rate of 8% per annum compounded continuously, compute the effective interest rate.

(b) Given an effective rate of 8% per annum, compute the nominal rate.

Solution (a) With the values $r = 0.08$ and $t = 1$, the formula $A = Pe^{rt}$ yields

$$A = Pe^{0.08(1)}$$

$$A \approx P(1.08329) \qquad \text{using a calculator}$$

This shows that the effective interest rate is approximately 8.33% per year.

(b) We now wish to compute the nominal rate r, given an effective rate of 8% per year. An effective rate of 8% means that the initial principal P grows to $P(1.08)$ by the end of the year. Thus, in the formula $A = Pe^{rt}$, we make the substitutions $A = P(1.08)$ and $t = 1$. This yields

$$P(1.08) = Pe^{r(1)}$$

Dividing both sides of this last equation by P, we have

$$1.08 = e^r$$

To solve this equation for r, we rewrite it in its equivalent logarithmic form:

$$r = \ln(1.08)$$

$$r \approx 0.07696 \qquad \text{using a calculator}$$

Thus, a nominal rate of about 7.70% per annum yields an effective rate of 8%. Table 4 summarizes these results. ■

TABLE 4

Comparison of Nominal and Effective Rates in Example 6

Nominal Rate (% per annum)	Effective Rate (% per annum)
8	8.33
7.70	8

Now we come to one of the remarkable and characteristic features of growth governed by the formula $A = Pe^{rt}$. By the **doubling time** we mean, as the name implies, the amount of time required for a given principal to double. The surprising fact here is that the doubling time does not depend on the principal P. To see why this is so, we begin with the formula

$$A = Pe^{rt}$$

We are interested in the time t at which $A = 2P$. Replacing A by $2P$ in the formula yields

$$2P = Pe^{rt}$$
$$2 = e^{rt}$$
$$rt = \ln 2$$
$$t = \frac{\ln 2}{r}$$

Denoting the doubling time by T_2, we have the following formula.

$$\text{Doubling time} = T_2 = \frac{\ln 2}{r}$$

As you can see, the formula for the doubling time T_2 does not involve P, but only r. Thus, at a given rate under continuous compounding, $2 and $2000 would both take the same amount of time to double. (This idea takes some getting used to.)

EXAMPLE 7 Compute the doubling time T_2 when a sum is invested at an interest rate of 4% per annum compounded continuously.

Solution $$T_2 = \frac{\ln 2}{r} = \frac{\ln 2}{0.04} \approx 17.3 \text{ years} \qquad \text{using a calculator}$$

There is a convenient approximation that allows us easily to estimate doubling times. Using a calculator, we see that

$$\ln 2 \approx 0.7$$

Using this approximation, we have the following rule of thumb for estimating doubling time.

$$T_2 \approx \frac{0.7}{r}$$

Let's use this rule to rework Example 7. With $r = 0.04$, we obtain

$$T_2 \approx \frac{0.7}{0.04} = \frac{70}{4} = 17.5 \text{ years}$$

Notice that this estimation is quite close to the actual doubling time obtained in Example 7.

The idea of doubling time is useful in graphing the function $A = Pe^{rt}$. In this discussion, we assume that P and r are constants, so that the amount A is a function of the time t. Suppose, for example, that a principal of $1000 is invested at 10% per annum compounded continuously. Then the function we wish to graph is

$$A = 1000e^{0.1t}$$

Now, the doubling time in this situation is

$$T_2 \approx \frac{0.7}{r} = \frac{0.7}{0.1} = 7 \text{ years}$$

TABLE 5

t (years)	A (dollars)
0	1000
7	2000
14	4000
21	8000
28	16000

Table 5 shows the results of doubling a principal of $1000 every 7 years.

We'll use the data in Table 5 to graph the function $A = 1000e^{0.1t}$. We'll mark off units on the t-axis in multiples of 7; on the A-axis, we'll use multiples of 2000. Figure 1 shows the result of plotting the points from Table 5 and then joining them with a smooth curve. Notice that the domain in this context is $[0, \infty)$.

You may have already noticed that aside from the arbitrary choice of letters, there is no difference between the functions $A = Pe^{rt}$ and $N = N_0 e^{kt}$. Both formulas have the same form. Thus, this single function serves as a model for phenomena as diverse as continuous compounding of interest, population growth, and radioactive decay.

In newspapers and in everyday speech, the term *exponential growth* is used rather loosely to describe any situation involving rapid growth. In the sciences, however, **exponential growth** refers specifically to growth governed by func-

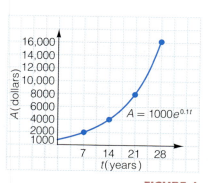

FIGURE 1
Exponential growth function

tions of the form $N = N_0 e^{kt}$ with $k > 0$. And since the function $A = Pe^{rt}$ has this form, we say that money grows exponentially under continuous compounding of interest. Similarly, in the sciences, **exponential decay** refers specifically to decay governed by functions of the form $N = N_0 e^{kt}$ with $k < 0$.

In the next example we predict the population of the world in the year 2000, assuming that the population grows exponentially. In this context, it can be shown that the constant k in the formula $N = N_0 e^{kt}$ represents the **relative growth rate** of the population N.* Algebraically, we will work with k just as we did in the calculations with compound interest, where k represented an interest rate.

EXAMPLE 8   Statistical projections from the United Nations indicate that the relative growth rate for the world's population during the years 1990–2000 will be approximately 1.6% per year. (This is down from the all-time high of 2% per year in 1970.) Use the formula $N = N_0 e^{kt}$ to predict the world population in the year 2000, given that the population in 1990 was 5.321 billion.

Solution   Let $t = 0$ correspond to the year 1990. Then the year 2000 corresponds to the value $t = 10$. Our given data therefore are $k = 0.016$, $t = 10$, and $N_0 = 5.321$ (in units of one billion). Using these values in the formula $N = N_0 e^{kt}$, we obtain

$$N = 5.321 e^{(0.016)10}$$
$$N \approx 6.244$$

Thus, our prediction for the world's population in the year 2000 is 6.244 billion.

EXAMPLE 9   The following statements about the increasing levels of carbon dioxide in the atmosphere appear in the book *The Challenge of Global Warming,* edited by Dean Edwin Abrahamson (Washington, D.C.: Island Press, 1989).

> Carbon dioxide, the single most important greenhouse gas, accounts for about half of the warming that has been experienced as a result of past emissions† and also for half of the projected future warming. The present [1988] concentration is now about 350 parts per million (ppm) and is increasing about 0.4% ... per year. ...

Assuming that the concentration of carbon dioxide in the atmosphere continues to increase exponentially at 0.4% per year, estimate when the concentration might reach 600 ppm. (This would be roughly twice the level estimated to exist prior to the Industrial Revolution.)

Solution   Let $t = 0$ correspond to the year 1988. In the formula $N = N_0 e^{kt}$, we want to determine the time t when $N = 600$. Using the values $N_0 = 350$, $N = 600$, and $k = 0.004$, we have

$$600 = 350 e^{0.004t} \qquad \text{or} \qquad \frac{600}{350} = e^{0.004t}$$

*The *relative growth rate* is $(\Delta N/N)/(\Delta t)$; it is the fractional change in N per unit time. This is distinct from (but sometimes confused with) the *average growth rate*, $\Delta N/\Delta t$.
†The two principal sources of these emissions are the burning of fossil fuels and the burning of vegetation in the tropics.

and, consequently,

$$\ln \tfrac{600}{350} = 0.004t$$

$$t = \frac{\ln \tfrac{600}{350}}{0.004} \approx 135 \qquad \text{using a calculator and rounding off}$$

Now, 135 years beyond our base year of 1988 is 2123. Rounding off once more, we summarize our result this way: If the carbon dioxide levels continue to increase exponentially at 0.4% per year, then the concentration will reach 600 ppm by approximately 2125.

In previous sections, we used the function $N = N_0 e^{kt}$ with $k < 0$ to describe radioactive decay. Let's return to that idea now. Just as we used the idea of a doubling time to graph an exponential growth function, we can use the half-life concept, introduced in Section 6.2, to graph an exponential decay function. Consider, for example, the radioactive element iodine-131, which has a half-life of about 8 days. Table 6 shows what fraction of an initial amount remains at 8-day intervals. Using the data in this table, we can draw the graph for the decay function $N = N_0 e^{kt}$ for iodine-131 (see Figure 2). Note that we are able to construct this graph without specifically evaluating the decay constant k.

TABLE 6

t (days)	N (amount)
0	N_0
8	$\tfrac{1}{2}N_0$
16	$\tfrac{1}{4}N_0$
24	$\tfrac{1}{8}N_0$
32	$\tfrac{1}{16}N_0$

FIGURE 2

An exponential decay function

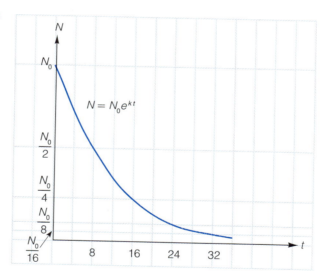

The next example deals with the subject of nuclear energy. The purpose of the example is not to present an argument for or against the use of nuclear power. Rather, the purpose is to show you that with an understanding of exponential decay, you will be better equipped to read about and to evaluate the issues.

EXAMPLE 10

An article on nuclear energy appeared in the January 1976 issue of *Scientific American*. The author of the article is physicist Hans Bethe, a Nobel Prize winner. At one point in the article, Professor Bethe discusses the disposal (through burial) of radioactive waste material from a nuclear reactor. The particular waste product under discussion is plutonium-239.

... Plutonium-239 has a half-life of nearly 25,000 years, and 10 half-lives are required to cut the radioactivity by a factor of 1000. Thus, the buried wastes must be kept out of the biosphere for 250,000 years.

(a) Supply the detailed calculations to support the statement that 10 half-lives are required before the radioactivity is reduced by a factor of 1000.
(b) Show how the figure of 10 half-lives can be obtained by estimation, as opposed to detailed calculation.

Solution **(a)** Let $\mathcal{N}_0$ denote the initial amount of plutonium-239 at time $t = 0$. Then the amount $\mathcal{N}$ present at time t is given by $\mathcal{N} = \mathcal{N}_0 e^{kt}$. We wish to determine t when $\mathcal{N} = \frac{1}{1000}\mathcal{N}_0$. First, we determine the value of k by using the half-life information. To say that the half-life is 25,000 years is to say that $\mathcal{N} = \frac{1}{2}\mathcal{N}_0$ when $t = 25,000$. Substituting these values into the formula $\mathcal{N} = \mathcal{N}_0 e^{kt}$ yields

$$\frac{1}{2}\mathcal{N}_0 = \mathcal{N}_0 e^{k(25,000)}$$
$$\frac{1}{2} = e^{25,000k} \qquad \textcolor{red}{\text{dividing by } \mathcal{N}_0}$$

Writing this last equation in its equivalent logarithmic form, we obtain

$$25,000k = \ln \tfrac{1}{2} \qquad \text{and therefore} \qquad k = \frac{\ln \tfrac{1}{2}}{25,000}$$

Now we can find t when $\mathcal{N} = \frac{1}{1000}\mathcal{N}_0$, as required. Substituting $\frac{1}{1000}\mathcal{N}_0$ for $\mathcal{N}$ in the decay law gives us

$$\frac{1}{1000}\mathcal{N}_0 = \mathcal{N}_0 e^{kt} \qquad \left(\text{where } k \text{ is} \frac{\ln \tfrac{1}{2}}{25,000}\right)$$

$$\frac{1}{1000} = e^{kt}$$

$$kt = \ln \frac{1}{1000}$$

$$t = \frac{\ln \tfrac{1}{1000}}{k} = \frac{\ln \tfrac{1}{1000}}{(\ln \tfrac{1}{2})/25,000} = \frac{\left(\ln \tfrac{1}{1000}\right)(25,000)}{\ln \tfrac{1}{2}}$$

Using a calculator, we obtain the value 249,144.6 years; however, given the time scale involved, it would be ludicrous to announce the answer in this form. Instead, we round off the answer to the nearest thousand years and say that after 249,000 years, the radioactivity will have decreased by a factor of 1000. Notice that this result confirms Professor Bethe's ballpark estimate of 10 half-lives, or 250,000 years.

(b) After 1 half-life: $\qquad \mathcal{N} = \dfrac{\mathcal{N}_0}{2}$

After 2 half lives: $\qquad \mathcal{N} = \dfrac{1}{2}\left(\dfrac{\mathcal{N}_0}{2}\right) = \dfrac{\mathcal{N}_0}{2^2}$

After 3 half-lives: $\qquad \mathcal{N} = \dfrac{1}{2}\left(\dfrac{\mathcal{N}_0}{2^2}\right) = \dfrac{\mathcal{N}_0}{2^3}$

Following this pattern, we see that after 10 half-lives, we should have

$$\mathcal{N} = \frac{\mathcal{N}_0}{2^{10}}$$

But as we noted in the first section of this chapter, 2^{10} is approximately 1000. Therefore, we have

$$\mathcal{N} \approx \frac{\mathcal{N}_0}{1000} \qquad \text{after 10 half-lives}$$

This is in agreement with Professor Bethe's statement.

EXERCISE SET 6.5

1. You invest $800 at 6% interest compounded annually. How much is in the account after 4 years, assuming that you make no subsequent withdrawal or deposit?

2. A sum of $1000 is invested at an interest rate of $5\frac{1}{2}\%$ compounded annually. How many years will it take before the sum exceeds $2500? (First find out when the amount equals $2500; then round off as in Example 2.)

3. At what interest rate (compounded annually) will a sum of $4000 grow to $6000 in 5 years?

4. A bank pays 7% interest compounded annually. What principal will grow to $10,000 in 10 years?

5. You place $500 in a savings account at 5% compounded annually. After 4 years you withdraw all your money and take it to a different bank, which advertises a rate of 6% compounded annually. What is the balance in this new account after 4 more years? (As usual, assume that no subsequent withdrawal or deposit is made.)

6. A sum of $3000 is placed in a savings account at 6% per annum. How much is in the account after 1 year if the interest is compounded (a) annually? (b) semiannually? (c) daily?

7. A sum of $1000 is placed in a savings account at 7% per annum. How much is in the account after 20 years if the interest is compounded (a) annually? (b) quarterly?

8. Your friend invests $2000 at $5\frac{1}{4}\%$ per annum compounded semiannually. You invest an equal amount at the same yearly rate, but compounded daily. How much larger is your account than your friend's after 8 years?

9. You invest $100 at 6% per annum compounded quarterly. How long will it take for your balance to exceed $120? (Round off your answer to the next quarter.)

10. A bank offers an interest rate of 7% per annum compounded daily. What is the effective rate?

11. What principal should you deposit at $5\frac{1}{2}\%$ per annum compounded semiannually so as to have $6000 after 10 years?

12. You place a sum of $800 in a savings account at 6% per annum compounded continuously. Assuming that you make no subsequent withdrawal or deposit, how much is in the account after 1 year? When will the balance reach $1000?

13. A bank offers an interest rate of $6\frac{1}{2}\%$ per annum compounded continuously. What principal will grow to $5000 in 10 years under these conditions?

14. Given a nominal rate of 6% per annum, compute the effective rate under continuous compounding of interest.

15. Suppose that under continuous compounding of interest, the effective rate is 6% per annum. Compute the nominal rate.

16. You have two savings accounts, each with an initial principal of $1000. The nominal rate on both accounts is $5\frac{1}{4}\%$ per annum. In the first account, interest is compounded semiannually. In the second account, interest is compounded continuously. How much more is in the second account after 12 years?

17. You want to invest $10,000 for 5 years, and you have a choice between two accounts. The first pays 6% per annum compounded annually. The second pays 5%

per annum compounded continuously. Which is the better investment?

18. Suppose that a certain principal is invested at 6% per annum compounded continuously.
(a) Use the rule $T_2 \approx 0.7/r$ to estimate the doubling time.
(b) Compute the doubling time using the formula $T_2 = (\ln 2)/r$.
(c) Do your answers in (a) and (b) differ by more than 2 months?

19. A sum of $1500 is invested at 5% per annum compounded continuously.
(a) Estimate the doubling time.
(b) Compute the actual doubling time.
(c) Let d_1 and d_2 denote the actual and estimated doubling times, respectively. Define d by

$$d = |d_1 - d_2|$$

What percent is d of the actual doubling time?

20. A sum of $5000 is invested at 10% per annum compounded continuously.
(a) Estimate the doubling time.
(b) Estimate the time required for the $5000 to grow to $40,000.

21. After carrying out the calculations in this problem, you'll see one of the reasons why the government imposes inheritance taxes and why laws are passed to prohibit savings accounts from being passed from generation to generation without restriction. Suppose that a family invests $1000 at 8% per annum compounded continuously. If this account were to remain intact, being passed from generation to

generation, for 300 years, how much would be in the account at the end of those 300 years?

22. A principal of $500 is invested at 7% per annum compounded continuously.
(a) Estimate the doubling time.
(b) Sketch a graph similar to the one in Figure 1, showing how the amount increases with time.

23. A principal of $7000 is invested at 5% per annum compounded continuously.
(a) Estimate the doubling time.
(b) Sketch a graph showing how the amount increases with time.

24. In one savings account, a principal of $1000 is deposited at 5% per annum. In a second account, a principal of $500 is deposited at 10% annum. Both accounts compound interest continuously.
(a) Estimate the doubling time for each account.
(b) On the same set of axes, sketch graphs showing the amount of money in each account over time. Give the (approximate) coordinates of the point where the two curves meet. In financial terms, what is the significance of this point? (In working this problem, assume that the initial deposits in each account were made at the same time.)
(c) During what period of time does the first account have the larger balance?

The statistics in Exercises 25–27 are taken from the 1990 World Population Data Sheet (Washington, D.C.: Population Reference Bureau, Inc., 1991). In Exercises 25, 26, and 27(a), complete the tables, assuming that the populations grow exponentially and that the indicated relative growth rates are valid through the year 2000.

25.

Region	1990 Population (billions)	Percent of Population in 1990	Relative Growth Rate (percent per year)	Year 2000 Population (billions)	Percent of World Population in 2000
World	5.321	100	1.8	?	?
More developed regions	1.214	?	0.5	?	?
Less developed regions	4.107	?	2.1	?	?

26.

Country	1990 Population (millions)	Relative Growth Rate (percent per year)	Year 2000 Population (millions)	Percent Increase in Population
United States	251.4	0.8	?	?
People's Republic of China	1119.9	1.4	?	?
Mexico	88.6	2.4	?	?

27. (a)

Country	1990 Population (millions)	Relative Growth Rate (percent per year)	Year 2000 Population (millions)	Percent Increase in Population
Iraq	18.8	3.9	?	?
United Kingdom	57.4	0.2	?	?

(b) Iraq has one of the highest population growth rates in the world, and United Kingdom has one of the lowest. Use the data given in part (a) to compute an estimate for the year in which the two populations could be equal. (Round off your final answer to the nearest five years.) *Hint:* You need to solve the equation $(18.8)e^{0.039t} = (57.4)e^{0.002t}$.

28. According to the United States Bureau of the Census, the population of the United States grew most rapidly during the period 1800–1810 and least rapidly during the period 1930–1940. Use the following data to compute the relative growth rate (percent per year) for each of these two periods. In 1800, the population was 5,308,483; in 1810, it was 7,239,881. In 1930, the population was 123,202,624; in 1940, it was 132,164,569.

29. According to the U.S. Bureau of the Census, the population of the United States in the year 1850 was 23,191,876; in 1900, the population was 62,947,714.
(a) Assuming that the population grew exponentially during this period, compute the growth constant k.
(b) Assuming continued growth at the same rate, predict the 1950 population.
(c) The actual population for 1950, according to the Bureau of the Census, was 150,697,361. How does this compare to your prediction in part (b)? Was the actual growth over the period 1900–1950 faster or slower than exponential growth with the growth constant k determined in part (a)?

In Exercises 30–32, use the population statistics given in the following table, provided by the U.S. Bureau of the Census.

State	1930	1940	1950
California	5,677,251	6,907,387	10,586,223
New York	12,588,066	13,479,142	14,830,192
North Dakota	680,845	641,935	619,636

30. (a) Assume that the population of California grew exponentially over the period 1930–1940. Compute the growth constant k and express it as a percent (per year).
(b) Assume that the population of California grew exponentially over the period 1940–1950. Compute the growth constant k and express it as a percent.

(c) What would the 1990 California population have been if the relative growth rate obtained in part (b) had remained in effect throughout the period 1950–1990? *Hint:* Let $t = 0$ correspond to 1950. Then find $\mathcal{N}$ when $t = 40$.
(d) How does your prediction in part (c) compare to the actual 1990 population of 29,279,000?

31. Repeat Exercise 30 for New York: *Note:* The 1990 population of New York was 17,627,000.

32. Repeat Exercise 30 for North Dakota. *Note:* The 1990 population of North Dakota was 634,000.

33. (a) The half-life of radium-226 is 1620 years. Draw a graph of the decay function for radium-226, similar to that shown in Figure 2.
(b) The half-life of radium-A is 3 min. Draw a graph of the decay function for radium-A.

34. (a) The half-life of thorium-232 is 1.4×10^{10} years. Draw a graph of the decay function.
(b) The half-life of thorium-A is 0.16 sec. Draw a graph of the decay function.

35. The half-life of plutonium-241 is 13 years.
(a) Compute the decay constant k.
(b) What fraction of an initial sample $\mathcal{N}_0$ would remain after 10 years? after 100 years?

36. The half-life of thorium-229 is 7340 years.
(a) Compute the time required for a given sample to be reduced by a factor of 1000. Show detailed calculations, as in Example 10(a).
(b) Express your answer in part (a) in terms of half-lives.
(c) As in Example 10(b), estimate the time required for a given sample of thorium-229 to be reduced by a factor of 1000. Compare your answer with that obtained in part (b).

37. Strontium-90, with a half-life of 28 years, is one of the waste products from nuclear fission reactors. One of the reasons great care is taken in the storage and disposal of this substance stems from the fact that strontium-90 is, in some chemical respects, similar to ordinary calcium. Thus, strontium-90 in the biosphere, entering the food chain via plants or animals, would eventually be absorbed into our bones. (In fact, everyone already has a measurable amount of strontium-90 in their bones as a result of fallout from atmospheric nuclear tests.)

(a) Compute the decay constant k for strontium-90.
(b) Compute the time required if a given quantity of strontium-90 is to be stored until the radioactivity is reduced by a factor of 1000.
(c) Using half-lives, estimate the time required for a given sample to be reduced by a factor of 1000. Compare your answer with that obtained in (b).

38. (a) Suppose that a certain country violates the ban against above-ground nuclear testing and, as a result, an island is contaminated with debris containing the radioactive substance iodine-131. A team of scientists from the United Nations wants to visit the island to look for clues in determining which country was involved. However, the level of radioactivity from the iodine-131 is estimated to be 30,000 times the safe level. Approximately how long must the team wait before it is safe to visit the island? The half-life of iodine-131 is 8 days.
(b) Rework part (a), assuming instead that the radioactive substance is strontium-90 rather than iodine-131. The half-life of strontium-90 is 28 years. Assume, as before, that the initial level of radioactivity is 30,000 times the safe level. (This exercise underscores the difference between a half-life of 8 days and one of 28 years.)

The statistics presented in Exercises 39(b)–42 are taken from the Global 2000 Report to the President. The following statement by then President Jimmy Carter provides some background on the report.

I am directing the Council on Environmental Quality and the Department of State, working in cooperation with ... other appropriate agencies, to make a one-year study of the probable changes in the world's population, natural resources, and environment through the end of the century.

President Jimmy Carter, May 23, 1977

39. In 1969 the United States National Academy of Sciences issued a report entitled *Resources and Man.* One conclusion in the report is that a world population of 10 billion "is close to (if not above) the maximum that an intensively managed world might hope to support with some degree of comfort and individual choice." (The figure "10 billion" is sometimes referred to as the *carrying capacity* of the Earth.)
(a) When the report was issued in 1969, the world population was about 3.6 billion, with a relative growth rate of 2% per year. Assuming continued exponential growth at this rate, estimate the year in which the Earth's carrying capacity of 10 billion might be reached.
(b) The data in the following table, from the *Global*

2000 Report, show the world population estimates for 1975. Use the "low" figures to estimate the year in which the Earth's carrying capacity of 10 billion might be reached.
(c) Use the "high" figures to estimate the year in which the Earth's carrying capacity of 10 billion might be reached.

Projection	Population (billions)	Relative Growth Rate (percent per year)
Low	4.043	1.5
High	4.134	2.0

40. *Depletion of Nonrenewable Energy Resources.* Suppose that the world population grows exponentially. Then, as a first approximation, it is reasonable to assume that the use of nonrenewable energy resources, such as petroleum and coal, also grows exponentially. Under these conditions, the following formula can be derived (using calculus):

$$A = \frac{A_0}{k}(e^{kT} - 1)$$

where A is the amount of the resource consumed from time $t = 0$ to $t = T$, the quantity A_0 is the amount of the resource consumed during the year $t = 0$, and k is the relative growth rate of annual consumption.
(a) Show that solving the formula for T yields

$$T = \frac{\ln[(Ak/A_0) + 1]}{k}$$

This formula gives the "life expectancy" T for a given resource. In the formula, A_0 and k are as previously defined, and A represents the total amount of the resource available.
(b) In the *Global 2000 Report to the President,* it is estimated that the 1976 worldwide consumption of oil was 21.7 billion barrels and that the total remaining oil resources ultimately available were 1661 billion barrels. Compute the "life expectancy" for oil if (i) the relative growth rate k is 1%/year; (ii) 2%/year; (iii) 3%/year. (This last value is, in fact, the predicted rate for the next decade.)
(c) In the *Global 2000 Report,* it is estimated that the 1976 worldwide consumption of natural gas was approximately 50 tcf (trillions of cubic feet), while the total remaining gas ultimately available was 8493 tcf. Compute the "life expectancy" for natural gas if $k = 2\%$/year.

41. In this exercise, the term *nonrenewable energy resources* refers collectively to petroleum, natural gas, coal, shale oil, and uranium. In the *Global 2000 Report,* it was

estimated that in 1976, the total worldwide consumption of these nonrenewable resources amounted to 250 quadrillion Btu. It was also estimated that in 1976, the total nonrenewable resources ultimately available were 161,241 quadrillion Btu. Use the formula for T in Exercise 40 to compute the life expectancy for these nonrenewable energy resources if
(a) $k = 1\%$ per year;   (b) $k = 2\%$ per year.

42.   The data in the following table, from the *Global 2000 Report,* show the world reserves for selected minerals. The term *world reserves* refers to known resources available with current technology. Use the formula for T derived in Exercise 40 to estimate the "depletion date" for each mineral under these conditions (depletion date = $T + 1976$).

Mineral	1976 World Reserves	1976 Consumption	Relative Growth Rate (percent per year)
Fluorine (million short tons)	37	2.1	4.58
Silver (million troy ounces)	6,100	305	2.33
Tin (thousand metric tons)	10,000	241	2.05
Copper (million short tons)	503	8.0	2.94
Phosphate rock (million metric tons)	25,732	107	5.17
Aluminum in bauxite (million short tons)	5,610	18	4.29

43.   In 1956, scientists B. Gutenberg and C. F. Richter developed a formula to estimate the amount of energy E released in an earthquake. The formula is

$$\log_{10} E = 11.4 + 1.5M$$

where E is the energy in units of ergs and M is the Richter magnitude.
(a) Solve the formula for E.
(b) The table shows the Richter magnitudes for two large earthquakes. If E_1 denotes the energy of the Alaskan quake and E_2 denotes the energy of the quake in San Francisco, compute the ratio E_1/E_2 to compare the energies of the two earthquakes. Round off your answer to the nearest integer.

Date	Location	M (Magnitude)
March 28, 1964	Alaska	8.6
October 17, 1989	San Francisco	7.1

44.   The age of some rocks can be estimated by measuring the ratio of the amounts of certain chemical elements within the rock. The method known as the *rubidium-strontium method* will be discussed here. This method has been used in dating the moon rocks brought back on the Apollo missions.

Rubidium-87 is a radioactive substance with a half-life of 4.7×10^{10} years. Rubidium-87 decays into the substance strontium-87, which is stable (nonradioactive). We are going to derive the following formula for the age of a rock:

$$T = \frac{\ln(N_s/N_r + 1)}{-k}$$

where T is the age of the rock, k is the decay constant for rubidium-87, N_s is the number of atoms of strontium-87 now present in the rock, and N_r is the number of atoms of rubidium-87 now present in the rock.
(a) Assume that initially, when the rock was formed, there were N_0 atoms of rubidium-87 and none of strontium-87. Then as time goes by, some of the rubidium atoms decay into strontium atoms, but the total number of atoms must still be N_0. Thus, after T years, we have

$$N_0 = N_r + N_s$$

or, equivalently,

$$N_s = N_0 - N_r \qquad (1)$$

However, according to the law of exponential decay for the rubidium-87, we must have $N_r = N_0 e^{kT}$. Solve this equation for N_0 and then use the result to eliminate N_0 from equation (1). Show that the result can be written

$$N_s = N_r e^{-kT} - N_r \qquad (2)$$

(b) Solve equation (2) for T to obtain the formula given at the beginning of this exercise.

45.   (Continuation of Exercise 44)   The half-life of rubidium-87 is 4.7×10^{10} years. Compute the decay constant k.

46.   (Continuation of Exercise 44)   Analysis of lunar rock samples taken on the Apollo 11 mission showed the strontium–rubidium ratio to be

$$\frac{N_s}{N_r} = 0.0588$$

Estimate the age of these lunar rocks.

47. (Continuation of Exercise 44)   Analysis of the so-called genesis rock sample taken on the Apollo 15 mission revealed a strontium–rubidium ratio of 0.0636. Estimate the age of this rock.

48. *Radiocarbon Dating.*   Because rubidium-87 decays so slowly, the technique of rubidium–strontium dating is generally considered effective only for objects older than 10 million years. By way of contrast, archeologists and geologists rely on the method of *radiocarbon dating* in assigning ages ranging from 500 to 50,000 years.

   Two types of carbon occur naturally in our environment—carbon-12, which is nonradioactive, and carbon-14, which has a half-life of 5730 years. All living plant and animal tissue contains both types of carbon, always in the same ratio. (The ratio is one part carbon-14 to 10^{12} parts carbon-12.) As long as the plant or animal is living, this ratio is maintained. When the organism dies, however, no new carbon-14 is absorbed, and the amount of carbon-14 begins to decrease exponentially. Since the amount of carbon-14 decreases exponentially, it follows that the level of radioactivity also must decrease exponentially. The formula describing this situation is

$$\mathcal{N} = \mathcal{N}_0 e^{kT}$$

where T is the age of the sample, $\mathcal{N}$ is the present level of radioactivity (in units of disintegrations per hour per gram of carbon), and $\mathcal{N}_0$ is the level of radioactivity T years ago, when the organism was alive. Given that the half-life of carbon-14 is 5730 years and that $\mathcal{N}_0 = 920$ disintegrations per hour per gram, show that the age T of a sample is given by

$$T = \frac{5730 \ln (\mathcal{N}/920)}{\ln \frac{1}{2}}$$

In Exercises 49–52, use the formula derived in Exercise 48 to estimate the age of each sample. Note: *Some technical complications arise in interpreting such results. Studies have shown that the ratio of carbon-12 to carbon-14 in the air, and*

therefore in living matter, has not in fact been constant over time. For instance, air pollution from factory smokestacks tends to increase the level of carbon-12. In the other direction, nuclear bomb testing increases the level of carbon-14.

49. Prehistoric cave paintings were discovered in the Lascaux cave in France. Charcoal from the site was analyzed and the level of radioactivity was found to be $\mathcal{N} = 141$ disintegrations per hour per gram. Estimate the age of the paintings.

50. Before radiocarbon dating was used, historians estimated that the tomb of Vizier Hemaka, in Egypt, was constructed about 4900 years ago. After radiocarbon dating became available, wood samples from the tomb were analyzed, and it was determined that the radioactivity level was 510 disintegrations per hour per gram. Estimate the age of the tomb on the basis of this reading and compare your answer to the figure already mentioned.

51. Analyses of the oldest campsites in the Western Hemisphere reveal a carbon-14 radioactivity level of $\mathcal{N} = 226$ disintegrations per hour per gram. Show that this implies an age of 11,500 years, to the nearest 500 years. (This would correspond to the last Ice Age. At that time, the sea level was significantly lower than today. It is believed that at the time of the last Ice Age, land stretched across what is now the Bering Strait, and humans first entered the Western Hemisphere by this route.)

52. The Dead Sea Scrolls are a collection of ancient manuscripts discovered in caves along the west bank of the Dead Sea. (The discovery occurred by accident when an Arab herdsman of the Taamireh tribe was searching for a stray goat.) When the linen wrappings on the scrolls were analyzed, the carbon-14 radioactivity level was found to be 723 disintegrations per hour per gram. Estimate the age of the scrolls using this information. Historical evidence suggests that some of the scrolls date back somewhere between 150 B.C. and A.D. 40. How do these dates compare with the estimate derived using radiocarbon dating?

CHAPTER SIX   SUMMARY OF PRINCIPAL TERMS AND FORMULAS

TERMS OR FORMULAS	PAGE REFERENCE	COMMENTS
1. Exponential function with base b	309	The exponential function with base b is defined by the equation $y = b^x$. It is understood here that the base b is a positive number other than 1.
2. Asymptote	309	A line is an asymptote for a curve if the separation distance between the curve and the line approaches zero as we move out farther and farther along the line.

TERMS OR FORMULAS	PAGE REFERENCE	COMMENTS
3. The number e	314	The irrational number e is one of the basic constants in mathematics, as is the irrational number π. To five decimal places, the value of e is 2.71828. In calculus e is the base most commonly used for exponential functions. The graph of the exponential function $y = e^x$ is shown in the box on page 316.
4. $\mathcal{N} = \mathcal{N}_0 e^{kt}$ $(k > 0)$	317, 349	In many instances throughout this chapter, we studied situations in which a quantity $\mathcal{N}$ grows or increases according to this formula, called the *growth law*. In this formula, the quantity $\mathcal{N}$ is a function of the time t, and $\mathcal{N}_0$ denotes the value of $\mathcal{N}$ at time $t = 0$. The number k is a constant, called the *growth constant*, that is related to the rate at which $\mathcal{N}$ increases.
5. $\mathcal{N} = \mathcal{N}_0 e^{kt}$ $(k < 0)$	318, 350	This is called the *decay law* for radioactive substances. In the formula, $\mathcal{N}$ denotes the amount of the substance present at time t, and $\mathcal{N}_0$ is the amount of the substance present at time $t = 0$. The negative number k is a constant, called the *decay constant*, that is related to the rate at which the decay takes place.
6. Half-life	319	The half-life of a radioactive substance is the time required for half of a given sample to disintegrate. Notice that the half-life is an amount of time, not an amount of the substance.
7. $\log_b x$	323	The expression $\log_b x$ denotes the exponent to which b must be raised to yield x. The equation $\log_b x = y$ is equivalent to $b^y = x$.
8. $\ln x$	327	The expression $\ln x$ means $\log_e x$. Logarithms to the base e are known as *natural logarithms*. For the graph of $y = \ln x$, see Figure 5 on page 327.
9. $\log_a x = \dfrac{\log_b x}{\log_b a}$	337	This is the change of base formula for converting logarithms from one base to another.
10. $A = P\left(1 + \dfrac{r}{n}\right)^{nt}$	344	This formula gives the amount A that accumulates after t years when a principal of P dollars is invested at an annual rate r that is compounded n times per year.
11. $A = Pe^{rt}$	346	This formula gives the amount A that accumulates after t years when a principal of P dollars is invested at an annual rate r that is compounded continuously.

WRITING MATHEMATICS

1. In Section 6.3, we defined logarithmic functions in terms of inverse functions.
   (a) Explain in general what is meant by a pair of inverse functions.
   (b) Assuming a knowledge of the function $f(x) = 2^x$, explain how to define and then graph the function $g(x) = \log_2 x$.

2. A student who was trying to simplify the expression $\dfrac{\ln(e^3)}{\ln(e^2)}$ wrote

$$\frac{\ln(e^3)}{\ln(e^2)} = \ln(e^3) - \ln(e^2) = 3 - 2 = 1$$

(a) What is the error here? What do you think is the most probable reason for making this error?

(b) Give the correct solution.

3. A student who wanted to simplify the expression $\dfrac{100}{\log_{10} 100}$ wrote

$$\frac{100}{\log_{10} 100} = \frac{\overset{1}{\cancel{100}}}{\log_{10} \underset{1}{\cancel{100}}}$$

$$= \frac{1}{\log_{10}}$$

Explain why this is nonsense, and then indicate the correct solution.

CHAPTER SIX   REVIEW EXERCISES

1. Which is larger, $\log_5 126$ or $\log_{10} 999$?

2. Graph the function $y = 3^{-x} - 3$. Specify the domain, range, intercept(s), and asymptote.

3. Suppose that the population of a colony of bacteria increases exponentially. If the population at the start of an experiment is 8000, and 4 hours later it is 10,000, how long (from the start of the experiment) will it take for the population to reach 12,000? (Express the answer in terms of base e logarithms.)

4. Express $\log_{10} 2$ in terms of base e logarithms.

5. Let f be the function defined by

$$f(x) = \begin{cases} 2^{-x} & \text{if } x < 0 \\ x^2 & \text{if } x \geq 0 \end{cases}$$

Sketch the graph of f and then use the horizontal line test to determine whether f is one-to-one.

6. Estimate 2^{60} in terms of an integral power of 10.

7. Solve for x: $\ln(x + 1) - 1 = \ln(x - 1)$

8. Suppose that $5000 is invested at 8% interest compounded annually. How many years will it take for the money to double? Make use of the approximations $\ln 2 \approx 0.7$ and $\ln 1.08 \approx 0.08$ to obtain a numerical answer.

9. On the same set of axes, sketch the graphs of $y = e^x$ and $y = \ln x$. Specify the domain and range for each function.

10. Solve for x: $xe^x - 2e^x = 0$.

11. Simplify: $\log_9\left(\frac{1}{27}\right)$.

12. Given that $\ln A = a$, $\ln B = b$, and $\ln C = c$, express $\ln\left[(A^2\sqrt{B})/C^3\right]$ in terms of a, b, and c.

13. The half-life of plutonium-241 is 13 years. What is the decay constant? Use the approximation $\ln \frac{1}{2} \approx -0.7$ to obtain a numerical answer.

14. Express as a single logarithm with a coefficient of 1: $3 \log_{10} x - \log_{10}(1 - x)$

15. Solve for x, leaving your answer in terms of base e logarithms: $5e^{2-x} = 12$

16. Let $f(x) = e^{x+1}$. Find a formula for $f^{-1}(x)$ and specify the domain of f^{-1}.

17. Suppose that in 1980 the population of a certain country was 2 million and increasing with a relative growth rate of 2% per year. Estimate the year in which the population will reach 3 million. (Use the approximation $\ln 1.5 \approx 0.40$ to obtain a numerical answer.)

18. Simplify: $\ln e + \ln\sqrt{e} + \ln 1 + \ln e^{\ln 10}$.

19. A principal of $1000 is deposited at 10% per annum compounded continuously. Estimate the doubling time and then sketch a graph showing how the amount increases with time.

20. Simplify: $\ln(\log_8 56 - \log_8 7)$

In Exercises 21–32, graph the function and specify the asymptote(s) and intercept(s).

21. $y = e^x$

22. $y = -e^{-x}$

23. $y = \ln x$

24. $y = \ln(x + 2)$

25. $y = 2^{x+1} + 1$

26. $y = \log_{10}(-x)$

27. $y = (1/e)^x$

28. $y = \left(\frac{1}{2}\right)^{-x}$

29. $y = e^{x+1} + 1$

30. $y = -\log_2(x + 1)$

31. $y = \ln(e^x)$

32. $y = e^{\ln x}$

In Exercises 33–49, solve the equation for x. (When logarithms appear in your answer, leave the answer in that form, rather than using a calculator.)

33. $\log_4 x + \log_4(x - 3) = 1$

34. $\log_3 x + \log_3(2x + 5) = 1$

35. $\ln x + \ln(x + 2) = \ln 15$

36. $\log_6 \dfrac{x + 4}{x - 1} = 1$

37. $\log_2 x + \log_2(3x + 10) - 3 = 0$

38. $2 \ln x - 1 = 0$ 39. $3 \log_9 x = \frac{1}{2}$

40. $e^{2x} = 6$ 41. $e^{1 - 5x} = 3\sqrt{e}$

42. $2^x = 100$

43. $\log_{10} x - 2 = \log_{10}(x - 2)$

44. $\log_{10}(x^2 - x - 10) = 1$

45. $\ln(x + 2) = \ln x + \ln 2$

46. $\ln(2x) = \ln 2 + \ln x$ 47. $\ln x^4 = 4 \ln x$

48. $(\ln x)^4 = 4 \ln x$ 49. $\log_{10} x = \ln x$

50. Solve for x: $(\ln x)/(\ln 3) = \ln x - \ln 3$. Use a calculator to evaluate your result and round off to the nearest integer.   *Answer:* 206765

In Exercises 51–66, simplify the expression without using a calculator.

51. $\log_{10}\sqrt{10}$ 52. $\log_7 1$

53. $\ln \sqrt[5]{e}$ 54. $\log_3 54 - \log_3 2$

55. $\log_{10} \pi - \log_{10} 10\pi$ 56. $\log_2 2$

57. 10^t, where $t = \log_{10} 16$ 58. $e^{\ln 5}$

59. $\ln(e^4)$ 60. $\log_{10}(10^{\sqrt{2}})$

61. $\log_{12} 2 + \log_{12} 18 + \log_{12} 4$

62. $(\log_{10} 8)/(\log_{10} 2)$ 63. $(\ln 100)/(\ln 10)$

64. $\log_5 2 + \log_{1/5} 2$ 65. $\log_2 \sqrt[7]{16\sqrt[3]{2\sqrt{2}}}$

66. $\log_{1/8}\left(\frac{1}{16}\right) + \log_5 0.02$

In Exercises 67–70, express the quantity in terms of a, b, and c, where $a = \log_{10} A$, $b = \log_{10} B$, and $c = \log_{10} C$.

67. $\log_{10} A^2 B^3 \sqrt{C}$ 68. $\log_{10} \sqrt[3]{AC/B}$

69. $16 \log_{10} \sqrt{A}\sqrt[4]{B}$ 70. $6 \log_{10}\left[B^{1/3}/(A\sqrt{C})\right]$

In Exercises 71–76 find consecutive integers n and n + 1 such that the given expression lies between n and n + 1. Do not use a calculator.

71. $\log_{10} 209$ 72. $\ln 2$ 73. $\log_6 100$

74. $\log_{10}\left(\frac{1}{12}\right)$ 75. $\log_{10} 0.003$ 76. $\log_3 244$

77. (a) On the same set of axes, graph the curves $y = \ln(x + 2)$ and $y = \ln(-x) - 1$. According to

your graph, in which quadrant do the two curves intersect?

(b) Find the x-coordinate of the intersection point.

78. The curve $y = ae^{bt}$ passes through the point $(3, 4)$ and has a y-intercept of 2. Find a and b.

79. A certain radioactive substance has a half-life of T years. Find the decay constant (in terms of T).

80. Find the half-life of a certain radioactive substance if it takes T years for one-third of a given sample to disintegrate. (Your answer will be in terms of T.)

81. A radioactive substance has a half-life of M min. What percentage of a given sample will remain after $4M$ min?

82. At the start of an experiment, the initial population is a bacteria in a colony. After b hours, c bacteria are present. Determine the population d hours after the start of the experiment.

83. The half-life of a radioactive substance is d days. If you begin with a sample weighing b g, how long will it be until c g remain?

84. Find the half-life of a radioactive substance if it takes D days for P percent of a given sample to disintegrate.

In Exercises 85–90, write the expression as a single logarithm with a coefficient of 1.

85. $\log_{10} 8 + \log_{10} 3 - \log_{10} 12$

86. $4 \log_{10} x - 2 \log_{10} y$

87. $\ln 5 - 3 \ln 2 + \ln 16$

88. $\ln(x^4 - 1) - \ln(x^2 - 1)$

89. $a \ln x + b \ln y$

90. $\ln(x^3 + 8) - \ln(x + 2) - 2 \ln(x^2 - 2x + 4)$

In Exercises 91–98, write the quantity using sums and differences of simpler logarithmic expressions. Express the answer so that logarithms of products, quotients, and powers do not appear.

91. $\ln \sqrt{(x - 3)(x + 4)}$ 92. $\log_{10} \dfrac{x^2 - 4}{x + 3}$

93. $\log_{10} \dfrac{3}{\sqrt{1 + x}}$ 94. $\ln \dfrac{x^2\sqrt{2x + 1}}{2x - 1}$

95. $\log_{10} \sqrt[3]{x/100}$ 96. $\log_{10} \sqrt{\dfrac{x^2 + 8}{(x^2 + 9)(x + 1)}}$

97. $\ln\left(\dfrac{1 + 2e}{1 - 2e}\right)^3$ 98. $\log_{10} \sqrt[3]{x\sqrt{1 + y^2}}$

99. Suppose that A dollars are invested at $R\%$ compounded annually. How many years will it take for the money to double?

100. A sum of $2800 is placed in a savings account at 9% per annum. How much is in the account after two years if the interest is compounded quarterly?

101. A bank offers an interest rate of 9.5% per annum compounded monthly. Compute the effective interest rate.

102. A sum of D dollars is placed in an account at $R\%$ per annum compounded continuously. When will the balance reach E dollars?

103. Compute the doubling time for a sum of D dollars invested at $R\%$ per annum compounded continuously.

104. Your friend invests D dollars at $R\%$ per annum compounded semiannually. You invest an equal amount at the same yearly rate, but compounded daily. How much larger is your account than your friend's after T years?

105. (a) You invest $660 at 5.5% per annum compounded quarterly. How long will it take for your balance to reach $1000? Round off your answer to the next quarter of a year.
(b) You invest D dollars at $R\%$ per annum compounded quarterly. How long will it take for your balance to reach nD dollars? (Assume that $n > 1$.)

In Exercises 106–110, find the domain of each function.

106. (a) $y = \ln x$
(b) $y = e^x$

107. (a) $y = \log_{10}\sqrt{x}$
(b) $y = \sqrt{\log_{10} x}$

108. (a) $f \circ g$ where $f(x) = \ln x$ and $g(x) = e^x$
(b) $g \circ f$ where f and g are as in part (a)

109. (a) $y = \log_{10}|x^2 - 2x - 15|$
(b) $y = \log_{10}(x^2 - 2x - 15)$

110. $y = \dfrac{2 + \ln x}{2 - \ln x}$

111. Find the range of the function defined by $y = \dfrac{e^x + 1}{e^x - 1}$.

In Exercises 112–122, use the graph and the properties of logarithms to estimate the quantity to the nearest tenth.

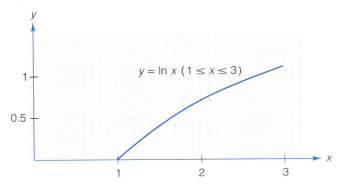

$y = \ln x \ (1 \le x \le 3)$

112. $\ln 2$

113. $\ln 0.5$

114. $\ln 3$

115. $\ln \frac{1}{9}$

116. $\ln 6$

117. $\ln 72$

118. $\ln 40.5$ *Hint:* $40.5 = \frac{81}{2}$

119. e *Hint:*   If $x = e$, then $\ln x = 1$.

120. $e^{1.1}$ *Hint:*   If $x = e^{1.1}$, then $\ln x = 1.1$.

121. $\log_2 3$

122. $\log_3 6$

123. Lester R. Brown, in his book *State of the World 1985* (New York: W. W. Norton and Company, 1985), makes the following statement:

> The projected [*population*] growth for North America, all of Europe, and the Soviet Union is less than the additions expected in either Bangladesh or Nigeria.

In this exercise, you are asked to carry out the type of calculations that could be used to support Lester Brown's projection for the period 1990–2025. The source of the data in the following table is the Population Division of the United Nations.

Region	1990 Population (millions)	Projected Relative Growth Rate (percent per year)	2025 Projected Population (millions)
North America	275.2	0.7	?
Soviet Union	291.3	0.7	?
Europe	499.5	0.2	?
Nigeria	113.3	3.1	?

(a) Complete the table, assuming that the populations grow exponentially and that the indicated growth rates are valid over the period 1990–2025.
(b) According to the projections in part (a), what will be the net increase in Nigeria's population over the period 1990–2025?
(c) According to the projections in part (a), what will be the net increase in the combined populations of North America, the Soviet Union, and Europe over the period 1990–2025?
(d) Compare your answers in parts (b) and (c). Do your results support or contradict Lester Brown's projection?

124. As of 1990, ten countries in the world had populations exceeding 100 million. These countries are listed (in order of population, from largest to smallest) in the table. According to *World Population Profile: 1987*, published by the United States Bureau of the Census, "The latest projections suggest that India's population may surpass China's in less than 60 years, or before today's youngsters in both countries reach old age." Using the data in the table, estimate the populations of the ten countries in the year 2050. (Assume that the indicated growth rates are valid over the period 1990–2050.) Tabulate your final answers, listing the countries in order of population, from largest to smallest. Do your results for China and India support the projection by the Census Bureau?

Country	1990 Population (millions)	Relative Growth Rate (percent per year)
1. China	1119.9	1.3
2. India	853.4	2.1
3. Soviet Union	291	0.9
4. United States	251.4	0.7
5. Indonesia	189.4	1.8
6. Brazil	150.4	1.9
7. Japan	123.6	0.4
8. Nigeria	118.8	2.9
9. Bangladesh	114.8	2.5
10. Pakistan	114.6	3.0

Source: *1990 World Population Data Sheet*, published by Population Reference Bureau, Inc., Washington, D.C.

125. The figure shows a portion of the graph of $y = e^x$, for $1.0970 \leq x \leq 1.1000$.
  (a) Use the graph to estimate $\ln 3$ to four decimal places.
  (b) Using your calculator, compute the percent error in the approximation in part (a).
  (c) Using the approximation in part (a), estimate the quantities $\ln \sqrt{3}$, $\ln 9$, and $\ln \frac{1}{3}$.

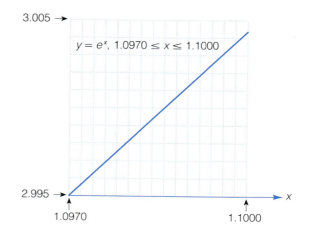

$y = e^x$, $1.0970 \leq x \leq 1.1000$

OPTIONAL TI-81 GRAPHING CALCULATOR EXERCISES

EXERCISES FOR CHAPTER SIX

1. As preparation, press ZOOM 6 and then make the following adjustments in the RANGE menu: $x_{min} = -2$, $x_{max} = 3$, $y_{min} = -1$, $y_{max} = 20$, and $y_{scl} = 2$.
  (a) Graph the functions $Y_1 = 2^x$, $Y_2 = 3^x$, and $Y_3 = 4^x$ on the same set of axes.
  (b) Where do you think the graph of $Y_4 = 75^x$ would fit into the picture that you obtained in part (a)? Add the graph of this function to the picture and see if you are right.

Instructions for Exercises 2–7:
(a) *Without using the calculator, determine the intercepts and asymptotes for the graph of each function.*
(b) *Use the calculator to graph the function. (Choose appropriate RANGE settings or ZOOM settings so that the essential features of the graph are clear.) Then check to see if your answers in part (a) are consistent with the graph.*

2. $y = -3^{x-2}$
3. $y = -3^{x-2} + 1$
4. $y = 4^{-x} - 4$
5. $y = 4 + \left(\frac{1}{4}\right)^x$
6. $y = 10^{x-1}$
7. $y = -10^{x-1} + 0.5$

8. If you add two quadratic functions, the result is again a quadratic (assuming that the x^2-terms don't add to zero). For example, if $f(x) = x^2 + 2x$ and $g(x) = 2x^2 - 1$, then the sum is $(f + g)(x) = 3x^2 + 2x - 1$, another quadratic. But in general, this is not the case with exponential functions. To explore a visual example, consider the function given by $y = 2^x + 2^{-x}$. (This is the sum of the two exponential functions 2^x and $\left(\frac{1}{2}\right)^x$.)
  (a) Graph the function $Y_1 = 2^x + 2^{-x}$. (Use ZOOM 6.) As you can see, the resulting graph is U-shaped.
  (b) Now add the graph of the function $Y_2 = x^2$ to your picture. Note (at this scale) how similar the graphs appear.
  (c) Actually, the graph of $Y_1 = 2^x + 2^{-x}$ rises much more steeply than $Y_2 = x^2$. To see this, make the following adjustments in the RANGE window and then press GRAPH: $y_{min} = 0$, $y_{max} = 100$, and $y_{scl} = 10$.

9. Follow parts (a) and (b) in Problem 8, but use the two functions $Y_1 = 2^x - 2^{-x}$ and $Y_2 = x^3$.

10. For this problem, first reread Example 2 in Section 6.1. (The example shows you how to find the approximate value of x for which $10^x = 2$.)

(a) On the same set of axes, graph the two functions $Y_1 = 10^x$ and $y = 2$. (Use the following settings in the RANGE window: $x_{min} = 0$, $x_{max} = 1$, $x_{scl} = 0.1$, $y_{min} = 0$, $y_{max} = 10$, and $y_{scl} = 1$.

(b) Explain how your graph in part (a) supports the answer $\frac{3}{10}$ that was obtained in Section 6.1.

11. For comparison, draw the graphs of 2^x, e^x, and 3^x on the same set of axes. Begin with the standard viewing rectangle. Then, to see the graphs more clearly when x is negative, use the following settings in the RANGE menu: $x_{min} = -1$, $x_{max} = 0$, $y_{min} = 0$, $y_{max} = 1$. To compare the graphs for positive values of x, use $x_{min} = 0$, $x_{max} = 2$, $y_{min} = 0$, $y_{max} = 10$. Note that the graph of e^x is bounded between the graphs of 2^x and 3^x, just as the number e is between 2 and 3. Also note in particular that we have

$$2^x < e^x < 3^x \qquad \text{when } x > 0$$

but

$$2^x > e^x > 3^x \qquad \text{when } x < 0$$

12. According to the text (p. 315), the line $y = x + 1$ is tangent to the graph of $y = e^x$ at the point $(0, 1)$. Confirm this visually by graphing the curve and the line on the same set of axes.

In Problems 13–18, first tell what translations or reflections are required to obtain the graph of the given function from that of $y = e^x$. Then graph the given function along with $y = e^x$, and check that the picture is consistent with your ideas.

13. $y = e^{-x}$ **14.** $y = -e^x$

15. $y = -e^{-x}$ **16.** (a) $y = e^x - 1$
                                                (b) $y = e^{-x} - 1$

17. (a) $y = e^{x-1}$ **18.** (a) $y = e^{x-1} - 1$
          (b) $y = e^{-x-1}$ (b) $y = e^{-x-1} - 1$

19. In Section 6.5 it was pointed out that the value of $[1 + (1/x)]^x$ approaches e as x gets larger and larger. In this exercise you will use the graphing calculator to reinforce this fact.

(a) Enter the functions $Y_1 = [1 + (1/x)]^x$ and $Y_2 = e$. (For e, use 2.71828.) Now press ZOOM 6 to see the graphs in the standard viewing rectangle. Note that the line $y = e$ appears to be an asymptote for the curve.

(b) To see more convincing evidence that the line $y = e$ is an asymptote, press the RANGE key and make the following changes: $x_{min} = 0$, $x_{max} = 100$, $x_{scl} = 10$, $y_{min} = 0$, $y_{max} = 3$. Press the GRAPH key. As you can see, the quantity $[1 + (1/x)]^x$ indeed seems to approach the value e as x gets larger and

larger. Complete the table in part (c) to see the numerical side of this.

(c)

x	5	10	20	50	100	10^3	10^4	10^5	10^6
$\left(1 + \dfrac{1}{x}\right)^x$									

20. (a) According to Section 6.3, what kind of symmetry would you see if you graphed the functions $Y_1 = e^x$ and $Y_2 = \ln x$ on the same set of axes?

(b) Graph those two functions on the same set of axes. Also, add the line $Y_3 = x$ to the picture. [Choose an appropriate viewing rectangle and then press ZOOM 5. (Why is the ZOOM 5 setting necessary?)]

21. Follow the steps in Exercise 20, but use the functions $Y_1 = 10^x$ and $Y_2 = \log_{10} x$, respectively.

22. Let $F(x) = e^x$, $G(x) = \ln x$, $H(x) = x - 2$, and $K(x) = x^2$. Specify the domain of each of the following functions and then check your answers by drawing the graphs. *Hint:* The answers for (c) and (d) are not the same.

(a) F (b) G (c) $G \circ F$ (d) $F \circ G$
(e) $G \circ H$ (f) $H \circ G$ (g) $G \circ K$ (h) $K \circ G$

23. Draw graphs to help answer the following questions. (You will need to experiment with the RANGE settings.)

(a) For which x-values do we have $\ln x < \log_{10} x$?
(b) For which x-values do we have $\ln x > \log_{10} x$?
(c) For which x-values do we have $e^x < 10^x$?
(d) For which x-values do we have $e^x > 10^x$?

24. This exercise uses the RANGE key and the TRACE key to demonstrate the very slow growth of the natural logarithm function $y = \ln x$. Begin by clearing all functions in the Y= menu and pressing ZOOM 6.

(a) Press the Y= key and enter the function $Y_1 = \ln x$. Press GRAPH and note how slowly the graph rises. We ask the question, How large must x be before the graph reaches a height of 10? Press the TRACE key and use the right-hand arrow key to move the cursor to the point on the graph where $x = 10$. As indicated on the screen display, the y-value is only about 2.3. Go on to part (b).

(b) Since we are trying to see when the graph of $Y_1 = \ln x$ reaches a height of 10, press the Y= key and add the graph of (the horizontal line) $Y_2 = 10$ to the picture obtained in part (a). Next, press the RANGE key and make the following changes: $x_{min} = 0$, $x_{max} = 100$, and $x_{scl} = 10$. Press the GRAPH key. Now press the TRACE key and move the cursor to the point on the graph where

$x = 100$. Note that the corresponding y-coordinate isn't even 5. Go on to part (c).

(c) Press the RANGE key and make the following changes: $x_{max} = 1000$ and $x_{scl} = 100$. Press the GRAPH key. Now press the TRACE key and, as before, determine the coordinates of the highest point shown on the graph. You'll find that at $x = 1000$, the value of y is almost 7. We are getting closer to 10.

(d) Repeat the process in part (c) using $x_{max} = 10000$ and $x_{scl} = 1000$. As you can see from the results, the curve has almost reached the height $y = 10$.

(e) The last step: In the RANGE menu, set $x_{max} = 100000$, $x_{scl} = 10000$, and $y_{max} = 15$. Press GRAPH. Now use the TRACE key and the arrow keys to estimate the value of x for which $\ln x = 10$. (*Question:* How would you find the *exact* value of x for which $\ln x = 10$?)

25. This exercise shows how to use the calculator to fit an exponential curve $y = ab^x$ to a data set. (In Chapter 5 we used linear functions, rather than exponential functions, to model a data set. But as explained in the introduction to Chapter 6, there are cases where the exponential model is more appropriate.) We will use the following population data for the United States. (This is the data used in the chapter introduction.)

Year	U.S. Population
1800	5,308,483
1810	7,239,881
1820	9,638,453
1830	12,866,020
1840	17,069,453
1850	23,191,876
1860	31,443,321
1870	38,558,371
1880	50,189,209
1890	62,947,714
1900	75,994,575

(a) Clear any existing statistical data by first pressing the "2nd" key, then the STAT key, the left-hand arrow key, and the "2" key. After that, press ENTER.

(b) To enter the data in the table, begin by pressing the "2nd" key, then the STAT key, and then the left-hand arrow key, but do not press ENTER yet. Instead, press the number "1" to select the edit option. As you can see, the screen display shows

$$\text{DATA}$$
$$x_1 =$$
$$y_1 = 1$$

For the x_1-value, type 1800, and then press ENTER to move the cursor down to the line for y_1. For the y_1-value, type 5308483, and then press ENTER. The remaining values are then entered in the same fashion.

(c) After you have entered all of the data, press the "2nd" key and then the STAT key. Now use the arrow to move the cursor to item 4 (which is the choice for exponential regression) and press ENTER. Then press ENTER again. You should see the following values displayed on your screen:

$$a = 4.3342717 \times 10^{-15}$$
$$b = 1.027380654$$

These are the values for the constants a and b in the regression curve $y = ab^x$. (There is also a value for r, a measure of how well the curve fits the data, but we do not need that here.)

(d) Press the Y= key and enter the function $Y_1 = ab^x$ using the values for a and b determined in part (c). Press the RANGE key and make the following adjustments to the standard viewing rectangle: $x_{min} = 1800$, $x_{max} = 1900$, $x_{scl} = 10$, $y_{min} = 0$, $y_{max} = 80,000,000$, $y_{scl} = 20,000,000$. Now, press the GRAPH key and check that your graph is consistent with Figure 1(b) in the chapter introduction.

CHAPTER SIX TEST

1. Graph the function $y = 2^{-x} - 3$. Specify the domain, range, intercepts, and asymptote.

2. Suppose that the population of a colony of bacteria increases exponentially. At the start of an experiment there are 6000 bacteria, and 1 hour later the population is 6200. How long (from the start of the experiment) will it take for the population to reach 10,000? Give two forms for the answer: one in terms of base e logarithms and the other a calculator approximation rounded off to the nearest hour.

3. Which is larger, $\log_2 17$ or $\log_3 80$? Explain the reasons for your answer.

4. Express $\log_2 15$ in terms of base e logarithms.

5. Estimate 2^{40} in terms of an integral power of 10.

6. Suppose that \$9500 is invested at 6% interest compounded annually. How long will it take for the amount in the account to reach \$12,000?

7. (a) For which values of x is the following identity valid?

$$e^{\ln x} = x$$

   (b) On the same set of axes, graph $y = e^x$ and $y = \ln x$.

8. Given that $\log_{10} A = a$ and $\log_{10} B = b$, express $\log_{10}(A^3/\sqrt{B})$ in terms of a and b.

9. Simplify each expression.

   (a) $\log_5 \dfrac{1}{\sqrt{5}}$

   (b) $\ln e^2 + \ln 1 - e^{\ln 3}$

10. Solve for x: $x^3 e^x - 4x e^x = 0$.

11. The half-life of a radioactive substance is four days.
    (a) Find the decay constant.
    (b) How much of an initial 2-g sample will remain after 10 days?

12. Express as a single logarithm with a coefficient of 1: $2 \ln x - \ln \sqrt[3]{x^2 + 1}$.

13. Solve for x, leaving your answer in terms of base e logarithms: $-2e^{3x-1} = 9$.

14. Let $g(x) = \log_{10}(x - 1)$. Find a formula for $g^{-1}(x)$ and specify the range of g^{-1}.

15. A principal of \$12,000 is deposited at 6% per annum compounded continuously.
    (a) Estimate the doubling time.
    (b) Sketch a graph showing how the amount increases with time.

16. (a) For which values of x is it true that $\ln (x^2) = 2 \ln x$?
    (b) For which values of x is it true that $(\ln x)^2 = 2 \ln x$?

17. Let $f(x) = \begin{cases} e^x & \text{if } x < 0, \\ \sqrt{1 - x^2} & \text{if } 0 \le x < 1. \end{cases}$

    (a) Sketch the graph of f.
    (b) Is the function f one-to-one?

18. Solve for x: $\frac{1}{2} - \log_{16}(x - 3) = \log_{16} x$.

TRIGONOMETRIC FUNCTIONS OF ANGLES

Trigonometry, as we know the subject today, is a branch of mathematics that is linked with algebra. As such it dates back only to the eighteenth century.

    When treated purely as a development of geometry, however, it goes back to the time of the great Greek mathematician–astronomers. . . .

Alfred Hooper in *Makers of Modern Mathematics* (New York: Random House, Inc., 1948)

It is certain that the division of the ecliptic into 360 degrees . . . [was] adopted by the Greeks from Babylon. . . . [But] It was Hipparchus [ca. 180–ca. 125 B.C.] who first divided the circle in general into 360 parts or degrees, and the introduction of this division coincides with his invention of trigonometry.

Sir Thomas Heath in *A History of Greek Mathematics,* vol. II (London: Oxford University Press, 1921)

The origins of trigonometry are obscure. There are some problems in the Rhind papyrus [ca. 1650 B.C.] that involve the cotangent of the dihedral angles at the base of a pyramid, and . . . the Babylonian cuneiform tablet Plimpton 322 essentially contains a remarkable table of secants. It may be that modern investigations into the mathematics of ancient Mesopotamia will reveal an appreciable development of practical trigonometry.

Howard Eves in *An Introduction to the History of Mathematics,* sixth edition (Philadelphia: Saunders College Publishing, 1990)

INTRODUCTION

In general, there are two approaches to trigonometry at the precalculus level. In a sense, these two approaches correspond to the two historical roots of the subject mentioned in the first of our opening quotations. One approach centers around the study of triangles. Indeed, the word "trigonometry" is derived from two Greek words, *trigonon,* meaning triangle, and *metria,* meaning measurement. It is with this triangle approach that we begin the chapter. The second approach, which is in a sense a generalization of the first, we will take up in Chapter 8. Before getting down to specifics, we offer the following bird's-eye view of Chapters 7 and 8 in terms of functions. As you'll see in subsequent chapters, a real familiarity with both approaches to trigonometry is necessary.

	INPUTS FOR THE FUNCTIONS	OUTPUTS	EMPHASIS
Chapter 7	angles	real numbers	The trigonometric functions are used to study triangles.
Chapter 8	real numbers	real numbers	The objects of study are the trigonometric functions themselves.

7.1 TRIGONOMETRIC FUNCTIONS OF ACUTE ANGLES

> . . . [the English clergyman and mathematician] *William Oughtred* [1574–1660] *introduced a vast array of characters into mathematics . . . In the* Trigonometria *(1657), he employed ∠ for angle . . . .*
>
> Florian Cajori in *A History of Mathematical Notations,* vol. I (La Salle, Illinois: The Open Court Publishing Company, 1928)

In elementary geometry, an **angle** is a figure formed by two rays with a common endpoint. As indicated in Figure 1, the common endpoint is called the **vertex** of the angle. Recall that there are several conventions used in naming angles; Figure 2 indicates some of these. In Figure 2, the symbol θ is the lowercase Greek letter *theta*. Greek letters are often used to name angles. For reference, the Greek alphabet is given in the endpapers at the back of this book. The ∠ symbol that you see in Figure 2 stands for the word "angle." When three letters are used in naming an angle, as with ∠ABC in Figure 2, the middle letter always indicates the vertex of the angle.

There are several ways to indicate the size of (i.e., the amount of rotation in) an angle. In this chapter we use the familiar units of *degrees.* Recall that 360 degrees is the measure of an angle obtained by rotating a ray through one complete circle. The symbol for degrees is °. For comparison, Figure 3 (on the next page) displays angles of various degrees. In this section we will not make any distinction whether the rotation generating an angle is clockwise or counterclockwise; in a subsequent section this distinction will be important. The angles in Figures 3(a) and 3(b) are **acute angles**; these are angles with degree measure θ satisfying $0° < \theta < 90°$. The angle in Figure 3(d) is an **obtuse angle**, that is, an angle with degree measure θ satisfying $90° < \theta < 180°$.

The degree is subdivided into two smaller units: the **minute**, which is $\frac{1}{60}$ of a degree; and the **second**, which is $\frac{1}{60}$ of a minute. The abbreviations for minutes and seconds are ′ and ″, respectively. Thus, we have

$$1° = 60' \qquad \text{There are 60 minutes in one degree.}$$

and

$$1' = 60'' \qquad \text{There are 60 seconds in one minute.}$$

Until modern handheld calculators became commonplace (in the 1970s and 1980s), these subdivisions of the degree were widely used. Nowadays, however,

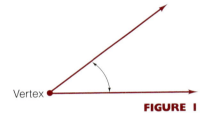

Vertex

FIGURE 1

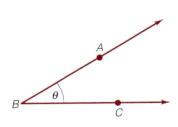

Notations for the angle at *B*:
∠*B*, θ, ∠*ABC*, ∠*CBA*

FIGURE 2

FIGURE 3

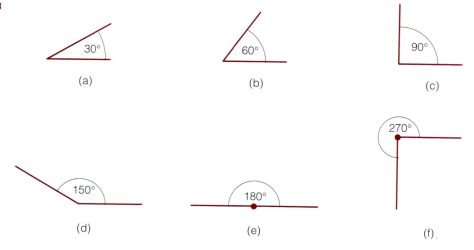

the convention in most scientific applications is to use ordinary decimal notation for the fractional parts of a degree, and this is the convention we follow in this text. (We mention in passing that minutes and seconds are still widely used in some portions of astronomy and in navigation.)

One more convention: Suppose, for example, that the measure of an angle, say angle B, is 70°. We can write this as

measure $\angle B = 70°$ or, more concisely, $m \angle B = 70°$

For ease of notation and speech, however, we will often simply write $\angle B = 70°$.

We are going to define six functions, the six *trigonometric functions*. As indicated by the title of this section, the inputs for these functions (in this section) will be acute angles. The outputs will be real numbers. In a later section we'll make a transition so that angles of any size can serve as inputs, and in Chapter 8 we'll make another transition, this time to real-number inputs. But for now, as we have said, the inputs will be acute angles.

For reasons that are more historical* than mathematical, the trigonometric functions have names that are words rather than single letters, such as f. The use of words rather than single letters for naming functions is by no means peculiar to trigonometry. For instance, in almost every computer-programming language, the absolute value function is ABS. So, for example, we have

ABS(5) = 5   and   ABS(−5) = 5

The names of the six trigonometric functions, along with their abbreviations, are as follows.

*For a discussion of the names of the trigonometric functions, see Howard Eves, *An Introduction to the History of Mathematics,* 6th ed., pp. 236–237 (Philadelphia: Saunders College Publishing, 1990).

NAME OF FUNCTION	ABBREVIATION
cosine	cos
sine	sin
tangent	tan
secant	sec
cosecant	csc
cotangent	cot

Now, let θ be an acute angle in a right triangle, as shown in Figure 4. We define the six trigonometric functions of θ in the box that follows.

FIGURE 4

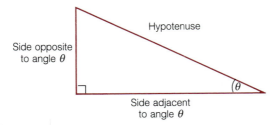

(Refer to Figure 4.)

$$\cos \theta = \frac{\text{length of side adjacent angle } \theta}{\text{length of hypotenuse}} \qquad \sec \theta = \frac{\text{length of hypotenuse}}{\text{length of side adjacent angle } \theta}$$

$$\sin \theta = \frac{\text{length of side opposite angle } \theta}{\text{length of hypotenuse}} \qquad \csc \theta = \frac{\text{length of hypotenuse}}{\text{length of side opposite angle } \theta}$$

$$\tan \theta = \frac{\text{length of side opposite angle } \theta}{\text{length of side adjacent angle } \theta} \qquad \cot \theta = \frac{\text{length of side adjacent angle } \theta}{\text{length of side opposite angle } \theta}$$

Our work for the next several sections will be devoted to exploring these definitions and their consequences. Two preliminary observations that will help you to understand and memorize the definitions are these.

1. An expression such as $\cos \theta$ really means $\cos(\theta)$, where cos or cosine is the name of the function and θ is an input. It is for historical rather than mathematical reasons that the parentheses are suppressed.
2. There are three pairs of reciprocals in the definitions: $\cos \theta$ and $\sec \theta$; $\sin \theta$ and $\csc \theta$; and $\tan \theta$ and $\cot \theta$.

The definitions that we have just given for the trigonometric functions involve ratios of sides in a right triangle. The fact that *ratios* are involved is important. We'll use two examples to explain and emphasize this. Figure 5 (on the next page) shows two right triangles; both contain an angle of 60 degrees. In Figure 5(a) all three sides of the triangle are specified, while in Figure 5(b), only one side is specified.

FIGURE 5

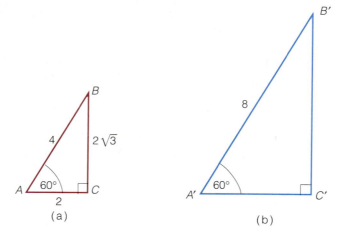

(a)

(b)

Using the data in Figure 5(a), we have

$$\tan 60° = \frac{\text{opposite}}{\text{adjacent}} = \frac{BC}{AC} = \frac{2\sqrt{3}}{2} = \sqrt{3}$$

So, from Figure 5(a), we have $\tan 60° = \sqrt{3}$. Now, the question is, will Figure 5(b) give us the same value for tan 60°? [It had better, otherwise tan will not qualify as a function. (Why?)] Indeed, Figure 5(b) does give us the same value for tan 60°, and it is not even necessary to use the fact that we are given a specific length in Figure 5(b). Elementary geometry* tells us that $\triangle ABC$ is similar to $\triangle A'B'C'$. Then, because corresponding sides of similar triangles are proportional, we have

$$\frac{BC}{AC} = \frac{B'C'}{A'C'} \tag{1}$$

Now look at the ratio on the right-hand side of this last equation; it represents tan 60° as we would calculate it in Figure 5(b). The left-hand side of the equation, as we've already seen from Figure 5(a), represents $\tan 60° = \sqrt{3}$. In summary, similar triangles and equation (1) tell us that we obtain the same value for tan 60° whether we use Figure 5(a) or 5(b).

The point of the example we've just concluded is this: For a given acute angle θ, the values of the trigonometric functions depend on the ratios of the lengths of the sides but not on the size of the particular right triangle in which θ resides. For emphasis, the next example makes this same point.

EXAMPLE 1 Figure 6 (on the next page) shows two right triangles. The first right triangle has sides 5, 12, and 13. The second right triangle is similar to the first (the angles are the same), but each side is 10 times longer than the corresponding side in the first triangle. Calculate and compare the values of sin θ, cos θ, and tan θ for both triangles.

*If two angles of one triangle are congruent to two corresponding angles of another triangle, then the triangles are similar.

FIGURE 6

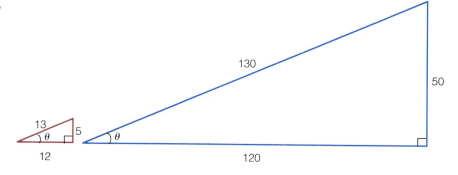

Solution SMALL TRIANGLE   LARGE TRIANGLE

$$\sin \theta = \frac{\text{opposite}}{\text{hypotenuse}} = \frac{5}{13} \qquad \sin \theta = \frac{50}{130} = \frac{5}{13}$$

$$\cos \theta = \frac{\text{adjacent}}{\text{hypotenuse}} = \frac{12}{13} \qquad \cos \theta = \frac{120}{130} = \frac{12}{13}$$

$$\tan \theta = \frac{\text{opposite}}{\text{adjacent}} = \frac{5}{12} \qquad \tan \theta = \frac{50}{120} = \frac{5}{12}$$

Observation The corresponding values for $\sin \theta$, $\cos \theta$, and $\tan \theta$ are the same for both triangles. ■■■

EXAMPLE 2 Let θ be the acute angle indicated in Figure 7. Determine the six quantities $\cos \theta$, $\sin \theta$, $\tan \theta$, $\sec \theta$, $\csc \theta$, and $\cot \theta$.

Solution We use the definitions:

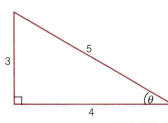

$$\cos \theta = \frac{\text{adjacent}}{\text{hypotenuse}} = \frac{4}{5} \qquad \sec \theta = \frac{\text{hypotenuse}}{\text{adjacent}} = \frac{5}{4}$$

$$\sin \theta = \frac{\text{opposite}}{\text{hypotenuse}} = \frac{3}{5} \qquad \csc \theta = \frac{\text{hypotenuse}}{\text{opposite}} = \frac{5}{3}$$

$$\tan \theta = \frac{\text{opposite}}{\text{adjacent}} = \frac{3}{4} \qquad \cot \theta = \frac{\text{adjacent}}{\text{opposite}} = \frac{4}{3}$$

FIGURE 7 Note the pairs of answers that are reciprocals. (As mentioned before, this helps in memorizing the definitions.) ■■■

EXAMPLE 3 Let β be the acute angle indicated in Figure 8 (on the next page). Find $\sin \beta$ and $\cos \beta$.

Solution In view of the definitions, we need to know the length of the hypotenuse in Figure 8. If we call this length h, then by the Pythagorean theorem we have

$$h^2 = 3^2 + 1^2 = 10$$
$$h = \sqrt{10}$$

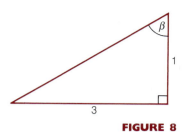

FIGURE 8

Therefore

$$\sin \beta = \frac{\text{opposite}}{\text{hypotenuse}} = \frac{3}{\sqrt{10}} = \frac{3\sqrt{10}}{10}$$

and

$$\cos \beta = \frac{\text{adjacent}}{\text{hypotenuse}} = \frac{1}{\sqrt{10}} = \frac{\sqrt{10}}{10}$$

In the next example we use a calculator to evaluate the trigonometric functions for a particular acute-angle input. All scientific calculators allow for two common units of angle measurement: the degree and the *radian*. Radian measure is discussed in Chapter 8; for the present chapter, though, we are using degree measure exclusively. So, before working through the next example on your own, you need to check that your calculator is set in the "degree mode" rather than the "radian mode." If necessary, consult the owner's manual for your calculator on this point.

One more calculator-related detail: For most calculators, when you want to evaluate a function, first you enter the input, then you press the appropriate function key. For example (assuming that your calculator is set in the degree mode), to evaluate sin 40°, first enter the input 40, then press the $\boxed{\text{SIN}}$ key. Try this now. If the resulting display on your calculator begins 0.6427 ..., then this was the appropriate sequence for your calculator. If not (but assuming that your calculator is indeed set in the degree mode), then you may have a calculator that requires you to enter the function *before* the input. (The Casio and Texas Instruments graphing calculators work that way, for example.) Again, consult your owner's manual on this point, as necessary.

EXAMPLE 4   Use a calculator to evaluate cos 40°, sin 40°, and tan 40°. Round off the results to three decimal places.

Solution   With the calculator set in the degree mode, the sequence of keystrokes and the results (rounded off to three decimal places) are as follows. You should verify each of these results for yourself.

EXPRESSION	KEYSTROKES	OUTPUT
cos 40°	40 $\boxed{\text{COS}}$	0.766
sin 40°	40 $\boxed{\text{SIN}}$	0.643
tan 40°	40 $\boxed{\text{TAN}}$	0.839

In the next example, we use a calculator to evaluate the three functions secant, cosecant, and cotangent for a particular input. These three functions are referred to as the **reciprocal functions**. This is because, from the definitions on page 369, it follows that

$$\sec \theta = \frac{1}{\cos \theta} \qquad \csc \theta = \frac{1}{\sin \theta} \qquad \cot \theta = \frac{1}{\tan \theta}$$

So, for example, although most calculators do not have a key labeled "sec," we can nevertheless evaluate the secant function by using the cosine key $\boxed{\text{COS}}$ and then the reciprocal key $\boxed{1/x}$.

EXAMPLE 5 Use a calculator to evaluate sec 10°, csc 10°, and cot 10°. Round off the results to three decimal places.

Solution As in the previous example, we first check that the calculator is in the degree mode. Then, the sequence of keystrokes and the results are as follows. (Again, you should verify each of these results for yourself.)

EXPRESSION	KEYSTROKES	OUTPUT
sec 10°	10 COS 1/x	1.015
csc 10°	10 SIN 1/x	5.759
cot 10°	10 TAN 1/x	5.671

There are certain angles for which we can evaluate the trigonometric functions without using a calculator. Three such angles are 30°, 45°, and 60°. We focus on these particular angles for now, not because they are somehow more fundamental than others, but because a ready knowledge of their trigonometric values will provide a useful source of examples in our subsequent work.

In order to obtain the trigonometric values for angles of 30° and 60°, we rely on the following theorem from geometry. (For a proof of this theorem, see Exercise 41.)

THEOREM   The 30°–60° Right Triangle

In a 30°–60° right triangle, the length of the side opposite the 30° angle is half the length of the hypotenuse. (See Figure 9.)

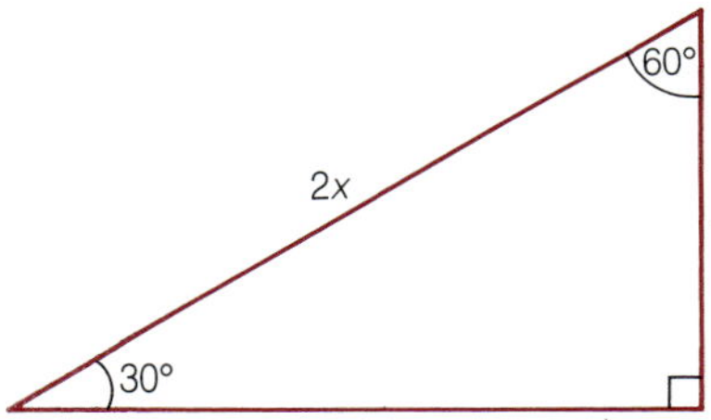

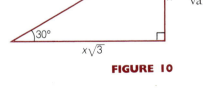

FIGURE 9

If we let y denote the length of the unmarked side in Figure 9, then (according to the Pythagorean theorem) we have $y^2 + x^2 = (2x)^2$ and, consequently,

$$y^2 = 4x^2 - x^2 = 3x^2$$
$$y = \sqrt{3}x$$

Our 30°–60° right triangle can now be labeled as in Figure 10. From this figure, the required trigonometric values can be written down by inspection. We will list only the values of the cosine, sine, and tangent, since the remaining three values are just the reciprocals of these.

FIGURE 10

$$\cos 30° = \frac{x\sqrt{3}}{2x} = \frac{\sqrt{3}}{2} \qquad \cos 60° = \frac{x}{2x} = \frac{1}{2}$$

$$\sin 30° = \frac{x}{2x} = \frac{1}{2} \qquad \sin 60° = \frac{x\sqrt{3}}{2x} = \frac{\sqrt{3}}{2}$$

$$\tan 30° = \frac{x}{x\sqrt{3}} = \frac{1}{\sqrt{3}} = \frac{\sqrt{3}}{3} \qquad \tan 60° = \frac{x\sqrt{3}}{x} = \sqrt{3}$$

There are two observations to be made here. First, the final answers do not involve x. Once again this shows that for a given angle, it is the *ratios* of the sides that determine the trigonometric values, not the particular triangle. Second, notice that

$$\sin 30° = \cos 60° \qquad \text{and} \qquad \sin 60° = \cos 30°$$

It is easy to see that this is no coincidence. The terms "opposite" and "adjacent" are relative ones. As a glance at Figure 10 shows, the side opposite to one acute angle is automatically adjacent to the other acute angle. We'll return to this point again in the next secton.

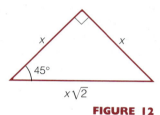

FIGURE 11

Another angle for which we can compute the trigonometric values without the aid of a calculator is 45°. Figure 11 shows a 45°–45° right triangle. Since the base angles of this triangle are equal, it follows from a theorem of geometry that the triangle is isosceles. That is, the triangle has two equal sides, as indicated by the use of the letter x twice in Figure 11. If we now use y to denote the length of the unmarked side in Figure 11, we have

$$y^2 = x^2 + x^2 = 2x^2 \qquad \text{and, consequently,} \qquad y = \sqrt{2}x$$

Our 45°–45° right triangle can now be labeled as in Figure 12, and the trigonometric values can then be obtained by inspection. As before, we list only the values for cosine, sine, and tangent, since the remaining three are just the reciprocals of these.

FIGURE 12

$$\cos 45° = \frac{x}{x\sqrt{2}} = \frac{1}{\sqrt{2}} = \frac{\sqrt{2}}{2}$$

$$\sin 45° = \frac{x}{x\sqrt{2}} = \frac{1}{\sqrt{2}} = \frac{\sqrt{2}}{2}$$

$$\tan 45° = \frac{x}{x} = 1$$

Table 1 summarizes the results we've now obtained for angles of 30, 45, and 60 degrees. These results should be memorized.

TABLE 1

θ	$\sin \theta$	$\cos \theta$	$\tan \theta$
30°	$\frac{1}{2}$	$\frac{\sqrt{3}}{2}$	$\frac{\sqrt{3}}{3}$
45°	$\frac{\sqrt{2}}{2}$	$\frac{\sqrt{2}}{2}$	1
60°	$\frac{\sqrt{3}}{2}$	$\frac{1}{2}$	$\sqrt{3}$

EXERCISE 7.1

A

In Exercises 1–4, use the definitions (as in Example 2) to evaluate the six trigonometric functions of (a) θ; (b) β. In cases where a radical occurs in a denominator, rationalize the denominator.

1.

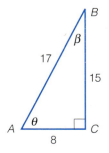

2.

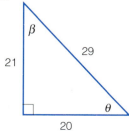

3.

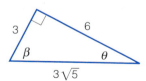

4.

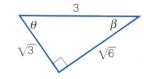

In Exercises 5–12, suppose that $\triangle ABC$ is a right triangle with $\angle C = 90°$.

5. If $AC = 3$ and $BC = 2$, find the following.
   (a) $\cos A$, $\sin A$, $\tan A$
   (b) $\sec B$, $\csc B$, $\cot B$

6. If $AC = 6$ and $BC = 2$, find the following.
   (a) $\cos A$, $\sin A$, $\tan A$
   (b) $\sec B$, $\csc B$, $\cot B$

7. If $AB = 13$ and $BC = 5$, compute the values of the six trigonometric functions of angle B.

8. If $AB = 3$ and $AC = 1$, compute the values of the six trigonometric functions of angle A.

9. If $AC = 1$ and $BC = \frac{3}{4}$, compute each quantity:
   (a) $\sin B$, $\cos A$;   (b) $\sin A$, $\cos B$;
   (c) $(\tan A)(\tan B)$.

10. If $AC = BC = 4$, compute the following:
    (a) $\sec A$, $\csc A$, $\cot A$;   (b) $\sec B$, $\csc B$, $\cot B$;
    (c) $(\cot A)(\cot B)$.

11. If $AB = 25$ and $AC = 24$, compute each of the required quantities:
    (a) $\cos A$, $\sin A$, $\tan A$;   (b) $\cos B$, $\sin B$, $\tan B$;
    (c) $(\tan A)(\tan B)$.

12. If $AB = 1$ and $BC = \sqrt{3}/2$, compute the following:
    (a) $\cos A$, $\sin B$;   (b) $\tan A$, $\cot B$;
    (c) $\sec A$, $\csc B$.

In Exercises 13–18, use a calculator to compute $\cos \theta$, $\sin \theta$, and $\tan \theta$ for the given value of θ. Round off each result to three decimal places.

13. $\theta = 65°$ 14. $\theta = 21°$ 15. $\theta = 38.5°$
16. $\theta = 12.4°$ 17. $\theta = 80.06°$ 18. $\theta = 0.99°$

In Exercises 19–24, use a calculator to compute $\sec \theta$, $\csc \theta$, and $\cot \theta$ (as in Example 5) for the given value of θ. Round off each result to three decimal places.

19. $\theta = 20°$ 20. $\theta = 40°$ 21. $\theta = 17.5°$
22. $\theta = 18.5°$ 23. $\theta = 1°$ 24. $\theta = 89.9°$

In Exercises 25–36, verify that each of the following equations is correct by evaluating each side. Do not use a calculator. The purpose of Exercises 25–36 is twofold. First, doing the problems will help you to review the values in Table 1 of this section. Second, the exercises serve as an algebra review. Note: *In the exercises, notation such as $\sin^2 \theta$ stands for $(\sin \theta)^2$. This common notational convention will be discussed in the next section.*

25. $\cos 60° = \cos^2 30° - \sin^2 30°$

26. $\cos 60° = 1 - 2 \sin^2 30°$

27. $\sin^2 30° + \sin^2 45° + \sin^2 60° = \frac{3}{2}$

28. $\sin 30° \cos 60° + \cos 30° \sin 60° = 1$

29. $2 \sin 30° \cos 30° = \sin 60°$

30. $2 \sin 45° \cos 45° = 1$

31. $\sin 30° = \sqrt{\frac{1}{2}(1 - \cos 60°)}$

32. $\cos 30° = \sqrt{\frac{1}{2}(1 + \cos 60°)}$

33. $\tan 30° = \dfrac{\sin 60°}{1 + \cos 60°}$

34. $\tan 30° = \dfrac{1 - \cos 60°}{\sin 60°}$

35. $1 + \tan^2 45° = \sec^2 45°$

36. $1 + \cot^2 60° = \csc^2 60°$

37. (a) Use your calculator to evaluate cos 30° and cos 45°. Give your answers to as many decimal places as your calculator allows.
    (b) In Table 1, there are expressions for cos 30° and cos 45°. Use your calculator to evaluate these expressions, and check to see that the results agree with those in part (a).

38. (a) Use your calculator to evaluate tan 30° and tan 60°. Give your answers to as many decimal places as your calculator allows.
    (b) In Table 1, there are expressions for tan 30° and tan 60°. Use your calculator to evaluate these expressions, and check to see that the results agree with those in part (a).

B

For Exercises 39 and 40, refer to the following figure. In each case, say which of the two given quantities is larger. If the quantities are equal, say so.

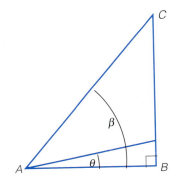

39. (a) $\cos \theta$, $\cos \beta$
    (b) $\sec \theta$, $\sec \beta$

40. (a) $\tan \theta$, $\tan \beta$
    (b) $\cot \theta$, $\cot \beta$

41. The following figure shows a 30°–60° right triangle, $\triangle ABC$. Prove that $AC = 2AB$. *Suggestion:* Construct $\triangle DBC$ as shown, congruent to $\triangle ABC$. Then note that $\triangle ADC$ is equilateral.

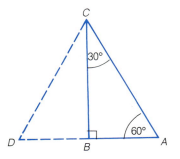

42. This exercise shows how to obtain radical expressions for sin 15° and cos 15°. In the figure, assume that $AB = BD = 2$.

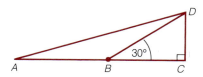

(a) In the right triangle *BCD,* note that $DC = 1$ because $\overline{DC}$ is opposite the 30° angle and $BD = 2$. Use the Pythagorean theorem to show that $BC = \sqrt{3}$.

(b) Use the Pythagorean theorem to show that
$$AD = 2\sqrt{2 + \sqrt{3}}.$$

(c) Show that the expression for *AD* in part (b) is equal to $\sqrt{6} + \sqrt{2}$. *Hint:* Two nonnegative quantities are equal if and only if their squares are equal.

(d) Explain why $\angle BAD = \angle BDA$.

(e) According to a theorem from geometry, an exterior angle of a triangle is equal to the sum of the two nonadjacent interior angles. Apply this to $\triangle ABD$ with exterior angle $DBC = 30°$, and show that $\angle BAD = 15°$.

(f) Using the figure and the values that you have obtained for the lengths, conclude that
$$\sin 15° = \frac{1}{\sqrt{6} + \sqrt{2}} \qquad \cos 15° = \frac{2 + \sqrt{3}}{\sqrt{6} + \sqrt{2}}$$

(g) Rationalize the denominators in part (f) to obtain
$$\sin 15° = \frac{\sqrt{6} - \sqrt{2}}{4} \qquad \cos 15° = \frac{\sqrt{6} + \sqrt{2}}{4}$$

(h) Use your calculator to check the results in part (g).

43. If an angle θ is an integral multiple of 3°, then the real number sin θ is either rational or expressible in terms of radicals. You've already seen examples of this with the angles 30°, 45°, and 60°. The accompanying table gives the values of sin θ for some multiples of 3°. Use your calculator to check the entries in the table. In Exercise Set 8.7, you will see how these results are derived. *Remark:* If θ is an integral number of degrees, but not a multiple of 3°, then the real number sin θ cannot be expressed in terms of radicals within the real-number system.

θ	$\sin \theta$
3°	$\frac{1}{16}\left[(\sqrt{6} + \sqrt{2})(\sqrt{5} - 1) - 2(\sqrt{3} - 1)\sqrt{5 + \sqrt{5}}\right]$
6°	$\frac{1}{8}(\sqrt{30 - 6\sqrt{5}} - \sqrt{5} - 1)$
9°	$\frac{1}{8}(\sqrt{10} + \sqrt{2} - 2\sqrt{5 - \sqrt{5}})$
12°	$\frac{1}{8}(\sqrt{10 + 2\sqrt{5}} - \sqrt{15} + \sqrt{3})$
15°	$\frac{1}{4}(\sqrt{6} - \sqrt{2})$
18°	$\frac{1}{4}(\sqrt{5} - 1)$

7.2   ALGEBRA AND THE TRIGONOMETRIC FUNCTIONS

Perhaps the first use of abbreviations for the trigonometric [functions] . . . goes back to the [Danish] physician and mathematician, Thomas Finck [1561–1646]. . . .

Florian Cajori in *A History of Mathematical Notations,* vol. II (Chicago: The Open Court Publishing Company, 1929)

Albert Girard (1595–1632), who seems to have lived chiefly in Holland, . . . interested himself in spherical trigonometry and trigonometry. In 1626, he published a treatise on trigonometry that contains the earliest use of our abbreviations sin, tan, *and* sec *for* sine, tangent, *and* secant.

Howard Eves in *An Introduction to the History of Mathematics,* sixth edition (Philadelphia: Saunders College Publishing, 1990)

In this section, we are going to concentrate on the algebra involved in working with the trigonometric functions. This will help to pave the way both for the applications in the next section and for some of the more analytical parts of trigonometry in the next chapter. We begin by listing some common notational conventions.

Notational Conventions

1. The quantity $(\sin \theta)^n$ is usually written $\sin^n \theta$. For example, $(\sin \theta)^2$ is written $\sin^2 \theta$. The same convention also applies to the other five trigonometric functions.*

2. Parentheses are often omitted in multiplication. For example, the product $(\sin \theta)(\cos \theta)$ is usually written $\sin \theta \cos \theta$. Similarly, $2(\sin \theta)$ is written $2 \sin \theta$.

3. An expression such as $\sin \theta$ really means $\sin(\theta)$, where sin or sine is the name of the function and θ is an input. It is for historical rather than mathematical reasons that the parentheses are suppressed. An exception to this, however, occurs in expressions such as $\sin(A + B)$; here the parentheses are necessary.

EXAMPLE 1 (a) Combine like terms: $3 \sin^2 \theta \cos \theta - 5 \sin^2 \theta \cos \theta$
(b) Carry out the multiplication: $(2 \sin \theta - 3 \cos \theta)^2$

Solution (a) We do this in the same way we would simplify the algebraic expression $3S^2C - 5S^2C$. Since

$$3S^2C - 5S^2C = -2S^2C$$

we have

$$3 \sin^2 \theta \cos \theta - 5 \sin^2 \theta \cos \theta = -2 \sin^2 \theta \cos \theta$$

Note If the original expression had been $3 \sin^2 \theta \cos \theta - 5 \sin \theta \cos^2 \theta$, then there would be no like terms to combine.

(b) We do this in the same way that we would expand $(2S - 3C)^2$. Since

$$(2S - 3C)^2 = 4S^2 - 12SC + 9C^2$$

we have

$$(2 \sin \theta - 3 \cos \theta)^2 = 4 \sin^2 \theta - 12 \sin \theta \cos \theta + 9 \cos^2 \theta$$

EXAMPLE 2 Factor: $\tan^2 A + 5 \tan A + 6$.

Solution **PRELIMINARY SOLUTION** To help us focus on the algebra that is actually involved, let's replace each occurrence of the quantity $\tan A$ by the letter T. Then

$$T^2 + 5T + 6 = (T + 3)(T + 2)$$

ACTUAL SOLUTION

$$\tan^2 A + 5 \tan A + 6 = (\tan A + 3)(\tan A + 2)$$

---

*A single exception to this convention occurs when $n = -1$. The meaning of expressions such as $\sin^{-1} x$ will be explained in Sections 8 and 9 in Chapter 8.

Note   After you are accustomed to working with trigonometric expressions, you should be able to eliminate the preliminary step.   ▮▮▮

EXAMPLE 3   Add $\sin \theta + \dfrac{1}{\cos \theta}$.

Solution   PRELIMINARY SOLUTION

$$S + \frac{1}{C} = \frac{S}{1} + \frac{1}{C} = \frac{S}{1} \cdot \frac{C}{C} + \frac{1}{C}$$

$$= \frac{SC}{C} + \frac{1}{C} = \frac{SC + 1}{C}$$

ACTUAL SOLUTION

$$\sin \theta + \frac{1}{\cos \theta} = \frac{\sin \theta}{1} + \frac{1}{\cos \theta} = \frac{\sin \theta}{1} \cdot \frac{\cos \theta}{\cos \theta} + \frac{1}{\cos \theta}$$

$$= \frac{\sin \theta \cos \theta}{\cos \theta} + \frac{1}{\cos \theta} = \frac{\sin \theta \cos \theta + 1}{\cos \theta}$$ ▮▮▮

There are numerous identities involving the trigonometric functions. Recall that an **identity** is an equation that is satisfied by all relevant values of the variables concerned. Two examples of identities are $(x - y)(x + y) = x^2 - y^2$ and $x^3 = x^4/x$. The first of these is true no matter what real numbers are used for x and y; the second is true for all real numbers except $x = 0$. For now, in the context of right-triangle trigonometry, we'll consider only a few of the most basic identities.

PROPERTY SUMMARY   BASIC RIGHT-TRIANGLE IDENTITIES

IDENTITY

1. $\sin^2 \theta + \cos^2 \theta = 1$

2. $\dfrac{\sin \theta}{\cos \theta} = \tan \theta$

3. $\sin(90° - \theta) = \cos \theta$; $\cos(90° - \theta) = \sin \theta$

4. $\sec \theta = \dfrac{1}{\cos \theta}$; $\csc \theta = \dfrac{1}{\sin \theta}$; $\cot \theta = \dfrac{1}{\tan \theta}$

EXAMPLES

1. $\sin^2 30° + \cos^2 30° = 1$
    $\sin^2 19° + \cos^2 19° = 1$

2. $\dfrac{\sin 60°}{\cos 60°} = \tan 60°$

3. $\sin 70° = \cos 20°$; $\sin 20° = \cos 70°$

4. $\cot 40° = \dfrac{1}{\tan 40°}$

The fourth set of identities in the box follows directly from the definitions on page 369. For the proofs of the other identities, we refer to Figure 1.

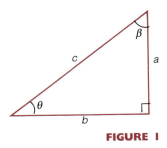

FIGURE I

PROOF THAT $\sin^2 \theta + \cos^2 \theta = 1$ Looking at Figure 1, we have $\sin \theta = a/c$ and $\cos \theta = b/c$. Thus

$$\sin^2 \theta + \cos^2 \theta = \left(\frac{a}{c}\right)^2 + \left(\frac{b}{c}\right)^2 = \frac{a^2}{c^2} + \frac{b^2}{c^2} = \frac{a^2 + b^2}{c^2}$$

$$= \frac{c^2}{c^2} \qquad \text{using } a^2 + b^2 = c^2, \text{ the Pythagorean theorem}$$

$$= 1 \qquad \text{as required}$$

PROOF THAT $(\sin \theta)/(\cos \theta) = \tan \theta$ Again with reference to Figure 1, we have, by definition, $\sin \theta = a/c$, $\cos \theta = b/c$ and $\tan \theta = a/b$. Therefore

$$\frac{\sin \theta}{\cos \theta} = \frac{a/c}{b/c} = \frac{a}{c} \times \frac{c}{b}$$

$$= \frac{a}{b} = \tan \theta \qquad \text{as required}$$

PROOF THAT $\sin(90° - \theta) = \cos \theta$ First of all, since the sum of the angles in any triangle is 180°, we have

$$\theta + \beta + 90° = 180°$$
$$\beta = 90° - \theta$$

Then $\sin(90° - \theta) = \sin \beta = b/c$. But also (by definition) $\cos \theta = b/c$. Thus

$$\sin(90° - \theta) = \cos \theta$$

since both expressions equal b/c. This is what we wanted to prove.

The proof that $\cos(90° - \theta) = \sin \theta$ is entirely similar and we shall omit it. We can conveniently summarize these last two results by recalling the notion of complementary angles. Two acute angles are said to be **complementary** provided their sum is 90°. Thus, the two angles θ and $90° - \theta$ are complementary. In view of this, we can restate the last two results as follows:

> If two angles are complementary, then the sine of (either) one equals the cosine of the other.

Incidentally, this result gives us an insight into the origin of the term *cosine*: it is a shortened form of the phrase *complement's sine*.

EXAMPLE 4   Suppose that B is an acute angle and $\cos B = \frac{2}{5}$. Find $\sin B$ and $\tan B$.

Solution   We'll show two methods. The first uses the identities $\sin^2 B + \cos^2 B = 1$ and $(\sin B)/(\cos B) = \tan B$. The second makes direct use of the Pythagorean theorem.

FIRST METHOD Replace $\cos B$ with $\frac{2}{5}$ in the identity $\sin^2 B + \cos^2 B = 1$. This yields

$$\sin^2 B + \left(\frac{2}{5}\right)^2 = 1$$

$$\sin^2 B = 1 - \frac{4}{25} = \frac{21}{25}$$

$$\sin B = \sqrt{\frac{21}{25}} = \frac{\sqrt{21}}{5}$$

Notice that we've chosen the positive square root here. This is because the values of the trigonometric functions of an acute angle are by definition positive. *Caution:* In a later section, we'll extend the definitions of the trigonometric functions to include angles that are not acute. In those cases, some of the trigonometric values will not be positive; we will then need to pay more attention to the matter of signs.

For $\tan B$, we have

$$\tan B = \frac{\sin B}{\cos B} = \frac{\sqrt{21}/5}{2/5} = \frac{\sqrt{21}}{2}$$

ALTERNATE METHOD Since $\cos B = \frac{2}{5} = \text{adjacent/hypotenuse}$, we can work with a right triangle labeled as in Figure 2. Using the Pythagorean theorem, we have $2^2 + x^2 = 5^2$, from which it follows that $x = \sqrt{21}$. Consequently,

$$\sin B = \frac{\text{opposite}}{\text{hypotenuse}} = \frac{x}{5} = \frac{\sqrt{21}}{5}$$

and

$$\tan B = \frac{\text{opposite}}{\text{adjacent}} = \frac{x}{2} = \frac{\sqrt{21}}{2} \qquad \text{as obtained previously}$$

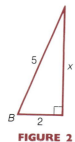

FIGURE 2

One of the many uses for the identities in this section is in simplifying trigonometric expressions. In the example that follows, we use the identity $\sin^2 A + \cos^2 A = 1$.

EXAMPLE 5 Combine and simplify: $\dfrac{\sin A}{\cos A} + \dfrac{\cos A}{\sin A}$.

Solution The common denominator is $\cos A \sin A$. Therefore we have

$$\frac{\sin A}{\cos A} + \frac{\cos A}{\sin A} = \frac{\sin A}{\cos A} \cdot \frac{\sin A}{\sin A} + \frac{\cos A}{\sin A} \cdot \frac{\cos A}{\cos A}$$

$$= \frac{\sin^2 A}{\cos A \sin A} + \frac{\cos^2 A}{\sin A \cos A}$$

$$= \frac{\sin^2 A + \cos^2 A}{\cos A \sin A}$$

$$= \frac{1}{\cos A \sin A}$$

This is the required result. If we wish, we can write this answer in an alternative form that doesn't involve fractions:

$$\frac{1}{\cos A \sin A} = \frac{1}{\cos A} \cdot \frac{1}{\sin A} = \sec A \csc A$$

One of the most useful techniques for simplifying a trigonometric expression is first to rewrite it in terms of sines and cosines and then to carry out the usual algebraic simplifications. This is demonstrated in Examples 6 through 8.

EXAMPLE 6 Simplify the expression $\csc A \tan A \cos^2 A$.

Solution We use the basic identities to express everything in terms of sines and cosines.

$$\csc A \tan A \cos^2 A = \frac{1}{\sin A} \cdot \frac{\sin A}{\cos A} \cdot \cos^2 A$$

$$= \frac{\cos^2 A}{\cos A} = \cos A$$

EXAMPLE 7 Express in terms of sines and cosines and simplify: $\dfrac{1 - \tan^2 B}{1 + \tan^2 B}$.

Solution
$$\frac{1 - \tan^2 B}{1 + \tan^2 B} = \frac{1 - \dfrac{\sin^2 B}{\cos^2 B}}{1 + \dfrac{\sin^2 B}{\cos^2 B}}$$

Since $\tan B = \dfrac{\sin B}{\cos B}$, it follows that $\tan^2 B = \dfrac{\sin^2 B}{\cos^2 B}$

$$= \frac{\cos^2 B \left(1 - \dfrac{\sin^2 B}{\cos^2 B} \right)}{\cos^2 B \left(1 + \dfrac{\sin^2 B}{\cos^2 B} \right)}$$

multiplying both the numerator and the denominator by $\cos^2 B$

$$= \frac{\cos^2 B - \sin^2 B}{\cos^2 B + \sin^2 B}$$

$$= \frac{\cos^2 B - \sin^2 B}{1} = \cos^2 B - \sin^2 B$$

EXAMPLE 8 Express in terms of sines and cosines and simplify: $\dfrac{\tan(90° - A)}{\cos A \sin A}$.

Solution
$$\frac{\tan(90° - A)}{\cos A \sin A} = \frac{\dfrac{\sin(90° - A)}{\cos(90° - A)}}{\cos A \sin A}$$

applying the identity $\tan \theta = \dfrac{\sin \theta}{\cos \theta}$

$$= \frac{\dfrac{\cos A}{\sin A}}{\cos A \sin A} \qquad \text{(Why?)}$$

$$= \frac{\cos A}{\sin A} \cdot \frac{1}{\cos A \sin A} = \frac{1}{\sin^2 A}$$

The given expression is therefore equal to $1/\sin^2 A$. We can also write this as $\csc^2 A$.

EXERCISE SET 7.2

A

In Exercises 1–12, carry out the indicated operations.

1. (a) $-SC + 12SC$
   (b) $-\sin\theta\cos\theta + 12\sin\theta\cos\theta$

2. (a) $10SC + 4SC - 16SC$
   (b) $10\sin\theta\cos\theta + 4\sin\theta\cos\theta - 16\sin\theta\cos\theta$

3. (a) $4C^3S - 12C^3S$
   (b) $4\cos^3\theta\sin\theta - 12\cos^3\theta\sin\theta$

4. (a) $-C^2S^2 + (2SC)^2$
   (b) $-\cos^2\theta\sin^2\theta + (2\sin\theta\cos\theta)^2$

5. (a) $(1 + T)^2$
   (b) $(1 + \tan\theta)^2$

6. (a) $(3 - 2T)^2$
   (b) $(3 - 2\tan\theta)^2$

7. (a) $(T + 3)(T - 2)$
   (b) $(\tan\theta + 3)(\tan\theta - 2)$

8. (a) $(S^2 - 3)(S^2 + 3)$
   (b) $(\sec^2\theta - 3)(\sec^2\theta + 3)$

9. (a) $\dfrac{S - C}{C - S}$
   (b) $\dfrac{\sin\theta - \cos\theta}{\cos\theta - \sin\theta}$

10. (a) $\dfrac{5 - 2T}{2T - 5}$
    (b) $\dfrac{5 - 2\tan\theta}{2\tan\theta - 5}$

11. (a) $C + \dfrac{2}{S}$
    (b) $\cos A + \dfrac{2}{\sin A}$

12. (a) $\dfrac{1}{S} - \dfrac{3}{C}$
    (b) $\dfrac{1}{\sin A} - \dfrac{3}{\cos A}$

In Exercises 13–18, factor each expression.

13. (a) $T^2 + 8T - 9$
    (b) $\tan^2\beta + 8\tan\beta - 9$

14. (a) $3S^2 + 2S - 8$
    (b) $3\sec^2\beta + 2\sec\beta - 8$

15. (a) $4C^2 - 1$
    (b) $4\cos^2 B - 1$

16. (a) $16S^3 - 9S^2$
    (b) $16\sin^3 B - 9\sin^2 B$

17. (a) $9S^2T^3 + 6ST^2$
    (b) $9\sec^2 B\tan^3 B + 6\sec B\tan^2 B$

18. (a) $5C^2c^2 - 15Cc$
    (b) $5\csc^2 B\cot^2 B - 15\csc B\cot B$

In Exercises 19–30, use the given information to determine the values of the remaining five trigonometric functions. (Assume that all of the angles are acute.) When radicals appear in a denominator, rationalize the denominator. In Exercises 19–24, use either of the methods shown in Example 4. In Exercises 25–30, use the second method shown in Example 4.

19. $\sin\theta = \frac{3}{4}$
20. $\sin\theta = \frac{2}{5}$
21. $\cos\beta = \sqrt{3}/5$
22. $\cos\beta = \sqrt{7}/3$
23. $\sin A = \frac{5}{13}$
24. $\cos A = \frac{8}{17}$
25. $\tan B = \frac{4}{3}$
26. $\tan B = 5$
27. $\sec C = \frac{3}{2}$
28. $\csc C = \sqrt{5}/2$
29. $\cot\alpha = \sqrt{3}/3$
30. $\cot\alpha = \sqrt{3}/2$

In Exercises 31–34, use a calculator to determine the values of the remaining five trigonometric functions. (In each case, θ denotes an acute angle.) Round off the final answers to three decimal places.

31. $\cos\theta = 0.4626$
32. $\sin\theta = 0.5917$
33. $\tan\theta = 1.1998$
34. $\sec\theta = 2.3283$

In Exercises 35–56, simplify each expression.

35. $\dfrac{\sin^2 A - \cos^2 A}{\sin A - \cos A}$ *Hint:* Factor the numerator.

36. $\dfrac{\sin^4 A - \cos^4 A}{\cos A - \sin A}$

37. $\sin^2\theta\cos\theta\csc^3\theta\sec\theta$

38. $\sin\theta\csc\theta\tan\theta$

39. $\cot B\sin^2 B\cot B$

40. $\dfrac{3\sin\theta + 6}{\sin^2\theta - 4}$

41. $\dfrac{\cos^2 A + \cos A - 12}{\cos A - 3}$

42. $\dfrac{\cos A - 2\sin A\cos A}{\cos^2 A - \sin^2 A + \sin A - 1}$

43. $\dfrac{\tan\theta}{\sec\theta - 1} + \dfrac{\tan\theta}{\sec\theta + 1}$

44. $\cot\theta + \dfrac{1 - 2\cos^2\theta}{\sin\theta\cos\theta}$

45. $\sec A\csc A - \tan A - \cot A$

46. $(\sec A + \tan A)(\sec A - \tan A)$

47. $\dfrac{\cot^2\theta}{\csc^2\theta} + \dfrac{\tan^2\theta}{\sec^2\theta}$

48. $\dfrac{\tan\theta + \tan\theta\sin\theta - \cos\theta\sin\theta}{\sin\theta\tan\theta}$

49. $\dfrac{\cos(90° - \theta)}{\cos\theta}$

50. $\sin^2(90° - \beta) + \cos^2(90° - \beta)$

51. $\dfrac{\cos^2(90° - A)}{\sin^2(90° - A)} - \dfrac{1}{\cos^2 A}$

52. $\dfrac{\cos(90° - A)}{\csc A} + \dfrac{\sin(90° - A)}{\sec A}$

53. $1 - \sin(90° - \theta)\cos\theta$

54. $\dfrac{\cos A \tan A}{\tan(90° - A)} - \dfrac{1}{\sin(90° - A)}$

55. $\dfrac{\dfrac{\cos\theta + 1}{\cos\theta} + 1}{\dfrac{\cos\theta - 1}{\cos\theta} - 1}$

56. $\dfrac{\dfrac{\cot\beta + 1}{\cot\beta} + 1}{\dfrac{\cot\beta - 1}{\cot\beta} - 1}$

B

57. Suppose that

$$A\sin\theta + \cos\theta = 1 \quad \text{and} \quad B\sin\theta - \cos\theta = 1$$

Show that $AB = 1$. *Hint:* Solve the first equation for A, the second for B, and then compute AB.

58. If $\sin\alpha + \cos\alpha = a$ and $\sin\alpha - \cos\alpha = b$, show that

$$\tan\alpha = \frac{a + b}{a - b}$$

59. If $a\sin^2\theta + b\cos^2\theta = 1$, show that

$$\sin^2\theta = \frac{1 - b}{a - b} \quad \text{and} \quad \tan^2\theta = \frac{b - 1}{1 - a}$$

60. This exercise shows how to obtain radical expressions for $\sin 18°$ and $\cos 18°$, using the following figure.

(a) Find $\angle B$, $\angle BDC$, and $\angle ADC$.
(b) Why does $AC = BC = BD$?
For the rest of this problem, assume $AD = 1$.
(c) Why does $CD = 1$?
(d) Let x denote the common lengths of $\overline{AC}$, $\overline{BC}$, and $\overline{BD}$. Use similar triangles to deduce that $x/(1 + x) = 1/x$. Then show that $x = \left(1 + \sqrt{5}\right)/2$.
(e) In $\triangle BDC$, draw an altitude from B to $\overline{DC}$, meeting $\overline{DC}$ at F. Use right triangle BFC to conclude that $\sin 18° = 1/(1 + \sqrt{5})$.
(f) Rationalize the denominator in part (e) to obtain $\sin 18° = \frac{1}{4}\left(\sqrt{5} - 1\right)$.
(g) Use the identity $\sin^2\theta + \cos^2\theta = 1$, along with part (f), to show that

$$\cos 18° = \tfrac{1}{4}\sqrt{10 + 2\sqrt{5}}$$

(h) Use your calculator to check the results in parts (f) and (g).

61. Suppose that β is an acute angle and

$$\sin\beta = \frac{m^2 - n^2}{m^2 + n^2} \quad (m > n > 0)$$

Show that

$$\cos\beta = \frac{2mn}{m^2 + n^2} \quad \text{and} \quad \tan\beta = \frac{m^2 - n^2}{2mn}$$

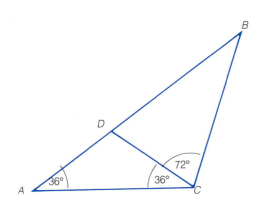

7.3 RIGHT-TRIANGLE APPLICATIONS

We began this chapter in Section 7.1 by defining the six trigonometric functions of an acute angle θ. Then, in Section 7.2, we looked at some of the algebra that is involved in working with the trigonometric functions. Now we are ready to apply these ideas in solving some basic problems involving right triangles.

EXAMPLE I   Use one of the trigonometric functions to find x in Figure 1.

Solution   Relative to the given 30° angle, x is the adjacent side. The length of the hypotenuse is 100 cm. Since the adjacent side and the hypotenuse are involved, we use the cosine function here:

$$\cos 30° = \frac{\text{adjacent}}{\text{hypotenuse}} = \frac{x}{100}$$

Consequently,

$$x = 100 \cos 30° = (100) \cdot \left(\frac{\sqrt{3}}{2}\right) = 50\sqrt{3}\text{ cm}$$

This is the result we are looking for.

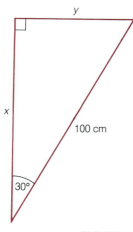

FIGURE I

We used the cosine function in Example 1 because the adjacent side and the hypotenuse were involved. We could instead use the secant. In that case, again with reference to Figure 1, the calculations would look like this:

$$\sec 30° = \frac{\text{hypotenuse}}{\text{adjacent}} = \frac{100}{x}$$

$$\frac{2}{\sqrt{3}} = \frac{100}{x}$$

$$2x = 100\sqrt{3}$$

$$x = \frac{100\sqrt{3}}{2} = 50\sqrt{3}\text{ cm} \qquad \text{as obtained previously}$$

EXAMPLE 2   Find y in Figure 1.

Solution   As you can see from Figure 1, the side of length y is opposite the 30° angle. Furthermore, we are given the length of the hypotenuse. Since the opposite side and the hypotenuse are involved, we use the sine function. This yields

$$\sin 30° = \frac{\text{opposite}}{\text{hypotenuse}} = \frac{y}{100}$$

and therefore

$$y = 100 \sin 30° = (100) \cdot \left(\tfrac{1}{2}\right) = 50\text{ cm}$$

This is the required answer. Actually, we could have obtained this particular result much faster by recalling that in the 30°–60° right triangle, the side opposite the 30° angle, namely y, is half the hypotenuse. That is, $y = \frac{100}{2} = 50$ cm, as obtained previously.

EXAMPLE 3   A ladder, which is leaning against the side of a building, forms an angle of 50° with the ground. If the foot of the ladder is 12 ft from the base of the building, how far up the side of the building does the ladder reach? See Figure 2.

Solution   In Figure 2 we have used y to denote the required distance. Notice that y is opposite the 50° angle, while the given side is adjacent to that angle. Since the opposite and adjacent sides are involved, we'll use the tangent function. (The cotangent function could also be used.) We have

$$\tan 50° = \frac{y}{12}$$

and therefore

$$y = 12 \tan 50°$$

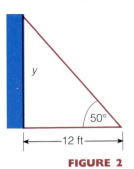

FIGURE 2

Without the use of a calculator or tables, this is our final answer. On the other hand, using a calculator we find that

$$y = 14 \text{ ft} \qquad \text{to the nearest foot}$$

In the next example we derive a useful formula for the area of a triangle. The formula can be used when we are given two sides and the included angle, but not the height.

EXAMPLE 4   Show that the area of the triangle in Figure 3(a) is given by $A = \frac{1}{2}ab \sin \theta$.

FIGURE 3

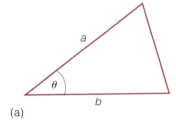
(a)

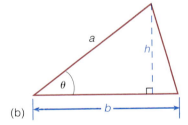
(b)

Solution   We draw an altitude, as shown in Figure 3(b), and we call the length of this altitude h. Then we have

$$\sin \theta = \frac{h}{a}$$

and therefore

$$h = a \sin \theta$$

This value for h can now be used in the usual formula for the area of a triangle:

$$A = \frac{1}{2}bh = \frac{1}{2}b(a \sin \theta) = \frac{1}{2}ab \sin \theta$$

The formula that we just derived in Example 4 is worth remembering. In words, the formula states that *the area of a triangle is equal to half the product of the lengths of two of the sides times the sine of the included acute angle.* In Section 7.4, we will see that the formula remains valid even when the included angle is not an acute angle.

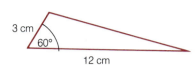

EXAMPLE 5 Find the area of the triangle in Figure 4.

Solution In the formula $A = \frac{1}{2}ab \sin \theta$, we let $a = 3$ cm, $b = 12$ cm, and $\theta = 60°$. Then

$$A = \frac{1}{2}(3)(12) \sin 60° = (18)\frac{\sqrt{3}}{2}$$
$$= 9\sqrt{3} \text{ cm}^2$$

FIGURE 4 This is the required area.

EXAMPLE 6 Figure 5 shows a regular pentagon inscribed in a circle of radius 2 in. (*Regular* means that all of the sides are equal and all of the angles are equal.) Find the area of the pentagon.

Solution The idea here is first to find the area of triangle *BOA* using the area formula from Example 4. Then, since the pentagon is composed of five such identical triangles, the area of the pentagon will be five times the area of triangle *BOA*. We will make use of the result from geometry that, in a regular *n*-sided polygon, the central angle is 360°/*n*. In our case, we therefore have

$$\angle BOA = \frac{360°}{5} = 72°$$

We can now find the area of triangle *BOA*:

$$\text{area} = \frac{1}{2}ab \sin \theta$$
$$= \frac{1}{2}(2)(2) \sin 72° = 2 \sin 72° \text{ in}^2$$

FIGURE 5

The area of the pentagon is five times this, or $10 \sin 72°$ in.^{2} (Using a calculator, this is about 9.51 in.^{2})

Now we introduce some terminology that will be used in the next two examples. Suppose that a surveyor sights an object at a point above the horizontal, as indicated in Figure 6(a). Then the angle between the line of sight and the horizontal is called the **angle of elevation**. The **angle of depression** is similarly defined for an object below the horizontal, as shown in Figure 6(b).

FIGURE 6

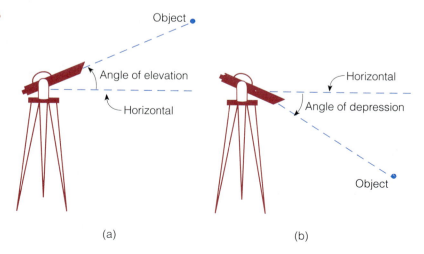

(a)   (b)

EXAMPLE 7 A helicopter hovers 800 ft directly above a small island that is off the California coast. From the helicopter, the pilot takes a sighting to a point P directly ashore on the mainland, at the water's edge. If the angle of depression is 35°, how far off the coast is the island? See Figure 7.

FIGURE 7

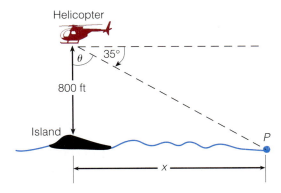

Helicopter

θ 35°

800 ft

Island

P

x

Solution   Let x denote the distance from the island to the mainland. Then, as you can see from Figure 7, we have $\theta + 35° = 90°$, from which it follows that $\theta = 55°$. Now we can write

$$\tan 55° = \frac{x}{800}$$

or

$$x = 800 \tan 55° \approx 1150 \text{ ft}$$ using a calculator and rounding to the nearest 50 feet

EXAMPLE 8 Two satellite tracking stations, located at points A and B in California's Mojave Desert, are 200 miles apart. At a prearranged time, both stations measure the angle of elevation of a satellite as it crosses the vertical plane containing A and B. (See Figure 8.) If the angles of elevation from A and from B are α and β, respectively, express the altitude h of the satellite in terms of α and β.

Solution   We want to express the length $h = SC$ in Figure 8 in terms of the angles α and β. Note that $\overline{SC}$ is a side of both of the right triangles SCA and SCB. Working first in the right triangle SCB, we have

$$\cot \beta = \frac{CA + 200}{h}$$

or

$$h \cot \beta = CA + 200 \tag{1}$$

We can eliminate CA from equation (1) as follows. Looking at right triangle SCA, we have

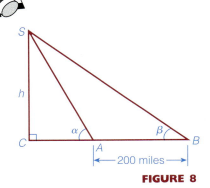

FIGURE 8

$$\cot \alpha = \frac{CA}{h} \quad \text{and thus} \quad CA = h \cot \alpha$$

Using this last equation to substitute for CA in equation (1), we obtain

$$h \cot \beta = h \cot \alpha + 200$$
$$h \cot \beta - h \cot \alpha = 200$$
$$h(\cot \beta - \cot \alpha) = 200$$

$$h = \frac{200}{\cot \beta - \cot \alpha} \text{ miles} \qquad \text{as required}$$

EXAMPLE 9 The arc shown in Figure 9 is a portion of the unit circle, $x^2 + y^2 = 1$. Express the following quantities in terms of θ.

(a) OA **(b)** AB **(c)** OC **(d)** Area of $\triangle OAC$

Solution

(a) In right triangle OAB:

$$\cos \theta = \frac{OA}{OB} = \frac{OA}{1}$$
$$OA = \cos \theta$$

(b) In right triangle OAB:

$$\sin \theta = \frac{AB}{OB} = \frac{AB}{1}$$
$$AB = \sin \theta$$

(c) In right triangle OAC, we have $\cos \theta = OC/OA$, and therefore

$$OC = OA \cos \theta = (\cos \theta)(\cos \theta) = \cos^2 \theta$$

(d) Area $\triangle OAC = \frac{1}{2}(OA)(OC)(\sin \theta)$ using the formula area $= \frac{1}{2}ab \sin \theta$

       $= \frac{1}{2}(\cos \theta)(\cos^2 \theta)(\sin \theta)$ using the results from parts (a) and (c)

       $= \frac{1}{2}\cos^3 \theta \sin \theta$

FIGURE 9

EXERCISE SET 7.3

A

For Exercises 1–6, refer to the following figure. (However, each problem is independent of the others.)

1. If $\angle A = 30°$ and $AB = 60$ cm, find AC and BC.

2. If $\angle A = 60°$ and $AB = 12$ cm, find AC and BC.

3. If $\angle B = 60°$ and $AC = 16$ cm, find BC and AB.

4. If $\angle B = 45°$ and $AC = 9$ cm, find BC and AB.

5. If $\angle B = 50°$ and $AB = 15$ cm, find BC and AC. (Round off your answers to one decimal place.)

6. If $\angle A = 25°$ and $AC = 100$ cm, find BC and AB. (Round off your answers to one decimal place.)

7. A ladder 18 ft long leans against a building. The ladder forms an angle of 60° with the ground.

(a) How high up the side of the building does the ladder reach? [Here and in part (b), give two forms for your answers: one with radicals and one (using a calculator) with decimals, rounded off to two places.]

(b) Find the horizontal distance from the foot of the ladder to the base of the building.

8. From a point level with and 1000 ft away from the base of the Washington Monument, the angle of elevation to the top of the monument is 29.05°. Determine the height of the monument to the nearest half foot.

9. Refer to the accompanying figure. At certain times, the planets Earth and Mercury line up in such a way that $\angle EMS$ is a right angle, as shown in the figure. At such times, $\angle SEM$ is found to be 21.16°. Use this information to estimate the distance MS of Mercury from the Sun. Assume that the distance from the Earth to the Sun is 93 million miles. (Round off your answer to the nearest million miles. Because Mercury's orbit is not exactly circular, the actual distance of Mercury from the Sun varies from about 28 million miles to 43 million miles.)

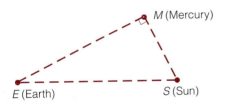

10. Determine the distance AB across the lake shown in the figure, using the following data: $AC = 400$ m, $\angle C = 90°$, and $\angle CAB = 40°$. Round off the answer to the nearest meter.

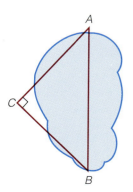

11. A building contractor wants to put a fence around the perimeter of a flat lot with the shape of a right triangle. One angle of the triangle is 41.4°, and the length of the hypotenuse is 58.5 m. Find the length of fencing required. Round off the answer to one decimal place.

12. Suppose that the contractor in Exercise 11 reviews his notes and finds that it is not the hypotenuse that is 58.5 m but rather the side opposite the 41.4° angle. Find the length of fencing required in this case. Again, round off the answer to one decimal place.

For Exercises 13 and 14, refer to following diagram for the roof of a house. In the figure, x is the length of a rafter measured from the top of a wall to the top of the roof; θ is the acute angle between a rafter and the horizontal; and h is the vertical distance from the top of the wall to the top of the roof.

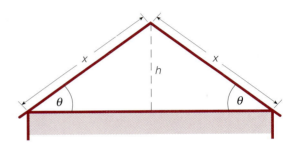

13. Suppose that $\theta = 39.4°$ and $x = 43.0$ ft.
   (a) Determine h. Round off the answer to one decimal place.
   (b) The *gable* is the triangular region bounded by the rafters and the attic ceiling. Find the area of the gable. Round off the final answer to one decimal place.

14. Suppose that $\theta = 34°$ and $h = 36.5$ ft.
   (a) Determine x. Round off the answer to one decimal place.
   (b) Find the area of the gable. Round off the final answer to one decimal place. [See Exercise 13(b) for the definition of gable.]

In Exercises 15 and 16, find the area of the triangle. In Exercise 16, use a calculator and round off the final answer to two decimal places.

15.

16.

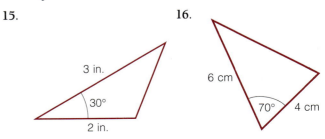

In Exercises 17–22, determine the area of the shaded region, given that the radius of the circle is 1 unit and the inscribed polygon is a regular polygon. Give two forms for each answer: an expression involving radicals or the trigonometric functions; a calculator approximation rounded off to three decimal places.

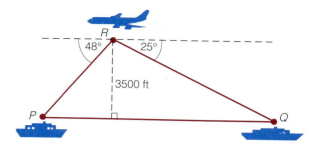

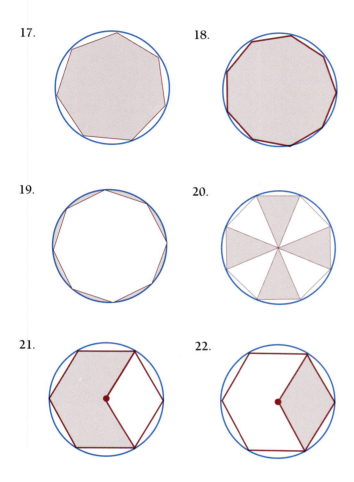

17.

18.

19.

20.

21.

22.

23. Show that the perimeter of the pentagon in Example 6 is 20 sin 36°.   *Hint:* In Figure 5, draw a perpendicular from O to $\overline{AB}$.

24. In triangle OAB, lengths $OA = OB = 6$ in. and $\angle AOB = 72°$. Find AB.   *Hint:* Draw a perpendicular from O to $\overline{AB}$. Round off the answer to one decimal place.

25. The accompanying figure shows two ships at points P and Q, which are in the same vertical plane with an airplane at point R. When the height of the airplane is 3500 ft, the angle of depression to P is 48° and that to Q is 25°. Find the distance between the two ships. Round off the answer to the nearest 10 feet.

26. An observer in a lighthouse is 66 ft above the surface of the water. The observer sees a ship and finds the angle of depression to be 0.7°. Estimate the distance of the ship from the base of the lighthouse. Round off the answer to the nearest 5 feet.

27. From a point on ground level, you measure the angle of elevation to the top of a mountain to be 38°. Then you walk 200 meters farther away from the mountain and find that the angle of elevation is now 20°. Find the height of the mountain. Round off the answer to the nearest meter.

28. A surveyor stands 30 yards from the base of a building. On top of the building is a vertical radio antenna. Let α denote the angle of elevation when the surveyor sights to the top of the building. Let β denote the angle of elevation when the surveyor sights to the top of the antenna. Express the length of the antenna in terms of the angles α and β.

29. In $\triangle ACD$, you are given $\angle C = 90°$, $\angle A = 60°$, and $AC = 18$ cm. If B is a point on $\overline{CD}$ and $\angle BAC = 45°$, find BD. Express the answer in terms of a radical (rather than using a calculator).

30. The radius of the circle in the following figure is 1 unit. Express the lengths OA, AB, and DC in terms of α.

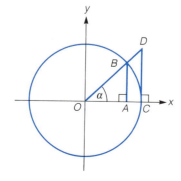

31. The arc in the next figure is a portion of the unit circle, $x^2 + y^2 = 1$.

(a) Express the following angles in terms of θ: $\angle BOA$, $\angle OAB$, $\angle BAP$, $\angle BPA$. (Assume θ is in degrees.)

(b) Express the following lengths in terms of $\sin \theta$ and $\cos \theta$: AO, AP, OB, BP.

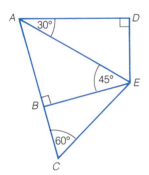

32. In the accompanying figure, suppose that $AD = 1$. Find the length of each of the other line segments in the figure. When radicals appear in an answer, leave the answer in that form, rather than using a calculator.

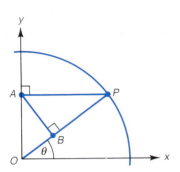

B

33. Refer to the figure. Express each of the following lengths as a function of θ.

(a) BC (b) AB (c) AC

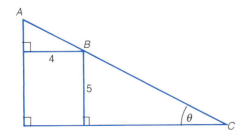

34. If $AB = 8$ in., express x as a function of θ.
Hint: First work Exercise 33.

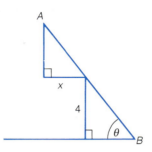

35. In the figure, line segment $\overline{BA}$ is tangent to the unit circle at A. Also, $\overline{CF}$ is tangent to the circle at F. Express the following lengths in terms of θ.

(a) DE (b) OE (c) CF
(d) OC (e) AB (f) OB

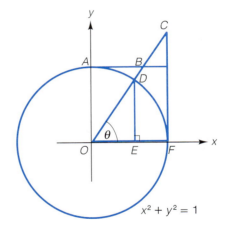

36. At point P on the earth's surface, the moon is observed to be directly overhead, while at the same time at point T, the moon is just visible. See the figure (on page 392).

(a) Show that $MP = \dfrac{OT}{\cos \theta} - OP$.

(b) Use a calculator and the following data to estimate the distance MP from the earth to the moon: $\theta = 89.05°$ and $OT = OP = 4000$ miles. Round off your answer to the nearest thousand miles. (Because the moon's orbit is not really circular, the actual distance varies from about 216,400 miles to 247,000 miles.)

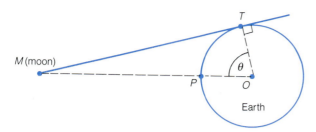

37. Refer to the figure. Let r denote the radius of the moon.

(a) Show that $r = \left(\dfrac{\sin\theta}{1-\sin\theta}\right)PS$.

(b) Use a calculator and the following data to estimate the radius r of the moon: $PS = 238{,}857$ miles and $\theta = 0.257°$. Round off your answer to the nearest 10 miles.

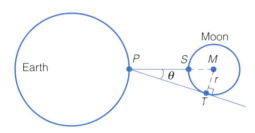

38. Figure A shows a regular hexagon inscribed in a circle of radius 1. Figure B shows a regular heptagon (seven-sided polygon) inscribed in a circle of radius 1. In Figure A, a line segment drawn from the center of the circle perpendicular to one of the sides is called an **apothem** of the polygon.

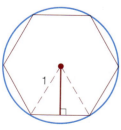

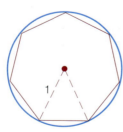

Figure A   Figure B

(a) Show that the length of the apothem in Figure A is $\sqrt{3}/2$.

(b) Show that the length of one side of the heptagon in Figure B is $2\sin(180°/7)$.

(c) Use a calculator to evaluate the expressions in parts (a) and (b). Round off each answer to four decimal places, and note how close the two values are. Approximately two thousand years ago, Heron of Alexandria made use of this coincidence when he used the length of the apothem of the hexagon to approximate the length of the side of the heptagon. (The apothem of the hexagon can be constructed with ruler and compass; the side of the regular heptagon cannot.)

39. (a) Show that the area of a regular n-gon inscribed in a circle of radius 1 unit is given by

$$A_n = \frac{n}{2}\sin\frac{360°}{n}$$

(b) Use a calculator and the formula in part (a) to complete the following table.

n	5	10	50	100	1000	5000	10,000
A_n							

(c) Explain why the successive values of A_n in your table get closer and closer to π.

40. The figure shows a regular pentagon and a regular hexagon, with a common side of length $2\,\text{cm}$. Compute the area within the hexagon but outside of the pentagon. Round off the answer to two decimal places.

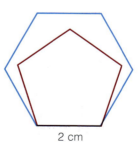

2 cm

C

41. In the accompanying figure, the smaller circle is tangent to the larger circle. Ray PQ is a common tangent and ray PR passes through the centers of both circles. If the radius of the smaller circle is a and the radius of the larger circle is b, show that $\sin\theta = (b-a)/(a+b)$. Then, using the identity $\sin^2\theta + \cos^2\theta = 1$, show that $\cos\theta = 2\sqrt{ab}/(a+b)$.

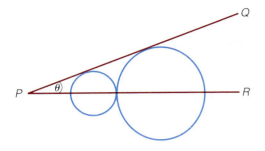

42. A vertical tower of height h stands on level ground. From a point P at ground level and due south of the tower, the angle of elevation to the top of the tower is θ. From a point Q at ground level and due west of the tower, the angle of elevation to the top of the tower is β. If d is the distance between P and Q, show that

$$h = \frac{d}{\sqrt{\cot^2 \theta + \cot^2 \beta}}$$

43. The following problem is taken from *An Elementary Treatise on Plane Trigonometry*, 8th ed., by R. D. Beasley (London: Macmillan and Co. 1884): The [angle of] elevation of a tower standing on a horizontal plane is observed; a feet nearer it is found to be 45°; b feet nearer still it is the complement of what it was at the first station; shew that the height of the tower is $ab/(a-b)$ feet.

7.4 TRIGONOMETRIC FUNCTIONS OF ANGLES

Be clear about the signs of the functions. Your calculator will show them to you correctly, but you still need to be able to figure them out for yourself, quadrant by quadrant.

David Halliday and Robert Resnick in *Fundamentals of Physics,* third edition. (New York: John Wiley and Sons, 1988)

Recall that the definitions of the trigonometric functions given in Section 7.1 apply only to acute angles. That is, as defined on page 369, the domains of trigonometric functions consist of all angles θ such that $0° < \theta < 90°$. In this section, we want to expand our definitions of the trigonometric functions to include angles of any size. To do this, we will need to be more explicit than we have been about angles and their size or measure.

For analytical purposes, we think of the two rays that form an angle as having been originally coincident; then while one ray is held fixed, the other is rotated to create the given angle. As Figure 1 indicates, the fixed ray is called the **initial side** of the angle, and the rotated ray is called the **terminal side.** By convention, we take the measure of an angle to be **positive** if the rotation is counterclockwise and **negative** if the rotation is clockwise. For example, the measure of the angle in Figure 2(a) is *positive* thirty degrees because the rotation is coun-

FIGURE I

FIGURE 2
Angles generated by a counterclockwise rotation have positive measure. Angles generated by a clockwise rotation have negative measure.

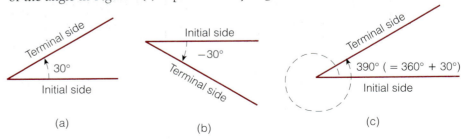

(a) (b) (c)

terclockwise, whereas the measure of the angle in Figure 2(b) is *negative* thirty degrees because the rotation is clockwise. Figure 2(c) shows an angle of 390°.

In our development of trigonometry, it will be convenient to have a *standard position* for angles. In a rectangular coordinate system, an angle is in **standard position** if the vertex is located at $(0, 0)$ and the initial side of the angle lies along the positive horizontal axis. Figure 3 shows examples of angles in standard position.

FIGURE 3
Examples of angles in standard position

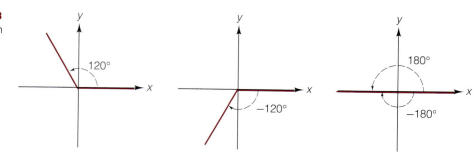

Now we are ready to extend our definitions of the trigonometric functions to accommodate angles of any size. We begin by placing the angle θ in standard position and drawing in the **unit circle** $x^2 + y^2 = 1$, as shown in Figure 4. (Recall from Chapter 3 that the equation $x^2 + y^2 = 1$ represents a circle of radius 1, with center at the origin.) Notice the notation $P(x, y)$ in Figure 4; this stands for the point P, with coordinates (x, y), where the terminal side of angle θ intersects the unit circle. With this notation, we define the six trigonometric functions of θ as follows.

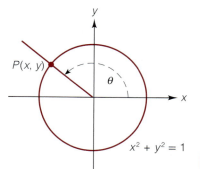

FIGURE 4
$P(x, y)$ denotes the point where the terminal side of angle θ intersects the unit circle.

DEFINITION   Trigonometric Functions of Angles

$$\cos \theta = x \qquad\qquad \sec \theta = \frac{1}{x} \quad (x \neq 0)$$

$$\sin \theta = y \qquad\qquad \csc \theta = \frac{1}{y} \quad (y \neq 0)$$

$$\tan \theta = \frac{y}{x} \quad (x \neq 0) \qquad \cot \theta = \frac{x}{y} \quad (y \neq 0)$$

Even before turning to some examples, we make three initial observations concerning our new definitions. The first two observations will help in memorizing the definitions. The third observation will help you to see why these new definitions are consistent with our previous work on right-triangle trigonometry.

1. $\cos \theta$ is the first coordinate of the point where the terminal side of angle θ intersects the unit circle; $\sin \theta$ is the second coordinate. (You can remember this by noting that, alphabetically, cosine comes before sine.)

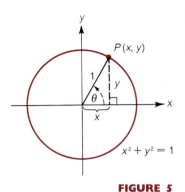

FIGURE 5

2. Just as with the right-triangle definitions, we have the same three pairs of reciprocals: cos θ and sec θ; sin θ and csc θ; tan θ and cot θ.

3. For the cases in which the angle θ is acute, these definitions are really equivalent to the original right-triangle definitions. Consider, for instance, the acute angle θ in Figure 5. According to our "new" definition of sine, we have

$$\sin \theta = y$$

On the other hand, applying the original right-triangle definition to Figure 5 yields

$$\sin \theta = \frac{\text{opposite}}{\text{hypotenuse}} = \frac{y}{1} = y$$

Thus, in both cases we obtain the same result. Using Figure 5, you should check for yourself that the same agreement also occurs for the other trigonometric functions.

EXAMPLE 1 Compute cos 90°, sin 90°, tan 90°, sec 90°, csc 90°, and cot 90°.

Solution We place the angle $\theta = 90°$ in standard position. Then, as Figure 6 indicates, the terminal side of the angle meets the unit circle at the point $(0, 1)$. Now we apply the definitions.

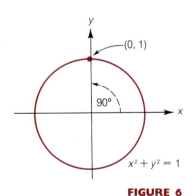

FIGURE 6

(0, 1)

By definition, cos 90° is this number. ⟶⌐ ⌐⟶ By definition, sin 90° is this number.

Thus,

$$\cos 90° = 0 \quad \text{and} \quad \sin 90° = 1$$

For the remaining trigonometric functions of 90°, we have

$$\tan 90° = \frac{y}{x} = \frac{1}{0}, \qquad \text{undefined}$$

$$\sec 90° = \frac{1}{x} = \frac{1}{0}, \qquad \text{undefined}$$

$$\csc 90° = \frac{1}{y} = \frac{1}{1} = 1$$

$$\cot 90° = \frac{x}{y} = \frac{0}{1} = 0$$

These are the required results.

EXAMPLE 2 Evaluate the trigonometric functions of $-180°$.

Solution The results can be read off from Figure 7 (on the next page).

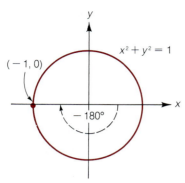

$$\cos(-180°) = x = -1 \qquad \sec(-180°) = \frac{1}{x} = \frac{1}{-1} = -1$$

$$\sin(-180°) = y = 0 \qquad \csc(-180°) = \frac{1}{y} = \frac{1}{0}, \qquad \text{undefined}$$

$$\tan(-180°) = \frac{y}{x} = \frac{0}{-1} = 0 \qquad \cot(-180°) = \frac{x}{y} = \frac{-1}{0}, \qquad \text{undefined}$$

FIGURE 7

In the two examples just concluded, we evaluated the trigonometric functions for $\theta = 90°$ and $\theta = -180°$. In the same manner, the trigonometric functions can be evaluated just as easily for any angle that is an integral multiple of $90°$. Table 1 shows the results of such calculations. Exercise 14 at the end of this section asks you to make these calculations for yourself.

TABLE I

θ	$\cos \theta$	$\sin \theta$	$\tan \theta$	$\sec \theta$	$\csc \theta$	$\cot \theta$
0°	1	0	0	1	undefined	undefined
90°	0	1	undefined	undefined	1	0
180°	−1	0	0	−1	undefined	undefined
270°	0	−1	undefined	undefined	−1	0
360°	1	0	0	1	undefined	undefined

EXAMPLE 3   Use Figure 8 (on the next page) to approximate the following trigonometric values to within successive tenths. Then use a calculator to check your answers. Round off the calculator values to two decimal places.

(a) $\cos 160°$ and $\sin 160°$ **(b)** $\cos(-40°)$ and $\sin(-40°)$

Solution   **(a)** Figure 8 shows that (the terminal side of) an angle of $160°$ in standard position meets the unit circle at a point in the second quadrant. Letting (x, y) denote the coordinates of that point, we have (from Figure 8)

$$-1.0 < x < -0.9 \qquad \text{and} \qquad 0.3 < y < 0.4$$

But, by definition, $\cos 160° = x$ and $\sin 160° = y$ and, consequently,

$$-1.0 < \cos 160° < -0.9 \qquad \text{and} \qquad 0.3 < \sin 160° < 0.4$$

The corresponding calculator results here are $\cos 160° \approx -0.94$ and $\sin 160° \approx 0.34$. Note that the estimations we obtained from Figure 8 are consistent with these calculator values. (When you check these calculator values for yourself, be sure that your calculator is set in the degree mode.)

(b) As you can verify using Figure 8, the terminal side of an angle of $-40°$ intersects the unit circle at the same point as does the terminal side of an angle of 320°. Letting (x, y) denote the coordinates of that point, we have (from Figure 8)

$$0.7 < x < 0.8 \quad \text{and} \quad -0.7 < y < -0.6$$

and, consequently,

$$0.7 < \cos(-40°) < 0.8 \quad \text{and} \quad -0.7 < \sin(-40°) < -0.6$$

The corresponding calculator values here are $\cos(-40°) \approx 0.77$ and $\sin(-40°) \approx -0.64$. Again, note that the estimations we obtain from Figure 8 are consistent with these calculator values. ▮▮▮

FIGURE 8

Degree measure on the unit circle for angles in standard position

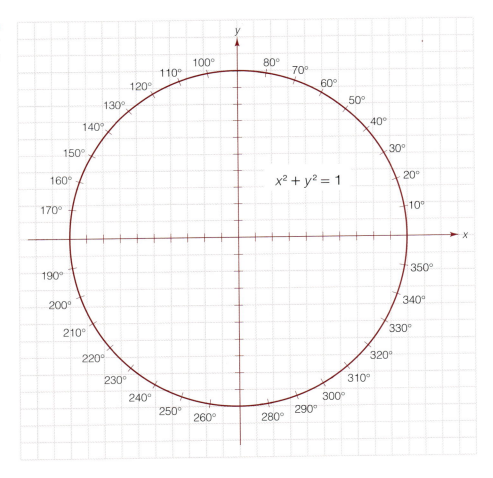

In evaluating the trigonometric functions, we'd like to take advantage, as much as possible, of the symmetry in Figure 8. To do this, we introduce the concept of a *reference angle*. (Even when we are using a calculator with the trigonometric functions, there are times, nevertheless, when we will need the reference angle concept. For instance, see Example 9 in Section 8.8.)

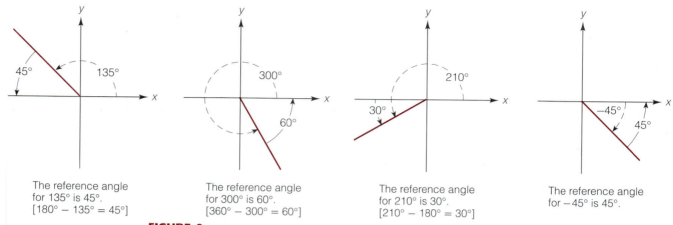

The reference angle for 135° is 45°. [180° − 135° = 45°]

The reference angle for 300° is 60°. [360° − 300° = 60°]

The reference angle for 210° is 30°. [210° − 180° = 30°]

The reference angle for −45° is 45°.

FIGURE 9

> **DEFINITION   The Reference Angle**
>
> Let θ be an angle in standard position, and suppose that θ is not a multiple of 90°. The **reference angle** associated with θ is the acute angle (with positive measure) formed by the x-axis and the terminal side of the angle θ.

In Figure 9 we show four examples of angles and their respective reference angles. The first part of Figure 9 shows how to find the reference angle for $\theta = 135°$. First we place the angle $\theta = 135°$ in standard position. Then we find the acute angle between the x-axis and the terminal side of θ. As you can see in this case, this acute angle is 45°; that is the reference angle associated with $\theta = 135°$. In the same way now, you should work through the remaining three parts of Figure 9.

Now let's look at an example to see how reference angles are used in evaluating the trigonometric functions. Suppose that we want to evaluate cos 150°. In Figure 10(a) we've placed the angle $\theta = 150°$ in standard position. As you can see, the reference angle for 150° is 30°. By definition, the value of cos 150° is the x-coordinate of the point P in Figure 10(a). To find this x-coordinate, we

FIGURE 10

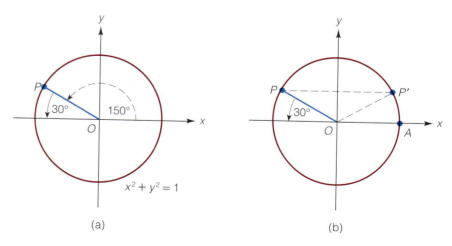

(a)

(b)

reflect the line segment $\overline{OP}$ in the y-axis; the reflected line segment is the segment $\overline{OP'}$ in Figure 10(b). Since $\angle P'OA = 30°$, the x-coordinate of the point P' is by definition cos 30°, or $\sqrt{3}/2$. The x-coordinate of P is then the negative of this. That is, the x-coordinate of P is $-\sqrt{3}/2$. It follows now, again by definition, that the value of cos 150° is $-\sqrt{3}/2$.

The same method that we have just used to evaluate cos 150° can be used to evaluate any of the trigonometric functions when the angles are not multiples of 90°. The following three steps summarize this method.

STEP 1 Determine the reference angle associated with the given angle.
STEP 2 Evaluate the given trigonometric function using the reference angle for the input.
STEP 3 Affix the appropriate sign to the number found in Step 2.

The next two examples illustrate this procedure.

EXAMPLE 4 Evaluate the following quantities.

(a) sin 135°   (b) cos 135°   (c) tan 135°

Solution   As Figure 11 indicates, the reference angle associated with 135° is 45°.

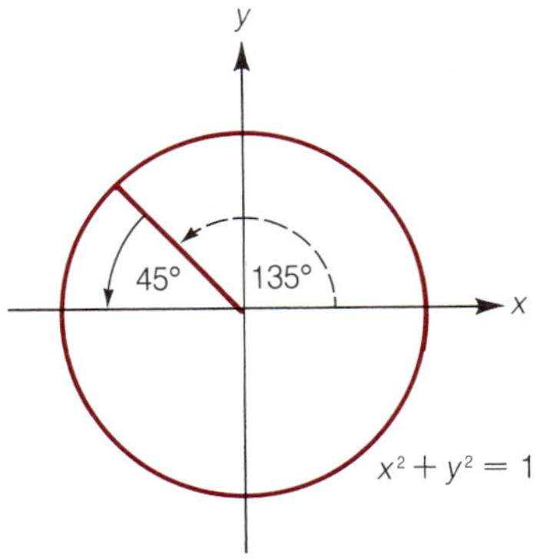

(a) STEP 1 The reference angle is 45°.
    STEP 2 sin 45° = $\sqrt{2}/2$
    STEP 3 sin θ is the y-coordinate. In the second quadrant, y-coordinates are positive. Thus sin 135° is positive, since the terminal side of $\theta = 135°$ lies in the second quadrant. We have therefore

$$\sin 135° = \frac{\sqrt{2}}{2}$$

FIGURE 11

(b) STEP 1 The reference angle is 45°.
    STEP 2 cos 45° = $\sqrt{2}/2$
    STEP 3 cos θ is the x-coordinate. In the second quadrant, x-coordinates are negative. Thus cos 135° is negative, since the terminal side of $\theta = 135°$ lies in the second quadrant. We have therefore

$$\cos 135° = -\frac{\sqrt{2}}{2}$$

(c) STEP 1 The reference angle is 45°.
    STEP 2 tan 45° = 1
    STEP 3 By definition, tan $\theta = y/x$. Now, the terminal side of $\theta = 135°$ lies in the second quadrant, in which y is positive and x is negative. Thus tan 135° is negative. We have therefore

$$\tan 135° = -1$$

EXAMPLE 5 Evaluate each of the following quantities.

(a) cos(−120°)   (b) cot(−120°)   (c) sec(−120°)

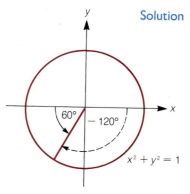

FIGURE 12

Solution   As Figure 12 shows, the reference angle for $-120°$ is $60°$.

(a) STEP 1 The reference angle for $-120°$ is $60°$.
STEP 2 $\cos 60° = \frac{1}{2}$
STEP 3 $\cos \theta$ is the x-coordinate. In the third quadrant, the x-coordinates are negative. Thus, $\cos(-120°)$ is negative, since the terminal side of $\theta = -120°$ lies in the third quadrant. It now follows that

$$\cos(-120°) = -\frac{1}{2}$$

(b) STEP 1 The reference angle for $-120°$ is $60°$.
STEP 2 $\cot 60° = \sqrt{3}/3$
STEP 3 By definition, $\cot \theta = x/y$. Now, the terminal side of $\theta = -120°$ lies in the third quadrant, in which both x and y are negative. So x/y is positive and we have

$$\cot(-120°) = \frac{\sqrt{3}}{3}$$

(c) We could follow our three-step procedure here, but in this case there is a faster method. In part (a) of this example, we found that $\cos(-120°) = -\frac{1}{2}$. Therefore $\sec(-120°) = -2$, because (according to the unit-circle definitions) $\sec \theta$ is the reciprocal of $\cos \theta$. ▮▮▮

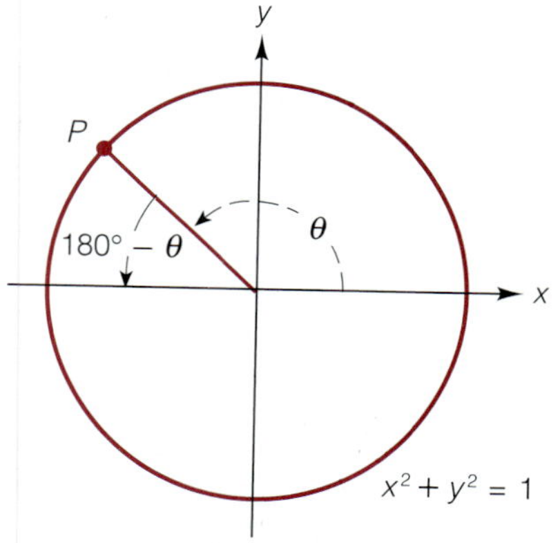

In this section we have used the unit circle to generalize our definitions of the trigonometric functions to accommodate angles of any size. We conclude this section now with an application that relies on both the unit-circle definitions and the right-triangle definitions. Recall that in the previous section we found that the area of a triangle is given by the formula $A = \frac{1}{2}ab \sin \theta$, where a and b are the lengths of two sides and θ is the angle included between those two sides. When we derived this result in the previous section, we were assuming that θ was an acute angle. In fact, however, the formula is still valid when θ is an obtuse angle (an angle between $90°$ and $180°$). To establish this result, we'll need to rely on the following identity.

$$\sin(180° - \theta) = \sin \theta$$

FIGURE 13

We will use Figure 13 to establish this identity for the case in which θ is an obtuse angle. (Exercise 53 shows you how to prove this identity when θ is an acute angle, and Section 8.6 discusses identities such as this from a more general point of view.) Figure 13 shows an obtuse angle θ in standard position. Let us apply our three-step procedure to determine $\sin \theta$. (As you'll see, the end result will be the required identity.)

STEP 1 As indicated in Figure 13, the reference angle for θ is $180° - \theta$.
STEP 2 The sine of the reference angle is $\sin(180° - \theta)$.
STEP 3 $\sin \theta$ is positive because the terminal side of θ lies in the second quadrant. Therefore, we have

$$\sin \theta = \sin(180° - \theta) \qquad \text{as required}$$

We are now prepared to show that the area A of the triangle in Figure 14(a) is given by $A = \frac{1}{2}ab \sin \theta$.

The area of any triangle is given by $A = \frac{1}{2}$(base)(height). Thus, referring to Figure 14(b), we have

$$A = \tfrac{1}{2}bh \qquad (1)$$

FIGURE 14

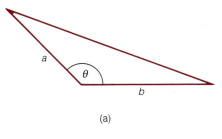

(a)

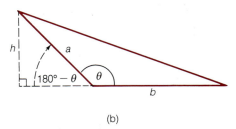

(b)

Also from Figure 14(b), we have

$$\sin(180° - \theta) = \frac{\text{opposite}}{\text{hypotenuse}} = \frac{h}{a}$$

$$h = a \sin(180° - \theta)$$

$$h = a \sin \theta \qquad \text{using the identity } \sin(180° - \theta) = \sin \theta \qquad (2)$$

Now we can use equation (2) to substitute for h in equation (1). This yields

$$A = \tfrac{1}{2}b(a \sin \theta) = \tfrac{1}{2}ab \sin \theta$$

The formula we've just derived is a useful one. We will use it in later work to prove the *law of sines*. For reference, we summarize the result in the box that follows.

Formula for the Area of a Triangle

If a and b are the lengths of two sides of a triangle and θ is the angle included between those two sides, then the area of the triangle is given by

$$\text{area} = \tfrac{1}{2}ab \sin \theta$$

In Words The area of a triangle equals half the product of the lengths of two sides times the sine of the included angle.

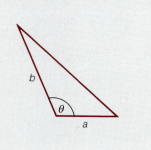

EXERCISE SET 7.4

A

In Exercises 1–8, sketch each angle in standard position and specify the reference angle.

1. (a) 110°
(b) −110°

2. (a) 160°
(b) −160°

3. (a) 200°
(b) −200°

4. (a) 225°
(b) −225°

5. (a) 300°
(b) −300°

6. (a) 325°
(b) −325°

7. (a) 60°
(b) −60°

8. (a) 460°
(b) −460°

In Exercises 9–13, use the definitions (not a calculator) to evaluate the six trigonometric functions of the angle. If a value is undefined, state this.

9. 270°   **10.** 450°   **11.** −270°

12. −630°   **13.** 810°

14. Use the definitions of the trigonometric functions to complete the table. (When you are finished, check your results against the values shown in Table 1.)

θ	$\cos \theta$	$\sin \theta$	$\tan \theta$	$\sec \theta$	$\csc \theta$	$\cot \theta$
0°						
90°						
180°						
270°						
360°						

In Exercises 15–28, use Figure 8 to approximate within successive tenths the given trigonometric values. Then use a calculator to approximate the values to the nearest hundredth.

15. sin 10° and sin(−10°)   **16.** cos 10° and cos(−10°)

17. cos 80° and cos(−80°)   **18.** sin 80° and sin(−80°)

19. sin 120° and sin(−120°)

20. cos 120° and cos(−120°)

21. sin 150° and sin(−150°)

22. cos 150° and cos(−150°)

23. cos 220° and cos(−220°)

24. sin 220° and sin(−220°)

25. cos 310° and cos(−310°)

26. sin 310° and sin(−310°)

27. sin(40° + 360°)

28. cos(40° + 360°)

In Exercises 29–38, evaluate each expression using the method shown in Examples 4 and 5.

29. (a) cos 315°   (b) cos(−315°)
       (c) sin 315°   (d) sin(−315°)

30. (a) cos 240°   (b) cos(−240°)
       (c) sin 240°   (d) sin(−240°)

31. (a) cos 300°   (b) cos(−300°)
       (c) sin 300°   (d) sin(−300°)

32. (a) cos 150°   (b) cos(−150°)
       (c) sin 150°   (d) sin(−150°)

33. (a) cos 210°   (b) cos(−210°)
       (c) sin 210°   (d) sin(−210°)

34. (a) cos 585°   (b) cos(−585°)
       (c) sin 585°   (d) sin(−585°)

35. (a) cos 390°   (b) cos(−390°)
       (c) sin 390°   (d) sin(−390°)

36. (a) cos 405°   (b) cos(−405°)
       (c) sin 405°   (d) sin(−405°)

37. (a) sec 600°   (b) csc(−600°)
       (c) tan 600°   (d) cot(−600°)

38. (a) sec 330°   (b) csc(−330°)
       (c) tan 330°   (d) cot(−330°)

In Exercises 39 and 40, complete the tables.

39.

θ	$\sin \theta$	$\cos \theta$	$\tan \theta$
0°			
30°			
45°			
60°			
90°			
120°			
135°			
150°			
180°			

40.

θ	$\sin \theta$	$\cos \theta$	$\tan \theta$
180°			
210°			
225°			
240°			
270°			
300°			
315°			
330°			
360°			

In Exercises 41–44, use the given information to determine the area of each triangle.

41. Two of the sides are 5 cm and 7 cm, and the angle between those sides is 120°.

42. Two of the sides are 3 m and 6 m, and the included angle is 150°.

43. Two of the sides are 21.4 cm and 28.6 cm, and the included angle is 98.5°. (Round off the final answer to one decimal place.)

44. Two of the sides are 5.98 cm and 8.05 cm, and the included angle is 107.1°. (Round off the answer to one decimal place.)

45. The figure shows an equilateral triangle inscribed in a circle of radius 12 cm. Use the method of Example 6 in Section 7.3 to compute the area of the triangle. Give two forms for your answer: one with a square root and the other a calculator approximation rounded off to two decimal places.

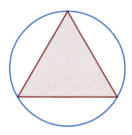

46. An equilateral triangle is inscribed in a circle of radius 8 cm. In each case, compute the area of the shaded region. Round off your final answers to three decimal places.

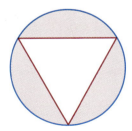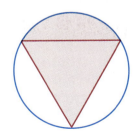

B

47. (a) Complete the table, using the words "positive" or "negative" as appropriate.

	Terminal Side of Angle θ Lies in			
	Quadrant I	Quadrant II	Quadrant III	Quadrant IV
cos θ and sec θ	positive	negative		
sin θ and csc θ				
tan θ and cot θ				

(b) The mnemonic (memory device) ASTC (all students *take calculus*) is sometimes used to recall the signs of the trigonometric values in each quadrant:

A . . . . . . All are positive in the first quadrant.

S . . . . . . Sine is positive in the second quadrant.

T . . . . . . Tangent is positive in third quadrant.

C . . . . . . Cosine is positive in the fourth quadrant.

Check the validity of this mnemonic against your chart in part (a).

48. The value of tan θ is undefined when θ = 90°. Use a calculator to complete the following tables to investigate the behavior of tan θ as θ approaches 90°. Round off each value to the nearest integer.

(a)

θ	89°	89.9°	89.99°	89.999°
tan θ				

(b)

θ	91°	90.1°	90.01°	90.001°
tan θ				

In Exercises 49–52, determine which of the two trigonometric values is the larger. Do not use a calculator. Hint: Sketch the angles in standard position or refer to Figure 8.

49. (a) sin 10° or sin 70°
(b) cos 10° or cos 70°

50. (a) tan 140° or tan 40°
(b) tan 130° or tan 230°

51. (a) cos 140° or cos 100°
(b) sin 140° or sin 100°

52. (a) sec 10° or sec 80°
(b) csc 190° or csc 250°

53. In the text we derived the identity $\sin(180° - \theta) = \sin \theta$ in the case when θ is an obtuse angle. Now we'll derive the identity when θ is an acute angle.   *Note:* If θ is an acute angle, then $180° - \theta$ is an obtuse angle.)
    (a) Sketch a figure showing the obtuse angle $180° - \theta$ in standard position. Note that the terminal side of the angle $180° - \theta$ lies in the second quadrant.
    (b) What is the reference angle for the angle $180° - \theta$?
    (c) Use steps (a) and (b) to conclude that $\sin(180° - \theta) = \sin \theta$.

54. Check that the identity $\sin(180° - \theta) = \sin \theta$ is valid in the three cases $\theta = 0°$, $\theta = 90°$, and $\theta = 180°$.

55. *Formula for* $\sin(\alpha + \beta)$ In the following figure, $\overline{AD} \perp \overline{BC}$ and $AD = 1$.

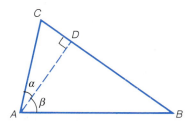

    (a) Show that $AC = \sec \alpha$ and $AB = \sec \beta$.
    (b) Show that

    $$\text{area } \triangle ADC = \tfrac{1}{2} \sec \alpha \sin \alpha$$
    $$\text{area } \triangle ADB = \tfrac{1}{2} \sec \beta \sin \beta$$
    $$\text{area } \triangle ABC = \tfrac{1}{2} \sec \alpha \sec \beta \sin(\alpha + \beta)$$

    (c) The sum of the areas of the two smaller triangles in part (b) equals the area of $\triangle ABC$. Use this fact and the expressions given in part (b) to show that

    $$\sin(\alpha + \beta) = \sin \alpha \cos \beta + \cos \alpha \sin \beta$$

    (d) Use the formula in part (c) to compute $\sin 75°$.
        Hint: $75° = 30° + 45°$
    (e) Show that $\sin 75° \neq \sin 30° + \sin 45°$.
    (f) Compute $\sin 105°$ and then check that $\sin 105° \neq \sin 45° + \sin 60°$.

For Exercises 56–59, let f, g, and h be the following three functions.

$$f(\theta) = \sin \theta \qquad g(\theta) = \cos \theta \qquad h(\theta) = \tan \theta$$

56. Evaluate each of the following (without using a calculator).
    (a) $f(30°)$ (b) $g(135°)$ (c) $h(-60°)$

57. If $\theta = 60°$, evaluate each of the following (without using a calculator).
    (a) $h(\theta)$ (b) $h(2\theta)$ (c) $h(\theta/2)$

58. If $\theta = 15°$, evaluate each of the following (without using a calculator).
    (a) $f(-2\theta)$ (b) $f(-4\theta)$
    (c) $[f(\theta)]^2 + [g(\theta)]^2$

59. Use a calculator to evaluate each of the following. Round off your answers to three decimal places.
    (a) $f(27°)/g(27°)$ (b) $h(27°)$
    (c) $[f(27°)]^2 + [g(27°)]^2$

C

60. Use the figure to prove the following theorem: The area of a quadrilateral is equal to half the product of the diagonals times the sine of the included angle.
    Hint: Find the areas of each of the four triangles that make up the quadrilateral. Show that the sum of those areas is $\tfrac{1}{2}(\sin \theta)[qs + rq + rp + ps]$. Then factor the quantity within the brackets.

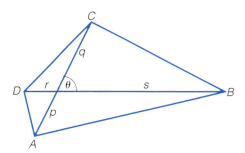

7.5   **TRIGONOMETRIC IDENTITIES**

A special name for the function which we call the sine is first found in the works of Āryabhata (c. 510) . . . .

It is further probable, from the efforts made to develop simple tables, that the Hindus were acquainted with the principles which we represent by the . . . [formula] $\sin^2 \phi + \cos^2 \phi = 1$ *. . . .*

David Eugene Smith in *History of Mathematics*, vol. II (New York: Ginn and Company, 1925)

As we mentioned in Section 7.2, there are many identities involving the trigonometric functions. In fact, that is one reason why these functions are so useful. In the box that follows, we list some of the same basic identities that we studied in Section 7.2. Keep in mind, however, that now the identities will be true for all angles for which the expressions are defined, rather than for acute angles only. This means that we will need to pay attention to the signs of the trigonometric values. (You'll see how this works in Examples 1 and 2.)

PROPERTY SUMMARY   BASIC TRIGONOMETRIC IDENTITIES

1. $\sin^2 \theta + \cos^2 \theta = 1$

2. $\dfrac{\sin \theta}{\cos \theta} = \tan \theta$

3. $\sec \theta = \dfrac{1}{\cos \theta}$; $\csc \theta = \dfrac{1}{\sin \theta}$; $\cot \theta = \dfrac{1}{\tan \theta}$

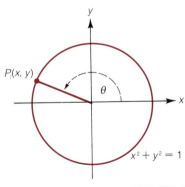

FIGURE I

The second and the third identities in the box are immediate consequences of the unit-circle definitions of the trigonometric functions. To see why the first identity in the box is valid, consider Figure 1. Since the point $P(x, y)$ lies on the unit circle, we have

$$x^2 + y^2 = 1$$

But, by definition, $x = \cos \theta$ and $y = \sin \theta$. Therefore, we have

$$\cos^2 \theta + \sin^2 \theta = 1$$

which is essentially what we wished to show. Incidentally, you should also become familiar with the equivalent forms of this identity:

$$\cos^2 \theta = 1 - \sin^2 \theta \qquad \text{and} \qquad \sin^2 \theta = 1 - \cos^2 \theta$$

EXAMPLE I Given that $\sin \theta = \frac{2}{3}$ and $90° < \theta < 180°$, find $\cos \theta$ and $\tan \theta$.

Solution Substituting $\sin \theta = \frac{2}{3}$ into the identity $\cos^2 \theta = 1 - \sin^2 \theta$ yields

$$\cos^2 \theta = 1 - \left(\frac{2}{3}\right)^2 = \frac{5}{9}$$

$$\cos \theta = \pm\sqrt{\frac{5}{9}} = \frac{\pm\sqrt{5}}{3}$$

To decide whether to choose the positive or negative value, note that the given inequality $90° < \theta < 180°$ tells us that the terminal side of θ lies in the second quadrant. Since in the second quadrant x-coordinates are negative, we choose the negative value here. Thus

$$\cos \theta = \frac{-\sqrt{5}}{3}$$

For tan θ, we have

$$\tan \theta = \frac{y}{x} = \frac{\sin \theta}{\cos \theta} = \frac{2/3}{-\sqrt{5}/3}$$

$$= -\frac{2}{\sqrt{5}} = -\frac{2\sqrt{5}}{5}$$

EXAMPLE 2   Suppose that

$$\cos \theta = t/2$$

where $270° < \theta < 360°$ and $t > 0$. Express the other five trigonometric values as functions of t.

Solution   Replacing $\cos \theta$ with the quantity $t/2$ in the identity $\sin^2 \theta = 1 - \cos^2 \theta$ yields

$$\sin^2 \theta = 1 - \left(\frac{t}{2}\right)^2 = 1 - \frac{t^2}{4} = \frac{4 - t^2}{4}$$

and, consequently,

$$\sin \theta = \pm\frac{\sqrt{4 - t^2}}{2}$$

To decide whether to choose the positive or the negative value here, note that the given inequality $270° < \theta < 360°$ tells us that the terminal side of θ lies in the fourth quadrant. Since in the fourth quadrant y-coordinates are negative, we choose the negative value here. Thus

$$\sin \theta = -\frac{\sqrt{4 - t^2}}{2}$$

To obtain $\tan \theta$, we use the identity $\tan \theta = (\sin \theta)/(\cos \theta)$. This yields

$$\tan \theta = \frac{-\sqrt{4 - t^2}/2}{t/2} = -\frac{\sqrt{4 - t^2}}{t}$$

The remaining three values can now be found simply by taking reciprocals:

$$\sec \theta = \frac{1}{\cos \theta} = \frac{2}{t}$$

$$\csc \theta = \frac{1}{\sin \theta} = -\frac{2}{\sqrt{4 - t^2}} = -\frac{2\sqrt{4 - t^2}}{4 - t^2} \qquad \text{rationalizing the denominator}$$

$$\cot \theta = \frac{1}{\tan \theta} = -\frac{t}{\sqrt{4 - t^2}} = -\frac{t\sqrt{4 - t^2}}{4 - t^2} \qquad \text{rationalizing the denominator}$$

In the next five examples, we are asked to show that certain trigonometric equations are, in fact, identities. The procedures here will be very much like those in our work on simplifying trigonometric expressions in Section 7.2, but now we have an advantage. In each case we are given an answer toward which to work. The identities in the next five examples should not be memorized; they are too specialized. Instead, you should concentrate on the proofs themselves, noting where the fundamental identities (such as $\sin^2 \theta + \cos^2 \theta = 1$) come into play.

EXAMPLE 3   Prove that the equation $\csc A \tan A \cos A = 1$ is an identity.

Solution   We begin with the left-hand side and express each factor in terms of sines or cosines;

$$\csc A \tan A \cos A = \frac{1}{\cancel{\sin A}} \cdot \frac{\cancel{\sin A}^{1}}{\cancel{\cos A}_{1}} \cdot \cancel{\cos A}^{1}$$

$$= 1 \qquad \text{as required}$$

EXAMPLE 4   Prove that $1 - \dfrac{\cot^2 \theta}{\csc^2 \theta} = \sin^2 \theta$.

Solution   We work with the left-hand side, expressing everything in terms of sines and cosines and then simplifying (as in Section 7.2).

$$1 - \frac{\cot^2 \theta}{\csc^2 \theta} = 1 - \frac{(\cos^2 \theta)/(\sin^2 \theta)}{1/\sin^2 \theta}$$

Since $\tan \theta = (\sin \theta)/(\cos \theta)$, it follows that $\tan^2 \theta = (\sin^2 \theta)/(\cos^2 \theta)$ and consequently $\cot^2 \theta = (\cos^2 \theta)/(\sin^2 \theta)$.

$$= 1 - \frac{\cos^2 \theta}{1}$$

Multiplying both numerator and denominator of the fraction by $\sin^2 \theta$

$$= \sin^2 \theta$$

EXAMPLE 5   Prove that $\cos^2 B - \sin^2 B = \dfrac{1 - \tan^2 B}{1 + \tan^2 B}$.

Solution   We begin with the right-hand side this time; it is the more complicated expression. As in the previous examples, we express everything in terms of sines and cosines.

$$\frac{1 - \tan^2 B}{1 + \tan^2 B} = \frac{1 - \dfrac{\sin^2 B}{\cos^2 B}}{1 + \dfrac{\sin^2 B}{\cos^2 B}} = \frac{\cos^2 B \left(1 - \dfrac{\sin^2 B}{\cos^2 B}\right)}{\cos^2 B \left(1 + \dfrac{\sin^2 B}{\cos^2 B}\right)}$$

$$= \frac{\cos^2 B - \sin^2 B}{\cos^2 B + \sin^2 B} = \frac{\cos^2 B - \sin^2 B}{1} = \cos^2 B - \sin^2 B$$

EXAMPLE 6   Prove that $\dfrac{\cos \theta}{1 - \sin \theta} = \dfrac{1 + \sin \theta}{\cos \theta}$.

Solution   The suggestions given in the previous examples are not applicable here. Everything is already in terms of sines and cosines. Furthermore, neither side appears more complicated than the other. A technique that does work here is to begin with the left-hand side and multiply numerator *and* denominator by the same quantity, namely, $1 + \sin\theta$. Doing so gives us

$$\frac{\cos\theta}{1-\sin\theta} = \frac{\cos\theta}{1-\sin\theta} \cdot \frac{1+\sin\theta}{1+\sin\theta}$$

$$= \frac{\cos\theta(1+\sin\theta)}{1-\sin^2\theta}$$

$$= \frac{\cos\theta(1+\sin\theta)}{\cos^2\theta} = \frac{1+\sin\theta}{\cos\theta}$$

The general strategy for each of the proofs in Examples 3 through 6 was the same. In each case we worked with one side of the given equation, and we transformed it (into equivalent expressions) until it was identical to the other side of the equation. This is not the only strategy that can be used. For instance, an alternate way to establish the identity in Example 6 is as follows. The given equation is equivalent to

$$\frac{\cos\theta}{1-\sin\theta} - \frac{1+\sin\theta}{\cos\theta} = 0 \qquad (1)$$

Now we show that equation (1) is an identity. To do this, we combine the two fractions on the left-hand side, using the common denominator $\cos\theta(1-\sin\theta)$:

$$\frac{\cos\theta}{1-\sin\theta} - \frac{1+\sin\theta}{\cos\theta} = \frac{\cos^2\theta - (1+\sin\theta)(1-\sin\theta)}{\cos\theta(1-\sin\theta)}$$

$$= \frac{\cos^2\theta - (1-\sin^2\theta)}{\cos\theta(1-\sin\theta)}$$

$$= \frac{\cos^2\theta + \sin^2\theta - 1}{\cos\theta(1-\sin\theta)} = \frac{1-1}{\cos\theta(1-\sin\theta)} = 0$$

This shows that equation (1) is an identity, and thus the equation given in Example 6 is an identity.

Another strategy that can be used in establishing trigonometric identities is to work independently with each side of the given equation until a common expression is obtained. This is the strategy used in Example 7.

EXAMPLE 7 Prove that $\dfrac{1}{1-\cos\beta} + \dfrac{1}{1+\cos\beta} = 2 + 2\cot^2\beta$.

Solution

$$\begin{aligned} \text{Left-hand} \atop \text{side} &= \frac{1(1+\cos\beta) + 1(1-\cos\beta)}{(1-\cos\beta)(1+\cos\beta)} \\ &= \frac{2}{1-\cos^2\beta} \end{aligned}$$

$$\begin{aligned} \text{Right-hand} \atop \text{side} &= 2 + \frac{2\cos^2\beta}{\sin^2\beta} \\ &= \frac{2\sin^2\beta + 2\cos^2\beta}{\sin^2\beta} \\ &= \frac{2(\sin^2\beta + \cos^2\beta)}{1-\cos^2\beta} \\ &= \frac{2}{1-\cos^2\beta} \end{aligned}$$

We've now established the required identity by showing that both sides are equal to the same quantity.

EXERCISE SET 7.5

A

In Exercises 1–8, use the given information to determine the remaining five trigonometric values. Rationalize any denominators that contain radicals.

1. (a) $\sin \theta = \frac{1}{5}$, $90° < \theta < 180°$
    (b) $\sin \theta = -\frac{1}{5}$, $180° < \theta < 270°$
2. (a) $\cos \theta = -\frac{3}{4}$, $90° < \theta < 180°$
    (b) $\cos \theta = \frac{3}{4}$, $270° < \theta < 360°$
3. (a) $\cos \theta = \frac{5}{13}$, $0° < \theta < 90°$
    (b) $\cos \theta = -\frac{5}{13}$, $180° < \theta < 270°$
4. (a) $\sin \theta = \sqrt{3}/6$, $0° < \theta < 90°$
    (b) $\sin \theta = -\sqrt{3}/6$, $270° < \theta < 360°$
5. $\csc A = -3$, $270° < A < 360°$ *Hint:* First find $\sin A$.
6. $\csc A = \frac{6}{5}$, $90° < A < 180°$
7. $\sec B = -\frac{3}{2}$, $180° < B < 270°$
8. $\sec B = \frac{25}{24}$, $270° < B < 360°$

In Exercises 9–14, use the given information to express the remaining five trigonometric values as functions of t or u. Assume that t and u are positive. Rationalize any denominators that contain radicals.

9. $\cos \theta = t/3$, $270° < \theta < 360°$
10. $\cos \theta = -2t/5$, $90° < \theta < 180°$
11. $\sin \theta = -3u$, $180° < \theta < 270°$
12. $\sin \theta = -u^3$, $270° < \theta < 360°$
13. $\cos \theta = u/\sqrt{3}$, $0° < \theta < 90°$
14. $\sin \theta = 2u/\sqrt{7}$, $0° < \theta < 90°$

In Exercises 15–36, prove that the equations are identities.

15. $\sin \theta \cos \theta \sec \theta \csc \theta = 1$
16. $\tan^2 A + 1 = \sec^2 A$
17. $(\sin \theta \sec \theta)/(\tan \theta) = 1$
18. $\tan \beta \sin \beta = \sec \beta - \cos \beta$
19. $(1 - 5 \sin x)/\cos x = \sec x - 5 \tan x$
20. $\dfrac{1}{\sin \theta} - \sin \theta = \cot \theta \cos \theta$
21. $\cos A(\sec A - \cos A) = \sin^2 A$

22. $\dfrac{\sin \theta}{\csc \theta} + \dfrac{\cos \theta}{\sec \theta} = 1$
23. $(1 - \sin \theta)(\sec \theta + \tan \theta) = \cos \theta$
24. $(\cos \theta - \sin \theta)^2 + 2 \sin \theta \cos \theta = 1$
25. $(\sec \alpha - \tan \alpha)^2 = \dfrac{1 - \sin \alpha}{1 + \sin \alpha}$
26. $\dfrac{\sin B}{1 + \cos B} + \dfrac{1 + \cos B}{\sin B} = 2 \csc B$
27. $\sin A + \cos A = \dfrac{\sin A}{1 - \cot A} - \dfrac{\cos A}{\tan A - 1}$
28. $(1 - \cos C)(1 + \sec C) = \tan C \sin C$
29. $\csc^2 \theta + \sec^2 \theta = \csc^2 \theta \sec^2 \theta$
30. $\cos^2 \theta - \sin^2 \theta = 1 - 2 \sin^2 \theta$
31. $\sin A \tan A = \dfrac{1 - \cos^2 A}{\cos A}$
32. $\dfrac{\cot A - 1}{\cot A + 1} = \dfrac{1 - \tan A}{1 + \tan A}$
33. $\cot^2 A + \csc^2 A = -\cot^4 A + \csc^4 A$
34. $\dfrac{\cot^2 A - \tan^2 A}{(\cot A + \tan A)^2} = 2 \cos^2 A - 1$
35. $\dfrac{\sin A - \cos A}{\sin A} + \dfrac{\cos A - \sin A}{\cos A} = 2 - \sec A \csc A$
36. $\tan A \tan B = \dfrac{\tan A + \tan B}{\cot A + \cot B}$

B
37. Only one of the following two equations is an identity. Decide which equation this is, and give a proof to show that it is, indeed, an identity. For the other equation, give an example showing that it is not an identity. (For example, to show that the equation $\sin \theta + \cos \theta = 1$ is not an identity, let $\theta = 30°$. Then the equation becomes $\frac{1}{2} + \frac{\sqrt{3}}{2} = 1$, which is false.)
    (a) $\dfrac{\csc^2 \alpha - 1}{\csc^2 \alpha} = \cos \alpha$
    (b) $(\sec^2 \alpha - 1)(\csc^2 \alpha - 1) = 1$

38. Follow the directions given in Exercise 37.

(a) $(\csc \beta - \cot \beta)^2 = \dfrac{1 + \cos \beta}{1 - \cos \beta}$

(b) $\cot \beta + \dfrac{\sin \beta}{1 + \cos \beta} = \csc \beta$

39. Prove the identity $\dfrac{\sin \theta}{1 - \cos \theta} = \dfrac{1 + \cos \theta}{\sin \theta}$ in two ways.

(a) Adapt the method of Example 6.
(b) Begin with the left-hand side and multiply numerator and denominator by $\sin \theta$.

In Exercises 40–44, prove that the equations are identities.

40. $\dfrac{2 \sin^3 \beta}{1 - \cos \beta} = 2 \sin \beta + 2 \sin \beta \cos \beta$

Hint: Write $\sin^3 \beta$ as $(\sin \beta)(\sin^2 \beta)$.

41. $\dfrac{\sec \theta - \csc \theta}{\sec \theta + \csc \theta} = \dfrac{\tan \theta - 1}{\tan \theta + 1}$

42. $1 - \dfrac{\sin^2 \theta}{1 + \cot \theta} - \dfrac{\cos^2 \theta}{1 + \tan \theta} = \sin \theta \cos \theta$

43. $(\sin^2 \theta)(1 + n \cot^2 \theta) = (\cos^2 \theta)(n + \tan^2 \theta)$

44. $(r \sin \theta \cos \phi)^2 + (r \sin \theta \sin \phi)^2 + (r \cos \theta)^2 = r^2$

45. (a) Factor the expression $\cos^3 \theta - \sin^3 \theta$.
(b) Prove the identity

$$\dfrac{\cos \phi \cot \phi - \sin \phi \tan \phi}{\csc \phi - \sec \phi} = 1 + \sin \phi \cos \phi$$

46. Prove the following identities. (These two identities, along with $\sin^2 \theta + \cos^2 \theta = 1$, are known as the *Pythagorean identities*. They will be discussed in the next chapter.)

(a) $\tan^2 \theta = \sec^2 \theta - 1$ (b) $\cot^2 \theta = \csc^2 \theta - 1$

47. If $\tan \alpha \tan \beta = 1$ and α and β are acute angles, show that $\sec \alpha = \csc \beta$. *Hint:* Make use of the identities in Exercise 46.

48. Refer to the figure (at the top of the next column). Express the slope m of the line as a function of θ. *Hint:* What are the coordinates of the point P?

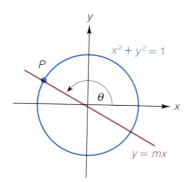

49. Suppose that $\sin \theta = \dfrac{p - q}{p + q}$, where p and q are positive and $90° < \theta < 180°$. Show that $\tan \theta = \dfrac{q - p}{2\sqrt{qp}}$.

50. In this exercise, you will use the unit-circle definitions of sine and cosine, along with the identity $\sin^2 \theta + \cos^2 \theta = 1$ to prove a surprising geometric result. In the figure, we show an equilateral triangle inscribed in the unit circle $x^2 + y^2 = 1$. The vertices of the equilateral triangle are $A\left(-\frac{1}{2}, -\frac{\sqrt{3}}{2}\right)$, $B(1, 0)$, and $C\left(-\frac{1}{2}, \frac{\sqrt{3}}{2}\right)$. Prove that for any point P on the unit circle, the sum of the squares of the distances from P to the three vertices is 6. *Hint:* Let the coordinates of P be $(\cos \theta, \sin \theta)$.

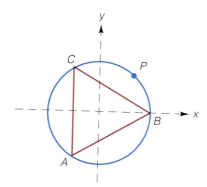

 CHAPTER SEVEN SUMMARY OF PRINCIPAL TERMS AND FORMULAS

FORMULAS OR NOTATIONS	PAGE REFERENCE	COMMENT
1. $\cos \theta$ $\sin \theta$ $\tan \theta$ $\sec \theta$	369	When θ is an acute angle in a right triangle, the six trigonometric functions of θ are defined as follows. (In these definitions, the word "adjacent" means "the length of the side adjacent to angle θ." The same convention applies to the word "opposite.")

FORMULAS OR NOTATIONS	PAGE REFERENCE	COMMENT
csc θ cot θ		$\cos \theta = \dfrac{\text{adjacent}}{\text{hypotenuse}}$ $\qquad$ $\sec \theta = \dfrac{\text{hypotenuse}}{\text{adjacent}}$ $\sin \theta = \dfrac{\text{opposite}}{\text{hypotenuse}}$ $\qquad$ $\csc \theta = \dfrac{\text{hypotenuse}}{\text{opposite}}$ $\tan \theta = \dfrac{\text{opposite}}{\text{adjacent}}$ $\qquad$ $\cot \theta = \dfrac{\text{adjacent}}{\text{opposite}}$
2. $\sin^n \theta$	377	$\sin^n \theta$ means $(\sin \theta)^n$. The same convention also applies to the other five trigonometric functions.
3. $\sin^2 \theta + \cos^2 \theta = 1$ $(\sin\theta)/(\cos \theta) = \tan \theta$ $\sin(90° - \theta) = \cos \theta$ $\cos(90° - \theta) = \sin \theta$ $\sec \theta = 1/\cos \theta$ $\csc \theta = 1/\sin \theta$ $\cot \theta = 1/\tan \theta$	378	These are some of the most fundamental trigonometric identities. (We'll see many more in the following chapter.) With the exception of the third and fourth identities in this list, we discussed these identities in the context of both right-triangle and unit-circle trigonometry. Although the third and fourth identities were discussed only in the context of right-triangles, you'll see in Chapter 8 that these two identities indeed hold for all angles.
4. $A = \frac{1}{2}ab \sin \theta$	385, 401	The area of a triangle equals half the product of the lengths of two sides times the sine of the included angle.
5. Initial side of an angle Terminal side of an angle	393	We think of the two rays that form an angle as having been originally coincident; then, while one ray is held fixed, the other is rotated to create the given angle. The fixed ray is the *initial side* of the angle, and the rotated ray is the *terminal side*. (See Figure 1 on page 393). The measure of an angle is *positive* if the rotation is counterclockwise and *negative* if the rotation is clockwise. (See Figure 2 on page 393.)
6. Standard position	394	In a rectangular coordinate system, an angle is in *standard position* if the vertex is located at $(0, 0)$ and the initial side of the angle lies along the positive horizontal axis. For examples, see Figure 3 on page 394.
7. $\cos \theta$ $\sin \theta$ $\tan \theta$ $\sec \theta$ $\csc \theta$ $\cot \theta$	394	If θ is an angle in standard position and $P(x, y)$ is the point where the terminal side of the angle meets the unit circle, then the six trigonometric functions of θ are defined as follows. $\cos \theta = x$ $\qquad$ $\sec \theta = \dfrac{1}{x}$ $(x \neq 0)$ $\sin \theta = y$ $\qquad$ $\csc \theta = \dfrac{1}{y}$ $(y \neq 0)$ $\tan \theta = \dfrac{y}{x}$ $(x \neq 0)$ $\qquad$ $\cot \theta = \dfrac{x}{y}$ $(y \neq 0)$
8. Reference angle	398	Let θ be an angle in standard position in an *x-y* coordinate system, and suppose that θ is not a multiple of 90°. The reference angle associated with θ is the acute angle (with positive measure) formed by the *x*-axis and the terminal side of the angle θ.

FORMULAS OR NOTATIONS	PAGE REFERENCE	COMMENT
9. The three-step procedure for evaluating the trigonometric functions.	399	The following three-step procedure can be used to evaluate the trigonometric functions for angles that are not multiples of 90°. STEP 1 Determine the reference angle associated with the given angle. STEP 2 Evaluate the given trigonometric function using the reference angle for the input. STEP 3 Affix the appropriate sign to the number found in Step 2. (See Examples 4 and 5 on pages 399–400.)

WRITING MATHEMATICS

1. A student who wanted to simplify the expression $\dfrac{30°}{\sin 30°}$ wrote

$$\frac{30°}{\sin 30°} = \frac{\cancel{30°}}{\sin \cancel{30°}} = \frac{1}{\sin}$$

Explain why this is nonsense, and then indicate the correct solution.

2. Determine if each statement is *true* or *false*. In each case, write out your reason or reasons in complete sentences. If you draw a diagram to accompany your writing, be sure that you clearly label any parts of the diagram to which you refer in the writing.
   (a) If θ is an angle in standard position and $180° < \theta < 270°$, then $\tan \theta < \sin \theta$.
   (b) If θ is an acute angle in a right triangle, then $\sin \theta < 1$.
   (c) If θ is an acute angle in a right triangle, then $\tan \theta < 1$.
   (d) For all angles θ, we have $\sin \theta = \sqrt{1 - \cos^2 \theta}$.
   (e) For all acute angles θ, we have $\sin \theta = \sqrt{1 - \cos^2 \theta}$.
   (f) The formula for the area of a triangle, $A = \frac{1}{2}ab \sin \theta$, is not valid when θ is a right angle.
   (g) Every angle θ satisfying $0° < \theta < 180°$ is an allowable input for the sine function.
   (h) Every angle θ satisfying $0° < \theta < 180°$ is an allowable input for the tangent function.

CHAPTER SEVEN REVIEW EXERCISES

For Exercises 1–16, evaluate each expression. (Don't use a calculator.)

1.	$\sin 135°$	**2.**	$\cos(-60°)$	**3.**	$\tan(-240°)$
4.	$\sin 450°$	**5.**	$\csc 210°$	**6.**	$\sec 225°$
7.	$\sin 270°$	**8.**	$\cot(-330°)$	**9.**	$\cos(-315°)$
10.	$\cos 180°$	**11.**	$\cos 1800°$	**12.**	$\sec 120°$
13.	$\csc 240°$	**14.**	$\csc(-45°)$	**15.**	$\sec 780°$
16.	$\sin^2 33° + \sin^2 57°$				

17. A student who was asked to simplify the expression $\sin^2 33° + \sin^2 57°$ wrote $\sin^2 33° + \sin^2 57° = \sin^2(33° + 57°) = \sin^2 90° = 1^2 = 1$.
    (a) Where is the error?
    (b) What is the correct answer?

18. If $\theta = 45°$, find each of the following.
    (a) $\cos \theta$ (b) $\cos^3 \theta$ (c) $\cos 2\theta$
    (d) $\cos 3\theta$ (e) $\cos(2\theta/3)$ (f) $(\cos 3\theta)/3$
    (g) $\cos(-\theta)$ (h) $\cos^3 5\theta$

In Exercises 19–31, the lengths of the three sides of a triangle are denoted a, b, and c; the angles opposite these sides are A, B, and C, respectively. In each exercise, use the given information to find the required quantities.

GIVEN	FIND
19. $B = 90°$, $A = 30°$ $b = 1$	a and c
20. $B = 90°$, $A = 60°$, $a = 1$	c and b
21. $B = 90°$, $\sin A = \frac{2}{5}$, $a = 7$	b
22. $B = 90°$, $\sec C = 4$, $c = \sqrt{2}$	b
23. $B = 90°$, $\cos A = \frac{3}{8}$	$\sin A$ and $\cot A$
24. $B = 90°$, $b = 1$, $\tan C = \sqrt{5}$	a
25. $b = 4$, $c = 5$, $A = 150°$	area of $\triangle ABC$
26. $A = 120°$, $b = 8$, area of $\triangle ABC = 12\sqrt{3}$	c
27. $c = 4$, $a = 2$, $B = 90°$	$\sin^2 A + \cos^2 B$
28. $B = 90°$, $2a = b$	A
29. $A = 30°$, $B = 120°$, $b = 16$	a
30. $a = 7$, $b = 8$, $\sin C = \frac{1}{4}$	area of $\triangle ABC$
31. $a = b = 5$, $\sin(\frac{1}{2}C) = \frac{9}{10}$	c and area of $\triangle ABC$

32. For the following figure, show that
$$y = x[\tan(\alpha + \beta) - \tan \beta]$$

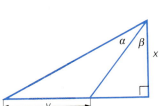

33. For the following figure, show that
$$y = \frac{x}{\cot \alpha - \cot \beta}$$

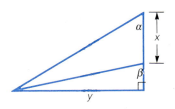

34. For the following figure, show that
$$\cot \theta = \frac{a}{b} + \cot \alpha$$

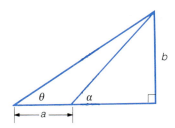

35. Suppose that θ is an acute angle in a right triangle and $\sin \theta = \dfrac{2p^2q^2}{p^4 + q^4}$. Find $\cos \theta$ and $\tan \theta$.

36. This problem is adapted from the text, *An Elementary Treatise on Plane Trigonometry*, by R. D. Beasley, first published in 1884.
(a) Prove the following identity, which will be used in part (b): $1 + \tan^2 \alpha = \sec^2 \alpha$.
(b) Suppose that α and θ are acute angles and $\tan \theta = \dfrac{1 + \tan \alpha}{1 - \tan \alpha}$. Express $\sin \theta$ and $\cos \theta$ in terms of α.   *Hint:*   Draw a right triangle, label one of the angles θ, and let the lengths of the sides opposite and adjacent to θ be $1 + \tan \alpha$ and $1 - \tan \alpha$, respectively.   *Answer:*

$$\sin \theta = \frac{1}{\sqrt{2}}(\cos \alpha + \sin \alpha)$$

$$\cos \theta = \frac{1}{\sqrt{2}}(\cos \alpha - \sin \alpha)$$

In Exercises 37–46, convert each expression into one involving only sines and cosines and then simplify. (Leave your answers in terms of sines and/or cosines.)

37. $\dfrac{\sin A + \cos A}{\sec A + \csc A}$

38. $\dfrac{\csc A \sec A}{\sec^2 A + \csc^2 A}$

39. $\dfrac{\sin A \sec A}{\tan A + \cot A}$

40. $\cos A + \tan A \sin A$

41. $\dfrac{\cos A}{1 - \tan A} + \dfrac{\sin A}{1 - \cot A}$

42. $\dfrac{1}{\sec A - 1} \div \dfrac{1}{\sec A + 1}$

43. $(\sec A + \csc A)^{-1}[(\sec A)^{-1} + (\csc A)^{-1}]$

44. $\dfrac{\dfrac{\tan^2 A - 1}{\tan^3 A + \tan A}}{\dfrac{\tan A + 1}{\tan^2 A + 1}}$

45. $\dfrac{\dfrac{\sin A + \cos A}{\sin A - \cos A} - \dfrac{\sin A - \cos A}{\sin A + \cos A}}{\dfrac{\sin A + \cos A}{\sin A - \cos A} + \dfrac{\sin A - \cos A}{\sin A + \cos A}}$

46. $\dfrac{\sin A \tan A - \cos A \cot A}{\sec A - \csc A}$

In Exercises 47–50, refer to the following figure. The figure shows a highly magnified view of the point P, where the terminal side of an angle of 10° (in standard position) meets the unit circle.

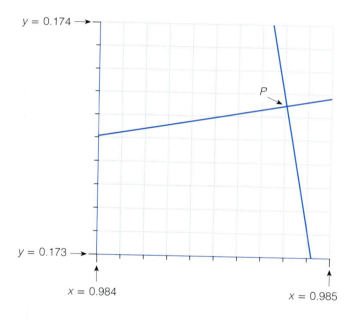

47. Using the figure, estimate the value of cos 10° to four decimal places. (Then use a calculator to check your estimate.)

48. Use your estimate in Exercise 47 to evaluate each of the following.
    (a) cos 170°   (b) cos 190°   (c) cos 350°

49. Using the figure, estimate the value of sin 10° to four decimal places. (Then use a calculator to check your estimate.)

50. Use your estimate in Exercise 49 to evaluate each of the following.
    (a) sin(−10°)   (b) sin(−190°)   (c) sin(−370°)

In Exercises 51–60, use the given information to find the required quantities.

	GIVEN	FIND
51.	$\cos\theta = \frac{3}{5}$, $0° < \theta < 90°$	$\sin\theta$ and $\tan\theta$
52.	$\sin\theta = -\frac{12}{13}$, $180° < \theta < 270°$	$\cos\theta$
53.	$\sec\theta = \frac{25}{7}$, $270° < \theta < 360°$	$\tan\theta$
54.	$\cot\theta = \frac{4}{3}$, $0° < \theta < 90°$	$\sin\theta$
55.	$\csc\theta = \frac{13}{12}$, $0° < \theta < 90°$	$\cot\theta$
56.	$\sin\theta = \frac{1}{5}$, $0° < \theta < 90°$	$\cos(90° - \theta)$
57.	$\cos\theta = 5t$, $0° < \theta < 90°$	$\tan(90° - \theta)$
58.	$\sin\theta = -\sqrt{5}\,u$, $90° < \theta < 270°$	$\sec\theta$
59.	$\tan\theta = \dfrac{\sqrt{2} - 1}{\sqrt{2} + 1}$, $0° < \theta < 90°$	$\sin\theta$
60.	$\cos\theta = \dfrac{2u}{u^2 + 1}$, $0° < \theta < 90°$	$\tan\theta$

61. Suppose θ is an acute angle and $\tan\theta + \cot\theta = 2$. Show that $\sin\theta + \cos\theta = \sqrt{2}$.

62. A 100-ft vertical antenna is on the roof of a building. From a point on the ground, the angles of elevation to the top and the bottom of the antenna are 51° and 37°, respectively. Find the height of the building.

63. In an isosceles triangle, the two base angles are each 35°, and the length of the base is 120 cm. Find the area of the triangle.

64. Find the perimeter and the area of a regular pentagon inscribed in a circle of radius 9 cm.

65. In triangle ABC, let h denote the length of the altitude from A to $\overline{BC}$. Show that $h = a/(\cot B + \cot C)$.

66. The length of each side of an equilateral triangle is $2a$. Show that the radius of the inscribed circle is $a/\sqrt{3}$ and the radius of the circumscribed circle is $2a/\sqrt{3}$.

67. From a helicopter h ft above the sea, the angle of depression to the pilot's horizon is θ. Show that $\cot \theta = R/\sqrt{2Rh + h^2}$, where R is the radius of the Earth.

68. In triangle ABC, angle C is a right angle. If a, b, and c denote the lengths of the sides opposite angles A, B, and C, respectively, show that

$$\frac{\sin A - \sin B}{\sin A + \sin B} = \frac{a - b}{a + b}$$

In Exercises 69–78, show that each equation is an identity.

69. $\sin^2(90° - \theta)\csc \theta - \tan^2(90° - \theta)\sin \theta = 0$

70. $\dfrac{1 - \sin \theta \cos \theta}{(\cos \theta)(\sec \theta - \csc \theta)} \times \dfrac{\sin^2 \theta - \cos^2 \theta}{\sin^3 \theta + \cos^3 \theta} = \sin \theta$

71. $\cos A \cot A = \csc A - \sin A$

72. $\sec A - 1 = (\sec A)(1 - \cos A)$

73. $\dfrac{\cot A - 1}{\cot A + 1} = \dfrac{\cos A - \sin A}{\cos A + \sin A}$

74. $\dfrac{\sin A}{\csc A - \cot A} = 1 + \cos A$

75. $\dfrac{1}{1 - \cos A} + \dfrac{1}{1 + \cos A} = 2 + 2\cot^2 A$

76. $\dfrac{1}{\csc A - \cot A} - \dfrac{1}{\csc A + \cot A} = 2\cot A$

77. $\dfrac{1}{1 + \sin^2 A} + \dfrac{1}{1 + \cos^2 A} + \dfrac{1}{1 + \sec^2 A}$
$\qquad\qquad + \dfrac{1}{1 + \csc^2 A} = 2$

78. $\dfrac{\sec A - \tan A}{\sec A + \tan A} = \left(\dfrac{\cos A}{1 + \sin A}\right)^2$

79. The radius of the circle in the following figure is 1 unit. The line segment $\overline{PT}$ is tangent to the circle at P, and $\overline{PN}$ is perpendicular to $\overline{OT}$. Express each of the following lengths in terms of θ.
(a) PN (b) ON (c) PT
(d) OT (e) NA (f) NT

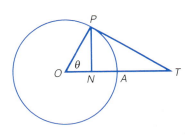

80. Refer to the figure accompanying Exercise 79.
(a) Show that the ratio of the area of triangle ONP to the area of triangle NPT is $\cot^2 \theta$.
(b) Show that the ratio of the area of triangle NPT to the area of triangle OPT is $\sin^2 \theta$.

81. Refer to the accompanying figure. From a point R, two tangent lines are drawn to a circle of radius 1. These tangents meet the circle at the points P and Q. At the midpoint of the circular arc $\overset{\frown}{PQ}$, a third tangent line is drawn, meeting the other two tangents at T and U, as shown. Express the area of triangle RTU as a function of θ (where θ is defined in the figure).

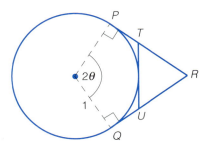

82. Suppose that in $\triangle ABC$, angle C is a right angle. Let a, b, and c denote the lengths of the sides opposite angles A, B, and C, respectively. Use the right-triangle definitions of sine and cosine to prove that

$$\frac{a^2\cos^2 B - b^2\cos^2 A}{a^2 - b^2} = 1$$

83. The following figure shows a semicircle with radius $AO = 1$.
(a) Use the figure to derive the formula

$$\tan \frac{\theta}{2} = \frac{\sin \theta}{1 + \cos \theta}$$

Hint: Show that $CD = \sin \theta$ and $OD = \cos \theta$. Then look at right triangle ADC to find $\tan(\theta/2)$.
(b) Use the formula developed in part (a) to show that

(i) $\tan 15° = \dfrac{1}{2 + \sqrt{3}} = 2 - \sqrt{3}$;

(ii) $\tan 22.5° = \sqrt{2} - 1$.

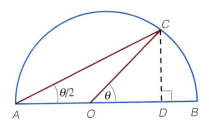

84. This exercise outlines geometric derivations of formulas for sin 2θ and cos 2θ. Refer to the figure, which shows a semicircle with diameter $\overline{SV}$ and radius $UV = 1$. In each case, supply the reasons for the given statement.

(a) $TU = \cos 2\theta$ (b) $TV = 1 + \cos 2\theta$

(c) $WT = \sin 2\theta$

(d) $WV = 2 \cos \theta$ *Hint:*   Use right triangle *SWV*.

(e) $SW = 2 \sin \theta$

(f) $\sin \theta = \dfrac{\sin 2\theta}{2 \cos \theta}$ *Hint:*   Use $\triangle WTV$ and the results in parts (c) and (d).

(g) $\sin 2\theta = 2 \sin \theta \cos \theta$

(h) $\cos \theta = \dfrac{1 + \cos 2\theta}{2 \cos \theta}$ *Hint:*   Use $\triangle WTV$.

(i) $\cos 2\theta = 2 \cos^2 \theta - 1$

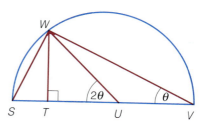

CHAPTER SEVEN TEST

1. Specify a value for each of the following expressions. (Exact values are required, not calculator approximations.)
   (a) $\tan 30°$ (b) $\sec 45°$ (c) $\sin^2 25° + \cos^2 25°$ (d) $\sin 53° - \cos 37°$

2. Suppose that angle θ is in standard position and the terminal side of the angle is in the third quadrant. Which of the following are positive?
   (a) $\sec \theta$ (b) $\csc \theta$ (c) $\cot \theta$

3. Evaluate each of the following.
   (a) $\sin(-270°)$ (b) $\cos 180°$ (c) $\tan 720°$

4. Factor: $2 \cot^2 \theta + 11 \cot \theta + 12$

5. Find the area of a triangle in which the lengths of two sides are 8 cm and 9 cm and the angle included between those two sides is 150°.

6. Evaluate each of the following. (Exact values are required, not calculator approximations.)
   (a) $\sin(-225°)$ (b) $\tan 330°$ (c) $\sec 120°$

7. If $\sin \theta = -\sqrt{5}/5$ and $180° < \theta < 270°$, find $\cos \theta$ and $\tan \theta$.

8. If $\sin \theta = -u/3$ and $270° < \theta < 360°$, express $\cos \theta$ and $\cot \theta$ as functions of u.

9. Simplify: $\dfrac{\dfrac{\cos \theta + 1}{\cos \theta} + 1}{\dfrac{\cos \theta - 1}{\cos \theta} - 1}$

10. A regular nine-sided polygon is inscribed in a circle of radius 3 m. Find the area of the polygon. Give two forms of the answer: one in terms of a trigonometric function and the other a calculator approximation rounded off to three decimal places.

11. Express each term using sines and cosines and simplify as much as possible:

    $\dfrac{\cot^2 \theta}{\csc^2 \theta} + \dfrac{\tan^2 \theta}{\sec^2 \theta}.$

12. In the figure, $\angle DAB = 25°$, $\angle CAB = 55°$, and $AB = 50$ cm. Find CD. Leave your answer in terms of the trigonometric functions (rather than using a calculator).

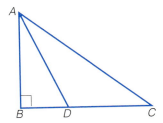

13. Without using a calculator, determine in each case which of the two trigonometric values is the larger. State your reasons. *Hint:* Sketch each pair of angles in standard position.
   (a) sin 5° or sin 85°   (b) sin 5° or cos 5°   (c) tan 175° or tan 185°

14. In △ABC (shown in the figure), ∠A = 20°, AB = 3.25 cm, and AC = 2.75 cm. Complete steps (a) and (b) to determine the length CB.
   (a) Find AD, DB, and CD. Leave your answers in terms of the trigonometric functions.
   (b) Find CB by applying the Pythagorean theorem in △CDB. For your final answer, use a calculator and round off to two decimal places.

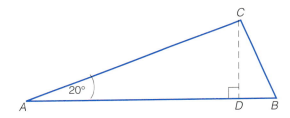

TRIGONOMETRIC FUNCTIONS OF REAL NUMBERS

Trigonometric functions were originally introduced and tabulated in calculations in astronomy and navigation and were used extensively in surveying. What is more important [now], they occur whenever we want to describe a periodic process, like the vibration of a string, the tides, planetary motion, alternating currents, or the emission of light by atoms.

From *The Mathematical Sciences,* edited by the Committee on Support of Research in the Mathematical Sciences with the collaboration of George Boehm (Cambridge, Mass.: The M.I.T. Press, 1969)

INTRODUCTION

As the opening quotation indicates, the trigonometric functions play an important role in many of the modern applications of mathematics. In these applications, the inputs for these functions are real numbers, rather than angles as in the previous chapter. In order to make the transition from angle inputs to real-number inputs, we introduce in Section 8.1 radian measure for angles. Then, in Section 8.2, we use radian measure to restate the definitions of the trigonometric functions in such a way that the domains are indeed sets of real numbers. Among other advantages, this allows us to analyze the trigonometric functions using the graphing techniques that we developed in Chapter 4.

We discuss the graphs of the trigonometric functions in Sections 8.3 through 8.5. In Section 8.6, we develop four fundamental trigonometric identities known as the *addition formulas for sine and cosine.* Then, in the section after this, we consider a number of identities that follow rather directly from these addition formulas. In Section 8.8, we use some of these identities (along with our earlier work) to solve trigonometric equations. In the course of solving these equations, we'll find it useful to introduce the *inverse trigonometric functions.* In this first exposure to these functions, our approach is computational, and we use a calculator to evaluate the functions. We approach the inverse trigonometric functions from a more conceptual point of view in the last section of this chapter, Section 8.9.

8.1 RADIAN MEASURE

Degree measure, traditionally used to measure the angles of a geometric figure, has one serious drawback. It is artificial. There is no intrinsic connection between a degree and the geometry of a circle. Why 360 degrees for one revolution? Why not 400? or 100?

Calculus, One and Several Variables, fifth edition, by S. L. Salas, Einar Hille, John T. Anderson (New York: John Wiley and Sons, Inc., 1986)

For the portion of trigonometry dealing with angles and geometric figures, the units of degrees are quite suitable for measuring angles. However, for the more analytical portions of trigonometry and for calculus, *radian* measure is used. The radian measure of an angle is defined as follows.

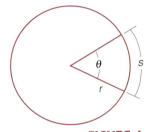

FIGURE I
The radian measure θ is defined by the equation $\theta = s/r$.

DEFINITION   The Radian Measure of an Angle

Place the vertex of the angle at the center of a circle of radius r. Let s denote the length of the arc intercepted by the angle, as indicated in Figure 1. The **radian measure** θ of the angle is the ratio of the arc length s to the radius r. In symbols,

$$\theta = \frac{s}{r} \tag{1}$$

In this definition, it is assumed that s and r have the same linear units.

EXAMPLE I   Determine the radian measure for each angle in Figure 2.

FIGURE 2

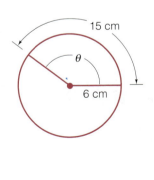

(a)

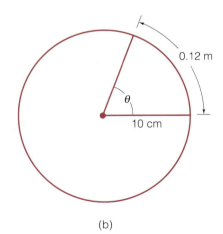

(b)

Solution   In Figure 2(a) we have $r = 6$ cm, $s = 15$ cm, and therefore

$$\theta = \frac{s}{r} = \frac{15 \text{ cm}}{6 \text{ cm}} = \frac{5}{2}$$

So, for the angle in Figure 2(a), we obtain $\theta = 5/2$ radians. In the calculations just completed, notice that although both s and r have the dimensions of length, the resulting radian measure s/r is simply a real number with no dimensions. (The dimensions "cancel out.")

Before carrying out the calculations for the radian measure in Figure 2(b), we have to arrange matters so that the same unit of length is used for the radius and the arc length. Converting the arc length s to centimeters, we have $s = 12$ cm, and consequently

$$\theta = \frac{s}{r} = \frac{12 \text{ cm}}{10 \text{ cm}} = \frac{6}{5}$$

Thus, in Figure 2(b), we have $\theta = 6/5$ radians. Again, note that the radian measure s/r is dimensionless. (Check for yourself that the same answer is obtained if we use meters rather than centimeters for the common unit of measurement.)   ▪▪▪

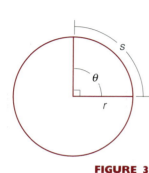

FIGURE 3

At first it may appear to you that the radian measure depends on the radius of the particular circle that we use. But as you will see, this is not the case. To gain some experience in working with the definition of radian measure, let's calculate the radian measure of the right angle in Figure 3. We begin with the formula $\theta = s/r$. Now, since θ is a right angle, the arc length s is one-quarter of the entire circumference. Thus,

$$s = \frac{1}{4}(2\pi r) = \frac{\pi r}{2} \tag{2}$$

Using equation (2) to substitute for s in equation (1), we get

$$\theta = \frac{\pi r/2}{r} = \frac{\pi r}{2} \times \frac{1}{r} = \frac{\pi}{2}$$

In other words,

$$90° = \frac{\pi}{2} \text{ radians} \tag{3}$$

Notice that the radius r does not appear in our answer.

For practical reasons, we would like to be able to convert rapidly between degree and radian measure. Multiplying both sides of equation (3) by 2 yields

$$180° = \pi \text{ radians} \tag{4}$$

Equation (4) is useful and should be memorized. For instance, dividing both sides of equation (4) by 6 yields

$$\frac{180°}{6} = \frac{\pi}{6}$$

or

$$30° = \frac{\pi}{6} \text{ radians}$$

Similarly, dividing both sides of equation (4) by 4 and 3, respectively, yields

$$45° = \frac{\pi}{4} \text{ radians} \quad \text{and} \quad 60° = \frac{\pi}{3} \text{ radians}$$

And, multiplying both sides of equation (4) by 2 gives us

$$360° = 2\pi \text{ radians}$$

Figure 4 summarizes some of these results.

FIGURE 4

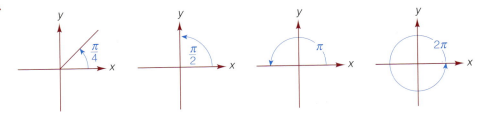

EXAMPLE 2 (a) Express 1° in radian measure.
  (b) Express 1 radian in terms of degrees.

Solution (a) Dividing both sides of equation (4) by 180 yields

$$1° = \frac{\pi}{180} \text{ radian} \quad (\approx 0.017 \text{ radian})$$

  (b) Dividing both sides of equation (4) by π yields

$$\frac{180°}{\pi} = 1 \text{ radian}$$

In other words, 1 radian is approximately 180°/3.14, or 57.3°. ▮▮▮

From the results in Example 2, we have the following rules for converting between radians and degrees.

> To convert from degrees to radians, multiply by $\pi/180°$. To convert from radians to degrees, multiply by $180°/\pi$.

EXAMPLE 3 Convert 150° to radians.

Solution $$150°\left(\frac{\pi}{180°}\right) = \frac{5\pi}{6} \qquad \text{reducing the fraction}$$

Thus,

$$150° = \tfrac{5\pi}{6} \text{ radians}$$

▮▮▮

EXAMPLE 4   Convert $11\pi/6$ radians to degrees.

Solution
$$\frac{11\pi}{6}\left(\frac{180°}{\pi}\right) = \frac{(11)(180°)}{6} = 11(30°) = 330°$$

So, $11\pi/6$ radians $= 330°$.

We saw in Example 2(b) that 1 radian is approximately 57°. It is also important to be able to visualize an angle of 1 radian without thinking in terms of degree measure. This is done as follows. In the equation $\theta = s/r$, we let $\theta = 1$. This yields $1 = s/r$ and, consequently, $r = s$. In other words, in a circle, 1 radian is the central angle that intercepts an arc equal in length to the radius of the circle. In the box we summarize this result. Figure 5 displays angles of 1, 2, and 3 radians.

FIGURE 5

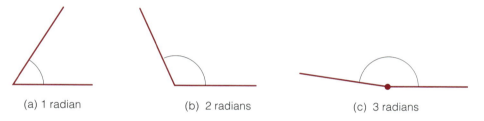

(a) 1 radian   (b) 2 radians   (c) 3 radians

PROPERTY SUMMARY   RADIAN MEASURE

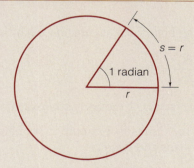

In a circle, 1 radian is the measure of the central angle
that intercepts an arc equal in length to the radius of the circle.

From now on, when we specify the measure of an angle, we will assume that the units are radians unless the degree symbol is explicitly used. (This convention is also used in calculus.) For instance, the equation $\theta = 2$ means that θ is 2 radians. Similarly, the expression $\sin \pi/6$ refers to the sine of $\pi/6$ radians.

EXAMPLE 5   Evaluate the following quantities: **(a)** $\sin\dfrac{\pi}{6}$;   **(b)** $\cos 2\pi$.

Solution   **(a)** $\dfrac{\pi}{6}$ radians $= 30°$, therefore,

$$\sin\frac{\pi}{6} = \sin 30° = \frac{1}{2} \qquad \text{because } \sin 30° = \frac{1}{2}$$

(b) 2π radians $= 360°$, therefore,

$$\cos 2\pi = \cos 360° = 1 \qquad \text{because } \cos 360° = 1$$

EXAMPLE 6 Without using a calculator, determine whether the following values are positive or negative: **(a)** $\cos 3$;   **(b)** $\sin 1$;   **(c)** $\tan 6$.

Solution   **(a)** Since π radians is $180°$, we estimate that 3 radians is slightly less than $180°$ (because 3 is slightly less than π). Thus, in standard position, the terminal side of an angle of 3 radians lies in the second quadrant. Therefore, $\cos 3$ is negative.

(b) Since 1 radian is approximately $60°$ [as we saw in Example 2(b) and in Figure 5], $\sin 1$ is positive.

(c) We know that 6 radians is slightly less than 2π radians, which is one revolution, or $360°$. Thus, in standard position, the terminal side of an angle of 6 radians lies in the fourth quadrant. Consequently, $\tan 6$ is negative.

As we mentioned in Chapter 7, all scientific calculators have keys with which you can indicate whether you are working in degrees or radians. In preparation for the next example, refer to the owner's manual for your calculator to make sure you know how it operates in this respect. (Here's a quick way to find out if you've properly set the calculator in the radian mode. Compute $\sin(\pi/6)$ and see if the calculator reads 0.5, as it indeed should.)

EXAMPLE 7 Use Figure 6 (on the next page) to approximate to within successive tenths the following trigonometric values. Then use a calculator to check your answers. Round off the calculator values to two decimal places.

(a) $\cos 3$ and $\sin 3$ **(b)** $\cos(-3)$ and $\sin(-3)$

Solution   **(a)** Figure 6 shows that (the terminal side of) an angle of 3 radians in standard position meets the unit circle at a point in the second quadrant. Letting (x, y) denote the coordinates of that point, we have (using the grid in Figure 6)

$$-1.0 < x < -0.9 \qquad \text{and} \qquad 0.1 < y < 0.2$$

But, by definition, $\cos 3 = x$ and $\sin 3 = y$ and, consequently,

$$-1.0 < \cos 3 < -0.9 \qquad \text{and} \qquad 0.1 < \sin 3 < 0.2$$

The corresponding calculator values here are

$$\cos 3 \approx -0.99 \qquad \text{and} \qquad \sin 3 \approx 0.14$$

Note that the estimates we obtained from Figure 6 are consistent with these calculator values. (When you verify these calculator values for yourself, be sure that the calculator is set in the radian mode.)

(b) In Figure 6, we want to know where the terminal side of an angle of -3 radians intersects the unit circle. There is no marker for this point in Figure 6. But we can locate the point, nevertheless, by using symmetry. Let (x, y) denote the point corresponding to 3 radians, determined in part (a). By re-

FIGURE 6

Radian measure on the unit circle for angles in standard position

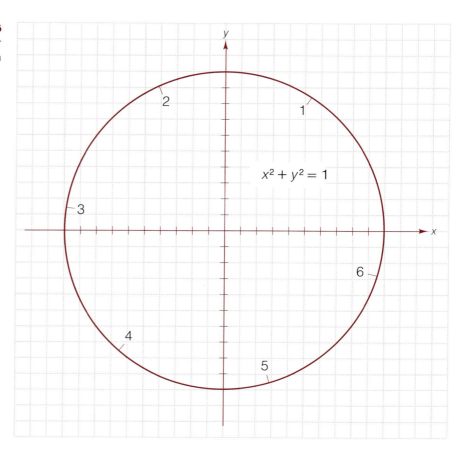

flecting this point in the x-axis, we obtain the point corresponding to -3 radians. This means that the required point will have the same x-coordinate as the point in part (a), while the y-coordinate will be just the negative of that in part (a). So, given that

$$-1.0 < \cos 3 < -0.9 \qquad \text{and} \qquad 0.1 < \sin 3 < 0.2$$

we conclude that

$$-1.0 < \cos(-3) < -0.9 \qquad \text{and} \qquad -0.2 < \sin(-3) < -0.1$$

For the corresponding calculator values here, we obtain

$$\cos(-3) \approx -0.99 \qquad \text{and} \qquad \sin(-3) \approx -0.14$$

Again, note that the estimates from Figure 6 are consistent with these calculator values. When you verify these calculator values for yourself, remember that the calculator should be in the radian mode. Also, you may need to consult your calculator owner's manual about using negative numbers as inputs for these functions. You may find that you cannot simply use the calculator's arithmetic subtraction sign for the input -3. Instead, you will first need to enter "3" and then press a "change sign" key. For instance, on some types of Casio and Texas Instruments calculators (set in the radian mode), the sequence of keystrokes for evaluating $\cos(-3)$ is

3 (+/−) (COS)

while on a Hewlett-Packard calculator, the sequence is

3 (CHS) (COS)

Let us now return to the defining equation for radian measure, $\theta = s/r$, and multiply both sides by r to obtain $s = r\theta$. This gives us a useful formula for arc length on a circle in terms of the radius r and the central angle θ subtended by the arc.

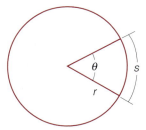

FIGURE 7

Formula for Arc Length

(Refer to Figure 7.) In a circle of radius r, the arc length s determined by a central angle of radian measure θ is given by

$$s = r\theta$$

In this formula, it is assumed that s and r have the same linear units.

EXAMPLE 8 Compute the indicated arc lengths in Figures 8(a) and 8(b). Express the answers both in terms of π and as decimal approximations rounded off to two decimal places.

FIGURE 8

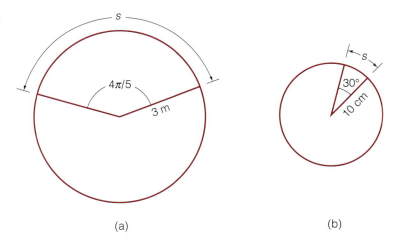

(a)   (b)

Solution   **(a)** We use the formula $s = r\theta$ with $r = 3$ m and $\theta = 4\pi/5$. This yields

$$s = (3\,\text{m})\left(\tfrac{4\pi}{5}\right) = \tfrac{12\pi}{5}\ \text{m}$$

Using a calculator, we find that this is approximately 7.54 m.

   (b) To apply the formula $s = r\theta$, the angle measure θ must be expressed in radians (because the formula is just a restatement of the definition of radian measure). We've seen previously that $30° = \pi/6$ radians. So we have

$$s = r\theta = (10\ \text{cm})\left(\tfrac{\pi}{6}\right) = \tfrac{5\pi}{3}\ \text{cm}$$

Before picking up the calculator, note that we can obtain a quick approximation here: since $\pi \approx 3$, we have $s \approx 5(3/3) = 5$ cm. With a calculator now, we obtain $s = \frac{5\pi}{3}$ cm ≈ 5.24 cm. ■■■

One of the advantages of using radian measure is that many formulas then take on particularly simple forms. The formulas $s = r\theta$ is one example of this. As another example, let us derive a formula for the area of a *sector* of a circle. (In Figure 9, the shaded region is the sector.) From geometry, we know that the area A of the sector is directly proportional to the measure θ of its central angle. In symbols,

$$A = k\theta \tag{5}$$

To determine the constant k, we use the fact that when $\theta = 2\pi$ radians, we have a complete circle. Thus, when $\theta = 2\pi$, the area must be πr^2. Substituting the values $\theta = 2\pi$ and $A = \pi r^2$ in equation (5) yields

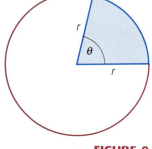

FIGURE 9

$$\pi r^2 = k(2\pi)$$

and therefore

$$\tfrac{1}{2}r^2 = k$$

Using this value for k in equation (5) gives us the following result:

Formula for the Area of a Sector

In a circle of radius r, the area of a sector with central angle of radian measure θ is given by

$$A = \frac{1}{2}r^2\theta$$

EXAMPLE 9   Compute the area of the sector in Figure 10.

Solution   We first convert 120° to radians:

$$120°\left(\frac{\pi}{180°}\right) = \frac{2\pi}{3} \text{ radians}$$

We then have

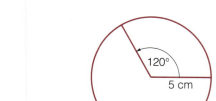

FIGURE 10

$$A = \tfrac{1}{2}r^2\theta$$
$$= \tfrac{1}{2}(5 \text{ cm})^2\left(\tfrac{2\pi}{3}\right) = \tfrac{25\pi}{3} \text{ cm}^2 \quad (\approx 26.18 \text{ cm}^2)$$ ■■■

The next example indicates how radian measure is used in the study of rotating objects. As a prerequisite for this example, we first define the terms *angular speed* and *linear speed*.

DEFINITION   Angular Speed and Linear Speed

Suppose that a wheel rotates about its axis at a constant rate.

1. Refer to Figure 11(a). If a radial line turns through an angle θ in time t, then the **angular speed** of the wheel (denoted by the Greek letter ω) is defined to be

$$\text{angular speed} = \omega = \frac{\theta}{t}$$

2. Refer to Figure 11(b). If a point P on the rotating wheel travels a distance d in time t, then the **linear speed** of P, denoted by v, is defined to be

$$\text{linear speed} = v = \frac{d}{t}$$

FIGURE 11

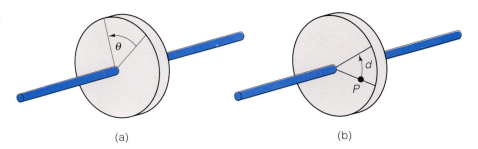

(a)   (b)

EXAMPLE 10   A circular gear in a motor rotates at the rate of 100 rpm (revolutions per minute).

(a) What is the angular speed of the gear in radians per minute?
(b) Find the linear speed of a point on the gear 4 cm from the center.

Solution   (a) Each revolution of the gear is 2π radians. So in 100 revolutions, there are

$$\theta = 100(2\pi) = 200\pi \text{ radians}$$

Consequently, we have

$$\omega = \frac{\theta}{t} = \frac{200\pi \text{ radians}}{1 \text{ min}} = 200\pi \text{ radians/min}$$

(b) We can use the formula $s = r\theta$ to find the distance traveled by the point in 1 minute. Using $r = 4$ cm and $\theta = 200\pi$, we obtain

$$s = r\theta = 4(200\pi) = 800\pi \text{ cm},$$

The linear speed therefore is

$$v = \frac{d}{t} = \frac{800\pi \text{ cm}}{1 \text{ min}}$$
$$= 800\pi \text{ cm/min} \quad (\approx 2513 \text{ cm/min})$$

EXERCISE SET 8.1

A

In Exercises 1–4, convert to radian measure. Express your answers both in terms of π and as decimal approximations rounded off to two decimal places.

1. (a) 45°   (b) 90°   (c) 135°
2. (a) 30°   (b) 150°   (c) 300°
3. (a) 0°   (b) 360°   (c) 450°
4. (a) 36°   (b) 35°   (c) 720°

In Exercises 5–8, convert the radian measures to degrees.

5. (a) π/12   (b) π/6   (c) π/4
6. (a) 3π   (b) 3π/2   (c) 2π
7. (a) π/3   (b) 5π/3   (c) 4π
8. (a) 5π/6   (b) 11π/6   (c) 0

In Exercises 9 and 10, convert the radian measures to degrees. Round off the answers to two decimal places.

9. (a) 2   (b) 3   (c) π^2
10. (a) 1.32   (b) 0.96   (c) 1/π

11. Suppose that the radian measure of an angle is $\frac{3}{2}$. Without using a calculator or tables, determine if this angle is larger or smaller than a right angle.
    Hint: What is the radian measure of a right angle?

12. Two angles in a triangle have radian measure π/5 and π/6. What is the radian measure of the third angle?

For Exercises 13 and 14, refer to the following figure, which shows all of the angles from 0° to 360° that are multiples of 30° or 45°.

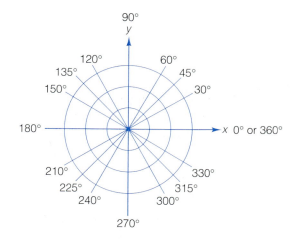

13. In the figure, relabel the angles in quadrants I and II using radian measure.

14. In the figure, relabel the angles in quadrants III and IV using radian measure.

In Exercises 15–18, use the definition $\theta = s/r$ to determine the radian measure of the indicated angle.

15.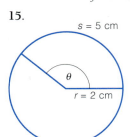

16.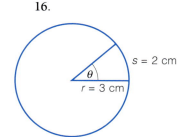

17.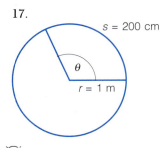

18.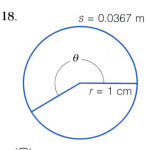

In Exercises 19–28, use Figure 6 to approximate the given trigonometric values to within successive tenths. Then use a calculator to compute the values to the nearest hundredth.

19. cos 1 and sin 1
20. cos 2 and sin 2
21. cos(−1) and sin(−1)
22. cos(−2) and sin(−2)
23. cos 4 and sin 4
24. cos 5 and sin 5
25. cos(−4) and sin(−4)
26. cos(−5) and sin(−5)
27. sin(1 + 2π)
28. sin(2 + 2π)

In Exercises 29–32, find the arc length s in each case.

29.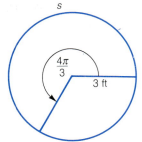

30.

31.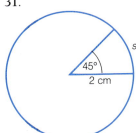

32.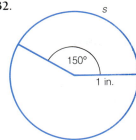

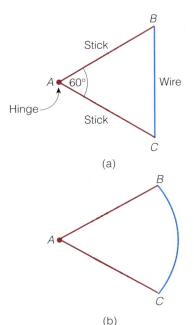

(a)

(b)

In Exercises 33 and 34, find the area of the sector determined by the given radius r and central angle θ. Express the answer both in terms of π and as a decimal approximation rounded off to two decimal places.

33. (a) $r = 6$ cm; $\theta = 2\pi/3$ (b) $r = 5$ m; $\theta = 80°$
     (c) $r = 24$ m; $\theta = \pi/20$ (d) $r = 1.8$ cm; $\theta = 144°$

34. (a) $r = 4$ cm; $\theta = \pi/10$ (b) $r = 16$ m; $\theta = 5°$
     (c) $r = 21$ ft; $\theta = 11\pi/6$ (d) $r = 4.2$ in.; $\theta = 170°$

35. In a circle of radius 1 cm, the area of a certain sector is $\pi/5$ cm^2. Find the radian measure of the central angle. Express the answer in terms of π rather than as a decimal approximation.

36. In a circle of radius 3 m, the area of a certain sector is 20 m^2. Find the degree measure of the central angle. Round off the answer to two decimal places.

In Exercises 37–42, you are given the rate of rotation of a wheel and its radius. In each case, determine the following:

(a) *the angular speed, in units of radians/sec;*
(b) *the linear speed, in units of cm/sec, of a point on the circumference of the wheel;*
(c) *the linear speed, in cm/sec, of a point halfway between the center of the wheel and the circumference.*

37. 6 revolutions/sec; $r = 12$ cm

38. 15 revolutions/sec; $r = 20$ cm

39. 1080°/sec; $r = 25$ cm   **40.** 2160°/sec; $r = 60$ cm

41. 500 rpm; $r = 45$ cm   **42.** 1250 rpm; $r = 10$ cm

B
43. Suppose that you have two sticks and a piece of wire, each of length 1 ft, fastened at the ends to form an equilateral triangle. See Figure (a). If side $\overline{BC}$ is bent out to form an arc of a circle with center A, then the angle at A will decrease from 60° to something less. See Figure (b). What is the measure of this new angle at A?

44. (a) When a clock reads 4:00, what is the radian measure of the (smaller) angle between the hour hand and minute hand?
   (b) When a clock reads 5:30, what is the radian measure of the (smaller) angle between the hour hand and minute hand?

In Exercises 45 and 46, suppose that a belt drives two wheels, of radii r and R, as indicated in the figure.

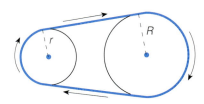

45. If $r = 6$ cm, $R = 10$ cm, and the angular speed of the larger wheel is 100 rpm, determine each of the following.
   (a) The angular speed of the larger wheel in radians per minute.
   (b) The linear speed of a point on the circumference of the larger wheel.
   (c) The angular speed of the smaller wheel in radians per minute. *Hint:* Because of the belt, the linear speed of a point on the circumference of the larger

wheel is equal to the linear speed of a point on the circumference of the smaller wheel.

(d) The angular speed of the smaller wheel in rpm.

46. Follow Exercise 45, assuming $r = 5$ cm, $R = 15$ cm, and the angular speed of the larger wheel is 1800 rpm.

The latitude of a point P on the surface of the Earth is specified by means of the angle θ in the figure. For instance, the latitude of Paris, France, is 48°52' N. The letter N is used here to indicate that the location is north of, rather than south of, the equator. (Recall that the notation 52' indicates $\frac{52}{60}$ of one degree.) In Exercises 47–52, use the arc length formula (and your calculator) to determine the distance $\overset{\frown}{PE}$ from the given location P to the equator. Assume that the Earth is a sphere with radius $OP = OE = 3960$ miles. Round off each answer to the nearest 10 miles.

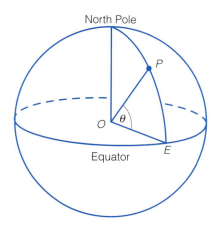

North Pole

Equator

47. Point Barrow, Alaska: 71°23' N

48. Singapore: 1°17' N

49. Honolulu: 21°19' N

50. Lagos: 6°27' N

51. Washington, D.C.:   38°54' N

52. Fairbanks, Alaska: 64°51' N

53. The accompanying figure shows a circular sector with radius r cm and central angle 2θ (radian measure). The perimeter of the sector is 12 cm.

(a) Express r as a function of θ.

(b) Express the area A of the sector as a function of θ.

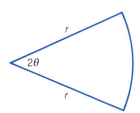

54. The following figure shows a semicircle of radius 1 unit and two adjacent sectors, AOC and COB.

(a) Show that the product P of the areas of the two sectors is given by

$$P = \frac{\pi\theta}{4} - \frac{\theta^2}{4}$$

Is this a quadratic function?

(b) For what value of θ is P a maximum?

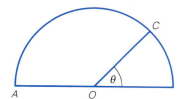

8.2   TRIGONOMETRIC FUNCTIONS OF REAL NUMBERS

The definitions of the trigonometric functions in Section 7.4 are based on the unit circle. Let us look at radian measure in this context. By definition, the radian measure θ is given by $\theta = s/r$. So when $r = 1$, we obtain

$$\theta = \frac{s}{1} = s$$

Thus, in the unit circle, the arc length *is* the radian measure of the angle (see Figure 1).

In view of this observation, we can now define the trigonometric functions in such a way that the domains are sets of real numbers, rather than angles.

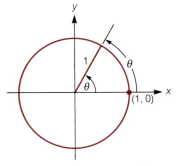

FIGURE 1

In the unit circle, the length of the intercepted arc is the radian measure of the central angle.

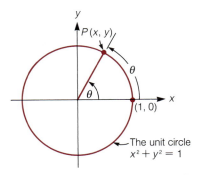

FIGURE 2

There are two reasons for doing this. First, for analytical work, the domains of the trigonometric functions are sets of real numbers. Second, if the domains are sets of real numbers, then the techniques for graphing that we developed in Chapter 4 can be used to analyze these functions.

In the definitions that follow, you may think of θ as either the measure of an arc length or the radian measure of an angle. But in both cases—and this is the point—θ denotes a real number. The conventions regarding the measurement of arc length on the unit circle are the same as those used previously for angles: We measure from the point $(1, 0)$ and we assume that the positive direction is counterclockwise.

Trigonometric Functions of Real Numbers

(Refer to Figure 2.) Let $P(x, y)$ denote the point on the unit circle such that its arc length from $(1, 0)$ is the real number θ. Or, equivalently, let $P(x, y)$ denote the point where the terminal side of the angle with radian measure θ intersects the unit circle. Then the six trigonometric functions are defined as follows.

$$\sin \theta = y \qquad\qquad \csc \theta = \frac{1}{y} \quad (y \neq 0)$$

$$\cos \theta = x \qquad\qquad \sec \theta = \frac{1}{x} \quad (x \neq 0)$$

$$\tan \theta = \frac{y}{x} \quad (x \neq 0) \qquad \cot \theta = \frac{x}{y} \quad (y \neq 0)$$

Let us immediately note that there is nothing new here as far as evaluating the trigonometric functions. For instance, $\sin \pi/6$ is still equal to $\frac{1}{2}$. [If this seems unfamiliar, review Example 5(a) on page 422]. Also, whether we are working in radians or degrees, our three-step procedure for evaluating the trigonometric functions is applicable. For instance, in Example 4(b) on page 399, we computed $\cos 135°$ ($\cos 3\pi/4$). Here is how that same procedure looks when radian measure is used. You should carefully compare each of the corresponding steps.

STEP 1 The reference angle for $3\pi/4$ is $\pi/4$.* $(\pi - 3\pi/4 = \pi/4)$
STEP 2 $\cos \pi/4 = \sqrt{2}/2$
STEP 3 $\cos \theta$ is the x-coordinate. In the second quadrant, x-coordinates are negative. Thus, $\cos 3\pi/4$ is negative, since the terminal side of $\theta = 3\pi/4$ lies in the second quadrant. (Equivalently, the arc of length $3\pi/4$ terminates in the second quadrant.) Thus we have

$$\cos \frac{3\pi}{4} = -\frac{\sqrt{2}}{2}$$

*Some texts refer to this as the reference *number,* instead of reference angle, to emphasize the fact that $3\pi/4$ is a real number.

For the remainder of this section we discuss some of the basic identities for the trigonometric functions of real numbers. First of all, there are five identities that are immediate consequences of the definitions:

$$\csc \theta = \frac{1}{\sin \theta} \qquad \tan \theta = \frac{\sin \theta}{\cos \theta}$$

$$\sec \theta = \frac{1}{\cos \theta} \qquad \cot \theta = \frac{\cos \theta}{\sin \theta}$$

$$\cot \theta = \frac{1}{\tan \theta}$$

The next identities that we consider are the three *Pythagorean identities*.

The Pythagorean Identities

$$\sin^2 \theta + \cos^2 \theta = 1$$
$$\tan^2 \theta + 1 = \sec^2 \theta$$
$$\cot^2 \theta + 1 = \csc^2 \theta$$

To establish the identity $\sin^2 \theta + \cos^2 \theta = 1$, we think of the real number θ, for the moment, as the radian measure of an angle in standard position, as indicated in Figure 2 on page 431. Now we proceed exactly as in the previous chapter, when we proved this identity in the context of angles. Since the point (x, y) in Figure 2 lies on the unit circle, we have

$$x^2 + y^2 = 1$$

Now, replacing x by $\cos \theta$ and y by $\sin \theta$ gives us

$$\cos^2 \theta + \sin^2 \theta = 1$$

which is what we wished to show. As we remarked in Chapter 7, you should also be familiar with two other ways of writing this identity:

$$\cos^2 \theta = 1 - \sin^2 \theta \qquad \text{and} \qquad \sin^2 \theta = 1 - \cos^2 \theta$$

To prove the second of the Pythagorean identities, we begin with $\sin^2 \theta + \cos^2 \theta = 1$ and divide both sides by the quantity $\cos^2 \theta$ to obtain

$$\frac{\sin^2 \theta}{\cos^2 \theta} + \frac{\cos^2 \theta}{\cos^2 \theta} = \frac{1}{\cos^2 \theta} \qquad \text{(assuming } \cos \theta \neq 0)$$

and, consequently,

$$\tan^2 \theta + 1 = \sec^2 \theta \qquad \text{as required}$$

Since the proof of the third Pythagorean identity is similar, we omit it here.

EXAMPLE 1 If $\sin \theta = \frac{2}{3}$ and $\frac{\pi}{2} < \theta < \pi$, compute $\cos \theta$ and $\tan \theta$.

Solution
$$\cos^2 \theta = 1 - \sin^2 \theta \qquad \text{using the first Pythagorean identity}$$
$$= 1 - \left(\frac{2}{3}\right)^2 \qquad \text{substituting}$$
$$= 1 - \frac{4}{9} = \frac{5}{9}$$

Consequently,
$$\cos \theta = \frac{\sqrt{5}}{3} \qquad \text{or} \qquad \cos \theta = \frac{-\sqrt{5}}{3}$$

Now, since $\frac{\pi}{2} < \theta < \pi$, it follows that $\cos \theta$ is negative. (Why?) Thus,
$$\cos \theta = \frac{-\sqrt{5}}{3} \qquad \text{as required}$$

To compute $\tan \theta$, we use the identity $\tan \theta = \sin \theta / \cos \theta$ to obtain
$$\tan \theta = \frac{2/3}{-\sqrt{5}/3} = \frac{2}{3} \times \frac{3}{-\sqrt{5}}$$
$$= -\frac{2}{\sqrt{5}}$$

If required, we can rationalize the denominator $\left(\text{by multiplying by } \sqrt{5}/\sqrt{5}\right)$ to obtain
$$\tan \theta = -\frac{2\sqrt{5}}{5}$$

EXAMPLE 2 If $\sec \theta = -\frac{5}{3}$ and $\pi < \theta < \frac{3\pi}{2}$, compute $\cos \theta$ and $\tan \theta$.

Solution Since $\cos \theta$ is the reciprocal of $\sec \theta$, we have $\cos \theta = -\frac{3}{5}$. We compute $\tan \theta$ using the second Pythagorean identity as follows.
$$\tan^2 \theta = \sec^2 \theta - 1$$
$$= \left(-\frac{5}{3}\right)^2 - 1 = \frac{25}{9} - \frac{9}{9} = \frac{16}{9}$$

Therefore,
$$\tan \theta = \frac{4}{3} \qquad \text{or} \qquad \tan \theta = -\frac{4}{3}$$

Since θ is between π and $3\pi/2$, the quantity $\tan \theta$ is positive. Thus,
$$\tan \theta = \frac{4}{3}$$

The next example shows how certain expressions containing radicals can be simplified through an appropriate trigonometric substitution. (This is often useful in calculus.)

EXAMPLE 3   In the expression $u/\sqrt{u^2 - 1}$, make the substitution $u = \sec\theta$ and show that the resulting expression is equal to $\csc\theta$. (Assume that $0 < \theta < \frac{\pi}{2}$.)

Solution   Replacing u by $\sec\theta$ in the given expression yields

$$\frac{u}{\sqrt{u^2 - 1}} = \frac{\sec\theta}{\sqrt{\sec^2\theta - 1}}$$

$$= \frac{\sec\theta}{\sqrt{\tan^2\theta}} \qquad \text{using the second Pythagorean identity}$$

$$= \frac{\sec\theta}{\tan\theta} = \frac{1/\cos\theta}{\sin\theta/\cos\theta} = \frac{1}{\cos\theta} \times \frac{\cos\theta}{\sin\theta}$$

$$= \frac{1}{\sin\theta} = \csc\theta$$

Question: Where did we use the condition $0 < \theta < \frac{\pi}{2}$?

The next three identities indicate the effects of replacing θ by $-\theta$ in the expressions $\sin\theta$, $\cos\theta$, and $\tan\theta$.

The Opposite-Angle Identities

$$\cos(-\theta) = \cos\theta$$

$$\sin(-\theta) = -\sin\theta$$

$$\tan(-\theta) = -\tan\theta$$

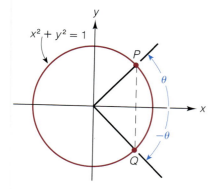

To see why the first two of these identities are true, consider Figure 3. (Although Figure 3 shows an arc of length θ terminating in the first quadrant, the same kind of argument we use will work when θ terminates in any quadrant.) By definition, the coordinates of P and Q are as follows:

P: $(\cos\theta, \sin\theta)$

Q: $(\cos(-\theta), \sin(-\theta))$

However, as you can see by looking at Figure 3, the x-coordinates of P and Q are the same, while the y-coordinates are negatives of each other. Thus,

$$\cos(-\theta) = \cos\theta$$

FIGURE 3   and

$$\sin(-\theta) = -\sin\theta \qquad \text{as we wished to show}$$

Now we can establish the third identity, involving $\tan(-\theta)$, as follows:

$$\tan(-\theta) = \frac{\sin(-\theta)}{\cos(-\theta)} = \frac{-\sin\theta}{\cos\theta}$$

$$= -\frac{\sin\theta}{\cos\theta} = -\tan\theta$$

EXAMPLE 4   (a) If $\sin t = -0.76$, find $\sin(-t)$.
(b) If $\cos s = -0.29$, find $\cos(-s)$.
(c) If $\sin u = 0.54$, find $\sin^2(-u) + \cos^2(-u)$.

Solution   (a) $\sin(-t) = -\sin t$
$= -(-0.76) = 0.76$

(b) $\cos(-s) = \cos s = -0.29$

(c) The opposite-angle identities are unnecessary here, and so too is the given value of $\sin u$. For, according to the first Pythagorean identity, we have

$$\sin^2(-u) + \cos^2(-u) = 1$$

The last set of identities that we are going to discuss in this section are simply consequences of the fact that the circumference C of the unit circle is 2π. (Substituting $r = 1$ in the formula $C = 2\pi r$ gives us $C = 2\pi$.) Thus, if we begin at any point P on the unit circle and travel a distance of 2π units along the perimeter, we return to the same point P. In other words, arc lengths of θ and $\theta + 2\pi$ [measured from $(1, 0)$, as usual] yield the same terminal point on the unit circle. Since the trigonometric functions are defined in terms of the coordinates of that point P, we obtain the following identities.

Periodicity of Sine and Cosine

$$\sin(\theta + 2\pi) = \sin \theta$$
$$\cos(\theta + 2\pi) = \cos \theta$$

These two identities are true for all real numbers θ. As you will see in the next section, they provide important information about the graphs of the sine and cosine functions; the graphs of both functions repeat themselves at intervals of 2π. Similar identities hold for the other trigonometric functions in their respective domains:

$$\tan(\theta + 2\pi) = \tan \theta \qquad \csc(\theta + 2\pi) = \csc \theta$$
$$\cot(\theta + 2\pi) = \cot \theta \qquad \sec(\theta + 2\pi) = \sec \theta$$

EXAMPLE 6   Evaluate $\sin \frac{5\pi}{2}$.

Solution   First, simply as a matter of arithmetic, we observe that $\frac{5\pi}{2} = \frac{\pi}{2} + 2\pi$. Thus, in view of our earlier remarks, we have

$$\sin \frac{5\pi}{2} = \sin \frac{\pi}{2} = 1$$

The preceding set of identities can be generalized as follows. If we start at a point P on the unit circle and make two complete counterclockwise revolutions, the arc length we travel is $2\pi + 2\pi = 4\pi$. Similarly, for three complete revolutions, the arc length traversed is $3(2\pi) = 6\pi$. And, in general, if k is an integer, the arc length for k complete revolutions is $2k\pi$. (When k is positive, the revolutions are counterclockwise; when k is negative, the revolutions are clockwise.) Consequently, we have the following identities.

> For any real number θ and any integer k, the following identities hold.
>
> $$\sin(\theta + 2k\pi) = \sin\theta \quad\text{and}\quad \cos(\theta + 2k\pi) = \cos\theta$$

EXAMPLE 7   Evaluate $\cos(-17\pi)$.

Solution $\cos(-17\pi) = \cos(\pi - 18\pi) = \cos\pi = -1$

EXERCISE SET 8.2

 A

1. Complete the following table.

θ	$\sin\theta$	$\cos\theta$	$\tan\theta$	$\csc\theta$	$\sec\theta$	$\cot\theta$
0						
$\pi/6$						
$\pi/4$						
$\pi/3$						
$\pi/2$						
$2\pi/3$						
$3\pi/4$						
$5\pi/6$						
π						

2. Complete the following table.

θ	$\sin\theta$	$\cos\theta$	$\tan\theta$	$\csc\theta$	$\sec\theta$	$\cot\theta$
π						
$7\pi/6$						
$5\pi/4$						
$4\pi/3$						
$3\pi/2$						
$5\pi/3$						
$7\pi/4$						
$11\pi/6$						

In Exercises 3–10, check that both sides of the identity are indeed equal for the given values of the variable t. For part (c) of each problem, use your calculator.

3. $\sin^2 t + \cos^2 t = 1$
   (a) $t = \pi/3$
   (b) $t = 5\pi/4$
   (c) $t = -53$

4. $\tan^2 t + 1 = \sec^2 t$
   (a) $t = 3\pi/4$
   (b) $t = -2\pi/3$
   (c) $t = \sqrt{5}$

5. $\cot^2 t + 1 = \csc^2 t$
   (a) $t = -\pi/6$
   (b) $t = 7\pi/4$
   (c) $t = 0.12$

6. $\cos(-t) = \cos t$
   (a) $t = \pi/6$
   (b) $t = -5\pi/3$
   (c) $t = -4$

7. $\sin(-t) = -\sin t$
   (a) $t = 3\pi/2$
   (b) $t = -5\pi/6$
   (c) $t = 13.24$

8. $\tan(-t) = -\tan t$
   (a) $t = -4\pi/3$
   (b) $t = \pi/4$
   (c) $t = 1000$

9. $\sin(t + 2\pi) = \sin t$
   (a) $t = 5\pi/3$
   (b) $t = -3\pi/2$
   (c) $t = \sqrt{19}$

10. $\cos(t + 2\pi) = \cos t$
    (a) $t = -5\pi/3$
    (b) $t = \pi$
    (c) $t = -\sqrt{3}$

In Exercises 11 and 12, show that the equation is not an identity by evaluating both sides using the given value of t and noting that the results are unequal.

11. $\cos 2t = 2\cos t$; $t = \pi/6$

12. $\sin 2t = 2\sin t$; $t = \pi/2$

13. If $\sin t = -\frac{3}{5}$ and $\pi < t < \frac{3\pi}{2}$, compute $\cos t$ and $\tan t$.

14. If $\cos t = \frac{5}{13}$ and $\frac{3\pi}{2} < t < 2\pi$, compute $\sin t$ and $\cot t$.

15. If $\sin t = \sqrt{3}/4$ and $\frac{\pi}{2} < t < \pi$, compute $\tan t$.

16. If $\sec s = -\sqrt{13}/2$ and $\sin s > 0$, compute $\tan s$.

17. If $\tan\alpha = \frac{12}{5}$ and $\cos\alpha > 0$, compute $\sec\alpha$, $\cos\alpha$, and $\sin\alpha$.

18. If $\cot\theta = -1/\sqrt{3}$ and $\cos\theta < 0$, compute $\csc\theta$ and $\sin\theta$.

19. In the expression $\sqrt{9 - x^2}$, make the substitution $x = 3 \sin \theta$ $(0 < \theta < \frac{\pi}{2})$, and show that the result is $3 \cos \theta$.

20. Make the substitution $u = 2 \cos \theta$ in the expression $1/\sqrt{4 - u^2}$, and simplify the result. (Assume that $0 < \theta < \pi$.)

21. In the expression $1/(u^2 - 25)^{3/2}$, make the substitution $u = 5 \sec \theta$ $(0 < \theta < \frac{\pi}{2})$, and show that the result is $(\cot^3 \theta)/125$.

22. In the expression $1/(x^2 + 5)^2$, replace x by $\sqrt{5} \tan \theta$, and show that the result is $(\cos^4 \theta)/25$.

23. In the expression $1/\sqrt{u^2 + 7}$, let $u = \sqrt{7} \tan \theta$, where $0 < \theta < \frac{\pi}{2}$, and simplify the result.

24. In the expression $\sqrt{x^2 - a^2}/x$ $(a > 0)$, let $x = a \sec \theta$ $(0 < \theta < \frac{\pi}{2})$, and simplify the result.

25. (a) If $\sin t = \frac{2}{3}$, find $\sin(-t)$.
(b) If $\sin \phi = -\frac{1}{4}$, find $\sin(-\phi)$.
(c) If $\cos \alpha = \frac{1}{5}$, find $\cos(-\alpha)$.
(d) If $\cos s = -\frac{1}{5}$, find $\cos(-s)$.

26. (a) If $\sin t = 0.35$, find $\sin(-t)$.
(b) If $\sin \phi = -0.47$, find $\sin(-\phi)$.
(c) If $\cos \alpha = 0.21$, find $\cos(-\alpha)$.
(d) If $\cos s = -0.56$, find $\cos(-s)$.

27. If $\cos t = -\frac{1}{3}$ $(\frac{\pi}{2} < t < \pi)$, compute the following:
(a) $\sin(-t) + \cos(-t)$; (b) $\sin^2(-t) + \cos^2(-t)$.

28. If $\sin(-s) = \frac{3}{5}$ $(\pi < s < \frac{3\pi}{2})$, compute the following:
(a) $\sin s$; (b) $\cos(-s)$; (c) $\cos s$;
(d) $\tan s + \tan(-s)$.

In Exercises 29 and 30, use one of the identities
$\cos(\theta + 2\pi k) = \cos \theta$ *or* $\sin(\theta + 2\pi k) = \sin \theta$ *to evaluate each expression.*

29. (a) $\cos\left(\frac{\pi}{4} + 2\pi\right)$ (b) $\sin\left(\frac{\pi}{3} + 2\pi\right)$
(c) $\sin\left(\frac{\pi}{2} - 6\pi\right)$

30. (a) $\sin\frac{17\pi}{4}$ (b) $\sin\left(\frac{-17\pi}{4}\right)$
(c) $\cos 11\pi$ (d) $\cos\frac{53\pi}{4}$
(e) $\tan\left(\frac{-7\pi}{4}\right)$ (f) $\cos\frac{7\pi}{4}$
(g) $\sec\left(\frac{11\pi}{6} + 2\pi\right)$ (h) $\csc\left(2\pi - \frac{\pi}{3}\right)$

In Exercises 31–34, use the Pythagorean identities to simplify the given expressions.

31. $\dfrac{\sin^2 t + \cos^2 t}{\tan^2 t + 1}$

32. $\dfrac{\sec^2 t - 1}{\tan^2 t}$

33. $\dfrac{\sec^2 \theta - \tan^2 \theta}{1 + \cot^2 \theta}$

34. $\dfrac{\csc^4 \theta - \cot^4 \theta}{\csc^2 \theta + \cot^2 \theta}$

In Exercises 35–42, prove that the equations are identities.

35. $\csc t = \sin t + \cot t \cos t$

36. $\sin^2 t - \cos^2 t = \dfrac{1 - \cot^2 t}{1 + \cot^2 t}$

37. $\dfrac{1}{1 + \sec s} + \dfrac{1}{1 - \sec s} = -2 \cot^2 s$

38. $\dfrac{1 + \tan s}{1 - \tan s} = \dfrac{\sec^2 s + 2 \tan s}{2 - \sec^2 s}$

39. $\cot \theta + \tan \theta + 1 = \dfrac{\cot \theta}{1 - \tan \theta} + \dfrac{\tan \theta}{1 - \cot \theta}$

40. $\dfrac{\sec s + \cot s \csc s}{\cos s} = \csc^2 s \sec^2 s$

41. $(\tan \theta)(1 - \cot^2 \theta) + (\cot \theta)(1 - \tan^2 \theta) = 0$

42. $(\cos \alpha \cos \beta - \sin \alpha \sin \beta) \times (\cos \alpha \cos \beta + \sin \alpha \sin \beta)$
$= \cos^2 \alpha - \sin^2 \beta$

43. If $\sec t = \frac{13}{5}$ and $\frac{3\pi}{2} < t < 2\pi$, evaluate
$$\dfrac{2 \sin t - 3 \cos t}{4 \sin t - 9 \cos t}$$

44. If $\sec t = (b^2 + 1)/2b$ and $\pi < t < \frac{3\pi}{2}$, find $\tan t$ and $\sin t$.

B

45. Use the accompanying figure to explain why the following four identities are valid. (The identities can be used to provide an algebraic foundation for the reference number technique that we've used to evaluate the trigonometric functions.)

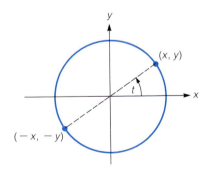

(i) $\sin(t + \pi) = -\sin t$
(ii) $\sin(t - \pi) = -\sin t$
(iii) $\cos(t + \pi) = -\cos t$
(iv) $\cos(t - \pi) = -\cos t$

46. In the equation $x^4 + 6x^2y^2 + y^4 = 32$, make the substitutions
$$x = X \cos\tfrac{\pi}{4} - Y \sin\tfrac{\pi}{4} \text{ and } y = X \sin\tfrac{\pi}{4} + Y \cos\tfrac{\pi}{4}$$
and show that the result simplifies to $X^4 + Y^4 = 16$.

47. Consider the equation $2 \sin^2 t - \sin t = 2 \sin t \cos t - \cos t$.
    (a) Evaluate each side of the equation when $t = \pi/6$.
    (b) Evaluate each side of the equation when $t = \pi/4$.
    (c) Is the given equation an identity?

48. Let $f(t) = (\sin t \cos t)(2 \sin t - 1)(2 \cos t - 1) \times (\tan t - 1)$.
    (a) Compute each of the following: $f(0)$, $f(\pi/6)$, $f(\pi/4)$, $f(\pi/3)$, and $f(\pi/2)$.
    (b) Is the equation $f(t) = 0$ an identity?

In Section 8.1, we pointed out that one of the advantages in using radian measure is that many formulas then take on particularly simple forms. Another reason for using radian measure is that the trigonometric functions can be closely approximated by very simple polynomial functions. To see examples of this, complete the tables in Exercises 49 and 50. Note: *The approximating polynomials in Exercises 49 and 50 are known as Taylor polynomials, after the English mathematician Brook Taylor (1685–1731). The theory of Taylor polynomials is developed in calculus.*

49.

θ	$1 - \dfrac{\theta^2}{2}$	$\cos \theta$
0.1		0.995004. . .
0.2		0.980066. . .
0.3		0.955336. . .

50.

θ	$\theta - \dfrac{\theta^3}{6}$	$\sin \theta$
0.1		0.099833. . .
0.2		0.198669. . .
0.3		0.295520. . .

8.3 GRAPHS OF THE SINE AND COSINE FUNCTIONS

Our focus in this section is on the sine and cosine functions. As preparation for the discussion, we want to understand what is meant by the term *periodic function*. By way of example, both of the functions in Figure 1 are **periodic**. That is, their graphs display patterns that repeat themselves at regular intervals.

In Figure 1(a), the graph of the function f repeats itself every six units. We say that *the period of f is 6*. Similarly, the period of g in Figure 1(b) is 2π. In both cases, notice that the period represents the minimum number of units that we must travel along the horizontal axis before the graph begins to repeat itself. Because of this, we can state the definition of a periodic function as follows.

FIGURE 1

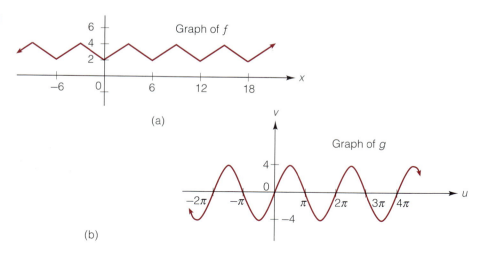

(a)

(b)

DEFINITION **A Periodic Function and Its Period**

A function f is said to be **periodic** if there is a number $p > 0$ such that

$$f(x + p) = f(x)$$

for all x in the domain of f. The smallest such number p is called the **period** of f.

We also want to define the term *amplitude* as it applies to periodic functions. For a function such as g in Figure 1(b), in which the graph is centered about the horizontal axis, the amplitude is simply the maximum height of the graph above the horizontal axis. Thus, the amplitude of g is 4. More generally, we define the amplitude of any periodic function as follows.

DEFINITION **Amplitude**

Let f be a periodic function and let m and M denote, respectively, the minimum and maximum values of the function. Then, by definition, the **amplitude** of f is the number

$$\frac{M - m}{2}$$

For the function g in Figure 1(b), this definition tells us that the amplitude is $\dfrac{4 - (-4)}{2} = 4$, which agrees with our previous value. Check for yourself now that the amplitude of the function in Figure 1(a) is 1.

Periodic functions are used throughout the sciences to analyze or describe a variety of phenomena ranging from the vibrations of an electron to the variations in the size of an animal population as it interacts with its environment. Figures 2 through 5 display some examples of periodic functions.

FIGURE 2

The surface temperature of the star Delta Cephei is a periodic function of time. The period is 5.367 days. The amplitude is $(6800 - 5400)/2 = 700K$.

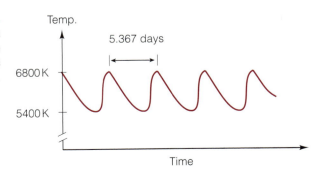

FIGURE 3

Electrical activity of the heart and blood pressure as periodic functions of time. The figure shows a typical ECG (electrocardiogram) and the corresponding graph of arterial blood pressure [adapted from *Physics for the Health Sciences* by C. R. Nave and B. C. Nave (Philadelphia: W. B. Saunders Company, 1975)]

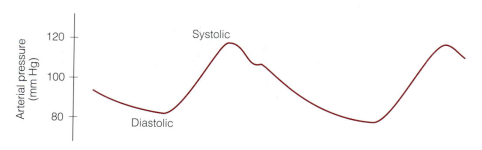

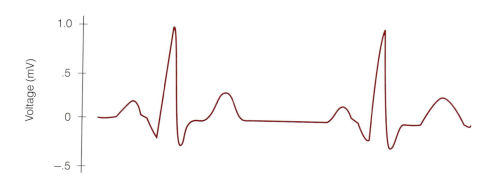

FIGURE 4

The sound wave generated by a note played on the bamboo flute is a periodic function. The sound wave is recorded on an oscilloscope. (Photograph by Professor Vern Ostdiek)

Let us now graph the sine function $f(\theta) = \sin\theta$. We are assuming that θ is in radians, so that the domain of the sine function is the set of all real numbers. Even before making any calculations, we can gain a strong intuitive insight into how the graph must look by carrying out the following experiment. After drawing the unit circle, $x^2 + y^2 = 1$, place your fingertip at the point $(1, 0)$ and then move your finger counterclockwise around the circle. As you do this, keep track

FIGURE 5
Sound waves

(a) The sound wave generated by a tuning fork vibrating at 440 vibrations per second
(On the piano, this note corresponds to the first A above middle C.)

(b) The sound wave generated by a tuning fork vibrating at 880 vibrations per second
[This note is an A one octave above that in part (a).]

(c) The sound wave generated by simultaneously striking the tuning forks in
parts (a) and (b).

of what happens to the y-coordinate of your fingertip. (The y-coordinate is sin θ).
Figure 6 (on the next page) indicates the general results.

Figure 6 tells us a great deal about the sine function: where the function is
increasing and decreasing; where the graph crosses the x-axis; and where the
high and low points of the graph occur. Furthermore, since at $\theta = 2\pi$ we've returned to our starting point $(1, 0)$, additional counterclockwise trips around the
unit circle will just result in repetitions of the pattern established in Figure 6. In
other words, the period of the function $y = \sin \theta$ is 2π.

As informative as Figure 6 is, however, there is not enough information
there to tell us the precise shape of the required graph. For instance, all three of
the curves in Figure 7 fit the specifications described in Figure 6.

Now let's set up a table so that we can accurately sketch the graph of
$y = \sin \theta$. Since the sine function is periodic (as we've just observed), with pe-
riod 2π, our table need only contain values of θ between 0 and 2π, as shown in
Table 1 (on the next page). This will establish the basic pattern for the graph.

In plotting the points obtained from Table 1, we use the approximation
$\sqrt{3}/2 \approx 0.87$. Rather than approximating π, however, we mark off units on the
horizontal axis in terms of π. The resulting graph is shown in Figure 8 (page
443). By continuing this same pattern to the left and right, we obtain the complete graph of $f(\theta) = \sin \theta$, as indicated in Figure 9.

Before going on to analyze the sine function or to study the graphs of the

FIGURE 6

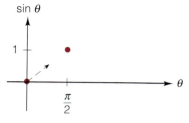

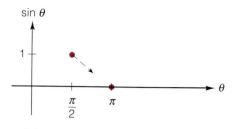

(a) As θ increases from 0 to $\pi/2$, the y-coordinate (sin θ) increases from 0 to 1.

(b) As θ increases from $\pi/2$ to π, the y-coordinate decreases from 1 back down to 0.

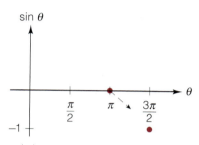

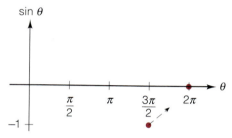

(c) As θ increases from π to $3\pi/2$, the y-coordinate decreases from 0 to -1.

(d) As θ increases from $3\pi/2$ to 2π, the y-coordinate increases from -1 back up to 0.

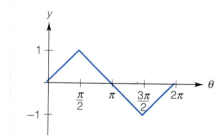

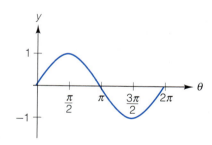

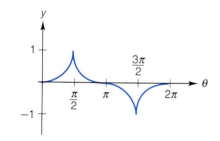

FIGURE 7

TABLE I

θ	0	$\dfrac{\pi}{6}$	$\dfrac{\pi}{3}$	$\dfrac{\pi}{2}$	$\dfrac{2\pi}{3}$	$\dfrac{5\pi}{6}$	π	$\dfrac{7\pi}{6}$	$\dfrac{4\pi}{3}$	$\dfrac{3\pi}{2}$	$\dfrac{5\pi}{3}$	$\dfrac{11\pi}{6}$	2π
$\sin \theta$	0	$\dfrac{1}{2}$	$\dfrac{\sqrt{3}}{2}$	1	$\dfrac{\sqrt{3}}{2}$	$\dfrac{1}{2}$	0	$-\dfrac{1}{2}$	$-\dfrac{\sqrt{3}}{2}$	-1	$-\dfrac{\sqrt{3}}{2}$	$-\dfrac{1}{2}$	0

FIGURE 8

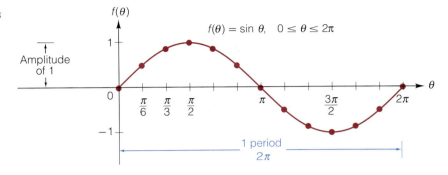

FIGURE 9

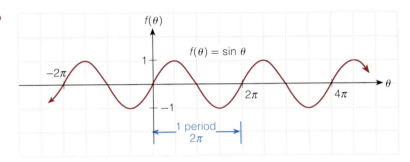

other trigonometric functions, we are going to make a slight change in the notation we have been using. To conform with common usage, we will use x instead of θ on the horizontal axis, and we will use y for the vertical axis. The sine function is then written simply as $y = \sin x$, where the real number x denotes the radian measure of an angle or, equivalently, the length of the corresponding arc on the unit circle. For reference, we redraw Figures 8 and 9 using this familiar x-y notation, as shown in Figure 10.

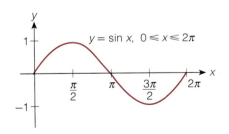

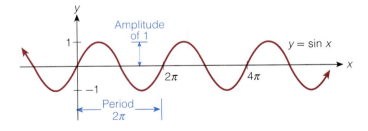

(a) One complete cycle of $y = \sin x$

(b) The period of $y = \sin x$ is 2π.

FIGURE 10

You should memorize the graph of the sine curve in Figure 10(b) so that you can sketch it without first setting up a table. Of course, once you know the shape and the location of the basic cycle shown in Figure 10(a), you automatically know the graph of the full sine curve in Figure 10(b).

We can use the graphs in Figure 10 to help us list some of the key properties of the sine function.

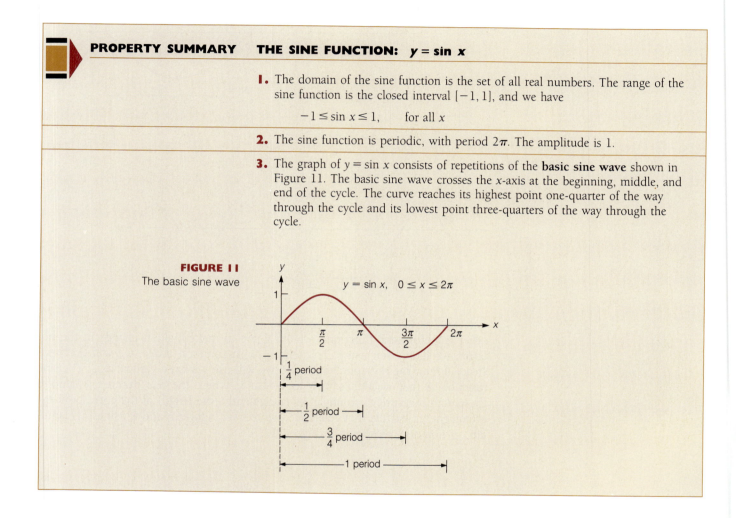

PROPERTY SUMMARY **THE SINE FUNCTION:** $y = \sin x$

1. The domain of the sine function is the set of all real numbers. The range of the sine function is the closed interval $[-1, 1]$, and we have

$$-1 \leq \sin x \leq 1, \quad \text{for all } x$$

2. The sine function is periodic, with period 2π. The amplitude is 1.

3. The graph of $y = \sin x$ consists of repetitions of the **basic sine wave** shown in Figure 11. The basic sine wave crosses the x-axis at the beginning, middle, and end of the cycle. The curve reaches its highest point one-quarter of the way through the cycle and its lowest point three-quarters of the way through the cycle.

FIGURE 11
The basic sine wave

$y = \sin x, \ 0 \leq x \leq 2\pi$

EXAMPLE 1 Figure 12 shows a portion of the graph of $y = \sin x$. What are the coordinates of the turning points A, B, and C? Give the x-coordinates both in terms of π and as decimal approximations rounded off to three places.

FIGURE 12

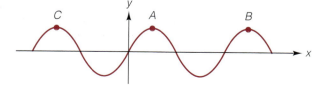

Solution From Figure 11 (which you should memorize), we know that the coordinates of the point A are $(\pi/2, 1)$. Therefore, since the period of the sine function is 2π, the coordinates of B and C are as follows.

Coordinates of B:	$\left(\frac{\pi}{2} + 2\pi, 1\right)$	or	$\left(\frac{5\pi}{2}, 1\right)$
Coordinates of C:	$\left(\frac{\pi}{2} - 2\pi, 1\right)$	or	$\left(-\frac{3\pi}{2}, 1\right)$

As you should check now for yourself, the calculator approximations here are

$$A(1.571, 1) \qquad B(7.854, 1) \qquad C(-4.712, 1)$$

There is an interesting fact about the graph of $y = \sin x$ that is useful in numerical work. If you look at the graph in the immediate vicinity of any of its x-intercepts, the graph closely resembles a straight line segment. For example, in Figure 13 we show the graph of $y = \sin x$ and the line $y = x$. Notice that the two graphs appear to be virtually identical in the vicinity of the origin. [Actually, the only point the two graphs have in common is $(0, 0)$. But near the origin, the distance between the two graphs is less than the thickness of the lines we draw—so the two graphs appear to be indistinguishable from one another.]

FIGURE 13

In the immediate vicinity of the origin, the graph of $y = \sin x$ closely resembles the line $y = x$.

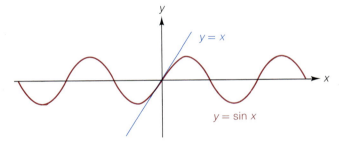

We summarize our observations as follows. (The symbol $\approx$ means "is approximately equal to.")

$$\sin x \approx x \qquad \text{when } |x| \text{ is close to } 0$$

As examples of this approximation, we have

$$\sin(0.07) \approx 0.07 \qquad \text{and} \qquad \sin(-0.0032) \approx -0.0032$$

You should check for yourself now just how close these approximations are to the values obtained using your calculator. (Remember to set the calculator in the radian mode.)

We could obtain the graph of the cosine function by setting up a table, just as we did with the sine function. A more interesting and informative way to proceed, however, is to use the identity

$$\cos x = \sin\left(x + \frac{\pi}{2}\right) \tag{1}$$

(Exercise 39 at the end of this section outlines a geometric proof of this identity. Also, after studying Section 8.6, you'll see a simple way to prove this identity algebraically.) From our work on graphing functions in Chapter 4, we can interpret equation (1) geometrically. Equation (1) tells us that the graph of $y = \cos x$ is obtained by translating the sine curve $\pi/2$ units to the left. The result is shown in Figure 14(a). Figure 14(b) displays one complete cycle of the cosine curve, from $x = 0$ to $x = 2\pi$.

FIGURE 14

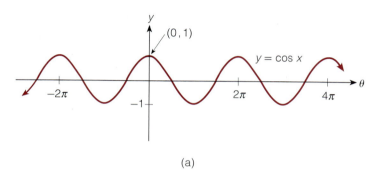

(a)

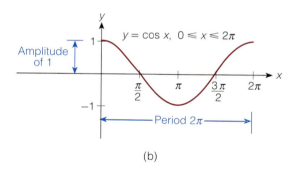

(b)

As with the sine function, you should memorize the graph and basic features of the cosine function. The key features of the cosine are summarized in the box that follows.

PROPERTY SUMMARY   THE COSINE FUNCTION: *y* = cos *x*

1. The domain of the cosine function is the set of all real numbers. The range of the cosine function is the closed interval $[-1, 1]$, and we have

$$-1 \le \cos x \le 1, \qquad \text{for all } x$$

2. The cosine function is periodic, with period 2π. The amplitude is 1.

3. The graph of $y = \cos x$ consists of repetitions of the **basic cosine wave** shown in Figure 15. At the beginning and end of the cycle, the basic cosine wave reaches its highest point. At the middle of the cycle, the curve reaches its lowest point. The *x*-intercepts occur one-quarter and three-quarters of the way through the cycle.

FIGURE 15
The basic cosine wave

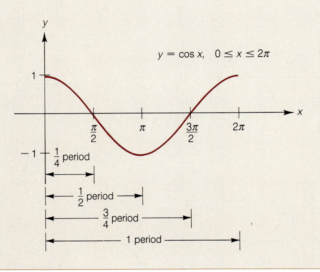

EXAMPLE 2   Figure 16 shows a portion of the graph of $y = \cos x$. What are the *x*-coordinates of the points *B* and *C*? Give the values both in terms of π and as decimal approximations rounded off to three places.

FIGURE 16

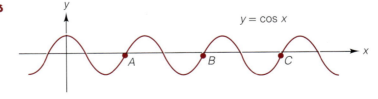

Solution   Comparing Figures 15 and 16, we know that the *x*-intercept at *A* is $3\pi/2$. Therefore, since the period of the cosine function is 2π, we have

$$x\text{-coordinate of } B: \quad \tfrac{3\pi}{2} + 2\pi = \tfrac{3\pi}{2} + \tfrac{4\pi}{2} = \tfrac{7\pi}{2}$$
$$x\text{-coordinate of } C: \quad \tfrac{3\pi}{2} + 4\pi = \tfrac{3\pi}{2} + \tfrac{8\pi}{2} = \tfrac{11\pi}{2}$$

Using a calculator now, we obtain

x-coordinate of B: $\frac{7\pi}{2} \approx 10.996$

x-coordinate of C: $\frac{11\pi}{2} \approx 17.279$

EXERCISE SET 8.3

A

In Exercises 1–8, specify the period and amplitude for each function.

1.
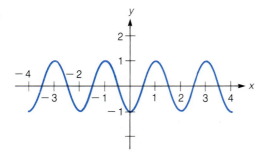

2.
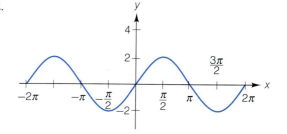

3.
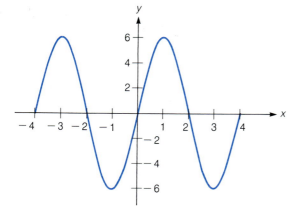

4.
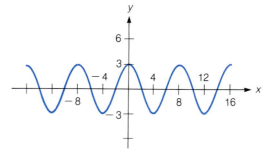

5.
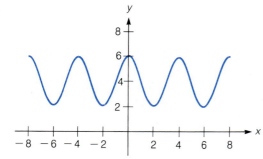

6.
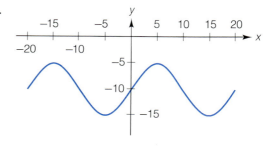

7.

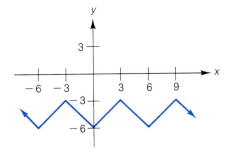

8.

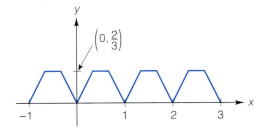

$\left(0, \frac{2}{3}\right)$

In Exercises 9–18, refer to the graph of $y = \sin x$ in the following figure. Specify the coordinates of the indicated points. Give the x-coodinates both in terms of π and as calculator approximations rounded off to three decimal places.

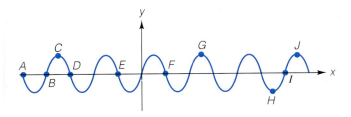

9.	C	10.	F	11.	G	12.	A
13.	B	14.	J	15.	D	16.	H
17.	E	18.	I

In Exercises 19–22, state whether the function $y = \sin x$ is increasing or decreasing on the given interval. (The terms increasing and decreasing are explained on page 189.)

19. $\frac{3\pi}{2} < x < 2\pi$ 20. $-\frac{\pi}{2} < x < \frac{\pi}{2}$

21. $\frac{5\pi}{2} < x < \frac{7\pi}{2}$ 22. $-\frac{5\pi}{2} < x < -2\pi$

In Exercises 23 and 24, use a calculator to complete the tables. Round off the answers to five decimal places. The results in the third column will tell you something about the accuracy of the approximation $\sin x \approx x$.

23.

x	sin x	$x - \sin x$
0.453		
0.253		
0.0253		
0.00253		

24.

x	sin x	$x - \sin x$
0.526		
0.126		
0.0126		
0.00126		

In Exercises 25–34, refer to the graph of $y = \cos x$ in the following figure. Specify the coordinates of the indicated points. Give the x-coordinates both in terms of π and as calculator approximations rounded off to three decimal places.

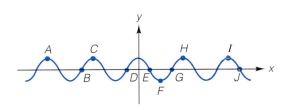

25.	J	26.	H	27.	A	28.	G
29.	E	30.	D	31.	I	32.	F
33.	B	34.	C

In Exercises 35–38, state whether the function $y = \cos x$ is increasing or decreasing on the given interval.

35. $0 < x < \pi$ 36. $6\pi < x < 7\pi$

37. $-\frac{\pi}{2} < x < 0$ 38. $-\frac{5\pi}{2} < x < -2\pi$

B

39. In the text we used the identity $\cos \theta = \sin(\theta + \pi/2)$ in obtaining the graph of $y = \cos x$ from that of $y = \sin x$. In this exercise you'll derive this identity. Refer to the following figure. (Although the figure shows the angle of radian measure θ in the first quadrant, the proof can be easily carried over for the other quadrants as well.)

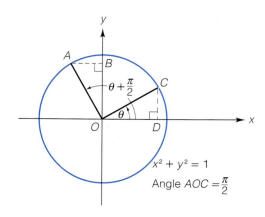

$x^2 + y^2 = 1$

Angle $AOC = \frac{\pi}{2}$

(a) What are the coordinates of C?
(b) Show that the triangles AOB and COD are congruent.
(c) Use the answers in parts (a) and (b) to show that the coordinates of A are $(-\sin \theta, \cos \theta)$.
(d) Since the radian measure of $\angle DOA$ is $\theta + \frac{\pi}{2}$, the coordinates of A (by definition) are $\left(\cos\left(\theta + \frac{\pi}{2}\right), \sin\left(\theta + \frac{\pi}{2}\right)\right)$. Now explain why

$$\cos \theta = \sin\left(\theta + \frac{\pi}{2}\right)$$

and

$$-\sin \theta = \cos\left(\theta + \frac{\pi}{2}\right)$$

8.4   GRAPHS OF $y = A \sin(Bx - C)$ AND $y = A \cos(Bx - C)$

The graphs of $y = \sin x$ and $y = \cos x$ are the building blocks we need for graphing functions of the form

$$y = A \sin(Bx - C) \qquad \text{and} \qquad y = A \cos(Bx - C)$$

As a first example, consider $y = 2 \sin x$. To obtain the graph of $y = 2 \sin x$ from that of $y = \sin x$, we multiply each y-coordinate on the graph of $y = \sin x$ by 2. As indicated in Figure 1, this changes the amplitude from 1 to 2, but it does not affect the period, which remains 2π. As a second example, Figure 2 shows the graphs of $y = \cos x$ and $y = \frac{1}{2} \cos x$. Note that the amplitude of $y = \frac{1}{2} \cos x$ is $\frac{1}{2}$ and the period is, again, 2π. More generally, graphs of functions of the form $y = A \sin x$ and $y = A \cos x$ always have an amplitude of $|A|$ and a period of 2π.

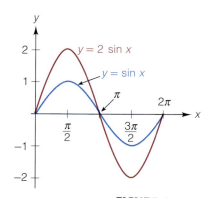

FIGURE 1
The amplitude of $y = 2 \sin x$ is 2.
Both $y = \sin x$ and $y = 2 \sin x$
have a period of 2π.

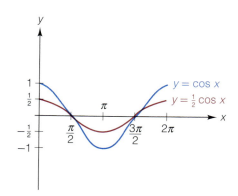

FIGURE 2
The amplitude of $y = \frac{1}{2} \cos x$ is $\frac{1}{2}$.
Both $y = \cos x$ and $y = \frac{1}{2} \cos x$
have a period of 2π.

EXAMPLE 1   Graph the function $y = -2 \sin x$ over one period. On which interval(s) is the function decreasing?

Solution   In Section 4.3 we saw that the graph of $y = -f(x)$ is obtained from that of $y = f(x)$ by reflection about the x-axis. Thus, we need only take the graph of $y = 2 \sin x$ from Figure 1 and reflect it about the x-axis. See Figure 3. Note that both functions have an amplitude of 2 and a period of 2π. From the graph in Figure 3, we can see that the function $y = -2 \sin x$ is decreasing on the intervals $(0, \pi/2)$ and $(3\pi/2, 2\pi)$.

FIGURE 3

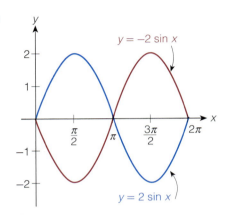

We have just seen that functions of the form $y = A \sin x$ and $y = A \cos x$ have an amplitude of $|A|$ and a period of 2π. The next two examples show how to analyze functions of the form $y = A \sin Bx$ and $y = A \cos Bx$ $(B > 0)$. As you'll see, these functions have an amplitude of $|A|$ and a period of $2\pi/B$.

EXAMPLE 2   Graph the function $y = \cos 3x$ over one period.

Solution   We know that the cosine curve $y = \cos x$ begins its basic pattern when $x = 0$ and completes that pattern when $x = 2\pi$. Thus, $y = \cos 3x$ will begin its basic pattern when $3x = 0$, and it will complete that pattern when $3x = 2\pi$. From the equation $3x = 0$ we conclude that $x = 0$, and from the equation $3x = 2\pi$ we conclude that $x = 2\pi/3$. Thus, the graph of $y = \cos 3x$ begins its basic pattern at $x = 0$ and completes the pattern at $x = 2\pi/3$. This tells us that the period is $2\pi/3$. Next, in preparation for drawing the graph, we divide the period into quarters, as shown in Figure 4(a) on the next page. In Figure 4(b) we've plotted the points with x-coordinates shown in Figure 4(a). (We've also plotted the point on the curve corresponding to $x = 0$, where the basic pattern is to begin.) From Figure 4(b), you can see that the amplitude is going to be 1. Now, with the points in Figure 4(b) as a guide, we can sketch one cycle of $y = \cos 3x$, as shown in Figure 4(c).

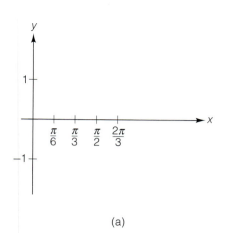

(a)

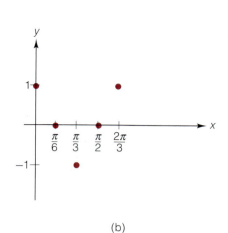

(b)

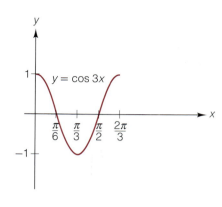

(c) The period of $y = \cos 3x$ is $2\pi/3$; the amplitude is 1.

FIGURE 4

EXAMPLE 3   Graph each function over one period: **(a)** $y = \frac{1}{2}\cos 3x$;   **(b)** $y = -\frac{1}{2}\cos 3x$.

Solution   **(a)** In Example 2 we graphed $y = \cos 3x$. To obtain the graph of $y = \frac{1}{2}\cos 3x$ from that of $y = \cos 3x$, we multiply each y-coordinate on the graph of $y = \cos 3x$ by $\frac{1}{2}$. As indicated in Figure 5(a), this changes the amplitude from 1 to $\frac{1}{2}$, but it does not affect the period, which remains $2\pi/3$.

(b) The graph of $y = -\frac{1}{2}\cos 3x$ is obtained by reflecting the graph of $y = \frac{1}{2}\cos 3x$ about the x-axis, as indicated in Figure 5(b). Both functions have a period of $2\pi/3$ and an amplitude of $\frac{1}{2}$.

FIGURE 5

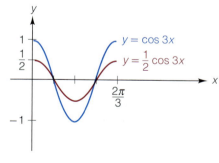

(a)

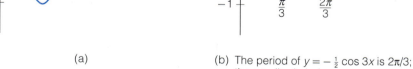

(b) The period of $y = -\frac{1}{2}\cos 3x$ is $2\pi/3$; the amplitude is $\frac{1}{2}$.

Before looking at more examples, let's take a moment to summarize where we are. Our work in Examples 2 and 3(a) shows how to graph $y = \frac{1}{2}\cos 3x$. The same technique we used in those examples can be applied to any function of the form $y = A \cos Bx$ or $y = A \sin Bx$. As indicated in the box that follows, for both functions, the amplitude is $|A|$ and the period is $2\pi/B$. (Exercise 39 asks you to use the method of Example 2 to show that the period is indeed $2\pi/B$.)

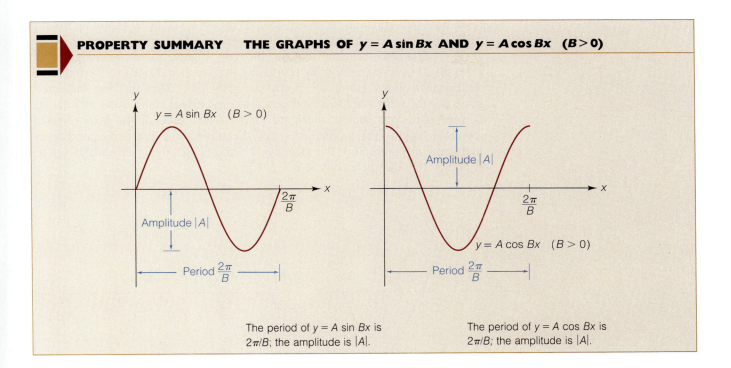

PROPERTY SUMMARY THE GRAPHS OF $y = A \sin Bx$ AND $y = A \cos Bx$ $(B > 0)$

The period of $y = A \sin Bx$ is $2\pi/B$; the amplitude is $|A|$.

The period of $y = A \cos Bx$ is $2\pi/B$; the amplitude is $|A|$.

EXAMPLE 4 In Figure 6, a function of the form $y = A \sin Bx$ $(B > 0)$ is graphed for one period. Determine the values of A and B.

Solution From the figure, we see that the amplitude is 4. Also from the figure, we know that three-fourths of the period is 9, so

$$\frac{3}{4}\left(\frac{2\pi}{B}\right) = 9$$

$$\frac{\pi}{2B} = 3$$

$$B = \frac{\pi}{6} \qquad \text{(Check the algebra in the last two lines.)}$$

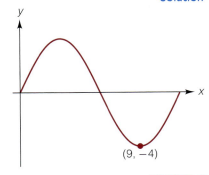

$(9, -4)$

FIGURE 6

In summary, we have $A = 4$ and $B = \pi/6$; the equation of the curve is $y = 4 \sin(\pi x/6)$. ▪▪▪

EXAMPLE 5 Graph the function $y = 4 \sin\left(2x - \frac{2\pi}{3}\right)$ over one period.

Solution The technique here is to factor the quantity within parentheses so that the coefficient of x is 1. We'll then be able to graph the function using a simple translation, as in Chapter 4. We have

$$y = 4 \sin\left(2x - \frac{2\pi}{3}\right)$$
$$= 4 \sin\left[2\left(x - \frac{\pi}{3}\right)\right] \qquad\qquad (1)$$

Now, note that equation (1) is obtained from $y = 4 \sin 2x$ by replacing x with $x - \frac{\pi}{3}$. Thus, the graph of equation (1) is obtained by translating the graph of $y = 4 \sin 2x$ a distance of $\pi/3$ units to the right. Figure 7(a) shows the graph of $y = 4 \sin 2x$ over one period. By translating this graph $\pi/3$ units to the right, we obtain the required graph, as shown in Figure 7(b). Note that the translated graph has the same amplitude and period as the original graph. Also, as a matter of arithmetic, you should check for yourself that each of the x-coordinates shown in Figure 7(b) is obtained simply by adding $\pi/3$ to a corresponding x-coordinate in Figure 7(a). For example, in Figure 7(a), the cycle ends at $x = \pi$; in Figure 7(b) the cycle ends at $\pi + \frac{\pi}{3} = \frac{4\pi}{3}$.

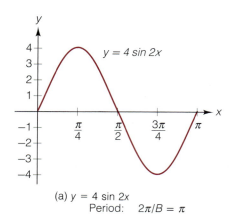

(a) $y = 4 \sin 2x$
Period: $2\pi/B = \pi$
Amplitude: $|A| = 4$

(b) $y = 4 \sin\left(2x - \frac{2\pi}{3}\right) = 4 \sin\left[2\left(x - \frac{\pi}{3}\right)\right]$

The graph is obtained by translating the graph of $y = 4 \sin 2x$ *a distance of $\pi/3$ units to the right. The period and amplitude are still π and 4, respectively.

FIGURE 7

In the example just completed, we used translation to graph the function $y = 4 \sin\left(2x - \frac{2\pi}{3}\right)$. In particular, we translated the graph of $y = 4 \sin 2x$ so that the starting point of the basic cycle was shifted from $x = 0$ to $x = \pi/3$. The number $\pi/3$ in this case is called the *phase shift* of the function. In the box on the next page, we define phase shift and we generalize the results of the graphing technique used in Example 5.

EXAMPLE 6   Specify the amplitude, period, and phase shift for each function.

    (a) $f(x) = 3 \cos(4x - 5)$ **(b)** $g(x) = -2 \cos\left(\pi x + \frac{2\pi}{3}\right)$

Solution   **(a)** By comparing the given equation with $y = A \cos(Bx - C)$, we see that $A = 3$, $B = 4$, and $C = 5$. Consequently, we have

$$\text{amplitude} = |A| = 3$$

$$\text{period} = \frac{2\pi}{B} = \frac{2\pi}{4} = \frac{\pi}{2}$$

$$\text{phase shift} = \frac{C}{B} = \frac{5}{4}$$

(Example 6 continued on page 455)

PROPERTY SUMMARY $y = A \sin(Bx - C)$ AND $y = A \cos(Bx - C)$ $(B > 0, C \neq 0)$

The graphs of $y = A \sin(Bx - C)$ and $y = A \cos(Bx - C)$ are obtained by horizontally translating the graphs of $y = A \sin Bx$ and $y = A \cos Bx$, respectively, so that the starting point of the basic cycle is shifted from $x = 0$ to $x = C/B$. The number C/B is called the **phase shift** for each of the functions $y = A \sin(Bx - C)$ and $y = A \cos(Bx - C)$. The amplitude and the period for these functions are $|A|$ and $2\pi/B$, respectively.

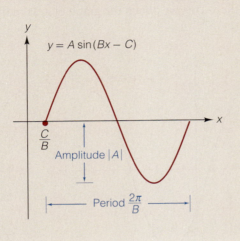 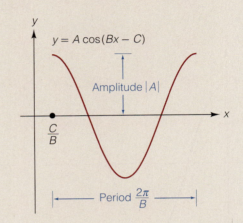

(*Example 6 continued.*)

For purposes of review, let's also calculate the phase shift without explicitly relying on the expression C/B. In the equation $f(x) = 3 \cos(4x - 5)$, we can factor a 4 out of the parentheses to obtain

$$f(x) = 3 \cos\left[4\left(x - \tfrac{5}{4}\right)\right]$$

This last equation tells us that we can obtain the graph of f by translating the graph of $y = 3 \cos 4x$. In particular, the translation shifts the starting point of the basic cycle from $x = 0$ to $x = \tfrac{5}{4}$. The number $\tfrac{5}{4}$ is the phase shift, as obtained previously.

(b) We have $A = -2$, $B = \pi$, and $C = -2\pi/3$, and therefore

$$\text{amplitude} = |A| = 2.$$

$$\text{period} = \frac{2\pi}{B} = \frac{2\pi}{\pi} = 2$$

$$\text{phase shift} = \frac{C}{B} = \frac{-2\pi/3}{\pi} = -\frac{2}{3}$$

EXAMPLE 7   Graph the following function over one period.

$$g(x) = -2 \cos\left(\pi x + \frac{2\pi}{3}\right)$$

Solution Our strategy is first to obtain the graph of $y = 2 \cos(\pi x + \frac{2\pi}{3})$. The graph of g can then be obtained by a reflection about the x-axis. We have

$$y = 2 \cos\left(\pi x + \frac{2\pi}{3}\right) = 2 \cos\left[\pi\left(x + \frac{2}{3}\right)\right]$$

Now, the graph of this last equation is obtained by translating the graph of $y = 2 \cos \pi x$ a distance of $\frac{2}{3}$ unit to the left. Figures 8(a) and 8(b) show the graphs of $y = 2 \cos \pi x$ and $y = 2 \cos[\pi(x + \frac{2}{3})]$. By reflecting the graph of this last equation about the x-axis, we obtain the graph of g, as required. See Figure 8(c).

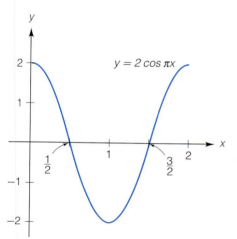

(a) $y = 2 \cos \pi x$
Amplitude: $|A| = 2$
Period: $2\pi/B = 2$

(b) $y = 2 \cos\left(\pi x + \frac{2\pi}{3}\right)$
Amplitude: 2
Period: 2
Phase shift: $-2/3$

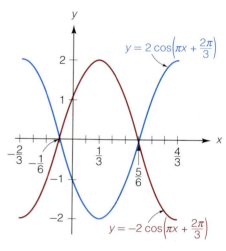

(c) $y = -2 \cos\left(\pi x + \frac{2\pi}{3}\right)$
Amplitude: 2
Period: 2
Phase shift: $-2/3$

FIGURE 8

EXERCISE SET 8.4

A

In Exercises 1–12, graph the functions for one period. In each case, specify the amplitude, period, x-intercepts, and interval(s) on which the function is increasing.

1. (a) $y = 2 \sin x$
    (b) $y = -\sin 2x$
2. (a) $y = 3 \sin x$
    (b) $y = \sin 3x$
3. (a) $y = \cos 2x$
    (b) $y = 2 \cos 2x$
4. (a) $y = \cos(x/2)$
    (b) $y = -\frac{1}{2} \cos(x/2)$
5. (a) $y = 3 \sin(\pi x/2)$
    (b) $y = -3 \sin(\pi x/2)$
6. (a) $y = 2 \sin \pi x$
    (b) $y = -2 \sin \pi x$

7. (a) $y = \cos 2\pi x$
    (b) $y = -4 \cos 2\pi x$
8. (a) $y = -2 \cos(x/4)$
    (b) $y = -2 \cos(\pi x/4)$
9. $y = 1 + \sin 2x$
10. $y = \sin(x/2) - 2$
11. $y = 1 - \cos(\pi x/3)$
12. $y = -2 - 2 \cos 3\pi x$

In Exercises 13–28, determine the amplitude, period, and phase shift for the given function. Graph the function over one period. Indicate the x-intercepts and the coordinates of the highest and lowest points on the graph.

13. $f(x) = \sin\left(x - \frac{\pi}{6}\right)$

14. $g(x) = \cos\left(x + \frac{\pi}{3}\right)$

15. $F(x) = -\cos\left(x + \dfrac{\pi}{4}\right)$

16. $G(x) = -\sin(x + 2)$

17. $y = \sin\left(2x - \dfrac{\pi}{2}\right)$

18. $y = \sin\left(3x + \dfrac{\pi}{2}\right)$

19. $y = \cos(2x - \pi)$

20. $y = \cos\left(x - \dfrac{\pi}{2}\right)$

21. $y = 3 \sin\left(\dfrac{1}{2}x + \dfrac{\pi}{6}\right)$

22. $y = -2 \sin(\pi x + \pi)$

23. $y = 4 \cos\left(3x - \dfrac{\pi}{4}\right)$

24. $y = \cos(x + 1)$

25. $y = \dfrac{1}{2} \sin\left(\dfrac{\pi x}{2} - \pi^2\right)$

26. $y = \cos\left(2x - \dfrac{\pi}{3}\right) + 1$

27. $y = 1 - \cos\left(2x - \dfrac{\pi}{3}\right)$

28. $y = 3 \cos\left(\dfrac{2x}{3} + \dfrac{\pi}{6}\right)$

In Exercises 29–34, determine whether the equation for the graph has the form $y = A \sin Bx$ or $y = A \cos Bx$ and then find the value of A and B.

29.

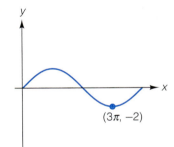

$(3\pi, -2)$

30.

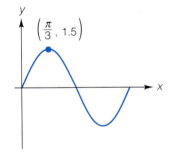

$\left(\dfrac{\pi}{3}, 1.5\right)$

31.

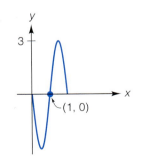

3

$(1, 0)$

32.

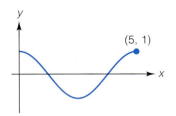

$(5, 1)$

33.

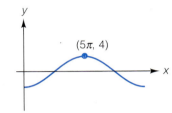

$(5\pi, 4)$

34.

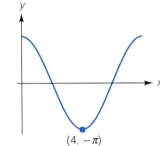

$(4, -\pi)$

B

35. In Section 8.7 you'll see the identity $\sin^2 x = \tfrac{1}{2} - \tfrac{1}{2} \cos 2x$. Use this identity to graph the function $y = \sin^2 x$ for one period.

36. In Section 8.7 you'll see the identity $\cos^2 x = \tfrac{1}{2} + \tfrac{1}{2} \cos 2x$. Use this identity to graph the function $y = \cos^2 x$ for one period.

37. In Section 8.7 we derive the identity $\sin 2x = 2 \sin x \cos x$. Use this to graph $y = \sin x \cos x$ for one period.

38. In Section 8.7 we derive the identity $\cos 2x = \cos^2 x - \sin^2 x$. Use this to graph $y = \cos^2 x - \sin^2 x$ for one period.

39. In Example 2 we showed that the period of $y = \cos 3x$ is $2\pi/3$. Use the same method to show that the period of $y = A \cos Bx$ is $2\pi/B$.

8.5 GRAPHS OF THE TANGENT AND THE RECIPROCAL FUNCTIONS

A third . . . function, the tangent of θ, or tan θ, is of secondary importance, in that it is not associated with wave phenomena. Nevertheless, it enters into the body of analysis so prominently that we cannot ignore it.

Samuel E. Urner and William B. Orange in *Elements of Mathematical Analysis* (Boston: Ginn and Company, 1950)

We have seen in the previous sections that the sine and cosine functions are periodic. The remaining four trigonometric functions are also periodic, but their graphs differ significantly from those of sine and cosine. The graphs of $y = \tan x$, $\cot x$, $\csc x$, and $\sec x$ all possess asymptotes.

We'll obtain the graph of the tangent function by a combination of both point-plotting and symmetry considerations. Table 1 displays a list of values for $y = \tan x$ using x-values in the interval $[0, \pi/2)$.

TABLE I

As *x* increases from 0 toward *π/2*, the values of tan *x* increase slowly at first, then more and more rapidly.

x	0	$\dfrac{\pi}{6}$	$\dfrac{\pi}{4}$	$\dfrac{\pi}{3}$	$\dfrac{5\pi}{12}$ (= 75°)	$\dfrac{17\pi}{36}$ (= 85°)	$\dfrac{89\pi}{180}$ (= 89°)	$\dfrac{\pi}{2}$
tan x	0	$\dfrac{\sqrt{3}}{3} \approx 0.58$	1	$\sqrt{3} \approx 1.73$	3.73	11.43	57.29	undefined

As indicated in Table 1, tan x is undefined when $x = \pi/2$. This follows from the identity

$$\tan x = \frac{\sin x}{\cos x} \tag{1}$$

When $x = \pi/2$, the denominator in this identity is zero. Indeed, when x is equal to any odd integral multiple of $\pi/2$ (e.g., $\pm 3\pi/2$, $\pm 5\pi/2$), the denominator in equation (1) will be zero and, consequently, tan x will be undefined.

Because tan x is undefined when $x = \pi/2$, we want to see how the graph behaves as x gets closer and closer to $\pi/2$. This is why the x-values $5\pi/12$, $17\pi/36$, and $89\pi/180$ are used in Table 1. In Figure 1, we've used the data in Table 1 to draw the graph of $y = \tan x$ for $0 \leq x < \pi/2$. As the figure indicates, the vertical line $x = \pi/2$ is an asymptote for the graph.

The graph of $y = \tan x$ can now be completed without further need for tables or a calculator. First, the identity $\tan(-x) = -\tan x$ (from Section 8.2) tells us that the graph of $y = \tan x$ is symmetric about the origin. So, after reflecting the graph in Figure 1 about the origin, we can draw the graph of $y = \tan x$ on the interval $(-\pi/2, \pi/2)$, as shown in Figure 2.

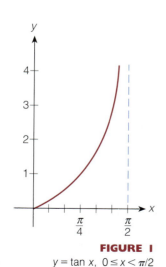

FIGURE I

$y = \tan x$, $0 \leq x < \pi/2$

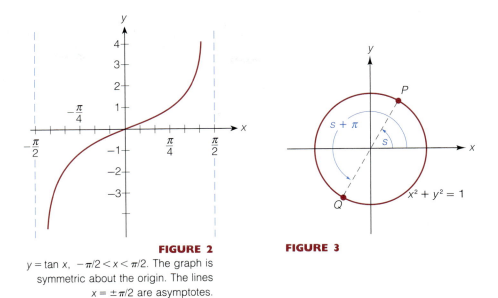

FIGURE 2

$y = \tan x$, $-\pi/2 < x < \pi/2$. The graph is symmetric about the origin. The lines $x = \pm \pi/2$ are asymptotes.

FIGURE 3

Now, to complete the graph of $y = \tan x$, we'll use the identity

$$\tan(s + \pi) = \tan s \qquad (2)$$

Looking at Figure 3, we can see why this identity is valid. By definition, the coordinates of P and Q are

$$P(\cos s, \sin s) \qquad \text{and} \qquad Q\big(\cos(s + \pi), \sin(s + \pi)\big)$$

On the other hand, the points P and Q are symmetric about the origin, and therefore the coordinates of Q are just the negatives of the coordinates of P. That is,

$$\cos(s + \pi) = -\cos s \qquad \text{and} \qquad \sin(s + \pi) = -\sin s$$

Consequently, we have

$$\tan(s + \pi) = \frac{\sin(s + \pi)}{\cos(s + \pi)} = \frac{-\sin s}{-\cos s}$$

$$= \frac{\sin s}{\cos s} = \tan s \qquad \text{as required}$$

[Although Figure 3 shows the angle with radian measure s terminating in the first quadrant, our proof is valid for the other quadrants as well. (Draw a figure for yourself and verify this.)]

Identity (2) tells us that the graph of $y = \tan x$ must repeat itself at intervals of length π. Taking this fact into account, along with Figure 2, we conclude that the period of $y = \tan x$ is exactly π. Our final graph of $y = \tan x$ is shown in the summary box that follows.

PROPERTY SUMMARY   THE TANGENT FUNCTION: $y = \tan x$

Domain: The set of all real numbers other than
         $\pm\pi/2,\ \pm3\pi/2,\ \pm5\pi/2,\ \ldots$

Range: $(-\infty, \infty)$

Period: π

Asymptotes: $x = \pm\pi/2,\ x = \pm3\pi/2,\ x = \pm5\pi/2,\ \ldots$

x-intercepts: $0,\ \pm\pi,\ \pm2\pi,\ \pm3\pi,\ \ldots$

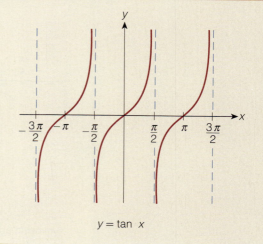

$y = \tan x$

EXAMPLE I   Graph the following functions for one period. In each case, specify the period, the asymptotes, and the intercepts.

(a) $y = \tan\left(x - \frac{\pi}{4}\right)$ (b) $y = -\tan\left(x - \frac{\pi}{4}\right)$

Solution   We begin with the graph of one period of $y = \tan x$, as shown in Figure 4(a). By translating this graph $\pi/4$ units to the right, we obtain the graph of $y = \tan\left(x - \frac{\pi}{4}\right)$, in Figure 4(b). With this translation, note that the left asymptote shifts from $x = -\frac{\pi}{2}$ to $x = -\frac{\pi}{2} + \frac{\pi}{4} = -\frac{\pi}{4}$; the right asymptote shifts from $x = \frac{\pi}{2}$ to $x = \frac{\pi}{2} + \frac{\pi}{4} = \frac{3\pi}{4}$; and the x-intercept shifts from 0 to $\frac{\pi}{4}$.

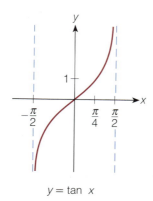

$y = \tan x$

Period: π
Asymptotes: $x = \pm\pi/2$
x-intercept: 0
y-intercept: 0

(a)

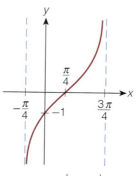

$y = \tan\left(x - \frac{\pi}{4}\right)$

Period: π
Asymptotes: $x = -\pi/4,\ x = 3\pi/4$
x-intercept: $\pi/4$
y-intercept: -1

(b)

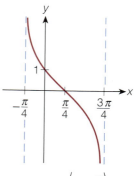

$y = -\tan\left(x - \frac{\pi}{4}\right)$

Period: π
Asymptotes: $x = -\pi/4,\ x = 3\pi/4$
x-intercept: $\pi/4$
y-intercept: 1

(c)

FIGURE 4

For the y-intercept of $y = \tan\left(x - \frac{\pi}{4}\right)$, replace x with 0 in the equation. This yields

$$y = \tan\left(0 - \frac{\pi}{4}\right) = \tan\left(-\frac{\pi}{4}\right) = -\tan\frac{\pi}{4} = -1$$

Finally, for the graph of $y = -\tan\left(x - \frac{\pi}{4}\right)$, we need only reflect the graph in Figure 4(b) about the x-axis; see Figure 4(c). ▮▮▮

EXAMPLE 2   Graph the function $y = \tan(x/2)$ for one period.

Solution   First we refer back to Figure 2, which shows the basic pattern for one period of $y = \tan x$. In this basic pattern, the asymptotes occur when the radian measure of the angle equals $-\pi/2$ or $\pi/2$. Consequently, for $y = \tan(x/2)$, the asymptotes occur when $x/2 = -\pi/2$ and when $x/2 = \pi/2$. From the equation $x/2 = -\pi/2$ we conclude that $x = -\pi$, and from the equation $x/2 = \pi/2$ we conclude that $x = \pi$. Thus, the asymptotes for $y = \tan(x/2)$ are $x = -\pi$ and $x = \pi$. The distance between these asymptotes, namely, 2π, is the period of $y = \tan(x/2)$. This is twice the period of $y = \tan x$. So basically, we want to draw a curve with the same general shape as $y = \tan x$ but twice as wide. See Figure 5. Note that the graph in Figure 5 passes through the origin, since when $x = 0$, we have

$$y = \tan(0/2) = \tan 0 = 0$$ ▮▮▮

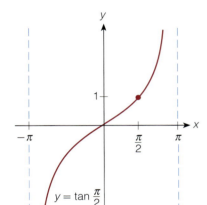

$y = \tan\dfrac{\pi}{2}$

FIGURE 5

The graph of the cotangent function can be obtained from that of the tangent function by means of the identity

$$\cot x = -\tan\left(x - \frac{\pi}{2}\right) \tag{3}$$

(Exercise 37 shows how to derive this identity.) According to identity (3), the graph of $y = \cot x$ can be obtained by first translating the graph of $y = \tan x$ to the right $\pi/2$ units and then reflecting the translated graph about the x-axis. When this is done, we obtain the graph shown in the box that follows.

PROPERTY SUMMARY   THE COTANGENT FUNCTION: $y = \cot x$

Domain: The set of all real numbers other than $0, \pm\pi, \pm 2\pi, \ldots$

Range: $(-\infty, \infty)$

Period: π (See Figure 6.)

Asymptotes: $x = 0, x = \pm\pi, x = \pm 2\pi, \ldots$

x-intercepts: $\pm\dfrac{\pi}{2}, \pm\dfrac{3\pi}{2}, \pm\dfrac{5\pi}{2}, \ldots$

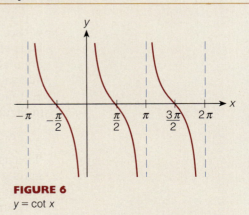

FIGURE 6
$y = \cot x$

EXAMPLE 3   Graph each of the following functions for one period.

(a) $y = \cot \pi x$ **(b)** $y = \frac{1}{2} \cot \pi x$ **(c)** $y = -\frac{1}{2} \cot \pi x$

Solution   **(a)** Looking at the graph of $y = \cot x$ in Figure 6, we see that one complete pattern or cycle of the graph is bounded by the asymptotes $x = 0$ and $x = \pi$. Now, for the function we are given, x has been replaced by πx. Thus, the corresponding asymptotes occur when $\pi x = 0$ and when $\pi x = \pi$; in other words, when $x = 0$ and when $x = 1$. See Figure 7(a).

(b) The graph of $y = \frac{1}{2} \cot \pi x$ will have the same general shape as that of $y = \cot \pi x$, but each y-coordinate on $y = \frac{1}{2} \cot \pi x$ will be half of the corresponding coordinate on $y = \cot \pi x$. See Figure 7(b).

(c) The graph of $y = -\frac{1}{2} \cot \pi x$ is obtained by reflecting the graph of $y = \frac{1}{2} \cot \pi x$ about the x-axis. See Figure 7(c).

FIGURE 7

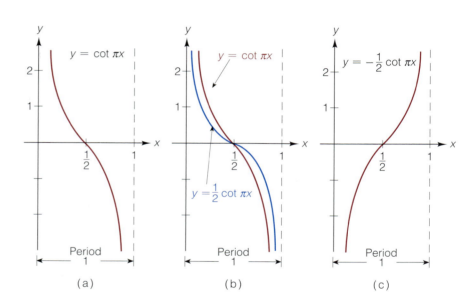

(a)   (b)   (c)

We conclude this section with a discussion of the graphs of $y = \csc x$ and $y = \sec x$. We will obtain these graphs in a series of easy steps, relying on the ideas of symmetry and translation. First consider the function $y = \csc x$. In Figure 8(a), we've set up a table and used it to sketch the graph of $y = \csc x$ on the interval $[\pi/2, \pi)$. Note that $\csc x$ is undefined when $x = \pi$. (Why?) As Figure 8(a) indicates, the vertical line $x = \pi$ is an asymptote for the graph. The graph in Figure 8(a) can be extended to the interval $(0, \pi)$ by means of the identity

$$\csc\left(\frac{\pi}{2} + s\right) = \csc\left(\frac{\pi}{2} - s\right)$$

(Exercise 38 shows you how to verify this identity.) This identity tells us that, starting at $x = \pi/2$, whether we travel a distance s to the right or a distance s to the left, the value of $y = \csc x$ is the same. In other words, the graph of $y = \csc x$ is symmetric about the line $x = \pi/2$. In view of this symmetry, we can sketch the graph of $y = \csc x$ on the interval $(0, \pi)$, as shown in Figure 8(b).

FIGURE 8

x	$\csc x$
$\pi/2$	1
$2\pi/3$	≈ 1.2
$5\pi/6$	2
$11\pi/12$	≈ 3.9

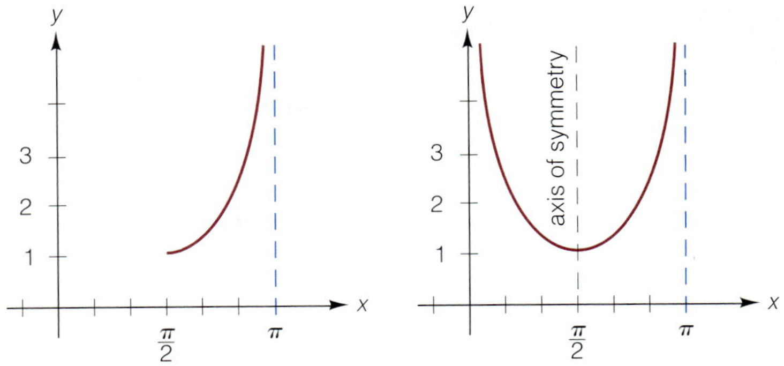

(a) $y = \csc x,\ \dfrac{\pi}{2} \le x < \pi$

(b) $y = \csc x,\ 0 < x < \pi$

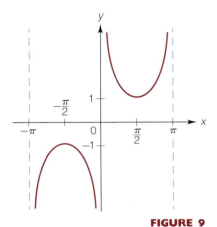

FIGURE 9

$y = \csc x,\ -\pi < x < \pi$

The graph is symmetric about the origin.

The next step in obtaining the graph of $y = \csc x$ is to use the fact that the graph is symmetric about the origin. To verify this, we need to check that $\csc(-x) = -\csc x$. We have

$$\csc(-x) = \frac{1}{\sin(-x)} = \frac{1}{-\sin x} = -\csc x \qquad \text{as required}$$

Now, taking into account this symmetry about the origin and the portion of the graph that we've already obtained in Figure 8(b), we can sketch the graph of $y = \csc x$ over the interval $(-\pi, \pi)$, as shown in Figure 9.

To complete the graph of $y = \csc x$, we observe that the values of $\csc x$ must repeat themselves at intervals of 2π. This is because $\csc x = 1/(\sin x)$, and the sine function has a period of 2π. In view of this, we can draw the graph of $y = \csc x$ as shown in the box that follows.

PROPERTY SUMMARY   THE COSECANT FUNCTION: $y = \csc x$

Domain: All real numbers other than $0, \pm\pi, \pm 2\pi, \ldots$

Range: $(-\infty, -1] \cup [1, \infty)$

Period: 2π (See Figure 10.)

Asymptotes: $x = 0, x = \pm\pi, x = \pm 2\pi, \ldots$

Intercepts: None

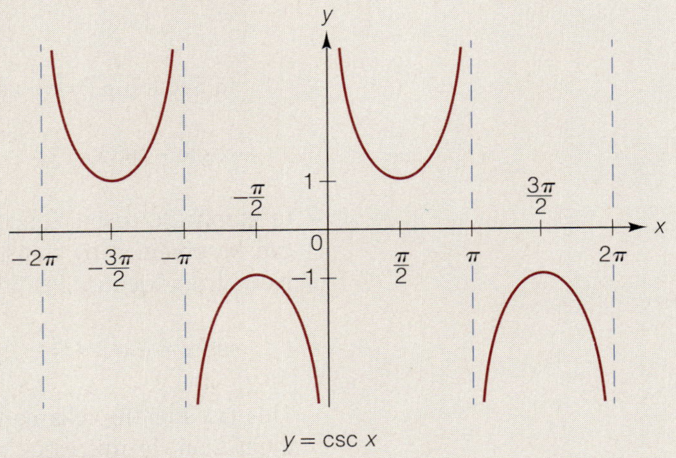

$y = \csc x$

FIGURE 10

EXAMPLE 4   Graph the function $y = \csc(x/3)$ for one period.

Solution   Since $\csc(x/3) = 1/\sin(x/3)$, it will be helpful first to graph one period of $y = \sin(x/3)$. This is done in Figure 11(a) (using the techniques of the previous section). Note that the period of $y = \sin(x/3)$ is 6π. This is also the period of $y = \csc(x/3)$, because $\csc(x/3)$ and $\sin(x/3)$ are just reciprocals. The asymptotes for $y = \csc(x/3)$, occur when $\sin(x/3) = 0$. From Figure 11(a), we see that $\sin(x/3) = 0$ when $x = 0$, when $x = 3\pi$, and when $x = 6\pi$. These asymptotes are sketched in Figure 11(b). The colored points in Figure 11(b) indicate where the value of $\sin(x/3)$ is 1 or -1. The graph of $y = \csc(x/3)$ must pass through these points. (Why?) Finally, using the points and the asymptotes in Figure 11(b), we can sketch the graph of $y = \csc(x/3)$, as shown in Figure 11(c).

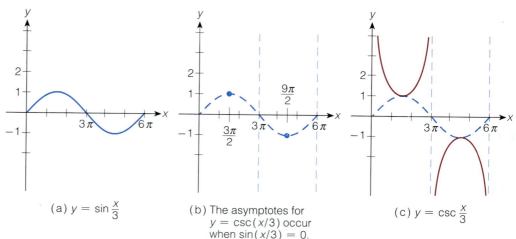

(a) $y = \sin \frac{x}{3}$

(b) The asymptotes for $y = \csc(x/3)$ occur when $\sin(x/3) = 0$.

(c) $y = \csc \frac{x}{3}$

FIGURE 11

In Section 8.3, we used the identity

$$\cos x = \sin\left(x + \tfrac{\pi}{2}\right) \tag{4}$$

to graph the cosine function. This identity tells us that the graph of $y = \cos x$ can be obtained by translating the graph of $y = \sin x$ to the left by $\pi/2$ units. Now, from identity (4), it follows that

$$\sec x = \csc\left(x + \tfrac{\pi}{2}\right)$$

This last identity tells us that the graph of $y = \sec x$ can also be obtained by a translation. In this case, by translating the graph of $y = \csc x$ a distance of $\pi/2$ units to the left, we obtain the graph of $y = \sec x$, as shown in the box that follows.

PROPERTY SUMMARY   THE SECANT FUNCTION: $y = \sec x$

Domain: All real numbers other than $\pm\pi/2$, $\pm 3\pi/2$, $\pm 5\pi/2$, . . .

Range: $(-\infty, -1] \cup [1, \infty)$

Period: 2π (See Figure 12.)

Asymptotes: $x = \pm\pi/2$, $x = \pm 3\pi/2$, $x = \pm 5\pi/2$, . . .

y-intercept: 1

x-intercept: None

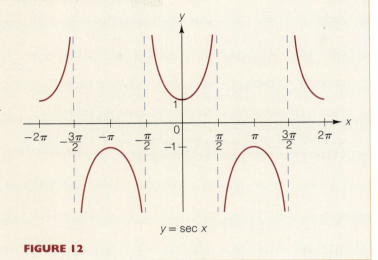

$y = \sec x$

FIGURE 12

EXERCISE SET 8.5

A

In Exercises 1–24, graph each function for one period, and show (or specify) the intercepts and asymptotes.

1. (a) $y = \tan\left(x + \frac{\pi}{4}\right)$
   (b) $y = -\tan\left(x + \frac{\pi}{4}\right)$

2. (a) $y = \tan\left(x - \frac{\pi}{3}\right)$
   (b) $y = -\tan\left(x - \frac{\pi}{3}\right)$

3. (a) $y = \tan(x/3)$
   (b) $y = -\tan(x/3)$

4. (a) $y = 2 \tan \pi x$
   (b) $y = -2 \tan \pi x$

5. $y = \frac{1}{2}\tan(\pi x/2)$

6. $y = -\frac{1}{2}\tan 2\pi x$

7. $y = \cot(\pi x/2)$

8. $y = \cot 2\pi x$

9. $y = -\cot\left(x - \frac{\pi}{4}\right)$

10. $y = \cot\left(x + \frac{\pi}{6}\right)$

11. $y = \frac{1}{2}\cot 2x$

12. $y = -\frac{1}{2}\cot(x/2)$

13. $y = \csc\left(x - \frac{\pi}{4}\right)$

14. $y = \csc\left(x - \frac{\pi}{6}\right)$

15. $y = -\csc(x/2)$

16. $y = 2 \csc x$

17. $y = \frac{1}{3}\csc \pi x$

18. $y = -\frac{1}{2}\csc 2\pi x$

19. $y = -\sec x$

20. $y = -2 \sec x$

21. $y = \sec(x - \pi)$

22. $y = \sec(x + 1)$

23. $y = 3 \sec(\pi x/2)$

24. $y = -2\sec(\pi x/3)$

In Exercises 25–28, graph each function for two periods. Specify the intercepts and the asymptotes.

25. (a) $y = 2\sin\left(3\pi x - \frac{\pi}{6}\right)$
    (b) $y = 2\csc\left(3\pi x - \frac{\pi}{6}\right)$

26. (a) $y = -\frac{1}{2}\sin\left(\pi x + \frac{\pi}{3}\right)$
    (b) $y = -\frac{1}{2}\csc\left(\pi x + \frac{\pi}{3}\right)$

27. (a) $y = -3\cos\left(2\pi x - \frac{\pi}{4}\right)$
    (b) $y = -3\sec\left(2\pi x - \frac{\pi}{4}\right)$

28. (a) $y = \cos\left(3x + \frac{\pi}{3}\right)$
    (b) $y = \sec\left(3x + \frac{\pi}{3}\right)$

B

For Exercises 29–32, six functions are defined as follows.

$$f(x) = \sin x \qquad g(x) = \csc x \qquad h(x) = \pi x - \frac{\pi}{6}$$
$$F(x) = \cos x \qquad G(x) = \sec x \qquad H(x) = \pi x + \frac{\pi}{4}$$

In each case, graph the indicated function for one period.

29. (a) $f \circ h$
    (b) $g \circ h$

30. (a) $F \circ H$
    (b) $G \circ H$

31. (a) $f \circ H$
    (b) $g \circ H$

32. (a) $F \circ h$
    (b) $G \circ h$

For Exercises 33–36, four functions are defined as follows.

$$f(x) = \csc x \qquad T(x) = \tan x$$
$$g(x) = \sec x \qquad A(x) = |x|$$

In each case, graph the indicated function over the interval $[-2\pi, 2\pi]$.

33. $A \circ T$

34. $A \circ g$

35. $A \circ f$

36. $f \circ A$

37. In the text, we used the identity

$$\cot s = -\tan\left(s - \tfrac{\pi}{2}\right)$$

to obtain the graph of $y = \cot x$ from that of $y = \tan x$. The following steps show one way to derive this identity. (Although the accompanying figure shows the angle with radian measure s terminating in the first quadrant, the proof is valid no matter in which quadrant the angle terminates.) In the figure, $PO \perp QO$.

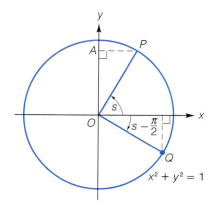

(a) Why are the coordinates of P and Q as follows?

$$P(\cos s, \sin s) \quad \text{and} \quad Q\!\left(\cos\!\left(s - \tfrac{\pi}{2}\right), \sin\!\left(s - \tfrac{\pi}{2}\right)\right)$$

(b) Using congruent triangles [and without reference to part (a)], explain why the y-coordinate of Q is the negative of the x-coordinate of P, and the x-coordinate of Q equals the y-coordinate of P.

(c) Use the results in parts (a) and (b) to conclude that

$$\sin\!\left(s - \tfrac{\pi}{2}\right) = -\cos s \quad \text{and} \quad \cos\!\left(s - \tfrac{\pi}{2}\right) = \sin s$$

(d) Use the result in part (c) to show that

$$\cot s = -\tan\!\left(s - \tfrac{\pi}{2}\right)$$

38. In this exercise we verify the identity $\csc\!\left(\tfrac{\pi}{2} + s\right) = \csc\!\left(\tfrac{\pi}{2} - s\right)$. Refer to the following figure, in which $\angle AOB = \angle BOC = s$ radians. [Although the figure shows s in the interval $0 < s < \pi/2$, a similar proof will work for other intervals as well. (A proof that does not depend upon a picture can be given using the formula for $\sin(s + t)$, which is developed in Section 8.6.)]

(a) Show that the triangles AOE and COD are congruent and, consequently, $AE = CD$.

(b) Explain why the y-coordinates of the points C and A are $\sin\!\left(\tfrac{\pi}{2} - s\right)$ and $\sin\!\left(\tfrac{\pi}{2} + s\right)$, respectively.

(c) Use parts (a) and (b) to conclude that $\sin\!\left(\tfrac{\pi}{2} + s\right) = \sin\!\left(\tfrac{\pi}{2} - s\right)$. It follows from this that $\csc\!\left(\tfrac{\pi}{2} + s\right) = \csc\!\left(\tfrac{\pi}{2} - s\right)$, as required.

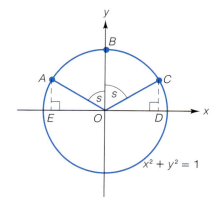

8.6 THE ADDITION FORMULAS FOR SINE AND COSINE

It has long been recognized that the addition formulas are the heart of trigonometry. Indeed, Professor Rademacher and others have shown that the entire body of trigonometry can be derived from the assumption that there exist functions S and C such that

1. $S(x - y) = S(x)C(y) - C(x)S(y)$

2. $C(x - y) = C(x)C(y) + S(x)S(y)$

3. $\displaystyle\lim_{x \to 0^+} \frac{S(x)}{x} = 1$

Professor Frederick H. Young in his article, "The Addition Formulas" from The Mathematics Teacher, vol. L (1957), pp. 45–48

For any real numbers r, s, and t, it is always true that $r(s + t) = rs + rt$. This is the so-called **distributive law** for real numbers. If f is a function, however, it is not true in general that $f(s + t) = f(s) + f(t)$. By way of example, consider the cosine function. It is not true in general that $\cos(s + t) = \cos s + \cos t$. For instance, with $s = \pi/6$ and $t = \pi/3$, we have

$$\cos\left(\frac{\pi}{6} + \frac{\pi}{3}\right) \stackrel{?}{=} \cos\frac{\pi}{6} + \cos\frac{\pi}{3}$$

$$\cos\frac{\pi}{2} \stackrel{?}{=} \cos\frac{\pi}{6} + \cos\frac{\pi}{3}$$

$$0 \stackrel{?}{=} \frac{\sqrt{3}}{2} + \frac{1}{2} \qquad \text{No!}$$

In this section, we will see just what $\cos(s + t)$ does equal. The correct formula for $\cos(s + t)$ is one of a group of four important trigonometric identities called the **addition formulas**.

The Addition Formulas

$$\sin(s + t) = \sin s \cos t + \cos s \sin t$$
$$\sin(s - t) = \sin s \cos t - \cos s \sin t$$
$$\cos(s + t) = \cos s \cos t - \sin s \sin t$$
$$\cos(s - t) = \cos s \cos t + \sin s \sin t$$

Our strategy for deriving these formulas will be as follows. First we'll prove the fourth formula—this takes some effort. The other three formulas are then relatively easy to derive from the fourth one. As background for our discussion, you will need to recall that the distance d between two points (x_1, y_1) and (x_2, y_2) is given by

$$d = \sqrt{(x_2 - x_1)^2 + (y_2 - y_1)^2}$$

To prove the fourth formula in the box, we use Figure 1 (on the next page). The idea behind the proof is as follows.* We begin in Figure 1(a) with the unit circle and the angles s, t, and $s - t$. Then we rotate $\triangle OPQ$ about the origin until the point P coincides with the point $(1, 0)$, as indicated in Figure 1(b). Although this rotation changes the coordinates for the points P and Q, it certainly has no effect upon the length of the line segment $\overline{PQ}$. Thus, whether we calculate PQ using the coordinates in Figure 1(a) or those in Figure 1(b), the results must be the same. As you'll see, by equating the two expressions for PQ, we will obtain the required formula.

*This idea for the proof can be traced back to the great French mathematician Augustin-Louis Cauchy (1789–1857).

FIGURE I

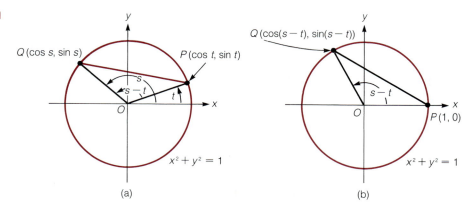

(a)   (b)

Applying the distance formula to Figure 1(a), we have

$$PQ = \sqrt{(\cos s - \cos t)^2 + (\sin s - \sin t)^2}$$

$$= \sqrt{\cos^2 s - 2 \cos s \cos t + \cos^2 t + \sin^2 s - 2 \sin s \sin t + \sin^2 t}$$

$$= \sqrt{2 - 2 \cos s \cos t - 2 \sin s \sin t} \qquad \text{(Why?)} \qquad (1)$$

Next, applying the distance formula to Figure 1(b), we have

$$PQ = \sqrt{[\cos(s - t) - 1]^2 + [\sin(s - t) - 0]^2}$$

$$= \sqrt{\cos^2(s - t) - 2 \cos(s - t) + 1 + \sin^2(s - t)}$$

The right-hand side of this last equation can be simplified using the fact that $\cos^2(s - t) + \sin^2(s - t) = 1$. We then have

$$PQ = \sqrt{2 - 2 \cos(s - t)} \qquad (2)$$

From equations (1) and (2), we conclude that

$$2 - 2 \cos(s - t) = 2 - 2 \cos s \cos t - 2 \sin s \sin t$$

$$-2 \cos(s - t) = -2 \cos s \cos t - 2 \sin s \sin t$$

$$\cos(s - t) = \cos s \cos t + \sin s \sin t$$

This completes the proof of the fourth addition formula. Before deriving the other three addition formulas, let's look at some applications of this result.

EXAMPLE I   Simplify the expression $\cos 2\theta \cos \theta + \sin 2\theta \sin \theta$.

Solution   According to the identity that we just proved, we have

$$\cos 2\theta \cos \theta + \sin 2\theta \sin \theta = \cos(2\theta - \theta) = \cos \theta$$

Thus, the given expression is equal to $\cos \theta$.

EXAMPLE 2   Simplify $\cos(\theta - \pi)$.

Solution   We use the formula for $\cos(s - t)$ with s and t replaced by θ and π, respectively. This yields

$$\cos(s - t) = \cos s \cos t + \sin s \sin t$$
$$\updownarrow \quad \updownarrow$$
$$\cos(\theta - \pi) = \cos \theta \cos \pi + \sin \theta \sin \pi$$
$$= (\cos \theta)(-1) + (\sin \theta)(0) = -\cos \theta$$

Thus, the required simplification is $\cos(\theta - \pi) = -\cos \theta$. ▮▮▮

In Example 2, we found that $\cos(\theta - \pi) = -\cos \theta$. This type of identity is often referred to as a **reduction formula**. The next example develops two basic reduction formulas that we will need to use later in this section.

EXAMPLE 3 Prove the following identities.

(a) $\cos\left(\frac{\pi}{2} - \alpha\right) = \sin \alpha$ (b) $\sin\left(\frac{\pi}{2} - \beta\right) = \cos \beta$

Solution (a) $\cos\left(\frac{\pi}{2} - \alpha\right) = \cos \frac{\pi}{2} \cos \alpha + \sin \frac{\pi}{2} \sin \alpha$
$$= (0) \cos \alpha + (1) \sin \alpha = \sin \alpha$$

This proves the identity. Incidentally, if we use degree measure instead of radian measure, this identity states that

$$\cos(90° - \alpha) = \sin \alpha$$

as we saw earlier, in Chapter 7.

(b) Since the identity $\cos\left(\frac{\pi}{2} - \alpha\right) = \sin \alpha$ holds for all values of α, we can simply replace α by the quantity $\frac{\pi}{2} - \beta$ to obtain

$$\cos\left[\frac{\pi}{2} - \left(\frac{\pi}{2} - \beta\right)\right] = \sin\left(\frac{\pi}{2} - \beta\right)$$
$$\cos\left(\frac{\pi}{2} - \frac{\pi}{2} + \beta\right) = \sin\left(\frac{\pi}{2} - \beta\right)$$
$$\cos \beta = \sin\left(\frac{\pi}{2} - \beta\right)$$

This proves the identity. ▮▮▮

The two identities in Example 3 are worth memorizing. For simplicity, we replace both α and β by the single letter θ to get

$$\cos\left(\frac{\pi}{2} - \theta\right) = \sin \theta$$
$$\sin\left(\frac{\pi}{2} - \theta\right) = \cos \theta$$

In Examples 2 and 3, radian measure was used. However, if you look back at Figure 1 and the derivation of the formula for $\cos(s - t)$, you will see that the derivation makes no reference, implicit or explicit, to a specific system of angle measurement. (For instance, the derivation does not involve the formula $s = r\theta$, which does require radian measure.) Thus, the formula for $\cos(s - t)$ is also valid when angles are measured in degrees. In the next example, we apply the formula in just such a case.

EXAMPLE 4   Use the formula for $\cos(s - t)$ to determine the exact value of $\cos 15°$.

Solution   First observe that $15° = 45° - 30°$. Then we have

$\cos 15° = \cos(45° - 30°)$

$\qquad = \cos 45° \cos 30° + \sin 45° \sin 30°$ using the formula for $\cos(s - t)$
with $s = 45°$ and $t = 30°$

$\qquad = \left(\dfrac{\sqrt{2}}{2}\right)\left(\dfrac{\sqrt{3}}{2}\right) + \left(\dfrac{\sqrt{2}}{2}\right)\left(\dfrac{1}{2}\right)$

$\qquad = \dfrac{\sqrt{6}}{4} + \dfrac{\sqrt{2}}{4} = \dfrac{\sqrt{6} + \sqrt{2}}{4}$

Thus, the exact value of $\cos 15°$ is $\frac{1}{4}(\sqrt{6} + \sqrt{2})$.

Now let's return to our derivations of the addition formulas. Using the fourth addition formula, we can easily derive the third formula as follows. In the formula

$\cos(s - t) = \cos s \cos t + \sin s \sin t$

we replace t by the quantity $-t$. This is permissible because the formula holds for all real numbers. We obtain

$\cos[s - (-t)] = \cos s \cos(-t) + \sin s \sin(-t)$

On the right-hand side of this equation, we can use the identities developed for $\cos(-t)$ and $\sin(-t)$ in Section 8.2. Doing this yields

$\cos(s + t) = (\cos s)(\cos t) + (\sin s)(-\sin t)$

which is equivalent to

$\cos(s + t) = \cos s \cos t - \sin s \sin t$

This is the third addition formula, as we wished to prove.

Next we derive the formula for $\sin(s + t)$. We have

$\sin(s + t) = \cos\left[\frac{\pi}{2} - (s + t)\right]$ replacing θ by $s + t$ in the identity
$\sin \theta = \cos\left[\frac{\pi}{2} - \theta\right]$

$\qquad = \cos\left[\left(\frac{\pi}{2} - s\right) - t\right]$

$\qquad = \cos\left(\frac{\pi}{2} - s\right) \cos t + \sin\left(\frac{\pi}{2} - s\right) \sin t$ (Why?)

$\qquad = \sin s \cos t + \cos s \sin t$ (Why?)

This proves the first addition formula.

Finally, we can use the first addition formula to prove the second as follows:

$\sin(s - t) = \sin[s + (-t)]$

$\qquad = \sin s \cos(-t) + \cos s \sin(-t)$

$\qquad = (\sin s)(\cos t) + (\cos s)(-\sin t)$

$\qquad = \sin s \cos t - \cos s \sin t$

This completes the proofs of the four addition formulas.

EXAMPLE 5   If $\sin s = \frac{3}{5}$, $0 < s < \frac{\pi}{2}$, and $\sin t = -\sqrt{3}/4$, $\pi < t < \frac{3\pi}{2}$, compute $\sin(s - t)$.

Solution $\sin(s - t) = \underbrace{\sin s}\ \cos t - \cos s\ \underbrace{\sin t}$ (3)

given as $\frac{3}{5}$ given as $-\sqrt{3}/4$

In view of equation (3), we need to find only $\cos t$ and $\cos s$. These can be determined using the Pythagorean identity $\cos^2 \theta = 1 - \sin^2 \theta$. We have

$$\cos^2 t = 1 - \sin^2 t$$
$$= 1 - \left(\frac{-\sqrt{3}}{4}\right)^2 = 1 - \frac{3}{16} = \frac{13}{16}$$

Therefore,

$$\cos t = \frac{\sqrt{13}}{4} \quad \text{or} \quad \cos t = \frac{-\sqrt{13}}{4}$$

We choose the negative value here for cosine, since it is given that $\pi < t < 3\pi/2$:

$$\cos t = \frac{-\sqrt{13}}{4}$$

Similarly, to find $\cos s$, we have

$$\cos^2 s = 1 - \sin^2 s = 1 - \left(\frac{3}{5}\right)^2 = \frac{16}{25}$$

Therefore,

$$\cos s = \frac{4}{5} \quad \cos s \text{ is positive, since } 0 < s < \pi/2$$

Finally, we substitute the values we've obtained for $\cos t$ and $\cos s$, along with the given data, back into equation (3). This yields

$$\sin(s - t) = \left(\frac{3}{5}\right)\left(\frac{-\sqrt{13}}{4}\right) - \left(\frac{4}{5}\right)\left(\frac{-\sqrt{3}}{4}\right)$$
$$= \frac{-3\sqrt{13}}{20} + \frac{4\sqrt{3}}{20} = \frac{-3\sqrt{13} + 4\sqrt{3}}{20}$$

EXERCISE SET 8.6

A

In Exercises 1–10, use the addition formulas to simplify the expression.

1. $\sin \theta \cos 2\theta + \cos \theta \sin 2\theta$
2. $\sin \frac{\pi}{6} \cos \frac{\pi}{3} + \cos \frac{\pi}{6} \sin \frac{\pi}{3}$
3. $\sin 3\theta \cos \theta - \cos 3\theta \sin \theta$

4. $\sin 110° \cos 20° - \cos 110° \sin 20°$
5. $\cos 2u \cos 3u - \sin 2u \sin 3u$
6. $\cos 2u \cos 3u + \sin 2u \sin 3u$
7. $\cos \frac{2\pi}{9} \cos \frac{\pi}{18} + \sin \frac{2\pi}{9} \sin \frac{\pi}{18}$
8. $\cos \frac{3\pi}{10} \cos \frac{\pi}{5} - \sin \frac{3\pi}{10} \sin \frac{\pi}{5}$

9. $\sin(A + B) \cos A - \cos(A + B) \sin A$

10. $\cos(s - t) \cos t - \sin(s - t) \sin t$

In Exercises 11–14, simplify each expression (as in Example 2).

11. $\sin\left(\theta - \frac{3\pi}{2}\right)$

12. $\cos\left(\frac{3\pi}{2} + \theta\right)$

13. $\cos(\theta + \pi)$

14. $\sin(\theta - \pi)$

15.   Expand $\sin(t + 2\pi)$ using the appropriate addition formula, and check to see that your answer agrees with the formula for $\sin(t + 2\pi)$ given on page 435.

16.   Follow the directions in Exercise 15, but use $\cos(t + 2\pi)$.

17.   Use the formula of $\cos(s + t)$ to compute the exact value of cos 75°.

18.   Use the formula for $\sin(s - t)$ to compute the exact value of $\sin(\pi/12)$.

19.   Use the formula for $\sin(s + t)$ to find $\sin(7\pi/12)$.

20.   Determine the exact value of (a) sin 105°; (b) cos 105°.

In Exercises 21–24, use the addition formulas to simplify each expression.

21. $\sin\left(\frac{\pi}{4} + s\right) - \sin\left(\frac{\pi}{4} - s\right)$

22. $\sin\left(t + \frac{\pi}{6}\right) - \sin\left(t - \frac{\pi}{6}\right)$

23. $\cos\left(\frac{\pi}{3} - \theta\right) - \cos\left(\frac{\pi}{3} + \theta\right)$

24. $\cos\left(\theta - \frac{\pi}{4}\right) + \cos\left(\theta + \frac{\pi}{4}\right)$

In Exercises 25–28, compute the indicated quantity using the following data.

$$\sin \alpha = \tfrac{12}{13}, \quad \pi/2 < \alpha < \pi$$
$$\cos \beta = -\tfrac{3}{5}, \quad \pi < \beta < 3\pi/2$$
$$\cos \theta = \tfrac{7}{25}, \quad -2\pi < \theta < -3\pi/2$$

25.   (a) $\sin(\alpha + \beta)$
      (b) $\cos(\alpha + \beta)$

26.   (a) $\sin(\alpha - \beta)$
      (b) $\cos(\alpha - \beta)$

27.   (a) $\sin(\theta - \beta)$
      (b) $\sin(\theta + \beta)$

28.   (a) $\cos(\alpha + \theta)$
      (b) $\cos(\alpha - \theta)$

29.   Suppose that $\sin \theta = \tfrac{1}{5}$ and $0 < \theta < \pi/2$.
      (a) Compute cos θ.
      (b) Compute sin 2θ.   *Hint:* $\sin 2\theta = \sin(\theta + \theta)$

30.   Suppose that $\cos \theta = \tfrac{12}{13}$ and $3\pi/2 < \theta < 2\pi$.
      (a) Compute sin θ.
      (b) Compute cos 2θ.   *Hint:* $\cos 2\theta = \cos(\theta + \theta)$.

31.   Given $\tan \theta = -\tfrac{2}{3}$, $\pi/2 < \theta < \pi$, and csc $\beta = 2$, $0 < \beta < \pi/2$, find $\sin(\theta + \beta)$ and $\cos(\beta - \theta)$.

32.   Given $\sec s = \tfrac{5}{4}$, sin $s < 0$, and cot $t = -1$, $\pi/2 < t < \pi$, find $\sin(s - t)$ and $\cos(s + t)$.

In Exercises 33–35, prove that each equation is an identity.

33. $\sin\left(t + \frac{\pi}{4}\right) = (\sin t + \cos t)/\sqrt{2}$

34. $\cos\left(t + \frac{\pi}{4}\right) = (\cos t - \sin t)/\sqrt{2}$

35. $\sin\left(t + \frac{\pi}{4}\right) + \cos\left(t + \frac{\pi}{4}\right) = \sqrt{2} \cos t$

36.   In the following figure, $AB = CD = 1$ and $BC = 2$. Find $\alpha + \beta$ and express your answer using radian measure. *Hint:*   Find $\cos(\alpha + \beta)$.

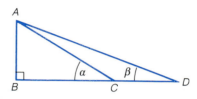

37.   (a) Prove that the following equation is an identity:

$$\cos x = \sin\left(x + \frac{\pi}{2}\right)$$

      (b) What does the identity in part (a) tell us about the graphs of sine and cosine? (We used this identity in Section 8.3 to obtain the graph of the cosine function.)

38.   Prove that the following equation is an identity:

$$\sin x = -\cos\left(x + \frac{\pi}{2}\right)$$

B

In Exercises 39–46, prove that each equation is an identity.

39. $\dfrac{\sin(s + t)}{\cos s \cos t} = \tan s + \tan t$

40. $\dfrac{\cos(s - t)}{\cos s \sin t} = \cot t + \tan s$

41. $\cos(A - B) - \cos(A + B) = 2 \sin A \sin B$

42. $\sin(A - B) + \sin(A + B) = 2 \sin A \cos B$

43. $\cos(A + B) \cos(A - B) = \cos^2 A - \sin^2 B$

44. $\sin(A + B) \sin(A - B) = \cos^2 B - \cos^2 A$

45. $\cos(\alpha + \beta) \cos \beta + \sin(\alpha + \beta) \sin \beta = \cos \alpha$

46. $\cos\left(\theta + \frac{\pi}{4}\right) + \sin\left(\theta - \frac{\pi}{4}\right) = 0$

47.   Let $f(t) = \cos^2 t + \cos^2\left(t + \frac{2\pi}{3}\right) + \cos^2\left(t - \frac{2\pi}{3}\right)$.
      (a) Complete the table. (Use a calculator.)

t	1	2	3	4
$f(t)$				

(b) On the basis of your results in part (a), make a conjecture about the function f. Prove that your conjecture is correct.

48. Let A, B, and C be the angles of a triangle, so that $A + B + C = \pi$.
   (a) Show that $\sin(A + B) = \sin C$.
   (b) Show that $\cos(A + B) = -\cos C$.
   (c) Show that $\tan(A + B) = -\tan C$.

49. Suppose that A, B, and C are the angles of a triangle, so that $A + B + C = \pi$. Show that

$$\cos^2 A + \cos^2 B + \cos^2 C + 2 \cos A \cos B \cos C = 1$$

50. Prove that

$$\frac{\sin(\alpha - \beta)}{\cos \alpha \cos \beta} + \frac{\sin(\beta - \gamma)}{\cos \beta \cos \gamma} + \frac{\sin(\gamma - \alpha)}{\cos \gamma \cos \alpha} = 0$$

51. Suppose that $a^2 + b^2 = 1$ and $c^2 + d^2 = 1$. Prove that $|ac + bd| \le 1$. *Hint:* Let $a = \cos \theta$, $b = \sin \theta$, $c = \cos \phi$, and $d = \sin \phi$.

In Exercises 52 and 53, simplify the expression.

52. $\cos\left(\frac{\pi}{6} + t\right) \cos\left(\frac{\pi}{6} - t\right) - \sin\left(\frac{\pi}{6} + t\right) \sin\left(\frac{\pi}{6} - t\right)$

   Hint: If your solution relies on four separate addition formulas, then you are doing this the hard way.

53. $\sin\left(\frac{\pi}{3} - t\right) \cos\left(\frac{\pi}{3} + t\right) + \cos\left(\frac{\pi}{3} - t\right) \sin\left(\frac{\pi}{3} + t\right)$

Exercises 54 and 55 outline simple geometric derivations of the formulas for $\sin(\alpha + \beta)$ and $\cos(\alpha + \beta)$ in the case where α and β are acute angles, with $\alpha + \beta < 90°$. The exercises rely on the accompanying figures, which are constructed as follows. Begin, in Figure A, with $\alpha = \angle GAD$, $\beta = \angle HAG$, and $AH = 1$. Then, from H, draw perpendiculars to $\overline{AD}$ and to $\overline{AG}$, as shown in Figure B. Finally, draw $\overline{FE} \perp \overline{BH}$ and $\overline{FC} \perp \overline{AD}$.

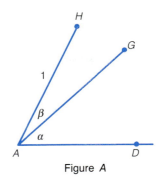

Figure A

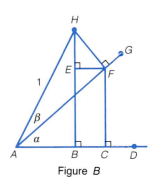

Figure B

54. *Formula for $\sin(\alpha + \beta)$* Supply the reasons or steps behind each statement.
   (a) $BH = \sin(\alpha + \beta)$ (b) $FH = \sin \beta$
   (c) $\angle BHF = \alpha$
   (d) $EH = \cos \alpha \sin \beta$
      Hint: Use $\triangle EFH$ and the result in part (b).
   (e) $AF = \cos \beta$ (f) $CF = \sin \alpha \cos \beta$
   (g) $\sin(\alpha + \beta) = \sin \alpha \cos \beta + \cos \alpha \sin \beta$
      Hint: $\sin(\alpha + \beta) = BH = EH + CF$

55. *Formula for $\cos(\alpha + \beta)$* Supply the reason or steps behind each statement
   (a) $\cos(\alpha + \beta) = AB$
   (b) $AC = \cos \alpha \cos \beta$
      Hint: Use $\triangle ACF$ and the result in Exercise 54(e).
   (c) $EF = \sin \alpha \sin \beta$
   (d) $\cos(\alpha + \beta) = \cos \alpha \cos \beta - \sin \alpha \sin \beta$
      Hint: $AB = AC - BC$

56. If triangle ABC is not a right triangle, and $\cos A = \cos B \cos C$, show that $\tan B \tan C = 2$.

8.7 FURTHER IDENTITIES

Ptolemy (c. 150) knew substantially the sine of half an angle . . . and it is probable that Hipparchus (c. 140 B.C.) and certain that Varahamihira (c. 505) knew the relation that we express as $\sin \frac{\phi}{2} = \sqrt{(1 - \cos \phi)/2}$.

David Eugene Smith in *History of Mathematics* (New York: Ginn and Company, 1925)

There are a number of basic identities that follow from the addition formulas for sine and cosine. In the following boxes, we summarize some of the most useful of these.

Addition Formulas for Tangent

a. $\tan(s + t) = \dfrac{\tan s + \tan t}{1 - \tan s \tan t}$

b. $\tan(s - t) = \dfrac{\tan s - \tan t}{1 + \tan s \tan t}$

Double-Angle Formulas

a. $\sin 2\theta = 2 \sin \theta \cos \theta$

b. $\cos 2\theta = \cos^2 \theta - \sin^2 \theta$

c. $\tan 2\theta = \dfrac{2 \tan \theta}{1 - \tan^2 \theta}$

Half-Angle Formulas

a. $\sin \dfrac{s}{2} = \pm \sqrt{\dfrac{1 - \cos s}{2}}$

b. $\cos \dfrac{s}{2} = \pm \sqrt{\dfrac{1 + \cos s}{2}}$

c. $\tan \dfrac{s}{2} = \dfrac{\sin s}{1 + \cos s}$

To prove the formula for $\tan(s + t)$, we begin with

$$\tan(s + t) = \frac{\sin(s + t)}{\cos(s + t)}$$

$$= \frac{\sin s \cos t + \cos s \sin t}{\cos s \cos t - \sin s \sin t} \tag{1}$$

Now we divide both numerator and denominator on the right-hand side of equation (1) by the quantity $\cos s \cos t$. This yields

$$\tan(s + t) = \frac{\dfrac{\sin s \cos t}{\cos s \cos t} + \dfrac{\cos s \sin t}{\cos s \cos t}}{\dfrac{\cos s \cos t}{\cos s \cos t} - \dfrac{\sin s \sin t}{\cos s \cos t}} = \frac{\tan s + \tan t}{1 - \tan s \tan t}$$

This proves the formula for $\tan(s + t)$. The formula for $\tan(s - t)$ can be deduced from this with the aid of the identity $\tan(-t) = -\tan t$, which was derived in Section 8.2. We have

$$\tan(s - t) = \tan[s + (-t)]$$

$$= \frac{\tan s + \tan(-t)}{1 - \tan s \tan(-t)} = \frac{\tan s + (-\tan t)}{1 - (\tan s)(-\tan t)}$$

$$= \frac{\tan s - \tan t}{1 + \tan s \tan t} \qquad \text{as required}$$

EXAMPLE 1 Compute $\tan \frac{7\pi}{12}$, using the fact that $\frac{7\pi}{12} = \frac{\pi}{4} + \frac{\pi}{3}$.

Solution $\tan \frac{7\pi}{12} = \tan\left(\frac{\pi}{4} + \frac{\pi}{3}\right)$

$$= \frac{\tan \frac{\pi}{4} + \tan \frac{\pi}{3}}{1 - \tan \frac{\pi}{4} \tan \frac{\pi}{3}} \qquad \begin{array}{l}\text{using the formula for } \tan(s + t) \\ \text{with } s = \frac{\pi}{4} \text{ and } t = \frac{\pi}{3}\end{array}$$

$$= \frac{1 + \sqrt{3}}{1 - (1)(\sqrt{3})}$$

Thus, $\tan \dfrac{7\pi}{12} = \dfrac{1 + \sqrt{3}}{1 - \sqrt{3}}$. We can write this answer in a more compact form by rationalizing the denominator. As you can check, the result is

$$\tan \tfrac{7\pi}{12} = -2 - \sqrt{3} \qquad\qquad \blacksquare\blacksquare\blacksquare$$

We turn now to the identities for $\sin 2\theta$, $\cos 2\theta$, and $\tan 2\theta$. All of these are derived in the same way: we replace 2θ by $(\theta + \theta)$ and use the appropriate addition formula. For instance, for $\sin 2\theta$ we have

$$\sin 2\theta = \sin(\theta + \theta) = \sin \theta \cos \theta + \cos \theta \sin \theta$$
$$= 2 \sin \theta \cos \theta$$

This establishes the formula for $\sin 2\theta$. (Exercise 37 asks you to carry out the corresponding derivations for $\cos 2\theta$ and $\tan 2\theta$.)

EXAMPLE 2 If $\sin \theta = \frac{4}{5}$ and $\pi/2 < \theta < \pi$, find $\cos \theta$, $\sin 2\theta$, and $\cos 2\theta$.

Solution We have

$$\cos^2 \theta = 1 - \sin^2 \theta = 1 - \left(\frac{4}{5}\right)^2 = \frac{9}{25}$$

Consequently, $\cos \theta = 3/5$ or $\cos \theta = -3/5$. We want the negative value for the cosine here, since $\pi/2 < \theta < \pi$. Thus,

$$\cos \theta = -\frac{3}{5}$$

Now that we know the values of $\cos \theta$ and $\sin \theta$, the double-angle formulas can be used to determine $\sin 2\theta$ and $\cos 2\theta$. We have

$$\sin 2\theta = 2 \sin \theta \cos \theta = 2\left(\frac{4}{5}\right)\left(-\frac{3}{5}\right) = -\frac{24}{25}$$

and

$$\cos 2\theta = \cos^2 \theta - \sin^2 \theta$$

$$= \left(-\frac{3}{5}\right)^2 - \left(\frac{4}{5}\right)^2 = \frac{9}{25} - \frac{16}{25} = -\frac{7}{25}$$

The required values are therefore $\cos \theta = -\frac{3}{5}$, $\sin 2\theta = -\frac{24}{25}$, and $\cos 2\theta = -\frac{7}{25}$.

EXAMPLE 3 If $x = 4 \sin \theta$, $0 < \theta < \pi/2$, express $\sin 2\theta$ in terms of x.

Solution The given equation is equivalent to $\sin \theta = x/4$, so we have

$$\sin 2\theta = 2 \sin \theta \cos \theta = 2\left(\frac{x}{4}\right)\cos \theta$$

$$= \frac{x}{2}\sqrt{1 - \sin^2 \theta} \qquad \text{(Why is the positive root appropriate?)}$$

$$= \frac{x}{2}\sqrt{1 - \frac{x^2}{16}} = \frac{x}{2}\sqrt{\frac{16 - x^2}{16}} = \frac{x\sqrt{16 - x^2}}{8}$$

EXAMPLE 4 Prove the following identity:

$$\cos 3\theta = 4 \cos^3 \theta - 3 \cos \theta$$

Solution

$$\cos 3\theta = \cos(2\theta + \theta)$$

$$= \underbrace{\cos 2\theta}_{(\cos^2 \theta - \sin^2 \theta)} \cos \theta - \underbrace{\sin 2\theta}_{(2 \sin \theta \cos \theta)} \sin \theta$$

$$= (\cos^2 \theta - \sin^2 \theta)\cos \theta - (2 \sin \theta \cos \theta)\sin \theta$$

$$= \cos^3 \theta - \sin^2 \theta \cos \theta - 2 \sin^2 \theta \cos \theta$$

Collecting like terms now gives us

$$\cos 3\theta = \cos^3 \theta - 3 \sin^2 \theta \cos \theta$$

Finally, we replace $\sin^2 \theta$ by the quantity $1 - \cos^2 \theta$. This yields

$$\cos 3\theta = \cos^3 \theta - 3(1 - \cos^2 \theta) \cos \theta$$

$$= \cos^3 \theta - 3 \cos \theta + 3 \cos^3 \theta$$

$$= 4 \cos^3 \theta - 3 \cos \theta \qquad \text{as required}$$

In the box that follows, we list several alternate ways of writing the formula for $\cos 2\theta$. Formulas (c) and (d) are quite useful in calculus.

Equivalent Forms of the Formula $\cos 2\theta = \cos^2 \theta - \sin^2 \theta$

a. $\cos 2\theta = 2 \cos^2 \theta - 1$ b. $\cos 2\theta = 1 - 2 \sin^2 \theta$

c. $\cos^2 \theta = \dfrac{1 + \cos 2\theta}{2}$ d. $\sin^2 \theta = \dfrac{1 - \cos 2\theta}{2}$

One way to prove identity (a) is as follows:

$$\cos 2\theta = \cos^2 \theta - \sin^2 \theta$$
$$= \cos^2 \theta - (1 - \cos^2 \theta)$$
$$= \cos^2 \theta - 1 + \cos^2 \theta$$
$$= 2 \cos^2 \theta - 1 \qquad \text{as required}$$

If, in this last equation, we add 1 to both sides and then divide by 2, the result is identity (c). (Verify this.) The proofs for (b) and (d) are similar; see Exercise 38.

EXAMPLE 5 Express $\cos^4 t$ in a form that does not involve powers of the trigonometric functions.

Solution

$$\cos^4 t = (\cos^2 t)^2 = \left(\frac{1 + \cos 2t}{2}\right)^2 \qquad \text{using the formula for } \cos^2 \theta$$

$$= \frac{1 + 2\cos 2t + \cos^2 2t}{4}$$

$$= \frac{1 + 2\cos 2t + \frac{1}{2}(1 + \cos 4t)}{4} \qquad \begin{array}{l}\text{using the formula for } \cos^2 \theta \\ \text{with } \theta = 2t\end{array}$$

An easy way to simplify this last expression is to multiply both the numerator and the denominator by 2. As you should check for yourself, the final result is

$$\cos^4 t = \frac{3 + 4\cos 2t + \cos 4t}{8}$$

The last three formulas we are going to prove in this section are the **half-angle formulas**:

$$\cos \frac{s}{2} = \pm\sqrt{\frac{1 + \cos s}{2}}$$

$$\sin \frac{s}{2} = \pm\sqrt{\frac{1 - \cos s}{2}}$$

$$\tan \frac{s}{2} = \frac{\sin s}{1 + \cos s}$$

To derive the formula for $\cos(s/2)$, we begin with one of the alternate forms of the cosine double-angle formula:

$$\cos^2 \theta = \frac{1 + \cos 2\theta}{2} \qquad \text{or} \qquad \cos \theta = \pm\sqrt{\frac{1 + \cos 2\theta}{2}}$$

Since this identity holds for all values of θ, we may replace θ by $s/2$ to obtain

$$\cos \frac{s}{2} = \pm\sqrt{\frac{1 + \cos 2(s/2)}{2}} = \pm\sqrt{\frac{1 + \cos s}{2}}$$

This is the required formula for $\cos(s/2)$. To derive the formula for $\sin(s/2)$, we follow exactly the same procedure, except that we begin with the identity $\sin^2 \theta = \frac{1}{2}(1 - \cos 2\theta)$. [Exercise 38(c) asks you to complete the proof.] In both formulas, the sign before the radical is determined by the quadrant in which the angle or arc $s/2$ terminates.

EXAMPLE 6 Evaluate $\cos 105°$ using a half-angle formula.

Solution

$$\cos 105° = \cos \frac{210°}{2} = \pm \sqrt{\frac{1 + \cos 210°}{2}} \qquad \text{using the formula for } \cos(s/2) \text{ with } s = 210°$$

$$= \pm \sqrt{\frac{1 + (-\sqrt{3}/2)}{2}}$$

$$= \pm \sqrt{\frac{1 - (\sqrt{3}/2)}{2} \cdot \frac{2}{2}} = \pm \sqrt{\frac{2 - \sqrt{3}}{4}}$$

$$= \frac{\pm \sqrt{2 - \sqrt{3}}}{2}$$

We want to choose the negative value here, since the terminal side of $105°$ lies in the second quadrant. Thus, we finally obtain

$$\cos 105° = \frac{-\sqrt{2 - \sqrt{3}}}{2}$$

Our last task is to establish the formula for $\tan(s/2)$. To do this, we first prove the equivalent identity, $\tan \theta = \dfrac{\sin 2\theta}{1 + \cos 2\theta}$.

PROOF THAT $\tan \theta = \dfrac{\sin 2\theta}{1 + \cos 2\theta}$

$$\frac{\sin 2\theta}{1 + \cos 2\theta} = \frac{2 \sin \theta \cos \theta}{2 \cos^2 \theta} \qquad \text{using the identity } \cos 2\theta = 2 \cos^2 \theta - 1 \text{ in the denominator}$$

$$= \frac{\sin \theta}{\cos \theta} = \tan \theta$$

If we now replace θ by $s/2$ in the identity $\tan \theta = \dfrac{\sin 2\theta}{1 + \cos 2\theta}$, the result is

$$\tan \frac{s}{2} = \frac{\sin s}{1 + \cos s}$$

This is the half-angle formula for the tangent.

We conclude this section with a summary of the principal trigonometric identities developed in this chapter. For completeness, the list also includes two sets of trigonometric identities that we did not discuss in the text but that are occasionally useful. These are the so-called **product-to-sum formulas** and **sum-to-product formulas**. Proofs and applications of these formulas are included in the exercises.

PROPERTY SUMMARY PRINCIPAL TRIGONOMETRIC IDENTITIES

I. Consequences of the Definitions

(a) $\csc\theta = \dfrac{1}{\sin\theta}$

(b) $\sec\theta = \dfrac{1}{\cos\theta}$

(c) $\cot\theta = \dfrac{1}{\tan\theta}$

(d) $\tan\theta = \dfrac{\sin\theta}{\cos\theta}$

(e) $\cot\theta = \dfrac{\cos\theta}{\sin\theta}$

II. Pythagorean Identities

(a) $\sin^2\theta + \cos^2\theta = 1$

(b) $\tan^2\theta + 1 = \sec^2\theta$

(c) $\cot^2\theta + 1 = \csc^2\theta$

III. Opposite-Angle Formulas

(a) $\sin(-\theta) = -\sin\theta$

(b) $\cos(-\theta) = \cos\theta$

(c) $\tan(-\theta) = -\tan\theta$

IV. Reduction Formulas

(a) $\sin(\theta + 2\pi k) = \sin\theta$

(b) $\cos(\theta + 2\pi k) = \cos\theta$

(c) $\sin\left(\dfrac{\pi}{2} - \theta\right) = \cos\theta$

(d) $\cos\left(\dfrac{\pi}{2} - \theta\right) = \sin\theta$

V. Addition Formulas

(a) $\sin(s + t) = \sin s \cos t + \cos s \sin t$

(b) $\sin(s - t) = \sin s \cos t - \cos s \sin t$

(c) $\cos(s + t) = \cos s \cos t - \sin s \sin t$

(d) $\cos(s - t) = \cos s \cos t + \sin s \sin t$

(e) $\tan(s + t) = \dfrac{\tan s + \tan t}{1 - \tan s \tan t}$

(f) $\tan(s - t) = \dfrac{\tan s - \tan t}{1 + \tan s \tan t}$

VI. Double-Angle Formulas

(a) $\sin 2\theta = 2 \sin\theta \cos\theta$

(b) $\cos 2\theta = \cos^2\theta - \sin^2\theta$

(c) $\tan 2\theta = \dfrac{2 \tan\theta}{1 - \tan^2\theta}$

VII. Half-Angle Formulas

(a) $\sin\dfrac{\theta}{2} = \pm\sqrt{\dfrac{1 - \cos\theta}{2}}$

(b) $\cos\dfrac{\theta}{2} = \pm\sqrt{\dfrac{1 + \cos\theta}{2}}$

(c) $\tan\dfrac{\theta}{2} = \dfrac{\sin\theta}{1 + \cos\theta}$

VIII. Product-to-Sum Formulas

(a) $\sin A \sin B = \frac{1}{2}[\cos(A - B) - \cos(A + B)]$

(b) $\sin A \cos B = \frac{1}{2}[\sin(A + B) + \sin(A - B)]$

(c) $\cos A \cos B = \frac{1}{2}[\cos(A + B) + \cos(A - B)]$

IX. Sum-to-Product Formulas

(a) $\sin\alpha + \sin\beta = 2 \sin\dfrac{\alpha + \beta}{2} \cos\dfrac{\alpha - \beta}{2}$

(b) $\sin\alpha - \sin\beta = 2 \cos\dfrac{\alpha + \beta}{2} \sin\dfrac{\alpha - \beta}{2}$

(c) $\cos\alpha + \cos\beta = 2 \cos\dfrac{\alpha + \beta}{2} \cos\dfrac{\alpha - \beta}{2}$

(d) $\cos\alpha - \cos\beta = -2 \sin\dfrac{\alpha + \beta}{2} \sin\dfrac{\alpha - \beta}{2}$

EXERCISE SET 8.7

A

In Exercises 1–12, use the given information to evaluate each expression.

1. $\tan t = \frac{3}{4}$, $\tan s = \frac{7}{24}$
   (a) $\tan(t + s)$ (b) $\tan(t - s)$
2. $\tan \theta = -\frac{3}{4}$
   (a) $\tan\left(\theta + \frac{1}{4}\pi\right)$ (b) $\tan\left(\theta - \frac{1}{4}\pi\right)$
3. $\tan x = \sqrt{2}$
   (a) $\tan\left(x + \frac{3}{4}\pi\right)$ (b) $\tan\left(x - \frac{5}{4}\pi\right)$
4. $\sin x = -\frac{5}{13}$, $3\pi/2 < x < 2\pi$
   (a) $\tan\left(x + \frac{1}{3}\pi\right)$ (b) $\tan\left(x - \frac{1}{3}\pi\right)$
5. $\cos \varphi = \frac{7}{25}$, $0° < \varphi < 90°$
   (a) $\sin 2\varphi$ (b) $\cos 2\varphi$ (c) $\tan 2\varphi$
6. $\cos \varphi = \frac{3}{5}$, $0° < \varphi < 90°$
   (a) $\sin 2\varphi$ (b) $\cos 2\varphi$ (c) $\tan 2\varphi$
7. $\tan u = -4$, $3\pi/2 < u < 2\pi$
   (a) $\sin 2u$ (b) $\cos 2u$ (c) $\tan 2u$
8. $\cot s = 2$, $\pi < s < 3\pi/2$
   (a) $\sin 2s$ (b) $\cos 2s$ (c) $\tan 2s$
9. $\sin \alpha = \sqrt{3}/2$, $0° < \alpha < 90°$
   (a) $\sin(\alpha/2)$ (b) $\cos(\alpha/2)$ (c) $\tan(\alpha/2)$
10. $\cos \beta = -\frac{1}{8}$, $180° < \beta < 270°$
    (a) $\sin(\beta/2)$ (b) $\cos(\beta/2)$ (c) $\tan(\beta/2)$
11. $\cos \theta = -\frac{7}{9}$, $\pi/2 < \theta < \pi$
    (a) $\sin(\theta/2)$ (b) $\cos(\theta/2)$ (c) $\tan(\theta/2)$
12. $\cos \theta = \frac{12}{13}$, $3\pi/2 < \theta < 2\pi$
    (a) $\sin(\theta/2)$ (b) $\cos(\theta/2)$ (c) $\tan(\theta/2)$

In Exercises 13–16, use the given information to compute each of the following: **(a)** $\sin 2\theta$; **(b)** $\cos 2\theta$; **(c)** $\sin(\theta/2)$; **(d)** $\cos(\theta/2)$.

13. $\sin \theta = \frac{3}{4}$ and $\pi/2 < \theta < \pi$
14. $\cos \theta = \frac{2}{5}$ and $3\pi/2 < \theta < 2\pi$
15. $\cos \theta = -\frac{1}{3}$ and $180° < \theta < 270°$
16. $\sin \theta = -\frac{1}{10}$ and $270° < \theta < 360°$

In Exercises 17–20, use an appropriate half-angle formula to evaluate each quantity.

17. (a) $\sin(\pi/12)$
    (b) $\cos(\pi/12)$
    (c) $\tan(\pi/12)$
18. (a) $\sin(\pi/8)$
    (b) $\cos(\pi/8)$
    (c) $\tan(\pi/8)$
19. (a) $\sin 105°$
    (b) $\cos 105°$
    (c) $\tan 105°$
20. (a) $\sin 165°$
    (b) $\cos 165°$
    (c) $\tan 165°$

In Exercises 21–28, refer to the two triangles and compute the quantities indicated.

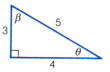 

21. (a) $\sin 2\theta$ (b) $\cos 2\theta$ (c) $\tan 2\theta$
22. (a) $\sin 2t$ (b) $\cos 2t$ (c) $\tan 2t$
23. (a) $\sin 2\beta$ (b) $\cos 2\beta$ (c) $\tan 2\beta$
24. (a) $\sin 2s$ (b) $\cos 2s$ (c) $\tan 2s$
25. (a) $\sin(\theta/2)$ (b) $\cos(\theta/2)$ (c) $\tan(\theta/2)$
26. (a) $\sin(s/2)$ (b) $\cos(s/2)$ (c) $\tan(s/2)$
27. (a) $\sin(\beta/2)$ (b) $\cos(\beta/2)$ (c) $\tan(\beta/2)$
28. (a) $\sin(t/2)$ (b) $\cos(t/2)$ (c) $\tan(t/2)$

In Exercises 29–32, use the given information to express $\sin 2\theta$ and $\cos 2\theta$ in terms of x.

29. $x = 5 \sin \theta$, $0 < \theta < \pi/2$
30. $x = \sqrt{2} \cos \theta$, $0 < \theta < \pi/2$
31. $x - 1 = 2 \sin \theta$, $0 < \theta < \pi/2$
32. $x + 1 = 3 \sin \theta$, $\pi/2 < \theta < \pi$

In Exercises 33–36, express each quantity in a form that does not involve powers of the trigonometric functions (as in Example 5).

33. $\sin^4 \theta$
34. $\cos^6 \theta$
35. $\sin^4(\theta/2)$
36. $\sin^6(\theta/4)$

37. Prove each of the following double-angle formulas. *Hint:* As in the text, replace 2θ with $\theta + \theta$, and use an appropriate addition formula.
    (a) $\cos 2\theta = \cos^2 \theta - \sin^2 \theta$
    (b) $\tan 2\theta = \dfrac{2 \tan \theta}{1 - \tan^2 \theta}$

38. (a) Beginning with the identity $\cos 2\theta = \cos^2 \theta - \sin^2 \theta$, prove that $\cos 2\theta = 1 - 2 \sin^2 \theta$.
    (b) Using the result in part (a), prove that $\sin^2 \theta = \frac{1}{2}(1 - \cos 2\theta)$.
    (c) Derive the formula for $\sin(s/2)$ as follows. Using the identity in part (b), replace θ with $s/2$, and then take square roots.

B

In Exercises 39–54, prove that the given equations are identities.

39. $\cos 2s = \dfrac{1 - \tan^2 s}{1 + \tan^2 s}$

40. $1 + \cos 2t = \cot t \sin 2t$

41. $\cos \theta = 2 \cos^2(\theta/2) - 1$

42. $\dfrac{\sin 2\theta}{\sin \theta} - \dfrac{\cos 2\theta}{\cos \theta} = \sec \theta$

43. $\sin^4 \theta = \dfrac{3 - 4 \cos 2\theta + \cos 4\theta}{8}$

44. $\sin 3\theta = 3 \sin \theta - 4 \sin^3 \theta$

45. $\sin 2\theta = \dfrac{2 \tan \theta}{1 + \tan^2 \theta}$

46. $2 \csc 2\theta = \dfrac{\csc^2 \theta}{\cot \theta}$

47. $\sin 2\theta = 2 \sin^3 \theta \cos \theta + 2 \sin \theta \cos^3 \theta$

48. $\cot \theta = \dfrac{1 + \cos 2\theta}{\sin 2\theta}$

49. $\dfrac{1 + \tan(\theta/2)}{1 - \tan(\theta/2)} = \tan \theta + \sec \theta$

50. $\tan \theta + \cot \theta = 2 \csc 2\theta$

51. $2 \sin^2(45° - \theta) = 1 - \sin 2\theta$

52. $(\sin \theta - \cos \theta)^2 = 1 - \sin 2\theta$

53. $1 + \tan \theta \tan 2\theta = \tan 2\theta \cot \theta - 1$

54. $\tan\left(\frac{\pi}{4} + \theta\right) - \tan\left(\frac{\pi}{4} - \theta\right) = 2 \tan 2\theta$

55. If $\cos(\alpha + \beta) = 0$, show that $\sin(\alpha + 2\beta) = \sin \alpha$.

56. If $\tan \alpha = \frac{1}{11}$ and $\tan \beta = \frac{5}{6}$, find $\alpha + \beta$, given that $0 < \alpha < \pi/2$ and $0 < \beta < \pi/2$. *Hint:* Compute $\tan(\alpha + \beta)$.

57. If $\sin \theta = \dfrac{a^2 - b^2}{a^2 + b^2}$, show that $\tan \dfrac{\theta}{2} = \dfrac{a - b}{a + b}$. Assume that $0 < \theta < \pi/2$ and a and b are positive.

58. Let $z = \tan \theta$. Show that

$$\cos 2\theta = \frac{1 - z^2}{1 + z^2} \qquad \text{and} \qquad \sin 2\theta = \frac{2z}{1 + z^2}$$

59. The following figure shows a semicircle with radius $AO = 1$.

(a) Use the figure to derive the formula

$$\tan \frac{\theta}{2} = \frac{\sin \theta}{1 + \cos \theta} \qquad \left(0 < \theta < \frac{\pi}{2}\right)$$

Hint: Show that $CD = \sin \theta$ and $OD = \cos \theta$. Then look at right triangle ADC to find $\tan(\theta/2)$.

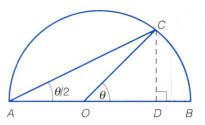

(b) Use the formula developed in part (a) to show that

(i) $\tan 15° = \dfrac{1}{2 + \sqrt{3}} = 2 - \sqrt{3}$;

(ii) $\tan(\pi/8) = \sqrt{2} - 1$.

60. In this exercise, we'll use the accompanying figure to prove the following identities:

$$\cos 2\theta = 2 \cos^2 \theta - 1$$
$$\sin 2\theta = 2 \sin \theta \cos \theta$$
$$\cos 3\theta = 4 \cos^3 \theta - 3 \cos \theta$$
$$\sin 3\theta = 3 \sin \theta - 4 \sin^3 \theta$$

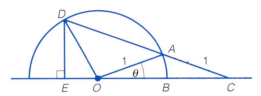

[The figure and technique in this exercise are adapted from the article by Wayne Dancer, "Geometric Proofs of Multiple Angle Formulas," in *American Mathematical Monthly* **44** (1937) 366–367.] The figure is constructed as follows. We start with $\angle AOB = \theta$ in standard position in the unit circle, as shown. The point C is chosen on the extended diameter such that $CA = 1$. Then $\overline{CA}$ is extended to meet the circle at D, and radius $\overline{DO}$ is drawn. Finally, from D, a perpendicular is drawn to the diameter, as shown.

Supply the reason or reasons that justify each of the following statements.

(a) $\angle ACO = \theta$

(b) $\angle DAO = 2\theta$ *Hint:* $\angle DAO$ is an exterior angle to $\triangle AOC$.

(c) $\angle ODA = 2\theta$

(d) $\angle DOE = 3\theta$

(e) From O, draw a perpendicular to $\overline{CD}$, meeting $\overline{CD}$ at F. From A, draw a perpendicular to $\overline{OC}$, meeting $\overline{OC}$ at G. Then $GC = OG = \cos \theta$ and $FA = DF = \cos 2\theta$.

(f) $\cos \theta = \dfrac{1 + \cos 2\theta}{2 \cos \theta}$ *Hint:* Use $\triangle CFO$.

(g) $\cos 2\theta = 2\cos^2\theta - 1$ *Hint:*   In the equation in part (f), solve for $\cos 2\theta$.

(h) $OF = \sin 2\theta$

(i) $\sin 2\theta = 2\sin\theta\cos\theta$ *Hint:*   Find $\sin\theta$ in $\triangle CFO$, and then solve the resulting equation for $\sin 2\theta$.

(j) $DC = 1 + 2\cos 2\theta$ and $EO = \cos 3\theta$

(k) $\cos\theta = \dfrac{2\cos\theta + \cos 3\theta}{1 + 2\cos 2\theta}$ *Hint:*   Compute $\cos\theta$ in $\triangle CDE$ and then use part (j).

(l) $\cos 3\theta = 4\cos^3\theta - 3\cos\theta$ *Hint:*   Use the results in parts (k) and (g).

(m) $DE = \sin 3\theta$

(n) $\sin 3\theta = 3\sin\theta - 4\sin^3\theta$ *Hint:*   Compute $\sin\theta$ in $\triangle CDE$.

61. Prove the following identities involving products of cosines. *Suggestion:* In each case, begin with the right-hand side and use the double-angle formula for the sine.

(a) $\cos\theta\cos 2\theta = \dfrac{\sin 4\theta}{4\sin\theta}$

(b) $\cos\theta\cos 2\theta\cos 4\theta = \dfrac{\sin 8\theta}{8\sin\theta}$

(c) $\cos\theta\cos 2\theta\cos 4\theta\cos 8\theta = \dfrac{\sin 16\theta}{16\sin\theta}$

62. Prove the following three formulas, known as the product-to-sum formulas.

(a) $\sin A\sin B = \frac{1}{2}[\cos(A - B) - \cos(A + B)]$

(b) $\sin A\cos B = \frac{1}{2}[\sin(A + B) + \sin(A - B)]$

(c) $\cos A\cos B = \frac{1}{2}[\cos(A + B) + \cos(A - B)]$

63. Express each product as a sum or difference of sines or cosines.

(a) $4\sin 6\theta\cos 2\theta$ (b) $2\sin A\sin 3A$

(c) $\cos 4\theta\cos 2\theta$

64. (a) Show that $\sin(\pi/4)\cos(\pi/12) = \frac{1}{4}(\sqrt{3} + 1)$.

(b) Evaluate $2(\sin 82.5°)(\cos 37.5°)$ without using a calculator.

65. Prove the following identities.

(a) $\sin(A + B)\sin(A - B) = \sin^2 A - \sin^2 B$

(b) $\cos(A + B)\cos(A - B) = \cos^2 A - \sin^2 B$

66. In the formula (from Exercise 62)

$$\sin A\cos B = \tfrac{1}{2}[\sin(A + B) + \sin(A - B)]$$

make the substitutions $A + B = \alpha$ and $A - B = \beta$ and deduce that

$$\sin\alpha + \sin\beta = 2\sin\frac{\alpha + \beta}{2}\cos\frac{\alpha - \beta}{2}$$

This is one of the sum-to-product formulas mentioned in the text.

67. Prove the following sum-to-product identities. *Hint:*   See the previous exercise.

(a) $\sin\alpha - \sin\beta = 2\cos\dfrac{\alpha + \beta}{2}\sin\dfrac{\alpha - \beta}{2}$

(b) $\cos\alpha + \cos\beta = 2\cos\dfrac{\alpha + \beta}{2}\cos\dfrac{\alpha - \beta}{2}$

(c) $\cos\alpha - \cos\beta = -2\sin\dfrac{\alpha + \beta}{2}\sin\dfrac{\alpha - \beta}{2}$

68. (a) Show that $\cos\frac{2\pi}{9} + \cos\frac{\pi}{9} = \sqrt{3}\cos\frac{\pi}{18}$.

(b) Show that $\sin 105° + \sin 15° = \sqrt{6}/2$.

69. Prove that $\dfrac{\sin\theta + \sin 3\theta}{\cos\theta + \cos 3\theta} = \tan 2\theta$.   *Hint:*   Use the sum-to-product formulas.

70. Prove that $\dfrac{\sin 7\theta - \sin 5\theta}{\cos 7\theta + \cos 5\theta} = \tan\theta$.

71. Prove that $\tan(x + y) = \dfrac{\sin 2x + \sin 2y}{\cos 2x + \cos 2y}$.

8.8   TRIGONOMETRIC EQUATIONS

In this section, we consider some techniques for solving equations involving the trigonometric functions. As usual, by a **solution** of an equation, we mean a value of the variable for which the equation becomes a true statement.

EXAMPLE I   Consider the trigonometric equation $\sin x + \cos x = 1$. Is $x = \pi/4$ a solution? Is $x = \pi/2$ a solution?

Solution To see if the value $x = \pi/4$ satisfies the given equation, we write

$$\sin \frac{\pi}{4} + \cos \frac{\pi}{4} \overset{?}{=} 1$$

$$\frac{\sqrt{2}}{2} + \frac{\sqrt{2}}{2} \overset{?}{=} 1$$

$$\sqrt{2} \overset{?}{=} 1 \qquad \text{No!}$$

Thus, $x = \pi/4$ is not a solution. In a similar fashion, we can check to see if $x = \pi/2$ is a solution:

$$\sin \frac{\pi}{2} + \cos \frac{\pi}{2} \overset{?}{=} 1$$

$$1 + 0 \overset{?}{=} 1 \qquad \text{Yes!}$$

Thus, $x = \pi/2$ is a solution. ▮▮▮

The example that we've just concluded serves to remind us of the difference between a *conditional equation* and an *identity*. An identity is true for all values of the variable in its domain. For example, the equation $\sin^2 t + \cos^2 t = 1$ is an identity: it is true for every real number t. In contrast to this, a conditional equation is true only for some (or perhaps even none) of the values of the variable. The equation in Example 1 is a conditional equation; we saw that it is false when $x = \pi/4$ and true when $x = \pi/2$. The equation $\sin t = 2$ is an example of a conditional equation that has no solution. (Why?) The equations that we are going to solve in this section are conditional equations that involve the trigonometric functions. In general, there is no single technique that can be used to solve every trigonometric equation. In the examples that follow, we illustrate some of the more common approaches to solving trigonometric equations.

EXAMPLE 2 Solve the equation $\sin \theta = \frac{1}{2}$.

Solution First of all, we know that $\sin(\pi/6) = \frac{1}{2}$. Thus, one solution is certainly $\theta = \pi/6$. To find another solution, we note that $\sin \theta$ is positive in the second quadrant, as well as in the first. In the second quadrant, the angle with a reference angle of $\pi/6$ is $5\pi/6$ (150°). Thus, $\sin(5\pi/6) = \frac{1}{2}$, and we conclude that $\theta = 5\pi/6$ is also a solution to the given equation. (See Figure 1 on the next page.)

Since $\sin \theta$ is negative in the third and fourth quadrants, we needn't look there for solutions. Nevertheless, there are other solutions, because the sine function is periodic, with period 2π. Therefore, since $\theta = \pi/6$ is a solution, we know that $\theta = (\pi/6) + 2\pi$ is another solution. So is $\theta = (\pi/6) + 4\pi$ a solution, and so on. Similarly, adding multiples of 2π to $\theta = 5\pi/6$ produces additional solutions. In general, the solutions of the equation $\sin \theta = \frac{1}{2}$ are given by

$$\frac{\pi}{6} + 2\pi k \qquad \text{and} \qquad \frac{5\pi}{6} + 2\pi k \qquad \text{where } k \text{ is any integer}$$

Figure 2 shows some of these solutions; they are the x-coordinates of the points where the sine curve intersects the line $y = \frac{1}{2}$. If required, we can express the solutions using degree measure. In that case, the solutions are given by

$$30° + 360k° \qquad \text{and} \qquad 150° + 360k° \qquad \text{where } k \text{ is any integer}$$

FIGURE 1

$$\sin\frac{\pi}{6} = \sin\frac{5\pi}{6} = \frac{1}{2}$$

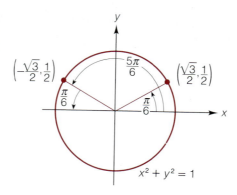

FIGURE 2

There are infinitely many real numbers x for which $\sin x = \frac{1}{2}$.

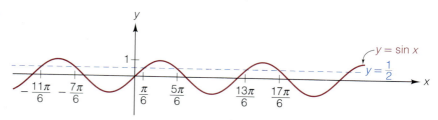

Conceptually, the next example is similar to the one we just completed, but now we'll need to use a calculator. In particular, we will be using the calculator to find a real number x such that $\cos x = 0.351$. After the example is completed, we'll indicate some of the theoretical points that are involved.

EXAMPLE 3   Find all solutions of the equation

$$\cos x = 0.351$$

in the open interval $(0, 2\pi)$. Round off the answers to three decimal places.

Solution   As indicated in Figure 3, there are two numbers x_1 and x_2 in the interval $(0, 2\pi)$ that satisfy the equation $\cos x = 0.351$. We can determine x_1 using a calculator set in the radian mode. With some types of Casio calculators, for example, the keystrokes are

$$0.351 \quad \boxed{\text{INV}} \quad \boxed{\text{COS}}$$

FIGURE 3

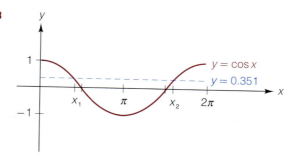

Or, for example, with a Texas Instruments TI-81 calculator, we have

(2nd) (COS) 0.351 (ENTER)

In any case, as you should check for yourself, the result (to three decimal places) is

$$x_1 = 1.212$$

For the value of x_2, we can reason as follows. From the graph in Figure 3, we see that x_2 lies in the interval $3\pi/2 < x < 2\pi$. Thinking in terms of angles, x_2 is the radian measure of an angle with terminal side in quadrant IV. Furthermore, x_1 is the measure of the reference angle for x_2. (How do we know, without relying on a calculator, that $x_1 < \pi/2$?) Therefore we have

$$x_2 = 2\pi - x_1 = 5.071 \qquad \text{using a calculator and rounding off}$$

In summary now, the two solutions in the interval $(0, 2\pi)$ are 1.212 and 5.071.
■■■

In the example just concluded, we used a calculator to find a number having a cosine of 0.351. The function that outputs such a number is called the **inverse cosine function.** There are two standard abbreviations for the name of this function, and a third abbreviation that is used on certain brands of calculators. The two standard abbreviations are $\cos^{-1}$ and arccos (read *arc cos*). And, as you saw in Example 3, the notation **INV COS** is used on some calculators. For now, we'll use the notation $\cos^{-1}$.

In the next section, we will analyze the inverse cosine function in some detail. But for our present purposes, you need only know the following definition:

$\cos^{-1}(x)$ denotes the number in the interval $[0, \pi]$ whose cosine is x.

As examples of this notation, we have

$$\cos^{-1}\left(\tfrac{1}{2}\right) = \tfrac{\pi}{3} \qquad \text{because } \cos\tfrac{\pi}{3} = \tfrac{1}{2} \text{ and } 0 < \tfrac{\pi}{3} < \pi$$
$$\cos^{-1}\left(\tfrac{1}{2}\right) \neq \tfrac{5\pi}{3} \qquad \text{because although } \cos\tfrac{5\pi}{3} = \tfrac{1}{2}, \text{ the number } \tfrac{5\pi}{3}$$
$$\text{is not in the required interval } [0, \pi]$$
$$\cos^{-1}\left(-\tfrac{1}{2}\right) = \tfrac{2\pi}{3} \qquad \text{because } \cos\tfrac{2\pi}{3} = -\tfrac{1}{2} \text{ and } 0 < \tfrac{2\pi}{3} < \pi$$
$$\cos^{-1}(0.351) \approx 1.212 \qquad \text{as we saw in Example 3}$$

In Example 3, Figure 3 tells us that the number x_1 is in the interval $[0, \pi]$ but x_2 is not. That is why the sequence of keystrokes

0.351 (COS⁻¹)

gave us the value for x_1 rather than x_2. (We then used the reference angle or reference number concept to obtain x_2.)

For some of the exercises in this section, you will need to use the *inverse sine function* or the *inverse tangent function,* rather than the inverse cosine function that we have just been discussing. For reference, we define the three functions in the box that follows. (All three functions are discussed at greater length in the next section.)

DEFINITION   Inverse Trigonometric Functions

NAME OF FUNCTION	ABBREVIATION	DEFINITION
Inverse cosine	$\cos^{-1}$	$\cos^{-1}(x)$ is the (unique) number in the interval $[0, \pi]$ whose cosine is x.
Inverse sine	$\sin^{-1}$	$\sin^{-1}(x)$ is the (unique) number in the interval $[-\pi/2, \pi/2]$ whose sine is x.
Inverse tangent	$\tan^{-1}$	$\tan^{-1}(x)$ is the (unique) number in the interval $(-\pi/2, \pi/2)$ whose tangent is x.

The next example shows how factoring can be used to solve a trigonometric equation.

EXAMPLE 4   Solve the equation $\cos^2 x + \cos x - 2 = 0$.

Solution   By factoring the expression on the left-hand side of the equation, we obtain

$$(\cos x + 2)(\cos x - 1) = 0$$

Therefore,

$$\cos x = -2 \quad \text{or} \quad \cos x = 1$$

We discard the result $\cos x = -2$, since the value of $\cos x$ is never less than -1. From the equation $\cos x = 1$, we conclude that $x = 0$ is one solution. After $x = 0$, the next time we have $\cos x = 1$ is when $x = 2\pi$. (You can see this by looking at the graph of $y = \cos x$ or by considering the unit circle definition of the cosine.) In general, then, the solutions of the equation $\cos^2 x + \cos x - 2 = 0$ are given by $x = 2\pi k$, where k is an integer.   ▪▪▪

In some equations, more than one trigonometric function is present. A common approach here is to express the various functions in terms of a single one. The next example demonstrates this technique.

EXAMPLE 5   Find all solutions of the equation $3 \tan^2 x - \sec^2 x - 5 = 0$.

Solution   We use the Pythagorean identity

$$\tan^2 x + 1 = \sec^2 x$$

to substitute for $\sec^2 x$ in the given equation. This gives us

$$3 \tan^2 x - (\tan^2 x + 1) - 5 = 0$$
$$2 \tan^2 x - 6 = 0$$
$$2 \tan^2 x = 6$$
$$\tan^2 x = 3$$
$$\tan x = \pm\sqrt{3}$$

Since the period of the tangent function is π, we need only find values of x between 0 and π that satisfy $\tan x = \pm\sqrt{3}$. The other solutions will then be obtained by adding multiples of π to these solutions. Now, in the first quadrant, the tangent is positive and we know that $\tan x = \sqrt{3}$ when $x = \pi/3$. In the second quadrant, $\tan x$ is negative and we know that $\tan 2\pi/3 = -\sqrt{3}$, since the reference angle for $2\pi/3$ is $\pi/3$. Thus, the solutions between 0 and π are $x = \pi/3$ and $x = 2\pi/3$. It follows that all of the solutions to the equation $3\tan^2 x - \sec^2 x - 5 = 0$ are given by

$$x = \tfrac{\pi}{3} + \pi k \qquad \text{and} \qquad x = \tfrac{2\pi}{3} + \pi k$$

where k is an integer.

The technique used in Example 5, of expressing the various functions in terms of a single function, is most useful when it does not involve introducing a radical expression. For instance, consider the equation $\sin s + \cos s = 1$. Although we could begin by replacing $\cos s$ by the expression $\pm\sqrt{1 - \sin^2 s}$, it turns out to be easier in this situation to begin by squaring both sides of the given equation. This is done in the next example.

EXAMPLE 6   Find all solutions of the equation $\sin s + \cos s = 1$ satisfying $0° \le s < 360°$.

Solution   Squaring both sides of the equation yields

$$(\sin s + \cos s)^2 = 1^2$$
$$\sin^2 s + 2\sin s \cos s + \cos^2 s = 1$$

$$\underset{\text{These add to 1}}{\underbrace{\qquad\qquad\qquad}}$$

Consequently, we have

$$2\sin s \cos s = 0$$

From this last equation, we conclude that $\sin s = 0$ or $\cos s = 0$. When $\sin s = 0$, we know that $s = 0°$ or $s = 180°$. And when $\cos s = 0$, we know that $s = 90°$ or $s = 270°$. Now we must go back and check which (if any) of these values is a solution to the *original* equation. This must be done whenever we square both sides in the process of solving an equation.

$s = 0°$: $\qquad\qquad \sin 0° + \cos 0° \overset{?}{=} 1$
$\qquad\qquad\qquad\qquad 0 + 1 \overset{?}{=} 1 \qquad$ True

$s = 90°$: $\qquad\qquad \sin 90° + \cos 90° \overset{?}{=} 1$
$\qquad\qquad\qquad\qquad 1 + 0 \overset{?}{=} 1 \qquad$ True

$s = 180°$: $\qquad \sin 180° + \cos 180° \overset{?}{=} 1$
$\qquad\qquad\qquad\qquad 0 + (-1) \overset{?}{=} 1 \qquad$ False

$s = 270°$: $\qquad \sin 270° + \cos 270° \overset{?}{=} 1$
$\qquad\qquad\qquad\qquad -1 + 0 \overset{?}{=} 1 \qquad$ False

We conclude that the only solutions of the equation $\sin s + \cos s = 1$ on the interval $0° \le s < 360°$ are $s = 0°$ and $s = 90°$.

In the example that follows, we consider an equation that involves a multiple of the unknown angle.

EXAMPLE 7   Solve the equation $\sin 3x = 1$ on the interval $0 \leq x \leq 2\pi$.

Solution   We know that $\sin(\pi/2) = 1$. Thus, one solution can be found by writing $3x = \pi/2$, from which we conclude that $x = \pi/6$. We can look for other solutions in the required interval by writing, more generally,

$$3x = \tfrac{\pi}{2} + 2\pi k$$

from which it follows that

$$x = \tfrac{\pi}{6} + \tfrac{2\pi k}{3}$$

Thus, when $k = 1$, we obtain

$$x = \tfrac{\pi}{6} + \tfrac{2\pi}{3} = \tfrac{\pi}{6} + \tfrac{4\pi}{6} = \tfrac{5\pi}{6}$$

When $k = 2$, we obtain

$$x = \tfrac{\pi}{6} + \tfrac{2\pi(2)}{3} = \tfrac{\pi}{6} + \tfrac{8\pi}{6} = \tfrac{3\pi}{2}$$

When $k = 3$, we obtain

$$x = \tfrac{\pi}{6} + \tfrac{2\pi(3)}{3} = \tfrac{\pi}{6} + 2\pi \qquad \text{which is greater than } 2\pi$$

We conclude that the solutions of $\sin 3x = 1$ on the interval $0 \leq x \leq 2\pi$ are $\pi/6$, $5\pi/6$, and $3\pi/2$. ∎

In Example 7, notice that we did not need to make use of a formula for $\sin 3x$, even though the expression $\sin 3x$ did appear in the given equation. In the next example, however, we do make use of the identity $\sin 2x = 2 \sin x \cos x$.

EXAMPLE 8   Solve the equation $\sin x \cos x = 1$.

Solution   We could begin by squaring both sides. (Exercise 55 will ask you to use this approach.) However, with the double-angle formula for sine in mind, we can proceed instead as follows. We multiply both sides of the given equation by 2. This yields

$$2 \sin x \cos x = 2$$

and, consequently,

$$\sin 2x = 2 \qquad \textcolor{red}{\text{using the double-angle formula}}$$

This last equation has no solution, since the value of the sine function never exceeds 1. Thus, the equation $\sin x \cos x = 1$ has no solution. ∎

For the last example in this section, we look at another case in which tables or a calculator are required.

EXAMPLE 9   Find all angles θ between $0°$ and $360°$ satisfying the equation $\sin \theta = 2 \cos \theta$.

Solution   We first want to rewrite the given equation using a single function. The easiest way to do this is to divide both sides by $\cos \theta$. (Nothing is lost here in assuming that $\cos \theta \neq 0$. If $\cos \theta$ were 0, then θ would be $90°$ or $270°$; but neither of those angles is a solution of the given equation.) Dividing through by cos θ, we obtain

$$\frac{\sin \theta}{\cos \theta} = \frac{2 \cos \theta}{\cos \theta} \qquad \text{or} \qquad \tan \theta = 2$$

From experience, we know that none of the angles with which we are familiar (the multiples of 30° and 45°) has a value of 2 for their tangent. Thus, a calculator or table is required at this point. Using a Casio calculator (set in the degree mode), we find an angle whose tangent is 2 by using the following keystrokes:

2 $\boxed{\text{TAN}^{-1}}$

By so doing, we find that

$$\theta \approx 63.4°$$

(Even if we were to retain all the decimals displayed by the calculator, we would still have only an approximate solution.) To find another value of θ, we note that the tangent is also positive in the third quadrant. Thus, a second (approximate) solution is given by

$$\theta = 180° + 63.4° = 243.4°$$

We conclude that the solutions between 0° and 360° to the equation $\sin \theta = 2 \cos \theta$ are approximately 63.4° and 243.4°. Note that the calculator provided only one value for θ; we still needed to work out the second value ourselves using the reference-angle concept. ▮▮▮

EXERCISE SET 8.8

A

1. Is $\theta = \pi/2$ a solution of the following equation?

$$2 \cos^2 \theta - 3 \cos \theta = 0$$

2. Is $x = 15°$ a solution of $(\sqrt{3}/3)\cos 2x + \sin 2x = 1$?

3. Is $x = 3\pi/4$ a solution of $\tan^2 x - 3 \tan x + 2 = 0$?

4. Is $t = 2\pi/3$ a solution of $2 \sin t + 2 \cos t = \sqrt{3} - 1$?

In Exercises 5–22, determine all solutions of the given equations. Express your answers using radian measure.

5. $\sin \theta = \sqrt{3}/2$

6. $\sin \theta = \sqrt{2}/2$

7. $\sin \theta = -\frac{1}{2}$

8. $\sin \theta + (\sqrt{2}/2) = 0$

9. $\cos \theta = -1$

10. $\cos \theta = \frac{1}{2}$

11. $\tan \theta = \sqrt{3}$

12. $\tan \theta + (\sqrt{3}/3) = 0$

13. $\tan x = 0$

14. $2 \sin^2 x - 3 \sin x + 1 = 0$

15. $2 \cos^2 \theta + \cos \theta = 0$

16. $\sin^2 x - \sin x - 6 = 0$

17. $\cos^2 t \sin t - \sin t = 0$

18. $\cos \theta + 2 \sec \theta = -3$

19. $2 \cos^2 x - \sin x - 1 = 0$

20. $2 \cot^2 x + \csc^2 x - 2 = 0$

21. $\sqrt{3} \sin t - \sqrt{1 + \sin^2 t} = 0$

22. $\sec \alpha + \tan \alpha = \sqrt{3}$

In Exercises 23–30, determine all of the solutions in the interval $0° \leq \theta < 360°$.

23. $\cos 3\theta = 1$

24. $\tan 2\theta = -1$

25. $\sin 3\theta = -\sqrt{2}/2$

26. $\sin(\theta/2) = \frac{1}{2}$

27. $\sin \theta = \cos(\theta/2)$ *Hint:* $\sin \theta = 2 \sin(\theta/2) \cos(\theta/2)$

28. $2 \sin^2 \theta - \cos 2\theta = 0$

29. $\sin 2\theta = \sqrt{3} \cos 2\theta$ *Hint:* Divide by $\cos 2\theta$.

30. $\sin 2\theta = -2 \cos \theta$

In Exercises 31–36, use a calculator where necessary to find all solutions in the interval $0° \leq \theta < 360°$.

31. $\sin \theta = \frac{1}{4}$

32. $3 \cos \theta = 5$

33. $2 \tan \theta = -4$

34. $3 \sin^2 \theta - 2 \sin \theta - 1 = 0$

35. $\cos^2 x - \cos x - 1 = 0$ 36. $2 \sin x = 3$

In Exercises 37–46, use a calculator to find all solutions in the interval $(0, 2\pi)$. Round off the answers to two decimal places.

37. $\cos x = 0.184$

38. $\cos t = -0.567$

39. $\sin x = 1/\sqrt{5}$

40. $\sin t = -0.301$

41. $\tan x = 6$

42. $\tan t = -5.25$

43. $\sin t = 5 \cos t$

44. $\sin x \cos x = 0.035$

45. $\sec t = 2.24$ *Hint:*   The equation is equivalent to $\cos t = 1/(2.24)$.

46. $\cot x = -3.27$

B

47. Find all solutions of the equation $\tan 3x - \tan x = 0$ in the interval $0 \le x < 2\pi$.   *Hint:*   Write $\tan 3x = \tan(2x + x)$ and use the addition formula for tangent.

48. Find all solutions of the equation $2 \sin x = 1 - \cos x$ in the interval $0° \le x < 360°$. Use a calculator and round off the answer(s) to one decimal place.

49. Find all solutions of the equation $\cos(x/2) = 1 + \cos x$ in the interval $0 \le x < 2\pi$.

50. Find all solutions of the equation

$$\sin 3x \cos x + \cos 3x \sin x = \sqrt{3}/2$$

in the interval $0 < x < 2\pi$.

51. Find all real numbers θ for which $\sec 4\theta + 2 \sin 4\theta = 0$.

52. Consider the equation $\sin^2 x - \cos^2 x = \frac{7}{25}$.
    (a) Solve the equation for $\cos x$.
    (b) Find all of the solutions of the equation satisfying $0 < x < \pi$.
    (c) Solve the original equation by means of a double-angle formula and use the result to check your answers in part (b).

53. Find a solution of the equation $4 \sin \theta - 3 \cos \theta = 2$ in the interval $0° < \theta < 90°$.   *Hint:*   Add $3 \cos \theta$ to both sides, and then square.

54. Find all solutions of the equation

$$\sin^3 \theta \cos \theta - \sin \theta \cos^3 \theta = -\tfrac{1}{4}$$

in the interval $0 < \theta < \pi$.   *Hint:*   Factor the left-hand side, and then use the double-angle formulas.

55. Consider the equation $\sin x \cos x = 1$.
    (a) Square both sides, and then replace $\cos^2 x$ by $1 - \sin^2 x$. Show that the resulting equation can be written $\sin^4 x - \sin^2 x + 1 = 0$.
    (b) Show that the equation $\sin^4 x - \sin^2 x + 1 = 0$ has no solutions. Conclude from this that the original equation has no real-number solutions.

56. The accompanying figure shows a portion of the graph of the periodic function

$$f(x) = \sin(1 + \sin x)$$

Find the x-coordinates of the turning points P, Q, and R. Round off the answers to three decimal places. *Hint:*   For Q, the x-coordinate is halfway between the x-coordinates of P and R.

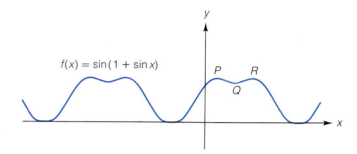

57. (a) Find the smallest solution of $\cos x = 0.412$ in the interval $(1000, \infty)$. (Round off the answer to three decimal places.)
    (b) Find the smallest solution of $\cos x = -0.412$ in the interval $(1000, \infty)$.

C

58. In this exercise, you will see how certain cubic equations can be solved by using the following identity (which we proved in Example 4 in Section 8.7):

$$4 \cos^3 \theta - 3 \cos \theta = \cos 3\theta \qquad (1)$$

For example, suppose that we wish to solve the equation

$$8x^3 - 6x - 1 = 0 \qquad (2)$$

To transform this equation into a form where the stated identity is useful, we make the substitution $x = a \cos \theta$, where a is a constant to be determined. With this substitution, equation (2) can be written

$$8a^3 \cos^3 \theta - 6a \cos \theta = 1 \qquad (3)$$

In equation (3), the coefficient of $\cos^3 \theta$ is $8a^3$. Since we want this coefficient to be 4 [as it is in equation (1)], we divide both sides of equation (3) by $2a^3$ to obtain

$$4 \cos^3 \theta - \frac{3}{a^2} \cos \theta = \frac{1}{2a^3} \qquad (4)$$

Next, a comparison of equations (4) and (1) leads us to require that $3/a^2 = 3$. Thus, $a = \pm 1$. For convenience, we choose $a = 1$; equation (4) then becomes

$$4 \cos^3 \theta - 3 \cos \theta = \tfrac{1}{2} \tag{5}$$

Comparing equation (5) with the identity in (1) leads us to the equation

$$\cos 3\theta = \tfrac{1}{2}$$

As you can check, the solutions here are of the form

$$\theta = 20° + 120k° \qquad \text{and} \qquad \theta = 100° + 120k°$$

Thus,

$$x = \cos(20° + 120k°) \qquad \text{and} \qquad x = \cos(100° + 120k°)$$

Now, however, as you can again check, only three of the angles yield distinct values for $\cos \theta$, namely, $\theta = 20°$, $\theta = 140°$, and $\theta = 260°$. Thus, the solutions of the equation $8x^3 - 6x - 1 = 0$ are given by $x = \cos 20°$, $x = \cos 140°$, and $x = \cos 260°$.

Use the method just described to solve the following equations.
(a) $x^3 - 3x + 1 = 0$
       Answers: 2 cos 40°, −2 cos 20°, 2 cos 80°
(b) $x^3 - 36x - 72 = 0$
(c) $x^3 - 6x + 4 = 0$ *Answers:*   2, $-1 \pm \sqrt{3}$
(d) $x^3 - 7x - 7 = 0$ (Round off your answers to three decimal places.)

8.9   THE INVERSE TRIGONOMETRIC FUNCTIONS

The notation $\cos^{-1} \theta$ must not be understood to signify $\dfrac{1}{\cos \theta}$.

John Herschel in *Philosophical Transactions of London,* 1813

In the previous section, we introduced the three inverse trigonometric functions $y = \sin^{-1} x$, $y = \cos^{-1} x$, and $y = \tan^{-1} x$. Because we were using these functions as tools to solve equations, our emphasis there was more computational than theoretical in nature. Now we want to take a second look at these functions, this time from a more conceptual point of view. As background for this work, you should be familiar with the material on inverse functions in Section 4.5.

As indicated in Figure 1(a) (on the next page), the sine function is not one-to-one. Therefore, there is no inverse function. However, let us now consider the **restricted sine function:**

$$y = \sin x, \qquad -\pi/2 \le x \le \pi/2$$

As indicated in Figure 1(b), the restricted sine function *is* one-to-one, and therefore the inverse function does exist in this case. We refer to the inverse of the restricted sine function as the **inverse sine function.**

Two notations are commonly used to denote the inverse sine function:

$$y = \sin^{-1} x \qquad \text{and} \qquad y = \arcsin x$$

Initially, at least, we will use the notation $y = \sin^{-1} x$. The graph of $y = \sin^{-1} x$ is easily obtained using the fact that the graph of a function and its inverse are reflections of one another about the line $y = x$. Figure 2(a) (on the next page) shows the graph of the restricted sine function and its inverse, $y = \sin^{-1} x$. Figure 2(b) shows the graph of $y = \sin^{-1} x$ alone. From Figure 2(b), we see that the domain of $y = \sin^{-1} x$ is the interval $[-1, 1]$, while the range is $[-\pi/2, \pi/2]$.

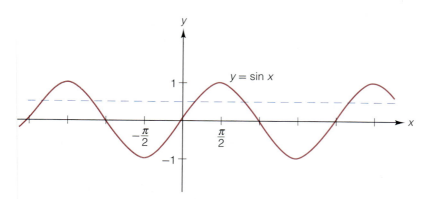

(a) The horizontal line test shows that the sine function is not one-to-one.

(b) By restricting the domain of the sine function to the closed interval $[-\pi/2, \pi/2]$, we obtain the restricted sine function. The horizontal line test shows that the restricted sine function is one-to-one.

FIGURE 1

FIGURE 2

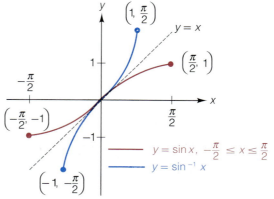

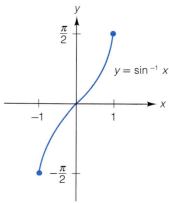

(a) The graphs of a function and its inverse are mirror images of one another about the line $y = x$.

(b) The inverse sine function

Values of $\sin^{-1} x$ are computed according to the following rule.

$\sin^{-1} x$ is that number in the interval $[-\pi/2, \pi/2]$ whose sine is x.

To see why this is so, we begin with the restricted sine function:

$$y = \sin x, \qquad -\pi/2 \le x \le \pi/2$$

As explained in Section 4.5, the inverse function is obtained by interchanging x and y (the inputs and the outputs). So, for the inverse function, we have

$$x = \sin y, \qquad -\pi/2 \le y \le \pi/2 \tag{1}$$

Equation (1) tells us that y is the number in the interval $[-\pi/2, \pi/2]$ whose sine is x. This is what we wished to show.

EXAMPLE 1   Evaluate each of the following: **(a)** $\sin^{-1}\left(\frac{1}{2}\right)$;   **(b)** $\sin^{-1}\left(-\frac{1}{2}\right)$.

Solution   **(a)** $\sin^{-1}\left(\frac{1}{2}\right)$ is that number in the interval $[-\pi/2, \pi/2]$ whose sine is $\frac{1}{2}$. Since $\sin(\pi/6) = \frac{1}{2}$, we conclude that $\sin^{-1}\left(\frac{1}{2}\right) = \pi/6$.

(b) Since $\sin(-\pi/6) = -\frac{1}{2}$ and $-\pi/6$ belongs to the interval $[-\pi/2, \pi/2]$, we conclude that $\sin^{-1}\left(-\frac{1}{2}\right) = -\pi/6$.   ▮▮▮

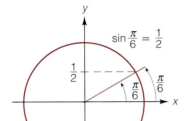

FIGURE 3

We can use the result in Example 1(a) to see why the notation *arcsin* is used for the inverse sine function. With the arcsin notation, the result in Example 1(a) can be written

$$\arcsin \frac{1}{2} = \frac{\pi}{6}$$

Now, as Figure 3 indicates, the arc length with a sine of $\frac{1}{2}$ is $\pi/6$. This is the idea behind the arcsin notation.

EXAMPLE 2   Evaluate $\arcsin \frac{3}{4}$.

Solution   The quantity $\arcsin \frac{3}{4}$ is that number (or arc length, or angle in radians) in the interval $[-\pi/2, \pi/2]$ whose sine is $\frac{3}{4}$. Since we are not familiar with an angle with a sine of $\frac{3}{4}$, we use a calculator:

$$0.75 \quad \boxed{\text{INV}} \quad \boxed{\text{SIN}}$$

The result is

$$\arcsin \frac{3}{4} \approx 0.85$$ ▮▮▮

EXAMPLE 3   Show that the following two expressions are not equal:

$$\sin^{-1} 0 \qquad \text{and} \qquad \frac{1}{\sin 0}$$

Solution   The quantity $\sin^{-1} 0$ is that number in the interval $[-\pi/2, \pi/2]$ whose sine is 0. Since $\sin 0 = 0$, we conclude that

$$\sin^{-1} 0 = 0$$

On the other hand, since $\sin 0 = 0$, the expression $1/(\sin 0)$ is not even defined. Thus, the two given expressions certainly are not equal.   ▮▮▮

If f and f^{-1} are any pair of inverse functions, then by definition,

$$f[f^{-1}(x)] = x \qquad \text{for every } x \text{ in the domain of } f^{-1}$$

and

$$f^{-1}[f(x)] = x \qquad \text{for every } x \text{ in the domain of } f$$

Applying these facts to the restricted sine function, $y = \sin x$, and its inverse, $y = \sin^{-1} x$, we obtain the following two identities.

$$\sin(\sin^{-1} x) = x \quad \text{for every } x \text{ in the inverval } [-1, 1]$$
$$\sin^{-1}(\sin x) = x \quad \text{for every } x \text{ in the interval } [-\pi/2, \pi/2]$$

The following example indicates that the domain restrictions accompanying these two identities cannot be ignored.

EXAMPLE 4   Compute each of the following quantities that is defined.

(a) $\sin^{-1}\left(\sin \frac{\pi}{4}\right)$ (b) $\sin^{-1}(\sin \pi)$

(c) $\sin(\sin^{-1} 2)$ (d) $\sin\left[\sin^{-1}(-1/\sqrt{5})\right]$

Solution   (a) Since $\frac{\pi}{4}$ lies in the domain of the restricted sine function, the identity $\sin^{-1}(\sin x) = x$ is applicable here. Thus,

$$\sin^{-1}\left(\sin \frac{\pi}{4}\right) = \frac{\pi}{4}$$

Check: $\sin \frac{\pi}{4} = \sqrt{2}/2$. Therefore $\sin^{-1}\left(\sin \frac{\pi}{4}\right) = \sin^{-1}(\sqrt{2}/2) = \frac{\pi}{4}$.

(b) The number π is not in the domain of the restricted sine function, so the identity $\sin^{-1}(\sin x) = x$ does not apply in this case. However, since $\sin \pi = 0$, we have

$$\sin^{-1}(\sin \pi) = \sin^{-1} 0 = 0$$

Thus, $\sin^{-1}(\sin \pi)$ is equal to 0, not π.

(c) The number 2 is not in the domain of the inverse sine function. Thus, the expression $\sin(\sin^{-1} 2)$ is undefined.

(d) The identity $\sin(\sin^{-1} x) = x$ is applicable here. (Why?) Thus,

$$\sin\left[\sin^{-1}(-1/\sqrt{5})\right] = -1/\sqrt{5}$$

EXAMPLE 5   If $\sin \theta = x/3$ and $0 < \theta < \pi/2$, express the quantity $\theta - \sin 2\theta$ as a function of x.

Solution   The given conditions tell us that $\theta = \sin^{-1}(x/3)$. That expresses θ in terms of x. To express $\sin 2\theta$ in terms of x, we have

$$\sin 2\theta = 2 \sin \theta \cos \theta = 2 \sin \theta \sqrt{1 - \sin^2 \theta}$$

$$= 2\left(\frac{x}{3}\right)\sqrt{1 - \frac{x^2}{9}} = \frac{2x}{3}\sqrt{\frac{9 - x^2}{9}}$$

$$= \frac{2x\sqrt{9 - x^2}}{9}$$

Now, combining the results, we have

$$\theta - \sin 2\theta = \sin^{-1}\left(\frac{x}{3}\right) - \frac{2x\sqrt{9 - x^2}}{9}$$

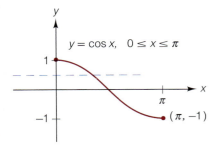

FIGURE 4

The restricted cosine function is one-to-one.

We turn now to the inverse cosine function. We'll use the same general procedure that we followed for the inverse sine to define the inverse cosine. We begin by defining the **restricted cosine function**:

$$y = \cos x, \qquad 0 \le x \le \pi$$

As indicated by the horizontal line test in Figure 4, this function is one-to-one. Since the restricted cosine function is one-to-one, the inverse function exists. We denote the **inverse cosine function** by

$$y = \cos^{-1} x \qquad \text{or} \qquad y = \arccos x$$

The graph of $y = \cos^{-1} x$ is obtained by reflecting the graph of the restricted cosine function about the line $y = x$. Figure 5(a) displays the graph of the restricted cosine function along with $y = \cos^{-1} x$. Figure 5(b) shows the graph of $y = \cos^{-1} x$ alone. From Figure 5(b), we see that the domain of $y = \cos^{-1} x$ is the interval $[-1, 1]$, while the range is $[0, \pi]$.

FIGURE 5

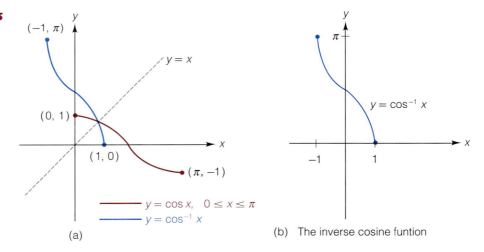

(a)

(b) The inverse cosine funtion

As we pointed out previously, the defining equations for inverse functions are

$$f(f^{-1}(x)) = x \qquad \text{for every } x \text{ in the domain of } f^{-1}$$

and

$$f^{-1}(f(x)) = x \qquad \text{for every } x \text{ in the domain of } f$$

In the particular case of the restricted cosine function and the inverse cosine function, these two identities read as follows.

$$\cos(\cos^{-1} x) = x \qquad \text{for every } x \text{ in the interval } [-1, 1]$$
$$\cos^{-1}(\cos x) = x \qquad \text{for every } x \text{ in the interval } [0, \pi]$$

As stated in the previous section, the values of $\cos^{-1}x$ are computed according to the following rule.

> $\cos^{-1}x$ is that number in the interval $[0, \pi]$ whose cosine is x.

To see why this is so, we begin with the restricted cosine function:

$$y = \cos x, \qquad 0 \le x \le \pi$$

Then for the inverse function, we interchange x and y (the inputs and the outputs) to obtain

$$x = \cos y, \quad 0 \le y \le \pi$$

This last equation tells us that y is the number in the interval $[0, \pi]$ whose cosine is x. This is what we wished to show.

EXAMPLE 6   Compute each of the following.

(a) $\cos^{-1}(0)$ (b) $\arccos\left(\cos\frac{2}{5}\right)$

Solution   (a) $\cos^{-1}(0)$ is that number in the interval $[0, \pi]$ whose cosine is 0. Since $\cos(\pi/2) = 0$, we have

$$\cos^{-1}(0) = \pi/2$$

(b) Since the number $\frac{2}{5}$ is in the domain of the restricted cosine function, the identity $\arccos(\cos x) = x$ is applicable. Thus,

$$\arccos\left(\cos\tfrac{2}{5}\right) = \tfrac{2}{5}$$

EXAMPLE 7   Show that $\sin(\cos^{-1}x) = \sqrt{1 - x^2}$ for $-1 \le x \le 1$.

Solution   We use the identity

$$\sin y = \sqrt{1 - \cos^2 y}, \qquad \text{which is valid for } 0 \le y \le \pi$$

Substituting $\cos^{-1}x$ for y in this identity, we obtain

$$\sin(\cos^{-1}x) = \sqrt{1 - [\cos(\cos^{-1}x)]^2}$$

or

$$\sin(\cos^{-1}x) = \sqrt{1 - x^2} \qquad \text{as required}$$

Before leaving this example, we point out an alternate method of solution that is useful when the restriction on x is $0 < x < 1$. In this case, we let $\theta = \cos^{-1}x$. Then θ is the radian measure of the acute angle whose cosine is x and we can sketch θ as shown in Figure 6. The sides of the triangle in Figure 6 have been labeled in such a way that $\cos \theta = x$. Then, by the Pythagorean theorem, we find that the third side of the triangle is $\sqrt{1 - x^2}$. We therefore have

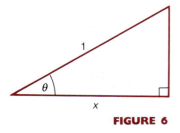

FIGURE 6

$$\sin \theta = \frac{\text{opposite}}{\text{hypotenuse}} = \frac{\sqrt{1 - x^2}}{1} = \sqrt{1 - x^2}$$

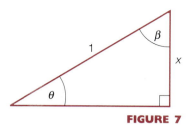

FIGURE 7

Just as there is a basic identity connecting $\sin x$ and $\cos x$, namely, $\sin^2 x + \cos^2 x = 1$, there is also an identity connecting $\sin^{-1} x$ and $\cos^{-1} x$:

$$\sin^{-1}x + \cos^{-1}x = \frac{\pi}{2} \qquad \text{for every } x \text{ in the closed interval } [-1, 1]$$

We can use Figure 7 to see why this identity is valid when $0 < x < 1$. (Exercise 45 shows you how to establish this identity for all values of x from -1 to 1.) In Figure 7, assume that θ and β are the radian measures of the indicated acute angles. Then we have

$$\sin \theta = \frac{x}{1} = x \qquad \text{and} \qquad \cos \beta = \frac{x}{1} = x$$

and therefore

$$\theta = \sin^{-1} x \qquad \text{and} \qquad \beta = \cos^{-1} x$$

But, from Figure 7, we know that $\theta + \beta = \pi/2$. Consequently, we have $\sin^{-1} x + \cos^{-1} x = \pi/2$, as required.

Now let us turn to the definition of the inverse tangent function. We begin by defining the **restricted tangent function**:

$$y = \tan x, \qquad -\pi/2 < x < \pi/2$$

Figure 8 shows the graph of the restricted tangent function. As you can check by applying the horizontal line test, the restricted tangent function is one-to-one. This tells us that the inverse function exists. The two common notations for this **inverse tangent function** are

$$y = \tan^{-1} x \qquad \text{or} \qquad y = \arctan x$$

The graph of $y = \tan^{-1} x$ is obtained in the same way we obtained the graphs of the inverse sine and the inverse cosine. That is, we use the fact that the graph of a function and its inverse are mirror images of one another about the line $y = x$. In Figure 9(a) (on the next page), we've reflected the restricted tangent function about the line $y = x$ to obtain the graph of $y = \tan^{-1} x$. In Figure 9(b) we show the graph of $y = \tan^{-1} x$ alone.

From the graph in Figure 9(b), we can see that the domain of the inverse tangent function is the set of all real numbers, and the range is the open interval $(-\pi/2, \pi/2)$. Furthermore, if we follow exactly the same line of reasoning used previously for the inverse sine and the inverse cosine, we obtain the following properties of the inverse tangent function.

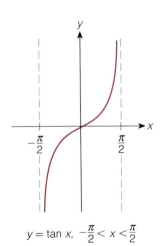

$y = \tan x, \ -\dfrac{\pi}{2} < x < \dfrac{\pi}{2}$

FIGURE 8
The restricted tangent function

$\tan^{-1} x$ is that number in the interval $(-\pi/2, \pi/2)$ whose tangent is x.

$$\tan(\tan^{-1} x) = x \qquad \text{for every real number } x$$
$$\tan^{-1}(\tan x) = x \qquad \text{for every } x \text{ in the open interval } (-\pi/2, \pi/2)$$

FIGURE 9

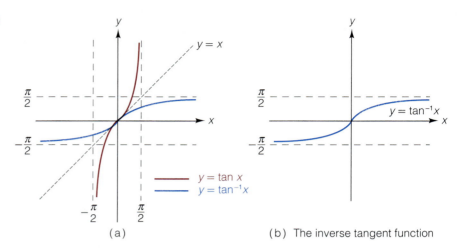

(a)   (b) The inverse tangent function

EXAMPLE 8   Compute the following: **(a)** $\tan^{-1}(-1)$ **(b)** $\tan\left(\tan^{-1}\sqrt{5}\right)$.

Solution   **(a)** The quantity $\tan^{-1}(-1)$ is that number in the interval $(-\pi/2, \pi/2)$ whose tangent is -1. Since $\tan(-\pi/4) = -1$, we have $\tan^{-1}(-1) = -\pi/4$.
(b) The identity $\tan(\tan^{-1} x) = x$ holds for all real numbers x. We therefore have $\tan\left(\tan^{-1}\sqrt{5}\right) = \sqrt{5}$.   ▌▌▌

EXAMPLE 9   If $\tan\theta = x/3$ and $0 < \theta < \pi/2$, express the quantity $\theta - \tan 2\theta$ as a function of x.

Solution   The given equation tells us that $\theta = \tan^{-1}(x/3)$. That expresses θ in terms of x. To express $\tan 2\theta$ in terms of x, we have

$$\tan 2\theta = \frac{2\tan\theta}{1 - \tan^2\theta} = \frac{2(x/3)}{1 - (x/3)^2}$$

$$= \frac{2x/3}{1 - (x^2/9)} = \frac{6x}{9 - x^2} \qquad \text{\color{red}multiplying numerator and denominator by 9}$$

Now, combining the results, we have

$$\theta - \tan 2\theta = \tan^{-1}\frac{x}{3} - \frac{6x}{9 - x^2}$$ ▌▌▌

EXAMPLE 10   Simplify the quantity $\sec(\tan^{-1} x)$, where $x > 0$.

Solution   We let $\theta = \tan^{-1} x$, or equivalently, $\tan\theta = x$. Now, as shown in Figure 10, we sketch a right triangle with an angle θ (in radians) whose tangent is x. The Pythagorean theorem tells us that the length of the hypotenuse in this triangle is $\sqrt{1 + x^2}$. Consequently, we have

$$\sec(\tan^{-1} x) = \sec\theta = \frac{\text{hypotenuse}}{\text{adjacent}} = \frac{\sqrt{1 + x^2}}{1} = \sqrt{1 + x^2}$$ ▌▌▌

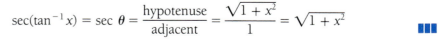

FIGURE 10
$\theta = \tan^{-1} x \ (x > 0)$ or $\tan\theta = x$

EXERCISE SET 8.9

A

In Exercises 1–20, evaluate each of the quantities that is defined, but do not use a calculator or tables. If a quantity is undefined, say so.

1. $\sin^{-1}(\sqrt{3}/2)$
2. $\cos^{-1}(-1)$
3. $\tan^{-1}\sqrt{3}$
4. $\arccos(-\sqrt{2}/2)$
5. $\arctan(-1/\sqrt{3})$
6. $\arcsin(-1)$
7. $\tan^{-1}1$
8. $\sin^{-1}0$
9. $\cos^{-1}2\pi$
10. $\arctan 0$
11. $\sin[\sin^{-1}(\frac{1}{4})]$
12. $\cos[\cos^{-1}(\frac{4}{3})]$
13. $\cos[\cos^{-1}(\frac{3}{4})]$
14. $\tan(\arctan 3\pi)$
15. $\arctan[\tan(-\pi/7)]$
16. $\sin(\arcsin 2)$
17. $\arcsin[\sin(\pi/2)]$
18. $\arccos[\cos(\pi/8)]$
19. $\arccos(\cos 2\pi)$
20. $\sin^{-1}[\sin(3\pi/2)]$

In Exercises 21–30, evaluate the given quantities without using a calculator or tables.

21. $\tan[\sin^{-1}(\frac{4}{5})]$
22. $\cos(\arcsin\frac{2}{7})$
23. $\sin(\tan^{-1}1)$
24. $\sin[\tan^{-1}(-1)]$
25. $\tan(\arccos\frac{5}{13})$
26. $\cos[\sin^{-1}(\frac{2}{3})]$
27. $\cos(\arctan\sqrt{3})$
28. $\sin[\cos^{-1}(\frac{1}{3})]$
29. $\sin[\arccos(-\frac{1}{3})]$
30. $\tan(\arcsin\frac{20}{29})$

31. Use a calculator to evaluate each of the following quantities. Express your answers to two decimal places; don't round off.

   (a) $\sin^{-1}(\frac{3}{4})$
   (b) $\cos^{-1}(\frac{2}{3})$
   (c) $\tan^{-1}\pi$
   (d) $\tan^{-1}(\tan^{-1}\pi)$

In Exercises 32–34, evaluate the given expressions without using a calculator or tables.

32. $\csc[\sin^{-1}(\frac{1}{2}) - \cos^{-1}(\frac{1}{2})]$
33. $\sec[\cos^{-1}(\sqrt{2}/2) + \sin^{-1}(-1)]$
34. $\cot[\cos^{-1}(-1/2) + \cos^{-1}(0) + \tan^{-1}(1/\sqrt{3})]$

35. Show that $\cos(\sin^{-1}x) = \sqrt{1 - x^2}$ for $-1 \le x \le 1$.
   Suggestion: Use the method of Example 7 in the text.

36. If $\sin\theta = 2x$ and $0 < \theta < \pi/2$, express $\theta + \cos 2\theta$ as a function of x.

37. If $\sin\theta = 3x/2$ and $0 < \theta < \pi/2$, express $\frac{1}{4}\theta - \sin 2\theta$ as a function of x.

38. If $\cos\theta = x - 1$ and $0 < \theta < \pi/2$, express $2\theta - \cos 2\theta$ as a function of x.

39. If $\tan\theta = \frac{1}{2}(x - 1)$ and $0 < \theta < \pi/2$, express $\theta - \cos\theta$ as a function of x.

40. If $\tan\theta = \frac{1}{3}(x + 1)$ and $0 < \theta < \pi/2$, express $2\theta + \tan 2\theta$ as a function of x.

B

41. Evaluate $\sin(2\tan^{-1}4)$.   *Hint:* $\sin 2\theta = 2\sin\theta\cos\theta$
42. Evaluate $\cos[2\sin^{-1}(\frac{5}{13})]$.
43. Evaluate $\sin(\arccos\frac{3}{5} - \arctan\frac{7}{13})$.   *Suggestion:* Use the formula for $\sin(x - y)$.
44. Show that $\sin[\sin^{-1}(\frac{1}{3}) + \sin^{-1}(\frac{1}{4})] = \dfrac{\sqrt{15} + 2\sqrt{2}}{12}$.

45. In the text we showed that $\sin^{-1}x + \cos^{-1}x = \pi/2$ for x in the open interval $(0, 1)$. Follow steps (a) and (b) to show that this identity actually holds for every x in the closed interval $[-1, 1]$.
   (a) Let $\alpha = \sin^{-1}x$ and $\beta = \cos^{-1}x$. Explain why $-\pi/2 \le \alpha + \beta \le 3\pi/2$.   *Hint:*   What are the ranges of the inverse sine and of the inverse cosine?
   (b) Use the addition formula for sine to show that $\sin(\alpha + \beta) = 1$. Conclude [with the help of part (a)] that $\alpha + \beta = \pi/2$, as required.

In Exercises 46–48 solve the given equations.

46. $\cos^{-1}t = \sin^{-1}t$ *Hint:*   Compute the cosine of both sides.
47. $\sin^{-1}(3t - 2) = \sin^{-1}t - \cos^{-1}t$
48. $\sin^{-1}(\sin^{-1}x) = \sin^{-1}(\cos^{-1}x)$

49. Show that $\arctan x + \arctan y = \arctan\dfrac{x + y}{1 - xy}$ when x and y are positive and $xy < 1$.

50. Use the identity in Exercise 49 to show that the following equations are correct.
   (a) $\arctan\frac{1}{2} + \arctan\frac{1}{3} = \pi/4$
   (b) $\arctan\frac{1}{4} + \arctan\frac{3}{5} = \pi/4$
   (c) $2\arctan\frac{1}{3} + \arctan\frac{1}{7} = \pi/4$
      Hint: $2\arctan\frac{1}{3} = \arctan\frac{1}{3} + \arctan\frac{1}{3}$
   (d) $\arctan\frac{1}{2} + \arctan\frac{1}{5} + \arctan\frac{1}{8} = \pi/4$

In Exercises 51–53, solve the equations.

51. $2\tan^{-1}x = \tan^{-1}\left(\dfrac{1}{4x}\right)$ Compute the tangent of both sides.

52. $\tan^{-1}x = \cos^{-1}x$

53. $2 \tan^{-1} \sqrt{t - t^2} = \tan^{-1} t + \tan^{-1}(1 - t)$

54. (a) As you can see in the accompanying figure, the graphs of $y = \cos^{-1} x$ and $y = \tan^{-1} x$ intersect at a point in the first quadrant. By solving the equation $\cos^{-1} x = \tan^{-1} x$, show that the x-coordinate of this intersection point is given by

$$x = \sqrt{\frac{\sqrt{5} - 1}{2}}$$

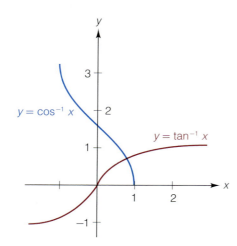

(b) Use the result in part (a) along with your calculator to specify the coordinates of the intersection point. Express both coordinates as decimals, rounded off to three places.

55. (a) The domain of the **restricted secant function** is $[0, \pi/2) \cup [\pi, 3\pi/2)$. Sketch the restricted secant function and note that it is one-to-one.

(b) The inverse secant function can be defined as follows:

$\sec^{-1} x$ is the unique number in the set $[0, \pi/2) \cup [\pi, 3\pi/2)$ whose secant is x.

Using the fact that this function is the inverse of the restricted secant function, sketch the graph of $y = \sec^{-1} x$.

(c) Evaluate $\sec^{-1}(2/\sqrt{3})$ and $\sec^{-1}(-2/\sqrt{3})$.

(d) Evaluate $\sec^{-1}(\sqrt{2})$ and $\sec^{-1}(-\sqrt{2})$.

(e) Evaluate $\sec(\sec^{-1} 2)$ and $\sec^{-1}(\sec 0)$.

56. Evaluate each of the following quantities. (The inverse secant function is defined in the previous exercise.)

(a) $\cos\left[\sec^{-1}\left(\frac{4}{3}\right)\right]$ (b) $\sin(\sec^{-1} 4)$

(c) $\tan[\sec^{-1}(-3)]$ (d) $\cot\left[\sec^{-1}\left(\frac{13}{12}\right)\right]$

In Exercises 57 and 58, we make use of some simple geometric figures to evaluate sums involving the inverse tangent function. The idea here is taken from the note by Professor Edward M. Harris entitled Behold! Sums of Arctan *(The College Mathematics Journal, vol. 18, no. 2, p. 141).*

57. In this exercise, we use the following figure to show that $\tan^{-1}\left(\frac{2}{3}\right) + \tan^{-1}\left(\frac{1}{5}\right) = \pi/4$.

(a) Show that $\angle ABC = \pi/2$.   *Hint:*   Compute the slopes of $\overline{AB}$ and $\overline{BC}$. (Assume that the grid lines are marked off at one-unit intervals.)

(b) Show that $AB = BC$ and conclude that $\angle BAC = \pi/4$,

(c) Now explain why $\alpha = \tan^{-1}\left(\frac{2}{3}\right)$, $\beta = \tan^{-1}\left(\frac{1}{5}\right)$, and $\tan^{-1}\left(\frac{2}{3}\right) + \tan^{-1}\left(\frac{1}{5}\right) = \pi/4$.

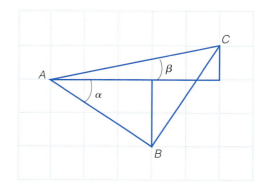

58. In this exercise, we use the accompanying figure to show that $\tan^{-1} 1 + \tan^{-1} 2 + \tan^{-1} 3 = \pi$.

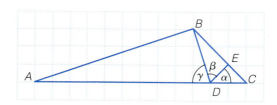

(a) By computing slopes, show that $\overline{DE} \perp \overline{BC}$ and $\overline{AB} \perp \overline{BD}$. (Assume that the grid lines are at one-unit intervals.)

(b) Determine the following lengths: DE, CE, BE, AB, and BD.

(c) Show that $\tan \alpha = 1$, $\tan \beta = 2$, and $\tan \gamma = 3$. Then explain why $\tan^{-1} 1 + \tan^{-1} 2 + \tan^{-1} 3 = \pi$.

CHAPTER EIGHT   SUMMARY OF PRINCIPAL FORMULAS AND TERMS

FORMULAS OR TERMS	PAGE REFERENCE	COMMENT
1. $\theta = \dfrac{s}{r}$	419	This equation defines the *radian measure* θ of an angle. We assume here that the vertex of the angle is placed at the center of a circle of radius r and that s is the length of the intercepted arc. See Figure 1 in Section 8.1.
2. $s = r\theta$	425	This formula expresses the arc length s on a circle in terms of the radius r and the radian measure θ of the central angle subtended by the arc.
3. $A = \frac{1}{2}r^2\theta$	426	This formula expresses the area of a sector of a circle in terms of the radius r and the radian measure θ of the central angle.
4. $\omega = \dfrac{\theta}{t}$ $v = \dfrac{d}{t}$	427	Suppose that a wheel rotates about its axis at a constant rate. If a radial line on the wheel turns through an angle of measure θ in time t, then the *angular speed* ω of the wheel is defined by $\omega = \theta/t$. If a point on the rotating wheel travels a distance d in time t, then the *linear speed* v of the point is defined by $v = d/t$.
5. $\cos t$ $\sin t$ $\tan t$ $\sec t$ $\csc t$ $\cot t$	431	Let $P(x, y)$ denote the point on the unit circle such that the arc length from $(1, 0)$ is t. Or equivalently, let $P(x, y)$ denote the point where the terminal side of the angle with radian measure t intersects the unit circle. Then the six trigonometric functions of the real number t are defined as follows. $\cos t = x$ $\qquad$ $\sec t = \dfrac{1}{x}$ $(x \neq 0)$ $\sin t = y$ $\qquad$ $\csc t = \dfrac{1}{y}$ $(y \neq 0)$ $\tan t = \dfrac{y}{x}$ $(x \neq 0)$ $\quad$ $\cot t = \dfrac{x}{y}$ $(y \neq 0)$
6. $\sec t = \dfrac{1}{\cos t}$ $\quad$ $\csc t = \dfrac{1}{\sin t}$ $\cot t = \dfrac{1}{\tan t}$ $\quad$ $\tan t = \dfrac{\sin t}{\cos t}$ $\cot t = \dfrac{\cos t}{\sin t}$	432	These five identities are direct consequences of the definitions of the trigonometric functions.
7. $\sin^2 t + \cos^2 t = 1$ $\tan^2 t + 1 = \sec^2 t$ $\cot^2 t + 1 = \csc^2 t$	432	These are the three *Pythagorean* identities.
8. $\cos(-t) = \cos t$ $\sin(-t) = -\sin t$ $\tan(-t) = -\tan t$	434	These three identities are sometimes referred to as the *opposite angle identities*. In each case, a trigonometric function of $-t$ is expressed in terms of that same function of t.
9. Periodic function	438, 439	A function f is periodic if there is a positive number p such that the equation $f(x + p) = f(x)$ holds for all x in the domain of f. The

FORMULAS OR TERMS	PAGE REFERENCE	COMMENT
		smallest such number p is called the *period* of f. Important examples: The period of $y = \sin x$ is 2π; the period of $y = \cos x$ is 2π; the period of $y = \tan x$ is π.
10. Amplitude	439	Let m and M denote the smallest and the largest values, respectively, of the periodic function f. Then the *amplitude* of f is defined to be the number $\frac{1}{2}(M - m)$. Important examples: The amplitude for both $y = \sin x$ and $y = \cos x$ is 1; amplitude is not defined for the function $y = \tan x$.
11. Period $= \dfrac{2\pi}{B}$	453, 455	This formula gives the period for the functions $y = A \sin(Bx - C)$ and $y = A \cos(Bx - C)$.
12. Phase shift $= \dfrac{C}{B}$	455	The phase shift serves as a guide in graphing functions of the form $y = A \sin(Bx - C)$ or $y = A \cos(Bx - C)$. For instance, to graph one complete cycle of $y = A \sin(Bx - C)$, first sketch one complete cycle of $y = A \sin Bx$, beginning at $x = 0$. Then draw a curve with exactly the same shape, but beginning at $x = C/B$ rather than $x = 0$. This will represent one cycle of $y = A \sin(Bx - C)$.
13. $\sin(s + t) = \sin s \cos t + \cos s \sin t$ $\sin(s - t) = \sin s \cos t - \cos s \sin t$ $\cos(s + t) = \cos s \cos t - \sin s \sin t$ $\cos(s - t) = \cos s \cos t + \sin s \sin t$	467	These identities are referred to as the *addition formulas* for sine and cosine.
14. $\cos(\frac{\pi}{2} - \theta) = \sin \theta$ $\sin(\frac{\pi}{2} - \theta) = \cos \theta$	469	These two identities, called *reduction formulas*, hold for all real numbers θ. In terms of angles, the identities state that the sine of an angle is equal to the cosine of its complement.
15. $\tan(s + t) = \dfrac{\tan s + \tan t}{1 - \tan s \tan t}$ $\tan(s - t) = \dfrac{\tan s - \tan t}{1 + \tan s \tan t}$	474	These two identities are referred to as the *addition formulas* for tangent.
16. $\sin 2\theta = 2 \sin \theta \cos \theta$ $\cos 2\theta = \cos^2 \theta - \sin^2 \theta$ $\tan 2\theta = \dfrac{2 \tan \theta}{1 - \tan^2 \theta}$	474	These are the *double-angle formulas*. There are four other forms of the double-angle formula for cosine. They appear in the box on page 476.
17. $\sin \dfrac{s}{2} = \pm\sqrt{\frac{1}{2}(1 - \cos s)}$ $\cos \dfrac{s}{2} = \pm\sqrt{\frac{1}{2}(1 + \cos s)}$ $\tan \dfrac{s}{2} = \dfrac{\sin s}{1 + \cos s}$	474	These three identities are referred to as the *half-angle formulas*. In the case of the half-angle formulas for sine and cosine, the sign before the radical is determined by the quadrant in which the angle or arc terminates.
18. Restricted sine function	491	The domain of the function $y = \sin x$ is the set of all real numbers. By allowing inputs only from the closed interval $[-\pi/2, \pi/2]$, we obtain the restricted sine function. (See Figure 1 on page 492.) The

FORMULAS OR TERMS	PAGE REFERENCE	COMMENT
		motivation for restricting the domain is that the restricted sine function is one-to-one, and therefore the inverse function exists.
19. $\sin^{-1} x$	486, 491	$\sin^{-1} x$ is that number in the interval $[-\pi/2, \pi/2]$ whose sine is x. An alternate but entirely equivalent form of this expression is arcsin x.
20. Restricted cosine function	495	The domain of the function $y = \cos x$ is the set of all real numbers. By allowing inputs only from the closed interval $[0, \pi]$, we obtain the restricted cosine function. (See Figure 4 on page 495.) The motivation for restricting the domain is that the restricted cosine function is one-to-one, and therefore the inverse function exists.
21. $\cos^{-1} x$	486, 495	$\cos^{-1} x$ is that number in the interval $[0, \pi]$ whose cosine is x. An alternate but entirely equivalent form of this expression is arccos x. The graph of the inverse cosine function is on page 495.
22. Restricted tangent function	497	The tangent function, $y = \tan x$, is not one-to-one. However, by allowing inputs only from the open interval $(-\pi/2, \pi/2)$, we obtain the restricted tangent function, which is one-to-one. (See Figure 8 on page 497.) Since the restricted tangent function is one-to-one, the inverse function exists.
23. $\tan^{-1} x$	486, 497	$\tan^{-1} x$ is that number in the interval $(-\pi/2, \pi/2)$ whose tangent is x. An alternate but entirely equivalent form of this expression is arctan x. The graph of the inverse tangent function is on page 498.

WRITING MATHEMATICS

Say whether the statement is *true* or *false*. Write out your reason or reasons in complete sentences. If you draw a diagram to accompany your writing, be sure that you clearly label any parts of the diagram to which you refer.

1. (In the accompanying figure, θ_1 and θ_2 are the radian measures of the indicated central angles.) Since the larger radius is twice the smaller radius, it follows that θ_2 is twice θ_1.

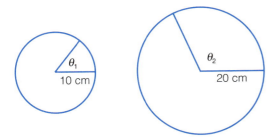

2. If t is a real number less than π, then $\sin t$ is positive.

3. If t is a real number such that $\pi < t < 3\pi/2$, then the product of $\sin t$ and $\cos t$ is positive.

4. In the immediate vicinity of the origin, the graph of $y = \sin x$ actually coincides with the line $y = x$.

5. $\dfrac{\sin \pi}{\pi} = \sin$

6. Between any two consecutive turning points, the graph of $y = \sin x$ is decreasing.

7. If t is a positive real number, then $\sin^2 t$ is a positive real number.

8. The equation $\tan^2 t + 1 = \sec^2 t$ is true for every real number t.

9. There is no real number x satisfying the equation $\cos(\pi/4 + x) = 2$.

10. There is no real number x satisfying the equation $\cos(\pi/4 + x) = \cos x$.

11. For every number x in the closed interval $[-1, 1]$, we have $\sin^{-1} x = 1/\sin x$.

12. There is no real number x for which $\sin^{-1} x = 1/\sin x$.   *Hint:*   Draw a careful sketch of the graphs of the inverse sine function and the cosecant function on the interval $0 < x \le 1$.

CHAPTER EIGHT REVIEW EXERCISES

1. Evaluate the following.
    (a) $\sin \frac{5\pi}{3}$ (b) $\cot \frac{11\pi}{6}$

2. Graph the function $y = 3 \cos 3\pi x$ over the interval $-\frac{1}{3} \le x \le \frac{1}{3}$. Specify the x-intercepts and the coordinates of the highest point on the graph.

3. Simplify the following expression:
    $$\cos t - \cos(-t) + \sin t - \sin(-t)$$

4. In the expression $1/\sqrt{4 - t^2}$, make the substitution $t = 2 \cos x$ and simplify the result. Assume that $0 < x < \pi$.

5. Graph the equation $y = \sec(2\pi x - 3)$ for one complete cycle.

6. If $\sec t = -\frac{5}{3}$ and $\pi < t < 3\pi/2$, compute $\cot t$.

7. Graph the function $y = -\sin(2x - \pi)$ for one complete period. Specify the amplitude, the period, and the phase shift.

8. Graph $y = -\frac{1}{2} \tan(\pi x/3)$ over one period.

In Exercises 9–20, evaluate each expression without using a calculator or tables.

9. $\cos \pi$

10. $\sin(-3\pi/2)$

11. $\csc(2\pi/3)$

12. $\tan(\pi/3)$

13. $\tan(11\pi/6)$

14. $\cos 0$

15. $\sin(\pi/6)$

16. $\sec(3\pi/4)$

17. $\cot(5\pi/4)$

18. $\tan(-7\pi/4)$

19. $\csc(-5\pi/6)$

20. $\sin^2(\pi/7) + \cos^2(\pi/7)$

Exercises 21–30 are calculator exercises. (Set your calculator to the radian mode.) In Exercises 21–26, where numerical answers are required, round off your results to three decimal places.

21. Evaluate $\sin 1$.

22. Evaluate $\cos 2$.

23. Evaluate $\sin(3\pi/2)$.

24. Evaluate $\sin(0.78)$.

25. Evaluate $\sin(\sin 0.0123)$.

26. Evaluate $\sin[\sin(\sin 0.0123)]$.

27. Verify that $\sin^2 1776 + \cos^2 1776 = 1$.

28. Verify that $\sin 14 = 2 \sin 7 \cos 7$.

29. Verify that $\cos(0.5) = \cos^2(0.25) - \sin^2(0.25)$.

30. Verify that $\cos(0.3) = \left[\frac{1}{2}(1 + \cos 0.6)\right]^{1/2}$.

31. In the expression $\sqrt{25 - x^2}$, make the substitution $x = 5 \sin \theta$ $(0 < \theta < \pi/2)$ and simplify the result.

32. In the expression $(49 + x^2)^{1/2}$, make the substitution $x = 7 \tan \theta$ $(0 < \theta < \pi/2)$ and simplify the result.

33. In the expression $(x^2 - 100)^{1/2}$, make the substitution $x = 10 \sec \theta$ $(0 < \theta < \pi/2)$ and simplify the result.

34. In the expression $(x^2 - 4)^{-3/2}$, make the substitution $x = 2 \sec \theta$ $(0 < \theta < \pi/2)$ and simplify the result.

35. In the expression $(x^2 + 5)^{-1/2}$, make the substitution $x = \sqrt{5} \tan \theta$ $(0 < \theta < \pi/2)$ and simplify the result.

36. If $\sin \theta = -\frac{5}{13}$ and $\pi < \theta < 3\pi/2$, compute $\cos \theta$.

37. If $\cos \theta = \frac{8}{17}$, $\sin \theta$ is negative, and $0 < \theta < 2\pi$, compute $\tan \theta$.

38. If $\sec \theta = -\frac{25}{7}$ and $\pi < \theta < 3\pi/2$, compute $\cot \theta$.

39. In the following figure, P is the center of the circle, which has radius $\sqrt{2}$ units. If the radian measure of

angle *BPA* is θ, express the area of the shaded region in terms of θ. Simplify your answer as much as possible.

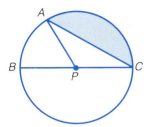

40. Express the area of the shaded region in the following figure in terms of r and θ. (Assume θ is in radians.)

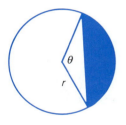

41. In the following figure, *ABCD* is a square, each side of which is 1 cm. The two arcs are portions of circles with radii of 1 cm and with centers *A* and *C*. Find the area of the shaded region. *Hint:* Draw $\overline{BD}$ and use your result from Exercise 40.

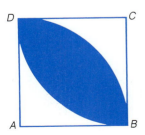

In Exercises 42–46, a function of the form $y = A \sin Bx$ or $y = A \cos Bx$ is graphed for one period. Determine the equation in each case. (Assume $B > 0$.)

42.

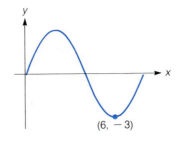

43.

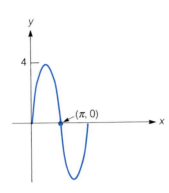

44.

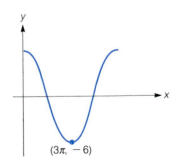

45.

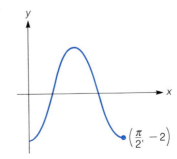

46.

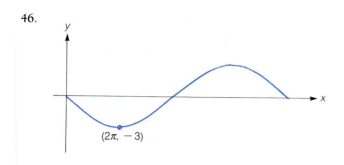

(2π, −3)

In Exercises 47–52, sketch the graph of each function for one complete cycle. In each case, specify the x-intercepts and the coordinates of the highest and lowest points on the graph.

47. $y = -3\cos 4x$
48. $y = 2\sin(3\pi x/4)$
49. $y = 2\sin(\frac{\pi x}{2} - \frac{\pi}{4})$
50. $y = -\sin(x - 1)$
51. $y = 3\cos(\frac{\pi x}{3} - \frac{\pi}{3})$
52. $y = -2\cos(x + \pi)$

In Exercises 53–56, sketch the graph of each function for one period.

53. (a) $y = \tan(\pi x/4)$
     (b) $y = \cot(\pi x/4)$
54. (a) $y = 2\cot 2x$
     (b) $y = 2\tan 2x$
55. (a) $y = 3\sec(x/4)$
     (b) $y = 3\csc(x/4)$
56. (a) $y = \sec \pi x$
     (b) $y = \csc \pi x$

In Exercises 57–66 simplify the expressions.

57. $\sin[x + (3\pi/2)]$
58. $\sin(\pi - x)$
59. $\cos(\pi - x)$
60. $\sin 10° \cos 80° + \cos 10° \sin 80°$
61. $\cos 175° \cos 25° + \sin 175° \sin 25°$
62. $\cos\frac{2\pi}{5}\cos\frac{\pi}{10} - \sin\frac{2\pi}{5}\sin\frac{\pi}{10}$
63. $\cos(x - \frac{2\pi}{3}) - \cos(x + \frac{2\pi}{3})$
64. $\sin(x + \frac{\pi}{6}) - \sin(x - \frac{\pi}{6})$
65. $\tan(x + 45°)\tan(x - 45°)$
66. $\tan x \tan y + \tan y \tan z + \tan z \tan x$, where $y = x + 60°$ and $z = x - 60°$
67. (a) Show that $\tan\frac{\pi}{12} = 2 - \sqrt{3}$.
         Hint: $\frac{\pi}{12} = \frac{\pi}{4} - \frac{\pi}{6}$
     (b) Show that $\cot\frac{\pi}{12} = 2 + \sqrt{3}$.
68. If x, y, and z are acute angles with sines $\frac{3}{5}$, $\frac{5}{13}$, and $\frac{8}{17}$, respectively, compute $\sin(x + y + z)$.

In Exercises 69–100, prove that the equations are identities.

69. $\cot(x + y) = \dfrac{\cot x \cot y - 1}{\cot x + \cot y}$

70. $\cos 2x = \dfrac{1 - \tan^2 x}{1 + \tan^2 x}$

71. $\sin 2x = \dfrac{2\tan x}{1 + \tan^2 x}$
72. $\sin^2 x - \sin^2 y = \sin(x + y)\sin(x - y)$
73. $\tan^2 x - \tan^2 y = \dfrac{\sin(x + y)\sin(x - y)}{\cos^2 x \cos^2 y}$
74. $2\csc 2x = \sec x \csc x$
75. $(\sin x)[\tan(x/2) + \cot(x/2)] = 2$
76. $\tan[x + (\pi/4)] = (1 + \tan x)/(1 - \tan x)$
77. $\tan[(\pi/4) + x] - \tan[(\pi/4) - x] = 2\tan 2x$
78. $\dfrac{\cot x - 1}{\cot x + 1} = \dfrac{1 - \sin 2x}{\cos 2x}$
79. $2\sin\left(\dfrac{\pi}{4} - \dfrac{x}{2}\right)\cos\left(\dfrac{\pi}{4} - \dfrac{x}{2}\right) = \cos x$
80. $\dfrac{\tan(x + y) - \tan y}{1 + \tan(x + y)\tan y} = \tan x$
81. $\dfrac{1 - \tan(\frac{1}{4}\pi - t)}{1 + \tan(\frac{1}{4}\pi - t)} = \tan t$
82. $\tan(\frac{1}{4}\pi + \frac{1}{2}t) = \tan t + \sec t$
83. $\dfrac{\tan(\alpha - \beta) + \tan\beta}{1 - \tan(\alpha - \beta)\tan\beta} = \tan\alpha$
84. $(1 + \tan\theta)[1 + \tan(\frac{1}{4}\pi - \theta)] = 2$
85. $\tan 3\theta = \dfrac{3t - t^3}{1 - 3t^2}$, where $t = \tan\theta$
86. $\sin(t + \frac{2}{3}\pi)\cos(t - \frac{2}{3}\pi) + \cos(t + \frac{2}{3}\pi)\sin(t - \frac{2}{3}\pi) = \sin 2t$
87. $\tan 2x + \sec 2x = \dfrac{\cos x + \sin x}{\cos x - \sin x}$
88. $\cos^4 x - \sin^4 x = \cos 2x$
89. $2\sin x + \sin 2x = 2\sin^3 x/(1 - \cos x)$
90. $1 + \tan x \tan(x/2) = \sec x$
91. $\tan\dfrac{x}{2} = \dfrac{1 - \cos x + \sin x}{1 + \cos x + \sin x}$
92. $\dfrac{\sin 3x}{\sin x} - \dfrac{\cos 3x}{\cos x} = 2$
93. $\sin(x + y)\cos y - \cos(x + y)\sin y = \sin x$
94. $\dfrac{\sin x + \sin 2x}{\cos x - \cos 2x} = \cot\dfrac{x}{2}$
95. $\dfrac{1 - \tan^2(x/2)}{1 + \tan^2(x/2)} = \cos x$
96. $4\sin(x/4)\cos(x/4)\cos(x/2) = \sin x$
97. $\sin 4x = 4\sin x \cos x - 8\sin^3 x \cos x$
98. $\cos 4x = 8\cos^4 x - 8\cos^2 x + 1$
99. $\sin 5x = 16\sin^5 x - 20\sin^3 x + 5\sin x$

100. $\cos 5x = 16 \cos^5 x - 20 \cos^3 x + 5 \cos x$

In Exercises 101–112, find all solutions of each equation in the interval $[0, 2\pi)$. In cases where a calculator is necessary, round off the answers to two decimal places.

101. $\tan x = 4.26$ **102.** $\tan x = -4.26$

103. $\csc x = 2.24$ **104.** $\sin(\sin x) = \pi/6$

105. $\tan^2 x - 3 = 0$ **106.** $\cot^2 x - \cot x = 0$

107. $\sin x - \cos 2x + 1 = 0$

108. $\sin x + \sin 2x = 0$

109. $3 \csc x - 4 \sin x = 0$

110. $2 \sin^2 x + \sin x - 1 = 0$

111. $2 \sin^4 x - 3 \sin^2 x + 1 = 0$

112. $4 \sin^2 2x + \cos 2x - 2 \cos^2 x - 2 = 0$

In Exercises 113–122, evaluate each expression (without using a calculator or tables).

113. $\cos^{-1}\left(-\sqrt{2}/2\right)$ **114.** $\arctan\left(\sqrt{3}/3\right)$

115. $\sin^{-1} 0$ **116.** $\arcsin \frac{1}{2}$

117. $\sin\left[\arccos\left(-\frac{1}{2}\right)\right]$ **118.** $\sin[\tan^{-1}(-1)]$

119. $\cot\left[\cos^{-1}\left(\frac{1}{2}\right)\right]$ **120.** $\sec\left[\cos^{-1}\left(\sqrt{2}/3\right)\right]$

121. $\sin\left(\frac{2\pi}{3} + \arccos \frac{3}{5}\right)$ **122.** $\tan\left[\frac{\pi}{4} + \sin^{-1}\left(\frac{5}{13}\right)\right]$

123. In this exercise we investigate the relationship between the variables x and θ in the accompanying figure. (Assume that θ is in radians.)
(a) Refer to the figure. Show that
$$\theta = \tan^{-1}(10/x) - \tan^{-1}(4/x).$$

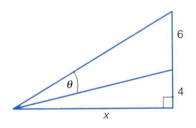

(b) Use your calculator to complete the following table. (Round off the results to two decimal places.) Which x-value in the table yields the largest value for θ?

x	0.1	1	2	3	10	100
θ						

(c) As indicated in the accompanying graph, the value of x that makes θ as large as possible is a number between 5 and 10, closer to 5 than to 10. Using calculus, it can be shown that this value of x is, in fact, $2\sqrt{10}$. Use your calculator to evaluate $2\sqrt{10}$; check that the result is consistent with the given graph. What is the corresponding value of θ in this case? Also, give the coordinates of the highest point on the accompanying graph. Round off both coordinates to two decimal places.

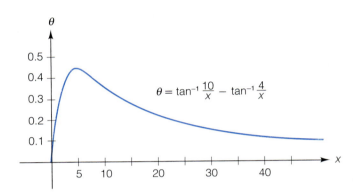

(d) What is the degree measure for the angle θ obtained in part (c)? Round off the answer to one decimal place.

In Exercises 124–128, show that each equation is an identity.

124. $\tan(\tan^{-1} x + \tan^{-1} y) = (x + y)/(1 - xy)$

125. $\tan^{-1}\left(x/\sqrt{1 - x^2}\right) = \sin^{-1} x$

126. $\sin(2 \arctan x) = 2x/(1 + x^2)$

127. $\cos(2 \cos^{-1} x) = 2x^2 - 1$

128. $\sin\left[\frac{1}{2} \sin^{-1}(x^2)\right] = \sqrt{\frac{1}{2} - \frac{1}{2}\sqrt{1 - x^4}}$

129. (a) Use a calculator to compute the quantity $\cos 20° \cos 40° \cos 60° \cos 80°$. Give your answer to as many decimal places as is shown on your calculator.
(b) Now use a product-to-sum formula to *prove* that the display on your calculator is the exact value of the given expression, not an approximation.

130. In this exercise, we will use the following figure to derive the half-angle formula for sine:
$$\sin \frac{\theta}{2} = \pm\sqrt{\frac{1}{2}(1 - \cos \theta)}$$

Our derivation will make use of the formula for the distance between two points and the identity

$\sin^2 t + \cos^2 t = 1$. However, we will not rely on an addition formula for sine, as we did in Section 8.7.

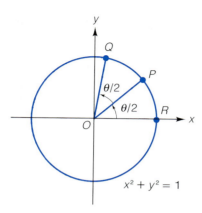

(a) Explain why the coordinates of P and Q are
$P\left(\cos\frac{\theta}{2}, \sin\frac{\theta}{2}\right)$, and $Q\left(\cos\theta, \sin\theta\right)$.

(b) Use the distance formula to show that
$$(PQ)^2 = 2 - 2\cos\frac{\theta}{2}\cos\theta - 2\sin\frac{\theta}{2}\sin\theta$$
and
$$(PR)^2 = 2 - \cos\frac{\theta}{2}$$

(c) Explain why $\triangle POR$ is congruent to $\triangle QOP$.

(d) From part (c), it follows that $(PQ)^2 = (PR)^2$. By equating the expressions for $(PQ)^2$ and $(PR)^2$ [obtained in part (b)], show that
$$\sin\frac{\theta}{2}\sin\theta = \left(\cos\frac{\theta}{2}\right)(1 - \cos\theta)$$

(e) Square both sides of the equation obtained in part (d); then replace $\cos^2\frac{\theta}{2}$ by $1 - \sin^2\frac{\theta}{2}$ and show that the resulting equation can be written
$$\left(\sin^2\frac{\theta}{2}\right)(2 - 2\cos\theta) = (1 - \cos\theta)^2$$

(f) Solve the equation in part (e) for the quantity $\sin\frac{\theta}{2}$. You should obtain
$$\sin\frac{\theta}{2} = \pm\sqrt{(1 - \cos\theta)/2}, \qquad \text{as required}$$

 ## OPTIONAL TI-81 GRAPHING-CALCULATOR EXERCISES
EXERCISES FOR SECTIONS 8.3 AND 8.4

As preparation for this exercise set, press the MODE key and check that the calculator is set for radians rather than degrees.

1. In this exercise, we use the TI-81 calculator to obtain several different views of $y = \sin x$ and $y = \cos x$.

    (a) Using the $Y =$ key, enter the function $Y_1 = \sin x$. Now press the ZOOM key followed by the 6 key. This yields the graph of $y = \sin x$ in the standard viewing window with $-10 \leq x \leq 10$ and $-10 \leq y \leq 10$.

    (b) For a better view of $y = \sin x$ (and to see how the RANGE key operates), press the RANGE key and enter the following settings. Then press the GRAPH key.

    RANGE
    $X_{min} = -8$
    $X_{max} = 8$
    $X_{scl} = 4$
    $Y_{min} = -8$
    $Y_{max} = 8$
    $Y_{scl} = 4$
    $X_{res} = 1$

You should obtain a picture similar to the following (except, of course, we've added some labels here).

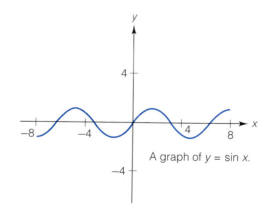

A graph of $y = \sin x$.

(c) As the graph shows, there is an x-intercept between 0 and 4. Using your knowledge of the sine function (and not using a calculator!), what is the exact x-coordinate for this intercept? Similarly, what is the exact coordinate for the x-intercept between 4 and 8?

(d) Now, using the Y= key, deactivate (but don't delete) $Y_1 = \sin x$ and enter $Y_2 = \cos x$. Graph $Y_2 = \cos x$. [Retain the RANGE key settings from part (b).] As you can see from the graph, there is an x-intercept between 0 and 4. What is the exact coordinate of that intercept?

(e) The TI-81 has a setting that allows us to mark off the x-axis in units of $\pi/2$. Press the ZOOM key followed by the 7 key. As you can see in the resulting graph of $Y_2 = \cos x$, the x-axis is indeed marked off in units of $\pi/2$. You can confirm this numerically by pressing the RANGE key and noting that the value for the x-scale is

$$X_{scl} = 1.570796327$$

This is the value of $\pi/2$ rounded off to nine decimal places.

(f) Press the Y= key and reactivate $Y_1 = \sin x$ (so that now both Y_1 and Y_2 are activated). Press the GRAPH key to see sin x and cos x on the same set of axes. Notice that the graphs are just translates of one another. By what distance would we have to shift the graph of $Y_1 = \sin x$ to the left in order for it to coincide with the graph of $Y_2 = \cos x$?

2. Press the Y= key and enter the following three functions:

$$Y_1 = \sin x \qquad Y_2 = 2 \sin x \qquad Y_3 = 3 \sin x$$

Now press the ZOOM key followed by the 7 key to obtain the three graphs. What are the amplitudes of the functions? What are the periods?

3. (a) Press the Y= key and enter the following three functions:

$$Y_1 = \cos x \qquad Y_2 = \cos 2x \qquad Y_3 = \cos 3x$$

Now press the ZOOM key followed by the 7 key to obtain the three graphs. Go on to part (b) for a better view of the graphs.

(b) In order to obtain a less cluttered view of the graphs, press the RANGE key and change the value of X_{min} to 0. Now press the GRAPH key to obtain the three graphs on the interval $0 \le x \le 2\pi$. What are the amplitudes of the functions? What are the periods?

Instructions for Problems 4–9.

(a) *First (without using the calculator), determine the amplitude, the period, and (where appropriate) the phase shift for the given function.*

(b) *Next, use the calculator to draw the graph of the function for two complete cycles. [In choosing an appropriate viewing rectangle, you will need to use the information obtained in part (a).]*

(c) *Specify the coordinates of the highest and lowest points on your graph.*

(d) *Use the TRACE key (and then the ARROW keys) on the calculator to check your answers in part (c).*

4. $y = -2.5 \cos(3x + 4)$ 5. $y = -2.5 \cos(3\pi x + 4)$
6. $y = \sin(0.5x - 0.75)$ 7. $y = \sin(0.5x + 0.75)$
8. $y = 9 \cos(11x - 8)$
9. $y = 0.02 \cos(100\pi x + 4\pi)$

10. Let $F(x) = \sin x$, $G(x) = x^2$, and $H(x) = x^3$. Which, if any, of the following four composite functions have graphs that do not go below the x-axis? First, try to answer without using the calculator, then use the calculator to check yourself. (You will learn more this way than if you were to draw the graphs immediately.)

$$y = G(F(x)) \qquad y = F(G(x))$$
$$y = F(H(x)) \qquad y = H(F(x))$$

11. (a) Enter the following two functions using the Y= key, but do not activate either of them yet.

$$Y_1 = \sin x \qquad \text{and} \qquad Y_2 = \sin(\sin x)$$

Press the ZOOM key and then the 7 key. Now press the RANGE key and set $Y_{min} = -1$ and $Y_{max} = 1$.

(b) Activate only the function $Y_1 = \sin x$ and draw the graph. Next, deactivate $Y_1 = \sin x$, activate $Y_2 = \sin(\sin x)$, and draw the graph. Are the two graphs identical?

(c) Activate both Y_1 and Y_2 and draw the graphs. As you can see, the two functions do have the same period, and their amplitudes are nearly equal.

(d) Estimate the amplitude of sin(sin x) as follows. With only Y_2 activated, regraph $Y_2 = \sin(\sin x)$. Now press the TRACE key and use the ARROW keys to move the cursor to the right until you reach what appears to be the highest point on the graph. What are the coordinates of this point (as indicated at the bottom of the calculator screen)? Is it the x-coordinate or the y-coordinate that gives the amplitude?

(e) Without a calculator, explain why the amplitude of $Y_2 = \sin(\sin x)$ is the number sin 1.

(f) Evaluate sin 1 and use the result to check your approximation in part (d).

EXERCISES FOR SECTION 8.5

1. Using the $Y=$ key, enter the function $Y_1 = \tan x$. Use the following settings. Press the mode key and check that your calculator is in the radian mode. Next, press the ZOOM key followed by the 7 key to mark off the axes in units of $\pi/2$. Check that the graph you obtain is consistent with the graph shown in the text on page 460.

2. In graphing the tangent function in the text, we used the identity $\tan(x + \pi) = \tan x$. Check this identity as follows. Retain the settings from Exercise 1 (including $Y_1 = \tan x$). Using the $Y=$ key, enter the function $Y_2 = \tan(x + \pi)$. Now, graph Y_1 and Y_2 and observe that the graphs indeed appear to be identical.

3. In the text, we obtained the graph of the cotangent function from that of the tangent function by means of the identity $\cot x = -\tan[x - (\pi/2)]$. Verify this identity graphically as follows. Using the $Y=$ key, enter the functions $Y_1 = \cot x \ (= 1 \div \tan x)$ and $Y_2 = -\tan[x - (\pi/2)]$. Graph Y_1 and Y_2 and observe that the graphs indeed appear to be identical.

4. In this exercise we compare the graphs of $Y_1 = \tan x$ and $Y_2 = x^3$ on the open interval $(-\pi/2, \pi/2)$. Use the $Y=$ key to enter these two functions. Next press ZOOM 7, and then (after waiting for the graphs to conclude) press the RANGE key and set $X_{min} = -1.5708$ and $X_{max} = 1.5708$. Now press the GRAPH key. Use the graphs to answer these questions:
   (a) Which of the two graphs appears to be horizontal as it passes through the origin?
   (b) For x-values in the interval $0 < x < \pi/2$, which quantity is larger, $\tan x$ or x^3?
   (c) For x-values in the interval $-\pi/2 < x < 0$, which quantity is larger, $\tan x$ or x^3?

In Exercises 5–10, graph each function. Adjust the range settings, where necessary, so that the graph is shown for at least two periods.

5. (a) $y = \tan(x/4)$
   (b) $y = \tan(4x)$

6. (a) $y = \cot(2x)$
   (b) $y = \cot(x/2)$

7. (a) $y = 0.5 \tan \pi x$
   (b) $y = 0.5 \tan[\pi x + (\pi/3)]$
   (c) $y = 0.5 \tan(\pi x + 1)$

8. (a) $y = 0.25 \cot x$
   (b) $y = 0.25 \cot[x + (\pi/4)]$
   (c) $y = -0.25 \cot[x + (\pi/4)]$

9. (a) $y = 0.4 \tan(x/2)$
   (b) $y = 0.4 \tan(x/3)$
   (c) $y = 0.4 \tan(x/5)$

10. (a) $y = 0.2 \cot(3x)$
    (b) $y = -0.2 \cot(4x)$
    (c) $y = -0.2 \cot(12x)$

11. Graph the functions $Y_1 = \sin x$ and $Y_2 = \csc x$ using the ZOOM 7 setting and a y-scale value of 1 in the RANGE menu. (To enter $\csc x$, use $1 \div \sin x$.) Observe that $|\sin x| \leq 1$ and $|\csc x| \geq 1$. At which points on the graph do we have $\sin x = \csc x$? At which points do we have $\sin x = -\csc x$?

12. Graph the functions $Y_1 = \cos x$ and $Y_2 = \sec x$ using the ZOOM 7 setting and a y-scale value of 1 in the RANGE menu. (To enter $\sec x$, use $1 \div \cos x$.) Observe that $|\cos x| \leq 1$ and $|\sec x| \geq 1$. At which points on the graph do we have $\cos x = \sec x$? At which points do we have $\cos x = -\sec x$?

In Exercises 13–16, graph each pair of functions on the same set of axes. Adjust the RANGE and/or ZOOM settings, where necessary, so that the graphs are displayed for at least two periods.

13. $Y_1 = 0.6 \sin(x/2)$
    $Y_2 = 0.6 \csc(x/2)$

14. $Y_1 = -1.2 \sin(\pi x/3)$
    $Y_2 = -1.2 \csc(\pi x/3)$

15. $Y_1 = 3 \cos[2x - (\pi/6)]$
    $Y_2 = 3 \sec[2x - (\pi/6)]$

16. $Y_1 = -1.5 \cos[\pi x - (\pi/6)]$
    $Y_2 = -1.5 \sec[\pi x - (\pi/6)]$

17. Using the ZOOM 7 setting, graph the two functions $Y_1 = \tan^2 x$ and $Y_2 = \sec^2 x - 1$. What do you observe? What does this demonstrate?

18. Using the ZOOM 7 setting, graph the two functions $Y_1 = \cot^2 x$ and $Y_2 = \csc^2 x - 1$. What do you observe? What does this demonstrate?

19. (a) Graph the two functions $Y_1 = \tan x$ and $Y_2 = x$ using the ZOOM 7 setting. Observe that $\tan x$ and x are very close to one another when x is close to 0. In fact, the approximation $\tan x \approx x$ is often used in science applications when it is known that x is close to 0.
    (b) To obtain a better view of $Y_1 = \tan x$ and $Y_2 = x$ near the origin, press the RANGE key and set $X_{min} = -1.57$ and $X_{max} = 1.57$. Then press the GRAPH key. Again, note that when x is close to 0, the values of $\tan x$ are indeed close to x. To see

numerical evidence of this, complete the following table.

x	0.000123	0.01	0.05	0.1	0.2	0.3	0.4	0.5
$\tan x$								

(c) If you study *Taylor polynomials* in calculus, you'll see that an even better approximation to tan x is

$$\tan x \approx x + \tfrac{1}{3}x^3 \qquad \text{when } x \text{ is close to } 0$$

Using the $Y=$ key, enter the function $Y_3 = x + \tfrac{1}{3}x^3$. Now, after deactivating $Y_2 = x$, press the GRAPH key to obtain the graphs of $Y_1 = \tan x$ and $Y_3 = x + \tfrac{1}{3}x^3$.

(d) To see numerical evidence of how well tan x is approximated by $x + \tfrac{1}{3}x^3$, add a third row to the table you worked out in part (b): in this third row, show the values for $x + \tfrac{1}{3}x^3$. When you've completed the table, note that the new values are much closer to tan x than were the values of x.

20. Using the $Y=$ key, enter the functions $Y_1 = \tan x$ and

$$Y_2 = \frac{\tan\left(x + \frac{\pi}{6}\right) - \tan\frac{\pi}{6}}{1 + \tan\left(x + \frac{\pi}{6}\right)\tan\frac{\pi}{6}}$$

Graph these two functions using the ZOOM 7 setting. What do you observe? *Remark:* After studying Section 8.7, you should be able to prove the identity $Y_1 = Y_2$.

21. *Finding a positive root of the equation* tan $x = x$. As preparation for this exercise, press the $Y=$ key and clear (delete) all functions. Next press ZOOM 6.

Finally, press the RANGE key and set $X_{min} = 0$ and $X_{max} = 6$.

(a) Using the $Y=$ key, enter the functions $Y_1 = \tan x$ and $Y_2 = x$. Press the GRAPH key. From the graph, estimate the smallest positive root of the equation tan $x = x$. (That is, estimate the x-coordinate of the point where $Y_1 = \tan x$ and $Y_2 = x$ intersect.) *Answer:* Something between 4 and 5, perhaps 4.5.

(b) We can obtain successively closer approximations to the root using the ZOOM key and the ARROW keys as follows. First press the ZOOM key and then the 2 key. If you now look carefully at the x-axis (at $x = 3$, actually), you will see a blinking cursor. Use the ARROW keys to move the cursor to the point where $Y_1 = \tan x$ tan $Y_2 = x$ intersect. Now press the ENTER key to obtain a magnified view of that intersection point. Next, use the ARROW keys to move the cursor back onto that intersection point. Record the x-value shown at the bottom of the calculator screen. This is our next approximation [after the one in part (a)] to the required root. By repeating this process, you can obtain better and better approximations to the required root. Continue the process until the first three decimal places of the intersection point remain the same when you progress to the next step.

22. (a) *Without using the graphing calculator:* In what general ways would you expect the graphs of $y = \sec^2 x$ and $y = |\sec x|$ to resemble one another?

(b) Use the calculator (with the ZOOM 7 setting) to draw the graphs of these two functions. Was your response in part (a) correct? In what way do the graphs differ?

EXERCISES FOR SECTION 8.8

1. In Example 3 in Section 8.8, we solved the equation cos $x = 0.351$ on the interval $(0, 2\pi)$. This exercise shows how to obtain the solutions graphically.

(a) With the calculator in the radian mode, enter the functions $Y_1 = \cos x$ and $Y_2 = 0.351$. Press ZOOM 7 to obtain the graphs with the x-axis marked off in units of $\pi/2$. Now, to focus on the interval from 0 to 2π (as in Example 3), press the RANGE key and set $X_{min} = 0$. Then press the GRAPH key.

(b) As you can see in the calculator display, there are two points where the line $Y_2 = 0.351$ meets the curve $Y_1 = \cos x$. The x-coordinates of these points

are solutions of the equation cos $x = 0.351$. In parts (c) through (e), we explain how to obtain these coordinates. But first, just to obtain a better overview of the situation, press the RANGE key and set $Y_{min} = -1$ and $Y_{max} = 1$. Now press the GRAPH key.

(c) The graph in part (b) shows that there are two solutions to the equation cos $x = 0.351$ in the interval $(0, 2\pi)$. Evidently, one solution is a little less than $\pi/2$ and the other is a little more than $3\pi/2$. We can obtain these solutions using the TRACE key and the ZOOM key. (We'll show how to look for the smaller of the two solutions; the other solution is obtained in a similar fashion.) Press the TRACE key. A blinking cursor will appear on the screen. By repeatedly pressing the left arrow key, move the cursor to the left (along the curve $Y_1 = \cos x$) until the cursor is positioned at the point where the curve $Y_1 = \cos x$ intersects the line $Y_2 = 0.351$. (If you overshoot the mark, use the right arrow key to back up.) As you can see from the read-out at the bottom of the screen, the x-coordinate of this point is approximately 1.19. This is a first approximation to the solution.

(d) To obtain a closer approximation, press ZOOM 2, and ENTER. This provides a magnified view of the region around the intersection point. Now press the TRACE key, and then use the arrow keys to move the cursor onto the apparent intersection point. From the read-out at the bottom of the screen, we have $x \approx 1.215$. This is our second approximation to the solution.

(e) By repeating the process in part (d), you can obtain closer and closer approximations to the required solution. Repeat the process until the first three decimal places of the approximation remain the same as you progress to the next step. You should obtain $x \approx 1.212$, in agreement with the result in Example 3.

2. In Example 9 in Section 8.8, we solved the equation $\sin \theta = 2 \cos \theta$ on the interval $0° \le \theta \le 360°$. This exercise shows how to obtain the solutions graphically.

(a) With the calculator set in the degree mode, enter the functions $Y_1 = \sin x$ and $Y_2 = 2 \cos x$. Next, press the RANGE key and set $X_{min} = 0$, $X_{max} = 360$,

$X_{scl} = 90$, $Y_{min} = -2$, and $Y_{max} = 2$. Now press the GRAPH key. Note that there are two intersection points, one in which x is less than 90° and one in which x is between 180° and 270°.

(b) Use the TRACE and ZOOM technique (explained in Exercise 1) to show that the two solutions are approximately 63.4° and 243.4°. (These are the two values that were obtained in Example 9.)

In Exercises 3–10, solve the equations on the interval $[0, 2\pi]$ using the TRACE and ZOOM technique described in Exercise 1. In each case, carry out the process far enough so that you are certain of the first two decimal places in each answer.

3. $\cos x = 0.623$

4. $\sin x = -0.438$

5. $\cos^2 x = 2 \sin x$

6. $\cos^3 x = 2 \sin x$

7. (a) $\cos 2x + 1 = \cos(2x + 1)$
   (b) $\cos 2x + 0.9 = \cos(2x + 1)$

8. $2 \sin x - 3 \cos x = \tan(x/4)$

9. $\sqrt{x} = \tan x$

10. (a) $\sin(\cos x) = \sin x$
    (b) $\cos(\sin x) = \sin x$

In Exercises 11–16, solve the equations on the interval $0° \le \theta \le 360°$ using the TRACE and ZOOM technique described in Exercises 1 and 2. In each case, carry out the process far enough so that you are certain of the first decimal place in each answer.

11. $\sin \theta = \cos(45° - \theta)$

12. $\sin(\theta/2) = \cos(\theta/3)$

13. $\sin^3 \theta + \cos^3 \theta = \frac{1}{2}$

14. $\sin^3 \theta - \cos^3 \theta = \frac{1}{2}$

15. $\tan \theta = \tan \sqrt{\theta}$

16. (a) $1 - \tan^2 \theta = 2 \sin(\theta/5)$
    (b) $1 - \tan^2 \theta = 2 \sin(\theta/6)$

EXERCISES FOR SECTION 8.9

1. To prepare the graphing screen for this exercise, press ZOOM 7. Then press the RANGE key and set $X_{min} = -1.57$ and $X_{max} = 1.57$. (The value of $\pi/2$ is approximately 1.57.)
   (a) Press the Y= key and define $Y_1 = x$ and $Y_2 = \sin^{-1}(\sin x)$. First graph only Y_2. Next, graph only Y_1. Finally graph both functions on the same screen. What do you observe?
   (b) What do the results in part (a) demonstrate?

2. Make the following adjustments to the RANGE settings from Exercise 1: $X_{min} = -1$ and $X_{max} = 1$. Next, press the Y= key. Retain the function $Y_1 = x$ (from Exercise

1), but redefine Y_2 as $Y_2 = \sin(\sin^{-1} x)$. Now graph the functions Y_1 and Y_2. What do you observe? What does this demonstrate?

3. To prepare the graphing screen for this exercise, press ZOOM 7. Then press the RANGE key and set $X_{min} = -1$ and $X_{max} = 1$. (These are the same RANGE settings used in Exercise 2.)
   (a) Press the Y= key and define $Y_1 = \sin^{-1} x + \cos^{-1} x$. Then, after deactivating any other functions in the menu, press the GRAPH key. What type of function does the resulting graph represent?
   (b) What is the *exact* value for the y-intercept of the

graph? What identity does the graph in part (a) demonstrate?

4. (a) In Example 7 in Section 8.9, we proved the identity $\sin(\cos^{-1}x) = \sqrt{1-x^2}$, for $-1 \le x \le 1$. Demonstrate this identity graphically by graphing the two functions $Y_1 = \sin(\cos^{-1}x)$ and $Y_2 = \sqrt{1-x^2}$.
    (b) Demonstrate the identity $\cos(\sin^{-1}x) = \sqrt{1-x^2}$, for $-1 \le x \le 1$, by graphing the functions $Y_2 = \sqrt{1-x^2}$ and $Y_3 = \cos(\sin^{-1}x)$ and noting that the graphs appear to be identical.

5. (a) Press the Y= key and clear or delete all functions. Enter the functions $Y_1 = \tan(\tan^{-1}x)$ and $Y_2 = x$. Press ZOOM 7 to obtain the graphs. What do you observe? What does this demonstrate?
    (b) Clear or delete the function Y_1 from part (a), and enter $Y_1 = \tan^{-1}(\tan x)$. Next, press the RANGE key and change the settings for X_{min} and X_{max} to $X_{min} = -1.57$ and $X_{max} = 1.57$. Now press the GRAPH key to obtain the graphs of $Y_1 = \tan^{-1}(\tan x)$ and $Y_2 = x$. What do you observe, and what does this demonstrate?
    (c) In part (b), why was it necessary to change the range settings?

6. In Example 10 in Section 8.9 we found that $\sec(\tan^{-1}x) = \sqrt{1+x^2}$, for $x > 0$. Actually, this

identity is valid for all real numbers. Demonstrate this graphically as follows. Using the Y= key, enter the functions $Y_1 = \sec(\tan^{-1}x) = 1/\cos(\tan^{-1}x)$ and $Y_2 = \sqrt{1+x^2}$. Press ZOOM 6 to obtain the graphs, and note that the graphs indeed appear identical.

In Exercises 7–15, solve the equations on the interval $[0, 1]$ using the TRACE and ZOOM technique described in Optional Exercise 1 on page 511. In each case, carry out the process far enough so that you are certain of the first two decimal places in each answer.

7. $\cos^{-1}x = \tan^{-1}x$

8. $x = \arccos x$

9. $\cos^{-1}x = x^2$

10. (a) $1.3\left(x - \tfrac{1}{2}\right)^2 = \cos^{-1}x$
    (b) $1.4\left(x - \tfrac{1}{2}\right)^2 = \cos^{-1}x$

11. $\tan^{-1}x = \sin 3x$

12. (a) $\arccos x = 2 \sin 3x$
    (b) $\arccos x = 2 \sin 4x$

13. (a) $\sin(2.3x) = \arctan x$
    (b) $\sin(2.2x) = \arctan x$

14. (a) $1/(\tan^{-1}x + \sin^{-1}x) = \sin 2x$
    (b) $1/(\tan^{-1}x + \sin^{-1}x) = \sin 3x$

15. (a) $1/(\sin^{-1}x + \cos^{-1}x) = 4x^3$
    (b) $1/(\sin^{-1}x + \cos^{-1}x) = 5x^3$

CHAPTER EIGHT TEST

1. Evaluate each expression without using a calculator or tables.
    (a) $\cos(4\pi/3)$ (b) $\csc(-5\pi/6)$ (c) $\sin^2(3\pi/4) + \cos^2(3\pi/4)$

2. In the expression $1/\sqrt{16 - t^2}$, make the substitution $t = 4 \sin u$ and simplify the result. Assume that $0 < u < \pi/2$.

3. Graph the function $y = 0.5 \sec(4\pi x - 1)$ for one complete cycle.

4. Graph the function $y = -\sin\left(3x - \tfrac{\pi}{4}\right)$ for one complete cycle. Specify the amplitude, period, and phase shift.

5. Graph the function $y = 3 \tan(\pi x/4)$ on the interval $0 \le x \le 4$.

6. (a) Convert $175°$ to radian measure.
    (b) Convert 5 radians to degree measure.

7. Two points B and C are on a circle of radius 5 cm. The center of the circle is A, and angle BAC is $75°$.
    (a) Find the length of the (shorter) arc of the circle from B to C.
    (b) Find the area of the (smaller) sector determined by angle BAC.

8. A wheel rotates about its axis with an angular speed of 25 revolutions/sec.
    (a) Find the angular speed of the wheel in radians/sec.
    (b) Find the linear speed of a point on the wheel that is 5 cm from the center.

9. Prove that the following equation is an identity:

$$\frac{\cot\theta}{1+\tan(-\theta)} + \frac{\tan\theta}{1+\cot(-\theta)} = \cot\theta + \tan\theta + 1$$

10. Use an appropriate addition formula to simplify the expression $\sin(\theta + 3\pi/2)$.

11. Compute $\cos 2t$ given that $\sin t = -2\sqrt{5}/5$ and $3\pi/2 < t < 2\pi$.

12. Use a calculator to find all solutions of the equation $\sin x = 3 \cos x$ in the interval $(0, 2\pi)$.

13. Find all solutions of the equation $2 \sin^2 x + 7 \sin x + 3 = 0$ on the interval $0 \le x \le 2\pi$.

14. If $\cos\alpha = 2/\sqrt{5}$ $(3\pi/2 < \alpha < 2\pi)$ and $\sin\beta = \frac{4}{5}$ $(\pi/2 < \beta < \pi)$, compute $\sin(\beta - \alpha)$.

15. Find all solutions of the equation $\sin(x + 30°) = \sqrt{3} \sin x$ on the interval $0° < x < 90°$.

16. If $\csc\theta = -3$ and $\pi < \theta < 3\pi/2$, compute $\sin(\theta/2)$.

17. Compute each of the following quantities:
    (a) $\sin^{-1}[\sin(\pi/10)]$;   (b) $\sin^{-1}(\sin 2\pi)$.

18. Compute $\cos(\arcsin\frac{3}{4})$.

19. Prove that the following equation is an identity:

$$\tan\left(\frac{\pi}{4} + \frac{\theta}{2}\right) = \frac{1 + \cos\theta + \sin\theta}{1 + \cos\theta - \sin\theta}$$

20. Simplify the following expression: $\sec(\arctan\sqrt{x^2 - 1})$ (assume that $x > 1$).

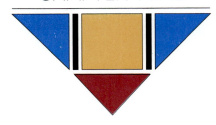

CHAPTER NINE

ADDITIONAL TOPICS IN TRIGONOMETRY

The subject of trigonometry is an excellent example of a branch of mathematics . . . which was motivated by both practical and intellectual interests — surveying, map-making, and navigation on the one hand, and curiosity about the size of the universe on the other. With it the Alexandrian mathematicians triangulated the universe and rendered precise their knowledge about the Earth and the heavens.

Morris Klein in *Mathematics in Western Culture* (New York: Oxford University Press, 1953)

INTRODUCTION

In this chapter we develop six topics that require a background in basic trigonometry. The first two of these topics, discussed in Section 9.1, are the law of sines and the law of cosines. These laws relate the angles and the lengths of the sides in any triangle. The next two sections, 9.2 and 9.3, introduce the important topic of vectors, first from a geometric standpoint, then from an algebraic standpoint. In Sections 9.4 and 9.5 we expand upon some of the ideas in Chapters 3 and 4 on graphs and equations. The two topics presented here are parametric equations and polar coordinates. The sixth and last topic that we introduce in this chapter is the trigonometric form for complex numbers. As you'll see, writing complex numbers in trigonometric form simplifies the processes of multiplication, division, and computation of roots.

9.1 THE LAW OF SINES AND THE LAW OF COSINES

The ratio of the sides of a triangle to each other is the same as the ratio of the sines of the opposite angles.

Bartholomaus Pitiscus (1561–1613) in his text, *Trigonometriae sive de dimensione triangulorum libri quinque.* This text was first published in Frankfort, Germany, in 1595, and according to several historians of mathematics, it was the first satisfactory textbook on trigonometry.

In this section we discuss two formulas relating the sides and the angles in any triangle: the *law of sines* and the *law of cosines*. These formulas can be used to determine an unknown side or angle using given information about the triangle. As you will see, which formula to apply in a particular case depends upon what data are initially given. To help you keep track of this, we'll use the following notations from elementary geometry.

NOTATION	EXPLANATION
SAA	One side and two angles are given.
SSA	Two sides and an angle opposite one of those sides are given.
SAS	Two sides and the included angle are given.
SSS	Three sides of the triangle are given.

Also, we will often follow the convention of denoting the angles of a triangle by A, B, and C and the lengths of the corresponding opposite sides by a, b, and c (see Figure 1). With this notation, we are ready to state the law of sines.

Law of Sines

In any triangle, the sines of the angles are proportional to the lengths of the opposite sides:

$$\frac{\sin A}{a} = \frac{\sin B}{b} = \frac{\sin C}{c}$$

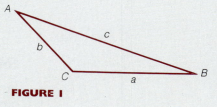

FIGURE 1

The proof of the law of sines is easy. We use the following result from Section 7.4: The area of any triangle is equal to half the product of two sides times the sine of the included angle. Thus, with reference to Figure 1, we have

$$\tfrac{1}{2}bc \sin A = \tfrac{1}{2}ac \sin B = \tfrac{1}{2}ab \sin C$$

since each of these three expressions equals the area of triangle ABC. Now we just multiply through by the quantity $2/abc$ to obtain

$$\frac{\sin A}{a} = \frac{\sin B}{b} = \frac{\sin C}{c}$$

which completes the proof.

EXAMPLE 1   (SAA)   Find the length x in Figure 2.

Solution   We can determine x directly by applying the law of sines. We have

$$\frac{\sin 30°}{20} = \frac{\sin 135°}{x}$$

Length of side opposite $\}$ ——↑   ↑—— $\{$ Length of side opposite
the 30° angle   the 135° angle

$$x \sin 30° = 20 \sin 135°$$

$$x\left(\frac{1}{2}\right) = 20\left(\frac{\sqrt{2}}{2}\right)$$

$$x = 20\sqrt{2} \text{ cm}$$

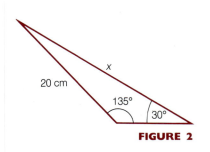

FIGURE 2

20 cm

x

135°

30°

EXAMPLE 2   (SAA)   Find the length y in Figure 3.

Solution   First we need to determine the angle θ in Figure 3. Since the sum of the angles in any triangle is 180°, we have

$$\theta = 180° - (75° + 62°) = 43°$$

Now, using the law of sines, we obtain

$$\frac{\sin 75°}{10.2} = \frac{\sin 43°}{y}$$

$$y \sin 75° = (10.2)\sin 43°$$

$$y = \frac{(10.2)\sin 43°}{\sin 75°}$$

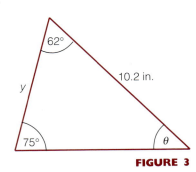

FIGURE 3

Without the use of a calculator or tables, this is the final form for the answer. On the other hand (as you should check for yourself), a calculator yields

$$y \approx 7.2 \text{ in.}$$

EXAMPLE 3   (SSA)   In triangle ABC, we are given $\angle C = 45°$, $b = 4\sqrt{2}$ ft, and $c = 8$ ft. Determine the remaining side and angles.

Solution   First let's draw a preliminary sketch conveying the given data; see Figure 4. (The sketch must be considered tentative. At the outset, we don't know whether the other angles are acute or even whether the given data are compatible.) To find angle B, we have (according to the law of sines)

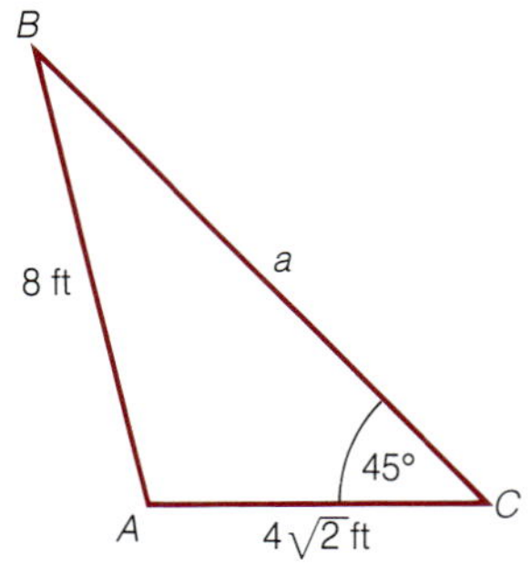

$$\frac{\sin B}{4\sqrt{2}} = \frac{\sin 45°}{8}$$

and therefore

$$8 \sin B = 4\sqrt{2} \sin 45° = 4\sqrt{2}\left(\frac{\sqrt{2}}{2}\right) = 4$$

$$\sin B = \frac{1}{2}$$

FIGURE 4

From our previous work, we know that one possibility for B is 30°, since $\sin 30° = \frac{1}{2}$. However, there is another possibility. Since the reference angle for 150° is 30°, we know that $\sin 150°$ is also equal to $\frac{1}{2}$. Which angle do we want? For the problem at hand, this is easy to answer. Since angle C is given as 45°, angle B cannot equal 150°, for the sum of 45° and 150° exceeds 180°. We conclude that

$$\angle B = 30°$$

Next, since $\angle B = 30°$ and $\angle C = 45°$, we have

$$\angle A = 180° - (30° + 45°) = 105°$$

Finally, we use the law of sines to find a. From the equation $\dfrac{\sin A}{a} = \dfrac{\sin C}{c}$, we have $a \sin C = c \sin A$, and therefore

$$a = \frac{c \sin A}{\sin C} = \frac{8 \sin 105°}{\sin 45°} = \frac{8 \sin 105°}{1/\sqrt{2}}$$

$$= 8\sqrt{2} \sin 105° \text{ ft} \approx 10.9 \text{ ft}$$

In the preceding example, two possibilities arose for the angle B: both 30° and 150°. However, it turned out that the value 150° was incompatible with the given information in the problem. In Exercise 13 at the end of this section, you will see a case in which both of two possibilities are compatible with the given data. This results in two distinct solutions to the problem. In contrast to this, Exercise 11(a) shows a case in which there is no triangle fulfilling the given conditions. For these reasons, the case SSA is sometimes referred to as the **ambiguous case**.

Now we turn to the law of cosines.

Law of Cosines

In any triangle, the square of the length of any side equals the sum of the squares of the lengths of the other two sides minus twice the product of the lengths of those other two sides times the cosine of their included angle.

$$a^2 = b^2 + c^2 - 2bc \cos A$$
$$b^2 = c^2 + a^2 - 2ca \cos B$$
$$c^2 = a^2 + b^2 - 2ab \cos C$$

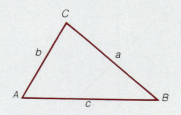

Before looking at a proof of this law, we make two preliminary comments. First, it is important to understand that the three equations in the box all follow the same pattern. For example, look at the first equation:

┌──── side and opposite angle ────┐
└──→ $a^2 = b^2 + c^2 - 2bc \cos A$ ←──┘
         ↑   ↑   ↑↑   ↑
sides that include this angle

Now check for yourself that the other two equations also follow this pattern. It is the pattern that is important here; after all, not every triangle is labeled *ABC*.

The second observation is that the law of cosines is a generalization of the Pythagorean theorem. In fact, look what happens to the equation

$$a^2 = b^2 + c^2 - 2bc \cos A$$

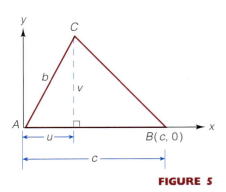

FIGURE 5

when angle A is a right angle:

$$a^2 = b^2 + c^2 - 2bc \underbrace{\cos 90°}_{0}$$

$$a^2 = b^2 + c^2 \qquad \text{which is the Pythagorean theorem}$$

Now let us prove the law of cosines:

$$a^2 = b^2 + c^2 - 2bc \cos A$$

(The other two equations can be proved in the same way. Indeed, just relabeling the figure would suffice.) The proof that we give uses coordinate geometry in a very nice way to complement the trigonometry. We begin by placing angle A in standard position, as indicated in Figure 5. (So, in the figure, angle A is then identified with angle CAB.) Then if u and v denote the lengths indicated in Figure 5, the coordinates of C are (u, v) and we have

$$\cos A = \frac{\text{adjacent}}{\text{hypotenuse}} = \frac{u}{b} \qquad \text{and therefore} \qquad u = b \cos A$$

Similarly, we have

$$\sin A = \frac{\text{opposite}}{\text{hypotenuse}} = \frac{v}{b} \qquad \text{and therefore} \qquad v = b \sin A$$

Thus, the coordinates of C are

$$(b \cos A, b \sin A)$$

(Exercise 44 at the end of this section asks you to check that these represent the coordinates of C even when angle A is not acute.) Now we use the distance formula,

$$d = \sqrt{(x_2 - x_1)^2 + (y_2 - y_1)^2}$$

to compute the required distance a between the points $C(b \cos A, b \sin A)$ and $B(c, 0)$. We have

$$
\begin{aligned}
a &= \sqrt{(b \cos A - c)^2 + (b \sin A - 0)^2} \\
&= \sqrt{b^2 \cos^2 A - 2bc \cos A + c^2 + b^2 \sin^2 A} \\
&= \sqrt{b^2 \underbrace{(\cos^2 A + \sin^2 A)}_{1} - 2bc \cos A + c^2} \\
&= \sqrt{b^2 + c^2 - 2bc \cos A}
\end{aligned}
$$

Squaring both sides of this last equation gives us the law of cosines, as we set out to prove.

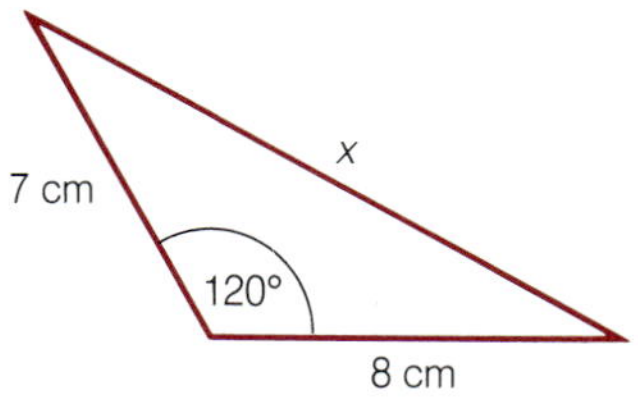

FIGURE 6

EXAMPLE 4 (SAS) Compute the length x in Figure 6.

Solution The law of cosines is directly applicable. We have

$$
\begin{aligned}
x^2 &= 7^2 + 8^2 - 2(7)(8)\cos 120° \\
&= 49 + 64 - 112\left(-\tfrac{1}{2}\right) = 169 \\
x &= \sqrt{169} = 13 \text{ cm}
\end{aligned}
$$

If the equation $a^2 = b^2 + c^2 - 2bc \cos A$ is solved for $\cos A$, the result is

$$\cos A = \frac{b^2 + c^2 - a^2}{2bc}$$

This expresses the cosine of an angle in a triangle in terms of the lengths of the sides. In a similar fashion, we obtain the corresponding formulas

$$\cos B = \frac{c^2 + a^2 - b^2}{2ca} \quad \text{and} \quad \cos C = \frac{a^2 + b^2 - c^2}{2ab}$$

These alternative forms for the law of cosines are used in the next example.

EXAMPLE 5   (SSS)   In triangle ABC, the sides are $a = 3$ units, $b = 5$ units, and $c = 7$ units. Find the angles.

Solution   Figure 7 summarizes the given data. We have

$$\cos A = \frac{b^2 + c^2 - a^2}{2bc} = \frac{5^2 + 7^2 - 3^2}{2(5)(7)} = \frac{65}{70} = \frac{13}{14}$$

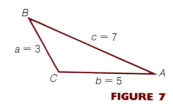

B

$c = 7$

$a = 3$

C $b = 5$ A

FIGURE 7

Now that we know $\cos A = \frac{13}{14}$, we can use a calculator (set in the degree mode) to find $\angle A$:

KEYSTROKES   **OUTPUT**

13 ÷ 14 = INV COS   21.7867893

Thus, $\angle A \approx 21.8°$. In a similar manner, we have

$$\cos B = \frac{c^2 + a^2 - b^2}{2ca} = \frac{7^2 + 3^2 - 5^2}{2(7)(3)} = \frac{33}{42} = \frac{11}{14}$$

So, $\cos B = \frac{11}{14}$ and a calculator then yields

$$\angle B \approx 38.2132107° \approx 38.2°$$

At this point, we can find $\angle C$ in either of two ways. Each has its advantage. The first way is to begin by computing $\cos C$ in the same way that we found $\cos A$ and $\cos B$. As you can check, the result is $\cos C = -\frac{1}{2}$. A calculator is not needed in this case. We know from previous work that $\angle C$ must be 120°. The second method that can be used relies on the fact that the sum of the angles in a triangle is 180°. Thus, we have

$$\angle C = 180° - \angle A - \angle B$$
$$\approx 180° - 21.7867893° - 38.2132107°$$
$$\approx 120.0°$$

This way is quicker than the first method. The disadvantage, however, is that we know that $\angle C$ is only approximately 120°. The first method (using the cosine law) is longer, but it tells us that $\angle C$ is exactly 120°. In summary, then, the three required angles are

$$\angle A \approx 21.8° \qquad \angle B \approx 38.2° \qquad \angle C = 120°$$

We conclude this section with an example indicating how the law of sines and the law of cosines are used in navigation. In this example, you'll see the term *bearing* used in specifying the location of one point relative to another. To explain this term, we refer to Figure 8.

FIGURE 8

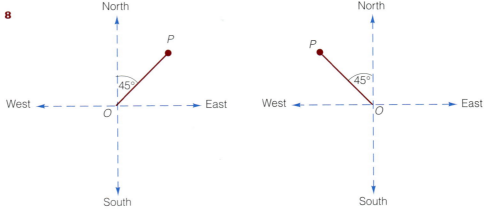

(a) The bearing of *P* from *O* is N45°E.   (b) The bearing of *P* from *O* is N45°W.

In Figure 8(a), the bearing of *P* from *O* is N45°E (read "north, 45° east"). This bearing tells us the *acute* angle between line segment $\overline{OP}$ and the north–south line through *O*. In Figure 8(b), the bearing of *P* from *O* is N45°W (read "north, 45° west"). Again, note that the bearing gives us the acute angle between line segment $\overline{OP}$ and the north–south line through *O*. Figure 9 provides additional examples of this terminology.

FIGURE 9
The bearing is specified by means of the acute angle measured from the north–south line.

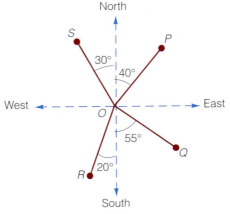

The bearing of *P* from *O* is N40° E.
The bearing of *Q* from *O* is S55° E.
The bearing of *R* from *O* is S20° W.
The bearing of *S* from *O* is N30° W.

EXAMPLE 6   A small fire is sighted from ranger stations *A* and *B*. The bearing of the fire from *A* is N35°E, and the bearing of the fire from *B* is N49°W. Station *A* is 1.3 miles due west of station *B*.

(a) How far is the fire from each ranger station?
(b) At fire station *C*, which is 1.5 miles from *A*, there is a helicopter that can be used to drop water on the fire. If the bearing of *C* from *A* is S42°E, find the distance from *C* to the fire, and find the bearing of the fire from *C*.

Solution (a) In Figure 10 we have sketched the situation involving ranger stations A and B and the fire (denoted by F). We compute the angles of $\triangle ABF$ as follows.

FIGURE 10

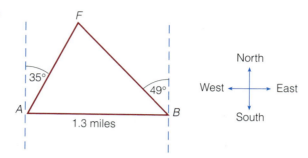

$$\angle FAB = 90° - 35° = 55° \qquad \angle FBA = 90° - 49° = 41°$$
$$\angle F = 180° - (55° + 41°) = 84°$$

We can now use the law of sines in $\triangle ABF$ to compute AF and BF.

$$\frac{\sin(\angle F)}{AB} = \frac{\sin(\angle FBA)}{AF} \qquad\qquad \frac{\sin(\angle F)}{AB} = \frac{\sin(\angle FAB)}{BF}$$

and therefore

$$\frac{\sin 84°}{1.3} = \frac{\sin 41°}{AF} \qquad\qquad \frac{\sin 84°}{1.3} = \frac{\sin 55°}{BF}$$

or

$$AF = \frac{(1.3)\sin 41°}{\sin 84°} \qquad\qquad BF = \frac{(1.3)\sin 55°}{\sin 84°}$$

We use a calculator to evaluate these expressions for AF and BF. As you should check for yourself, the results (rounded to the nearest tenth of a mile) are

$$AF \approx 0.9 \text{ miles} \qquad \text{and} \qquad BF \approx 1.1 \text{ miles}$$

(b) We draw a sketch of the situation, as shown in Figure 11.

In Figure 11, we can compute CF, the distance from the helicopter to the fire, using the law of cosines in $\triangle CAF$. First, note that $\angle CAF = 48° + 55° = 103°$. So we have

$$CF = \sqrt{AC^2 + AF^2 - 2 \cdot AC \cdot AF \cdot \cos 103°}$$
$$= \sqrt{1.5^2 + \left(\frac{1.3 \sin 41°}{\sin 84°}\right)^2 - 2(1.5)\left(\frac{1.3 \sin 41°}{\sin 84°}\right)\cos 103°}$$
$$\approx 1.9 \text{ miles} \qquad \text{using a calculator and rounding to one decimal place}$$

In order to find the bearing of the fire at F from the fire station at C, we need to determine the angle α in Figure 11. First we find the angle β. Using the law of sines in $\triangle CAF$, we have

$$\frac{\sin \beta}{AF} = \frac{\sin 103°}{CF}$$

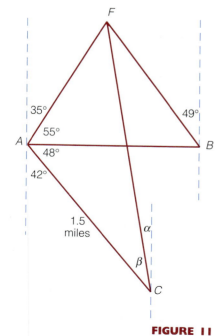

FIGURE 11

and therefore

$$\sin \beta = \frac{AF \cdot \sin 103°}{CF}$$

Before using the inverse sine function and our calculator to compute β, we note from Figure 11 that β is an acute angle (because β is an angle in $\triangle CAF$, and in that triangle, $\angle CAF$ is greater than 90°). So in this particular application of the law of sines, there is no ambiguity. We have then

$$\beta = \sin^{-1}\left(\frac{AF \cdot \sin 103°}{CF}\right)$$

Now, on the right-hand side of this last equation, we substitute the expressions we obtained previously for AF and CF; then, using a calculator set in the degree mode, we obtain

$$\beta \approx 26° \qquad \text{rounding off to the nearest degree}$$

(You should verify this calculator value for yourself.) Now that we know β, we can determine the bearing of the fire from station C. From Figure 11, we have

$$\alpha + \beta = 42° \qquad \text{(Why?)}$$

and therefore

$$\alpha = 42° - \beta = 42° - 26° = 16° \qquad \text{(to the nearest one degree)}$$

In summary now, fire station C is approximately 1.9 miles from the fire, and the bearing of the fire from station C is N16°W.

EXERCISE SET 9.1

A

In Exercises 1–8, assume that the vertices and the lengths of the sides of a triangle are labeled as in Figure 1 in the text. For Exercises 1–4, leave your answers in terms of radicals or the trigonometric functions; that is, don't use a calculator. In Exercises 5–8, use a calculator and round off your final answers to one decimal place.

1. If $\angle A = 60°$, $\angle B = 45°$, and $BC = 12$ cm, find AC.
2. If $\angle A = 30°$, $\angle B = 135°$, and $BC = 4$ cm, find AC.
3. If $\angle B = 100°$, $\angle C = 30°$, and $AB = 10$ cm, find BC.
4. If $\angle A = \angle B = 35°$, and $AB = 16$ cm, find AC and BC.
5. If $\angle A = 36°$, $\angle B = 50°$, and $b = 12.61$ cm, find a and c.
6. If $\angle B = 81°$, $\angle C = 55°$, and $b = 6.24$ cm, find c and a.
7. If $a = 29.45$ cm, $b = 30.12$ cm, and $\angle B = 66°$, find the remaining side and angles of the triangle.

8. If $a = 52.15$ cm, $c = 42.90$ cm, and $\angle A = 125°$, find the remaining side and angles of the triangle.

In Exercises 9 and 10, use degree measure for your answers. In parts (c) and (d), use a calculator and round off the results to one decimal place.

9. (a) In $\triangle ABC$, $\sin B = \sqrt{2}/2$. What are the possible values for $\angle B$?
   (b) In $\triangle DEF$, $\cos E = \sqrt{2}/2$. What are the possible values for $\angle E$?
   (c) In $\triangle GHI$, $\sin H = \frac{1}{4}$. What are the possible values for $\angle H$?
   (d) In $\triangle JKL$, $\cos K = -\frac{2}{3}$. What are the possible values for $\angle K$?

10. (a) In $\triangle ABC$, $\sin B = \sqrt{3}/2$. What are the possible values for $\angle B$?
    (b) In $\triangle DEF$, $\cos E = -\sqrt{3}/2$. What are the possible values for $\angle E$?

(c) In $\triangle GHI$, $\sin H = \frac{2}{9}$. What are the possible values for $\angle H$?

(d) In $\triangle JKL$, $\cos K = \frac{2}{3}$. What are the possible values for $\angle K$?

11. (a) Show that there is no triangle satisfying the conditions $a = 2.0$ ft, $b = 6.0$ ft, and $\angle A = 23.1°$.
    Hint: Try computing $\sin B$ using the law of sines.

    (b) If $a = 2.0$ ft, $b = 3.0$ ft, $\angle A = 23.1°$, and $\angle B$ is obtuse, show that $c = 1.1$ ft.

12. (a) Show that there is no triangle with $a = 2$, $b = 3$, and $\angle A = 42°$.

    (b) Is there any triangle in which $a = 2$, $b = 3$, and $\angle A = 41°$?

13. Let $b = 1$, $a = \sqrt{2}$, and $\angle B = 30°$.

    (a) Use the law of sines to show that $\sin A = \sqrt{2}/2$. Conclude that $\angle A = 45°$ or $\angle A = 135°$.

    (b) Assuming that $\angle A = 45°$, determine the remaining parts of $\triangle ABC$.

    (c) Assuming that $\angle A = 135°$, determine the remaining parts of $\triangle ABC$.

    (d) Find the area of the two triangles.

14. Let $a = 30$, $b = 36$, and $\angle A = 20°$.

    (a) Show that $\angle B = 24.23°$ or $\angle B = 155.77°$ (rounding off to two decimal places).

    (b) Determine the remaining parts for each of the two possible triangles. Round off your final results to two decimal places. [However, in your calculations, do *not* work with values that have been rounded off. (Why?)]

    (c) Find the areas of the two triangles.

15. Find the lengths a, b, c, and d in the following figure. Leave your answers in terms of trigonometric functions (rather than using a calculator).

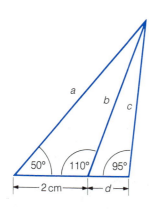

16. In the figure, $\overline{PQR}$ is a straight line segment. Find the distance PR. Round off your final answer to two decimal places.

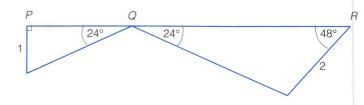

17. Two points P and Q are on opposite sides of a river (see the sketch). From P to another point R on the same side is 300 ft. Angles PRQ and RPQ are found to be 20° and 120°, respectively. Compute the distance from P to Q, across the river. (Round off your answer to the nearest foot.)

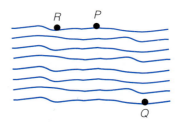

18. Determine the angle θ in the accompanying figure. Round off your answer to two decimal places.

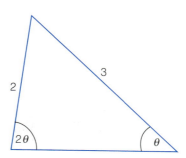

In Exercises 19–22, use the law of cosines to determine the length x in each figure. For Exercises 19 and 20, leave your answers in terms of radicals. In Exercises 21 and 22, use a calculator and round off the answers to one decimal place.

19. (a)

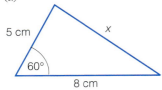

(b)

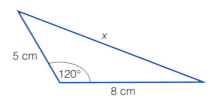

20. (a)

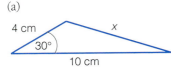

(b)

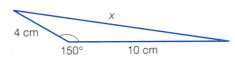

21. (a)

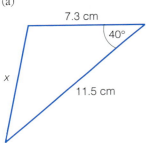

(b)

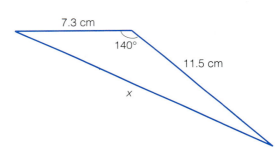

22. (a)

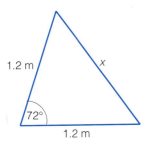

(b)

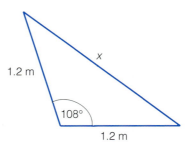

In Exercises 23 and 24, refer to the following figure.

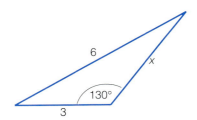

23. In applying the law of cosines to the figure, a student incorrectly writes $x^2 = 3^2 + 6^2 - 2(3)(6)\cos 130°$. Why is this incorrect? What is the correct equation?

24. In applying the law of cosines to the figure, a student writes $6^2 = 3^2 + x^2 + 6x \cos 50°$. Why is this correct?

In Exercises 25 and 26, use the given information to find the cosine of each angle in △ABC.

25. $a = 6$ cm, $b = 7$ cm, $c = 10$ cm

26. $a = 17$ cm, $b = 8$ cm, $c = 15$ cm (For this particular triangle, you can check your answers, because there is an alternate method of solution that does not require the law of cosines.)

In Exercises 27–30, compute each angle of the given triangle. Where necessary, use a calculator and round off to one decimal place.

27. $a = 7$, $b = 8$, $c = 13$
28. $a = 33$, $b = 7$, $c = 37$
29. $a = b = 2/\sqrt{3}$, $c = 2$
30. $a = 36$, $b = 77$, $c = 85$

In Exercises 31–34, round off each answer to one decimal place.

31. A regular pentagon is inscribed in a circle of unit radius. Find the perimeter of the pentagon. *Hint:* First find the length of a side using the law of cosines.

32. Find the perimeter of a regular nine-sided polygon inscribed in a circle of radius 4 cm. (See the hint for Exercise 31.)

33. In triangle ABC, $\angle A = 40°$, $b = 6.1$ cm, and $c = 3.2$ cm.
     (a) Find a using the law of cosines.
     (b) Find angle C using the law of sines.
     (c) Find angle B.

34. In parallelogram $ABCD$, you are given $AB = 6$ in., $AD = 4$ in., and $\angle A = 40°$. Find the length of each diagonal.

35. Town B is 26 miles from town A at a bearing of $S15°W$. Town C is 54 miles from town A at a bearing of $S7°E$. Compute the distance from town B to town C. Round off your final answer to the nearest mile.

36. Town C is 5 miles due east of town D. Town E is 12 miles from town C at a bearing (from C) of $N52°E$.
     (a) How far apart are towns D and E? (Round off to the nearest one-half mile.)
     (b) Find the bearing of town E from town D. (Round off the angle to the nearest degree.)

37. An airplane crashes in a lake and is spotted by observers at lighthouses A and B along the coast. Lighthouse B is 1.50 miles due east of lighthouse A. The bearing of the airplane from lighthouse A is $S20°E$; the bearing of the plane from lighthouse B is $S42°W$. Find the distance from each lighthouse to the crash site. (Round off your final answer to two decimal places.)

38. (Continuation of Exercise 37) A rescue boat is in the lake, three-fourths of a mile from lighthouse B, and at a bearing of $S35°E$ from lighthouse B.
     (a) Find the distance from the rescue boat to the airplane. Express your answer using miles and feet, with the portion in feet rounded off to the nearest ten feet.

(b) Find the bearing of the plane from the rescue boat. (Your answer should have the form of $S\theta°W$. Round off θ to two decimal places.)

39. (Refer to the figure.) When the Sun is viewed from the Earth, it subtends an angle of $\theta = 32'$ $(= \frac{32}{60}$ degree). Assuming that the distance d from the Earth to the Sun is 92,690,000 miles, compute the diameter D of the Sun. Round off the answer to the nearest ten thousand miles.

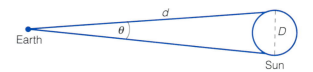

40. Compute the lengths CD and CE in the accompanying figure. Round off the final answers to two decimal places.

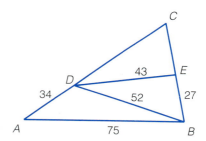

B

41. (a) Let m and n be positive numbers, with $m > n$. Furthermore, suppose that in triangle ABC the lengths a, b, and c are given by

$$a = 2mn + n^2 \qquad b = m^2 - n^2$$
$$c = m^2 + n^2 + mn$$

Show that $\cos C = -\frac{1}{2}$ and conclude that $\angle C = 120°$.

(b) Give an example of a triangle in which the lengths of the sides are whole numbers and one of the angles is 120°. (Specify the three sides; you needn't find the other angles.)

42. If the lengths of two adjacent sides of a parallelogram are a and b, and if the acute angle formed by these two sides is θ, show that the product of the lengths of the two diagonals is given by the expression

$$\sqrt{(a^2 + b^2)^2 - 4a^2b^2 \cos^2 \theta}$$

43. Two trains leave the railroad station at noon. The first train travels along a straight track at 90 mph. The second train travels at 75 mph along another straight track that makes an angle of 130° with the first track. At what time are the trains 400 miles apart? Round off your answer to the nearest minute.

44. In this exercise, you will complete a detail mentioned in the text in the proof of the law of cosines. Let the positive numbers u and v denote the lengths indicated in the following figure, so that the coordinates of C are $(-u, v)$. Show that $u = -b \cos A$ and $v = b \sin A$. Conclude from this that the coordinates of C are

$$(b \cos A, b \sin A)$$

Hint: Use the right-triangle definitions for cosine and sine along with the addition formulas for $\cos(180° - \theta)$ and $\sin(180° - \theta)$.

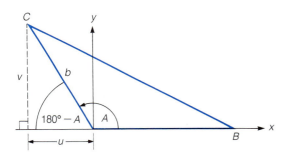

45. The following figure shows a quadrilateral with sides a, b, c, and d, inscribed in a circle.

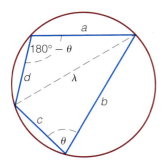

If λ denotes the length of the diagonal indicated in the figure, prove that

$$\lambda^2 = \frac{(ab + cd)(ac + bd)}{bc + ad}$$

This result is known as **Brahmagupta's theorem**. It is named after its discoverer, a seventh-century Hindu mathematician. *Hint:* Assume as given the theorem

from geometry stating that when a quadrilateral is inscribed in a circle, the opposite angles are supplementary. Apply the law of cosines in both of the triangles in the figure to obtain expressions for λ^2. Then eliminate $\cos \theta$ from one equation.

46. Prove the following identity for $\triangle ABC$:

$$\frac{\cos A}{a} + \frac{\cos B}{b} + \frac{\cos C}{c} = \frac{a^2 + b^2 + c^2}{2abc}$$

Suggestion: Use the law of cosines to substitute for a^2, for b^2, and for c^2 in the numerator of the expression on the right-hand side.

47. In $\triangle ABC$, suppose that $a^4 + b^4 + c^4 = 2(a^2 + b^2)c^2$. Find $\angle C$. (There are two answers.) *Hint:* Solve the given equation for c^2.

48. In this section we have seen that the cosines of the angles in a triangle can be expressed in terms of the lengths of the sides. For instance, for $\cos A$ in $\triangle ABC$, we obtained $\cos A = (b^2 + c^2 - a^2)/2bc$. This exercise shows how to derive corresponding expressions for the sines of the angles. For ease of notation in this exercise, let us agree to use the letter T to denote the following quantity:

$$T = 2(a^2b^2 + b^2c^2 + c^2a^2) - (a^4 + b^4 + c^4)$$

Then the sines of the angles in $\triangle ABC$ are given by

$$\sin A = \frac{\sqrt{T}}{2bc} \qquad \sin B = \frac{\sqrt{T}}{2ac} \qquad \sin C = \frac{\sqrt{T}}{2ab}$$

In the steps that follow, we'll derive the first of these three formulas, the derivations for the other two being entirely similar.

(a) In $\triangle ABC$, why is the positive root always appropriate in the formula $\sin A = \sqrt{1 - \cos^2 A}$?

(b) In the formula in part (a), replace $\cos A$ by $(b^2 + c^2 - a^2)/2bc$ and show that the result can be written

$$\sin A = \frac{\sqrt{4b^2c^2 - (b^2 + c^2 - a^2)^2}}{2bc}$$

(c) On the right-hand side of the equation in part (b), carry out the indicated multiplication. After combining like terms, you should obtain $\sin A = \sqrt{T}/2bc$, as required.

49. In the two easy steps that follow, we derive the law of sines by using the formulas obtained in Exercise 48. (Since the formulas in Exercise 48 were obtained using the law of cosines, we are, in essence, showing how to derive the law of sines from the law of cosines.)

(a) Use the formulas in Exercise 48 to check that each of the three fractions $(\sin A)/a$, $(\sin B)/b$, and $(\sin C)/c$ is equal to $\sqrt{T}/2abc$.

(b) Conclude from part (a) that $(\sin A)/a = (\sin B)/b = (\sin C)/c$.

C

50. *Heron's formula*   Approximately 2000 years ago, Heron of Alexandria derived a formula for the area of a triangle in terms of the lengths of the sides. A more modern derivation of Heron's formula is indicated in the steps that follow.

(a) Use the expression for $\sin A$ in Exercise 48(b) to show that

$$\sin^2 A = \frac{(a - b + c)(a + b - c)(b + c - a)(b + c + a)}{4b^2c^2}$$

Hint: Use difference-of-squares factoring repeatedly.

(b) Let s denote one-half of the perimeter of $\triangle ABC$. That is, let $s = \frac{1}{2}(a + b + c)$. Using this notation (which is due to Euler), verify that

(i) $a + b + c = 2s$

(ii) $-a + b + c = 2(s - a)$
(iii) $a - b + c = 2(s - b)$
(iv) $a + b - c = 2(s - c)$

Then, using this notation and the result in part (a), show that

$$\sin A = \frac{2\sqrt{s(s - a)(s - b)(s - c)}}{bc}$$

Note: Since $\sin A$ is positive (why?), the positive root is appropriate here.

(c) Use the result in part (b) and the formula area $\triangle ABC = \frac{1}{2}bc \sin A$ to conclude that

$$area \ \triangle ABC = \sqrt{s(s - a)(s - b)(s - c)}$$

This is Heron's formula. For historical background and a purely geometric proof, see *An Introduction to the History of Mathematics,* 6th ed., by Howard Eves (Philadelphia: Saunders College Publishing, 1990), pp. 178 and 194.

9.2 VECTORS IN THE PLANE, A GEOMETRIC APPROACH

The idea of a parallelogram of velocities may be found in various ancient Greek authors, and the concept of a parallelogram of forces was not uncommon in the sixteenth and seventeenth centuries.

Michael J. Crowe in *A History of Vector Analysis* (Notre Dame, Ind.: University of Notre Dame Press, 1967)

Vector notation is compact. If we can express a law of physics in vector form we usually find it easier to understand and to manipulate mathematically.

David Halliday and Robert Resnick in *Fundamentals of Physics,* third edition (New York: John Wiley and Sons, 1988)

Certain quantities, such as temperature, length, and mass, can be specified by means of a single number (assuming that a system of units has been agreed on). We call these quantities **scalars.** On the other hand, quantities such as force and velocity are characterized by both a *magnitude* (a number) and a *direction.* We call these quantities **vectors.**

Geometrically, a vector is a directed line segment or arrow. The vector in Figure 1, for instance, represents a wind velocity of 5 mph from the west. The length of this vector represents the magnitude of the wind velocity, while the direction of the vector indicates the direction of the wind velocity. As another example, the vector in Figure 2 represents a force acting on an object: the mag-

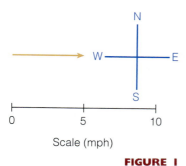

FIGURE 1
A vector representing a wind velocity of 5 mph from the west.

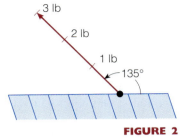

FIGURE 2

A vector representing a force of 3 lb acting at an angle of 135° with the horizontal.

FIGURE 3

nitude of the force is 3 pounds, and the force acts at an angle of 135° with the horizontal.

In a moment, we are going to discuss the important concept of vector addition, but first let us agree on some matters of notation. Suppose that we have a vector drawn from a point P to a point Q, as shown in Figure 3. The point P in Figure 3 is called the **initial point** of the vector, and Q is the **terminal point.** We can denote this vector by

$$\overrightarrow{PQ}$$

The length of the vector $\overrightarrow{PQ}$ is denoted by $|\overrightarrow{PQ}|$. On the printed page, vectors are often indicated by boldface letters, such as **a**, **A**, and **v**.

A word about notation. As you know, the notation (a, b) can denote either a point in the x-y plane or an open interval on the number line. In each instance, however, there is usually no danger of confusion; the context makes it clear which meaning is intended. Now we have a similar situation with the notation for the length of a vector. As has perhaps already occurred to you, the same vertical bars that we are using to denote the length of a vector are also used, in another context, to denote the absolute value of a real number. Again, it will be clear from the context which meaning is intended. Some books avoid this situation by using double bars to indicate the length of a vector: $\|\mathbf{v}\|$. In this text, we use the notation $|\mathbf{v}|$ simply because that is the one found in most calculus books.

If two vectors **a** and **b** have the same length and the same direction, we say that they are *equal* and we write **a** = **b** (see Figure 4). Notice that our definition for vector equality involves magnitude and direction, but not location. Thus, when it is convenient to do so, we are free to move a given vector to another location, provided we do not alter the magnitude or direction.

FIGURE 4

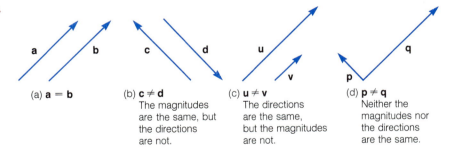

(a) **a** = **b**

(b) **c** ≠ **d**
The magnitudes are the same, but the directions are not.

(c) **u** ≠ **v**
The directions are the same, but the magnitudes are not.

(d) **p** ≠ **q**
Neither the magnitudes nor the directions are the same.

As motivation for the definition of vector addition, let us suppose that an object moves from a point P to a point Q. Then we can represent this *displacement* by the vector $\overrightarrow{PQ}$. (Indeed, the word *vector* is derived from the Latin *vectus*, meaning "carried.") Now suppose that after moving from P to Q, the object moves from Q to R. Then, as you can see in Figure 5, the net effect is a displacement from P to R. We say in this case that the vector $\overrightarrow{PR}$ is the **sum** or **resultant** of the vectors $\overrightarrow{PQ}$ and $\overrightarrow{QR}$, and we write

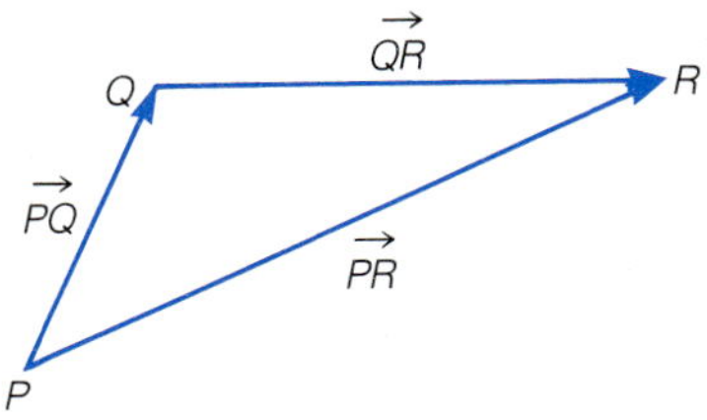

$$\overrightarrow{PQ} + \overrightarrow{QR} = \overrightarrow{PR}$$

FIGURE 5

These ideas are formalized in the definition that follows.

DEFINITION Vector Addition

Let **u** and **v** be two vectors. Position **v** (without changing its magnitude or direction) so that its initial point coincides with the terminal point of **u**, as in Figure 6(a). Then, as indicated in Figure 6(b), the vector **u** + **v** is the directed line segment from the initial point of **u** to the terminal point of **v**. The vector **u** + **v** is called the **sum** or **resultant** of **u** and **v**.

FIGURE 6

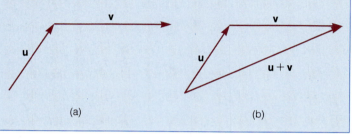

(a)   (b)

EXAMPLE I Referring to Figure 7, **(a)** determine the initial and terminal points of **u** + **v**; **(b)** compute $|\mathbf{u} + \mathbf{v}|$.

FIGURE 7

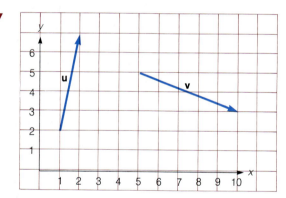

Solution **(a)** According to the definition, we first need to move **v** (without changing its length or direction) so that its initial point coincides with the terminal point of **u**. From Figure 7, we see that this can be accomplished by moving each point of **v** three units in the negative x-direction and two units in the positive y-direction. Figure 8(a) shows the new location of **v**, and Figure 8(b) indicates the sum **u** + **v**. From Figure 8(b), we see that the initial and terminal points of **u** + **v** are $(1, 2)$ and $(7, 5)$, respectively.

(b) We can use the distance formula to determine $|\mathbf{u} + \mathbf{v}|$. Using the points $(1, 2)$ and $(7, 5)$ that were obtained in part (a), we have

$$|\mathbf{u} + \mathbf{v}| = \sqrt{(7 - 1)^2 + (5 - 2)^2} = \sqrt{45} = \sqrt{9 \cdot 5} = 3\sqrt{5}$$

FIGURE 8

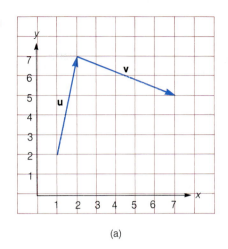

(a)

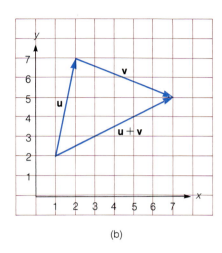

(b)

One important consequence of our definition for vector addition is that this operation is commutative. That is, for any two vectors **u** and **v**, we have

u + **v** = **v** + **u**

Figure 9 indicates why this is so.

In view of Figure 9, vector addition can also be carried out by using the **parallelogram law.** To determine **u** + **v**, position **u** and **v** so that their initial points coincide. Then, as indicated in Figure 10, the vector **u** + **v** is the directed diagonal of the parallelogram determined by **u** and **v**.

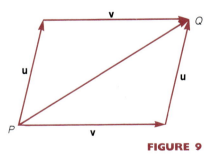

FIGURE 9

The upper triangle shows the sum **u** + **v**, whereas the lower triangle shows the sum **v** + **u**. Since in both cases the resultant is $\overrightarrow{PQ}$, it follows that

u + **v** = **v** + **u.**

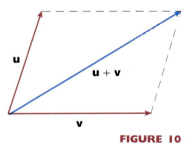

FIGURE 10

The parallelogram law for vector addition.

It is a fact—and it has been verified experimentally—that if two forces **F** and **G** act on an object, the net effect is the same as if just the resultant **F** + **G** acted on the object. In Example 2, we use the parallelogram law to compute the resultant of two forces. Note that the units of force used in this example are **newtons** (N), where $1\,\text{N} \approx 0.2248$ lb.

EXAMPLE 2   Two forces **F** and **G** act on an object. As indicated in Figure 11, the force **G** acts horizontally to the right with a magnitude of 12 N, while **F** acts vertically upward with a magnitude of 16 N. Determine the magnitude and direction of the resultant force.

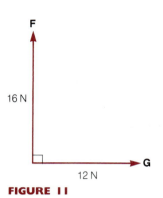

FIGURE 11

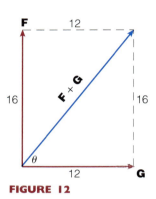

FIGURE 12

Solution   We complete the parallelogram, as shown in Figure 12. Now we need to calculate the length of **F** + **G** and the angle θ. Applying the Pythagorean theorem in Figure 12, we have

$$|\mathbf{F} + \mathbf{G}| = \sqrt{12^2 + 16^2} = \sqrt{144 + 256} = \sqrt{400} = 20$$

Also from Figure 12, we have

$$\tan \theta = \frac{16}{12} = \frac{4}{3}$$

Consequently,

$$\theta = \tan^{-1}\frac{4}{3} \approx 53.1° \qquad \text{using a calculator set in the degree mode}$$

Now we can summarize our results. The magnitude of **F** + **G** is 20 N, and the angle θ between **F** + **G** and the horizontal is (approximately) 53.1°.

    In Example 2, we determined the resultant for two perpendicular forces. The next example shows how to compute the resultant when the forces are not perpendicular. Our calculations will make use of both the law of sines and the law of cosines.

EXAMPLE 3   Determine the resultant of the two forces in Figure 13. (Round off the answers to one decimal place.)

Solution   As in the previous example, we complete the parallelogram. In Figure 14 (on the next page), the angle in the lower right-hand corner of the parallelogram is 140°. This is because the sum of two adjacent angles in any parallelogram is al-

FIGURE 13

ways 180°. Letting d denote the length of the diagonal in Figure 14, we can use the law of cosines to write

$$d^2 = 15^2 + 5^2 - 2(15)(5)\cos 140°$$
$$d = \sqrt{250 - 150 \cos 140°} = \sqrt{250 + 150 \cos 40°} \qquad \text{(Why?)}$$
$$= \sqrt{25(10 + 6 \cos 40°)} = 5\sqrt{10 + 6 \cos 40°}$$
$$\approx 19.1 \qquad \text{using a calculator}$$

FIGURE 14

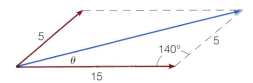

So the magnitude of the resultant is 19.1 N (to one decimal place). To specify the direction of the resultant, we need to determine the angle θ in Figure 14. Using the law of sines, we have

$$\frac{\sin \theta}{5} = \frac{\sin 140°}{d}$$

and, consequently,

$$\sin \theta = \frac{5 \sin 40°}{d} \qquad \text{(Why?)}$$

$$= \frac{5 \sin 40°}{5\sqrt{10 + 6 \cos 40°}} = \frac{\sin 40°}{\sqrt{10 + 6 \cos 40°}}$$

Using a calculator now, we obtain

$$\theta = \sin^{-1}\left(\frac{\sin 40°}{\sqrt{10 + 6 \cos 40°}}\right) \approx 9.7°$$

In summary, the magnitude of the resultant force is about 19.1 N, and the angle θ between the resultant and the 15-N force is approximately 9.7°. ∎

As background for the next example, we introduce the notion of *components* of a vector. (You will see this concept again in the next section, but in a more algebraic context.) Suppose that the initial point of a vector **v** is located at the origin of a rectangular coordinate system, as shown in Figure 15. Now suppose we draw perpendiculars from the terminal point of **v** to the axes, as indicated in Figure 16 (on the next page). Then the coordinates v_x and v_y in Figure 16 are called the **components** of the vector **v** in the x- and y-directions, respectively. For an example involving components, refer back to Figure 12. The horizontal component of the vector **F** + **G** is 12 N, and the vertical component is 16 N.

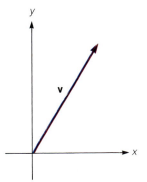

FIGURE 15

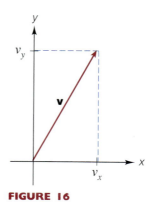

FIGURE 16

EXAMPLE 4 Determine the horizontal and vertical components of the velocity vector **v** in Figure 17.

Solution From Figure 17, we can write

$$\cos 30° = \frac{\text{adjacent}}{\text{hypotenuse}} = \frac{v_x}{70}$$

and, consequently,

$$v_x = (\cos 30°)(70) = \frac{\sqrt{3}}{2}(70)$$

$$= 35\sqrt{3} \approx 61 \text{ cm/sec} \qquad \text{(to two significant digits)}$$

Similarly, we have

$$\sin 30° = \frac{v_y}{70}$$

and therefore,

$$v_y = (\sin 30°)(70) = 35 \text{ cm/sec}$$

In summary, now, the x-component of the velocity is about 61 cm/sec, and the y-component is 35 cm/sec.

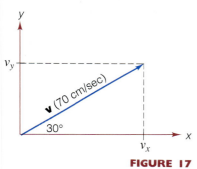

FIGURE 17

Our last example in this section will indicate how vectors are used in navigation. First, however, let us introduce some terminology. Suppose that an airplane has a **heading** of due east. This means that the airplane is pointed due east, and if there were no wind effects, the plane would indeed travel due east with respect to the ground. The **air speed** is the speed of the airplane relative to the air, whereas the **ground speed** is the plane's speed relative to the ground. Again, if there were no wind effects, then the air speed and the ground speed would be equal. Now suppose that the heading and air speed of an airplane are represented by the velocity vector **V** in Figure 18. (The direction of **V** is the heading; the magnitude of **V** is the air speed.) Also suppose that the wind velocity is represented by the vector **W** in Figure 18. Then the vector sum **V** + **W** represents the actual velocity of the plane with respect to the ground. The direc-

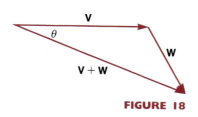

FIGURE 18

FIGURE 19

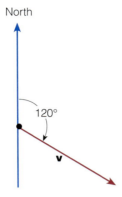

North

120°

v

tion of **V** + **W** is called the **course**; it is the direction in which the airplane is moving with respect to the ground. The magnitude of **V** + **W** is the ground speed (which was defined previously). The angle θ in Figure 18 is called, naturally enough, the **drift angle**.

In navigation, directions are given in terms of the angle measured clockwise from true north. For example, the direction of the velocity vector **v** in Figure 19 is 120°. We say in this case that the **bearing** of **v** is 120°.

EXAMPLE 5 [Refer to Figure 20(a).] The heading and air speed of an airplane are 60° and 250 mph, respectively. If the wind is 40 mph from 150°, find the ground speed, the drift angle, and the course.

FIGURE 20

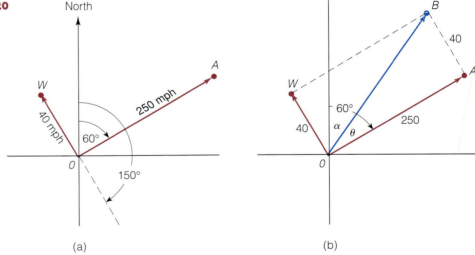

(a)   (b)

Solution In Figure 20(a), the vector $\overrightarrow{OA}$ represents the air speed of 250 mph and the heading of 60°. The vector $\overrightarrow{OW}$ represents a wind of 40 mph from 150°. Also, angle $WOA = 90°$. (Why?) In Figure 20(b), we have completed the parallelogram to obtain the vector sum $\overrightarrow{OA} + \overrightarrow{OW} = \overrightarrow{OB}$. The length of $\overrightarrow{OB}$ represents the ground speed, θ is the drift angle, and α is the bearing of the course. Because triangle BOA is a right triangle, we have

$$\tan \theta = \frac{40}{250} = \frac{4}{25} \qquad \left|\overrightarrow{OB}\right| = \sqrt{250^2 + 40^2}$$
$$\theta = 9.1° \qquad\qquad \approx 253.2$$

From these calculations, we conclude that the ground speed is approximately 253.2 mph and the drift angle is about 9.1°. We still need to compute the bearing of the course. From Figure 20(b), we have

$$\alpha = 60° - \theta \approx 60° - 9.1° = 50.9°$$

Thus, the bearing of the course is 50.9° (to one decimal place). ▪▪▪

EXERCISE SET 9.2

A

In Exercises 1–26, assume that the coordinates of the points P, Q, R, S, and O are as follows:

    $P(-1, 3)$ $Q(4, 6)$ $R(4, 3)$
    $S(5, 9)$ $O(0, 0)$

For each exercise, draw the indicated vector (using graph paper) and compute its magnitude. In Exercises 7–20, compute the sums using the definition given on page 530. In Exercises 21–26, use the parallelogram law to compute the sums.

1. $\overrightarrow{PQ}$
2. $\overrightarrow{QP}$
3. $\overrightarrow{SQ}$
4. $\overrightarrow{QS}$
5. $\overrightarrow{OP}$
6. $\overrightarrow{PO}$
7. $\overrightarrow{PQ} + \overrightarrow{QS}$
8. $\overrightarrow{SQ} + \overrightarrow{QP}$
9. $\overrightarrow{OP} + \overrightarrow{PQ}$
10. $\overrightarrow{OS} + \overrightarrow{SQ}$
11. $(\overrightarrow{OS} + \overrightarrow{SQ}) + \overrightarrow{QP}$
12. $(\overrightarrow{OS} + \overrightarrow{SP}) + \overrightarrow{PR}$
13. $\overrightarrow{OP} + \overrightarrow{QS}$
14. $\overrightarrow{QS} + \overrightarrow{PO}$
15. $\overrightarrow{SR} + \overrightarrow{PO}$
16. $\overrightarrow{OS} + \overrightarrow{QO}$
17. $\overrightarrow{OP} + \overrightarrow{RQ}$
18. $\overrightarrow{OP} + \overrightarrow{QR}$
19. $\overrightarrow{SQ} + \overrightarrow{RO}$
20. $\overrightarrow{SQ} + \overrightarrow{OR}$
21. $\overrightarrow{OP} + \overrightarrow{OR}$
22. $\overrightarrow{OP} + \overrightarrow{OQ}$
23. $\overrightarrow{RP} + \overrightarrow{RS}$
24. $\overrightarrow{QP} + \overrightarrow{QR}$
25. $\overrightarrow{SO} + \overrightarrow{SQ}$
26. $\overrightarrow{SQ} + \overrightarrow{SR}$

In Exercises 27–32, the vectors **F** and **G** denote two forces that act on an object: **G** acts horizontally to the right, and **F** acts vertically upward. In each case, use the information that is given to compute $|\mathbf{F} + \mathbf{G}|$ and θ, where θ is the angle between **G** and the resultant.

27. $|\mathbf{F}| = 4\,\text{N}, |\mathbf{G}| = 5\,\text{N}$
28. $|\mathbf{F}| = 15\,\text{N}, |\mathbf{G}| = 6\,\text{N}$
29. $|\mathbf{F}| = |\mathbf{G}| = 9\,\text{N}$
30. $|\mathbf{F}| = 28\,\text{N}, |\mathbf{G}| = 1\,\text{N}$
31. $|\mathbf{F}| = 3.22\,\text{N}, |\mathbf{G}| = 7.21\,\text{N}$
32. $|\mathbf{F}| = 4.06\,\text{N}, |\mathbf{G}| = 26.83\,\text{N}$

In Exercises 33–38, the vectors **F** and **G** represent two forces acting on an object, as indicated in the following figure. In each case, use the given information to compute (to two decimal places) the magnitude and direction of the resultant. (Give the direction of the resultant by specifying the angle θ between **F** and the resultant.)

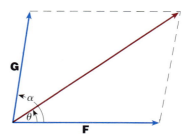

33. $|\mathbf{F}| = 5\,\text{N}, |\mathbf{G}| = 4\,\text{N}, \alpha = 80°$
34. $|\mathbf{F}| = 8\,\text{N}, |\mathbf{G}| = 10\,\text{N}, \alpha = 60°$
35. $|\mathbf{F}| = 16\,\text{N}, |\mathbf{G}| = 25\,\text{N}, \alpha = 35°$
36. $|\mathbf{F}| = 4.24\,\text{N}, |\mathbf{G}| = 9.01\,\text{N}, \alpha = 45°$

37. $|\mathbf{F}| = 50$ N, $|\mathbf{G}| = 25$ N, $\alpha = 130°$

38. $|\mathbf{F}| = 1.26$ N, $|\mathbf{G}| = 2.31$ N, $\alpha = 160°$

In Exercises 39–46, the initial point for each vector is the origin, and θ denotes the angle (measured counterclockwise) from the x-axis to the vector. In each case, compute the horizontal and vertical components of the given vector. (Round off your answers to two decimal places.)

39. The magnitude of $\mathbf{V}$ is 16 cm/sec, and $\theta = 30°$.

40. The magnitude of $\mathbf{V}$ is 40 cm/sec, and $\theta = 60°$.

41. The magnitude of $\mathbf{F}$ is 14 N, and $\theta = 75°$.

42. The magnitude of $\mathbf{F}$ is 23.12 N, and $\theta = 52°$.

43. The magnitude of $\mathbf{V}$ is 1 cm/sec, and $\theta = 135°$.

44. The magnitude of $\mathbf{V}$ is 12 cm/sec, and $\theta = 120°$.

45. The magnitude of $\mathbf{F}$ is 1.25 N, and $\theta = 145°$.

46. The magnitude of $\mathbf{F}$ is 6.34 N, and $\theta = 175°$.

In Exercises 47–50, use the given flight data to compute the ground speed, the drift angle, and the course. (Round off your answers to two decimal places.)

47. The heading and air speed are 30° and 300 mph, respectively; the wind is 25 mph from 120°.

48. The heading and air speed are 45° and 275 mph, respectively; the wind is 50 mph from 135°.

49. The heading and air speed are 100° and 290 mph, respectively; the wind is 45 mph from 190°.

50. The heading and air speed are 90° and 220 mph, respectively; the wind is 80 mph from
(a) 180°; (b) 90°.

B

51. A block weighing 12 lb rests on an inclined plane, as indicated in the following figure. Determine the components of the weight perpendicular to and parallel to the plane. Round off your answers to two decimal places. *Hint:* In Figure (b), the component of the weight perpendicular to the plane is $|\overrightarrow{OR}|$; the component parallel to the plane is $|\overrightarrow{OP}|$. Why does angle QOR equal 35°?

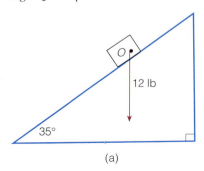

(a)

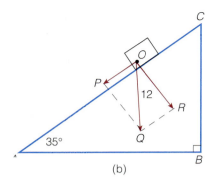

(b)

In Exercises 52 and 53, you are given the weight of a block on an inclined plane, along with the angle θ that the inclined plane makes with the horizontal. In each case, determine the components of the weight perpendicular to and parallel to the plane. (Round off your answers to two decimal places where necessary.)

52. 15 lb; $\theta = 30°$ **53.** 12 lb; $\theta = 10°$

54. A block rests on an inclined plane that makes an angle of 20° with the horizontal. The component of the weight parallel to the plane is 34.2 lb.
   (a) Determine the weight of the block. (Round off your answer to one decimal place.)
   (b) Determine the component of the weight perpendicular to the plane. (Round off your answer to one decimal place.)

55. In Section 9.3, we will see that vector addition is associative. That is, for any three vectors $\mathbf{A}$, $\mathbf{B}$, and $\mathbf{C}$, we have $(\mathbf{A} + \mathbf{B}) + \mathbf{C} = \mathbf{A} + (\mathbf{B} + \mathbf{C})$. In this exercise, you are going to check that this property holds in a particular case. Let $\mathbf{A}$, $\mathbf{B}$, and $\mathbf{C}$ be the vectors with initial and terminal points as follows:

Vector	Initial Point	Terminal Point
A	$(-1, 2)$	$(2, 4)$
B	$(1, 2)$	$(3, 0)$
C	$(6, 2)$	$(4, -3)$

   (a) Use the definition of vector addition on page 530 to determine the initial and terminal points of $(\mathbf{A} + \mathbf{B}) + \mathbf{C}$. *Suggestion:* Use graph paper.
   (b) Use the definition of vector addition to determine the initial and terminal points of $\mathbf{A} + (\mathbf{B} + \mathbf{C})$. [Your answers should agree with those in part (a).]

9.3 VECTORS IN THE PLANE, AN ALGEBRAIC APPROACH

A great many of the mathematical ideas that apply to physics and engineering are collected in the concept of vector spaces. This branch of mathematics has applications in such practical problems as calculating the vibrations of bridges and airplane wings. Logical extensions to spaces of infinitely many dimensions are widely used in modern theoretical physics as well as in many branches of mathematics itself.

From *The Mathematical Sciences,* edited by the Committee on Support of Research in the Mathematical Sciences with the collaboration of George Boehm (Cambridge, Mass.: The MIT Press, 1969)

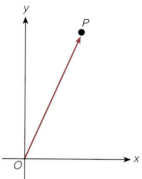

FIGURE 1

The geometric concept of a vector in the plane can be recast in an algebraic setting. This is useful both for computational purposes and (as our opening quotation implies) for more advanced work.

Consider an *x-y* coordinate system and a vector $\overrightarrow{OP}$ with initial point the origin, as shown in Figure 1. We call $\overrightarrow{OP}$ the **position vector** (or **radius vector**) of the point *P*. Most of our work in this section will involve such position vectors. There is no loss of generality in focusing on these types of vectors, for, as indicated in Figure 2, each vector **v** in the plane is equal to a unique position vector $\overrightarrow{OP}$.

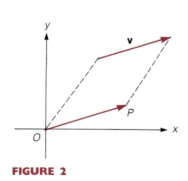

FIGURE 2

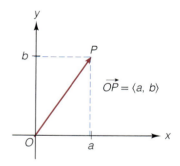

FIGURE 3
The vector $\overrightarrow{OP}$ is denoted by $\langle a, b \rangle$.
The coordinates *a* and *b* are the components of the vector.

If the coordinates of the point *P* are (a, b), we call *a* and *b* the **components** of the vector $\overrightarrow{OP}$, and we use the notation

$$\langle a, b \rangle$$

to denote this vector (see Figure 3). The number *a* is the **horizontal component** or **x-component** of the vector; *b* is the **vertical component** or **y-component**.

In the previous section, we said that two vectors are equal provided they have the same length and the same direction. For vectors $\langle a, b \rangle$ and $\langle c, d \rangle$, this implies that

$$\langle a, b \rangle = \langle c, d \rangle \qquad \text{if and only if} \qquad a = c \quad \text{and} \quad b = d$$

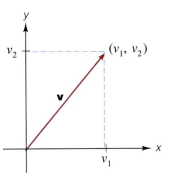

FIGURE 4

(Notice the similarity here to equality of complex numbers. Two complex numbers are equal provided their corresponding real and imaginary parts are equal; two vectors are equal provided their corresponding components are equal.)

It's easy to calculate the length of a vector **v** when its components are given. Suppose that $\mathbf{v} = \langle v_1, v_2 \rangle$. Applying the Pythagorean theorem in Figure 4, we have

$$|\mathbf{v}|^2 = v_1^2 + v_2^2$$

and, consequently,

$$|\mathbf{v}| = \sqrt{v_1^2 + v_2^2}$$

Although Figure 4 shows the point (v_1, v_2) in the first quadrant, you can check for yourself that the same formula results when (v_1, v_2) is located in any of the other three quadrants. We therefore have the following general formula.

The Length of a Vector

If $\mathbf{v} = \langle v_1, v_2 \rangle$, then

$$|\mathbf{v}| = \sqrt{v_1^2 + v_2^2}$$

EXAMPLE 1 Compute the length of a vector $\mathbf{v} = \langle 2, -4 \rangle$.

Solution $|\mathbf{v}| = \sqrt{v_1^2 + v_2^2} = \sqrt{2^2 + (-4)^2} = \sqrt{20} = 2\sqrt{5}$ ■■■

Even if the initial point of a vector **v** is not the origin, we can still find the *components* of **v** by determining the components of the equivalent position vector. The formula that we are going to derive for this is as follows.

If the coordinates of the points P and Q are $P(x_1, y_1)$ and $Q(x_2, y_2)$, then

$$\overrightarrow{PQ} = \langle x_2 - x_1, y_2 - y_1 \rangle$$

To derive this formula, we first construct the right triangle shown in Figure 5(a).

FIGURE 5

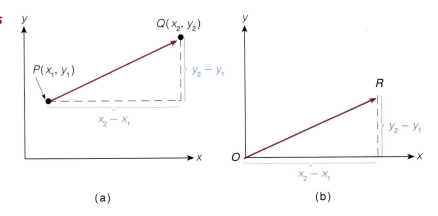

(a)   (b)

Now we let R denote the point $(x_2 - x_1, y_2 - y_1)$ and draw the position vector $\overrightarrow{OR}$ shown in Figure 5(b). Since the right triangles in Figures 5(a) and 5(b) are congruent and have corresponding legs that are parallel, we have

$$\overrightarrow{PQ} = \overrightarrow{OR} = \langle x_2 - x_1, y_2 - y_1 \rangle$$

as required. We can summarize this result as follows. For any vector $\mathbf{v}$, we have

x-component of $\mathbf{v}$:   (x-coordinate of terminal point) − (x-coordinate of initial point)

y-component of $\mathbf{v}$:   (y-coordinate of terminal point) − (y-coordinate of initial point)

EXAMPLE 2   Let P and Q be the points $P(3, 1)$ and $Q(7, 3)$. Find the components of $\overrightarrow{PQ}$.

Solution x-component of $\overrightarrow{PQ} = x$-coordinate of $Q - x$-coordinate of P
$$= 7 - 3 = 4$$

y-component of $\overrightarrow{PQ} = y$-coordinate of $Q - y$-coordinate of P
$$= 3 - 1 = 2$$

Consequently,

$$\overrightarrow{PQ} = \langle 4, 2 \rangle$$

Vector addition is particularly simple to carry out when the vectors are in component form. Indeed, we can use the parallelogram law to verify the following result.

THEOREM   Vector Addition

If $\mathbf{u} = \langle u_1, u_2 \rangle$ and $\mathbf{v} = \langle v_1, v_2 \rangle$, then $\mathbf{u} + \mathbf{v} = \langle u_1 + v_1, u_2 + v_2 \rangle$.

This theorem tells us that vector addition can be carried out *componentwise*; in other words, to add two vectors, just add the corresponding components. For example,

$$\langle 1, 2 \rangle + \langle 3, 7 \rangle = \langle 4, 9 \rangle$$

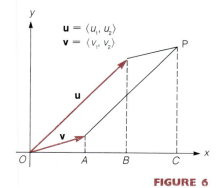

FIGURE 6

To see why this theorem is valid, consider Figure 6. (*Caution:* In reading the derivation that follows, don't confuse notation such as OA with $\overrightarrow{OA}$; recall that OA denotes the length of the line segment $\overline{OA}$.) Since the x-component of $\mathbf{u}$ is u_1, we have

$$OB = u_1$$

Also,

$$BC = v_1 \qquad \text{(Why?)}$$

Therefore,

$$OC = OB + BC = u_1 + v_1$$

But OC is the x-component of the vector $\overrightarrow{OP}$ and, by the parallelogram law, $\overrightarrow{OP} = \mathbf{u} + \mathbf{v}$. In other words, the x-component of $\mathbf{u} + \mathbf{v}$ is $u_1 + v_1$, as we wished to show. The fact that the y-component of $\mathbf{u} + \mathbf{v}$ is $u_2 + v_2$ is proved in a similar fashion.

The vector $\langle 0, 0 \rangle$ is called the **zero vector**, and it is denoted by $\mathbf{0}$. Notice that for any vector $\mathbf{v} = \langle v_1, v_2 \rangle$, we have

$$\mathbf{v} + \mathbf{0} = \mathbf{v} \qquad \text{because} \qquad \langle v_1, v_2 \rangle + \langle 0, 0 \rangle = \langle v_1, v_2 \rangle$$

and

$$\mathbf{0} + \mathbf{v} = \mathbf{v} \qquad \text{because} \qquad \langle 0, 0 \rangle + \langle v_1, v_2 \rangle = \langle v_1, v_2 \rangle$$

So for the operation of vector addition, the zero vector plays the same role as does the real number zero in addition of real numbers. There are, in fact, several other ways in which vector addition resembles ordinary addition of real numbers. We'll return to this point again near the end of this section.

In the box that follows, we define an operation called **scalar multiplication**, in which a vector is "multiplied" by a real number (a **scalar**) to obtain another vector

---

DEFINITION Scalar Multiplication

	EXAMPLES
For each real number k and each vector $\mathbf{v} = \langle x, y \rangle$, we define a vector $k\mathbf{v}$ by the equation $$k\mathbf{v} = k\langle x, y \rangle = \langle kx, ky \rangle$$	If $\mathbf{v} = \langle 2, 1 \rangle$, then $$2\mathbf{v} = \langle 4, 2 \rangle$$ $$3\mathbf{v} = \langle 6, 3 \rangle$$ $$-1\mathbf{v} = \langle -2, -1 \rangle$$

In geometric terms, the length of $k\mathbf{v}$ is $|k|$ times the length of $\mathbf{v}$. The vectors $\mathbf{v}$ and $k\mathbf{v}$ have the same direction if $k > 0$ and opposite directions if $k < 0$. For example, let $\mathbf{v} = \langle 2, 1 \rangle$. In Figure 7(a) we show the vectors $\mathbf{v}$ and $3\mathbf{v}$, while in Figure 7(b) we show $\mathbf{v}$ and $-2\mathbf{v}$.

FIGURE 7

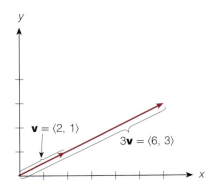

(a) The vectors **v** and 3**v** have the same direction. The length of 3**v** is three times that of **v**.

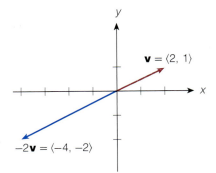

(b) The vectors **v** and −2**v** have opposite directions. The length of −2**v** is twice that of **v**.

EXAMPLE 3 Let $\mathbf{v} = \langle 3, 4 \rangle$ and $\mathbf{w} = \langle -1, 2 \rangle$. Compute each of the following.

(a) $\mathbf{v} + \mathbf{w}$ (b) $-2\mathbf{v} + 3\mathbf{w}$ (c) $|-2\mathbf{v} + 3\mathbf{w}|$

Solution (a) $\mathbf{v} + \mathbf{w} = \langle 3, 4 \rangle + \langle -1, 2 \rangle = \langle 3-1, 4 + 2 \rangle = \langle 2, 6 \rangle$

(b) $-2\mathbf{v} + 3\mathbf{w} = -2\langle 3, 4 \rangle + 3\langle -1, 2 \rangle = \langle -6, -8 \rangle + \langle -3, 6 \rangle$
$$= \langle -9, -2 \rangle$$

(c) $|-2\mathbf{v} + 3\mathbf{w}| = |\langle -9, -2 \rangle| = \sqrt{(-9)^2 + (-2)^2} = \sqrt{85}$ ∎

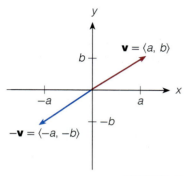

FIGURE 8

For each vector $\mathbf{v}$, we define a vector $-\mathbf{v}$, called the **negative** of $\mathbf{v}$, by the equation

$$-\mathbf{v} = -1\mathbf{v}$$

Thus, if $\mathbf{v} = \langle a, b \rangle$, then $-\mathbf{v} = \langle -a, -b \rangle$. As indicated in Figure 8, the vectors $\mathbf{v}$ and $-\mathbf{v}$ have the same length but opposite directions.

We can use the ideas in the preceding paragraph to define **vector subtraction**. Given two vectors $\mathbf{u}$ and $\mathbf{v}$, we define a vector $\mathbf{u} - \mathbf{v}$ by the equation

$$\mathbf{u} - \mathbf{v} = \mathbf{u} + (-\mathbf{v}) \tag{1}$$

First, let's see what equation (1) is saying in terms of components. Then we will indicate a simple geometric interpretation of vector subtraction.

If $\mathbf{u} = \langle u_1, u_2 \rangle$ and $\mathbf{v} = \langle v_1, v_2 \rangle$, then equation (1) tells us that

$$\mathbf{u} - \mathbf{v} = \langle u_1, u_2 \rangle + \langle -v_1, -v_2 \rangle$$
$$= \langle u_1 + (-v_1), u_2 + (-v_2) \rangle$$

That is,

$$\mathbf{u} - \mathbf{v} = \langle u_1 - v_1, \ u_2 - v_2 \rangle \tag{2}$$

In other words, to subtract two vectors, just subtract the corresponding components.

Vector subtraction can be interpreted geometrically. According to the formula in the second box on page 539, the right-hand side of equation (2) represents a vector drawn from the terminal point of $\mathbf{v}$ to the terminal point of $\mathbf{u}$.

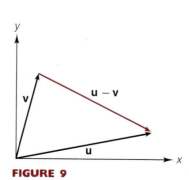

FIGURE 9

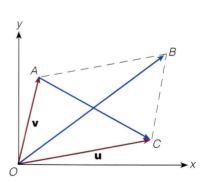

FIGURE 10

$\mathbf{u} + \mathbf{v}$ and $\mathbf{u} - \mathbf{v}$ are the directed diagonals of parallelogram $OABC$:
$\mathbf{u} + \mathbf{v} = \overrightarrow{OB}$ and $\mathbf{u} - \mathbf{v} = \overrightarrow{AC}$.

Figure 9 summarizes this fact, and Figure 10 provides a geometric comparison of vector addition and vector subtraction.

EXAMPLE 4 Let $\mathbf{u} = \langle 5, 3\rangle$ and $\mathbf{v} = \langle -1, 2\rangle$. Compute $3\mathbf{u} - \mathbf{v}$.

Solution

$$\begin{aligned} 3\mathbf{u} - \mathbf{v} &= 3\langle 5, 3\rangle - \langle -1, 2\rangle \\ &= \langle 15, 9\rangle - \langle -1, 2\rangle = \langle 16, 7\rangle \end{aligned}$$

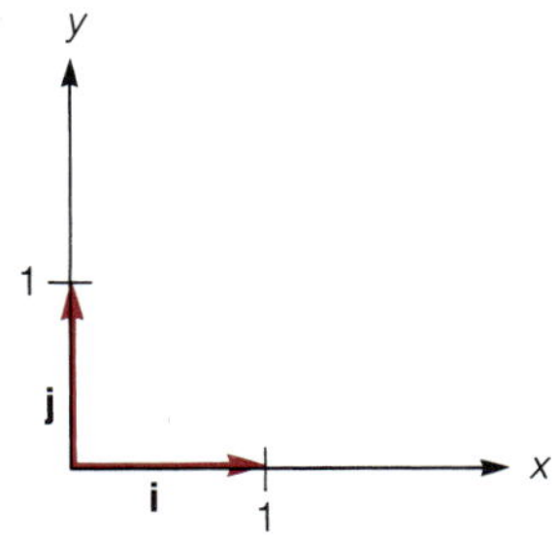

FIGURE 11

The next three examples deal with unit vectors. By definition, any vector with a length of 1 is called a **unit vector.** Two particularly useful unit vectors are

$$\mathbf{i} = \langle 1, 0\rangle \qquad \text{and} \qquad \mathbf{j} = \langle 0, 1\rangle$$

These are shown in Figure 11.

Any vector $\mathbf{v} = \langle x, y\rangle$ can be uniquely expressed in terms of the unit vectors $\mathbf{i}$ and $\mathbf{j}$ as follows:

$$\langle x, y\rangle = x\mathbf{i} + y\mathbf{j} \tag{3}$$

To verify equation (3), we have

$$\begin{aligned} x\mathbf{i} + y\mathbf{j} &= x\langle 1, 0\rangle + y\langle 0, 1\rangle \\ &= \langle x, 0\rangle + \langle 0, y\rangle = \langle x, y\rangle \end{aligned}$$

EXAMPLE 5

(a) Express the vector $\langle 3, -7\rangle$ in terms of the unit vectors $\mathbf{i}$ and $\mathbf{j}$.
(b) Express the vector $\mathbf{v} = -4\mathbf{i} + 5\mathbf{j}$ in component form.

Solution

(a) Using equation (3), we can write

$$\langle 3, -7\rangle = 3\mathbf{i} + (-7)\mathbf{j} = 3\mathbf{i} - 7\mathbf{j}$$

(b)
$$\begin{aligned} \mathbf{v} &= -4\mathbf{i} + 5\mathbf{j} = -4\langle 1, 0\rangle + 5\langle 0, 1\rangle \\ &= \langle -4, 0\rangle + \langle 0, 5\rangle = \langle -4, 5\rangle \end{aligned}$$

Thus, the component form of $\mathbf{v}$ is $\langle -4, 5\rangle$.

EXAMPLE 6 Find a unit vector $\mathbf{u}$ that has the same direction as the vector $\mathbf{v} = \langle 3, 4\rangle$.

Solution First, let us determine the length of $\mathbf{v}$:

$$|\mathbf{v}| = \sqrt{3^2 + 4^2} = \sqrt{25} = 5$$

So we want a vector whose length is one-fifth that of $\mathbf{v}$ and whose direction is the same as that of $\mathbf{v}$. Such a vector is

$$\mathbf{u} = \tfrac{1}{5}\mathbf{v} = \tfrac{1}{5}\langle 3, 4\rangle = \langle \tfrac{3}{5}, \tfrac{4}{5}\rangle$$

(You should check for yourself now that the length of this vector is 1.)

EXAMPLE 7 The angle from the positive x-axis to the unit vector $\mathbf{u}$ is $\pi/3$, as indicated in Figure 12 (on the next page). Determine the components of $\mathbf{u}$.

Solution   Let P denote the terminal point of $\mathbf{u}$. Since P lies on the unit circle, the coordinates of P are by definition $(\cos(\pi/3), \sin(\pi/3))$. We therefore have

$$\mathbf{u} = \left\langle \frac{1}{2}, \frac{\sqrt{3}}{2} \right\rangle$$

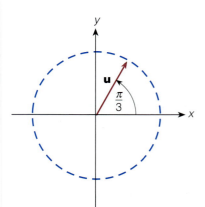

FIGURE 12

It was mentioned earlier that there are several ways in which vector addition resembles ordinary addition of real numbers. In the previous section, for example, we saw that vector addition is commutative. That is, for any two vectors $\mathbf{u}$ and $\mathbf{v}$,

$$\mathbf{u} + \mathbf{v} = \mathbf{v} + \mathbf{u}$$

By using components, we can easily verify that this property holds. (In the previous section, we used a geometric argument to establish this property.) We begin by letting $\mathbf{u} = \langle u_1, u_2 \rangle$ and $\mathbf{v} = \langle v_1, v_2 \rangle$. Then we have

$$\begin{aligned}
\mathbf{u} + \mathbf{v} &= \langle u_1, u_2 \rangle + \langle v_1, v_2 \rangle \\
&= \langle u_1 + v_1, u_2 + v_2 \rangle \\
&= \langle v_1 + u_1, v_2 + u_2 \rangle \qquad \text{Addition of real numbers is commutative.} \\
&= \langle v_1, v_2 \rangle + \langle u_1, u_2 \rangle \\
&= \mathbf{v} + \mathbf{u}
\end{aligned}$$

which is what we wanted to show.

There are a number of other properties of vector addition and scalar multiplication that can be proved in a similar fashion. In the following box, we list a particular collection of these properties, known as the **vector space properties**. (Exercises 55–58 ask that you verify these properties by using components, just as we did for the commutative property.)

Properties of Vector Addition and Scalar Multiplication

For all vectors $\mathbf{u}$, $\mathbf{v}$, and $\mathbf{w}$, and for all scalars (real numbers) a and b, the following properties hold.

1. $\mathbf{u} + (\mathbf{v} + \mathbf{w}) = (\mathbf{u} + \mathbf{v}) + \mathbf{w}$ **5.** $a(\mathbf{u} + \mathbf{v}) = a\mathbf{u} + a\mathbf{v}$

2. $\mathbf{0} + \mathbf{v} = \mathbf{v} + \mathbf{0} = \mathbf{v}$ **6.** $(a + b)\mathbf{v} = a\mathbf{v} + b\mathbf{v}$

3. $\mathbf{v} + (-\mathbf{v}) = \mathbf{0}$ **7.** $(ab)\mathbf{v} = a(b\mathbf{v})$

4. $\mathbf{u} + \mathbf{v} = \mathbf{v} + \mathbf{u}$ **8.** $1\mathbf{v} = \mathbf{v}$

EXERCISE SET 9.3

A

In Exercises 1–6, sketch each vector in an x-y coordinate system, and compute the length of the vector.

1. $\langle 4, 3 \rangle$ **2.** $\langle 5, 12 \rangle$ **3.** $\langle -4, 2 \rangle$

4. $\langle -6, -6 \rangle$ **5.** $\langle \frac{3}{4}, -\frac{1}{2} \rangle$ **6.** $\langle -3, 0 \rangle$

In Exercises 7–12, the coordinates of two points P and Q are given. In each case, determine the components of the vector $\overrightarrow{PQ}$. Write your answers in the form $\langle a, b \rangle$.

7. $P(2, 3)$ and $Q(3, 7)$ **8.** $P(5, 1)$ and $Q(4, 9)$

9. $P(-2, -3)$ and $Q(-3, -2)$

10. $P(0, -4)$ and $Q(0, -8)$

11. $P(-5, 1)$ and $Q(3, -4)$ 12. $P(1, 0)$ and $Q(0, 1)$

*In Exercises 13–32, assume that the vectors **a**, **b**, **c**, and **d** are defined as follows:*

$$\mathbf{a} = \langle 2, 3 \rangle \qquad \mathbf{b} = \langle 5, 4 \rangle \qquad \mathbf{c} = \langle 6, -1 \rangle \qquad \mathbf{d} = \langle -2, 0 \rangle$$

Compute each of the indicated quantities.

13. $\mathbf{a} + \mathbf{b}$ 14. $\mathbf{c} + \mathbf{d}$

15. $2\mathbf{a} + 4\mathbf{b}$ 16. $-2\mathbf{c} + 2\mathbf{d}$

17. $|\mathbf{b} + \mathbf{c}|$ 18. $|5\mathbf{b} + 5\mathbf{c}|$

19. $|\mathbf{a} + \mathbf{c}| - |\mathbf{a}| - |\mathbf{c}|$ 20. $1/|\mathbf{d}|$

21. $\mathbf{a} + (\mathbf{b} + \mathbf{c})$ 22. $(\mathbf{a} + \mathbf{b}) + \mathbf{c}$

23. $3\mathbf{a} + 4\mathbf{a}$ 24. $|4\mathbf{b} + 5\mathbf{b}|$

25. $\mathbf{a} - \mathbf{b}$ 26. $\mathbf{b} - \mathbf{c}$

27. $3\mathbf{b} - 4\mathbf{d}$ 28. $\dfrac{1}{|3\mathbf{b} - 4\mathbf{d}|}(3\mathbf{b} - 4\mathbf{a})$

29. $\mathbf{a} - (\mathbf{b} + \mathbf{c})$ 30. $(\mathbf{a} - \mathbf{b}) - \mathbf{c}$

31. $|\mathbf{c} + \mathbf{d}|^2 - |\mathbf{c} - \mathbf{d}|^2$

32. $|\mathbf{a} + \mathbf{b}|^2 + |\mathbf{a} - \mathbf{b}|^2 - 2|\mathbf{a}|^2 - 2|\mathbf{b}|^2$

*In Exercises 33–38, express each vector in terms of the unit vectors **i** and **j**.*

33. $\langle 3, 8 \rangle$ 34. $\langle 4, -2 \rangle$

35. $\langle -8, -6 \rangle$ 36. $\langle -9, 0 \rangle$

37. $3\langle 5, 3 \rangle + 2\langle 2, 7 \rangle$

38. $|\langle 12, 5 \rangle|\langle 3, 4 \rangle + |\langle 3, 4 \rangle|\langle 12, 5 \rangle$

In Exercises 39–42, express each vector in the form $\langle a, b \rangle$.

39. $\mathbf{i} + \mathbf{j}$ 40. $\mathbf{i} - 2\mathbf{j}$

41. $5\mathbf{i} - 4\mathbf{j}$ 42. $\dfrac{1}{|\mathbf{i} + \mathbf{j}|}(\mathbf{i} + \mathbf{j})$

In Exercises 43–48, find a unit vector having the same direction as the given vector.

43. $\langle 4, 8 \rangle$ 44. $\langle -3, 3 \rangle$ 45. $\langle 6, -3 \rangle$

46. $\langle -12, 5 \rangle$ 47. $8\mathbf{i} - 9\mathbf{j}$ 48. $\langle 7, 3 \rangle - \mathbf{i} + \mathbf{j}$

In Exercises 49–54, you are given an angle θ measured counterclockwise from the positive x-axis to a unit vector $\mathbf{u} = \langle u_1, u_2 \rangle$. In each case, determine the components u_1 and u_2.

49. $\theta = \pi/6$ 50. $\theta = \pi/4$ 51. $\theta = 2\pi/3$

52. $\theta = 3\pi/4$ 53. $\theta = 5\pi/6$ 54. $\theta = 3\pi/2$

B

In Exercises 55–58, let $\mathbf{u} = \langle u_1, u_2 \rangle$, $\mathbf{v} = \langle v_1, v_2 \rangle$, and $\mathbf{w} = \langle w_1, w_2 \rangle$. Refer to the box on page 544.

55. Verify properties 1 and 2.

56. Verify properties 3 and 4.

57. Verify properties 5 and 6.

58. Verify properties 7 and 8.

*In Exercises 59–75, we study the dot product of two vectors. Given two vectors $\mathbf{A} = \langle x_1, y_1 \rangle$ and $\mathbf{B} = \langle x_2, y_2 \rangle$, we define the **dot product** $\mathbf{A} \cdot \mathbf{B}$ as follows.*

$$\mathbf{A} \cdot \mathbf{B} = x_1 x_2 + y_1 y_2$$

*For example, if $\mathbf{A} = \langle 3, 4 \rangle$ and $\mathbf{B} = \langle -2, 5 \rangle$, then $\mathbf{A} \cdot \mathbf{B} = (3)(-2) + (4)(5) = 14$. Notice that the dot product of two vectors is a real number. For this reason, the dot product is also known as the **scalar product**. For Exercises 59–61, the vectors $\mathbf{u}$, $\mathbf{v}$, and $\mathbf{w}$ are defined as follows:*

$$\mathbf{u} = \langle -4, 5 \rangle \qquad \mathbf{v} = \langle 3, 4 \rangle \qquad \mathbf{w} = \langle 2, -5 \rangle$$

59. (a) Compute $\mathbf{u} \cdot \mathbf{v}$ and $\mathbf{v} \cdot \mathbf{u}$.
     (b) Compute $\mathbf{v} \cdot \mathbf{w}$ and $\mathbf{w} \cdot \mathbf{v}$.
     (c) Show that for any two vectors $\mathbf{A}$ and $\mathbf{B}$, we have $\mathbf{A} \cdot \mathbf{B} = \mathbf{B} \cdot \mathbf{A}$. That is, show that the dot product is commutative.   *Hint:* Let $\mathbf{A} = \langle x_1, y_1 \rangle$ and let $\mathbf{B} = \langle x_2, y_2 \rangle$.

60. (a) Compute $\mathbf{v} + \mathbf{w}$.
     (b) Compute $\mathbf{u} \cdot (\mathbf{v} + \mathbf{w})$.
     (c) Compute $\mathbf{u} \cdot \mathbf{v} + \mathbf{u} \cdot \mathbf{w}$.
     (d) Show that for any three vectors $\mathbf{A}$, $\mathbf{B}$, and $\mathbf{C}$, we have $\mathbf{A} \cdot (\mathbf{B} + \mathbf{C}) = \mathbf{A} \cdot \mathbf{B} + \mathbf{A} \cdot \mathbf{C}$.

61. (a) Compute $\mathbf{v} \cdot \mathbf{v}$ and $|\mathbf{v}|^2$.
     (b) Compute $\mathbf{w} \cdot \mathbf{w}$ and $|\mathbf{w}|^2$.

62. Show that for any vector $\mathbf{A}$, we always have $|\mathbf{A}|^2 = \mathbf{A} \cdot \mathbf{A}$. That is, the square of the length of a vector is equal to the dot product of the vector with itself.   *Hint:* Let $\mathbf{A} = \langle x, y \rangle$.

Let θ (where $0 \le \theta \le \pi$) denote the angle between the two nonzero vectors $\mathbf{A}$ and $\mathbf{B}$. Then it can be shown that the cosine of θ is given by the formula

$$\cos \theta = \frac{\mathbf{A} \cdot \mathbf{B}}{|\mathbf{A}| \, |\mathbf{B}|}$$

(See Exercise 75 for the derivation of this result.) In Exercises 63–68, use this formula to find the cosine of the angle between the given pair of vectors. Also, in each case use a calculator to compute the angle. Express the angle using degrees and using radians. Round off the values to two decimal places.

63. $\mathbf{A} = \langle 4, 1 \rangle$ and $\mathbf{B} = \langle 2, 6 \rangle$

64. $\mathbf{A} = \langle 3, -1 \rangle$ and $\mathbf{B} = \langle -2, 5 \rangle$

65. $\mathbf{A} = \langle 5, 6 \rangle$ and $\mathbf{B} = \langle -3, -7 \rangle$

66. $\mathbf{A} = \langle 3, 0 \rangle$ and $\mathbf{B} = \langle 1, 4 \rangle$

67. (a) $\mathbf{A} = \langle -8, 2 \rangle$ and $\mathbf{B} = \langle 1, -3 \rangle$
       (b) $\mathbf{A} = \langle -8, 2 \rangle$ and $\mathbf{B} = \langle -1, 3 \rangle$

68. (a) $\mathbf{A} = \langle 7, 12 \rangle$ and $\mathbf{B} = \langle 1, 2 \rangle$
       (b) $\mathbf{A} = \langle 7, 12 \rangle$ and $\mathbf{B} = \langle -1, -2 \rangle$

69. (a) Compute the cosine of the angle between the vectors $\langle 2, 5 \rangle$ and $\langle -5, 2 \rangle$.
       (b) What can you conclude from your answer in part (a)?
       (c) Draw a sketch to check your conclusion in part (b).

70. Follow Exercise 69, but use the vectors $\langle 6, -8 \rangle$ and $\langle -4, -3 \rangle$.

71. Suppose that $\mathbf{A}$ and $\mathbf{B}$ are nonzero vectors and $\mathbf{A} \cdot \mathbf{B} = 0$. Explain why $\mathbf{A}$ and $\mathbf{B}$ are perpendicular.

72. Find a value for t such that the vectors $\langle 15, -3 \rangle$ and $\langle -4, t \rangle$ are perpendicular.

73. Find a unit vector that is perpendicular to the vector $\langle -12, 5 \rangle$. (There are two answers.)

C

74. Suppose that $\mathbf{A}$ and $\mathbf{B}$ are nonzero vectors such that $|\mathbf{A} + \mathbf{B}|^2 = |\mathbf{A}|^2 + |\mathbf{B}|^2$. Show that $\mathbf{A}$ and $\mathbf{B}$ are perpendicular. *Hint:* Let $\mathbf{A} = \langle x_1, y_1 \rangle$ and let $\mathbf{B} = \langle x_2, y_2 \rangle$. After making some calculations, you should be able to apply the result in Exercise 71.

75. Refer to the accompanying figure. In this exercise we are going to derive the following formula for the cosine of the angle θ between two nonzero vectors $\mathbf{A}$ and $\mathbf{B}$:

$$\cos \theta = \frac{\mathbf{A} \cdot \mathbf{B}}{|\mathbf{A}|\,|\mathbf{B}|}$$

(a) Let $\mathbf{A} = \langle x_1, y_1 \rangle$ and $\mathbf{B} = \langle x_2, y_2 \rangle$. Compute the length of the vector $\mathbf{C}$ in the figure using the fact that $\mathbf{C} = \langle x_2 - x_1, y_2 - y_1 \rangle$.

(b) Using your result in part (a), show that
$$|\mathbf{C}|^2 = |\mathbf{A}|^2 + |\mathbf{B}|^2 - 2\mathbf{A} \cdot \mathbf{B}$$

(c) According to the law of cosines, we have
$$|\mathbf{C}|^2 = |\mathbf{A}|^2 + |\mathbf{B}|^2 - 2|\mathbf{A}|\,|\mathbf{B}| \cos \theta$$

Set this expression for $|\mathbf{C}|^2$ equal to the expression obtained in part (b), and then solve for $\cos \theta$ to obtain the required formula.

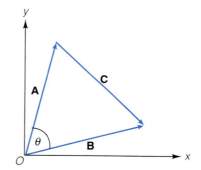

9.4 PARAMETRIC EQUATIONS

We will introduce the idea of parametric equations through a simple example. Suppose that we have a point $P(x, y)$ that moves in the x-y plane, and that the x- and y-coordinates of P at time t (in seconds) are given by the following pair of equations:

$$x = 2t \qquad \text{and} \qquad y = \tfrac{1}{2}t^2 \qquad (t \geq 0)$$

This pair of *parametric equations* tells us the location of the point P at any time t. For instance, when $t = 1$, we have

$$x = 2t = 2(1) = 2 \qquad \text{and} \qquad y = \tfrac{1}{2}t^2 = \tfrac{1}{2}(1)^2 = \tfrac{1}{2}$$

In other words, after one second, the coordinates of P are $\left(2, \tfrac{1}{2}\right)$. Let's see where P is after two seconds. We have

$$x = 2t = 2(2) = 4 \qquad \text{and} \qquad y = \tfrac{1}{2}t^2 = \tfrac{1}{2}(2)^2 = 2$$

TABLE I

t	x	y
0	0	0
1	2	$\frac{1}{2}$
2	4	2
3	6	$\frac{9}{2}$
4	8	8
5	10	$\frac{25}{2}$

So, after two seconds, the location of the point P is $(4, 2)$. In Table 1, we have computed the values of x and y (and hence the location of P) for integral values of t running from 0 to 5. Figure 1(a) shows the points that are determined. In Figure 1(b) we have joined the points with a smooth curve. [If we were not certain of the pattern emerging in Figure 1(a), we could have computed additional points using fractional values for t.]

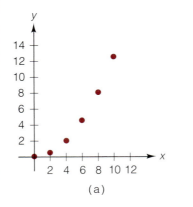

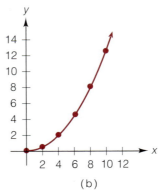

FIGURE I

$x = 2t, \; y = \frac{1}{2}t^2 \quad (t \geq 0)$

(a)   (b)

The curve in Figure 1(b) appears to be a portion of a parabola. We can confirm this as follows. From the equation $x = 2t$, we obtain $t = x/2$. Now we use this result to substitute for t in the equation $y = \frac{1}{2}t^2$. This yields

$$y = \frac{1}{2}\left(\frac{x}{2}\right)^2 = \frac{1}{8}x^2$$

We conclude from this that the curve in Figure 1(b) is a portion of the parabola $y = \frac{1}{8}t^2$. The equations $x = 2t$ and $y = \frac{1}{2}t^2$ (with the restriction $t \geq 0$) are the **parametric equations** for the curve. The variable t is called the **parameter**. In our initial example, the parameter t represented time. In other cases the parameter might have an interpretation as a slope of a certain line, or as an angle. And in still other cases, there may be no immediate physical interpretation for the parameter.

Before looking at additional examples, we point out two advantages in using parametric equations to describe curves. First, by restricting the values of the parameter (as we did in our initial example), we can focus on specific portions of a curve. Second, parametric equations let us think of a curve as a path traced out by a moving point; as the parameter t increases, a definite direction of motion is established.

In physics, parametric equations are often used to describe the motion of an object. Suppose, for example, that a ball is thrown from a height of 6 feet, with an initial speed of 88 ft/s and at an angle of 35° with the horizontal, as shown in Figure 2. Then (neglecting air resistance and spin), it can be shown that the parametric equations for the path of the ball are

$$x = (88 \cos 35°)t \tag{1}$$

$$y = 6 + (88 \sin 35°)t - 16t^2 \tag{2}$$

In these equations, x and y are measured in feet, and t is in seconds, with $t = 0$ corresponding to the instant the ball is thrown.

FIGURE 2

The path of a ball thrown from a height of 6 ft, with an initial speed of 88 ft/s, at an angle of 35° with the horizontal.

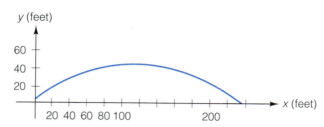

With equations (1) and (2) we can calculate the location of the ball at any time t. For instance, to determine the location when $t = 1$ second, we substitute $t = 1$ in the equations to obtain

$$x = (88 \cos 35°)(1) \approx 72.1 \text{ ft}$$

and

$$y = 6 + (88 \sin 35°)(1) - 16(1)^2 \approx 40.5 \text{ ft}$$

So, after one second, the ball has traveled a horizontal distance of approximately 72.1 ft, and the height of the ball is approximately 40.5 ft. In Figure 3 we display this information along with the results for similar calculations corresponding to $t = 2$ and $t = 3$ seconds. [You should use equations (1) and (2) and your calculator to check the results in Figure 3 for yourself.]

FIGURE 3

The location of the ball after 1, 2, and 3 seconds.

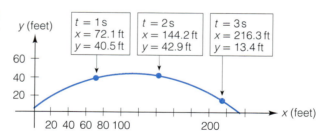

From Figure 3, we can see that the total horizontal distance traveled by the ball in flight is something between 220 ft and 240 ft. Using the parametric equations for the path of the ball, we can determine this distance exactly. We can also find out how long the ball is in the air. When the ball does hit the ground, its y-coordinate is zero. Replacing y by zero in equation (2), we have

$$0 = 6 + (88 \sin 35°)t - 16t^2$$

This is a quadratic equation in the variable t. To solve for t, we use the quadratic formula:

$$t = \frac{-b \pm \sqrt{b^2 - 4ac}}{2a}$$

with the following values for a, b, and c:

$$a = -16 \qquad b = 88 \sin 35° \qquad c = 6$$

As Exercise 19 asks you to check, the results are $t \approx 3.27$ and $t \approx -0.11$. We discard the negative root here because in Figure 3, the values of t are nonnegative. Now, using $t = 3.27$, we can compute x:

$$x = (88 \cos 35°)t \approx (88 \cos 35°)(3.27) \approx 235.7 \text{ ft}$$

In summary, the ball is in the air for about 3.3 seconds, and the total horizontal distance is approximately 236 feet.

Just as in Figure 1, the curve in Figures 2 and 3 is a parabola. Exercise 20 asks you to verify this by solving equation (1) for t in terms of x and then substituting the result in equation (2). This process of obtaining an explicit equation relating x and y from the parametric equations is referred to as *eliminating the parameter*. The next example shows a technique that is often useful in this context. As background for this example, we point out that the graph of an equation of the form $\dfrac{x^2}{a^2} + \dfrac{y^2}{b^2} = 1$ is an *ellipse*, as indicated in Figure 4. (We will study this curve in detail in Chapter 12.)

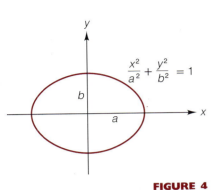

FIGURE 4

The ellipse $(x^2/a^2) + (y^2/b^2) = 1$ is symmetric about both coordinate axes. The intercepts of the ellipse are $x = \pm a$ and $y = \pm b$.

EXAMPLE 1   The parametric equations of an ellipse are

$$x = 6 \cos t \qquad \text{and} \qquad y = 3 \sin t \qquad (0 \le t \le 2\pi)$$

(a) Eliminate the parameter t to obtain an x-y equation for the curve.
(b) Graph the ellipse and indicate the points corresponding to $t = 0$, $\pi/2$, π, $3\pi/2$, and 2π. As t increases, what is the direction of travel along the curve?

Solution   **(a)**   So that we can apply the identity $\cos^2 t + \sin^2 t = 1$, we divide the first equation by 6 and the second by 3. This gives us

$$\frac{x}{6} = \cos t \qquad \text{and} \qquad \frac{y}{3} = \sin t$$

Squaring and then adding these two equations yields

$$\left(\frac{x}{6}\right)^2 + \left(\frac{y}{3}\right)^2 = \cos^2 t + \sin^2 t = 1$$

or

$$\frac{x^2}{6^2} + \frac{y^2}{3^2} = 1$$

This is the x-y equation for the ellipse.

(b) Figure 5 (on the next page) shows the graph of the ellipse $(x^2/6^2) + (y^2/3^2) = 1$. When $t = 0$, the parametric equations yield

$$x = 6 \cos 0 = 6 \qquad \text{and} \qquad y = 3 \sin 0 = 0$$

This, with $t = 0$, we obtain the point $(6, 0)$ on the ellipse, as indicated in Figure 5. The points corresponding to $t = \pi/2$, π, $3\pi/2$, and 2π are obtained similarly. Note that as t increases, the direction of travel around the ellipse is counterclockwise, and that when $t = 2\pi$, we are back to the starting point $(6, 0)$.

FIGURE 5

Parametric equations for this ellipse are $x = 6 \cos t$ and $y = 3 \sin t$. The x-y equation is $(x^2/6^2) + (y^2/3^2) = 1$.

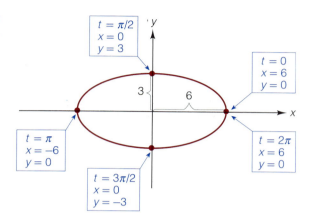

The parametric equations $x = 6 \cos t$ and $y = 3 \sin t$ that we graphed in Figure 5 are by no means the only parametric equations yielding that ellipse. Indeed, there are numerous parametric equations that give rise to the same ellipse. One such pair is

$$x = 6 \sin t \quad \text{and} \quad y = 3 \cos t \quad (0 \le t \le 2\pi)$$

As you can check for yourself, these equations lead to the same x-y equation as before, namely, $(x^2/6^2) + (y^2/3^2) = 1$. The difference now is that in tracing out the curve from the parametric equations, we start (when $t = 0$) with the point $(0, 3)$ on the ellipse and we travel clockwise rather than counterclockwise. Similarly, the equations

$$x = 6 \sin 2t \quad \text{and} \quad y = 3 \cos 2t$$

also produce the ellipse. In this case, as you can check for yourself, as t runs from 0 to 2π, we make two complete trips around the ellipse. If we think of t as time, then the parametric equations $x = 6 \sin 2t$ and $y = 3 \cos 2t$ describe a point traveling around the ellipse $(x^2/6^2) + (y^2/3^2) = 1$ twice as fast as would be the case with the equations $x = 6 \sin t$ and $y = 3 \cos t$.

In Example 1 we saw that the parametric equations $x = 6 \cos t$ and $y = 3 \sin t$ represent an ellipse. The same method we used in Example 1 can be used to establish the following general results.

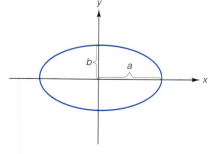

$x = a \cos t$ and $y = b \sin t$
or
$x = a \sin t$ and $y = b \cos t$

FIGURE 6

Parametric Equations for the Ellipse

Let a and b be positive constants. Then the parametric equations

$$x = a \cos t \quad \text{and} \quad y = b \sin t \tag{3}$$

represent the ellipse shown in Figure 6. As t increases, we move around the ellipse in a counterclockwise direction. The parametric equations

$$x = a \sin t \quad \text{and} \quad y = b \cos t$$

also represent the ellipse in Figure 6. In this case, the direction of motion is clockwise.

There is a particular case involving the parametric equations $x = a \cos t$ and $y = b \sin t$ that deserves mention. When a and b are equal, we have $x = a \cos t$ and $y = a \sin t$, from which we deduce that

$$x^2 + y^2 = a^2 \cos^2 t + a^2 \sin^2 t$$
$$= a^2(\cos^2 t + \sin^2 t) = a^2$$

Thus, with $a = b$, the parametric equations describe a circle of radius a. And specializing further still, if $a = b = 1$, we obtain $x = \cos t$ and $y = \sin t$ as parametric equations for the unit circle. This agrees with our unit circle definitions for sine and cosine back in Section 7.4.

EXAMPLE 2 The position of a point $P(x, y)$ at time t is given by the parametric equations

$$x = 1 + 2t \quad \text{and} \quad y = 2 + 4t \quad (t \geq 0)$$

Find the x-y equation for the path traced out by the point P.

Solution From the parametric equation for x, we obtain $t = (x - 1)/2$. Using this to substitute for t in the second parametric equation, we have

$$y = 2 + 4\left(\frac{x - 1}{2}\right) = 2 + 2(x - 1) = 2x$$

This tells us that the point P moves along the line $y = 2x$. However, the entire line is not traced out, but only a portion of it. This is because of the original restriction $t \geq 0$. To see which portion of the line is described by the equations, we can successively let $t = 0$, 1, and 2 in the parametric equations. As you can check, the points obtained are $(1, 2)$, $(3, 6)$, and $(5, 10)$. So, as t increases, we move to the right along the line $y = 2x$, starting from the point $(1, 2)$. See Figure 7.

One of the recurrent techniques that we have used for analyzing parametric equations in this section has been that of eliminating the parameter. It is important to note, however, that it is not always a simple matter to eliminate the parameter. And indeed, in some cases it may not even be possible. When this occurs we have several techniques available. We can set up a table with t, x, and y and plot points; we can use the techniques of calculus; or we can use a graphing calculator or computer to obtain the graph. Even when we use a graphing calculator, however, the techniques of calculus are helpful in understanding why the graph looks as it does.

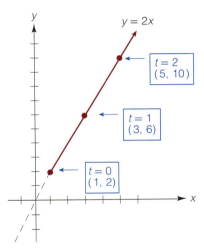

FIGURE 7

The parametric equations $x = 1 + 2t$ and $y = 2 + 4t$, with the restriction $t \geq 0$, describe the portion of the line $y = 2x$ to the right of and including the point $(1, 2)$.

EXERCISE SET 9.4

A

In Exercises 1–6, you are given the parametric equations of a curve and a value for the parameter t. Find the coordinates of the point on the curve corresponding to the given value of t.

1. $x = 2 - 4t$, $y = 3 - 5t$; $t = 0$

2. $x = 3 - t^2$, $y = 4 + t^3$; $t = -1$

3. $x = 5 \cos t$, $y = 2 \sin t$; $t = \pi/6$

4. $x = 4 \cos 2t$, $y = 6 \sin 2t$; $t = \pi/3$

5. $x = 3 \sin^3 t$, $y = 3 \cos^3 t$; $t = \pi/4$

6. $x = \sin t - \sin 2t$, $y = \cos t + \cos 2t$; $t = 2\pi/3$

In Exercises 7–16, graph the parametric equations after eliminating the parameter t. Specify the direction on the curve corresponding to increasing values of t.

7. $x = t + 1,\ y = t^2$

8. $x = 2t - 1,\ y = t^2 - 1$

9. $x = t^2 - 1,\ y = t + 1$

10. $x = t - 4,\ y = |t|$

11. $x = 5 \cos t,\ y = 2 \sin t$

12. $x = 2 \sin t,\ y = 3 \cos t$

13. $x = 4 \cos 2t,\ y = 6 \sin 2t$

14. $x = 2 \cos(t/2),\ y = \sin(t/2)$

15. (a) $x = 2 \cos t,\ y = 2 \sin t$
    (b) $x = 4 \cos t,\ y = 2 \sin t$

16. (a) $x = 3 \sin t,\ y = 3 \cos t$
    (b) $x = 5 \sin t,\ y = 3 \cos t$

17. The accompanying figure shows the parametric equations and the path for a ball thrown from a height of 5 ft, with an initial speed of 100 ft/s and at an angle of 70° with the horizontal.
    (a) Compute the *x*- and *y*-coordinates of the ball when $t = 1, 2,$ and 3 seconds. (Round off the answers to one decimal place.)
    (b) How long is the ball in flight? (Round off the answer to two decimal places.) What is the total horizontal distance traveled by the ball before it lands? (Round off to the nearest foot.) Check that your answer is consistent with the figure.

18. The figure at the bottom of this page shows the parametric equations and the path for a ball thrown from a height of 5 ft, with an initial speed of 100 ft/s and at an angle of 45° with the horizontal. (So except for the initial angle, the data is the same as in Exercise 17.)
    (a) Compute the *x*- and *y*-coordinates of the ball when $t = 1, 2,$ and 3 seconds. (Round off the answers to one decimal place.)
    (b) How long is the ball in flight? (Round off the answer to two decimal places.) What is the total horizontal distance traveled by the ball before it lands? (Round off to the nearest foot.)

19. In the text we said that the solutions of the quadratic equation $0 = 6 + (88 \sin 35°)t - 16t^2$ are $t \approx 3.27$ and $t \approx -0.11$. Use the quadratic formula and your calculator to verify these results.

20. Refer to parametric equations (1) and (2) in the text. By eliminating the parameter *t*, show that the equations describe a parabola. (That is, show that the resulting *x*-*y* equation has the form $y = ax^2 + bx + c$.)

B

21. The curve in the accompanying figure is called an *astroid*. (It is also known as a *hypocycloid of four cusps*.) A pair of parametric equations for the curve is $x = \cos^3 t,\ y = \sin^3 t$. By eliminating the parameter *t*, find the *x*-*y* equation for the curve.   *Hint:*   In each equation, raise both sides to the two-thirds power.

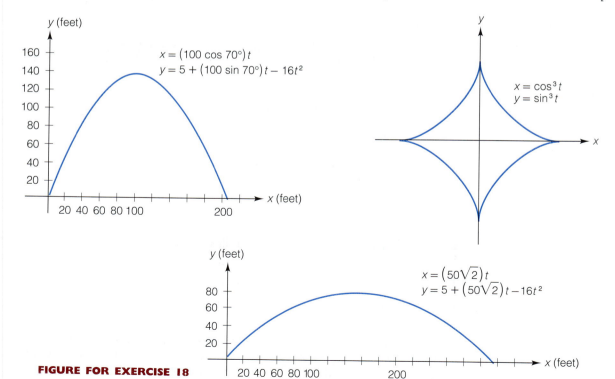

$x = \cos^3 t$
$y = \sin^3 t$

$x = (100 \cos 70°)\,t$
$y = 5 + (100 \sin 70°)\,t - 16t^2$

$x = (50\sqrt{2})\,t$
$y = 5 + (50\sqrt{2})\,t - 16t^2$

FIGURE FOR EXERCISE 18

22. The curve in the accompanying figure is called the *Folium of Descartes*. The *x-y* equation for the curve is $x^3 + y^3 = 3xy$. By substituting $y = xt$ in the equation of the Folium, obtain the following parametric equations for the curve:

$$x = \frac{3t}{1 + t^3} \quad \text{and} \quad y = \frac{3t^2}{1 + t^3}$$

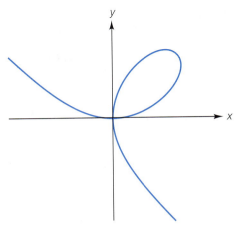

The Folium of Descartes: $x^3 + y^3 = 3xy$

9.5 INTRODUCTION TO POLAR COORDINATES

[Jakob Bernoulli (1654–1705)] *was one of the first to use polar coordinates in a general manner, and not simply for spiral shaped curves.*

Florian Cajori in *A History of Mathematics,* fourth edition (New York: Chelsea Publishing Company, 1985)

FIGURE I
This polar coordinate graph was discovered by Henri Berger, a student in one of the author's mathematics classes at U.C.L.A. in Spring of 1988.

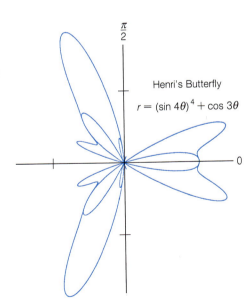

Henri's Butterfly

$r = (\sin 4\theta)^4 + \cos 3\theta$

Up until now, we have always specified the location of a point in the plane by means of a rectangular coordinate system. In this section, we introduce another coordinate system that can be used to locate points in the plane. This is the sys-

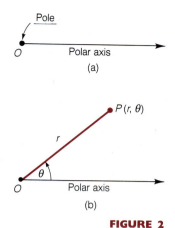

FIGURE 2

tem of **polar coordinates**. We begin by drawing a half-line or ray emanating from a fixed point O. The fixed point O is called the **pole** or **origin**, and the half-line is called the **polar axis**. As a matter of convention, the polar axis is usually depicted as being horizontal and extending to the right, as indicated in Figure 2(a). Now let P be any point in the plane. As indicated in Figure 2(b), we initially let r denote the distance from O to P and we let θ denote the angle measured from the polar axis counterclockwise to $\overline{OP}$. Then the ordered pair $(r, θ)$ serves to locate the point P with respect to the pole and the polar axis. We refer to r and θ as the polar coordinates of P, and we write $P(r, θ)$ to indicate that P is the point with polar coordinates $(r, θ)$.

Plotting points in polar coordinates is facilitated by the use of polar coordinate graph paper, such as that shown in Figure 3(a). Figure 3(b) shows the points with polar coordinates $A(2, π/6)$, $B(3, 2π/3)$, and $C(4, 0)$.

FIGURE 3

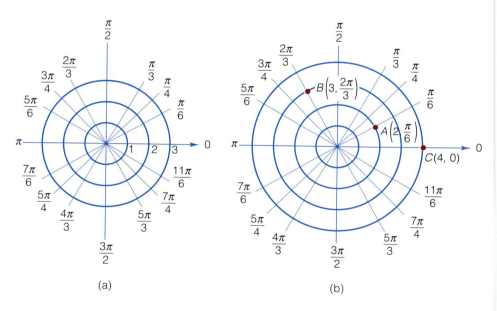

(a)   (b)

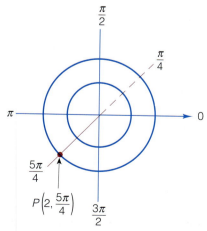

FIGURE 4

There is a minor complication that arises in using polar rather than rectangular coordinates. Consider, for example, the point $C(4, 0)$ in Figure 3(b). This point could just as well have been labeled with the coordinates $(4, 2π)$, or $(4, 2kπ)$ for any integral value of k. Similarly, the coordinates $(r, θ)$ and $(r, θ + 2kπ)$ represent the same point for all integral values of k. This is in marked contrast to the situation with rectangular coordinates, where the coordinate representation of each point is unique. Also, what polar coordinates should we assign to the origin? For if $r = 0$ in Figure 2(b), we cannot really define an angle θ. The convention agreed on to cover this case is that the coordinates $(0, θ)$ denote the origin for all values of θ. Finally, we point out that in working with polar coordinates, it is sometimes useful to let r take on negative values. For example, consider the point $P(2, 5π/4)$ in Figure 4. The coordinates $(2, 5π/4)$ indicate that to reach P from the origin, we go 2 units in the direction

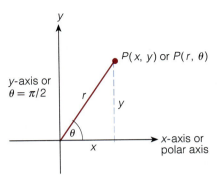

FIGURE 5

$5\pi/4$. Alternately, we can describe this as -2 units in the $\pi/4$ direction. (Refer again to Figure 4.) For reasons such as this, we will adhere to the convention that the polar coordinates (r, θ) and $(-r, \theta + \pi)$ represent the same point. (Since r can now, in fact, be positive, negative, or zero, r is referred to as a *directed distance*.)

It is often useful to consider simultaneously both rectangular and polar coordinates. To do this, we draw the two coordinate systems so that the origins coincide and the positive x-axis coincides with the polar axis (see Figure 5). Suppose now that a point P, other than the origin, has rectangular coordinates (x, y) and polar coordinates (r, θ), as indicated in Figure 5. We wish to find equations relating the two sets of coordinates. From Figure 5, we see that

$$x^2 + y^2 = r^2 \qquad \sin \theta = \frac{y}{r} \qquad \cos \theta = \frac{x}{r} \qquad \tan \theta = \frac{y}{x}$$

Although Figure 5 displays the point P in the first quadrant, it can be shown that these same equations hold when P is in any quadrant. For reference we summarize these equations as follows.

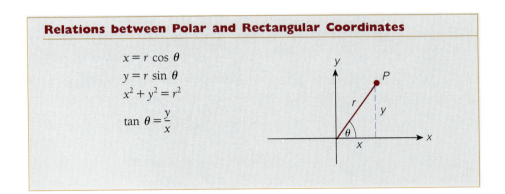

Relations between Polar and Rectangular Coordinates

$$x = r \cos \theta$$
$$y = r \sin \theta$$
$$x^2 + y^2 = r^2$$
$$\tan \theta = \frac{y}{x}$$

EXAMPLE I   The polar coordinates of a point are $(5, \pi/6)$. What are the rectangular coordinates?

Solution   We are given that $r = 5$ and $\theta = \pi/6$. Thus,

$$x = r \cos \theta = 5 \cos \frac{\pi}{6} = 5\left(\frac{\sqrt{3}}{2}\right)$$

and

$$y = r \sin \theta = 5 \sin \frac{\pi}{6} = 5\left(\frac{1}{2}\right)$$

The rectangular coordinates are therefore $\left(\frac{5}{2}\sqrt{3}, \frac{5}{2}\right)$.

The definition of the graph of an equation in polar coordinates is similar to the corresponding definition for rectangular coordinates. The **graph** of an equation in polar coordinates is the set of all points (r, θ) with coordinates that satisfy the given equation. It is often the case that the equation of a curve is simpler in one coordinate system than in another. The next two examples show instances of this.

EXAMPLE 2   Convert each polar equation to rectangular form.

(a) $r = \cos\theta + 2\sin\theta$ (b) $r^2 = \sin 2\theta$

Solution   (a) In view of the transformation equations $x = r\cos\theta$ and $y = r\sin\theta$, we multiply both sides of the given equation by r to obtain

$$r^2 = r\cos\theta + 2r\sin\theta$$

and therefore

$$x^2 + y^2 = x + 2y \qquad \text{or} \qquad x^2 - x + y^2 - 2y = 0$$

This is the rectangular form of the given equation.
   Question for review: What is the graph of this last equation?

(b) Using the double-angle formula for $\sin 2\theta$, we have

$$r^2 = 2\sin\theta\cos\theta$$

Now, in order to obtain the expressions $r\sin\theta$ and $r\cos\theta$ on the right-hand side of the equation, we multiply both sides by r^2. This yields

$$r^4 = 2(r\sin\theta)(r\cos\theta)$$

and, consequently,

$$(x^2 + y^2)^2 = 2yx$$

or

$$x^4 + 2x^2y^2 + y^4 - 2xy = 0$$

This is the rectangular form of the given equation. Notice how much simpler the equation is in its polar coordinate form.   ■■■

EXAMPLE 3   Convert the rectangular equation $x^2 + y^2 + ax = a\sqrt{x^2 + y^2}$ to polar form, expressing r as a function of θ. Assume that a is a constant.

Solution   Using the relations $x^2 + y^2 = r^2$ and $x = r\cos\theta$, we obtain

$$r^2 + ar\cos\theta = ar$$

Notice that this equation is satisfied by $r = 0$. In other words, the graph of this equation will pass through the origin. This is consistent with the fact that the original equation is satisfied when x and y are both zero. Now assume for the moment that $r \neq 0$. Then we can divide both sides of the last equation by r to obtain

$$r + a\cos\theta = a \qquad \text{or} \qquad r = a - a\cos\theta$$

This expresses r as a function of θ, as required. Notice that when $\theta = 0$, we obtain

$$r = a - a(1) = 0$$

That is, nothing has been lost in dividing through by r; the graph will still pass through the origin. ▪▪▪

The three examples that we've just completed dealt with the algebraic aspects of polar coordinates. Now let us take a more geometric approach. We are going to graph curves defined by polar equations. As a first example, consider the equation

$$r = 2 \cos \theta$$

We need to keep in mind at this stage that r and θ are polar, not rectangular, coordinates. Thus, we should not expect the graph of $r = 2 \cos \theta$ to be the familiar cosine wave.

First we set up a table using convenient values of θ, as shown in Table 1.

TABLE 1

Values for $r = 2 \cos \theta$

θ	0	$\dfrac{\pi}{6}$	$\dfrac{\pi}{4}$	$\dfrac{\pi}{3}$	$\dfrac{\pi}{2}$	$\dfrac{2\pi}{3}$	$\dfrac{3\pi}{4}$	$\dfrac{5\pi}{4}$	π
$r = 2 \cos \theta$	2	$\sqrt{3} \approx 1.7$	$\sqrt{2} \approx 1.4$	1	0	-1	$-\sqrt{2} \approx -1.4$	$-\sqrt{3} \approx -1.7$	-2

As you can check, values of θ beyond π in this case merely lead to points already listed. For instance, with $\theta = 4\pi/3$, we have

$$r = 2 \cos \frac{4\pi}{3} = 2\left(-\frac{1}{2}\right) = -1$$

However, the point $(-1, 4\pi/3)$ is the same point as $(1, \pi/3)$, according to our convention regarding negative values of r. In Figure 6, we have plotted the points obtained in Table 1 and connected them with a smooth curve. (The curve is, in fact, a circle, as we will show.)

FIGURE 6
$r = 2 \cos \theta$

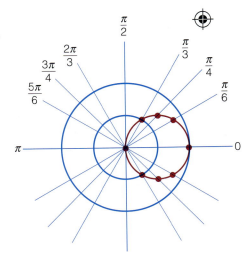

As a check on our work in the example we just completed, we can convert the equation $r = 2 \cos \theta$ to rectangular form. Multiplying both sides by r, we have

$$r^2 = 2r \cos \theta$$

and, consequently,

$$x^2 + y^2 = 2x$$

or

$$x^2 - 2x + y^2 = 0$$

Completing the square now, we find that

$$x^2 - 2x + 1 + y^2 = 1 \qquad \text{or} \qquad (x - 1)^2 + y^2 = 1$$

This represents a circle with center at $(1, 0)$ and with radius 1, in agreement with Figure 6.

EXAMPLE 4 Graph the equation $r = 2 + 2 \cos \theta$.

Solution Again we need to keep in mind that we are using polar rather than rectangular coordinates. Thus, even though we have just seen that $r = 2 \cos \theta$ represents a circle, we cannot expect the graph of $r = 2 + 2 \cos \theta$ to be simply a translate of the circle. As before, we begin by setting up a table. This time, however, we find that values of θ beyond π yield new points on the graph. Thus, we continue the table to $\theta = 2\pi$. In Figure 7, we have plotted the points obtained from Table 2

TABLE 2

Values for $r = 2 + 2 \cos \theta$

θ	0	$\dfrac{\pi}{6}$	$\dfrac{\pi}{3}$	$\dfrac{\pi}{2}$	$\dfrac{2\pi}{3}$	$\dfrac{5\pi}{6}$	π	$\dfrac{7\pi}{6}$	$\dfrac{4\pi}{3}$	$\dfrac{3\pi}{2}$	$\dfrac{5\pi}{3}$	$\dfrac{11\pi}{6}$	2π
$r = 2 + 2 \cos \theta$	4	3.7	3	2	1	0.3	0	0.3	1	2	3	3.7	4

FIGURE 7
$r = 2 + 2 \cos \theta$

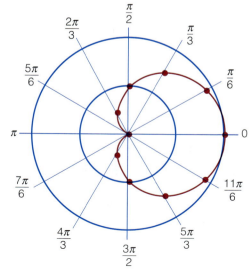

and connected them with a smooth curve. The resulting heart-shaped curve is known as a **cardioid**.

We saw in Chapter 3 that symmetry considerations can often be used to lessen the amount of work involved in graphing equations. This is also true for polar equations. In the box that follows, we list four tests for symmetry in polar coordinates.

Symmetry Tests in Polar Coordinates

If the following substitution yields an equivalent equation:	then the graph is symmetric about:
1. replacing θ with $-\theta$,	the x-axis;
2. replacing r and θ with $-r$ and $-\theta$, respectively,	the y-axis;
3. replacing θ with $\pi - \theta$,	the y-axis;
4. replacing r with $-r$,	the origin.

The validity of the first test follows from the fact that the points (r, θ) and $(r, -\theta)$ are reflections of each other about the x-axis [see Figure 8(a)]. (This first test could have been used to reduce the labor involved in graphing the cardioid $r = 2 + 2 \cos \theta$ in Figure 7.) The validity of test 2 follows from the fact that the points (r, θ) and $(-r, -\theta)$ are reflections of each other about the y-axis, as indicated in Figure 8(b). The other tests can be justified in a similar manner.

FIGURE 8

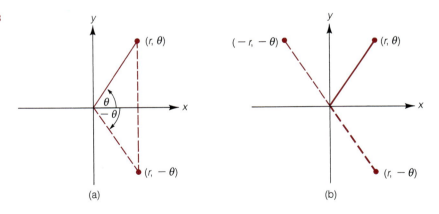

(a)   (b)

EXAMPLE 5   Graph the equation $r^2 = 4 \cos 2\theta$.

Solution   As θ varies from 0 to π, the values of 2θ run from 0 to 2π, and hence $\cos 2\theta$ runs through one complete cycle of values. Thus, in setting up a table to graph this equation, we do not need to consider values of θ beyond π. Furthermore, we do not need to consider values of θ in the interval $\pi/4 < \theta < 3\pi/4$ because $\cos 2\theta$ is negative there, whereas r^2 is always nonnegative. We begin by setting up a table, using a calculator as necessary. See Table 3. If we plot the points in Table 3 and join them with a smooth curve, we obtain the graph in Figure 9(a).

TABLE 3

θ	0	$\pi/12$	$\pi/8$	$\pi/6$	$\pi/4$
$r = \pm 2\sqrt{\cos 2\theta}$	± 2	± 1.86	± 1.68	± 1.41	0

Rather than set up another table with values of θ running from $3\pi/4$ to π, we can rely on symmetry to complete the graph. According to the first two symmetry tests, the curve is symmetric about both the x-axis and the y-axis. Thus, we obtain the graph shown in Figure 9(b). This curve is called a **lemniscate**. ■■■

FIGURE 9

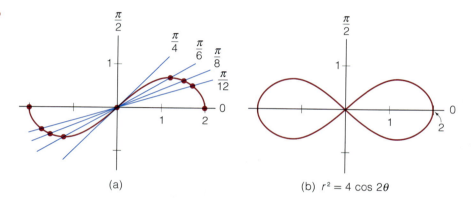

(a)

(b) $r^2 = 4 \cos 2\theta$

All the polar equations we have graphed so far have involved the trigonometric functions. As the next example indicates, this need not always be the case.

EXAMPLE 6 Graph the equation $r = \theta/\pi$ for $\theta \geq 0$.

Solution The equation shows that as θ increases, so does r. We set up a table of values as in the previous examples. Using the entries in Table 4, we construct the graph shown in Figure 10. The curve is known as the **spiral of Archimedes**. ■■■

TABLE 4

θ	0	$\dfrac{\pi}{4}$	$\dfrac{\pi}{2}$	$\dfrac{3\pi}{4}$	π	$\dfrac{5\pi}{4}$	$\dfrac{3\pi}{2}$	$\dfrac{7\pi}{4}$	2π	$\dfrac{5\pi}{2}$	3π	4π
r	0	$\dfrac{1}{4}$	$\dfrac{1}{2}$	$\dfrac{3}{4}$	1	$\dfrac{5}{4}$	$\dfrac{3}{2}$	$\dfrac{7}{4}$	2	$\dfrac{5}{2}$	3	4

FIGURE 10
$r = \theta/\pi$ $(\theta \geq 0)$

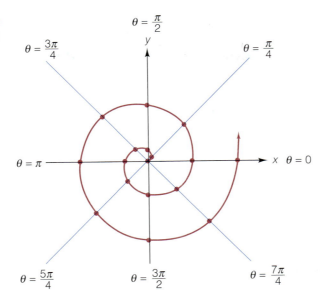

EXERCISE SET 9.5

A

In Exercises 1–3, convert the given polar coordinates to rectangular coordinates.

1. (a) $(3, 2\pi/3)$ (b) $(4, 11\pi/6)$ (c) $(4, -\pi/6)$

2. (a) $(5, \pi/4)$ (b) $(-5, \pi/4)$ (c) $(-5, -\pi/4)$

3. (a) $(1, \pi/2)$ (b) $(1, 5\pi/2)$ (c) $(-1, \pi/8)$

In Exercises 4–6, convert the given rectangular coordinates to polar coordinates. Express your answers in such a way that r is nonnegative and $0 \leq \theta < 2\pi$.

4. $(3, \sqrt{3})$ 5. $(-1, -1)$ 6. $(0, -2)$

In Exercises 7–16, convert the polar equations to rectangular form.

7. $r = 2 \cos \theta$
8. $2 \sin \theta - 3 \cos \theta = r$
9. $r = \tan \theta$
10. $r = 4$
11. $r = 3 \cos 2\theta$
12. $r = 4 \sin 2\theta$
13. $r^2 = 8/(2 - \sin^2 \theta)$
14. $r^2 = 1/(3 + \cos^2 \theta)$
15. $r \cos \left(\theta - \frac{\pi}{6}\right) = 2$
16. $r \sin \left(\theta + \frac{\pi}{4}\right) = 6$

In Exercises 17–24, convert the rectangular equations to polar form.

17. $3x - 4y = 2$
18. $x^2 + y^2 = 25$
19. $y^2 = x^3$
20. $y = x^2$
21. $2xy = 1$
22. $x^2 + 4x + y^2 + 4y = 0$
23. $9x^2 + y^2 = 9$
24. $x^2(x^2 + y^2) = y^2$

25. Notice that the rectangular form of the polar equation $r \cos \theta = a$ is $x = a$. Thus, the graph of $r \cos \theta = a$ is a vertical line with an x-intercept of a. Use this observation to graph the following polar equations.
(a) $r \cos \theta = 3$ (b) $r \cos \theta = -2$

26. Notice that the rectangular form of the polar equation $r \sin \theta = a$ is $y = a$. Thus, the graph of $r \sin \theta = a$ is a horizontal line with a y-intercept of a. Use this observation to graph the following polar equations.
(a) $r \sin \theta = 4$ (b) $r \sin \theta = -1$

In Exercises 27–50, graph the given equations.

27. $r \cos \theta = 5$ *Hint:*   See Exercise 25.
28. $r \sin \theta = \pi$
29. $r = 3 \cos \theta$
30. $r = 2 \sin \theta$
31. $r = 1 - \cos \theta$
32. $r = 1 + \cos \theta$
33. $r^2 = 2 \sin 2\theta$
34. $r^2 = 2 \cos 2\theta$
35. $r = 1$
36. $r = 2\theta/\pi, \ \theta \geq 0$
37. $\theta = 1$
38. $\theta = \pi/6$
39. $r = 4 \sin \theta + 2 \cos \theta$ *Hint:*   First convert to rectangular form.

40. $r(2 \sin \theta + \cos \theta) = 1$ (Use the hint in the previous exercise.)

41. $r = 2 \sin 2\theta$ (four-leafed rose)

42. $r = \sin 3\theta$ (three-leafed rose)

43. $r = 2 + \sin \theta$ (limaçon)

44. $r = 1 + 2 \sin \theta$ (limaçon with an inner loop)

45. $r = 1 - 2 \cos \theta$ (limaçon with an inner loop)

46. $r = 2 + 2 \sin \theta$ (cardioid)

47. $r^2 = 4 \sin 2\theta$ (lemniscate)

48. $r = 8 \tan \theta$ (kappa curve)

49. $r = \csc \theta + 2$ (conchoid of Nicomedes)

50. $r = e^\theta$ (logarithmic spiral)

In Exercises 51 and 52, match one of the graphs with each equation.   *Hint:*   *Compute the values of r when $\theta = 0$, $\pi/2$, π, and $3\pi/2$.*

51. (a) $r = 3 + 3 \sin \theta$ (b) $r = 3 - 3 \sin \theta$
    (c) $r = 3 - 3 \cos \theta$ (d) $r = 3 + 3 \cos \theta$

A 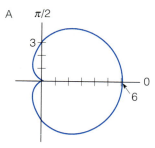   B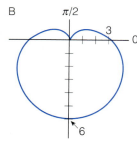

C 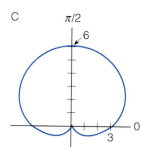   D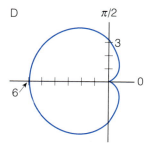

52. (a) $r = 3 + 2 \cos \theta$ (b) $r = 2 + 3 \cos \theta$
    (c) $r = 2 - 3 \cos \theta$ (d) $r = 3 - 2 \cos \theta$

A 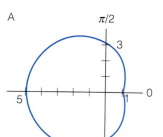   B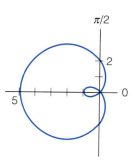

C 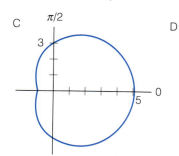   D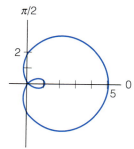

B

53. Show that the distance between the two points $P_1(r_1, \theta_1)$ and $P_2(r_2, \theta_2)$ is given by
$$d = \sqrt{r_1^2 + r_2^2 - 2r_1 r_2 \cos(\theta_1 - \theta_2)}$$

54. By converting the polar equation
$$r = a \cos \theta + b \sin \theta$$
to rectangular form, show that the graph is a circle, and find the center and the radius.

55. Show that the rectangular form of the equation $r = a \sin 3\theta$ is $(x^2 + y^2)^2 = ay(3x^2 - y^2)$.

56. Show that the rectangular form of the equation
$$r = ab/(1 - a \cos \theta) \quad (a < 1)$$
is $(1 - a^2)x^2 + y^2 - 2a^2bx - a^2b^2 = 0$.

57. Show that the polar form of the equation $(x^2/a^2) - (y^2/b^2) = 1$ is $r^2 = a^2b^2/(b^2 \cos^2 \theta - a^2 \sin^2 \theta)$.

58. Let k denote a positive constant, and let F_1 and F_2 denote the points with rectangular coordinates $(-k, 0)$ and $(k, 0)$, respectively. A curve known as the *lemniscate of Bernoulli* is defined as the set of points $P(x, y)$ such that $(F_1P) \times (F_2P) = k^2$.
    (a) Show that the rectangular equation of the curve is $(x^2 + y^2)^2 = 2k^2(x^2 - y^2)$.
    (b) Show that the polar equation is $r^2 = 2k^2 \cos 2\theta$.
    (c) Graph the equation $r^2 = 2k^2 \cos 2\theta$.   *Hint:*   See Example 5 in the text.

9.6 TRIGONOMETRIC FORM FOR COMPLEX NUMBERS

Go to Mr. DeMoivre; he knows these things better than I do.

Isaac Newton [according to Boyer's *A History of Mathematics* (New York: John Wiley, and Sons, Inc., 1968)]

In this section, we explore one of the many important connections between trigonometry and the complex number system. We begin with an observation that was first made in the year 1797 by the Norwegian surveyor and mathematician Caspar Wessel. Wessel realized, essentially, that the complex numbers could be visualized as points in the x-y plane, the complex number $a + bi$ being identified with the point (a, b). In this context, we often refer to the x-y plane as the **complex plane**, and we refer to the complex numbers as points in this plane.

EXAMPLE 1 Plot the point $2 + 3i$ in the complex plane.

Solution The complex number $2 + 3i$ is identified with the point $(2, 3)$. See Figure 1.

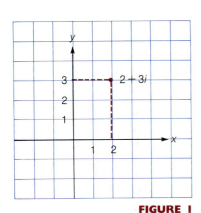

FIGURE 1

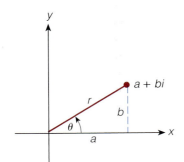

FIGURE 2

As indicated in Figure 2, the distance from the origin to the point $a + bi$ is denoted by r. We call the distance r the **modulus** of the complex number $a + bi$. The angle θ in Figure 2 (measured counterclockwise from the positive x-axis) is referred to as the **argument** of the complex number $a + bi$. (Using the terminology of Section 9.5, r and θ are the **polar coordinates** of the point $a + bi$.)

From Figure 2, we have the following three equations relating the quantities a, b, r, and θ. (Although Figure 2 shows $a + bi$ in the first quadrant, the equations remain valid for the other quadrants as well.)

$$r = \sqrt{a^2 + b^2} \tag{1}$$
$$a = r \cos \theta \tag{2}$$
$$b = r \sin \theta \tag{3}$$

If we have a complex number $z = a + bi$, we can use equations (2) and (3) to write

$$z = (r \cos \theta) + (r \sin \theta)i = r(\cos \theta + i \sin \theta)$$

That is,

$$z = r(\cos \theta + i \sin \theta) \tag{4}$$

The expression that appears on the right-hand side of equation (4) is called the **trigonometric** (or **polar**) **form** of the complex number z. In contrast to this, the expression $a + bi$ is referred to as the **rectangular form** of the complex number z.

EXAMPLE 2   Express the complex number $z = 3\left(\cos \frac{\pi}{3} + i \sin \frac{\pi}{3}\right)$ in rectangular form.

Solution
$$z = 3\left(\cos \tfrac{\pi}{3} + i \sin \tfrac{\pi}{3}\right)$$
$$= 3\left(\tfrac{1}{2} + \tfrac{1}{2}\sqrt{3}i\right) = \tfrac{3}{2} + \tfrac{3}{2}\sqrt{3}i$$

The rectangular form is therefore $\tfrac{3}{2} + \tfrac{3}{2}\sqrt{3}i$.

EXAMPLE 3   Find the trigonometric form for the complex number $-\sqrt{2} + i\sqrt{2}$.

Solution   We are asked to write the given number in the form $r(\cos \theta + i \sin \theta)$, so we need to find r and θ. Using equation (1) and the values $a = -\sqrt{2}$, $b = \sqrt{2}$, we have

$$r = \sqrt{\left(-\sqrt{2}\right)^2 + \left(\sqrt{2}\right)^2} = \sqrt{2 + 2} = \sqrt{4} = 2$$

Now that we know r, we can use equations (2) and (3) to determine θ. From equation (2), we obtain

$$\cos \theta = \frac{a}{r} = \frac{-\sqrt{2}}{2} \tag{5}$$

Similarly, equation (3) gives us

$$\sin \theta = \frac{b}{r} = \frac{\sqrt{2}}{2} \tag{6}$$

One angle satisfying both of equations (5) and (6) is $\theta = 3\pi/4$. (There are other angles, and we'll return to this point in a moment.) In summary, then, we have $r = 2$ and $\theta = 3\pi/4$, so the required trigonometric form is

$$2\left(\cos \tfrac{3\pi}{4} + i \sin \tfrac{3\pi}{4}\right)$$

In the example we just completed, we noted that $\theta = 3\pi/4$ was only one angle satisfying the conditions $\cos \theta = -\sqrt{2}/2$ and $\sin \theta = \sqrt{2}/2$. Another such angle is $\frac{3\pi}{4} + 2\pi$. Indeed, any angle of the form $\frac{3\pi}{4} + 2\pi k$, where k is an integer, would do just as well. The upshot of this is that θ, the argument of a complex number, is not uniquely determined. In Example 3, we followed a common convention in converting to trigonometric form: we picked θ in the interval $0 \le \theta < 2\pi$. Furthermore, although it won't cause us any difficulties in this section, you might also note that the argument θ is undefined for the complex number $0 + 0i$. (Why?)

We are now ready to derive a formula that will make it easy to multiply two complex numbers in trigonometric form. Suppose that the two complex numbers are

$$r(\cos \alpha + i \sin \alpha) \qquad \text{and} \qquad R(\cos \beta + i \sin \beta)$$

Then their product is

$$rR[(\cos \alpha + i \sin \alpha)(\cos \beta + i \sin \beta)]$$
$$= rR[(\cos \alpha \cos \beta - \sin \alpha \sin \beta) + i(\sin \alpha \cos \beta + \cos \alpha \sin \beta)]$$
$$= rR[\cos(\alpha + \beta) + i \sin(\alpha + \beta)] \qquad \text{using the addition formulas from Section 8.6}$$

Notice that the modulus of the product is rR, which is the product of the two original moduli. Also, the argument is $\alpha + \beta$, which is the *sum* of the two original arguments. So, to multiply two complex numbers, just multiply the moduli and add the arguments. There is a similar rule for obtaining the quotient of two complex numbers: Divide their moduli and subtract the arguments. These two rules are stated more precisely in the box that follows. (For a proof of the division rule, see Exercise 75.)

Let $z = r(\cos \alpha + i \sin \alpha)$ and $w = R(\cos \beta + i \sin \beta)$. Then

$$zw = rR[\cos(\alpha + \beta) + i \sin(\alpha + \beta)]$$

Also, if $R \neq 0$, then

$$\frac{z}{w} = \frac{r}{R}[\cos(\alpha - \beta) + i \sin(\alpha - \beta)]$$

EXAMPLE 4

Let $z = 8\left(\cos \frac{5\pi}{3} + i \sin \frac{5\pi}{3}\right)$ and $w = 4\left(\cos \frac{2\pi}{3} + i \sin \frac{2\pi}{3}\right)$. Compute **(a)** zw; **(b)** z/w. Express each answer in both trigonometric and rectangular form.

Solution

(a) $zw = (8)(4)\left[\cos\left(\frac{5\pi}{3} + \frac{2\pi}{3}\right) + i \sin\left(\frac{5\pi}{3} + \frac{2\pi}{3}\right)\right]$

$= 32\left(\cos \frac{7\pi}{3} + i \sin \frac{7\pi}{3}\right) = 32\left(\cos \frac{\pi}{3} + i \sin \frac{\pi}{3}\right)$ trigonometric form

$= 32\left(\frac{1}{2} + \frac{1}{2}\sqrt{3}i\right) = 16 + 16\sqrt{3}i$ rectangular form

(b) $\dfrac{z}{w} = \dfrac{8}{4}\left[\cos\left(\frac{5\pi}{3} - \frac{2\pi}{3}\right) + i \sin\left(\frac{5\pi}{3} - \frac{2\pi}{3}\right)\right]$

$= 2(\cos \pi + i \sin \pi)$ trigonometric form

$= 2(-1 + i \cdot 0) = -2$ rectangular form

EXAMPLE 5

Compute z^2, where $z = r(\cos \theta + i \sin \theta)$.

Solution

$z^2 = [r(\cos \theta + i \sin \theta)][r(\cos \theta + i \sin \theta)]$

$= r^2[\cos(\theta + \theta) + i \sin(\theta + \theta)]$

$= r^2(\cos 2\theta + i \sin 2\theta)$

The result in Example 5 is a particular case of an important theorem attributed to Abraham DeMoivre (1667–1754). In the box that follows, we state *DeMoivre's theorem*. (The theorem can be proved using mathematical induction, which is discussed in Chapter 13.)

DeMoivre's Theorem

Let n be a natural number. Then

$$[r(\cos \theta + i \sin \theta)]^n = r^n(\cos n\theta + i \sin n\theta)$$

EXAMPLE 6 Use DeMoivre's theorem to compute $\left(-\sqrt{2} + i\sqrt{2}\right)^5$. Express your answer in rectangular form.

Solution In Example 3, we saw that the trigonometric form of $-\sqrt{2} + i\sqrt{2}$ is given by

$$-\sqrt{2} + i\sqrt{2} = 2\left(\cos \tfrac{3\pi}{4} + i \sin \tfrac{3\pi}{4}\right)$$

Therefore,

$$\begin{aligned}
\left(-\sqrt{2} + i\sqrt{2}\right)^5 &= 2^5\left(\cos \tfrac{15\pi}{4} + i \sin \tfrac{15\pi}{4}\right) \\
&= 32\left[\tfrac{1}{2}\sqrt{2} + i\left(-\tfrac{1}{2}\sqrt{2}\right)\right] \\
&= 16\sqrt{2} - 16\sqrt{2}i
\end{aligned}$$

The next two examples show how DeMoivre's theorem is used in computing roots. If n is a natural number and $z^n = w$, then we say that z is an **nth root** of w. The work in the examples also relies on the following observation about equality between nonzero complex numbers in trigonometric form: If $r(\cos \theta + i \sin \theta) = R(\cos A + i \sin A)$, then $r = R$ and $\theta = A + 2\pi k$, where k is an integer.

EXAMPLE 7 Find the cube roots of $8i$.

Solution First we express $8i$ in trigonometric form. As can be seen from Figure 3,

$$8i = 8\left(\cos \tfrac{\pi}{2} + i \sin \tfrac{\pi}{2}\right)$$

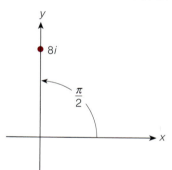

Now we let $z = r(\cos \theta + i \sin \theta)$ denote a cube root of $8i$. Then the equation $z^3 = 8i$ becomes

$$r^3(\cos 3\theta + i \sin 3\theta) = 8\left(\cos \tfrac{\pi}{2} + i \sin \tfrac{\pi}{2}\right) \tag{7}$$

From equation (7), we conclude that $r^3 = 8$ and, consequently, that $r = 2$. Also from equation (7), we have

$$3\theta = \tfrac{\pi}{2} + 2\pi k$$

or

$$\theta = \tfrac{\pi}{6} + \tfrac{2\pi k}{3} \qquad \text{dividing by 3} \tag{8}$$

FIGURE 3

If we let $k = 0$, equation (8) yields $\theta = \pi/6$. Thus, one of the cube roots of $8i$ is

$$2\left(\cos \tfrac{\pi}{6} + i \sin \tfrac{\pi}{6}\right) = 2\left(\tfrac{1}{2}\sqrt{3} + i \cdot \tfrac{1}{2}\right) = \sqrt{3} + i$$

Next we let $k = 1$ in equation (8). As you can check, this yields $\theta = 5\pi/6$. So another cube root of $8i$ is

$$2\left(\cos \tfrac{5\pi}{6} + i \sin \tfrac{5\pi}{6}\right) = 2\left(-\tfrac{1}{2}\sqrt{3} + i \cdot \tfrac{1}{2}\right) = -\sqrt{3} + i$$

Similarly, using $k = 2$ in equation (8) yields $\theta = 3\pi/2$. (Verify this.) Consequently, a third cube root is

$$2\left(\cos \tfrac{3\pi}{2} + i \sin \tfrac{3\pi}{2}\right) = 2[0 + i(-1)] = -2i$$

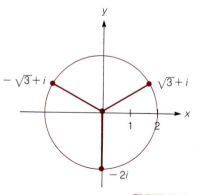

FIGURE 4

We have now found three distinct cube roots of $8i$. If we were to continue the process, using $k = 3$, for example, we would find that no additional roots are obtained in this manner. We therefore conclude that there are exactly three cube roots of $8i$. We have plotted these cube roots in Figure 4. Notice that the points lie equally spaced on a circle of radius 2 about the origin. ∎

When we used DeMoivre's theorem in Example 7, we found that the number $8i$ had three distinct cube roots. Along these same lines, it is true in general that any nonzero number $a + bi$ possesses exactly n distinct nth roots. This is true even if $b = 0$. In the next example, for instance, we compute the five fifth roots of 2.

EXAMPLE 8   Compute the five fifth roots of 2.

Solution   We will follow the procedure we used in Example 7. In trigonometric form, the number 2 becomes $2(\cos 0 + i \sin 0)$. Now we let $z = r(\cos \theta + i \sin \theta)$ denote a fifth root of 2. Then the equation $z^5 = 2$ becomes

$$r^5(\cos 5\theta + i \sin 5\theta) = 2(\cos 0 + i \sin 0) \tag{9}$$

From equation (9), we see that $r^5 = 2$ and, consequently, $r = 2^{1/5}$. Also from equation (9), we have

$$5\theta = 0 + 2\pi k \qquad \text{or} \qquad \theta = \tfrac{2\pi k}{5}$$

Using the values $k = 0$, 1, 2, 3, and 4 in succession, we obtain the following results:

k	0	1	2	3	4
θ	0	$2\pi/5$	$4\pi/5$	$6\pi/5$	$8\pi/5$

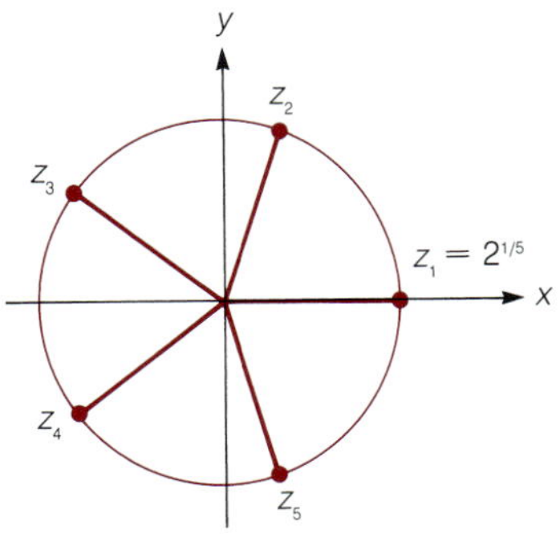

The five fifth roots of 2 are therefore

$$z_1 = 2^{1/5}(\cos 0 + i \sin 0) = 2^{1/5} \qquad z_4 = 2^{1/5}\left(\cos \tfrac{6\pi}{5} + i \sin \tfrac{6\pi}{5}\right)$$
$$z_2 = 2^{1/5}\left(\cos \tfrac{2\pi}{5} + i \sin \tfrac{2\pi}{5}\right) \qquad z_5 = 2^{1/5}\left(\cos \tfrac{8\pi}{5} + i \sin \tfrac{8\pi}{5}\right)$$
$$z_3 = 2^{1/5}\left(\cos \tfrac{4\pi}{5} + i \sin \tfrac{4\pi}{5}\right)$$

In Figure 5, we have plotted these five fifth roots. Notice that the points are equally spaced on a circle of radius $2^{1/5}$ about the origin. ∎

FIGURE 5

EXERCISE SET 9.6

A

In Exercises 1–8, plot each point in the complex plane.

1. $4 + 2i$ 2. $4 - 2i$ 3. $-5 + i$
4. $-3 - 5i$ 5. $1 - 4i$ 6. i
7. $-i$ 8. $1(= 1 + 0i)$

In Exercises 9–18, convert each complex number to rectangular form.

9. $2\left(\cos \tfrac{1}{4}\pi + i \sin \tfrac{1}{4}\pi\right)$ 10. $6\left(\cos \tfrac{5}{3}\pi + i \sin \tfrac{5}{3}\pi\right)$
11. $4\left(\cos \tfrac{5}{6}\pi + i \sin \tfrac{5}{6}\pi\right)$ 12. $3\left(\cos \tfrac{3}{2}\pi + i \sin \tfrac{3}{2}\pi\right)$
13. $\sqrt{2}(\cos 225° + i \sin 225°)$ 14. $\tfrac{1}{2}(\cos 240° + i \sin 240°)$
15. $\sqrt{3}\left(\cos \tfrac{1}{2}\pi + i \sin \tfrac{1}{2}\pi\right)$ 16. $5(\cos \pi + i \sin \pi)$
17. $4(\cos 75° + i \sin 75°)$ *Hint:*   Use the addition formulas from Section 8.6 to evaluate $\cos 75°$ and $\sin 75°$.

18. $2\left(\cos \frac{1}{8}\pi + i \sin \frac{1}{8}\pi\right)$ *Hint:* Use the half-angle formulas from Section 8.7 to evaluate $\cos(\pi/8)$ and $\sin(\pi/8)$.

In Exercises 19–28, convert from rectangular to trigonometric form. (In each case, choose an argument θ such that $0 \le \theta < 2\pi$.)

19. $\frac{1}{2}\sqrt{3} + \frac{1}{2}i$

20. $\sqrt{2} + \sqrt{2}i$

21. $-1 + \sqrt{3}i$

22. -4

23. $-2\sqrt{3} - 2i$

24. $-3\sqrt{2} - 3\sqrt{2}i$

25. $-6i$

26. $-4 - 4\sqrt{3}i$

27. $\frac{1}{4}\sqrt{3} - \frac{1}{4}i$

28. 16

In Exercises 29–54, carry out the indicated operations. Express your results in rectangular form for those cases in which the trigonometric functions are readily evaluated without tables or a calculator.

29. $2(\cos 22° + i \sin 22°) \times 3(\cos 38° + i \sin 38°)$

30. $4(\cos 5° + i \sin 5°) \times 6(\cos 130° + i \sin 130°)$

31. $\sqrt{2}\left(\cos \frac{1}{3}\pi + i \sin \frac{1}{3}\pi\right) \times \sqrt{2}\left(\cos \frac{4}{3}\pi + i \sin \frac{4}{3}\pi\right)$

32. $\left(\cos \frac{1}{5}\pi + i \sin \frac{1}{5}\pi\right) \times \left(\cos \frac{1}{20}\pi + i \sin \frac{1}{20}\pi\right)$

33. $3\left(\cos \frac{1}{7}\pi + i \sin \frac{1}{7}\pi\right) \times \sqrt{2}\left(\cos \frac{1}{7}\pi + i \sin \frac{1}{7}\pi\right)$

34. $\sqrt{3}(\cos 3° + i \sin 3°) \times \sqrt{3}(\cos 38° + i \sin 38°)$

35. $6(\cos 50° + i \sin 50°) \div 2(\cos 5° + i \sin 5°)$

36. $\sqrt{3}(\cos 140° + i \sin 140°) \div 3(\cos 5° + i \sin 5°)$

37. $2^{4/3}\left(\cos \frac{5}{12}\pi + i \sin \frac{5}{12}\pi\right) \div 2^{1/3}\left(\cos \frac{1}{4}\pi + i \sin \frac{1}{4}\pi\right)$

38. $\sqrt{6}\left(\cos \frac{16}{9}\pi + i \sin \frac{16}{9}\pi\right) \div \sqrt{2}\left(\cos \frac{1}{9}\pi + i \sin \frac{1}{9}\pi\right)$

39. $\left(\cos \frac{2}{5}\pi + i \sin \frac{2}{5}\pi\right) \div \left(\cos \frac{2}{5}\pi + i \sin \frac{2}{5}\pi\right)$

40. $\left(\cos \frac{2}{5}\pi + i \sin \frac{2}{5}\pi\right) \div \left(\cos \frac{1}{10}\pi + i \sin \frac{1}{10}\pi\right)$

41. $\left[3\left(\cos \frac{1}{3}\pi + i \sin \frac{1}{3}\pi\right)\right]^5$

42. $\left[\sqrt{2}\left(\cos \frac{5}{6}\pi + i \sin \frac{5}{6}\pi\right)\right]^4$

43. $\left[\frac{1}{2}\left(\cos \frac{1}{24}\pi + i \sin \frac{1}{24}\pi\right)\right]^6$

44. $\left[\sqrt{3}(\cos 70° + i \sin 70°)\right]^3$

45. $\left[2^{1/5}(\cos 63° + i \sin 63°)\right]^{10}$

46. $\left[2\left(\cos \frac{1}{5}\pi + i \sin \frac{1}{5}\pi\right)\right]^3$

47. $2(\cos 200° + i \sin 200°) \times \sqrt{2}(\cos 20° + i \sin 20°)$ $\times \frac{1}{2}(\cos 5° + i \sin 5°)$

48. $\left[\dfrac{\cos(\pi/8) + i \sin(\pi/8)}{\cos(-\pi/8) + i \sin(-\pi/8)}\right]^5$

49. $\left[\frac{1}{2}(1 - \sqrt{3}i)\right]^5$ *Hint:* Convert to trigonometric form.

50. $(1 - i)^3$

51. $(-2 - 2i)^5$

52. $\left(-\frac{1}{2} + \frac{1}{2}\sqrt{3}i\right)^6$

53. $(-2\sqrt{3} - 2i)^4$

54. $(1 + i)^{16}$

In Exercises 55–61, use DeMoivre's theorem to find the indicated roots. Express the results in rectangular form.

55. Cube roots of $-27i$

56. Cube roots of 2

57. Eighth roots of 1

58. Square roots of i

59. Cube roots of 64

60. Square roots of $-\frac{1}{2} - \frac{1}{2}\sqrt{3}i$

61. Sixth roots of 729

Use a calculator to complete Exercises 62–65.

62. Compute $(9 + 9i)^6$.

63. Compute $(7 - 7i)^8$.

64. Compute the cube roots of $1 + 2i$. Express your answers in rectangular form, with the real and imaginary parts rounded off to two decimal places.

65. Compute the fifth roots of i. Express your answers in rectangular form, with the real and imaginary parts rounded off to two decimal places.

B

In Exercises 66–68, find the indicated roots. Express the results in rectangular form.

66. Find the fourth roots of i.   *Hint:* Use the half-angle formulas from Section 8.7.

67. Find the fourth roots of $8 - 8\sqrt{3}i$.   *Hint:* Use the addition formulas or the half-angle formulas.

68. Find the square roots of $7 + 24i$.   *Hint:* You'll need to use the half-angle formulas from Section 8.7.

69. (a) Compute the three cube roots of 1.
    (b) Let z_1, z_2, and z_3 denote the three cube roots of 1. Verify that $z_1 + z_2 + z_3 = 0$ and also that $z_1 z_2 + z_2 z_3 + z_3 z_1 = 0$.

70. (a) Compute the four fourth roots of 1.
    (b) Verify that the sum of these four fourth roots is 0.

71. Evaluate $\left(-\frac{1}{2} + \frac{1}{2}\sqrt{3}i\right)^5 + \left(-\frac{1}{2} - \frac{1}{2}\sqrt{3}i\right)^5$.   *Hint:* Use DeMoivre's theorem.

72. Show that $\left(-\frac{1}{2} + \frac{1}{2}\sqrt{3}i\right)^6 + \left(-\frac{1}{2} - \frac{1}{2}\sqrt{3}i\right)^6 = 2$.

73. Compute $(\cos \theta + i \sin \theta)(\cos \theta - i \sin \theta)$.

74. In the identity $(\cos \theta + i \sin \theta)^2 = \cos 2\theta + i \sin 2\theta$, carry out the actual multiplication on the left-hand side of the equation. Then equate the corresponding real parts and the corresponding imaginary parts from each side of the equation that results. What do you obtain?

75. Show that
$$\frac{r(\cos \alpha + i \sin \alpha)}{R(\cos \beta + i \sin \beta)} = \frac{r}{R}[\cos(\alpha - \beta) + i \sin(\alpha - \beta)]$$

Suggestion: Begin with the quantity on the left side and multiply it by $(\cos \beta - i \sin \beta)/(\cos \beta - i \sin \beta)$.

76. Show that
$$1 + \cos \theta + i \sin \theta = 2 \cos\left(\tfrac{\theta}{2}\right)\left[\cos\left(\tfrac{\theta}{2}\right) + i \sin\left(\tfrac{\theta}{2}\right)\right]$$

77. If $z = r(\cos\theta + i\sin\theta)$, show that

$$\frac{1}{z} = \frac{1}{r}(\cos\theta - i\sin\theta)$$

Hint: $1/z = [1(\cos 0 + i\sin 0)]/[r(\cos\theta + i\sin\theta)]$.

78. Show that $\dfrac{1 + \sin\theta + i\cos\theta}{1 + \sin\theta - i\cos\theta} = \sin\theta + i\cos\theta$.

Assume that $\theta \neq \frac{3}{2}\pi + 2\pi k$, where k is an integer.
Hint: Work with the left-hand side; first "rationalize" the denominator by multiplying by the quantity
$$\frac{(1 + \sin\theta) + i\cos\theta}{(1 + \sin\theta) + i\cos\theta}.$$

▌▌ CHAPTER NINE SUMMARY OF PRINCIPAL FORMULAS AND NOTATION

FORMULAS OR NOTATIONS	PAGE REFERENCE	COMMENT
1. $\dfrac{\sin A}{a} = \dfrac{\sin B}{b} = \dfrac{\sin C}{c}$	516	This is the *law of sines.* In words, it states that in any triangle, the sines of the angles are proportional to the lengths of the opposite sides.
2. $a^2 = b^2 + c^2 - 2bc\cos A$	518	This is the *law of cosines.* We can use it to calculate the third side of a triangle when we are given two sides and the included angle. The formula can also be used to calculate the angles of a triangle when we know the three sides, as in Example 5 on page 520.
3. Vector	528 538	Geometrically, a *vector* in the plane is a directed line segment. Vectors can be used to represent quantities that have both magnitude and direction, such as force and velocity. For examples, see Figures 1 and 2 in Section 9.2. Algebraically, a vector in the plane is an ordered pair of real numbers, denoted by $\langle x, y\rangle$. (See Figure 3 in Section 9.3.) The numbers x and y are called the *components* of the vector $\langle x, y\rangle$.
4. Vector equality	529 538	Geometrically, two vectors are said to be equal provided they have the same length and the same direction. In terms of components, this means that two vectors are equal if and only if their corresponding components are equal.
5. Vector addition	530 531 540	Geometrically, two vectors can be added by using the parallelogram law, as in Figure 10 in Section 9.2. Algebraically, this is equivalent to the following componentwise formula for vector addition: $$\langle a, b\rangle + \langle c, d\rangle = \langle a + c, b + d\rangle$$
6. $\lvert\mathbf{v}\rvert = \sqrt{v_1^2 + v_2^2}$	529 539	The expression on the left-hand side of the equation denotes the *length* or *magnitude* of the vector $\mathbf{v}$. The expression on the right-hand side of the equation tells us how to compute the length of $\mathbf{v}$ in terms of its respective x- and y-components, v_1 and v_2.
7. $\overrightarrow{PQ} = \langle x_2 - x_1, y_2 - y_1\rangle$	539	This formula gives the components of the vector $\overrightarrow{PQ}$, where P and Q are the points $P(x_1, y_1)$ and $Q(x_2, y_2)$. See Figure 5 in Section 9.3 for a derivation of this formula.
8. The zero vector	541	The *zero vector*, denoted by $\mathbf{0}$, is the vector $\langle 0, 0\rangle$. The zero vector plays the same role in vector addition as does the real number zero in addition of real numbers. For any vector $\mathbf{v}$, we have $$\mathbf{0} + \mathbf{v} = \mathbf{v} + \mathbf{0} = \mathbf{v}$$

FORMULAS OR NOTATIONS	PAGE REFERENCE	COMMENT
9. Scalar multiplication	541	For each real number k and each vector $\mathbf{v} = \langle x, y \rangle$, the vector $k\mathbf{v}$ is defined by the equation $$k\mathbf{v} = \langle kx, ky \rangle$$ Geometrically, the length of $k\mathbf{v}$ is $\lvert k \rvert$ times the length of $\mathbf{v}$. The operation that forms the vector $k\mathbf{v}$ from the scalar k and the vector $\mathbf{v}$ is called *scalar multiplication*.
10. $\mathbf{i} = \langle 1, 0 \rangle$ $\quad\;\; \mathbf{j} = \langle 0, 1 \rangle$	543	These two equations define the *unit vectors* $\mathbf{i}$ and $\mathbf{j}$. See Figure 11 in Section 9.3. Any vector $\langle x, y \rangle$ can be expressed in terms of the unit vectors $\mathbf{i}$ and $\mathbf{j}$ as follows: $$\langle x, y \rangle = x\mathbf{i} + y\mathbf{j}$$
11. Parametric equations	546 547	Suppose the x- and y-coordinates of the points on a curve are expressed as functions of another variable t: $$x = f(t) \quad \text{and} \quad y = g(t)$$ These equations are parametric equations of the curve, and t is called a *parameter*. An example: A pair of parametric equations for the ellipse $(x^2/a^2) + (y^2/b^2) = 1$ is $x = a \cos t$ and $y = b \sin t$.
12. $x = r \cos \theta$ $\quad\;\; y = r \sin \theta$ $\quad\;\; x^2 + y^2 = r^2$ $\quad\;\; \tan \theta = \dfrac{y}{x}$	555	These formulas relate the rectangular coordinates of a point (x, y) to its polar coordinates (r, θ). See the boxed figure on page 555.
13. The complex plane	563	Refer to Figure 2 on page 563. A complex number $a + bi$ can be identified with the point (a, b) in the x-y plane. In this context, the x-y plane is referred to as the *complex plane*. The distance r in the figure is the *modulus* of the complex number $a + bi$; the angle θ is an *argument* of $a + bi$.
14. $z = r(\cos \theta + i \sin \theta)$	563	The expression on the right-hand side is the *trigonometric form* of the complex number $z = a + bi$. (The expression $a + bi$ is the *rectangular form* of the complex number z.) The trigonometric form and the rectangular form are related by the following equations: $$r = \sqrt{a^2 + b^2} \qquad a = r \cos \theta \qquad b = r \sin \theta$$
15. DeMoivre's theorem	565	Let n be a natural number and let $z = r(\cos \theta + i \sin \theta)$. Then DeMoivre's theorem states that $$z^n = r^n(\cos n\theta + i \sin n\theta)$$ For applications of this theorem, see Examples 6, 7, and 8 in Section 9.6.

WRITING MATHEMATICS

A regular feature in *Mathematics Magazine* is the "Proof Without Words" section. The idea is to present a proof of a well-known theorem in such a way that the entire proof is "obvious" from a well-chosen picture and at most several equations. Of course, what is obvious to a mathematician may not be so obvious to her or his students (as you well know).

The following clever proof without words for the law of cosines was developed by Professor Sidney H. Kung of Jacksonville University. (It was published in *Mathematics Magazine*, vol. 63, no. 5, December 1990.) Study the proof (or discuss it with friends) until you see how it works. Then, in a paragraph or two, write out the details of the proof as if you were explaining it to a classmate. (To facilitate your explanation, you will probably need to label some of the points in the given figure.)

Proof without words for
the law of cosines, $\theta < 90°$
(by Professor Sidney H. Kung)

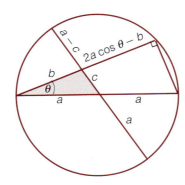

$$(2a\cos\theta - b)\,b = (a - c)(a + c)$$
$$c^2 = a^2 + b^2 - 2ab\cos\theta$$

 ## CHAPTER NINE REVIEW EXERCISES

In Exercises 1–8, the lengths of the three sides of a triangle are denoted by a, b, and c; the angles opposite these sides are A, B, and C, respectively. In each exercise, use the given data to find the remaining sides and angles. Use a calculator and round off the final answers to one decimal place.

1. $\angle A = 40°$, $\angle B = 85°$, $c = 16$ cm

2. $\angle C = 84°$, $\angle B = 16°$, $a = 9$ cm

3. (a) $a = 8$ cm, $b = 9$ cm, $\angle A = 52°$, $\angle B < 90°$
   (b) Use the data in part (a), but assume $\angle B > 90°$.

4. (a) $a = 6.25$ cm, $b = 9.44$ cm, $\angle A = 12°$, $\angle B < 90°$
   (b) Use the data in part (a), but assume $\angle B > 90°$.

5. $a = 18$ cm, $b = 14$ cm, $\angle C = 24°$

6. $a = 32.16$ cm, $b = 50.12$ cm, $\angle C = 156°$

7. $a = 4$ cm, $b = 7$ cm, $c = 9$ cm

8. $a = 12.61$ cm, $b = 19.01$ cm, $c = 14.14$ cm

In Exercises 9–18, refer to the following figure. In each case, determine the indicated quantity. Use a calculator, and round off your result to two decimal places.

9. BE

10. BC

11. area of $\triangle BCE$

12. BD

13. area of $\triangle ABD$

14. DE

15. CD

16. AD

17. AC

18. area of $\triangle DEC$

Figure for Exercises 9–18

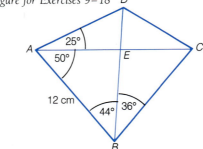

In Exercises 19 and 20, refer to the following figure, in which $\overline{BD}$ bisects $\angle ABC$.

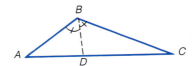

19. If $AB = 24$ cm, $BC = 40$ cm, and $AC = 56$ cm, find the length of the angle bisector $\overline{BD}$.

20. If $AB = 105$ cm, $BC = 120$ cm, and $AC = 195$ cm, find the length of the angle bisector $\overline{BD}$.

In Exercises 21 and 22, refer to the following figure, in which AD = DC.

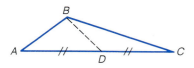

21. If $AB = 12$ cm, $BC = 26$ cm, and $AD = DC = 17$ cm, find BD.

22. If $AB = 16$ cm, $BC = 22$ cm, and $AD = DC = 17$ cm, find BD.

23. In $\triangle ABC$, $a = 4$, $b = 5$, and $c = 6$.
    (a) Find $\cos A$ and $\cos C$.
    (b) Using the results in part (a), show that angle C is twice angle A.

24. In $\triangle ABC$, suppose that angle C is twice angle A. Show that $ab = c^2 - a^2$.

In Exercises 25–30, refer to the framework shown in the accompanying figure. (Assume that the figure is drawn to scale.) In the figure, AC is parallel to $\overline{ED}$ and $BE = BD$. In each case, compute the indicated quantity. Round off the final answers to two decimal places.

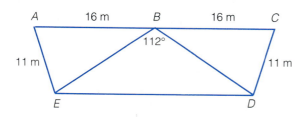

25. ∠BED   26. ∠ABE   27. ∠AEB
28. ∠BAE   29. BE   30. DE

31. The given figure shows a regular pentagon inscribed in a circle. The radius of the circle is 10 cm. Determine the following lengths, rounding off the answers to two decimal places:
    (a) one side of the pentagon;   (b) AC.

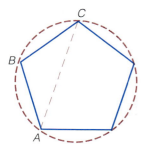

32. A pilot on a training flight is supposed to leave the airport and fly for 100 miles in the direction N40°E. Then he is supposed to turn around and fly directly back to the airport. By error, however, he makes the 100-mile return trip in the direction S25°W. How far is he now from the airport, and in what direction must he fly? (Round off the answers to one decimal place.)

33. The following figures show the same circle of diameter D inscribed first in an equilateral triangle and then in a regular hexagon. Let a denote the length of a side of the triangle and let b denote the length of a side of the hexagon. Show that $ab = D^2$.

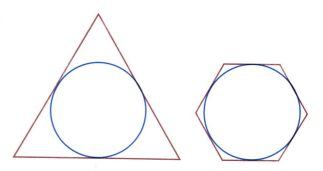

34. Suppose that triangle ABC is a right triangle with the right angle at C. Use the law of cosines to prove the following statements.
    (a) The square of the distance from C to the midpoint of the hypotenuse is equal to one-fourth the square of the hypotenuse.
    (b) The sum of the squares of the distances from C to the two points that trisect the hypotenuse is equal to five-ninths the square of the hypotenuse.
    (c) Let P, Q, and R be points on the hypotenuse such that $AP = PQ = QR = RB$. Derive a result similar to your results in parts (a) and (b) for the sum of the squares of the distances from C to P, Q, and R.

35. The perimeter and the area of the triangle in the figure are 20 cm and $10\sqrt{3}$ cm², respectively. Find a and b. (Assume $a < b$.)

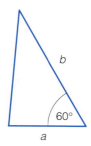

36. In $\triangle ABC$, suppose that $\angle B - \angle A = 90°$.
  (a) Show that $c^2 = (b^2 - a^2)^2/(a^2 + b^2)$. *Hint:* Let P be the point on $\overline{AC}$ such that $\angle CBP = 90°$ and $\angle PBA = \angle A$. Show that $CP = (a^2 + b^2)/2b$.
  (b) Show that $\dfrac{2}{c^2} = \dfrac{1}{(b + a)^2} + \dfrac{1}{(b - a)^2}$.

37. Two forces **F** and **G** act on an object, **G** horizontally to the right with a magnitude of 15 N, and **F** vertically upward with a magnitude of 20 N. Determine the magnitude and direction of the resultant force. (Use a calculator to determine the angle between the horizontal and the resultant; round off the result to one decimal place.)

38. Determine the resultant of the two forces in the accompanying figure. (Use a calculator, and round off the values you obtain for the magnitude and direction to one decimal place.)

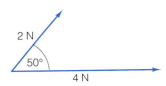

39. Determine the horizontal and vertical components of the velocity vector **v** in the following figure. (Use a calculator, and round off your answers to one decimal place.)

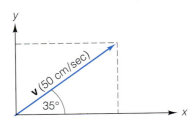

40. The heading and air speed of an airplane are 50° and 220 mph, respectively. If the wind is 60 mph from 140°, find the ground speed, the drift angle, and the course. (Use a calculator, and round off your answers to one decimal place.)

41. A block rests on an inclined plane that makes an angle of 24° with the horizontal. The component of the weight parallel to the plane is 14.8 lb. Determine the weight of the block and the component of the weight perpendicular to the plane. (Use a calculator, and round off your answers to one decimal place.)

42. Find the length of the vector $\langle 20, 99 \rangle$.

43. For which values of b will the vectors $\langle 2, 6 \rangle$ and $\langle -5, b \rangle$ have the same length?

44. The coordinates of the points A and B are $(2, 6)$ and $(-7, 4)$, respectively. Find the components of the following vectors. Write your answers in the form $\langle x, y \rangle$.
  (a) $\overrightarrow{AB}$ (b) $\overrightarrow{BA}$ (c) $3\overrightarrow{AB}$ (d) $\dfrac{1}{|\overrightarrow{AB}|}\overrightarrow{AB}$

*In Exercises 45–56, compute each of the indicated quantities, given that the vectors **a**, **b**, **c**, and **d** are defined as follows:*

$$\mathbf{a} = \langle 3, 5 \rangle \qquad \mathbf{b} = \langle 7, 4 \rangle \qquad \mathbf{c} = \langle 2, -1 \rangle \qquad \mathbf{d} = \langle 0, 3 \rangle$$

45. $\mathbf{a} + \mathbf{b}$ **46.** $\mathbf{b} - \mathbf{d}$

47. $3\mathbf{c} + 2\mathbf{a}$ **48.** $|\mathbf{a}|$

49. $|\mathbf{b} + \mathbf{d}|^2 - |\mathbf{b} - \mathbf{d}|^2$ **50.** $\mathbf{a} + (\mathbf{b} + \mathbf{c})$

51. $(\mathbf{a} + \mathbf{b}) + \mathbf{c}$ **52.** $\mathbf{a} - (\mathbf{b} - \mathbf{c})$

53. $(\mathbf{a} - \mathbf{b}) - \mathbf{c}$

54. $|\mathbf{a} + \mathbf{b}|^2 + |\mathbf{a} - \mathbf{b}|^2 - 2|\mathbf{a}|^2 - 2|\mathbf{b}|^2$

55. $4\mathbf{c} + 2\mathbf{a} - 3\mathbf{b}$ **56.** $|4\mathbf{c} + 2\mathbf{a} - 3\mathbf{b}|$

57. Express the vector $\langle 7, -6 \rangle$ in terms of the unit vectors **i** and **j**.

58. Express the vector $4\mathbf{i} - 6\mathbf{j}$ in the form $\langle a, b \rangle$.

59. Find a unit vector having the same direction as $\langle 6, 4 \rangle$.

60. Find two unit vectors that are perpendicular to $\langle \cos\theta, \sin\theta \rangle$. Simplify the components as much as possible.

In Exercises 61–68, graph the polar equations.

61. (a) $r = 2 - 2\cos\theta$ **62.** (a) $r^2 = 4\cos 2\theta$
      (b) $r = 2 - 2\sin\theta$ (b) $r^2 = 4\sin 2\theta$

63. (a) $r = 2\cos\theta - 1$ **64.** (a) $r = 2\cos\theta$
      (b) $r = 2\sin\theta - 1$ (b) $r = 2\cos\theta + 1$

65. (a) $r = 4\sin 2\theta$ **66.** (a) $r = 4\sin 3\theta$
      (b) $r = 4\cos 2\theta$ (b) $r = 4\cos 3\theta$

67. (a) $r = 1 + 2\sin(\theta/2)$ **68.** (a) $r = (1.5)^\theta \ (\theta \geq 0)$
      (b) $r = 1 - 2\cos(\theta/2)$ (b) $r = (1.5)^\theta \ (\theta \leq 0)$

In Exercises 69–74, graph the curve (or line) determined by the parametric equations. Indicate the direction of travel along the curve as t increases.

69. $x = 3 - 5t, \ y = 1 + t$ **70.** $x = t - 1, \ y = 3t$

71. $x = 3\sin t, \ y = 6\cos t$

72. $x = 2\cos t, \ y = 2\sin t$

73. $x = 4\sec t, \ y = 3\tan t$

74. $x = 1 + 3\cos t, \ y = 2 + 4\sin t$

In Exercises 75–78, convert each complex number to rectangular form.

75. $3\left(\cos \frac{1}{3}\pi + i \sin \frac{1}{3}\pi\right)$

76. $\cos \frac{1}{6}\pi + i \sin \frac{1}{6}\pi$

77. $2^{1/4}\left(\cos \frac{7}{4}\pi + i \sin \frac{7}{4}\pi\right)$

78. $5\left[\cos\left(-\frac{1}{4}\pi\right) + i \sin\left(-\frac{1}{4}\pi\right)\right]$

In Exercises 79–82, express the complex numbers in trigonometric form.

79. $\frac{1}{2}\left(1 + \sqrt{3}i\right)$

80. $3i$

81. $-3\sqrt{2} - 3\sqrt{2}i$

82. $2\sqrt{3} - 2i$

In Exercises 83–90, carry out the indicated operations. Express your results in rectangular form for those cases in which the trigonometric functions are readily evaluated without tables or a calculator.

83. $5\left(\cos \frac{1}{7}\pi + i \sin \frac{1}{7}\pi\right) \times 2\left(\cos \frac{3}{28}\pi + i \sin \frac{3}{28}\pi\right)$

84. $4\left(\cos \frac{1}{12}\pi + i \sin \frac{1}{12}\pi\right) \times 3\left(\cos \frac{1}{12}\pi + i \sin \frac{1}{12}\pi\right)$

85. $8\left(\cos \frac{1}{12}\pi + i \sin \frac{1}{12}\pi\right) \div 4\left(\cos \frac{1}{3}\pi + i \sin \frac{1}{3}\pi\right)$

86. $4(\cos 32° + i \sin 32°) \div 2^{1/2}(\cos 2° + i \sin 2°)$

87. $\left[3^{1/4}\left(\cos \frac{1}{36}\pi + i \sin \frac{1}{36}\pi\right)\right]^{12}$

88. $\left[2\left(\cos \frac{2}{15}\pi + i \sin \frac{2}{15}\pi\right)\right]^{5}$

89. $\left(\sqrt{3} + i\right)^{10}$

90. $\left(\sqrt{2} - \sqrt{2}i\right)^{15}$

In Exercises 91–94, use DeMoivre's theorem to find the indicated roots. Express your results in rectangular form.

91. Sixth roots of 1

92. Cube roots of $-64i$

93. Square roots of $\sqrt{2} - \sqrt{2}i$

94. Fourth roots of $1 + \sqrt{3}i$

95. Find the five fifth roots of $1 + i$. Use a calculator to express the roots in rectangular form. (Round off each decimal to two places in the final answer.)

96. If $z = r(\cos \theta + i \sin \theta)$, show that

$$r[\cos(\theta + \pi) + i \sin(\theta + \pi)] = -z$$

OPTIONAL TI-81 GRAPHING CALCULATOR EXERCISES
EXERCISES FOR SECTIONS 9.4 AND 9.5

1. In Figure 1 of Section 9.4 we graphed the parametric equations $x = 2t$, $y = \frac{1}{2}t^2$ for $t \geq 0$. This exercise shows how to use the calculator to obtain the graph. As preparation press ZOOM 6 for the standard viewing rectangle.
   (a) Press the MODE key and choose the parametric mode (rather than the function mode). Then press ENTER. Next, press the Y= key and enter $X_1 = 2t$, $Y_1 = \frac{1}{2}t^2$. (The parameter t is typed using the same X|T key that you have always used for graphing ordinary x-y equations.) Finally, press GRAPH.
   (b) To see which values of the parameter t were used in generating the graph in part (a), press the RANGE key. As you can then see, t runs from a minimum of 0 to a maximum of about 6.28 (which is approximately 2π). Change the setting for t_{min} to read $t_{min} = -6.28$. Now press GRAPH. How is the graph changed?
   (c) Press the RANGE key and set $t_{min} = 2$ and $t_{max} = 3$. Now press GRAPH.
   (d) To determine the coordinates of the left-hand endpoint of your graph in part (c), press the TRACE key and then use the left ARROW key to move the cursor to the left-hand endpoint of the graph. The display at the bottom of the screen tells

us that the endpoint has the coordinates $(4, 2)$ and that the corresponding value of t is 2.

2. In this exercise we use the calculator to obtain the graph shown in Figure 2 in Section 9.4. The parametric equations are $x = (88 \cos 35°)t$ and $y = 6 + (88 \sin 35°)t - 16t^2$. Before entering these two equations in the Y= menu, press the MODE key and select degree measure rather than radian measure. Now press the Y= key and enter the two given parametric equations. Press ZOOM 6 and then make the following RANGE adjustments: $t_{min} = 0$, $t_{max} = 4$, $t_{step} = 0.5$, $x_{min} = 0$, $x_{max} = 240$, $x_{scl} = 20$, $y_{min} = 0$, $y_{max} = 60$, $y_{scl} = 20$. Finally, press the GRAPH key. Check that your result is consistent with the graph shown in Figure 2 in Section 9.4. (If you want to improve the appearance of your graph, press the RANGE key and set $t_{step} = 0.01$. Then press GRAPH.)

In Exercises 3–10, graph the parametric equations for the given values of the parameter t. As preparation, press the MODE key and select the radian mode. Press the Y= key and clear all of the functions. Press ZOOM 6 so that you are beginning with the standard viewing rectangle. (In each case, you are on your own as far as making suitable adjustments in the RANGE menu. However, as a general rule, try to choose settings that will utilize as much of the viewing screen as possible. For instance, in graphing the circle given by $x = \cos t$, $y = \sin t$, it would be natural to choose the settings $x_{min} = -1$, $x_{max} = 1$, $y_{min} = -1$, and $y_{max} = 1$. Also, after choosing these settings, the ZOOM 5 command would be useful to ensure that you are seeing the correct proportions in the graph.)

3. $x = 4 \cos t$, $y = 3 \sin t$, $0 \le t \le 2\pi$ (an ellipse)

4. $x = 4 \cos t$, $y = -3 \sin t$, $0 \le t \le 2\pi$ (the same as the ellipse in Exercise 3, but traced out in the opposite direction)

5. $x = 4 \cos t$, $y = 3 \sin t$, $0 \le t \le \pi/2$ (one-quarter of an ellipse)

6. $x = 4 \cos^3 t$, $y = 4 \sin^3 t$, $0 \le t \le 2\pi$ [This curve is known as the *astroid* or *hypocycloid of four cusps*. It was first studied by the Danish astronomer Olof Roemer (1644–1710) and by the Swiss mathematician James Bernoulli (1654–1705).]

7. $x = 2 \cos t + \cos 2t$, $y = 2 \sin t - \sin 2t$, $0 \le t \le 2\pi$. [This curve is the *deltoid*. It was first studied by the Swiss mathematician Leonhard Euler (1707–1783).]

8. $x = \dfrac{\cos t}{1 + \sin^2 t}$, $y = \dfrac{\sin t \cos t}{1 + \sin^2 t}$, $0 \le t \le 2\pi$ (the *lemniscate of Bernoulli*)

9. $x = 2 \tan t$, $y = 2 \cos^2 t$, $0 \le t \le 2\pi$ [This is the *witch of Agnesi*, named after the Italian mathematician and scientist Maria Gaetana Agnesi (1718–1799). For background on how the curve got its name, see the text by Howard Eves, *An Introduction to the History of Mathematics*, 6th ed., p. 442 (Philadelphia: Saunders College Publishing, 1990).]

10. $x = \dfrac{t^2 - 1}{3t^2 + 1}$, $y = \dfrac{t(t^2 - 1)}{3t^2 + 1}$, $-\infty < t < \infty$ (the *folium of Descartes*) *Hint:* Use $t_{min} = -10$, $t_{max} = 10$, $t_{step} = 0.05$.

In Exercises 11–22 we graph polar equations. A polar equation of the form $r = f(\theta)$ can always be expressed in parametric form by means of the equations

$$x = r \cos \theta \qquad (1)$$
$$y = r \sin \theta \qquad (2)$$

In equations (1) and (2), we replace r by $f(\theta)$. This yields

$$x = f(\theta) \cos \theta \qquad (3)$$
$$y = f(\theta) \sin \theta \qquad (4)$$

Equations (3) and (4) form a set of parametric equations; the graph of these equations will be the polar curve $r = f(\theta)$.

As preparation for Exercises 11–22, press the MODE key and check that the calculator is set in the radian mode and in the parametric mode. Press the Y= key and clear all functions that

are present. Finally press ZOOM 6 so that you are starting with the standard viewing rectangle.

11. In Section 9.5, we graphed the polar equation $r = 2 \cos \theta$ by setting up a table and plotting points. To graph this same equation using the graphing calculator, we use equations (3) and (4) to write

$$x = f(\theta) \cos \theta = 2 \cos \theta \cos \theta = 2 \cos^2 \theta$$

and

$$y = f(\theta) \sin \theta = 2 \cos \theta \sin \theta$$

(a) Press the Y= key and enter the parametric equations $X_1 = 2 \cos^2 t$ and $Y_1 = 2 \cos t \sin t$. Now press the GRAPH key. Go on to part (b) for a better view of the graph.

(b) Use the zoom box (ZOOM 1) to obtain a closer look at the graph. (The precise coordinates of the corners of the box are unimportant; what matters is that the entire curve be contained within the box.) After you have obtained a larger picture of the graph, press ZOOM 5 so that the true proportions of the curve are shown. You should obtain a circle, as shown in Figure 6 in Section 9.5.

In Exercises 12–18, graph the polar equations using the technique explained in Exercise 11.

12. (a) $r = \cos 2\theta$ *(four-leafed rose)*
    (b) $r = \sin 2\theta$

13. (a) $r = \cos 4\theta$ *(eight-leafed rose)*
    (b) $r = \sin 4\theta$

14. (a) $r = \cos 3\theta$ *(three-leafed rose)*
    (b) $r = \sin 3\theta$

15. (a) $r = \cos 5\theta$ *(five-leafed rose)*
    (b) $r = \sin 5\theta$

16. $r = 4(1 + \cos \theta)$ *(cardioid)*

17. $r = \theta$ *(spiral of Archimedes)* *Suggestion:* Experiment with different values for t_{max} in the RANGE menu.

18. $r^2 = 4 \cos 2\theta$ *(lemniscate of Bernoulli)* *Hints:* The equation is equivalent to the two equations $r = \pm 2\sqrt{\cos 2\theta}$. You'll find, however, that you need to consider only one of these. Also, you can improve the appearance of your final graph if you set $t_{step} = 0.01$. (Don't make this adjustment until the end; the graph will be drawn very slowly.)

19. The *limaçon of Pascal* is defined by the equation $r = a \cos \theta + b$. [The curve is named after Étienne Pascal (1588–1640), the father of Blaise Pascal. According to Howard Eves, however, the curve is misnamed; it appears earlier in the writings of Albrecht Dürer (1471–1528).] When $a = b$, the curve is a

cardioid; when $a = 0$ the curve is a circle. Graph the following limaçons.

(a) $r = 2 \cos \theta + 1$ (b) $r = 2 \cos \theta - 1$
(c) $r = \cos \theta + 2$ (d) $r = \cos \theta - 2$
(e) $r = 2 \cos \theta + 2$ (f) $r = \cos \theta$

20. Graph the *cissoid of Diocles*, $r = 10 \sin \theta \tan \theta$. [The Greek mathematician Diocles (ca. 180 B.C.) used the geometric properties of this curve to solve the problem of duplicating the cube. For an explanation of the problem and historical background, see the text by Howard Eves, *An Introduction to the History of Mathematics*, 6th ed., pp. 109–127 (Philadelphia: Saunders College Publishing, 1990).]

21. Graph the *logarithmic spiral* (or *spiral of Descartes*), $r = e^{\theta/10}$ using the standard viewing rectangle (modified

with ZOOM 5) and the following settings.
(a) $t_{max} = 10$ (b) $t_{max} = 20$ (c) $t_{max} = 25$

22. The curve $r = \cos^4(\theta/4)$ has a surprising property that is very difficult to observe without the aid of a graphing calculator or a computer. After entering the parametric equations for this curve, press ZOOM 6. Next, make the following adjustments in the RANGE menu: $t_{max} = 13$, $x_{min} = -1$, $x_{max} = 1$, $y_{min} = -1$, $y_{max} = 1$. Then press ZOOM 5. As you can see, the graph appears to have a simple inner loop to the left of the origin. This is not the case, however. Use the ZOOM 2 command to zoom in on the origin several times. What do you find?

CHAPTER NINE TEST

1. In triangle *ABC*, let $A = 120°$, $b = 5$ cm, and $c = 3$ cm. Find a.

2. The sides of a triangle are 2 cm, 3 cm, and 4 cm. Determine the cosine of the angle opposite the longest side. On the basis of your answer, explain whether or not the angle opposite the longest side is an acute angle.

3. Two of the angles in a triangle are 30° and 45°. If the side opposite the 45° angle is $20\sqrt{2}$ cm, find the side opposite the 30° angle.

4. Each side of the square *STUV* is 8 cm long. The point *P* lies on diagonal $\overline{SU}$ such that $SP = 2$ cm. Find the distance from *P* to *V*.

5. In $\triangle ABC$, $b = 5.8$ cm, $c = 3.2$ cm, and $A = 27°$. Find the remaining sides and angles of the triangle.

6. Two forces **F** and **G** act on an object. The force **G** acts horizontally with a magnitude of 2 N, and **F** acts vertically upward with a magnitude of 4 N.
    (a) Find the magnitude of the resultant.
    (b) Find $\tan \theta$, where θ is the angle between **G** and the resultant.

7. Two forces act on an object, as shown in the following figure.
    (a) Find the magnitude of the resultant. (Leave your answer in terms of radicals and the trigonometric functions.)
    (b) Find $\sin \theta$, where θ is the angle between the 12-N force and the resultant.

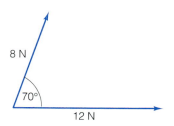

8. The heading and air speed of an airplane are 40° and 300 mph, respectively. If the wind is 50 mph from 130°, find the ground speed and the tangent of the

drift angle. (Leave your answer in terms of radicals and the trigonometric functions.)

9. Let $\mathbf{A} = \langle 2, 4 \rangle$, $\mathbf{B} = \langle 3, -1 \rangle$, and $\mathbf{C} = \langle 4, -4 \rangle$.
   (a) Find $2\mathbf{A} + 3\mathbf{B}$.
   (b) Find $|2\mathbf{A} + 3\mathbf{B}|$.
   (c) Express $\mathbf{C} - \mathbf{B}$ in terms of $\mathbf{i}$ and $\mathbf{j}$.

10. Let P and Q be the points $(4, 5)$ and $(-7, 2)$, respectively. Find a unit vector having the same direction as $\overrightarrow{PQ}$.

11. Convert the polar equation $r^2 = \cos 2\theta$ to rectangular form.

12. Graph the equation $r = 2(1 - \cos \theta)$.

13. Given the parametric equations $x = 4 \sin t$ and $y = 2 \cos t$, eliminate the parameter t and sketch the graph.

14. Find the rectangular form for the complex number $z = 2(\cos \frac{2}{3}\pi + i \sin \frac{2}{3}\pi)$.

15. Find the trigonometric form of the complex number $\sqrt{2} - \sqrt{2}i$.

16. Complete the statement: According to DeMoivre's theorem, if n is a natural number, then

    $$[r(\cos \theta + i \sin \theta)]^n = \underline{\hspace{3cm}}$$

17. Let $z = 3(\cos \frac{2\pi}{9} + i \sin \frac{2\pi}{9})$ and $w = 5(\cos \frac{\pi}{9} + i \sin \frac{\pi}{9})$. Compute the product zw and express your answer in rectangular form.

18. Compute the cube roots of $64i$.

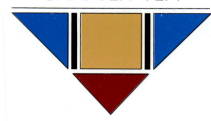

CHAPTER TEN

SYSTEMS OF EQUATIONS

Many problems in a variety of disciplines give rise to ... [linear] systems. For example, in physics, in order to find the currents in an electrical circuit containing known voltages and resistances, a system of linear equations must be solved. In chemistry, the balancing of chemical equations requires the solution of a system of linear equations. And in economics, the Leontief input–output model reduces problems concerning the production and consumption of goods to systems of linear equations.

Leslie Hogben in *Elementary Linear Algebra* (St. Paul: West Publishing Company, 1987)

INTRODUCTION

In this chapter we consider systems of equations. Roughly speaking, a system of equations is just a collection of equations with a common set of unknowns. In solving such systems, we try to find values for the unknowns that simultaneously satisfy each of the equations in the system. In Section 10.1 we review two techniques that are often presented in intermediate algebra for solving systems involving two linear equations in two unknowns. An important technique for solving larger systems of equations is developed in Section 10.2; this technique is known as *Gaussian elimination*. After that, in Section 10.3, we introduce matrices as a tool for reducing the amount of bookkeeping involved in Gaussian elimination. Two additional techniques for solving systems of linear equations are explained in Sections 10.4 and 10.5. The technique in Section 10.4 uses the idea of an inverse matrix; the technique in Section 10.5 involves determinants and Cramer's rule. Section 10.6 contains a brief discussion of nonlinear systems of equations. In the last section of this chapter, Section 10.7, we consider systems of inequalities. As an application of this topic and of the earlier work in the chapter, we conclude Section 10.7 with an introduction to linear programming.

10.1 SYSTEMS OF TWO LINEAR EQUATIONS IN TWO UNKNOWNS

Both in theory and in applications, it's often necessary to solve two equations in two unknowns. You may have been introduced to the idea of simultaneous equations in a previous course in algebra; however, to put matters on a firm foundation, we begin here with the basic definitions. By a **linear equation in two variables** we mean an equation of the form

$$ax + by = c$$

where the constants a and b are not both zero. The two variables needn't always be denoted by the letters x and y, of course; it is the *form* of the equation that matters. Table 1 displays some examples.

An ordered pair of numbers (x_0, y_0) is said to be a **solution of the linear equation** $ax + by = c$ provided we obtain a true statement when we replace x and y in the equation by x_0 and y_0, respectively. For example, the ordered pair $(3, 2)$ is a solution of the equation $x - y = 1$, since $3 - 2 = 1$. On the other hand, $(2, 3)$ is not a solution of $x - y = 1$, since $2 - 3 \neq 1$.

Now consider a **system** of two linear equations in two unknowns:

$$\begin{cases} ax + by = c \\ dx + ey = f \end{cases}$$

If we can find an ordered pair that is a solution to both equations, then we say that ordered pair is a **solution of the system**. Sometimes, to emphasize the fact that a solution must satisfy both equations, we refer to the system as a pair of **simultaneous equations**. A system that has at least one solution is said to be **consistent**. If there are no solutions, the system is **inconsistent**.

TABLE 1

Equation in Two Variables	Is It Linear?	
	Yes	No
$3x - 8y = 12$	✓	
$-s + 4t = 0$	✓	
$2x - 3y^2 = 1$		✓
$y = 4 - 2x$	✓	
$\dfrac{4}{u} + \dfrac{5}{v} = 3$		✓

EXAMPLE 1 Consider the system

$$\begin{cases} x + y = 2 \\ 2x - 3y = 9 \end{cases}$$

(a) Is $(1, 1)$ a solution of the system?
(b) Is $(3, -1)$ a solution of the system?

Solution (a) Although $(1, 1)$ is a solution of the first equation, it is not a solution of the system because it does not satisfy the second equation. (Check this for yourself.)

(b) $(3, -1)$ satisfies the first equation:

$$3 + (-1) \overset{?}{=} 2 \qquad \text{True}$$

$(3, -1)$ satisfies the second equation:

$$2(3) - 3(-1) \overset{?}{=} 9$$
$$6 + 3 \overset{?}{=} 9 \qquad \text{True}$$

Since $(3, -1)$ satisfies both equations, it is a solution of the system.

We can gain an important perspective on systems of linear equations by looking at Example 1 in graphical terms. Table 2 shows how each of the statements in that example can be rephrased using geometric ideas with which we are already familiar.

TABLE 2

Algebraic Idea	Corresponding Geometric Idea
1. The ordered pair $(1, 1)$ is a solution of the equation $x + y = 2$.	1. The point $(1, 1)$ lies on the line $x + y = 2$. See Figure 1.
2. The ordered pair $(1, 1)$ is not a solution of the equation $2x - 3y = 9$.	2. The point $(1, 1)$ does not lie on the line $2x - 3y = 9$. See Figure 1.
3. The ordered pair $(3, -1)$ is a solution of the system $$\begin{cases} x + y = 2 \\ 2x - 3y = 9 \end{cases}$$	3. The point $(3, -1)$ lies on both of the lines $x + y = 2$ and $2x - 3y = 9$. See Figure 1.

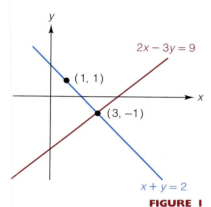

FIGURE 1

In Example 1, we verified that $(3, -1)$ is a solution of the system

$$\begin{cases} x + y = 2 \\ 2x - 3y = 9 \end{cases}$$

Are there any other solutions of this particular system? No; Figure 1 shows us that there are no other solutions, since $(3, -1)$ is clearly the only point common to both lines. In a moment we'll look at two important methods for solving systems of linear equations in two unknowns. But even before we consider these methods, we can say something about the solutions of linear systems.

■▶ PROPERTY SUMMARY   POSSIBILITIES FOR SOLUTIONS OF LINEAR SYSTEMS

Given a system of two linear equations in two unknowns, exactly one of the following cases must occur.

CASE 1 The graphs of the two linear equations intersect in exactly one point. Thus, there is exactly one solution to the system. See Figure 2.

CASE 2 The graphs of the two linear equations are parallel lines. Therefore, the lines do not intersect, and the system has no solution. See Figure 3.

CASE 3 The two equations actually represent the same line. Thus, there are infinitely many points of intersection and correspondingly infinitely many solutions. See Figure 4.

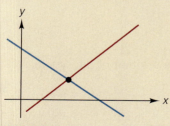

FIGURE 2
A consistent system with exactly one solution.

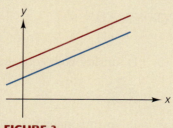

FIGURE 3
An inconsistent system has no solution.

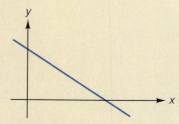

FIGURE 4
A consistent system with infinitely many solutions.

We are going to review two methods from basic algebra for solving systems of two linear equations in two unknowns. These methods are the **substitution method** and the **addition–subtraction method**. We'll begin by demonstrating the substitution method. Consider the system

$$\begin{cases} 3x + 2y = 17 & (1) \\ 4x - 5y = -8 & (2) \end{cases}$$

We first choose one of the two equations and then use it to express one of the variables in terms of the other. In the case at hand, neither equation appears particularly simpler than the other, so let's just start with the first equation and solve for x in terms of y. We have

$$3x = 17 - 2y$$
$$x = \tfrac{1}{3}(17 - 2y) \qquad (3)$$

Now we use equation (3) to substitute for x in the equation that we have not used yet, namely, equation (2). This yields

$$4\left[\tfrac{1}{3}(17 - 2y)\right] - 5y = -8$$
$$4(17 - 2y) - 15y = -24 \qquad \text{multiplying by 3}$$
$$-23y = -92$$
$$y = 4$$

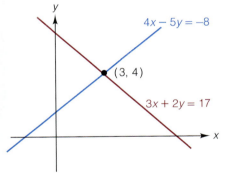

FIGURE 5

The value $y = 4$ that we have just obtained can be used in equation (3) to find x. Replacing y with 4 in equation (3) yields

$$x = \tfrac{1}{3}[17 - 2(4)] = \tfrac{1}{3}(9) = 3$$

We have now found that $x = 3$ and $y = 4$. As you can easily check, this pair of values indeed satisfies both of the original equations. We write our solution as the ordered pair $(3, 4)$. Figure 5 summarizes the situation. It shows that the system is consistent and that $(3, 4)$ is the only solution.

Generally speaking, it is not necessary to graph the equations in a given system in order to decide whether the system is consistent. Rather, this information will emerge as you attempt to follow an algebraic method of solution. Examples 2 and 3 will illustrate this.

EXAMPLE 2 Solve the system

$$\begin{cases} \tfrac{3}{2}x - 3y = -9 \\ x - 2y = 4 \end{cases}$$

Solution We use the substitution method. Since it is easy to solve the second equation for x, we begin there:

$$x - 2y = 4$$
$$x = 4 + 2y$$

Now we substitute this result in the first equation of our system to obtain

$$\tfrac{3}{2}(4 + 2y) - 3y = -9$$
$$6 + 3y - 3y = -9$$
$$6 = -9 \qquad \text{False}$$

Since the substitution process leads us to this obviously false statement, we conclude that the given system has no solution; that is, the system is inconsistent. *Question:* What can you say about the graphs of the two given equations? ∎

EXAMPLE 3 Solve the system

$$\begin{cases} 3x + 4y = 12 \\ 2y = 6 - \tfrac{3}{2}x \end{cases}$$

Solution We use the method of substitution. Since it is easy to solve the second equation for y, we begin there:

$$2y = 6 - \tfrac{3}{2}x \qquad \text{and therefore} \qquad y = 3 - \tfrac{3}{4}x$$

Now we use this result to substitute for y in the first equation of the original system. The result is

$$3x + 4\left(3 - \tfrac{3}{4}x\right) = 12$$
$$3x + 12 - 3x = 12$$
$$3x - 3x = 12 - 12$$
$$0 = 0 \qquad \text{Always true}$$

This last identity imposes no restrictions on x. Graphically speaking, this says that our two lines intersect for *every* value of x. In other words, the two lines coincide. We could have foreseen this initially had we solved both equations for y. As you can verify, the result in both cases is

$$y = -\tfrac{3}{4}x + 3$$

Every point on this line yields a solution to our system of equations. In summary then, our system is consistent and the solutions to the system have the form $\left(x, -\tfrac{3}{4}x + 3\right)$, where x can be any real number. For instance, when $x = 0$, we obtain the solution $(0, 3)$. When $x = 1$, we obtain the solution $\left(1, \tfrac{9}{4}\right)$. The idea here is that for *each* value of x, we obtain a solution; thus, there are infinitely many solutions. ∎

Now let us turn to the addition–subtraction method of solving systems of equations. By way of example, consider the system

$$\begin{cases} 2x + 3y = 5 \\ 4x - 3y = -1 \end{cases}$$

Notice that if we add these two equations, the result will be an equation involving only the unknown x. Carrying this out gives us

$$6x = 4$$
$$x = \tfrac{4}{6} = \tfrac{2}{3}$$

There are now several ways in which the corresponding value of y may be obtained. As you can easily check, substituting the value $x = \frac{2}{3}$ in either of the original equations leads to the result $y = \frac{11}{9}$.

Another way to find y is by multiplying both sides of the first equation by -2. (You'll see why in a moment.) We display the work this way:

$$2x + 3y = 5 \qquad \xrightarrow{\text{Multiply by } -2} \qquad -4x - 6y = -10$$

$$4x - 3y = -1 \qquad \xrightarrow{\text{No change}} \qquad 4x - 3y = -1$$

Adding the last two equations then gives us

$$-9y = -11$$
$$y = \tfrac{11}{9}$$

The required solution is therefore $\left(\frac{2}{3}, \frac{11}{9}\right)$.

In the previous example, we were able to find x directly by adding the two equations. As the next example shows, it may be necessary first to multiply both sides of each equation by an appropriate constant.

EXAMPLE 4 Solve the system

$$\begin{cases} 5x - 3y = 4 \\ 2x + 4y = 1 \end{cases}$$

Solution To eliminate x, we could multiply the second equation by $\frac{5}{2}$ and then subtract the resulting equation from the first equation. However, to avoid working with fractions, we proceed as follows.

$$5x - 3y = 4 \qquad \xrightarrow{\text{Multiply by 2}} \qquad 10x - 6y = 8 \tag{4}$$

$$2x + 4y = 1 \qquad \xrightarrow{\text{Multiply by 5}} \qquad 10x + 20y = 5 \tag{5}$$

Subtracting equation (5) from equation (4) then yields

$$-26y = 3$$
$$y = -\tfrac{3}{26}$$

To find x, we return to the original system and work in a similar manner:

$$5x - 3y = 4 \qquad \xrightarrow{\text{Multiply by 4}} \qquad 20x - 12y = 16$$

$$2x + 4y = 1 \qquad \xrightarrow{\text{Multiply by 3}} \qquad 6x + 12y = 3$$

Upon adding the last two equations, we obtain

$$26x = 19$$
$$x = \tfrac{19}{26}$$

The solution of the given system of equations is therefore $\left(\frac{19}{26}, -\frac{3}{26}\right)$. ∎

We conclude this section with some problems that can be solved using simultaneous equations.

EXAMPLE 5 Determine constants b and c so that the parabola $y = x^2 + bx + c$ passes through the points $(-3, 1)$ and $(1, -2)$.

Solution Since the point $(-3, 1)$ lies on the curve $y = x^2 + bx + c$, the coordinates must satisfy the equation. Thus, we have

$$1 = (-3)^2 + b(-3) + c$$
$$-8 = -3b + c \tag{6}$$

This gives us one equation in two unknowns. We need another equation involving b and c. Since the point $(1, -2)$ also lies on the graph of $y = x^2 + bx + c$, we must have

$$-2 = 1^2 + b(1) + c$$
$$-3 = b + c \tag{7}$$

Rewriting equations (6) and (7), we have the system

$$\begin{cases} -3b + c = -8 & (8) \\ b + c = -3 & (9) \end{cases}$$

Subtracting equation (9) from (8) then yields

$$-4b = -5 \quad \text{and therefore} \quad b = \tfrac{5}{4}$$

One way to obtain the corresponding value of c is to replace b by $\tfrac{5}{4}$ in equation (9). This yields

$$\tfrac{5}{4} + c = -3$$
$$c = -3 - \tfrac{5}{4} = -\tfrac{17}{4}$$

The required values of b and c are therefore

$$b = \tfrac{5}{4}, \qquad c = -\tfrac{17}{4}$$

(Exercise 58 at the end of this section asks you to check that the parabola $y = x^2 + \tfrac{5}{4}x - \tfrac{17}{4}$ indeed passes through the given points.) ∎

EXAMPLE 6 In Figure 6 (on the next page), find the area of the triangular region bounded by the lines $y = 3x$, $y = -\tfrac{1}{2}x + 7$, and the x-axis.

Solution To compute the area of the triangle, we will need to know the base and the height. To determine the base, we first compute the x-intercept of the line $y = -\tfrac{1}{2}x + 7$:

$$0 = -\tfrac{1}{2}x + 7$$
$$\tfrac{1}{2}x = 7$$
$$x = 14$$

The height of the triangle is the y-coordinate of the intersection point of the lines $y = 3x$ and $y = -\tfrac{1}{2}x + 7$ (see Figure 6). The substitution method can be used to solve this pair of simultaneous equations for y. From the equation

FIGURE 6

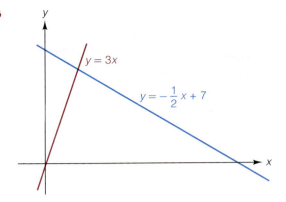

$y = 3x$, we obtain $x = \frac{1}{3}y$. Then substituting this in the equation $y = -\frac{1}{2}x + 7$ yields

$$y = -\tfrac{1}{2}\left(\tfrac{1}{3}y\right) + 7$$
$$6y = -y + 42$$
$$7y = 42$$
$$y = 6$$

Now that we know that the base of the triangle is 14 and the height is 6, we can compute the required area:

$$A = \tfrac{1}{2}bh$$
$$= \tfrac{1}{2}(14)(6) = 42 \text{ square units}$$

In the next example, we solve a mixture problem using a system of two equations in two unknowns.

EXAMPLE 7 Suppose that a chemistry student can obtain two acid solutions from the stockroom. The first solution is 20% acid and the second solution is 45% acid. (The percentages are by volume.) How many milliliters of each solution should the student mix together to obtain 100 ml of a 30% acid solution?

Solution We begin by assigning letters to denote the required quantities.

Let x denote the number of milliliters of the 20% solution to be used.

Let y denote the number of milliliters of the 45% solution to be used.

Then the data can be summarized as in Table 3 (on the next page). Since the final mixture must total 100 ml, we have the equation

$$x + y = 100 \tag{10}$$

This gives us one equation in two unknowns. However, we need a second equation. Looking at the data in the right-hand column of Table 3, we can write

$$\underbrace{0.20x}_{\substack{\text{Amount of acid in} \\ x \text{ ml of the 20\%} \\ \text{solution}}} + \underbrace{0.45y}_{\substack{\text{Amount of acid in} \\ y \text{ ml of the 45\%} \\ \text{solution}}} = \underbrace{(0.30)(x + y)}_{\substack{\text{Amount of acid in} \\ \text{the final mixture}}}$$

TABLE 3

Type of Solution	Number of ml	Percent of Acid	Total Acid (ml)
First solution (20% acid)	x	20	$(0.20)x$
Second solution (45% acid)	y	45	$(0.45)y$
Mixture	$x + y$	30	$(0.30)(x + y)$

Thus,

$$0.20x + 0.45y = 0.30(x + y)$$
$$20x + 45y = 30(x + y)$$
$$4x + 9y = 6(x + y) = 6x + 6y$$
$$-2x + 3y = 0 \tag{11}$$

Equations (10) and (11) can be solved by either the substitution method or the addition method. As Exercise 59 at the end of this section asks you to show, the results are

$$x = 60 \text{ ml} \quad \text{and} \quad y = 40 \text{ ml}$$

EXERCISE SET 10.1

A

1. Which of the following are linear equations in two variables?
   (a) $3x + 3y = 10$ (b) $2x + 4xy + 3y = 1$
   (c) $u - v = 1$ (d) $x = 2y + 6$

2. Which of the following are linear equations in two variables?
   (a) $y = x$ (b) $y = x^2$
   (c) $\dfrac{4}{x} - \dfrac{3}{y} = -1$ (d) $2w + 8z = -4w + 3$

3. Is $(5, 1)$ a solution of the following system?
$$\begin{cases} 2x - 8y = 2 \\ 3x + 7y = 22 \end{cases}$$

4. Is $(14, -2)$ a solution of the following system?
$$\begin{cases} x + y = 12 \\ x - y = 4 \end{cases}$$

5. Is $(0, -4)$ a solution of the following system?
$$\begin{cases} \frac{1}{6}x + \frac{1}{2}y = -2 \\ \frac{2}{3}x + \frac{3}{4}y = 2 \end{cases}$$

6. Is $(12, -8)$ a solution of the system in Exercise 5?

7. Is $(3, -2)$ a solution of the following system?
$$\begin{cases} \frac{2}{7}x - \frac{1}{5}y = \frac{44}{35} \\ \frac{1}{3}x - \frac{5}{4}y = \frac{7}{2} \end{cases}$$

In Exercises 8–18, use the substitution method to find all solutions of each system.

8. $\begin{cases} 4x - y = 7 \\ -2x + 3y = 9 \end{cases}$ 9. $\begin{cases} 3x - 2y = -19 \\ x + 4y = -4 \end{cases}$

10. $\begin{cases} 6x - 2y = -3 \\ 5x + 3y = 4 \end{cases}$ 11. $\begin{cases} 4x + 2y = 3 \\ 10x + 4y = 1 \end{cases}$

12. $\begin{cases} \frac{3}{2}x - 5y = 1 \\ x + \frac{3}{4}y = -1 \end{cases}$ 13. $\begin{cases} 13x - 8y = -3 \\ -7x + 2y = 0 \end{cases}$

14. $\begin{cases} 4x + 6y = 3 \\ -6x - 9y = -\frac{9}{2} \end{cases}$ 15. $\begin{cases} -\frac{2}{5}x + \frac{1}{4}y = 3 \\ \frac{1}{4}x - \frac{2}{5}y = -3 \end{cases}$

16. $\begin{cases} 0.02x - 0.03y = 1.06 \\ 0.75x + 0.50y = -0.01 \end{cases}$

17. $\begin{cases} \sqrt{2}x - \sqrt{3}y = \sqrt{3} \\ \sqrt{3}x - \sqrt{8}y = \sqrt{2} \end{cases}$ 18. $\begin{cases} 7x - 3y = -12 \\ \frac{14}{3}x - 2y = 2 \end{cases}$

In Exercises 19–28, use the addition–subtraction method to find all solutions of each system of equations.

19. $\begin{cases} 5x + 6y = 4 \\ 2x - 3y = -3 \end{cases}$

20. $\begin{cases} -8x + y = -2 \\ 4x - 3y = 1 \end{cases}$

21. $\begin{cases} 4x + 13y = -5 \\ 2x - 54y = -1 \end{cases}$

22. $\begin{cases} 16x - 3y = 100 \\ 16x + 10y = 10 \end{cases}$

23. $\begin{cases} \frac{1}{4}x - \frac{1}{3}y = 4 \\ \frac{2}{7}x - \frac{1}{7}y = \frac{1}{10} \end{cases}$

Suggestion: First clear both equations of fractions.

24. $\begin{cases} 2.1x - 3.5y = 1.2 \\ 1.4x + 2.6y = 1.1 \end{cases}$

25. $\begin{cases} 8x + 16y = 5 \\ 2x + 5y = \frac{5}{4} \end{cases}$

26. $\begin{cases} 8x + 16y = 5 \\ 2x + 4y = 1 \end{cases}$

27. $\begin{cases} 125x - 40y = 45 \\ \frac{1}{10}x + \frac{1}{10}y = \frac{3}{10} \end{cases}$

28. $\begin{cases} \sqrt{6}x - \sqrt{3}y = 3\sqrt{2} - \sqrt{3} \\ \sqrt{2}x - \sqrt{5}y = \sqrt{6} + \sqrt{5} \end{cases}$

29. Find b and c, given that the parabola $y = x^2 + bx + c$ passes through $(0, 4)$ and $(2, 14)$.

30. Determine the constants a and b, given that the parabola $y = ax^2 + bx + 1$ passes through $(-1, 11)$ and $(3, 1)$.

31. Determine the constants A and B, given that the line $Ax + By = 2$ passes through the points $(-4, 5)$ and $(7, -9)$.

32. Find the area of the triangular region in the first quadrant bounded by the lines $y = 5x$ and $y = -3x + 6$ and the x-axis.

33. Find the area of the triangular region in the first quadrant bounded by the x-axis and the lines $y = 2x - 5$ and $y = -\frac{1}{2}x + 3$.

34. Find the area of the triangular region in the first quadrant bounded by the y-axis and the lines $y = 2x + 2$ and $y = -\frac{3}{2}x + 9$.

35. A student in a chemistry laboratory has access to two acid solutions. The first solution is 10% acid and the second is 35% acid. (The percentages are by volume.) How many cubic centimeters of each should she mix together to obtain 200 cm³ of a 25% acid solution?

36. One salt solution is 15% salt and another is 20% salt. How many cubic centimeters of each solution must be mixed to obtain 50 cm³ of a 16% salt solution?

37. A shopkeeper has two types of coffee beans on hand. One type sells for $5.20/lb, the other for $5.80/lb. How many pounds of each type must be mixed to produce 16 lb of a blend that sells for $5.50/lb?

38. A certain alloy contains 10% tin and 30% copper. (The percentages are by weight.) How many pounds of tin and how many pounds of copper must be melted with

1000 lb of the given alloy to yield a new alloy containing 20% tin and 35% copper?

B

39. Find x and y in terms of a and b:

$$\begin{cases} \dfrac{x}{a} + \dfrac{y}{b} = 1 \\[2mm] \dfrac{x}{b} + \dfrac{y}{a} = 1 \end{cases}$$

Does your solution impose any conditions on a and b?

40. Solve the following system for x and y in terms of a and b, where $a \neq b$:

$$\begin{cases} ax + by = 1/a \\ b^2x + a^2y = 1 \end{cases}$$

41. Solve the following system for x and y in terms of a and b, where $a \neq b$:

$$\begin{cases} ax + a^2y = 1 \\ bx + b^2y = 1 \end{cases}$$

Does your solution impose any additional conditions on a and b?

42. Solve the following system for s and t:

$$\begin{cases} \dfrac{3}{s} - \dfrac{4}{t} = 2 \\[2mm] \dfrac{5}{s} + \dfrac{1}{t} = -3 \end{cases}$$

Hint: Make the substitutions $1/s = x$ and $1/t = y$ in order to obtain a system of two linear equations.

43. Solve the following system for s and t:

$$\begin{cases} \dfrac{1}{2s} - \dfrac{1}{2t} = -10 \\[2mm] \dfrac{2}{s} + \dfrac{3}{t} = 5 \end{cases}$$

(Use the hint in Exercise 42.)

In Exercises 44–53, find all solutions of the given systems. For Exercises 47–52, use a calculator and round off the final answers to two decimal places.

44. $\begin{cases} 0.5x - 0.8y = 0.3 \\ 0.4x - 0.1y = 0.9 \end{cases}$

45. $\begin{cases} \dfrac{2w - 1}{3} + \dfrac{z + 2}{4} = 4 \\[2mm] \dfrac{w + 3}{2} - \dfrac{w - z}{3} = 3 \end{cases}$

46. $\dfrac{x - y}{2} = \dfrac{x + y}{3} = 1$

47. $\begin{cases} 1.03x - 2.54y = 5.47 \\ 3.85x + 4.29y = -1.84 \end{cases}$

48. $\begin{cases} 2.39x + 8.16y = -2.83 \\ 1.01x + 2.98y = 4.41 \end{cases}$

49. $\begin{cases} 2 \ln x - 5 \ln y = 11 \\ \ln x + \ln y = -5 \end{cases}$

50. $\begin{cases} 3 \ln x + \ln y = 3 \\ 4 \ln x - 6 \ln y = -7 \end{cases}$

Hint: Let $u = \ln x$ and $v = \ln y$

51. $\begin{cases} e^x - 3e^y = 2 \\ 3e^x + e^y = 16 \end{cases}$

52. $\begin{cases} e^x + 2e^y = 4 \\ \frac{1}{2}e^x - e^y = 0 \end{cases}$

Hint: Let $u = e^x$ and $v = e^y$

53. $\begin{cases} 4\sqrt{x^2 - 3x} - 3\sqrt{y^2 + 6y} = -4 \\ \frac{1}{2}\left(\sqrt{x^2 - 3x} + \sqrt{y^2 + 6y}\right) = 3 \end{cases}$

Hint: Let $u = \sqrt{x^2 - 3x}$ and $v = \sqrt{y^2 + 6y}$.

54. The sum of two numbers is 64. Twice the larger number plus five times the smaller number is 20. Find the two numbers. (Let x denote the larger number and let y denote the smaller number.)

55. In a two-digit number, the sum of the digits is 14. Twice the tens digit exceeds the units digit by one. Find the number.

56. The sum of the digits in a two-digit number is 14. Furthermore, the number itself is 2 greater than 11 times the tens digit. Find the number.

57. The perimeter of a rectangle is 34 in. If the length is 2 in. more than twice the width, find the length and the width.

58. Verify that the parabola $y = x^2 + \frac{5}{4}x - \frac{17}{4}$ indeed passes through $(-3, 1)$ and $(1, -2)$, as stated in the text.

59. Consider the following system from Example 7.

$$\begin{cases} x + y = 100 \\ -2x + 3y = 0 \end{cases}$$

(a) Solve this system using the method of substitution.
Answer: $(60, 40)$

(b) Solve the system using the addition–subtraction method.

60. Consider the following system:

$$\begin{cases} 2x - 5y = -2 \\ 3x + 4y = 5 \end{cases}$$

(a) Solve the system using the substitution method.

(b) Solve the system using the addition–subtraction method. *Answer:* $\left(\frac{17}{23}, \frac{16}{23}\right)$

(c) Verify that your solution indeed satisfies both equations of the system.

61. Solve the following system for x and y:

$$\begin{cases} by = x + ab \\ cy = x + ac \end{cases}$$

(Assume a, b, and c are constants and that $b \neq c$.)

62. Solve for x and y in terms of a, b, c, d, e, and f:

$$\begin{cases} ax + by = c \\ dx + ey = f \end{cases}$$

(Assume that $ae - bd \neq 0$.)

63. Solve the following system for x and y in terms of a and b, where $a \neq b$:

$$\begin{cases} \dfrac{a}{bx} + \dfrac{b}{ay} = a + b \\ \dfrac{b}{x} + \dfrac{a}{y} = a^2 + b^2 \end{cases}$$

C

64. Solve the following system for x and y in terms of a and b, where $ab \neq -1$:

$$\begin{cases} \dfrac{x + y - 1}{x - y + 1} = a \\ \dfrac{y - x + 1}{x - y + 1} = ab \end{cases}$$

Answer: $\left(\dfrac{a + 1}{ab + 1}, \dfrac{a(b + 1)}{ab + 1}\right)$

65. Given that the lines $7x + 5y = 4$, $x + ky = 3$, and $5x + y + k = 0$ are concurrent (pass through a common point), what are the possible values for k?

66. Solve the following system for x and y in terms of a and b:

$$\begin{cases} (a + b)x + (a^2 + b^2)y = a^3 + b^3 \\ (a - b)x + (a^2 - b^2)y = a^3 - b^3 \end{cases}$$

(Assume that a and b are nonzero and that $a \neq \pm b$.)

67. The vertices of triangle ABC are $A(2a, 0)$, $B(2b, 0)$, and $C(0, 2)$.

(a) Show that the equations of the sides are $x + by = 2b$, $x + ay = 2a$, and $y = 0$.

(b) Show that the equations of the medians are $x + (2a - b)y = 2a$, $x + (2b - a)y = 2b$, and $2x + (a + b)y = 2(a + b)$.

(c) Show that all three medians intersect at the point $\left(\frac{2}{3}(a + b), \frac{2}{3}\right)$.

(d) Show that the equations of the altitudes are $bx - y = 2ab$, $ax - y = 2ab$, and $x = 0$.

(e) Show that all three altitudes intersect at the point $(0, -2ab)$.

(f) Show that the equations of the perpendicular bisectors of the sides are $x = a + b$, $bx - y = b^2 - 1$, and $ax - y = a^2 - 1$.

(g) Show that all three perpendicular bisectors intersect at the point $(a + b, ab + 1)$.

(h) Show that the three points found in parts (c), (e), and (g) all lie on a line. This is the **Euler line** of the triangle.

10.2 GAUSSIAN ELIMINATION

A method of solution is perfect if we can foresee from the start, and even prove, that following that method we shall obtain our aim.

Gottfried Wilhelm von Leibniz (1646–1716)

... the first electronic computer, the ABC, named after its designers Atanasoff and Berry, was built specifically to solve systems of 29 equations in 29 unknowns, a formidable task without the aid of a computer.

Angela B. Shiflet in *Discrete Mathematics for Computer Science* (St. Paul: West Publishing Company, 1987)

In the previous section, we solved systems of linear equations in two unknowns. In this section, we introduce the technique known as Gaussian elimination for solving systems of linear equations in which there are more than two unknowns.*

As a first example, consider the following system of three linear equations in the three unknowns x, y, and z:

$$\begin{cases} 3x + 2y - z = -3 \\ 5y - 2z = 2 \\ 5z = 20 \end{cases}$$

This system is easy to solve using the process of *back-substitution*. Dividing both sides of the third equation by 5 yields $z = 4$. Then substituting $z = 4$ back into the second equation gives us

$$5y - 2(4) = 2$$
$$5y = 10$$
$$y = 2$$

Finally, substituting the values $z = 4$ and $y = 2$ back into the first equation yields

$$3x + 2(2) - 4 = -3$$
$$3x = -3$$
$$x = -1$$

We have now found that $x = -1$, $y = 2$, and $z = 4$. If you go back and check, you will find that these values indeed satisfy all three equations in the given system. Furthermore, the algebra we've just carried out shows that these are the only possible values for x, y, and z satisfying all three equations. We summarize by saying that the **ordered triple** $(-1, 2, 4)$ is the solution of the given system.

The system that we just considered was easy to solve (using back-

---

*The technique is named after Carl Friedrich Gauss (1777–1855). Early in the nineteenth century, Gauss used this technique (and introduced the method of least squares for minimizing errors) in analyzing the orbit of the asteroid Pallas. However, the essentials of Gaussian elimination were in existence long before Gauss's time. Indeed, a version of the method appears in the Chinese text *Chui-Chang Suan-Shu* ("Nine Chapters on the Mathematical Art"), written approximately two thousand years ago.

substitution) because of the special form in which it was written. This form is called **upper-triangular form**. Although the following definition of upper-triangular form refers to systems with three unknowns, the same type of definition can be given for systems with any number of unknowns. Table 1 displays examples of systems in upper-triangular form.

Upper-Triangular Form (Three Variables)

A system of linear equations in x, y, and z is said to be in **upper-triangular** form provided x appears in no equation after the first and y appears in no equation after the second. (It is possible that y may not even appear in the second equation.)

TABLE 1

Examples of Systems in Upper-Triangular Form

2 Unknowns: x, y	3 Unknowns: x, y, z	4 Unknowns: x, y, z, t
$\begin{cases} 3x + 5y = 7 \\ 8y = 5 \end{cases}$	$\begin{cases} 4x - 3y + 2z = -5 \\ 7y + z = 9 \\ -4z = 3 \end{cases}$	$\begin{cases} x - y + z - 4t = 1 \\ 3y - 2z + t = -1 \\ 3z - 5t = 4 \\ 6t = 7 \end{cases}$
	$\begin{cases} 15x - 2y + z = 1 \\ 3z = -8 \end{cases}$	$\begin{cases} 2x + y + 2z - t = -3 \\ 4z + 3t = 1 \\ 5t = 6 \end{cases}$
		$\begin{cases} 8x + 3y - z + t = 2 \\ 2y + z - 4t = 1 \end{cases}$

When we were solving linear systems of equations with two unknowns in the previous section, we observed that there were three possibilities: a unique solution, infinitely many solutions, and no solution. As the next three examples indicate, the situation is similar when dealing with larger systems.

EXAMPLE 1   Find all solutions of the system

$$\begin{cases} x + y + 2z = 2 \\ 3y - 4z = -5 \\ 6z = 3 \end{cases}$$

Solution   The system is in upper-triangular form, and we can use back-substitution. Dividing the third equation by 6 yields $z = \frac{1}{2}$. Substituting this value for z back into the second equation then yields

$$3y - 4\left(\tfrac{1}{2}\right) = -5$$
$$3y = -3$$
$$y = -1$$

Now, substituting the values $z = \frac{1}{2}$ and $y = -1$ back in the first equation, we obtain

$$x + (-1) + 2\left(\tfrac{1}{2}\right) = 2$$
$$x = 2$$

As you can easily check, the values $x = 2$, $y = -1$, and $z = \frac{1}{2}$ indeed satisfy all three equations. Furthermore, the algebra we've just carried out shows that these are the only possible values for x, y, and z satisfying all three equations. We summarize by saying that the unique solution to our system is the ordered triple $(2, -1, \frac{1}{2})$. ∎

EXAMPLE 2   Find all solutions of the system

$$\begin{cases} -2x + y + 3z = 6 \\ \qquad\qquad 2z = 10 \end{cases}$$

Solution   Again, the system is in upper-triangular form, and we use back-substitution. Solving the second equation for z yields $z = 5$. Then replacing z by 5 in the first equation gives us

$$-2x + y + 3(5) = 6$$
$$-2x + y = -9$$
$$y = 2x - 9$$

At this point, we've made use of both equations in the given system. There is no third equation to provide additional restrictions on x, y, or z. We know from the previous section that the equation $y = 2x - 9$ has infinitely many solutions, all of the form

$$(x, 2x - 9) \qquad \text{where } x \text{ is a real number}$$

It follows, then, that there are infinitely many solutions to the given system and they may be written

$$(x, 2x - 9, 5) \qquad \text{where } x \text{ is a real number}$$

For instance, choosing in succession $x = 0$, $x = 1$, and $x = 2$ yields the solutions $(0, -9, 5)$, $(1, -7, 5)$, and $(2, -5, 5)$. (We remark in passing that any linear system in upper-triangular form in which the number of unknowns exceeds the number of equations will always have infinitely many solutions.) ∎

EXAMPLE 3   Find all solutions of the system

$$\begin{cases} 4x - 7y + 3z = 1 \\ 3x + \ y - 2z = 4 \\ 4x - 7y + 3z = 6 \end{cases}$$

(Note that the system is not in upper-triangular form.)

Solution   Look at the left-hand sides of the first and third equations: they are identical. Thus, if there were values for x, y, and z that satisfied both equations, it would follow that $1 = 6$, which is clearly impossible. We conclude that the given system has no solutions. ∎

As Examples 1 and 2 have demonstrated, systems in upper-triangular form can be readily solved. In view of this, it would be useful to have a technique for converting a given system into an equivalent system in upper-triangular form. (The expression **equivalent system** means a system with exactly the same set of solutions as the original system.) **Gaussian elimination** is one such technique.

We will demonstrate this technique in Examples 4, 5, and 6. In using Gaussian elimination, we will rely on what are called the three **elementary operations**, listed in the box that follows. These are operations that can be performed on an equation in a system without altering the set of solutions of the system.

The Elementary Operations

1. Multiply both sides of an equation by a nonzero constant.

2. Interchange the order in which two equations of a system are listed.

3. To one equation add a multiple of another equation in the system.

EXAMPLE 4 Find all solutions of the system

$$\begin{cases} x + 2y + z = 3 \\ 2x + y + z = 16 \\ x + y + 2z = 9 \end{cases}$$

Solution First we want to eliminate x from the second and third equations. To eliminate x from the second equation, we add to it -2 times the first equation. The result is the equivalent system

$$\begin{cases} x + 2y + z = 3 \\ -3y - z = 10 \\ x + y + 2z = 9 \end{cases}$$

To eliminate x from the third equation, we add to it -1 times the first equation. The result is the equivalent system

$$\begin{cases} x + 2y + z = 3 \\ -3y - z = 10 \\ -y + z = 6 \end{cases}$$

Now to bring the system into upper-triangular form, we need to eliminate y from the third equation. We could do this by adding $-\frac{1}{3}$ times the second equation to the third equation. However, to avoid working with fractions as long as possible, we proceed instead to interchange the second and third equations to obtain the equivalent system

$$\begin{cases} x + 2y + z = 3 \\ -y + z = 6 \\ -3y - z = 10 \end{cases}$$

Now we add -3 times the second equation to the last equation to obtain the equivalent system

$$\begin{cases} x + 2y + z = 3 \\ -y + z = 6 \\ -4z = -8 \end{cases}$$

The system is now in upper-triangular form, and back-substitution yields, in turn, $z = 2$, $y = -4$, and $x = 9$. (Check this for yourself.) The required solution is therefore $(9, -4, 2)$. ▌▌▌

In Table 2 we list some convenient abbreviations that are used in describing the elementary operations. You should plan on using these abbreviations yourself: they'll make it simpler for you (and your instructor) to check your work. In Table 2, the notation E_i stands for the ith equation in a system. For instance, for the initial system in Example 4, the symbol E_1 denotes the first equation: $x + 2y + z = 3$.

TABLE 2

Abbreviations for the Elementary Operations

Abbreviation	Explanation
1. cE_i	Multiply both sides of the ith equation by c.
2. $E_i \leftrightarrow E_j$	Interchange the ith and jth equations.
3. $cE_i + E_j$	To the jth equation, add c times the ith equation.

EXAMPLE 5 Solve the system

$$\begin{cases} 4x - 3y + 2z = 40 \\ 5x + 9y - 7z = 47 \\ 9x + 8y - 3z = 97 \end{cases}$$

Solution
$$\begin{cases} 4x - 3y + 2z = 40 \\ 5x + 9y - 7z = 47 \\ 9x + 8y - 3z = 97 \end{cases} \xrightarrow{(-1)E_2 + E_1} \begin{cases} -x - 12y + 9z = -7 \\ 5x + 9y - 7z = 47 \\ 9x + 8y - 3z = 97 \end{cases}$$

adding -1 times the second equation to the first equation

$$\xrightarrow[9E_1 + E_3]{5E_1 + E_2} \begin{cases} -x - 12y + 9z = -7 \\ -51y + 38z = 12 \\ -100y + 78z = 34 \end{cases}$$

$$\xrightarrow{\frac{1}{2}E_3} \begin{cases} -x - 12y + 9z = -7 \\ -51y + 38z = 12 \\ -50y + 39z = 17 \end{cases}$$

$$\xrightarrow{(-1)E_3 + E_2} \begin{cases} -x - 12y + 9z = -7 \\ -y - z = -5 \\ -50y + 39z = 17 \end{cases}$$

to allow working with smaller but integral coefficients

$$\xrightarrow{-50E_2 + E_3} \begin{cases} -x - 12y + 9z = -7 \\ -y - z = -5 \\ 89z = 267 \end{cases}$$

The system is now in upper-triangular form. Solving the third equation, we obtain $z = 3$. Substituting this value back into the second equation yields $y = 2$. (Check this for yourself.) Finally, substituting $z = 3$ and $y = 2$ back into the first equation yields $x = 10$. (Again, check this for yourself.) The solution to the system is therefore $(10, 2, 3)$. ∎

EXAMPLE 6 Solve the system

$$\begin{cases} x + 2y + 4z = 0 \\ x + 3y + 9z = 0 \end{cases}$$

Solution This system is similar to the one in Example 2 in that there are fewer equations than there are unknowns. By subtracting the first equation from the second, we readily obtain an equivalent system in upper-triangular form:

$$\begin{cases} x + 2y + 4z = 0 \\ y + 5z = 0 \end{cases}$$

Although the system is now in upper-triangular form, notice that the second equation does not determine y or z uniquely; that is, there are infinitely many number pairs (y, z) satisfying the second equation. We can solve the second equation for y in terms of z; the result is $y = -5z$. Now we replace y with $-5z$ in the first equation to obtain

$$x + 2(-5z) + 4z = 0 \qquad \text{or} \qquad x = 6z$$

At this point, we've used both of the equations in the system to express x and y in terms of z. Furthermore, there is no third equation in the system to provide additional restrictions on x, y, or z. We therefore conclude that the given system has infinitely many solutions. These solutions have the form

$$(6z, -5z, z) \qquad \text{where } z \text{ is any real number}$$ ∎

EXAMPLE 7 Solve the system

$$\begin{cases} x - 4y + z = 3 \\ 3x + 5y - 2z = -1 \\ 7x + 6y - 3z = 2 \end{cases}$$

Solution $$\begin{cases} x - 4y + z = 3 \\ 3x + 5y - 2z = -1 \\ 7x + 6y - 3z = 2 \end{cases} \xrightarrow[-7E_1 + E_3]{-3E_1 + E_2} \begin{cases} x - 4y + z = 3 \\ 17y - 5z = -10 \\ 34y - 10z = -19 \end{cases}$$

$$\xrightarrow{-2E_2 + E_3} \begin{cases} x - 4y + z = 3 \\ 17y - 5z = -10 \\ 0 = 1 \end{cases}$$

From the third equation in this last system, we conclude that this system, and consequently the original system, has no solution. (Reason: If there *were* values for x, y, and z satisfying the original system, then it would follow that $0 = 1$, which is clearly impossible.) ∎

In the next two examples, we introduce the subject of **partial fractions**. The basic goal here is to write a given fractional expression as a sum or difference of two or more *simpler* fractions. For instance, in Example 8, we will be given the fraction $1/(x-1)(x+1)$ and we will be asked to find constants A and B so that

$$\frac{1}{(x-1)(x+1)} = \frac{A}{x-1} + \frac{B}{x+1}$$

When A and B are determined, we refer to the right-hand side of the preceding equation as the **partial fraction decomposition** of the given fraction. Detailed techniques for working with partial fractions are developed as the need arises in calculus courses. The two examples and the exercises that follow will provide you with ample background for that later work.

EXAMPLE 8 Determine constants A and B so that the following equation is an identity (i.e., the equation is to hold for all values of x for which the denominators are not zero):

$$\frac{1}{(x-1)(x+1)} = \frac{A}{x-1} + \frac{B}{x+1} \tag{1}$$

Solution First, to clear equation (1) of fractions, we multiply both sides of the equation by the quantity $(x-1)(x+1)$. This yields

$$1 = (x+1)A + (x-1)B = Ax + A + Bx - B$$
$$1 = (A+B)x + (A-B) \tag{2}$$

Since this last equation is to be an identity, we now reason as follows:

$$\begin{bmatrix} \text{coefficient of } x \text{ on the left-} \\ \text{hand side of equation (2)} \end{bmatrix} = \begin{bmatrix} \text{coefficient of } x \text{ on the right-} \\ \text{hand side of equation (2)} \end{bmatrix}$$

Therefore,

$$0 = A + B \tag{3}$$

Similarly, we have

$$\begin{bmatrix} \text{constant term on the left-} \\ \text{hand side of equation (2)} \end{bmatrix} = \begin{bmatrix} \text{constant term on the right-} \\ \text{hand side of equation (2)} \end{bmatrix}$$

Therefore,

$$1 = A - B \tag{4}$$

Rewriting equations (3) and (4) together gives us a system of two linear equations in the unknowns A and B:

$$\begin{cases} A + B = 0 \\ A - B = 1 \end{cases}$$

Adding these two equations yields

$$2A = 1 \qquad \text{and therefore} \qquad A = \tfrac{1}{2}$$

Subtracting the two equations gives us

$$B - (-B) = 0 - 1 \qquad \text{and therefore} \qquad B = -\tfrac{1}{2}$$

We have now found that $A = \tfrac{1}{2}$ and $B = -\tfrac{1}{2}$, as required. The partial fraction decomposition of $1/(x-1)(x+1)$ is therefore

$$\frac{1}{(x-1)(x+1)} = \frac{\tfrac{1}{2}}{x-1} + \frac{-\tfrac{1}{2}}{x+1}$$

$$= \frac{1}{2(x-1)} - \frac{1}{2(x+1)}$$

You should check for yourself that the result of combining the two fractions on the right-hand side of this last equation is indeed $1/(x-1)(x+1)$. ■■■

In Example 8, we were able to find the partial fraction decomposition by solving a system of two linear equations. There is a shortcut that is sometimes useful in such problems. As before, we multiply both sides of equation (1) by $(x-1)(x+1)$ to obtain

$$1 = (x+1)A + (x-1)B \tag{5}$$

Now, equation (5) is to hold for all values of x; in particular, it must hold when x is -1. Replacing x with -1 in the equation then yields

$$1 = -2B$$
$$B = -\tfrac{1}{2} \qquad \text{as obtained previously}$$

The value of A can be obtained in a similar way. We replace x by 1 in equation (5). As you can check, this readily yields $A = \tfrac{1}{2}$, which again agrees with the result we obtained previously.

EXAMPLE 9 Determine the constants A, B, and C so that the following equation is an identity:

$$\frac{2x+1}{(x+2)(x-3)^2} = \frac{A}{x+2} + \frac{B}{(x-3)} + \frac{C}{(x-3)^2}$$

Solution First, to clear the given equation of fractions, we multiply both sides by the quantity $(x+2)(x-3)^2$. After dividing out the common factors, we have

$$2x + 1 = A(x-3)^2 + B(x+2)(x-3) + C(x+2)$$
$$= A(x^2 - 6x + 9) + B(x^2 - x - 6) + C(x+2)$$
$$= Ax^2 - 6Ax + 9A + Bx^2 - Bx - 6B + Cx + 2C$$
$$2x + 1 = (A+B)x^2 + (-6A - B + C)x + 9A - 6B + 2C$$

Equating the coefficients of x^2 on both sides of this identity yields

$$A + B = 0$$

Similarly, equating the x-coefficients on both sides yields

$$-6A - B + C = 2$$

Finally, equating the constant terms on both sides gives us

$$9A - 6B + 2C = 1$$

We now have a system of three equations in the three unknowns A, B, and C:

$$\begin{cases} A + B &= 0 \\ -6A - B + C &= 2 \\ 9A - 6B + 2C &= 1 \end{cases}$$

Exercise 12 at the end of this section asks you to verify that the required values here are $A = -\frac{3}{25}$, $B = \frac{3}{25}$, and $C = \frac{7}{5}$. The partial fraction decomposition is therefore

$$\frac{2x + 1}{(x + 2)(x - 3)^2} = \frac{-3}{25(x + 2)} + \frac{3}{25(x - 3)} + \frac{7}{5(x - 3)^2}$$

Here is an alternative method for working Example 9; it relies on the short-cut mentioned earlier. Again, we first multiply the given equation by $(x + 2)(x - 3)^2$ to obtain

$$2x + 1 = A(x - 3)^2 + B(x + 2)(x - 3) + C(x + 2) \tag{6}$$

Now we just replace x by 3 in this equation. This yields

$$7 = 5C \qquad \text{or} \qquad C = \tfrac{7}{5} \qquad \text{as obtained previously}$$

The value of A can be obtained in a similar manner. We replace x by -2 in equation (6). As you can check, this readily yields $A = -\frac{3}{25}$, which again agrees with the value obtained previously. We've now computed A and C using the short method. But as you'll see if you try, the value of B cannot be obtained in the same way. One way to find B here is by substituting the values found for A and C in equation (6) to obtain

$$2x + 1 = -\tfrac{3}{25}(x - 3)^2 + B(x + 2)(x - 3) + \tfrac{7}{5}(x + 2)$$

This equation is to hold for all values of x; in particular, it must hold for the convenient value $x = 0$. Substituting $x = 0$ in this last equation yields

$$1 = -\tfrac{3}{25}(-3)^2 + B(2)(-3) + \tfrac{7}{5}(2)$$
$$25 = -27 - 150B + 70 \qquad \text{\color{red}multiplying by 25}$$
$$150B = 18 \qquad \text{\color{red}collecting like terms}$$
$$B = \tfrac{18}{150} = \tfrac{3}{25} \qquad \text{\color{red}dividing by 150 and simplifying}$$

We have now found that $B = \frac{3}{25}$, in agreement with the value obtained previously.

EXERCISE SET 10.2

A

In Exercises 1–10, the systems of linear equations are in upper-triangular form. Find all solutions of each system.

1. $\begin{cases} 2x + y + z = -9 \\ 3y - 2z = -4 \\ 8z = -8 \end{cases}$

2. $\begin{cases} -3x + 7y + 2z = -19 \\ y + z = 1 \\ -2z = -2 \end{cases}$

3. $\begin{cases} 8x + 5y + 3z = 1 \\ 3y + 4z = 2 \\ 5z = 3 \end{cases}$

4. $\begin{cases} 2x + 7z = -4 \\ 5y - 3z = 6 \\ 6z = 18 \end{cases}$

5. $\begin{cases} -4x + 5y = 0 \\ 3y + 2z = 1 \\ 3z = -1 \end{cases}$

6. $\begin{cases} 3x - 2y + z = 4 \\ 3z = 9 \end{cases}$

7. $\begin{cases} -x + 8y + 3z = 0 \\ 2z = 0 \end{cases}$

8. $\begin{cases} -x + y + z + w = 9 \\ 2y - z - w = 9 \\ 3z + 2w = 1 \\ 11w = 22 \end{cases}$

9. $\begin{cases} 2x + 3y + z + w = -6 \\ y + 3z - 4w = 23 \\ 6z - 5w = 31 \\ -2w = 10 \end{cases}$

10. $\begin{cases} 7x - y - z + w = 3 \\ 2y - 3z - 4w = -2 \\ 3w = 6 \end{cases}$

In Exercises 11–30, find all solutions of each system.

11. $\begin{cases} x + y + z = 12 \\ 2x - y - z = -1 \\ 3x + 2y + z = 22 \end{cases}$

12. $\begin{cases} A + B = 0 \\ -6A - B + C = 2 \\ 9A - 6B + 2C = 1 \end{cases}$
    Answer: $\left(-\frac{3}{25}, \frac{3}{25}, \frac{7}{5}\right)$

13. $\begin{cases} 2x - 3y + 2z = 4 \\ 4x + 2y + 3z = 7 \\ 5x + 4y + 2z = 7 \end{cases}$

14. $\begin{cases} x + 2z = 5 \\ y - 30z = -16 \\ x - 2y + 4z = 8 \end{cases}$

15. $\begin{cases} 3x + 3y - 2z = 13 \\ 6x + 2y - 5z = 13 \\ 7x + 5y - 3z = 26 \end{cases}$

16. $\begin{cases} 2x + 5y - 3z = 4 \\ 4x - 3y + 2z = 9 \\ 5x + 6y - 2z = 18 \end{cases}$

17. $\begin{cases} x + y + z = 1 \\ -2x + y + z = -2 \\ 3x + 6y + 6z = 5 \end{cases}$

18. $\begin{cases} 7x + 5y - 7z = -10 \\ 2x + y + z = 7 \\ x + y - 3z = -8 \end{cases}$

19. $\begin{cases} 2x - y + z = -1 \\ x + 3y - 2z = 2 \\ -5x + 6y - 5z = 5 \end{cases}$

20. $\begin{cases} -2x + 2y - z = 0 \\ 3x - 4y + z = 1 \\ 5x - 8y + z = 4 \end{cases}$

21. $\begin{cases} 2x - y + z = 4 \\ x + 3y + 2z = -1 \\ 7x + 5z = 11 \end{cases}$

22. $\begin{cases} 3x + y - z = 10 \\ 8x - y - 6z = -3 \\ 5x - 2y - 5z = 1 \end{cases}$

23. $\begin{cases} x + y + z + w = 4 \\ x - 2y - z - w = 3 \\ 2x - y + z - w = 2 \\ x - y + 2z - 2w = -7 \end{cases}$

24. $\begin{cases} x + y - 3z + 2w = 0 \\ -2x - 2y + 6z + w = -5 \\ -x + 3y + 3z + 3w = -5 \\ 2x + y - 3z - w = 4 \end{cases}$

25. $\begin{cases} 2x + 3y + 2z = 5 \\ x + 4y - 3z = 1 \end{cases}$

26. $\begin{cases} 4x - y - 3z = 2 \\ 6x + 5y - z = 0 \end{cases}$

27. $\begin{cases} x - 2y - 2z + 2w = -10 \\ 3x + 4y - z - 3w = 11 \\ -4x - 3y - 3z + 8w = -21 \end{cases}$

28. $\begin{cases} 2x + y + z + w = 1 \\ x + 3y - 3z - 3w = 0 \\ -3x - 4y + 2z + 2w = -1 \end{cases}$

29. $\begin{cases} 4x - 2y + 3z = -2 \\ 6y - 4z = 6 \end{cases}$

30. $\begin{cases} 6x - 2y - 5z + w = 3 \\ 5x - y + z + 5w = 4 \\ -4y - 7z - 9w = -5 \end{cases}$

In Exercises 31–40, determine the constants (denoted by capital letters) so that each equation is an identity.

31. (a) $\dfrac{1}{(x - 2)(x + 2)} = \dfrac{A}{x - 2} + \dfrac{B}{x + 2}$

    (b) $\dfrac{x}{(x - 2)(x + 2)} = \dfrac{A}{x - 2} + \dfrac{B}{x + 2}$

32. (a) $\dfrac{2}{(x + 3)(x + 2)} = \dfrac{A}{x + 3} + \dfrac{B}{x + 2}$

    (b) $\dfrac{x - 9}{(x + 3)(x + 2)} = \dfrac{A}{x + 3} + \dfrac{B}{x + 2}$

33. $\dfrac{3x}{(x + 4)(x - 5)} = \dfrac{A}{x + 4} + \dfrac{B}{x - 5}$

34. $\dfrac{4x + 26}{(x - 1)(x + 5)} = \dfrac{A}{x - 1} + \dfrac{B}{x + 5}$

35. (a) $\dfrac{3}{(x - 4)(x + 4)} = \dfrac{A}{x - 4} + \dfrac{B}{x + 4}$

    (b) $\dfrac{3}{(x - 4)(x + 4)^2} = \dfrac{A}{x - 4} + \dfrac{B}{x + 4} + \dfrac{C}{(x + 4)^2}$

36. (a) $\dfrac{1}{(x + 1)(x + 6)} = \dfrac{A}{x + 1} + \dfrac{B}{x + 6}$

    (b) $\dfrac{1}{(x + 1)(x + 6)^2} = \dfrac{A}{x + 1} + \dfrac{B}{x + 6} + \dfrac{C}{(x + 6)^2}$

37. $\dfrac{1}{(x + 1)(x^2 - x + 1)} = \dfrac{A}{x + 1} + \dfrac{Bx + C}{x^2 - x + 1}$

38. $\dfrac{1}{(x + 1)^2(x^2 - x + 1)} = \dfrac{A}{x + 1} + \dfrac{B}{(x + 1)^2} + \dfrac{Cx + D}{x^2 - x + 1}$

39. $\dfrac{4}{x(1 - x)} = \dfrac{A}{x} + \dfrac{B}{1 - x}$

40. $\dfrac{4}{x^2(1 - x)} = \dfrac{A}{x} + \dfrac{B}{x^2} + \dfrac{C}{1 - x}$

B

41. Determine the constants A, B, C, and D so that the following equation is an identity:

$$\frac{1}{(x^2 + x + 1)(x^2 - x + 1)} = \frac{Ax + B}{x^2 + x + 1} + \frac{Cx + D}{x^2 - x + 1}$$

42. Suppose that the height of an object as a function of time is given by $f(t) = at^2 + bt + c$, where t is time in

seconds, $f(t)$ is the height in feet at time t, and a, b, and c are certain constants. If after 1, 2, and 3 seconds the corresponding heights are 184 ft, 136 ft, and 56 ft, find the time at which the object is at ground level (height = 0 ft).

43. Find the equation of the circle passing through the points $(-1, -5)$, $(1, 2)$, and $(-2, 0)$. Write your answer in the form $Ax^2 + Ay^2 + Bx + Cy + D = 0$.

44. Find the equation of a circle that passes through the origin and the points $(8, -4)$ and $(7, -1)$. Specify the center and radius.

45. Solve the following system for x, y, and z:

$$\begin{cases} e^x + e^y - 2e^z = 2a \\ e^x + 2e^y - 4e^z = 3a \\ \tfrac{1}{2}e^x - 3e^y + e^z = -5a \end{cases}$$

(Assume that $a > 0$.)   *Hint:*   Let $A = e^x$, $B = e^y$, and $C = e^z$. Solve the resulting linear system for A, B, and C.

46. Solve the following system for α, β, and γ.

$$\begin{cases} \ln \alpha - \ln \beta - \ln \gamma = 2 \\ 3 \ln \alpha + 5 \ln \beta - 2 \ln \gamma = 1 \\ 2 \ln \alpha - 4 \ln \beta + \ln \gamma = 2 \end{cases}$$

47. The following figure displays three circles that are mutually tangent. The line segments joining the centers have lengths a, b, and c, as shown. Let r_1, r_2, and r_3 denote the radii of the circles, as indicated in the figure. Show that

$$r_1 = \frac{a + c - b}{2} \qquad r_2 = \frac{a + b - c}{2}$$

$$r_3 = \frac{b + c - a}{2}$$

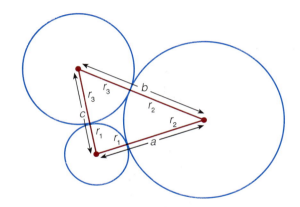

48. Consider the system

$$\begin{cases} \lambda x + y + z = a \\ x + \lambda y + z = b \\ x + y + \lambda z = c \end{cases}$$

(a) Assuming that the value of the constant λ is neither 1 nor -2, find all solutions of the system.

(b) If $\lambda = -2$, how must a, b, and c be related for the system to have solutions? Find these solutions.

(c) If $\lambda = 1$, how must a, b, and c be related for the system to have solutions? Find these solutions.

In Exercises 49 and 50, the lowercase letters a, b, p, and q denote given constants. In each case, determine the values of A and B so that the equation is an identity.

49. (a) $\dfrac{1}{(x - a)(x - b)} = \dfrac{A}{x - a} + \dfrac{B}{x - b}$ $(a \neq b)$

    (b) $\dfrac{px + q}{(x - a)(x - b)} = \dfrac{A}{x - a} + \dfrac{B}{x - b}$ $(a \neq b)$

50. (a) $\dfrac{1}{(x - a)(x + a)} = \dfrac{A}{x - a} + \dfrac{B}{x + a}$ $(a \neq 0)$

    (b) $\dfrac{px + q}{(x - a)(x + a)} = \dfrac{A}{x - a} + \dfrac{B}{x + a}$ $(a \neq 0)$

C

51. Express A, B, and C in terms of a, b, c, p, and q so that the following is an identity. (Assume a, b, and c are all unequal.)

$$\frac{x^2 + px + q}{(x - a)(x - b)(x - c)} = \frac{A}{x - a} + \frac{B}{x - b} + \frac{C}{x - c}$$

52. Express A, B, and C in terms of a, b, and c so that the following equation is an identity. (Assume a, b, and c are all unequal).

$$\frac{1}{(1 - ax)(1 - bx)(1 - cx)} = \frac{A}{1 - ax} + \frac{B}{1 - bx} + \frac{C}{1 - cx}$$

The following exercise appears in Algebra for Colleges and Schools *by H. S. Hall and S. R. Knight, revised by F. L. Sevenoak (New York: The Macmillan Company, 1906).*

53. A, B, and C are three towns forming a triangle. A man has to walk from one to the next, ride thence to the next, and drive thence to his starting point. He can walk, ride, and drive a mile in a, b, and c minutes, respectively. If he starts from B he takes $a + c - b$ hours, if he starts from C he takes $b + a - c$ hours, and if he starts from A he takes $c + b - a$ hours. Find the length of the circuit. [Assume that the circuit from A to B to C is counterclockwise.]

10.3 MATRICES

Arthur Cayley (1821–1895) and James Joseph Sylvester (1814–1897), two English mathematicians, invented the matrix ... in the 1850s.... The operations of addition and multiplication of matrices were later defined and the algebra of matrices was then developed. In 1925, Werner Heisenberg, a German physicist, used matrices in developing his theory of quantum mechanics, extending the role of matrices from algebra to the area of applied mathematics.

John K. Luedeman and Stanley M. Lukawecki in *Elementary Linear Algebra* (St. Paul: West Publishing Co., 1986)

Recall that in Gaussian Elimination, row operations are used to change the coefficient matrix to an upper triangular matrix. The solution is then found by back substitution, starting from the last equation in the reduced system.

Steven C. Althoen and Renate McLaughlin in "Gauss–Jordan Reduction: A Brief History," *American Mathematical Monthly* **94** (1987):130–142.

As you saw in the previous section, there can be a good deal of bookkeeping involved in using Gaussian elimination to solve systems of equations. We can lighten the load somewhat by using matrices, where a **matrix** is simply a rectangular array of numbers, enclosed in parentheses or brackets. Here are three examples:

$$\begin{pmatrix} 2 & 3 \\ -5 & 4 \end{pmatrix} \qquad \begin{pmatrix} -6 & 0 & 1 & \frac{1}{4} \\ \frac{2}{3} & 1 & 5 & 8 \end{pmatrix} \qquad \begin{pmatrix} \pi & 0 & 0 \\ 0 & 1 & 9 \\ -1 & -2 & 3 \\ -4 & 8 & 6 \end{pmatrix}$$

The particular numbers constituting a matrix are called its **entries** or **elements**. For instance, the entries in the matrix $\begin{pmatrix} 2 & 3 \\ -5 & 4 \end{pmatrix}$ are the four numbers 2, 3, -5, and 4. In this section, the entries will always be real numbers. However, it is also possible to consider matrices in which some or all of the entries are nonreal complex numbers.

It is convenient to agree on a standard system for labeling the rows and columns of a matrix. The rows are numbered from top to bottom and the columns from left to right, as indicated in the following example:

$$\begin{array}{c} \text{column 1} \quad \text{column 2} \\ \begin{array}{c} \text{row 1} \longrightarrow \\ \text{row 2} \longrightarrow \\ \text{row 3} \longrightarrow \end{array} \begin{pmatrix} 5 & -3 \\ 0 & 2 \\ -1 & 16 \end{pmatrix} \end{array}$$

We express the **size**, or **dimension**, of a matrix by specifying the number of rows and the number of columns, in that order. For instance, we would say that the matrix

$$\begin{pmatrix} 5 & -3 \\ 0 & 2 \\ -1 & 16 \end{pmatrix}$$

is a 3 × 2 (read "3 by 2") matrix, not 2 × 3. The following example will help fix in your mind the terminology that we have introduced.

EXAMPLE I   Consider the matrix

$$\begin{pmatrix} 1 & 3 & 5 \\ 7 & 9 & 11 \end{pmatrix}$$

(a) List the entries.
(b) What is the size of the matrix?
(c) Which element is in the second row, third column?

Solution   **(a)** The entries are 1, 3, 5, 7, 9, and 11.
(b) Since there are two rows and three columns, this is a 2 × 3 matrix.
(c) To locate the element in the second row, third column, draw lines through the second row and the third column and see where they intersect:

$$\begin{pmatrix} 1 & 3 & 5 \\ 7 & 9 & 11 \end{pmatrix}$$

Thus, the entry in row 2, column 3 is 11.

There is a natural way to use matrices to describe and solve systems of linear equations. Consider, for example, the following system of linear equations in **standard form** (with the x, y, and z terms lined up on the left-hand side and the constant terms on the right-hand side of each equation):

$$\begin{cases} x + 2y - 3z = 4 \\ 3x \quad\quad + z = 5 \\ -x - 3y + 4z = 0 \end{cases}$$

The **coefficient matrix** of this system is the matrix

$$\begin{pmatrix} 1 & 2 & -3 \\ 3 & 0 & 1 \\ -1 & -3 & 4 \end{pmatrix}$$

As the name implies, the coefficient matrix of the system is the matrix whose entries are the coefficients of x, y, and z, written in the same relative positions as they appear in the system. Notice the zero appearing in the second row, second column of the matrix. It is there because the coefficient of y in the second equation is in fact zero. The **augmented matrix** of the system of equations considered here is

$$\begin{pmatrix} 1 & 2 & -3 & 4 \\ 3 & 0 & 1 & 5 \\ -1 & -3 & 4 & 0 \end{pmatrix}$$

As you can see, the augmented matrix is formed by *augmenting* the coefficient matrix with the column of constant terms taken from the right-hand side of the given system of equations. To help relate the augmented matrix to the original system of equations, we will write the augmented matrix this way:

$$\left(\begin{array}{ccc|c} 1 & 2 & -3 & 4 \\ 3 & 0 & 1 & 5 \\ -1 & -3 & 4 & 0 \end{array}\right)$$

EXAMPLE 2 Write the coefficient matrix and the augmented matrix for the system

$$\begin{cases} 8x - 2y + z = 1 \\ 3x - 4z + y = 2 \\ 12y - 3z - 6 = 0 \end{cases}$$

Solution First write the system in standard form, with the x, y, and z, terms lined up and the constant terms on the right. This yields

$$\begin{cases} 8x - 2y + z = 1 \\ 3x + y - 4z = 2 \\ 12y - 3z = 6 \end{cases}$$

The coefficient matrix is then

$$\begin{pmatrix} 8 & -2 & 1 \\ 3 & 1 & -4 \\ 0 & 12 & -3 \end{pmatrix}$$

while the augmented matrix is

$$\left(\begin{array}{ccc:c} 8 & -2 & 1 & 1 \\ 3 & 1 & -4 & 2 \\ 0 & 12 & -3 & 6 \end{array}\right)$$

In the previous section we used the three elementary operations in solving systems of linear equations. In Table 1 we express these operations in the language of matrices. The matrix operations are called the **elementary row operations**.

TABLE 1

Elementary Operations for a System of Linear Equations	Corresponding Elementary Row Operation for a Matrix
1. Multiply both sides of an equation by a nonzero constant.	1′. Multiply each entry in a given row by a nonzero constant.
2. Interchange two equations.	2′. Interchange two rows.
3. To one equation, add a multiple of another equation.	3′. To one row, add a multiple of another row.

Table 2 displays an example of each elementary row operation. In the table, notice that the notation for describing these operations is essentially the same as that introduced in the previous section. For example, in the previous section, $10E_1$ indicated that the first equation in a system was multiplied by 10. Now, $10R_1$ indicates that each entry in the first *row* of a matrix is multiplied by 10.

We are now ready to use matrices to solve systems of equations. In the example that follows, we'll use the same system of equations used in Example 5 of the previous section. *Suggestion:* After reading the next example, carefully compare each step with the corresponding one taken in Example 5 of the previous section.

TABLE 2

Examples of the Elementary Row Operations		Comments
$\begin{pmatrix} 1 & 2 & 3 \\ 4 & 5 & 6 \\ 7 & 8 & 9 \end{pmatrix} \xrightarrow{\;10R_1\;} \begin{pmatrix} 10 & 20 & 30 \\ 4 & 5 & 6 \\ 7 & 8 & 9 \end{pmatrix}$		Multiply each entry in row 1 by 10.
$\begin{pmatrix} 1 & 2 & 3 \\ 4 & 5 & 6 \\ 7 & 8 & 9 \end{pmatrix} \xrightarrow{\;R_2 \leftrightarrow R_3\;} \begin{pmatrix} 1 & 2 & 3 \\ 7 & 8 & 9 \\ 4 & 5 & 6 \end{pmatrix}$		Interchange rows 2 and 3.
$\begin{pmatrix} 1 & 2 & 3 \\ 4 & 5 & 6 \\ 7 & 8 & 9 \end{pmatrix} \xrightarrow{\;-4R_1 + R_2\;} \begin{pmatrix} 1 & 2 & 3 \\ 0 & -3 & -6 \\ 7 & 8 & 9 \end{pmatrix}$		To each entry in row 2, add -4 times the corresponding entry in row 1.

EXAMPLE 3 Solve the system

$$\begin{cases} 4x - 3y + 2z = 40 \\ 5x + 9y - 7z = 47 \\ 9x + 8y - 3z = 97 \end{cases}$$

Solution

$$\left(\begin{array}{ccc|c} 4 & -3 & 2 & 40 \\ 5 & 9 & -7 & 47 \\ 9 & 8 & -3 & 97 \end{array}\right) \xrightarrow{(-1)R_2 + R_1} \left(\begin{array}{ccc|c} -1 & -12 & 9 & -7 \\ 5 & 9 & -7 & 47 \\ 9 & 8 & -3 & 97 \end{array}\right)$$

$$\xrightarrow[\;9R_1 + R_3\;]{5R_1 + R_2} \left(\begin{array}{ccc|c} -1 & -12 & 9 & -7 \\ 0 & -51 & 38 & 12 \\ 0 & -100 & 78 & 34 \end{array}\right)$$

$$\xrightarrow{\frac{1}{2}R_3} \left(\begin{array}{ccc|c} -1 & -12 & 9 & -7 \\ 0 & -51 & 38 & 12 \\ 0 & -50 & 39 & 17 \end{array}\right)$$

$$\xrightarrow{(-1)R_3 + R_2} \left(\begin{array}{ccc|c} -1 & -12 & 9 & -7 \\ 0 & -1 & -1 & -5 \\ 0 & -50 & 39 & 17 \end{array}\right)$$

$$\xrightarrow{-50R_2 + R_3} \left(\begin{array}{ccc|c} -1 & -12 & 9 & -7 \\ 0 & -1 & -1 & -5 \\ 0 & 0 & 89 & 267 \end{array}\right)$$

This last augmented matrix represents a system of equations in upper-triangular form:

$$\begin{cases} -x - 12y + 9z = -7 \\ -y - z = -5 \\ 89z = 267 \end{cases}$$

As you should now check for yourself, this yields the values $z = 3$, then $y = 2$, then $x = 10$. The solution of the original system is therefore $(10, 2, 3)$. ∎

For the remainder of this section (and in part of the next) we are going to study matrices without reference to systems of equations. As motivation for this, we point out that matrices are essential tools in many fields of study. For example, a knowledge of matrices and their properties is needed for work in computer graphics. To begin, we need to say what it means for two matrices to be equal.

DEFINITION   Equality of Matrices

Two matrices are equal provided they have the same size (same number of rows, same number of columns) and the corresponding entries are equal.

EXAMPLES

$$\begin{pmatrix} 2 & 3 \\ 4 & 5 \end{pmatrix} = \begin{pmatrix} 2 & 3 \\ 4 & 5 \end{pmatrix}$$

$$\begin{pmatrix} 2 & 3 & 0 \\ 4 & 5 & 0 \end{pmatrix} \neq \begin{pmatrix} 2 & 3 \\ 4 & 5 \end{pmatrix}$$

$$\begin{pmatrix} 2 & 3 \\ 4 & 5 \end{pmatrix} \neq \begin{pmatrix} 2 & 3 \\ 5 & 4 \end{pmatrix}$$

Now we can define matrix addition and subtraction. These operations are defined only between matrices of the same size.

DEFINITION   Matrix Addition

EXAMPLES

To add (or subtract) two matrices of the same size, add (or subtract) the corresponding entries.

$$\begin{pmatrix} 2 & 3 \\ -1 & 4 \end{pmatrix} + \begin{pmatrix} 6 & 1 \\ 0 & -4 \end{pmatrix} = \begin{pmatrix} 8 & 4 \\ -1 & 0 \end{pmatrix}$$

$$\begin{pmatrix} 2 & 3 \\ -1 & 4 \\ 9 & 10 \end{pmatrix} - \begin{pmatrix} 6 & 1 \\ 0 & -4 \\ 3 & 2 \end{pmatrix} = \begin{pmatrix} -4 & 2 \\ -1 & 8 \\ 6 & 8 \end{pmatrix}$$

Many properties of the real numbers also apply to matrices. For instance, matrix addition is *commutative:*

$$A + B = B + A \qquad \text{where } A \text{ and } B \text{ are matrices of the same size}$$

Matrix addition is also *associative:*

$$A + (B + C) = (A + B) + C \qquad \text{where } A, B, \text{ and } C \text{ are matrices of the same size}$$

In the next example, we verify these properties in two specific instances.

EXAMPLE 4   Let $A = \begin{pmatrix} 1 & 2 \\ 3 & 4 \end{pmatrix}$, $B = \begin{pmatrix} 0 & -5 \\ 8 & -1 \end{pmatrix}$, and $C = \begin{pmatrix} 6 & 7 \\ 8 & 9 \end{pmatrix}$.

(a) Show that $A + C = C + A$.
(b) Show that $A + (B + C) = (A + B) + C$.

Solution   **(a)** $A + C = \begin{pmatrix} 1 & 2 \\ 3 & 4 \end{pmatrix} + \begin{pmatrix} 6 & 7 \\ 8 & 9 \end{pmatrix} = \begin{pmatrix} 7 & 9 \\ 11 & 13 \end{pmatrix}$

$\phantom{\text{Solution   (a) }} C + A = \begin{pmatrix} 6 & 7 \\ 8 & 9 \end{pmatrix} + \begin{pmatrix} 1 & 2 \\ 3 & 4 \end{pmatrix} = \begin{pmatrix} 7 & 9 \\ 11 & 13 \end{pmatrix}$

This shows that $A + C = C + A$, since both $A + C$ and $C + A$ represent the matrix $\begin{pmatrix} 7 & 9 \\ 11 & 13 \end{pmatrix}$.

(b) First we compute $A + (B + C)$:

$$A + (B + C) = \begin{pmatrix} 1 & 2 \\ 3 & 4 \end{pmatrix} + \left[\begin{pmatrix} 0 & -5 \\ 8 & -1 \end{pmatrix} + \begin{pmatrix} 6 & 7 \\ 8 & 9 \end{pmatrix} \right]$$

$$= \begin{pmatrix} 1 & 2 \\ 3 & 4 \end{pmatrix} + \begin{pmatrix} 6 & 2 \\ 16 & 8 \end{pmatrix}$$

$$= \begin{pmatrix} 7 & 4 \\ 19 & 12 \end{pmatrix}$$

Next we compute $(A + B) + C$:

$$(A + B) + C = \left[\begin{pmatrix} 1 & 2 \\ 3 & 4 \end{pmatrix} + \begin{pmatrix} 0 & -5 \\ 8 & -1 \end{pmatrix} \right] + \begin{pmatrix} 6 & 7 \\ 8 & 9 \end{pmatrix}$$

$$= \begin{pmatrix} 1 & -3 \\ 11 & 3 \end{pmatrix} + \begin{pmatrix} 6 & 7 \\ 8 & 9 \end{pmatrix}$$

$$= \begin{pmatrix} 7 & 4 \\ 19 & 12 \end{pmatrix}$$

We conclude from these calculations that $A + (B + C) = (A + B) + C$, since both sides of that equation represent the matrix $\begin{pmatrix} 7 & 4 \\ 19 & 12 \end{pmatrix}$. ▬▬▬

We will now define an operation on matrices called **scalar multiplication**. First of all, the word *scalar* here just means *real number*. So we are talking about multiplying a matrix by a real number. (In more advanced work, nonreal complex scalars are also considered.)

DEFINITION   Scalar Multiplication

EXAMPLES

To multiply a matrix by a scalar, multiply each entry in the matrix by that scalar.

$$2 \begin{pmatrix} 5 & 9 & 0 \\ -1 & 2 & 3 \end{pmatrix} = \begin{pmatrix} 10 & 18 & 0 \\ -2 & 4 & 6 \end{pmatrix}$$

$$1 \begin{pmatrix} 1 & 2 \\ 3 & 4 \end{pmatrix} = \begin{pmatrix} 1 & 2 \\ 3 & 4 \end{pmatrix}$$

There are two simple but useful properties of scalar multiplication that are worth noting at this point. We'll omit the proofs of these two properties; however, Example 5 does ask us to verify them for a particular case.

PROPERTY SUMMARY   PROPERTIES OF SCALAR MULTIPLICATION

1. $c(kM) = (ck)M$ for all scalars c and k and any matrix M

2. $c(M + \mathcal{N}) = cM + c\mathcal{N}$,   where c is any scalar and M and $\mathcal{N}$ are any matrices of the same size

EXAMPLE 5   Let $c = 2$, $k = 3$, $M = \begin{pmatrix} 1 & 2 \\ 3 & 4 \end{pmatrix}$, and $N = \begin{pmatrix} 5 & 6 \\ 7 & 8 \end{pmatrix}$.

(a) Show that $c(kM) = (ck)M$.
(b) Show that $c(M + N) = cM + cN$.

Solution   (a) $c(kM) = 2\left[3\begin{pmatrix} 1 & 2 \\ 3 & 4 \end{pmatrix} \right] = 2\begin{pmatrix} 3 & 6 \\ 9 & 12 \end{pmatrix} = \begin{pmatrix} 6 & 12 \\ 18 & 24 \end{pmatrix}$

$(ck)M = (2 \cdot 3)\begin{pmatrix} 1 & 2 \\ 3 & 4 \end{pmatrix} = 6\begin{pmatrix} 1 & 2 \\ 3 & 4 \end{pmatrix} = \begin{pmatrix} 6 & 12 \\ 18 & 24 \end{pmatrix}$

Thus, $c(kM) = (ck)M$, since in both cases the result is $\begin{pmatrix} 6 & 12 \\ 18 & 24 \end{pmatrix}$.

(b) $c(M + N) = 2\left[\begin{pmatrix} 1 & 2 \\ 3 & 4 \end{pmatrix} + \begin{pmatrix} 5 & 6 \\ 7 & 8 \end{pmatrix} \right]$

$= 2\begin{pmatrix} 6 & 8 \\ 10 & 12 \end{pmatrix} = \begin{pmatrix} 12 & 16 \\ 20 & 24 \end{pmatrix}$

$cM + cN = 2\begin{pmatrix} 1 & 2 \\ 3 & 4 \end{pmatrix} + 2\begin{pmatrix} 5 & 6 \\ 7 & 8 \end{pmatrix}$

$= \begin{pmatrix} 2 & 4 \\ 6 & 8 \end{pmatrix} + \begin{pmatrix} 10 & 12 \\ 14 & 16 \end{pmatrix} = \begin{pmatrix} 12 & 16 \\ 20 & 24 \end{pmatrix}$

Thus, $c(M + N) = cM + cN$, since both sides equal $\begin{pmatrix} 12 & 16 \\ 20 & 24 \end{pmatrix}$.

A matrix with zeros for all of its entries plays the same role in matrix addition as does the number zero in ordinary addition of real numbers. For instance, in the case of 2×2 matrices, we have

$$\begin{pmatrix} a & b \\ c & d \end{pmatrix} + \begin{pmatrix} 0 & 0 \\ 0 & 0 \end{pmatrix} = \begin{pmatrix} a & b \\ c & d \end{pmatrix}$$

and

$$\begin{pmatrix} 0 & 0 \\ 0 & 0 \end{pmatrix} + \begin{pmatrix} a & b \\ c & d \end{pmatrix} = \begin{pmatrix} a & b \\ c & d \end{pmatrix}$$

for all real numbers a, b, c, and d. The matrix $\begin{pmatrix} 0 & 0 \\ 0 & 0 \end{pmatrix}$ is called the **additive identity** for 2×2 matrices. Similarly, any matrix with all zero entries is the additive identity for matrices of that size. It is sometimes convenient to denote an additive identity matrix by a boldface zero: **0**. With this notation, we can write

$$A + 0 = 0 + A = A \qquad \text{for any matrix } A$$

For this matrix equation, it is understood that the size of the matrix **0** is the same as the size of A. With this notation, we also have

$$A - A = 0 \qquad \text{for any matrix } A$$

Our last topic in this section is matrix multiplication. We will begin with the simplest case and then work up to the more general situation. By convention, a matrix with only one row is called a **row vector**. Examples of row vectors are

$$(2 \quad 13), \qquad (-1 \quad 4 \quad 3), \qquad \text{and} \qquad (0 \quad 0 \quad 0 \quad 1)$$

Similarly, a matrix with only one column is called a **column vector**. Examples are

$$\begin{pmatrix} 2 \\ 13 \end{pmatrix}, \qquad \begin{pmatrix} -1 \\ 4 \\ 3 \end{pmatrix}, \qquad \text{and} \qquad \begin{pmatrix} 0 \\ 0 \\ 0 \\ 1 \end{pmatrix}$$

The following definition tells us how to multiply a row vector and a column vector when they have the same number of entries.

The Inner Product of a Row Vector and a Column Vector

EXAMPLES

Let A be a row vector and B a column vector, and assume that the number of columns in A is the same as the number of rows in B. Then the **inner product** $A \cdot B$ is defined to be the number obtained by multiplying the corresponding entries and then adding the products.

$$(1 \quad 2 \quad 3) \cdot \begin{pmatrix} 4 \\ 5 \\ 6 \end{pmatrix} = 1 \cdot 4 + 2 \cdot 5 + 3 \cdot 6$$
$$= 32$$

$$(1 \quad 2) \cdot \begin{pmatrix} 4 \\ 5 \\ 6 \end{pmatrix} \quad \text{is not defined}$$

$$(1 \quad 2 \quad 3) \cdot \begin{pmatrix} 4 \\ 5 \end{pmatrix} \quad \text{is not defined}$$

An important observation here is that the end result of taking the inner product is always just a number. The definition of matrix product that we now give depends on this observation.

The Product of Two Matrices

Let A and B be two matrices, and assume that the number of columns in A is the same as the number of rows in B. Then the product matrix AB is computed according to the following rule.

   The entry in the ith row and the jth column of AB is the inner product of the ith row of A with the jth column of B.

The matrix AB will have as many rows as A and as many columns as B.

As an example of matrix multiplication, we will compute the product AB, where $A = \begin{pmatrix} 1 & 2 \\ 3 & 4 \end{pmatrix}$ and $B = \begin{pmatrix} 5 & 6 & 0 \\ 7 & 8 & 1 \end{pmatrix}$. In other words, we will compute $\begin{pmatrix} 1 & 2 \\ 3 & 4 \end{pmatrix}\begin{pmatrix} 5 & 6 & 0 \\ 7 & 8 & 1 \end{pmatrix}$. Before we attempt to carry out the calculations of any matrix multiplication, however, we should check on two points.

1. Is the product defined? That is, does the number of columns in A equal the number of rows in B? In this case, yes; the common number is 2.

2. What is the size of the product? According to the definition, the product AB will have as many rows as A and as many columns as B. Thus, the size of AB will be 2×3.

Schematically, then, the situation looks like this:

$$\begin{pmatrix} 1 & 2 \\ 3 & 4 \end{pmatrix} \begin{pmatrix} 5 & 6 & 0 \\ 7 & 8 & 1 \end{pmatrix} = \begin{pmatrix} ? & ? & ? \\ ? & ? & ? \end{pmatrix}$$

We have six positions to fill. The computations are presented in Table 3.

TABLE 3

Position	How to Compute	Computation
row 1, column 1	inner product of row 1 and column 1 $$\begin{pmatrix} 1 & 2 \\ 3 & 4 \end{pmatrix} \begin{pmatrix} 5 & 6 & 0 \\ 7 & 8 & 1 \end{pmatrix}$$	$1 \cdot 5 + 2 \cdot 7 = 19$
row 1, column 2	inner product of row 1 and column 2 $$\begin{pmatrix} 1 & 2 \\ 3 & 4 \end{pmatrix} \begin{pmatrix} 5 & 6 & 0 \\ 7 & 8 & 1 \end{pmatrix}$$	$1 \cdot 6 + 2 \cdot 8 = 22$
row 1, column 3	inner product of row 1 and column 3 $$\begin{pmatrix} 1 & 2 \\ 3 & 4 \end{pmatrix} \begin{pmatrix} 5 & 6 & 0 \\ 7 & 8 & 1 \end{pmatrix}$$	$1 \cdot 0 + 2 \cdot 1 = 2$
row 2, column 1	inner product of row 2 and column 1 $$\begin{pmatrix} 1 & 2 \\ 3 & 4 \end{pmatrix} \begin{pmatrix} 5 & 6 & 0 \\ 7 & 8 & 1 \end{pmatrix}$$	$3 \cdot 5 + 4 \cdot 7 = 43$
row 2, column 2	inner product of row 2 and column 2 $$\begin{pmatrix} 1 & 2 \\ 3 & 4 \end{pmatrix} \begin{pmatrix} 5 & 6 & 0 \\ 7 & 8 & 1 \end{pmatrix}$$	$3 \cdot 6 + 4 \cdot 8 = 50$
row 2, column 3	inner product of row 2 and column 3 $$\begin{pmatrix} 1 & 2 \\ 3 & 4 \end{pmatrix} \begin{pmatrix} 5 & 6 & 0 \\ 7 & 8 & 1 \end{pmatrix}$$	$3 \cdot 0 + 4 \cdot 1 = 4$

Reading from the table, we see that our result is

$$AB = \begin{pmatrix} 1 & 2 \\ 3 & 4 \end{pmatrix} \begin{pmatrix} 5 & 6 & 0 \\ 7 & 8 & 1 \end{pmatrix} = \begin{pmatrix} 19 & 22 & 2 \\ 43 & 50 & 4 \end{pmatrix}$$

EXAMPLE 6   Let $A = \begin{pmatrix} 1 & 2 \\ 3 & 4 \end{pmatrix}$ and $B = \begin{pmatrix} 5 & 6 \\ 7 & 8 \end{pmatrix}$. By computing AB and then BA, show that $AB \neq BA$. This shows that, in general, matrix multiplication is not commutative.

Solution
$$AB = \begin{pmatrix} 1 & 2 \\ 3 & 4 \end{pmatrix} \begin{pmatrix} 5 & 6 \\ 7 & 8 \end{pmatrix} = \begin{pmatrix} 1 \cdot 5 + 2 \cdot 7 & 1 \cdot 6 + 2 \cdot 8 \\ 3 \cdot 5 + 4 \cdot 7 & 3 \cdot 6 + 4 \cdot 8 \end{pmatrix} = \begin{pmatrix} 19 & 22 \\ 43 & 50 \end{pmatrix}$$

$$BA = \begin{pmatrix} 5 & 6 \\ 7 & 8 \end{pmatrix} \begin{pmatrix} 1 & 2 \\ 3 & 4 \end{pmatrix} = \begin{pmatrix} 5 \cdot 1 + 6 \cdot 3 & 5 \cdot 2 + 6 \cdot 4 \\ 7 \cdot 1 + 8 \cdot 3 & 7 \cdot 2 + 8 \cdot 4 \end{pmatrix} = \begin{pmatrix} 23 & 34 \\ 31 & 46 \end{pmatrix}$$

Comparing the two matrices AB and BA, we conclude that $AB \neq BA$.

EXERCISE SET 10.3

A

In Exercises 1–4, specify the size of each matrix.

1. (a) $\begin{pmatrix} -4 & 0 & 5 \\ 2 & 8 & -1 \end{pmatrix}$ (b) $\begin{pmatrix} 7 & 1 \\ 4 & -3 \\ 0 & 0 \end{pmatrix}$

2. (a) $\begin{pmatrix} 1 & 0 \\ 0 & -1 \end{pmatrix}$ (b) $\begin{pmatrix} 1 \\ 6 \\ 8 \\ 1 \end{pmatrix}$

3. $\begin{pmatrix} 1 & a & b & c \\ a & 1 & 0 & a \\ b & 0 & 1 & b \\ c & a & b & 1 \\ 0 & 0 & 0 & 1 \end{pmatrix}$ 4. $(-3 \quad 1 \quad 6 \quad 0)$

In Exercises 5–8, write the coefficient matrix and the augmented matrix for each system.

5. $\begin{cases} 2x + 3y + 4z = 10 \\ 5x + 6y + 7z = 9 \\ 8x + 9y + 10z = 8 \end{cases}$ 6. $\begin{cases} 5x - y + z = 0 \\ 4y + 2z = 1 \\ 3x + y + z = -1 \end{cases}$

7. $\begin{cases} x \quad\;\; + z + w = -1 \\ x + y \quad\;\; + 2w = 0 \\ \quad\;\; y + z + w = 1 \\ 2x - y - z \quad\;\; = 2 \end{cases}$ 8. $\begin{cases} 8x - 8y \quad\;\; = 5 \\ x - y + z = 1 \end{cases}$

In Exercises 9–22, use matrices to solve each system of equations.

9. $\begin{cases} x - y + 2z = 7 \\ 3x + 2y - z = -10 \\ -x + 3y + z = -2 \end{cases}$ 10. $\begin{cases} 2x - 3y + 4z = 14 \\ 3x - 2y + 2z = 12 \\ 4x + 5y - 5z = 16 \end{cases}$

11. $\begin{cases} x \quad\;\; + z = -2 \\ -3x + 2y \quad\;\; = 17 \\ x - y - z = -9 \end{cases}$ 12. $\begin{cases} 5x + y + 10z = 23 \\ 4x + 2y - 10z = 76 \\ 3x - 4y \quad\;\; = 18 \end{cases}$

13. $\begin{cases} x + y + z = -4 \\ 2x - 3y + z = -1 \\ 4x + 2y - 3z = 33 \end{cases}$ 14. $\begin{cases} 2x + 3y - 4z = 7 \\ x - y + z = -\frac{3}{2} \\ 6x - 5y - 2z = -7 \end{cases}$

15. $\begin{cases} 3x - 2y + 6z = 0 \\ x + 3y + 20z = 15 \\ 10x - 11y - 10z = -9 \end{cases}$ 16. $\begin{cases} 3A - 3B + C = 4 \\ 6A + 9B - 3C = -7 \\ A - 2B - 2C = -3 \end{cases}$

17. $\begin{cases} 4x - 3y + 3z = 2 \\ 5x + y - 4z = 1 \\ 9x - 2y - z = 3 \end{cases}$ 18. $\begin{cases} 6x + y - z = -1 \\ -3x + 2y + 2z = 2 \\ 5y + 3z = 1 \end{cases}$

19. $\begin{cases} x - y + z + w = 6 \\ x + y - z + w = 4 \\ x + y + z - w = -2 \\ -x + y + z + w = 0 \end{cases}$

20. $\begin{cases} x + 2y - z - 2w = 5 \\ 2x + y + 2z + w = -7 \\ -2x - y - 3z - 2w = 10 \\ z + w = -3 \end{cases}$

21. $\begin{cases} 15A + 14B + 26C = 1 \\ 18A + 17B + 32C = -1 \\ 21A + 20B + 38C = 0 \end{cases}$

22. $\begin{cases} A + B + C + D + E = -1 \\ 3A - 2B - 2C + 3D + 2E = 13 \\ 3C + 4D - 4E = 7 \\ 5A - 4B \quad\quad\;\; + E = 30 \\ C \quad\quad\;\; - 2E = 3 \end{cases}$

In Exercises 23–50, the matrices A, B, C, D, E, F, and G are defined as follows:

$$A = \begin{pmatrix} 2 & 3 \\ -1 & 4 \end{pmatrix} \quad B = \begin{pmatrix} 1 & -1 \\ 3 & 0 \end{pmatrix} \quad C = \begin{pmatrix} 1 & 0 \\ 0 & 1 \end{pmatrix}$$

$$D = \begin{pmatrix} -1 & 2 & 3 \\ 4 & 0 & 5 \end{pmatrix} \quad E = \begin{pmatrix} 2 & 1 \\ 8 & -1 \\ 6 & 5 \end{pmatrix}$$

$$F = \begin{pmatrix} 5 & -1 \\ -4 & 0 \\ 2 & 3 \end{pmatrix} \quad G = \begin{pmatrix} 0 & 0 \\ 0 & 0 \\ 0 & 0 \end{pmatrix}$$

In each exercise, carry out the indicated matrix operations if they are defined. If an operation is not defined, say so.

23. $A + B$	24. $A - B$	25. $2A + 2B$
26. $2(A + B)$	27. AB	28. BA
29. AC	30. CA	31. $3D + E$
32. $E + F$	33. $2F - 3G$	34. DE
35. ED	36. DF	37. FD
38. $A + D$	39. $G + A$	40. DG
41. GD	42. $(A + B) + C$	43. $A + (B + C)$
44. CD	45. DC	46. $5E - 3F$
47. $A^2(=AA)$	48. A^2A	49. AA^2
50. C^2		

51. Let

$$A = \begin{pmatrix} -1 & 3 & 4 \\ 3 & 2 & -3 \\ 9 & 1 & 6 \end{pmatrix} \quad B = \begin{pmatrix} 7 & 0 & 1 \\ 0 & 0 & 3 \\ -1 & 2 & 4 \end{pmatrix}$$

$$C = \begin{pmatrix} 4 & 6 & 1 \\ 2 & 1 & 3 \\ -1 & -1 & 2 \end{pmatrix}$$

(a) Compute $A(B + C)$. (b) Compute $AB + AC$.
(c) Compute $(AB)C$ (d) Compute $A(BC)$.

B

52. Let $A = \begin{pmatrix} 1 & 2 \\ 3 & 4 \end{pmatrix}$ and $B = \begin{pmatrix} 5 & 6 \\ 7 & 8 \end{pmatrix}$. Let A^2 and B^2 denote the matrix products AA and BB, respectively. Compute each of the following.
    (a) $(A + B)(A + B)$ (b) $A^2 + 2AB + B^2$
    (c) $A^2 + AB + BA + B^2$

53. Let $A = \begin{pmatrix} 3 & 5 \\ 7 & 9 \end{pmatrix}$ and $B = \begin{pmatrix} 2 & 4 \\ 6 & 8 \end{pmatrix}$. Compute each of the following.
    (a) $A^2 - B^2$ (b) $(A - B)(A + B)$
    (c) $(A + B)(A - B)$ (d) $A^2 + AB - BA - B^2$

54. Let
$$A = \begin{pmatrix} 1 & 0 \\ 0 & 1 \end{pmatrix} \qquad B = \begin{pmatrix} 1 & 0 \\ 0 & -1 \end{pmatrix}$$
$$C = \begin{pmatrix} -1 & 0 \\ 0 & 1 \end{pmatrix} \qquad D = \begin{pmatrix} -1 & 0 \\ 0 & -1 \end{pmatrix}$$

Complete the following multiplication table.

	A	B	C	D
A				
B			D	
C				
D				

Hint: In the second row, third column, D is the proper entry because (as you can check) $BC = D$.

55. In this exercise, let us agree to write the coordinates (x, y) of a point in the plane as the 2×1 matrix $\begin{pmatrix} x \\ y \end{pmatrix}$.
    (a) Let $A = \begin{pmatrix} 1 & 0 \\ 0 & -1 \end{pmatrix}$ and $Z = \begin{pmatrix} x \\ y \end{pmatrix}$. Compute the matrix AZ. After computing AZ, observe that it represents the point obtained by reflecting $\begin{pmatrix} x \\ y \end{pmatrix}$ about the x-axis.
    (b) Let $B = \begin{pmatrix} -1 & 0 \\ 0 & 1 \end{pmatrix}$ and $Z = \begin{pmatrix} x \\ y \end{pmatrix}$. Compute the matrix BZ. After computing BZ, observe that it represents the point obtained by reflecting $\begin{pmatrix} x \\ y \end{pmatrix}$ about the y-axis.

(c) Let A, B, and Z represent the matrices defined in parts (a) and (b). Compute the matrix $(AB)Z$, and then interpret it in terms of reflection about the axes.

56. If $A = \begin{pmatrix} a & b \\ c & d \end{pmatrix}$ and $I = \begin{pmatrix} 1 & 0 \\ 0 & 1 \end{pmatrix}$, show that
$$AI = IA = A$$
(The matrix I is called the *identity matrix* of order 2.)

57. A function f is defined as follows. The domain of f is the set of all 2×2 matrices (with real entries). If $A = \begin{pmatrix} a & b \\ c & d \end{pmatrix}$, then $f(A) = ad - bc$.
    (a) Let $A = \begin{pmatrix} 1 & 2 \\ 3 & 4 \end{pmatrix}$ and $B = \begin{pmatrix} 3 & -1 \\ 5 & 8 \end{pmatrix}$. Compute $f(A)$, $f(B)$, and $f(AB)$. Is it true, in this case, that $f(A) \cdot f(B) = f(AB)$?
    (b) Let $A = \begin{pmatrix} a & b \\ c & d \end{pmatrix}$ and $B = \begin{pmatrix} e & f \\ g & h \end{pmatrix}$. Show that $f(A) \cdot f(B) = f(AB)$.

58. The **trace** of a 2×2 matrix $\begin{pmatrix} a & b \\ c & d \end{pmatrix}$ is defined by
$$\mathrm{tr}\begin{pmatrix} a & b \\ c & d \end{pmatrix} = a + d$$
    (a) If $A = \begin{pmatrix} 1 & 2 \\ 3 & 4 \end{pmatrix}$ and $B = \begin{pmatrix} 5 & 6 \\ 7 & 8 \end{pmatrix}$, verify that $\mathrm{tr}(A + B) = \mathrm{tr}\,A + \mathrm{tr}\,B$.
    (b) If $A = \begin{pmatrix} a & b \\ c & d \end{pmatrix}$ and $B = \begin{pmatrix} e & f \\ g & h \end{pmatrix}$, show that $\mathrm{tr}(A + B) = \mathrm{tr}\,A + \mathrm{tr}\,B$.

59. Let $A = \begin{pmatrix} a & b \\ c & d \end{pmatrix}$. The **transpose** of A is the matrix denoted by A^T and defined by $A^\mathrm{T} = \begin{pmatrix} a & c \\ b & d \end{pmatrix}$. In other words, A^T is obtained by switching the columns and rows of A. Show that the following equations hold for all 2×2 matrices A and B.
    (a) $(A + B)^\mathrm{T} = A^\mathrm{T} + B^\mathrm{T}$ (b) $(A^\mathrm{T})^\mathrm{T} = A$
    (c) $(AB)^\mathrm{T} = B^\mathrm{T}A^\mathrm{T}$

60. Find an example of two 2×2 matrices A and B for which $AB = \mathbf{0}$ but neither A nor B is $\mathbf{0}$.

10.4 THE INVERSE OF A SQUARE MATRIX

We have noted Cayley's work in analytic geometry, especially in connection with the use of determinants; but [Arthur] Cayley [1821–1895] also was one of the first men to study matrices, another instance of the British concern for form and structure in algebra.

Carl B. Boyer in *A History of Mathematics*, 2d ed., revised by Uta C. Merzback (New York: John Wiley and Sons, 1991)

A matrix in which there are as many rows as there are columns is called a **square matrix**. So, two examples of square matrices are

$$A = \begin{pmatrix} 1 & 2 \\ 3 & 4 \end{pmatrix} \quad \text{and} \quad B = \begin{pmatrix} -5 & 6 & 7 \\ \frac{1}{2} & 0 & 1 \\ 8 & 4 & -3 \end{pmatrix}$$

The matrix A is said to be a square matrix of **order two** (or, more simply, a 2×2 matrix); B is a square matrix of **order three** (that is, a 3×3 matrix). We will first present the concepts and techniques of this section in terms of square matrices of order two. After that, we'll show how the ideas carry over to larger square matrices.

We begin by defining a special matrix I_2:

$$I_2 = \begin{pmatrix} 1 & 0 \\ 0 & 1 \end{pmatrix}$$

The matrix I_2 has the following property. For every 2×2 matrix $A = \begin{pmatrix} a & b \\ c & d \end{pmatrix}$, we have (as you can easily verify)

$$\begin{pmatrix} a & b \\ c & d \end{pmatrix}\begin{pmatrix} 1 & 0 \\ 0 & 1 \end{pmatrix} = \begin{pmatrix} a & b \\ c & d \end{pmatrix} \quad \text{and} \quad \begin{pmatrix} 1 & 0 \\ 0 & 1 \end{pmatrix}\begin{pmatrix} a & b \\ c & d \end{pmatrix} = \begin{pmatrix} a & b \\ c & d \end{pmatrix}$$

In other words,

$$AI_2 = A \quad \text{and} \quad I_2A = A$$

This shows that the matrix I_2 plays the same role in the multiplication of 2×2 matrices as does the number 1 in the multiplication of real numbers. For this reason, I_2 is called the **multiplicative identity** for square matrices of order two.

If we have two real numbers a and b such that $ab = 1$, then we say that a and b are (multiplicative) inverses. In the box that follows we apply this same terminology to matrices.

DEFINITION   The Inverse of a 2 × 2 Matrix

If A and B are 2×2 matrices such that

$$AB = I_2 \quad \text{and} \quad BA = I_2$$

then A and B are said to be **inverses** of one another.

EXAMPLE 1   Find the inverse of $A = \begin{pmatrix} 1 & 2 \\ 3 & 4 \end{pmatrix}$.

Solution   We need to find numbers a, b, c, and d such that the following two matrix equations are valid.

$$\begin{pmatrix} 1 & 2 \\ 3 & 4 \end{pmatrix}\begin{pmatrix} a & b \\ c & d \end{pmatrix} = \begin{pmatrix} 1 & 0 \\ 0 & 1 \end{pmatrix} \tag{1}$$

$$\begin{pmatrix} a & b \\ c & d \end{pmatrix}\begin{pmatrix} 1 & 2 \\ 3 & 4 \end{pmatrix} = \begin{pmatrix} 1 & 0 \\ 0 & 1 \end{pmatrix} \tag{2}$$

From equation (1), we have

$$\begin{pmatrix} a + 2c & b + 2d \\ 3a + 4c & 3b + 4d \end{pmatrix} = \begin{pmatrix} 1 & 0 \\ 0 & 1 \end{pmatrix}$$

For this last equation to be valid, the corresponding entries in the two matrices must be equal. (Why?) Consequently, we obtain four equations, two involving a and c, and two involving b and d:

$$\begin{cases} a + 2c = 1 \\ 3a + 4c = 0 \end{cases} \qquad \begin{cases} b + 2d = 0 \\ 3b + 4d = 1 \end{cases}$$

As Exercise 32(a) asks you to show, the solution to the first system is $a = -2$ and $c = \frac{3}{2}$; and for the second system, $b = 1$ and $d = -\frac{1}{2}$. Furthermore (as you can check), these same values are obtained for a, b, c, and d if we begin with matrix equation (2) rather than (1). Thus, the required inverse matrix is

$$\begin{pmatrix} -2 & 1 \\ \frac{3}{2} & -\frac{1}{2} \end{pmatrix}$$

Exercise 32(b) asks you to carry out the matrix multiplication to confirm that we indeed have

$$\begin{pmatrix} 1 & 2 \\ 3 & 4 \end{pmatrix}\begin{pmatrix} -2 & 1 \\ \frac{3}{2} & -\frac{1}{2} \end{pmatrix} = \begin{pmatrix} -2 & 1 \\ \frac{3}{2} & -\frac{1}{2} \end{pmatrix}\begin{pmatrix} 1 & 2 \\ 3 & 4 \end{pmatrix}$$
$$= \begin{pmatrix} 1 & 0 \\ 0 & 1 \end{pmatrix}$$

In Example 1 we found the inverse of a square matrix by solving systems of equations. However, as you've seen in previous sections, there are systems of equations that do not have solutions. In the present context, this implies that there are matrices that do not have inverses. For example (as Exercise 33 asks you to show), the matrix $\begin{pmatrix} 2 & 5 \\ 6 & 15 \end{pmatrix}$ does not have an inverse. A matrix that does not have an inverse is called a **singular matrix**. If a matrix does possess an inverse, we say that the matrix is **nonsingular**. So, according to the result in Example 1, the matrix $\begin{pmatrix} 1 & 2 \\ 3 & 4 \end{pmatrix}$ is nonsingular.

It can be proven that if a square matrix has an inverse, then that inverse is unique. (In other words, you can never find two different inverses for the same matrix.) The inverse of a nonsingular matrix A is denoted by A^{-1} (read A *inverse*).

There are several methods that can be used to compute the inverse of a matrix (or to show that there is no inverse). One method involves working with systems of equations, as in Example 1. A more efficient method, however, is the following. (We'll demonstrate the method with an example, but we won't give a formal proof of its validity.)

Suppose that we want to compute the inverse (if it exists) for the matrix

$$A = \begin{pmatrix} 6 & 2 \\ 13 & 4 \end{pmatrix}$$

We begin by writing a larger matrix, formed by merging A and I_2 as follows.

$$\begin{pmatrix} 6 & 2 & \vdots & 1 & 0 \\ 13 & 4 & \vdots & 0 & 1 \end{pmatrix}$$

A —————↑   ↑————— I_2

A convenient abbreviation for this matrix is $(A \mid I_2)$. Now we carry out elementary row operations on this matrix until one of the following two situations occurs:

1. Either the matrix takes on the form

$$\begin{pmatrix} 1 & 0 & \vdots & a & b \\ 0 & 1 & \vdots & c & d \end{pmatrix}$$

I_2 —————↑

In this case, $A^{-1} = \begin{pmatrix} a & b \\ c & d \end{pmatrix}$.

2. Or, one of the rows to the left of the dashed line consists entirely of zeros. In this case, A^{-1} does not exist.

The calculations for our example then run as follows.

$$\begin{pmatrix} 6 & 2 & \vdots & 1 & 0 \\ 13 & 4 & \vdots & 0 & 1 \end{pmatrix} \xrightarrow{-2R_1 + R_2} \begin{pmatrix} 6 & 2 & \vdots & 1 & 0 \\ 1 & 0 & \vdots & -2 & 1 \end{pmatrix}$$

$$\xrightarrow{R_1 \leftrightarrow R_2} \begin{pmatrix} 1 & 0 & \vdots & -2 & 1 \\ 6 & 2 & \vdots & 1 & 0 \end{pmatrix}$$

$$\xrightarrow{-6R_1 + R_2} \begin{pmatrix} 1 & 0 & \vdots & -2 & 1 \\ 0 & 2 & \vdots & 13 & -6 \end{pmatrix}$$

$$\xrightarrow{\frac{1}{2}R_2} \begin{pmatrix} 1 & 0 & \vdots & -2 & 1 \\ 0 & 1 & \vdots & \frac{13}{2} & -3 \end{pmatrix}$$

The inverse matrix can now be read off:

$$A^{-1} = \begin{pmatrix} -2 & 1 \\ \frac{13}{2} & -3 \end{pmatrix}$$

(You should verify for yourself that we indeed have $AA^{-1} = I_2$ and $A^{-1}A = I_2$.)

Inverse matrices can be used in solving certain systems of equations in which the number of unknowns is the same as the number of equations. Before explaining how this works, we describe how a system of equations can be written in matrix form. Consider the system

$$\begin{cases} x + 2y = 8 \\ 3x + 4y = 6 \end{cases} \tag{3}$$

As defined in the previous section, the coefficient matrix A for this system is

$$A = \begin{pmatrix} 1 & 2 \\ 3 & 4 \end{pmatrix}$$

Now we define two matrices X and B:

$$X = \begin{pmatrix} x \\ y \end{pmatrix} \qquad B = \begin{pmatrix} 8 \\ 6 \end{pmatrix}$$

Then system (3) can be written as a single matrix equation:

$$AX = B \qquad (4)$$

To see why this is so, we expand equation (4) to obtain

$$\begin{pmatrix} 1 & 2 \\ 3 & 4 \end{pmatrix}\begin{pmatrix} x \\ y \end{pmatrix} = \begin{pmatrix} 8 \\ 6 \end{pmatrix} \qquad \text{\color{red}using the definitions of } A, X, \text{ and } B$$

$$\begin{pmatrix} x + 2y \\ 3x + 4y \end{pmatrix} = \begin{pmatrix} 8 \\ 6 \end{pmatrix} \qquad \text{\color{red}carrying out the matrix multiplication}$$

By equating the corresponding entries of the matrices in this last equation, we obtain $x + 2y = 8$ and $3x + 4y = 6$, as given initially in system (3).

EXAMPLE 2 Use an inverse matrix to solve the system

$$\begin{cases} x + 2y = 8 \\ 3x + 4y = 6 \end{cases}$$

Solution As explained just prior to this example, the matrix form for this system is

$$AX = B \qquad (5)$$

where

$$A = \begin{pmatrix} 1 & 2 \\ 3 & 4 \end{pmatrix} \qquad X = \begin{pmatrix} x \\ y \end{pmatrix} \qquad B = \begin{pmatrix} 8 \\ 6 \end{pmatrix}$$

From Example 1, we know that

$$A^{-1} = \begin{pmatrix} -2 & 1 \\ \frac{3}{2} & -\frac{1}{2} \end{pmatrix}$$

Now we multiply both sides of equation (5) by A^{-1} to obtain

$$A^{-1}(AX) = A^{-1}B$$
$$(A^{-1}A)X = A^{-1}B \qquad \text{\color{red}Matrix multiplication is associative.}$$
$$I_2 X = A^{-1}B$$
$$X = A^{-1}B \qquad \text{(Why?)}$$

Substituting the actual matrices into this last equation, we have

$$\begin{pmatrix} x \\ y \end{pmatrix} = \begin{pmatrix} -2 & 1 \\ \frac{3}{2} & -\frac{1}{2} \end{pmatrix}\begin{pmatrix} 8 \\ 6 \end{pmatrix} = \begin{pmatrix} -10 \\ 9 \end{pmatrix}$$

Therefore $x = -10$ and $y = 9$, as required.

All of the ideas we have discussed for square matrices of order two can be carried over rather directly to larger square matrices. It is easy to check that the following matrices, I_3 and I_4, are the multiplicative identities for 3×3 and 4×4 matrices, respectively.

$$I_3 = \begin{pmatrix} 1 & 0 & 0 \\ 0 & 1 & 0 \\ 0 & 0 & 1 \end{pmatrix} \qquad I_4 = \begin{pmatrix} 1 & 0 & 0 & 0 \\ 0 & 1 & 0 & 0 \\ 0 & 0 & 1 & 0 \\ 0 & 0 & 0 & 1 \end{pmatrix}$$

These identity matrices are described by saying that they have ones down the **main diagonal** and zeros everywhere else. (Larger identity matrices can be defined following this same pattern.) Sometimes, when it is clear from the context or as a matter of convenience, we'll omit the subscript and denote the appropriately sized identity matrix simply by I. (This is done in the box that follows.)

PROPERTY SUMMARY   THE INVERSE OF A SQUARE MATRIX

1. Suppose that A and B are square matrices of the same size, and let I denote the identity matrix of the same size. Then A and B are said to be **inverses** of one another provided

$$AB = I \quad \text{and} \quad BA = I$$

2. It can be shown that every square matrix has at most one inverse.

3. A square matrix that has an inverse is said to be **nonsingular** (or **invertible**). A square matrix that does not have an inverse is called **singular**.

4. If the matrix A is nonsingular, then the inverse of A is denoted by A^{-1}. In this case, we have

$$AA^{-1} = I \quad \text{and} \quad A^{-1}A = I$$

EXAMPLE 3   Let $A = \begin{pmatrix} 5 & 0 & 2 \\ 2 & 2 & 1 \\ -3 & 1 & -1 \end{pmatrix}$. Use the elementary row operations to compute A^{-1}, if it exists.

Solution   Following the method that we described for the 2×2 case, we first write down the matrix $(A \mid I_3)$ formed by merging A and I_3:

$$\begin{pmatrix} 5 & 0 & 2 & \vdots & 1 & 0 & 0 \\ 2 & 2 & 1 & \vdots & 0 & 1 & 0 \\ -3 & 1 & -1 & \vdots & 0 & 0 & 1 \end{pmatrix}$$

Now we carry out the elementary row operations, trying to obtain I_3 to the left of the dashed line. We have

$$\begin{pmatrix} 5 & 0 & 2 & \vdots & 1 & 0 & 0 \\ 2 & 2 & 1 & \vdots & 0 & 1 & 0 \\ -3 & 1 & -1 & \vdots & 0 & 0 & 1 \end{pmatrix} \xrightarrow{-2R_2 + R_1} \begin{pmatrix} 1 & -4 & 0 & \vdots & 1 & -2 & 0 \\ 2 & 2 & 1 & \vdots & 0 & 1 & 0 \\ -3 & 1 & -1 & \vdots & 0 & 0 & 1 \end{pmatrix}$$

$$\xrightarrow[\substack{-2R_1 + R_2 \\ 3R_1 + R_3}]{} \begin{pmatrix} 1 & -4 & 0 & \vdots & 1 & -2 & 0 \\ 0 & 10 & 1 & \vdots & -2 & 5 & 0 \\ 0 & -11 & -1 & \vdots & 3 & -6 & 1 \end{pmatrix}$$

$$\xrightarrow{1R_3 + R_2} \begin{pmatrix} 1 & -4 & 0 & \vdots & 1 & -2 & 0 \\ 0 & -1 & 0 & \vdots & 1 & -1 & 1 \\ 0 & -11 & -1 & \vdots & 3 & -6 & 1 \end{pmatrix} \xrightarrow{(-1)R_2} \begin{pmatrix} 1 & -4 & 0 & \vdots & 1 & -2 & 0 \\ 0 & 1 & 0 & \vdots & -1 & 1 & -1 \\ 0 & -11 & -1 & \vdots & 3 & -6 & 1 \end{pmatrix}$$

$$\xrightarrow[\substack{4R_2 + R_1 \\ 11R_2 + R_3}]{} \begin{pmatrix} 1 & 0 & 0 & \vdots & -3 & 2 & -4 \\ 0 & 1 & 0 & \vdots & -1 & 1 & -1 \\ 0 & 0 & -1 & \vdots & -8 & 5 & -10 \end{pmatrix} \xrightarrow{(-1)R_3} \begin{pmatrix} 1 & 0 & 0 & \vdots & -3 & 2 & -4 \\ 0 & 1 & 0 & \vdots & -1 & 1 & -1 \\ 0 & 0 & 1 & \vdots & 8 & -5 & 10 \end{pmatrix}$$

The required inverse is therefore

$$A^{-1} = \begin{pmatrix} -3 & 2 & -4 \\ -1 & 1 & -1 \\ 8 & -5 & 10 \end{pmatrix}$$

EXERCISE SET 10.4

A

In Exercises 1–4, the matrices A, B, C, and D are defined as follows.

$$A = \begin{pmatrix} 4 & -1 \\ -5 & 2 \end{pmatrix} \qquad B = \begin{pmatrix} \frac{1}{2} & 5 \\ 3 & 1 \end{pmatrix}$$

$$C = \begin{pmatrix} 3 & 0 & -2 \\ 0 & 5 & 6 \\ 1 & 4 & -7 \end{pmatrix} \qquad D = \begin{pmatrix} 1 & 2 & 3 \\ 4 & 5 & 6 \\ 7 & 8 & 9 \end{pmatrix}$$

1. Compute AI_2 and I_2A to verify that $AI_2 = I_2A = A$.
2. Compute BI_2 and I_2B to verify that $BI_2 = I_2B = B$.
3. Compute CI_3 and I_3C to verify that $CI_3 = I_3C = C$.
4. Compute DI_3 and I_3D to verify that $DI_3 = I_3D = D$.

In Exercises 5–12, compute A^{-1}, if it exists, using the method of Example 1.

5. $A = \begin{pmatrix} 7 & 9 \\ 4 & 5 \end{pmatrix}$

6. $A = \begin{pmatrix} 3 & -8 \\ 2 & -5 \end{pmatrix}$

7. $A = \begin{pmatrix} -3 & 1 \\ 5 & 6 \end{pmatrix}$

8. $A = \begin{pmatrix} -4 & 0 \\ 9 & 3 \end{pmatrix}$

9. $A = \begin{pmatrix} -2 & 3 \\ -4 & 6 \end{pmatrix}$

10. $A = \begin{pmatrix} \frac{5}{3} & -2 \\ -\frac{2}{3} & 1 \end{pmatrix}$

11. $A = \begin{pmatrix} \frac{1}{3} & \frac{1}{3} \\ -\frac{1}{9} & \frac{2}{9} \end{pmatrix}$

12. $A = \begin{pmatrix} -3 & 7 \\ 12 & -28 \end{pmatrix}$

In Exercises 13–26, compute the inverse matrix, if it exists, using elementary row operations (as shown in Example 3).

13. $\begin{pmatrix} 2 & 1 \\ 3 & 2 \end{pmatrix}$

14. $\begin{pmatrix} -6 & 5 \\ 18 & -15 \end{pmatrix}$

15. $\begin{pmatrix} 0 & -11 \\ 1 & 6 \end{pmatrix}$

16. $\begin{pmatrix} -2 & 13 \\ -4 & 25 \end{pmatrix}$

17. $\begin{pmatrix} \frac{2}{3} & -\frac{1}{4} \\ -8 & 3 \end{pmatrix}$

18. $\begin{pmatrix} -\frac{2}{5} & \frac{1}{3} \\ -6 & 5 \end{pmatrix}$

19. $\begin{pmatrix} -5 & 4 & -3 \\ 10 & -7 & 6 \\ 8 & -6 & 5 \end{pmatrix}$

20. $\begin{pmatrix} 1 & 0 & -2 \\ -3 & -1 & 6 \\ 2 & 1 & -5 \end{pmatrix}$

21. $\begin{pmatrix} 1 & 2 & -1 \\ 0 & 3 & 0 \\ -4 & 0 & 5 \end{pmatrix}$

22. $\begin{pmatrix} 1 & -4 & -8 \\ 1 & 2 & 5 \\ 1 & 1 & 3 \end{pmatrix}$

23. $\begin{pmatrix} -7 & 5 & 3 \\ 3 & -2 & -2 \\ 3 & -2 & -1 \end{pmatrix}$

24. $\begin{pmatrix} 2 & -1 & -1 \\ 1 & 0 & -1 \\ -2 & 1 & 2 \end{pmatrix}$

25. $\begin{pmatrix} 1 & 2 & 3 \\ 4 & 5 & 6 \\ 7 & 8 & 9 \end{pmatrix}$

26. $\begin{pmatrix} 2 & 1 & 3 \\ 4 & 5 & -7 \\ 2 & 1 & 3 \end{pmatrix}$

27. If $A = \begin{pmatrix} 3 & 8 \\ 4 & 11 \end{pmatrix}$, then $A^{-1} = \begin{pmatrix} 11 & -8 \\ -4 & 3 \end{pmatrix}$. Use this fact and the method of Example 2 to solve the following systems.

(a) $\begin{cases} 3x + 8y = 5 \\ 4x + 11y = 7 \end{cases}$ 　(b) $\begin{cases} 3x + 8y = -12 \\ 4x + 11y = 0 \end{cases}$

28. If $A = \begin{pmatrix} 3 & -7 \\ 4 & -9 \end{pmatrix}$, then $A^{-1} = \begin{pmatrix} -9 & 7 \\ -4 & 3 \end{pmatrix}$. Use this fact and the method of Example 2 to solve the following systems.

(a) $\begin{cases} 3x - 7y = 30 \\ 4x - 9y = 39 \end{cases}$ 　(b) $\begin{cases} 3x - 7y = -45 \\ 4x - 9y = -71 \end{cases}$

29. The inverse of the matrix $A = \begin{pmatrix} 3 & 2 & 6 \\ 1 & 1 & 2 \\ 2 & 2 & 5 \end{pmatrix}$ is

$$A^{-1} = \begin{pmatrix} 1 & 2 & -2 \\ -1 & 3 & 0 \\ 0 & -2 & 1 \end{pmatrix}.$$ Use this fact to solve the following systems.

(a) $\begin{cases} 3x + 2y + 6z = 28 \\ x + y + 2z = 9 \\ 2x + 2y + 5z = 22 \end{cases}$

(b) $\begin{cases} 3x + 2y + 6z = -7 \\ x + y + 2z = -2 \\ 2x + 2y + 5z = -6 \end{cases}$

30. The inverse of the matrix $A = \begin{pmatrix} 1 & -1 & 1 \\ 2 & -3 & 2 \\ -4 & 6 & 1 \end{pmatrix}$ is

$$A^{-1} = \begin{pmatrix} 3 & -\frac{7}{5} & -\frac{1}{5} \\ 2 & -1 & 0 \\ 0 & \frac{2}{5} & \frac{1}{5} \end{pmatrix}.$$ Use this fact to solve the following systems.

(a) $\begin{cases} x - y + z = 5 \\ 2x - 3y + 2z = -15 \\ -4x + 6y + z = 25 \end{cases}$

(b) $\begin{cases} x - y + z = 1 \\ 2x - 3y + 2z = -2 \\ -4x + 6y + z = 0 \end{cases}$

B

31. Let $A = \begin{pmatrix} 1 & -6 & 3 \\ 2 & -7 & 3 \\ 4 & -12 & 5 \end{pmatrix}$.

(a) Compute the matrix product AA. What do you observe?
(b) Use the result in part (a) to solve the following system.

$\begin{cases} x - 6y + 3z = \frac{19}{2} \\ 2x - 7y + 3z = 11 \\ 4x - 12y + 5z = 19 \end{cases}$

32. (a) Solve the following two systems and then check to see that your results agree with those given in Example 1.

$\begin{cases} a + 2c = 1 \\ 3a + 4c = 0 \end{cases}$ $\begin{cases} b + 2d = 0 \\ 3b + 4d = 1 \end{cases}$

(b) At the end of Example 1, it is asserted that

$$\begin{pmatrix} 1 & 2 \\ 3 & 4 \end{pmatrix}\begin{pmatrix} -2 & 1 \\ \frac{3}{2} & -\frac{1}{2} \end{pmatrix} = \begin{pmatrix} -2 & 1 \\ \frac{3}{2} & -\frac{1}{2} \end{pmatrix}\begin{pmatrix} 1 & 2 \\ 3 & 4 \end{pmatrix}$$
$$= \begin{pmatrix} 1 & 0 \\ 0 & 1 \end{pmatrix}$$

Carry out the indicated matrix multiplications to verify that these equations are valid.

33. Use the technique in Example 1 to show that the matrix $\begin{pmatrix} 2 & 5 \\ 6 & 15 \end{pmatrix}$ does not have an inverse.

34. Use the elementary row operations (as in Example 3) to find the inverse of the following matrix.

$$\begin{pmatrix} 1 & 1 & 1 & 1 \\ 1 & 2 & 3 & 4 \\ 1 & 3 & 6 & 10 \\ 1 & 4 & 10 & 20 \end{pmatrix}$$

35. Let $A = \begin{pmatrix} 2 & 3 \\ 4 & 5 \end{pmatrix}$ and $B = \begin{pmatrix} 7 & 8 \\ 6 & 7 \end{pmatrix}$.

(a) Compute A^{-1}, B^{-1}, and $B^{-1}A^{-1}$.
(b) Compute $(AB)^{-1}$. What do you observe?

36. Let $A = \begin{pmatrix} a & b \\ c & d \end{pmatrix}$. Compute A^{-1}. (Assume that $ad - bc \neq 0$.)

10.5 DETERMINANTS AND CRAMER'S RULE

The idea of the determinant...dates back essentially to Leibniz (1693), the Swiss mathematician Gabriel Cramer (1750), and Lagrange (1773); the name is due to Cauchy (1812). Y. Mikami has pointed out that the Japanese mathematician Seki Kōwa had the idea of a determinant sometime before 1683.

Dirk J. Struik in *A Concise History of Mathematics*, 4th rev. ed. (New York: Dover Publications, 1987)

As you saw in the previous section, square matrices and their inverses can be used to solve certain systems of equations. In this section, we are going to associate a number with each square matrix. This number is called the **determinant** of the matrix. As you'll see, this too has an application in solving systems of equations.

The determinant of a matrix A can be denoted by det A or $|A|$. Determinants are also denoted simply by replacing the parentheses of matrix notation with vertical lines. Thus, three examples of determinants are

$$\begin{vmatrix} 1 & 2 \\ 3 & 4 \end{vmatrix} \qquad \begin{vmatrix} 1 & 2 & 3 \\ 4 & 5 & 6 \\ 7 & 8 & 9 \end{vmatrix} \qquad \begin{vmatrix} 3 & 7 & 8 & 9 \\ 5 & 6 & 4 & 3 \\ -9 & 9 & 0 & 1 \\ 1 & 3 & -2 & 1 \end{vmatrix}$$

A determinant with n rows and n columns is said to be an **nth-order determinant**. Therefore, the determinants we've just written are, respectively, second-, third-, and fourth-order determinants. As with matrices, we speak of the numbers in a determinant as its **entries**. We also number the rows and the columns of a determinant as we do with matrices. However, unlike matrices, each determinant has a definite value. The value of a second-order determinant is defined as follows.

DEFINITION   2 × 2 Determinant

$$\begin{vmatrix} a & b \\ c & d \end{vmatrix} = ad - bc$$

Table 1 illustrates how this definition is used to evaluate, or *expand*, a second-order determinant.

TABLE 1

Determinant	Value of Determinant
$\begin{vmatrix} 3 & 7 \\ 5 & 10 \end{vmatrix}$	$\begin{vmatrix} 3 & 7 \\ 5 & 10 \end{vmatrix} = 3(10) - 7(5) = 30 - 35 = -5$
$\begin{vmatrix} 3 & -7 \\ 5 & 10 \end{vmatrix}$	$\begin{vmatrix} 3 & -7 \\ 5 & 10 \end{vmatrix} = 3(10) - (-7)(5) = 30 + 35 = 65$
$\begin{vmatrix} a & a^3 \\ 1 & a^2 \end{vmatrix}$	$\begin{vmatrix} a & a^3 \\ 1 & a^2 \end{vmatrix} = a(a^2) - a^3(1) = a^3 - a^3 = 0$

In general, the value of an nth-order determinant $(n > 2)$ is defined in terms of certain determinants of order $n - 1$. For instance, the value of a third-order determinant is defined in terms of second-order determinants. We'll use the following example to introduce the necessary terminology here:

$$\begin{vmatrix} 8 & 3 & 5 \\ 2 & 4 & 6 \\ 9 & 1 & 7 \end{vmatrix}$$

Pick a given entry—say, 8—and imagine crossing out all entries occupying the same row or the same column as 8.

$$\begin{vmatrix} 8 & 3 & 5 \\ 2 & 4 & 6 \\ 9 & 1 & 7 \end{vmatrix}$$

Then we are left with the second-order determinant $\begin{vmatrix} 4 & 6 \\ 1 & 7 \end{vmatrix}$. This second-order determinant is called the **minor** of the entry 8. Similarly, to find the minor of the entry 6 in the original determinant, imagine crossing out all entries that occupy the same row or the same column as 6.

$$\begin{vmatrix} 8 & 3 & 5 \\ 2 & 4 & 6 \\ 9 & 1 & 7 \end{vmatrix}$$

We are left with the second-order determinant $\begin{vmatrix} 8 & 3 \\ 9 & 1 \end{vmatrix}$, which by definition is the minor of the entry 6. In the same manner, *the minor of any element is the determinant obtained by crossing out the entries occupying the same row or column as the given element.*

Closely related to the minor of an entry in a determinant is the cofactor of that entry. The **cofactor** of an entry is defined as the minor multiplied by $+1$ or -1, according to the scheme displayed in Figure 1.

$$\begin{vmatrix} + & - & + \\ - & + & - \\ + & - & + \end{vmatrix}$$

FIGURE 1

After looking at an example, we'll give a more formal rule for computing cofactors, one that will not rely on a figure and that will also apply to larger determinants.

EXAMPLE 1   Consider the determinant

$$\begin{vmatrix} 1 & 2 & 3 \\ 4 & 5 & 6 \\ 7 & 8 & 9 \end{vmatrix}$$

Compute the minor and the cofactor of the entry 4.

Solution   By definition, we have

$$\text{minor of } 4 = \begin{vmatrix} 2 & 3 \\ 8 & 9 \end{vmatrix} = 18 - 24 = -6$$

Thus, the minor of the entry 4 is -6. To compute the cofactor of 4, we first notice that 4 is located in the second row and first column of the given determinant. Upon checking the corresponding position in Figure 1, we see a negative sign. Therefore, we have by definition

$$\text{cofactor of } 4 = (-1)(\text{minor of } 4)$$
$$= (-1)(-6) = 6$$

The cofactor of 4 is therefore 6. ∎

The following rule tells us how cofactors can be computed without relying on Figure 1. For reference, we also restate the definition of a minor. (You should verify for yourself that this rule yields results that are consistent with Figure 1.)

DEFINITION   Minors and Cofactors

The **minor** of an entry b in a determinant is the determinant formed by suppressing the entries in the row and in the column in which b appears.

Suppose that the entry b is in the ith row and the jth column. Then the **cofactor** of b is given by the expression

$$(-1)^{i+j}(\text{minor of } b)$$

We are now prepared to state the definition that tells us how to evaluate a third-order determinant. Actually, as you'll see later, the definition is quite general and may be applied to determinants of any size.

DEFINITION   The Value of a Determinant

Multiply each entry in the first row of the determinant by its cofactor and then add the results. The value of the determinant is defined to be this sum.

To see how this definition is used, let us evaluate the determinant

$$\begin{vmatrix} 1 & 2 & 3 \\ 4 & 5 & 6 \\ 7 & 8 & 9 \end{vmatrix}$$

The definition tells us to multiply each entry in the first row by its cofactor and then add the results. Carrying out this procedure, we have

$$\begin{vmatrix} 1 & 2 & 3 \\ 4 & 5 & 6 \\ 7 & 8 & 9 \end{vmatrix} = 1\begin{vmatrix} 5 & 6 \\ 8 & 9 \end{vmatrix} - 2\begin{vmatrix} 4 & 6 \\ 7 & 9 \end{vmatrix} + 3\begin{vmatrix} 4 & 5 \\ 7 & 8 \end{vmatrix}$$

$$= 1(45 - 48) - 2(36 - 42) + 3(32 - 35)$$

$$= 0 \quad \text{(Check the arithmetic!)}$$

So the value of this particular determinant is zero. The procedure we've used here is referred to as **expanding the determinant along its first row**. The following theorem (stated here without proof) tells us that the value of a determinant can be obtained by expanding along any row or along any column; the results are the same in all cases.

Theorem

Select any row or any column in a determinant and multiply each element in that row or column by its cofactor. Then add the results. The number obtained will be the value of the determinant. (In other words, the number obtained will be the same as that obtained by expanding the determinant along its first row.)

According to this theorem, we could have evaluated the determinant

$$\begin{vmatrix} 1 & 2 & 3 \\ 4 & 5 & 6 \\ 7 & 8 & 9 \end{vmatrix}$$

by expanding it along any row or any column. Let's expand it along the second column and check to see that the result agrees with the value obtained earlier. Expanding along the second column, we have

$$\begin{vmatrix} 1 & 2 & 3 \\ 4 & 5 & 6 \\ 7 & 8 & 9 \end{vmatrix} = -2\begin{vmatrix} 4 & 6 \\ 7 & 9 \end{vmatrix} + 5\begin{vmatrix} 1 & 3 \\ 7 & 9 \end{vmatrix} - 8\begin{vmatrix} 1 & 3 \\ 4 & 6 \end{vmatrix}$$

$$= -2(36 - 42) + 5(9 - 21) - 8(6 - 12)$$
$$= -2(-6) + 5(-12) - 8(-6)$$
$$= 12 - 60 + 48 = 0 \qquad \text{as obtained previously}$$

We would have obtained the same result had we chosen to begin with any other row or column. (Exercise 13 at the end of this section asks you to verify this.)

There are three basic rules that make it easier to evaluate determinants. These are summarized in the box that follows. (Suggestions for proving these can be found in the exercises.)

PROPERTY SUMMARY   RULES FOR MANIPULATING DETERMINANTS

EXAMPLES

1. If each entry in a given row is multiplied by the constant k, then the value of the determinant is multiplied by k. This is true for columns, also.

$$10\begin{vmatrix} 1 & 3 & 4 \\ 1 & 2 & 3 \\ 4 & 5 & 6 \end{vmatrix} = \begin{vmatrix} 10 & 30 & 40 \\ 1 & 2 & 3 \\ 4 & 5 & 6 \end{vmatrix}$$

$$k\begin{vmatrix} a & b & c \\ d & e & f \\ g & h & i \end{vmatrix} = \begin{vmatrix} a & kb & c \\ d & ke & f \\ g & kh & i \end{vmatrix}$$

2. If a multiple of one row is added to another row, the value of the determinant is not changed. This applies to columns, also.

$$\begin{vmatrix} a & b & c \\ d & e & f \\ g & h & i \end{vmatrix} = \begin{vmatrix} a & b & c \\ d + ka & e + kb & f + kc \\ g & h & i \end{vmatrix}$$

3. If two rows are interchanged, then the value of the determinant is multiplied by -1. This applies to columns, also.

$$\begin{vmatrix} 1 & 2 & 3 \\ 4 & 5 & 6 \\ a & b & c \end{vmatrix} = -\begin{vmatrix} 4 & 5 & 6 \\ 1 & 2 & 3 \\ a & b & c \end{vmatrix}$$

EXAMPLE 2   Evaluate the determinant $\begin{vmatrix} 15 & 14 & 26 \\ 18 & 17 & 32 \\ 21 & 20 & 42 \end{vmatrix}$.

Solution

$$\begin{vmatrix} 15 & 14 & 26 \\ 18 & 17 & 32 \\ 21 & 20 & 42 \end{vmatrix} = (3 \times 2)\begin{vmatrix} 5 & 14 & 13 \\ 6 & 17 & 16 \\ 7 & 20 & 21 \end{vmatrix}$$ using rule 1 to factor 3 from the first column and 2 from the third column

$$= 6\begin{vmatrix} 5 & 1 & 13 \\ 6 & 1 & 16 \\ 7 & -1 & 21 \end{vmatrix}$$ using rule 2 to subtract the third column from the second column

$$= 6\begin{vmatrix} 12 & 0 & 34 \\ 13 & 0 & 37 \\ 7 & -1 & 21 \end{vmatrix}$$ using rule 2 to add the third row to the first and second rows

$$= 6\left[-\left(-1\begin{vmatrix} 12 & 34 \\ 13 & 37 \end{vmatrix}\right)\right]$$ expanding the determinant along the second column

$$= 6\begin{vmatrix} 12 & 34 \\ 1 & 3 \end{vmatrix}$$ using rule 2 to subtract the first row from the second row

$$= 6(36 - 34) = 12$$

The value of the given determinant is therefore 12. Notice the general strategy. We use rules 1 and 2 until one column (or one row) contains two zeros. At that point, it is a simple matter to expand the determinant along that column (or row).

EXAMPLE 3   Show that

$$\begin{vmatrix} 1 & 1 & 1 \\ a & b & c \\ a^2 & b^2 & c^2 \end{vmatrix} = (b - a)(c - a)(c - b)$$

Solution

$$\begin{vmatrix} 1 & 1 & 1 \\ a & b & c \\ a^2 & b^2 & c^2 \end{vmatrix} = \begin{vmatrix} 1 & 0 & 0 \\ a & b - a & c - a \\ a^2 & b^2 - a^2 & c^2 - a^2 \end{vmatrix}$$ using rule 2 to subtract the first column from the second and the third columns

$$= \begin{vmatrix} 1 & 0 & 0 \\ a & b - a & c - a \\ a^2 & (b-a)(b+a) & (c-a)(c+a) \end{vmatrix}$$

$$= (b - a)(c - a)\begin{vmatrix} 1 & 0 & 0 \\ a & 1 & 1 \\ a^2 & b + a & c + a \end{vmatrix}$$ using rule 1 to factor $(b - a)$ from the second column and $(c - a)$ from the third column

$$= (b - a)(c - a)[(c + a) - (b + a)]$$ expanding the determinant along the first row

$$= (b - a)(c - a)(c - b)$$

The definition that we gave for third-order determinants can also be applied to define fourth-order (or larger) determinants. Consider, for example, the fourth-order determinant A given by

$$A = \begin{vmatrix} 25 & 40 & 5 & 10 \\ 9 & 0 & 3 & 6 \\ -2 & 3 & 11 & -17 \\ -3 & 4 & 7 & 2 \end{vmatrix}$$

By definition, then, we could evaluate this determinant by selecting the first row, multiplying each entry by its cofactor, and then adding the results. This yields

$$A = 25\begin{vmatrix} 0 & 3 & 6 \\ 3 & 11 & -17 \\ 4 & 7 & 2 \end{vmatrix} - 40\begin{vmatrix} 9 & 3 & 6 \\ -2 & 11 & -17 \\ -3 & 7 & 2 \end{vmatrix}$$

$$+ 5\begin{vmatrix} 9 & 0 & 6 \\ -2 & 3 & -17 \\ -3 & 4 & 2 \end{vmatrix} - 10\begin{vmatrix} 9 & 0 & 3 \\ -2 & 3 & 11 \\ -3 & 4 & 7 \end{vmatrix}$$

The problem is now reduced to evaluating four third-order determinants. If we had instead expanded down the second column, the ensuing work would be somewhat less because of the zero in that column. Nevertheless, there would still be three third-order determinants to evaluate.

Because of the amount of computation involved, we in fact rarely evaluate fourth-order determinants directly from the definition. Instead, we use the three rules (given just before Example 2) to simplify matters first. In the example that follows, we show how this is done. (We will accept the fact that the three rules are applicable in evaluating determinants of any size.)

EXAMPLE 4   Evaluate the determinant A given by

$$A = \begin{vmatrix} 25 & 40 & 5 & 10 \\ 9 & 0 & 3 & 6 \\ -2 & 3 & 11 & -17 \\ -3 & 4 & 7 & 2 \end{vmatrix}$$

Solution $A = 5 \times 3 \begin{vmatrix} 5 & 8 & 1 & 2 \\ 3 & 0 & 1 & 2 \\ -2 & 3 & 11 & -17 \\ -3 & 4 & 7 & 2 \end{vmatrix} = 15 \begin{vmatrix} 2 & 8 & 0 & 0 \\ 3 & 0 & 1 & 2 \\ -35 & 3 & 0 & -39 \\ -24 & 4 & 0 & -12 \end{vmatrix}$

Factor 5 from the first row and 3 from the second row.

Subtract the second row from the first. Subtract 11 times the second row from the third. Subtract 7 times the second row from the fourth.

$= 15 \times 2 \times 4 \begin{vmatrix} 1 & 4 & 0 & 0 \\ 3 & 0 & 1 & 2 \\ -35 & 3 & 0 & -39 \\ -6 & 1 & 0 & -3 \end{vmatrix} = 120 \cdot (-1) \begin{vmatrix} 1 & 4 & 0 \\ -35 & 3 & -39 \\ -6 & 1 & -3 \end{vmatrix}$

Factor 2 from the first row and 4 from the fourth row.

Expand the determinant along the third column.

$= -120 \cdot (-3) \begin{vmatrix} 1 & 4 & 0 \\ -35 & 3 & 13 \\ -6 & 1 & 1 \end{vmatrix} = 360 \begin{vmatrix} 1 & 0 & 0 \\ -35 & 143 & 13 \\ -6 & 25 & 1 \end{vmatrix}$

Factor (-3) from the third column.

Subtract 4 times the first column from the second.

$= 360 \begin{vmatrix} 143 & 13 \\ 25 & 1 \end{vmatrix} = 360 \begin{vmatrix} -7 & 7 \\ 25 & 1 \end{vmatrix}$

Expand the determinant along the first row.

Subtract 6 times the second row from the first row.

$= 360(-7 - 175) = -65520$

Do the arithmetic!

The value of the given fourth-order determinant is therefore -65520. ■

Determinants can be used to solve certain systems of linear equations in which there are as many unknowns as there are equations. In the box that follows, we state **Cramer's rule** for solving a system of three linear equations in three unknowns.* A more general, but entirely similar, version of Cramer's rule holds for n equations in n unknowns.

Cramer's Rule

Consider the system

$$\begin{cases} a_1x + b_1y + c_1z = d_1 \\ a_2x + b_2y + c_2z = d_2 \\ a_3x + b_3y + c_3z = d_3 \end{cases}$$

Let the four determinants D, D_x, D_y, and D_z be defined as follows:

$$D = \begin{vmatrix} a_1 & b_1 & c_1 \\ a_2 & b_2 & c_2 \\ a_3 & b_3 & c_3 \end{vmatrix} \qquad D_x = \begin{vmatrix} d_1 & b_1 & c_1 \\ d_2 & b_2 & c_2 \\ d_3 & b_3 & c_3 \end{vmatrix}$$

$$D_y = \begin{vmatrix} a_1 & d_1 & c_1 \\ a_2 & d_2 & c_2 \\ a_3 & d_3 & c_3 \end{vmatrix} \qquad D_z = \begin{vmatrix} a_1 & b_1 & d_1 \\ a_2 & b_2 & d_2 \\ a_3 & b_3 & d_3 \end{vmatrix}$$

Then if $D \neq 0$, the unique values of x, y, and z satisfying the system are given by

$$x = \frac{D_x}{D} \qquad y = \frac{D_y}{D} \qquad z = \frac{D_z}{D}$$

[If $D = 0$, the solutions (if any) can be found using Gaussian elimination.]

Notice that the determinant D in Cramer's rule is just the determinant of the coefficient matrix of the given system. If you replace the first column of D with the column of numbers on the right side of the given system, you obtain D_x. The determinants D_y and D_z are obtained similarly.

Before actually proving Cramer's rule, let's take a look at how the rule is applied.

EXAMPLE 5   Use Cramer's rule to find all solutions of the following system of equations:

$$\begin{cases} 2x + 2y - 3z = -20 \\ x - 4y + z = 6 \\ 4x - y + 2z = -1 \end{cases}$$

Solution   First, we list the determinants D, D_x, D_y, and D_z:

$$D = \begin{vmatrix} 2 & 2 & -3 \\ 1 & -4 & 1 \\ 4 & -1 & 2 \end{vmatrix} \qquad D_x = \begin{vmatrix} -20 & 2 & -3 \\ 6 & -4 & 1 \\ -1 & -1 & 2 \end{vmatrix}$$

$$D_y = \begin{vmatrix} 2 & -20 & -3 \\ 1 & 6 & 1 \\ 4 & -1 & 2 \end{vmatrix} \qquad D_z = \begin{vmatrix} 2 & 2 & -20 \\ 1 & -4 & 6 \\ 4 & -1 & -1 \end{vmatrix}$$

*The rule is named after one of its discoverers, the Swiss mathematician Gabriel Cramer (1704–1752).

The calculations for evaluating D begin as follows:

$$D = \begin{vmatrix} 2 & 2 & -3 \\ 1 & -4 & 1 \\ 4 & -1 & 2 \end{vmatrix} = \begin{vmatrix} 0 & 10 & -5 \\ 1 & -4 & 1 \\ 0 & 15 & -2 \end{vmatrix}$$

Subtract twice the second row from the first.
Subtract 4 times the second row from the third.

Since we now have two zeros in the first column, it is an easy matter to expand D along that column to obtain

$$D = -1 \begin{vmatrix} 10 & -5 \\ 15 & -2 \end{vmatrix}$$

$$= -1[-20 - (-75)] = -1(55) = -55$$

The value of D is therefore -55. (Since this value is nonzero, Cramer's rule does apply.) As Exercise 31 at the end of this section asks you to verify, the values of the other three determinants are

$$D_x = 144 \qquad D_y = 61 \qquad D_z = -230$$

By Cramer's rule, then, the unique values of x, y, and z that satisfy the system are

$$x = \frac{D_x}{D} = \frac{144}{-55} = -\frac{144}{55} \qquad y = \frac{D_y}{D} = \frac{61}{-55} = -\frac{61}{55} \qquad z = \frac{D_z}{D} = \frac{-230}{-55} = \frac{46}{11}$$

∎

One way we can prove Cramer's rule is to use Gaussian elimination to solve the system

$$\begin{cases} a_1 x + b_1 y + c_1 z = d_1 \\ a_2 x + b_2 y + c_2 z = d_2 \\ a_3 x + b_3 y + c_3 z = d_3 \end{cases} \tag{1}$$

A much shorter and simpler proof, however, has been found by D. E. Whitford and M. S. Klamkin.* This is the proof we give here; it makes effective use of the rules employed in this section for manipulating determinants.

Consider the system of equations (1) and assume that $D \neq 0$. We will show that *if* x, y, and z satisfy the system, then in fact $x = D_x/D$, with similar equations giving y and z. (Exercise 64 at the end of this section then shows how to check that these values indeed satisfy the given system.) We have

$$D_x = \begin{vmatrix} d_1 & b_1 & c_1 \\ d_2 & b_2 & c_2 \\ d_3 & b_3 & c_3 \end{vmatrix} \qquad \text{by definition}$$

$$= \begin{vmatrix} (a_1 x + b_1 y + c_1 z) & b_1 & c_1 \\ (a_2 x + b_2 y + c_2 z) & b_2 & c_2 \\ (a_3 x + b_3 y + c_3 z) & b_3 & c_3 \end{vmatrix} \qquad \begin{matrix} \text{using the equations in (1)} \\ \text{to substitute for } d_1, d_2, \text{ and } d_3 \end{matrix}$$

*The proof was published in the *American Mathematical Monthly* **60** (1953), 186–187.

$$= \begin{vmatrix} a_1x & b_1 & c_1 \\ a_2x & b_2 & c_2 \\ a_3x & b_3 & c_3 \end{vmatrix}$$ subtracting y times the second column as well as z times the third column from the first column

$$= x \begin{vmatrix} a_1 & b_1 & c_1 \\ a_2 & b_2 & c_2 \\ a_3 & b_3 & c_3 \end{vmatrix}$$ factoring x out of the first column

$$= xD$$ by definition

We now have $D_x = xD$, which is equivalent to $x = D_x/D$, as required. The formulas for y and z are obtained similarly.

EXERCISE SET 10.5

A

In Exercises 1–6, evaluate the determinants.

1. (a) $\begin{vmatrix} 2 & -17 \\ 1 & 6 \end{vmatrix}$ (b) $\begin{vmatrix} 1 & 6 \\ 2 & -17 \end{vmatrix}$

2. (a) $\begin{vmatrix} 1 & 0 \\ 0 & 1 \end{vmatrix}$ (b) $\begin{vmatrix} 0 & 1 \\ 0 & 1 \end{vmatrix}$

3. (a) $\begin{vmatrix} 5 & 7 \\ 500 & 700 \end{vmatrix}$ (b) $\begin{vmatrix} 5 & 500 \\ 7 & 700 \end{vmatrix}$

4. (a) $\begin{vmatrix} -8 & -3 \\ 4 & -5 \end{vmatrix}$ (b) $\begin{vmatrix} -3 & -8 \\ -5 & 4 \end{vmatrix}$

5. $\begin{vmatrix} \sqrt{2}-1 & \sqrt{2} \\ \sqrt{2} & \sqrt{2}+1 \end{vmatrix}$ 6. $\begin{vmatrix} \sqrt{3}+\sqrt{2} & 1+\sqrt{5} \\ 1-\sqrt{5} & \sqrt{3}-\sqrt{2} \end{vmatrix}$

In Exercises 7–12, refer to the following determinant:

$$\begin{vmatrix} -6 & 3 & 8 \\ 5 & -4 & 1 \\ 10 & 9 & -10 \end{vmatrix}$$

7. Evaluate the minor of the entry 3.
8. Evaluate the cofactor of the entry 3.
9. Evaluate the minor of -10.
10. Evaluate the cofactor of -10.
11. (a) Multiply each entry in the first row by its minor and find the sum of the results.
    (b) Multiply each entry in the first row by its cofactor and find the sum of the results.
    (c) Does the answer in part (a) or part (b) give you the value of the determinant?
12. (a) Multiply each entry in the first column by its cofactor and find the sum of the results.
    (b) Follow the same instructions as in part (a), but use the second column.

(c) Follow the same instructions as in part (a), but use the third column.

13. Let $A = \begin{vmatrix} 1 & 2 & 3 \\ 4 & 5 & 6 \\ 7 & 8 & 9 \end{vmatrix}$. Evaluate A by expanding it along the indicated row or column: (a) second row; (b) third row; (c) first column; (d) third column.

In Exercises 14–21, evaluate the determinants.

14. $\begin{vmatrix} 1 & 2 & -1 \\ 2 & -1 & 1 \\ 4 & 0 & 2 \end{vmatrix}$ 15. $\begin{vmatrix} 5 & 10 & 15 \\ 1 & 2 & 3 \\ -9 & 11 & 7 \end{vmatrix}$

16. $\begin{vmatrix} 8 & 4 & 2 \\ 3 & 9 & 3 \\ -2 & 8 & 6 \end{vmatrix}$ 17. $\begin{vmatrix} 1 & 2 & -3 \\ 4 & 5 & -9 \\ 0 & 0 & 1 \end{vmatrix}$

18. $\begin{vmatrix} 3 & 0 & 0 \\ 0 & 19 & 0 \\ 0 & 0 & 10 \end{vmatrix}$ 19. $\begin{vmatrix} -6 & -8 & 18 \\ 25 & 12 & 15 \\ -9 & 4 & 13 \end{vmatrix}$

20. $\begin{vmatrix} 23 & 0 & 47 \\ -37 & 0 & 18 \\ 14 & 0 & 25 \end{vmatrix}$ 21. $\begin{vmatrix} 16 & 0 & -64 \\ -8 & 15 & -12 \\ 30 & -20 & 10 \end{vmatrix}$

22. Use the method illustrated in Example 3 in the text to show that

$$\begin{vmatrix} 1 & 1 & 1 \\ a & b & c \\ a^3 & b^3 & c^3 \end{vmatrix} = (b-a)(c-a)(c^2-b^2+ac-ab)$$

$$= (b-a)(c-a)(c-b)(a+b+c)$$

23. Use the method shown in Example 3 to express the determinant $\begin{vmatrix} 1 & x & x^2 \\ 1 & y & y^2 \\ 1 & z & z^2 \end{vmatrix}$ as a product.

24. Show that
$$\begin{vmatrix} 1 & 1 & 1 \\ a^2 & b^2 & c^2 \\ a^3 & b^3 & c^3 \end{vmatrix} = (b - a)(c - a)(bc^2 - b^2c + ac^2 - ab^2)$$
$$= (b - a)(c - a)(c - b)(bc + ac + ab)$$

25. Simplify the determinant $\begin{vmatrix} 1 & 1 & 1 \\ 1 & 1 + x & 1 \\ 1 & 1 & 1 + y \end{vmatrix}$.

26. Use the method shown in Example 3 to express the following determinant as a product of three factors:
$$\begin{vmatrix} 1 & 1 & 1 \\ a & b & c \\ bc & ca & ab \end{vmatrix}$$

In Exercises 27–30, evaluate the determinants.

27. $\begin{vmatrix} 1 & -1 & 0 & 2 \\ 0 & 1 & -1 & 0 \\ 2 & 1 & 0 & -1 \\ -2 & 2 & 1 & 1 \end{vmatrix}$

28. $\begin{vmatrix} 3 & -2 & 3 & 4 \\ 1 & 4 & -3 & 2 \\ 6 & 3 & -6 & -3 \\ -1 & 0 & 1 & 5 \end{vmatrix}$

29. $\begin{vmatrix} 2 & 0 & 0 & 0 \\ 0 & 3 & 0 & 0 \\ 0 & 0 & 4 & 0 \\ 0 & 0 & 0 & 5 \end{vmatrix}$

30. $\begin{vmatrix} 7 & -8 & 1 & 2 \\ 21 & 4 & 3 & -1 \\ -35 & 8 & 3 & -2 \\ 14 & 16 & 0 & 1 \end{vmatrix}$

31. Verify the following statements (from Example 5).

    (a) $\begin{vmatrix} -20 & 2 & -3 \\ 6 & -4 & 1 \\ -1 & -1 & 2 \end{vmatrix} = 144$

    (b) $\begin{vmatrix} 2 & -20 & -3 \\ 1 & 6 & 1 \\ 4 & -1 & 2 \end{vmatrix} = 61$

    (c) $\begin{vmatrix} 2 & 2 & -20 \\ 1 & -4 & 6 \\ 4 & -1 & -1 \end{vmatrix} = -230$

32. Consider the following system:
$$\begin{cases} x + y + 3z = 0 \\ x + 2y + 5z = 0 \\ x - 4y - 8z = 0 \end{cases}$$

    (a) Without doing any calculations, find one obvious solution of this system.
    (b) Calculate the determinant D.
    (c) List all solutions of this system.

In Exercises 33–42, use Cramer's rule to solve those systems for which $D \neq 0$. In cases where $D = 0$, use Gaussian elimination or matrix methods.

33. $\begin{cases} 3x + 4y - z = 5 \\ x - 3y + 2z = 2 \\ 5x - 6z = -7 \end{cases}$

34. $\begin{cases} 3A - B - 4C = 3 \\ A + 2B - 3C = 9 \\ 2A - B + 2C = -8 \end{cases}$

35. $\begin{cases} 3x + 2y - z = -6 \\ 2x - 3y - 4z = -11 \\ x + y + z = 5 \end{cases}$

36. $\begin{cases} 5x - 3y - z = 16 \\ 2x + y - 3z = 5 \\ 3x - 2y + 2z = 5 \end{cases}$

37. $\begin{cases} 2x + 5y + 2z = 0 \\ 3x - y - 4z = 0 \\ x + 2y - 3z = 0 \end{cases}$

38. $\begin{cases} 4u + 3v - 2w = 14 \\ u + 2v - 3w = 6 \\ 2u - v + 4w = 2 \end{cases}$

39. $\begin{cases} 12x - 11z = 13 \\ 6x + 6y - 4z = 26 \\ 6x + 2y - 5z = 13 \end{cases}$

40. $\begin{cases} 3x + 4y + 2z = 1 \\ 4x + 6y + 2z = 7 \\ 2x + 3y + z = 11 \end{cases}$

41. $\begin{cases} x + y + z + w = -7 \\ x - y + z - w = -11 \\ 2x - 2y - 3z - 3w = 26 \\ 3x + 2y + z - w = -9 \end{cases}$

42. $\begin{cases} 2A - B - 3C + 2D = -2 \\ A - 2B + C - 3D = 4 \\ 3A - 4B + 2C - 4D = 12 \\ 2A + 3B - C - 2D = -4 \end{cases}$

43. Find all values of x for which
$$\begin{vmatrix} x - 4 & 0 & 0 \\ 0 & x + 4 & 0 \\ 0 & 0 & x + 1 \end{vmatrix} = 0$$

44. Find all values of x for which
$$\begin{vmatrix} 1 & x & x^2 \\ 1 & 1 & 1 \\ 4 & 5 & 0 \end{vmatrix} = 0$$

45. By expanding the determinant $\begin{vmatrix} a & b & c \\ a & b & c \\ d & e & f \end{vmatrix}$ down the first column, show that its value is zero.

46. By expanding the determinant $\begin{vmatrix} ka & kb & kc \\ d & e & f \\ g & h & i \end{vmatrix}$ along its first row, show that it is equal to $k\begin{vmatrix} a & b & c \\ d & e & f \\ g & h & i \end{vmatrix}$.

B

47. Show that
$$\begin{vmatrix} a_1 + A_1 & b_1 & c_1 \\ a_2 + A_2 & b_2 & c_2 \\ a_3 + A_3 & b_3 & c_3 \end{vmatrix} = \begin{vmatrix} a_1 & b_1 & c_1 \\ a_2 & b_2 & c_2 \\ a_3 & b_3 & c_3 \end{vmatrix} + \begin{vmatrix} A_1 & b_1 & c_1 \\ A_2 & b_2 & c_2 \\ A_3 & b_3 & c_3 \end{vmatrix}$$

48. Show that

$$\begin{vmatrix} a_1 & b_1 & c_1 \\ a_2 & b_2 & c_2 \\ a_3 & b_3 & c_3 \end{vmatrix} = \begin{vmatrix} a_1 + kb_1 & b_1 & c_1 \\ a_2 + kb_2 & b_2 & c_2 \\ a_3 + kb_3 & b_3 & c_3 \end{vmatrix}$$

49. By expanding each determinant along a row or column, show that

$$\begin{vmatrix} a_1 & b_1 & c_1 \\ a_2 & b_2 & c_2 \\ a_3 & b_3 & c_3 \end{vmatrix} = - \begin{vmatrix} a_2 & b_2 & c_2 \\ a_1 & b_1 & c_1 \\ a_3 & b_3 & c_3 \end{vmatrix}$$

50. Solve for x in terms of a, b, and c:

$$\begin{vmatrix} a & a & x \\ c & c & c \\ b & x & b \end{vmatrix} = 0 \qquad (c \neq 0)$$

51. Show that

$$\begin{vmatrix} 1+a & 1 & 1 & 1 \\ 1 & 1+b & 1 & 1 \\ 1 & 1 & 1+c & 1 \\ 1 & 1 & 1 & 1+d \end{vmatrix}$$

$$= abcd\left(\frac{1}{a} + \frac{1}{b} + \frac{1}{c} + \frac{1}{d} + 1\right)$$

52. Show that

$$\begin{vmatrix} 1 & a & a^2 \\ a^2 & 1 & a \\ a & a^2 & 1 \end{vmatrix} = (a^3 - 1)^2$$

53. Consider the determinant $D = \begin{vmatrix} a_1 & b_1 & c_1 \\ a_2 & b_2 & c_2 \\ a_3 & b_3 & c_3 \end{vmatrix}$. Let A_1 denote the cofactor of a_1, let B_1 denote the cofactor of b_1, and so on. Prove that

$$\begin{vmatrix} B_2 & C_2 \\ B_3 & C_3 \end{vmatrix} = a_1 D$$

54. Show that

$$\begin{vmatrix} 1 & bc & b+c \\ 1 & ca & c+a \\ 1 & ab & a+b \end{vmatrix} = (b-c)(c-a)(a-b)$$

55. Find the value of each determinant.

(a) $\begin{vmatrix} 1 & 0 & 0 \\ x & 1 & 0 \\ x & y & 1 \end{vmatrix}$ (b) $\begin{vmatrix} 1 & 0 & 0 & 0 \\ x & 1 & 0 & 0 \\ x & y & 1 & 0 \\ x & y & z & 1 \end{vmatrix}$

56. Evaluate the determinant.

$$\begin{vmatrix} a & b & c \\ a & a+b & a+b+c \\ a & 2a+b & 3a+2b+c \end{vmatrix}$$

57. Show that

$$\begin{vmatrix} 1 & a & a & a \\ 1 & b & a & a \\ 1 & a & b & a \\ 1 & a & a & b \end{vmatrix} = (b-a)^3$$

58. Show that

$$\begin{vmatrix} a & 1 & 1 & 1 \\ 1 & a & 1 & 1 \\ 1 & 1 & a & 1 \\ 1 & 1 & 1 & a \end{vmatrix} = (a-1)^3(a+3)$$

59. Solve the following system for x, y, and z:

$$\begin{cases} ax + by + cz = k \\ a^2x + b^2y + c^2z = k^2 \\ a^3x + b^3y + c^3z = k^3 \end{cases}$$

(Assume that the values of a, b, and c are all distinct and all nonzero.)

60. Show that the equation

$$\begin{vmatrix} x & y & 1 \\ x_1 & y_1 & 1 \\ 1 & m & 0 \end{vmatrix} = 0$$

represents a line with slope m, passing through (x_1, y_1).

For Exercise 61, use the fact that the equation of a line passing through (x_1, y_1) and (x_2, y_2) can be written

$$\begin{vmatrix} x & y & 1 \\ x_1 & y_1 & 1 \\ x_2 & y_2 & 1 \end{vmatrix} = 0$$

For Exercise 62, use the fact that the equation of a circle passing through (x_1, y_1), (x_2, y_2), and (x_3, y_3) can be written

$$\begin{vmatrix} x^2+y^2 & x & y & 1 \\ x_1^2+y_1^2 & x_1 & y_1 & 1 \\ x_2^2+y_2^2 & x_2 & y_2 & 1 \\ x_3^2+y_3^2 & x_3 & y_3 & 1 \end{vmatrix} = 0$$

61. Find the equation of the line passing through $(-3, -1)$ and $(2, 9)$. Write the answer in the form $y = mx + b$.

62. Find the equation of the circle passing through $(7, 0)$, $(5, -6)$, and $(-1, -4)$. Write the answer in the form $(x-h)^2 + (y-k)^2 = r^2$.

C

63. Show that the area of the triangle in Figure A is $\frac{1}{2}\begin{vmatrix} a & b \\ c & d \end{vmatrix}$. *Hint:* Figure B indicates how the required area can be found using a rectangle and three right triangles.

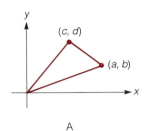

A

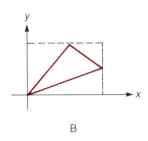

B

(c) Show that the equation in part (b) can be written

$$\begin{vmatrix} a_1 & b_1 & c_1 & d_1 \\ a_1 & b_1 & c_1 & d_1 \\ a_2 & b_2 & c_2 & d_2 \\ a_3 & b_3 & c_3 & d_3 \end{vmatrix} = 0$$

(d) Now explain why the equation in (c) indeed holds.

65. Let D denote the determinant of the matrix $\begin{pmatrix} a & b \\ c & d \end{pmatrix}$.

(a) Show that the inverse of this matrix is
$\frac{1}{D}\begin{pmatrix} d & -b \\ -c & a \end{pmatrix}$. Assume that $ad - bc \neq 0$.

(b) Use the result in part (a) to find the inverse of the
matrix $\begin{pmatrix} -6 & 7 \\ 1 & 9 \end{pmatrix}$.

66. Assuming that $p + q + r = 0$, solve the following
equation for x:

$$\begin{vmatrix} p - x & r & q \\ r & q - x & p \\ q & p & r - x \end{vmatrix} = 0$$

Hint: After adding certain rows or columns, the
quantity $p + q + r$ will appear.

64. This exercise completes the derivation of Cramer's rule
given in the text. Using the same notation, and
assuming $D \neq 0$, we need to show that the values
$x = D_x/D$, $y = D_y/D$, and $z = D_z/D$ satisfy the equations
in (1) on page 625. We will show that these values
satisfy the first equation in (1), the verification for the
other equations being entirely similar.

(a) Check that substituting the values $x = D_x/D$,
$y = D_y/D$, and $z = D_z/D$ in the first equation of (1)
yields an equation equivalent to

$$a_1 D_x + b_1 D_y + c_1 D_z - d_1 D = 0$$

(b) Show that the equation in part (a) can be written

$$a_1 \begin{vmatrix} b_1 & c_1 & d_1 \\ b_2 & c_2 & d_2 \\ b_3 & c_3 & d_3 \end{vmatrix} - b_1 \begin{vmatrix} a_1 & c_1 & d_1 \\ a_2 & c_2 & d_2 \\ a_3 & c_3 & d_3 \end{vmatrix}$$
$$+ c_1 \begin{vmatrix} a_1 & b_1 & d_1 \\ a_2 & b_2 & d_2 \\ a_3 & b_3 & d_3 \end{vmatrix} - d_1 \begin{vmatrix} a_1 & b_1 & c_1 \\ a_2 & b_2 & c_2 \\ a_3 & b_3 & c_3 \end{vmatrix} = 0$$

10.6 NONLINEAR SYSTEMS OF EQUATIONS

In the previous sections of this chapter, we looked at several techniques for
solving systems of linear equations. In particular, we saw that if solutions exist,
they can always be found by Gaussian elimination. In the present section, we
consider **nonlinear systems** of equations, that is, systems in which at least one
of the equations is not linear. There is no single technique that serves to solve all
nonlinear systems. However, simple substitution will often suffice. The work in
this section focuses on examples showing some of the more common approaches. In all the examples and in the exercises, we will be concerned exclusively with solutions (x, y) in which both x and y are real numbers. In Example
1, the system consists of one linear equation and one quadratic equation. Such a
system can always be solved by substitution.

EXAMPLE 1 Find all solutions (x, y) of the following system, where x and y are real numbers:

$$\begin{cases} 2x + y = 1 \\ y = 4 - x^2 \end{cases}$$

Solution   We use the second equation to substitute for y in the first equation. This will yield an equation with only one unknown:

$$2x + (4 - x^2) = 1$$
$$-x^2 + 2x + 3 = 0$$
$$x^2 - 2x - 3 = 0$$
$$(x - 3)(x + 1) = 0$$

$$x - 3 = 0 \quad | \quad x + 1 = 0$$
$$x = 3 \quad | \quad x = -1$$

The values $x = 3$ and $x = -1$ can now be substituted back into either of the original equations. Substituting $x = 3$ in the equation $y = 4 - x^2$ yields $y = -5$. Similarly, substituting $x = -1$ in the equation $y = 4 - x^2$ gives us $y = 3$. We have now determined the two ordered pairs $(3, -5)$ and $(-1, 3)$. As you can easily check, both of these are solutions of the given system. Figure 1 displays the graphical interpretation of this result. The line $2x + y = 1$ intersects the parabola $y = 4 - x^2$ at the points $(-1, 3)$ and $(3, -5)$.

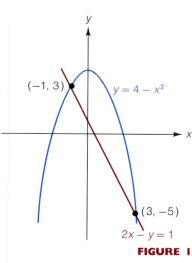

FIGURE 1

EXAMPLE 2   Where do the graphs of the parabola $y = x^2$ and the circle $x^2 + y^2 = 1$ intersect? See Figure 2.

Solution   The system that we wish to solve is

$$\begin{cases} y = x^2 & (1) \\ x^2 + y^2 = 1 & (2) \end{cases}$$

In view of equation (1), we can replace the x^2-term of equation (2) by y. Doing this yields

$$y + y^2 = 1$$

or

$$y^2 + y - 1 = 0$$

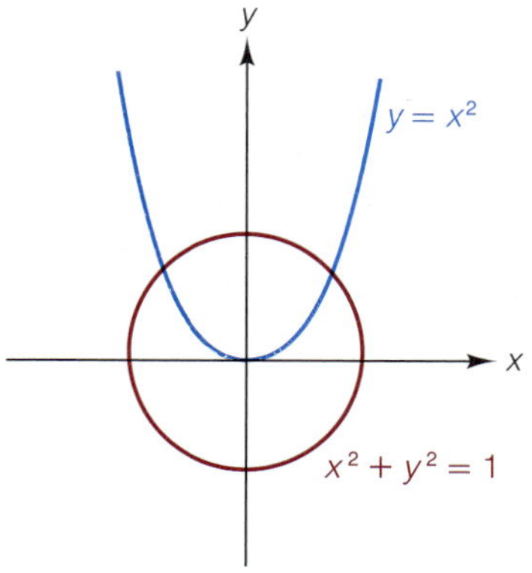

This last equation can be solved by using the quadratic formula with $a = 1$, $b = 1$, and $c = -1$. As you can check, the results are

$$y = \frac{-1 + \sqrt{5}}{2} \qquad \text{and} \qquad y = \frac{-1 - \sqrt{5}}{2}$$

FIGURE 2

However, from Figure 2, it is clear that the y-coordinate at each intersection point is positive. Therefore, we discard the negative number $y = -\frac{1}{2}(1 + \sqrt{5})$ from further consideration in this context. Substituting the positive number $y = \frac{1}{2}(-1 + \sqrt{5})$ back in the equation $y = x^2$ then gives us

$$x^2 = \frac{1}{2}(-1 + \sqrt{5})$$

Therefore,

$$x = \pm\sqrt{\tfrac{1}{2}(-1 + \sqrt{5})}$$

By choosing the positive square root, we obtain the x-coordinate for the intersection point in the first quadrant. That point is therefore

$$\left(\sqrt{\tfrac{1}{2}(-1 + \sqrt{5})}, \tfrac{1}{2}(-1 + \sqrt{5})\right) \approx (0.79, 0.62) \qquad \text{using a calculator}$$

Similarly, the negative square root yields the x-coordinate for the intersection point in the second quadrant. That point is

$$\left(-\sqrt{\tfrac{1}{2}(-1 + \sqrt{5})},\ \tfrac{1}{2}(-1 + \sqrt{5})\right) \approx (-0.79, 0.62) \qquad \text{using a calculator}$$

We have now found the two intersection points, as required.   ■■■

EXAMPLE 3   Find all solutions (x, y) of the following system, where x and y are real numbers:

$$\begin{cases} xy = 1 & (3) \\ y = 3x + 1 & (4) \end{cases}$$

Solution   Since these equations are easy to graph, we do so, because that will tell us something about the required solutions. As Figure 3 indicates, there are two intersection points, one in the first quadrant, the other in the third quadrant. One way to begin now would be to solve equation (3) for one unknown in terms of the other. However, to avoid introducing fractions at the outset, let us use equation (4) to substitute for y in equation (3). This yields

$$x(3x + 1) = 1$$

or

$$3x^2 + x - 1 = 0$$

This last equation can be solved by using the quadratic formula. As you should verify, the solutions are

$$x = \frac{-1 + \sqrt{13}}{6} \qquad \text{and} \qquad x = \frac{-1 - \sqrt{13}}{6}$$

The corresponding y-values can now be obtained by substituting for x in either of the given equations. We will substitute in equation (4). (Exercise 23 at the end of this section asks you to substitute in equation (3) as well and then to show that the very different-looking answers obtained in that way are in fact equal to those found here.) Substituting $x = \tfrac{1}{6}(-1 + \sqrt{13})$ in the equation $y = 3x + 1$ gives us

$$y = 3\left(\frac{-1 + \sqrt{13}}{6}\right) + 1$$

$$= \frac{-1 + \sqrt{13}}{2} + \frac{2}{2} = \frac{1 + \sqrt{13}}{2}$$

Thus, one of the intersection points is

$$\left(\tfrac{1}{6}(-1 + \sqrt{13}),\ \tfrac{1}{2}(1 + \sqrt{13})\right)$$

Notice that this must be the first-quadrant point of intersection, since both coordinates are positive. The intersection point in the third quadrant is obtained in exactly the same manner. As you can check, substituting the value $x = -\tfrac{1}{6}(1 + \sqrt{13})$ in the equation $y = 3x + 1$ yields $y = \tfrac{1}{2}(1 - \sqrt{13})$. Thus, the other intersection point is

$$\left(-\tfrac{1}{6}(1 + \sqrt{13}),\ \tfrac{1}{2}(1 - \sqrt{13})\right)$$

As Figure 3 indicates, there are no other solutions.   ■■■

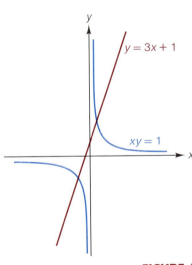

$y = 3x + 1$

$xy = 1$

FIGURE 3

EXAMPLE 4 Find all real numbers x and y satisfying the system of equations

$$\begin{cases} y = \sqrt{x} \\ (x + 2)^2 + y^2 = 1 \end{cases}$$

Solution We use the first equation to substitute for y in the second equation. This yields

$$(x + 2)^2 + \left(\sqrt{x}\right)^2 = 1$$
$$x^2 + 4x + 4 + x = 1$$
$$x^2 + 5x + 3 = 0$$

Using the quadratic formula to solve this last equation, we have

$$x = \frac{-5 \pm \sqrt{5^2 - 4(1)(3)}}{2(1)} = \frac{-5 \pm \sqrt{13}}{2}$$

The two values of x are thus

$$-\tfrac{1}{2}(5 - \sqrt{13}) \qquad \text{and} \qquad -\tfrac{1}{2}(5 + \sqrt{13})$$

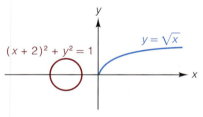

$y = \sqrt{x}$

$(x + 2)^2 + y^2 = 1$

FIGURE 4

However, notice that both of these quantities are negative. (The second is obviously negative; without using a calculator, can you explain why the first is negative?) Thus, neither of these quantities is an appropriate x-input in the equation $y = \sqrt{x}$, since we are looking for y-values that are real numbers. We conclude from this that there are no pairs of real numbers x and y satisfying the given system. In geometric terms, this means that the two graphs do not intersect. See Figure 4.

In the next example, we look at a system that can be reduced to a linear system through appropriate substitutions.

EXAMPLE 5 Solve the system

$$\begin{cases} \dfrac{2}{x^2} - \dfrac{3}{y^2} = -6 \\[2mm] \dfrac{3}{x^2} + \dfrac{4}{y^2} = 59 \end{cases}$$

Solution Let $u = 1/x^2$ and $v = 1/y^2$, so that the system becomes

$$\begin{cases} 2u - 3v = -6 \\ 3u + 4v = 59 \end{cases}$$

This is now a linear system. As Exercise 46 asks you to show, the solution is $u = 9$, $v = 8$. In view of the definitions of u and v, then, we have

$$\frac{1}{x^2} = 9 \qquad\qquad \frac{1}{y^2} = 8$$
$$x^2 = 1/9 \qquad\qquad y^2 = 1/8$$
$$x = \pm 1/3 \qquad\qquad y = \pm 1/\sqrt{8} = \pm 1/(2\sqrt{2})$$
$$= \pm \sqrt{2}/4$$

This gives us four possible solutions for the original system:

$$\left(\frac{1}{3}, \frac{\sqrt{2}}{4}\right) \quad \left(\frac{1}{3}, -\frac{\sqrt{2}}{4}\right) \quad \left(-\frac{1}{3}, \frac{\sqrt{2}}{4}\right) \quad \left(-\frac{1}{3}, -\frac{\sqrt{2}}{4}\right)$$

As you can check, all four of these pairs satisfy the given system. ▪▪▪

EXAMPLE 6 Determine all solutions (x, y) of the following system, where x and y are real numbers:

$$\begin{cases} y = 3^x & (5) \\ y = 3^{2x} - 2 & (6) \end{cases}$$

Solution We'll use the substitution method. First we rewrite equation (6) as

$$y = (3^x)^2 - 2 \tag{7}$$

Now, in view of equation (5), we can replace 3^x with y in equation (7) to obtain

$$\begin{aligned} y &= y^2 - 2 \\ 0 &= y^2 - y - 2 \\ &= (y + 1)(y - 2) \end{aligned}$$

From this last equation, we see that $y = -1$ or $y = 2$. With $y = -1$, equation (5) becomes $-1 = 3^x$, contrary to the fact that 3^x is positive for all real numbers x. Thus, we discard the case where $y = -1$. On the other hand, if $y = 2$, equation (5) becomes

$$2 = 3^x$$

We can solve this exponential equation by taking the logarithm of both sides. Using base e logarithms, we have

$$\ln 2 = \ln 3^x = x \ln 3$$

and, consequently,

$$x = \frac{\ln 2}{\ln 3} \qquad \text{Caution: } \frac{\ln 2}{\ln 3} \neq \ln 2 - \ln 3$$

We've now found that $x = (\ln 2)/(\ln 3)$ and $y = 2$. Figure 5 displays a graphical interpretation of this result. (Using a calculator for the x-coordinate, we find that $x \approx 0.6$, which is consistent with Figure 5.) ▪▪▪

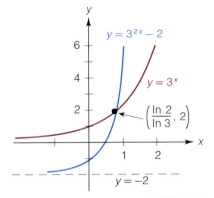

FIGURE 5

EXERCISE SET 10.6

A

In Exercises 1–22, find all solutions (x, y) of the given systems, where x and y are real numbers.

1. $\begin{cases} y = 3x \\ y = x^2 \end{cases}$

2. $\begin{cases} y = x + 3 \\ y = 9 - x^2 \end{cases}$

3. $\begin{cases} x^2 + y^2 = 25 \\ 24y = x^2 \end{cases}$

4. $\begin{cases} 3x + 4y = 12 \\ x^2 - y + 1 = 0 \end{cases}$

5. $\begin{cases} xy = 1 \\ y = -x^2 \end{cases}$

6. $\begin{cases} x + 2y = 0 \\ xy = -2 \end{cases}$

7. $\begin{cases} 2x^2 + y^2 = 17 \\ x^2 + 2y^2 = 22 \end{cases}$

8. $\begin{cases} x - 2y = 1 \\ y^2 - x^2 = 3 \end{cases}$

9. $\begin{cases} y = 1 - x^2 \\ y = x^2 - 1 \end{cases}$

10. $\begin{cases} xy = 4 \\ y = 4x \end{cases}$

11. $\begin{cases} xy = 4 \\ y = 4x + 1 \end{cases}$

12. $\begin{cases} \dfrac{2}{x^2} + \dfrac{5}{y^2} = 3 \\ \dfrac{3}{x^2} - \dfrac{2}{y^2} = 1 \end{cases}$

13. $\begin{cases} \dfrac{1}{x^2} - \dfrac{3}{y^2} = 14 \\ \dfrac{2}{x^2} + \dfrac{1}{y^2} = 35 \end{cases}$

14. $\begin{cases} y = -\sqrt{x} \\ (x - 3)^2 + y^2 = 4 \end{cases}$

15. $\begin{cases} y = -\sqrt{x - 1} \\ (x - 3)^2 + y^2 = 4 \end{cases}$

16. $\begin{cases} y = -\sqrt{x - 6} \\ (x - 3)^2 + y^2 = 4 \end{cases}$

17. $\begin{cases} y = 2^x \\ y = 2^{2x} - 12 \end{cases}$

18. $\begin{cases} y = e^{4x} \\ y = e^{2x} + 6 \end{cases}$

19. $\begin{cases} 2(\log_{10} x)^2 - (\log_{10} y)^2 = -1 \\ 4(\log_{10} x)^2 - 3(\log_{10} y)^2 = -11 \end{cases}$

20. $\begin{cases} y = \log_2(x + 1) \\ y = 5 - \log_2(x - 3) \end{cases}$

21. $\begin{cases} 2^x \cdot 3^y = 4 \\ x + y = 5 \end{cases}$

22. $\begin{cases} a^{2x} + a^{2y} = 10 \\ a^{x+y} = 4 \end{cases}$ $(a > 0)$

Hint: Use the substitutions $a^x = t$ and $a^y = u$.

23. Let $x = \frac{1}{6}(-1 + \sqrt{13})$. Using this x-value, show that the equations $y = 3x + 1$ and $y = 1/x$ yield the same y-value.

24. A sketch shows that the line $y = 100x$ intersects the parabola $y = x^2$ at the origin. Are there any other intersection points? If so, find them. If not, explain why not.

25. Solve the following system for x and y:

$$\begin{cases} ax + by = 2 \\ abxy = 1 \end{cases}$$

(Assume that neither a nor b is zero.)

B

26. Let a, b, and c be constants (with $a \neq 0$), and consider the system

$$\begin{cases} y = ax^2 + bx + c \\ y = k \end{cases}$$

For which value of k (in terms of a, b, and c) will the system have exactly one solution? What is that solution? What is the relationship between the solution you've found and the graph of $y = ax^2 + bx + c$?

27. Find all solutions of the system

$$\begin{cases} x^3 + y^3 = 3473 \\ x + y = 23 \end{cases}$$

28. Solve the following system for x, y, and z:

$$\begin{cases} yz = p^2 \\ zx = q^2 \\ xy = r^2 \end{cases}$$

(Assume that p, q, and r are positive constants.)

29. If the diagonal of a rectangle has length d and the perimeter of the rectangle is $2p$, express the lengths of the sides in terms of d and p.

30. Solve the following system for x and y:

$$\begin{cases} xy + pq = 2px \\ x^2y^2 + p^2q^2 = 2q^2y^2 \end{cases} \quad (pq \neq 0)$$

Hint: Square the first equation, then subtract the second. This results in an equation that can be written $(2px + qy)(px - qy) = 0$.

31. Solve the following system for u and v:

$$\begin{cases} u^2 - v^2 = 9 \\ \sqrt{u + v} + \sqrt{u - v} = 4 \end{cases}$$

Hint: Square the second equation.

32. Solve the given system for x and y in terms of p and q:

$$\begin{cases} q^{\ln x} = p^{\ln y} \\ (px)^{\ln a} = (qy)^{\ln b} \end{cases}$$

(Assume all constants and variables are positive.)

33. The accompanying figure shows the graph of a function $N = N_0 e^{kt}$. In each case, determine the constants N_0 and k so that the graph passes through the two given points.

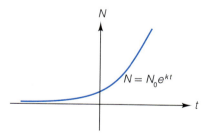

(a) $(2, 3)$ and $(8, 24)$
(b) $\left(\frac{1}{2}, 1\right)$ and $(4, 10)$

34. [As background for this exercise, first work Exercise 33(a).] Measurements (begun in 1958, by Charles D. Keeling of the Scripps Institution of Oceanography) show that the concentration of carbon dioxide in the atmosphere has increased exponentially over the years 1958–1988. In 1958, the average carbon dioxide

concentration was $N = 315$ ppm (parts per million), and in 1988 it was $N = 351$ ppm.

(a) Use this data to determine the constants k and N_0 in the growth law $N = N_0 e^{kt}$. *Hint:* The curve $N = N_0 e^{kt}$ passes through the two points $(1958, 315)$ and $(1988, 351)$. Use this to obtain a system of two equations in two unknowns.

(b) Stephen H. Schneider in his article "The Changing Climate" [*Scientific American* **261** (1989), 70–79] estimates that, under certain circumstances, the concentration of carbon dioxide in the atmosphere could reach 600 ppm by the year 2080. Use your results in part (a) to make a projection for the year 2080. Is your projection higher or lower than Dr. Schneider's?

35. Solve the following system for x, y, and z in terms of p, q, and r.

$$\begin{cases} x(x + y + z) = p^2 \\ y(x + y + z) = q^2 \\ z(x + y + z) = r^2 \end{cases}$$

(Assume that p, q, and r are nonzero.) *Hint:* Denote $x + y + z$ by w; then add the three equations.

36. (a) Find the points where the line $y = -2x - 2$ intersects the parabola $y = \frac{1}{2}x^2$.

(b) On the same set of axes, sketch the line $y = -2x - 2$ and the parabola $y = \frac{1}{2}x^2$. Be certain that your sketch is consistent with the results obtained in part (a).

37. The area of a right triangle is 180 cm^2, and the hypotenuse is 41 cm. Find the lengths of the two legs.

38. The sum of two numbers is 8, while their product is -128. What are the two numbers?

39. The perimeter of a rectangle is 46 cm, and the area is 60 cm^2. Find the length and the width.

40. Find all right triangles for which the perimeter is 24 units and the area is 24 square units.

41. Solve the following system for x and y using the substitution method:

$$\begin{cases} x^2 + y^2 = 5 & (1) \\ xy = 2 & (2) \end{cases}$$

42. The substitution method in Exercise 41 leads to a quadratic equation. Here is an alternative approach to solving that system; this approach leads to linear equations. Multiply equation (2) by 2 and add the resulting equation to equation (1). Now take square roots to conclude that $x + y = \pm 3$. Next multiply equation (2) by 2 and subtract the resulting equation from equation (1). Take square roots to conclude that

$x - y = \pm 1$. You now have the following four linear systems, each of which can be solved (with almost no work) by the addition–subtraction method:

$$\begin{cases} x + y = 3 \\ x - y = 1 \end{cases} \qquad \begin{cases} x + y = 3 \\ x - y = -1 \end{cases}$$

$$\begin{cases} x + y = -3 \\ x - y = 1 \end{cases} \qquad \begin{cases} x + y = -3 \\ x - y = -1 \end{cases}$$

Solve these systems and compare your results with those obtained in Exercise 41.

43. Solve the following system using the method explained in Exercise 42:

$$\begin{cases} x^2 + y^2 = 7 \\ xy = 3 \end{cases}$$

44. Solve the following system using the substitution method:

$$\begin{cases} 3xy - 4x^2 = 2 \\ -5x^2 + 3y^2 = 7 \end{cases}$$

(Begin by solving the first equation for y.)

45. Here is an alternative approach for solving the system in Exercise 44. Let $y = mx$, where m is a constant to be determined. Replace y with mx in both equations of the system to obtain the following pair of equations:

$$x^2(3m - 4) = 2 \qquad (1)$$
$$x^2(-5 + 3m^2) = 7 \qquad (2)$$

Now divide equation (2) by equation (1). After clearing fractions and simplifying, you can write the resulting equation as $2m^2 - 7m + 6 = 0$. Solve this last equation by factoring. The values of m can then be used in equation (1) to determine values for x. In each case, the corresponding y-values are determined by the equation $y = mx$.

46. Solve the following system. (You should obtain $u = 9$ and $v = 8$, as stated in Example 5.)

$$\begin{cases} 2u - 3v = -6 \\ 3u + 4v = 59 \end{cases}$$

47. Solve the following system for x and y:

$$\begin{cases} xy + pq = 2px \\ x^2 y^2 + p^2 q^2 = 2q^2 y^2 \end{cases}$$

Hint: Square the first equation, then subtract the second. This results in an equation that can be factored.

C

48. Solve the following system for x and y:

$$\begin{cases} \dfrac{1}{x^2} + \dfrac{1}{xy} = \dfrac{1}{a^2} \\[2mm] \dfrac{1}{y^2} + \dfrac{1}{xy} = \dfrac{1}{b^2} \end{cases}$$

(Assume that a and b are positive.)

49. Solve the following system for x and y:

$$\begin{cases} x^4 = y^6 \\[2mm] \ln \dfrac{x}{y} = \dfrac{\ln x}{\ln y} \end{cases}$$

10.7   SYSTEMS OF INEQUALITIES. LINEAR PROGRAMMING

... the variety of problems in business, science, and industry that can be solved by using the linear programming model is impressive, to say the least. Problems in transportation, scheduling, capital budgeting, production planning, inventory control, advertising, and many more areas can be solved by the linear programming approach.

Stanley J. Farlow and Gary M. Haggard in *Finite Mathematics and Its Applications* (New York: Random House, 1988)

In the first section of this chapter, we solved systems of equations in two unknowns. Now let us consider systems of inequalities in two unknowns. The techniques we develop are used in calculus when discussing functions of two variables and in business and economics in the study of linear programming.

Let a, b, and c denote real numbers, and assume that a and b are not both zero. Then all of the following are called **linear inequalities**:

$$ax + by + c < 0 \qquad ax + by + c > 0$$
$$ax + by + c \le 0 \qquad ax + by + c \ge 0$$

An ordered pair of numbers (x_0, y_0) is said to be a **solution** of a given inequality (linear or not) provided we obtain a true statement upon substituting x_0 and y_0 for x and y, respectively. For instance, $(-1, 1)$ is a solution of the inequality $2x + 3y - 6 < 0$, since substitution yields

$$2(-1) + 3(1) - 6 < 0$$
$$-5 < 0 \qquad \text{True}$$

Like a linear equation in two unknowns, a linear inequality has infinitely many solutions. For this reason, we often represent the solutions graphically. When we do this, we say that we are **graphing the inequality**.

For example, let us graph the inequality $y < 2x$. First we observe that the coordinates of the points on the line $y = 2x$ do not, by definition, satisfy this inequality. So it remains to consider points above the line and points below the line. We will in fact show that the required graph consists of all points that lie below the line. Take any point $P(x_0, y_0)$ not on the line $y = 2x$. Let $Q(x_0, 2x_0)$ be the point on $y = 2x$ with the same first coordinate as P. Then, as indicated in Figure 1, the point P lies below the line if and only if the y-coordinate of P is less than the y-coordinate of Q. In other words, (x_0, y_0) lies below the line if and

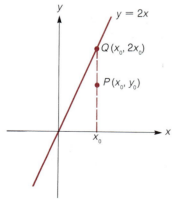

FIGURE I

only if $y_0 < 2x_0$. This last statement is equivalent to saying that (x_0, y_0) satisfies the inequality $y < 2x$. This shows that the graph of $y < 2x$ consists of all points below the line $y = 2x$ (see Figure 2). The broken line in Figure 2 indicates that the points on $y = 2x$ are not part of the required graph. If the original inequality had been $y \leq 2x$, then we would use a solid line rather than a broken one, indicating that the line is included in the graph. And if the original inequality had been $y > 2x$, the graph would be the region above the line.

Just as the graph of $y < 2x$ is the region below the line $y = 2x$, it is true in general that the graph of $y < f(x)$ is the region below the graph of the function f. For example, Figures 3 and 4 display the graphs of $y < x^2$ and $y \geq x^2$.

Example 1 summarizes the technique developed so far for graphing an inequality. Following this example, we will point out a useful alternative method.

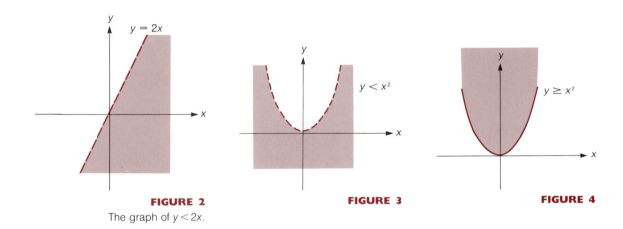

FIGURE 2
The graph of $y < 2x$.

FIGURE 3

FIGURE 4

EXAMPLE 1   Graph the inequality $4x - 3y \leq 12$.

Solution   The graph will include the line $4x - 3y = 12$ and either the region above or the region below the line. To decide which region, solve the inequality for y:

$$4x - 3y \leq 12$$
$$-3y \leq -4x + 12$$
$$y \geq \frac{4}{3}x - 4 \qquad \text{multiplying or dividing by a negative number reverses an inequality}$$

This last inequality tells us that we want the region above the line, as well as the line itself. Figure 5 displays the required graph.

There is another method that we can use in Example 1 to determine which side of the line we want. This method involves a **test point**. We pick any convenient point that is not on the line $4x - 3y = 12$. Then we test to see if this point satisfies the given inequality. For example, let us pick the point $(0, 0)$. Substituting these coordinates in the inequality $4x - 3y \leq 12$ yields

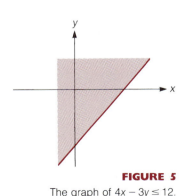

FIGURE 5
The graph of $4x - 3y \leq 12$.

$$4(0) - 3(0) \overset{?}{\leq} 12$$
$$0 \overset{?}{\leq} 12 \qquad \text{True}$$

We conclude from this that the required side of the line is that on which the point $(0,0)$ resides. In agreement with Figure 5, we see this is the region above the line.

Next, we discuss systems of inequalities in two unknowns. As with systems of equations, a **solution** of a system of inequalities is an ordered pair (x_0, y_0) that satisfies all of the inequalities in the system. As a first example, let us graph the points that satisfy the following nonlinear system:

$$\begin{cases} y - x^2 \geq 0 \\ x^2 + y^2 < 1 \end{cases}$$

By writing the first inequality as $y \geq x^2$, we see that it describes the set of points on or above the parabola $y = x^2$. For the second inequality, we must decide whether it describes the points within or outside the circle $x^2 + y^2 = 1$. One way to do this is to choose $(0,0)$ as a test point. Substituting the values $x = 0$ and $y = 0$ in the inequality $x^2 + y^2 < 1$ yields the true statement $0^2 + 0^2 < 1$. Since $(0,0)$ lies within the circle and satisfies the inequality, we conclude that the inequality $x^2 + y^2 < 1$ describes the set of all points within the circle. Now we put our information together. We wish to graph the points that lie on or above the parabola $y = x^2$ but within the circle $x^2 + y^2 = 1$. This is the shaded region shown in Figure 6.

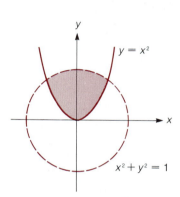

FIGURE 6

EXAMPLE 2   Graph the system

$$\begin{cases} -x + 3y \leq 12 \\ x + y \leq 8 \\ x \geq 0 \\ y \geq 0 \end{cases}$$

Solution   Solving the first inequality for y, we have

$$3y \leq x + 12$$
$$y \leq \tfrac{1}{3}x + 4$$

Thus, the first inequality is satisfied by the points on or below the line $-x + 3y = 12$. Similarly, by solving the second inequality for y, we see that it describes the set of points on or below the line $x + y = 8$. The third inequality, $x \geq 0$, describes the points on or to the right of the y-axis. Similarly, the fourth inequality, $y \geq 0$, describes the points on or above the x-axis. We summarize these statements in Figure 7(a). The arrows indicate which side of each line we wish to consider.

Finally, Figure 7(b) shows the graph of the given system. The coordinates of each point in the shaded region satisfy all four of the given inequalities. (Exercise 25 at the end of this section asks you to show that the coordinates listed in the figure are correct.)

We can use Figure 7(b) to introduce some terminology that is useful in describing sets of points in the plane. The **vertices** of a region are the corners, or points, where the adjacent bounding sides meet. Thus, the vertices of the shaded region in Figure 7(b) are the four points $(0,0)$, $(8,0)$, $(3,5)$, and $(0,4)$. The shaded region in Figure 7(b) is **convex.** This means that given any two

FIGURE 7

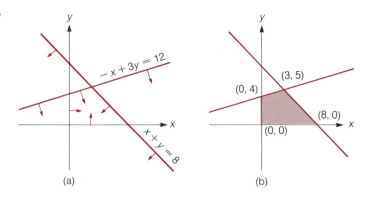

(a)   (b)

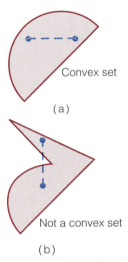

Convex set

(a)

Not a convex set

(b)

FIGURE 8

points in that region, the straight line segment joining these two points lies wholly within the region. Figure 8(a) displays another example of a convex set, while Figure 8(b) shows a set that is not convex. The shaded region in Figure 8(b) is also an example of a **bounded region**. By this we mean that the region can be wholly contained within some (sufficiently large) circle. Perhaps the simplest example of a region that is not bounded is the entire x-y plane itself. Figure 9 shows another example of an unbounded region.

We are prepared now to consider a typical problem in linear programming.

Problem Let $C = x + 3y$. Find the values of x and y that yield the largest possible value of C, given that x and y satisfy the following set of inequalities.

$$\begin{cases} -x + 3y \le 12 \\ x + y \le 8 \\ x \ge 0 \\ y \ge 0 \end{cases}$$

The function $C = x + 3y$ in this problem is known as the **objective function**. The given inequalities are called the **constraints**. In analyzing this problem we will take advantage of the fact that we've already graphed this system of inequalities in Example 2. For convenience, we repeat the graph here (Figure 10). Since the coordinates of each point in the shaded region of Figure 10 satisfy the constraints of the present problem, we refer to the region as the set of **feasible solutions**. The problem then is to find that point among the feasible solutions

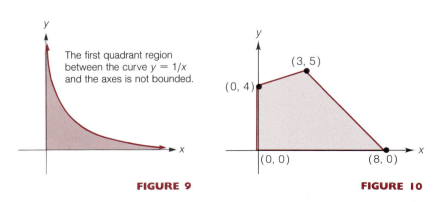

The first quadrant region between the curve $y = 1/x$ and the axes is not bounded.

FIGURE 9   **FIGURE 10**

that yields the largest possible value for C. A feasible solution (x_0, y_0) that yields the largest value for C is called an **optimal solution**. Some linear programming problems ask for a smallest rather than a largest value of the objective function. A feasible solution yielding the smallest value for C is also called an optimal solution.

We will see subsequently that there is a general method for solving this type of linear programming problem, but as a start, let us experiment. How large can the value of C be, subject to the given constraints? For instance, can C equal 30? With $C = 30$, we have

$$x + 3y = 30$$

Figure 11 shows the graph of this line. Since the line has no points in common with the set of feasible solutions, we conclude that C cannot equal 30 under the conditions of the problem.

Still experimenting, in Figure 12 we have graphed the lines $x + 3y = C$ for various values of C. Observe that the lines are all parallel. To see why this is so, solve the equation $x + 3y = C$ for y. This yields

$$y = -\frac{1}{3}x + \frac{1}{3}C$$

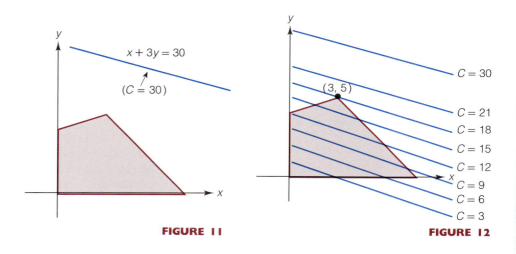

FIGURE 11   FIGURE 12

As you can see, the slope here is $m = -\frac{1}{3}$, independent of C. Since the lines all have the same slope, they are parallel. On the other hand, the equation $y = -\frac{1}{3}x + \frac{1}{3}C$ shows how the value of C does influence the graph. The y-intercept of the line is $C/3$. So as C increases, the line moves *up*, but the slope is always $-\frac{1}{3}$.

Now, as you can see in Figure 12, with $C = 18$ the line $x + 3y = 18$ intersects the set of feasible solutions at the vertex $(3, 5)$. However, it is also evident from the figure and the remarks in the previous paragraph that when C is greater than 18, the line $x + 3y = C$ will not intersect the set of feasible solutions. We conclude therefore that the vertex $(3, 5)$ is the optimal solution. That is, subject to the given constraints, the value of C is largest when $x = 3$ and $y = 5$.

In the example just concluded, we found that the optimal solution to the linear programming problem occurred at a vertex. The following theorem (stated here without proof) tells us that the optimal solution, when it exists, always occurs at a vertex. (In the statement of the theorem, the word "nonstrict" refers to inequalities that involve the symbols $\leq$ or $\geq$ rather than $<$ or $>$.)

Linear Programming Theorem

Consider the objective function $C = ax + by + c$, where the nonnegative variables x and y are subject to constraints specified by nonstrict linear inequalities. Then if C assumes a maximum (or minimum) value, this occurs at a vertex. Furthermore, if the set of feasible solutions is both convex and bounded, then a maximum and minimum always exist.

In view of this theorem, we can solve linear programming problems by first finding the vertices of the feasibility set and then computing and comparing the values of the objective function at those vertices. This is done in Example 3, which follows. Technically speaking, the theorem we've stated guarantees a maximum and a minimum only when the feasibility set is convex and bounded. However, this is not as great a restriction as it might at first seem. As it turns out, the graph of any set of linear inequalities forms a convex set. (For a look at a case in which the feasibility set is not bounded, see Exercise 44.)

EXAMPLE 3 Let $C = 3x - 2y$. Find the maximum and the minimum value of C subject to the following set of constraints.

$$\begin{cases} 6y - x \geq 5 \\ 4x + 3y \leq 34 \\ 2y - x \leq 8 \\ 4x - y \geq 3 \end{cases}$$

What are the optimal solutions in each case?

Solution In Figure 13 (on the next page), we have graphed the given system of inequalities using the method discussed in Example 2. Now we need to determine the coordinates of the four vertices. To find the coordinates of E, we solve the system

$$\begin{cases} 6y - x = 5 & \text{(1)} \\ 4x - y = 3 & \text{(2)} \end{cases}$$

Solving equation (1) for x, we have

$$6y - 5 = x \qquad \text{(3)}$$

Now we use equation (3) to substitute for x in equation (2). This yields

$$4(6y - 5) - y = 3$$
$$24y - 20 - y = 3$$
$$23y = 23$$
$$y = 1$$

Finally, substituting $y = 1$ in equation (3) gives us $x = 6(1) - 5 = 1$.

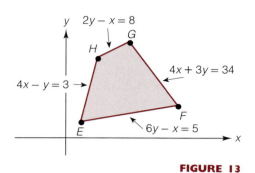

FIGURE 13

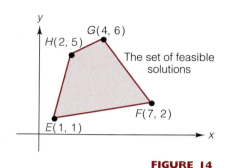

FIGURE 14

TABLE 1

Vertex	$C = 3x - 2y$
$E(1, 1)$	$3(1) - 2(1) = 1$
$F(7, 2)$	$3(7) - 2(2) = 17$
$G(4, 6)$	$3(4) - 2(6) = 0$
$H(2, 5)$	$3(2) - 2(5) = -4$

We have now found that the coordinates of the vertex E are $(1, 1)$. The coordinates of F, G, and H can be found in a similar fashion. The results are displayed in Figure 14. Notice that the set of feasible solutions forms a convex and bounded set. Thus the objective function attains a maximum and a minimum value at certain vertices. We compute the value of C at each vertex and tabulate the results as shown in Table 1.

The largest value of C in this table is 17. This is the required maximum. The corresponding optimal solution is $(7, 2)$. Similarly, the smallest value of C is -4. This is the required minimum. The corresponding optimal solution is $(2, 5)$. ∎

The next example deals with a type of linear programming problem known as the **transportation problem**. In this type of problem we are required to minimize the cost of shipping goods from warehouses to retail outlets.

EXAMPLE 4   A motorcycle manufacturer must fill orders from two dealers. The first dealer, D_1, has ordered 20 motorcycles, while the second dealer, D_2, has ordered 30 motorcycles. The manufacturer has the motorcycles stored in two warehouses, W_1 and W_2. There are 40 motorcycles in W_1 and 15 in W_2. The shipping costs, per motorcycle, are as follows: \$15 from W_1 to D_1; \$13 from W_1 to D_2; \$14 from W_2 to D_1; \$16 from W_2 to D_2. Under these conditions, find the number of motorcycles to be shipped from each warehouse to each dealer if the total shipping cost is to be a minimum. What is this minimum cost?

Solution   The first step is to summarize the data. We do this as follows.

Shipping Cost Per Motorcycle
W_1 to D_1:   \$15
W_1 to D_2:   \$13
W_2 to D_1:   \$14
W_2 to D_2:   \$16

Supply and Demand	
W_1 (40 motorcycles in stock)	W_2 (15 motorcycles in stock)
D_1 (wants 20 motorcycles)	D_2 (wants 30 motorcycles)

The problem asks how many motorcycles should be shipped from each warehouse to each dealer.

1. Let x denote the number of motorcycles to be shipped from W_1 to D_1.

2. Then $20 - x$ denotes the number of motorcycles to be shipped from W_2 to D_1. (Why?)

3. Let y denote the number of motorcycles to be shipped from W_1 to D_2.

4. Then $30 - y$ denotes the number of motorcycles to be shipped from W_2 to D_2.

If C denotes the total shipping cost, we then have

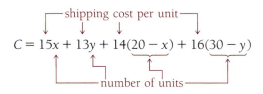

As you can check, this equation simplifies to

$$C = x - 3y + 760$$

This is the objective function that we must minimize.

Now let us examine the constraints of the problem. The number of items shipped from the warehouse certainly must be a nonnegative number in each case. Thus we must have

$$x \geq 0$$
$$y \geq 0$$
$$20 - x \geq 0$$
$$30 - y \geq 0$$

Furthermore, since at most 40 motorcycles can be shipped from W_1, we must have

$$x + y \leq 40$$

Similarly, since at most 15 motorcycles can be shipped from W_2, we have

$$(20 - x) + (30 - y) \leq 15$$

or

$$50 - x - y \leq 15$$
$$-x - y \leq -35$$
$$x + y \geq 35$$

In summary, then, we have the following set of constraints:

$$\begin{cases} x \geq 0 \\ y \geq 0 \\ 20 - x \geq 0 \\ 30 - y \geq 0 \\ x + y \leq 40 \\ x + y \geq 35 \end{cases}$$

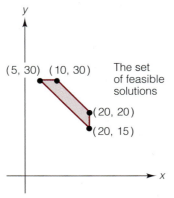

FIGURE 15

Figure 15 shows the graph of the region described by this set of constraints. (Exercise 33 at the end of this section asks you to verify that the graph and the coordinates of the vertices are as indicated.) Since the region is both convex and bounded, we know that the objective function assumes its minimum value at one of the vertices. We compute the value of C at each vertex.

Vertex	$C = x - 3y + 760$
$(20, 15)$	$C = 20 - 3(15) + 760 = 735$
$(20, 20)$	$C = 20 - 3(20) + 760 = 720$
$(10, 30)$	$C = 10 - 3(30) + 760 = 680$
$(5, 30)$	$C = 5 - 3(30) + 760 = 675$

As the table shows, the minimum shipping cost is $675. The corresponding vertex is $(5, 30)$. Therefore, the shipping cost will be minimized if the deliveries are made as follows.

$$x = 5:\qquad \text{ship 5 motorcycles from } W_1 \text{ to } D_1$$
$$20 - x = 15:\qquad \text{ship 15 motorcycles from } W_2 \text{ to } D_1$$
$$y = 30:\qquad \text{ship 30 motorcycles from } W_1 \text{ to } D_2$$
$$30 - y = 0:\qquad \text{ship no motorcycles from } W_2 \text{ to } D_2$$

The linear programming problems that we have considered have involved only two unknowns, x and y. Because there were only two unknowns, we were able to utilize graphical methods in solving the problems. As you might guess, linear programming problems actually arising in business and industry may involve dozens or even thousands of unknowns. For such cases, there is a purely algebraic method of solution called the **simplex method**, which makes use of matrices. Furthermore, this method can be programmed for computer solution. Several commercial firms now market computer packages for solving linear programming problems.*

*For an excellent discussion of the simplex method as well as methods of computer solution, see *An Introduction to Management Science*, 6th ed., by David R. Anderson et al. (St. Paul: West Publishing Co., 1991).

EXERCISE SET 10.7

A

In Exercises 1 and 2, decide whether or not the ordered pairs are solutions of the given inequality.

1. $4x - 6y + 3 \geq 0$ (a) $(1, 2)$ (b) $\left(0, \frac{1}{2}\right)$
2. $5x + 2y < 1$ (a) $(-1, 3)$ (b) $(0, 0)$

In Exercises 3–16, graph the given inequalities.

3. $2x - 3y > 6$
4. $2x - 3y < 6$
5. $2x - 3y \geq 6$
6. $2x - 3y \leq 6$
7. $x - y < 0$
8. $y \leq \frac{1}{2}x - 1$
9. $x \geq 1$
10. $y < 0$
11. $x > 0$
12. $y \leq \sqrt{x}$
13. $y > x^3 + 1$
14. $y \leq |x - 2|$
15. $x^2 + y^2 \geq 25$
16. $y \geq e^x - 1$

In Exercises 17–22, graph the systems of inequalities.

17. $\begin{cases} y \leq x^2 \\ x^2 + y^2 \leq 1 \end{cases}$

18. $\begin{cases} y < x \\ x^2 + y^2 < 1 \end{cases}$

19. $\begin{cases} y \geq 1 \\ y \leq |x| \end{cases}$

20. $\begin{cases} x \geq 0 \\ y \geq 0 \\ y < \sqrt{x} \\ x \leq 4 \end{cases}$

21. $\begin{cases} x \geq 0 \\ y \geq 0 \\ y \leq 1 - x^2 \end{cases}$

22. $\begin{cases} y < 2x \\ y > \frac{1}{2}x \end{cases}$

In Exercises 23–34, graph the systems of linear inequalities. In each case specify the vertices. Is the region convex? Is the region bounded?

23. $\begin{cases} y \leq x + 5 \\ y \leq -2x + 14 \\ x \geq 0 \\ y \geq 0 \end{cases}$

24. $\begin{cases} y \geq x + 5 \\ y \geq -2x + 14 \\ x \geq 0 \\ y \geq 0 \end{cases}$

25. $\begin{cases} -x + 3y \leq 12 \\ x + y \leq 8 \\ x \geq 0 \\ y \geq 0 \end{cases}$

26. $\begin{cases} y \geq 2x \\ y \geq -x + 6 \end{cases}$

27. $\begin{cases} 0 \leq 2x - y + 3 \\ x + 3y \leq 23 \\ 5x + y \leq 45 \end{cases}$

28. $\begin{cases} 0 \leq 2x - y + 3 \\ x + 3y \leq 23 \\ 5x + y \leq 45 \\ x \geq 0 \\ y \geq 0 \end{cases}$

29. $\begin{cases} 5x + 6y < 30 \\ y > 0 \end{cases}$

30. $\begin{cases} 5x + 6y < 30 \\ x > 0 \end{cases}$

31. $\begin{cases} 5x + 6y < 30 \\ x > 0 \\ y > 0 \end{cases}$

32. $\begin{cases} 2x + 3y \geq 6 \\ 2x + 3y \leq 12 \end{cases}$

33. $\begin{cases} x \geq 0 \\ y \geq 0 \\ 20 - x \geq 0 \\ 30 - y \geq 0 \\ x + y \leq 40 \\ x + y \geq 35 \end{cases}$

34. $\begin{cases} x \geq 0 \\ y \geq 0 \\ 3x - y + 1 \geq 0 \\ 0 \leq x - y + 3 \\ y \leq 5 \\ x \leq \frac{1}{3}(17 - y) \\ x \leq \frac{1}{2}(y + 8) \end{cases}$

In Exercises 35–43, find the maximum and the minimum values of the objective function subject to the given constraints. Specify the corresponding optimal solutions in each case.

35. Objective function: $C = 3y - x$

Constraints: $\begin{cases} y \leq x + 1 \\ y \leq -2x + 10 \\ x \geq 0 \\ y \geq 0 \end{cases}$

36. Objective function: $C = 2x - y + 8$

Constraints: $\begin{cases} 4x + 3y \geq 12 \\ 6x + 5y \leq 30 \\ x \geq 0 \\ y \geq 0 \end{cases}$

37. Objective function: $C = 2x + y$

Constraints: $\begin{cases} x \geq 0 \\ x \leq 1 \\ y \geq 0 \\ 2y - x \leq 3 \end{cases}$

38. Objective function: $C = 2x - y$

Constraints: Use the same constraints given in Exercise 37.

39. Objective function: $C = 10y + 9x - 1$

Constraints: $\begin{cases} x + y \geq 3 \\ y - x + 1 \geq 0 \\ y \leq -x + 12 \\ y \leq 6 \\ x \geq 1 \end{cases}$

40. Objective function: $C = 10y + 11x - 1$

Constraints: Use the same constraints given in Exercise 39.

41. Objective function: $C = 10y + 3x + 100$

Constraints: $\begin{cases} 0 \leq 4x - y + 4 \\ x + 5y \leq 41 \\ x + 3y \leq 27 \\ 2x + y \leq 24 \\ x \geq 0 \\ y \geq 0 \end{cases}$

42. Objective function: $C = 20y + 7x + 100$

Constraints: Use the same constraints given in Exercise 41.

43. (a) Objective function: $C = 19x + 100y$

Constraints: Use the same constraints given in Exercise 41.

(b) Objective function: $C = 21x + 100y$

Constraints: Use the same constraints given in Exercise 41.

(This exercise shows that a slight change in the objective function may yield a relatively large change in the optimal solution.)

44. Consider the objective function $C = 4y - x$ subject to the constraints

$$\begin{cases} y \geq -\frac{3}{2}x + 4 \\ 2y \geq x \\ y \geq 0 \\ x \geq 0 \end{cases}$$

(a) Given that there is an optimal solution yielding a minimum value of C, find that solution and the corresponding value of C.

(b) Find a feasible solution for which $C = 100$.

(c) Find a feasible solution for which $C = 1000$.
(d) Find a feasible solution for which $C = 10^6$.

45. An automobile manufacturer must fill orders from two dealers. The first dealer, D_1, has ordered 40 cars, while the second dealer, D_2, has ordered 25 cars. The manufacturer has the cars stored in two locations, W_1 and W_2. There are 30 cars in W_1 and 50 cars in W_2. The shipping costs per car are as follows: $180 from W_1 to D_1; $150 from W_1 to D_2; $160 from W_2 to D_1; $170 from W_2 to D_2. Under these conditions, how many cars should be shipped from each storage location to each dealer so as to minimize the total shipping cost? What is this minimum cost?

46. A television manufacturer must fill orders from two retailers. The first retailer, R_1 has ordered 55 television sets, while the second retailer, R_2, has ordered 75 sets. The manufacturer has the television sets stored in two warehouses, W_1 and W_2. There are 100 sets in W_1 and 120 sets in W_2. The shipping costs, per television set, are as follows: $8 from W_1 to R_1; $12 from W_1 to R_2; $13 from W_2 to R_1; $7 from W_2 to R_2. Under these conditions, find the number of television sets to be shipped from each warehouse to each retailer if the total shipping cost is to be a minimum. What is this minimum cost?

47. A factory produces two products, A and B. The production time (in hours) for each unit is as follows:

Product	Assembly	Inspection and Packing
A	0.40	0.20
B	0.50	0.40

Each day at most 500 worker-hours of assembly time and 300 worker-hours of inspection/packing time are available. The profit realized on each unit of A is $0.60; on each unit of B it is $0.80. Find the maximum daily profit under these conditions.
Hints: Let x denote the number of units of A produced each day and let y denote the number of units of B produced each day. After graphing the constraints you'll find that the coordinates of one vertex are not whole numbers. Round these off to the nearest whole number. (Why?)

48. Due to increased energy costs, the owner of the factory described in Exercise 47 decides to limit production. In particular, the owner decides that no more than 800 units of A and 500 units of B should be produced daily. Under these additional constraints, how many units of A and B should be produced daily to maximize profits? What is the maximum profit?

49. A farmer can allocate at most 600 acres for growing cherry tomatoes and regular tomatoes. The labor requirements and the profits are as follows.

	Cherry Tomatoes	Regular Tomatoes
Labor (hours per acre)	3	2
Profit (dollars per acre)	50	36

What is the farmer's maximum profit if 1350 hr of labor are available? How many acres should be devoted to each crop to achieve the maximum profit?

50. Consider the situation described in Exercise 49. Now additionally suppose that due to market considerations, the farmer decides to devote at most 550 acres to regular tomatoes and at most 100 acres to cherry tomatoes. What is the farmer's maximum profit in this case?

51. Suppose that you inherit $12,000 but that there are stipulations attached. Some (or all) of the money must be invested in two stocks, A and B. Stock A is low-risk and B is medium-risk. The stipulations require that at most $6,000 be invested in A and at least $2,000 be invested in B. Furthermore, the amount invested in B is not to exceed twice the amount invested in A. The expected returns on the stocks are $0.06 per dollar for A and $0.08 per dollar for B. Under these conditions, how much should you invest in each stock so as to maximize your expected returns?

52. You have $10,000 to invest in three stocks, A, B, and C. At least $1,000 must be invested in each stock. The combined investment in A and B must be at least $5,000, while the investment in B cannot exceed five times that in A. The expected returns on the stocks are as follows: $0.04 per dollar for A; $0.05 per dollar for B; $0.06 per dollar for C. Under these conditions, how much should you invest in each stock so as to maximize the expected returns? *Hint:* Let x and y denote the dollar amounts invested in A and B, respectively. Then $10,000 - x - y$ denotes the number of dollars invested in C.

53. In Exercise 52, suppose instead that the expected returns are $0.05 per dollar on A and $0.04 per dollar on B. The return on C remains at $0.06 per dollar. Now how much should you invest in each stock so as to maximize the expected returns?

54. In Exercise 52, suppose instead that the expected return on B is $0.07 per dollar. The expected returns on A and C remain the same. Now how much should you invest to maximize the expected returns?

55. A veterinarian wishes to prepare a supply of dog food by mixing two commercially available foods, A and B. The mixture is to contain at least 50 units of carbohydrates, at least 36 units of protein, and at least 40 units of fat. The specifications for the two commercially available products are listed in the following table.

Product	Units of Carbohydrates (per pound)	Units of Protein (per pound)	Units of Fat (per pound)
A	3	2	2
B	5	4	8

The veterinarian can purchase food A for $0.44 per pound and B for $0.80 per pound. How many pounds of each food should be mixed to meet the dietary requirements at the least cost?

B

56. Graph the following system of inequalities and specify the vertices.

$$\begin{cases} x \geq 0 \\ y \geq e^x \\ y \leq e^{-x} + 1 \end{cases}$$

A formula such as

$$f(x, y) = \sqrt{2x - y + 1} \tag{1}$$

defines a function of two variables. The inputs for such a function are ordered pairs (x, y) of real numbers. For example, using the ordered pair $(3, 5)$ as an input, we have

$$f(3, 5) = \sqrt{2(3) - 5 + 1} = \sqrt{2}$$

So the input $(3, 5)$ yields an output of $\sqrt{2}$. We define the domain for this function just as we did in Chapter 4: The domain is the set of all inputs that yield real-number outputs. For instance, the ordered pair $(1, 4)$ is not in the domain of the function we have been discussing, because (as you should check for yourself) $f(1, 4) = \sqrt{-1}$, which is not a real number. We can determine the domain of the function in equation (1) by requiring that the quantity under the radical sign be nonnegative. Thus, we require that $2x - y + 1 \geq 0$ and, consequently, $y \leq 2x + 1$. (Check this.) The following figure shows the graph of this inequality; the domain of our function is the set of ordered pairs making up the graph. In Exercises 57–62, follow a similar procedure and sketch the domain of the given function.

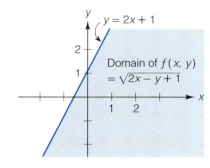

57. $f(x, y) = \sqrt{x + y + 2}$ **58.** $f(x, y) = \sqrt{x^2 + y^2 - 1}$

59. $g(x, y) = \sqrt{25 - x^2 - y^2}$ **60.** $g(x, y) = \ln(x^2 - y)$

61. $h(x, y) = \ln(xy)$ **62.** $h(x, y) = \sqrt{x} + \sqrt{y}$

 CHAPTER TEN   SUMMARY OF PRINCIPAL TERMS

TERMS	PAGE REFERENCE	COMMENT
1. Linear equation in two variables	579	A linear equation in two variables is an equation of the form $ax + by = c$, where a, b, and c are constants and x and y are variables or unknowns. Similarly, a linear equation in three variables is an equation of the form $ax + by + cz = d$.
2. Solution of a linear equation	579	A solution of the linear equation $ax + by = c$ is an ordered pair of numbers (x_0, y_0) such that $ax_0 + by_0 = c$. Similarly, a solution of the linear equation $ax + by + cz = d$ is an ordered triple of numbers (x_0, y_0, z_0) such that $ax_0 + by_0 + cz_0 = d$.
3. Consistent system; inconsistent system	579	A system of equations is *consistent* if it has at least one solution; otherwise, the system is *inconsistent*. See Figures 2, 3, and 4 in Section 10.1 for a geometric interpretation of these terms.

TERMS	PAGE REFERENCE	COMMENT
4. Upper-triangular form (three variables)	590	A system of linear equations in x, y, and z is said to be in *upper-triangular form* if x appears in no equation after the first, and y appears in no equation after the second. This definition can be extended to include systems with any number of unknowns. See Table 1 in Section 10.2 for examples of systems that are in upper-triangular form. When a system is in upper-triangular form, it is a simple matter to obtain the solutions; see, for instance, Examples 1 and 2 in Section 10.2.
5. Gaussian elimination	591, 592	This is a technique for converting a system of equations to upper-triangular form. See Examples 4 and 5 in Section 10.2.
6. Elementary operations	592	These are operations that can be performed on an equation in a system without altering the solution set. See the box on page 592 for a list of these operations.
7. Matrix	600	A *matrix* is a rectangular array of numbers, enclosed in parentheses or brackets. The numbers constituting the rectangular array are called the *entries* or the *elements* in the matrix. The *size or dimension* of a matrix is expressed by specifying the number of rows and the number of columns, in that order. For examples of this terminology, see Example 1 in Section 10.3.
8. Matrix equality	604	Two matrices are said to be *equal* provided they are the same size and their corresponding entries are equal.
9. Matrix addition and subtraction	604	To add two matrices of the same size, add the corresponding entries. Similarly, to subtract two matrices of the same size, subtract the corresponding entries.
10. Matrix multiplication	607	Let A and B be two matrices. The matrix product AB is defined only when the number of columns in A is the same as the number of rows in B. In this case, the matrix AB will have as many rows as A and as many columns as B. The entry in the ith row and jth column of AB is the number formed as follows: Multiply the corresponding entries in the ith row of A and the jth column of B; then add the results.
11. Square matrix	611	A matrix in which there are as many rows as there are columns is called a *square matrix*. An $n \times n$ square matrix is said to be an nth-order square matrix.
12. Multiplicative identity matrix	611, 614	For the set of $n \times n$ matrices, the multiplicative identity matrix I_n is the $n \times n$ matrix with ones down the main diagonal (upper left corner to bottom right corner) and zeros everywhere else. For example, $$I_2 = \begin{pmatrix} 1 & 0 \\ 0 & 1 \end{pmatrix} \quad \text{and} \quad I_3 = \begin{pmatrix} 1 & 0 & 0 \\ 0 & 1 & 0 \\ 0 & 0 & 1 \end{pmatrix}$$ For every $n \times n$ matrix A, we have $AI_n = A$ and $I_n A = A$.
13. Inverse matrix	611, 615	Given an $n \times n$ matrix A, the inverse matrix (if it exists) is denoted by A^{-1}. The defining equations for A^{-1} are $AA^{-1} = I$ and $A^{-1}A = I$. Here, I stands for the multiplicative identity matrix that is the same size as A. If a square matrix has an inverse, it is said to be *nonsingular*; if there is no inverse, then the matrix is *singular*.

TERMS	PAGE REFERENCE	COMMENT
14. Minor	618, 619	The *minor* of an entry b in a determinant is the determinant formed by suppressing the entries in the row and column in which b appears.
15. Cofactor	619	If an entry b appears in the ith row and the jth column of a determinant, then the *cofactor* of b is computed by multiplying the minor of b times the number $(-1)^{i+j}$.
16. Determinant	617, 618, 620	The value of the 2×2 determinant $\begin{vmatrix} a & b \\ c & d \end{vmatrix}$ is defined to be $ad - bc$. For larger determinants, the value can be found as follows. Select any row or column and multiply each entry in that row or column by its cofactor. Then add the results. The resulting sum is the value of the determinant. It can be shown that this value is independent of the particular row or column that is chosen.
17. Cramer's rule	624	This rule yields the solutions of certain systems of linear equations in terms of determinants. For a statement of the rule, see page 624. Example 5 on pages 624–625 shows how the rule is applied. The proof of Cramer's rule begins on page 625.
18. Linear inequality in two variables	636	A linear inequality in two variables is any one of the four types of inequalities that result when the equal sign in the equation $ax + by = c$ is replaced by one of the four symbols $<, \leq, >, \geq$. For any of these linear inequalities, a *solution* is an ordered pair of numbers (x_0, y_0) with the property that a true statement is obtained when x and y (in the inequality) are replaced by x_0 and y_0, respectively.
19. Vertices	638	The *vertices* of a region in the x-y plane are the corners or points where the adjacent bounding sides meet.
20. Convex region	638	A region in the x-y plane is *convex* if, given any two points in the region, the line segment joining those points lies wholly within the region.
21. Bounded region	639	A region in the x-y plane is *bounded* if it can be wholly contained within some (sufficiently large) circle.
22. Objective function and constraints	639	In linear programming problems, the function to be maximized or minimized is called the *objective function*. The inequalities restricting the variables in the objective function are referred to as the *constraints*.
23. Feasible solutions	639	In a linear programming problem, the set of points obtained by graphing the constraints is called the *set of feasible solutions*.
24. Optimal solution	640	In linear programming, a feasible solution that yields a largest or smallest value for the objective function is called an *optimal solution*.
25. Linear programming theorem	641	Consider an objective function $C = ax + by + c$, where the nonnegative variables x and y are subject to constraints specified by nonstrict inequalities. Then if C assumes a maximum (or a minimum) value, this occurs at a vertex. Furthermore, if the set of feasible solutions is both convex and bounded, then both a maximum and a minimum always exist.

WRITING MATHEMATICS

On your own or with a group of classmates, complete the following exercise. Then (strictly on your own) write out a detailed solution. This will involve a combination of English composition and algebra (much like the exposition in this textbook). At each stage, be sure to tell the reader (in complete sentences) where you are headed and why each of the main steps is necessary.

Three integers are said to form a *Pythagorean triple* if the square of the largest is equal to the sum of the squares of the other two. For example, 3, 4, and 5 form a Pythagorean triple because $3^2 + 4^2 = 5^2$. Do you know other Pythagorean triples? This exercise shows how to develop expressions for generating an infinite number of Pythagorean triples.

(a) Refer to the figure. Suppose that the coordinates of the point P are $(a/c, b/c)$, where a, b, and c are integers. Explain why a, b, and c form a Pythagorean triple.

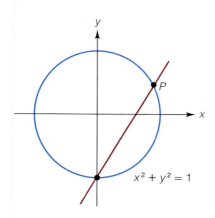

(b) Let m denote the slope of the line passing through P and $(0, -1)$. Using the methods of Section 10.6 show that the coordinates of the point P are

$$\left(\frac{2m}{m^2 + 1}, \frac{m^2 - 1}{m^2 + 1} \right)$$

(c) Use the results in parts (a) and (b) to explain why the three numbers $2m$, $m^2 - 1$, and $m^2 + 1$ form a Pythagorean triple. Then complete the following table to obtain examples of Pythagorean triples.

m	$2m$	$m^2 - 1$	$m^2 + 1$
2			
4			
6			
8			

CHAPTER TEN   REVIEW EXERCISES

In Exercises 1–38, solve each system of equations. If there are no solutions in a particular case, say so. In cases where there are literal (rather than numerical) coefficients, specify any restrictions that your solutions impose on those coefficients.

1. $\begin{cases} x + y = -2 \\ x - y = 8 \end{cases}$

2. $\begin{cases} x - y = 1 \\ x + y = 5 \end{cases}$

3. $\begin{cases} 2x + y = 2 \\ x + 2y = 7 \end{cases}$

4. $\begin{cases} 3x + 2y = 6 \\ 5x + 4y = 4 \end{cases}$

5. $\begin{cases} 7x + 2y = 9 \\ 4x + 5y = 63 \end{cases}$

6. $\begin{cases} \dfrac{x}{2} + \dfrac{y}{3} = 9 \\ \dfrac{x}{5} - \dfrac{y}{2} = -4 \end{cases}$

7. $\begin{cases} 2x - \dfrac{y}{2} = -8 \\ \dfrac{x}{3} + \dfrac{y}{8} = -1 \end{cases}$

8. $\begin{cases} 3x - 14y - 1 = 0 \\ -6x + 28y - 3 = 0 \end{cases}$

9. $\begin{cases} 3x + 5y - 1 = 0 \\ 9x - 10y - 8 = 0 \end{cases}$

10. $\begin{cases} 9x + 15y - 1 = 0 \\ 6x + 10y + 1 = 0 \end{cases}$

11. $\begin{cases} \dfrac{2}{3}x = -\dfrac{1}{2}y - 12 \\ \dfrac{x}{2} = y + 2 \end{cases}$

12. $\begin{cases} 0.1x + 0.2y = -5 \\ -0.2x - 0.5y = 13 \end{cases}$

13. $\begin{cases} \dfrac{1}{x} + \dfrac{1}{y} = -1 \\ \dfrac{2}{x} + \dfrac{5}{y} = -14 \end{cases}$

14. $\begin{cases} \dfrac{2}{x} + \dfrac{15}{y} = -9 \\ \dfrac{1}{x} + \dfrac{10}{y} = -2 \end{cases}$

15. $\begin{cases} ax + (1 - a)y = 1 \\ (1 - a)x + y = 0 \end{cases}$

16. $\begin{cases} ax - by - 1 = 0 \\ (a - 1)x + by + 2 = 0 \end{cases}$

17. $\begin{cases} 2x - y = 3a^2 - 1 \\ 2y + x = 2 - a^2 \end{cases}$

18. $\begin{cases} 3ax + 2by = 3a^2 - ab + 2b^2 \\ 3bx + 2ay = 2a^2 + 5ab - 3b^2 \end{cases}$

19. $\begin{cases} 5x - y = 4a^2 - 6b^2 \\ 2x + 3y = 5a^2 + b^2 \end{cases}$

20. $\begin{cases} \dfrac{2b}{x} - \dfrac{3}{y} = 7ab \\ \dfrac{4a}{x} + \dfrac{5a}{by} = 3a^2 \end{cases}$

21. $\begin{cases} px - qy = q^2 \\ qx + py = p^2 \end{cases}$

22. $\begin{cases} x - y = \dfrac{a - b}{a + b} \\ x + y = 1 \end{cases}$

23. $\begin{cases} \dfrac{4a}{x} - \dfrac{3b}{y} = a - 7b \\ \dfrac{3a^2}{x} - \dfrac{2b^2}{y} = (3a + b)(a - 2b) \end{cases}$

24. $\begin{cases} 6b^2 x^{-1/2} - 5a^2 y^{-1/2} - a^2 b^2 = 0 \\ 2b^2 x^{-1/2} + a^2 y^{-1/2} - 3a^2 b^2 = 0 \end{cases}$
   Hint: Let $u = x^{-1/2}$ and $v = y^{-1/2}$.

25. $\begin{cases} x + y + z = 9 \\ x - y - z = -5 \\ 2x + y - 2z = -1 \end{cases}$

26. $\begin{cases} x - 4y + 2z = 9 \\ 2x + y + z = 3 \\ 3x - 2y - 3z = -18 \end{cases}$

27. $\begin{cases} 4x - 4y + z = 4 \\ 2x + 3y + 3z = -8 \\ x + y + z = -3 \end{cases}$

28. $\begin{cases} x - 8y + z = 1 \\ 5x + 16y + 3z = 3 \\ 4x - 4y - 4z = -4 \end{cases}$

29. $\begin{cases} -2x + y + z = 1 \\ x - 2y + z = -2 \\ x + y - 2z = 4 \end{cases}$

30. $\begin{cases} -x + y + z = 1 \\ x - y + z = -1 \\ x + y - z = 1 \end{cases}$

31. $\begin{cases} 4x + 2y - 3z = 15 \\ 2x + y + 3z = 3 \end{cases}$

32. $\begin{cases} 3x + 2y + 17z = 1 \\ x + 2y + 3z = 3 \end{cases}$

33. $\begin{cases} x + 2y - 3z = -2 \\ 2x - y + z = 1 \\ 3x - 4y + 5z = 1 \end{cases}$

34. $\begin{cases} 9x + y + z = 0 \\ -3x + y - z = 0 \\ 3x - 5y + 3z = 0 \end{cases}$

35. $\begin{cases} x + y + z = a + b \\ 2x - y + 2z = -a + 5b \\ x - 2y + z = -2a + 4b \end{cases}$

36. $\begin{cases} ax + by - 2az = 4ab + 2b^2 \\ x + y + z = 4a + 2b \\ bx + ay + 4az = 5a^2 + b^2 \end{cases}$

37. $\begin{cases} x + y + z + w = 8 \\ 3x + 3y - z - w = 20 \\ 4x - y - z + 2w = 18 \\ 2x + 5y + 5z - 5w = 8 \end{cases}$

38. $\begin{cases} x - 2y + 3z + w = 1 \\ x + y + z + w = 5 \\ 2x + 3y + 2z - w = 3 \\ 3x + y - z + 2w = 4 \end{cases}$

For Exercises 39–50, determine the constants (denoted by capital letters) so that each equation is an identity.

39. $\dfrac{1}{(x - 10)(x + 10)} = \dfrac{A}{x - 10} + \dfrac{B}{x + 10}$

40. $\dfrac{x}{(x - 5)(x + 5)} = \dfrac{A}{x - 5} + \dfrac{B}{x + 5}$

41. $\dfrac{2x}{(x + 1)^2} = \dfrac{A}{x + 1} + \dfrac{B}{(x + 1)^2}$

42. $\dfrac{x + 2}{(x - 1)^2} = \dfrac{A}{x - 1} + \dfrac{B}{(x - 1)^2}$

43. $\dfrac{5}{x(x - 4)} = \dfrac{A}{x} + \dfrac{B}{x - 4}$

44. $\dfrac{6x}{(x + 2)(x + 3)} = \dfrac{A}{x + 2} + \dfrac{B}{x + 3}$

45. $\dfrac{1}{(x - 1)(x + 3)^2} = \dfrac{A}{x - 1} + \dfrac{B}{x + 3} + \dfrac{C}{(x + 3)^2}$

46. $\dfrac{x^2}{(x + 2)(x - 2)^2} = \dfrac{A}{x + 2} + \dfrac{B}{x - 2} + \dfrac{C}{(x - 2)^2}$

47. $\dfrac{4x^2 + 2x + 15}{(x - 1)(x^2 + x + 5)} = \dfrac{A}{x - 1} + \dfrac{Bx + C}{x^2 + x + 5}$

48. $\dfrac{1}{x(x^2 + 16)} = \dfrac{A}{x} + \dfrac{Bx + C}{x^2 + 16}$

49. $\dfrac{1}{x^3 + 64} = \dfrac{A}{x + 4} + \dfrac{Bx + C}{x^2 - 4x + 16}$

50. $\dfrac{1}{x^3 + 3x^2 + 3x + 1} = \dfrac{A}{x + 1} + \dfrac{B}{(x + 1)^2} + \dfrac{C}{(x + 1)^3}$

In Exercises 51–54, the lowercase letters a and b denote nonzero constants, with $a \neq b$. In each case, determine the values of A and B (and C, if applicable) so that the equation is an identity.

51. $\dfrac{x}{(x - a)^3} = \dfrac{A}{x - a} + \dfrac{B}{(x - a)^2} + \dfrac{C}{(x - a)^3}$

52. $\dfrac{ax(1 - a)}{(x - a)(x + a^2)} = \dfrac{A}{x - a} + \dfrac{B}{x + a^2}$

53. $\dfrac{(a - b)(a + b - x)}{(x - a)(x - b)} = \dfrac{A}{x - a} + \dfrac{B}{x - b}$

54. $\dfrac{4a^2x + 6a^3}{(x + a)(x + 2a)(x + 3a)} = \dfrac{A}{x + a} + \dfrac{B}{x + 2a} + \dfrac{C}{x + 3a}$

In Exercises 55–62, evaluate each of the determinants.

55. $\begin{vmatrix} 1 & 5 \\ -6 & 4 \end{vmatrix}$

56. $\begin{vmatrix} \frac{1}{6} & 1 \\ 2 & 12 \end{vmatrix}$

57. $\begin{vmatrix} 4 & 0 & 3 \\ -2 & 1 & 5 \\ 0 & 2 & -1 \end{vmatrix}$

58. $\begin{vmatrix} 2 & 6 & 4 \\ 6 & 18 & 24 \\ 15 & 5 & -10 \end{vmatrix}$

59. $\begin{vmatrix} 1 & 5 & 7 \\ 1 & 5 & 7 \\ 17 & 19 & 21 \end{vmatrix}$

60. $\begin{vmatrix} 0 & 2 & 4 & 0 \\ 4 & 0 & 6 & 2 \\ 0 & 0 & 1 & 1 \\ 14 & 7 & 1 & 0 \end{vmatrix}$

61. $\begin{vmatrix} 1 & 0 & 0 & 0 \\ 0 & 2 & 0 & 0 \\ 0 & 0 & 3 & 0 \\ 0 & 0 & 0 & 4 \end{vmatrix}$

62. $\begin{vmatrix} 1 & a & b & c \\ 0 & 2 & d & e \\ 0 & 0 & 3 & f \\ 0 & 0 & 0 & 4 \end{vmatrix}$

63. Show that $\begin{vmatrix} a & b & c \\ b & c & a \\ c & a & b \end{vmatrix} = 3abc - a^3 - b^3 - c^3$.

64. Show that $\begin{vmatrix} 1 & 1 & 1 \\ 1 & 1 + x & 1 \\ 1 & 1 & 1 + x^2 \end{vmatrix} = x^3$.

65. Show that
$$\begin{vmatrix} a^2 + x & b & c & d \\ -b & 1 & 0 & 0 \\ -c & 0 & 1 & 0 \\ -d & 0 & 0 & 1 \end{vmatrix} = a^2 + b^2 + c^2 + d^2 + x$$

66. In a two-digit number, the sum of the digits is 11. Four times the units digit exceeds the tens digit by 4. Find the number.

67. Determine constants a and b so that the parabola $y = ax^2 + bx - 1$ passes through the points $(-2, 5)$ and $(2, 9)$.

68. Find two numbers whose sum and difference are 52 and 10, respectively.

69. The vertices of triangle ABC are $A(-2, 0)$, $B(4, 0)$, and $C(0, 6)$. Let A_1, B_1, and C_1 denote the midpoints of sides $\overline{BC}$, $\overline{AC}$, and $\overline{AB}$, respectively.
(a) Find the point where the line segments $\overline{AA_1}$ and $\overline{BB_1}$ intersect. *Answer:* $\left(\frac{2}{3}, 2\right)$
(b) Follow the instructions given in part (a) using $\overline{BB_1}$ and $\overline{CC_1}$.
(c) Follow the instructions given in part (a) using $\overline{AA_1}$ and $\overline{CC_1}$.
(d) Let P denote the point $\left(\frac{2}{3}, 2\right)$ that you found in part (a). Compute each of the following ratios: AP/PA_1, BP/PB_1, CP/PC_1. What do you observe?

70. An *altitude* of a triangle is a line segment drawn from a vertex perpendicular to the opposite side. Suppose that the vertices of triangle ABC are as given in Exercise 69. Find the intersection point for each pair of altitudes. What do you observe about the three answers?

71. This exercise appears in *Plane and Solid Analytic Geometry*, by W. F. Osgood and W. G. Graustein (New York: Macmillan, 1920): Let P be any point (a, a) of the line $x - y = 0$, other than the origin. Through P draw two lines, of arbitrary slopes m_1 and m_2, intersecting the x-axis in A_1 and A_2, and the y-axis in B_1 and B_2, respectively. Prove that the lines $\overline{A_1B_2}$ and $\overline{A_2B_1}$ will, in general, meet on the line $x + y = 0$.

72. Determine constants h, k, and r so that the circle $(x - h)^2 + (y - k)^2 = r^2$ passes through the three points $(0, 0)$, $(0, 1)$, and $(1, 0)$.

73. The vertices of a triangle are the points of intersection of the lines $y = x - 1$, $y = -x - 2$, and $y = 2x + 3$. Find the equation of the circle passing through these three intersection points.

74. The vertices of triangle ABC are $A(0, 0)$, $B(3, 0)$, and $C(0, 4)$.
(a) Find the center and the radius of the circle that passes through A, B, and C. This circle is called the **circumcircle** for triangle ABC.

(b) The figure shows the **inscribed circle** for triangle *ABC*; this is the circle that is tangent to all three sides of the triangle. Find the center and the radius of this circle using the following two facts.
   (i) The center of the inscribed circle is the common intersection point of the three angle bisectors of the triangle.
   (ii) The line that bisects angle *B* has slope $-\frac{1}{2}$.

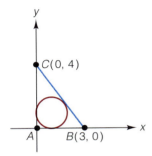

(c) Verify the following statement for triangle *ABC*. (The statement actually holds for any triangle; the result is known as *Euler's theorem*.) Let *R* and *r* denote the radii of the circles in parts (a) and (b), respectively. Then the distance *d* between the centers of the circles satisfies the equation

$$d^2 = R^2 - 2rR$$

(d) Verify the following statement for triangle *ABC*. (The statement actually holds for any triangle.) The area of triangle *ABC* is equal to one-half the product of the perimeter and the radius of the inscribed circle.

(e) Verify the following statement for triangle *ABC*. (The statement actually holds for any triangle.) The sum of the reciprocals of the lengths of the altitudes is equal to the reciprocal of the radius of the inscribed circle.

In Exercises 75–98, the matrices A, B, C, D, E, and F are defined as follows.

$$A = \begin{pmatrix} 3 & -2 \\ 1 & 5 \end{pmatrix} \qquad B = \begin{pmatrix} 2 & 1 \\ 1 & 8 \end{pmatrix} \qquad C = \begin{pmatrix} -1 & 0 \\ 0 & -1 \end{pmatrix}$$

$$D = \begin{pmatrix} -4 & 0 & 6 \\ 1 & 3 & 2 \end{pmatrix} \qquad E = \begin{pmatrix} 3 & -1 \\ 4 & 1 \\ -5 & 9 \end{pmatrix} \qquad F = \begin{pmatrix} 2 & 6 \\ 5 & 3 \\ 5 & 8 \end{pmatrix}$$

In each exercise, carry out the indicated matrix operations if they are defined. If an operation is not defined, say so.

75. $2A + 2B$

76. $2(A + B)$

77. $4B$

78. $B + 4$

79. AB

80. BA

81. $AB - BA$

82. $B + E$

83. $B + C$

84. $A(B + C)$

85. $AB + AC$

86. $(B + C)A$

87. $BA + CA$

88. $D + E$

89. DE

90. $(EE)D$

91. $E(ED)$

92. $E + F$

93. EF

94. $3E - 2F$

95. $(A + B) + C$

96. $A + (B + C)$

97. $(AB)C$

98. $A(BC)$

99. For a square matrix *A*, the notation A^2 means *AA*. Similarly, A^3 means *AAA*. If $A = \begin{pmatrix} 1 & 1 \\ 0 & 1 \end{pmatrix}$, verify that

$$A^2 = \begin{pmatrix} 1 & 2 \\ 0 & 1 \end{pmatrix} \qquad \text{and} \qquad A^3 = \begin{pmatrix} 1 & 3 \\ 0 & 1 \end{pmatrix}$$

100. Let *A* be the matrix $\begin{pmatrix} 0 & 0 & 0 \\ a & 0 & 0 \\ b & c & 0 \end{pmatrix}$. Compute A^2 and A^3.

For Exercises 101–104, in each case compute the inverse of the matrix in part (a), and then use that inverse to solve the system of equations in part (b).

101. (a) $\begin{pmatrix} 1 & 5 \\ 2 & 9 \end{pmatrix}$ (b) $\begin{cases} x + 5y = 3 \\ 2x + 9y = -4 \end{cases}$

102. (a) $\begin{pmatrix} 5 & -4 \\ 14 & -11 \end{pmatrix}$ (b) $\begin{cases} 5x - 4y = 2 \\ 14x - 11y = 5 \end{cases}$

103. (a) $\begin{pmatrix} 1 & -2 & 3 \\ 2 & -5 & 10 \\ -1 & 2 & -2 \end{pmatrix}$

   (b) $\begin{cases} x - 2y + 3z = -2 \\ 2x - 5y + 10z = -3 \\ -x + 2y - 2z = 6 \end{cases}$

104. (a) $\begin{pmatrix} -1 & 4 & 2 \\ -3 & 10 & 5 \\ 0 & 3 & 1 \end{pmatrix}$

   (b) $\begin{cases} -x + 4y + 2z = 8 \\ -3x + 10y + 5z = 0 \\ 3y + z = -12 \end{cases}$

In Exercises 105 and 106, find the inverse of each matrix.

105. $\begin{pmatrix} 5 & 3 & 6 & -7 \\ 3 & -4 & 0 & -9 \\ 0 & 1 & -1 & -1 \\ 2 & 2 & 3 & -2 \end{pmatrix}$

106. $\begin{pmatrix} 1 & -1 & 0 & -3 \\ 5 & -2 & 3 & -11 \\ 2 & 3 & 2 & -3 \\ 4 & 5 & 5 & -5 \end{pmatrix}$

In Exercises 107–114, compute D, D_x, D_y, D_z (and D_w where appropriate) for each system of equations. Use Cramer's rule to solve the systems in which $D \neq 0$. If $D = 0$, solve the system using Gaussian elimination or matrix methods.

107. $\begin{cases} 2x - y + z = 1 \\ 3x + 2y + 2z = 0 \\ x - 5y - 3z = -2 \end{cases}$

108. $\begin{cases} x + 2y - z = -1 \\ 2x - 3y + 3z = 3 \\ 2x + 3y + z = 1 \end{cases}$

109. $\begin{cases} x + 2y + 3z = -1 \\ 4x + 5y + 6z = 2 \\ 7x + 8y + 9z = -3 \end{cases}$

110. $\begin{cases} 3x + 2y - 2z = 0 \\ 2x + 3y - z = 0 \\ 8x + 7y - 5z = 0 \end{cases}$

111. $\begin{cases} 3x + 2y - 2z = 1 \\ 2x + 3y - z = -2 \\ 8x + 7y - 5z = 0 \end{cases}$

112. $\begin{cases} x + y + z + w = 5 \\ x - y - z + w = 3 \\ 2x + 3y + 3z + 2w = 21 \\ 4z - 3w = -7 \end{cases}$

113. $\begin{cases} 2x - y + z + 3w = 15 \\ x + 2y + 2w = 12 \\ 3y + 3z + 4w = 12 \\ -4x + y - 4z = -11 \end{cases}$

114. $\begin{cases} x + y + z = (a + b)^2 \\ \dfrac{bx}{a} + \dfrac{ay}{b} - z = 0 \\ x + y - z = (a - b)^2 \end{cases}$

In Exercises 115–128, find all solutions (x, y) for each system, where x and y are real numbers.

115. $\begin{cases} y = 6x \\ y = x^2 \end{cases}$

116. $\begin{cases} y = 4x \\ y = x^3 \end{cases}$

117. $\begin{cases} y = 9 - x^2 \\ y = x^2 - 9 \end{cases}$

118. $\begin{cases} x^3 - y = 0 \\ xy - 16 = 0 \end{cases}$

119. $\begin{cases} x^2 - y^2 = 9 \\ x^2 + y^2 = 16 \end{cases}$

120. $\begin{cases} 2x + 3y = 6 \\ y = \sqrt{x + 1} \end{cases}$

121. $\begin{cases} x^2 + y^2 = 1 \\ y = \sqrt{x} \end{cases}$

122. $\begin{cases} \dfrac{x}{11} + \dfrac{y}{12} = 2 \\ \dfrac{xy}{132} = 1 \end{cases}$

123. $\begin{cases} x^2 + y^2 = 1 \\ y = 2x^2 \end{cases}$

124. $\begin{cases} x^2 - 3xy + y^2 = -11 \\ 2x^2 + xy - y^2 = 8 \end{cases}$

125. $\begin{cases} x^2 + 2xy + 3y^2 = 68 \\ 3x^2 - xy + y^2 = 18 \end{cases}$

126. $\begin{cases} \dfrac{x^2}{a^2} + \dfrac{y^2}{b^2} = 1 \\ \dfrac{x^2}{b^2} + \dfrac{y^2}{a^2} = 1 \quad (a > b > 0) \end{cases}$

127. $\begin{cases} 2(x - 3)^2 - (y + 1)^2 = -1 \\ -3(x - 3)^2 + 2(y + 1)^2 = 6 \end{cases}$
   Hint: Let $u = x - 3$ and $v = y + 1$.

128. $\begin{cases} x^4 = y - 1 \\ y - 3x^2 + 1 = 0 \end{cases}$

Exercises 129–134 appear (in German) in an algebra text by Leonhard Euler, first published in 1770. The English versions given here are taken from the translated version, Elements of Algebra, *5th ed., by Leonhard Euler (London: Longman, Orme, and Co., 1840). [This, in turn, has been reprinted by Springer-Verlag (New York, 1984).]*

129. Required two numbers, whose sum may be s, and their proportion as a to b.

$$\text{\textit{Answer:}} \quad \frac{as}{a + b} \quad \text{and} \quad \frac{bs}{a + b}$$

130. The sum $2a$, and the sum of the squares $2b$, of two numbers being given; to find the numbers.
$$\text{\textit{Answer:}} \quad a - \sqrt{b - a^2} \quad \text{and} \quad a + \sqrt{b - a^2}$$

131. To find three numbers, so that [the sum of] one-half of the first, one-third of the second, and one-quarter of the third, shall be equal to 62; one-third of the first, one-quarter of the second, and on-fifth of the third, equal to 47; and one-quarter of the first, one-fifth of the second, and one-sixth of the third, equal to 38. *Answer:* 24, 60, 120

132. Required two numbers, whose product may be 105, and whose squares [when added] may together make 274.

133. Required two numbers, whose product may be m, and the sum of the squares n $(n \geq 2m)$.

134. Required two numbers such that their sum, their product, and the difference of their squares may all be equal.

In Exercises 135–140, graph each system of inequalities and specify whether the region is convex or bounded.

135. $\begin{cases} x^2 + y^2 \geq 1 \\ y - 4x \leq 0 \\ y - x \geq 0 \\ x \geq 0 \\ y \geq 0 \end{cases}$

136. $\begin{cases} x^2 + y^2 \leq 1 \\ x \geq 0 \\ y \geq 0 \end{cases}$

137. $\begin{cases} y - \sqrt{x} \leq 0 \\ y \geq 0 \\ x \geq 1 \\ x - 4 \leq 0 \end{cases}$

138. $\begin{cases} y - |x| \leq 0 \\ x + 1 \geq 0 \\ x - 1 \leq 0 \\ y + 1 \geq 0 \end{cases}$

139. $\begin{cases} y \le 1/x \\ y \ge 0 \\ x \ge 1 \end{cases}$ **140.** $\begin{cases} y - 100x \le 0 \\ y - x^2 \ge 0 \\ x \ge 0 \end{cases}$

In Exercises 141 and 142, find the maximum and the minimum values of the objective function subject to the given constraints. Specify the optimal solution in each case.

141. Objective function: $C = 7x + 6y + 6$
Constraints: $2y - 5x \le 10, \quad 3y - x \le 28$
$2x + y \le 32, \quad y + 5x \le 70$
$x \ge 0$
$y \ge 0$

142. Objective function: $C = 7y - 5x + 80$
Constraints: $3x + y \ge 13, \quad y - 2x \le 3$
$7x - 4y \le 5, \quad 2y - x \le 15$

OPTIONAL TI-81 GRAPHING CALCULATOR EXERCISES

EXERCISES FOR CHAPTER TEN

1. The graphing calculator can be used to determine (or at least approximate) the solution of many types of systems of equations in two variables. Consider, for instance, the following system, discussed in Example 1 in the first section of this chapter.

$$\begin{cases} x + y = 2 \\ 2x - 3y = 9 \end{cases}$$

(a) Solve each equation for y. Then press the $Y=$ key and enter the results as Y_1 and Y_2. If you've done the algebra correctly, you should have $Y_1 = 2 - x$ and $Y_2 = \frac{2}{3}x - 3$. Press ZOOM 6 to see the graphs in the standard viewing rectangle.

(b) As you can see, the lines intersect at a point in the fourth quadrant. Press the TRACE key and then use the ARROW keys to move the cursor to what appears to be the intersection point. The readout at the bottom of the screen tells you that the point is close to $(3, -1)$. [From the text, we know that $(3, -1)$ is the precise intersection point.]

(c) To obtain a closer approximation to the intersection point, press ZOOM 2. The blinking cursor will still be on the approximate intersection point. Now, press ENTER to zoom in on that point. Next press the TRACE key and then use the ARROW keys to move the cursor to the intersection point. Note the readout at the bottom of the screen. The values will be closer than they were in part (b) to $x = 3$ and $y = -1$. (Repeating the process will yield a still closer approximation.)

In Exercises 2–10, use the procedure described in Exercise 1 to determine or approximate the solutions of each system. Repeat the procedure until you are sure of the second decimal place in each coordinate of the intersection point.

2. $\begin{cases} 3x - 8y = 63 \\ 2x + y = 4 \end{cases}$ **3.** $\begin{cases} x - y - 1 = 0 \\ y = 2\sqrt{x + 3} \end{cases}$

4. $\begin{cases} y = e^x \\ y = x^2 \end{cases}$ **5.** $\begin{cases} y = e^{x/2} \\ y = x^2 \end{cases}$

6. $\begin{cases} y = \ln x \\ y = e^{-x} \end{cases}$ **7.** $\begin{cases} x^2 - 2x + y = 0 \\ y = 3x^3 - x^2 - 10x \end{cases}$

8. $\begin{cases} y = \sqrt{x} - 1 \\ y = \ln x \end{cases}$ *Hint:* There are two solutions. You will need a viewing rectangle that is larger than the standard viewing rectangle to see that there is a second solution.

9. $\begin{cases} y = x^3 \\ y = \frac{1}{2}(e^x + e^{-x}) \end{cases}$ *Hint:* There are two solutions.

10. $\begin{cases} y = x^3 \\ y = \frac{1}{2}(e^x - e^{-x}) \end{cases}$ *Hint:* There are four solutions. You can cut the amount of work in half by using the fact that both graphs are symmetric about the origin.

Exercise 11 shows how to use the calculator to carry out some basic computations with matrices and determinants. As examples, we will use the following two matrices:

$$A = \begin{pmatrix} 5 & 0 & 2 \\ 2 & 2 & 1 \\ -3 & 1 & -1 \end{pmatrix} \qquad B = \begin{pmatrix} -2 & 1 & -9 \\ 0 & 7 & 3 \\ -1 & 6 & 4 \end{pmatrix}$$

To enter matrix A, press the MATRX key and then the right (or the left) ARROW key. The menu for editing matrices now appears on the screen. Since we wish to edit the matrix A, press the number 1. Next, type 3 ENTER and then type 3 ENTER again. As you can see on the screen, the effect is to change the default setting from six rows and six columns to three rows and three columns.

The blinking cursor is now located on the default value zero for the matrix element a_{11}. Since we want a_{11} to be 5, type 5 ENTER. The blinking cursor is now located on the default value zero for the matrix element a_{12}. Since we want a_{12} to be 0, type

0 ENTER. Continue this process for the remaining entries in A. After you have entered the last element, $a_{33} = -1$, press the MATRX key to return to the matrix screen; then press the right ARROW key to go to the edit menu.

Type 2 so that you can edit the B matrix. Now follow the same procedure that you used in defining A. After you have entered the last element of B, namely $b_{33} = 4$, press the 2nd key and then the CLEAR key; this takes you back to the home screen.

Before doing any computations, we want to check for typing errors in entering the elements of A and B. To see the matrix A displayed, press the 2nd key, then the [A] key (which is the same as the 1 key), and then the ENTER key. As you can see, the matrix A is now displayed on the screen. If you detect typing errors, press the MATRX key, then the right ARROW key, then 1. You can now retype any entries that need to be fixed. When you are finished, or if there were no errors to correct, exit to the home screen by pressing the 2nd key and then the CLEAR key. Repeat the same procedure for checking the entries in B.

11. (a) Use the following sequence of keystrokes to compute the matrix product AB.

$$\boxed{\text{2nd}}\ \boxed{\text{[A]}}\ \boxed{\times}\ \boxed{\text{2nd}}\ \boxed{\text{[B]}}\ \boxed{\text{ENTER}}$$

(b) Use the following sequence of keystrokes to compute the matrix product BA.

$$\boxed{\text{2nd}}\ \boxed{\text{[B]}}\ \boxed{\times}\ \boxed{\text{2nd}}\ \boxed{\text{[A]}}\ \boxed{\text{ENTER}}$$

(c) Compute A^{-1} using the following sequence of keystrokes.

$$\boxed{\text{2nd}}\ \boxed{\text{[A]}}\ \boxed{x^{-1}}\ \boxed{\text{ENTER}}$$

Check the answer that the calculator gives you by referring to Example 3 in Section 3 of this chapter.

(d) Compute B^{-1} using the following sequence of keystrokes.

$$\boxed{\text{2nd}}\ \boxed{\text{[B]}}\ \boxed{x^{-1}}\ \boxed{\text{ENTER}}$$

The resulting display will show the first column of B^{-1}. Press the right ARROW key to see the other two columns. As you can see, the values that are

displayed are ten-place decimal approximations. To see the entries of B^{-1} rounded off to three decimal places, use the following sequence of keystrokes. (In the sequence, the command following the ALPHA key is a comma; this is the same as the decimal point key.)

$$\boxed{\text{MATH}}\ \boxed{\blacktriangleright}\ \boxed{1}\ \boxed{\text{2nd}}\ \boxed{\text{[B]}}\ \boxed{x^{-1}}$$
$$\boxed{\text{ALPHA}}\ \boxed{,}\ \boxed{3}\ \boxed{)}\ \boxed{\text{ENTER}}$$

(e) Use the following keystrokes to compute the determinant of the matrix A.

$$\boxed{\text{MATRX}}\ \boxed{5}\ \boxed{\text{2nd}}\ \boxed{\text{[A]}}\ \boxed{\text{ENTER}}$$

For Exercises 12–16, enter the following matrices and then use the calculator to carry out the indicated operations. In cases where the results involve decimals, round off each entry to three decimal places.

$$A = \begin{pmatrix} 3 & 2 & 6 \\ 1 & 1 & 2 \\ 2 & 2 & 5 \end{pmatrix} \qquad B = \begin{pmatrix} 5 & 4 & 1 \\ -1 & 2 & 9 \\ 0 & 8 & -6 \end{pmatrix} \qquad C = \begin{pmatrix} -21 \\ 17 \\ 8 \end{pmatrix}$$

12. (a) AB (b) AC (c) A^{-1}
(d) B^{-1} (e) $(BA)^{-1}$ (f) $A^{-1}B^{-1}$
Hint: In part (f), first recalculate A^{-1}. [The calculator will automatically store the result in the (temporary) ANS file.] Then use the following keystrokes.

$$\boxed{\text{2nd}}\ \boxed{\text{ANS}}\ \boxed{\times}\ \boxed{\text{2nd}}\ \boxed{\text{[B]}}\ \boxed{x^{-1}}\ \boxed{\text{ENTER}}$$

13. (a) $\det(A)$ (b) $\det(B)$ (c) $\det(AB)$
(d) $[\det(A)] \times [\det(B)]$

14. (a) $\det(A^{-1})$ (b) $1/\det(A)$

15. (a) $\det(B^{-1})$ (b) $1/\det(B)$

16. (a) A^2 (This indicates the matrix product AA.)
(b) A^3 *Hint:* Assuming that your result from part (a) is still displayed, use the keystrokes $\boxed{\times}\ \boxed{\text{2nd}}$ $\boxed{\text{[A]}}\ \boxed{\text{ENTER}}$.
(c) A^4

CHAPTER TEN TEST

1. Determine all solutions of the system

$$\begin{cases} 3x + 4y = 12 \\ y = x^2 + 2x + 3 \end{cases}$$

2. Find all solutions of the system

$$\begin{cases} x - 2y = 13 \\ 3x + 5y = -16 \end{cases}$$

3. (a) Find all solutions of the following system using Gaussian elimination:

$$\begin{cases} x + 4y - z = 0 \\ 3x + y + z = -1 \\ 4x - 4y + 5z = -7 \end{cases}$$

   (b) Compute D, D_x, D_y, and D_z for the system in part (a). [Then check your answer in part (a) using Cramer's rule.]

4. Suppose that the matrices A and B are defined as follows:

$$A = \begin{pmatrix} 1 & -3 \\ 2 & -1 \end{pmatrix} \qquad B = \begin{pmatrix} 0 & 4 \\ 1 & 3 \end{pmatrix}$$

   (a) Compute $2A - B$. (b) Compute BA.

5. Determine the area of the triangular region in the first quadrant that is bounded by the x-axis and the lines $y = 2x$ and $y = -x + 6$.

6. Find all solutions of the system

$$\begin{cases} \dfrac{1}{2x} + \dfrac{1}{3y} = 10 \\ -\dfrac{5}{x} - \dfrac{4}{y} = -4 \end{cases}$$

7. Specify the coefficient matrix for the system

$$\begin{cases} x + y - z = -1 \\ 2x - y + 2z = 11 \\ x - 2y + z = 10 \end{cases}$$

   Also specify the augmented matrix for this system.

8. Use matrix methods to find all solutions of the system displayed in the previous problem.

9. Find the equation of a line that passes through the point of intersection of the lines $x + y = 11$ and $3x + 2y = 7$ and that is perpendicular to the line $2x - 4y = 7$.

10. Determine the constants A, B, and C such that the following equation is an identity:

$$\frac{x - 2}{(x + 1)(x - 1)^2} = \frac{A}{x + 1} + \frac{B}{x - 1} + \frac{C}{(x - 1)^2}$$

11. Consider the determinant

$$\begin{vmatrix} 2 & 3 & -1 \\ 0 & 1 & 4 \\ 5 & -2 & 6 \end{vmatrix}$$

   (a) What is the minor of the entry in the third row, second column?
   (b) What is the cofactor of the entry in the third row, second column?

12. Evaluate the determinant

$$\begin{vmatrix} 4 & -5 & 0 \\ -8 & 10 & 7 \\ 16 & 20 & 14 \end{vmatrix}$$

13. Find all solutions of the system

$$\begin{cases} x^2 + y^2 = 15 \\ xy = 5 \end{cases}$$

14. Find the solutions of the system

$$\begin{cases} A + 2B + 3C = 1 \\ 2A - B - C = 2 \end{cases}$$

15. (a) Determine the inverse of the following matrix.

$$\begin{pmatrix} 10 & -2 & 5 \\ 6 & -1 & 4 \\ 1 & 0 & 1 \end{pmatrix}$$

   (b) Use the inverse matrix determined in part (a) to solve the following system.

$$\begin{cases} 10u - 2v + 5w = -1 \\ 6u - v + 4w = -2 \\ u + w = 3 \end{cases}$$

16. Graph the inequality $5x - 6y \geq 30$.

17. Determine the constants P and Q, given that the parabola $y = Px^2 + Qx - 5$ passes through the two points $(-2, -1)$ and $(-1, -2)$.

18. Graph the inequality $x^2 - 4x + y^2 + 3 > 0$. Is the solution set bounded? Is it convex?

19. Graph the following system of inequalities and specify the vertices:

$$\begin{cases} x \geq 0 \\ y \geq 0 \\ 2y - x \leq 14 \\ x + 3y \leq 36 \\ 9x + y \leq 99 \end{cases}$$

20. A factory assembles two types of motors, small and large. The cost of material is $15 for a small motor and $30 for a large motor. Two hours of labor are required to assemble a small motor and 6 hours are needed for a large motor. The cost of labor is $8/hr. The owner of the factory can allocate at most $1500 per day for material and $2000 per day for labor. Find the largest daily profit that the owner can make if the small motors are sold for $40 each and the large motors for $104 each. Assume that every motor produced is sold. *Hint:* Profit equals revenue minus cost.

CHAPTER ELEVEN

ROOTS OF POLYNOMIAL EQUATIONS

It is necessary that I make some general statements concerning the nature of equations, ...

René Descartes (1596–1650) in *La Géométrie* (1637)

INTRODUCTION

In this chapter we continue the work begun in Chapter 2 on solving polynomial equations. We begin in Section 11.1 with a second look at the long division process for polynomials. In this section you'll see how synthetic division is used to abbreviate the long division process. Section 11.2 presents two theorems about polynomials: the remainder theorem and the factor theorem. The remainder theorem and the factor theorem are then applied repeatedly in the next section (Section 11.3) to develop some fundamental results regarding polynomial equations and their solutions. You can view much of the material in this section as a kind of generalization of what you already know about quadratic equations. The last two sections of this chapter present additional results that are useful in actually solving polynomial equations or in determining the nature of their solutions.

11.1 MORE ON DIVISION OF POLYNOMIALS

In Section 1.8 we studied the long division process for polynomials. There is a theorem, commonly referred to as the *division algorithm,* that summarizes rather nicely the key results of that long division process. We state the theorem here without proof.

> ### The Division Algorithm
>
> Let $p(x)$ and $d(x)$ be polynomials, and assume that $d(x)$ is not the zero polynomial. Then there are unique polynomials $q(x)$ and $R(x)$ such that
>
> $$p(x) = d(x) \cdot q(x) + R(x)$$
>
> where either $R(x)$ is the zero polynomial or the degree of $R(x)$ is less than the degree of $d(x)$.

The polynomials $p(x)$, $d(x)$, $q(x)$, and $R(x)$ are referred to, respectively, as the **dividend, divisor, quotient,** and **remainder.** When $R(x) = 0$, we have $p(x) = d(x) \cdot q(x)$ and we say that $d(x)$ and $q(x)$ are **factors** of $p(x)$. Also, since $d(x)$ is not the zero polynomial, notice that the equation $p(x) = d(x) \cdot q(x) + R(x)$ implies that the degree of $q(x)$ is less than or equal to the degree of $p(x)$. (Why?)

EXAMPLE 1   Let $p(x) = x^3 + 2x^2 - 4$ and $d(x) = x - 3$. Use the long division process to find the polynomials $q(x)$ and $R(x)$ such that

$$p(x) = d(x) \cdot q(x) + R(x)$$

where either $R(x) = 0$ or the degree of $R(x)$ is less than the degree of $d(x)$.

Solution   After inserting the term $0x$ in the dividend $p(x)$, we use long division to divide $p(x)$ by $d(x)$:

$$
\begin{array}{r}
x^2 + 5x + 15 \\
x - 3 \overline{)\, x^3 + 2x^2 + 0x - 4} \\
\underline{x^3 - 3x^2} \\
5x^2 + 0x \\
\underline{5x^2 - 15x} \\
15x - 4 \\
\underline{15x - 45} \\
41
\end{array}
$$

We now have

$$\underbrace{x^3 + 2x^2 - 4}_{p(x)} = \underbrace{(x - 3)}_{d(x)}\underbrace{(x^2 + 5x + 15)}_{q(x)} + \underbrace{41}_{R(x)}$$

Thus, $q(x) = x^2 + 5x + 15$ and $R(x) = 41$. Notice that the degree of $R(x)$ is less than the degree of $d(x)$.

The long division procedure for polynomials can be streamlined when the divisor is of the form $x - r$. This shortened version, known as **synthetic division,** will be useful in subsequent sections when we are solving polynomial equations.

We can explain the idea behind synthetic division by using the long division carried out in Example 1. The basic idea is that in the long division process, it is the *coefficients* of the various polynomials that carry all the necessary information. In our example, for instance, the quotient and remainder can be abbreviated by writing down a sequence of four numbers:

$$1 \quad 5 \quad 15 \quad 41$$

By studying the long division process, you will find that these numbers are obtained through the following four steps:

STEP 1   Write down the first coefficient of the dividend. This will be the first coefficient of the quotient.

Result $\boxed{1}$

STEP 2   Multiply the 1 obtained in the previous step by the -3 in the divisor. Then subtract this from the second coefficient of the dividend:

$$-3 \times 1 = -3 \qquad 2 - (-3) = 5$$

Result $\boxed{5}$

STEP 3 Multiply the 5 obtained in the previous step by the -3 in the divisor. Then subtract this from the third coefficient of the dividend:

$$-3 \times 5 = -15 \qquad 0 - (-15) = 15 \qquad\qquad Result \quad \boxed{15}$$

STEP 4 Multiply the 15 obtained in the previous step by the -3 in the divisor. Then subtract this from the fourth coefficient of the dividend:

$$-3 \times 15 = -45 \qquad -4 - (-45) = 41 \qquad\qquad Result \quad \boxed{41}$$

A convenient format for setting up this process involves writing the constant term of the divisor and the coefficients of the dividend as follows:

$$-3 \underline{\big|} \quad 1 \quad\quad 2 \quad\quad 0 \quad -4$$

Now, using this format, let us again go through the four steps we have just described:

STEP 1 Bring down the 1.

$$
\begin{array}{r|rrrr}
-3 & 1 & 2 & 0 & -4 \\
\hline
& 1 & & &
\end{array}
$$

STEP 2 $\quad -3 \times 1 = -3$
$\qquad\quad 2 - (-3) = 5$

$$
\begin{array}{r|rrrr}
-3 & 1 & 2 & 0 & -4 \\
& & -3 & & \\
\hline
& 1 & 5 & &
\end{array}
$$

STEP 3 $\quad -3 \times 5 = -15$
$\qquad\quad 0 - (-15) = 15$

$$
\begin{array}{r|rrrr}
-3 & 1 & 2 & 0 & -4 \\
& & -3 & -15 & \\
\hline
& 1 & 5 & 15 &
\end{array}
$$

STEP 4 $\quad -3 \times 15 = -45$
$\qquad\quad -4 - (-45) = 41$

$$
\begin{array}{r|rrrr}
-3 & 1 & 2 & 0 & -4 \\
& & -3 & -15 & -45 \\
\hline
& 1 & 5 & 15 & 41
\end{array}
$$

Although we have now obtained the required sequence of numbers, 1 5 15 41, there is one further simplification that can be made. In steps 2 through 4, we can add instead of subtract if we use 3 instead of -3 in the initial format. (You will see the motivation for this in the next section when we discuss the remainder theorem.) With this change, let us now summarize the technique of synthetic division using the example with which we've been working (See the box on page 662.) The method is applicable for any polynomial division in which the form of the divisor is $x - a$.

EXAMPLE 2 Use synthetic division to divide $x^3 - 6x + 4$ by $x - 2$.

Solution

$$
\begin{array}{r|rrrr}
2 & 1 & 0 & -6 & 4 \\
& & 2 & 4 & -4 \\
\hline
& 1 & 2 & -2 & 0
\end{array}
$$

Looking at the third row of numbers in the synthetic division we've carried out, we see that the quotient is $x^2 + 2x - 2$ and the remainder is 0. In other words, both $x - 2$ and $x^2 + 2x - 2$ are factors of $x^3 - 6x + 4$, and we have

$$x^3 - 6x + 4 = (x - 2)(x^2 + 2x - 2)$$

To Divide $x^3 + 2x^2 - 4$ by $x - 3$ Using Synthetic Division

COMMENTS

Format

$$\begin{array}{r|rrrr} 3 & 1 & 2 & 0 & -4 \\ \hline \end{array}$$

Since the divisor is $x - 3$, the format begins with 3. The coefficients from the dividend are written in the order corresponding to decreasing powers of x. A zero coefficient is inserted as a place holder.

Procedure

$$\begin{array}{r|rrrr} 3 & 1 & 2 & 0 & -4 \\ & & 3 & 15 & 45 \\ \hline & 1 & 5 & 15 & 41 \end{array}$$

Step 1: Bring down the 1.
Step 2: $3 \times 1 = 3$; $2 + 3 = 5$
Step 3: $3 \times 5 = 15$; $0 + 15 = 15$
Step 4: $3 \times 15 = 45$; $-4 + 45 = 41$

Answer

Quotient:
$x^2 + 5x + 15$
Remainder: 41

The degree of the first term in the quotient is one less than the degree of the first term of the dividend.

EXAMPLE 3   Use synthetic division to divide $x^5 - a^5$ by $x - a$.

Solution

$$\begin{array}{r|rrrrrr} a & 1 & 0 & 0 & 0 & 0 & -a^5 \\ & & a & a^2 & a^3 & a^4 & a^5 \\ \hline & 1 & a & a^2 & a^3 & a^4 & 0 \end{array}$$

As before, we read off the quotient and remainder from the third row of numbers. The quotient is

$$x^4 + ax^3 + a^2x^2 + a^3x + a^4$$

and the remainder is zero. So, we have

$$\underbrace{x^5 - a^5}_{\text{dividend}} = \underbrace{(x - a)}_{\text{divisor}}\underbrace{(x^4 + ax^3 + a^2x^2 + a^3x + a^4)}_{\text{quotient}} + \underbrace{0}_{\text{remainder}}$$

or

$$x^5 - a^5 = (x - a)(x^4 + ax^3 + a^2x^2 + a^3x + a^4)$$

The last equation in Example 3 tells us how to factor $x^5 - a^5$. One factor is $x - a$. Notice the pattern in the second factor $x^4 + ax^3 + a^2x^2 + a^3x + a^4$.

first term: a^0x^{5-1}
second term: a^1x^{5-2}
third term: a^2x^{5-3}
fourth term: a^3x^{5-4}
fifth term: a^4x^{5-5}

In the same way that we've found a factorization for $x^5 - a^5$, we can find a factorization for $x^n - a^n$ for any positive integer $n \geq 2$. In each case, the first factor is $x - a$, while the second factor follows the same pattern just described for $x^5 - a^5$. We state the general result in the box that follows. The result can be proved by using *mathematical induction* (discussed in a later chapter) or by using the *remainder theorem* (discussed in the next section).

Factorization of $x^n - a^n$

$$x^n - a^n = (x - a)(x^{n-1} + ax^{n-2} + a^2x^{n-3} + \cdots + a^{n-1})$$

In our development of synthetic division earlier in this section, we assumed that the form of the divisor was $x - r$. The next example shows what to do when the form of the divisor is $x + r$.

EXAMPLE 4   Use synthetic division to divide $x^4 - 2x^3 + 5x^2 - 4x + 3$ by $x + 1$.

Solution   We first need to write the divisor $x + 1$ in the form $x - r$. We have

$$x + 1 = x - (-1)$$

In other words, r is -1, and this is the value we use to set up the synthetic division format. The format then is

$$\underline{-1\,|}\quad 1 \quad -2 \quad\quad 5 \quad -4 \quad\quad 3$$

Now we carry out the synthetic division procedure:

$$
\begin{array}{r|rrrrr}
-1 & 1 & -2 & 5 & -4 & 3 \\
   &   & -1 & 3 & -8 & 12 \\
\hline
   & 1 & -3 & 8 & -12 & 15
\end{array}
$$

The quotient is therefore $x^3 - 3x^2 + 8x - 12$, and the remainder is 15. We can summarize this result by writing

$$x^4 - 2x^3 + 5x^2 - 4x + 3 = (x + 1)(x^3 - 3x^2 + 8x - 12) + 15$$

(Notice that the degree of the remainder is less than the degree of the divisor, in agreement with the division algorithm.) ∎

EXERCISE SET 11.1

A

In Exercises 1–20, use synthetic division to find the quotients and remainders. Also, in each case, write the result of the division in the form $p(x) = d(x) \cdot q(x) + R(x)$, as in equation (2) in the text.

1. $\dfrac{x^2 - 6x - 2}{x - 5}$

2. $\dfrac{3x^2 + 4x - 1}{x - 1}$

3. $\dfrac{4x^2 - x - 5}{x + 1}$

4. $\dfrac{x^2 - 1}{x + 2}$

5. $\dfrac{6x^3 - 5x^2 + 2x + 1}{x - 4}$

6. $\dfrac{x^4 - 4x^3 + 6x^2 - 4x + 1}{x - 1}$

7. $\dfrac{x^3 - 1}{x - 2}$

8. $\dfrac{x^3 - 8}{x - 2}$

9. $\dfrac{x^5 - 1}{x + 2}$

10. $\dfrac{x^3 - 8x^2 - 1}{x + 3}$

11. $\dfrac{x^4 - 6x^3 + 2}{x + 4}$

12. $\dfrac{3x^3 - 2x^2 + x + 1}{x - \frac{1}{2}}$

13. $\dfrac{x^3 - 4x^2 - 3x + 6}{x - 10}$

14. $\dfrac{1 + 3x + 3x^2 + x^3}{x + 1}$

15. $\dfrac{x^3 - x^2}{x + 5}$

16. $\dfrac{5x^4 - 4x^3 + 3x^2 - 2x + 1}{x + \frac{1}{2}}$

17. $\dfrac{14 - 27x - 27x^2 + 54x^3}{x - \frac{2}{3}}$

18. $\dfrac{14 - 27x - 27x^2 + 54x^3}{x + \frac{2}{3}}$

19. $\dfrac{x^4 + 3x^2 + 12}{x - 3}$

20. (a) $\dfrac{x^4 - 16}{x - 2}$ (b) $\dfrac{x^4 + 16}{x + 2}$

In Exercises 21–24, each expression has the form $x^n - a^n$. Write each expression as a product of two factors (as in the box on page 663).

21. $x^5 - 32$ 22. $y^6 - 1$

23. $z^4 - 81$ 24. $x^7 - y^7$

B

In Exercises 25–28, use synthetic division to determine the quotient $q(x)$ and the remainder $R(x)$ in each case.

25. $\dfrac{6x^2 - 8x + 1}{3x - 4}$ *Hint:* Divide both numerator and denominator by 3. (Why?)

26. $\dfrac{4x^3 + 6x^2 - 6x - 5}{2x - 3}$ 27. $\dfrac{6x^3 + 1}{2x + 1}$

28. $\dfrac{5x^3 - 3x^2 + 1}{3x + 1}$

29. When $x^3 + kx + 1$ is divided by $x + 1$, the remainder is -4. Find k.

30. (a) Show that when $x^3 + kx + 6$ is divided by $x + 3$, the remainder is $-21 - 3k$.

(b) Determine a value of k such that $x + 3$ will be a factor of $x^3 + kx + 6$.

31. When $x^2 + 2px - 3q^2$ is divided by $x - p$, the remainder is zero. Show that $p^2 = q^2$.

32. Given that $x - 3$ is a factor of $x^3 - 2x^2 - 4x + 3$, solve the equation $x^3 - 2x^2 - 4x + 3 = 0$.

The process of synthetic division applies equally well when some or all of the coefficients are nonreal complex numbers. In Exercises 33–36, use synthetic division to determine the quotient $q(x)$ and the remainder $R(x)$ in each case.

33. $\dfrac{x^2 - 4x + 1}{x - i}$ 34. $\dfrac{x^3 - 2x^2 - 4}{x - 3i}$

35. $\dfrac{x^2 - 2x + 2}{x - (1 + i)}$ 36. $\dfrac{x^3 - x^2 + 4x - 4}{x + 2i}$

37. Given: the identity $f(x) = d(x)q(x) + R(x)$ holds for the following polynomials.

$$f(x) = 2x^5 + 5x^4 - 8x^3 + 7x^2 - 9 \qquad d(x) = x^2 - 3$$
$$q(x) = 2x^3 + 5x^2 - 2x + 22 \qquad R(x) = -6x + 57$$

Evaluate $f(\sqrt{3})$.   *Hint (of sorts):* There's an easy way and a tedious way.

38. Given: the identity $f(t) = d(t)q(t) + R(t)$ holds for the following polynomials.

$$f(t) = t^5 - 3t^4 + 2t^3 - 5t^2 + 6t - 7 \qquad d(t) = t - 4$$
$$q(t) = t^4 + t^3 + 6t^2 + 19t + 82 \qquad R(t) = 321$$

Evaluate $f(4)$.

39. Find the remainder when $t^5 - 5a^4t + 4a^5$ is divided by $t - a$.

40. When $f(x)$ is divided by $(x - a)(x - b)$, the remainder is $Ax + B$. Apply the division algorithm to show that

$$A = \dfrac{f(a) - f(b)}{a - b} \quad \text{and} \quad B = \dfrac{bf(a) - af(b)}{b - a}$$

ROOTS OF POLYNOMIAL EQUATIONS: THE REMAINDER THEOREM AND THE FACTOR THEOREM

11.2

Descartes recommended [in his La Géométrie of 1637] that all terms [of an equation] should be taken to one side and equated with zero. Though he was not the first to suggest this, he was the earliest writer to realize the advantage to be gained. He pointed out that a polynomial $f(x)$ was divisible by $(x - a)$ if and only if a was a root of $f(x)$.

David M. Burton in *The History of Mathematics, An Introduction*, 2d ed. (Dubuque, Iowa: Wm. C. Brown Publishers, 1991)

The techniques for solving polynomial equations of degree 2 were discussed in Section 2.3. Now we want to extend those ideas. Our focus in this section and in the remainder of the chapter is on solving polynomial equations of any degree, that is, equations of the form

$$f(x) = a_n x^n + a_{n-1} x^{n-1} + \cdots + a_1 x + a_0 = 0 \qquad (1)$$

Here, as in Chapter 2, a **root** or **solution** of equation (1) is a number r that when substituted for x leads to a true statement. Thus, r is a root of equation (1) provided that $f(r) = 0$. We also refer to the number r in this case as a **zero** of the function f.

EXAMPLE 1
(a) Is -3 a zero of the function f defined by $f(x) = x^4 + x^2 - 6$?
(b) Is $\sqrt{2}$ a root of the equation $x^4 + x^2 - 6 = 0$?

Solution
(a) By definition, -3 will be a zero of f if $f(-3) = 0$. We have

$$f(-3) = (-3)^4 + (-3)^2 - 6 = 81 + 9 - 6 = 84 \neq 0$$

Thus, -3 is not a zero of the function f.
(b) To check if $\sqrt{2}$ is a root of the given equation, we have

$$(\sqrt{2})^4 + (\sqrt{2})^2 - 6 \overset{?}{=} 0$$
$$4 + 2 - 6 \overset{?}{=} 0$$
$$0 \overset{?}{=} 0 \qquad \text{True}$$

Thus, $\sqrt{2}$ is a root of the equation $x^4 + x^2 - 6 = 0$.

There are cases in which a root of an equation is what we call a **repeated root**. Consider, for instance, the equation $x(x - 1)(x - 1) = 0$. We have

$$x(x - 1)(x - 1) = 0$$

| $x = 0$ | $x - 1 = 0$ | $x - 1 = 0$ |
| | $x = 1$ | $x = 1$ |

The roots of the equation are therefore 0, 1, and 1. The repeated root here is $x = 1$. We say in this case that 1 is a **double root** or, equivalently, that 1 is a **root of multiplicity 2**. More generally, if a root is repeated k times, we call it a **root of multiplicity k**.

EXAMPLE 2 State the multiplicity of each root of the equation

$$(x - 4)^2 (x - 5)^3 = 0$$

Solution We have $(x - 4)(x - 4)(x - 5)(x - 5)(x - 5) = 0$. By setting each factor equal to zero, we obtain the roots 4, 4, 5, 5, and 5. From this we see that the root 4 has multiplicity 2, while the root 5 has multiplicity 3. Notice that it is not really necessary to write out all the factors as we did here; the exponents of the factors in the original equation give us the required multiplicities.

There are two simple but important theorems that will form the basis for much of our subsequent work with polynomials: the *remainder theorem* and the *factor theorem*. We begin with a statement of the remainder theorem.

> ### The Remainder Theorem
>
> When a polynomial $f(x)$ is divided by $x - r$, the remainder is $f(r)$.

Before turning to a proof of the remainder theorem, let's see what the theorem is saying in two particular cases. First, suppose that we divide the polynomial $f(x) = 2x^2 - 3x + 4$ by $x - 1$. Then according to the remainder theorem, the remainder in this case should be the number $f(1)$. Let's check:

$$
\begin{array}{r|rrr}
1 & 2 & -3 & 4 \\
  &   & 2 & -1 \\
\hline
  & 2 & -1 & 3
\end{array}
\qquad
\begin{aligned}
f(x) &= 2x^2 - 3x + 4 \\
f(1) &= 2(1)^2 - 3(1) + 4 \\
f(1) &= 3
\end{aligned}
$$

As the calculations show, the remainder is indeed equal to $f(1)$. As a second example, let us divide the polynomial $g(x) = ax^2 + bx + c$ by $x - r$. According to the remainder theorem, the remainder should be $g(r)$. Again, let us check:

$$
\begin{array}{r|ccc}
r & a & b & c \\
  &   & ar & ar^2 + br \\
\hline
  & a & ar + b & ar^2 + br + c
\end{array}
\qquad
\begin{aligned}
g(x) &= ax^2 + bx + c \\
g(r) &= ar^2 + br + c
\end{aligned}
$$

The calculations show that the remainder is equal to $g(r)$, as we wished to check.

In the example just concluded, we verified that the remainder theorem holds for any quadratic polynomial $g(x) = ax^2 + bx + c$. A general proof of the remainder theorem can easily be given along these same lines. The only drawback is that it becomes slightly cumbersome to carry out the synthetic division process when the dividend is

$$a_n x^n + a_{n-1} x^{n-1} + \cdots + a_1 x + a_0$$

For this reason, mathematicians often prefer to base the proof of the remainder theorem on the division algorithm given in the previous section. This is the path we will follow here.

To prove the remainder theorem, we must show that when the polynomial $f(x)$ is divided by $x - r$, the remainder is $f(r)$. Now according to the division algorithm, we can write

$$f(x) = (x - r) \cdot q(x) + R(x) \tag{2}$$

In this identity, either $R(x)$ is the zero polynomial or the degree of $R(x)$ is less than that of $x - r$. But since the degree of $x - r$ is 1, we must have in this case that the degree of $R(x)$ is zero. Thus, in *either* case, the remainder $R(x)$ is a constant. Denoting this constant by c, we can rewrite equation (2) as

$$f(x) = (x - r) \cdot q(x) + c$$

Now if we set $x = r$ in this identity, we obtain

$$f(r) = (r - r) \cdot q(r) + c = c$$

We have now shown that $f(r) = c$. But by definition, c is the remainder $R(x)$. Thus, $f(r) = R(x)$. This proves the remainder theorem.

EXAMPLE 3   Let $f(x) = 2x^3 - 5x^2 + x - 6$.

(a) Use the remainder theorem to evaluate $f(3)$.
(b) Is $x - 3$ a factor of $f(x) = 2x^3 - 5x^2 + x - 6$?

Solution   (a) According to the remainder theorem, $f(3)$ is the remainder when $f(x)$ is divided by $x - 3$. Using synthetic division, we have

$$3|\begin{array}{cccc} 2 & -5 & 1 & -6 \\ & 6 & 3 & 12 \\ \hline 2 & 1 & 4 & 6 \end{array}$$

The remainder is 6, and therefore $f(3) = 6$.

(b) By definition, $x - 3$ is a factor of $f(x)$ if we obtain a zero remainder when $f(x)$ is divided by $x - 3$. But from our work in part (a), we know that the remainder is 6, not 0. So, $x - 3$ is not a factor of $f(x)$.

EXAMPLE 4   In the previous section it was stated, but not proved, that $x - a$ is a factor of $f(x) = x^n - a^n$. Use the remainder theorem to prove this fact.

Solution   We need to show that when $f(x) = x^n - a^n$ is divided by $x - a$, the remainder is 0. By the remainder theorem, the remainder is equal to $f(a)$, which is easy to find:

$$f(a) = a^n - a^n = 0$$

Thus, the remainder is zero, and $x - a$ is a factor of $x^n - a^n$, as required.

From our experience with quadratic equations, we know that there is a close connection between factoring a quadratic polynomial $f(x)$ and solving the polynomial equation $f(x) = 0$. The factor theorem states this relationship between roots and factors in a precise form. Furthermore, the factor theorem tells us that this relationship holds for polynomials of all degrees, not just quadratics.

> **The Factor Theorem**
>
> Let $f(x)$ be a polynomial. If $f(r) = 0$, then $x - r$ is a factor of $f(x)$. Conversely, if $x - r$ is a factor of $f(x)$, then $f(r) = 0$.

In terms of roots, we can summarize the factor theorem by saying that r is a root of the equation $f(x) = 0$ if and only if $x - r$ is a factor of $f(x)$. To prove the factor theorem, let us begin by assuming that $f(r) = 0$. We want to show that $x - r$ is a factor of $f(x)$. Now, according to the remainder theorem, if $f(x)$ is divided by $x - r$, the remainder is $f(r)$. So we can write

$$f(x) = (x - r) \cdot q(x) + f(r) \qquad \text{for some polynomial } q(x)$$

But since $f(r)$ is zero, this equation becomes

$$f(x) = (x - r) \cdot q(x)$$

This last equation tells us that $x - r$ is a factor of $f(x)$, as we wished to prove.

Now, conversely, let us assume that $x - r$ is a factor of $f(x)$. We want to show that $f(r) = 0$. Since $x - r$ is a factor of $f(x)$, we can write

$$f(x) = (x - r) \cdot q(x) \qquad \text{for some polynomial } q(x)$$

If we now let $x = r$ in this last equation, we obtain

$$f(r) = (r - r) \cdot q(r) = 0$$

as we wished to show.

The example that follows indicates how the factor theorem can be used to solve equations.

EXAMPLE 5 Solve the equation $x^3 - 2x + 1 = 0$, given that one root is $x = 1$.

Solution Since $x = 1$ is a root, the factor theorem tells us that $x - 1$ is a factor of $x^3 - 2x + 1$. In other words,

$$x^3 - 2x + 1 = (x - 1) \cdot q(x) \qquad \text{for some polynomial } q(x)$$

To determine this other factor $q(x)$, we divide $x^3 - 2x + 1$ by $x - 1$:

$$
\begin{array}{r|rrrr}
1 & 1 & 0 & -2 & 1 \\
  &   & 1 & 1 & -1 \\
\hline
  & 1 & 1 & -1 & 0
\end{array}
$$

Thus, $q(x) = x^2 + x - 1$ and we have the factorization

$$x^3 - 2x + 1 = (x - 1)(x^2 + x - 1)$$

Using this identity, the original equation becomes

$$(x - 1)(x^2 + x - 1) = 0$$

Now the problem is reduced to solving the linear equation $x - 1 = 0$ and the quadratic equation $x^2 + x - 1 = 0$. The linear equation yields $x = 1$, but we already knew that 1 was a root. So if we are to find any additional roots, they must come from the equation $x^2 + x - 1 = 0$. Using the quadratic formula, we obtain

$$x = \frac{-1 \pm \sqrt{(1)^2 - 4(1)(-1)}}{2(1)} = \frac{-1 \pm \sqrt{5}}{2}$$

We now have three roots of the cubic equation $x^3 - 2x + 1 = 0$. They are 1, $\frac{1}{2}(-1 + \sqrt{5})$, and $\frac{1}{2}(-1 - \sqrt{5})$. As you will see in the next section, a cubic equation can have at most three roots. So in this case, we have determined all the roots; that is, we have solved the equation. ▮▮▮

Before going on to other examples, let's take a moment to summarize the technique we used in Example 5. We want to solve a polynomial equation $f(x) = 0$, given that one root is $x = r$. Since r is a root, the factor theorem tells us that $x - r$ is a factor of $f(x)$. Then, with the aid of synthetic division, we obtain a factorization

$$f(x) = (x - r) \cdot q(x) \qquad \text{for some polynomial } q(x)$$

This gives rise to the two equations, $x - r = 0$ and $q(x) = 0$. Since the first of these only reasserts that $x = r$ is a root, we try to solve the second equation,

$q(x) = 0$. We refer to the equation $q(x) = 0$ as the **reduced equation.** Example 5 showed you the idea behind this terminology; the degree of $q(x)$ is one less than that of $f(x)$. If, as in Example 5, the reduced equation happens to be a quadratic equation, then we can always determine the remaining roots by factoring or by the quadratic formula. In subsequent sections, we will look at techniques that are helpful in cases where $q(x)$ is not quadratic.

EXAMPLE 6   Solve the equation $x^4 + 2x^3 - 7x^2 - 20x - 12 = 0$, given that $x = 3$ and $x = -2$ are roots.

Solution   Since $x = 3$ is a root, the factor theorem tells us that $x - 3$ is a factor of the polynomial $x^4 + 2x^3 - 7x^2 - 20x - 12$. That is,

$$x^4 + 2x^3 - 7x^2 - 20x - 12 = (x - 3) \cdot q(x) \qquad \text{for some polynomial } q(x)$$

As in Example 5, we can find $q(x)$ by synthetic division. We have

$$\begin{array}{r|rrrrr}
3 & 1 & 2 & -7 & -20 & -12 \\
  &   & 3 & 15 & 24 & 12 \\
\hline
  & 1 & 5 & 8 & 4 & 0
\end{array}$$

Thus, $q(x) = x^3 + 5x^2 + 8x + 4$, and our original equation is equivalent to

$$(x - 3)(x^3 + 5x^2 + 8x + 4) = 0$$

Now, $x = -2$ is also a root of this equation. But $x = -2$ surely is not a root of the equation $x - 3 = 0$. Therefore, $x = -2$ must be a root of the reduced equation $x^3 + 5x^2 + 8x + 4 = 0$. Again the factor theorem is applicable. Since $x = -2$ is a root of the reduced equation, $x + 2$ must be a factor of $x^3 + 5x^2 + 8x + 4$. That is,

$$x^3 + 5x^2 + 8x + 4 = (x + 2) \cdot p(x) \qquad \text{for some polynomial } p(x)$$

We can use synthetic division to determine $p(x)$:

$$\begin{array}{r|rrrr}
-2 & 1 & 5 & 8 & 4 \\
   &   & -2 & -6 & -4 \\
\hline
   & 1 & 3 & 2 & 0
\end{array}$$

Thus, $p(x) = x^2 + 3x + 2$, and our reduced equation can be written

$$(x + 2)(x^2 + 3x + 2) = 0$$

This gives rise to a second reduced equation.

$$x^2 + 3x + 2 = 0$$

In this case, the roots are readily obtained by factoring. We have

$$x^2 + 3x + 2 = 0$$
$$(x + 2)(x + 1) = 0$$
$$x + 2 = 0 \quad | \quad x + 1 = 0$$
$$x = -2 \quad | \quad x = -1$$

Now, of the two roots we've just found, the -2 happens to be one of the roots that we were initially given in the statement of the problem. The -1, on the other hand, is a distinct additional root. In summary, then, we have three distinct roots: 3, −2, and −1. The root −2 has multiplicity 2. As you will see in the next section, a fourth-degree equation can have at most four (not necessarily distinct) roots. So in the case at hand, we have found all the roots of the given equation.

EXAMPLE 7   In each case, find a polynomial equation $f(x) = 0$ satisfying the given conditions. If there is no such equation, say so.

(a) The numbers −1, 4, and 5 are roots.
(b) A factor of $f(x)$ is $x − 3$, and −4 is a root of multiplicity 2.
(c) The degree of f is 4, the number −5 is a root of multiplicity 3, and 6 is a root of multiplicity 2.

Solution   (a) If $f(x)$ is any polynomial containing the factors $(x + 1)$, $(x − 4)$, and $(x − 5)$, then the equation $f(x) = 0$ will certainly be satisfied when $x = −1$, $x = 4$, or $x = 5$. The simplest polynomial equation in this case is therefore

$$(x + 1)(x − 4)(x − 5) = 0$$

This is a polynomial equation with the required roots. If required, we can carry out the multiplication on the left-hand side of the equation. As you can check, this yields $x^3 − 8x^2 + 11x + 20 = 0$.

(b) According to the factor theorem, since −4 is a root, $x + 4$ must be a factor of $f(x)$. In fact, since −4 is a root of multiplicity 2, the quantity $(x + 4)^2$ must be a factor of $f(x)$. The following equation therefore satisfies the given conditions:

$$(x − 3)(x + 4)^2 = 0$$

(c) We are given two roots, one with multiplicity 3, the other with multiplicity 2. Thus the degree of $f(x)$ must be at least $3 + 2 = 5$. (Why?) But this then contradicts the given condition that the degree of f should be 4. Consequently, there is no polynomial equation that satisfies the given conditions.

EXERCISE SET 11.2

A

In Exercises 1–6, determine whether the given value for the variable is a root of the equation.

1. $12x − 8 = 112$; $x = 10$
2. $12x^2 − x − 20 = 0$; $x = \frac{5}{4}$
3. $x^2 − 2x − 4 = 0$; $x = 1 − \sqrt{5}$
4. $1 − x + x^2 − x^3 = 0$; $x = −1$
5. $2x^2 − 3x + 1 = 0$; $x = \frac{1}{2}$
6. $(x − 1)(x − 2)(x − 3) = 0$; $x = 4$

In Exercises 7–14, determine whether the given value is a zero of the function.

7. $f(x) = 3x − 2$; $x = \frac{2}{3}$
8. $g(x) = 1 + x^2$; $x = −1$
9. $h(x) = 5x^3 − x^2 + 2x + 8$; $x = −1$
10. $F(x) = −2x^5 + 3x^4 + 8x^3$; $x = 0$
11. $f(t) = 1 + 2t + t^3 − t^5$; $t = 2$
12. $f(t) = 1 + 2t + t^3 − t^5$; $t = \sqrt{2}$

13. $f(x) = 2x^3 - 3x + 1$; (a) $x = \frac{1}{2}(\sqrt{3} - 1)$;
    (b) $x = \frac{1}{2}(\sqrt{3} + 1)$

14. $g(x) = x^4 + 8x^3 + 9x^2 - 8x - 10$; (a) $x = 1$;
    (b) $x = \sqrt{6} - 4$; (c) $x = \sqrt{6} + 4$

15. List the distinct roots of each of the following equations. In the case of a repeated root, give its multiplicity.
    (a) $(x - 1)(x - 2)^3(x - 3) = 0$
    (b) $(x - 1)(x - 1)(x - 1) = 0$
    (c) $(x - 5)^6(x + 1)^4 = 0$ (d) $x^5(x - 1) = 0$

16. In this exercise, we verify that the remainder theorem is valid for the cubic polynomial

    $$g(x) = ax^3 + bx^2 + cx + d$$

    (a) Compute $g(r)$.
    (b) Using synthetic division, divide $g(x)$ by $x - r$. Check that the remainder you obtain is the same as the answer in part (a).

In Exercises 17–24, use the remainder theorem (as in Example 3) to evaluate $f(x)$ for the given value of x.

17. $f(x) = 4x^3 - 6x^2 + x - 5$; $x = -3$
18. $f(x) = 2x^3 - x - 4$; $x = 4$
19. $f(x) = 6x^4 + 5x^3 - 8x^2 - 10x - 3$; $x = \frac{1}{2}$
20. $f(x) = x^5 - x^4 - x^3 - x^2 - x - 1$; $x = -2$
21. $f(x) = x^2 + 3x - 4$; $x = -\sqrt{2}$
22. $f(x) = -3x^3 + 8x^2 - 12$; $x = 5$
23. $f(x) = \frac{1}{2}x^3 - 5x^2 - 13x - 10$; $x = 12$
24. $f(x) = x^7 - 7x^6 + 5x^4 + 1$; $x = -3$

In Exercises 25–38, you are given a polynomial equation and one or more roots. Solve each equation using the method shown in Examples 5 and 6. To help you decide if you have found all the roots in each case, you may rely on the following theorem, discussed in the next section: A polynomial equation of degree n has at most n (not necessarily distinct) roots.

25. $x^3 - 4x^2 - 9x + 36 = 0$; -3 is a root.
26. $x^3 + 7x^2 + 11x + 5 = 0$; -1 is a root.
27. $x^3 + x^2 - 7x + 5 = 0$; 1 is a root.
28. $x^3 + 8x^2 - 3x - 24 = 0$; -8 is a root.
29. $3x^3 - 5x^2 - 16x + 12 = 0$; -2 is a root.
30. $2x^3 - 5x^2 - 46x + 24 = 0$; 6 is a root.
31. $2x^3 + x^2 - 5x - 3 = 0$; $-\frac{3}{2}$ is a root.
32. $6x^4 - 19x^3 - 25x^2 + 18x + 8 = 0$; 4 and $-\frac{1}{3}$ are roots.
33. $x^4 - 15x^3 + 75x^2 - 125x = 0$; 5 is a root.
34. $2x^3 + 5x^2 - 8x - 20 = 0$; 2 is a root.
35. $x^4 + 2x^3 - 23x^2 - 24x + 144 = 0$; -4 and 3 are roots.

36. $6x^5 + 5x^4 - 29x^3 - 25x^2 - 5x = 0$; $\sqrt{5}$ and $-\frac{1}{3}$ are roots.
37. $x^3 + 7x^2 - 19x - 9 = 0$; -9 is a root.
38. $4x^5 - 15x^4 + 8x^3 + 19x^2 - 12x - 4 = 0$; 1, 2, and -1 are roots.

In Exercises 39 and 40, use the given information to solve each equation.

39. $f(x) = 4x^4 - 12x^3 + 5x^2 + 6x + 1 = 0$; $2x^2 - 3x - 1$ is a factor of $f(x)$.

40. $f(x) = 3x^5 - 3x^4 - x^3 - 24x^2 + 24x + 8 = 0$; $3x^2 - 3x - 1$ is a factor of $f(x)$.

For Exercises 41 and 42, refer to the following computer-generated tables for the functions f and g.

$f(t) = t^3 - 4t + 3$		$g(t) = t^5 + 2t^4 + t^3 + 2t^2 - t - 2$	
t	$f(t)$	t	$g(t)$
0.500	1.125000	-3.00	-89.0
0.750	0.421875	-2.50	-22.15625
1.000	0.000000	-2.25	-7.4228515625
1.250	-0.046875	-2.00	0.0
1.500	0.375000	-1.75	2.8603515625

41. (a) What is the remainder when $f(t)$ is divided by $t - \frac{1}{2}$?
    (b) What is the remainder when $f(t)$ is divided by $t - 1.25$?
    (c) Specify a linear factor of $f(t)$.
    (d) Solve the equation $f(t) = 0$.

42. (a) What is the remainder when $g(t)$ is divided by $t + 3$?
    (b) What is the remainder when $g(t)$ is divided by $t + \frac{5}{2}$?
    (c) Specify a linear factor of $g(t)$.

In Exercises 43–50, find a polynomial equation $f(x) = 0$ satisfying the given conditions. If no such equation is possible, state this.

43. Degree 3; the coefficient of x^3 is 1; three roots are 3, -4, and 5.

44. Degree 3; the coefficients are integers; $\frac{1}{2}$, $\frac{2}{5}$, and $-\frac{3}{4}$ are roots.

45. Degree 3; -1 is a root of multiplicity two; $x + 6$ is a factor of $f(x)$.

46. Degree 3; 4 is a root of multiplicity two; -1 is a root of multiplicity two.

47. Degree 4; $\frac{1}{2}$ is a root of multiplicity three; $x^2 - 3x - 4$ is a factor of $f(x)$.

48. Degree 3; -2 and -3 are roots.

49. Degree 4; $x^2 - 3x + 1$ and $x + 6$ are factors of $f(x)$.

50. Degree 4; the coefficients are integers; $\frac{1}{2}$ is a root of multiplicity two; $2x^2 - 4x - 1$ is a factor of $f(x)$.

In Exercises 51–54, use the remainder theorem (as in Example 3) to evaluate $f(x)$ for the given value of x. Use a calculator, and round off your answers to two decimal places.

51. $f(x) = x^3 - 3x^2 + 12x + 9$; $x = 1.16$

52. $f(x) = x^3 - 2x - 5$;
 (a) $x = 2.09$
 (b) $x = 2.094$
 (c) $x = 2.0945$

53. $f(x) = x^3 - 5x - 2$;
 (a) $x = 2.41$
 (b) $x = 2.42$

54. $f(x) = x^4 - 2x^3 - 5x^2 + 10x - 3$;
 (a) $x = -2.3$
 (b) $x = -2.302$
 (c) $x = -2.30277$

55. A computer was used to generate the following information about the function $f(x) = x^3 + x^2 - 18x + 10$.

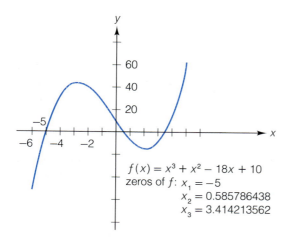

$f(x) = x^3 + x^2 - 18x + 10$
zeros of f: $x_1 = -5$
$x_2 = 0.585786438$
$x_3 = 3.414213562$

 (a) Determine the exact values for the zeros x_2 and x_3.
 (b) Use your calculator to evaluate the answers in part (a), and check that they are consistent with the computer values given in the figure.

56. A computer was used to generate the following information about the function
$g(x) = -2x^3 + 7x^2 - 2x - 6$.

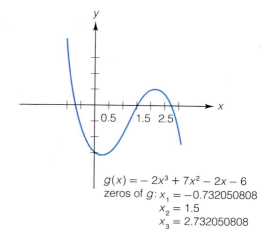

$g(x) = -2x^3 + 7x^2 - 2x - 6$
zeros of g: $x_1 = -0.732050808$
$x_2 = 1.5$
$x_3 = 2.732050808$

 (a) Determine the exact values for the zeros x_1 and x_3.
 (b) Use your calculator to evaluate the answers in part (a), and check that they are consistent with the computer values given in the figure.

B

In Exercises 57–60, determine whether the given value is a zero of the function.

57. $f(x) = \frac{1}{2}x^2 + bx + c$; $x = -b - \sqrt{b^2 - 2c}$

58. $Q(x) = ax^2 + bx + c$; $x = \left(-b + \sqrt{b^2 - 4ac}\right)/2a$
 Hint: Look before you leap!

59. $F(x) = 2x^4 + 4x + 1$; $x = \frac{1}{2}\left(-\sqrt{2} + \sqrt{2\sqrt{2} - 2}\right)$

60. $f(x) = x^3 - 3x^2 + 3x - 3$; (a) $x = \sqrt[3]{2} - 1$;
 (b) $x = \sqrt[3]{2} + 1$

61. Determine values for a and b such that $x - 1$ is a factor of both $x^3 + x^2 + ax + b$ and $x^3 - x^2 - ax + b$.

62. Determine a quadratic equation with the given roots.
 (a) a/b, $-b/a$ (b) $-a + 2\sqrt{2b}$, $-a - 2\sqrt{2b}$

63. One root of the equation $x^2 + bx + 1 = 0$ is twice the other; find b. (There are two answers.)

64. Determine a value for a such that one root of the equation $ax^2 + x - 1 = 0$ is five times the other.

C

65. Solve the equation $x^3 - 12x + 16 = 0$, given that one of the roots has multiplicity 2.

11.3 THE FUNDAMENTAL THEOREM OF ALGEBRA

One of the most intriguing problems was the question of the number of roots of an equation, which brought in negative and imaginary quantities, and led to the conclusion, in the work of Girard and Descartes, that an equation of degree n can have no more than n roots. The more precise statement, that an equation of degree n always has one root, and hence always has n roots (allowing for multiple roots), became known as the fundamental theorem of algebra. After several attempts by D'Alembert, Euler, and others, the proof was finally given by Gauss in 1799.

From *A Source Book in Mathematics, 1200–1800,* D. J. Struik, ed. (Princeton: Princeton University Press, 1986)

Every equation of algebra has as many solutions as the exponent of the highest term indicates.

Albert Girard, 1629, as quoted by David M. Burton in *The History of Mathematics, An Introduction,* 2d ed., (Dubuque, Iowa: Wm. C. Brown Publishers, 1991)

… we arrive from all the investigations explained above at the rigorous proof of the theorem that every integral rational algebraic function of one variable can be decomposed into real factors of the first and second degree.

Carl Friedrich Gauss in his doctoral dissertation (1799)

Does every polynomial equation (of degree at least 1) have a root? Or, to put the question another way, is it possible to write a polynomial equation that has no solution? Certainly, if we consider only real roots (i.e., roots that are real numbers), then it is easy to specify an equation with no real root:

$$x^2 = -1$$

This equation has no real root because the square of a real number is never negative. On the other hand, if we expand our base of operations from the real number system to the complex number system, then $x^2 = -1$ does indeed have a root. In fact, both i and $-i$ are roots in this case.

As it turns out, the situation just described for the equation $x^2 = -1$ holds quite generally. That is, within the complex number system, every polynomial equation of degree at least 1 has at least one root. This is the substance of a remarkable theorem that was first proved by the great mathematician Carl Friedrich Gauss in 1799. (Gauss was only 22 years old at the time.) Although there are many fundamental theorems in algebra, Gauss's result has come to be known as *the* **fundamental theorem of algebra.**

The Fundamental Theorem of Algebra

Every polynomial equation of the form

$$a_n x^n + a_{n-1} x^{n-1} + \cdots + a_1 x + a_0 = 0 \qquad (n \geq 1,\ a_n \neq 0)$$

has at least one root among the complex numbers. (This root may be a real number.)

Although most of this section deals with polynomials with real coefficients, Gauss's theorem still applies in cases in which some or all of the coefficients are nonreal complex numbers. The proof of Gauss's theorem is usually given in the post-calculus course called complex variables.

The fundamental theorem of algebra asserts that every polynomial equation of degree at least 1 has a root. There are two initial observations to be made here. First, notice that the theorem says nothing about actually finding the root. Second, notice that the theorem deals only with polynomial equations. Indeed, it is easy to specify a nonpolynomial equation that does not have a root. Such an equation is $1/x = 0$. The expression on the left-hand side of this equation can never be zero because the numerator is 1.

EXAMPLE I   Which of the following equations has at least one root?

(a) $x^3 - 17x^2 + 6x - 1 = 0$ (b) $\sqrt{2}x^{47} - \pi x^{25} + \sqrt{3} = 0$
(c) $x^2 - 2ix + (3 + i) = 0$

Solution   All three equations are polynomial equations. So, according to the fundamental theorem of algebra, each equation has at least one root.   ∎

In Chapter 2 we used factoring as a tool for solving quadratic equations. The next theorem, a consequence of the fundamental theorem of algebra, tells us that (in principle, at least) any polynomial of degree n can be factored into a product of n linear factors. In proving the **linear factors theorem,** we will need to use the factor theorem. If you reread the proof of that theorem in the previous section, you will see that it makes no difference whether the number r appearing in the factor $x - r$ is a real number or a nonreal complex number. Thus, the factor theorem is valid in either case.

The Linear Factors Theorem

Let $f(x) = a_n x^n + a_{n-1} x^{n-1} + \cdots + a_1 x + a_0$ ($n \geq 1$, $a_n \neq 0$). Then $f(x)$ can be expressed as a product of n linear factors:

$$f(x) = a_n(x - r_1)(x - r_2) \cdots (x - r_n)$$

(The complex numbers r_k appearing in these factors are not necessarily all distinct, and some or all of the r_k may be real numbers.)

PROOF OF THE LINEAR FACTORS THEOREM   According to the fundamental theorem of algebra, the equation $f(x) = 0$ has a root. Let us call this root r_1. By the factor theorem, $x - r_1$ is a factor of $f(x)$, and we can write

$$f(x) = (x - r_1) \cdot Q_1(x)$$

for some polynomial $Q_1(x)$ that has degree $n - 1$ and leading coefficient a_n. If the degree of $Q_1(x)$ happens to be zero, we are done. On the other hand, if the degree of $Q_1(x)$ is at least 1, another application of the fundamental theorem of algebra followed by the factor theorem gives us

$$Q_1(x) = (x - r_2) \cdot Q_2(x)$$

where the degree of $Q_2(x)$ is $n - 2$ and the leading coefficient of $Q_2(x)$ is a_n. We now have

$$f(x) = (x - r_1)(x - r_2) \cdot Q_2(x)$$

We continue this process until the quotient is $Q_n(x) = a_n$. As a result, we obtain

$$f(x) = (x - r_1)(x - r_2) \cdots (x - r_n)a_n$$
$$= a_n(x - r_1)(x - r_2) \cdots (x - r_n)$$

as we wished to show.

The linear factors theorem tells us that any polynomial can be expressed as a product of linear factors. The theorem gives us no information, however, as to how those factors can actually be obtained. The next example demonstrates a case in which the factors are readily obtainable; this is always the case with quadratic polynomials.

EXAMPLE 2 Express each of the following second-degree polynomials in the form $a_n(x - r_1)(x - r_2)$.

(a) $3x^2 - 5x - 2$ **(b)** $x^2 - 4x + 5$

Solution **(a)** A factorization for $3x^2 - 5x - 2$ can be found by simple trial and error. We have

$$3x^2 - 5x - 2 = (3x + 1)(x - 2)$$

We now write the factor $3x + 1$ as $3\left(x + \frac{1}{3}\right)$. This, in turn, can be written $3\left[x - \left(-\frac{1}{3}\right)\right]$. The final factorization is then

$$3x^2 - 5x - 2 = 3\left[x - \left(-\frac{1}{3}\right)\right](x - 2)$$

(b) From the factor theorem, or from our more elementary work with quadratic equations, we know that if r_1 and r_2 are the roots of the equation $x^2 - 4x + 5 = 0$, then $x - r_1$ and $x - r_2$ are the factors of $x^2 - 4x + 5$. That is, $x^2 - 4x + 5 = (x - r_1)(x - r_2)$. The values for r_1 and r_2 in this case are readily obtained by using the quadratic formula. As you can check, the results are

$$r_1 = 2 + i \quad \text{and} \quad r_2 = 2 - i$$

The required factorization is therefore

$$x^2 - 4x + 5 = [x - (2 + i)][x - (2 - i)]$$

Using the linear factors theorem, we can show that every polynomial equation of degree n ($n \geq 1$) has exactly n roots. To help you follow the reasoning, we make two preliminary comments. First, we agree that a root of multiplicity k will be counted as k roots. For example, although the third-degree equation $(x - 1)(x - 4)^2 = 0$ has only two distinct roots, namely, 1 and 4, it has three roots *if* we agree to count the repeated root 4 two times. The second preliminary comment concerns the *zero-product property,* which states that $pq = 0$ if and only if $p = 0$ or $q = 0$. When we stated this in Section 2.3, we were working within the real number system. However, the property is also valid within the complex number system. Now let's state and prove our theorem.

> **Theorem**
>
> Every polynomial equation of degree $n \geq 1$ has exactly n roots, where a root of multiplicity k is counted k times.

PROOF OF THEOREM   Using the linear factors theorem, we can write the nth-degree polynomial equation $f(x) = a_n x^n + a_{n-1} x^{n-1} + \cdots + a_0 = 0$ as

$$f(x) = a_n(x - r_1)(x - r_2) \cdots (x - r_n) = 0 \qquad (a_n \neq 0) \tag{1}$$

By the factor theorem, each of the numbers $r_1, r_2, \ldots, r_n$ is a root. Some of these numbers may in fact be equal; in other words, we may have repeated roots in this list. In any case, if we agree to count a root of multiplicity k as k roots, then we obtain exactly n roots from the list $r_1, r_2, \ldots, r_n$. Furthermore, the equation $f(x) = 0$ can have no other roots, as we now show. Suppose that r is any number distinct from all the numbers $r_1, r_2, \ldots, r_n$. Replacing x with r in equation (1) yields

$$f(r) = a_n(r - r_1)(r - r_2) \cdots (r - r_n)$$

But the expression on the right-hand side of this last equation cannot be zero, because none of the factors is zero. Thus, $f(r)$ is not zero, and so r is not a root. This completes the proof of the theorem.

EXAMPLE 3   Find a polynomial $f(x)$ with leading coefficient 1 such that the equation $f(x) = 0$ has only those roots specified in Table 1. What is the degree of this polynomial?

Solution   The expressions $(x - 3)^2$, $(x + 2)$, and $(x - 0)^2$ all must appear as factors of $f(x)$ and, furthermore, no other linear factor can appear. The form of $f(x)$ is therefore

$$f(x) = a_n(x - 3)^2(x + 2)(x - 0)^2$$

Since the leading coefficient a_n is to be 1, we can rewrite this last equation as

$$f(x) = x^2(x - 3)^2(x + 2)$$

This is the required polynomial. The degree here is 5. This can be seen either by multiplying out the factors or by simply adding the multiplicities of the roots in Table 1.

TABLE 1

Root	Multiplicity
3	2
−2	1
0	2

EXAMPLE 4   Find a quadratic function f that has zeros of 3 and 5 and a graph that passes through the point $(2, -9)$.

Solution   The general form of a quadratic function with 3 and 5 as zeros is $f(x) = a_n(x - 3)(x - 5)$. Since the graph passes through $(2, -9)$, we have

$$-9 = a_n(2 - 3)(2 - 5) = a_n(3)$$
$$-3 = a_n$$

The required function is therefore

$$f(x) = -3(x - 3)(x - 5)$$

If we wish, we can carry out the multiplication and rewrite this as

$$f(x) = -3x^2 + 24x - 45$$

The next example shows how we can use the factored form of a polynomial $f(x)$ to determine the relationships between the coefficients of the polynomial and the roots of the equation $f(x) = 0$.

EXAMPLE 5   Let r_1 and r_2 be the roots of the equation $x^2 + bx + c = 0$. Show that

$$r_1r_2 = c \qquad \text{and} \qquad r_1 + r_2 = -b$$

Solution   Since r_1 and r_2 are the roots of the equation $x^2 + bx + c = 0$, we have the identity

$$x^2 + bx + c = (x - r_1)(x - r_2)$$

After multiplying out the right-hand side, we can rewrite this identity as

$$x^2 + bx + c = x^2 - (r_1 + r_2)x + r_1r_2$$

By equating coefficients, we readily obtain $r_1r_2 = c$ and $r_1 + r_2 = -b$, as required.

The technique used in Example 5 can be used to obtain similar relationships between the roots and the coefficients of polynomial equations of any given degree. In Table 2, for instance, we show the relationships obtained in Example 5, along with the corresponding relationships that can be derived for a cubic equation. (Exercise 31 at the end of this section asks you to verify the results for the cubic equation.)

TABLE 2

Equation	Roots	Relationships Between Roots and Coefficients
$x^2 + bx + c = 0$	r_1, r_2	$r_1 + r_2 = -b$ $r_1r_2 = c$
$x^3 + bx^2 + cx + d = 0$	r_1, r_2, r_3	$r_1 + r_2 + r_3 = -b$ $r_1r_2 + r_2r_3 + r_3r_1 = c$ $r_1r_2r_3 = -d$

We conclude this section with some remarks concerning the solving of polynomial equations by formulas. You know that the roots of the quadratic equation $ax^2 + bx + c = 0$ are given by the formula

$$x = \frac{-b \pm \sqrt{b^2 - 4ac}}{2a}$$

The question is, are there similar formulas for the solutions of higher-degree equations? By "similar" we mean a formula involving the coefficients and radicals. To answer this question, we look at a bit of history. As early as 1700 B.C., Babylonian mathematicians were able to solve quadratic equations. This is clear from the study of the clay tablets with cuneiform numerals that archeologists have found. The ancient Greeks also were able to solve quadratic equations. Like the Babylonians, the Greeks worked without the aid of algebra as we know it. The mathematicians of ancient Greece used geometric constructions to solve equations. Of course, since all quantities were interpreted geometrically, negative roots were never considered.

The general quadratic formula was known to the Moslem mathematicians sometime before A.D. 1000. For the next 500 years, mathematicians searched for, but did not discover, a formula to solve the general cubic equation. Indeed, in 1494, Luca Pacioli stated in his text *Summa di Arithmetica* that the general cubic equation could not be solved by the algebraic techniques then available. All of this was to change, however, within the next several decades.

Around 1515, the Italian mathematician Scipione del Ferro solved the cubic equation $x^3 + px + q = 0$ using algebraic techniques. This essentially constituted a solution of the seemingly more general equation $x^3 + bx^2 + cx + d = 0$. The reason for this is that if we make the substitution $x = y - b/3$ in the latter equation, the result is a cubic equation with no y^2-term. By 1540, the Italian mathematician Ludovico Ferrari had solved the general fourth-degree equation. Actually, at that time in Renaissance Italy, there was considerable controversy as to exactly who discovered the various formulas first. Details of the dispute can be found in any text on the history of mathematics. But for our purposes here, the point is simply that by the middle of the sixteenth century, all polynomial equations of degree 4 or less could be solved by the formulas that had been discovered. The common feature of these formulas was that they involved the coefficients, the four basic operations of arithmetic, and various radicals. For example, a formula for the solution of the equation $x^3 + px + q = 0$ is as follows:

$$x = \sqrt[3]{\frac{-q}{2} + \sqrt{\frac{q^2}{4} + \frac{p^3}{27}}} + \sqrt[3]{\frac{-q}{2} - \sqrt{\frac{q^2}{4} + \frac{p^3}{27}}}$$

To get some idea of the practical difficulties inherent in computing with this formula, try using it to show that $x = -2$ is a root of the cubic equation $x^3 + 4x + 16 = 0$.

For more than 200 years after the cubic and quartic (fourth-degree) equations had been solved, mathematicians continued to search for a formula that would yield the solutions of the general fifth-degree equation. The first breakthrough, if it can be called that, occurred in 1770, when the French mathematician Joseph Louis Lagrange found a technique that served to unify and summarize all of the previous methods used for the equations of degrees 2, 3, and 4. However, Lagrange then showed that his technique could not work in the case of the general fifth-degree equation. While we will not describe the details of Lagrange's work here, it is worth pointing out that he relied on the types of relationships between roots and coefficients that we looked at in Table 2 of this section.

Finally, in 1828, the Norwegian mathematician Niels Henrik Abel proved that for the general polynomial equation of degree 5 or higher, there could be no formula yielding the solutions in terms of the coefficients and radicals. This is not to say that such equations do not possess solutions. In fact they must, as we saw earlier in this section. It is just that we cannot in every case express the solutions in terms of the coefficients and radicals. For example, it can be shown that the equation $x^5 - 6x + 3 = 0$ has a real root between 0 and 1, but this number cannot be expressed in terms of the coefficients and radicals. (We can, however, compute the root to as many decimal places as we wish, as you will see in the next section.) In 1830, the French mathematician Evariste Galois completed matters by giving conditions for determining exactly which polynomial equations can be solved in terms of coefficients and radicals.

EXERCISE SET 11.3

A

According to the fundamental theorem of algebra, which of the equations in Exercises 1 and 2 have at least one root?

1. (a) $x^5 - 14x^4 + 8x + 53 = 0$
   (b) $(4.17)x^3 + (2.06)x^2 + (0.01)x + 1.23 = 0$
   (c) $ix^2 + (2 + 3i)x - 17 = 0$
   (d) $x^{2.1} + 3x^{0.3} + 1 = 0$
2. (a) $\sqrt{3}x^{17} + \sqrt{2}x^{13} + \sqrt{5} = 0$
   (b) $17x^{\sqrt{3}} + 13x^{\sqrt{2}} + \sqrt{5} = 0$
   (c) $1/(x^2 + 1) = 0$ (d) $2^{3x} - 2^x - 1 = 0$

In Exercises 3–10, express each polynomial in the form $a_n(x - r_1)(x - r_2) \cdots (x - r_n)$.

3. $x^2 - 2x - 3$
4. $x^3 - 2x^2 - 3x$
5. $4x^2 + 23x - 6$
6. $6x^2 + x - 12$
7. $x^2 - 5$
8. $x^2 + 5$
9. $x^2 - 10x + 26$
10. $x^3 + 2x^2 - 3x - 6$

In Exercises 11–16, find a polynomial $f(x)$ with leading coefficient 1 such that the equation $f(x) = 0$ has the given roots and no others. If the degree of $f(x)$ is 7 or more, express $f(x)$ in factored form; otherwise, express $f(x)$ in the form $a_nx^n + a_{n-1}x^{n-1} + \cdots + a_1x + a_0$.

11.

Root	1	−3
Multiplicity	2	1

12.

Root	0	4
Multiplicity	2	1

13.

Root	2	−2	2i	−2i
Multiplicity	1	1	1	1

14.

Root	2 + i	2 − i
Multiplicity	1	1

15.

Root	$\sqrt{3}$	$-\sqrt{3}$	4i	−4i
Multiplicity	2	2	1	1

16.

Root	5	1	1 − i	1 + i
Multiplicity	2	3	1	1

In Exercises 17–20, express the polynomial $f(x)$ in the form $a_nx^n + a_{n-1}x^{n-1} + \cdots + a_1x + a_0$.

17. Find a quadratic function that has zeros −4 and 9 and a graph that passes through the point $(3, 5)$.

18. Find a quadratic function that has a maximum value of 2 and that has −2 and 4 as zeros.

19. Find a third-degree polynomial function that has zeros −5, 2, and 3 and a graph that passes through the point $(0, 1)$.

20. Find a fourth-degree polynomial function that has zeros $\sqrt{2}$, $-\sqrt{2}$, 1, and −1 and a graph that passes through $(2, -20)$.

In Exercises 21–28, find a quadratic equation with the given roots and no others. Write your answers in the form $Ax^2 + Bx + C = 0$. Suggestion: Make use of Table 2.

21. $r_1 = -i$, $r_2 = -\sqrt{3}$
22. $r_1 = 1 + i\sqrt{3}$, $r_2 = 1 - i\sqrt{3}$
23. $r_1 = 9$, $r_2 = -6$ 24. $r_1 = 5$, $r_2 = \frac{3}{4}$
25. $r_1 = 1 + \sqrt{5}$, $r_2 = 1 - \sqrt{5}$
26. $r_1 = 6 - 5i$, $r_2 = 6 + 5i$
27. $r_1 = a + \sqrt{b}$, $r_2 = a - \sqrt{b}$ $(b > 0)$
28. $r_1 = a + bi$, $r_2 = a - bi$

B

29. Express the polynomial $x^4 + 64$ as a product of four linear factors. *Hint:* First add and subtract the quantity $16x^2$ so that you can use the difference-of-squares factoring formula.

30. Suppose that p and q are positive integers with $p > q$. Find a quadratic equation with integer coefficients whose roots are $\sqrt{p}/(\sqrt{p} \pm \sqrt{p-q})$.

31. Let r_1, r_2, and r_3 be the roots of the equation $x^3 + bx^2 + cx + d = 0$. Use the method shown in Example 5 to verify the following relationships:

$$r_1 + r_2 + r_3 = -b$$
$$r_1 r_2 + r_2 r_3 + r_3 r_1 = c$$
$$r_1 r_2 r_3 = -d$$

32. Let r_1, r_2, r_3, and r_4 be the roots of the equation $x^4 + bx^3 + cx^2 + dx + e = 0$. Use the method shown in Example 5 to prove the following facts:

$$r_1 + r_2 + r_3 + r_4 = -b$$
$$r_1 r_2 + r_2 r_3 + r_3 r_4 + r_4 r_1 + r_2 r_4 + r_3 r_1 = c$$
$$r_1 r_2 r_3 + r_2 r_3 r_4 + r_3 r_4 r_1 + r_4 r_1 r_2 = -d$$
$$r_1 r_2 r_3 r_4 = e$$

33. Solve the equation $x^3 - 4x^2 - 9x + 36 = 0$, given that the sum of two of the roots is 0. *Suggestion:* Use Table 2.

34. Solve the equation $x^3 - 75x + 250 = 0$, given that two of the roots are equal. *Suggestion:* Use Table 2.

35. Let α and β be the roots of the polynomial equation $x^2 + bx + c = 0$.
(a) Show that $\alpha^2 + \beta^2 = b^2 - 2c$. *Hint:* Use Table 2 along with the identity $\alpha^2 + \beta^2 = (\alpha + \beta)^2 - 2\alpha\beta$.
(b) Show that $\dfrac{1}{\alpha^2} + \dfrac{1}{\beta^2} = \dfrac{b^2 - 2c}{c^2}$.
(c) Show that $\alpha^3 + \beta^3 = -b(b^2 - 3c)$.

C

36. (a) Let r_1, r_2, r_3, and r_4 be four real roots of the equation $x^4 + ax^2 + bx + c = 0$. Show that $r_1 + r_2 + r_3 + r_4 = 0$. *Hint:* Use the first formula in Exercise 32.
(b) Suppose a circle intersects the parabola $y = x^2$ in the points $(x_1, y_1), \ldots, (x_4, y_4)$. Show that $x_1 + x_2 + x_3 + x_4 = 0$. *Hint:* Use the result in part (a).

37. (a) Let r_1, r_2, and r_3 be three distinct numbers that are roots of the equation $f(x) = Ax^2 + Bx + C = 0$. Show that $f(x) = 0$ for all values of x. *Hint:* You need to show that $A = B = C = 0$. First show that both A and B are zero as follows. If either A or B were nonzero, then the equation $f(x) = 0$ would be a polynomial equation of degree at most 2 with three distinct roots. Why is that impossible?
(b) Use the result in part (a) to prove the following identity.

$$\frac{a^2 - x^2}{(a-b)(a-c)} + \frac{b^2 - x^2}{(b-c)(b-a)}$$
$$+ \frac{c^2 - x^2}{(c-a)(c-b)} - 1 = 0$$

Hint: Let $f(x)$ denote the quadratic expression on the left-hand side of the equation. Compute $f(a)$, $f(b)$, and $f(c)$.

38. Prove that the following equation is an identity:

$$\frac{(x-a)(x-b)c^2}{(c-a)(c-b)} + \frac{(x-b)(x-c)a^2}{(a-b)(a-c)}$$
$$+ \frac{(x-c)(x-a)b^2}{(b-c)(b-a)} - x^2 = 0$$

Hint: Use the result in Exercise 37(a).

11.4 RATIONAL AND IRRATIONAL ROOTS

Jacques Peletier (1517–1582), a French man of letters, poet, and mathematician, had observed as early as 1558, that the root of an equation is a divisor of the last term.

Florian Cajori in *A History of Mathematics*, 4th ed. (New York: Chelsea Publishing Co., 1985)

Of the two leading commentators on Descartes in the 17th century, the first (1649) was his warm personal friend, Florimond de Beaune, an officeholder at Blois. He also wrote on algebra, being one of the first to treat scientifically of the superior and inferior limits of the roots of a numerical equation.

David Eugene Smith in *History of Mathematics*, vol. I (New York: Ginn and Co., 1923)

As we saw in the previous section, not every polynomial equation has a real root. Furthermore, even if a polynomial equation does possess a real root, that root isn't necessarily a rational number. (The equation $x^2 = 2$ provides a simple example.) If a polynomial equation with integer coefficients does have a rational root, however, we can find that root by applying the **rational roots theorem**, which we now state.

The Rational Roots Theorem

Consider the polynomial equation
$$a_n x^n + a_{n-1} x^{n-1} + \cdots + a_1 x + a_0 = 0 \qquad (n \geq 1, \, a_n \neq 0)$$
and suppose that all the coefficients are integers. Let p/q be a rational number, where p and q have no common factors other than ± 1. If p/q is a root of the equation, then p is a factor of a_0 and q is a factor of a_n.

A proof of the rational roots theorem is outlined in Exercise 39 at the end of this section. For the moment, though, let's just see why the theorem is plausible. Suppose that the two rational numbers a/b and c/d are the roots of a certain quadratic equation. Then, from our experience with quadratics (or by the linear factors theorem), we know that the equation can be written in the form

$$k\left(x - \frac{a}{b}\right)\left(x - \frac{c}{d}\right) = 0 \tag{1}$$

where k is a constant. Now, as Exercise 28 will ask you to check, if we carry out the multiplication and clear of fractions, equation (1) becomes

$$(kbd)x^2 - (kad + kbc)x + kac = 0 \tag{2}$$

Observe that a and c (the numerators of the two roots) are factors of the constant term kac in equation (2), just as the rational roots theorem asserts. Furthermore, b and d (the denominators of the roots) are factors of the coefficient of the x^2-term in equation (2), again as the theorem asserts.

The following example shows how the rational roots theorem can be used to solve a polynomial equation.

EXAMPLE 1 Find the rational roots (if any) of the equation $2x^3 - x^2 - 9x - 4 = 0$. Then solve the equation.

Solution First we list the factors of a_0, the factors of a_n, and the possibilities for rational roots:

> factors of $a_0 = -4$: $\pm 1, \pm 2, \pm 4$
>
> factors of $a_3 = 2$: $\pm 1, \pm 2$
>
> possible rational roots: $\pm \dfrac{1}{1}, \pm \dfrac{1}{2}, \pm \dfrac{2}{1}, \pm \dfrac{4}{1}$

Now we can use synthetic division to test whether or not any of these possibilities is a root. (A zero remainder will tell us that we have a root.) As you can check, the first three possibilities $\left(1, -1, \text{and } \tfrac{1}{2}\right)$ are not roots. However, using $-\tfrac{1}{2}$, we have

$$
\begin{array}{r|rrrr}
-1/2 & 2 & -1 & -9 & -4 \\
      &   & -1 & 1 & 4 \\
\hline
      & 2 & -2 & -8 & 0 \\
\end{array}
$$

Thus, $x = -\tfrac{1}{2}$ is a root. We could now continue to check the remaining possibilities in this same manner. At this point, however, it is simpler to consider the reduced equation $2x^2 - 2x - 8 = 0$, or $x^2 - x - 4 = 0$. Since this is a quadratic equation, it can be solved directly. We have

$$
x = \frac{-(-1) \pm \sqrt{(-1)^2 - 4(1)(-4)}}{2(1)} = \frac{1 \pm \sqrt{17}}{2}
$$

We have now determined three distinct roots. Since the degree of the original equation is 3, there can be no other roots. We conclude that $x = -\tfrac{1}{2}$ is the only rational root. The three roots of the equation are $-\tfrac{1}{2}$ and $\left(1 \pm \sqrt{17}\right)/2$. ∎ ■■■

As Example 1 indicates, the number of possibilities for rational roots can be relatively large, even for rather simple equations. The next theorem we develop allows us to reduce the number of possibilities. We say that a real number B is an **upper bound** for the roots of an equation if every real root is less than or equal to B. Similarly, a real number b is a **lower bound** if every real root is greater than or equal to b. The following theorem tells us how synthetic division can be used in determining upper and lower bounds for roots.

The Upper and Lower Bound Theorem for Real Roots

Consider the polynomial equation

$$
f(x) = a_n x^n + a_{n-1} x^{n-1} + \cdots + a_1 x + a_0 = 0
$$

where all of the coefficients are real numbers and a_n is positive.

1. If we use synthetic division to divide $f(x)$ by $x - B$, where $B > 0$, and we obtain a third row containing no negative numbers, then B is an upper bound for the real roots of $f(x) = 0$.

2. If we use synthetic division to divide $f(x)$ by $x - b$, where $b < 0$, and we obtain a third row in which the numbers are alternately positive and negative, then b is a lower bound for the real roots of $f(x) = 0$. (In determining whether the signs alternate in the third row, zeros may be suitably denoted by $+0$ or -0.)

We will prove the first part of this theorem. A proof of the second part can be developed along similar lines. To prove the first part of the theorem, we use the division algorithm to write

$$
f(x) = (x - B) \cdot Q(x) + R \tag{3}
$$

The remainder R here is a constant that may be zero. To show that B is an upper bound, we must show that any number greater than B is not a root. Toward this end, let p be a number that is greater than B. Note that p must be positive, since B is positive. Then with $x = p$, equation (3) becomes

$$f(p) = (p - B) \cdot Q(p) + R \tag{4}$$

We are now going to show that the right-hand side of equation (4) is positive. This will tell us that p is not a root. First, look at the factor $(p - B)$. This is positive, since p is greater than B. Next consider $Q(p)$. By hypothesis, the coefficients of $Q(x)$ are all nonnegative. Furthermore, the leading coefficient of $Q(x)$ is a number a_n that is positive. Since p is also positive, it follows that $Q(p)$ must be positive. Finally, the number R is nonnegative because, in the synthetic division of $f(x)$ by $x - B$, all the numbers in the third row are nonnegative. It now follows that the right-hand side of equation (4) is positive. Consequently, $f(p)$ is not zero and p is not a root of the equation $f(x) = 0$. This is what we wished to show.

EXAMPLE 2   Determine the rational roots, or show that none exist, for the equation

$$\frac{1}{4}x^4 - \frac{3}{4}x^3 + \frac{17}{4}x^2 + 4x + 5 = 0$$

Solution   We will use the rational roots theorem along with the upper and lower bound theorem. First of all, if we are to apply the rational roots theorem, then our equation must have integer coefficients. In view of this, we multiply both sides of the given equation by 4 to obtain

$$x^4 - 3x^3 + 17x^2 + 16x + 20 = 0$$

As in the previous example, we list the factors of a_0, the factors of a_n, and the possibilities for rational roots:

> factors of $a_0 = 20$: $\pm 1, \pm 2, \pm 4, \pm 5, \pm 10, \pm 20$
>
> factors of $a_4 = 1$: ± 1
>
> possible rational roots: $\pm 1, \pm 2, \pm 4, \pm 5, \pm 10, \pm 20$

Our strategy here will be to first check for positive roots, beginning with 1 and working upward. The checks for $x = 1$, $x = 2$, and $x = 4$ are as follows:

```
1| 1  -3   17   16   20      2| 1  -3   17   16    20      4| 1  -3   17    16    20
        1   -2   15   31              2   -2   30    92              4    4    84   400
     1  -2   15   31   51         1  -1   15   46   112         1   1   21   100   420
```

As you can see, none of the remainders here is zero. However, notice that in the division corresponding to $x = 4$, all the numbers appearing in the third row are nonnegative. It therefore follows that 4 is an upper bound for the roots of the given equation. In view of this, we needn't bother to check the remaining values $x = 5$, $x = 10$, and $x = 20$, since none of those can be roots. At this point, we can conclude that the given equation has no positive rational roots.

Next we check for negative rational roots, beginning with -1 and working downward (if necessary). Checking $x = -1$, we have

```
-1| 1  -3   17   16   20
         -1    4  -21    5
      1  -4   21   -5   25
```

Two conclusions can be drawn from this synthetic division. First, $x = -1$ is not a root of the equation. Second, -1 is a lower bound for the roots because the signs in the third row of the synthetic division alternate. This means that we needn't bother to check whether any of the numbers -2, -4, -5, -15, or -20 are roots; none of them can be roots, since they are all less than -1.

Let us summarize our results. We have shown that the given equation has no positive rational roots and no negative rational roots. Furthermore, by inspection we see that zero is not a root of the equation. Thus, the given equation possesses no rational roots.

We conclude this section by demonstrating a method for approximating irrational roots. The method depends on the **location theorem**.

> ### The Location Theorem
>
> Let $f(x)$ be a polynomial, all of whose coefficients are real numbers. If a and b are real numbers such that $f(a)$ and $f(b)$ have opposite signs, then the equation $f(x) = 0$ has at least one real root between a and b.

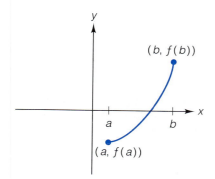

FIGURE 1

Figure 1 indicates why this theorem is plausible. If the point $(a, f(a))$ lies below the x-axis, and $(b, f(b))$ lies above the x-axis, then it certainly seems that the graph of f must cross the x-axis at some point x_0 between a and b. At this intercept, we have $f(x_0) = 0$; that is, x_0 is a root of the equation $f(x) = 0$. (The location theorem is a special case of the so-called intermediate value theorem; the proof is usually discussed in calculus courses.)

Our technique for approximating (or "locating") irrational roots uses the **method of successive approximations**. We will demonstrate this method in Example 3.

EXAMPLE 3

The following equation has exactly one positive root. Locate this root between successive hundredths.

$$f(x) = x^3 + 2x - 4 = 0$$

Solution

First we need to find two numbers a and b such that $f(a)$ and $f(b)$ have opposite signs. By inspection (or by trial and error), we find that $f(1) = -1$ and $f(2) = 8$. Thus, according to the location theorem, the equation $f(x) = 0$ has a real root in the interval $(1, 2)$.

Now that we have located the root between successive integers, we can locate it between successive tenths. We compute $f(1.0)$, $f(1.1)$, $f(1.2)$, and so on, up through $f(2.0)$ if necessary, until we find a sign change. As you can verify, using synthetic division and a calculator, the results are as follows:

$$f(1.0) = -1 \qquad f(1.1) = -0.469$$
$$f(1.2) = 0.128 \quad \longleftarrow \text{ sign change}$$

This shows that the root lies between 1.1 and 1.2.

Having located the root between successive tenths, we follow a similar process to locate the root between successive hundredths. Using synthetic division and a calculator, we obtain these results:

$$f(1.10) = -0.469 \qquad f(1.13) \approx -0.297 \qquad f(1.16) \approx -0.119$$
$$f(1.11) \approx -0.412 \qquad f(1.14) \approx -0.238 \qquad f(1.17) \approx -0.058$$
$$f(1.12) \approx -0.355 \qquad f(1.15) \approx -0.179 \qquad f(1.18) \approx 0.003$$

←—— sign change

This shows that the root lies between 1.17 and 1.18, so we've located the root between successive hundredths, as required. ∎

The procedure described in Example 3 could be continued to yield closer and closer approximations for the required root. We note in passing that such calculations are easily handled with the aid of a programmable calculator.

EXERCISE SET 11.4

A

In Exercises 1–6, list the possibilities for rational roots.

1. $4x^3 - 9x^2 - 15x + 3 = 0$
2. $x^4 - x^3 + 10x^2 - 24 = 0$
3. $8x^5 - x^2 + 9 = 0$
4. $18x^4 - 10x^3 + x^2 - 4 = 0$
5. $\frac{2}{3}x^3 - x^2 - 5x + 2 = 0$
6. $\frac{1}{2}x^4 - 5x^3 + \frac{4}{3}x^2 + 8x - \frac{1}{3} = 0$

In Exercises 7–12, show that each equation has no rational roots.

7. $x^3 - 3x + 1 = 0$ 8. $x^3 + 8x^2 - 1 = 0$
9. $x^3 + x^2 - x + 1 = 0$
10. $x^4 + 4x^3 + 4x^2 - 16 = 0$
11. $12x^4 - x^2 - 6 = 0$
12. $4x^5 - x^4 - x^3 - x^2 + x - 8 = 0$

For Exercises 13–25, find the rational roots of each equation and then solve the equation. (Use the rational roots theorem and the upper and lower bound theorem, as in Example 2.)

13. $x^3 + 3x^2 - x - 3 = 0$
14. $2x^3 - 5x^2 - 3x + 9 = 0$
15. $4x^3 + x^2 - 20x - 5 = 0$
16. $3x^3 - 16x^2 + 17x - 4 = 0$
17. $9x^3 + 18x^2 + 11x + 2 = 0$
18. $4x^3 - 10x^2 - 25x + 4 = 0$
19. $x^4 + x^3 - 25x^2 - x + 24 = 0$
20. $10x^4 + 107x^3 + 301x^2 + 171x + 23 = 0$
21. $x^4 - 4x^3 + 6x^2 - 4x + 1 = 0$

22. $24x^3 - 46x^2 + 29x - 6 = 0$
23. $x^3 - \frac{5}{2}x^2 - 23x + 12 = 0$
24. $x^3 - \frac{17}{3}x^2 - \frac{10}{3}x + 8 = 0$
25. $2x^4 - \frac{9}{10}x^3 - \frac{29}{10}x^2 + \frac{27}{20}x - \frac{3}{20} = 0$

In Exercises 26 and 27, determine integral upper and lower bounds for the real roots of the equations. (Follow the method used within the solution of Example 2.)

26. (a) $x^3 + 2x^2 - 5x + 20 = 0$
    (b) $x^5 - 3x^2 + 100 = 0$
27. (a) $5x^4 - 10x - 12 = 0$
    (b) $3x^4 - 4x^3 + 5x^2 - 2x - 4 = 0$
    (c) $2x^4 - 7x^3 - 5x^2 + 28x - 12 = 0$

28. Referring to equation (1) in this section, multiply out the left-hand side and then clear the equation of fractions. Check that your result agrees with equation (2).

In Exercises 29–34, each equation has exactly one positive root. In each case, locate the root between successive hundredths. Use a calculator.

29. $x^3 + x - 1 = 0$ 30. $x^3 - 2x - 5 = 0$
31. $x^5 - 200 = 0$ 32. $x^3 - 3x^2 + 3x - 26 = 0$
33. $x^3 - 8x^2 + 21x - 22 = 0$
34. $2x^4 - x^3 - 12x^2 - 16x - 8 = 0$

In Exercises 35–38, each equation has exactly one negative root. In each case, use a calculator to locate the root between successive hundredths.

35. $x^3 + x^2 - 2x + 1 = 0$ 36. $x^5 + 100 = 0$
37. $x^3 + 2x^2 + 2x + 101 = 0$
38. $x^4 + 4x^3 - 6x^2 - 8x - 3 = 0$

B

39. This exercise outlines a proof of the rational roots theorem. At one point in the proof, we will need to rely on the following fact, which is proved in courses on number theory.

FACT FROM NUMBER THEORY Suppose that A, B, and C are integers and that A is a factor of the number BC. Then if A has no factor in common with C (other than ± 1), A must be a factor of B.

(a) Let $A = 2$, $B = 8$, and $C = 5$. Verify that the fact from number theory is correct here.

(b) Let $A = 20$, $B = 8$, and $C = 5$. Note that A is a factor of BC, but A is not a factor of B. Why doesn't this contradict the fact from number theory?

(c) Now we're ready to prove the rational roots theorem. We begin with a polynomial equation with integer coefficients:

$$a_n x^n + a_{n-1} x^{n-1} + \cdots + a_1 x + a_0 = 0$$
$$(n \geq 1,\ a_n \neq 0)$$

We assume that the rational number p/q is a root of the equation and that p and q have no common factors other than 1. Why is the following equation now true?

$$a_n\left(\frac{p}{q}\right)^n + a_{n-1}\left(\frac{p}{q}\right)^{n-1} + \cdots + a_1\left(\frac{p}{q}\right) + a_0 = 0$$

(d) Show that the last equation in part (c) can be written

$$p\left(a_n p^{n-1} + a_{n-1} q p^{n-2} + \cdots + a_1 q^{n-1}\right) = -a_0 q^n$$

Since p is a factor of the left-hand side of this last equation, p must also be a factor of the right-hand side. That is, p must be a factor of $a_0 q^n$. But since p and q have no common factors, neither do p and q^n. Our fact from number theory now tells us that p must be a factor of a_0, as we wished to show. (The proof that q is a factor of a_n is carried out in a similar manner.)

40. The location theorem asserts that the polynomial equation $f(x) = 0$ has a root in the open interval (a, b) whenever $f(a)$ and $f(b)$ have unlike signs. If $f(a)$ and $f(b)$ have the same sign, can the equation $f(x) = 0$ have a root between a and b? *Hint:* Look at the graph of $f(x) = x^2 - 2x + 1$ with $a = 0$ and $b = 2$.

Use a calculator for Exercises 41–43. Round off your answers to two decimal places.

41. On the same set of axes, sketch the graphs of $y = x^3$ and $y = 1 - 3x$. Find the x-coordinate of the point where the graphs intersect.

42. On the same set of axes, sketch the graphs of $y = x^3$ and $y = x + 1$. Find the x-coordinate of the point where the graphs intersect.

43. Find the x-coordinate of the point where the curves $y = x^2 - 1$ and $y = \sqrt{x}$ meet.

44. In a note that appeared in *The Two-Year College Mathematics Journal* (vol. 12, 1981, pp. 334–336), Professors Warren Page and Leo Chosid explain how the process of testing for rational roots can be shortened. In essence, their result is as follows. Suppose that we have a polynomial with integer coefficients and we are testing for a possible root p/q. Then, if a noninteger is generated at any point in the synthetic division process, p/q cannot be a root of the polynomial. For example, suppose we want to know if $\frac{4}{3}$ is a root of $6x^4 - 10x^3 + 2x^2 - 9x + 8 = 0$. The first few steps of the synthetic division are as follows.

$$
\begin{array}{r|rrrrr}
\frac{4}{3} & 6 & -10 & 2 & -9 & 8 \\
 & & 8 & -\frac{8}{3} & & \\
\hline
 & 6 & -2 & & &
\end{array}
$$

Since the noninteger $-\frac{8}{3}$ has been generated in the synthetic division process, the process can be stopped; $\frac{4}{3}$ is not a root of the polynomial. Use this idea to shorten your work in testing to see if the numbers $\frac{3}{4}$, $\frac{1}{8}$, and $-\frac{3}{2}$ are roots of the equation $8x^5 - 5x^4 + 3x^2 - 2x - 6 = 0$.

45. In a note that appeared in *The College Mathematics Journal* (vol. 20, 1989, pp. 139–141), Professor Don Redmond proved the following interesting result.

Consider the polynomial equation $f(x) = a_n x^n + a_{n-1} x^{n-1} + \cdots + a_1 x + a_0 = 0$, and suppose that the degree of $f(x)$ is at least two and that all of the coefficients are integers. If the three numbers a_0, a_n, and $f(1)$ are all odd, then the given equation has no rational roots.

Use this result to show that the following equations have no rational roots.
(a) $9x^5 - 8x^4 + 3x^2 - 2x + 27 = 0$
(b) $5x^5 + 5x^4 - 11x^2 - 3x - 25 = 0$

C

In Exercises 46–50, you need to know that a prime number is a positive integer greater than 1 with no factors other than itself and 1. Thus, the first seven prime numbers are 2, 3, 5, 7, 11, 13, and 17.

46. Find all prime numbers p for which the equation $x^2 + x - p = 0$ has a rational root.

47. Find all prime numbers p for which the equation $x^3 + x^2 + x - p = 0$ has at least one rational root. For each value of p that you find, find the corresponding *real* roots of the equation.

48. Consider the equation $x^3 + px - q = 0$, where p and q are prime numbers. Observe that there are only four possible rational roots here: $1, -1, q,$ and $-q$.
  (a) Show that if $x = 1$ is a root, then we must have $q = 3$ and $p = 2$. What are the remaining roots in this case?
  (b) Show that none of the numbers $-1, q,$ and $-q$ can be a root of the equation. *Hint:* For each case, assume the contrary and deduce a contradiction.

49. Consider the equation $x^2 + x - pq = 0$, where p and q are prime numbers. If this equation has rational roots, show that these roots must be -3 and 2. *Suggestion:*

The possible rational roots are $\pm 1, \pm p, \pm q,$ and $\pm pq$. In each case, assume that the given number is a root and see where that leads.

50. If p and q are prime numbers, show that the equation $x^3 + px - pq = 0$ has no rational roots.

51. Find all integral values of b for which the equation $x^3 - b^2 x^2 + 3bx - 4 = 0$ has a rational root.

52. Let $P(a, b)$ be a point on the first-quadrant portion of the curve $y = x^2$ such that the distance of P from the origin is equal to ab. (Assume that $a \neq 0$.)
  (a) Use a calculator and the method demonstrated in this section to find the value of a; round off your answer to two decimal places.
  (b) Find the exact value of a.

11.5 CONJUGATE ROOTS AND DESCARTES'S RULE OF SIGNS

We now proceed to investigate a remarkable theorem, implicit in the work of [Thomas] Harriot [1560–1621] but first used explicitly by Descartes (1637), which limits the number of positive or negative roots of an equation....

Remarkable as the Harriot–Descartes's Rule of Signs is, it still leaves uncertainty as to the exact number of real roots in an equation: it only gives an upper limit to them. The problem of finding an exact test ... was finally solved in 1829 by [Jacques Charles François] Sturm.

H. W. Turnbull in *Theory of Equations* (Edinburgh: Oliver and Boyd, 1939)

An equation can have as many true [positive] roots as it contains changes of sign, from plus to minus or from minus to plus; and as many false [negative] roots as the number of times two plus signs or two minus signs are found in succession.

René Descartes (1637)

As you know from earlier work involving quadratic equations with real coefficients, when nonreal complex roots occur, they occur in conjugate pairs. For instance, as you can check by means of the quadratic formula, the roots of the equation $x^2 - 2x + 5 = 0$ are $1 + 2i$ and $1 - 2i$. The **conjugate roots theorem** tells us that the situation is the same for all polynomial equations with real coefficients.

The Conjugate Roots Theorem

Let $f(x)$ be a polynomial, all of whose coefficients are real numbers. Suppose that $a + bi$ is a root of the equation $f(x) = 0$, where a and b are real and $b \neq 0$. Then $a - bi$ is also a root of the equation.

To prove the conjugate roots theorem, we use four of the properties of complex conjugates listed in Chapter 1:

Property 1: $\overline{z_1 z_2} = \overline{z_1}\,\overline{z_2}$

Property 2: $(\overline{z})^m = \overline{z^m}$

Property 3: $\overline{r} = r$ for every real number r

Property 4: $\overline{z_1} + \overline{z_2} = \overline{z_1 + z_2}$

Using these properties, we can prove the theorem as follows. We begin with a polynomial with real coefficients:

$$f(x) = a_n x^n + a_{n-1} x^{n-1} + \cdots + a_1 x + a_0$$

We must show that if $z = a + bi$ is a root of $f(x) = 0$, then $\overline{z} = a - bi$ is also a root. We have

$$
\begin{aligned}
f(\overline{z}) &= a_n \overline{z}^n + a_{n-1}\overline{z}^{n-1} + \cdots + a_1 \overline{z} + a_0 \\
&= \overline{a_n}\,\overline{z}^n + \overline{a_{n-1}}\,\overline{z}^{n-1} + \cdots + \overline{a_1}\overline{z} + \overline{a_0} && \text{Properties 3 and 2} \\
&= \overline{a_n z^n} + \overline{a_{n-1}z^{n-1}} + \cdots + \overline{a_1 z} + \overline{a_0} && \text{Property 1} \\
&= \overline{a_n z^n + a_{n-1}z^{n-1} + \cdots + a_1 z + a_0} && \text{Property 4} \\
&= \overline{f(z)} = \overline{0} && f(z) = 0, \text{ since } z \text{ is a root} \\
&= 0 && \text{Property 3}
\end{aligned}
$$

We have now shown that $f(\overline{z}) = 0$, given that $f(z) = 0$. Thus, $\overline{z}$ is a root, as we wished to show.

Although the conjugate roots theorem concerns nonreal complex roots, it can nevertheless be used to obtain information about real roots, as the next two examples demonstrate.

EXAMPLE 1 Solve the equation $f(x) = 2x^4 - 3x^3 + 12x^2 + 22x - 60 = 0$, given that one root is $1 + 3i$.

Solution Since all the coefficients of $f(x)$ are real numbers, we know that the conjugate of $1 + 3i$ must also be a root. Thus, $1 + 3i$ and $1 - 3i$ are roots, from which it follows that $[x - (1 + 3i)]$ and $[x - (1 - 3i)]$ are factors of $f(x)$. As you can check, the product of these two factors is $x^2 - 2x + 10$. Thus, we must have

$$f(x) = (x^2 - 2x + 10) \cdot Q(x)$$

for some polynomial $Q(x)$. We compute $Q(x)$ using long division:

$$
\require{enclose}
\begin{array}{r}
2x^2 + x - 6 \\[-3pt]
x^2 - 2x + 10 \enclose{longdiv}{2x^4 - 3x^3 + 12x^2 + 22x - 60} \\
\underline{2x^4 - 4x^3 + 20x^2} \\
x^3 - 8x^2 + 22x \\
\underline{x^3 - 2x^2 + 10x} \\
-6x^2 + 12x - 60 \\
\underline{-6x^2 + 12x - 60} \\
0
\end{array}
$$

Thus, $Q(x) = 2x^2 + x - 6$, and the original equation becomes

$$f(x) = (x^2 - 2x + 10)(2x^2 + x - 6) = 0$$

We can now find any additional roots by solving the equation $2x^2 + x - 6 = 0$. We have

$$2x^2 + x - 6 = 0$$
$$(2x - 3)(x + 2) = 0$$

$$2x - 3 = 0 \quad | \quad x + 2 = 0$$
$$x = \frac{3}{2} \quad | \quad x = -2$$

We now have four distinct roots of the original equation: $1 + 3i$, $1 - 3i$, $\frac{3}{2}$, and -2. Since the degree of the equation is 4, there can be no other roots. ▮▮▮

EXAMPLE 2 Show that the equation $x^3 - 2x^2 + x - 1 = 0$ has at least one irrational root.

Solution Allowing for multiple roots, the given equation has three roots. Therefore, since nonreal complex roots occur in pairs, the equation can have either two nonreal complex roots or no nonreal complex roots. In either case, then, the remaining root or roots must be real. Furthermore, the equation has no rational roots; this is easily checked using the rational roots theorem. It follows now that the equation must have at least one irrational root. (The reasoning here relies on the fact that each real number is either rational or irrational, but not both.) ▮▮▮

There is a theorem, similar to the conjugate roots theorem, that tells us about irrational roots of the form $a + b\sqrt{c}$. As background for this theorem, let us look at two preliminary examples. First, we consider the equation $x^2 - 2x - 5 = 0$. As you can check, the roots in this case are $1 + \sqrt{6}$ and $1 - \sqrt{6}$. However, it is not true in general that irrational roots such as these always occur in pairs. Consider as a second example the quadratic equation

$$(x + 2)(x - \sqrt{3}) = 0$$

or

$$x^2 + (2 - \sqrt{3})x - 2\sqrt{3} = 0$$

Here, one of the roots is $\sqrt{3}$, yet $-\sqrt{3}$ is not a root. This type of behavior can occur in polynomial equations where not all of the coefficients are rational. On the other hand, when the coefficients are all rational, we do have the following theorem. (See Exercise 41 at the end of this section for a proof.)

Theorem

Let $f(x)$ be a polynomial in which all the coefficients are rational. Suppose that $a + b\sqrt{c}$ is a root of the equation $f(x) = 0$, where a, b, and c are rational and $\sqrt{c}$ is irrational. Then $a - b\sqrt{c}$ is also a root of the equation.

EXAMPLE 3 Find a quadratic equation with rational coefficients and a leading coefficient of 1 such that one of the roots is $r_1 = 4 + 5\sqrt{3}$.

Solution   If one root is $r_1 = 4 + 5\sqrt{3}$, then the other is $r_2 = 4 - 5\sqrt{3}$. We denote the required equation by $x^2 + bx + c = 0$. Thus, according to Table 2 in Section 11.3, we have

$$b = -(r_1 + r_2) = -\left[(4 + 5\sqrt{3}) + (4 - 5\sqrt{3})\right] = -8$$

and

$$c = r_1 r_2 = (4 + 5\sqrt{3})(4 - 5\sqrt{3}) = 16 - 75 = -59$$

The required equation is therefore $x^2 - 8x - 59 = 0$. This answer can also be obtained without using the table. Since the roots are $4 \pm 5\sqrt{3}$, we can write the required equation as $\left[x - (4 + 5\sqrt{3})\right]\left[x - (4 - 5\sqrt{3})\right] = 0$. As you can now check by multiplying out the two factors, this equation is equivalent to $x^2 - 8x - 59 = 0$, as obtained previously. ∎

    We conclude this section with a discussion of **Descartes's rule of signs**. This rule, published by Descartes in 1637, provides us with information about the types of roots an equation can have, even before we attempt to solve the equation. In order to state Descartes's rule of signs, we first explain what is meant by a variation in sign in a polynomial with real coefficients. Suppose that $f(x)$ is a polynomial with real coefficients, written in descending (or ascending) powers of x. For example, let $f(x) = 2x^3 - 4x^2 - 3x + 1$. Then we say that there is a **variation in sign** if two successive coefficients have opposite signs. In the case of $f(x) = 2x^3 - 4x^2 - 3x + 1$, there are two variations in sign, the first occurring as we go from 2 to -4 and the second occurring as we go from -3 to 1. In looking for variations in sign, we ignore terms with zero coefficients. Table 1 shows a few more examples of how we count variations in sign.

    We now state Descartes's rule of signs and look at some examples. Although the proof of this theorem is not difficult, it is rather lengthy, and we shall omit it here.

TABLE 1

Polynomial	Number of Variations in Sign
$x^2 + 4x$	0
$-3x^5 + x^2 + 1$	1
$x^3 + 3x^2 - x + 6$	2

Descartes's Rule of Signs

Let $f(x)$ be a polynomial, all of whose coefficients are real numbers, and consider the equation $f(x) = 0$. Then:

a. The number of positive roots either is equal to the number of variations in sign of $f(x)$ or is less than that by an even integer.

b. The number of negative roots either is equal to the number of variations in sign of $f(-x)$ or is less than that by an even integer.

EXAMPLE 4   Use Descartes's rule of signs to obtain information regarding the roots of the equation $x^3 + 8x + 5 = 0$.

Solution   Let $f(x) = x^3 + 8x + 5$. Then, since there are no variations in sign for $f(x)$, we see from part (a) of Descartes's rule that the given equation has no positive roots. Next we compute $f(-x)$ to learn about the possibilities for negative roots: we have $f(-x) = -x^3 - 8x + 5$. So $f(-x)$ has one sign change, and consequently [by part (b) of Descartes's rule] the original equation has one negative root. Furthermore, notice that zero is not a root of the equation. Thus, the equation has only one real root, a negative root. Since the equation has a total of three roots, we can conclude that we have one negative root and two nonreal complex roots. The two nonreal roots will be complex conjugates. ■■■

EXAMPLE 5 Use Descartes's rule to obtain information regarding the roots of the equation $x^4 + 3x^2 - 7x - 5 = 0$.

Solution Let $f(x) = x^4 + 3x^2 - 7x - 5$. Then $f(x)$ has one variation in sign. So according to part (a) of Descartes's rule, the equation has one positive root. That leaves us three roots still to account for, since the degree of the equation is 4. We have $f(-x) = x^4 + 3x^2 + 7x - 5$. Since $f(-x)$ has one sign change, we know from part (b) of Descartes's rule that the equation has one negative root. Noting now that zero is not a root, we conclude that the two remaining roots must be nonreal complex roots. In summary, then, the equation has one positive root, one negative root, and two nonreal complex (conjugate) roots. ■■■

EXAMPLE 6 Use Descartes's rule to obtain information regarding the roots of the equation $f(x) = x^3 - x^2 + 3x + 2 = 0$.

Solution Since $f(x)$ has two variations in sign, the given equation has either two positive roots or no positive roots. To see how many negative roots are possible, we compute

$$f(-x) = (-x)^3 - (-x)^2 + 3(-x) + 2$$
$$= -x^3 - x^2 - 3x + 2$$

Since $f(-x)$ has one variation in sign, we conclude from part (b) of Descartes's rule that the equation has exactly one negative root. In summary, then, there are two possibilities:

either: one negative root and two positive roots;

or: one negative root and two nonreal complex roots. ■■■

By using Descartes's rule in Examples 4 and 5, we were able to determine the exact numbers of positive roots and negative roots for the given equations. As Example 6 indicates, however, there are cases in which a direct application of Descartes's rule provides several distinct possibilities for the types of roots, rather than a single definitive result. (Exercises 55 and 56 in the Review Exercises for this chapter illustrate a technique that is sometimes useful in gaining additional information from Descartes's rule. In particular, Exercise 56 will show you that the equation in Example 6 has no positive roots.)

EXERCISE SET 11.5

A

In Exercises 1–16, an equation is given, followed by one or more roots of the equation. In each case, determine the remaining roots.

1. $x^2 - 14x + 53 = 0$; $x = 7 - 2i$
2. $x^2 - x - \frac{1535}{4} = 0$; $x = \frac{1}{2} + 8\sqrt{6}$
3. $x^3 - 13x^2 + 59x - 87 = 0$; $x = 5 + 2i$
4. $x^4 - 10x^3 + 30x^2 - 10x - 51 = 0$; $x = 4 + i$
5. $x^4 + 10x^3 + 38x^2 + 66x + 45 = 0$; $x = -2 + i$
6. $2x^3 + 11x^2 + 30x - 18 = 0$; $x = -3 - 3i$
7. $4x^3 - 47x^2 + 232x + 61 = 0$; $x = 6 - 5i$

8. $9x^4 + 18x^3 + 20x^2 - 32x - 64 = 0$; $x = -1 + \sqrt{3}\,i$

9. $4x^4 - 32x^3 + 81x^2 - 72x + 162 = 0$; $x = 4 + \sqrt{2}\,i$

10. $2x^4 - 17x^3 + 137x^2 - 57x - 65 = 0$; $x = 4 - 7i$

11. $x^4 - 22x^3 + 140x^2 - 128x - 416 = 0$; $x = 10 + 2i$

12. $4x^4 - 8x^3 + 24x^2 - 20x + 25 = 0$; $x = \frac{1}{2}(1 + 3i)$

13. $15x^3 - 16x^2 + 9x - 2 = 0$; $x = \frac{1}{3}(1 + \sqrt{2}i)$

14. $x^5 - 5x^4 + 30x^3 + 18x^2 + 92x - 136 = 0$;
$x = -1 + i\sqrt{3},\ x = 3 - 5i$

15. $x^7 - 3x^6 - 4x^5 + 30x^4 + 27x^3 - 13x^2 - 64x + 26 = 0$;
$x = 3 - 2i,\ x = -1 + i,\ x = 1$

16. $x^6 - 2x^5 - 2x^4 + 2x^3 + 2x + 1 = 0$; $x = 1 + \sqrt{2}$

In Exercises 17–20, find a quadratic equation with rational coefficients, one of whose roots is the given number. Write your answer so that the coefficient of x^2 is 1. Use either of the methods shown in Example 3.

17. $r_1 = 1 + \sqrt{6}$ 18. $r_1 = 2 - \sqrt{3}$

19. $r_1 = \frac{1}{3}(2 + \sqrt{10})$ 20. $r_1 = \frac{1}{2} + \frac{1}{4}\sqrt{5}$

In Exercises 21–36, use Descartes's rule of signs to obtain information regarding the roots of the equations.

21. $x^3 + 5 = 0$ 22. $x^4 + x^2 + 1 = 0$

23. $2x^5 + 3x + 4 = 0$ 24. $x^3 + 8x - 3 = 0$

25. $5x^4 + 2x - 7 = 0$ 26. $x^3 - 4x^2 + x - 1 = 0$

27. $x^3 - 4x^2 - x - 1 = 0$

28. $x^8 + 4x^6 + 3x^4 + 2x^2 + 5 = 0$

29. $3x^8 + x^6 - 2x^2 - 4 = 0$

30. $12x^4 - 5x^3 - 7x^2 - 4 = 0$

31. $x^9 - 2 = 0$ 32. $x^9 + 2 = 0$

33. $x^8 - 2 = 0$ 34. $x^8 + 2 = 0$

35. $x^6 + x^2 - x - 1 = 0$ 36. $x^7 + x^2 - x - 1 = 0$

B

37. Consider the equation $x^4 + cx^2 + dx - e = 0$, where c, d, and e are positive. Show that the equation has one positive root, one negative root, and two nonreal complex roots.

38. Consider the equation $x^n - 1 = 0$.
(a) Show that the equation has $n - 2$ nonreal complex roots when n is even.

(b) How many nonreal complex roots are there when n is odd?

39. Find the polynomial $f(x)$ of lowest degree with rational coefficients and with leading coefficient 1, such that $\sqrt{3} + 2i$ is a root of the equation $f(x) = 0$.

40. Find a quadratic polynomial $f(x)$ with integer coefficients, such that $(2 + \sqrt{3})/(2 - \sqrt{3})$ is a root of the equation $f(x) = 0$. *Hint:* First rationalize the given root.

C

41. Let $f(x)$ be a polynomial, with rational coefficients. Suppose that $a + b\sqrt{c}$ is a root of $f(x) = 0$, where a, b, and c are rational and $\sqrt{c}$ is irrational. Complete the following steps to prove that $a - b\sqrt{c}$ is also a root of the equation $f(x) = 0$.
(a) If $b = 0$, we're done. Why?
(b) (From now on we'll assume that $b \neq 0$.) Let
$d(x) = [x - (a + b\sqrt{c})][x - (a - b\sqrt{c})]$. Explain why $d(a + b\sqrt{c}) = 0$.
(c) Verify that $d(x) = (x - a)^2 - b^2 c$. Thus, $d(x)$ is a quadratic polynomial with rational coefficients.
(d) Now suppose that we use the long division process to divide the polynomial $f(x)$ by the quadratic polynomial $d(x)$. We'll obtain a quotient $Q(x)$ and a remainder. Since the degree of $d(x)$ is 2, our remainder will be of degree 1 or less. In other words, the general form of this remainder will be $Cx + D$. Furthermore, C and D will have to be rational, because all of the coefficients in $f(x)$ and in $d(x)$ are rational. In summary, we have the identity

$$f(x) = d(x) \cdot Q(x) + (Cx + D)$$

Now make the substitution $x = a + b\sqrt{c}$ in this identity, and conclude that $C = D = 0$.
(e) Using the result in part (d), we have

$$f(x) = [x - (a + b\sqrt{c})][x - (a - b\sqrt{c})] \cdot Q(x)$$

Let $x = a - b\sqrt{c}$ in this last identity and conclude that $a - b\sqrt{c}$ is a root of $f(x) = 0$, as required.

42. Find a polynomial $f(x)$ with integer coefficients, such that one root of $f(x) = 0$ is $x = \sqrt{2} + \sqrt[3]{2}$.

CHAPTER ELEVEN   SUMMARY OF PRINCIPAL TERMS AND THEOREMS

TERMS OR THEOREM	PAGE REFERENCE	COMMENT
1. Division algorithm	659	This theorem summarizes the results of the long division process for polynomials. Suppose that $p(x)$ and $d(x)$ are polynomials and $d(x)$ is not the zero polynomial. Then according to the *division algorithm,* there are unique polynomials $q(x)$ and $R(x)$ such that $$p(x) = d(x) \cdot q(x) + R(x)$$ where either $R(x)$ is the zero polynomial or the degree of $R(x)$ is less than the degree of $d(x)$.
2. Root, solution, zero	665	A *root,* or *solution,* of a polynomial equation $f(x) = 0$ is a number r such that $f(r) = 0$. The root r is also called a *zero* of the function f.
3. The remainder theorem	666	The *remainder theorem* asserts that when a polynomial $f(x)$ is divided by $x - r$, the remainder is $f(r)$. Example 3 in Section 11.2 shows how this theorem can be used with synthetic division to evaluate a polynomial.
4. The factor theorem	667	The *factor theorem* makes two statements about a polynomial $f(x)$. First, if $f(r) = 0$, then $x - r$ is a factor of $f(x)$. And second, if $x - r$ is a factor of $f(x)$, then $f(r) = 0$.
5. The fundamental theorem of algebra	673	Let $f(x)$ be a polynomial of degree 1 or greater. The *fundamental theorem of algebra* asserts that the equation $f(x) = 0$ has at least one root among the complex numbers. (See Example 1 in Section 11.3.)
6. The linear factors theorem	674	Let $f(x) = a_n x^n + a_{n-1} x^{n-1} + \cdots + a_0$, where $n \geq 1$ and $a_n \neq 0$. Then this theorem asserts that $f(x)$ can be expressed as a product of n linear factors: $$f(x) = a_n(x - r_1)(x - r_2) \cdots (x - r_n)$$ (The complex numbers $r_1, r_2, \ldots, r_n$ are not necessarily all distinct, and some or all of them may be real numbers.) On page 676, the linear factors theorem is used to prove that every polynomial equation of degree $n \geq 1$ has exactly n roots, where a root of multiplicity k is counted k times.
7. The rational roots theorem	681	Given a polynomial equation, this theorem tells us which rational numbers are candidates for roots of the equation. For a statement of the theorem, see page 681. A proof of the theorem is outlined in Exercise 39, Exercise Set 11.4. For an example of how the theorem is applied, see Example 1 in Section 11.4.
8. Upper bound (for roots); lower bound (for roots)	682	A real number B is an upper bound for the roots of an equation if every real root is less than or equal to B. Similarly, a real number b is a lower bound for the roots if every real root is greater than or equal to b.

TERMS OR THEOREM	PAGE REFERENCE	COMMENT
9. The upper and lower bound theorem for roots	682	This theorem tells how synthetic division can be used in determining upper and lower bounds for roots of equations. For the statement and proof of the theorem, see page 682 in Section 11.4. For a demonstration of how the theorem is applied, see Example 2 in Section 11.4.
10. The location theorem	684	Let $f(x)$ be a polynomial with real coefficients. If a and b are real numbers such that $f(a)$ and $f(b)$ have opposite signs, then the equation $f(x) = 0$ has at least one root between a and b. To see how this theorem is applied, see Example 3 in Section 11.4.
11. The conjugate roots theorem	687	Let $f(x)$ be a polynomial with real coefficients, and suppose that $a + bi$ is a root of the equation $f(x) = 0$, where a and b are real numbers and $b \neq 0$. Then this theorem asserts that $a - bi$ is also a root of the equation. (In other words, for polynomial equations in which all of the coefficients are real numbers, when complex nonreal roots occur, they occur in conjugate pairs.) For illustrations of how this theorem is applied, see Examples 1 and 2 in Section 11.5.
12. Variation in sign	690	Suppose that $f(x)$ is a polynomial with real coefficients, written in descending or ascending powers of x. Then a variation in sign occurs whenever two successive coefficients have opposite signs. For examples, see Table 1 in Section 11.5.
13. Descartes's rule of signs	690	Let $f(x)$ be a polynomial, all of whose coefficients are real numbers, and consider the equation $f(x) = 0$. Then, according to Descartes's rule: **a.** The number of positive roots either is equal to the number of variations in sign of $f(x)$ or is less than that by an even integer. **b.** The number of negative roots either is equal to the number of variations in sign of $f(-x)$ or is less than that by an even integer. Examples 4–6 in Section 11.5 show how Descartes's rule is applied.

WRITING MATHEMATICS

Discuss each of the following statements with a classmate and decide whether it is true or false. Then (on your own), write out the reason (or reasons) for each decision.

1.   Every equation has a root.
2.   Every polynomial equation of degree 4 has four distinct roots.
3.   No cubic equation can have a root of multiplicity 4.
4.   The degree of the polynomial $x(x - 1)(x - 2)(x - 3)$ is 3.
5.   The degree of the polynomial $6(x + 1)^2(x - 5)^4$ is 6.

6. Every polynomial of degree n, where $n \geq 1$, can be written in the form
$(x - r_1)(x - r_2) \cdots (x - r_n)$.

7. The sum of the roots of the polynomial equation $x^2 - px + q = 0$ is p.

8. The product of the roots of the polynomial equation $2x^3 - x^2 + 3x - 1 = 0$ is 1.

9. Although a polynomial equation of degree n may have n distinct roots, the fundamental theorem of algebra tells us how to find only one of the roots.

10. Every polynomial equation of degree $n \geq 1$ has at least one real root.

11. If all the coefficients in a polynomial equation are real, then at least one of the roots must be real.

12. Every cubic equation whose roots are $\sqrt{5}$, $\sqrt{6}$, and $\sqrt{7}$ can be written in the form

$$a_n(x - \sqrt{5})(x - \sqrt{6})(x - \sqrt{7}) = 0$$

13. Every polynomial equation of degree at least 1 has at least one real root.

14. According to the rational roots theorem, there are four possibilities for the rational roots of the equation $x^5 + 6x^2 - 2 = 0$.

15. According to the rational roots theorem, there are only two possibilities for the rational roots of the equation $\frac{1}{3}x^3 - 5x + 1 = 0$.

16. The sum of the roots of the equation $x^2 - 12x + 16 = 0$ is -12.

17. According to the location theorem, if $f(x)$ is a polynomial and $f(a)$ and $f(b)$ have the same sign, then the equation $f(x) = 0$ has at least one root between a and b.

18. According to Descartes's rule, the equation $x^7 - 4x + 3 = 0$ has two positive roots.

CHAPTER ELEVEN   REVIEW EXERCISES

In Exercises 1 and 2, you are given polynomials $p(x)$ and $d(x)$. In each case, use synthetic division to determine polynomials $q(x)$ and $R(x)$ such that

$$p(x) = d(x) \cdot q(x) + R(x)$$

where either $R(x) = 0$ or the degree of $R(x)$ is less than the degree of $d(x)$.

1. $p(x) = x^4 + 3x^3 - x^2 - 5x + 1$; $d(x) = x + 2$

2. $p(x) = 4x^4 + 2x + 1$; $d(x) = x - 2$

In Exercises 3–8, use synthetic division to find the quotients and the remainders.

3. $\dfrac{x^4 - 2x^2 + 8}{x - 3}$

4. $\dfrac{x^3 - 1}{x - 2}$

5. $\dfrac{2x^3 - 5x^2 - 6x - 3}{x + 4}$

6. $\dfrac{x^3 + x - 3\sqrt{2}}{x - \sqrt{2}}$

7. $\dfrac{5x^2 - 19x - 4}{x + 0.2}$

8. $\dfrac{x^3 - 3a^2x^2 - 4a^4x + 9a^6}{x - a^2}$

In Exercises 9–16, use synthetic division and the remainder theorem to find the indicated values of the functions. Use a calculator for Exercises 15 and 16.

9. $f(x) = x^5 - 10x + 4$; $f(10)$

10. $f(x) = x^4 + 2x^3 - x$; $f(-2)$

11. $f(x) = x^3 - 10x^2 + x - 1$; $f\left(\frac{1}{10}\right)$

12. $f(x) = x^4 - 2a^2x^2 + 3a^3x - a^4$; $f(-a)$

13. $f(x) = x^3 + 3x^2 + 3x + 1$; $f(a - 1)$

14. $f(x) = x^3 - 1$; $f(1.1)$

15. $f(x) = x^4 + 4x^3 - 6x^2 - 8x - 2$;
    (a) $f(-0.3)$ (Round off the result to two decimal places.)
    (b) $f(-0.39)$ (Round off the result to three decimal places.)
    (c) $f(-0.394)$ (Round off the result to five decimal places.)

16. $f(-4.907378)$, where f is the function in Exercise 15. (Round off the result to three decimal places.)

17. Find a value for a such that 3 is a root of the equation $x^3 - 4x^2 - ax - 6 = 0$.

18. For which values of b will -1 be a root of the equation $x^3 + 2b^2x^2 + x - 48 = 0$?

19. For which values of a will $x - 1$ be a factor of the polynomial $a^2x^3 + 3ax^2 + 2$?

20. Use synthetic division to verify that $\sqrt{2} - 1$ is a root of the equation $x^6 + 14x^3 - 1 = 0$.

21. Let $f(x) = ax^3 + bx^2 + cx + d$ and suppose that r is a root of the equation $f(x) = 0$.
    (a) Show that $r - h$ is a root of the equation $f(x + h) = 0$.
    (b) Show that $-r$ is a root of the equation $f(-x) = 0$.
    (c) Show that kr is a root of the equation $f(x/k) = 0$.

22. Suppose that r is a root of the equation $a_2x^2 + a_1x + a_0 = 0$. Show that mr is a root of the quadratic equation $a_2x^2 + ma_1x + m^2a_0 = 0$.

In Exercises 23–28, list the possibilities for the rational roots of the equations.

23. $x^5 - 12x^3 + x - 18 = 0$

24. $x^5 - 12x^3 + x - 17 = 0$

25. $2x^4 - 125x^3 + 3x^2 - 8 = 0$

26. $\frac{3}{5}x^3 - 8x^2 - \frac{1}{2}x + \frac{3}{2} = 0$

27. $x^3 + x - p = 0$, where p is a prime number.

28. $x^3 + x - pq = 0$, where both p and q are prime numbers.

In Exercises 29–36, each equation has at least one rational root. Solve the equations. *Suggestion:* Use the upper and lower bound theorem to eliminate some of the possibilities for rational roots.

29. $2x^3 + x^2 - 7x - 6 = 0$

30. $x^3 + 6x^2 - 8x - 7 = 0$

31. $2x^3 - x^2 - 14x + 10 = 0$

32. $2x^3 + 12x^2 + 13x + 15 = 0$

33. $\frac{3}{2}x^3 + \frac{1}{2}x^2 + \frac{1}{2}x - 1 = 0$

34. $x^4 - 2x^3 - 13x^2 + 38x - 24 = 0$

35. $x^5 + x^4 - 14x^3 - 14x^2 + 49x + 49 = 0$

36. $8x^5 + 12x^4 + 14x^3 + 13x^2 + 6x + 1 = 0$

37. Solve the equation $x^3 - 9x^2 + 24x - 20 = 0$, using the fact that one of the roots has multiplicity 2.

38. One root of the equation $x^2 + kx + 2k = 0$ ($k \neq 0$) is twice the other. Find k and find the roots of the equation.

39. State each of the following theorems.
    (a) The division algorithm
    (b) The remainder theorem

(c) The factor theorem
(d) The fundamental theorem of algebra

40. Find a quadratic equation with roots $a - \sqrt{a^2 - 1}$ and $a + \sqrt{a^2 - 1}$ ($a > 1$).

In Exercises 41–44, write each polynomial in the form
$$a_n(x - r_1)(x - r_2) \cdots (x - r_n)$$

41. $6x^2 + 7x - 20$

42. $x^2 + x - 1$

43. $x^4 - 4x^3 + 5x - 20$

44. $x^4 - 4x^2 - 5$

Each of Exercises 45–48 gives an equation, followed by one or more roots. Solve the equation.

45. $x^3 - 7x^2 + 25x - 39 = 0$, $x = 2 - 3i$

46. $x^3 + 6x^2 - 24x + 160 = 0$; $x = 2 + 2i\sqrt{3}$

47. $x^4 - 2x^3 - 4x^2 + 14x - 21 = 0$; $x = 1 + i\sqrt{2}$

48. $x^5 + x^4 - x^3 + x^2 + x - 1 = 0$; $x = \frac{1}{2}(1 + i\sqrt{3})$, $x = \frac{1}{2}(-1 - \sqrt{5})$

In Exercises 49–54, use Descartes's rule of signs to obtain information regarding the roots of the equations.

49. $x^3 + 8x - 7 = 0$

50. $3x^4 + x^2 + 4x - 2 = 0$

51. $x^3 + 3x + 1 = 0$

52. $2x^6 + 3x^2 + 6 = 0$

53. $x^4 - 10 = 0$

54. $x^4 + 5x^2 - x + 2 = 0$

55. Consider the equation $x^3 + x^2 + x + 1 = 0$.
    (a) Use Descartes's rule to show that the equation has either one or three negative roots.
    (b) Now show that the equation cannot have three negative roots. *Hint:* Multiply both sides of the equation by $x - 1$. Then simplify the left-hand side and reapply Descartes's rule to the new equation.
    (c) Actually, the original equation can be solved using only the basic algebraic techniques discussed in Chapters 1 and 2. Solve the equation in this manner.

56. Use Descartes's rule to show that the equation $x^3 - x^2 + 3x + 2 = 0$ has no positive roots. *Hint:* Multiply both sides of the equation by $x + 1$ and apply Descartes's rule to the resulting equation.

57. Let P be the point in the first quadrant where the curve $y = x^3$ intersects the circle $x^2 + y^2 = 1$. Using a calculator, locate the x-coordinate of P within successive hundredths.

58. Let P be the point in the first quadrant where the parabola $y = 4 - x^2$ intersects the curve $y = x^3$. Using a calculator, locate the x-coordinate of P within successive hundredths.

59. Consider the equation $x^3 - 36x - 84 = 0$.
    (a) Use Descartes's rule to check that this equation has exactly one positive root.

(b) Use the upper and lower bound theorem to show that 7 is an upper bound for the positive root.

(c) Using a calculator, locate the positive root within successive hundredths,

60. Consider the equation $x^3 - 3x + 1 = 0$.

(a) Use Descartes's rule to check that this equation has exactly one negative root.

(b) Use the upper and lower bound theorem to show that -2 is a lower bound for the negative root.

(c) Using a calculator, locate the negative root within successive hundredths.

In Exercises 61–64, find polynomial equations that have integer coefficients and the given values as roots.

61. $4 - \sqrt{5}$

62. $a + b$ and $a - b$ (a and b are integers)

63. $6 - 2i$ and $\sqrt{5}$

64. $\dfrac{5 + \sqrt{6}}{5 - \sqrt{6}}$ *Hint:*   First rationalize the expression.

65. Find a fourth-degree polynomial equation with integer coefficients, such that $x = 1 + \sqrt{2} + \sqrt{3}$ is a root.

Hint: Begin by writing the given relationship as $x - 1 = \sqrt{2} + \sqrt{3}$; then square both sides.

66. Find a cubic equation with integer coefficients, such that $x = 1 + \sqrt[3]{2}$ is a root. Is $1 - \sqrt[3]{2}$ also a root of the equation?

In Exercises 67–71, first determine the zeros of each function; then sketch the graph.

67. $y = x^3 - 2x^2 - 3x$
68. $y = x^4 + 3x^3 + 3x^2 + x$
69. $y = x^4 - 4x^2$
70. $y = x^3 + 6x^2 + 5x - 12$
71. $y = x^5 - 8x^4 + 25x^3 - 38x^2 + 28x - 8$

In Exercises 72 and 73, you are given a binomial expression of the form $x + r$ or $x - r$, followed by a polynomial $f(x)$. In each case, determine whether the binomial expression is a factor of $f(x)$.

72. (a) $x - 1$; $f(x) = x^4 + 3x^3 - x^2 + 3x - 18$
    (b) $x + 3$; $f(x)$ as in part (a)

73. (a) $x - \sqrt{3}$; $f(x) = x^6 - 3x^4 - 16x^2 + 48$
    (b) $x + \sqrt{3}$; $f(x)$ as in part (a)
    (c) $x - 2$; $f(x)$ as in part (a)
    (d) $x + 2$; $f(x)$ as in part (a)

OPTIONAL TI-81 GRAPHING CALCULATOR EXERCISES

EXERCISES FOR CHAPTER ELEVEN

In Exercises 1–9, you are given a polynomial equation $f(x) = 0$. In parentheses following each equation is the example number and the section in which that equation was considered. For each equation $f(x) = 0$, (a) graph the corresponding function $y = f(x)$, and (b) use the ZOOM and TRACE keys to determine (or at least approximate) the x-intercepts of the graph. These x-intercepts are the roots of the equation $f(x) = 0$. In each case, check that your values are consistent with the results in the text.

1. $x^4 + x^2 - 6 = 0$ (Example 1 in 11.2)
2. $2x^3 - 5x^2 + x - 6 = 0$ (Example 3 in 11.2)
3. $x^3 - 2x + 1 = 0$ (Example 5 in 11.2)
4. $x^4 + 2x^3 - 7x^2 - 20x - 12 = 0$ (Example 6 in 11.2)
5. $2x^3 - x^2 - 9x - 4 = 0$ (Example 1 in 11.4)
6. $\frac{1}{4}x^4 - \frac{3}{4}x^3 + \frac{17}{4}x^2 + 4x + 5 = 0$ (Example 2 in 11.4)
7. $x^3 + 2x - 4 = 0$ (Example 3 in 11.4)
8. $2x^4 - 3x^3 + 12x^2 + 22x - 60 = 0$ (Example 1 in 11.5)
9. $x^3 - 2x^2 + x - 1 = 0$ (Example 2 in 11.5)

10. According to Descartes's rule, the equation $x^3 + 8x + 5 = 0$ has one negative root and two nonreal complex roots. (See Example 4 in Section 11.5.) Verify this result by graphing the function $f(x) = x^3 + 8x + 5$.

11. According to Descartes's rule, the equation $x^4 + 3x^2 - 7x - 5 = 0$ has one negative root, one positive root, and two nonreal complex roots. (See Example 5 in Section 11.5.) Verify this result by graphing the function $f(x) = x^4 + 3x^2 - 7x - 5$.

12. Consider the equation $x^3 - x^2 + 3x + 2 = 0$. In Example 6 of Section 11.5 we found that there were two possible cases for the roots:

    Case 1:   one negative root and two positive roots

    Case 2:   one negative root and two nonreal complex roots

(a) By graphing the function $f(x) = x^3 - x^2 + 3x + 2$, determine which of the two cases, in fact, describes the roots.

(b) Use the ZOOM and the TRACE keys to approximate the real root(s).

CHAPTER ELEVEN TEST

1. Let $f(x) = 6x^4 - 5x^3 + 7x^2 - 2x - 2$. Make use of the remainder theorem and synthetic division to compute $f(\frac{1}{2})$.

2. Solve the equation $x^3 + x^2 - 11x - 15 = 0$, given that one of the roots is -3.

3. List the possibilities for the rational roots of the equation $2x^5 - 4x^3 + x - 6 = 0$.

4. Find a quadratic function with zeros 1 and -8 and with a y-intercept of -24.

5. Use synthetic division to divide $4x^3 + x^2 - 8x + 3$ by $x + 1$.

6. (a) State the factor theorem.
   (b) State the fundamental theorem of algebra.

7. (a) The equation $x^3 - 2x^2 - 1 = 0$ has just one positive root. Use the upper and lower bound theorem to determine the smallest integer that is an upper bound for that root.
   (b) Locate the root between successive tenths. (Use a calculator.)

8. Solve the equation $x^5 - 6x^4 + 11x^3 + 16x^2 - 50x + 52 = 0$, given that two of the roots are $1 + i$ and $3 - 2i$.

9. Let $p(x) = x^4 + 2x^3 - x + 6$ and $d(x) = x^2 + 1$. Find polynomials $q(x)$ and $R(x)$ such that $p(x) = d(x) \cdot q(x) + R(x)$.

10. Express the polynomial $2x^2 - 6x + 5$ in the factored form $a_n(x - r_1)(x - r_2)$.

11. Consider the equation $x^4 - x^3 + 24 = 0$.
    (a) List the possibilities for rational roots.
    (b) Use the upper and lower bound theorem to show that 2 is an upper bound for the roots.
    (c) In view of parts (a) and (b), what possibilities now remain for positive rational roots?
    (d) Which (if any) of the possibilities in part (c) are actually roots?

12. (a) Find the rational roots of the cubic equation $2x^3 - x^2 - x - 3 = 0$.
    (b) Find all solutions of the equation in part (a).

13. Use Descartes's rule of signs to obtain information regarding the roots of the following equation: $3x^4 + x^2 - 5x - 1 = 0$.

14. Find a cubic polynomial $f(x)$ with integer coefficients, such that $1 - 3i$ and -2 are roots of the equation $f(x) = 0$.

15. Find a polynomial $f(x)$ with leading coefficient 1, such that the equation $f(x) = 0$ has the following roots and no other:

Root	Multiplicity
2	1
3i	3
$1 + \sqrt{2}$	2

Write your answer in the form

$$a_n(x - r_1)(x - r_2) \cdots (x - r_n)$$

THE CONIC SECTIONS

The Greeks knew the properties of the curves given by cutting a cone with a plane — the ellipse, the parabola and hyperbola. Kepler discovered by analysis of astronomical observations, and Newton proved mathematically . . . that the planets move in ellipses. The geometry of Ancient Greece thus became the cornerstone of modern astronomy.

John Lighton Synge (1897 – 1987)

Navigator, lay in a conic section flight path to the cloud's center.

Captain James T. Kirk in *Star Trek, The Motion Picture*

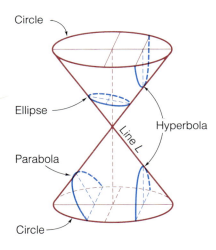

The **conic sections** are the curves formed when a plane intersects the surface of a right circular cone. As indicated in the figure in the margin, these curves are the circle, the ellipse, the hyperbola, and the parabola.

The study of the conics dates back over 2000 years to Ancient Greece, where Apollonius of Perga (262 – 190 B.C.) wrote an eight-volume treatise on the subject. However, until the seventeenth century, the conic sections were studied only as a portion of pure (as opposed to applied) mathematics. Then, in the seventeenth century, it was discovered that the conic sections were crucial in expressing some of the most important laws of nature. This is essentially the observation that is made in the opening quotation by the physicist J. L. Synge.

12.1 THE PARABOLA

In Section 5.2, we saw that the graph of a quadratic function $y = ax^2 + bx + c$ is a symmetric U-shaped curve called a parabola. In this section, we give a more general definition of the parabola, a definition that emphasizes the geometric properties of the curve.

The Parabola

A **parabola** is the set of all points in the plane equally distant from a fixed line and a fixed point not on the line. the fixed line is called the **directrix**. The fixed point is called the **focus**.

Let us initially suppose that the focus of the parabola is the point $(0, p)$ and the directrix is the line $y = -p$. We will assume throughout this section that p is positive. To understand the geometric content of the definition of the parabola, the special graph paper displayed in Figure 1(a) is useful. The common center of the concentric circles in Figure 1(a) is the focus $(0, p)$. Thus, all the points on a given circle are at a fixed distance from the focus. The radii of the circles increase in increments of p units. Similarly, the broken horizontal lines in the figure are drawn at intervals that are multiples of p units from the directrix $y = -p$. By considering the points where the circles intersect the horizontal lines, we can find a number of points equally distant from the focus $(0, p)$ and the directrix $y = -p$ [see Figure 1(b)].

FIGURE 1

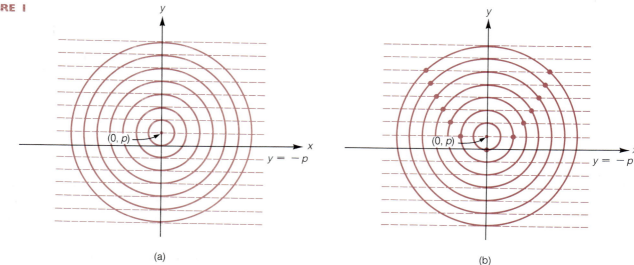

(a)

(b)

The graphical method just described lets us locate many points equally distant from the focus $(0, p)$ and the directrix $y = -p$. Figure 1(b) shows that the points on the parabola are symmetric about a line, in this case the y-axis. Also, by studying the figure, you should be able to convince yourself that in this case there can be no points below the x-axis that satisfy the stated condition. However, to completely describe the required set of points, and to show that our new definition is consistent with the old, Figure 1(b) is inadequate. We need to bring algebraic methods to bear on the problem. Thus, let d_1 denote the distance from the point $P(x, y)$ to the focus $(0, p)$ and let d_2 denote the distance from $P(x, y)$ to the directrix $y = -p$, as shown in Figure 2. The distance d_1 is then

$$d_1 = \sqrt{(x - 0)^2 + (y - p)^2} = \sqrt{x^2 + y^2 - 2py + p^2}$$

The distance d_2 in Figure 2 is just the vertical distance between the points P and Q. Thus,

$$d_2 = (y\text{-coordinate of } P) - (y\text{-coordinate of } Q)$$
$$= y - (-p) = y + p$$

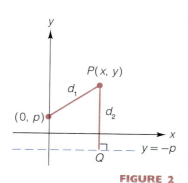

FIGURE 2

(Absolute value signs are unnecessary here for, as noted earlier, P cannot lie below the x-axis.) The condition that P be equally distant from the focus $(0, p)$ and the directrix $y = -p$ can be expressed by the equation

$$d_1 = d_2$$

Using the expressions we've found for d_1 and d_2, this last equation becomes

$$\sqrt{x^2 + y^2 - 2py + p^2} = y + p$$

We can obtain a simpler but equivalent equation by squaring both sides. (Two nonnegative quantities are equal if and only if their squares are equal.) Thus,

$$x^2 + y^2 - 2py + p^2 = y^2 + 2py + p^2$$

After combining like terms, this equation becomes

$$x^2 = 4py$$

This is the equation of a parabola with focus $(0, p)$ and directrix $y = -p$. In the box that follows, we summarize the properties of the parabola $x^2 = 4py$. As Figure 3(b) indicates, the terminology we've introduced applies equally well to an arbitrary parabola for which the axis of symmetry is not necessarily parallel to one of the coordinate axes and the vertex is not necessarily the origin.

PROPERTY SUMMARY   THE PARABOLA

1. A **parabola** is the set of points equidistant from a fixed line called the **directrix** and a fixed point, not on the line, called the **focus**.

2. The **axis** of a parabola is the line drawn through the focus and perpendicular to the directrix.

3. The **vertex** of a parabola is the point where the parabola intersects its axis. The vertex is located halfway between the focus and the directrix. See Figure 3.

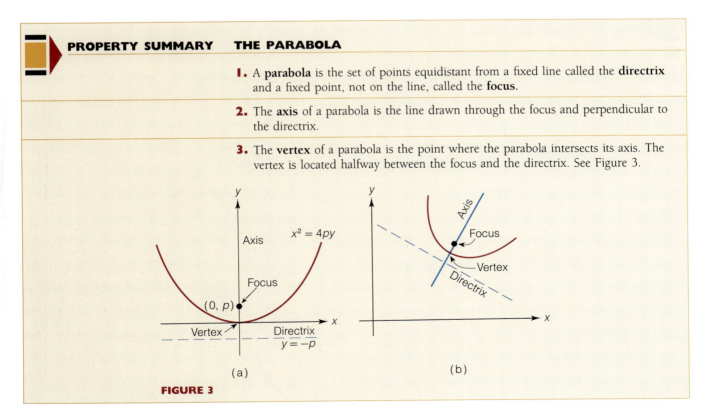

(a)

(b)

FIGURE 3

EXAMPLE 1   Determine the equation of the parabola in Figure 4, given that the curve passes through the point $(3, 5)$. Specify the focus and the directrix.

Solution   We have just seen that the general equation for a parabola in this position is $x^2 = 4py$. Since the point $(3, 5)$ lies on the curve, its coordinates must satisfy the equation $x^2 = 4py$. Thus,

$$3^2 = 4p(5)$$
$$9 = 20p$$
$$\frac{9}{20} = p$$

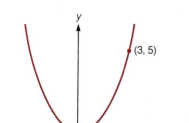

FIGURE 4

With $p = \frac{9}{20}$, the equation $x^2 = 4py$ becomes

$$x^2 = 4\left(\frac{9}{20}\right)y = \frac{9}{5}y \qquad \text{or} \qquad x^2 = \frac{9}{5}y \qquad \text{as required}$$

Furthermore, since $p = \frac{9}{20}$, the focus is $\left(0, \frac{9}{20}\right)$ and the directrix is $y = -\frac{9}{20}$.

We have seen that the equation of a parabola with focus $(0, p)$ and directrix $y = -p$ is $x^2 = 4py$. By following the same method, we can obtain general equations for parabolas with other orientations. The basic results are summarized in Figure 5.

PROPERTY SUMMARY   BASIC EQUATIONS FOR THE PARABOLA

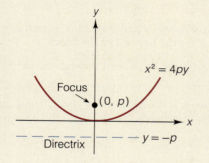

(a) $x^2 = 4py$

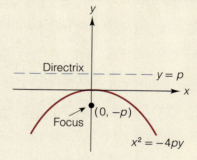

(b) $x^2 = -4py$

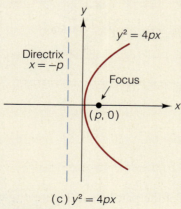

(c) $y^2 = 4px$

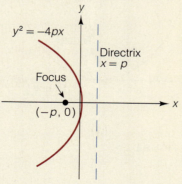

(d) $y^2 = -4px$

FIGURE 5

By a **chord** of a parabola, we mean a straight line segment joining any two points on the curve. If the chord passes through the focus, it is called a **focal chord**. For purposes of graphing, it is useful to know the length of the focal chord perpendicular to the axis of the parabola. This is the length of the horizontal line segment $\overline{AB}$ in Figure 6 and the vertical line segment $\overline{A'B'}$ in Figure 7. We will call this length the **focal width**.

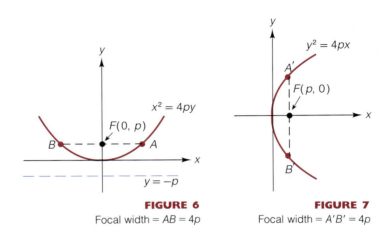

FIGURE 6
Focal width = $AB = 4p$

FIGURE 7
Focal width = $A'B' = 4p$

In Figure 6, the distance from A to F is the same as the distance from A to the line $y = -p$. (Why?) But the distance from A to the line $y = -p$ is $2p$. Therefore, $AF = 2p$ and AB is twice this, or $4p$. We have shown that the focal width of the parabola is $4p$. In other words, given a parabola $x^2 = 4py$, the width at its focus is $4p$, the coefficient of y. In the same way, the length of the focal chord $\overline{A'B'}$ in Figure 7 is also $4p$, the coefficient of x in that case.

EXAMPLE 2   Find the focus and directrix of the parabola $y^2 = -4x$, and sketch the graph.

Solution   Comparing the basic equation $y^2 = -4px$ [in Figure 5(d)] with the equation at hand, we see that $4p = 4$ and thus $p = 1$. The focus is therefore $(-1, 0)$, and the directrix is $x = 1$. The basic form of the graph will be as in Figure 5(d). For purposes of graphing, we note that the focal width is 4 (the absolute value of the coefficient of x). This, along with the fact that the vertex is $(0, 0)$, gives us enough information to draw the graph (see Figure 8).

In Examples 1 and 2 and in Figure 5, the vertex of each parabola is located at the origin. Now we want to consider parabolas that are translated (shifted) from this standard position. As background for this, we review and generalize the results about translation in Section 4.3. Consider, as an example, the two equations

$$y = (x - 1)^2 \qquad \text{and} \qquad y = x^2 + 1$$

As we saw in Section 4.3, the graphs of these two equations can each be obtained by translating the graph of $y = x^2$. To graph $y = (x - 1)^2$, translate the graph of $y = x^2$ one unit to the right; to graph $y = x^2 + 1$, translate $y = x^2$ one unit up.

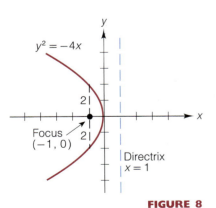

FIGURE 8

To see the underlying pattern here, rewrite the equation $y = x^2 + 1$ as $y - 1 = x^2$. Then the situation is this:

EQUATION	HOW GRAPH OBTAINED
$y = (x - 1)^2$	translation in positive x-direction
$y - 1 = x^2$	translation in positive y-direction

Observation   *In the equation $y = x^2$, the effect of replacing x with $x - 1$ is to translate the graph one unit in the positive x-direction. Similarly, replacing y with $y - 1$ translates the graph one unit in the positive y-direction.*

As a second example before we generalize, consider the equations

$$y = (x + 1)^2 \qquad \text{and} \qquad y = x^2 - 1$$

The first equation involves a translation in the negative x-direction; the second involves a translation in the negative y-direction. As before, to see the underlying pattern, we rewrite the second equation as $y + 1 = x^2$. Now we have the following situation.

EQUATION	HOW GRAPH OBTAINED
$y = (x + 1)^2$	translation in negative x-direction
$y + 1 = x^2$	translation in negative y-direction

Observation   *In the equation $y = x^2$, the effect of replacing x with $x + 1$ is to translate the graph one unit in the negative x-direction. Similarly, replacing y with $y + 1$ translates the graph one unit in the negative y-direction.*

Both of the examples that we have just considered are specific instances of the following basic result. (The result is valid whether or not the given equation and graph represent a function.)

PROPERTY SUMMARY   TRANSLATION AND COORDINATES

Suppose that we have an equation that determines a graph in the x-y plane, and let h and k denote positive numbers. Then, replacing x with $x - h$ or $x + h$, or replacing y with $y - k$ or $y + k$, has the following effects on the graph of the original equation.

REPLACEMENT	RESULTING TRANSLATION
1. x replaced with $x - h$	h units in the positive x-direction
2. y replaced with $y - k$	k units in the positive y-direction
3. x replaced with $x + h$	h units in the negative x-direction
4. y replaced with $y + k$	k units in the negative y-direction

---

EXAMPLE 3   Graph the parabola $(y + 1)^2 = -4(x - 2)$. Specify the vertex, the focus, the directrix, and the axis of symmetry.

Solution   The given equation is obtained from $y^2 = -4x$ (which we graphed in the previous example) by replacing x and y with $x - 2$ and $y + 1$, respectively. So the required graph is obtained by translating the parabola in Figure 8 two units to the right and one unit down. In particular, this means that the vertex moves from $(0,0)$ to $(2,-1)$; the focus moves from $(-1,0)$ to $(1,-1)$; the directrix moves from $x = 1$ to $x = 3$; and the axis of symmetry moves from $y = 0$ (which is the x-axis) to $y = -1$. The required graph is shown in Figure 9. ■■■

FIGURE 9

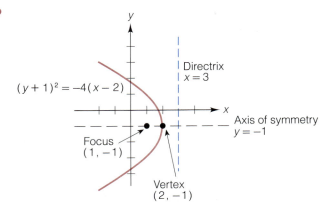

Directrix
$x = 3$

$(y + 1)^2 = -4(x - 2)$

Axis of symmetry
$y = -1$

Focus
$(1, -1)$

Vertex
$(2, -1)$

EXAMPLE 4   Graph the parabola $2y^2 - 4y - x + 5 = 0$, and specify each of the following: vertex, focus, directrix, axis of symmetry, and focal width.

Solution   Just as we did in Section 5.2, we use the technique of completing the square:

$$2(y^2 - 2y \quad) = x - 5$$
$$2(y^2 - 2y + 1) = x - 5 + 2 \qquad \text{adding 2 to both sides}$$
$$(y - 1)^2 = \tfrac{1}{2}(x - 3)$$

The graph of this last equation is obtained by translating the graph of $y^2 = \tfrac{1}{2}x$ "right 3, up 1." This moves the vertex from $(0,0)$ to $(3,1)$. Now, for $y^2 = \tfrac{1}{2}x$, the focus and directrix are determined by setting $4p = \tfrac{1}{2}$. Therefore $p = \tfrac{1}{8}$, and consequently the focus and directrix of $y^2 = \tfrac{1}{2}x$ are $\left(\tfrac{1}{8},0\right)$ and $x = -\tfrac{1}{8}$, respectively. Thus, the focus and directrix of the translated curve are $\left(3\tfrac{1}{8},1\right)$ and $x = 2\tfrac{7}{8}$, respectively. Figure 10(a) shows the graph of $y^2 = \tfrac{1}{2}x$, and Figure 10(b) shows the

FIGURE 10

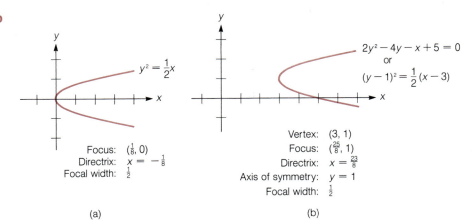

$y^2 = \tfrac{1}{2}x$

Focus: $\left(\tfrac{1}{8}, 0\right)$
Directrix: $x = -\tfrac{1}{8}$
Focal width: $\tfrac{1}{2}$

(a)

$2y^2 - 4y - x + 5 = 0$
or
$(y - 1)^2 = \tfrac{1}{2}(x - 3)$

Vertex: $(3, 1)$
Focus: $\left(\tfrac{25}{8}, 1\right)$
Directrix: $x = \tfrac{23}{8}$
Axis of symmetry: $y = 1$
Focal width: $\tfrac{1}{2}$

(b)

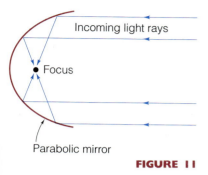

Incoming light rays

Focus

Parabolic mirror

FIGURE 11

translated graph. (You should check for yourself that the information accompanying Figure 10(b) is correct.)

Many applications of the parabola in the sciences are due to its *reflection property*. Figure 11 shows a cross section of a parabolic mirror in a telescope. As indicated in Figure 11, light rays coming in parallel to the axis of the parabola are reflected through the focus. (The word *focus* comes from a Latin word meaning "fireplace.") Parabolic reflectors are also used in communication and surveillance systems, in radio telescopes, and in automobile headlights. In an automobile headlight, the bulb is located at the focus. To diagram the parabolic mirror of an automobile headlight, just reverse the directions of the arrows in Figure 11.

EXERCISE SET 12.1

A

In Exercises 1–22, graph the parabolas. In each case, specify the focus, the directrix, and the focal width. For Exercises 13–22, also specify the vertex.

1. $x^2 = 4y$
2. $x^2 = 16y$
3. $y^2 = -8x$
4. $y^2 = 12x$
5. $x^2 = -20y$
6. $x^2 - y = 0$
7. $y^2 + 28x = 0$
8. $4y^2 + x = 0$
9. $x^2 = 6y$
10. $y^2 = -10x$
11. $4x^2 = 7y$
12. $3y^2 = 4x$
13. $y^2 - 6y - 4x + 17 = 0$
14. $y^2 + 2y + 8x + 17 = 0$
15. $x^2 - 8x - y + 18 = 0$
16. $x^2 + 6y + 18 = 0$
17. $y^2 + 2y - x + 1 = 0$
18. $2y^2 - x + 1 = 0$
19. $2x^2 - 12x - y + 18 = 0$
20. $y + \sqrt{2} = (x - 2\sqrt{2})^2$
21. $2x^2 - 16x - y + 33 = 0$
22. $\frac{1}{4}y^2 - y - x + 1 = 0$

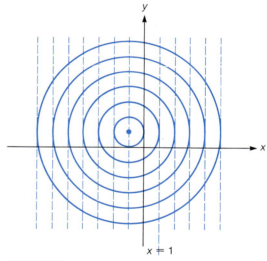

$x = 1$

FIGURE A

B

For Exercises 23 and 24, refer to the following figure.

23. Make a photocopy of Figure A. Then in your copy of Figure A, indicate (by drawing dots) eleven points that are equidistant from the point $(-1, 1)$ and the line $x = 1$. What is the equation of the line of symmetry for the set of dots?

24. The eleven dots that you located in Exercise 23 are part of a parabola. Find the equation of that parabola and sketch its graph. Specify the vertex, the focus and directrix, and the focal width.

For Exercises 25–30, find the equation of the parabola satisfying the given conditions. In each case, assume that the vertex is at the origin.

25. The focus is $(0, 3)$.
26. The directrix is $y - 8 = 0$.
27. The directrix is $x + 32 = 0$.
28. The focus lies on the y-axis, and the parabola passes through the point $(7, -10)$.
29. The parabola is symmetric about the x-axis, the x-coordinate of the focus is negative, and the length of the focal chord perpendicular to the x-axis is 9.

30. The focus is the smaller of the two x-intercepts of the circle $x^2 - 8x + y^2 - 6y + 9 = 0$.

31. Let P denote the point $(8, 8)$ on the parabola $x^2 = 8y$, and let $\overline{PQ}$ be a focal chord.
    (a) Find the equation of the line through the point $(8, 8)$ and the focus.
    (b) Find the coordinates of Q.
    (c) Find the length of $\overline{PQ}$.
    (d) Find the equation of the circle with this focal chord as a diameter.
    (e) Show that the circle determined in part (d) intersects the directrix of the parabola in only one point. Conclude from this that the directrix is tangent to the circle. Draw a sketch of the situation.

32. The following figure shows a parabolic cross section of a reflecting mirror. Find the distance from the vertex of the parabola to the focus. *Hint:* Choose a convenient coordinate system.

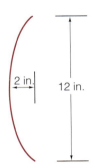

2 in.   12 in.

33. An arch is in the shape of a parabola with a vertical axis. The arch is 15 ft high at the center and 40 ft wide at the base. At what height above the base is the width 20 ft? *Hint:* Choose a convenient coordinate system in which the vertex of the parabola is at the origin.

34. The following figure depicts the cable of a suspension bridge. The cable is in the form of a parabola with a vertical axis. The horizontal highway on the bridge is 300 ft long. The longest of the vertical supporting wires is 100 ft; the shortest is 40 ft. Find the length of a supporting wire that is 50 ft from the middle.

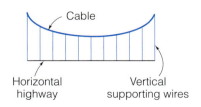

Cable

Horizontal   Vertical
highway   supporting wires

35. The segments $\overline{AA'}$ and $\overline{BB'}$ are focal chords of the parabola $x^2 = 2y$. The coordinates of A and B are $(4, 8)$ and $(-2, 2)$, respectively.
    (a) Find the equation of the line through A and B'.
    (b) Find the equation of the line through B and A'.
    (c) Show that the two lines you have found intersect at a point on the directrix.

36. Let $\overline{PQ}$ be the horizontal focal chord of the parabola $x^2 = 8y$. Let R denote the point where the directrix of the parabola meets the y-axis. Show that $\overline{PR}$ is perpendicular to $\overline{QR}$.

37. Suppose $\overline{PQ}$ is a focal chord of the parabola $y = x^2$ such that the coordinates of P are $(2, 4)$.
    (a) Find the coordinates of Q.
    (b) Find the coordinates of M, the midpoint of $\overline{PQ}$.
    (c) A perpendicular is drawn from M to the y-axis, meeting the y-axis at S. Also, a line perpendicular to the focal chord is drawn through M, meeting the y-axis at T. Find ST and verify that it is equal to one-half the focal width of the parabola.

38. Let F be the focus of the parabola $x^2 = 8y$, and let P denote the point on the parabola with coordinates $(8, 8)$. Let $\overline{PQ}$ be a focal chord. If V denotes the vertex of the parabola, verify that
$$PF \cdot FQ = VF \cdot PQ$$

39. In the following figure, triangle OAB is equilateral and $\overline{AB}$ is parallel to the x-axis. Find the length of a side and the area of triangle OAB.

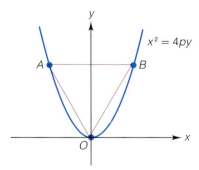

$x^2 = 4py$

40. If $\overline{PQ}$ is a focal chord of the parabola $x^2 = 4py$ and the coordinates of P are (x_0, y_0), show that the coordinates of Q are
$$\left(\frac{-4p^2}{x_0}, \frac{p^2}{y_0} \right)$$

41. If $\overline{PQ}$ is a focal chord of the parabola $y^2 = 4px$ and the coordinates of P are (x_0, y_0), show that the coordinates of Q are

$$\left(\frac{p^2}{x_0}, \frac{-4p^2}{y_0} \right)$$

42. Let F and V denote the focus and the vertex, respectively, of the parabola $x^2 = 4py$. If $\overline{PQ}$ is a focal chord of the parabola, show that

$$PF \cdot FQ = VF \cdot PQ$$

43. Let $\overline{PQ}$ be a focal chord of the parabola $y^2 = 4px$, and let M be the midpoint of $\overline{PQ}$. A perpendicular is drawn from M to the x-axis, meeting the x-axis at S. Also from M, a line segment is drawn that is perpendicular to $\overline{PQ}$ and that meets the x-axis at T. Show that the length of $\overline{ST}$ is one-half the focal width of the parabola.

44. Let $\overline{AB}$ be a chord (not necessarily a focal chord) of the parabola $y^2 = 4px$, and suppose that $\overline{AB}$ subtends a right angle at the vertex. (In other words, $\angle AOB = 90°$, where O is the origin in this case.) Find the x-intercept of the segment $\overline{AB}$. What is surprising about the result?   *Hint:*   Begin by writing the coordinates of A and B as $A\left(\frac{a^2}{4p}, a \right)$ and $B\left(\frac{b^2}{4p}, b \right)$.

12.2   THE ELLIPSE

We have here apparently [in the work of Anthemius of Tralles (a sixth-century Greek architect and mathematician)] the first mention of the construction of an ellipse by means of a string stretched tight round the foci.

Sir Thomas Heath in *A History of Greek Mathematics*, vol. II (Oxford: The Clarendon Press, 1921)

The true orbit of Mars was even less of a circle than the Earth's. It took almost two years for Kepler to realize that its orbital shape is that of an ellipse. An ellipse is the shape of a circle when viewed at an angle.

Phillip Flower in *Understanding the Universe* (St. Paul: West Publishing Company, 1990)

The heavenly motions are nothing but a continuous song for several voices, to be perceived by the intellect, not by the ear.

Johannes Kepler (1571–1630)

In this section, we discuss the symmetric oval-shaped curve known as the *ellipse*. As Kepler discovered, and Newton later proved, this is the curve described by the planets in their motions around the sun.

> ### The Ellipse
>
> An **ellipse** is the set of all points in the plane, the sum of whose distances from two fixed points is constant. Each fixed point is called a **focus** (plural: **foci**).

Subsequently, we will derive an equation describing the ellipse, just as we found an equation for the parabola in the previous section. But first let us consider some rather immediate consequences of the definition. In fact, we can learn a great deal about the ellipse even before we derive its equation.

There is a simple mechanical method for constructing an ellipse that arises

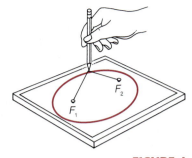

FIGURE 1

directly from the definition of the curve. Mark the given foci, say F_1 and F_2, on a drawing board and insert thumbtacks at those points. Now take a piece of string whose length is greater than the distance from F_1 to F_2 and tie the ends of the string to the tacks. Next, pull the string taut with a pencil point. Then if you move the pencil while keeping the string taut, the curve traced out will be an ellipse, as indicated in Figure 1. The reason the curve is an ellipse is that for each point on the curve, the sum of the distances from the foci is constant, the constant being the length of the string. By actually carrying out this construction for yourself several times, each time varying the distance between the foci or the length of the string, you can learn a great deal about the ellipse. For instance, when the distance between the foci is small compared to the length of the string, the ellipse begins to resemble a circle, as in Figure 2. On the other hand, when the distance between the foci is nearly equal to the length of the string, the ellipse becomes relatively flat, as in Figure 3.

We now derive one of the standard forms for the equation of an ellipse. As indicated in Figure 4, we assume that the foci are $F_1\ (-c, 0)$ and $F_2(c, 0)$ and that the sum of the distances from a point $P(x, y)$ on the ellipse to the foci is $2a$. Since the point P lies on the ellipse, we have

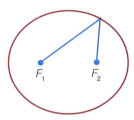

FIGURE 2
When the distance between the foci is small compared to the length of the string, the ellipse resembles a circle.

$$F_1P + F_2P = 2a$$

and therefore

$$\sqrt{(x + c)^2 + (y - 0)^2} + \sqrt{(x - c)^2 + (y - 0)^2} = 2a$$

or

$$\sqrt{(x - c)^2 + y^2} = 2a - \sqrt{(x + c)^2 + y^2} \tag{1}$$

Now, by following a straightforward but lengthy process of squaring and simplifying (as outlined in detail in Exercise 35), we find that equation (1) becomes

$$(a^2 - c^2)x^2 + a^2y^2 = a^2(a^2 - c^2) \tag{2}$$

To write equation (2) in a more symmetric form, we define the positive number b by the equation

$$b^2 = a^2 - c^2 \tag{3}$$

FIGURE 3
When the distance between the foci is nearly equal to the length of the string, the ellipse is relatively flat.

Note For this definition to make sense, we need to know that the right-hand side of equation (3) is positive. (See Exercise 36 at the end of this section for details.)

FIGURE 4

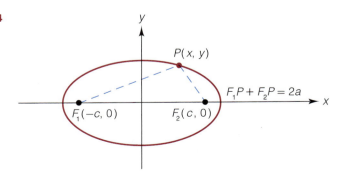

Finally, using equation (3) to substitute for $a^2 - c^2$ in equation (2), we obtain

$$b^2x^2 + a^2y^2 = a^2b^2$$

which can be written

$$\frac{x^2}{a^2} + \frac{y^2}{b^2} = 1 \qquad (a > b) \tag{4}$$

We have now shown that the coordinates of each point on the ellipse satisfy equation (4). Conversely, it can be shown that if x and y satisfy equation (4), then the point (x, y) indeed lies on the ellipse. We refer to equation (4) as the **standard form** for the equation of an ellipse with foci $(-c, 0)$ and $(c, 0)$. For an ellipse in this form, it will always be the case that a is greater than b; this follows from equation (3).

For purposes of graphing, we want to know the intercepts of the ellipse. To find the x-intercepts, we set $y = 0$ in equation (4) to obtain

$$\frac{x^2}{a^2} = 1$$

$$x^2 = a^2 \qquad \text{or} \qquad x = \pm a$$

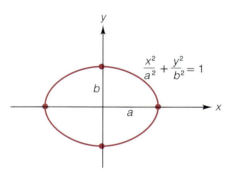

FIGURE 5
The intercepts of the ellipse $(x^2/a^2) + (y^2/b^2) = 1$ are $x = \pm a$ and $y = \pm b$

The x-intercepts are therefore a and $-a$. In a similar fashion, you can check that the y-intercepts are b and $-b$. Also (according to the symmetry tests in Section 3.2), note that the graph of equation (4) must be symmetric about both coordinate axes. Figure 5 shows the graph of the ellipse $(x^2/a^2) + (y^2/b^2) = 1$. (Calculator exercises at the end of this section will help convince you that the general shape of the curve in Figure 5 is correct.)

In the box that follows (at the top of page 711), we define several terms that are useful in describing and analyzing the ellipse.

The eccentricity (as defined in the box) provides a numerical measure of how much the ellipse deviates from being a circle. As Figure 7 indicates, the closer the eccentricity is to zero, the more the ellipse resembles a circle. In the other direction, as the eccentricity approaches 1, the ellipse becomes increasingly flat.

In the box at the bottom of page 711, we summarize our discussion up to this point. (In the box, note that we've used the letter e to denote the eccentricity. This is the conventional choice, even though the same letter is used with a very different meaning in connection with exponential functions and logarithms.)

EXAMPLE 1 Find the lengths of the major and minor axes of the ellipse $9x^2 + 16y^2 = 144$. Specify the coordinates of the foci and the eccentricity. Graph the ellipse.

Solution To convert the equation $9x^2 + 16y^2 = 144$ to standard form, divide both sides by 144. This yields

$$\frac{9x^2}{144} + \frac{16y^2}{144} = \frac{144}{144}$$

$$\frac{x^2}{16} + \frac{y^2}{9} = 1$$

$$\frac{x^2}{4^2} + \frac{y^2}{3^2} = 1$$

DEFINITION Terminology for the Ellipse

1. The **focal axis** is the line passing through the foci of the ellipse.

2. The **center** is the point midway between the foci. This is the point C in Figure 6.

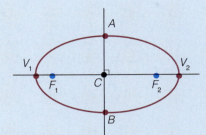

FIGURE 6

3. The **vertices** (singular: **vertex**) are the two points where the focal axis meets the ellipse. These are the points V_1 and V_2 in Figure 6.

4. The **major axis** is the line segment joining the vertices. This is the line segment $\overline{V_1 V_2}$ in Figure 6. The **minor axis** is the line segment through the center of the ellipse, perpendicular to the major axis and with endpoints on the ellipse. This is the segment $\overline{AB}$ in Figure 6.

5. The **eccentricity** is the ratio $\dfrac{c}{a} = \dfrac{\sqrt{a^2 - b^2}}{a}$.

FIGURE 7

Eccentricity is a number between 0 and 1 that describes the shape of an ellipse. The narrowest ellipse in this figure has the same proportions as the orbit of Halley's comet. By way of contrast, the eccentricity of Earth's orbit is 0.0017; if an ellipse with this eccentricity were included in Figure 7, it would appear indistinguishable from a circle.

Eccentricity = 0.4

Eccentricity = 0.8

Eccentricity = 0.967

PROPERTY SUMMARY THE ELLIPSE $\dfrac{x^2}{a^2} + \dfrac{y^2}{b^2} = 1$ $\quad (a > b)$

Foci: $(\pm c, 0)$, where $c^2 = a^2 - b^2$

Center: $(0, 0)$

Vertices: $(\pm a, 0)$

Length of major axis: $2a$

Length of minor axis: $2b$

Eccentricity: $e = \dfrac{c}{a} = \dfrac{\sqrt{a^2 - b^2}}{a}$

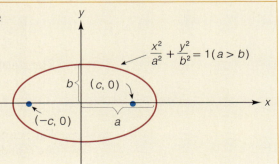

This is the standard form. Comparing this equation with $(x^2/a^2) + (y^2/b^2) = 1$, we see that $a = 4$ and $b = 3$. Thus, the major and minor axes are 8 and 6 units, respectively. Next, to determine the foci, we use the equation $c^2 = a^2 - b^2$. We have

$$c^2 = 4^2 - 3^2 = 7$$
$$c = \sqrt{7}$$

(We choose the positive square root because $c > 0$.) It follows that the coordinates of the foci are $\left(-\sqrt{7}, 0\right)$ and $\left(\sqrt{7}, 0\right)$. We now calculate the eccentricity using the formula $e = c/a$. This yields $e = \sqrt{7}/4$. Figure 8 shows the graph along with the required information. ∎

FIGURE 8

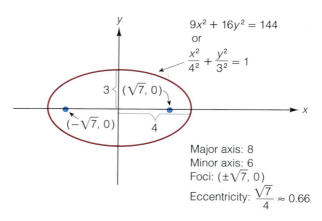

$9x^2 + 16y^2 = 144$
or
$$\frac{x^2}{4^2} + \frac{y^2}{3^2} = 1$$

Major axis: 8
Minor axis: 6
Foci: $(\pm\sqrt{7}, 0)$
Eccentricity: $\dfrac{\sqrt{7}}{4} \approx 0.66$

EXAMPLE 2   The foci of an ellipse are $(-1, 0)$ and $(1, 0)$. The eccentricity is $\frac{1}{3}$. Find the equation of the ellipse (in standard form), and specify the lengths of the major and minor axes.

Solution   Since the foci are $(\pm 1, 0)$, we have $c = 1$. Using the equation $e = c/a$ with $e = \frac{1}{3}$ and $c = 1$, we obtain

$$\frac{1}{3} = \frac{1}{a} \qquad \text{and therefore} \qquad a = 3$$

Recall now that b^2 is defined by the equation $b^2 = a^2 - c^2$. In view of this, we have

$$b^2 = 3^2 - 1^2 = 8$$
$$b = \sqrt{8} = 2\sqrt{2}$$

The equation of the ellipse in standard form is therefore

$$\frac{x^2}{3^2} + \frac{y^2}{(2\sqrt{2})^2} = 1$$

Furthermore, since $a = 3$ and $b = 2\sqrt{2}$, the lengths of the major and minor axes are 6 and $4\sqrt{2}$ units, respectively. ∎

EXAMPLE 3 The eccentricity of an ellipse is $\frac{4}{5}$, and the sum of the distances from a point P on the ellipse to the foci is 10 units. Compute the distance between the foci F_1 and F_2. See Figure 9.

Solution We are required to find the distance between the foci F_1 and F_2. By definition, this is the quantity $2c$. Since the sum of the distances from a point on the ellipse to the foci is 10 units, we have

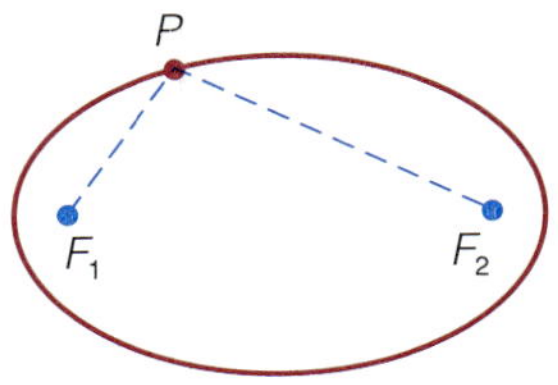

$$2a = 10 \qquad \text{by definition of } 2a$$
$$a = 5$$

Now we substitute the values $a = 5$ and $e = \frac{4}{5}$ in the formula $e = c/a$:

$$\frac{4}{5} = \frac{c}{5} \qquad \text{and therefore} \qquad c = 4$$

It now follows that the required distance $2c$ is 8 units.

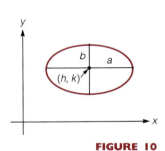

$$e = \frac{4}{5}$$
$$F_1P + F_2P = 10$$

FIGURE 9

Suppose now that we translate the graph of $(x^2/a^2) + (y^2/b^2) = 1$ by h units in the x-direction and k units in the y-direction, as shown in Figure 10. Then the equation of the translated ellipse is

$$\frac{(x - h)^2}{a^2} + \frac{(y - k)^2}{b^2} = 1 \tag{5}$$

Equation (5) is another **standard form** for the equation of an ellipse. As indicated in Figure 10, the center of this ellipse is the point (h, k). In the next example, we use the technique of completing the square to convert an equation for an ellipse to standard form. Once the equation is in standard form, the graph is readily obtained.

FIGURE 10

EXAMPLE 4 Determine the center, foci, and eccentricity of the ellipse

$$4x^2 + 9y^2 - 8x - 54y + 49 = 0$$

Graph the ellipse.

Solution We will convert the given equation to standard form by using the technique of completing the square. We have

$$4x^2 - 8x + 9y^2 - 54y = -49$$
$$4(x^2 - 2x) + 9(y^2 - 6y) = -49$$
$$4(x^2 - 2x + 1) + 9(y^2 - 6y + 9) = -49 + (4)(1) + 9(9)$$
$$4(x - 1)^2 + 9(y - 3)^2 = 36$$

Dividing this last equation by 36, we obtain

$$\frac{(x - 1)^2}{9} + \frac{(y - 3)^2}{4} = 1$$

or

$$\frac{(x - 1)^2}{3^2} + \frac{(y - 3)^2}{2^2} = 1$$

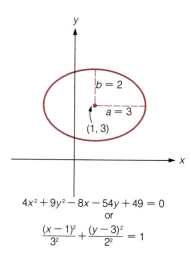

$4x^2 + 9y^2 - 8x - 54y + 49 = 0$

or

$$\frac{(x-1)^2}{3^2} + \frac{(y-3)^2}{2^2} = 1$$

FIGURE 11

This last equation represents an ellipse with center at $(1, 3)$ and with $a = 3$ and $b = 2$. We can calculate c using the formula $c^2 = a^2 - b^2$:

$$c^2 = 3^2 - 2^2 = 5 \qquad \text{and therefore} \qquad c = \sqrt{5}$$

Since the center of this ellipse is $(1, 3)$, the foci are therefore $\left(1 + \sqrt{5}, 3\right)$ and $\left(1 - \sqrt{5}, 3\right)$. Finally the eccentricity is c/a, which is $\sqrt{5}/3$. Figure 11 shows the graph of this ellipse. ∎

In developing the equation $(x^2/a^2) + (y^2/b^2) = 1$, we assumed that the foci of the ellipse were located on the x-axis at the points $(-c, 0)$ and $(c, 0)$. If, instead, the foci are located on the y-axis at the points $(0, c)$ and $(0, -c)$, then the same method we used in the previous case will show the equation of the ellipse to be

$$\frac{x^2}{b^2} + \frac{y^2}{a^2} = 1 \qquad (a > b)$$

We still assume that $2a$ represents the sum of the distances from a point on the ellipse to the foci. In the box that follows, we summarize the situation for the ellipse with foci $(0, c)$ and $(0, -c)$.

PROPERTY SUMMARY   **THE ELLIPSE** $\dfrac{x^2}{b^2} + \dfrac{y^2}{a^2} = 1$ $(a > b)$

Foci: $(0, \pm c)$, where $c^2 = a^2 - b^2$

Center: $(0, 0)$

Vertices: $(0, \pm a)$

Length of major axis: $2a$

Length of minor axis: $2b$

Eccentricity: $e = \dfrac{c}{a} = \dfrac{\sqrt{a^2 - b^2}}{a}$

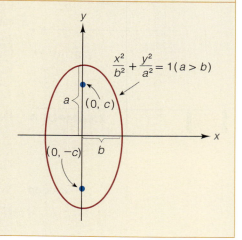

We now have two standard forms for the ellipse:

Foci on the x-axis at $(\pm c, 0)$: $\dfrac{x^2}{a^2} + \dfrac{y^2}{b^2} = 1$ $(a > b)$

Foci on the y-axis at $(0, \pm c)$: $\dfrac{x^2}{b^2} + \dfrac{y^2}{a^2} = 1$ $(a > b)$

Because a is greater than b in both standard forms, it is always easy to determine by inspection whether the foci lie on the x-axis or the y-axis. Consider, for instance, the equation $(x^2/5^2) + (y^2/6^2) = 1$. In this case, since $6 > 5$, we have $a = 6$. And since 6^2 appears under y^2, we conclude that the foci lie on the y-axis.

EXAMPLE 5   The point $(5, 3)$ lies on an ellipse with vertices $\left(0, \pm 2\sqrt{21}\right)$. Find the equation of the ellipse. Write the answer in both standard form and the form $Ax^2 + By^2 = C$.

Solution   Since the vertices are $\left(0, \pm 2\sqrt{21}\right)$, the standard form in this case is $(x^2/b^2) + (y^2/a^2) = 1$. Furthermore, in view of the coordinates of the vertices, we have $a = 2\sqrt{21}$. Therefore,

$$\frac{x^2}{b^2} + \frac{y^2}{84} = 1$$

Now, since the point $(5, 3)$ lies on the ellipse, its coordinates must satisfy this last equation. We then have

$$\frac{5^2}{b^2} + \frac{3^2}{84} = 1$$

$$\frac{25}{b^2} + \frac{3}{28} = 1$$

$$\frac{25}{b^2} = \frac{25}{28}$$

From this last equation, we see that $b^2 = 28$ and therefore

$$b = \sqrt{28} = 2\sqrt{7}$$

Now that we've determined a and b, we can write the equation of the ellipse in standard form. It is

$$\frac{x^2}{\left(2\sqrt{7}\right)^2} + \frac{y^2}{\left(2\sqrt{21}\right)^2} = 1$$

As you should verify for yourself, this equation can also be written in the equivalent form

$$3x^2 + y^2 = 84$$ ■■■

We conclude this section by mentioning a few of the applications of the ellipse. Each planet in the solar system moves in an elliptical orbit with the sun at one focus. Some gears in machines are elliptical rather than circular. (In certain brands of racing bikes, one of the gears in front is elliptical rather than circular.) Like the parabola, the ellipse has a *reflection property* that is related to the foci. As indicated in Figure 12, a light ray or sound emitted from one focus of an ellipse is always reflected through the other focus. This property is used in the design of "whispering galleries." These are rooms (with elliptical cross sections) in which a person standing at focus F_2 can hear a whisper from focus F_1 while others closer to F_1 might hear nothing. Statuary Hall in the Capitol building in Washington, D.C. is a whispering gallery. This idea is used in a modern medical device known as the *lithotripter,* in which high-energy sound waves are used to break up kidney stones. The patient is positioned with the kidney stone at one focus in an elliptical water bath while sound waves are emitted from the other focus.

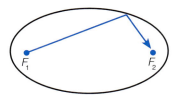

FIGURE 12
The reflection property of the ellipse. A light ray or sound emitted from one focus is reflected through the other focus.

EXERCISE SET 12.2

A

In Exercises 1–24, graph the ellipses. In each case, specify the lengths of the major and minor axes, the foci, and the eccentricity. For Exercises 13–24, also specify the center of the ellipse.

1. $4x^2 + 9y^2 = 36$
2. $4x^2 + 25y^2 = 100$
3. $x^2 + 16y^2 = 16$
4. $9x^2 + 25y^2 = 225$
5. $x^2 + 2y^2 = 2$
6. $2x^2 + 3y^2 = 3$
7. $16x^2 + 9y^2 = 144$
8. $25x^2 + y^2 = 25$
9. $15x^2 + 3y^2 = 5$
10. $9x^2 + y^2 = 4$
11. $2x^2 + y^2 = 4$
12. $36x^2 + 25y^2 = 400$
13. $\dfrac{(x-5)^2}{5^2} + \dfrac{(y+1)^2}{3^2} = 1$
14. $\dfrac{(x-1)^2}{2^2} + \dfrac{(y+4)^2}{3^2} = 1$
15. $\dfrac{(x-1)^2}{1^2} + \dfrac{(y-2)^2}{2^2} = 1$
16. $\dfrac{x^2}{4^2} + \dfrac{(y-3)^2}{2^2} = 1$
17. $\dfrac{(x+3)^2}{3^2} + \dfrac{y^2}{1^2} = 1$
18. $\dfrac{(x-2)^2}{2^2} + \dfrac{(y-2)^2}{2^2} = 1$
19. $3x^2 + 4y^2 - 6x + 16y + 7 = 0$
20. $16x^2 + 64x + 9y^2 - 54y + 1 = 0$
21. $5x^2 + 3y^2 - 40x - 36y + 188 = 0$
22. $x^2 + 16y^2 - 160y + 384 = 0$
23. $16x^2 + 25y^2 - 64x - 100y + 564 = 0$
24. $4x^2 + 4y^2 - 32x + 32y + 127 = 0$

In Exercises 25–32, find the equation of the ellipse satisfying the given conditions. Write the answer both in standard form and in the form $Ax^2 + By^2 = C$.

25. Foci $(\pm 3, 0)$; vertices $(\pm 5, 0)$
26. Foci $(0, \pm 1)$; vertices $(0, \pm 4)$
27. Vertices $(\pm 4, 0)$; eccentricity $\frac{1}{4}$
28. Foci $(0, \pm 2)$; endpoints of minor axis $(\pm 5, 0)$
29. Foci $(0, \pm 2)$; endpoints of major axis $(0, \pm 5)$
30. Endpoints of major axis $(\pm 10, 0)$; endpoints of minor axis $(0, \pm 4)$
31. Center at the origin; vertices on the x-axis; length of major axis twice the length of minor axis; $\left(1, \sqrt{2}\right)$ lies on ellipse
32. Eccentricity $\frac{3}{5}$; one endpoint of minor axis $(-8, 0)$; center at the origin
33. Solve the equation $(x^2/3^2) + (y^2/2^2) = 1$ for y to obtain $y = \pm\frac{1}{3}\sqrt{36 - 4x^2}$. Then complete the following table and use the results to graph the given equation. (Use the fact that the graph must be symmetric about the y-axis.)

x	0	0.5	1.0	1.5	2.0	2.5	3
y	± 2						0

34. Solve the equation $(x^2/1^2) + (y^2/4^2) = 1$ for y to obtain $y = \pm 4\sqrt{1 - x^2}$. Then complete the following table and use the results to graph the given equation. (Use the fact that the graph must be symmetric about the y-axis.)

x	0	0.1	0.2	0.3	0.4	0.5	0.6
y	± 4						

x	0.7	0.8	0.9	1.0
y				0

35. This exercise outlines the steps needed to complete the derivation of the equation $(x^2/a^2) + (y^2/b^2) = 1$.
    (a) Square both sides of equation (1) on page 709. After simplifying, you should obtain
    $$a\sqrt{(x+c)^2 + y^2} = a^2 + xc$$
    (b) Square both sides of the equation in part (a). Show that the result can be written
    $$a^2x^2 - x^2c^2 + a^2y^2 = a^4 - a^2c^2$$
    (c) Verify that the equation in part (b) is equivalent to
    $$(a^2 - c^2)x^2 + a^2y^2 = a^2(a^2 - c^2)$$
    (d) Using the equation in part (c), replace the quantity $a^2 - c^2$ with b^2. Then show that the resulting equation can be rewritten $(x^2/a^2) + (y^2/b^2) = 1$, as required.

36. In the text we defined the positive number b^2 by the equation $b^2 = a^2 - c^2$. For this definition to make sense, we need to show that the quantity $a^2 - c^2$ is indeed positive. This can be done as follows. First, recall that in any triangle, the sum of the lengths of two sides is always greater than the length of the third side. Now apply this fact to triangle F_1PF_2 in Figure 4 to show that $2a > 2c$. Conclude from this that $a^2 - c^2 > 0$, as required.

37. As you know, there is a simple expression for the circumference of a circle of radius a, namely, $2\pi a$. However, there is no similar type of elementary expression for the circumference of an ellipse. (The circumference of an ellipse can be computed to as many decimal places as required using the methods of calculus.) Nevertheless, there are some interesting elementary formulas that allow us to approximate the circumference of an ellipse quite closely. Three such formulas follow, along with the names of their discoverers and the dates of discovery. Each formula yields an approximate value for the circumference of the ellipse $(x^2/a^2) + (y^2/b^2) = 1$.

$$C_1 = \pi\left[a + b + \tfrac{1}{2}(\sqrt{a} - \sqrt{b})^2\right]$$
Giuseppe Peano, 1887

$$C_2 = \pi\left[3(a + b) - \sqrt{(a + 3b)(3a + b)}\right]$$
Srinivasa Ramanujan, 1914

$$C_3 = \frac{\pi}{2}\left[a + b + \sqrt{2(a^2 + b^2)}\right]$$
R. A. Johnson, 1930

Use these formulas to complete the following table of approximations for the circumference of the ellipse $(x^2/5^2) + (y^2/3^2) = 1$. Round off the values of C_1, C_2, and C_3 to six decimal places. In order to complete the right-hand column of the table, you need two facts. First, the actual circumference of the ellipse, rounded off to six decimal places, is 25.526999. Second, percentage error in an approximation is given by

$$\frac{|(\text{true value}) - (\text{approximation})|}{\text{true value}} \times 100$$

Round off the percentage errors to two significant digits. Which of the three approximations is the best?

	Approximation Obtained	Percentage Error
C_1		
C_2		
C_3		

C

38. In Section 12.1, we defined the parabola in terms of a focus–directrix property. This exercise shows that the ellipse also has a type of focus–directrix property.

Refer to the accompanying figure. Let P be a point on the ellipse $(x^2/a^2) + (y^2/b^2) = 1$, and suppose that the coordinates of P are (x, y).

(a) Show that $F_1P = a + ex$. *Hint:* If you try to do this from scratch, it can involve a rather lengthy calculation. Begin instead with the equation $a\sqrt{(x + c)^2 + y^2} = a^2 + xc$ [which appears in Exercise 35(a)], and divide both sides by a.

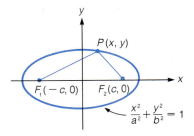

(b) Show that $F_2P = a - ex$. *Hint:* Make use of the result in part (a), along with the fact that the sum of F_1P and F_2P is, by definition, $2a$.

(c) In the accompanying figure, $P(x, y)$ denotes an

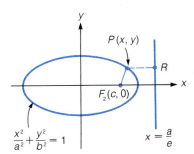

arbitrary point on the ellipse $(x^2/a^2) + (y^2/b^2) = 1$, and R is the foot of the perpendicular drawn from P to the line $x = a/e$. Show that $F_2P/PR = e$. (This is the *focus–directrix property* of the ellipse. The line $x = a/e$ is the directrix in this case.) *Hint:* This result is easy to prove if you make use of the formula for F_2P in part (b).

12.3 THE HYPERBOLA

> *... Menaechmus [c. 350 B.C.] is reputed to have discovered the curves that were later known as the ellipse, the parabola, and the hyperbola.*

Carl B. Boyer in *A History of Mathematics*, 2d ed., revised by Uta C. Merzbach (New York: John Wiley and Sons, 1991)

The ellipse is the general shape of any closed orbit.... It is also possible to have orbits that are not closed, whose shapes are represented by open curves. Even if the two bodies are not bound together by their mutual gravitational attraction, their gravitational attraction for each other still affects their relative motion. Then the general shape of their orbit is a hyperbola.

Theodore P. Snow in *The Dynamic Universe: An Introduction to Astronomy*, 4th ed. (St. Paul: West Publishing Company, 1990)

In the previous section, we defined an ellipse as the set of points P such that the sum of the distances from P to two fixed points is constant. By considering the difference instead of the sum, we are led to the definition of the hyperbola.

The Hyperbola

A **hyperbola** is the set of all points in the plane, the difference of whose distances from two fixed points is a positive constant. Each fixed point is called a **focus**.

As with the ellipse, we label the foci F_1 and F_2. Before obtaining an equation for the hyperbola, we can see the general features of the curve by using two sets of concentric circles, with centers F_1 and F_2, to locate points satisfying the definition of a hyperbola. In Figure 1, we've plotted a number of points P such that either $F_1P - F_2P = 3$ or $F_2P - F_1P = 3$. By joining these points, we obtain the graph of the hyperbola shown in Figure 1.

FIGURE 1

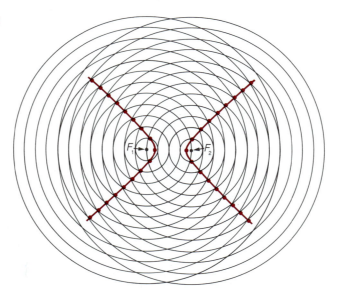

Unlike the parabola or the ellipse, the hyperbola is composed of two distinct parts, or **branches.** As you can check, the left branch in Figure 1 corresponds to the equation $F_2P - F_1P = 3$, while the right branch corresponds to the equation $F_1P - F_2P = 3$. Figure 1 also reveals that the hyperbola possesses two types of symmetry. First, it is symmetric about the line passing through the two foci F_1 and F_2; this line is referred to as the **focal axis** of the hyperbola. Second, the hyperbola is symmetric about the line that is the perpendicular bisector of the segment $\overline{F_1F_2}$.

To derive an equation for the hyperbola, let us initially assume that the foci are located at the points $F_1(-c, 0)$ and $F_2(c, 0)$, as indicated in Figure 2. We will

FIGURE 2

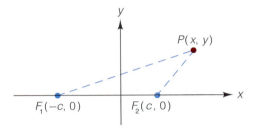

use $2a$ to denote the positive constant referred to in the definition of the hyperbola. By definition, then, $P(x, y)$ lies on the hyperbola if and only if

$$|F_1P - F_2P| = 2a$$

or, equivalently,

$$F_1P - F_2P = \pm 2a$$

If we use the formula for the distance between two points, this last equation becomes

$$\sqrt{(x + c)^2 + (y - 0)^2} - \sqrt{(x - c)^2 + (y - 0)^2} = \pm 2a$$

or

$$\sqrt{(x + c)^2 + y^2} - \sqrt{(x - c)^2 + y^2} = \pm 2a$$

We can simplify this equation by carrying out the same procedure that we used for the ellipse in the previous section. As Exercise 42 asks you to verify, the resulting equation is

$$(c^2 - a^2)x^2 - a^2y^2 = a^2(c^2 - a^2) \tag{1}$$

Before further simplifying equation (1), we point out that the quantity $c^2 - a^2$ [which appears twice in equation (1)] is positive. To see why this is so, refer back to Figure 2. In triangle F_1F_2P (as in any triangle), the length of any side is less than the sum of the lengths of the other two sides. Thus

$$F_1P < F_1F_2 + F_2P \quad \text{and} \quad F_2P < F_1F_2 + F_1P$$

and therefore

$$F_1P - F_2P < F_1F_2 \quad \text{and} \quad F_2P - F_1P < F_1F_2$$

These last two equations tell us that

$$|F_1P - F_2P| < F_1F_2$$

Therefore, in view of the definitions of $2a$ and $2c$, we have

$$0 < 2a < 2c \qquad \text{or} \qquad 0 < a < c$$

This last inequality tells us that $c^2 - a^2$ is positive, as we wished to show.

Now, since $c^2 - a^2$ is positive, we may define the positive number b by the equation

$$b^2 = c^2 - a^2$$

With this notation, equation (1) becomes

$$b^2 x^2 - a^2 y^2 = a^2 b^2$$

Dividing by $a^2 b^2$, we obtain

$$\frac{x^2}{a^2} - \frac{y^2}{b^2} = 1 \qquad\qquad (2)$$

We have now shown that the coordinates of every point on the hyperbola satisfy equation (2). Conversely, it can be shown that if the coordinates of a point satisfy equation (2), then the point satisfies the original definition of the hyperbola. Equation (2) is the **standard form** for the equation of a hyperbola with foci $F_1(-c, 0)$ and $F_2(c, 0)$.

The intercepts of the hyperbola are readily obtained from equation (2). To find the x-intercepts, we set y equal to zero to obtain

$$\frac{x^2}{a^2} = 1$$

$$x^2 = a^2 \qquad \text{or} \qquad x = \pm a$$

Thus, the hyperbola crosses the x-axis at the points $(-a, 0)$ and $(a, 0)$. On the other hand, the curve does not cross the y-axis, for if we set x equal to zero in equation (2), we obtain $-y^2/b^2 = 1$ or

$$y^2 = -b^2 \qquad (b > 0)$$

Since the square of any real number y is nonnegative, this last equation has no solution. Therefore, the graph does not cross the y-axis. Finally, let us note that (according to the symmetry tests in Section 3.2) the graph of equation (2) must be symmetric about both coordinate axes.

Before graphing the hyperbola $(x^2/a^2) - (y^2/b^2) = 1$, we point out the important fact that the two lines $y = (b/a)x$ and $y = -(b/a)x$ are asymptotes for the curve. We can see why as follows. First we solve equation (2) for y:

$$\frac{x^2}{a^2} - \frac{y^2}{b^2} = 1$$

$$b^2 x^2 - a^2 y^2 = a^2 b^2 \qquad \text{\color{red}{multiplying by } } a^2 b^2$$

$$- a^2 y^2 = a^2 b^2 - b^2 x^2$$

$$y^2 = \frac{b^2 x^2 - a^2 b^2}{a^2} = \frac{b^2(x^2 - a^2)}{a^2}$$

$$y = \pm \frac{b}{a} \sqrt{x^2 - a^2} \qquad\qquad (3)$$

TABLE 1

x	$\sqrt{x^2 - 5^2}$
100	99.875
1000	999.987
10000	9999.999

Now, as x grows arbitrarily large, the value of the quantity $\sqrt{x^2 - a^2}$ becomes closer and closer to x itself. Table 1 provides some empirical evidence for this statement in the case when $a = 5$. (A formal proof of the statement properly belongs to calculus.) In summary, then, we have the approximation $\sqrt{x^2 - a^2} \approx x$ as x grows arbitrarily large. So, in view of equation (3), we have

$$y = \pm \frac{b}{a} \sqrt{x^2 - a^2} \approx \pm \frac{b}{a} x \qquad \text{as } x \text{ grows arbitrarily large}$$

In other words, the two lines $y = \pm (b/a)x$ are asymptotes for the hyperbola.

A simple way to sketch the two asymptotes and then graph the hyperbola is as follows. First draw the rectangle with vertices (a, b), $(-a, b)$, $(-a, -b)$, and $(a, -b)$, as indicated in Figure 3(a). The slopes of the diagonals in this rectangle are b/a and $-b/a$. Thus by extending these diagonals as in Figure 3(b), we obtain the two asymptotes $y = \pm (b/a)x$. Now, since the x-intercepts of the hyperbola are a and $-a$, we can sketch the curve as shown in Figure 3(c).

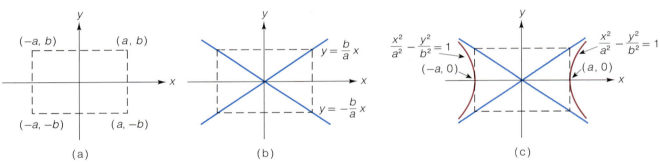

FIGURE 3
Steps in graphing the hyperbola
$(x^2/a^2) - (y^2/b^2) = 1$ and its asymptotes.

For the hyperbola in Figure 3(c), the two points $(\pm a, 0)$, where the curve meets the x-axis, are referred to as **vertices**. The midpoint of the line segment joining the two vertices is called the **center** of the hyperbola. (Equivalently, we can define the center as the point of intersection of the two asymptotes.) For the hyperbola in Figure 3(c), the center coincides with the origin. The line segment joining the vertices of a hyperbola is the **transverse axis** of the hyperbola. For reference, in the box that follows (on page 722), we summarize our work up to this point on the hyperbola. *Note:* Several new terms describing the hyperbola are also given in the box.

In general, if A, B, and C are positive numbers, then the graph of an equation of the form

$$Ax^2 - By^2 = C$$

will be a hyperbola of the type shown in Figure 4. Example 1 shows why this is so.

EXAMPLE 1 Graph the hyperbola $16x^2 - 9y^2 = 144$ after determining the following: vertices, foci, eccentricity, lengths of the transverse and conjugate axes, and asymptotes.

PROPERTY SUMMARY   THE HYPERBOLA $\dfrac{x^2}{a^2} - \dfrac{y^2}{b^2} = 1$

1. The **foci** are the points $F_1(-c, 0)$ and $F_2(c, 0)$. The hyperbola is the set of points P such that $|F_1P - F_2P| = 2a$.

2. The **focal axis** is the line passing through the foci.

3. The **vertices** are the points at which the hyperbola intersects its focal axis. In Figure 4, these are the two points $V_1(-a, 0)$ and $V_2(a, 0)$.

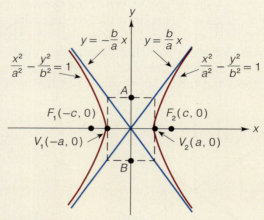

FIGURE 4

4. The **center** is the point on the focal axis midway between the foci. The center of the hyperbola in Figure 4 is the origin.

5. The **transverse axis** is the line segment joining the two vertices. In Figure 4, the length of the transverse axis $\overline{V_1V_2}$ is $2a$.

6. The **conjugate axis** is the line segment perpendicular to the transverse axis, passing through the center and extending a distance b on either side of the center. In Figure 4, this is the segment $\overline{AB}$.

7. The **eccentricity** e is defined by $e = c/a$, where $c^2 = a^2 + b^2$

Solution   First convert the given equation to standard form by dividing through by 144. This yields

$$\frac{x^2}{9} - \frac{y^2}{16} = 1 \qquad \text{or} \qquad \frac{x^2}{3^2} - \frac{y^2}{4^2} = 1$$

By comparing this with the equation $(x^2/a^2) - (y^2/b^2) = 1$, we see that $a = 3$ and $b = 4$. The value of c can be determined by using the equation $c^2 = a^2 + b^2$. We have

$$c^2 = 3^2 + 4^2 = 25 \quad \text{and therefore} \quad c = 5$$

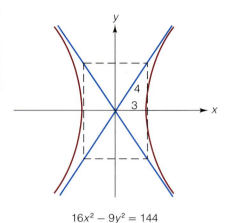

$16x^2 - 9y^2 = 144$
or
$\dfrac{x^2}{3^2} - \dfrac{y^2}{4^2} = 1$

FIGURE 5

Now that we know the values of a, b, and c, we can list the required information:

Vertices: $(\pm 3, 0)$	Length of transverse axis ($= 2a$): 6
Foci: $(\pm 5, 0)$	Length of conjugate axis ($= 2b$): 8
Eccentricity: $\dfrac{5}{3}$	Asymptotes: $y = \pm\dfrac{4}{3}x$

The graph of the hyperbola is shown in Figure 5. ▮▮▮

The same method that is used to derive the equation for the hyperbola with foci $(\pm c, 0)$ can be used when the foci are instead located on the y-axis at the points $(0, \pm c)$. The equation of the hyperbola in this case is

$$\frac{y^2}{a^2} - \frac{x^2}{b^2} = 1$$

We summarize the basic properties of this hyperbola in the following box.

PROPERTY SUMMARY   THE HYPERBOLA $\dfrac{y^2}{a^2} - \dfrac{x^2}{b^2} = 1$

Foci: $(0, \pm c)$, where $c^2 = a^2 + b^2$

Vertices: $(0, \pm a)$

Asymptotes: $y = \pm (a/b)x$

Length of transverse axis: $2a$

Length of conjugate axis: $2b$

Eccentricity: $e = c/a$

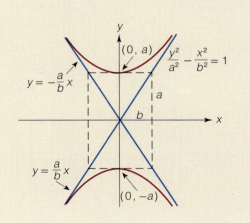

EXAMPLE 2   Use the technique of completing the square to show that the graph of the following equation is a hyperbola.

$$9y^2 - 54y - 25x^2 + 200x - 544 = 0$$

Graph the hyperbola and specify the center, the vertices, the foci, the length of the transverse axis, and the equations of the asymptotes.

Solution
$$9(y^2 - 6y \quad) - 25(x^2 - 8x \quad) = 544 \qquad \text{factoring}$$
$$9(y^2 - 6y + 9) - 25(x^2 - 8x + 16) = 544 + 81 - 400 \qquad \text{completing the squares}$$
$$9(y - 3)^2 - 25(x - 4)^2 = 225$$
$$\frac{(y-3)^2}{5^2} - \frac{(x-4)^2}{3^2} = 1 \qquad \text{dividing by 225}$$

Now, the graph of this last equation is obtained by translating the graph of the equation

$$\frac{y^2}{5^2} - \frac{x^2}{3^2} = 1 \tag{4}$$

four units to the right and three units up. So first we analyze the graph of equation (4). The general form of the graph is shown in the box just prior to this example. In our case, we have $a = 5$, $b = 3$, and

$$c = \sqrt{a^2 + b^2} = \sqrt{5^2 + 3^2} = \sqrt{34} \ (\approx 5.8)$$

Consequently, the vertices [for equation (4)] are $(0, \pm 5)$; the foci are $\left(0, \pm\sqrt{34}\right)$; and the asymptotes are $y = \pm\frac{5}{3}x$. Figure 6 shows the graph of this hyperbola. Finally, by translating the graph in Figure 6 to the right four units and up three units, we obtain the graph of the original equation, as shown in Figure 7. You should verify for yourself that the information accompanying Figure 7 is correct.

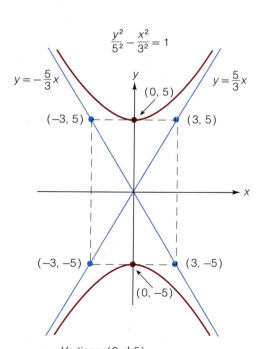

Vertices: $(0, \pm 5)$
Foci: $(0, \pm\sqrt{34})$
Length of transverse axis: 10
Asymptotes: $y = \pm\frac{5}{3}x$

FIGURE 6

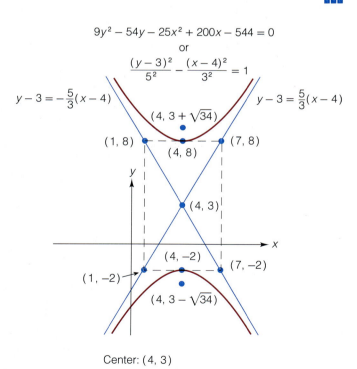

Center: $(4, 3)$
Vertices: $(4, 8)$, $(4, -2)$
Foci: $(4, 3 \pm\sqrt{34})$
Length of transverse axis: 10
Asymptotes: $y - 3 = \pm\frac{5}{3}(x - 4)$

FIGURE 7

If you reread the example just completed, you'll see that it was not necessary to know in advance that the given equation represented a hyperbola. Rather, this fact emerged naturally after we completed the square. Indeed, completing the square is a useful technique for identifying the graph of any equation of the form

$$Ax^2 + Cy^2 + Dx + Ey + F = 0$$

EXAMPLE 3 Identify the graph of the equation

$$4x^2 - 32x - y^2 + 2y + 63 = 0$$

Solution As before, we complete the squares:

$$4(x^2 - 8x) - (y^2 - 2y) = -63$$
$$4(x^2 - 8x + 16) - (y^2 - 2y + 1) = -63 + 4(16) - 1$$
$$4(x - 4)^2 - (y - 1)^2 = 0$$

Since the right-hand side of this last equation is 0, dividing both sides by 4 will not bring the equation into one of the standard forms. Indeed, if we factor the left-hand side of the equation as a difference of two squares, we obtain

$$[2(x - 4) - (y - 1)][2(x - 4) + (y - 1)] = 0$$
$$(2x - y - 7)(2x + y - 9) = 0$$

$$
\begin{array}{c|c}
2x - y - 7 = 0 & 2x + y - 9 = 0 \\
-y = -2x + 7 & y = -2x + 9 \\
y = 2x - 7 &
\end{array}
$$

Thus the given equation is equivalent to the two linear equations $y = 2x - 7$ and $y = -2x + 9$. These two lines, taken together, constitute the graph. See Figure 8.

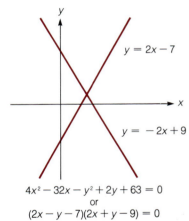

$y = 2x - 7$

$y = -2x + 9$

$4x^2 - 32x - y^2 + 2y + 63 = 0$
or
$(2x - y - 7)(2x + y - 9) = 0$

FIGURE 8

The two lines that we graphed in Figure 8 are actually the asymptotes for the hyperbola $4(x - 4)^2 - (y - 1)^2 = 1$. (Verify this for yourself.) For that reason, the graph in Figure 8 is referred to as a **degenerate hyperbola**. There are other cases similar to this that can arise in graphing equations of the form $Ax^2 + Cy^2 + Dx + Ey + F = 0$. For instance, as you can check for yourself by completing the squares, the graph of the equation

$$x^2 - 2x + 4y^2 - 16y + 17 = 0$$

consists of the single point $(1, 2)$. We refer to the graph in this case as a **degenerate ellipse**. Similarly, as you can check by completing the squares, the equation $x^2 - 2x + 4y^2 - 16y + 18 = 0$ has no graph; there are no points with coordinates that satisfy the equation.

We conclude this section by listing several applications of the hyperbola. Some comets have hyperbolic orbits. Unlike Halley's comet, which has an elliptical orbit, these comets pass through the solar system once and never return. The Cassegrain telescope (invented by the Frenchman Sieur Cassegrain in 1672) uses both a hyperbolic mirror and a parabolic mirror. The Hubble Space Telescope, launched into orbit in April 1990, utilizes a Cassegrain telescope. The hyperbola is also used in some navigation systems. In the LORAN (LOng RAnge Navigation) system, an airplane or a ship at a point P receives radio signals that were transmitted simultaneously from two locations, F_1 and F_2. The time difference between the two signals is converted to a difference in distances: $F_1P - F_2P$. This locates the ship along one branch of a hyperbola. Then, using data from a second set of signals, the ship is located along a second hyperbola. The intersection of the two hyperbolas then determines the location of the airplane or ship.

EXERCISE SET 12.3

A

For Exercises 1–24, graph the hyperbolas. In each case in which the hyperbola is nondegenerate, specify the following: vertices, foci, lengths of transverse and conjugate axes, eccentricity, and equations of the asymptotes. In Exercises 11–24, also specify the centers.

1. $x^2 - 4y^2 = 4$
2. $y^2 - x^2 = 1$
3. $y^2 - 4x^2 = 4$
4. $25x^2 - 9y^2 = 225$
5. $16x^2 - 25y^2 = 400$
6. $9x^2 - y^2 = 36$
7. $2y^2 - 3x^2 = 1$
8. $x^2 - y^2 = 9$
9. $4y^2 - 25x^2 = 100$
10. $x^2 - 3y^2 = 3$
11. $\dfrac{(x-5)^2}{5^2} - \dfrac{(y+1)^2}{3^2} = 1$
12. $\dfrac{(x-5)^2}{3^2} - \dfrac{(y+1)^2}{5^2} = 1$
13. $\dfrac{(y-2)^2}{2^2} - \dfrac{(x-1)^2}{1^2} = 1$
14. $\dfrac{(y-3)^2}{2^2} - \dfrac{x^2}{1^2} = 1$
15. $\dfrac{(x+3)^2}{4^2} - \dfrac{(y-4)^2}{4^2} = 1$
16. $\dfrac{(x+1)^2}{5^2} - \dfrac{(y+2)^2}{3^2} = 1$
17. $x^2 - y^2 + 2y - 5 = 0$
18. $16x^2 - 32x - 9y^2 + 90y - 353 = 0$
19. $x^2 - y^2 - 4x + 2y - 6 = 0$
20. $x^2 - 8x - y^2 + 8y - 25 = 0$
21. $y^2 - 25x^2 + 8y - 9 = 0$
22. $9y^2 - 18y - 4x^2 - 16x - 43 = 0$
23. $x^2 + 7x - y^2 - y + 12 = 0$
24. $9x^2 + 9x - 16y^2 + 4y + 2 = 0$
25. Let $P(x, y)$ be a point in the first quadrant on the hyperbola $(x^2/2^2) - (y^2/1^2) = 1$. Let Q be the point in the first quadrant with the same x-coordinate as P and lying on an asymptote to the hyperbola. Show that $PQ = \frac{1}{2}\left(x - \sqrt{x^2 - 4}\right)$.

26. The distance PQ in Exercise 25 represents the vertical distance between the hyperbola and the asymptote. Complete the following table to see numerical evidence that this separation distance approaches zero as x gets larger and larger. (Round off each entry to one significant digit.)

x	10	50	100	500	1000	10000
PQ						

In Exercises 27–36, determine the equation of the hyperbola satisfying the given conditions. Write each answer in the form $Ax^2 - By^2 = C$ or in the form $Ay^2 - Bx^2 = C$.

27. Foci $(\pm 4, 0)$; vertices $(\pm 1, 0)$
28. Foci $(0, \pm 5)$; vertices $(0, \pm 3)$
29. Asymptotes $y = \pm\frac{1}{2}x$; vertices $(\pm 2, 0)$
30. Asymptotes $y = \pm x$; foci $(0, \pm 1)$
31. Asymptotes $y = \pm\sqrt{10}x/5$; foci $\left(\pm\sqrt{7}, 0\right)$
32. Length of transverse axis 6; eccentricity $\frac{4}{3}$; center $(0, 0)$; focal axis horizontal
33. Vertices $(0, \pm 7)$; passing through the point $(1, 9)$
34. Eccentricity 2; foci $(\pm 1, 0)$
35. Length of transverse axis 6; length of conjugate axis 2; foci on the y-axis; center at the origin
36. Asymptotes $y = \pm 2x$; passing through $\left(1, \sqrt{3}\right)$
37. Show that the two asymptotes of the hyperbola $x^2 - y^2 = 16$ are perpendicular to each other.
38. (a) Verify that the point $P\left(6, 4\sqrt{3}\right)$ lies on the hyperbola $16x^2 - 9y^2 = 144$.
    (b) In Example 1, we found that the foci of this hyperbola were $F_1(-5, 0)$ and $F_2(5, 0)$. Compute the lengths F_1P and F_2P, where P is the point $\left(6, 4\sqrt{3}\right)$.
    (c) Verify that $|F_1P - F_2P| = 2a$.
39. (a) Verify that the point $P(5, 6)$ lies on the hyperbola $5y^2 - 4x^2 = 80$.

(b) Find the foci.
(c) Compute the lengths of the line segments $\overline{F_1P}$ and $\overline{F_2P}$, where P is the point $(5, 6)$.
(d) Verify that $|F_1P - F_2P| = 2a$.

B

40. (a) Let e_1 denote the eccentricity of the hyperbola $(x^2/4^2) - (y^2/3^2) = 1$, and let e_2 denote the eccentricity of the hyperbola $(x^2/3^2) - (y^2/4^2) = 1$. Verify that $e_1^2 e_2^2 = e_1^2 + e_2^2$.
    (b) Let e_1 and e_2 denote the eccentricities of the hyperbolas $(x^2/A^2) - (y^2/B^2) = 1$ and $(y^2/B^2) - (x^2/A^2) = 1$, respectively. Verify that $e_1^2 e_2^2 = e_1^2 + e_2^2$.

41. (a) If the hyperbola $(x^2/a^2) - (y^2/b^2) = 1$ has perpendicular asymptotes, show that $a = b$. What is the eccentricity in this case?
    (b) Show that the asymptotes of the hyperbola $(x^2/a^2) - (y^2/a^2) = 1$ are perpendicular. What is the eccentricity of this hyperbola?

42. Derive equation (1) in this section from the equation that precedes it.

43. Let $P(x, y)$ be a point on the right-hand branch of the hyperbola $(x^2/a^2) - (y^2/b^2) = 1$. As usual, let F_2 denote the focus located at $(c, 0)$. The following steps outline a proof of the fact that the length of the line segment $\overline{F_2P}$ in this case is given by $F_2P = xe - a$.
    (a) Explain why
    $$\sqrt{(x + c)^2 + y^2} - \sqrt{(x - c)^2 + y^2} = 2a$$
    (b) In the preceding equation, add the quantity $\sqrt{(x - c)^2 + y^2}$ to both sides and then square both sides. Show that the result can be written
    $$xc - a^2 = a\sqrt{(x - c)^2 + y^2}$$
    or
    $$xc - a^2 = a(F_2P)$$

(c) Divide both sides of the preceding equation by a to show that $xe - a = F_2P$, as required.

44. Let $P(x, y)$ be a point on the right-hand branch of the hyperbola $(x^2/a^2) - (y^2/b^2) = 1$. As usual, let F_1 denote the focus located at $(-c, 0)$. Show that $F_1P = xe + a$.
    Hint: Use the result in Exercise 43 along with the fact that the right-hand branch is defined by the equation $F_1P - F_2P = 2a$.

45. Let P be a point on the right-hand branch of the hyperbola $x^2 - y^2 = k^2$. If d denotes the distance from P to the center of the hyperbola, show that
    $$d^2 = (F_1P)(F_2P)$$

46. Let $P(x, y)$ be a point on the right-hand branch of the hyperbola $(x^2/a^2) - (y^2/b^2) = 1$. From P, a line segment is drawn perpendicular to the line $x = a/e$, meeting this line at D. If F_2 denotes (as usual) the focus located at $(c, 0)$, show that
    $$\frac{F_2P}{PD} = e$$

    The line $x = a/e$ is called the *directrix* of the hyperbola corresponding to the focus F_2. *Hint:* Use the expression for F_2P developed in Exercise 43.

47. Consider the hyperbola $(x^2/a^2) - (y^2/b^2) = 1$, and let D be the point where the asymptote $y = (b/a)x$ meets the directrix $x = a/e$. If F denotes the focus corresponding to this directrix, show that the line segment $\overline{FD}$ is perpendicular to the asymptote.

48. Consider the hyperbola $(x^2/a^2) - (y^2/b^2) = 1$ and suppose that a perpendicular is drawn from the focus $F_2(c, 0)$ to the asymptote $y = (b/a)x$, meeting the asymptote at A.
    (a) Show that $F_2A = b$.
    (b) Show that the distance from the center of the hyperbola to A is equal to a.

12.4 **THE FOCUS–DIRECTRIX PROPERTY OF CONICS**

Here then [in the Conics, written by Apollonius of Perga (ca. 262–ca. 190 B.C.)] we have the properties of the three curves expressed in the precise language of the Pythagorean application of areas, and the curves are named accordingly: parabola ($\pi\alpha\rho\alpha\beta o\lambda\eta$) where the rectangle is exactly applied [equal], hyperbola ($\upsilon\pi\epsilon\rho\beta o\lambda\eta$) where it exceeds, and ellipse ($\epsilon\lambda\lambda\epsilon\iota\psi\iota\zeta$) where it falls short.

Sir Thomas Heath in *A History of Greek Mathematics*, vol. II (Oxford: The Clarendon Press, 1921)

The most important contribution which Pappus [ca. 300] made to our knowledge of the conics was his publication of the focus, directrix, eccentricity theorem.

Julian Lowell Coolidge in *A History of the Conic Sections and Quadric Surfaces* (London: Oxford University Press, 1946)

Parabola

Focus

Directrix

FIGURE 1

We begin by recalling the focus–directrix property that we used in defining the parabola. A point P is on a parabola if and only if the distance from P to the focus is equal to the distance from P to the directrix. For the parabola in Figure 1, this means that for each point P on the parabola, we have $FP = PD$ or, equivalently,

$$\frac{FP}{PD} = 1$$

The ellipse and the hyperbola also can be characterized by focus–directrix properties. We'll begin with the ellipse. To help us in subsequent computations, we need to know the lengths of the line segments $\overline{F_1P}$ and $\overline{F_2P}$ in Figure 2.

FIGURE 2
Line segments drawn from a point on the ellipse to the foci are called focal radii. Here $\overline{F_1P}$ and $\overline{F_2P}$ are the focal radii.

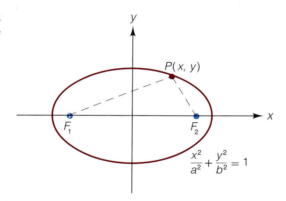

These line segments joining a point on the ellipse to the foci are called **focal radii.** In the box that follows, we give a formula for the length of each focal radius. In the formulas, $e\,(=c/a)$ denotes the eccentricity of the ellipse. (For the derivation of these formulas, see Exercise 15 at the end of this section.)

The Focal Radii of the Ellipse $\dfrac{x^2}{a^2} + \dfrac{y^2}{b^2} = 1$

As indicated in Figure 2, let $P(x, y)$ be a point on the ellipse. Then the lengths of the focal radii are

$$F_1P = a + ex \qquad \text{and} \qquad F_2P = a - ex$$

EXAMPLE 1   Figure 3 shows the ellipse $x^2 + 3y^2 = 28$. Compute the lengths of the focal radii drawn to the point $P(-1, 3)$.

FIGURE 3

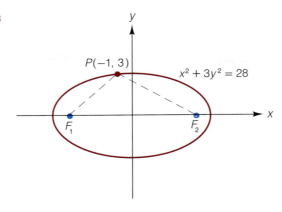

Solution To apply the formulas in the box, we need to know the values of a and e for the given ellipse. (We already know that $x = -1$; that is given.) As you should verify for yourself using the techniques of Section 12.2, we have $a = 2\sqrt{7}$ and $e = \sqrt{6}/3$. Therefore

$$F_1P = a + ex$$
$$= 2\sqrt{7} + \left(\frac{\sqrt{6}}{3}\right)(-1) = \frac{6\sqrt{7} - \sqrt{6}}{3}$$

and

$$F_2P = a - ex$$
$$= 2\sqrt{7} - \left(\frac{\sqrt{6}}{3}\right)(-1) = \frac{6\sqrt{7} + \sqrt{6}}{3}$$

◼◼◼

In Figure 4 we show the ellipse $(x^2/a^2) + (y^2/b^2) = 1$ and the horizontal line $x = a/e$. The line $x = a/e$ is a **directrix** of the ellipse. In Figure 4, F_2P is the distance from the focus F_2 to a point P on the ellipse, and PD is the distance from

FIGURE 4

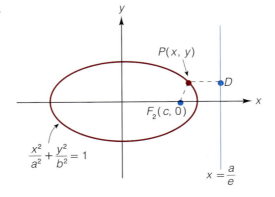

P to the directrix $x = a/e$. We are going to prove the following remarkable *focus–directrix property of the ellipse:* For any point $P(x, y)$ on the ellipse, the ratio of F_2P to PD is equal to e, the eccentricity of the ellipse. To prove this, we proceed as follows:

$$\frac{F_2P}{PD} = \frac{a - ex}{PD}$$ using our formula for F_2P

$$= \frac{a - ex}{\dfrac{a}{e} - x}$$ using Figure 4

$$= \frac{e(a - ex)}{a - ex}$$ multiplying both numerator and denominator by e

$$= e$$

So, $F_2P/PD = e$, as we wished to show.

The directrix $x = a/e$ is associated with the focus $F_2(c, 0)$. What about the focus $F_1(-c, 0)$? From the symmetry of the ellipse, it follows that the line $x = -a/e$ is the directrix associated with this focus. More specifically, referring to Figure 5, not only do we have $F_2P/PD = e$, we also have $F_1P/PE = e$.

FIGURE 5
For every point P on the ellipse, we have
$F_1P/PE = e$ and $F_2P/PD = e$.

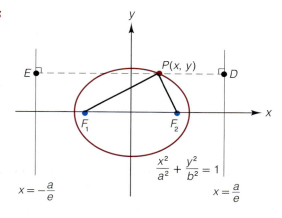

We have seen that each point P on the ellipse in Figure 5 satisfies the following two equations:

$$\frac{F_1P}{PE} = e \tag{1}$$

$$\frac{F_2P}{PD} = e \tag{2}$$

Now, conversely, suppose that the point $P(x, y)$ in Figure 6 satisfies equations (1) and (2). We are going to show that $P(x, y)$ must, in fact, lie on the ellipse $(x^2/a^2) + (y^2/b^2) = 1$.*

*The short proof given here was communicated to the author by Professor Ray Redheffer. For a proof using equation (2) but not equation (1), see Exercise 16.

FIGURE 6

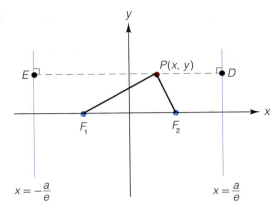

In Figure 6, the distance from E to D is $2a/e$. Therefore

$$PE + PD = 2a/e$$

and, consequently,

$$\frac{F_1P}{e} + \frac{F_2P}{e} = \frac{2a}{e} \qquad \text{using equations (1) and (2) to substitute for } PE \text{ and } PD$$

or

$$F_1P + F_2P = 2a \qquad \text{multiplying by } e$$

Thus, by definition, P lies on the ellipse $(x^2/a^2) + (y^2/b^2) = 1$, as we wished to show.

In the box that follows, we summarize the focus–directrix properties of the ellipse $(x^2/a^2) + (y^2/b^2) = 1$. For reference, we also include the original defining property of the ellipse from Section 12.2.

Focus and Focus–Directrix Properties of the Ellipse $\dfrac{x^2}{a^2} + \dfrac{y^2}{b^2} = 1$

Refer to Figure 7 (on the next page).

1. A point P is on the ellipse if and only if the sum of the distances from P to the foci $F_1(-c, 0)$ and $F_2(c, 0)$ is $2a$.

2. A point P is on the ellipse if and only if

$$\frac{F_2P}{PD} = e$$

where F_2P is the distance from P to the focus F_2, PD is the distance from P to the directrix $x = a/e$, and $e\ (= c/a)$ is the eccentricity of the ellipse.

3. A point P is on the ellipse if and only if

$$\frac{F_1P}{PE} = e$$

where F_1P is the distance from P to the focus F_1 and PE is the distance from P to the directrix $x = -a/e$.

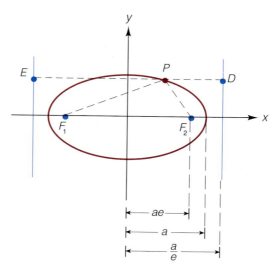

FIGURE 7
The ellipse $(x^2/a^2) + (y^2/b^2) = 1$ with foci
$F_1(-c, 0)$ and $F_2(c, 0)$ and directrices
$x = \pm a/e$.

EXAMPLE 2 The foci of an ellipse are $(\pm 3, 0)$ and the directrix corresponding to the focus $(3, 0)$ is $x = 5$. Find the equation of the ellipse. Write the answer in the form $Ax^2 + By^2 = C$.

Solution Using the given data (and referring to Figure 7), we have

$$ae = 3 \qquad \text{and} \qquad a/e = 5$$

From the first equation we obtain $e = 3/a$. Substituting this value of e in the second equation yields

$$\frac{a}{3/a} = 5 \qquad \text{and therefore} \qquad a^2 = 15$$

Now we can calculate b^2 by using the relation $b^2 = a^2 - c^2$ and the given information $c = 3$:

$$b^2 = a^2 - c^2 = 15 - 9 = 6$$

Substituting the values we've obtained for a^2 and b^2 in the equation $(x^2/a^2) + (y^2/b^2) = 1$ yields

$$\frac{x^2}{15} + \frac{y^2}{6} = 1$$

or

$$2x^2 + 5y^2 = 30 \qquad \text{multiplying by 30}$$

The focus–directrix property can be developed for the hyperbola in just the same way that we have proceeded for the ellipse. In fact, the algebra is so similar that we shall omit the details here and simply summarize the results. The hyperbola $(x^2/a^2) - (y^2/b^2) = 1$ has two directrices: they are the vertical lines $x = a/e$ in Figure 8(a) and $x = -a/e$ in Figure 8(b). Notice that these equations are identical to those for the directrices of the ellipse. In the box following Figure 8, we summarize the focus and the focus–directrix properties of the hyperbola.

FIGURE 8
The hyperbola $(x^2/a^2) - (y^2/b^2) = 1$ has two directrices. The directrix $x = a/e$ corresponds to the focus $F_2(c, 0)$. The directrix $x = -a/e$ corresponds to the focus $F_1(-c, 0)$.

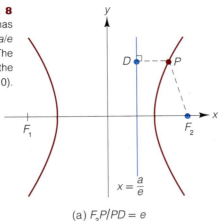

(a) $F_2P/PD = e$

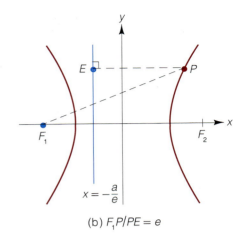

(b) $F_1P/PE = e$

Focus and Focus–Directrix Properties of the Hyperbola
$$\frac{x^2}{a^2} - \frac{y^2}{b^2} = 1$$

Refer to Figure 8.

1. A point P is on the hyperbola if and only if the difference of the distances from P to the foci $F_1(-c, 0)$ and $F_2(c, 0)$ is $2a$.

2. A point P is on the hyperbola if and only if

$$\frac{F_2P}{PD} = e$$

where F_2P is the distance from P to the focus F_2, PD is the distance from P to the directrix $x = a/e$, and $e\,(= c/a)$ is the eccentricity of the hyperbola.

3. A point P is on the hyperbola if and only if

$$\frac{F_1P}{PE} = e$$

where F_1P is the distance from P to the focus F_1 and PE is the distance from P to the directrix $x = -a/e$.

EXAMPLE 3   Determine the foci, the eccentricity, and the directrices for the hyperbola $9x^2 - 16y^2 = 144$.

Solution   To convert the given equation to standard form, we divide both sides by 144. This yields

$$\frac{x^2}{4^2} - \frac{y^2}{3^2} = 1$$

So, we have $a = 4$, $b = 3$, and

$$c = \sqrt{a^2 + b^2} = \sqrt{4^2 + 3^2} = 5$$

The foci are therefore $(0, \pm 5)$, and the eccentricity is

$$e = \frac{c}{a} = \frac{5}{4}$$

The directrices are

$$x = \pm\frac{a}{e} = \pm\frac{4}{5/4} = \pm\frac{16}{5}$$

EXAMPLE 4   Figure 9 shows the graph of the hyperbola $9x^2 - 16y^2 = 144$. (This is the hyperbola discussed in the previous example.)

(a) Verify that the point $P\left(-5, \frac{9}{4}\right)$ lies on this hyperbola.
(b) Compute F_2D and PD, and verify that

$$\frac{F_2P}{PD} = e$$

FIGURE 9

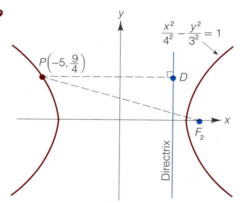

Solution   (a) Substituting the values $x = -5$ and $y = \frac{9}{4}$ in the equation of the hyperbola yields

$$9(-5)^2 - 16\left(\tfrac{9}{4}\right)^2 = 144$$

or

$$225 - 81 = 144$$

Since this last equation is correct, we conclude that the point $P\left(-5, \frac{9}{4}\right)$ indeed lies on the hyperbola.

(b) From the previous example, we know that the coordinates of the focus F_2 are $(5, 0)$. So, using the distance formula, we have

$$F_2P = \sqrt{(5 - (-5))^2 + \left(0 - \frac{9}{4}\right)^2}$$

$$= \sqrt{100 + \frac{81}{16}} = \sqrt{\frac{1681}{16}} = \frac{41}{4}$$

Next, we use the fact (from Example 3) that the equation of the directrix in Figure 9 is $x = \frac{16}{5}$. Thus, in Figure 9,

$$PD = |-5| + \tfrac{16}{5} = \tfrac{41}{5}$$

Finally, we compute the ratio F_2P/PD:

$$\frac{F_2P}{PD} = \frac{41/4}{41/5} = \frac{5}{4}$$

This is the same number that we obtained for the eccentricity in Example 3. So, we have verified in this case that the ratio of F_2P to PD is equal to the eccentricity.

The focus–directrix property provides a unified approach to the parabola, the ellipse, and the hyperbola. For both the ellipse and the hyperbola, we've seen that for any point P on the curve, we have

$$\frac{\text{distance from } P \text{ to a focus}}{\text{distance from } P \text{ to the corresponding directrix}} = e \qquad (3)$$

where e is a positive constant, the eccentricity of the curve. For the ellipse we have $e < 1$, and for the hyperbola we have $e > 1$. Equation (3) also holds for the parabola if we agree on the convention that the eccentricity of a parabola is 1. This is because with $e = 1$, equation (3) tells us that the distance from P to the focus equals the distance from P to the directrix, which is the defining condition for the parabola. The following theorem summarizes these remarks.

Theorem

Refer to Figure 10. Let $\mathscr{L}$ be a fixed line, F a fixed point, and e a positive constant. Consider the set of points P satisfying the condition

$$\frac{FP}{PD} = e$$

Then

(a) If $e = 1$, the set of points is a *parabola* with focus F and directrix $\mathscr{L}$;

(b) If $0 < e < 1$, the set of points is an *ellipse* with focus F, corresponding directrix $\mathscr{L}$, and eccentricity e;

(c) If $e > 1$, the set of points is a *hyperbola* with focus F, corresponding directrix $\mathscr{L}$, and eccentricity e.

FIGURE 10

Note In our development in this section, for simplicity, we've always considered cases in which the directrix is vertical. However, the preceding theorem is valid for any orientation of the line $\mathcal{L}$.

EXERCISE SET 12.4

A

In Exercises 1–4, you are given an ellipse and a point P on the ellipse. Find F_1P and F_2P, the lengths of the focal radii.

1. $x^2 + 3y^2 = 76$; $P(-8, 2)$
2. $x^2 + 3y^2 = 57$; $P(3, -4)$
3. $(x^2/15^2) + (y^2/5^2) = 1$; $P(9, 4)$
4. $2x^2 + 3y^2 = 14$; $P(-1, -2)$

In Exercises 5–10, determine the foci, the eccentricity, and the directrices for each ellipse and hyperbola.

5. (a) $(x^2/4^2) + (y^2/3^2) = 1$
    (b) $(x^2/4^2) - (y^2/3^2) = 1$
6. (a) $x^2 + 4y^2 = 1$
    (b) $x^2 - 4y^2 = 1$
7. (a) $12x^2 + 13y^2 = 156$
    (b) $12x^2 - 13y^2 = 156$
8. (a) $x^2 + 2y^2 = 2$
    (b) $x^2 - 2y^2 = 2$
9. (a) $25x^2 + 36y^2 = 900$
    (b) $25x^2 - 36y^2 = 900$
10. (a) $4x^2 + 25y^2 = 100$
    (b) $4x^2 - 25y^2 = 100$

In Exercises 11 and 12, use the given information to find the equation of the ellipse. Write the answer in the form $Ax^2 + By^2 = C$.

11. The foci are $(\pm 1, 0)$ and the directrices are $x = \pm 4$.
12. The foci are $(\pm\sqrt{3}, 0)$ and the eccentricity is $\frac{2}{3}$.

In Exercises 13 and 14, use the given information to find the equation of the hyperbola. Write the answer in the form $Ax^2 - By^2 = C$.

13. The foci are $(\pm 2, 0)$ and the directrices are $x = \pm 1$.
14. The foci are $(\pm 3, 0)$ and the eccentricity is 2.

B

15. In this exercise we show that the focal radii of the ellipse $(x^2/a^2) + (y^2/b^2) = 1$ are $F_1P = a + ex$ and $F_2P = a - ex$. The method used here, which avoids the use of radicals, appears in an eighteenth-century text, *Traité analytique des sections coniques* (Paris: 1707), by the Marquis de l'Hospital [1661–1704]. (For another method, one that does use radicals, see Exercise 38 in Section 12.2.)

    For convenience, let $d_1 = F_1P$ and $d_2 = F_2P$, as indicated in the accompanying figure.

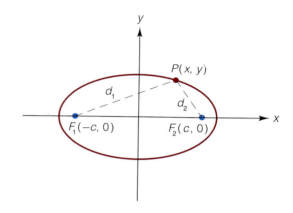

(a) Using the distance formula, verify that

$$d_1^2 = (x + c)^2 + y^2 \quad \text{and} \quad d_2^2 = (x - c)^2 + y^2$$

(b) Use the two equations in part (a) to show that $d_1^2 - d_2^2 = 4cx$.
(c) Explain why $d_1 + d_2 = 2a$.
(d) Factor the left-hand side of the equation in part (b) and substitute for one of the factors using the equation in part (c). Show that the result can be written

$$d_1 - d_2 = 2cx/a$$

(e) Add the equations in parts (c) and (d). Show that the resulting equation can be written $d_1 = a + ex$, as required.
(f) Use the equation in part (c) and the result in part (e) to show that $d_2 = a - ex$.

16. In this exercise we show that if the point $P(x, y)$ in the accompanying figure satisfies the condition

$$\frac{F_2P}{PD} = e$$

then, in fact, $P(x, y)$ lies on the ellipse $(x^2/a^2) + (y^2/b^2) = 1$.

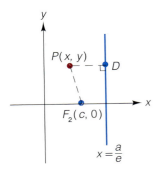

(a) From the given equation we have $(F_2P)^2 = e^2(PD)^2$. Use the distance formula and the figure to deduce from this equation that

$$(x - c)^2 + y^2 = e^2\left(\frac{a}{e} - x\right)^2$$

(b) In the equation in part (a), replace e with c/a. After carrying out the indicated operations and simplifying, show that the equation can be written

$$a^2x^2 - c^2x^2 + a^2y^2 = a^4 - a^2c^2$$

(c) The equation in part (b) is equivalent to $(a^2 - c^2)x^2 + a^2y^2 = a^2(a^2 - c^2)$. Now replace the quantity $a^2 - c^2$ by b^2 and show that the resulting equation can be written $(x^2/a^2) + (y^2/b^2) = 1$; thus, P lies on the ellipse, as we wished to show.

CHAPTER TWELVE SUMMARY OF PRINCIPAL TERMS AND FORMULAS

TERMS AND FORMULAS	PAGE REFERENCE	COMMENT
1. Conic sections	699	These are the curves that are formed when a plane intersects the surface of a right circular cone. As indicated in the chapter introduction, these curves are the circle, the ellipse, the hyperbola, and the parabola.
2. Parabola	699	A parabola is the set of all points in the plane equally distant from a fixed line and a fixed point not on the line. The fixed line is called the **directrix**, and the fixed point is called the **focus**.
3. Axis (of a parabola)	701	This is the line drawn through the focus of the parabola, perpendicular to the directrix.
4. Vertex (of a parabola)	701	This is the point at which the parabola intersects its axis.
5. Focal chord (of a parabola)	703	A focal chord of a parabola is a line segment passing through the focus, with endpoints on the parabola.
6. Focal width (of a parabola)	703	The focal width of a parabola is the length of the focal chord that is perpendicular to the axis of the parabola. For a given value of p, the two parabolas $x^2 = 4py$ and $y^2 = 4px$ have the same focal width; it is $4p$.
7. Ellipse	708	An ellipse is the set of all points in the plane, the sum of whose distances from two fixed points is constant. Each fixed point is called a **focus** of the ellipse.
8. Eccentricity (of an ellipse)	711	The eccentricity is a number that measures how much the ellipse deviates from being a circle. See Figure 6 in Section 12.2. The eccentricity e is defined by the formula $e = c/a$, where c and a are defined by the following conventions. The distance between the foci is denoted by $2c$. The sum of the distances from a point on the ellipse to the two foci is denoted by $2a$.

TERMS AND FORMULAS	PAGE REFERENCE	COMMENT
9. Center (of an ellipse)	711	This is the midpoint of the line segment joining the foci.
10. Major and minor axes (of an ellipse)	711	In Figure 5, Section 12.2, the major and minor axes are the line segments $\overline{V_1V_1}$ and $\overline{AB}$, respectively. For the ellipse $(x^2/a^2) + (y^2/b^2) = 1$, the lengths of the major and minor axes are $2a$ and $2b$, respectively.
11. $\dfrac{x^2}{a^2} + \dfrac{y^2}{b^2} = 1 \qquad (a > b)$	711	This is the standard form for the equation of an ellipse with foci $(\pm c, 0)$. The specifications for this ellipse are summarized in the box on page 711.
$\dfrac{x^2}{b^2} + \dfrac{y^2}{a^2} = 1 \qquad (a > b)$	714	This is the standard form for the equation of an ellipse with foci $(0, \pm c)$. The specifications for this ellipse are summarized in the box on page 714.
12. Hyperbola	718	A hyperbola is the set of all points in the plane, the difference of whose distances from two fixed points is a positive constant. The two fixed points are the **foci**, and the line passing through the foci is the **focal axis**.
13. Focal axis (of a hyperbola)	719	This is the line passing through the foci.
14. Asymptotes for the hyperbola	720, 721	As defined in earlier chapters, a line is an **asymptote** for a curve if the distance between the line and the curve approaches zero as we move farther and farther out along the line. The asymptotes for the hyperbola $\dfrac{x^2}{a^2} - \dfrac{y^2}{b^2} = 1$ are the two lines $y = \pm\dfrac{b}{a}x$. For the hyperbola $\dfrac{y^2}{a^2} - \dfrac{x^2}{b^2} = 1$, the asymptotes are $y = \pm\dfrac{a}{b}x$.
15. Vertices (of a hyperbola)	721	The two points at which the hyperbola intersects its focal axis are called **vertices**. See Figure 4 on page 722 and the boxed figure on page 723.
16. Center (of a hyperbola)	721	This is the point on the focal axis midway between the foci. See Figure 4 on page 722 and the boxed figure on page 723.
17. Transverse axis	721	This is the line segment joining the two vertices of a hyperbola. See Figure 4 on page 722 and the boxed figure on page 723.
18. Conjugate axis	722	This is the line segment perpendicular to the transverse axis of the hyperbola, passing through the center and extending a distance b on either side of the center. See Figure 4 on page 722 and the boxed figure on page 723.
19. Eccentricity (of a hyperbola)	722	For both of the standard forms for the hyperbola, the eccentricity e is defined by $e = c/a$.
20. $\dfrac{x^2}{a^2} - \dfrac{y^2}{b^2} = 1$	722	This is the standard form for the equation of a hyperbola with foci $(\pm c, 0)$.
$\dfrac{y^2}{a^2} - \dfrac{x^2}{b^2} = 1$	723	This is the standard form for the equation of a hyperbola with foci $(0, \pm c)$.

TERMS AND FORMULAS	PAGE REFERENCE	COMMENT
21. Focus–directrix property of conics	735	Refer to Figure 10 on page 735. Let $\mathcal{L}$ be a fixed line, F a fixed point, and e a positive constant. Consider the set of points P satisfying the condition $\dfrac{FP}{PD} = e$. Then: **(a)** If $e = 1$ the set of points is a *parabola* with focus F and directrix $\mathcal{L}$; **(b)** If $0 < e < 1$, the set of points is an *ellipse* with focus F, corresponding directrix $\mathcal{L}$, and eccentricity e; **(c)** If $e > 1$, the set of points is a *hyperbola* with focus F, corresponding directrix $\mathcal{L}$, and eccentricity e.

WRITING MATHEMATICS

Write out your answers to the following questions in complete sentences. If you draw a diagram to accompany your writing, or if you use equations, be sure that you clearly label any elements to which you refer.

1. Refer to Figure 1 on page 709. If the thumbtacks are 3 in. apart, what length of string should be used to produce an ellipse with eccentricity $\frac{2}{3}$?

2. Refer to Figure 1 on page 709 to explain each of the following.
   (a) When the eccentricity of an ellipse is close to 1, the ellipse is very flat.
   (b) When the eccentricity of an ellipse is close to 0, the ellipse resembles a circle.

3. Investigate the geometric significance of the eccentricity of a hyperbola by completing the three steps that follow. Then, write a report telling what you have done, what patterns you have observed, and what relationship you have found between the eccentricity and the shape of the hyperbola.
   (a) Use the definition of the eccentricity e to show that $e = \sqrt{1 + (b/a)^2}$. (This shows that the eccentricity is always greater than 1.)
   (b) Compute the eccentricity for each of the following hyperbolas. (Use a calculator.)
   (i) $(0.0201)x^2 - y^2 = 0.0201$
   (ii) $3x^2 - y^2 = 3$
   (iii) $8x^2 - y^2 = 8$
   (iv) $15x^2 - y^2 = 15$
   (v) $99x^2 - y^2 = 99$
   (c) On the same set of axes, sketch the first-quadrant portion of each of the hyperbolas in part (b).

CHAPTER TWELVE   REVIEW EXERCISES

In Exercises 1–4, find the equation of the parabola satisfying the given conditions. In each case, assume that the vertex is $(0, 0)$.

1. (a) The focus is $(4, 0)$.   (b) The focus is $(0, 4)$.

2. The focus lies on the x-axis, and the curve passes through the point $(3, 1)$.

3. The parabola is symmetric about the y-axis, the y-coordinate of the focus is positive, and the length of the focal chord perpendicular to the y-axis is 12.

4. The focus of the parabola is the center of the circle $x^2 - 8x + y^2 + 15 = 0$.

In Exercises 5–7, find the equation of the ellipse satisfying the given conditions. Write your answer in the form $Ax^2 + By^2 = C$.

5. Foci $(\pm 2, 0)$; endpoints of major axis $(\pm 8, 0)$

6. Foci $(0, \pm 1)$; endpoints of minor axis $(\pm 4, 0)$

7. Eccentricity $\frac{4}{5}$; one end of minor axis $(-6, 0)$; center at the origin

8. For any point P on an ellipse, the sum of the distances from $(1, 2)$ and $(-1, -2)$ is 12. Find the equation of the ellipse. *Hint:* Use the distance formula and the definition of an ellipse.

In Exercises 9–12, find the equation of the hyperbola satisfying the given conditions. Write each answer in the form $Ax^2 - By^2 = C$ or in the form $Ay^2 - Bx^2 = C$.

9. Foci $(\pm 6, 0)$; vertices $(\pm 2, 0)$

10. Asymptotes $y = \pm 2x$; foci $(0, \pm 3)$

11. Eccentricity 4; foci $(\pm 3, 0)$

12. Length of transverse axis 3; eccentricity $\frac{5}{4}$; center $(0, 0)$; focal axis horizontal

In Exercises 13–18, graph the parabolas, and in each case specify the vertex, the focus, the directrix, and the focal width.

13. $x^2 = 10y$

14. $x^2 = 5y$

15. $x^2 = -12(y - 3)$

16. $x^2 = -8(y + 1)$

17. $(y - 1)^2 = -4(x - 1)$

18. $(y + 3)^2 = 2(x - 1)$

In Exercises 19–24, graph the ellipses, and in each case specify the center, the foci, the lengths of the major and minor axes, and the eccentricity.

19. $x^2 + 2y^2 = 4$

20. $4x^2 + 9y^2 = 144$

21. $49x^2 + 9y^2 = 441$

22. $9x^2 + y^2 = 9$

23. $\dfrac{(x - 1)^2}{5^2} + \dfrac{(y + 2)^2}{3^2} = 1$

24. $\dfrac{(x + 3)^2}{3^2} + \dfrac{y^2}{3^2} = 1$

In Exercises 25–30, graph the hyperbolas. In each case specify the center, the vertices, the foci, the equations of the asymptotes, and the eccentricity.

25. $x^2 - 2y^2 = 4$

26. $4x^2 - 9y^2 = 144$

27. $49y^2 - 9x^2 = 441$

28. $9y^2 - x^2 = 9$

29. $\dfrac{(x - 1)^2}{5^2} - \dfrac{(y + 2)^2}{3^2} = 1$

30. $\dfrac{(y + 3)^2}{3^2} - \dfrac{x^2}{3^2} = 1$

In Exercises 31–44, use the technique of completing the square to graph the given equation. If the graph is a parabola, specify the vertex, axis, focus, and directrix. If the graph is an ellipse, specify the center, foci, and lengths of the major and minor axes. If the graph is a hyperbola, specify the center, vertices, foci, and

equations of the asymptotes. Finally, if the equation has no graph, say so.

31. $3x^2 + 4y^2 - 6x + 16y + 7 = 0$

32. $y^2 - 16x - 8y + 80 = 0$

33. $y^2 + 4x + 2y - 15 = 0$

34. $16x^2 + 64x + 9y^2 - 54y + 1 = 0$

35. $16x^2 - 32x - 9y^2 + 90y - 353 = 0$

36. $x^2 + 6x - 12y + 33 = 0$

37. $5x^2 + 3y^2 - 40x - 36y + 188 = 0$

38. $x^2 - y^2 - 4x + 2y - 6 = 0$

39. $9x^2 - 90x - 16y^2 + 32y + 209 = 0$

40. $x^2 + 2y - 12 = 0$

41. $y^2 - 25x^2 + 8y - 9 = 0$

42. $x^2 + 16y^2 - 160y + 384 = 0$

43. $16x^2 + 25y^2 - 64x - 100y + 564 = 0$

44. $16x^2 - 25y^2 - 64x + 100y - 36 = 0$

45. Let F_1 and F_2 denote the foci of the hyperbola $5x^2 - 4y^2 = 80$.
    (a) Verify that the point P with coordinates $(6, 5)$ lies on the hyperbola.
    (b) Compute the quantity $(F_1P - F_2P)^2$.

46. Show that the coordinates of the vertex of the parabola $Ax^2 + Dx + Ey + F = 0$ are given by
$$x = -\frac{D}{2A} \quad \text{and} \quad y = \frac{D^2 - 4AF}{4AE}$$

47. If the equation $Ax^2 + Cy^2 + Dx + Ey + F = 0$ represents an ellipse or a hyperbola, show that the center is the point
$$\left(-\frac{D}{2A}, -\frac{E}{2C} \right)$$

48. The figure shows the parabola $x^2 = 4py$ and a circle with center at the origin and diameter $3p$. If V and F denote the vertex and focus of the parabola, respectively, show that the common chord of the circle and parabola bisects the line segment $\overline{VF}$.

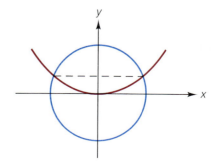

49. The following figure shows an ellipse and a parabola. As indicated in the figure, the curves are symmetric about the x-axis and they both have an x-intercept of 5. Find the equation of the ellipse and the parabola, given that the point $(3, 0)$ is a focus for both curves.

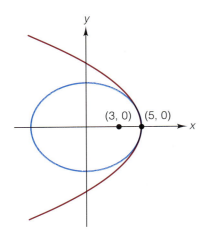

In Exercises 50 and 51, (a) compute the eccentricity; (b) determine each focus–directrix pair; (c) sketch the graph and show the foci and the directrices.

50. $x^2 - 6x + 4y^2 - 32y + 69 = 0$

51. $x^2 - 6x - 4y^2 + 32y - 59 = 0$

In Exercises 52–54, find the equation of the conic satisfying the given conditions.

52. eccentricity $= \frac{2}{3}$; directrices are $x = \pm 3$

53. eccentricity $= \frac{3}{2}$; directrices are $x = \pm 3$

54. eccentricity $= 1$; directrix is $y = 3$; focus is $(0, 5)$

OPTIONAL TI-81 GRAPHING CALCULATOR EXERCISES

EXERCISES FOR CHAPTER TWELVE

In Exercises 1–4, graph the parabolas. These graphs provide examples of the four basic orientations shown in Figure 5 on page 702.

1. $x^2 = 8y$ *Hint:* Enter this as $\frac{1}{8}x^2$.

2. $x^2 = -8y$

3. $y^2 = 8x$ *Hint:* Enter this as two functions $\sqrt{8x}$ and $-\sqrt{8x}$.

4. $y^2 = -8x$

5. In this exercise we graph the parabola $4x + y^2 + 2y - 7 = 0$.
   (a) By completing the square, show that the equation can be written $(y + 1)^2 = -4(x - 2)$.
   (b) Solving the equation in part (a) for y gives us the two functions $y = -1 \pm \sqrt{-4(x - 2)}$. Enter these two functions in the Y= menu and press ZOOM 6. Now check that your graph is consistent with Figure 9 on page 705.

6. (a) Complete the square to determine the vertex of the parabola $y = 2x^2 - 16x + 33$.
   (b) Use the calculator to graph this function. Check that the general appearance of the graph you obtain is consistent with your result in part (a).

(c) Use the ZOOM and TRACE keys to estimate the coordinates of the vertex. How close are the calculator values to the values you found in part (a)?

Exercises 7–10 deal with lines that are tangent to a parabola. To determine the equations for the tangent lines in these problems, use the following result, which was known (in a purely geometric form) to Archimedes.

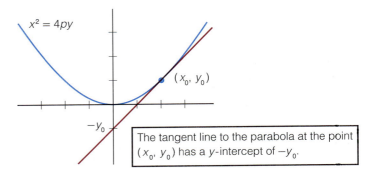

The tangent line to the parabola at the point (x_0, y_0) has a y-intercept of $-y_0$.

7. As a first example, consider the parabola $x^2 = 8y$. Note that the point $(4, 2)$ lies on this parabola. (Why?)

According to the statement accompanying the figure, the tangent to the parabola at $(4, 2)$ will have a y-intercept of $y_0 = -2$. Use this fact to find the equation of the tangent line. Then use the calculator to graph the parabola and the tangent line.

8. Graph the parabola $x^2 = 2y$ and the tangent to the curve at the point $(-2, 2)$.

9. Graph the parabola $x^2 = y$ and the tangent to the curve at the point $(-3, 9)$.

10. In this exercise, we use the graphing calculator to verify a particular case of the following general result:
Suppose $\overline{PQ}$ is a focal chord of a parabola. Then the tangents at P and Q are perpendicular to each other, and they intersect at a point on the directrix.
(a) The line segment $\overline{PQ}$ is a focal chord of the parabola $x^2 = 8y$ and the coordinates of P are $(8, 8)$. Find the coordinates of Q. Also find the equation of the directrix.
(b) Find the equations of the tangent lines at P and Q.
(c) Graph the parabola, the two tangent lines, and the directrix. (Use ZOOM 6, then ZOOM 5.) If your calculations are correct, the tangent lines will meet at a point on the directrix and they will be perpendicular.

11. On page 710 in the text was the statement that calculator exercises at the end of that section would help to convince you that the general shape of the ellipse shown in Figure 5 (on page 710) is correct. Now let's make that "graphing calculator exercises" rather than merely "calculator exercises."
(a) Take the equation $(x^2/3^2) + (y^2/2^2) = 1$, as given in Exercise 33 on page 716, and solve for y to obtain
$$y = \pm\tfrac{1}{3}\sqrt{36 - 4x^2}$$
Press the Y= key, enter these two functions, press ZOOM 6, and then ZOOM 5. Which types of symmetry does the ellipse possess?
(b) Take the equation $(x^2/1^2) + (y^2/4^2) = 1$, as given in Exercise 34 on page 716, and solve for y to obtain
$$y = \pm 4\sqrt{1 - x^2}$$
Press the Y= key, enter these two functions, press ZOOM 6, then ZOOM 5. Which types of symmetry does the ellipse possess?

12. (a) Solve the equation $4x^2 + 9y^2 = 36$ for y, and then graph this ellipse (as in Exercise 11).
(b) If we translate the ellipse in part (a) right one unit and up three units, what is the equation of the new ellipse?
(c) Check that if you solve the equation obtained in part (b) for y, the result can be written
$$y = \pm\tfrac{1}{3}\sqrt{36 - 4(x - 1)^2} + 3$$

(d) Press the Y= key and enter the two functions given in part (c); then press ZOOM 6 to see the graph in the standard viewing rectangle. What is the center of this ellipse?

13. In this exercise we graph four ellipses. Each ellipse will have the same major axis, but the eccentricities will vary from 0 to 0.8. (Actually, we'll only graph the top half of each ellipse, the portion corresponding to $y \geq 0$, because we can enter at most four functions in the Y= menu.)
(a) The equation $b^2x^2 + 16y^2 = 16b^2$ represents an ellipse. Show that the length of the horizontal axis of this ellipse is 8.
(b) Check that the following values for b^2 yield the indicated eccentricities.

b^2	16	13.44	10.24	5.76
eccentricity	0	0.4	0.6	0.8

(c) Solve the equation $b^2x^2 + 16y^2 = 16b^2$ for y. Show that the result can be written
$$y = \pm(0.25)\sqrt{b^2(16 - x^2)}$$
Now use the calculator to graph the (top half) of the four ellipses corresponding to the four values of b^2 in the table. Use the following settings: first press ZOOM 6; then press the RANGE key and set $x_{min} = -4$, $x_{max} = 4$, $y_{min} = 0$, and $y_{max} = 4$; finally, press ZOOM 5 so that the same scale is used on both axes. [Without this last step, the circle (that is, the ellipse with eccentricity 0) will not look like a circle.] If you have done things correctly, your graphs should be similar to the following figure.

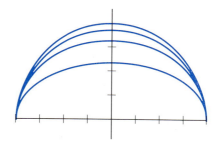

14. The *auxiliary circle* of an ellipse is defined to be the circle with diameter the same as the major axis of the ellipse. Determine the equation of the auxiliary circle for the ellipse $9x^2 + 25y^2 = 225$. Then use the calculator to graph the ellipse and the circle.

For Exercises 15 and 16, use the following fact:
The equation of the tangent line to the ellipse $(x^2/a^2) + (y^2/b^2) = 1$ at the point (x_0, y_0) on the ellipse is

$$\frac{x_0 x}{a^2} + \frac{y_0 y}{b^2} = 1$$

(Note that this last equation is a linear equation.) In each exercise, determine the equation of the tangent to the ellipse at the given point; then graph the ellipse and the tangent line.

15. $3x^2 + y^2 = 12$; $(1, -3)$

16. $x^2 + 3y^2 = 28$; $(4, 2)$

17. (As background for this exercise, you need to have completed Exercises 14 and 15.) In this exercise you're going to illustrate an interesting result concerning an ellipse and its auxiliary circle.
    (a) Graph the ellipse $x^2 + 3y^2 = 12$.
    (b) Find the values of a, b, and c for this ellipse.
    (c) Find the equation of the auxiliary circle for this ellipse. Add the graph of this circle to your picture from part (a).
    (d) Verify that the point $P(3, 1)$ lies on the ellipse. Then find the equation of the tangent line to the ellipse at P. Next, we want to add this tangent line to the picture in part (c), but we already have four functions defined in the Y= menu. Delete the two functions in the Y= menu that correspond to the bottom halves of the ellipse and the circle. Now you can add the tangent line to the picture.
    (e) In part (b) you determined the value of c. Find the equation of the line passing through $(-c, 0)$ and perpendicular to the tangent in part (d). Then add the graph of this line to your picture. If you've done things correctly, you should obtain a figure similar to the following one. The figure provides an example of this general result: The line through the focus and perpendicular to the tangent meets the tangent at a point on the auxiliary circle.

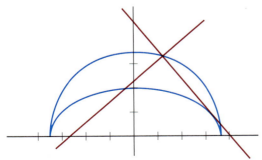

FIGURE FOR PROBLEM 17
The line through the focus and perpendicular to a tangent meets the tangent at a point on the auxiliary circle.

18. In this exercise you are going to look at the graph of the hyperbola $16x^2 - 9y^2 = 144$ from two perspectives. As preparation, clear all functions from the Y= menu and press ZOOM 6 for the standard viewing rectangle.
    (a) Solve the equation $16x^2 - 9y^2 = 144$ for y, enter the two functions that you obtain in the Y= menu, and press GRAPH.
    (b) Determine the equations of the two asymptotes for this hyperbola. Then enter the equations for the asymptotes in the Y= menu and press GRAPH.
    (c) Looking at the graph obtained in part (b), you can see that the hyperbola seems to be drawing closer to its asymptotes as $|x|$ gets large. To see more dramatic evidence of this, press ZOOM 3 and ENTER. At this scale, the hyperbola is virtually indistinguishable from its asymptotes. (Use the RANGE key to see what scale is involved.)

19. In this exercise we graph the hyperbola

$$\frac{(y - 3)^2}{5^2} - \frac{(x - 4)^2}{3^2} = 1$$

   (a) Show that the equation is equivalent to

$$y = 3 \pm 5\sqrt{1 + (x - 4)^2/9}$$

   Enter these two functions in the Y= menu and press ZOOM 6 to see the hyperbola in the standard viewing rectangle. Then, for a better view, make the following adjustments in the RANGE menu: $x_{min} = -20$, $x_{max} = 20$, $x_{scl} = 2$, $y_{min} = -20$, $y_{max} = 20$, and $y_{scl} = 2$.
   (b) Show that the equations of the asymptotes are

$$y = \pm \tfrac{5}{3}(x - 4) + 3$$

   Enter these two equations in the Y= menu and press GRAPH. Check to see that your result is consistent with the graph shown in Figure 7 on page 724.

20. In this exercise we graph a hyperbola in which the axes of the curve are not parallel to the coordinate axes. The equation is $x^2 + 4xy - 2y^2 = 6$.
    (a) Use the quadratic formula to solve the equation for y in terms of x. Show that the result can be written

$$y = x \pm \tfrac{1}{2}\sqrt{6x^2 - 12}$$

   (b) Enter these two functions in the Y= menu. Press ZOOM 6 and then make the following adjustments in the RANGE menu: $x_{min} = -6$, $x_{max} = 6$, $x_{scl} = 0.5$. Finally, press GRAPH.

21. In this exercise we graph an ellipse in which the axes of the curve are not parallel to the coordinate axes. The equation is $x^2 - 3x + xy + y^2 = 1$.

(a) Use the quadratic formula to solve the equation for y in terms of x. Show that the result can be written

$$y = \frac{-x \pm \sqrt{-3x^2 + 12x + 4}}{2}$$

(b) Press the $Y=$ key and enter the two functions that you obtained in part (a). Press ZOOM 6. Then, for a better view of the curve, press the RANGE key and set $x_{min} = -1$, $x_{max} = 5$, $y_{min} = -4$, and $y_{max} = 2$.

(c) Looking at the graph in part (b), make a rough estimate for the coordinates of the center of the ellipse.

(d) In general, the graph of an equation of the form

$$Ax^2 + Bxy + Cy^2 + Dx + Ey + F = 0$$

is a conic section. Furthermore, in cases where the equation represents an ellipse or a hyperbola, the coordinates (h, k) of the center of the curve are given by

$$h = \frac{2CD - BE}{B^2 - 4AC} \quad \text{and} \quad k = \frac{2AE - BD}{B^2 - 4AC}$$

Use these formulas to compute the center of the ellipse $x^2 - 3x + xy + y^2 = 1$. How close was your estimate in part (c)?

CHAPTER TWELVE TEST

1. Find the focus and the directrix of the parabola $y^2 = -12x$, and sketch the graph.

2. Graph the hyperbola $x^2 - 4y^2 = 4$. Specify the foci and the asymptotes.

3. The foci of an ellipse are $(0, \pm 2)$, and the eccentricity is $\frac{1}{2}$. Determine the equation of the ellipse. Write your answer in standard form.

4. Determine the equation of the hyperbola with foci $(\pm 2, 0)$ and with asymptotes $y = \pm(1/\sqrt{3})x$. Write your answer in standard form.

5. Let F_1 and F_2 denote the foci of the hyperbola $5x^2 - 4y^2 = 80$.
   (a) Verify that the point P with coordinates $(6, 5)$ lies on the hyperbola.
   (b) Compute the quantity $(F_1P - F_2P)^2$.

6. Graph the ellipse $4x^2 + 25y^2 = 100$. Specify the foci, the directrices, and the lengths of the major and minor axes.

7. Graph the equation $16x^2 + y^2 - 64x + 2y + 65 = 0$.

8. Graph the equation $\dfrac{(x + 4)^2}{3^2} - \dfrac{(y - 4)^2}{1^2} = 1$.

9. Graph the parabola $(x - 1)^2 = 8(y - 2)$. Specify the focal width and the vertex.

For Problems 10–12, refer to the following figure.

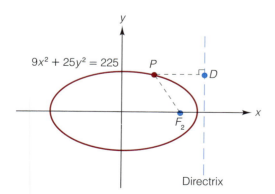

10. Find the equation of the directrix shown in the figure.

11. Suppose that the y-coordinate of the point P is $\frac{12}{5}$.

    (a) Find the x-coordinate of P.

    (b) Find the (horizontal) distance from P to the directrix.

    (c) Find the focus F_2. Then use the distance formula to compute F_2P.

    (d) Check your answer in part (c) by using one of the formulas for the length of a focal radius.

12. Using the results in Problem 11(b) and (c), verify that the ratio F_2P/PD is equal to the eccentricity of the ellipse.

CHAPTER THIRTEEN

ADDITIONAL TOPICS IN ALGEBRA

... in the ordinary treatise on the elements of algebra, these [additional] topics are either completely omitted or treated carelessly. For this reason, I am certain that the material I have gathered in this book is quite sufficient to remedy that defect.

Leonhard Euler (1707–1783) in his classic text, *Introductio in analysis infinitorum* (1748)

INTRODUCTION

A strong background in algebra is an important prerequisite for courses in calculus and in probability and statistics. In this final chapter we develop several additional topics that help provide a foundation in those areas of study. We begin in Section 13.1 with the principle of mathematical induction. This gives us a framework for proving statements about the natural numbers. In Section 13.2 we discuss the binomial theorem, which is used to analyze and expand expressions of the form $(a + b)^n$. As you'll see, the proof of the binomial theorem uses mathematical induction. Section 13.3 introduces the related (but distinct) concepts of sequences and series. Then in the next two sections (Sections 13.4 and 13.5), we study arithmetic and geometric sequences and series. The last topic introduced in Section 13.5 concerns finding sums of infinite geometric series. In a sense, this is an appropriate topic for our last chapter, for it is closely related to the idea of a *limit,* which is the starting point for calculus. Finally, in the last two sections of this chapter (Sections 13.6 and 13.7), we introduce the important concepts of permutations, combinations, and probability.

13.1 MATHEMATICAL INDUCTION

Mathematical induction is not a method of discovery but a technique of proving rigorously what has already been discovered.

David M. Burton in *The History of Mathematics, An Introduction* (Boston: Allyn and Bacon, 1985)

In Pascal's treatise [Traité du triangle arithmétique, written in 1653] ... appears one of the earliest acceptable statements of the method of mathematical induction.

Howard Eves in *An Introduction to the History of Mathematics,* 5th ed. (Philadelphia: Saunders, 1983)

Is mathematics an experimental science? The answer to this question is both yes and no, as the following example illustrates. Consider the problem of determining a formula for the sum of the first n odd natural numbers:

$$1 + 3 + 5 + \cdots + (2n - 1)$$

We begin by doing some calculations in the hope that this may shed some light on the problem. Table 1 shows the results of calculating the sum of the first n odd natural numbers for values of n ranging from 1 to 5. Upon inspecting the table, we observe that each sum in the right-hand column is the square of the corresponding entry in the left-hand column. For instance, for $n = 5$, we see that

$$\underbrace{1 + 3 + 5 + 7 + 9}_{\text{five terms}} = 5^2$$

Now let us try the next case, where $n = 6$, and see if the pattern persists. That is, we want to know if it is true that

$$\underbrace{1 + 3 + 5 + 7 + 9 + 11}_{\text{six terms}} = 6^2$$

As you can easily check, this last equation is true. Thus, based on the experimental (or empirical) evidence, we are led to the following conjecture:

CONJECTURE The sum of the first n odd natural numbers is n^2. That is, $1 + 3 + 5 + \cdots + (2n - 1) = n^2$, for each natural number n.

At this point, the "law" we've discovered is indeed really only a conjecture. After all, we've checked it only for values of n ranging from 1 to 6. It is conceivable at this point (although we may feel it is unlikely) that the conjecture is false for certain values of n. For the conjecture to be useful, we must be able to prove that it holds without exception for *all* natural numbers n. In fact, we will subsequently prove that this conjecture is valid. But before explaining the method of proof to be used, let us look at one more example.

Again, let n denote a natural number. Then consider the following question: Which quantity is the larger, 2^n or $(n + 1)^2$? As before, we begin by doing some calculations. This is the experimental stage of our work. According to Table 2, the quantity $(n + 1)^2$ is larger than 2^n for each value of n up through $n = 5$. Thus, we make the following conjecture:

CONJECTURE $(n + 1)^2 > 2^n$ for all natural numbers n.

Again, we note that this is only a conjecture at this point. Indeed, if we try the case where $n = 6$, we find that the pattern does not persist. That is, when $n = 6$, we find that 2^n is 64, while $(n + 1)^2$ is only 49. So in this example, the conjecture is not true in general; we have found a value of n for which it fails.

The preceding examples show that experimentation does have a place in mathematics, but we must be careful with the results. When experimentation leads to a conjecture, proof is required before the conjecture can be viewed as a valid law. For the remainder of this section, we shall discuss one such method of proof—*mathematical induction*.

In order to state the principle of mathematical induction, we first introduce

TABLE 1

n	$1 + 3 + 5 + \cdots + (2n - 1)$
1	1 $\qquad\qquad = 1$
2	$1 + 3$ $\qquad\quad = 4$
3	$1 + 3 + 5$ $\qquad = 9$
4	$1 + 3 + 5 + 7$ $\quad = 16$
5	$1 + 3 + 5 + 7 + 9 = 25$

TABLE 2

n	2^n	$(n + 1)^2$
1	2	4
2	4	9
3	8	16
4	16	25
5	32	36

some notation. Suppose that for each natural number n we have a statement P_n to be proved. Consider, for instance, the first conjecture we arrived at:

$$1 + 3 + 5 + \cdots + (2n - 1) = n^2$$

If we denote this statement by P_n, then we have, for example, that

P_1 is the statement that $1 = 1^2$
P_2 is the statement that $1 + 3 = 2^2$
P_3 is the statement that $1 + 3 + 5 = 3^2$

With this notation, we can now state the **principle of mathematical induction.**

Principle of Mathematical Induction

Suppose that for each natural number n, we have a statement P_n for which the following two conditions hold:

1. P_1 is true.

2. For each natural number k, if P_k is true, then P_{k+1} is true.

Then all the statements are true; that is, P_n is true for all natural numbers n.

The idea behind mathematical induction is a simple one. Think of each statement P_n as the rung of a ladder to be climbed. Then we make the analogy shown in Table 3.

TABLE 3

Mathematical Induction		Ladder Analogy	
Hypotheses	1. P_1 is true. 2. If P_k is true, then P_{k+1} is true, for any k.	Hypotheses	1'. You can reach the first rung. 2'. If you are on the kth rung, you can reach the $(k+1)$st rung, for any k.
Conclusion	3. P_n is true for all n.	Conclusion	3'. You can climb the entire ladder.

According to the principle of mathematical induction, we can prove that a statement or formula P_n is true for all n if we carry out the following two steps:

STEP 1 Show that P_1 is true.

STEP 2 Assume that P_k is true, and on the basis of this assumption, show that P_{k+1} is true.

In Step 2, the assumption that P_k is true is referred to as the **induction hypothesis.** (In computer science, Step 1 is sometimes referred to as the **initialization step.**) Let us now turn to some examples of proof by mathematical induction.

EXAMPLE 1   Use mathematical induction to prove that

$$1 + 3 + 5 + \cdots + (2n - 1) = n^2$$

for all natural numbers n.

Solution   Let P_n denote the statement that $1 + 3 + 5 + \cdots + (2n - 1) = n^2$. Then we want to show that P_n is true for all natural numbers n.

STEP 1   We must check that P_1 is true. But P_1 is just the statement that $1 = 1^2$, which is true.

STEP 2   Assuming that P_k is true, we must show that P_{k+1} is true. Thus, we assume that

$$1 + 3 + 5 + \cdots + (2k - 1) = k^2 \qquad (1)$$

That is the induction hypothesis. We must now show that

$$1 + 3 + 5 + \cdots + (2k - 1) + [2(k + 1) - 1] = (k + 1)^2 \qquad (2)$$

In order to derive equation (2) from equation (1), we add the quantity $[2(k + 1) - 1]$ to both sides of equation (1). (The motivation for this stems from the observation that the left-hand sides of equations (1) and (2) differ only by the quantity $[2(k + 1) - 1]$.) We obtain

$$1 + 3 + 5 + \cdots + (2k - 1) + [2(k + 1) - 1] = k^2 + [2(k + 1) - 1]$$
$$= k^2 + 2k + 1$$
$$= (k + 1)^2$$

This last equation is what we wanted to show. And having now carried out Steps 1 and 2, we conclude by the principle of mathematical induction that P_n is true for all natural numbers n.

EXAMPLE 2   Use mathematical induction to prove that

$$2^3 + 4^3 + 6^3 + \cdots + (2n)^3 = 2n^2(n + 1)^2$$

for all natural numbers n.

Solution   Let P_n denote the statement that

$$2^3 + 4^3 + 6^3 + \cdots + (2n)^3 = 2n^2(n + 1)^2$$

Then we want to show that P_n is true for all natural numbers n.

STEP 1   We must check that P_1 is true. P_1 is the statement that

$$2^3 = 2(1^2)(1 + 1)^2 \qquad \text{or} \qquad 8 = 8$$

Thus, P_1 is true.

STEP 2   Assuming that P_k is true, we must show that P_{k+1} is true. Thus, we assume that

$$2^3 + 4^3 + 6^3 + \cdots + (2k)^3 = 2k^2(k + 1)^2 \qquad (3)$$

We must now show that

$$2^3 + 4^3 + 6^3 + \cdots + (2k)^3 + [2(k + 1)]^3 = 2(k + 1)^2(k + 2)^2 \qquad (4)$$

Adding $[2(k + 1)]^3$ to both sides of equation (3) yields

$$2^3 + 4^3 + 6^3 + \cdots + (2k)^3 + [2(k + 1)]^3 = 2k^2(k + 1)^2 + [2(k + 1)]^3$$
$$= 2k^2(k + 1)^2 + 8(k + 1)^3$$
$$= 2(k + 1)^2[k^2 + 4(k + 1)]$$
$$= 2(k + 1)^2(k^2 + 4k + 4)$$
$$= 2(k + 1)^2(k + 2)^2$$

We have now derived equation (4) from equation (3), as we wished to do. Having carried out Steps 1 and 2, we conclude by the principle of mathematical induction that P_n is true for all natural numbers n. ▮▮▮

EXAMPLE 3   As indicated in Table 4, the number 3 is a factor of $2^{2n} - 1$ when $n = 1, 2, 3,$ and 4. Use mathematical induction to show that 3 is a factor of $2^{2n} - 1$ for all natural numbers n.

Solution   Let P_n denote the statement that 3 is a factor of $2^{2n} - 1$. We want to show that P_n is true for all natural numbers n.

TABLE 4

n	$2^{2n} - 1$
1	$3\ (= 3 \cdot 1)$
2	$15\ (= 3 \cdot 5)$
3	$63\ (= 3 \cdot 21)$
4	$255\ (= 3 \cdot 85)$

STEP 1   We must check that P_1 is true. But P_1 in this case is just the statement that 3 is a factor of $2^{2(1)} - 1$; that is, 3 is a factor of 3, which is surely true.

STEP 2   Assuming that P_k is true, we must show that P_{k+1} is true. Thus, we assume that

$$3 \text{ is a factor of } 2^{2k} - 1 \tag{5}$$

and we must show that

$$3 \text{ is a factor of } 2^{2(k+1)} - 1$$

The strategy here will be to rewrite the expression $2^{2(k+1)} - 1$ in such a way that the induction hypothesis, statement (5), can be applied. We have

$$2^{2(k+1)} - 1 = 2^{2k+2} - 1$$
$$= 2^2 \cdot 2^{2k} - 1$$
$$= 4 \cdot 2^{2k} - 4 + 3$$
$$= 4(2^{2k} - 1) + 3 \tag{6}$$

Now, look at the right-hand side of equation (6). By the induction hypothesis, 3 is a factor of $2^{2k} - 1$. Thus, 3 is a factor of $4(2^{2k} - 1)$, from which it certainly follows that 3 is a factor of $4(2^{2k} - 1) + 3$. In summary, then, 3 is a factor of the right-hand side of equation (6). Consequently, 3 must be a factor of the left-hand side of equation (6), which is what we wished to show.

Having now completed Steps 1 and 2, we conclude by the principle of mathematical induction that P_n is true for all natural numbers n. In other words, 3 is a factor of $2^{2n} - 1$ for all natural numbers n. ▮▮▮

There are instances in which a given statement P_n is false for certain initial values of n, but true thereafter. An example of this is provided by the statement

$$2^n > (n + 1)^2$$

As you can easily check, this statement is false for $n = 1, 2, 3, 4,$ and 5. But, as Example 4 shows, the statement is true for $n \geq 6$. In Example 4, we adapt the principle of mathematical induction by beginning in Step 1 with a consideration of P_6 rather than P_1.

EXAMPLE 4 Use mathematical induction to prove that

$$2^n > (n + 1)^2 \qquad \text{for all natural numbers } n \geq 6$$

Solution STEP 1 We must first check that P_6 is true. But P_6 is simply the assertion that

$$2^6 > (6 + 1)^2 \qquad \text{or} \qquad 64 > 49$$

Thus, P_6 is true.

STEP 2 Assuming that P_k is true, where $k \geq 6$, we must show that P_{k+1} is true. Thus, we assume that

$$2^k > (k + 1)^2 \qquad \text{where } k \geq 6 \tag{7}$$

We must show that

$$2^{k+1} > (k + 2)^2$$

Multiplying both sides of inequality (7) by 2 gives us

$$2(2^k) > 2(k + 1)^2 = 2k^2 + 4k + 2$$

This can be rewritten

$$2^{k+1} > k^2 + 4k + (k^2 + 2)$$

However, since $k \geq 6$, it is certainly true that

$$k^2 + 2 > 4$$

We therefore have

$$2^{k+1} > k^2 + 4k + 4 \qquad \text{or} \qquad 2^{k+1} > (k + 2)^2$$

as we wished to show.

Having now completed Steps 1 and 2, we conclude that P_n is true for all natural numbers $n \geq 6$.

EXERCISE SET 13.1

A

In Exercises 1–18, use the principle of mathematical induction to show that the statements are true for all natural numbers.

1. $1 + 2 + 3 + \cdots + n = \frac{1}{2}n(n + 1)$
2. $2 + 4 + 6 + \cdots + 2n = n(n + 1)$
3. $1 + 4 + 7 + \cdots + (3n - 2) = \frac{1}{2}n(3n - 1)$
4. $5 + 9 + 13 + \cdots + (4n + 1) = n(2n + 3)$
5. $1^2 + 2^2 + 3^2 + \cdots + n^2 = \frac{1}{6}n(n + 1)(2n + 1)$

6. $2^2 + 4^2 + 6^2 + \cdots + (2n)^2 = \frac{2}{3}n(n + 1)(2n + 1)$
7. $1^2 + 3^2 + 5^2 + \cdots + (2n - 1)^2 = \frac{1}{3}n(2n - 1)(2n + 1)$
8. $2 + 2^2 + 2^3 + \cdots + 2^n = 2^{n+1} - 2$
9. $3 + 3^2 + 3^3 + \cdots + 3^n = \frac{1}{2}(3^{n+1} - 3)$
10. $e^x + e^{2x} + e^{3x} + \cdots + e^{nx} = \dfrac{e^{(n+1)x} - e^x}{e^x - 1} \quad (x \neq 0)$
11. $1^3 + 2^3 + 3^3 + \cdots + n^3 = \left[\frac{1}{2}n(n + 1)\right]^2$
12. $2^3 + 4^3 + 6^3 + \cdots + (2n)^3 = 2n^2(n + 1)^2$

13. $1^3 + 3^3 + 5^3 + \cdots + (2n-1)^3 = n^2(2n^2 - 1)$

14. $1 \cdot 2 + 3 \cdot 4 + 5 \cdot 6 + \cdots + (2n-1)(2n)$
$$= \tfrac{1}{3}n(n+1)(4n-1)$$

15. $1 \cdot 3 + 3 \cdot 5 + 5 \cdot 7 + \cdots + (2n-1)(2n+1)$
$$= \tfrac{1}{3}n(4n^2 + 6n - 1)$$

16. $\dfrac{1}{1 \times 3} + \dfrac{1}{2 \times 4} + \dfrac{1}{3 \times 5} + \cdots + \dfrac{1}{n(n+2)}$
$$= \dfrac{n(3n+5)}{4(n+1)(n+2)}$$

17. $1 + \dfrac{3}{2} + \dfrac{5}{2^2} + \dfrac{7}{2^3} + \cdots + \dfrac{2n-1}{2^{n-1}} = 6 - \dfrac{2n+3}{2^{n-1}}$

18. $1 + 2 \cdot 2 + 3 \cdot 2^2 + 4 \cdot 2^3 + \cdots + n \cdot 2^{n-1}$
$$= (n-1)2^n + 1$$

19. Show that $n \le 2^{n-1}$ for all natural numbers n.

20. Show that 3 is a factor of $n^3 + 2n$ for all natural numbers n.

21. Show that $n^2 + 4 < (n+1)^2$ for all natural numbers $n \ge 2$.

22. Show that $n^3 > (n+1)^2$ for all natural numbers $n \ge 3$.

In Exercises 23–26, prove that the statement is true for all natural numbers in the specified range. Use a calculator to carry out Step 1.

23. $(1.5)^n > 2n, \ n \ge 7$

24. $(1.25)^n > n, \ n \ge 11$

25. $(1.1)^n > n, \ n \ge 39$

26. $(1.1)^n > 5n, \ n \ge 60$

B

27. Let $f(n) = \dfrac{1}{1 \cdot 2} + \dfrac{1}{2 \cdot 3} + \dfrac{1}{3 \cdot 4} + \cdots + \dfrac{1}{n(n+1)}$.

(a) Complete the following table.

n	1	2	3	4	5
$f(n)$					

(b) On the basis of the results in the table, what would you guess to be the value of $f(6)$? Compute $f(6)$ to see if this is correct.

(c) Make a conjecture about the value of $f(n)$, and prove it using mathematical induction.

28. Let $f(n) = \dfrac{1}{1 \times 3} + \dfrac{1}{3 \times 5} + \dfrac{1}{5 \times 7} + \cdots$
$$+ \dfrac{1}{(2n-1)(2n+1)}.$$

(a) Complete the following table.

n	1	2	3	4
$f(n)$				

(b) On the basis of the results in the table, what would you guess to be the value of $f(5)$? Compute $f(5)$ to see if your guess is correct.

(c) Make a conjecture about the value of $f(n)$, and prove it using mathematical induction.

29. Suppose that a function f satisfies the following conditions:

$$f(1) = 1$$
$$f(n) = f(n-1) + 2\sqrt{f(n-1)} + 1 \qquad \text{for } n \ge 2$$

(a) Complete the table.

n	1	2	3	4	5
$f(n)$					

(b) On the basis of the results in the table, what would you guess to be the value of $f(6)$? Compute $f(6)$ to see if your guess is correct.

(c) Make a conjecture about the value of $f(n)$ when n is a natural number, and prove the conjecture using mathematical induction.

30. This exercise demonstrates the necessity of carrying out both Step 1 and Step 2 before considering an induction proof valid.

(a) Let P_n denote the statement that $n^2 + 1$ is even. Check that P_1 is true. Then give an example showing that P_n is not true for all n.

(b) Let Q_n denote the statement that $n^2 + n$ is odd. Show that Step 2 of an induction proof can be completed in this case, but not Step 1.

31. A *prime number* is a natural number that has no factors other than itself and 1. For technical reasons, 1 is not considered a prime. Thus, the list of the first seven primes looks like this: 2, 3, 5, 7, 11, 13, 17. Let P_n be the statement that $n^2 + n + 11$ is prime. Check that P_n is true for all values of n less than 10. Check that P_{10} is false.

32. Prove that if $x \ne 1$,

$$1 + 2x + 3x^2 + \cdots + nx^{n-1} = \dfrac{1 - x^n}{(1-x)^2} - \dfrac{nx^n}{1-x}$$

for all natural numbers n.

33. If $r \ne 1$, show that $1 + r + r^2 + \cdots + r^{n-1} = \dfrac{(r^n - 1)}{(r-1)}$

for all natural numbers n.

34. Use mathematical induction to show that

$$x^n - 1 = (x - 1)(1 + x + x^2 + \cdots + x^{n-1})$$

for all natural numbers n.

35. Prove that 5 is a factor of $n^5 - n$ for all natural numbers $n \geq 2$.

36. Prove that 4 is a factor of $5^n + 3$ for all natural numbers n.

37. Prove that 5 is a factor of $2^{2n+1} + 3^{2n+1}$ for all nonnegative integers n.

38. Prove that 8 is a factor of $3^{2n} - 1$ for all natural numbers n.

39. Prove that 3 is a factor of $2^{n+1} + (-1)^n$ for all nonnegative integers n.

40. Prove that 6 is a factor of $n^3 + 3n^2 + 2n$ for all natural numbers n.

41. Use mathematical induction to show that $x - y$ is a factor of $x^n - y^n$ for all natural numbers n. *Suggestion for Step 2:* Verify and then use the fact that $x^{k+1} - y^{k+1} = x^k(x - y) + (x^k - y^k)y$.

In Exercises 42 and 43, use mathematical induction to prove that the formulas hold for all natural numbers n.

42. $\log_{10}(a_1 a_2 \cdots a_n) = \log_{10} a_1 + \log_{10} a_2 + \cdots + \log_{10} a_n$

43. $(1 + p)^n \geq 1 + np$, where $p > -1$

13.2 THE BINOMIAL THEOREM

A mathematician, like a painter or a poet, is a maker of patterns.

G. H. Hardy (1877–1947)

[Blaise Pascal] *made numerous discoveries relating to this array and set them forth in his* Traité du triangle arithmétique, *published posthumously in 1665, and among these was essentially our present Binomial Theorem for positive integral exponents.*

David Eugene Smith in *History of Mathematics,* vol. II (New York: Ginn and Co., 1925)

If you look back at Section 1.8, you will see that two of the special products listed there are

$$(a + b)^2 = a^2 + 2ab + b^2$$

and

$$(a + b)^3 = a^3 + 3a^2b + 3ab^2 + b^3$$

Our present goal is to develop a general formula, known as the *binomial theorem,* for expanding any product of the form $(a + b)^n$, when n is a natural number.

We begin by looking for patterns in the expansion of $(a + b)^n$. To do this, let's list the expansions of $(a + b)^n$ for $n = 1, 2, 3, 4,$ and 5. (Exercises 1 and 2 ask you to verify these results simply by repeated multiplication.)

$$(a + b)^1 = a + b$$
$$(a + b)^2 = a^2 + 2ab + b^2$$
$$(a + b)^3 = a^3 + 3a^2b + 3ab^2 + b^3$$
$$(a + b)^4 = a^4 + 4a^3b + 6a^2b^2 + 4ab^3 + b^4$$
$$(a + b)^5 = a^5 + 5a^4b + 10a^3b^2 + 10a^2b^3 + 5ab^4 + b^5$$

After surveying these results, we note the following patterns.

> **PROPERTY SUMMARY**   **PATTERNS OBSERVED IN $(a + b)^n$ for $n = 1, 2, 3, 4, 5$**

GENERAL STATEMENT	EXAMPLE
There are $n + 1$ terms.	There are 4 ($= 3 + 1$) terms in the expansion of $(a + b)^3$.
The expansion begins with a^n and ends with b^n.	$(a + b)^3$ begins with a^3 and ends with b^3.
The sum of the exponents in each term is n.	The sum of the exponents in each term of $(a + b)^3$ is 3.
The exponents of a decrease by 1 from term to term.	$(a + b)^3 = a^③ + 3a^②b + 3a^①b^2 + a^⓪b^3$
The exponents of b increase by 1 from term to term.	$(a + b)^3 = a^3b^⓪ + 3a^2b^① + 3ab^② + b^③$
When n is even, the coefficients are symmetric about the middle term.	The sequence of coefficients for $(a + b)^4$ is 1, 4, 6, 4, 1.
When n is odd, the coefficients are symmetric about the two middle terms.	The sequence of coefficients for $(a + b)^5$ is 1, 5, 10, 10, 5, 1.

The patterns we have just observed for $(a + b)^n$ persist for all natural numbers n. (This will follow from the binomial theorem, which is proved at the end of this section.) Thus, for example, the form of $(a + b)^6$ must be as follows:

$$(a + b)^6 = a^6 + \underline{?}a^5b + \underline{?}a^4b^2 + \underline{?}a^3b^3 + \underline{?}a^2b^4 + \underline{?}ab^5 + b^6$$

The problem now is to find the proper coefficient for each term. To do this, we need to discover additional patterns in the expansion of $(a + b)^n$.

We have already written out the expansions of $(a + b)^n$ for values of n ranging from 1 to 5. Let us now write only the coefficients appearing in those expansions. The resulting triangular array of numbers is known as **Pascal's triangle.**[*] For reasons of symmetry, we begin with $(a + b)^0$ rather than $(a + b)^1$.

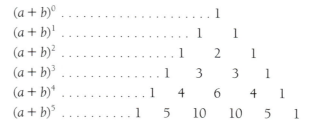

$(a + b)^0$ 1
$(a + b)^1$ 1   1
$(a + b)^2$ 1   2   1
$(a + b)^3$ 1   3   3   1
$(a + b)^4$ 1   4   6   4   1
$(a + b)^5$ 1   5   10   10   5   1

The key observation regarding Pascal's triangle is this: *Each entry in the array (other than the 1's along the sides) is the sum of the two numbers diagonally above it.* For instance, the 6 that appears in the fifth row is the sum of the two 3's diagonally above it. Using this observation, we can form as many additional rows as we please. The coefficients for $(a + b)^n$ will then appear in the $(n + 1)$st

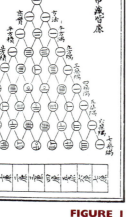

FIGURE 1
"Pascal's" triangle,
by Chu-Shi-Kie,
A.D. 1303

*The array is named after Blaise Pascal, a seventeenth-century French mathematician and philosopher. However, as Figure 1 indicates, the Pascal triangle was known to Chinese mathematicians centuries earlier.

row of the array.† For instance, to obtain the row corresponding to $(a + b)^6$, we have

sixth row, $(a + b)^5$:   1   5   10   10   5   1

seventh row, $(a + b)^6$:   1   6   15   20   15   6   1

Thus, the sequence of coefficients for $(a + b)^6$ is 1, 6, 15, 20, 15, 6, 1. This answers the question raised earlier about the expansions of $(a + b)^6$. We have

$$(a + b)^6 = a^6 + 6a^5b + 15a^4b^2 + 20a^3b^3 + 15a^2b^4 + 6ab^5 + b^6$$

For analytical work or for larger values of the exponent n, it is inefficient to rely on Pascal's triangle. For this reason, we point out another pattern in the expansions of $(a + b)^n$.

> In the expansion of $(a + b)^n$, the coefficient of any term after the first can be generated as follows. In the *previous* term, multiply the coefficient by the exponent of a and then divide by the number of that previous term.

To see how this observation is used, let's compute the second, third, and fourth coefficients in the expansion of $(a + b)^6$. To compute the coefficient of the second term, we go back to the first term, which is a^6. We have

$$\text{coefficient of second term} = \frac{1 \cdot 6}{1} = 6$$

where the 1 in the numerator is the coefficient of first term, the 6 is the exponent of a in first term, and the 1 in the denominator is the number of first term.

Thus, the second term is $6a^5b$ and, consequently, we have

$$\text{coefficient of third term} = \frac{6 \cdot 5}{2} = 15$$

where the 6 in the numerator is the coefficient of second term, the 5 is the exponent of a in second term, and the 2 in the denominator is the number of second term.

Continuing now with this method, you should check for yourself that the coefficient of the fourth term in the expansion of $(a + b)^6$ is 20.

Note   We now know that the first four coefficients are 1, 6, 15, and 20. By symmetry, it follows that the complete sequence of coefficients for this expansion is 1, 6, 15, 20, 15, 6, 1. No additional calculation for the coefficients is necessary.

---

†That these numbers actually are the appropriate coefficients follows from the binomial theorem, which is proved at the end of this section.

EXAMPLE I Expand $(2x - y^2)^7$.

Solution First we write the expansion of $(a + b)^7$ using the method explained just prior to this example, or using Pascal's triangle. As you should check for yourself, the expansion is

$$(a + b)^7 = a^7 + 7a^6b + 21a^5b^2 + 35a^4b^3 + 35a^3b^4 + 21a^2b^5 + 7ab^6 + b^7$$

Now we make the substitutions $a = 2x$ and $b = -y^2$. This yields

$$[2x + (-y^2)]^7 = (2x)^7 + 7(2x)^6(-y^2) + 21(2x)^5(-y^2)^2$$
$$+ 35(2x)^4(-y^2)^3 + 35(2x)^3(-y^2)^4$$
$$+ 21(2x)^2(-y^2)^5 + 7(2x)(-y^2)^6 + (-y^2)^7$$
$$= 128x^7 - 448x^6y^2 + 672x^5y^4 - 560x^4y^6 + 280x^3y^8$$
$$- 84x^2y^{10} + 14xy^{12} - y^{14}$$

This is the required expansion. Notice how the signs alternate in the final answer; this is characteristic of all expansions of the form $(a - b)^n$. ∎

In preparation for the binomial theorem, we introduce two notations that are used not only in connection with the binomial theorem, but in many other areas of mathematics as well. The first of these notations is $n!$ (read "n factorial").

DEFINITION The Factorial Symbol

EXAMPLES

$n! = 1 \cdot 2 \cdot 3 \cdots n$

where n is a natural number

$0! = 1$

$3! = 1 \cdot 2 \cdot 3 = 6$

$\dfrac{6!}{4!} = \dfrac{6 \cdot 5 \cdot 4 \cdot 3 \cdot 2 \cdot 1}{4 \cdot 3 \cdot 2 \cdot 1}$

$= 6 \times 5 = 30$

EXAMPLE 2 Simplify the expression $\dfrac{(n + 1)!}{(n - 1)!}$.

Solution $\dfrac{(n + 1)!}{(n - 1)!} = \dfrac{(n + 1) \cdot n \cdot (n - 1)!}{(n - 1)!} = (n + 1) \cdot n = n^2 + n$ ∎

The second notation that we introduce in preparation for the binomial theorem is $\dbinom{n}{k}$. This notation is read "n choose k" (because it can be shown that $\dbinom{n}{k}$ is equal to the number of ways of choosing a subset of k elements from a set with n elements).

> **DEFINITION   The Binomial Coefficient** $\binom{n}{k}$
>
> Let n and k be nonnegative integers with $k \leq n$. Then the binomial coefficient $\binom{n}{k}$ is defined by
> $$\binom{n}{k} = \frac{n!}{k!(n-k)!}$$
>
> **EXAMPLE**
> $$\binom{5}{2} = \frac{5!}{2!(5-2)!} = \frac{5!}{2!3!}$$
> $$= \frac{5 \cdot 4 \cdot 3 \cdot 2 \cdot 1}{(2 \cdot 1)(3 \cdot 2 \cdot 1)}$$
> $$= \frac{5 \cdot 4}{2 \cdot 1} = 10$$

The binomial coefficients are so named because they are indeed the coefficients in the expansion of $(a + b)^n$. More precisely, the relationship is this:

The coefficients in the expansion of $(a + b)^n$ are the $n + 1$ numbers

$$\binom{n}{0}, \binom{n}{1}, \binom{n}{2}, \cdots, \binom{n}{n}$$

Subsequently, we will see why this statement is true. For now, however, let us look at an example. Consider the binomial coefficients $\binom{3}{0}, \binom{3}{1}, \binom{3}{2}$, and $\binom{3}{3}$. According to our statement, these four quantities should be the coefficients in the expansion of $(a + b)^3$. Let us check:

$$\binom{3}{0} = \frac{3!}{0!(3-0)!} = \frac{3!}{1(3!)} = 1$$
$$\binom{3}{1} = \frac{3!}{1!(3-1)!} = \frac{3 \cdot 2 \cdot 1}{1(2 \cdot 1)} = 3$$
$$\binom{3}{2} = \frac{3!}{2!(3-2)!} = \frac{3 \cdot 2 \cdot 1}{(2 \cdot 1)1} = 3$$
$$\binom{3}{3} = \frac{3!}{3!(3-3)!} = \frac{3!}{3!0!} = 1$$

The values of $\binom{3}{0}, \binom{3}{1}, \binom{3}{2}$, and $\binom{3}{3}$ are thus 1, 3, 3, and 1, respectively. But these last four numbers are indeed the coefficients in the expansion of $(a + b)^3$, as we wished to check.

We are now in a position to state the binomial theorem, after which we will look at several applications. Finally, at the end of this section, we will use mathematical induction to prove the theorem. In the statement of the theorem that follows, we are assuming that the exponent n is a natural number.

> **Binomial Theorem**
> $$(a + b)^n = \binom{n}{0}a^n + \binom{n}{1}a^{n-1}b + \binom{n}{2}a^{n-2}b^2 + \cdots + \binom{n}{n-1}ab^{n-1} + \binom{n}{n}b^n$$

One of the uses of the binomial theorem is in identifying specific terms in an expansion without computing the entire expansion. This is particularly helpful when the exponent n is relatively large. Looking back at the statement of the binomial theorem, there are three observations we can make. First, the coefficient of the rth term is $\binom{n}{r-1}$. For instance, the coefficient of the third term is $\binom{n}{3-1} = \binom{n}{2}$. The second observation is that the exponent for a in the rth term is $n - (r-1)$. For instance, the exponent for a in the third term is $n - (3-1) = n - 2$. Finally, we observe that the exponent for b in the rth term is $r - 1$, the same quantity that appears in the lower position of the corresponding binomial coefficient. For instance, the exponent for b in the third term is $r - 1 = 3 - 1 = 2$. We summarize these three observations with the following statement.

> The rth term in the expansion of $(a + b)^n$ is
> $$\binom{n}{r-1} a^{n-r+1} b^{r-1}$$

EXAMPLE 3 Find the fifteenth term in the expansion of $\left(x^2 - \dfrac{1}{x}\right)^{18}$.

Solution Using the values $r = 15$, $n = 18$, $a = x^2$, and $b = -1/x$, we have

$$\binom{n}{r-1} a^{n-r+1} b^{r-1} = \binom{18}{15-1}(x^2)^{18-15+1}\left(\frac{-1}{x}\right)^{15-1}$$

$$= \binom{18}{14} x^8 \cdot \frac{1}{x^{14}}$$

$$= \frac{18 \times 17 \times 16 \times 15 \times (14!)}{14!(4 \times 3 \times 2 \times 1)} x^{-6}$$

$$= \frac{18 \times 17 \times 16 \times 15}{4 \times 3 \times 2 \times 1} x^{-6}$$

$$= 3060x^{-6} \qquad \text{as required}$$

EXAMPLE 4 Find the coefficient of the term containing x^4 in the expression of $(x + y^2)^{30}$.

Solution Again we use the fact that the rth term in the expansion of $(a + b)^n$ is $\binom{n}{r-1} a^{n-r+1} b^{r-1}$. In this case, n is 30 and x plays the role of a. The exponent for x is then $n - r + 1$ or $30 - r + 1$. To see when this exponent is 4, we write

$$30 - r + 1 = 4 \qquad \text{and therefore} \qquad r = 27$$

The required coefficient is therefore $\binom{30}{27-1}$. We then have

$$\binom{30}{26} = \frac{30!}{26!(30-26)!} = \frac{30 \times 29 \times 28 \times 27}{4 \times 3 \times 2 \times 1}$$

After carrying out the indicated arithmetic, we find that $\binom{30}{26} = 27{,}405$. This is the required coefficient.   ▪▪▪

EXAMPLE 5   Find the coefficient of the term containing a^9 in the expansion of $(a + 2\sqrt{a})^{10}$.

Solution   The rth term in this expansion is

$$\binom{10}{r-1} a^{10-r+1} (2\sqrt{a})^{r-1}$$

We can rewrite this as

$$\binom{10}{r-1} a^{10-r+1} (2^{r-1})(a^{1/2})^{r-1}$$

or

$$2^{r-1} \binom{10}{r-1} a^{10-r+1+(r-1)/2}$$

This shows that the general form of the coefficient we wish to find is $2^{r-1} \binom{10}{r-1}$. We now need to determine r when the exponent of a is 9. Thus, we require that

$$10 - r + 1 + \frac{r-1}{2} = 9$$

$$-r + \frac{r-1}{2} = -2$$

$$-2r + r - 1 = -4 \qquad \text{multiplying by 2}$$

$$r = 3$$

The required coefficient is now obtained by substituting $r = 3$ in the expression $2^{r-1} \binom{10}{r-1}$. Thus, the required coefficient is

$$2^2 \binom{10}{2} = \frac{4 \cdot 10!}{2!(10-2)!} = 2 \cdot 10 \cdot 9 = 180 \qquad \text{▪▪▪}$$

There are three simple identities involving the binomial coefficients that will simplify our proof of the binomial theorem.

IDENTITY 1 $\binom{r}{0} = 1$ for all nonnegative integers r

IDENTITY 2 $\binom{r}{r} = 1$ for all nonnegative integers r

IDENTITY 3 $\binom{k}{r} + \binom{k}{r-1} = \binom{k+1}{r}$ for all natural numbers k and r with $r \le k$

All three of these identities can be proved directly from the definitions of the binomial coefficients, without the need for mathematical induction. The proofs of the first two are straightforward, and we omit them here. The proof of identity 3 runs as follows:

$$\binom{k}{r} + \binom{k}{r-1} = \frac{k!}{r!(k-r)!} + \frac{k!}{(r-1)!(k-r+1)!}$$

$$= \frac{k!}{r(r-1)!(k-r)!} + \frac{k!}{(r-1)!(k-r+1)(k-r)!}$$

Now, the common denominator on the right-hand side of the last equation is $r(r-1)!(k-r+1)(k-r)!$ Thus, we have

$$\binom{k}{r} + \binom{k}{r-1} = \frac{k!(k-r+1)}{r(r-1)!(k-r+1)(k-r)!} + \frac{k!r}{r(r-1)!(k-r+1)(k-r)!}$$

$$= \frac{k!(k-r+1) + k!r}{r(r-1)!(k-r+1)(k-r)!}$$

$$= \frac{k!(k-r+1+r)}{r(r-1)!(k-r+1)(k-r)!} = \frac{k!(k+1)}{r!(k-r+1)!}$$

$$= \frac{(k+1)!}{r![(k+1)-r]!}$$

$$= \binom{k+1}{r} \qquad \text{as required}$$

Taken together, the three identities show why the $(n+1)$st row of Pascal's triangle consists of the numbers $\binom{n}{0}, \binom{n}{1}, \binom{n}{2}, \ldots, \binom{n}{n}$. Identities 1 and 2 tell us that this row of numbers begins and ends with 1. Identity 3 is then just a statement of the fact that each entry in the row, other than the initial and final 1, is generated by adding the two entries diagonally above it.

We conclude this section by using mathematical induction to prove the binomial theorem. The statement P_n that we wish to prove for all natural numbers n is this:

$$(a+b)^n = \binom{n}{0}a^n + \binom{n}{1}a^{n-1}b + \cdots + \binom{n}{n-1}ab^{n-1} + \binom{n}{n}b^n$$

First, we check that P_1 is true. The statement P_1 asserts that

$$(a+b)^1 = \binom{1}{0}a^1 + \binom{1}{1}a^0 b$$

However, in view of identities 1 and 2, this last equation becomes

$$(a+b)^1 = 1 \cdot a + 1 \cdot b$$

which is surely true. Now let us assume that P_k is true and, on the basis of this assumption, show that P_{k+1} is true. The statement P_k is

$$(a+b)^k = \binom{k}{0}a^k + \binom{k}{1}a^{k-1}b + \cdots + \binom{k}{k-1}ab^{k-1} + \binom{k}{k}b^k$$

Multiplying both sides of this equation by the quantity $(a + b)$ yields

$$(a + b)^{k+1} = (a + b)\left[\binom{k}{0}a^k + \binom{k}{1}a^{k-1}b + \cdots + \binom{k}{k-1}ab^{k-1} + \binom{k}{k}b^k\right]$$

$$= a\left[\binom{k}{0}a^k + \binom{k}{1}a^{k-1}b + \cdots + \binom{k}{k-1}ab^{k-1} + \binom{k}{k}b^k\right]$$

$$\quad + b\left[\binom{k}{0}a^k + \binom{k}{1}a^{k-1}b + \cdots + \binom{k}{k-1}ab^{k-1} + \binom{k}{k}b^k\right]$$

$$= \binom{k}{0}a^{k+1} + \binom{k}{1}a^k b + \cdots + \binom{k}{k-1}a^2 b^{k-1} + \binom{k}{k}ab^k$$

$$\quad + \binom{k}{0}a^k b + \binom{k}{1}a^{k-1}b^2 + \cdots + \binom{k}{k-1}ab^k + \binom{k}{k}b^{k+1}$$

$$= \binom{k}{0}a^{k+1} + \left[\binom{k}{1} + \binom{k}{0}\right]a^k b + \cdots + \left[\binom{k}{k} + \binom{k}{k-1}\right]ab^k + \binom{k}{k}b^{k+1}$$

We can now make some substitutions on the right-hand side of this last equation. The initial binomial coefficient $\binom{k}{0}$ can be replaced by $\binom{k+1}{0}$, for both are equal to 1 according to identity 1. Similarly, the binomial coefficient $\binom{k}{k}$ appearing at the end of the equation can be replaced by $\binom{k+1}{k+1}$, since both are equal to 1 according to identity 2. Finally, we can use identity 3 to simplify each of the sums in the brackets. We obtain

$$(a + b)^{k+1} = \binom{k+1}{0}a^{k+1} + \binom{k+1}{1}a^k b + \cdots + \binom{k+1}{k}ab^k + \binom{k+1}{k+1}b^{k+1}$$

But this last equation is just the statement P_{k+1}; that is, we have derived P_{k+1} from P_k, as we wished to do. The induction proof is now complete.

EXERCISE SET 13.2

A

In Exercises 1 and 2, verify each statement directly, without using the techniques developed in this section.

1. (a) $(a + b)^2 = a^2 + 2ab + b^2$
   (b) $(a + b)^3 = a^3 + 3a^2 b + 3ab^2 + b^3$
   Hint: $(a + b)^3 = (a + b)(a + b)^2$

2. (a) $(a + b)^4 = a^4 + 4a^3 b + 6a^2 b^2 + 4ab^3 + b^4$
   Hint: Use the result in Exercise 1(b) and the fact that $(a + b)^4 = (a + b)(a + b)^3$.
   (b) $(a + b)^5 = a^5 + 5a^4 b + 10a^3 b^2 + 10a^2 b^3 + 5ab^4 + b^5$

In Exercises 3–28, carry out the indicated expansions.

3. $(a + b)^9$

4. $(a - b)^9$

5. $(2A + B)^3$

6. $(1 + 2x)^6$

7. $(1 - 2x)^6$

8. $(3x^2 - y)^5$

9. $(\sqrt{x} + \sqrt{y})^4$

10. $(\sqrt{x} - \sqrt{y})^4$

11. $(x^2 + y^2)^5$

12. $(5A - B^2)^3$

13. $[1 - (1/x)]^6$

14. $(3x + y^2)^4$

15. $[(x/2) - (y/3)]^3$

16. $(1 - z^2)^7$

17. $(ab^2 + c)^7$

18. $[x - (1/x)]^8$

19. $(x + \sqrt{2})^8$

20. $(4A - \frac{1}{2})^5$

21. $(\sqrt{2} - 1)^3$

22. $(1 + \sqrt{5})^4$

23. $(\sqrt{2} + \sqrt{3})^5$

24. $(\frac{1}{2} - 2a)^6$

25. $(2\sqrt[3]{2} - \sqrt[3]{4})^3$

26. $(x + y + 1)^4$
   Suggestion: Rewrite the expression as $[(x + y) + 1]^4$.

27. $(x^2 - 2x - 1)^5$

Suggestion: Rewrite the expression as $[x^2 - (2x + 1)]^5$.

28. $[x^2 - 2x - (1/x)]^6$

In Exercises 29–38, evaluate or simplify each expression.

29. $5!$

30. (a) $3! + 2!$
(b) $(3 + 2)!$

31. $\binom{7}{3}\binom{3}{2}$

32. $\dfrac{20!}{18!}$

33. (a) $\binom{5}{3}$ (b) $\binom{5}{4}$

34. (a) $\binom{7}{7}$ (b) $\binom{7}{0}$

35. $\dfrac{(n + 2)!}{n!}$

36. $\dfrac{n[(n - 2)!]}{(n + 1)!}$

37. $\binom{6}{4} + \binom{6}{3} - \binom{7}{4}$

38. $(3!)! + (3!)^2$

39. Find the fifteenth term in the expansion of $(a + b)^{16}$.

40. Find the third term in the expansion of $(a - b)^{30}$.

41. Find the one hundredth term in the expansion of $(1 + x)^{100}$.

42. Find the twenty-third term in the expansion of $[x - (1/x^2)]^{25}$.

43. Find the coefficient of the term containing a^4 in the expansion of $\left(\sqrt{a} - \sqrt{x}\right)^{10}$.

44. Find the coefficient of the term containing a^4 in the expansion of $(3a - 5x)^{12}$.

45. Find the coefficient of the term containing y^8 in the expansion of $[(x/2) - 4y]^9$.

46. Find the coefficient of the term containing x^6 in the expansion of $[x^2 + (1/x)]^{12}$.

47. Find the coefficient of the term containing x^3 in the expansion of $\left(1 - \sqrt{x}\right)^8$.

48. Find the coefficient of the term containing a^8 in the expansion of $\left[a - (2/\sqrt{a})\right]^{14}$.

49. Find the term that does not contain A in the expansion of $[(1/A) + 3A^2]^{12}$.

50. Find the coefficient of B^{-10} in the expansion of $[(B^2/2) - (3/B^3)]^{10}$.

B

51. Show that the coefficient of x^n in the expansion of $(1 + x)^{2n}$ is $(2n)!/(n!)^2$.

52. Find n so that the coefficients of the eleventh and thirteenth terms in $(1 + x)^n$ are the same.

53. (a) Complete the following table.

k	0	1	2	3	4	5	6	7	8
$\binom{8}{k}$									

(b) Use the results in part (a) to verify that
$$\binom{8}{0} + \binom{8}{1} + \binom{8}{2} + \cdots + \binom{8}{8} = 2^8.$$

(c) By taking $a = b = 1$ in the expansion of $(a + b)^n$, show that $\binom{n}{0} + \binom{n}{1} + \binom{n}{2} + \cdots + \binom{n}{n} = 2^n$.

C

54. Two real numbers A and B are defined by $A = \sqrt[99]{99!}$ and $B = \sqrt[100]{100!}$. Which number is larger, A or B? *Hint:* Compare A^{9900} and B^{9900}.

55. This exercise outlines a proof of the identity
$$\binom{n}{0}^2 + \binom{n}{1}^2 + \binom{n}{2}^2 + \cdots + \binom{n}{n}^2 = \binom{2n}{n}$$

(a) Verify that
$$(1 + x)^n\left(1 + \frac{1}{x}\right)^n = \frac{(1 + x)^{2n}}{x^n} \qquad (1)$$

(This requires only basic algebra, not the binomial theorem.)

(b) Show that the coefficient of the term independent of x on the right side of equation (1) is $\binom{2n}{n}$.

(c) Use the binomial theorem to expand $(1 + x)^n$. Then show that the coefficient of the term independent of x on the left side of equation (1) is
$$\binom{n}{0}^2 + \binom{n}{1}^2 + \binom{n}{2}^2 + \cdots + \binom{n}{n}^2$$

13.3 INTRODUCTION TO SEQUENCES AND SERIES

Students of calculus do not always understand that infinite series are primarily tools for the study of functions.

George F. Simmons in *Calculus with Analytic Geometry* (New York-McGraw-Hill, 1985)

This section and the next two sections in this chapter deal with numerical sequences. We will begin with a somewhat informal definition of this concept.

Then, after looking at some examples and terminology, we will present a more formal definition. A **numerical sequence** is an ordered list of numbers. Here are four examples:

example A: $1, \sqrt{2}, 10$

example B: $2, 4, 6, 8, \ldots$

example C: $1, \frac{1}{2}, \frac{1}{4}, \frac{1}{8}, \ldots$

example D: $1, 1, 1, 1, \ldots$

The individual entries in a numerical sequence are called the **terms** of the sequence. In this chapter, the terms in each sequence will always be real numbers, so for convenience, we will drop the adjective *numerical* and refer simply to *sequences*. (It is worth pointing out, however, that in more advanced courses, sequences are studied in which the individual terms are functions.) Any sequence possessing only a finite number of terms is called a **finite sequence**. Thus, the sequence in example A is a finite sequence. On the other hand, examples B, C, and D are examples of what we call **infinite sequences**; each contains infinitely many terms. As example D indicates, it is not necessary that all the terms in a sequence be distinct. In this chapter, all the sequences we discuss will be infinite sequences.

In examples B, C, and D, the three dots are read "and so on." In using this notation, we are assuming that it is clear what the subsequent terms of the sequence are. Toward this end, we often specify a formula for the *n*th term in a sequence. Example B in this case would appear this way:

$$2, 4, 6, 8, \ldots, 2n, \ldots$$

A letter with subscripts is often used to denote the various terms in a sequence. For instance, if we denote the sequence in example B by $a_1, a_2, a_3, \ldots$, then we have $a_1 = 2$, $a_2 = 4$, $a_3 = 6$, and, in general, $a_n = 2n$.

EXAMPLE 1   Consider the sequence $a_1, a_2, a_3, \ldots$, in which the *n*th term a_n is given by

$$a_n = \frac{n}{n+1}$$

Compute the first three terms of the sequence, as well as the one thousandth term.

Solution   To obtain the first term, we replace *n* by 1 in the given formula. This yields

$$a_1 = \frac{1}{1+1} = \frac{1}{2}$$

The other terms are similarly obtained. We have

$$a_2 = \frac{2}{2+1} = \frac{2}{3}$$

$$a_3 = \frac{3}{3+1} = \frac{3}{4}$$

$$a_{1000} = \frac{1000}{1000+1} = \frac{1000}{1001}$$

In the example just concluded, we were given an explicit formula for the nth term of the sequence. The next example shows a different way of specifying a sequence, in which each term except the first is defined in terms of the previous term. This is an example of a **recursive definition**. (Recursive definitions are particularly useful in computer programming.)

EXAMPLE 2   Compute the first three terms of the sequence $b_1, b_2, b_3, \ldots$, which is defined recursively by

$$b_1 = 4$$
$$b_n = 2(b_{n-1} - 1) \qquad \text{for } n \geq 2$$

Solution   We are given the first term: $b_1 = 4$. To find b_2, we replace n by 2 in the formula $b_n = 2(b_{n-1} - 1)$ to obtain

$$b_2 = 2(b_1 - 1) = 2(4 - 1) = 6$$

Thus, $b_2 = 6$. Next we use this value of b_2 in the formula $b_n = 2(b_{n-1} - 1)$ to obtain b_3. Replacing n by 3 in this formula yields

$$b_3 = 2(b_2 - 1) = 2(6 - 1) = 10$$

We have now found the first three terms of the sequence: $b_1 = 4$, $b_2 = 6$, and $b_3 = 10$. ∎

If you think about the central idea behind the concept of a sequence, you can see that a function is involved: For each input n, we have an output a_n. For this reason, the formal definition of a sequence is phrased as follows.

DEFINITION   Sequence

A **sequence** is a function whose domain is the set of natural numbers.

If we denote the function by f for the moment, then $f(1)$ would denote what we have been calling a_1, the first term of the sequence. Similarly, $f(2) = a_2$, $f(3) = a_3$, and so on.

EXAMPLE 3   Consider the sequence with nth term

$$a_n = \frac{1}{2^n}$$

Graph this sequence for $n = 1, 2, 3,$ and 4.

Solution   First, we compute the required values of a_n, as shown in the following table.

n	1	2	3	4
a_n	$\frac{1}{2}$	$\frac{1}{4}$	$\frac{1}{8}$	$\frac{1}{16}$

Now, locating the inputs n on the horizontal axis and the outputs a_n on the vertical axis, we obtain the graph shown in Figure 1. ∎

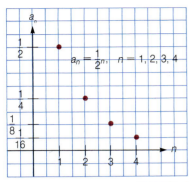

FIGURE 1

The graph of the sequence

$a_n = \dfrac{1}{2^n}$ for $n = 1, 2, 3, 4$

We will often be interested in the sum of certain terms of a sequence. Consider, for example, the sequence

$$10, 20, 30, 40, \ldots$$

in which the nth term is $10n$. The sum of the first four terms in this sequence is

$$10 + 20 + 30 + 40 = 100$$

More generally, the sum of the first n terms in this sequence is indicated by

$$10 + 20 + 30 + \cdots + 10n$$

This expression is an example of a **finite series**, which simply means a sum of a finite number of terms.

We can indicate the sum of the first n terms of the sequence a_1, a_2, a_3, ... by

$$a_1 + a_2 + a_3 + \cdots + a_n$$

Another way to indicate this sum uses what is called **sigma notation**, which we now introduce. The capital Greek letter sigma is written Σ. We define the notation $\displaystyle\sum_{k=1}^{n} a_k$ by the equation

$$\sum_{k=1}^{n} a_k = a_1 + a_2 + a_3 + \cdots + a_n$$

For example, $\displaystyle\sum_{k=1}^{3} a_k$ stands for the sum $a_1 + a_2 + a_3$, the idea in this case being to replace the subscript k successively by 1, 2, and 3 and then add the results.

For a more concrete example, let's evaluate the expression $\displaystyle\sum_{k=1}^{4} k^2$. We have

$$\sum_{k=1}^{4} k^2 = 1^2 + 2^2 + 3^3 + 4^2$$

$$= 1 + 4 + 9 + 16 = 30$$

There is nothing special about the choice of the letter k in the expression $\displaystyle\sum_{k=1}^{4} k^2$. For instance, we could equally well write

$$\sum_{j=1}^{4} j^2 = 1^2 + 2^2 + 3^2 + 4^2 = 30$$

The letter k in the expression $\displaystyle\sum_{k=1}^{4} k^2$ is called the **index of summation**. Similarly, the letter j appearing in $\displaystyle\sum_{j=1}^{4} j^2$ is the index of summation in that case. As we have seen, the choice of the letter used for the index of summation has no effect on the value of the indicated sum. For this reason, the index of summation is referred to as a *dummy variable*. The next two examples provide further practice with the sigma notation.

EXAMPLE 4   Express each of the following sums without sigma notation.

(a) $\displaystyle\sum_{k=1}^{3} (3k - 2)^2$ (b) $\displaystyle\sum_{i=1}^{4} ix^{i-1}$ (c) $\displaystyle\sum_{j=1}^{5} (a_{j+1} - a_j)$

Solution   (a) The notation $\displaystyle\sum_{k=1}^{3} (3k - 2)^2$ directs us to replace k successively by 1, 2, and 3 in the expression $(3k - 2)^2$ and then add the results. We thus obtain

$$\sum_{k=1}^{3} (3k - 2)^2 = 1^2 + 4^2 + 7^2 = 66$$

(b) The notation $\displaystyle\sum_{i=1}^{4} ix^{i-1}$ directs us to replace i successively by 1, 2, 3, and 4 in the expression ix^{i-1} and then add the results. We have

$$\sum_{i=1}^{4} ix^{i-1} = 1x^0 + 2x^1 + 3x^2 + 4x^3$$

$$= 1 + 2x + 3x^2 + 4x^3$$

(c) To expand $\displaystyle\sum_{j=1}^{5} (a_{j+1} - a_j)$, we replace j successively by 1, 2, 3, 4, and 5 in the expression $(a_{j+1} - a_j)$ and then add. We obtain

$$\sum_{j=1}^{5} (a_{j+1} - a_j) = (a_2 - a_1) + (a_3 - a_2) + (a_4 - a_3) + (a_5 - a_4) + (a_6 - a_5)$$

$$= a_2 - a_1 + a_3 - a_2 + a_4 - a_3 + a_5 - a_4 + a_6 - a_5$$

Combining like terms, we have

$$\sum_{j=1}^{5} (a_{j+1} - a_j) = a_6 - a_1$$

Sums such as $\displaystyle\sum_{j=1}^{5} (a_{j+1} - a_j)$ are known as **collapsing** or **telescoping sums**.

EXAMPLE 5   Use sigma notation to rewrite each sum.

(a) $\dfrac{x}{1!} + \dfrac{x^2}{2!} + \dfrac{x^3}{3!} + \cdots + \dfrac{x^{12}}{12!}$ (b) $\dfrac{x}{2!} + \dfrac{x^2}{3!} + \dfrac{x^3}{4!} + \cdots + \dfrac{x^n}{(n+1)!}$

Solution   (a) Since the exponents on x run from 1 to 12, we choose a dummy variable, say k, running from 1 to 12. Also, we notice that if the numerator of a given term in the sum is x^k, then the corresponding denominator is $k!$. Consequently, the sum can be written

$$\frac{x}{1!} + \frac{x^2}{2!} + \frac{x^3}{3!} + \cdots + \frac{x^{12}}{12!} = \sum_{k=1}^{12} \frac{x^k}{k!}$$

(b) Since the exponents on x run from 1 to n, we choose a dummy variable, say k, running from 1 to n. (Note that both of the letters n and x would be inappropriate here as dummy variables.) Also, we notice that if the numerator of a given term in the sum is x^k, then the corresponding denominator is $(k+1)!$. Thus, the given sum can be written

$$\frac{x}{2!} + \frac{x^2}{3!} + \frac{x^3}{4!} + \cdots + \frac{x^n}{(n+1)!} = \sum_{k=1}^{n} \frac{x^k}{(k+1)!}$$

EXERCISE SET 13.3

A

In Exercises 1–20, compute the first five terms in each sequence. (In Exercises 1–12, where the nth term is given by an explicit formula, for instance, $a_n = n/(n+1)$, assume that the equation holds for all natural numbers n.)

1. $a_n = \dfrac{n}{n+1}$

2. $a_n = n^3$

3. $a_n = (n-1)^2$

4. $b_n = n!$

5. $b_n = [1 + (1/n)]^n$

6. $b_n = \dfrac{n^n}{n!}$

7. $u_n = (-1)^n$

8. $a_n = (-1)^n \sqrt{n^n}$

9. $a_n = (-1)^{n+1}/n!$

10. $S_n = 1/[(n+1)!]$

11. $a_n = 3n$

12. $b_n = 3$

13. $a_1 = 1;\ a_n = (1 + a_{n-1})^2,\ n \ge 2$

14. $a_1 = 2;\ a_n = \sqrt{a_{n-1}^2 + 1},\ n \ge 2$

15. $a_1 = 2;\ a_2 = 2;\ a_n = a_{n-1}a_{n-2},\ n \ge 3$

16. $F_1 = 1;\ F_2 = 1;\ F_n = F_{n-1} + F_{n-2},\ n \ge 3$

17. $a_1 = 1;\ a_{n+1} = na_n,\ \ge 1$

18. $a_1 = 1;\ a_2 = 2;\ a_n = a_{n-1}/a_{n-2},\ n \ge 3$

19. $a_1 = 0;\ a_n = 2^{a_{n-1}},\ n \ge 2$

20. $a_1 = 0;\ a_2 = 1;\ a_n = \frac{1}{2}(a_{n-1} + a_{n-2}),\ n \ge 3$

In Exercises 21–26, graph the sequences for the indicated values of n.

21. $a_n = (-1)^n;\ n = 1, 2, 3, 4$

22. $a_n = 2n - 1;\ n = 1, 2, 3$

23. $a_n = 1/n;\ n = 1, 2, 3, 4$

24. $a_n = (-1)^n/n,\ n = 1, 2, 3, 4$

25. $a_n = (n-1)/(n+1);\ n = 1, 2, 3$

26. $a_n = [1 + (1/n)]^n;\ n = 1, 2, 3$

In Exercises 27–32, find the sum of the first five terms in the sequence with an nth term as given.

27. $a_n = 2^n$

28. $b_n = 2^{-n}$

29. $a_n = n^2 - n$

30. $b_n = (n-1)!$

31. $a_n = (-1)^n/n!$

32. $a_1 = 1;\ a_n = \dfrac{1}{n-1} - \dfrac{1}{n+1},\ n \ge 2$

33. Find the sum of the first five terms of the sequence that is defined recursively by $a_1 = 1;\ a_2 = 2;$ $a_n = a_{n-1}^2 + a_{n-2}^2,\ n \ge 3$.

34. Find the sum of the first four terms of the sequence defined recursively by $a_1 = 2;\ a_n = n/a_{n-1},\ n \ge 2$.

35. Find the sum of the first four terms of the sequence defined recursively by $a_1 = 2;\ a_n = (a_{n-1})^2,\ n \ge 2$.

36. Find the sum of the first six terms of the sequence defined recursively by $a_1 = 1;\ a_2 = 2;\ a_{n+1} = a_n a_{n-1},$ $n \ge 2$.

In Exercises 37–48, express each of the sums without using sigma notation. Simplify your answers where possible.

37. $\displaystyle\sum_{k=1}^{3} (k - 1)$

38. $\displaystyle\sum_{k=1}^{5} k$

39. $\displaystyle\sum_{k=4}^{5} k^2$

40. $\displaystyle\sum_{k=2}^{6} (1-2k)$

41. $\displaystyle\sum_{n=1}^{3} x^n$

42. $\displaystyle\sum_{n=1}^{3} (n-1)x^{n-2}$

43. $\displaystyle\sum_{n=1}^{4} \frac{1}{n}$

44. $\displaystyle\sum_{n=0}^{4} 3^n$

45. $\displaystyle\sum_{j=1}^{9} \log_{10} \frac{j}{j+1}$

46. $\displaystyle\sum_{j=2}^{5} \log_{10} j$

47. $\displaystyle\sum_{j=1}^{6} \left(\frac{1}{j} - \frac{1}{j+1}\right)$

48. $\displaystyle\sum_{j=1}^{5} (x^{j+1} - x^j)$

In Exercises 49–58, rewrite the sums using sigma notation.

49. $5 + 5^2 + 5^3 + 5^4$

50. $5 + 5^2 + 5^3 + \cdots + 5^n$

51. $x + x^2 + x^3 + x^4 + x^5 + x^6$

52. $x + 2x^2 + 3x^3 + 4x^4 + 5x^5 + 6x^6$

53. $\frac{1}{1} + \frac{1}{2} + \frac{1}{3} + \cdots + \frac{1}{12}$

54. $\frac{1}{1} + \frac{1}{2} + \frac{1}{3} + \cdots + \frac{1}{n}$

55. $2 - 2^2 + 2^3 - 2^4 + 2^5$

56. $\dbinom{10}{3} + \dbinom{10}{4} + \dbinom{10}{5} + \cdots + \dbinom{10}{10}$

57. $1 - 2 + 3 - 4 + 5$

58. $\frac{1}{2} - \frac{1}{4} + \frac{1}{8} - \frac{1}{16} + \frac{1}{32} - \frac{1}{64} + \frac{1}{128}$

B

59. The *Fibonacci numbers* $F_1, F_2, F_3, \ldots$ are defined as follows:

$$F_1 = 1; \quad F_2 = 1; \quad F_{n+2} = F_n + F_{n+1} \quad (n \geq 1)$$

(a) Verify that the first ten Fibonacci numbers are as follows.

F_1	F_2	F_3	F_4	F_5	F_6	F_7	F_8	F_9	F_{10}
1	1	2	3	5	8	13	21	34	55

(b) Complete the following table.

n	$F_1 + F_2 + F_3 + \cdots + F_n$	$F_{n+2} - 1$
1		
2		
3		
4		
5		

(c) Use mathematical induction to prove that

$$F_1 + F_2 + F_3 + \cdots + F_n = F_{n+2} - 1$$

for all natural numbers n.

(d) Use mathematical induction to prove that

$$F_n \geq n \qquad \text{for all natural numbers } n \geq 5$$

(e) Use mathematical induction to show that

$$F_1^2 + F_2^2 + F_3^2 + \cdots + F_n^2 = F_n F_{n+1}$$

for all natural numbers n.

(f) Use mathematical induction to show that

$$F_{n+1}^2 = F_n F_{n+2} + (-1)^n$$

for all natural numbers n. *Hint for Step 2:* Add $F_{k+1} F_{k+2}$ to both sides of the equation in the induction hypothesis. Then factor F_{k+1} from the left-hand side and factor F_{k+2} from the first two terms on the right-hand side.

(g) This part of the exercise requires a knowledge of matrix multiplication. Show that

$$\begin{pmatrix} 1 & 1 \\ 1 & 0 \end{pmatrix}^n = \begin{pmatrix} F_{n+1} & F_n \\ F_n & F_{n-1} \end{pmatrix} \qquad \text{for } n \geq 2$$

13.4 ARITHMETIC SEQUENCES AND SERIES

One of the most natural ways to generate a sequence is to begin with a fixed number a and then repeatedly add a fixed constant d. This yields the sequence

$$a, a+d, a+2d, a+3d, \ldots$$

Such a sequence is called an **arithmetic sequence** or **arithmetic progression**. Notice that the difference between any two consecutive terms is the constant d.

We call d the **common difference**. Here are several examples of arithmetic sequences:

example A: 1, 2, 3, ...
example B: 3, 7, 11, 15, ...
example C: 10, 5, 0, −5, ...

In example A, the first term is $a = 1$ and the common difference is $d = 1$. For example B, we have $a = 3$. The value of d in this example is found by subtracting any two consecutive terms; thus, $d = 4$. Finally, in example C, we have $a = 10$ and $d = -5$. Notice when the common difference is negative, the terms of the sequence decrease.

There is a simple formula for the nth term in an arithmetic sequence

$$a, a + d, a + 2d, a + 3d, \ldots$$

In this sequence, notice that

$$a_1 = a + 0d$$
$$a_2 = a + 1d$$
$$a_3 = a + 2d$$
$$a_4 = a + 3d$$

Following this pattern, it appears that the formula for a_n should be

$$a_n = a + (n - 1)d$$

Indeed, this is the correct formula, and Exercise 32 asks you to verify it using mathematical induction.

---

nth Term of an Arithmetic Sequence

The nth term of an arithmetic sequence $a, a + d, a + 2d, \ldots$ is given by

$$a_n = a + (n - 1)d$$

EXAMPLE 1   Determine the one hundredth term of the arithmetic sequence

7, 10, 13, 16, ...

Solution   The first term is $a = 7$, and the common difference is $d = 3$. Substituting these values in the formula $a_n = a + (n - 1)d$ yields

$$a_n = 7 + (n - 1)3 = 3n + 4$$

To find the 100th term, we replace n by 100 in this last equation to obtain

$$a_{100} = 3(100) + 4 = 304 \qquad \text{as required}$$

EXAMPLE 2   Determine the arithmetic sequence in which the second term is −2 and the eighth term is 40.

Solution   We are given that the second term is -2. Using this information in the formula $a_n = a + (n - 1)d$, we have

$$-2 = a + (2 - 1)d = a + d$$

This gives us one equation in two unknowns. We are also given that the eighth term is 40. Therefore,

$$40 = a + (8 - 1)d = a + 7d$$

We now have a system of two equations in two unknowns:

$$\begin{cases} -2 = a + d \\ 40 = a + 7d \end{cases}$$

Subtracting the first equation from the second gives us

$$42 = 6d \qquad \text{or} \qquad 7 = d$$

To find a, we replace d by 7 in the first equation of the system. This yields

$$-2 = a + 7 \qquad \text{or} \qquad -9 = a$$

We have now determined the sequence, since we know that the first term is -9 and the common difference is 7. The first four terms of the sequence are -9, -2, 5, and 12.   ▮▮▮

Next we would like to derive a formula for the sum of the first n terms of an arithmetic sequence. Such a sum is referred to as an **arithmetic series**. If we use S_n to denote the required sum, we have

$$S_n = a + (a + d) + \cdots + [a + (n - 2)d] + [a + (n - 1)d] \qquad (1)$$

Of course, we must obtain the same sum if we add the terms from right to left rather than left to right. That is, we must have

$$S_n = [a + (n - 1)d] + [a + (n - 2)d] + \cdots + (a + d) + a \qquad (2)$$

Let us now add equations (1) and (2). Adding the left-hand sides is easy; we obtain $2S_n$. Now we add the corresponding terms on the right-hand sides. For the first terms, we have

$$\underset{\substack{\uparrow \\ \text{first term} \\ \text{in equation (1)}}}{a} + \underset{\substack{\text{first term} \\ \text{in equation (2)}}}{\underline{[a + (n - 1)d]}} = 2a + (n - 1)d$$

Next we add the second terms:

$$\underset{\substack{\uparrow \\ \text{second term} \\ \text{in equation (1)}}}{(a + d)} + \underset{\substack{\text{second term} \\ \text{in equation (2)}}}{\underline{[a + (n - 2)d]}} = 2a + d + (n - 2)d = 2a + (n - 1)d$$

Notice that the sum of the second terms is again $2a + (n - 1)d$, the same quantity we arrived at with the first terms. As you can check, this pattern continues all the way through to the last terms. For instance,

$$\underset{\substack{\text{last term} \\ \text{in equation (1)}}}{\underline{[a + (n - 1)d]}} + \underset{\substack{\uparrow \\ \text{last term} \\ \text{in equation (2)}}}{a} = 2a + (n - 1)d$$

We conclude from these observations that by adding the right-hand sides of equations (1) and (2), the quantity $2a + (n - 1)d$ is added a total of n times. Therefore,

$$2S_n = n[2a + (n - 1)d]$$

$$S_n = \frac{n}{2}[2a + (n - 1)d]$$

This gives us the desired formula for the sum of the first n terms in an arithmetic sequence. There is an alternate form of this formula, which now follows rather quickly:

$$S_n = \frac{n}{2}[2a + (n - 1)d]$$

$$= \frac{n}{2}\{a + [a + (n - 1)d]\}$$

$$= \frac{n}{2}(a + a_n) = n\left(\frac{a + a_n}{2}\right)$$

This last equation is easy to remember. It says that the sum of an arithmetic series is obtained by averaging the first and last terms and then multiplying this average by n, the number of terms. For reference, we summarize both formulas as follows.

Formulas for the Sum of an Arithmetic Series

$$S_n = \frac{n}{2}[2a + (n - 1)d]$$

$$S_n = n\left(\frac{a + a_n}{2}\right)$$

EXAMPLE 3 Find the sum of the first 30 terms of the arithmetic sequence 2, 6, 10, 14, ...

Solution We have $a = 2$, $d = 4$, and $n = 30$. Substituting these values in the formula $S_n = (n/2)[2a + (n - 1)d]$ then yields

$$S_{30} = \frac{30}{2}[2(2) + (30 - 1)4]$$

$$= 15[4 + 29(4)]$$

$$= 1800 \qquad \text{(Check the arithmetic)}$$

EXAMPLE 4 In a certain arithmetic sequence, the first term is 6 and the fortieth term is 71. Find the sum of the first 40 terms and also the common difference for the sequence.

Solution We have $a = 6$ and $a_{40} = 71$. Using these values in the formula $S_n = n(a + a_n)/2$ yields

$$S_{40} = \frac{40}{2}(6 + 71) = 20(77) = 1540$$

The sum of the first 40 terms is thus 1540. The value of d can now be found by using the formula $a_n = a + (n - 1)d$. We have $a_{40} = a + (40-1)d$, and therefore

$$71 = 6 + 39d$$
$$39d = 65$$
$$d = \frac{65}{39} = \frac{5}{3}$$

The required value of d is therefore $\frac{5}{3}$. ▮▮▮

EXAMPLE 5 Show that $\displaystyle\sum_{k=1}^{50} (3k - 2)$ represents an arithmetic series, and compute the sum.

Solution There are two different ways to see that $\displaystyle\sum_{k=1}^{50} (3k - 2)$ is an arithmetic series. One way is simply to write out the first few terms and look at the pattern. We then have

$$\sum_{k=1}^{50} (3k - 2) = 1 + 4 + 7 + 10 + \cdots + 148$$

From this it is clear that we are indeed summing the terms in an arithmetic sequence in which $d = 3$ and $a = 1$.

Another, more formal way to show that $\displaystyle\sum_{k=1}^{50} (3k - 2)$ represents an arithmetic series is to prove that the difference between successive terms in the indicated sum is a constant. Now, the form of a typical term in this sum is $(3k - 2)$. Thus, the form of the next term must be $[3(k + 1) - 2]$. The difference between these terms is then

$$[3(k + 1) - 2] - (3k - 2) = 3k + 3 - 2 - 3k + 2 = 3$$

The difference therefore is constant, as we wished to show.

To evaluate $\displaystyle\sum_{k=1}^{50} (3k - 2)$, we can use either of the two formulas for the sum of an arithmetic series. Using the formula $S_n = (n/2)[2a + (n - 1)d]$, we obtain

$$S_{50} = \frac{50}{2}[2(1) + 49(3)] = 25(149) = 3725$$

Thus, the required sum is 3725. You should check for yourself that the same value is obtained using the formula $S_n = n(a + a_n)/2$. ▮▮▮

EXERCISE SET 13.4

A

1. Find the common difference d for each of the following arithmetic sequences.
   (a) $1, 3, 5, 7, \ldots$ (b) $10, 6, 2, -2, \ldots$
   (c) $\frac{2}{3}, 1, \frac{4}{3}, \frac{5}{3}, \ldots$
   (d) $1, 1 + \sqrt{2}, 1 + 2\sqrt{2}, 1 + 3\sqrt{2}, \ldots$

2. Which of the following are arithmetic sequences?
   (a) $2, 4, 8, 16, \ldots$ (b) $5, 9, 13, 17, \ldots$
   (c) $3, \frac{11}{5}, \frac{7}{5}, \frac{3}{5}, \ldots$ (d) $-1, -1, -1, -1, \ldots$
   (e) $-1, 1, -1, 1, \ldots$

In Exercises 3–8, find the indicated term in each sequence.

3. $10, 21, 32, 43, \ldots ; a_{12}$ 4. $7, 2, -3, -8, \ldots ; a_{20}$

5. $6, 11, 16, 21, \ldots ; a_{100}$ 6. $\frac{2}{5}, \frac{4}{5}, \frac{6}{5}, \frac{8}{5}, \ldots ; a_{30}$

7. $-1, 0, 1, 2, \ldots ; a_{1000}$

8. $42, 1, -40, -81, \ldots ; a_{15}$

9. The 4th term in an arithmetic sequence is -6, and the 10th term is 5. Find the common difference and the first term.

10. The 5th term in an arithmetic sequence is $\frac{1}{2}$, and the 20th term is $\frac{7}{8}$. Find the first three terms of the sequence.

11. The 60th term in an arithmetic sequence is 105, and the common difference is 5. Find the first term.

12. Find the common difference in an arithmetic sequence in which $a_{10} - a_{20} = 70$.

13. Find the common difference in an arithmetic sequence in which $a_{15} - a_7 = -1$.

14. Find the sum of the first 16 terms in the sequence 2, 11, 20, 29, ...

15. Find the sum of the first 1000 terms in the sequence 1, 2, 3, 4, ...

16. Find the sum of the first 50 terms in an arithmetic series that has 1st term -8 and 50th term 139.

17. Find the sum: $\dfrac{\pi}{3} + \dfrac{2\pi}{3} + \pi + \dfrac{4\pi}{3} + \cdots + \dfrac{13\pi}{3}$.

18. Find the sum: $\dfrac{1}{e} + \dfrac{3}{e} + \dfrac{5}{e} + \cdots + \dfrac{21}{e}$.

19. Determine the first term of an arithmetic sequence in which the common difference is 5 and the sum of the first 38 terms is 3534.

20. The sum of the first 12 terms in an arithmetic sequence is 156. What is the sum of the first and twelfth terms?

21. In a certain arithmetic sequence, the 1st term is 4 and the 16th term is -100. Find the sum of the first 16 terms and also the common difference for the sequence.

22. The 5th and 50th terms of an arithmetic sequence are 3 and 30, respectively. Find the sum of the first 10 terms.

23. The 8th term in an arithmetic sequence is 5, and the sum of the first 10 terms is 20. Find the common difference and the first term of the sequence.

In Exercises 24–26, find each sum.

24. $\displaystyle\sum_{i=1}^{10} (2i - 1) = 1 + 3 + 5 + \cdots + 19$

25. $\displaystyle\sum_{k=1}^{20} (4k + 3)$ 26. $\displaystyle\sum_{n=5}^{100} (2n - 1)$

27. The sum of three consecutive terms in an arithmetic sequence is 30, and their product is 360. Find the three terms. *Suggestion:* Let x denote the *middle* term and d the common difference.

28. The sum of three consecutive terms in an arithmetic sequence is 21, and the sum of their squares is 197. Find the three terms.

29. The sum of three consecutive terms in an arithmetic sequence is 6, and the sum of their cubes is 132. Find the three terms.

30. In a certain arithmetic sequence, $a = -4$ and $d = 6$. If $S_n = 570$, find n.

31. Let $a_1 = \dfrac{1}{1 + \sqrt{2}}$, $a_2 = -1$, and $a_3 = \dfrac{1}{1 - \sqrt{2}}$.
    (a) Show that $a_2 - a_1 = a_3 - a_2$.
    (b) Find the sum of the first six terms in the arithmetic sequence
    $$\dfrac{1}{1 + \sqrt{2}}, -1, \dfrac{1}{1 - \sqrt{2}}, \ldots$$

32. Using mathematical induction, prove that the nth term of the sequence $a, a + d, a + 2d, \ldots$ is given by
    $$a_n = a + (n - 1)d$$

B

33. Let b denote a positive constant. Find the sum of the first n terms in the sequence.
    $$\dfrac{1}{1 + \sqrt{b}}, \dfrac{1}{1 - b}, \dfrac{1}{1 - \sqrt{b}}, \ldots$$

34. The sum of the first n terms in a certain arithmetic sequence is given by $S_n = 3n^2 - n$. Show that the rth term is given by $a_r = 6r - 4$.

35. Let $a_1, a_2, a_3, \ldots$ be an arithmetic sequence, and let S_k denote the sum of the first k terms. If $S_n/S_m = n^2/m^2$, show that

$$\frac{a_n}{a_m} = \frac{2n - 1}{2m - 1}$$

36. If the common difference in an arithmetic sequence is twice the first term, show that

$$\frac{S_n}{S_m} = \frac{n^2}{m^2}$$

37. The lengths of the sides of a right triangle form three consecutive terms in an arithmetic sequence. Show that the triangle is similar to the 3-4-5 right triangle.

38. Suppose that $1/a$, $1/b$, and $1/c$ are three consecutive terms in an arithmetic sequence. Show that:

(a) $\dfrac{a}{c} = \dfrac{a - b}{b - c}$ (b) $b = \dfrac{2ac}{a + c}$

39. Suppose that a, b, and c are three positive numbers with $a > c > 2b$. If $1/a$, $1/b$, and $1/c$ are consecutive terms in an arithmetic sequence, show that

$$\ln(a + c) + \ln(a - 2b + c) = 2\ln(a - c)$$

13.5 GEOMETRIC SEQUENCES AND SERIES

A **geometric sequence** or **geometric progression** is a sequence of the form

$$a, \ ar, \ ar^2, \ ar^3, \ldots \qquad \text{where } a \text{ and } r \text{ are constants}$$

As you can see, each term after the first in a geometric sequence is obtained by multiplying the previous term by r. The number r is called the **common ratio,** since the ratio of any term to the previous one is always r. For instance, the ratio of the fourth term to the third is $ar^3/ar^2 = r$. Here are two examples of geometric sequences:

$$1, \ \frac{1}{2}, \ \frac{1}{4}, \ \frac{1}{8}, \ \ldots$$

$$10, \ -100, \ 1000, \ -10{,}000, \ \ldots$$

In the first example, we have $a = 1$ and $r = \frac{1}{2}$; in the second example, we have $a = 10$ and $r = -10$.

EXAMPLE 1 In a certain geometric sequence, the first term is 2, the third term is 3, and the common ratio is negative. Find the second term.

Solution Let x denote the second term, so that the sequence begins

$$2, \ x, \ 3, \ \ldots$$

By definition, the ratios $3/x$ and $x/2$ must be equal. Thus, we have

$$\frac{3}{x} = \frac{x}{2}$$
$$x^2 = 6$$
$$x = \pm\sqrt{6}$$

Now, the second term must be negative, since the first term is positive and the common ratio is negative. Thus, the second term is $x = -\sqrt{6}$.

TABLE I

n	a_n
1	ar^0
2	ar^1
3	ar^2
4	ar^3
$\vdots$	$\vdots$

The formula for the nth term of a geometric sequence is easily deduced by considering Table 1. The table indicates that the exponent on r is one less than the value of n in each case. On the basis of this observation, it appears that the nth term must be given by $a_n = ar^{n-1}$. Indeed, it can be shown by mathematical induction that this formula does hold for all natural numbers n. (Exercise 30 asks you to carry out the proof.) We summarize this result as follows:

---

nth Term of a Geometric Sequence

The nth term of the geometric sequence $a, ar, ar^2, \ldots$ is given by

$$a_n = ar^{n-1}$$

EXAMPLE 2 Find the seventh term in the geometric sequence 2, 6, 18, ...

Solution We can find the common ratio r by dividing the second term by the first. Thus, $r = 3$. Now, using $a = 2$, $r = 3$, and $n = 7$ in the formula $a_n = ar^{n-1}$, we have

$$a_7 = 2(3)^6 = 2(729) = 1458$$

The seventh term of the sequence is therefore 1458. ∎

Suppose that we begin with a geometric sequence $a, ar, ar^2, \ldots$, in which $r \neq 1$. If we add the first n terms and denote the sum by S_n, we have

$$S_n = a + ar + ar^2 + \cdots + ar^{n-2} + ar^{n-1} \tag{1}$$

This sum is called a **finite geometric series**. We would like to find a formula for S_n. To do this, we multiply equation (1) by r to obtain

$$rS_n = ar + ar^2 + ar^3 + \cdots + ar^{n-1} + ar^n \tag{2}$$

We now subtract equation (2) from equation (1). This yields (after combining like terms)

$$S_n - rS_n = a - ar^n$$
$$S_n(1 - r) = a(1 - r^n)$$
$$S_n = \frac{a(1 - r^n)}{1 - r} \qquad (r \neq 1)$$

This is the formula for the sum of a finite geometric series. We summarize this result in the box that follows.

Formula for the Sum of a Geometric Series

Let S_n denote the sum $a + ar + ar^2 + \cdots + ar^{n-1}$, and assume that $r \neq 1$.

$$S_n = \frac{a(1 - r^n)}{1 - r}$$

EXAMPLE 3 Evaluate the sum $\dfrac{1}{2^1} + \dfrac{1}{2^2} + \dfrac{1}{2^3} + \cdots + \dfrac{1}{2^{10}}$.

Solution This is a finite geometric series with $a = \frac{1}{2}$, $r = \frac{1}{2}$, and $n = 10$. Using these values in the formula for S_n yields

$$S_{10} = \frac{\frac{1}{2}\left[1 - \left(\frac{1}{2}\right)^{10}\right]}{1 - \frac{1}{2}} = \frac{\frac{1}{2}\left[1 - \frac{1}{1024}\right]}{\frac{1}{2}}$$

$$= 1 - \frac{1}{1024} = \frac{1023}{1024}$$

We would now like to attach a meaning to certain expressions of the form

$$a + ar + ar^2 + \cdots$$

Such an expression is called an **infinite geometric series.** The three dots indicate (intuitively at least) that the additions are to be carried out indefinitely, without end. To see how to proceed here, let us look at some examples involving finite geometric series. In particular, we will consider the series

$$\frac{1}{2^1} + \frac{1}{2^2} + \frac{1}{2^3} + \cdots + \frac{1}{2^n}$$

for increasing values of n. The idea is to look for a pattern as n grows ever larger. Let $S_1 = \dfrac{1}{2}$, $S_2 = \dfrac{1}{2^1} + \dfrac{1}{2^2}$, $S_3 = \dfrac{1}{2^1} + \dfrac{1}{2^2} + \dfrac{1}{2^3}$, and, in general,

$$S_n = \frac{1}{2^1} + \frac{1}{2^2} + \cdots + \frac{1}{2^n}$$

TABLE 2

n	$S_n = \dfrac{1}{2} + \dfrac{1}{2^2} + \cdots + \dfrac{1}{2^n}$
1	0.5
2	0.75
5	0.96875
10	0.999023437 ...
15	0.999969482 ...
20	0.999999046 ...
25	0.999999970 ...

Then we can compute S_n for any given value of n by means of the formula for the sum of a finite geometric series. From Table 2, which displays the results of such calculations, it seems clear that as n grows larger and larger, the value of S_n grows ever closer to 1. More precisely (but leaving the details for calculus), it can be shown that the value of S_n can be made as close to 1 as we please, provided only that n is sufficiently large. For this reason, we say that the *sum of the infinite geometric series* $\dfrac{1}{2^1} + \dfrac{1}{2^2} + \dfrac{1}{2^3} + \cdots$ is 1. That is,

$$\frac{1}{2^1} + \frac{1}{2^2} + \frac{1}{2^3} + \cdots = 1$$

We can arrive at this last result another way. First we compute the sum of the finite geometric series $\dfrac{1}{2^1} + \dfrac{1}{2^2} + \cdots + \dfrac{1}{2^n}$. As you can check, the result is

$$S_n = 1 - \left(\frac{1}{2}\right)^n$$

Now, as n grows larger and larger, the value of $\left(\frac{1}{2}\right)^n$ gets closer and closer to zero. Thus, as n grows ever larger, the value of S_n will more and more resemble $1 - 0$, or 1.

Now let us repeat our reasoning to obtain a formula for the sum of the infinite geometric series

$$a + ar + ar^2 + \cdots \qquad \text{where } |r| < 1$$

First we consider the finite geometric series

$$a + ar + ar^2 + \cdots + ar^{n-1}$$

The sum S_n in this case is

$$S_n = \frac{a(1 - r^n)}{1 - r}$$

We want to know how S_n behaves as n grows ever larger. This is where the assumption $|r| < 1$ is crucial. Just as $\left(\frac{1}{2}\right)^n$ approaches zero as n grows larger and larger, so will r^n approach zero as n grows larger and larger. Thus, as n grows ever larger, the sum S_n will more and more resemble

$$\frac{a(1 - 0)}{1 - r} = \frac{a}{1 - r}$$

For this reason, we say that the sum of the infinite geometric series is $\frac{a}{1 - r}$. We will make free use of this result in the subsequent examples. However, a more rigorous development of infinite series properly belongs to calculus.

Formula for the Sum of an Infinite Geometric Series

Suppose that $|r| < 1$. Then the sum S of the infinite geometric series $a + ar + ar^2 + \cdots$ is given by

$$S = \frac{a}{1 - r}$$

EXAMPLE 4 Find the sum of the infinite geometric series $1 + \frac{2}{3} + \frac{4}{9} + \cdots$

Solution In this case, we have $a = 1$ and $r = \frac{2}{3}$. Thus,

$$S = \frac{a}{1 - r} = \frac{1}{1 - \frac{2}{3}} = \frac{1}{\frac{1}{3}} = 3$$

The sum of the series is 3.

EXAMPLE 5 Find a fraction equivalent to the repeating decimal $0.2\overline{35}$.

Solution Let $S = 0.2\overline{35}$. Then we have

$$S = 0.2353535 \ldots$$

$$= \frac{2}{10} + \frac{35}{1000} + \frac{35}{100{,}000} + \frac{35}{10{,}000{,}000} + \cdots$$

Now, the expression following $\frac{2}{10}$ on the right-hand side of this last equation is an infinite geometric series in which $a = \frac{35}{1000}$ and $r = \frac{1}{100}$. Thus,

$$S = \frac{2}{10} + \frac{a}{1-r}$$

$$= \frac{2}{10} + \frac{\frac{35}{1000}}{1 - \frac{1}{100}} = \frac{233}{990} \qquad \text{(Check the arithmetic!)}$$

The given decimal is therefore equivalent to 233/990.

EXERCISE SET 13.5

A

1. Find the second term in a geometric sequence in which the first term is 9, the third term is 4, and the common ratio is positive.

2. Find the fifth term in a geometric sequence in which the fourth term is 4, the sixth term is 6, and the common ratio is negative.

3. The product of the first three terms in a geometric sequence is 8000. If the first term is 4, find the second and third terms.

In Exercises 4–8, find the indicated term of the geometric sequence given.

4. $9, 81, 729, \ldots ; a_7$

5. $-1, 1, -1, 1, \ldots ; a_{100}$

6. $\frac{1}{2}, \frac{1}{4}, \frac{1}{8}, \ldots ; a_9$

7. $\frac{2}{3}, \frac{4}{9}, \frac{8}{27}, \ldots ; a_8$

8. $1, -\sqrt{2}, 2, \ldots ; a_6$

9. Find the common ratio in a geometric sequence in which the first term is 1 and the seventh term is 4096.

10. Find the first term in a geometric sequence in which the common ratio is $\frac{4}{3}$ and the tenth term is $\frac{16}{9}$.

11. Find the sum of the first ten terms of the sequence $7, 14, 28, \ldots$.

12. Find the sum of the first five terms of the sequence $-\frac{1}{2}, \frac{3}{10}, -\frac{9}{50}, \ldots$.

13. Find the sum: $1 + \sqrt{2} + 2 + \cdots + 32$.

14. Find the sum of the first 12 terms in the sequence $-4, -2, -1, \ldots$.

In Exercises 15–17, evaluate each sum.

15. $\displaystyle\sum_{k=1}^{6} \left(\frac{3}{2}\right)^k$

16. $\displaystyle\sum_{k=1}^{6} \left(\frac{2}{3}\right)^{k+1}$

17. $\displaystyle\sum_{k=2}^{6} \left(\frac{1}{10}\right)^k$

In Exercises 18–22, determine the sum of each infinite geometric series.

18. $\dfrac{1}{4} + \dfrac{1}{4^2} + \dfrac{1}{4^3} + \cdots$

19. $\frac{2}{3} - \frac{4}{9} + \frac{8}{27} - \cdots$

20. $\frac{9}{10} + \frac{9}{100} + \frac{9}{1000} + \cdots$

21. $1 + \dfrac{1}{1.01} + \dfrac{1}{(1.01)^2} + \cdots$

22. $-1 - \dfrac{1}{\sqrt{2}} - \dfrac{1}{2} - \cdots$

In Exercises 23–27, express each repeating decimal as a fraction.

23. $0.555\ldots$

24. $0.\overline{47}$

25. $0.1\overline{23}$

26. $0.050505\ldots$

27. $0.\overline{432}$

B

28. The lengths of the sides in a right triangle form three consecutive terms of a geometric sequence. Find the common ratio of the sequence. (There are two distinct answers.)

29. The product of three consecutive terms in a geometric sequence is -1000, and their sum is 15. Find the common ratio. (There are two answers.)
Suggestion: Denote the terms by a/r, a and ar.

30. Use mathematical induction to prove that the nth term of the geometric sequence $a, ar, ar^2, \ldots$ is ar^{n-1}.

31. Show that the sum of the following infinite geometric series is $\frac{3}{2}$.

$$\frac{\sqrt{3}}{\sqrt{3}+1} + \frac{\sqrt{3}}{\sqrt{3}+3} + \cdots$$

32. Let A_1 denote the area of an equilateral triangle, each side of which is 1 unit long. A second equilateral triangle is formed by joining the midpoints of the sides

of the first triangle. Let A_2 denote the area of this second triangle. This process is then repeated to form a third triangle with area A_3; and so on. Find the sum of the areas: $A_1 + A_2 + A_3 + \cdots$.

33. Let $a_1, a_2, a_3, \ldots$ be a geometric sequence such that $r \neq 1$. Let $S = a_1 + a_2 + a_3 + \cdots + a_n$, and let $T = \dfrac{1}{a_1} + \dfrac{1}{a_2} + \cdots + \dfrac{1}{a_n}$. Show that $\dfrac{S}{T} = a_1 a_n$.

34. Suppose that a, b, and c are three consecutive terms in a geometric sequence. Show that $\dfrac{1}{a + b}$, $\dfrac{1}{2b}$, and $\dfrac{1}{c + b}$ are three consecutive terms in an arithmetic sequence.

35. A ball is dropped from a height of 6 ft. Assuming that on each bounce the ball rebounds to one-third of its previous height, find the total distance traveled by the ball.

13.6 PERMUTATIONS AND COMBINATIONS

I have all the words. What I am seeking is the perfect order of the words in the sentence.

James Joyce (1882–1941)

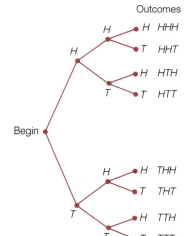

Outcomes

FIGURE 1

Suppose that you flip a coin twice, keeping track of the result each time. How many possible outcomes are there for this experiment? Using H and T to denote heads and tails, respectively, we can list the possible outcomes this way:

$$HH \qquad HT \qquad TH \qquad TT$$

For instance, HH describes the experiment in which we obtain two heads in succession. Looking at this list then, we can say that there are four possible outcomes for the experiment.

Now let's ask the same question for an experiment in which we flip a coin three times in succession. How many possible outcomes are there? One way to organize the counting procedure uses the *tree diagram* in Figure 1. A tree diagram shows us every possibility at each stage in the experiment. Each path through the tree (from left to right) corresponds to a possible outcome for the experiment. For instance, the topmost path corresponds to the case in which we obtain three heads in succession (HHH). As you can see in Figure 1, there are eight possible outcomes for our experiment, corresponding to the eight branches on the right side of the tree.

EXAMPLE 1 There are four roads from town A to town B and three roads from town B to town C, as indicated schematically in Figure 2. Draw a tree diagram to find out how many routes there are from A to C via B.

FIGURE 2

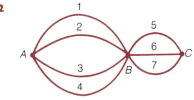

Solution Our tree diagram in Figure 3 has 12 branches on the right side. (Count them!) So there are 12 routes from A to C via B. ∎

FIGURE 3

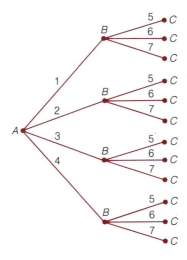

Rather than actually counting all 12 branches in Figure 3, we could have reasoned as follows. There are four possible ways to go from A to B, and for each of these ways, there are then three ways to go from B to C. Thus there are 4 *times* 3, or 12, ways to go from A to C. The *multiplication principle,* which we now state, formalizes this idea.

The Multiplication Principle

Suppose that a task involves two consecutive operations. If the first operation can be performed in M ways, and the second operation can be performed in N ways, then the entire task can be performed in MN ways.

EXAMPLE 2 How many different two-digit numbers can be formed using the digits 1, 3, 5, 7, and 9 if **(a)** no digit is repeated (within a given number)? **(b)** repetitions are allowed?

Solution **(a)** We can think of the task of forming a two-digit number as composed of two consecutive operations: first choose a tens digit, then choose a units digits. There are five choices for the tens digit. Then, after choosing a tens digit (assuming that no repetitions are allowed), there will be four choices left for the units digit. It follows now from the multiplication principle that there are $5 \times 4 = 20$ possible choices. That is, there are 20 different two-digit numbers that can be formed.

(b) With repetitions allowed, there are five choices for the tens digit and five choices for the units digit. Therefore (using the multiplication principle again) there are $5 \times 5 = 25$ different two-digit numbers that can be formed. ∎

The multiplication principle can be extended to tasks involving three or more consecutive operations. This idea is used in the next example.

EXAMPLE 3   You buy three books at a book sale: one by Hemingway, one by Joyce, and one by Fitzgerald. How many ways are there to arrange these on a (single) book-shelf?

Solution   There are three positions we have to fill on the shelf: left, middle, and right. Starting with the left position, there are three choices. Then, once we make a choice for that left position, there are two choices remaining for the middle position. And finally, after choosing a book for that middle position, we'll have but one choice for the right position. Thus, according to the multiplication principle, there are $3 \times 2 \times 1 = 6$ ways to arrange the three books.

Let's list the six possible arrangements from Example 3. Using *H* for Hemingway, *J* for Joyce, and *F* for Fitzgerald, the six arrangements are:

HJF	*JHF*	*FJH*
HFJ	*JFH*	*FHJ*

Each of the six arrangements here is called a *permutation* of the set $\{H, J, F\}$. In general, by a **permutation** of a set of *n* distinct objects we just mean a listing or arrangement of those objects in a definite order. For instance, if we take the set $\{a, b, c, d\}$, then two (of the many possible) permutations of this set are *abcd* and *dcba*.

EXAMPLE 4   How many permutations are there of the set $\{a, b, c, d\}$?

Solution   To form a permutation, we have to assign a letter to each of four positions:

—— —— —— ——

We have four choices for the first position. Then, after filling that position, we'll have three letters left from which to choose for the second position. After carrying that out, there will be only two letters from which to choose for the third position. And finally, with the first three positions filled, there will be but one letter left for the fourth position. In summary, we have

4 choices   3 choices   2 choices   1 choice

Therefore (by the multiplication principle) the total number of permutations or arrangements is $4 \times 3 \times 2 \times 1 = 24$. Notice that this is 4!. (Recall from Section 13.2 that *n*! stands for the product of the first *n* natural numbers.)

In the example just concluded, we saw that there were 4! permutations of a set with four elements. By following the same kind of reasoning (and using mathematical induction), it can be shown that there are *n*! permutations of a set with *n* elements. For reference we state this fact in the box that follows.

> The number of permutations of a set with *n* elements is *n*!

EXAMPLE 5   A regular deck of playing cards contain 52 cards. How many arrangements (permutations) of the deck are there?

Solution   The number of permutations of a set with 52 elements is 52!. With the aid of a calculator we find that

$$52! \approx 8 \times 10^{67}$$

We've said that a permutation of a set with n elements is a listing or arrangement of all n of those elements in some definite order. Sometimes, though, we want to consider arrangements that use less than all n of the elements. For example, let's take the set $\{H, J, F\}$ that we used earlier and this time list the possible arrangements using only two elements at a time:

> HJ   JH   FH
> HF   JF   FJ

Each arrangement here is called a **permutation of three objects taken two at a time**. The number of such permutations is denoted by $P(3, 2)$. So we have $P(3, 2) = 6$, since there are six arrangements. More generally, we have the following definition.

DEFINITION   Permutation of *n* Objects Taken *r* at a Time

By a **permutation of *n* objects taken *r* at a time**, we mean a listing or an arrangement of r of the objects in a definite order, where $r \leq n$. The number of such arrangements is denoted by

$$P(n, r)$$

EXAMPLE 6   Using the set $\{a, b, c, d\}$, list the permutations of these four letters taken two at a time. Evaluate $P(4, 2)$.

Solution

Using a and b:	ab, ba	Using b and c:	bc, cb
Using a and c:	ac, ca	Using b and d:	bd, db
Using a and d:	ad, da	Using c and d:	cd, dc

We've listed 12 permutations of the four given letters taken two at a time. (Convince yourself that we've listed all possibilities.) So $P(4, 2) = 12$.

We could have evaluated $P(4, 2)$ in Example 6 by reasoning as follows. To form a permutation of four letters taken two at a time, we have to fill two positions:

First letter   Second letter

There are four choices for the first letter. Then, after making a choice, we're left with three choices for the second letter. Therefore (according to the multiplication principle) there are $4 \times 3 = 12$ possible choices, in agreement with the result in Example 6. By using this type of reasoning, we can obtain the following general expression for the number of permutations of n objects taken r at a time. (Exercise 54 asks you to prove that the two expressions for $P(n, r)$ in the box are equal.)

$$P(n, r) = n(n - 1) \cdots (n - r + 1) = \frac{n!}{(n - r)!}$$

EXAMPLE 7 (a) Evaluate $P(8, 5)$, the number of permutations of eight objects taken five at a time.

(b) Evaluate $P(8, 8)$.

Solution (a) In the first expression for $P(n, r)$ in the box, let $n = 8$ and $r = 5$. This yields

$$P(8, 5) = 8 \cdot (7) \cdots (8 - 5 + 1)$$
$$= 8 \cdot (7)(6)(5)(4) = 6720$$

So there are 6720 permutations of eight objects taken five at a time.

(b) $P(8, 8) = 8 \cdot (7) \cdots (8 - 8 + 1)$
$$= 8 \cdot (7) \cdots (1) = 8! = 40,320$$

So there are 8! or 40,320 permutations of eight objects taken eight at a time. (Note that a permutation of eight objects taken eight at a time is just a permutation of those eight objects.)

EXAMPLE 8 From a set of ten pictures, five are to be selected and hung in a row. How many arrangements are possible?

Solution We want the number of permutations of ten objects taken five at a time. This number is

$$P(10, 5) = (10)(9) \cdots (10 - 5 + 1)$$
$$= (10)(9)(8)(7)(6)$$
$$= 30,240$$

There are, therefore, 30,240 possible arrangements.

EXAMPLE 9 How many possible arrangements are there of the letters $\{a, b, c, d, e\}$ in which a and b are next to each other?

Solution First think of a and b as a single object and count the number of arrangements of the *four* objects (ab), c, d, and e. This number is

$$P(4, 4) = 4! \qquad \text{(Why?)}$$
$$= 4 \cdot (3)(2)(1) = 24$$

Now, in each of these 24 arrangements, the letters a and b can appear in one of two ways, either ab or ba. Consequently (by the multiplication principle), the number of possible arrangements of $\{a, b, c, d, e\}$ in which a and b are adjacent is $24 \times 2 = 48$.

EXAMPLE 10 How many possible arrangements are there of the letters $\{a, b, c, d, e\}$ in which a and b are not next to each other?

Solution In this type of counting problem, the direct approach is not the simplest. Instead, we reason as follows. The number of arrangements in which a and b are not adjacent equals the total number of possible arrangements of the five letters, minus the number of arrangements of the five letters in which a and b *are* adjacent. So (using the result in Example 9), the required number of arrangements is

$$P(5, 5) - 48 = 5 \cdot (4)(3)(2)(1) - 48$$
$$= 120 - 48 = 72$$

Therefore there are 72 ways to arrange the five given letters such that a and b are not adjacent. ▪▪▪

In working with permutations, we've seen that the order in which the objects are listed is an essential factor. Now we want to introduce the idea of a *combination,* in which the order of the objects is not a consideration. Our work will be based on the following definition.

DEFINITION   Combination of *n* Objects Taken *r* at a Time

Suppose that r objects are selected from a set of n objects, where $r \leq n$. Then this selection or subset of r objects is called a **combination of the n objects taken r at a time**. The number of such combinations is denoted

$$C(n, r)$$

To illustrate this definition, let's take the set $\{H, J, F\}$. As we saw earlier, there are six permutations of these three letters taken two at a time:

Permutations involving H and J: HJ, JH
Permutations involving J and F: JF, FJ
Permutations involving H and F: HF, FH

Now, the two permutations HJ and JH use exactly the same letters. So, corresponding to these two permutations, we have the single combination or set of letters $\{H, J\}$. Also, whether we write this set as $\{H, J\}$ or $\{J, H\}$, it's still the same combination of letters. Similarly, corresponding to the two permutations JF and FJ we have the single combination $\{J, F\}$. And finally, corresponding to the two permutations HF and FH, we have the combination $\{H, F\}$. Thus, although we have six permutations of the three letters taken two at a time, there are only three combinations of the letters taken two at a time. We can summarize these results by writing

$$P(3, 2) = 6 \qquad \text{and} \qquad C(3, 2) = 3$$

EXAMPLE 11   Using the set $\{1, 2, 3, 4\}$, list both the combinations and the permutations of these four numbers taken three at a time. Evaluate $C(4, 3)$ and $P(4, 3)$.

Solution

COMBINATION	PERMUTATIONS GENERATED FROM THE COMBINATION					
$\{1, 2, 3\}$	123	132	213	231	312	321
$\{1, 2, 4\}$	124	142	241	214	412	421
$\{1, 3, 4\}$	134	143	341	314	413	431
$\{2, 3, 4\}$	234	243	342	324	423	432

We've found four combinations. (Convince yourself that there are no other distinct combinations using three of the four given digits.) Thus $C(4, 3) = 4$. Next, let's evaluate $P(4, 3)$. Assuming that the listing we've made is complete, a direct count shows that $P(4, 3) = 24$. You can check this result for yourself using the formula for $P(n, r)$ that was given earlier. Actually, there is yet another way to calculate $P(4, 3)$, and we'll show it here because it will help you to follow the reasoning in the next paragraph. Each of our combinations contains three numbers. So, corresponding to each combination there are 3! or six permutations. Thus, by the multiplication principle, we have $P(4, 3) = 6 \times C(4, 3) = 6 \times 4 = 24$.

We can derive a formula for $C(n, r)$ as follows. First of all, the number of permutations of n objects taken r at a time is

$$P(n, r) = n(n - 1) \cdots (n - r + 1)$$

We can think of the task of forming any one of these permutations as composed of two consecutive operations:

1. Select r of the n objects.
2. Then arrange these r objects in some definite order.

Since operation 1 can be performed in $C(n, r)$ ways, and then operation 2 can be performed in $r!$ ways, the multiplication principle tells us that

$$P(n, r) = C(n, r) \cdot r!$$

Therefore

$$C(n, r) = \frac{P(n, r)}{r!} = \frac{n(n - 1) \cdots (n - r + 1)}{r!}$$

We can simplify this last fraction. The technique is to multiply both the numerator and the denominator by the quantity $(n - r)!$. This gives us

$$C(n, r) = \frac{n(n - 1) \cdots (n - r + 1)[(n - r)!]}{r![(n - r)!]}$$

The numerator of this last fraction is just $n!$, so we have

$$C(n, r) = \frac{n!}{r!(n - r)!}$$

Do you recognize the fraction on the right-hand side of this last equation? It's the binomial coefficient $\binom{n}{r}$ that we defined in connection with our earlier work on the binomial theorem. In summary, then, we have the following two expressions for $C(n, r)$. (*Note:* In the second expression, there are r factors in the numerator and r factors in the denominator.)

$$C(n, r) = \binom{n}{r} = \frac{n!}{r!(n - r)!}$$
$$= \frac{n(n - 1) \cdots (n - r + 1)}{r!}$$

EXAMPLE 12   There are 100 senators in the United States Senate. How many 4-member committees are possible?

Solution   Since we are interested in the members of a committee rather than the order in which they are chosen or seated, we use combinations rather than permutations. We want the number of combinations of the 100 senators taken 4 at a time. This number is

$$C(100, 4) = \binom{100}{4} = \frac{100!}{4!(100 - 4)!}$$

$$= \frac{100(99)(98)(97)(96!)}{4!(96!)}$$

$$= \frac{100(99)(98)(97)}{4!}$$

$$= 3,921,225 \qquad \text{(using a calculator)}$$

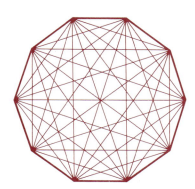

FIGURE 4   So there are 3,921,225 possible 4-member committees.

EXAMPLE 13   Figure 4 shows a decagon (a ten-sided figure) with all its diagonals. How many diagonals are there?

Solution   Each pair of vertices determines either a diagonal or a side of the decagon. So the total number of diagonals and sides is the same as the number of pairs of vertices. This latter number is $C(10, 2)$, the number of combinations of the ten vertices taken two at a time. Evaluating $C(10, 2)$, we have

$$C(10, 2) = \binom{10}{2} = \frac{10!}{2!(10 - 2)!} = \frac{10(9)(8!)}{2(8!)} = 45$$

Thus the total number of sides and diagonals is 45. Since 10 of these are sides, we're left with $45 - 10 = 35$ diagonals. (For a different approach to this problem, see Exercise 53.)

EXAMPLE 14   On a certain committee there are eight Republicans and seven Democrats. How many possibilities are there for a five-member subcommittee that consists of three Republicans and two Democrats?

Solution   Think of the task of forming a five-member subcommittee as composed of two consecutive operations: First choose three Republicans; then choose two Democrats. The number of ways to choose three Republicans is $C(8, 3)$, and the number of ways to choose two Democrats is $C(7, 2)$. Therefore, according to the multiplication principle, the total number of subcommittees here is

$$C(8, 3) \cdot C(7, 2) = \binom{8}{3} \cdot \binom{7}{2}$$

$$= \frac{8!}{3!(8 - 3)!} \cdot \frac{7!}{2!(7 - 2)!}$$

$$= \frac{8(7)(6)}{3(2)} \cdot \frac{7(6)}{2} = 1176 \qquad \text{(Check the arithmetic)}$$

So there are 1176 possible subcommittees consisting of three Republicans and two Democrats.

EXERCISE SET 13.6

A

1. There are three roads from town A to town B and two roads from town B to town C. By drawing a tree diagram, find out how many routes there are from A to C via B. (Check your answer using the multiplication principle.)

2. There are five roads from town A to town B and three roads from town B to town C. Draw a tree diagram to find out how many routes there are from A to C via B. (Check your answer using the multiplication principle.)

3. There are three roads from town A to town B, two roads from town B to town C, and two roads from town C to town D. Draw a tree diagram to find out how many routes there are from town A to town D via B and C.

4. At the beginning of this section we noted that if a coin is flipped twice, there are four possible outcomes. Draw a tree diagram illustrating this.

5. How many different two-digit numbers can be formed using the digits 2, 4, 6, and 8 if no digit is repeated (within a given number)?

6. Follow Exercise 5, but allow repetitions.

7. How many three-digit numbers can be formed using the digits 1, 2, 3, 4, and 5 if no digit can be repeated (within a given number)?

8. How many two-digit numbers that are even can be formed using the digits 2, 4, 5, 7, and 9 if no digit is repeated (within a given number)?

9. How many three-digit numbers that are less than 600 can be formed using the digits 3, 4, 5, 6, 7, and 8 if repetitions are allowed?

10. In a certain state, the license plates contain three digits followed by three letters. (Repetitions of digits or letters on a given plate are allowed.) How many different license plates are possible?

11. Follow Exercise 10, but assume that repetitions are not allowed.

12. In how many ways can five books be arranged on a bookshelf?

13. In how many ways can ten books be arranged on a bookshelf?

14. How many permutations are there of the set $\{a, e, i, o, u\}$?

15. If the number of permutations of a certain set of letters is 24, how many letters must there be in the set?

16. Five framed photographs are going to be hung on the wall in one row. If one particular photograph must

always be in the middle, how many arrangements are possible?

For Exercises 17–20, evaluate each expression.

17. $P(8, 1)$ 18. $P(8, 2)$

19. $P(6, 3)$ 20. $P(11, 3)$

21. Show that $P(n, n - 1) = n!$.

22. Show that $P(n, 1) = n$.

23. Which is larger, $P(100, 50)$ or $P(101, 51)$?

24. Which is larger, $P(100, 50)$ or $P(101, 52)$?

25. How many possible arrangements are there of the six letters $\{a, b, c, d, e, f\}$ in which b and c are next to each other? (Use the method shown in Example 9.)

26. How many arrangements are there of the six letters $\{a, b, c, d, e, f\}$ in which b and c are not next to each other? (Use the method shown in Example 10.)

27. How many four-digit numbers can be formed using the digits $\{2, 7, 1, 8\}$ such that 7 and 1 are next to each other? (No repetitions.)

28. How many four-digit numbers can be formed using the digits $\{2, 7, 1, 8\}$ such that 7 and 1 are not next to each other? (No repetitions.)

29. How many three-digit numbers can be formed from the ten digits $\{0, 1, 2, \ldots, 9\}$ if **(a)** repetitions are allowed? (Don't forget, the first digit cannot be 0.) **(b)** repetitions are not allowed?

30. How many three-digit numbers can be formed using the digits $\{1, 2, 3, 4, 5, 6\}$ if **(a)** repetitions are not allowed and each number ends with 32? **(b)** repetitions are not allowed and the number formed in each case is odd? **(c)** repetitions are not allowed and each number formed is divisible by 5?

31. How many three-letter code words can be formed using the letters in the word *equations* if **(a)** repetitions are allowed? **(b)** repetitions are not allowed? **(c)** repetitions are not allowed and each code word begins with q? **(d)** repetitions are not allowed and each code word begins with a vowel and ends with a consonant?

32. Using the set $\{a, b, c, d\}$, list both the combinations and the permutations of these four letters taken two at a time. Evaluate $C(4, 2)$ and $P(4, 2)$.

For Exercises 33–38, evaluate each expression.

33. (a) $C(12, 11)$ 34. (a) $C(15, 14)$
    (b) $C(12, 1)$ (b) $C(15, 1)$

35. (a) $C(8, 3)$ 36. (a) $C(10, 8)$
    (b) $C(8, 5)$ (b) $C(10, 2)$

37. (a) $C(12, 5)$
    (b) $C(12, 7)$

38. (a) $C(15, 13)$
    (b) $C(15, 2)$

39. Simplify $C(n, n - 2)$.

40. Simplify $C(n, 1)$.

41. There are eight people on a certain committee. How many three-member subcommittees are possible?

42. How many five-member subcommittees are possible from a committee with ten members?

43. There are ten people in a certain club.
    (a) How many different six-member committees are possible?
    (b) Of the ten people in the club, suppose that four are men and six are women. How many six-member committees are possible if each committee must be composed of three men and three women?

44. Suppose you are one of 12 people in a certain club.
    (a) How many five-member committees can be formed that include you?
    (b) How many five-member committees can be formed that exclude you?

45. Use the method shown in Example 13 to find out how many diagonals can be drawn in a hexagon. (A hexagon is a six-sided polygon.)

46. Use the method shown in Example 13 to count the number of diagonals that can be drawn in a regular 14-sided polygon.

47. Suppose that we are given eight points in a plane, and no three of the points lie on the same straight line. How many different triangles can be formed in which the vertices are chosen from these eight points?

48. Suppose that we are given 12 points lying on a certain circle. How many triangles can be formed if the vertices are chosen from the 12 points?

B

49. You have one penny, one nickel, one dime, and one quarter. How many different sums of money can you form using one or more of these coins? *Hint:* You need to compute $C(4, 1) + C(4, 2) + C(4, 3) + C(4, 4)$.

50. You have a $1 bill, a $5 bill, a $10 bill, a $20 bill, and a $100 bill. How many different sums of money can you form using one or more of these bills? (Study the hint in the previous exercise.)

51. A group consists of nine Republicans and eight Democrats. How many different eight-member committees can be formed that contain **(a)** six Republicans and two Democrats? **(b)** at least six Republicans?

52. (a) From a group of 12 people, a committee with eight members will be formed. Then, from those eight, a subcommittee with three people will be chosen. How many ways are there to carry out these two consecutive operations?
    (b) From a group of 12 people, a committee of three will be chosen. Later, five additional people are picked to increase the committee size to eight. How many ways are there to carry out these two consecutive operations?
    (c) Use parts (a) and (b) to explain why
    $$\binom{12}{8}\binom{8}{3} = \binom{12}{3}\binom{9}{5}$$
    (d) Verify the equality in part (c) by direct calculation (involving ratios of factorials).
    (e) Show that $\binom{n}{k}\binom{k}{r} = \binom{n}{r}\binom{n - r}{k - r}$, where $n \geq k \geq r$.
    Suggestion: You can do this either by direct calculation or by repeating the reasoning used in parts (a), (b), and (c) with the quantities 12, 8, and 3 replaced by n, k, and r, respectively.

53. In Example 13 we computed the number of diagonals in a decagon by considering combinations. Here is a different way to approach the problem. First observe that there are seven diagonals emanating from each of the 10 vertices. Why does this imply that the total number of diagonals is $\dfrac{7 \cdot 10}{2} = 35$?

54. Show that $n(n - 1) \cdots (n - r + 1) = \dfrac{n!}{(n - r)!}$.

13.7   INTRODUCTION TO PROBABILITY

The true logic of this world is in the calculus of probabilities.

James Clerk Maxwell

In order to define the mathematical notion of probability, we first introduce the idea of a *sample space* for an experiment. As a particularly simple first example, we can think of flipping a coin as an experiment in which there are two possible *outcomes*: heads (*H*) or tails (*T*). The set {*H, T*} consisting of all possible outcomes for this experiment is called a **sample space**. Intuitively, at least, we believe that each outcome in this sample space is equally likely to occur.

As another example of this terminology, let's suppose that an experiment consists of tossing a single die and then noting the number of dots showing on the top face. Then a sample space for this experiment is the set

$$\{1, 2, 3, 4, 5, 6\}$$

Here, as in the previous example, our sample space consists of all possible outcomes for the experiment. And again, intuitively at least, we believe that each outcome is equally likely to occur. In this section, all the sample spaces that we consider are finite; that is, they contain finitely many outcomes.

Sometimes we wish to focus on a particular outcome, or a particular set of outcomes, in an experiment. For instance, in tossing the die we might be interested in whether the die shows an even number of dots. In this context, we refer to the set {2, 4, 6} as an *event* in the sample space {1, 2, 3, 4, 5, 6}. More generally, any subset of a given sample space is referred to as an **event**. We shall use the notation $n(E)$ to denote the number of outcomes in an event E. For example, if E is the event "obtaining an even number" that we've just discussed, then $n(E) = 3$, because there are three outcomes in the set $E = \{2, 4, 6\}$.

EXAMPLE I   Use set notation to specify each of the following events in the die-tossing experiment using the sample space {1, 2, 3, 4, 5, 6}.

Event *A*:   obtaining an odd number

Event *B*:   obtaining a number less than 3

Event *C*:   obtaining a 6

Event *D*:   obtaining a number less than 7

Event *E*:   obtaining a 7

Solution   Event *A*:   {1, 3, 5}   Event *D*:   {1, 2, 3, 4, 5, 6}
Event *B*:   {1, 2}   Event *E*:   ∅
Event *C*:   {6}

Two of the events we've listed here deserve special mention. First, notice that event *D* is the entire sample space, simply because every outcome in the sample space is a number less than 7. At the other extreme we have event *E*, "obtaining a 7." There are no outcomes in the sample space corresponding to this event. So (remembering that an event is a subset of the sample space), we say that event *E* is the so-called **empty set**, that is, the set with no elements. By convention, the empty set is denoted by the symbol ∅. ∎

EXAMPLE 2   Using the events *A, B, C, D,* and *E* in Example 1, compute $n(A)$, $n(B)$, $n(C)$, $n(D)$, and $n(E)$.

Solution   We need only count the number of outcomes in each set. The results are

$n(A) = 3$ $n(D) = 6$

$n(B) = 2$ $n(E) = 0$

$n(C) = 1$

For each event E in a finite sample space, we are going to define a number $P(E)$ called the *probability* of the event. Informally speaking, the number $P(E)$ is a rating of how likely it is that the event E will occur in any trial of the experiment. As you'll see, the number $P(E)$ always satisfies the condition $0 \leq P(E) \leq 1$.

DEFINITION   Probability of an Event

Let S be a finite sample space in which each outcome is equally likely to occur. If E is an event in S, then the **probability** of E, denoted by $P(E)$, is defined by

$$P(E) = \frac{n(E)}{n(S)} = \frac{\text{number of outcomes in } E}{\text{number of outcomes in } S}$$

EXAMPLE 3   Find the probability for each of the events A, B, C, D, and E discussed in Examples 1 and 2.

Solution $P(A) = P(\text{obtaining an odd number}) = \dfrac{n(A)}{n(S)} = \dfrac{3}{6} = \dfrac{1}{2}$

$P(B) = P(\text{obtaining a number less than 3}) = \dfrac{n(B)}{n(S)} = \dfrac{2}{6} = \dfrac{1}{3}$

$P(C) = P(\text{obtaining a 6}) = \dfrac{n(C)}{n(S)} = \dfrac{1}{6}$

$P(D) = P(\text{obtaining a number less than 7}) = \dfrac{n(D)}{n(S)} = \dfrac{6}{6} = 1$

$P(E) = P(\text{obtaining a 7}) = \dfrac{n(E)}{n(S)} = \dfrac{0}{6} = 0$

Notice that each of the probabilities that we computed in Example 3 is a number between 0 and 1, as mentioned previously. Also note that a probability of 0 is assigned to the "impossible" event, whereas a probability of 1 is assigned to the "certain" event.

EXAMPLE 4   Two (distinct) letters are selected at random from the set $\{a, b, c, d, e\}$. What is the probability that one of the letters is c?

Solution   Let the sample space S consist of all combinations of the five given letters taken two at a time. For instance, ab, ac, ad, and ae are four of the outcomes in this sample space. We needn't list every outcome in S in order to find $n(S)$; for, from our work in the previous section, we know that

$$n(S) = C(5, 2) = \frac{5!}{2!\,3!} = 10$$

Now we let E be the event "c is selected." We can specify E as follows:

$E = \{ac, bc, dc, ec\}$

Thus $n(E) = 4$ and, consequently,

$$P(E) = \frac{n(E)}{n(S)} = \frac{4}{10} = \frac{2}{5}$$

So the probability that c is one of the two letters selected is $\frac{2}{5}$. This is also expressed by saying there is a 40% chance that one of the two letters is c. ▮▮▮

EXAMPLE 5 Suppose that five faulty transistors are accidentally packaged with ten reliable transistors. What is the probability that 3 transistors chosen at random from this package all will be reliable?

Solution Let the sample space S consist of the combinations of the 15 transistors taken three at a time. Then we have

$$n(S) = C(15, 3) = \frac{15!}{3!12!} = 455$$

Next let E denote the event "choose three reliable transistors." Since these three must be chosen from the ten reliable ones, we have

$$n(E) = C(10, 3) = \frac{10!}{3!7!} = 120$$

Thus

$$P(E) = \frac{n(E)}{n(S)} = \frac{120}{455} = \frac{24}{91} \qquad \text{as required}$$ ▮▮▮

To continue our development of probability a bit further, it will be convenient to recall two basic ideas from set theory. Suppose that A and B are two sets. Then the **union** of these two sets, written $A \cup B$, is defined to be the set of elements that are in A *or* in B (or in both). For example, if

$$A = \{2, 3, 4, 5\} \qquad \text{and} \qquad B = \{4, 5, 6, 7\}$$

then

$$A \cup B = \{2, 3, 4, 5, 6, 7\}$$

The **intersection** of two sets A and B, written $A \cap B$, is defined as the set of elements that belong to both A *and* B. For example, using A and B as just defined, we have

$$A \cap B = \{4, 5\}$$

because 4 and 5 are the only elements common to both sets. As another example, if $E = \{1, 2\}$ and $F = \{3, 4\}$, then $E \cap F = \varnothing$, because there are no elements common to the two sets. Two sets that have no elements in common are said to be **disjoint.**

If A and B are events (remember, that means subsets of a sample space), then $A \cup B$ is the event "A or B," whereas $A \cap B$ is the event "A and B." Here's an example. Let S be the sample space $\{1, 2, 3, 4, 5, 6\}$ that we've used for the die-tossing experiment. And let the events A and B be as follows:

   A: obtaining an even number

   B: obtaining a number greater than four

Then we have

$$A \cup B = \text{"obtaining an even number" } or$$
$$\text{"obtaining a number greater than four"}$$
$$= \{2, 4, 6\} \cup \{5, 6\}$$
$$= \{2, 4, 5, 6\}$$

whereas

$$A \cap B = \text{"obtaining an even number" } and$$
$$\text{"obtaining a number greater than four"}$$
$$= \{2, 4, 6\} \cap \{5, 6\}$$
$$= \{6\}$$

EXAMPLE 6   A coin is flipped three times and the result is recorded each time. Let A, B, and C denote the following events.

   A: obtaining exactly two heads

   B: obtaining no heads

   C: obtaining more than one tail

Compute the following probabilities:

(a) $P(A)$, $P(B)$, and $P(C)$ **(b)** $P(A \text{ or } B)$ **(c)** $P(B \text{ or } C)$

Solution   Our sample space S will be as follows. (See Figure 1 in the previous section if you're not certain how we've obtained the eight outcomes in S.)

$$S = \{HHH, HHT, HTH, HTT, THH, THT, TTH, TTT\}$$

(a) Using the definitions given for A, B, and C, we have

$$A = \{HHT, HTH, THH\}$$
$$B = \{TTT\}$$
$$C = \{HTT, THT, TTH, TTT\}$$

Thus

$$P(A) = \frac{n(A)}{n(S)} = \frac{3}{8} \qquad P(B) = \frac{n(B)}{n(S)} = \frac{1}{8} \qquad P(C) = \frac{n(C)}{n(S)} = \frac{4}{8} = \frac{1}{2}$$

(b) The event "A or B" is $A \cup B$, so we have

$$A \cup B = \{HHT, HTH, THH, TTT\}$$

and, consequently,

$$P(A \cup B) = \frac{n(A \cup B)}{n(S)} = \frac{4}{8} = \frac{1}{2}$$

(c) Likewise, the event "B or C" is $B \cup C$, so we have

$$B \cup C = \{TTT, HTT, THT, TTH\}$$

and, consequently,

$$P(B \cup C) = \frac{n(B \cup C)}{n(S)} = \frac{4}{8} = \frac{1}{2} \qquad \text{as required}$$

We can use the results in Example 6 to introduce a rule that simplifies certain probability computations. If you review the results in Example 6, you'll see that $P(A \cup B) = \frac{1}{2}$, $P(A) = \frac{3}{8}$, and $P(B) = \frac{1}{8}$. Thus we have

$$P(A \cup B) = P(A) + P(B) \qquad \text{because } \tfrac{1}{2} = \tfrac{3}{8} + \tfrac{1}{8}$$

That is,

$$P(A \text{ or } B) = P(A) + P(B) \tag{1}$$

So (in this case at least), we find that the probability of the event "A or B" is the sum of the individual probabilities of A and B. Furthermore, it can be shown that equation (1) remains valid for any two events (in a given sample space) that have no outcomes in common. *Terminology:* Two events with no outcomes in common are said to be **mutually exclusive.** For instance, the events A and B in Example 6 are mutually exclusive. (Why?)

The remarks in the previous paragraph can be summarized this way: If E and F are mutually exclusive, then $P(E \text{ or } F) = P(E) + P(F)$. This formula is *not* valid, however, in cases in which the two events fail to be mutually exclusive. To see an example of this, let's take the events B and C in Example 6. In that example we found that $P(B \cup C) = \frac{1}{2}$, $P(B) = \frac{1}{8}$, and $P(C = \frac{1}{2})$. Consequently we have

$$P(B \cup C) \neq P(B) + P(C) \qquad \text{because } \tfrac{1}{2} \neq \tfrac{1}{8} + \tfrac{1}{2}$$

The difference here, as we've said, has to do with the fact that B and C are not mutually exclusive. As you can check, B and C do have an outcome in common, namely TTT. Because of this, when we compute $P(B \cup C)$, the outcome TTT is counted but once; but when we compute $P(B) + P(C)$, the outcome TTT is counted twice (once with B and once again with C). This is essentially the reason why we obtain different values for the quantities $P(B \cup C)$ and $P(B) + P(C)$. By following this same type of reasoning (in which we keep track of how many times an outcome is counted) the following general formula can be derived. Notice that the formula agrees with equation (1) in the case when $A \cap B = \varnothing$ (that is, when A and B are mutually exclusive).

The Addition Theorem for Probabilities

If A and B are events in a sample space S, then

$$P(A \cup B) = P(A) + P(B) - P(A \cap B)$$

EXAMPLE 7   Three balls are drawn at random from a bag containing eight red and seven yellow balls. Compute the probability for each of the following events:

(a) drawing three red balls;   **(b)** drawing three yellow balls;
(c) drawing either three red or three yellow balls.

Solution   **(a)** The total number of possible outcomes in the experiment is $C(15, 3)$, the number of ways of choosing three balls from a set of 15. We have

$$C(15, 3) = \frac{15!}{3!12!} = \frac{15 \cdot 14 \cdot 13}{3 \cdot 2} = 455$$

Next, we need to know how many outcomes are in the event "drawing three red balls." This number is

$$C(8, 3) = \frac{8!}{3!5!} = \frac{8 \cdot 7 \cdot 6}{3 \cdot 2} = 56$$

Consequently we have

$$P(\text{drawing 3 red balls}) = \frac{56}{455} = \frac{8}{65}$$

(b) Following the reasoning used in part (a), we find that

$$P(\text{drawing 3 yellow balls}) = \frac{C(7, 3)}{C(15, 3)} = \frac{35}{455} \qquad \text{(Check the arithmetic)}$$

$$= \frac{1}{13}$$

(c) Let A and B denote the events described in parts (a) and (b), respectively. Then $A \cup B$ is the event "drawing either three red or three yellow balls," and we have

$$P(A \cup B) = P(A) + P(B) - P(A \cap B)$$
$$= \frac{8}{65} + \frac{1}{13} - P(\varnothing)$$
$$= \frac{8}{65} + \frac{5}{65} - 0 = \frac{13}{65} = \frac{1}{5}$$

So the probability of drawing either three red or three yellow balls is $\frac{1}{5}$.

EXAMPLE 8   A card is selected from a well-shuffled deck of 52 cards. What is the probability that it is a jack or a heart?

Solution In this experiment there are 52 possible outcomes. Let A denote the event "selecting a jack," and let B denote the event "selecting a heart." Then $P(A \cup B)$ is the probability of the event "selecting a jack or a heart," and we have

$$P(A \cup B) = P(A) + P(B) - P(A \cap B)$$

$$= \frac{4}{52} + \frac{13}{52} - \frac{1}{52} \quad \text{(Why?)}$$

$$= \frac{16}{52} = \frac{4}{13} \quad \text{as required}$$

EXERCISE SET 13.7

A

Exercises 1–20 refer to an experiment that consists of tossing a single die and then noting the number of dots showing on the top face. As in the text, assume that the sample space for this experiment is

$$S = \{1, 2, 3, 4, 5, 6\}$$

Let A, B, C, D, E, F, and G be the following events:

A: *obtaining an even number*
B: *obtaining a number greater than 3*
C: *obtaining a 1*
D: *obtaining a number greater than 1*
E: *obtaining an odd number*
F: *obtaining a number less than 8*
G: *obtaining an 8*

1. Use set notation to specify the following events.
   (a) A (b) B (c) $A \cup B$ (d) $A \cap B$

2. Use set notation to specify the following events.
   (a) C (b) D (c) $C \cup D$ (d) $C \cap D$

3. Use set notation to specify the following events.
   (a) E (b) F (c) $E \cup A$ (d) $E \cap A$

4. Use set notation to specify the following events.
   (a) G (b) $G \cup B$ (c) $G \cap B$ (d) $G \cup F$

5. Are events A and B mutually exclusive?

6. Are events A and C mutually exclusive?

7. Are events C and D mutually exclusive?

8. Are events B and E mutually exclusive?

For Exercises 9–20, compute each of the indicated quantities.

9. (a) $n(A)$ (b) $n(B)$ (c) $n(A \cup B)$

10. (a) $n(C)$ (b) $n(D)$ (c) $n(C \cup D)$

11. (a) $n(E)$ (b) $n(F)$ (c) $n(G)$

12. $n(B \cap E)$

13. (a) $P(A)$ (b) $P(B)$

14. (a) $P(C)$ (b) $P(D)$

15. (a) $P(E)$ (b) $P(F)$ 16. $P(G)$

17. (a) $P(A \cup B)$ (b) $P(A) + P(B)$

18. (a) $P(B \cup C)$ (b) $P(B) + P(C)$

19. (a) $P(B \cup E)$ (b) $P(B) + P(E)$

20. (a) $P(D) + P(E) - P(D \cap E)$ (b) $P(D \cup E)$

Exercises 21–40 refer to the experiment described in Example 6, in which a coin is flipped three times and the result is recorded each time. As in Example 6, let the sample space S be

$$S = \{HHH, HHT, HTH, HTT, THH, THT, TTH, TTT\}$$

Let A, B, C, D, E, and F be the following events:

A: *obtaining exactly two tails*
B: *obtaining two heads on the first two tosses*
C: *obtaining heads on the first toss*
D: *obtaining no heads*
E: *obtaining tails on both the second and third tosses*
F: *obtaining tails on the first toss*

Compute each of the following probabilities.

21. $P(A)$ 22. $P(B)$ 23. $P(C)$

24. $P(D)$ 25. $P(E)$ 26. $P(F)$

27. $P(A \cup B)$ 28. $P(A \cap B)$ 29. $P(A \cup C)$

30. $P(A \cap C)$ 31. $P(C \cup F)$ 32. $P(C \cap F)$

33. $P(E \cup F)$ 34. $P(E \cap F)$ 35. $P(B \cup C)$

36. $P(B \cap C)$ 37. $P(D \cap A)$ 38. $P(D \cup A)$

39. $P(D \cup E)$ 40. $P(D \cap E)$

41. Two letters are selected at random (and without replacement) from the set $\{a, b, c, d, e, f\}$. What is the probability that one of the letters is d?

42. The numbers 1, 2, 3, and 4 are arranged in random order to form a four-digit number. What is the probability that the number is greater than 4000?

43. Suppose that two faulty transistors are accidentally packaged with a dozen reliable ones.
    (a) If one transistor is chosen at random from the package, what is the probability that it will be reliable?
    (b) If five transistors are chosen at random from the package, what is the probability that they are all reliable?

44. Suppose that five cartons of spoiled milk are inadvertently placed on a grocery shelf along with 15 cartons of fresh milk.
    (a) If three cartons are selected at random, what is the probability that they all contain unspoiled milk?
    (b) What is the probability that three cartons selected at random all contain spoiled milk?

In Exercises 45–48, use the addition theorem to compute the required probabilities.

45. If $P(A) = \frac{1}{3}$, $P(B) = \frac{1}{4}$, and $A \cap B = \varnothing$, compute $P(A \cup B)$.

46. If $P(A) = \frac{2}{5}$, $P(B) = \frac{3}{10}$, and $P(A \cap B) = \frac{1}{5}$, compute $P(A \cup B)$.

47. If $P(A \cup B) = \frac{7}{9}$, $P(A) = \frac{5}{9}$, and $P(B) = \frac{4}{9}$, compute $P(A \cap B)$.

48. If $P(A) = \frac{1}{3}$, $P(A \cup B) = \frac{2}{3}$, and $P(A \cap B) = \frac{1}{6}$, compute $P(B)$.

49. Two balls are drawn at random from a bag containing six red and ten black balls. Compute the probability for each of the following events:
    (a) Drawing two red balls
    (b) Drawing two black balls
    (c) Drawing either two red or two black balls

50. A card is selected at random from a deck of 52 cards. Compute the probability of each of the following events:
    (a) Selecting a 4 or 2
    (b) Selecting a red card or an ace
    (c) Selecting an ace or a picture card

51. Your mathematics instructor claims that the probabilities of your getting A's on the first exam, on the second exam, and on both exams are $\frac{1}{2}$, $\frac{3}{10}$, and $\frac{1}{10}$, respectively. According to the figures, what is the probability that you will get an A on at least one exam? *Hint:* Let G and H denote the events "A on first exam" and "A on second exam," respectively. Then you want to compute $P(G \text{ or } H)$.

52. Two cards are drawn at random from a regular 52-card deck. Find the probability that either both are black or both are 2's.

53. Five cards are drawn at random from a regular deck of 52 cards. Compute the probability that all five cards are clubs.

B

54. An experiment consists of selecting at random one of the 100 points in the x-y plane shown in the accompanying figure. Compute the probability for each of the following events.
    (a) The point lies on the line $y = x$.
    (b) The point lies on the line $y = 2x$.
    (c) The point lies on the circle $x^2 + y^2 = 25$.
    (d) The point lies within (but not on) the circle $x^2 + y^2 = 25$.
    (e) The point lies on or within the circle $x^2 + y^2 = 25$.
    (f) The point lies outside the circle $x^2 + y^2 = 25$.
    (g) The point lies inside the circle $x^2 + y^2 = 25$ but not on the graph of $y = |x - 3|$.
    (h) The point lies either on the circle $x^2 + y^2 = 25$ or on the line $y = x$.
    (i) The point lies on the x-axis or on the circle $x^2 + y^2 = 25$.
    (j) The sum of the x- and y-coordinates of the point is 18.
    (k) The sum of the x- and y-coordinates of the point is 21.
    (l) The y-coordinate of the point is greater than the square of the x-coordinate.

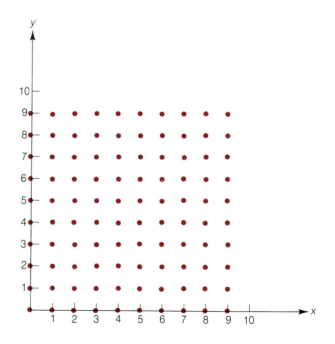

CHAPTER THIRTEEN   SUMMARY OF PRINCIPAL TERMS AND FORMULAS

TERMS OR FORMULAS	PAGE REFERENCE	COMMENT
1. Principle of mathematical induction	748	This principle can be stated as follows. Suppose that for each natural number n we have a statement P_n. Suppose P_1 is true. Also suppose that P_{k+1} is true whenever P_k is true. Then according to the **principle of mathematical induction**, all the statements are true; that is, P_n is true for all natural numbers n. The principle of mathematic induction has the status of an axiom; that is, we accept its validity without proof.
2. Pascal's triangle	754	Pascal's triangle refers to the triangular array of numbers displayed on page 754. Additional rows can be added to the triangle according to the following rule: Each entry in the array (other than the 1's along the sides) is the sum of the two numbers diagonally above it. The numbers in the nth row of Pascal's triangle are the coefficients of the terms in the expansion of $(a+b)^{n-1}$.
3. $n!$ (read: n factorial)	756	This denotes the product of the first n natural numbers. For example, $4! = (4)(3)(2)(1) = 24$.
4. $\binom{n}{k}$ (read: n choose k)	757	Let n and k be nonnegative integers with $k \le n$. Then the **binomial coefficient** $\binom{n}{k}$ is defined by $$\binom{n}{k} = \frac{n!}{k!(n-k)!}$$
5. Binomial theorem	757	The **binomial theorem** is a formula that allows us to analyze and expand expressions of the form $(a+b)^n$. If we use the sigma notation, then the statement of the binomial theorem that appears on page 757 can be abbreviated to read as follows. $$(a+b)^n = \sum_{k=0}^{n} \binom{n}{k} a^{n-k} b^k$$
6. Sequence	763, 764	In the context of the present chapter, a **sequence** is an ordered list of real numbers: $a_1, a_2, a_3, \ldots$ The numbers a_k are called the **terms** of the sequence. A sequence can also be defined as a *function* whose domain is the set of natural numbers.
7. $\sum\limits_{k=1}^{n} a_k$	765	This expression stands for the sum $a_1 + a_2 + a_3 + \cdots + a_n$. The letter k in the expression is referred to as the **index of summation**.

TERMS OR FORMULAS	PAGE REFERENCE	COMMENT
8. Arithmetic sequence	768	A sequence in which the successive terms differ by a constant is called an **arithmetic sequence**. The general form of an arithmetic sequence is $$a, a+d, a+2d, a+3d, \ldots$$ In this sequence, d is referred to as the **common difference**.
9. $a_n = a + (n-1)d$	769	This is the formula for the nth term of an arithmetic sequence.
10. $S_n = \dfrac{n}{2}[2a + (n-1)d]$ $S_n = n\left(\dfrac{a + a_n}{2}\right)$	771	These are the formulas for the sum of the first n terms of an arithmetic sequence.
11. Geometric sequence	774	A sequence in which the ratio of successive terms is constant is called a **geometric sequence**. The general form of a geometric sequence is $$a, ar, ar^2, \ldots$$ The number r is referred to as the **common ratio**.
12. $a_n = ar^{n-1}$	775	This is the formula for the nth term of a geometric sequence.
13. $S_n = \dfrac{a(1 - r^n)}{1 - r}$	775	This is the formula for the sum of the first n terms of a geometric sequence.
14. $S = \dfrac{a}{1 - r}$	777	This is the formula for the sum of the infinite geometric series $a + ar + ar^2 + \cdots$, where $|r| < 1$.
15. Multiplication principle	780	Suppose that a task involves two consecutive operations. If the first operation can be performed in M ways, and the second operation can be performed in N ways, then the entire task can be performed in MN ways.
16. Permutation	781	A **permutation** of a set of n objects is a listing or arrangement of those objects in a definite order.
17. $P(n, r) = n(n-1) \cdots (n - r + 1)$ $= \dfrac{n!}{(n-r)!}$	782, 783	$P(n, r)$ denotes the number of permutations of n objects taken r at a time $(r \leq n)$.
18. $C(n, r) = \dfrac{n!}{r!(n-r)!}$ $= \dfrac{n(n-1) \cdots (n - r + 1)}{r!}$	784, 785	$C(n, r)$ denotes the number of ways of selecting r objects from a set of n objects $(r \leq n)$. Each selection or subset of r objects is called a **combination** of the n objects taken r at a time.
19. Sample space; event	789	The sample space for an experiment is the set of all possible outcomes of the experiment. A subset of the sample space is called an **event**. The notation $n(E)$ is used to denote the number of outcomes in an event E.

TERMS OR FORMULAS	PAGE REFERENCE	COMMENT
20. $P(E)$	790	Let S be a finite sample space in which each outcome is equally likely to occur. If E is an event in S, then the probability of E, denoted by $P(E)$, is defined by $$P(E) = \frac{n(E)}{n(S)} = \frac{\text{number of outcomes in } E}{\text{number of outcomes in } S}$$
21. The addition theorem for probabilities	793	If A and B are events in a sample space S, then $$P(A \cup B) = P(A) + P(B) - P(A \cap B)$$

WRITING MATHEMATICS

1. Look up the word *anagram* in a dictionary. Then explain which concept from this chapter is illustrated by the following special anagram:

   TELEGRAPH $\leftrightarrow$ GREAT HELP

2. **(a)** Write down a formula for the number of permutations of five objects taken r at a time.
   (b) Does the formula define a function? If your answer is no, explain your reasons. If your answer is *yes*, explain your reasons and describe the domain, the range, and the graph.

3. The following exercise appears in the text *Essentials of Finite Mathematics: Matrices, Linear Programming, Probability, Markov Chains* by Robert F. Brown and Brenda W. Brown (New York: Ardsley House Publishers, 1990). Working with a classmate or your instructor, solve the problem. Then, on your own, write out the complete solution. Be sure to let the reader know what you are doing and which principles you are applying.

   In a chemical experiment, six different chemicals will be mixed by adding one at a time to the mixture. There are two chemicals, out of the six, that cannot be added one after the other (in either order) without causing an undesirable reaction. In how many ways can the experiment be performed? (*Hint:* In how many ways can the experiment be performed by adding the two reacting chemicals one after the other?)

 CHAPTER THIRTEEN REVIEW EXERCISES

In Exercises 1–10, use the principle of mathematical induction to show that the statements are true for all natural numbers.

1. $5 + 10 + 15 + \cdots + 5n = \frac{5}{2}n(n + 1)$

2. $10 + 10^2 + 10^3 + \cdots + 10^n = \frac{10}{9}(10^n - 1)$

3. $1 \cdot 2 + 2 \cdot 3 + 3 \cdot 4 + \cdots + n(n + 1) = \frac{1}{3}n(n + 1)(n + 2)$

4. $\frac{1}{2} + \frac{2}{2^2} + \frac{3}{2^3} + \cdots + \frac{n}{2^n} = 2 - \frac{2 + n}{2^n}$

5. $1 + 3 \cdot 2 + 5 \cdot 2^2 + 7 \cdot 2^3 + \cdots$
   $+ (2n - 1) \cdot 2^{n-1} = 3 + (2n - 3) \cdot 2^n$

6. $\frac{1}{1 \cdot 4} + \frac{1}{4 \cdot 7} + \frac{1}{7 \cdot 10} + \cdots$
   $+ \frac{1}{(3n - 2)(3n + 1)} = \frac{n}{3n + 1}$

7. $1 + 2^2 \cdot 2 + 3^2 \cdot 2^2 + 4^2 \cdot 2^3 + \cdots + n^2 \cdot 2^{n-1}$
   $= (n^2 - 2n + 3)2^n - 3$

8. 9 is a factor of $n^3 + (n + 1)^3 + (n + 2)^3$.

9. 3 is a factor of $7^n - 1$.

10. 8 is a factor of $9^n - 1$.

In Exercises 11–20, expand the given expressions.

11. $(3a + b^2)^4$

12. $(5a - 2b)^3$

13. $(x + \sqrt{x})^4$

14. $(1 - \sqrt{3})^6$

15. $(x^2 - 2y^2)^5$

16. $\left(\dfrac{1}{a} + \dfrac{2}{b}\right)^3$

17. $\left(1 + \dfrac{1}{x}\right)^5$

18. $\left(x^3 + \dfrac{1}{x^2}\right)^6$

19. $(a\sqrt{b} - b\sqrt{a})^4$

20. $(x^{-2} + y^{5/2})^8$

21. Find the fifth term in the expansion of $(3x + y^2)^5$.

22. Find the eighth term in the expansion of $(2x - y)^9$.

23. Find the coefficient of the term containing a^5 in the expansion of $(a - 2b)^7$.

24. Find the coefficient of the term containing b^8 in the expansion of $\left(2a - \dfrac{b}{3}\right)^{10}$

25. Find the coefficient of the term containing x^3 in the expansion of $(1 + \sqrt{x})^8$.

26. Expand $(1 + \sqrt{x} + x)^6$. *Suggestion:* Rewrite the expression as $[(1 + \sqrt{x}) + x]^6$.

In Exercises 27–32, verify each assertion by computing the indicated binomial coefficients.

27. $\dbinom{2}{0}^2 + \dbinom{2}{1}^2 + \dbinom{2}{2}^2 = \dbinom{4}{2}$

28. $\dbinom{3}{0}^2 + \dbinom{3}{1}^2 + \dbinom{3}{2}^2 + \dbinom{3}{3}^2 = \dbinom{6}{3}$

29. $\dbinom{4}{0}^2 + \dbinom{4}{1}^2 + \dbinom{4}{2}^2 + \dbinom{4}{3}^2 + \dbinom{4}{4}^2 = \dbinom{8}{4}$

30. $\dbinom{2}{0} + \dbinom{2}{1} + \dbinom{2}{2} = 2^2$

31. $\dbinom{3}{0} + \dbinom{3}{1} + \dbinom{3}{2} + \dbinom{3}{3} = 2^3$

32. $\dbinom{4}{0} + \dbinom{4}{1} + \dbinom{4}{2} + \dbinom{4}{3} + \dbinom{4}{4} = 2^4$

In Exercises 33–38, compute the first four terms in each sequence. Also, in Exercises 33–35, graph the sequences for n = 1, 2, 3, and 4.

33. $a_n = \dfrac{2n}{n + 1}$

34. $a_n = \dfrac{3n - 2}{3n + 2}$

35. $a_n = (-1)^n\left(1 - \dfrac{1}{n + 1}\right)$

36. $a_0 = 4;\ a_n = 2a_{n-1},\ n \geq 1$

37. $a_0 = -3;\ a_n = 4a_{n-1},\ n \geq 1$

38. $a_0 = 1;\ a_1 = 2;\ a_n = 3a_{n-1} + 2a_{n-2},\ n \geq 2$

In Exercises 39 and 40, evaluate each sum.

39. (a) $\displaystyle\sum_{k=1}^{3} (-1)^k(2k + 1)$ (b) $\displaystyle\sum_{k=0}^{8}\left(\dfrac{1}{k + 1} - \dfrac{1}{k + 2}\right)$

40. (a) $\displaystyle\sum_{j=1}^{4} \dfrac{(-1)^j}{j}$ (b) $\displaystyle\sum_{n=1}^{5}\left(\dfrac{1}{n} - \dfrac{1}{n + 1}\right)$

In Exercises 41 and 42, rewrite each sum using sigma notation.

41. $5/3 + 5/3^2 + 5/3^3 + 5/3^4 + 5/3^5$

42. $1/2 - 2/2^2 + 3/2^3 - 4/2^4 + 5/2^2 - 6/2^6$

In Exercises 43–46, find the indicated term in each sequence.

43. $5, 9, 13, 17, \ldots;\ a_{18}$

44. $5, \dfrac{9}{2}, 4, \dfrac{7}{2}, \ldots;\ a_{20}$

45. $10, 5, \dfrac{5}{2}, \dfrac{5}{4}, \ldots;\ a_{12}$

46. $\sqrt{2} + 1, 1, \sqrt{2} - 1, 3 - 2\sqrt{2}, \ldots;\ a_{10}$

47. Determine the sum of the first 12 terms of an arithmetic sequence in which the first term is 8 and the twelfth term is $\dfrac{43}{2}$.

48. Find the sum of the first 45 terms in the sequence $10, \dfrac{29}{3}, \dfrac{28}{3}, 9, \ldots$.

49. Find the sum of the first 10 terms in the sequence 7, 70, 700, ...

50. Find the sum of the first 12 terms in the sequence $\dfrac{1}{3}, -\dfrac{2}{9}, \dfrac{4}{27}, -\dfrac{8}{81}, \ldots$.

51. In a certain geometric sequence, the third term is 4 and the fifth term is 10. Find the sixth term, given that the common ratio is negative.

52. For a certain infinite geometric series, the first term is 2 and the sum is $\dfrac{18}{11}$. Find the common ratio r.

In Exercises 53–56, find the sum of each infinite geometric series.

53. $\dfrac{3}{5} + \dfrac{3}{25} + \dfrac{3}{125} + \cdots$

54. $\dfrac{7}{10} + \dfrac{7}{100} + \dfrac{7}{1000} + \cdots$

55. $\dfrac{1}{9} - \dfrac{1}{81} + \dfrac{1}{729} - \cdots$

56. $1 + \dfrac{1}{1 + \sqrt{2}} + \dfrac{1}{(1 + \sqrt{2})^2} + \cdots$

57. Find a fraction equivalent to $0.\overline{45}$.

58. Find a fraction equivalent to $0.2\overline{13}$.

In Exercises 59–62, verify each equation using the formula for the sum of an arithmetic series:

$$S_n = \dfrac{n}{2}[2a + (n - 1)d]$$

(The formulas given in Exercises 59–62 appear in Elements of Algebra *by Leonhard Euler, first published in 1770.)*

59. $1 + 2 + 3 + \cdots + n = n + n(n - 1)/2$

60. $1 + 3 + 5 + \cdots$ (to n terms) $= n + 2n(n - 1)/2$

61. $1 + 4 + 7 + \cdots$ (to n terms) $= n + 3n(n - 1)/2$

62. $1 + 5 + 9 + \cdots$ (to n terms) $= n + 4n(n - 1)/2$

63. In this exercise, you will use the following (remarkably simple) formula for approximating sums of powers of integers:

$$1^k + 2^k + 3^k + \cdots + n^k \approx \frac{\left(n + \tfrac{1}{2}\right)^{k+1}}{k + 1} \qquad (1)$$

[This formula appears in the article, "Sums of Powers of Integers" by B. L. Burrows and R. F. Talbot, published in the *American Mathematical Monthly* **91** (1984):394.]

(a) Use formula (1) to estimate the sum $1^2 + 2^2 + 3^2 + \cdots + 50^2$. Round off your answer to the nearest integer.

(b) Compute the exact value of the sum in part (a) using the formula $\displaystyle\sum_{k=1}^{n} k^2 = \frac{1}{6}n(n + 1)(2n + 1)$.

(This formula can be proved using mathematical induction.) Then compute the percent error for the approximation obtained in part (a). The percent error is given by

$$\frac{|\text{actual value} - \text{approximate value}|}{\text{actual value}} \times 100$$

(c) Use formula (1) to estimate the sum $1^4 + 2^4 + 3^4 + \cdots + 200^4$. Round off your answer to six significant digits.

(d) The following formula for the sum $1^4 + 2^4 + \cdots + n^4$ can be proved using mathematical induction:

$$\sum_{k=1}^{n} k^4 = \frac{n(n + 1)(2n + 1)(3n^2 + 3n - 1)}{30}$$

Use this formula to compute the sum in part (c). Round off your answer to six significant digits. Then use this result to compute the percent error for the approximation in part (c).

64. According to *Stirling's formula* [named after James Stirling (1692–1770)], the quantity $n!$ can be approximated as follows:

$$n! \approx \sqrt{2\pi n}\left(\frac{n}{e}\right)^n$$

In this formula, e is the constant $2.718\ldots$ (discussed in Section 6.2). Use a calculator to complete the following table. Round off your answers to five significant digits. As you will see, the numbers in the right-hand column approach 1 as n increases. This shows that, in a certain sense, the approximation improves as n increases.

n	$n!$	$\sqrt{2\pi n}(n/e)^n$	$\dfrac{n!}{\sqrt{2\pi n}(n/e)^n}$
10			
20			
30			
40			
50			
60			
65			

65. How many different two-digit numbers can be formed using the digits 2, 4, 6, and 8? (Repetitions are allowed.)

66. How many permutations are there of the set $\{A, B, C, D, E\}$?

67. Using the set $\{a, b, c, d\}$, list the permutations of these four letters taken three at a time. Evaluate $P(4, 3)$.

68. Follow Exercise 67, using combinations rather than permutations, and evaluate $C(4, 3)$.

69. From a collection of 12 pictures, eight are to be selected and hung in a row. How many arrangements are possible?

70. How many possible arrangements are there of the letters a, b, c, d, e, f in which a, b, and c appear consecutively (but not necessarily in that order)?

71. How many possible arrangements are there of the letters a, b, c, d, e, f in which e and f are not next to each other?

72. How many four-member committees are possible in a club with 15 members?

73. On a certain committee there are nine males and ten females. How many possibilities are there for a four-member subcommittee consisting of two females and two males?

74. Given ten points in a plane, no three of which lie on the same line, find the number of different triangles that can be formed when the vertices are selected from the given points.

75. Suppose that three faulty batteries are accidentally shipped along with nine reliable batteries. What is the probability that two batteries selected at random from this shipment will both be reliable?

76. The accompanying figure shows a 12-sided figure with all its diagonals. How many diagonals are there?

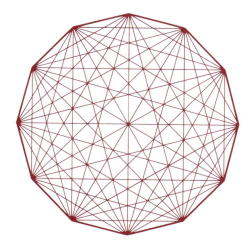

77. Two balls are selected at random from a bag containing five red and seven green balls. Compute the probability for each of the following events:
    (a) drawing two red balls;
    (b) drawing two green balls;
    (c) drawing either two red or two green balls.

For Exercises 78–88, consider an experiment in which two dice are tossed. Assume that one die is red and the other is white. Then the outcomes can be recorded as ordered pairs. Let us agree that the first element in any ordered pair corresponds to the number of dots showing on the red die and the second element corresponds to the number of dots showing on the white die. For example, (5, 4) denotes the outcome "5 on red and 4 on white." Compute the probabilities for the following events.

78. Obtaining a sum of 2
79. Obtaining a sum of 3
80. Obtaining a sum of 5
81. Obtaining a sum of 7
82. Obtaining a sum of 8
83. Obtaining a sum of either 7 or 11
84. Obtaining "doubles" (i.e., the same number of dots on both dice)
85. Obtaining a sum of at least 10
86. Obtaining a sum that is either even or less than 5
87. Obtaining any sum other than 7 and 8
88. Obtaining either doubles or a sum greater than 9

89. An integer between 1 and 15, inclusive, is selected at random. Compute the probabilities for the following events.
    (a) Selecting an even number
    (b) Selecting an odd number
    (c) Selecting either a number greater than 10 or an even number
    (d) Selecting either a multiple of 2 or a multiple of 3

90. (a) How many three-digit numbers are there with no repeated digits? (The first digit cannot be zero.)
    (b) A three-digit number is selected at random. What is the probability that it has no repeated digits?

91. (a) How many three-digit numbers are there with one or more repeated digits?
    (b) What is the probability that a three-digit number selected at random will contain one or more repeated digits?

92. The sum of three consecutive terms in a geometric sequence is 13, and the sum of the reciprocals is $\frac{13}{9}$. What are the possible values for the common ratio? (There are four answers.)

93. The nonzero numbers a, b, and c are consecutive terms in a geometric sequence, and $a + b + c = 70$. Furthermore, $4a$, $5b$, and $4c$ are consecutive terms in an arithmetic sequence. Find a, b, and c.

94. The nonzero numbers a, b, and c are consecutive terms in a geometric sequence, and a, $2b$, and c are consecutive terms in an arithmetic sequence. Show that the common ratio in the geometric sequence must be either $2 + \sqrt{3}$ or $2 - \sqrt{3}$.

95. If the numbers $\frac{1}{b + c}$, $\frac{1}{c + a}$, and $\frac{1}{a + b}$ are consecutive terms in an arithmetic sequence, show that a^2, b^2, and c^2 are also consecutive terms in an arithmetic sequence.

96. If a, b, and c are consecutive terms in a geometric sequence, prove that $\frac{1}{a + b}$, $\frac{1}{2b}$, and $\frac{1}{c + b}$ are consecutive terms in an arithmetic sequence.

97. (a) Find a value for x such that $3 + x$, $4 + x$, and $5 + x$ are consecutive terms in a geometric sequence.
    (b) Given three numbers a, b, and c, find a value for x (in terms of a, b, and c) such that $a + x$, $b + x$, and $c + x$ are consecutive terms in a geometric sequence.

98. If $\ln(A + C) + \ln(A + C - 2B) = 2 \ln(A - C)$, show that $1/A$, $1/B$, and $1/C$ are consecutive terms in an arithmetic sequence.

99. Let a_1, a_2, a_3, ... be an arithmetic sequence with common difference d, and let $r\,(\neq 1)$ be a real number. In this exercise we develop a formula for the sum of the series

$$a_1 + ra_2 + r^2a_3 + \cdots + r^{n-1}a_n$$

The method we use here is essentially the same as the method used in the text to derive the formula for the sum of a finite geometric series. So, as background for this exercise, you should review the derivation on page 563 for the sum of a geometric series.
(a) Let S denote the required sum. Show that

$$S - rS = a_1 + rd + r^2d + \cdots + r^{n-1}d - r^n a_n$$
$$= a_1 + \frac{d(r - r^n)}{1 - r} - r^n a_n$$

(b) Show that

$$S = \frac{a_1 - r^n a_n}{1 - r} + \frac{d(r - r^n)}{(1 - r)^2}$$

100. Use your calculator and the formula in Exercise 99(b) to find the sum of each of the following series.
(a) $1 + 2 \times 2 + 2^2 \times 3 + 2^3 \times 4 + \cdots + 2^{13} \times 14$
(b) $2 + 4 \times 5 + 4^2 \times 8 + 4^3 \times 11 + \cdots + 4^6 \times 20$
(c) $3 - \dfrac{1}{2} \cdot 5 + \dfrac{1}{2^2} \cdot 7 - \dfrac{1}{2^3} \cdot 9 + \cdots +$ (to 10 terms)

101. Let a_1, a_2, a_3, a_4, $\cdots$ be an arithmetic sequence in which all of the terms are positive. Show that the following equations are valid.
(a) $\dfrac{1}{\sqrt{a_1} + \sqrt{a_2}} + \dfrac{1}{\sqrt{a_2} + \sqrt{a_3}} = \dfrac{2}{\sqrt{a_1} + \sqrt{a_3}}$
(b) $\dfrac{1}{\sqrt{a_1} + \sqrt{a_2}} + \dfrac{1}{\sqrt{a_2} + \sqrt{a_3}} + \dfrac{1}{\sqrt{a_3} + \sqrt{a_4}}$
$$= \dfrac{3}{\sqrt{a_1} + \sqrt{a_4}}$$
(c) $\dfrac{1}{\sqrt{a_1} + \sqrt{a_2}} + \dfrac{1}{\sqrt{a_2} + \sqrt{a_3}} +$
$$\cdots + \dfrac{1}{\sqrt{a_{n-1}} + \sqrt{a_n}} = \dfrac{n - 1}{\sqrt{a_1} + \sqrt{a_n}}$$

CHAPTER THIRTEEN TEST

1. Use the principle of mathematical induction to show that the following formula is valid for all natural numbers n: $1^2 + 2^2 + 3^2 + \cdots + n^2 = \dfrac{n(n + 1)(2n + 1)}{6}$.

2. Express each of the following sums without using sigma notation, and then evaluate each sum.

(a) $\displaystyle\sum_{k=0}^{2} (10k - 1)$ (b) $\displaystyle\sum_{k=1}^{3} (-1)^k k^2$

3. (a) Write the formula for the sum S_n of a finite geometric series.
(b) Evaluate the sum $\dfrac{3}{2} + \dfrac{3^2}{2^2} + \dfrac{3^3}{2^3} + \cdots + \dfrac{3^{10}}{2^{10}}$.

4. (a) Determine the coefficient of the term containing a^3 in the expansion of $(a - 2b^3)^{11}$.
(b) Find the fifth term of the expansion in part (a).

5. Expand the expression $(3x^2 + y^3)^5$.

6. Determine the sum of the first 12 terms of an arithmetic sequence in which the first term is 8 and the twelfth term is $\frac{43}{2}$.

7. Find the sum of the following infinite geometric series: $\frac{7}{10} + \frac{7}{100} + \frac{7}{1000} + \cdots$.

8. A sequence is defined recursively as follows: $a_1 = 1$, $a_2 = 1$, and $a_n = (a_{n-1})^2 + a_{n-2}$ for $n \geq 3$. Determine the fourth and the fifth terms in this sequence.

9. In a certain geometric sequence, the third term is 4 and the fifth term is 10. Find the sixth term, given that the common ratio is negative.

10. What is the twentieth term in the arithmetic sequence -61, -46, -31, $\ldots$?

11. A telephone area code for the United States, Canada, and the Caribbean is a three-digit number. The first digit cannot be 0 or 1, the second digit *must* be 0 or 1, and the third digit cannot be 0. Subject to these constraints, repetitions are allowed within a number. (For example, the area codes for New York City and Hawaii are 212 and 808, respectively.) In 1991, there were only five area codes not yet assigned. How many area codes were in use in 1991?

12. From a collection of 14 pictures, eight are to be selected and hung in a row. How many arrangements are possible?

13. Suppose that three faulty batteries are accidentally shipped along with twelve reliable batteries. What is the probability that two batteries selected at random from this shipment will both be reliable?

14. Two balls are selected at random from a bag containing three red and seven green balls. Compute the probability for each of the following events: (a) drawing two red balls;   (b) drawing two green balls;   (c) drawing either two red or two green balls.

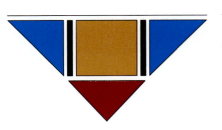

APPENDIX

A.1   SIGNIFICANT DIGITS AND CALCULATORS

Many of the numbers that we use in scientific work and in daily life are approximations. In some cases the approximations arise because the numbers were obtained through measurements or experiments. Consider, for example, the following statement from an astronomy textbook:

> The diameter of the Moon is 3476 km

We interpret this statement as meaning that the actual diameter D is closer to 3476 km than it is to either 3475 km or 3477 km. In other words,

> $3475.5 \text{ km} \leq D \leq 3476.5 \text{ km}$

The interval [3475.5, 3476.5] in this example provides information about the accuracy of the measurement. Another way to indicate accuracy in an approximation is by specifying the number of *significant digits* it contains. The measurement 3476 km has four significant digits. In general, the number of significant digits in a given number is found as follows.

Significant Digits		
The number of significant digits in a given number is determined by counting the digits from left to right, beginning with the left-most nonzero digit.	EXAMPLES	
	Number	Number of Significant Digits
	1.43	3
	0.52	2
	0.05	1
	4837	4
	4837.0	5

Numbers obtained through measurements are not the only source of approximations in scientific work. For example, to five significant digits, we have the following approximation for the irrational number π:

> $\pi \approx 3.1416$

This statement tells use that π is closer to 3.1416 than it is to either 3.1415 or 3.1417. In other words,

$$3.14155 \le \pi \le 3.14165$$

Table 1 provides some additional examples of the ideas we've introduced.

TABLE 1

Number	Number of Significant Digits	Range of Measurement
37	2	[36.5, 37.5]
37.0	3	[36.95, 37.05]
268.1	4	[268.05, 268.15]
1.036	4	[1.0355, 1.0365]
0.036	2	[0.0355, 0.0365]

There is an ambiguity involving zero that can arise in counting significant digits. Suppose that someone measures the width w of a rectangle and reports the result as 30 cm. How many significant digits are there? If the value 30 cm was obtained by measuring to the nearest 10 cm, then only the digit 3 is significant, and we can conclude only that the width w lies in the range 25 cm $\le w \le$ 35 cm. On the other hand, if the 30 cm was obtained by measuring to the nearest 1 cm, then both the digits 3 and 0 are significant and we have 29.5 cm $\le w \le$ 30.5 cm.

By using **scientific notation** we can avoid the type of ambiguity discussed in the previous paragraph. A number written in the form

$$b \times 10^n \qquad \text{where } 1 \le b < 10 \text{ and } n \text{ is an integer}$$

is said to be expressed in scientific notation. For the example in the previous paragraph, then, we would write

$$w = 3 \times 10^1 \text{ cm} \qquad \text{if the measurement were to the nearest 10 cm}$$

and

$$w = 3.0 \times 10^1 \text{ cm} \qquad \text{if the measurement were to the nearest 1 cm}$$

As the figures in Table 2 indicate, for a number $b \times 10^n$ in scientific notation, the number of significant digits is just the number of digits in b. (This is one of the advantages in using scientific notation; the number of significant digits, and hence the accuracy of the measurement, is readily apparent.)

TABLE 2

Measurement	Number of Significant Digits	Range of Measurement
Mass of the earth:		
6×10^{27} g	1	[5.5×10^{27} g, 6.5×10^{27} g]
6.0×10^{27} g	2	[5.95×10^{27} g, 6.05×10^{27} g]
5.974×10^{27} g	4	[5.9735×10^{27} g, 5.9745×10^{27} g]
Mass of a proton:		
1.67×10^{-24} g	3	[1.665×10^{-24} g, 1.675×10^{-24} g]

There are a number of exercises in this text in which a calculator either is required or is extremely useful. Although you could work many of these exercises using tables instead of a calculator, this author recommends that you purchase a scientific calculator. *Scientific calculator* is a generic term describing a calculator with (at least) the following features or functions beyond the usual arithmetic functions:

1. Memory
2. Scientific notation
3. Powers and roots
4. Logarithms (base ten and base *e*)
5. Trigonometric functions
6. Inverse functions

Because of the variety and differences in the calculators that are available, no specific instructions are provided in this section of the appendix for operating a calculator. When you buy a calculator, read the owner's manual carefully and work through some of the examples in it. In general, learn to use the memory capabilities of the calculator so that, as far as possible, you don't need to write down the results of the intermediate steps in a given calculation.

Many of the numerical exercises in the text ask that you round off the answers to a specified number of decimal places. Our rules for rounding off are as follows.

Rules for Rounding Off a Number (With More Than *n* Decimal Places) to *n* Decimal Places

1. If the digit in the $(n + 1)$st decimal place is greater than 5, increase the digit in the nth place by 1. If the digit in the $(n + 1)$st place is less than 5, leave the nth digit unchanged.

2. If the digit in the $(n + 1)$st decimal place is 5 and there is at least one nonzero digit to the right of this 5, increase the digit in the nth decimal place by 1.

3. If the digit in the $(n + 1)$st decimal place is 5 and there are no nonzero digits to the right of this 5, then increase the digit in the nth decimal place by 1 only if this results in an even digit.

The examples in Table 3 illustrate the use of these rules.

TABLE 3

Number	Rounded to One Decimal Place	Rounded to Three Decimal Places
4.3742	4.4	4.374
2.0515	2.1	2.052
2.9925	3.0	2.992

These same rules can be adapted for rounding off a result to a specified number of significant digits. As examples of this, we have

2347 rounded off to two significant digits is $2300 = 2.3 \times 10^3$

2347 rounded off to three significant digits is $2350 = 2.35 \times 10^3$

975 rounded off to two significant digits is $980 = 9.8 \times 10^2$

0.985 rounded off to two significant digits is $0.98 = 9.8 \times 10^{-1}$

In calculator exercises that ask you to round off your answers, it's important that you postpone rounding until the final calculation is carried out. For example, suppose that you are required to determine the hypotenuse x of the right triangle in Figure 1 to two significant digits. Using the Pythagorean theorem, we have

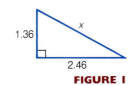

1.36

x

2.46

FIGURE 1

$$x = \sqrt{(1.36)^2 + (2.46)^2}$$

$$= 2.8 \qquad \text{(using a calculator and rounding off the final result to two significant digits)}$$

On the other hand, if we first round off each of the given lengths to two significant digits, we obtain

$$x = \sqrt{(1.4)^2 + (2.5)^2}$$

$$= 2.9 \qquad \text{(to two significant digits)}$$

This last result is inappropriate and we can see why as follows. As Table 4 shows, the maximum possible values for the sides are 1.365 and 2.465, respectively.

Thus, the maximum possible value for the hypotenuse must be

$$\sqrt{(1.365)^2 + (2.465)^2} = 2.817 \ldots \qquad \text{(calculator display)}$$

$$= 2.8 \qquad \text{(to two significant digits)}$$

This shows that the value 2.9 is indeed inappropriate, as we stated previously.

An error often made by people working with calculators and approximations is to report a final answer with a greater degree of accuracy than the data warrant. Consider, for example, the right triangle in Figure 2. Using the Pythagorean theorem and a calculator with an eight-digit display, we obtain

$$h = 3.6055513 \text{ cm}$$

This value for h is inappropriate, since common sense tells us that the answer should be no more accurate than the data used to obtain that answer. In particular, since the given sides of the triangle apparently were measured only to the nearest tenth of a centimeter, we certainly should not expect any improvement in accuracy for the resulting value of the hypotenuse. An appropriate form for the value of h here would be $h = 3.6$ cm.

TABLE 4

Number	Range of Measurement
1.36	[1.355, 1.365]
2.46	[2.455, 2.465]

FIGURE 2

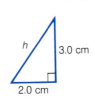

h

3.0 cm

2.0 cm

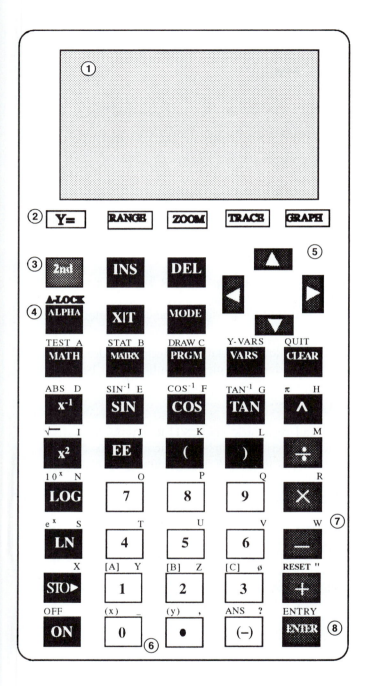

A.2 USING THE TI-81

Prepared by M. G. Settle © 1991 MR Creations

In the coming weeks, you will find that graphing functions greatly increases your ability to learn concepts in mathematics. The TI-81 Graphics Calculator can increase your efficiency in producing these graphs. Before learning how to use the TI-81, take a moment to learn its layout and basic features.

The Display Screen   At the top of the TI-81 is an 8-line-by-16-character display screen. (See area ①.) On this screen, computations are entered, and the results are displayed. Additionally, this screen is used to display the graph of the function.

The Graph Keys   Just below the display screen of the TI-81 are five keys. (See area ②.) These keys are used to define, display, manipulate, and interpret the graph of a function.

The 2nd Key   The top leftmost key below the Graph Keys is the **2nd** key. (See area ③.) This key is used to access the functions displayed on the left above some keys (such as "10^x" and "$\sqrt{}$").

The ALPHA Key   Just below the **2nd** key is the **ALPHA** key. (See area ④.) This key is used to access the characters displayed on the right above some keys (such as "A" and "?").

Directional Arrow Keys   The four blue keys at the top right (see area ⑤) are used to move the cursor during editing and to trace along the curve of a graph.

The Function Keys   The remaining black keys are used to access various mathematical functions ("SIN", "LOG", "x^2", etc.), to edit entries (the **DEL** and **INS** keys), and to display menus (the **MATH**, **MATRX**, **PRGM**, and **VARS** keys).

The Numeric Keypad   The gray keys are used to type numeric values. (See area ⑥.)

The Operation Keys   On the lower right are the four operation keys (**+** , **−** , **×** , **÷**). (See area ⑦.)

The ENTER Key   At the bottom right is the **ENTER** key. Pressing this key informs the TI-81 to execute the command (instruction) that you have typed. (See area ⑧.)

A-5

The opening screen
and the cursor. ⑨

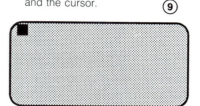

Comment

If you have trouble seeing
the display, press the 2nd
key and then hold down the
down-arrow key (▼) or the
up-arrow key (▲) to change
the contrast.

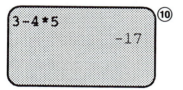

Order of operations yields −17.

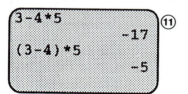

Use parentheses to override
the order of operations.

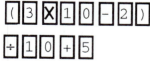

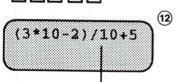

Place the cursor here, using
the left-arrow key.

GETTING STARTED

With the TI-81 in hand, find the **ON** key (at the left of the bottom row) and press it to turn on the TI-81. The TI-81 responds by presenting the cursor (a blinking square character). If you have trouble seeing the screen, try tilting the calculator slightly. If you still can't see the cursor, press the **2nd** key and hold the down-arrow key (▼) to set the contrast. (See frame ⑨.)

DOING CALCULATIONS

Because the TI-81 is different from a standard calculator, we will take a moment to observe some common computations on the TI-81.

Example: Compute $3 - 4 \times 5$.

Solution: Type $3 - 4 \times 5$ and then press the **ENTER** key. As a result, the multiplication is performed to yield $3 - 20$, and then the subtraction is performed. The number -17 is displayed to the right of the computation entered. (See frame ⑩.)

As you would expect, the TI-81 follows the standard algebraic order of operations (precedence) rules. To override these rules, parentheses can be entered. The parentheses keys are located just above the gray numeric keypad section.

Example: Compute $(3 - 4) \times 5$.

Solution: Type $(3 - 4) \times 5$ and then press the **ENTER** key. As a result, the computation in the parentheses is performed first (subtraction in this case), and then the multiplication is performed to yield -5. (See frame ⑪.)

EDITING COMPUTATIONS: THE INS KEY

In the above examples, the TI-81 operates like a calculator. However, the TI-81 is different from a standard scientific calculator. The first major difference is the TI-81's ability to edit computations. After a computation is typed, it can be edited (changed) before pressing the **ENTER** key. Consider the following example.

Example: Evaluate $(3x - 2)/(x + 5)$ for x equal to 10.

Solution: Type $(3 \times 10 - 2) \div 10 + 5$ but *do not* press the **ENTER** key yet! As typed, the computations in the parentheses will be performed, divided by 10, and then 5 will be added to the results. This is *not* what we want. We want the computation in the parentheses divided by the *sum* of 10 and 5 (not just 10). To edit (change) the computation entered, the directional-arrow keys are used.

Press the left-arrow key (◄) until the cursor is under the 1 of the second 10, as shown in frame ⑫. Now, press the **INS** key to enter the Insert mode. The cursor changes to a blinking underscore character.

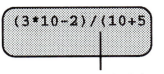

Press the left parenthesis key, and the (character is inserted.

Press the left parenthesis key, (, to insert the left parenthesis. (See frame ⑬.)

Press the right-arrow key (▶) to move the cursor to the right of the 5.

Press the right parenthesis key,), to insert the right parenthesis. (See frame ⑭.)

Finally, press the **ENTER** key to execute the computation and display the results. (See frame ⑮.)

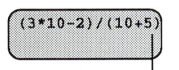

Place the cursor using the right-arrow key. Press the right parenthesis key, and the) character is inserted.

(3*10-2)/(10+5)
        1.866666667 ⑮

After the **ENTER** key is pressed, the computation is performed and the results are displayed.

EDITING COMPUTATIONS: THE DEL KEY

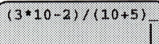

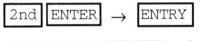

With the **ENTRY** command (the **2nd** key followed by the **ENTER** key), the last entry is displayed and the cursor is placed at the right end of the computation. You can now edit the displayed computation.

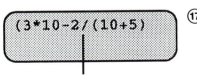

Pressing the **DEL** key removes the) character.

As just demonstrated, text can be inserted in a computation by use of the **INS** key (the parentheses in the preceding example). Text can also be deleted from a computation by using the **DEL** key. In the following example, you will use the **replay** feature to display the preceding computation for editing. Next, you will use the **DEL** key to delete the inside parentheses. Finally, you will use the **ENTER** key to execute and display the resulting computation.

Example: Compute $(3 \times 10 - 2 \div 10 + 5)$.

Solution: To do this computation, we can simply edit the existing computation, $(3 \times 10 - 2) \div (10 + 5)$, by deleting the inside parentheses.

Press the **2nd** key (the cursor changes to a blinking up-arrow) followed by the **ENTER** key (the ENTRY command) to **replay** the last command entered (the computation above). (See frame ⑯.)

Press the left-arrow key to move the cursor to the) character after the 2.

Press the **DEL** key, and the) character is deleted. (See frame ⑰.)

Now move the cursor to the (character after the / character. Press the **DEL** key to delete the (character.

Finally, press the **ENTER** key to execute and display the results of the new computation. (See frame ⑱.)

Comment

A shortcut for the **ENTRY** command is the up-arrow key. That is, press the up-arrow key and the last command executed is displayed.

| CLEAR | This key erases the screen. |

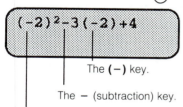

After pressing the **(−)** key and the **2** key.

After pressing the **x²** key and the **ENTER** key.

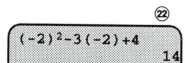

The **(−)** key.

The − (subtraction) key.

The **(−)** key.

Comment

The TI-81 understands juxtaposition, as in 3(−2), to mean multiplication. You could have written 3 * −2 instead of 3(−2).

EXERCISE SET I

1. Do the following computations using TI-81.
   (a) 5 − 14/8
   (b) 2/3 + 1/5
   (c) 1.56 − (3 + 13/5)

2. Evaluate $x^2 - 4x + 5$ for $x = 3$.

3. Type the computation $(5 - 3 \times 6) \div (12 + 3 \times 8)$.
   a. Press the **ENTER** key and record the results.
   b. Use the ENTRY command to redisplay the computation. Edit the computation to $(5 - 3) * 6/(12 + 3 * 8)$. Press the **ENTER** key and record the results.
   c. Use the ENTRY command to redisplay the computation. Edit the computation to $5 - 3 * (6/12 + 3 * 8)$. Press the **ENTER** key and record the results.

COMPUTATIONS: THE (−), ANS, AND CLEAR KEYS

To clear the TI-81 screen, the **CLEAR** key is used. That is, press the **CLEAR** key, and the screen is erased. To redisplay the *results* of the last computation, the **ANS** key [the **2nd** key followed by the (−) key] followed by the **ENTER** key can be pressed. Actually, **ANS** is a storage location in the memory of the TI-81. Pressing this key displays the characters "Ans" on the screen. Pressing the **ENTER** key results in the TI-81 displaying the contents of the **ANS** memory location. As you will see, **ANS** can be used in a computation. Finally, *to enter a negative number, the* **(−)** *key must be used*. To see how to use these keys, consider the following example.

Example: For $f(x) = x^2 - 3x + 4$, find $f(-2)$.

Solution: First, $f(-2)$ means evaluate $x^2 - 3x + 4$ for x equal to -2. Before doing this evaluation, you need to learn how to square a negative number. To do this, start by pressing the **CLEAR** key to clear the display.

Now, press the (−) key, followed by the **2** key. (See frame ⑲.)

Press the **x²** key and then press the **ENTER** key. (See frame ⑳.) As you can see, the (−) key produces the opposite of a value. Order of operations resulted in the computation $-1 \times 2 \times 2$, which is not what we want. To square -2, we need to type $(-2)^2$. Now, to find $f(-2)$ for $f(x) = x^2 - 3x + 4$, do the following.

Press the (key, the (−) key, the **2** key, the) key, and the **x²** key to display $(-2)^2$.

Press the − key, the **3** key, the (key, the (−) key, the **2** key, and the) key to display $(-2)^2 - 3(-2)$.

Press the + key and the **4** key to finish the entry. (See frame ㉑.)

Finally, press the **ENTER** key to compute and then display the results. (See frame ㉒.) Thus, $f(-2)$ is 14.

To demonstrate the use of the **ANS** key, consider the following example.

Example: For $f(x) = x^2 - 3x + 4$, find $3f(-2) - 6$.

Solution: Rather than find $f(-2)$ again, we can use the **ANS** key, because the value of $f(-2)$ is currently stored in the **ANS** memory location, which holds (stores) the value of the last calculation, even after the calculator is turned off.

Press the **3** key, the **×** key, the **ANS** key, (**2nd** key followed by **(−)** key), the **−** key, the **6** key, and then the **ENTER** key. (See frame ㉓.) So, $3f(-2) - 6$ equals 36.

Comment If you now pressed the **ANS** key followed by the **x²** key and then pressed the **ENTER** key, the displayed results would be 1296. This is because 36 is currently stored in the **ANS** memory location, and 36 squared is 1296.

EXERCISE SET 2

Match the following:

A. ENTRY	F. ENTER
B. 2nd	G. INS
C. ANS	H. DEL
D. DEL	I. CLEAR
E. (−)	J. ▼

_____ 1. A key pressed to compute and display a computation entered.

_____ 2. A key pressed to indicate the opposite of a number.

_____ 3. A key pressed to erase the display.

_____ 4. A key pressed to move the cursor on the screen.

_____ 5. A key pressed to remove (delete) a character from a computation displayed on the screen.

_____ 6. A key pressed to access the function listed on the top of a key (such as ENTRY or the 10^x function).

_____ 7. A key pressed to provide access to the most recently displayed computed results.

8. If you enter the computation $3 \times 5 - 8$, press the **ENTER** key, and then press the **2nd** key followed by the **(−)** key (the **ANS** key), followed by the **x²** key and the **ENTER** key, what number will be displayed?

9. For $f(x) = 3 - \dfrac{x}{x + 5}$, find $f(-1)$.

10. For $f(x) = -x^2 + 3x - 1$, find $f(5)$.

㉓

3XAns−6

36

3 | 2nd | (−) | − | 6 | ENTER

ANS

Comment

To turn off the TI-81, press the **2nd** key followed by the **ON** key. If no key is pressed for five minutes, the TI-81 automatically turns off. Press the **ON** key to return to where you were when the TI-81 was turned off.

Press the **RANGE** key.

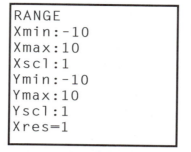

```
RANGE
Xmin:-10
Xmax:10
Xscl:1
Ymin:-10
Ymax:10
Yscl:1
Xres=1
```

The default-range display.

Press the **GRAPH** key.

The graph display.

Use the **QUIT** key to return to the text display.

Press the **Y=** key. Type

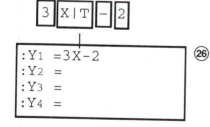

```
:Y1 =3X-2
:Y2 =
:Y3 =
:Y4 =
```

Press the **GRAPH** key.

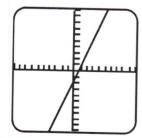

The graph of $f(x) = 3x - 2$.

A FIRST GRAPH

㉔ A major feature of the TI-81 is its ability to display the graph of a function. In the following examples, you will learn to use the **RANGE** key (to specify the x-axis interval and the y-axis interval), the **Y=** key (to display a prompt for entering the rule for a function), the **GRAPH** key (to switch to a graph display), and the QUIT command (to return to the text display).

The Right Range   Before displaying a first graph, press the **RANGE** key and check that the values displayed are as shown in frame ㉔.

Comment   If the values displayed are different than those shown in ㉔, press the **ZOOM** key followed by the **6** key and then the **CLEAR** key to return to the default values shown in frame ㉔. Later, you will learn to set any range values you desire. Press the **RANGE** key to bring the display back to the range display shown in frame ㉔.

The Graph Screen   Before inserting the rule for a function to be graphed, we will view the graph screen. Press the **GRAPH** key and you will see a set of axes with the default range settings displayed. (See frame ㉕.) The leftmost mark on the x-axis represents -10, and the rightmost mark on the x-axis represents 10. Likewise, the topmost mark on the y-axis represents 10, and the bottommost mark on the y-axis represents -10. The scale for both the x-axis and the y-axis is 1. That is, each mark represents 1 unit. Use the QUIT command (the **2nd** key followed by the **CLEAR** key) to bring the display back to the text display.

Graphing a Function   We will now graph the linear function $f(x) = 3x - 2$. The graph of this function is a straight line crossing the y-axis at $(0, -2)$. Let's work through the following example.

Example: Graph $f(x) = 3x - 2$.

Solution: With the TI-81 in text display, press the **Y=** key. The screen clears, and the prompt ":Y1=" is displayed.

Press the **3** key.

Press the **X|T** key to place the X on the display.

Press the **−** key.

Press the **2** key. (See frame ㉖.)

Press the **GRAPH** key to have the graph drawn and displayed on the graph screen. The TI-81 display changes to the graph screen, and the ordered pairs defined by the function rule $f(x) = 3x - 2$ are plotted on the axis. (See frame ㉗.)

Use the QUIT command (the **2nd** key followed by the **CLEAR** key) to switch back to the home screen (the text display).

Clearing the function definition(s) clears the graphics screen.

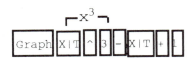

Graph Y₁=X^3-X+1   ㉘

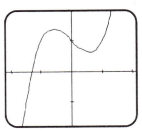

The graph of
$f(x) = x^3 - x + 1$.

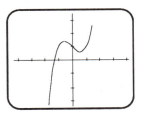

The ZOOM menu.

The graph of
$f(x) = x^3 - x + 1$ after zooming in (ZOOM 2).

Checking the Displays   At this point, you can press the **RANGE** key and display the current ranges for the x-axis and y-axis. Pressing the **GRAPH** key again toggles the display back to the graph screen. Also, the **Y=** key can be pressed to display the function definition (rule). The QUIT command can be used to switch back to the home screen. Finally, you can clear the home screen by pressing the **CLEAR** key. To clear the graphics screen, press the **Y=** key, and then with the cursor on the function rule, press the **CLEAR** key.

As another example of graphing a function, consider the following.

Example: Graph $f(x) = x^3 - x + 1$.

Solution: First, press the **Y=** key followed by the **CLEAR** key to clear the graphics screen.

Press the **X|T** key to place X on the screen.

Press the ^ key followed by the **3** key to get **x³**. [The screen displays "Y1 = X^3", which means $f(x) = x^3$.]

㉙   Press the **−** key to get "Y1 = X^3−".

Press the **X|T** key to get "Y1 = X^3 − X".

Press the **+** key and then the **1** key to get "Y1 = X^3−X+1". (See frame ㉘.)

Press the **GRAPH** key to display the graph of the function $f(x) = x^3 - x + 1$. (See frame ㉙.)

ZOOM IN, ZOOM OUT

A major feature TI-81 graphics feature is its ability to enlarge (*zoom in on*) or reduce (*zoom out on*) a graph. To zoom in on a graph, the **ZOOM** key is pressed, followed by the **2** key. To zoom out on a graph, the **ZOOM** key is pressed, followed by the **3** key.

Example: Graph $f(x) = x^3 - x + 1$ with $-2.25 < x < 2.25$.

Solution: For the graph in this example, the function domain is $-10 < x < 10$, the default-range values. If we zoom in on the graph of this function, the new x-values will be approximately $-2.25 < x < 2.25$ (one-fourth of -10 and one-fourth of 10).

With the graph of $f(x) = x^3 - x + 1$ on the screen, press the **ZOOM** key.

A menu of options is presented. (See frame ㉚.)

㉛   Press the **2** key to select the Zoom In option. You are returned to the graph screen with a blinking dot in the center of the screen. The blowup will be centered on this blinking dot, which can be moved using the directional-arrow keys.

Press the **ENTER** key. The new graph appears as shown in frame ㉛.

A-11

```
RANGE
Xmin:-2.36315789
Xmax:2.3736842107
Xscl:1
Ymin:-2.14285714
Ymax:2.87142858
Yscl:1
Xres=1
```

The new range display after
zooming in on the graph of
$f(x) = x^3 - x + 1$.

```
:Y₁=√(X-4)
:Y₂=
:Y₃=
:Y₄=
```

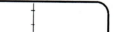

The graph of
$f(x) = \sqrt{x - 4}$.

③ Press the **RANGE** key to check that the x-values for the function $f(x) = x^3 - x + 1$ are approximately between -2.25 and 2.25. (See frame ③.) As you can see, one-fourth of the x- and y-ranges were computed based on the location of the blinking dot [which the TI-81 didn't consider to be exactly at $(0,0)$]. To set x between -2.25 and 2.25 exactly, do the following.

With the cursor on the $-$ in the XMin line, type -2.25 and then press the **ENTER** key.

With the cursor on the 2 in the XMax line, type 2.25 and then press the **ENTER** key.

Press the **GRAPH** key to display the graph of $f(x) = x^3 - x + 1$ with this set of values for x.

Use the **QUIT** command to return to the text screen.

Comment If you now zoom out (press the **ZOOM** key followed by the **3** key), the x-values will be approximately between -10 and 10. If you zoom out again, the x-values will be approximately between -40 and 40. Remember, ZOOM 6 (the Standard option on the ZOOM menu) returns you to the default-range settings.

EXERCISE SET 3

Use the TI-81 to display the following graphs, and then transfer the information to graph paper.
1. Graph $f(x) = x^3 + x^2 - 2$ for $-5 \le x \le 5$ and $-5 \le y \le 5$.
2. Graph $f(x) = x^2 - 2x + 1$ for $-2.25 \le x \le 2.25$ and $-1 \le y \le 5$.
3. Graph $f(x) = -x^2 - 2x + 2$ for $-10 \le x \le 10$ and $-10 \le y \le 10$.

THE ZOOM BOX

When you use the Zoom In and Zoom Out options on the ZOOM menu, the TI-81 automatically adjusts the range of values displayed. These values are computed based on the location of the blinking dot on the screen and the four factor (one-fourth for Zoom In and four times for Zoom Out). This factor can be changed by selecting the Set Factors option from the ZOOM menu. Of course, you can set exactly the ranges you want by entering them on the RANGE display. One of the best features of the TI-81 is the Box option on the ZOOM menu. To see for yourself how this feature operates, work through the following ③ example.

Example: Graph $f(x) = \sqrt{(x - 4)}$.

Solution: Press the **Y=** key, and then use the **CLEAR** key to clear the function definiton for Y1 (and thus clear the graphics screen). Use the Standard option on the ZOOM menu to reset the range values to their default settings.

Enter the function $f(x) = \sqrt{(x - 4)}$ by pressing the **Y=** key, the $\sqrt{}$ key, the **(** key, the **X|T** key, the **−** key, the **4** key, and the **)** key. (See frame ③.) Press the **GRAPH** key. (See frame ③.)

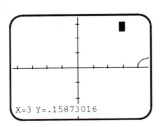

X=3 Y=.15873016

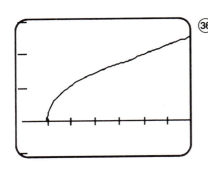

35 With the graph of $f(x) = \sqrt{(x-4)}$ on the screen, press the **ZOOM** key and select the box option. (Press the **ENTER** key or the **1** key.) The message X=.10526316 Y=.15873016 appears at the bottom of the graph screen. A blinking dot appears at the origin. Use the directional-arrow keys to move the blinking dot as close as possible to the point (3, 3). Press the **ENTER** key. Notice that the dot changes shape. (See frame **35**.)

Now, press the right-arrow key until the X= message displays X = 10.

Press the down-arrow key until the Y= message reads Y = −1.111111. Press the **ENTER** key. Notice the box drawn on the screen. This is known as a *zoom box*.

36 Press the **ENTER** key, and the graph is enlarged to fill the zoom box. (See frame **36**.)

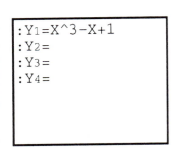

EXERCISE SET 4

Use the TI-81 to graph the following functions over the given intervals. Start with the default ranges, then use the zoom box feature.

1. Graph $f(x) = x^2 - 2x - 1$ for $-2 \le x \le 4$ and $-3 \le y \le 8$.
2. Graph $f(x) = \sqrt{x} - 4$ for $-1 \le x \le 8$ and $-4 \le y \le 1$.
3. Graph $f(x) = \sqrt{(4-x)}$ for $-5 \le x \le 5$ and $-1 \le y \le 4$.
4. Graph $f(x) = (3x - 1)/(x + 2)$ for $-5 \le x \le 5$ and $-5 \le y \le 8$.

THE TRACE FEATURE

```
: Y1=X^3−X+1
: Y2=
: Y3=
: Y4=
```

37 Often you will want to find the coordinates of a point on a graph drawn by the TI-81. To do so, you can use the *trace* feature as in the next example.

Example: Find the relative maximum for the function $f(x) = x^3 - x + 1$ over the interval $-1 \le x \le 1$.

Solution: Although you could graph the function and visually sight the point where the graph is at its highest point (where the function value or y-coordinate is greater than that of any other point in the specified interval), using the trace feature provides more accuracy.

38 Press the **ZOOM** key and select the Standard option to set the range values to their default settings.

Press the **Y=** key, and then press the **CLEAR** key to erase the function definition for Y1 (and thus the graphics screen).

Enter the function $f(x) = x^3 - x + 1$. (See frame **37**.)

Press the **GRAPH** key to display the graph of $f(x) = x^3 - x + 1$.

Use the zoom box to display the portion of the graph from $-1 \le x \le 1$ and $-0.5 \le y \le 2$. (See frame **38**.)

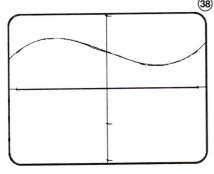

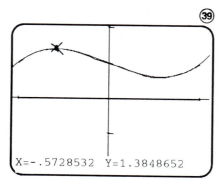

X=-.5728532 Y=1.3848652

Solve: $\begin{cases} 3x - 5y = 10 \\ 2x + y = 3 \end{cases}$

Solution: (▢, ▢)

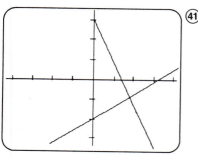

The graphs of
$y = \frac{3}{5}x - 2$ and $y = -2x + 3$.

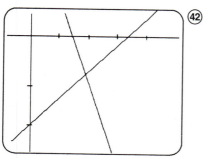

A blowup of the graphs of
$y = \frac{3}{5}x - 2$ and $y = -2x + 3$.

(39) Press the **TRACE** key. A blinking dot will appear on the curve.

Press the left-arrow key until the X= message displays X = −0.5728532. (See frame **(39)**.) Thus, the function $f(x) = x^3 - x + 1$ over the interval $-1 \leq x \leq 1$ reaches its relative maximum at −0.5728532, and the maximum value is 1.3848652 (approximately).

Comment If you now use the Zoom In option, the enlargement will center on the current location of the blinking dot displayed by the trace feature. Pressing the **TRACE** key again will improve the accuracy of the approximation of a point's coordinates.

Solving a System of Equations As another example of the use of the trace feature of the TI-81, we now consider solving a *system* of linear equations, such as the one shown in frame **(40)**. This example will also demonstrate how to display two graphs on the TI-81 graphics screen.

Example: Solve the system of equations

$$3x - 5y = 10 \quad \text{and} \quad 2x + y = 3$$

Solution: To solve the system graphically on the TI-81, the equations need to be rewritten in function form (or slope and y-intercept form). The equation $3x - 5y = 10$ becomes $y = \frac{3}{5}x - 2$, and the equation $2x + y = 3$ becomes $y = -2x + 3$.

Press the **Y=** key and use the **CLEAR** key to clear the Y1 definition.

Use the **RANGE** key to set the ranges to the values Xmin = −4, Xmax = 4, Ymin = −3.2, Ymax = 3.2.

Press the **Y=** key and type (3 ÷ 5)X − 2 for Y1; then press the **ENTER** key.

Type −2X + 3 for Y2.

Press the **GRAPH** key to display the graphs of the two linear functions. (See frame **(41)**.)

With the graph of the two linear functions on the screen, press the **TRACE** key. The message "X = .04210526 Y = −1.974737" will appear at the bottom of the screen.

Press the right-arrow key until the blinking dot appears at the intersection of the two lines. (The screen should display "X = 1.8947368 Y = −.8631579".)

Zoom in on the graph by pressing the **ZOOM** key followed by the **2** key and then the **ENTER** key. (See frame **(42)**.)

Press the **TRACE** key. The message "X = 1.9052632 Y = −.8568421" will appear on the screen.

Press the right-arrow key once to place the blinking dot at the intersection of the two lines (X = 1.9263158). Thus, using graphics, the solution to the system is (1.9263158, −0.8442105).

Comment   The actual solution to the system of equations is $(\frac{25}{13}, -\frac{11}{13})$ or (1.9231, −0.8461), accurate to four decimal places. Continuing this process of Trace and Zoom will yield an *x*-coordinate of 1.9230 and a *y*-coordinate of −0.8462.

EXERCISE SET 5

1. For $f(x) = x^3 - x^2 + 1$, use the trace and zoom features to find the minimal value in the interval $0 \le x \le 1$.
2. Find the point of intersection of the graph of $f(x) = x^2 - 3x - 1$ and that of $f(x) = 0.3x - 2$ with an *x*-coordinate that is between 0 and 1.
3. Solve: $3x - y = 4$ and $2x + 3y = 3$.
4. Solve: $y = x^2 - x$ and $y = -2x + 1$.

SUMMARY

When the TI-81 is turned on (using the **ON** key at the bottom left of the calculator), the screen is blank, with the cursor at the top right. If no entry is made for six minutes, the calculator automatically turns off. Power can be restored by pressing the **ON** key. To turn the TI-81 off, use the OFF command. (Press the **2nd** key followed by the **ON** key.)

Computation   Order-of-operations rules are followed in all computations. To override these rules, parentheses can be used.

To enter a negative value, the **(−)** key must be used. To enter the subtraction operation, the **−** key must be used.

```
ERROR 06 SYNTAX
1:Goto Error
2:Quit
```

If an error message (menu) appears on the screen, respond by pressing the **ENTER** key to select option 1 (try to edit the offending command), or press the **2** key to QUIT (stop trying to execute the command).

An entry can be edited (before pressing the **ENTER** key) by using the directional-arrow keys, the **DEL** key, and the **INS** key.

The ENTRY command (the **2nd** key followed by the **ENTER** key or the up-arrow key) is used to *replay* the last command entered.

The **CLEAR** key is used to clear the text screen.

The **ENTER** key is used to carry out a command (usually to compute and then to display the results of a computation).

Graphs   The graph screen is cleared by clearing the function definitions.

The **RANGE** key is used to display the current range settings (the portion of the *x*-axis, the portion of the *y*-axis, and the scales for the axes). These range values can be altered to the values desired.

The Standard option of the ZOOM menu is used to reset the range values to their default settings (− 10, 10, 1, − 10, 10, 1).

The **Y =** key is used to enter the function to be graphed. The prompt :Y1 = is displayed. At this point, you enter the function rule that will determine the ordered pairs to be graphed. Pressing the **GRAPH** key draws the graph using the current range settings.

To type the independent variable x when entering a function rule, press the **X|T** key.

The **2nd** key followed by the **CLEAR** key (the QUIT command) is used to switch the view back to the text screen.

With a graph displayed on the screen, the **ZOOM** key can be used to zoom in (reduce the graph viewing area) or zoom out (enlarge the viewing area).

With a graph displayed on the screen, the **TRACE** key can be used to place a blinking dot (pixel) on the graphed curve. The x-coordinate and y-coordinate of the dot are displayed at the bottom of the display. Pressing the left/right directional-arrow keys moves the dot along the graph, changing the coordinates displayed.

YOU MIGHT WANT TO KNOW

Q1 Is there a way to have computed values reported to a fixed number of decimal places (for example, accurate to hundredths)?

A1 Yes. Press the **MODE** key. The menu shown in frame ④③ is displayed. Press the down-arrow key once to make the word Float start flashing. Press the right-arrow key to place the cursor on 2 (for 2 decimal places). Press the **ENTER** key. The 2 starts flashing. (See frame ④④.) Use the QUIT command (the **2nd** key followed by the **CLEAR** key) to return to the text screen.

Q2 If I am to graph the function $f(t) = t^2 - 3t + 2$, do I use the **ALPHA** key and the **4** key to type T^2 − 3T + 2 in response to the :Y1 = prompt?

A2 No. You must use the **X|T** key to type X^2 − 3X + 2.

Q3 What is the fifth root of 27, and how do I find it?

A3 You want a number such that five times itself yields 27. To use the TI-81 to estimate that number, type 27^(1 ÷ 5), and then press the **ENTER** key. The number 1.933182045 is displayed. Remember from algebra, $X^{p/q}$ means $\sqrt[q]{X^p}$. Thus, $27^{1/5}$ is $\sqrt[5]{27}$.

   If you were looking for the third root of 27, you could press the **MATH** key to display the MATH menu. (See frame ④⑤.) Press the **4** key to select option 4 (the $\sqrt[3]{}$). You are returned to the text screen with the $\sqrt[3]{}$ symbol on the screen. Type 27, and then press the **ENTER** key. The number 3 is displayed, because (3)(3)(3) is 27.

Q4 How can I use the TI-81 to graph a line through the points $(-3, 5)$ and $(4, -1)$?

A4 By using the DRAW menu: first, use the **ZOOM** key and select the Standard option to set the ranges to their default values $(-10, 10, 1, -10, 10, 1)$. Also, you might want to press the **Y=** key and then use the **CLEAR** key to erase any existing function definitions and clear the graphics screen.

   Make sure you are on the home screen (use QUIT). See the comment in the margin.

④③

```
Norm Sci Eng
Float 0123456789
Rad Deg
Function Param
Connected Dot
Sequence Simul
Grid Off Grid On
Rect Polar
```

④④

```
Norm Sci Eng
Float 0123456789
Rad Deg
Function Param
Connected Dot
Sequence Simul
Grid Off Grid On
Rect Polar
```

④⑤

```
MATH NUM HYP PRB
1:R P(
2:P R(
3:3
4:3√
5:!
6:°
7↓r
```

Comment

If you are on the graphic screen when you select the DRAW menu, the points $(-3, 5)$ and $(4, -1)$ are entered as if you were defining a zoom box.

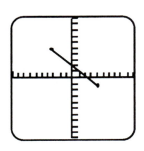

```
DRAW
1:ClrDraw
2:Line(
3:PT-On(
4:PT-Off(
5:PT-CHg(
6:DrawF
7:Shade(
```

The output of the command Line (−3, 5, 4, −1).

```
TEST
1:=
2:≠
3:>
4:≥
5:<
6:≤
```

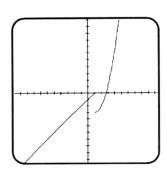

Press the **2nd** key followed by the **PRGM** key, and the DRAW menu is presented. (See frame ㊻.)

Press the **2** key to select the Line option.

The prompt Line(is presented on the text screen. Press the **(−)** key and the **3** key to get Line(−3. Now, press the **ALPHA** key followed by the **.** key to get Line(−3,. Press the **5** key, the comma key, the **4** key, the comma key, the **(−)** key, the **1** key, and then the **)** key to get Line (−3, 5, 4, −1). Press the **ENTER** key. (See frame ㊼.)

As you can see, the line segment from (−3, 5) to (4, −1) was drawn. You may now use a zoom box to zoom in on the area of interest. However, this may be difficult. When the TI-81 *draws* points on the screen (using the DRAW menu options), these points are not handled like function definitions (from the **Y=** key). If you zoom in on a line segment, the graph screen is redrawn, and any points from the DRAW options are lost. But if you use the ENTRY command (press the **2nd** key followed by the **ENTER** key, or press the up-arrow key) to *replay* the last command, the line segment will be redrawn on the graph screen with the new range values.

To clear any points placed on the graph screen as a result of the DRAW options, you can select Option 1 (ClrDraw) from the DRAW menu.

Q5 Is there a way to have the TI-81 graph a piecewise-defined function such as the following?

$$f(x) = \begin{cases} x - 1 & \text{for } x \le 1 \\ x^2 - 2x - 2 & \text{for } x > 1 \end{cases}$$

A5 Yes. First, set the range values to default using the Standard option from the ZOOM menu. Also, press the **MODE** key and select the Dot mode. (Move the flashing rectangle cursor down to line 5, then over to Dot, press the **ENTER** key, and then use the QUIT command.) Press the **Y=** key and type the following for the definition of function Y1:

$$:Y1 = (X - 1)(X \le 1) + (X^2 - 2X - 2)(X > 1)$$

To get the inequalities ≤ and >, use the TEST command (the **2nd** key followed by the **MATH** key) and select the one you want from the TEST menu (see frame ㊽.) With the graph on the screen (see frame ㊾), you can use a zoom box to look at the portion of interest of the graph. When you are finished with the graph, change the MODE setting back to Connected.

Q6 I tried to graph the function $f(x) = \sin x$, but all I got was a straight (horizontal) line below the negative x-axis. When I zoomed in on the graph, I got a horizontal line just above the positive x-axis. What's going on?

A6 You have graphed the sine function from −10 degrees to 10 degrees. To correct this, press the **MODE** key. Make sure the Rad option is selected in line 3. (If not, use the directional-arrow keys to place the flashing rectangle on Rad and then press the **ENTER** key.) Press the **RANGE** key and set the values: −6.28, 6.28 1, −2, 2, 1. Press the **GRAPH** key.

Note: If you wish to graph the function $f(x) = \sin x$ using degree measure for x, select the Deg option on line 3 of the MODE menu, but set the x-range values from −360 to 360.

A.3 SOME PROOFS FOR SECTION 1.2

Some of the fundamental algebraic properties of the real number system are listed on page 6 in Section 1.2. In this appendix we use those properties to prove the following theorem.

Theorem

Let a and b be real numbers. Then

(a) $a \cdot 0 = 0$ (d) $a(-b) = -ab$

(b) $-a = (-1)a$ (e) $(-a)(-b) = ab$

(c) $-(-a) = a$

PROOF OF PART (a)

$$a \cdot 0 = a \cdot (0 + 0) \qquad \text{additive identity property}$$
$$= a \cdot 0 + a \cdot 0 \qquad \text{distributive property}$$

Now since $a \cdot 0$ is a real number, it has an additive inverse $-(a \cdot 0)$. Adding this to both sides of the last equation, we obtain

$$a \cdot 0 + [-(a \cdot 0)] = (a \cdot 0 + a \cdot 0) + [-(a \cdot 0)]$$
$$a \cdot 0 + [-(a \cdot 0)] = a \cdot 0 + \{a \cdot 0 + [-(a \cdot 0)]\} \qquad \text{associative property of addition}$$
$$0 = a \cdot 0 + 0 \qquad \text{additive inverse property}$$
$$0 = a \cdot 0 \qquad \text{additive identity property}$$

Thus $a \cdot 0 = 0$, as we wished to show.

PROOF OF PART (b)

$$0 = 0 \cdot a \qquad \text{using part (a) and the commutative property of multiplication}$$
$$= [1 + (-1)]a \qquad \text{additive inverse property}$$
$$= 1 \cdot a + (-1)a \qquad \text{distributive property}$$
$$= a + (-1)a \qquad \text{multiplicative identity property}$$

Now, by adding $-a$ to both sides of this last equation, we obtain

$$-a + 0 = -a + [a + (-1)a]$$
$$-a = (-a + a) + (-1)a \qquad \text{additive identity property and associative property of addition}$$
$$= 0 + (-1)a \qquad \text{additive inverse property}$$
$$= (-1)a \qquad \text{additive identity property}$$

This last equation asserts that $-a = (-1)a$, as we wished to show.

PROOF OF PART (c)

$$-(-a) + (-a) = 0 \qquad \text{additive inverse property}$$

By adding a to both sides of this last equation, we obtain

$$[-(-a) + (-a)] + a = 0 + a$$
$$-(-a) + (-a + a) = a \qquad \text{associative property of addition and additive identity property}$$
$$-(-a) + 0 = a \qquad \text{additive inverse property}$$
$$-(-a) = a \qquad \text{additive identity property}$$

This last equation states that $-(-a) = a$, as we wished to show.

PROOF OF PART (d)

$$a(-b) = a[(-1)b] \qquad \text{using part (b)}$$
$$= [a(-1)]b \qquad \text{associative property of multiplication}$$
$$= [(-1)a]b \qquad \text{commutative property of multiplication}$$
$$= (-1)(ab) \qquad \text{associative property of multiplication}$$
$$= -(ab) \qquad \text{using part (b)}$$

Thus $a(-b) = -ab$, as we wished to show.

PROOF OF PART (e)

$$(-a)(-b) = -[(-a)b] \qquad \text{using part (d)}$$
$$= -[b(-a)] \qquad \text{commutative property of multiplication}$$
$$= -[-(ba)] \qquad \text{using part (d)}$$
$$= ba \qquad \text{using part (c)}$$
$$= ab \qquad \text{commutative property of multiplication}$$

We've now shown that $(-a)(-b) = ab$, as required.

A.4 $\sqrt{2}$ IS IRRATIONAL

The proof is by reductio ad absurdum, *and* reductio ad absurdum, *which Euclid loved so much, is one of a mathematician's finest weapons.*

G. H. Hardy (1877–1947)

We will use an *indirect proof* to show that the square root of two is an irrational number. The strategy is as follows.

1. We suppose that $\sqrt{2}$ is a rational number.
2. Using (1) and the usual rules of logic and algebra, we derive a contradiction.
3. On the basis of the contradiction in (2), we conclude that the supposition in (1) is untenable; that is, we conclude that $\sqrt{2}$ is irrational.

In carrying out the proof, we'll assume as known the following three statements.

If x is an even natural number, then $x = 2k$, for some natural number k.

Any rational number can be written in the form a/b, where the integers a and b have no common integral factors other than ± 1. (In other words, any fraction can be reduced to lowest terms.)

If x is a natural number and x^2 is even, then x is even.

Our indirect proof now proceeds as follows. Suppose that $\sqrt{2}$ were a rational number. Then we would be able to write

$$\sqrt{2} = \frac{a}{b} \qquad \text{where } a \text{ and } b \text{ are natural numbers with no} \atop \text{common factor other than } 1 \qquad\qquad (1)$$

Since both sides of equation (1) are positive, we can square both sides to obtain the equivalent equation

$$2 = \frac{a^2}{b^2}$$

or

$$2b^2 = a^2 \qquad\qquad (2)$$

Since the left-hand side of equation (2) is an even number, the right-hand side must be even. But if a^2 is even then a is even, and so

$$a = 2k, \text{ for some natural number } k$$

Using this last equation to substitute for a in equation (2), we have

$$2b^2 = (2k)^2 = 4k^2$$

or

$$b^2 = 2k^2$$

Hence (reasoning as before) b^2 is even, and therefore b is even. But then we have that both b and a are even, contrary to our hypothesis that b and a have no common factor other than 1. We conclude from this that equation (1) cannot hold; that is, there is no rational number a/b such that $\sqrt{2} = a/b$. Thus $\sqrt{2}$ is irrational, as we wished to prove.

TABLES

TABLE I

Squares and Square Roots

NO.	SQ.	SQ. RT.	NO.	SQ.	SQ. RT.
1	1	1.000	51	2,601	7.141
2	4	1.414	52	2,704	7.211
3	9	1.732	53	2.809	7.280
4	16	2.000	54	2,916	7.348
5	25	2.236	55	3.025	7.416
6	36	2.449	56	3,136	7.483
7	49	2.646	57	3,249	7.550
8	64	2.828	58	3,364	7.616
9	81	3.000	59	3,481	7.681
10	100	3.162	60	3,600	7.746
11	121	3.317	61	3,721	7.810
12	144	3.464	62	3,844	7.874
13	169	3.606	63	3.969	7.937
14	196	3.742	64	4,096	8.000
15	225	3.873	65	4,225	8.062
16	256	4.000	66	4,356	8.124
17	289	4.123	67	4,489	8.185
18	324	4.243	68	4,624	8.246
19	361	4.359	69	4,761	8.307
20	400	4.472	70	4,900	8.367
21	441	4.583	71	5,041	8.426
22	484	4.690	72	5,184	8.485
23	529	4.796	73	5,329	8.544
24	576	4.899	74	5,476	8.602
25	625	5.000	75	5,625	8.660
26	676	5.099	76	5,776	8.718
27	729	5.196	77	5,929	8.775
28	784	5.292	78	6,084	8.832
29	841	5.385	79	6,241	8.888
30	900	5.477	80	6,400	8.944
31	961	5.568	81	6,561	9.000
32	1,024	5.657	82	6,724	9.055
33	1,089	5.745	83	6,889	9.110
34	1,156	5.831	84	7,056	9.165
35	1,225	5.916	85	7,225	9.220
36	1,296	6.000	86	7,396	9.274
37	1,369	6.083	87	7,569	9.327
38	1,444	6.164	88	7,744	9.381
39	1,521	6.245	89	7,921	9.434
40	1,600	6.325	90	8,100	9.487
41	1,681	6.403	91	8,281	9.539
42	1,764	6.481	92	8,464	9.592
43	1,849	6.557	93	8,649	9.644
44	1,936	6.633	94	8,836	9.695
45	2,025	6.708	95	9,025	9.747
46	2,116	6.782	96	9,216	9.798
47	2,209	6.856	97	9,409	9.849
48	2,304	6.928	98	9,604	9.899
49	2,401	7.000	99	9,801	9.950
50	2,500	7.071	100	10,000	10.000

TABLE 2

Exponential Functions

x	e^x	e^{-x}	x	e^x	e^{-x}
0.00	1.0000	1.0000	1.5	4.4817	0.2231
0.01	1.0101	0.9901	1.6	4.9530	0.2019
0.02	1.0202	0.9802	1.7	5.4739	0.1827
0.03	1.0305	0.9704	1.8	6.0496	0.1653
0.04	1.0408	0.9608	1.9	6.6859	0.1496
0.05	1.0513	0.9512	2.0	7.3891	0.1353
0.06	1.0618	0.9418	2.1	8.1662	0.1225
0.07	1.0725	0.9324	2.2	9.0250	0.1108
0.08	1.0833	0.9231	2.3	9.9742	0.1003
0.09	1.0942	0.9139	2.4	11.023	0.0907
0.10	1.1052	0.9048	2.5	12.182	0.0821
0.11	1.1163	0.8958	2.6	13.464	0.0743
0.12	1.1275	0.8869	2.7	14.880	0.0672
0.13	1.1388	0.8781	2.8	16.445	0.0608
0.14	1.1503	0.8694	2.9	18.174	0.0550
0.15	1.1618	0.8607	3.0	20.086	0.0498
0.16	1.1735	0.8521	3.1	22.198	0.0450
0.17	1.1853	0.8437	3.2	24.533	0.0408
0.18	1.1972	0.8353	3.3	27.113	0.0369
0.19	1.2092	0.8270	3.4	29.964	0.0334
0.20	1.2214	0.8187	3.5	33.115	0.0302
0.21	1.2337	0.8106	3.6	36.598	0.0273
0.22	1.2461	0.8025	3.7	40.447	0.0247
0.23	1.2586	0.7945	3.8	44.701	0.0224
0.24	1.2712	0.7866	3.9	49.402	0.0202
0.25	1.2840	0.7788	4.0	54.598	0.0183
0.30	1.3499	0.7408	4.1	60.340	0.0166
0.35	1.4191	0.7047	4.2	66.686	0.0150
0.40	1.4918	0.6703	4.3	73.700	0.0136
0.45	1.5683	0.6376	4.4	81.451	0.0123
0.50	1.6487	0.6065	4.5	90.017	0.0111
0.55	1.7333	0.5769	4.6	99.484	0.0101
0.60	1.8221	0.5488	4.7	109.95	0.0091
0.65	1.9155	0.5220	4.8	121.51	0.0082
0.70	2.0138	0.4966	4.9	134.29	0.0074
0.75	2.1170	0.4724	5.0	148.41	0.0067
0.80	2.2255	0.4493	5.5	244.69	0.0041
0.85	2.3396	0.4274	6.0	403.43	0.0025
0.90	2.4596	0.4066	6.5	665.14	0.0015
0.95	2.5857	0.3867	7.0	1096.6	0.0009
1.0	2.7183	0.3679	7.5	1808.0	0.0006
1.1	3.0042	0.3329	8.0	2981.0	0.0003
1.2	3.3201	0.3012	8.5	4914.8	0.0002
1.3	3.6693	0.2725	9.0	8103.1	0.0001
1.4	4.0552	0.2466	10.0	22026	0.00005

TABLE 3

Logarithms to the Base Ten

x	0	1	2	3	4	5	6	7	8	9
1.0	.0000	.0043	.0086	.0128	.0170	.0212	.0253	.0294	.0334	.0374
1.1	.0414	.0453	.0492	.0531	.0569	.0607	.0645	.0682	.0719	.0755
1.2	.0792	.0828	.0864	.0899	.0934	.0969	.1004	.1038	.1072	.1106
1.3	.1139	.1173	.1206	.1239	.1271	.1303	.1335	.1367	.1399	.1430
1.4	.1461	.1492	.1523	.1553	.1584	.1614	.1644	.1673	.1703	.1732
1.5	.1761	.1790	.1818	.1847	.1875	.1093	.1931	.1959	.1987	.2014
1.6	.2041	.2068	.2095	.2122	.2148	.2175	.2201	.2227	.2253	.2279
1.7	.2304	.2330	.2355	.2380	.2405	.2430	.2455	.2480	.2504	.2529
1.8	.2553	.2577	.2601	.2625	.2648	.2672	.2695	.2718	.2742	.2765
1.9	.2788	.2810	.2833	.2856	.2878	.2900	.2923	.2945	.2967	.2989
2.0	.3010	.3032	.3054	.3075	.3096	.3118	.3139	.3160	.3181	.3201
2.1	.3222	.3243	.3263	.3284	.3304	.3324	.3345	.3365	.3385	.3404
2.2	.3424	.3444	.3464	.3483	.3502	.3522	.3541	.3560	.3579	.3598
2.3	.3617	.3636	.3655	.3674	.3692	.3711	.3729	.3747	.3766	.3784
2.4	.3802	.3820	.3838	.3856	.3874	.3892	.3909	.3927	.3945	.3962
2.5	.3979	.3997	.4014	.4031	.4048	.4065	.4082	.4099	.4116	.4133
2.6	.4150	.4166	.4183	.4200	.4216	.4232	.4249	.4265	.4281	.4298
2.7	.4314	.4330	.4346	.4362	.4378	.4393	.4409	.4425	.4440	.4456
2.8	.4472	.4487	.4502	.4518	.4533	.4548	.4564	.4579	.4594	.4609
2.9	.4624	.4639	.4654	.4669	.4683	.4698	.4713	.4728	.4742	.4757
3.0	.4771	.4786	.4800	.4814	.4829	.4843	.4857	.4871	.4886	.4900
3.1	.4914	.4928	.4942	.4955	.4969	.4983	.4997	.5011	.5024	.5038
3.2	.5051	.5065	.5079	.5092	.5105	.5119	.5132	.5145	.5159	.5172
3.3	.5185	.5198	.5211	.5224	.5237	.5250	.5263	.5276	.5289	.5302
3.4	.5315	.5328	.5340	.5353	.5366	.5378	.5391	.5403	.5416	.5428
3.5	.5441	.5453	.5465	.5478	.5490	.5502	.5514	.5527	.5539	.5551
3.6	.5563	.5575	.5587	.5599	.5611	.5623	.5635	.5647	.5658	.5670
3.7	.5682	.5694	.5705	.5717	.5729	.5740	.5752	.5763	.5775	.5786
3.8	.5798	.5809	.5821	.5832	.5843	.5855	.5866	.5877	.5888	.5899
3.9	.5911	.5922	.5933	.5944	.5955	.5966	.5977	.5988	.5999	.6010
4.0	.6021	.6031	.6042	.6053	.6064	.6075	.6085	.6096	.6107	.6117
4.1	.6128	.6138	.6149	.6160	.6170	.6180	.6191	.6201	.6212	.6222
4.2	.6232	.6243	.6253	.6263	.6274	.6284	.6294	.6304	.6314	.6325
4.3	.6335	.6345	.6355	.6365	.6375	.6385	.6395	.6405	.6415	.6425
4.4	.6435	.6444	.6454	.6464	.6474	.6484	.6493	.6503	.6513	.6522
4.5	.6532	.6542	.6551	.6561	.6571	.6580	.6590	.6599	.6609	.6618
4.6	.6628	.6637	.6646	.6656	.6665	.6675	.6684	.6693	.6702	.6712
4.7	.6721	.6730	.6739	.6749	.6758	.6767	.6776	.6785	.6794	.6803
4.8	.6812	.6821	.6830	.6839	.6848	.6857	.6866	.6875	.6884	.6893
4.9	.6902	.6911	.6920	.6928	.6937	.6946	.6955	.6964	.6972	.6981
5.0	.6990	.6998	.7007	.7016	.7024	.7033	.7042	.7050	.7059	.7067
5.1	.7076	.7084	.7093	.7101	.7110	.7118	.7126	.7135	.7143	.7152
5.2	.7160	.7168	.7177	.7185	.7193	.7202	.7210	.7218	.7226	.7235
5.3	.7243	.7251	.7259	.7267	.7275	.7284	.7292	.7300	.7308	.7316
5.4	.7324	.7332	.7340	.7348	.7356	.7364	.7372	.7380	.7388	.7396
x	0	1	2	3	4	5	6	7	8	9

TABLE 3

Logarithms to the Base Ten *(continued)*

x	0	1	2	3	4	5	6	7	8	9
5.5	.7404	.7412	.7419	.7427	.7435	.7443	.7451	.7459	.7466	.7474
5.6	.7482	.7490	.7497	.7505	.7513	.7520	.7528	.7536	.7543	.7551
5.7	.7559	.7566	.7574	.7582	.7589	.7597	.7604	.7612	.7619	.7627
5.8	.7634	.7642	.7649	.7657	.7664	.7672	.7679	.7686	.7694	.7701
5.9	.7709	.7716	.7723	.7731	.7738	.7745	.7752	.7760	.7767	.7774
6.0	.7782	.7789	.7796	.7803	.7810	.7818	.7825	.7832	.7839	.7846
6.1	.7853	.7860	.7868	.7875	.7882	.7889	.7896	.7903	.7910	.7917
6.2	.7924	.7931	.7938	.7945	.7952	.7959	.7966	.7973	.7980	.7987
6.3	.7993	.8000	.8007	.8014	.8021	.8028	.8035	.8041	.8048	.8055
6.4	.8062	.8069	.8075	.8082	.8089	.8096	.8102	.8109	.8116	.8122
6.5	.8129	.8136	.8142	.8149	.8156	.8162	.8169	.8176	.8182	.8189
6.6	.8195	.8202	.8209	.8215	.8222	.8228	.8235	.8241	.8248	.8254
6.7	.8261	.8267	.8274	.8280	.8287	.8293	.8299	.8306	.8312	.8319
6.8	.8325	.8331	.8338	.8344	.8351	.8357	.8363	.8370	.8376	.8382
6.9	.8388	.8395	.8401	.8407	.8414	.8420	.8426	.8432	.8439	.8445
7.0	.8451	.8457	.8463	.8470	.8476	.8482	.8488	.8494	.8500	.8506
7.1	.8513	.8519	.8525	.8531	.8537	.8543	.8549	.8555	.8561	.8567
7.2	.8573	.8579	.8585	.8591	.8597	.8603	.8609	.8615	.8621	.8627
7.3	.8633	.8639	.8645	.8651	.8657	.8663	.8669	.8675	.8681	.8686
7.4	.8692	.8698	.8704	.8710	.8716	.8722	.8727	.8733	.8739	.8745
7.5	.8751	.8756	.8762	.8768	.8774	.8779	.8785	.8791	.8797	.8802
7.6	.8808	.8814	.8820	.8825	.8831	.8837	.8842	.8848	.8854	.8859
7.7	.8865	.8871	.8876	.8882	.8887	.8893	.8899	.8904	.8910	.8915
7.8	.8921	.8927	.8932	.8938	.8943	.8949	.8954	.8960	.8965	.8971
7.9	.8976	.8982	.8987	.8993	.8998	.9004	.9009	.9015	.9020	.9025
8.0	.9031	.9036	.9042	.9047	.9053	.9058	.9063	.9069	.9074	.9079
8.1	.9085	.9090	.9096	.9101	.9106	.9112	.9117	.9122	.9128	.9133
8.2	.9138	.9143	.9149	.9154	.9159	.9165	.9170	.9175	.9180	.9186
8.3	.9191	.9196	.9201	.9206	.9212	.9217	.9222	.9227	.9232	.9238
8.4	.9243	.9248	.9253	.9258	.9263	.9269	.9274	.9279	.9284	.9289
8.5	.9294	.9299	.9304	.9309	.9315	.9320	.9325	.9330	.9335	.9340
8.6	.9345	.9350	.9355	.9360	.9365	.9370	.9375	.9380	.9385	.9390
8.7	.9395	.9400	.9405	.9410	.9415	.9420	.9425	.9430	.9435	.9440
8.8	.9445	.9450	.9455	.9460	.9465	.9469	.9474	.9479	.9484	.9489
8.9	.9494	.9499	.9504	.9509	.9513	.9518	.9523	.9528	.9533	.9538
9.0	.9542	.9547	.9552	.9557	.9562	.9566	.9571	.9576	.9581	.9586
9.1	.9590	.9595	.9600	.9605	.9609	.9614	.9619	.9624	.9628	.9633
9.2	.9638	.9643	.9647	.9652	.9657	.9661	.9666	.9671	.9675	.9680
9.3	.9685	.9689	.9694	.9699	.9703	.9708	.9713	.9717	.9722	.9727
9.4	.9731	.9736	.9741	.9745	.9750	.9754	.9759	.9763	.9768	.9773
9.5	.9777	.9782	.9786	.9791	.9795	.9800	.9805	.9809	.9814	.9818
9.6	.9823	.9827	.9832	.9836	.9841	.9845	.9850	.9854	.9859	.9863
9.7	.9868	.9872	.9877	.9881	.9886	.9890	.9894	.9899	.9903	.9908
9.8	.9912	.9917	.9921	.9926	.9930	.9934	.9939	.9943	.9948	.9952
9.9	.9956	.9961	.9965	.9969	.9974	.9978	.9983	.9987	.9991	.9996
x	0	1	2	3	4	5	6	7	8	9

TABLE 4

Natural Logarithms

n	$\ln n$	n	$\ln n$	n	$\ln n$
		4.5	1.5041	9.0	2.1972
0.1	−2.3026	4.6	1.5261	9.1	2.2083
0.2	−1.6094	4.7	1.5476	9.2	2.2192
0.3	−1.2040	4.8	1.5686	9.3	2.2300
0.4	−0.9163	4.9	1.5892	9.4	2.2407
0.5	−0.6931	5.0	1.6094	9.5	2.2513
0.6	−0.5108	5.1	1.6292	9.6	2.2618
0.7	−0.3567	5.2	1.6487	9.7	2.2721
0.8	−0.2231	5.3	1.6677	9.8	2.2824
0.9	−0.1054	5.4	1.6864	9.9	2.2925
1.0	0.0000	5.5	1.7047	10	2.3026
1.1	0.0953	5.6	1.7228	11	2.3979
1.2	0.1823	5.7	1.7405	12	2.4849
1.3	0.2624	5.8	1.7579	13	2.5649
1.4	0.3365	5.9	1.7750	14	2.6391
1.5	0.4055	6.0	1.7918	15	2.7081
1.6	0.4700	6.1	1.8083	16	2.7726
1.7	0.5306	6.2	1.8245	17	2.8332
1.8	0.5878	6.3	1.8405	18	2.8904
1.9	0.6419	6.4	1.8563	19	2.9444
2.0	0.6931	6.5	1.8718	20	2.9957
2.1	0.7419	6.6	1.8871	25	3.2189
2.2	0.7885	6.7	1.9021	30	3.4012
2.3	0.8329	6.8	1.9169	35	3.5553
2.4	0.8755	6.9	1.9315	40	3.6889
2.5	0.9163	7.0	1.9459	45	3.8067
2.6	0.9555	7.1	1.9601	50	3.9120
2.7	0.9933	7.2	1.9741	55	4.0073
2.8	1.0296	7.3	1.9879	60	4.0943
2.9	1.0647	7.4	2.0015	65	4.1744
3.0	1.0986	7.5	2.0149	70	4.2485
3.1	1.1314	7.6	2.0281	75	4.3175
3.2	1.1632	7.7	2.0412	80	4.3820
3.3	1.1939	7.8	2.0541	85	4.4427
3.4	1.2238	7.9	2.0669	90	4.4998
3.5	1.2528	8.0	2.0794	95	4.5539
3.6	1.2809	8.1	2.0919	100	4.6052
3.7	1.3083	8.2	2.1041		
3.8	1.3350	8.3	2.1163		
3.9	1.3610	8.4	2.1282		
4.0	1.3863	8.5	2.1401		
4.1	1.4110	8.6	2.1518		
4.2	1.4351	8.7	2.1633		
4.3	1.4586	8.8	2.1748		
4.4	1.4816	8.9	2.1861		

TABLE 5

Values of Trigonometric Functions

Angle θ									
Degrees	Radians	sin θ	csc θ	tan θ	cot θ	sec θ	cos θ		
0°00'	.0000	.0000	No value	.0000	No value	1.000	1.0000	1.5708	90°00'
10	029	029	343.8	029	343.8	000	000	679	50
20	058	058	171.9	058	171.9	000	000	650	40
30	087	087	114.6	087	114.6	000	1.0000	621	30
40	116	116	85.95	116	85.94	000	.9999	592	20
50	145	145	68.76	145	68.75	000	999	563	10
1°00'	.0175	.0175	57.30	.0175	57.29	1.000	.9998	1.5533	89°00'
10	204	204	49.11	204	49.10	000	998	504	50
20	233	233	42.98	233	42.96	000	997	475	40
30	262	262	38.20	262	38.19	000	997	446	30
40	291	291	34.38	291	34.37	000	996	417	20
50	320	320	31.26	320	31.24	001	995	388	10
2°00'	.0349	.0349	28.65	.0349	28.64	1.001	.9994	1.5359	88°00'
10	378	378	26.45	378	26.43	001	993	330	50
20	407	407	24.56	407	24.54	001	992	301	40
30	436	436	22.93	437	22.90	001	990	272	30
40	465	465	21.49	466	21.47	001	989	243	20
50	495	494	20.23	495	20.21	001	988	213	10
3°00'	.0524	.0523	19.11	.0524	19.08	1.001	.9986	1.5184	87°00'
10	553	552	18.10	553	18.07	002	985	155	50
20	582	581	17.20	582	17.17	002	983	126	40
30	611	610	16.38	612	16.35	002	981	097	30
40	640	640	15.64	641	15.60	002	980	068	20
50	669	669	14.96	670	14.92	002	978	039	10
4°00'	.0698	.0698	14.34	.0699	14.30	1.002	.9976	1.5010	86°00'
10	727	727	13.76	729	13.73	003	974	981	50
20	756	756	13.23	758	13.20	003	971	952	40
30	785	785	12.75	787	12.71	003	969	923	30
40	814	814	12.29	816	12.25	003	967	893	20
50	844	843	11.87	846	11.83	004	964	864	10
5°00'	.0873	.0872	11.47	.0875	11.43	1.004	.9962	1.4835	85°00'
10	902	901	11.10	904	11.06	004	959	806	50
20	931	929	10.76	934	10.71	004	957	777	40
30	960	958	10.43	963	10.39	005	954	748	30
40	.0989	.0987	10.13	.0992	10.08	005	951	719	20
50	.1018	.1016	9.839	.1022	9.788	005	948	690	10
6°00'	.1047	.1045	9.567	.1051	9.514	1.006	.9945	1.4661	84°00'
10	076	074	9.309	080	9.255	006	942	632	50
20	105	103	9.065	110	9.010	006	939	603	40
30	134	132	8.834	139	8.777	006	936	573	30
40	164	161	8.614	169	8.556	007	932	544	20
50	193	190	8.405	198	8.345	007	929	515	10
7°00'	.1222	.1219	8.206	.1228	8.144	1.008	.9925	1.4486	83°00'
		cos θ	sec θ	cot θ	tan θ	csc θ	sin θ	Radians	Degrees
									Angle θ

TABLE 5

Values of Trigonometric Functions (continued)

Angle θ									
Degrees	Radians	sin θ	csc θ	tan θ	cot θ	sec θ	cos θ		
7°00′	.1222	.1219	8.206	.1228	8.144	1.008	.9925	1.4486	83°00′
10	251	248	8.016	257	7.953	008	922	457	50
20	280	276	7.834	287	7.770	008	918	428	40
30	309	305	7.661	317	7.596	009	914	399	30
40	338	334	7.496	346	7.429	009	911	370	20
50	367	363	7.337	376	7.269	009	907	341	10
8°00′	.1396	.1392	7.185	.1405	7.115	1.010	.9903	1.4312	82°00′
10	425	421	7.040	435	6.968	010	899	283	50
20	454	449	6.900	465	827	011	894	254	40
30	484	478	765	495	691	011	890	224	30
40	513	507	636	524	561	012	886	195	20
50	542	536	512	554	435	012	881	166	10
9°00′	.1571	.1564	6.392	.1584	6.314	1.012	.9877	1.4137	81°00′
10	600	593	277	614	197	013	872	108	50
20	629	622	166	644	6.084	013	868	079	40
30	658	650	6.059	673	5.976	014	863	050	30
40	687	679	5.955	703	871	014	858	1.4021	20
50	716	708	855	733	769	015	853	1.3992	10
10°00′	.1745	.1736	5.759	.1763	5.671	1.015	.9848	1.3963	80°00′
10	774	765	665	793	576	016	843	934	50
20	804	794	575	823	485	016	838	904	40
30	833	822	487	853	396	017	833	875	30
40	862	851	403	883	309	018	827	846	20
50	891	880	320	914	226	018	822	817	10
11°00′	.1920	.1908	5.241	.1944	5.145	1.019	.9816	1.3788	79°00′
10	949	937	164	.1974	5.066	019	811	759	50
20	.1978	965	089	.2004	4.989	020	805	730	40
30	.2007	.1994	5.016	035	915	020	799	701	30
40	036	.2022	4.945	065	843	021	793	672	20
50	065	051	876	095	773	022	787	643	10
12°00′	.2094	.2079	4.810	.2126	4.705	1.022	.9781	1.3614	78°00′
10	123	108	745	156	638	023	775	584	50
20	153	136	682	186	574	024	769	555	40
30	182	164	620	217	511	024	763	526	30
40	211	193	560	247	449	025	757	497	20
50	240	221	502	278	390	026	750	468	10
13°00′	.2269	.2250	4.445	.2309	4.331	1.026	.9744	1.3439	77°00′
10	298	278	390	339	275	027	737	410	50
20	327	306	336	370	219	028	730	381	40
30	356	334	284	401	165	028	724	352	30
40	385	363	232	432	113	029	717	323	20
50	414	391	182	462	061	030	710	294	10
14°00′	.2443	.2419	4.134	.2493	4.011	1.031	.9703	1.3265	76°00′
		cos θ	sec θ	cot θ	tan θ	csc θ	sin θ	Radians	Degrees
								Angle θ	

TABLE 5

Values of Trigonometric Functions (continued)

Angle θ									
Degrees	Radians	sin θ	csc θ	tan θ	cot θ	sec θ	cos θ		
14°00'	.2443	.2419	4.134	.2493	4.011	1.031	.9703	1.3265	76°00'
10	473	447	086	524	3.962	031	696	235	50
20	502	476	4.039	555	914	032	689	206	40
30	531	504	3.994	586	867	033	681	177	30
40	560	532	950	617	821	034	674	148	20
50	589	560	906	648	776	034	667	119	10
15°00'	.2618	.2588	3.864	.2679	3.732	1.035	.9659	1.3090	75°00'
10	647	616	822	711	689	036	652	061	50
20	676	644	782	742	647	037	644	032	40
30	705	672	742	773	606	038	636	1.3003	30
40	734	700	703	805	566	039	628	1.2974	20
50	763	728	665	836	526	039	621	945	10
16°00'	.2793	.2756	3.628	.2867	3.487	1.040	.9613	1.2915	74°00'
10	822	784	592	899	450	041	605	886	50
20	851	812	556	931	412	042	596	857	40
30	880	840	521	962	376	043	588	828	30
40	909	868	487	.2944	340	044	580	799	20
50	938	896	453	.3026	305	045	572	770	10
17°00'	.2967	.2924	3.420	.3057	3.271	1.046	.9563	1.2741	73°00'
10	.2996	952	388	089	237	047	555	712	50
20	.3025	.2979	357	121	204	048	546	683	40
30	054	.3007	326	153	172	048	537	654	30
40	083	035	295	185	140	049	528	625	20
50	113	062	265	217	108	050	520	595	10
18°00'	.3142	.3090	3.236	.3249	3.078	1.051	.9511	1.2566	72°00'
10	171	118	207	281	047	052	502	537	50
20	200	145	179	314	3.018	053	492	508	40
30	229	173	152	346	2.989	054	483	479	30
40	258	201	124	378	960	056	474	450	20
50	287	228	098	411	932	057	465	421	10
19°00'	.3316	.3256	3.072	.3443	2.904	1.058	.9455	1.2392	71°00'
10	345	283	046	476	877	059	446	363	50
20	374	311	3.021	508	850	060	436	334	40
30	403	338	2.996	541	824	061	426	305	30
40	432	365	971	574	798	062	417	275	20
50	462	393	947	607	773	063	407	246	10
20°00'	.3491	.3420	2.924	.3640	2.747	1.064	.9397	1.2217	70°00'
10	520	448	901	673	723	065	387	188	50
20	549	475	878	706	699	066	377	159	40
30	578	502	855	739	675	068	367	130	30
40	607	529	833	772	651	069	356	101	20
50	636	557	812	805	628	070	346	072	10
21°00'	.3665	.3584	2.790	.3839	2.605	1.071	.9336	1.2043	69°00'
	cos θ	sec θ	cot θ	tan θ	csc θ	sin θ	Radians	Degrees	

| Angle θ | | | | | | | | | |

TABLE 5

Values of Trigonometric Functions (continued)

Angle θ								
Degrees	Radians	sin θ	csc θ	tan θ	cot θ	sec θ	cos θ	
21°00′	.3665	.3584	2.790	.3839	2.605	1.071	.9336	1.2043	69°00′
10	694	611	769	872	583	072	325	1.2014	50
20	723	638	749	906	560	074	315	1.1985	40
30	752	665	729	939	539	075	304	956	30
40	782	692	709	.3973	517	076	293	926	20
50	811	719	689	.4006	496	077	283	897	10
22°00′	.3840	.3746	2.669	.4040	2.475	1.079	.9272	1.1868	68°00′
10	869	773	650	074	455	080	261	839	50
20	898	800	632	108	434	081	250	810	40
30	927	827	613	142	414	082	239	781	30
40	956	854	595	176	394	084	228	752	20
50	985	881	577	210	375	085	216	723	10
23°00′	.4014	.3907	2.559	.4245	2.356	1.086	.9205	1.1694	67°00′
10	043	934	542	279	337	088	194	665	50
20	072	961	525	314	318	089	182	636	40
30	102	.3987	508	348	300	090	171	606	30
40	131	.4014	491	383	282	092	159	577	20
50	160	041	475	417	264	093	147	548	10
24°00′	.4189	.4067	2.459	.4452	2.246	1.095	.9135	1.1519	66°00′
10	218	094	443	487	229	096	124	490	50
20	247	120	427	522	211	097	112	461	40
30	276	147	411	557	194	099	100	432	30
40	305	173	396	592	177	100	088	403	20
50	334	200	381	628	161	102	075	374	10
25°00′	.4363	.4226	2.366	.4663	2.145	1.103	.9063	1.1345	65°00′
10	392	253	352	699	128	105	051	316	50
20	422	279	337	734	112	106	038	286	40
30	451	305	323	770	097	108	026	257	30
40	480	331	309	806	081	109	013	228	20
50	509	358	295	841	066	111	.9001	199	10
26°00′	.4538	.4384	2.281	.4877	2.050	1.113	.8988	1.1170	64°00′
10	567	410	268	913	035	114	975	141	50
20	596	436	254	950	020	116	962	112	40
30	625	462	241	.4986	2.006	117	949	083	30
40	654	488	228	.5022	1.991	119	936	054	20
50	683	514	215	059	977	121	923	1.1025	10
27°00′	. .4712	.4540	2.203	.5095	1.963	1.122	.8910	1.0996	63°00′
	cos θ	sec θ	cot θ	tan θ	csc θ	sin θ	Radians	Degrees
								Angle θ

TABLE 5

Values of Trigonometric Functions *(continued)*

Angle θ									
Degrees	Radians	sin θ	csc θ	tan θ	cot θ	sec θ	cos θ		
27°00′	.4712	.4540	2.203	.5095	1.963	1.122	.8910	1.0996	63°00′
10	741	566	190	132	949	124	897	966	50
20	771	592	178	169	935	126	884	937	40
30	800	617	166	206	921	127	870	908	30
40	829	643	154	243	907	129	857	879	20
50	858	669	142	280	894	131	843	850	10
28°00′	.4887	.4695	2.130	.5317	1.881	1.133	.8829	1.0821	62°00′
10	916	720	118	354	868	134	816	792	50
20	945	746	107	392	855	136	802	763	40
30	.4974	772	096	430	842	138	788	734	30
40	.5003	797	085	467	829	140	774	705	20
50	032	823	074	505	816	142	760	676	10
29°00′	.5061	.4848	2.063	.5543	1.804	1.143	.8746	1.0647	61°00′
10	091	874	052	581	792	145	732	617	50
20	120	899	041	619	780	147	718	588	40
30	149	924	031	658	767	149	704	559	30
40	178	950	020	696	756	151	689	530	20
50	207	.4975	010	735	744	153	675	501	10
30°00′	.5236	.5000	2.000	.5774	1.732	1.155	.8660	1.0472	60°00
10	265	025	1.990	812	720	157	646	443	50
20	294	050	980	851	709	159	631	414	40
30	323	075	970	890	698	161	616	385	30
40	352	100	961	930	686	163	601	356	20
50	381	125	951	.5969	675	165	587	327	10
31°00′	.5411	.5150	1.942	.6009	1.664	1.167	.8572	1.0297	59°00′
10	440	175	932	048	653	169	557	268	50
20	469	200	923	088	643	171	542	239	40
30	498	225	914	128	632	173	526	210	30
40	527	250	905	168	621	175	511	181	20
50	556	275	896	208	611	177	496	152	10
32°00′	.5585	.5299	1.887	.6249	1.600	1.179	.8480	1.0123	58°00′
10	614	324	878	289	590	181	465	094	50
20	643	348	870	330	580	184	450	065	40
30	672	373	861	371	570	186	434	036	30
40	701	398	853	412	560	188	418	1.0007	20
50	730	422	844	453	550	190	403	.9977	10
33°00′	.5760	.5446	1.836	.6494	1.540	1.192	.8387	.9948	57°00′
		cos θ	sec θ	cot θ	tan θ	csc θ	sin θ	Radians	Degrees
								Angle θ	

TABLE 5

Values of Trigonometric Functions (continued)

Angle θ									
Degrees	Radians	sin θ	csc θ	tan θ	cot θ	sec θ	cos θ		
33°00'	.5760	.5446	1.836	.6494	1.540	1.192	.8387	.9948	57°00'
10	789	471	828	536	530	195	371	919	50
20	818	495	820	577	520	197	355	890	40
30	847	519	812	619	511	199	339	861	30
40	876	544	804	661	501	202	323	832	20
50	905	568	796	703	492	204	307	803	10
34°00'	.5934	.5592	1.788	.6745	1.483	1.206	.8290	.9774	56°00'
10	963	616	781	787	473	209	274	745	50
20	.5992	640	773	830	464	211	258	716	40
30	.6021	644	766	873	455	213	241	687	30
40	050	688	758	916	446	216	225	657	20
50	080	712	751	.6959	437	218	208	628	10
35°00'	.6109	.5736	1.743	.7002	1.428	1.221	.8192	.9599	55°00'
10	138	760	736	046	419	223	175	570	50
20	167	783	729	089	411	226	158	541	40
30	196	807	722	133	402	228	141	512	30
40	225	831	715	177	393	231	124	483	20
50	254	854	708	221	385	233	107	454	10
36°00'	.6283	.5878	1.701	.7265	1.376	1.236	.8090	.9425	54°00'
10	312	901	695	310	368	239	073	396	50
20	341	925	688	355	360	241	056	367	40
30	370	948	681	400	351	244	039	338	30
40	400	972	675	445	343	247	021	308	20
50	429	.5995	668	490	335	249	.8004	279	10
37°00'	.6458	.6018	1.662	.7536	1.327	1.252	.7986	.9250	53°00'
10	487	041	655	581	319	255	969	221	50
20	516	065	649	627	311	258	951	192	40
30	545	088	643	673	303	260	934	163	30
40	574	111	636	720	295	263	916	134	20
50	603	134	630	766	288	266	898	105	10
38°00'	.6632	.6157	1.624	.7813	1.280	1.269	.7880	.9076	52°00'
10	661	180	618	860	272	272	862	047	50
20	690	202	612	907	265	275	844	.9018	40
30	720	225	606	.7954	257	278	826	.8988	30
40	749	248	601	.8002	250	281	808	959	20
50	778	271	595	050	242	284	790	930	10
39°00'	.6807	.6293	1.589	.8098	1.235	1.287	.7771	.8901	51°00'
	cos θ	sec θ	cot θ	tan θ	csc θ	sin θ	Radians	Degrees	

| Angle θ |

TABLE 5

Values of Trigonometric Functions (*continued*)

Angle θ									
Degrees	Radians	sin θ	csc θ	tan θ	cot θ	sec θ	cos θ		
39°00'	.6807	.6293	1.589	.8098	1.235	1.287	.7771	.8901	51°00'
10	836	316	583	146	228	290	753	872	50
20	865	338	578	195	220	293	735	843	40
30	894	361	572	243	213	296	716	814	30
40	923	383	567	292	206	299	698	785	20
50	952	406	561	342	199	302	679	756	10
40°00'	.6981	.6428	1.556	.8391	1.192	1.305	.7660	.8727	50°00'
10	.7010	450	550	441	185	309	642	698	50
20	039	472	545	491	178	312	623	668	40
30	069	494	540	541	171	315	604	639	30
40	098	517	535	591	164	318	585	610	20
50	127	539	529	642	157	322	566	581	10
41°00'	.7156	.6561	1.524	.8693	1.150	1.325	.7547	.8552	49°00'
10	185	583	519	744	144	328	528	523	50
20	214	604	514	796	137	332	509	494	40
30	243	626	509	847	130	335	490	465	30
40	272	648	504	899	124	339	470	436	20
50	301	670	499	.8952	117	342	451	407	10
42°00'	.7330	.6691	1.494	.9004	1.111	1.346	.7431	.8378	48°00'
10	359	713	490	057	104	349	412	348	50
20	389	734	485	110	098	353	392	319	40
30	418	756	480	163	091	356	373	290	30
40	447	777	476	217	085	360	353	261	20
50	476	799	471	271	079	364	333	232	10
43°00'	.7505	.6820	1.466	.9325	1.072	1.367	.7314	.8203	47°00'
10	534	841	462	380	066	371	294	174	50
20	563	862	457	435	060	375	274	145	40
30	592	884	453	490	054	379	254	116	30
40	621	905	448	545	048	382	234	087	20
50	650	926	444	601	042	386	214	058	10
44°00'	.7679	.6947	1.440	.9657	1.036	1.390	.7193	.8029	46°00'
10	709	967	435	713	030	394	173	.7999	50
20	738	.6988	431	770	024	398	153	970	40
30	767	.7009	427	827	018	402	133	941	30
40	796	030	423	884	012	406	112	912	20
50	825	050	418	.9942	006	410	092	883	10
45°00'	.7854	.7071	1.414	1.000	1.000	1.414	.7071	.7854	45°00'
		cos θ	sec θ	cot θ	tan θ	csc θ	sin θ	Radians	Degrees
								Angle θ	

ANSWERS TO SELECTED EXERCISES

CHAPTER ONE

Exercise Set 1.1

1. x^5y^2 **3.** $(x+1)^3$ **5. a.** $4x$
b. $3(x^2+1)$ or $3x^2+3$ **7. a.** $3a+2b$
b. $3a^2+2b^2$ **9. a.** $(2a+1)^3(2b+1)^2$
b. $(2a+1)^3(2b+1)^2$ **11.** 22
13. 25 **15.** 36 **17.** 21
19. 42 **21.** 2 **23.** 6
25. a. 256 **b.** 512 **c.** 81
27. 2^{3^4} is larger **29. a.** 390,625
b. 1,953,125 **31. a.** 2^{2^5} is larger
b. $2^{2^5}=4,294,967,295$; $5^{2^2}=625$
33. $2+x$ or $x+2$ **35.** $4(a+b^2)$
37. x^2+y^2 **39.** $x+2x^2$
41. $\frac{1}{3}(x+y+z)$ **43.** $\frac{1}{3}(x^2+y^2+z^2)$
45. $2xy-1$ **47. a.** 1000 **b.** $\frac{5}{8}$
49. a. (i) 6^7 is larger; (ii) 7^8 is larger;
(iii) 8^9 is larger **b.** $a=1$ or $a=2$

Exercise Set 1.2

1. 832 **3.** 4410
5. a. $A^2+2AB+B^2$ **b.** A^2-B^2
c. $A^2-2AB+B^2$ **7. a.** $x^2+4xy+4y^2$
b. x^2-4y^2 **c.** $x^2-4xy+4y^2$
9. a. Ax^2-Ax+A **b.** x^3+1
11. commutative property of addition
13. additive inverse property
15. associative property of addition
17. associative property of multiplication
19. identity property of addition
21. closure property with respect to
addition **23.** distributive property
25. $b=1$, $c=1$ (other answers are
possible) **27.** $x=1$, $y=2$ (other
answers are possible) **29.** $x=2$,
$y=1$ (other answers are possible)

31. $x=2$, $y=3$ (other answers are
possible) **33.** $a=5$, $b=3$, $c=1$
(other answers are possible) **35.** -23
37. 2 **39.** -1 **41.** 2 **43.** $\frac{31}{15}$
45. $-5/x$ **47.** $2y^2/(x+y)$
49. $-\frac{49}{36}$ **51.** $\frac{6}{35}$ **53.** $\frac{3}{5}$
55. a. $\frac{22}{37}$ **b.** -1 **57.** -1 **59.** $\frac{5}{7}$
61. 16 **63.** 0 **65.** 1 **67.** 145
73. $y \approx -1.6$
75. a.

x	1	1.2500	1.5000	1.7500	2.0000	2.2500
y	0	0.2880	0.3704	0.3848	0.3750	0.3566

b. $x=1.750$ **c.** $y \approx 0.3849$. This is close
to the table value corresponding to
$x=1.750$.

Exercise Set 1.3

1. a. natural number, integer, rational
number **b.** integer, rational number
3. a. rational number **b.** irrational
number **5. a.** natural number,
integer, rational number **b.** rational
number **7. a.** rational number
b. rational number **9.** irrational
number **11.** natural number, integer,
rational number **13. a.** $\frac{27}{5}$ **b.** $\frac{49}{9}$
15. a. $\frac{99}{100}$ **b.** 1 **17. a.** $\frac{17}{10}$ **b.** $\frac{16}{9}$
19. a.

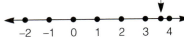

b.

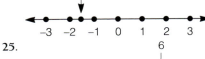

21.

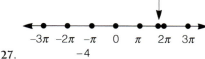

23.

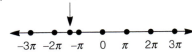

25.

27.

29. false **31.** true **33.** false
35. $2<x<5$

37. $1 \le x \le 4$

39. $0 \le x < 3$

41. $x>-3$

43. $x \ge -1$

45. $x < 1$

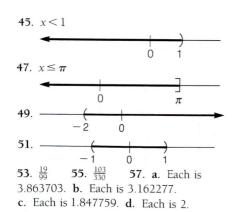

47. $x \leq \pi$

49.

51.

53. $\frac{19}{99}$ **55.** $\frac{103}{330}$ **57. a.** Each is 3.863703. **b.** Each is 3.162277. **c.** Each is 1.847759. **d.** Each is 2.

Exercise Set 1.4

1. 3   **3.** 6   **5.** 2   **7.** 0
9. 0   **11.** 17   **13.** 1   **15.** 0
17. 25   **19.** -1 **21.** 0   **23.** 3
25. $\sqrt{2} - 2$ **27.** $x - 3$
29. $t^2 + 1$ **31.** $\sqrt{3} + 4$
33. $-2x + 7$ **35.** 1   **37.** $3x + 11$
39. $|x - 4| = 8$ **41.** $|x - 1| = \frac{1}{2}$
43. $|x - 1| \geq \frac{1}{2}$ **45.** $|y + 4| < 1$
47. $|y| < 3$ **49.** $|x^2 - a^2| < M$

51.

53.

55.

57.

59.

61.

63. a.

b.

c. The interval in (b) does not include 2.

Exercise Set 1.5

1. -10 **3.** $\frac{2}{3}$ **5.** 2   **7.** $(x + y)^2$
9. $(x + y)^2 + 3$ **11.** $\left[\frac{1}{2}(x^2 - 2y^3)\right]^2$
13. $|x - 1|^3$ **15. a.** a^{15} **b.** $(a + 1)^{15}$
c. $(a + 1)^{15}$ **17. a.** y^{11} **b.** $(y + 1)^{11}$
c. $(y + 1)^{18}$ **19. a.** $x^2 + 3$
b. $1/(x^2 + 3)$ **c.** 12   **21. a.** t^6 **b.** $1/t^6$

c. $(t^2 + 3)^6$ **23. a.** x^4/y^5 **b.** y^5/x^4
c. $|y^{10}/x^8|$ **25. a.** $4x^6$ **b.** $16x^6$ **c.** 4
27. a. 1 **b.** 1 **c.** 1   **29. a.** $\frac{11}{100}$
b. $\frac{100}{11}$ **c.** 1000   **31.** 7   **33.** $\frac{6}{13}$
35. a. $\frac{1}{20}$ **b.** $\frac{1}{225}$ **37.** $1/a^6b^3$
39. a^4b^2/c^6 **41.** y^{12}/x^6z^{12}
43. y^{20}/x^6z^{16} **45.** $1/a^8b^{12}c^{16}$ **47.** y^6

49. b **51.** x^{2p} **53.** $\dfrac{a^{6q}}{b^{3p}}$

55. 576   **57.** 12
59. 9.29×10^7 miles
61. 6.68×10^4 mph
63. 2.5×10^{19} miles
65. a. 8.6688×10^1 days
b. 1.0604772×10^4 days
c. 8.9424×10^4 days
67. a. 1.0×10^{-9} seconds
b. 1.0×10^{-18} seconds
c. 1.0×10^{-24} seconds   **69.** 1
71. x^2 **73.** 8 min

Exercise Set 1.6

1. false   **3.** true   **5.** true
7. true   **9.** true   **11. a.** -4
b. undefined   **13. a.** $\frac{2}{5}$ **b.** $-\frac{2}{5}$
15. a. undefined **b.** undefined
17. a. $\frac{4}{3}$ **b.** $-\frac{3}{5}$ **19. a.** -2 **b.** 2
21. a. $3\sqrt{2}$ **b.** $3\sqrt[3]{2}$ **23. a.** $7\sqrt{2}$
b. $-2\sqrt[5]{2}$ **25. a.** $\frac{5}{2}$ **b.** $\frac{2}{5}$
27. a. $3\sqrt{2}$ **b.** $3\sqrt[3]{2}$ **29. a.** $-4\sqrt{2}$
b. $5\sqrt[4]{2}$ **31. a.** 0.3 **b.** 0.2
33. $-14\sqrt{6}$ **35.** $2\sqrt{2}$
37. a. $6x$ **b.** $-6y$ **39. a.** $ab\sqrt{ab}$
b. a^2b^2 **41.** $6ab^2c^2\sqrt{2ac}$
43. $2ab\sqrt[4]{b}$ **45.** $3ab\sqrt{2a}$
47. $2a^4\sqrt[3]{2b^2}/c^3$ **49.** $a^2\sqrt[6]{5}/b$
51. $4\sqrt{7}/7$ **53.** $\sqrt{2}/4$
55. $(\sqrt{5} - 1)/4$ **57.** $-2 - \sqrt{3}$
59. $61\sqrt{5}/5$ **61.** $\sqrt[3]{5}/5$
63. $\sqrt[4]{27}$ **65.** $\sqrt[4]{8a^3b^3}/2ab^2$
67. $3\sqrt[5]{2ab}/2ab^2$
69. $x(\sqrt{x} + 2)/(x - 4)$
71. $(x - 2\sqrt{ax} + a)/(x - a)$
73. $1/(\sqrt{x} + \sqrt{5})$
75. $1/(\sqrt{2 + h} + \sqrt{2})$ **77.** $a = 9$,
$b = 16$ (other answers are possible)
79. $u = 1$, $v = 8$ (other answers are
possible)   **81.** Both values are
1.645751 (approximately).

83. $(a + b)/(a - b)$
85. $\frac{1}{4}(\sqrt{2} + 2 - \sqrt{6})$

Exercise Set 1.7

1. $\sqrt[5]{a^3} = \left(\sqrt[5]{a}\right)^3$ **3.** $\sqrt[3]{5^2} = \left(\sqrt[3]{5}\right)^2$
5. $\sqrt[4]{(x^2 + 1)^3} = \left(\sqrt[4]{x^2 + 1}\right)^3$
7. $\sqrt[3]{2^{xy}} = \left(\sqrt[3]{2}\right)^{xy}$ **9.** $p^{2/3}$
11. $(1 + u)^{4/7}$ **13.** $\left(a^2 + b^2\right)^{3/p}$
15. 4   **17.** $\frac{1}{6}$ **19.** undefined
21. 5   **23.** 2   **25.** 4   **27.** -2
29. -10 **31.** $\frac{1}{7}$ **33.** undefined
35. $\frac{1}{216}$ **37.** 25   **39.** -1
41. $\frac{255}{16}$ **43.** $-\frac{28976}{243}$ **45.** $6a^{7/12}$
47. $2^{3/2}a^{1/12}$ **49.** $x^2 + 1$
51. a. $2^{1/3}3^{5/6}$ **b.** $\sqrt[6]{972}$
53. a. $2^{7/12}3^{1/3}$ **b.** $\sqrt[12]{10368}$
55. a. $x^{2/3}y^{4/5}$ **b.** $\sqrt[15]{x^{10}y^{12}}$
57. a. $x^{(a+b)/3}$ **b.** $\sqrt[3]{x^{a+b}}$
59. $(x + 1)^{2/3}$ **61.** $(x + y)^{2/5}$
63. $2x^{1/6}$ **65.** $x^{1/6}y^{1/8}$ **67.** $9^{10/9}$
69. $a = 9$, $b = 16$ (other answers are
possible)   **71.** $u = 1$, $v = 8$ (other
answers are possible)   **73.** $x = 4$,
$m = 2$ (other answers are possible)
75.

n	2	5	10	100
$n^{1/n}$	1.4142	1.3797	1.2589	1.0471

n	10^3	10^4	10^5	10^6
$n^{1/n}$	1.0069	1.0009	1.0001	1.0000

77. a. $2^{3/2}$ is larger **b.** $5^{1/2}$ is larger
c. $2^{1/2}$ is larger **d.** $\left(\frac{1}{2}\right)^{1/3}$ is larger
e. $10^{1/10}$ is larger
79. $(0.5)^{1/3}$ is closer to zero;
$(0.5)^{1/3} \approx 0.7937$; $(0.5)^{1/4} \approx 0.8409$

Exercise Set 1.8

1. a. the set of all real numbers **b.** the
set of all nonnegative real numbers
3. a. the set of all real numbers **b.** the
set of all real numbers   **5. a.** the set
of all nonnegative real numbers **b.** the
set of all positive real numbers except 1
7. a. degree: 3; coefficients: 4, -2, -6,
-1 **b.** degree: 2; coefficients: a, b, c
9. a. degree: 1; coefficient: 6

b. degree: 3; coefficients: $-1, 1, 6$
11. $20x^2 + 2x + 1$ **13.** $-2x - 2$
15. $-4x^2 - 12x - 8$
17. $4x^2 - 19x - 14$ **19.** $-ax^2 + 4bx$
21. $2x^3 - 8x^2 - 10x$ **23.** $x^2 - 3x + 2$
25. $2x^2 + 6x + 4$ **27.** $x^4 + 4x^3 + 3x^2$
29. $4x^2y^2 - 8xy + 3$ **31.** $x^2 + 2 + 1/x^2$
33. $a + b + 4\sqrt{a + b} + 3$
35. $x - 3x^{1/2} + 2$
37. $y^3 - y^2 - 11y - 10$
39. $x^2 - 4y^2 + 4yz - z^2$
41. a. $x^2 - y^2$ **b.** $x^4 - 25$
43. a. $A^2 - 16$ **b.** $a^2 + 2ab + b^2 - 16$
45. a. $x^2 - 16x + 64$ **b.** $4x^4 - 20x^2 + 25$
47. a. $x + 2\sqrt{xy} + y$
b. $2x + y - 2\sqrt{x^2 + xy}$
49. a. $a^3 + 3a^2 + 3a + 1$
b. $27x^6 - 54x^4a^2 + 36x^2a^4 - 8a^6$
51. a. $x^3 + 1$ **b.** $x^6 + 1$
53. quotient: $x - 5$; remainder: -11;
thus $x^2 - 8x + 4 = (x - 3)(x - 5) - 11$
55. quotient: $x - 11$; remainder: 53; thus
$x^2 - 6x - 2 = (x + 5)(x - 11) + 53$
57. quotient: $3x^2 - \frac{3}{2}x - \frac{1}{4}$; remainder: $\frac{13}{4}$;
thus $6x^3 - 2x + 3 =$
$(2x + 1)\left(3x^2 - \frac{3}{2}x - \frac{1}{4}\right) + \frac{13}{4}$
59. quotient: $t^4 - 3t^3 + 9t^2 - 27t + 81$;
remainder: -241; thus $t^5 + 2 =$
$(t + 3)(t^4 - 3t^3 + 9t^2 - 27t + 81) - 241$
61. quotient: $u^5 + 2u^4 + 4u^3 + 8u^2 +$
$16u + 32$; remainder: 0;
thus $u^6 - 64 = (u - 2)(u^5 + 2u^4 + 4u^3 +$
$8u^2 + 16u + 32) + 0$
63. quotient: $5x^2 + 15x + 17$;
remainder: $-24x - 83$;
thus $5x^4 - 3x^2 + 2 =$
$(x^2 - 3x + 5)(5x^2 + 15x + 17) +$
$(-24x - 83)$
65. quotient: $t^2 - 2t - 4$; remainder: 0;
thus $t^4 - 4t^3 + 4t^2 - 16 =$
$(t^2 - 2t + 4)(t^2 - 2t - 4) + 0$
67. quotient: $z^4 + z^3 + z^2 + z + 1$;
remainder: 0; thus $z^5 - 1 =$
$(z - 1)(z^4 + z^3 + z^2 + z + 1) + 0$
69. $a^2 + 2ab + b^2 - 9$
71. $a^2 + 2ab + b^2 - x^2 - 2xy - y^2$
73. $a + 2\sqrt{ab} + b$ **75.** $x - y$

77. a.

x	$\dfrac{x^2 - 16}{x - 4}$	x	$\dfrac{x^2 - 16}{x - 4}$
3.9	7.9	4.1	8.1
3.99	7.99	4.01	8.01
3.999	7.999	4.001	8.001
3.9999	7.9999	4.0001	8.0001
3.99999	7.99999	4.00001	8.00001

b. 8 **c.** Since $\dfrac{x^2 - 16}{x - 4} = x + 4$, note that
$x + 4$ approaches 8 as x approaches 4.
83. a. $ar^2 + br + c$ **b.** $ar^3 + br^2 + cr + d$
c. $ar^4 + br^3 + cr^2 + dr + e$
85. a.

n	1	2	3	4	5	6
$\dfrac{n^2}{n + 1}$	$\dfrac{1}{2}$	$\dfrac{4}{3}$	$\dfrac{9}{4}$	$\dfrac{16}{5}$	$\dfrac{25}{6}$	$\dfrac{36}{7}$

b. no

Exercise Set 1.9

1. a. $(x + 8)(x - 8)$ **b.** $7x^2(x^2 + 2)$
c. $z(11 + z)(11 - z)$ **d.** $(ab + c)(ab - c)$
3. a. $(x + 3)(x - 1)$ **b.** $(x - 3)(x + 1)$
c. irreducible **d.** $(-x + 3)(x + 1)$ or
$-(x - 3)(x + 1)$
5. a. $(x + 1)(x^2 - x + 1)$
b. $(x + 6)(x^2 - 6x + 36)$
c. $8(5 - x^2)(25 + 5x^2 + x^4)$
d. $(4ax - 5)(16a^2x^2 + 20ax + 25)$
7. a. $(12 + x)(12 - x)$ **b.** irreducible
c. $(141 + y)(147 - y)$
9. a. $h^3(1 + h)(1 - h)$
b. $h^3(10 + h)(10 - h)$
c. $(h + 1)^3(11 + h)(9 - h)$
11. a. $(x - 8)(x - 5)$ **b.** irreducible
13. a. $(x + 9)(x - 4)$ **b.** $(x - 9)(x - 4)$
15. a. $(3x + 2)(x - 8)$ **b.** irreducible
17. a. $(3x - 1)(2x + 5)$
b. $(6x + 5)(x - 1)$ **19. a.** $(t^2 + 1)^2$
b. $(t + 1)^2(t - 1)^2$ **c.** irreducible
21. a. $x(4x^2 - 20x - 25)$ **b.** $x(2x - 5)^2$
23. a. $(a - c)(b + a)$

b. $(u + v - y)(x + u + v)$
25. $(xz + t)(xz + y)$ **27.** $\left(a^2 - 2b^2c^2\right)^2$
29. irreducible
31. $(x + 4)(x^2 - 4x + 16)$
33. $x(x^2 + 3xy + 3y^2)$
35. $(x - y)(x^2 + xy + y^2 + 1)$
37. a. $(p^2 + 1)(p + 1)(p - 1)$
b. $(p^4 + 1)(p^2 + 1)(p + 1)(p - 1)$
39. $(x + 1)^3$ **41.** irreducible
43. $\left(\frac{5}{4} + c\right)\left(\frac{5}{4} - c\right)$
45. $(z^2 + \frac{9}{4})(z + \frac{3}{2})(z - \frac{3}{2})$
47. $\left(\dfrac{5}{mn} - 1\right)\left(\dfrac{25}{m^2n^2} + \dfrac{5}{mn} + 1\right)$
49. $\left(\frac{1}{2}x + y\right)^2$
51. $(x - a)(8x - 8a + 1)(8x - 8a - 1)$
53. $(x - y - a)(x - y + a)$
55. $x(7x - 3)(3x + 13)$
57. $(5 - 2x + 3y)(5 + 2x - 3y)$
59. $(ax + b)(x + 1)$ **61.** $-x(x + 1)^{1/2}$
63. $x(x + 1)^{-3/2}$ **65.** $-\frac{2}{3}x(2x + 3)^{1/2}$
67. a. 199 **b.** 296 **c.** 1999
69. $(A + B)^3$
71. $x\left(a^2 + x^2\right)^{-3/2}(2a^2 + x^2)$
73. $(y^2 - pq)(y^2 + pq - py - qy)$
75. $2(u - t)(ut - 1)$
77. $(x^2 + 4x + 8)(x^2 - 4x + 8)$
79. $2(x - t)(x + y + z + t)$
81. $3(b - a)(c - a)(c - b)$

Exercise Set 1.10

1. $x - 3$ **3.** $1/(x - 2)(x^2 + 4)$
5. $1/(x - 2)$ **7.** $3b/2a$ **9.** $\dfrac{a^2 + 1}{a - 1}$
11. $\dfrac{x^2 + xy + y^2}{(x - y)^2}$ **13.** 2
15. $\dfrac{x^2 + x}{(x + 4)(x - 2)}$ **17.** $\dfrac{x^2 - xy + y^2}{(x - y)(x + y)}$
19. $\dfrac{x - 3y}{x + 3y}$ **21.** $\dfrac{4x - 2}{x^2}$
23. $\dfrac{36 - a^2}{6a}$ **25.** $\dfrac{4x + 11}{(x + 3)(x + 2)}$
27. $\dfrac{3x^2 + 6x - 6}{(x - 2)(x + 2)}$
29. $\dfrac{6ax^2 - 4ax + a}{(x - 1)^3}$
31. $\dfrac{2x^2 - 3}{(x - 3)(x + 3)(x - 2)}$ **33.** $\dfrac{8}{x - 5}$
35. $\dfrac{2a}{a + b}$ **37.** $\dfrac{-9}{(x + 5)(x - 4)^2}$

39. $\dfrac{-p^2 - pq - q^2}{p(2p+q)(p-5q)}$ **41.** 0

43. 0   **45.** $\dfrac{-1}{x+a}$ **47.** $\dfrac{1+x}{1-x}$

49. $-\dfrac{1}{ax}$ **51.** $\dfrac{x+4}{3-2x}$ **53.** $a-1$

55. $-1/(4+2h)$ **57.** $(a+x)/(a^2x^2)$

59. $x/(1+2x)$ **61.** $1/(a+a^2)$

63. $(x^2+y^2)/(x^2-y^2)$ **65.** $(t^2+1)/t$

67. 0   **69.** $(a^2-b^2)/2$ **71.** $2/x^3$

73. $(px+q)/(x-a)(x-b)$ **75.** 0

77. -1 **79. a.** Both are $\frac{1}{42}$.

Exercise Set 1.11

1.

i^2	i^3	i^4	i^5	i^6	i^7	i^8
-1	$-i$	1	i	-1	$-i$	1

3. a. real part: 4; imaginary part: 5
b. real part: 4; imaginary part: -5
c. real part: $\frac{1}{2}$; imaginary part: -1
d. real part: 0; imaginary part: 16
5. $c=4, d=-3$ **7. a.** $14-4i$
b. $-4-8i$ **9. a.** $19-17i$
b. $19-17i$ **c.** $\frac{11}{26}-\frac{23}{26}i$ **d.** $\frac{11}{25}+\frac{23}{25}i$
11. a. $11-i$ **b.** $11-7i$ **c.** 4
13. $4-2i$ **15.** $30+19i$ **17.** 13
19. $-191-163i$ **21.** $19-4i$
23. $-70+84i$ **25.** $539+1140i$
27. $-46+9i$ **29.** $\frac{6}{97}+\frac{35}{97}i$
31. $\frac{6}{97}-\frac{35}{97}i$ **33.** $-\frac{5}{13}+\frac{12}{13}i$
35. -4 **37.** $\frac{1}{26}+\frac{5}{26}i$ **39.** $-i$
41. $12i$ **43.** $-3\sqrt{5}i$ **45.** -35
47. $12\sqrt{2}i$ **49. a.** $(a+c)+(b+d)i$
b. $(a-c)+(b-d)i$
c. $(ac-bd)+(bc+ad)i$
d. $\dfrac{ac+bd}{c^2+d^2}+\dfrac{bc-ad}{c^2+d^2}i$

57. real part: $\dfrac{2a^2-2b^2}{a^2+b^2}$;
imaginary part: 0   **59.** 0

Chapter One Review Exercises

1. $(a-4b)(a+4b)$
3. $(1-a)(7+4a+a^2)$ **5.** $x(ax+b)^2$
7. $(4x+1)(2x+1)$
9. $a^4x^4(1-xa)(1+xa)(1+x^2a^2)$
11. $(2+a)^3$ **13.** $z^3(4xy+1)(xy-1)$

15. $(1-x)(1+x+x^2)(1+x)(1-x+x^2)$
17. $(ax+b-2abx)(ax+b+2abx)$
19. $(a-b)(a+b+c+ab)$ **21.** 8
23. 1   **25.** $\frac{1}{16}$ **27.** $\frac{1}{9}$ **29.** ab^3c^4
31. $2a^2b^4$ **33.** $5\sqrt[3]{2}$ **35.** $5ab\sqrt{6b}$
37. $|t|$ **39.** $2|x|$ **41. a.** $x^{13/12}$
b. $x\sqrt[12]{x}$ **43. a.** $t^{17/30}$ **b.** $\sqrt[30]{t^{17}}$
45. $2\sqrt{3}$ **47.** $\frac{1}{3}(\sqrt{6}+\sqrt{3})$
49. $(a^2+\sqrt{a^4-x^4})/x^2$
51. $\frac{1}{12}(2\sqrt{3}+3\sqrt{2}-\sqrt{30})$
53. $1/(\sqrt{t}+a)$
55. $1/(\sqrt[3]{x^2}+2\sqrt[3]{x}+4)$
57. $9x^2+6x+1$
59. $9x^4+6x^2y^2+y^4$
61. $8a^3+36a^2+54a+27$
63. $1-27a^3$ **65.** $x-y$
67. $x+3x^{2/3}y^{1/3}+3x^{1/3}y^{2/3}+y$
69. $x^4-6x^3+11x^2-6x+1$
71. $5+6i$ **73.** 6
75. $\frac{1}{2}(1-\sqrt{3}i)$ **77.** $3\sqrt{2}-4\sqrt{2}i$
81. a. Both are $2\sqrt{10}$. **b.** 3
83. distributive property, associative
property of addition   **85.** distributive
property   **87.** $|x-6|=2$
89. $|a-b|=3$ **91.** $|x|>10$
93. $\sqrt{6}-2$ **95.** x^4+x^2+1
97. a. $-2x+5$ **b.** 1 **c.** $2x-5$
99. a. true **b.** true **c.** true **d.** false
101. [number line: 3 to 5]
103. [number line: arrow to 4]
105. 1.4×10^{-3} **107.** 1.2001×10^1
109. -37 **111.** -12
113. x^3+2x^2-x-2
115. $3x+2$
117. quotient: x^2-x; remainder: 4
119. quotient: x^2-2; remainder: 3
121. [number line: 3 to 9]
123. [number line: -2 to 0]
125. a. [number line: -1, 4, 9]
b. [number line: -1 to 9]
127. $\frac{1259}{99}$

129.

x	$1+\dfrac{x}{2}$	$\sqrt{1+x}$
0.1	1.05	1.048808
0.01	1.005	1.004987
0.001	1.0005	1.000499

131. a. 0.447214 **b.** 0.447214
e. $\sqrt{5}/5$ **137.** x^2

Chapter One Test

1. a. $(6-x)(12+x^2)$ **b.** irreducible
2. a. irreducible **b.** $z(3xy-1)(2xy+1)$
3. $-3x(1-x^2)^{-3/2}(2x^2+x-2)$
4. $(3a+2b)(3x-2y)$ **5.** $13\sqrt[3]{2}$
6. a. $x^{29/35}$ **b.** $\sqrt[35]{x^{29}}$
7. $x^2+\sqrt{x^4-1}$
8. $8a^3+36a^2b^2+54ab^4+27b^6$
9. $21-i$ **10. a.** distributive property
b. commutative property of
multiplication, associative property of
multiplication   **11.** 1
12. a. 3.96×10^9 light-years
b. 6.750×10^{-7} meters
13. quotient: $x^2-\frac{3}{2}x+\frac{1}{4}$; remainder: $-\frac{39}{4}$
14. $\frac{281}{9}$ **15. a.** $\left(\frac{39}{10},\frac{41}{10}\right)$ **b.** $[-2,\infty)$

CHAPTER TWO

Exercise Set 2.1

1. a. The problem asks us to find two
numbers x and y.
b.

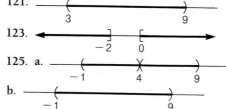

in English	in algebraic notation
Find two numbers.	x and y
Their sum is -9.	$x+y=-9$
The sum of their squares is 153.	$x^2+y^2=153$

c. $x^2+9x-36=0$
3. a. The problem asks us to find two
numbers x and y.

b.

in English	in algebraic notation
Find two numbers.	x and y
Their sum is 17.	$x + y = 17$
Their product is 52.	$xy = 52$

c. $x^2 - 17x + 52 = 0$

5. a. The problem asks us to find two consecutive integers x and $x + 1$.

b.

in English	in algebraic notation
Find two consecutive integers.	x and $x + 1$
The sum of their cubes is 341.	$x^3 + (x + 1)^3 = 341$

c. $2x^3 + 3x^2 + 3x - 340 = 0$

7. a. The problem asks us to find two numbers x and y.

b.

in English	in algebraic notation
Find two numbers	x and y
Their average is 4.	$\frac{1}{2}(x + y) = 4$
The average of their squares is 80.	$\frac{1}{2}(x^2 + y^2) = 80$

c. $x^2 - 8x - 48 = 0$

9. a. The problem asks us to find the length l and width of w of a rectangle.

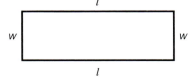

b.

in English	in algebraic notation
Find the width and length of a rectangle	w and l
The perimeter is 40 cm.	$2w + 2l = 40$
The area is 64 cm².	$lw = 64$

c. $w^2 - 20w + 64 = 0$

11. a. The problem asks us to find the number of years t until the total interest earned is \$8,760.

b. $(3000)(0.07)t + (5000)(0.08)t + (10000)(0.085)t = 8760$

13. $7400\left(1 + \dfrac{r}{10}\right) + 12600\left(1 + \dfrac{1 + r}{10}\right) = 31260$

15. a. $h = 4r$ **b.** $V = 4\pi r^3$ **c.** $V = \frac{1}{16}\pi h^3$

17. a. $r + h = 4$ **b.** $V = \pi r^2(4 - r)$

c.

r	1.0	1.5	2.0	2.5	3.0	3.5
V	9.42	17.67	25.13	29.45	28.27	19.24

d. 2.5 cm **e.** 2.6 cm or 2.7 cm **f.** no

19. a. $r^2 h = 3$

c.

r	0.25	0.50	0.75	1.00
S	37.700	18.866	12.690	9.935

r	1.25	1.50	1.75	2.00
S	8.997	9.457	11.026	13.421

d. 1.25 cm **e.** 1.30 cm **f.** yes, 1.29 cm (other answers are possible)

21. $A = \frac{17}{144}x^2 - \frac{1}{9}x + \frac{1}{18}$

Exercise Set 2.2

1. -1 **3.** 3 **5.** -11 **7.** 12
9. 3 **11.** -6 **13.** -7
15. -3 **17.** -15 **19.** 9

21. $-\frac{13}{7}$ **23.** 8
25. the set of all real numbers except ± 5
27. no solution **29.** no solution
31. 3 **33. a.** no solution **b.** $\frac{2}{7}$
c. $-\frac{7}{3}$ **35. a.** 1 **b.** 1 **c.** no solution
37. 6, 4 **39.** $-\frac{11}{2}, -\frac{13}{2}$
41. $5, -\frac{10}{3}$ **43.** 2.60 **45.** 0.04
47. $x = (b + 1)/a$ **49.** $x = 1$
51. $x = 1/(a + b)$ **53.** $x = ab/(b + c)$
55. $x = (a^2 - b^2)/2a$ **57. a.** $x = a/bd$
b. $x = (a - cd)/bd$ **59. a.** $-1/15x$
b. $\frac{1}{30}$ **61. a.** $(8x + 5)/(5x^2)$ **b.** $-\frac{10}{31}$
63. -3 **65.** $x = (a^2 + ab + b^2)/(a - b)$
67. $x = b$ **69.** $x = (a + b)/2$
71. $r = d/(1 - dt)$
73. $h = (S - 2\pi r^2)/(2\pi r)$
75. $r = (Sl - a)/(S - l)$
79. $x = 2pq/(p + q)$

Exercise Set 2.3

1. yes **3.** no **5.** no **7.** $6, -1$
9. ± 10 **11.** $\frac{6}{5}$ **13.** $-\frac{1}{5}, \frac{3}{2}$
15. $1, -3$ **17.** $\frac{1}{2}, 6$ **19.** $-13, 12$
21. ± 7 **23.** $\pm\frac{1}{2}$ **25. a.** $-1, -9$
b. $-1, -9$ **27. a.** $\frac{1}{2}\left(1 \pm \sqrt{21}\right)$
b. $\frac{1}{2}\left(1 \pm \sqrt{21}\right)$ **29. a.** $\frac{1}{4}\left(-3 \pm \sqrt{41}\right)$
b. $\frac{1}{4}\left(-3 \pm \sqrt{41}\right)$ **31. a.** $-\frac{1}{6}, -\frac{5}{2}$
b. $-\frac{1}{6}, -\frac{5}{2}$ **33. a.** $\frac{1}{4}\left(1 \pm \sqrt{41}\right)$
b. $\frac{1}{4}\left(1 \pm \sqrt{41}\right)$ **35. a.** $\frac{1}{6}\left(6 \pm \sqrt{42}\right)$
b. $\frac{1}{6}\left(6 \pm \sqrt{42}\right)$ **37.** $-9 \pm 9\sqrt{2}$
39. $1 \pm i\sqrt{5}$ **41.** $\frac{5}{8}, -2$
43. $\frac{1}{12}\left(5 \pm i\sqrt{119}\right)$ **45.** $\pm 2\sqrt{6}$
47. $\frac{1}{2}\left(1 \pm \sqrt{5}\right)$ **49.** $\frac{1}{3}\left(3 \pm \sqrt{15}\right)$
51. $\sqrt{5}/2, -2\sqrt{5}/5$
53. $x = 1/2y, x = 1/y$ **55.** $x = 0,$
$x = p + q$ **57.** $x = -5a/4, x = 4a/3$
59. $r = \frac{1}{2}\left(-h \pm \sqrt{h^2 + 40}\right)$
61. $t = 0, t = v_0/16$ **63.** $x = 0,$
$x = -2b, x = b$ **65.** two **67.** two
69. one **71.** two **73.** 36
75. $\pm 2\sqrt{5}$ **77.** $-1.47, -1.53$
79. $\frac{5}{2}, -4$ **81.** $\frac{1}{3}, \frac{1}{2}$ **83.** 1
85. $-3 \pm 2\sqrt{2}$ **87. b.** $k = -\frac{1}{2}$
c. $\frac{1}{2}\left(-1 \pm \sqrt{5}\right)$
89. $A = 0$ and $B = 0$; $A = 1$ and $B = -2$
95. $x = \dfrac{a + b}{ab}, x = \dfrac{-2}{a + b}$

Exercise Set 2.4

1. 12 and -5 3. 13 and 4
5. 21   7. 8   9. 5
11. 12 and 6   13. 78   15. 44
17. 5 cm, 12 cm, and 13 cm
19. 2.5 cm   21. 20 cm
23. 4.8 in. and 5.2 in.   25. 10 cm, 30 cm, and 30 cm   27. 8 years
29. 10 years   31. 6.6%
33. \$8000 account at 7.5%, \$9000 account at 8.5%   35. 80 cm^3 of 10% acid, 120 cm^3 of 35% acid
37. 15 lb   39. 8 tons   41. 17 cm
43. $\frac{1}{2}(3 - \sqrt{5})$ 45. $\frac{1}{2}(1 + \sqrt{5})$
47. a. 6 sec b. 1 sec (rising), 5 sec (falling)   49. $12\frac{3}{8}$ min
51. 55 mph   53. 350 mph
55. 15 miles   57. \$5117

59. $\frac{1}{24}$ mile/min   61. $\dfrac{2v_1 v_2}{v_1 + v_2}$ mph

63. $\dfrac{100(d + D)}{dr + DR}$ years

65. $\dfrac{a^2 - 2ab + 2b^2}{2(a - b)}$ in.

67. $\dfrac{b\sqrt{a^2 - 1}}{a^2 - 1}$ units

69. 4.62 cm   71. $\dfrac{ab}{a + b}$

Exercise Set 2.5

1. 0 and 16   3. 5
5. 0, 2, and -2 7. 1, 15, and -15
9. 0 and $\frac{5}{2}$ 11. 0, -3, and 1
13. 0, $\frac{1}{4}$, and -6 15. $\pm\sqrt{3}$
17. ± 1 19. $\pm\sqrt{6}/3$
21. $2 + \sqrt[3]{5} \approx 3.71$
23. $-4 - \sqrt[5]{16} \approx -5.74$
25. a. $3 + \sqrt[4]{30} \approx 5.34$ and $3 - \sqrt[4]{30} \approx 0.66$ b. no solution
27. ± 1 29. $\pm\sqrt{6}/3$
31. 0, $\pm\sqrt{6}$ and ± 2

33. $\pm\sqrt{\dfrac{-1 + \sqrt{5}}{2}}$

35. $\pm\sqrt{\dfrac{-3 + \sqrt{17}}{2}}$ 37. -2 and 1

39. $\frac{1}{3}$ and $\frac{1}{4}$ 41. $\frac{3}{5}$ and 4   43. $\pm\frac{1}{3}$
45. ± 27 47. ± 3 49. $1 + \sqrt[3]{7}$
51. -4 53. ± 1 and $\pm\frac{1}{27}$
55. -1 57. 9   59. -1
61. $-\frac{1}{2}$ 63. ± 2 and ± 3

Exercise Set 2.6

1. $(-\infty, -1)$ 3. $\left[\frac{1}{3}, \infty\right)$
5. $\left(-\infty, -\frac{9}{2}\right)$ 7. $[-6, \infty)$
9. $\left(-\infty, \frac{5}{2}\right)$ 11. $\left[-\frac{17}{23}, \infty\right)$
13. $[4, 6]$ 15. $\left[-\frac{1}{2}, 1\right]$
17. $(3.98, 3.998)$ 19. a. $\left[-\frac{1}{2}, \frac{1}{2}\right]$
b. $\left(-\infty, -\frac{1}{2}\right) \cup \left(\frac{1}{2}, \infty\right)$
21. a. $(-\infty, 0) \cup (0, \infty)$ b. no solution
23. a. $(-\infty, 3)$ b. $(1, 3)$
c. $(-\infty, 1) \cup (3, \infty)$ 25. a. $[-4, \infty)$
b. $[-4, 6]$ c. $(-\infty, -4) \cup (6, \infty)$
27. a. $(a - c, \infty)$ b. $(a - c, a + c)$
c. $(-\infty, a - c] \cup [a + c, \infty)$
29. $(-10, 14)$ 31. $(-11, 1)$
33. a. $(-2h, h)$ b. $(h, -2h)$
35. $(0.49, 0.51)$ 37. \$2000
39. $[-13°, 887°]$
41. $[9200°, 11700°]$
43. a.

a	b	$\sqrt{ab}$ (G.M.)	$\frac{1}{2}(a + b)$ (A.M.)	Which is larger, G.M. or A.M.?
1	2	1.4142	1.5	A.M.
1	3	1.7320	2.0	A.M.
1	4	2.0000	2.5	A.M.
2	3	2.4495	2.5	A.M.
3	4	3.4641	3.5	A.M.
5	10	7.0711	7.5	A.M.
9	10	9.4868	9.5	A.M.
99	100	99.4987	99.5	A.M.
999	1000	999.4999	999.5	A.M.

65. -4 and $-\frac{20}{9}$ 67. no solution
69. 2   71. ± 1.80 73. 0.47
75. $x = a + b$ 77. $x = a^2 b^2/(2a + b)^2$
79. $\frac{1}{2}(1 \pm \sqrt{21})$ 81. $x = -a/15$
85. $\frac{1}{2}(\sqrt{3} - 9)$
87. 0.21, -1.21, 1.56, and -2.56

89. $x = 0$ and $x = \dfrac{a(2^m - 1)}{2^m + 1}$

45. a.

x	y	$\dfrac{x}{y} + \dfrac{y}{x}$	True or False? $\dfrac{x}{y} + \dfrac{y}{x} \geq 2$
1	1	2.0000	true
2	3	2.1667	true
3	5	2.2667	true
4	7	2.3214	true
5	9	2.3556	true
9	10	2.0111	true
49	50	2.0004	true
99	100	2.0001	true

51. c. $x = 5$, dimensions are 5 ft by 5 ft

Exercise Set 2.7

1. $(-3, 2)$ 3. $(-\infty, 2) \cup (9, \infty)$
5. $(-\infty, 4] \cup [5, \infty)$
7. $(-\infty, -4] \cup [4, \infty)$ 9. no solution
11. $(-7, -6) \cup (0, \infty)$
13. $[-15, 0] \cup [15, \infty)$ 15. $(-\infty, \infty)$
17. $\left(-\infty, -\frac{3}{4}\right) \cup \left(-\frac{2}{3}, 0\right)$
19. $\left(-\infty, \dfrac{-1 - \sqrt{5}}{2}\right) \cup \left(\dfrac{-1 + \sqrt{5}}{2}, \infty\right)$
21. $\left[4 - \sqrt{14}, 4 + \sqrt{14}\right]$
23. $[-4, -3] \cup [1, \infty)$
25. $(-\infty, -6) \cup (-5, -4)$
27. $\left(-\infty, -\frac{1}{3}\right) \cup \left(\frac{1}{3}, 2\right) \cup (2, \infty)$
29. $\left(-\infty, -\frac{2}{3}\right] \cup \left\{3, -\frac{1}{2}\right\}$
31. $\left(-\infty, -\sqrt{5}\right] \cup [-2, 2] \cup \left[\sqrt{5}, \infty\right)$
33. $(-3, 3) \cup (4, \infty)$
35. $\left(-\sqrt{14}, -2\sqrt{2}\right) \cup \left(2\sqrt{2}, \sqrt{14}\right)$
37. $\left(-\infty, \dfrac{3 - \sqrt{11}}{2}\right) \cup \left(\dfrac{3 + \sqrt{11}}{2}, 4\right)$
39. $(-2, -1) \cup (1, \infty)$ 41. $(-1, 1]$
43. $\left(-\infty, \frac{3}{2}\right) \cup [2, \infty)$
45. $(-\infty, -1) \cup (0, 9)$
47. $\left(-\frac{7}{2}, -\frac{4}{3}\right) \cup (-1, 0) \cup (1, \infty)$
49. $(-\infty, -1)$ 51. $(-1, 0)$
53. $(-5, \infty)$ 55. $[-1, 1) \cup (2, 4]$
57. $\left(-1, 2 - \sqrt{5}\right) \cup \left(1, 2 + \sqrt{5}\right)$
59. $\left(-\frac{3}{2}, 0\right)$ 61. $(-\infty, -1) \cup (0, \infty)$

63. a. $(-\infty, -1] \cup [5, \infty)$
b. $(-\infty, -1) \cup (5, \infty)$
65. $(-\infty, -2) \cup [2, \infty)$
67. $\left(-\infty, 0\right) \cup \left(\frac{1}{2}, \infty\right)$ **69.** $\left(\frac{1}{5}, \frac{4}{5}\right)$
71. $(0, 2)$
73. a.

x	$x^2 + 1000$	$2x^2 + x$	$\dfrac{x^2 + 1000}{2x^2 + x}$
1	1001	3	333.67
2	1004	10	100.4
5	1025	55	18.64
10	1100	210	5.24
100	11,000	20,100	0.55
200	41,000	80,200	0.51
10,000	100,001,000	200,010,000	0.50

b. $\left(-\frac{1}{2}, 0\right) \cup (2000, \infty)$ **c.** 2001
75. $\left(-\infty, \dfrac{3a + 5b}{8}\right)$

Chapter Two Review Exercises

1. T   **3.** F   **5.** F   **7.** T
9. F   **11.** F   **13.** $\frac{1}{3}$ **15.** $-\frac{37}{11}$
17. $-\frac{11}{3}$ **19.** 8   **21.** -2 and -6
23. $\frac{1}{3}$ and $-\frac{1}{2}$ **25.** -6 and 4
27. 10 and -1 **29.** $\frac{2}{3}$ and 3
31. $\frac{1}{2}\left(-1 \pm \sqrt{3}\right)$ **33.** $\pm\sqrt{3}$
35. 0 and $\pm\sqrt{1 + \sqrt{3}}$
37. -6 and -8 **39.** 6561 and 256
41. -7 **43.** $\frac{1}{4}$ **45.** 16   **47.** 10
49. 2   **51.** $x = a + b$ **53.** $x = a$
55. $x = 1/2y$ **57.** $x = (a + b)/ab$ and $x = -2/a$ **59.** $x = -a$ and $x = -b$
61. $x = \left(\dfrac{a - b}{a + b}\right)^2$ and $x = \left(\dfrac{a + b}{a - b}\right)^2$
63. $(-2, 1)$ **65.** $\left[-\frac{1}{2}, \frac{1}{2}\right]$
67. $\left(-\frac{41}{10}, -\frac{39}{10}\right)$
69. $(-\infty, -2] \cup [3, \infty)$ **71.** $(-8, 5)$
73. $\left(3 - \sqrt{10}, 3 + \sqrt{10}\right)$
75. $(-5, -3) \cup (3, 5)$ **77.** $(-2, \infty)$
79. $(-\infty, -1) \cup [1, 9]$
81. $\left(-\infty, -\frac{1}{2}\right) \cup \left[\frac{1}{6}, \infty\right)$ **83.** $(0, 25]$
85. $\left[-\frac{1}{2}, \frac{1}{2}\right]$
87. $\left(-\infty, 3 - 2\sqrt{2}\right] \cup \left[3 + 2\sqrt{2}, \infty\right)$

89. 47   **91.** $-1 + \sqrt{2}$ cm
93. $(16\text{ cm}, \infty)$ **95.** 21, 22, and 23
97. 7.5 mph, 1 hr 17 min
99. $\frac{1}{4}\left[-(a + b) + \sqrt{a^2 + 6ab + b^2}\right]$
101. the ball thrown from a height of 100 ft   **103.** $\sqrt{2}/2$

Chapter Two Test

1. 5   **2.** $\frac{14}{11}$ and -1
3. a. -5 and 1 **b.** $-2 \pm \sqrt{5}$
4. ± 2 and $\pm\sqrt{3}$ **5.** ± 1 and $\pm\frac{27}{8}$
6. 2   **7.** 1 and $-\frac{1}{3}$
8. $x = (de - b)/(a - ce)$ **9.** $x = a + b$
10. $(-\infty, 3]$ **11.** $\left(\frac{27}{10}, \frac{31}{10}\right)$
12. $\left[\frac{7}{3}, 3\right]$ **13.** $[-8, \infty)$
14. $\left(-2, \dfrac{-3 - \sqrt{3}}{3}\right] \cup \left(-1, \dfrac{-3 + \sqrt{3}}{3}\right] \cup (0, \infty)$
15. 168 lb of Kona beans and 112 lb of Columbian beans   **16.** 12:20 P.M.
17. $\left[-\frac{3}{2}, \frac{3}{2}\right]$ **18. a.** $S = \frac{3}{2}\pi h^2$
b. $h = \sqrt{2S/3\pi}$

CHAPTER THREE

Exercise Set 3.1

1.

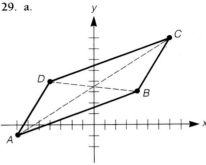

3. a.

b. 6 square units   **5. a.** 5 **b.** 13
7. a. 10 **b.** 9   **9.** 4   **11. a.** $\left(4, \frac{1}{2}\right)$
b. $(-6, 7)$ **13. a.** yes **b.** yes **c.** no
15. 0; the three points are collinear.
17. a. $(6, 5)$ **b.** $\left(\frac{1}{2}, -\frac{3}{2}\right)$ **c.** $(1, -4)$
19. $(11, -8)$ **21.** $10x - 8y + 11 = 0$
23. $x^2 + y^2 = 25$
25. $3x^2 + 10x + 3y^2 + 3 = 0$
27. $(x + 6)^2 + (y - 5)^2 = 1$
29. a.

b. Both are $\left(0, \frac{7}{2}\right)$. **c.** The diagonals bisect each other.   **31. a.** 128 **b.** 96
33. $a = \sqrt{2}$, $b = \sqrt{3}$, $c = 2$, $d = \sqrt{5}$, $e = \sqrt{6}$, $f = \sqrt{7}$, $g = 2\sqrt{2}$
39. a. $5x^2 + 9y^2 = 45$
b. $5x^2 + 4xy + 8y^2 = 36$
41. 7 and -3 **43.** No, the closest point is 2 units away.

Exercise Set 3.2

1. no   **3.** no   **5.** yes
7. a.

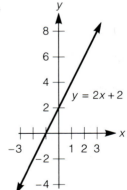

b. $\left(\frac{7}{5}, \frac{24}{5}\right)$ and $(-3, -4)$

9. a.

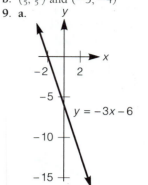

$y = -3x - 6$

b. $\left(-\frac{6}{5}, -\frac{12}{5}\right)$ **11.** x-intercept: 4; y-intercept: 3

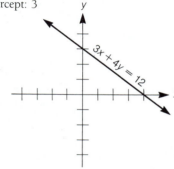

$3x + 4y = 12$

13. x-intercept: 2; y-intercept: -4

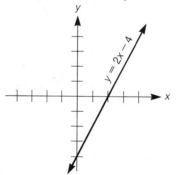

$y = 2x - 4$

15. x-intercept: 1; y-intercept: 1

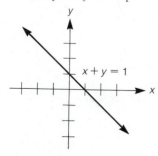

$x + y = 1$

17. a. x-intercepts: $-1, -2$; y-intercept: 2 **b.** no x-intercept; y-intercept: 3

19. a. x-intercepts: $\frac{1}{2}\left(-1 \pm \sqrt{5}\right)$; y-intercept: -1 **b.** no x-intercept; y-intercept: 1

21. x-intercept: 2; y-intercept: 3

23. x-intercept: $\frac{10}{3}$; y-intercept: -2

25. x-intercept: 2; y-intercept: -8

27. x-intercept: 5; no y-intercept

29. x-intercepts: 0, $-1 + 2\sqrt{3} \approx 2.46$, $-1 - 2\sqrt{3} \approx -4.46$; y-intercept: 0

31. x-intercepts: $\pm\sqrt{3} \approx \pm 1.73$; y-intercept: -3

33. a.

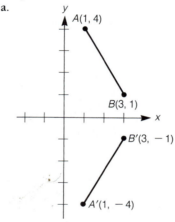

$A(1, 4)$
$B(3, 1)$
$B'(3, -1)$
$A'(1, -4)$

b.

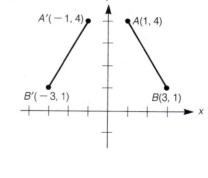

$A'(-1, 4)$
$A(1, 4)$
$B'(-3, 1)$
$B(3, 1)$

c.

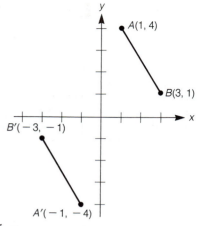

$A(1, 4)$
$B(3, 1)$
$B'(-3, -1)$
$A'(-1, -4)$

35. a.

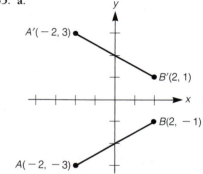

$A'(-2, 3)$
$B'(2, 1)$
$A(-2, -3)$
$B(2, -1)$

b.

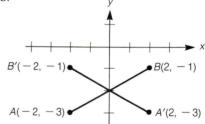

$B'(-2, -1)$
$B(2, -1)$
$A(-2, -3)$
$A'(2, -3)$

c.

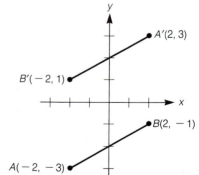

$A'(2, 3)$
$B'(-2, 1)$
$B(2, -1)$
$A(-2, -3)$

37. a. x-axis: no; y-axis: yes; origin: no
b. x-axis: yes; y-axis: no; origin: no
39. a. x-axis: no; y-axis: yes; origin: no
b. x-axis: no; y-axis: no; origin: no
c. x-axis: yes; y-axis: no; origin: no
d. x-axis: no; y-axis: no; origin: yes
41. x-axis: yes; y-axis: yes; origin: yes
43. x-axis: no; y-axis: no; origin: no
45. x-axis: no; y-axis: no; origin: no
47. x- and y-intercept: 0; symmetry: y-axis

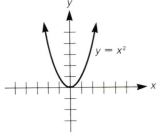

49. No x- or y-intercept; symmetry: origin

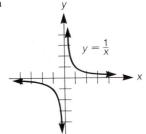

51. x- and y-intercept: 0; symmetry: y-axis

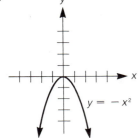

53. No x- or y-intercept; symmetry: origin

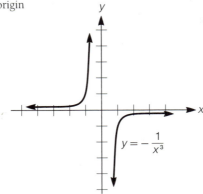

55. x- and y-intercept: 0; symmetry: y-axis

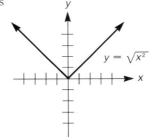

57. x- and y-intercept: 1; symmetry: none

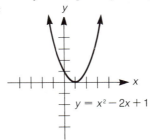

59. x-intercept: 2; no y-intercept; symmetry: x-axis

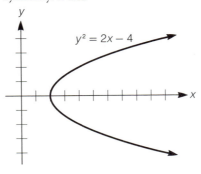

61. x-intercepts; $\frac{1}{4}\left(-1 \pm \sqrt{33}\right)$; y-intercept: -4; symmetry: none

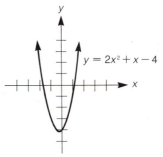

63. a. x-intercept: 2; y-intercept: -6; symmetry: none

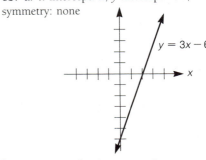

b. x-intercept: 2; y-intercept: 6; symmetry: none

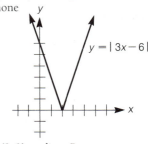

65. center: $(3, 1)$; radius: 5; x-intercepts: $3 \pm 2\sqrt{6}$; y-intercepts: 5, -3 **67.** center: $(0, 0)$; radius: $\sqrt[4]{2}$; x-intercepts: $\pm\sqrt[4]{2}$; y-intercepts: $\pm\sqrt[4]{2}$ **69.** center: $(-4, 3)$; radius: 1; no x- or y-intercepts
71. center: $\left(-3, \frac{1}{3}\right)$; radius: $\sqrt{2}$; x-intercepts: $-3 \pm \frac{1}{3}\sqrt{17}$; no y-intercepts   **73.** area = πc^2; circumference = $2\pi c$
75. $\left(x + \frac{1}{2}\right)^2 + (y - 2)^2 = \frac{169}{4}$
77. $(x + 3)^2 + (y - 2)^2 = 64$

79. a.

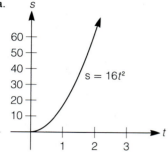

b. $0 \leq s \leq 16$

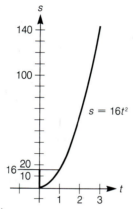

c. $16 \leq s \leq 64$

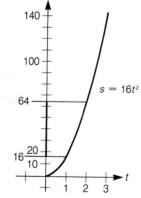

81.

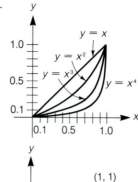

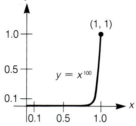

83. a. 500 bacteria **b.** 1.5 hours
c. $t = 3.5$ hours **d.** between $t = 3$ and
$t = 4$ **85.** x-intercepts: $0, \pm 3\sqrt{3}$;
y-intercept: 0; symmetry: origin

x	0	0.5	1	1.5	2	2.5	3
y	0	-13.4	-26	-37.1	-46	-51.9	-54

x	3.5	4	4.5	5	5.5	6
y	-51.6	-44	-30.4	-10	17.9	54

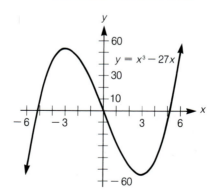

1. a. -2 **b.** 3 **c.** $-\frac{7}{3}$ **d.** 3
3. a. 1

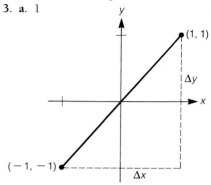

b. 0

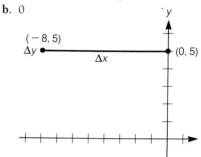

c. -1

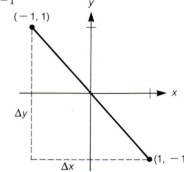

d. -1

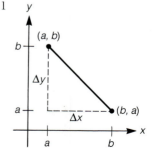

A-44

5. $m_3 < m_2 < m_4 < m_1$ **7.** no
9. yes   **11. a.** $y = -5x - 9$
b. $y = 4x - 20$ **c.** $y = \frac{1}{3}x + \frac{4}{3}$
d. $y = -x + 1$ **13. a.** $y = 2x$
b. $y = -2x - 4$ **c.** $y = \frac{1}{7}x - \frac{11}{7}$
15. $x = -3$ **17.** $y = 4$ **19.** y-axis
21. a. $y = -4x + 7$ **b.** $y = 2x + \frac{3}{2}$
c. $y = -\frac{4}{3}x + 14$ **23. a.** $y = 4x + 11$

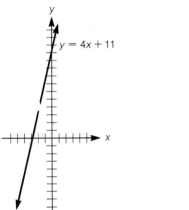

b. $y = \frac{1}{2}x - \frac{5}{4}$

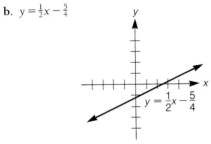

c. $y = -\frac{5}{6}x + 5$

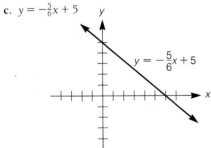

d. $y = \frac{3}{4}x + \frac{3}{2}$

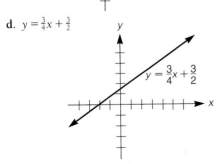

e. $y = 4x - 2$

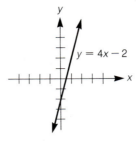

25. $y = x + 2$

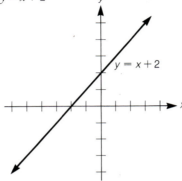

27. $y - 4 = 0$

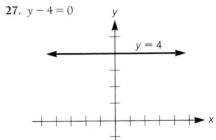

29. a. x-intercept: 5; y-intercept: 3;
area: $\frac{15}{2}$; perimeter: $8 + \sqrt{34}$
b. x-intercept: 5; y-intercept: -3;
area: $\frac{15}{2}$; perimeter: $8 + \sqrt{34}$
31. a. neither **b.** parallel
c. perpendicular **d.** perpendicular
e. neither **f.** parallel
33. $y = \frac{2}{5}x + \frac{12}{5}$, $2x - 5y + 12 = 0$
35. $y = -\frac{4}{3}x + \frac{16}{3}$, $4x + 3y - 16 = 0$
37. $y = \frac{3}{5}x + 11$; $3x - 5y + 55 = 0$

39. a.

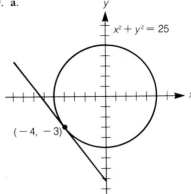

b. $y = -\frac{4}{3}x - \frac{25}{3}$
47.

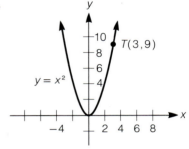

x	2.5	2.9	2.99	2.999	2.9999
y	6.25	8.41	8.9401	8.994001	8.99940001
Δx	0.5	0.1	0.01	0.001	0.0001
Δy	2.75	0.59	0.0599	0.005999	0.00059999
m	5.5	5.9	5.99	5.999	5.9999

The slope of PT approaches 6, so the
tangent slope is 6.
49. $\left(-\frac{1}{2}, -\frac{1}{8}\right)$ **51.** $\left(8, \frac{1}{8}\right)$
53. 44.1 square units
55. a.

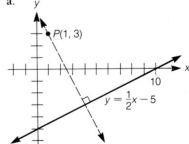

b. $y = -2x + 5$ **c.** $(4, -3)$ **d.** $3\sqrt{5}$
59. a. $x = 410 - \frac{2}{3}p$ **b.** 230 units
c. \$307.50 **63. a.** $y = -mx - b$
b. $y = -mx + b$ **c.** $y = mx - b$
65. a. $3x^2 + 3y^2 + 3x - 11y = 0$
b. $3x^2 + 3y^2 + 3x - 11y = 0$ **c.** $\left(0, \frac{4}{3}\right)$
h. $\left(-\frac{2}{3}, 2\right)$

Chapter Three Review Exercises

1. $y = -2x - 6$ **3.** $y = \frac{1}{4}x - \frac{5}{2}$
5. $y = 2x + 8$ **7.** $y = -2$
9. $y = x + 1$ **11.** $9x - 4y + 14 = 0$
13. $2x + y = 0$ **15.** $3x - 4y = 0$
17. $x + y - 1 = 0$ or $x - 2y - 4 = 0$
19. 13 **21.** $(15, 6)$
23. x-axis: no; y-axis: yes; origin: no
25. x-axis: no; y-axis: no; origin: yes
27. x-axis: no; y-axis: no; origin: yes
29. x-axis: no; y-axis: no, origin: yes
31. x-axis: no; y-axis: no; origin: yes
33. x-axis: yes; y-axis: no; origin: no
35. x-axis: yes; y-axis: yes; origin: yes
37. x-axis: yes; y-axis: no; origin: no
39. x-axis: yes; y-axis: no; origin: no
41. $\left(-\frac{27}{13}, \frac{31}{13}\right)$ and $(3, -1)$
43. $(-2, 4)$ and $(1, 1)$
45.

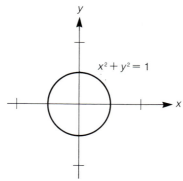

47.

49.

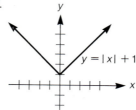

51.

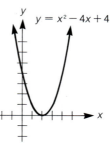

53.

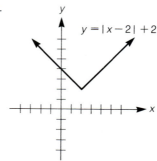

55.

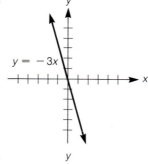

57.

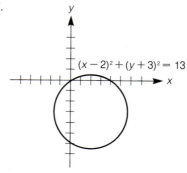

59.

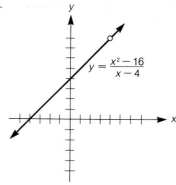

61.

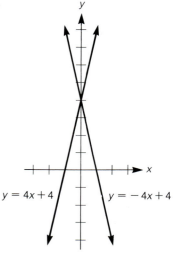

63. 19 **65.** $m = x^2 - 2x + 4$
67. $MA = MB = MC = \sqrt{b^2 + c^2}$; they are
all equal. **69. a.** $\sqrt{130}$ **b.** $-\frac{11}{3}$
c. $\left(\frac{7}{2}, -\frac{1}{2}\right)$ **71. a.** 2 **b.** $\sqrt{3}/3$
c. $(0, 0)$ **73.** $\frac{49}{6}$ **75. a.** They all
equal $\left(\frac{5}{3}, \frac{11}{3}\right)$. **b.** They all equal
$\left(\frac{2}{3}a + \frac{2}{3}b, \frac{2}{3}c\right)$. The medians of the triangle
intersect at a point that is $\frac{2}{3}$ of the
distance from each vertex to the midpoint
on the opposite side. **77.** $6\sqrt{2}$
81. $13\sqrt{5}/5$ **83.** $17\sqrt{13}/13$

1. no symmetry

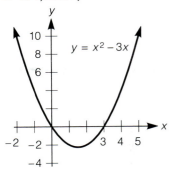

3. no symmetry

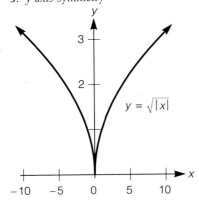

5. y-axis symmetry

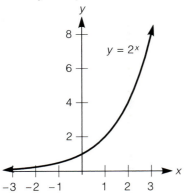

7. origin symmetry

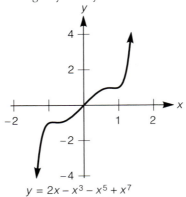

$$y = 2x - x^3 - x^5 + x^7$$

9.

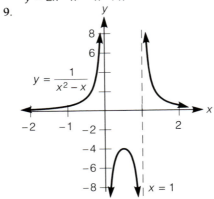

11. a. origin symmetry

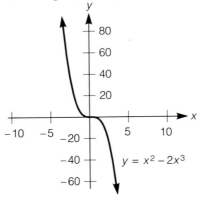

b.

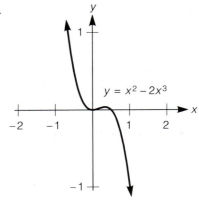

c. no symmetry

13.

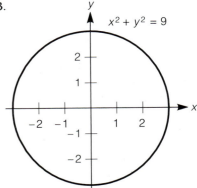

15.

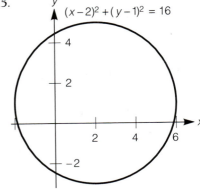

17. a.

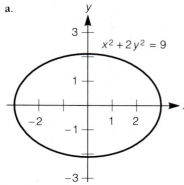

$x^2 + 2y^2 = 9$

b. $(-3, 0)$ and $(3, 0)$

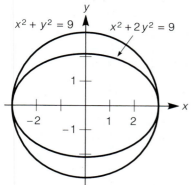

$x^2 + y^2 = 9$ $x^2 + 2y^2 = 9$

19. a.

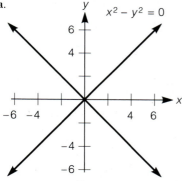

$x^2 - y^2 = 0$

b.

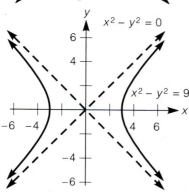

$x^2 - y^2 = 0$

$x^2 - y^2 = 9$

21. c.

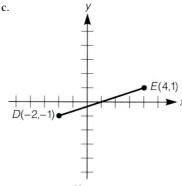

$E(4,1)$

$D(-2,-1)$

d.

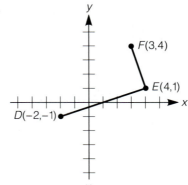

$F(3,4)$

$E(4,1)$

$D(-2,-1)$

e.

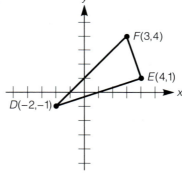

$F(3,4)$

$E(4,1)$

$D(-2,-1)$

23. $y = -x$, $y = -4x$, $y = -8x$

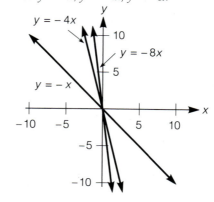

$y = -4x$

$y = -8x$

$y = -x$

25.

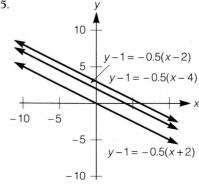

$y - 1 = -0.5(x - 2)$

$y - 1 = -0.5(x - 4)$

$y - 1 = -0.5(x + 2)$

27. a. x-intercept: 2; y-intercept: 3

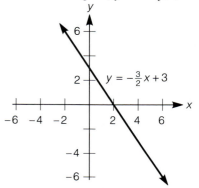

$y = -\frac{3}{2}x + 3$

b. x-intercept: -2; y-intercept: -3

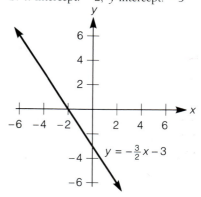

$y = -\frac{3}{2}x - 3$

c. x-intercept: 6; y-intercept: 5

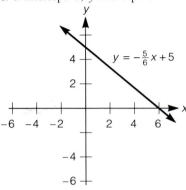

d. x-intercept: −6; y-intercept: −5

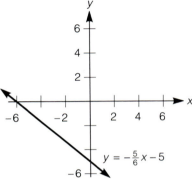

e. It is a line with x-intercept a and y-intercept b.

Chapter Three Test

1. $y = 5x - 7$ **2.** $y = \frac{6}{5}x - \frac{17}{5}$
3. $y = -3x$ **4.** $y = \frac{4}{5}x$ **5.** $(3, 9)$
6. a. x-axis: no; y-axis: no; origin: yes
b. x-axis: no; y-axis: yes; origin: no
c. x-axis: yes; y-axis: no; origin: no
7. x-intercept: 5; y-intercept: −3

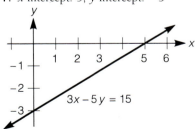

8. x-intercepts: $\frac{1}{2}(5 \pm \sqrt{59})$;
y-intercepts: $\frac{1}{2}(-3 \pm \sqrt{43})$

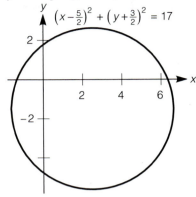

9. no x-intercept; y-intercept: 3

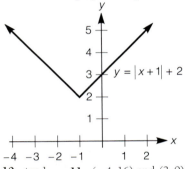

10. $t = 1$ **11.** $(-4, 16)$ and $(3, 9)$
12. yes **13.** $5x + 7y = 31$ **14.** $\frac{121}{5}$

CHAPTER FOUR

Exercise Set 4.1

1. a. $(-\infty, \infty)$ **b.** $\left(-\infty, \frac{1}{5}\right]$
c. $\left(-\infty, \frac{1}{5}\right) \cup \left(\frac{1}{5}, \infty\right)$
3. a. $(-\infty, 4) \cup (4, \infty)$
b. $(-\infty, -2) \cup (-2, 2) \cup (2, \infty)$
c. $(-\infty, -2) \cup (2, \infty)$
5. a. $(-\infty, -1] \cup [5, \infty)$
b. $(-\infty, -1) \cup (5, \infty)$ **c.** $(-\infty, \infty)$
7. a. all real numbers except 1
b. all real numbers except 1
9. a. $[4, \infty)$ **b.** $(-\infty, \infty)$
11. $f, g, F,$ and H are functions
13. a. range $f = \{1, 2, 3\}$,
range $g = \{2, 3\}$, range $F = \{1\}$,
range $H = \{1, 2\}$ **b.** range $g = \{i, j\}$,
range $F = \{i, j\}$, range $G = \{k\}$
15. a. $y = (x - 3)^2$ **b.** $y = x^2 - 3$
c. $y = (3x)^2$ **d.** $y = 3x^2$ **17. a.** -1

b. 1 **c.** 5 **d.** $-\frac{5}{4}$ **e.** $z^2 - 3z + 1$
f. $x^2 - x - 1$ **g.** $a^2 - a - 1$
h. $x^2 + 3x + 1$ **i.** 1 **j.** $4 - 3\sqrt{3}$
k. $1 - \sqrt{2}$ **l.** 2 **19. a.** $12x^2$ **b.** $6x^2$
c. $3x^4$ **d.** $9x^4$ **e.** $\frac{3}{4}x^2$ **f.** $\frac{3}{2}x^2$
21. a. 1 **b.** -7 **c.** -3 **d.** $-\frac{7}{18}$
e. $-2x^2 - 4x - 1$ **f.** $1 - 2x^2 - 4xh - 2h^2$
g. $-4xh - 2h^2$ **h.** $-4x - 2h$
23. a. domain: all real numbers except
2; range: all real numbers except 2 **b.** $\frac{1}{2}$
c. 0 **d.** 1 **e.** $\dfrac{2x^2 - 1}{x^2 - 2}$ **f.** $\dfrac{2 - x}{1 - 2x}$
g. $\dfrac{2a - 1}{a - 2}$ **h.** $\dfrac{2x - 3}{x - 3}$
25. a. $d(1) = 80; D\left(\frac{3}{2}\right) = 108$;
$d(2) = 128; d(t_0) = -16t_0^2 + 96t_0$ **b.** $0, 6$
c. $\frac{1}{4}(12 \pm \sqrt{143})$ **27.** $g(3) = 1$;
$g(x + 4) = |x|$ **29. a.** $2x + h$
b. $4x + 2h - 3$ **c.** $3x^2 + 3xh + h^2$
31. a. $-1/(x - 1)(a - 1)$ **b.** $-1/2(x - 1)$
c. $-1/[(x - 1)(x + h - 1)]$
d. $-1/2(2 + h)$ **33. a.** $\frac{11}{5}$ **b.** ± 2
c. $0, 1$ **d.** $\frac{3}{2}, -1$ **35. a.** \$125.51
b. \$363.76
37. a.

n	2	3	4	5
$g(n)$	1.4142	1.4422	1.4142	1.3797

n	6	7	8
$g(n)$	1.3480	1.3205	1.2968

b. 15 **39. a.** $f(a) = 0; f(2a) = \frac{1}{3}$;
$f(3a) = \frac{1}{2}$; no **43.** $a = \frac{1}{2}, b = -\frac{3}{2}$
49. $-\frac{1}{3}$ **51. a.** 0 **b.** 1 **c.** 2 **d.** 6
e. -1 **f.** -2 **g.** -6 **h.** $\frac{1}{2}$ **53.** 0
55. a. x **b.** $\frac{22}{7}$ **57.** G is not a
function, because it assigns more than
one value to $x = 4$. Let $G(x)$ be the rule
that assigns to each positive integer the
nearest prime less than or equal to x.
59. $P(1) = 17; P(2) = 19; P(3) = 23$;
$P(4) = 29$; yes: $x = 17$ (other answers are
possible)

Exercise Set 4.2

1. domain: $[-4, 2]$; range: $[-3, 3]$
3. domain: $[-4, -1) \cup (-1, 4]$;
range $[-2, 3)$ **5.** domain: $[-4, 4]$;
range: $[-2, 2)$ **7.** domain: $[-4, 3]$;

range: {2}   **9. a.** 1 **b.** −3 **c.** no
d. $x = 2$ **e.** −1   **11. a.** positive
b. $f(-2) = 4, f(1) = 1, f(2) = 2, f(3) = 0$
c. $f(2)$ **d.** −3 **e.** 3 **f.** domain: $[-2, 4]$;
range: $[-2, 4]$ **13. a.** $g(-2)$ **b.** 5
c. $f(2) - g(2)$ **d.** −2, 3 **e.** f only
15. range f: $[0, 4]$; range g: $[-3, 3]$
17.

Function	$|x|$	x^2	x^3
Turning Point	$(0, 0)$	$(0, 0)$	none
Maximum Value	none	none	none
Minimum Value	0	0	none
Interval(s) where Increasing	$[0, \infty)$	$[0, \infty)$	$(-\infty, \infty)$
Interval(s) where Decreasing	$(-\infty, 0]$	$(-\infty, 0]$	none

19. a. $[-1, 1]$ **b.** 1 **c.** −1
d. $[0, 1]$ and $[3, 4]$ **e.** $[1, 3]$
21. a. $[-3, 0]$ **b.** 0 **c.** −3 **d.** $[2, 4]$
e. $[0, 2]$ **23. a.** function **b.** not a
function **c.** not a function **d.** function
25. 7

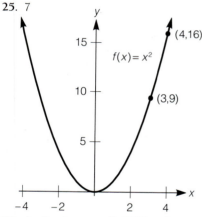

27. the line between $(1, T(1))$ and
$(4, T(4))$
29.

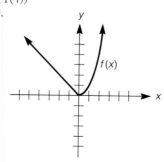

31.

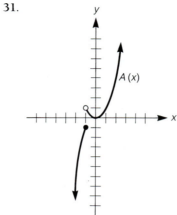

33.

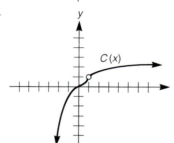

35. a.

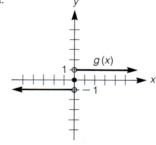

b.

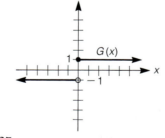

37.

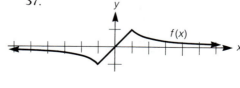

39. $t = 5$

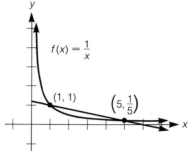

41. $(x + h, f(x + h))$
43. a.

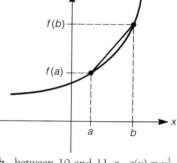

b. between 10 and 11 **c.** $g(x) = x^3$
d. (i) 2.1; (ii) 2.01; (iii) 2.001;
target value = 2
45. a. 48

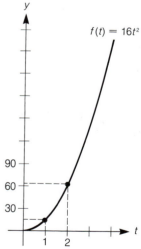

b. 80

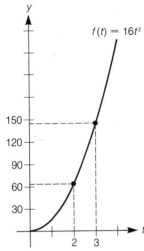

$f(t) = 16t^2$

c. slope = 64 ft/sec

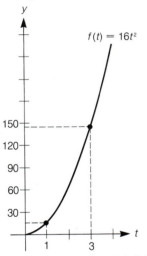

$f(t) = 16t^2$

d. (i) 33.6; (ii) 32.016; (iii) 32.00016; target value = 32

Exercise Set 4.3

1. a. C; **b.** F; **c.** I; **d.** A; **e.** J; **f.** K; **g.** D; **h.** B; **i.** E; **j.** H; **k.** G

3.

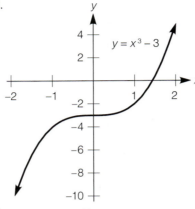

$y = x^3 - 3$

5.

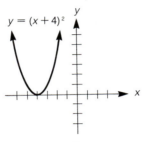

$y = (x+4)^2$

7.

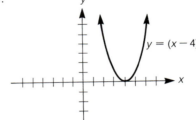

$y = (x-4)^2$

9.

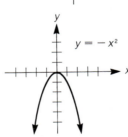

$y = -x^2$

11.

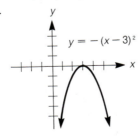

$y = -(x-3)^2$

13.

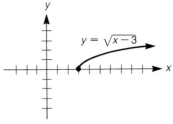

$y = \sqrt{x-3}$

15.

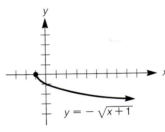

$y = -\sqrt{x+1}$

17.

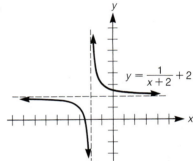

$y = \dfrac{1}{x+2} + 2$

19.

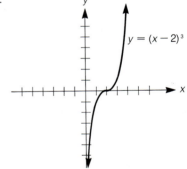

$y = (x-2)^3$

21.

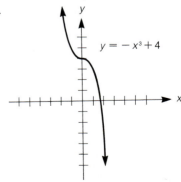

$y = -x^3 + 4$

23. a.

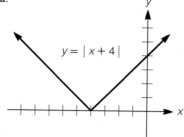

$y = |x + 4|$

b.

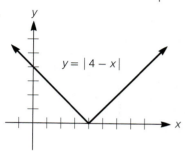

$y = |4 - x|$

c.

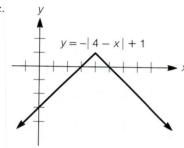

$y = -|4 - x| + 1$

25.

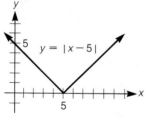

$y = |x - 5|$

27.

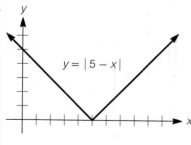

$y = |5 - x|$

29.

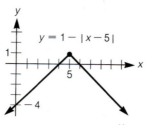

$y = 1 - |x - 5|$

31.

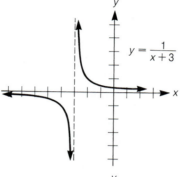

$y = \dfrac{1}{x + 3}$

33.

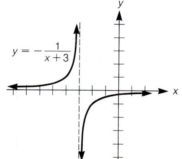

$y = -\dfrac{1}{x + 3}$

35.

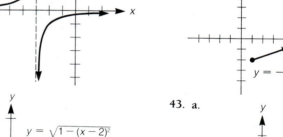

$y = \sqrt{1 - (x - 2)^2}$

37.

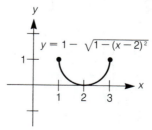

$y = 1 - \sqrt{1 - (x - 2)^2}$

39.

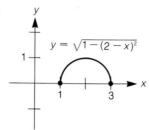

$y = \sqrt{1 - (2 - x)^2}$

41. a.

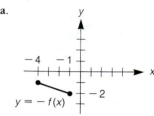

$y = -f(x)$

b.

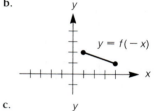

$y = f(-x)$

c.

$y = -f(-x)$

43. a.

$y = g(-x)$

b.

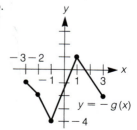

$y = -g(x)$

c.

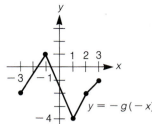

$y = -g(-x)$

45. a.

x	x^2	$x^2 - 1$	$x^2 + 1$
0	0	-1	1
± 1	1	0	2
± 2	4	3	5
± 3	9	8	10

b. They are translations of ± 1 unit.

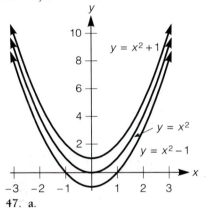

$y = x^2 + 1$
$y = x^2$
$y = x^2 - 1$

47. a.

x	$\sqrt{x}$	$-\sqrt{x}$
0	0.0	0.0
1	1.0	-1.0
2	1.4	-1.4
3	1.7	-1.7
4	2.0	-2.0
5	2.2	-2.2

b. They are reflections across the x-axis.

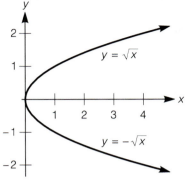

$y = \sqrt{x}$

$y = -\sqrt{x}$

49. a. $(a + 3, b)$; **b.** $(a, b - 3)$;
c. $(a + 3, b - 3)$; **d.** $(a, -b)$; **e.** $(-a, b)$;
f. $(-a, -b)$; **g.** $(-a + 3, b)$;
h. $(-a + 3, -b + 1)$

51.

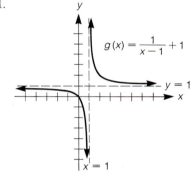

$g(x) = \dfrac{1}{x - 1} + 1$

$y = 1$

$x = 1$

53. $[1, 2)$
57. a. $(-\infty, -4) \cup (-2, \infty)$
b. $(-\infty, 2) \cup (4, \infty)$

Exercise Set 4.4

1. a. $x^2 - x - 7$ **b.** $-x^2 + 5x + 5$ **c.** 5
3. a. $x^2 - 2x - 8$ **b.** $-x^2 + 2x + 8$
5. a. $4x - 2$ **b.** $4x - 2$ **c.** -4
7. a. $\dfrac{-x^4 + 22x^2 - 4x - 80}{2x^3 - x^2 - 18x + 9}$ **b.** $-\dfrac{80}{9}$
9. a. $-x^5 + 9x^3 + 2x^2 - 18$
b. $-x^5 + 9x^3 + 2x^2 - 18$ **c.** -24
11. a. $-6x - 14$ **b.** -74 **c.** $-6x - 7$
d. -67 **13. a.** $(f \circ g)(x) = -x$;
$(f \circ g)(-2) = 2$; $(g \circ f)(x) = 2 - x$;
$(g \circ f)(-2) = 4$
b. $(f \circ g)(x) = 9x^2 - 3x - 6$;
$(f \circ g)(-2) = 36$;
$(g \circ f)(x) = -3x^2 + 9x + 14$;
$(g \circ f)(-2) = -16$

c. $(f \circ g)(x) = \frac{1}{3}(1 - x^4)$;
$(f \circ g)(-2) = -5$; $(g \circ f)(x) = 1 - \frac{1}{81}x^4$;
$(g \circ f)(-2) = \frac{65}{81}$ **d.** $(f \circ g)(x) = 2^{x^2 + 1}$;
$(f \circ g)(-2) = 32$; $(g \circ f)(x) = 2^{2x} + 1$;
$(g \circ f)(-2) = \frac{17}{16}$ **e.** $(f \circ g)(x) = 3x^5 - 4x^2$;
$(f \circ g)(-2) = -112$; $(g \circ f)(x) = 3x^5 - 4x^2$;
$(g \circ f)(-21) = -112$
f. $(f \circ g)(x) = x$; $(f \circ g)(-2) = -2$;
$(g \circ f)(x) = x$; $(g \circ f)(-2) = -2$
15. a. $(-x + 7)/6x$ **b.** $(-t + 7)/6t$ **c.** $\frac{5}{12}$
d. $(1 - 6x)/7$ **e.** $(1 - 6y)/7$ **f.** $-\frac{11}{7}$
17. a. $M(7) = \frac{13}{5}$, $M[M(7)] = 7$ **b.** x
c. 7 **19. a.** 1 **b.** -3 **c.** -1 **d.** 2
e. 2 **f.** -3
21. a. Both are $4x^3 - 3x^2 + 6x - 1$.
b. Both are $ax^2 + bx + c$.
c. $(f \circ I)(x) = (I \circ f)(x) = f(x)$ for *any*
function $f(x)$.
23.

x	0	1	2	3	4
$(f \circ g)(x)$	1	3	2	undef.	2

x	-1	0	1	2	3	4
$(g \circ f)(x)$	0	0	3	4	2	undef.

25. a. $(f \circ g)(x) = 6x - 7$

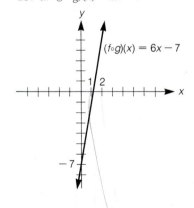

$(f \circ g)(x) = 6x - 7$

b. $(g \circ f)(x) = 6x - 1$

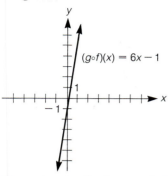

27. a. domain: $[0, \infty)$; range: $[-3, \infty)$

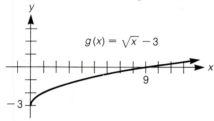

b. domain: $(-\infty, \infty)$; range $(-\infty, \infty)$

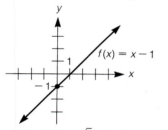

c. $(f \circ g)(x) = \sqrt{x} - 4$; domain: $[0, \infty)$; range: $[-4, \infty)$

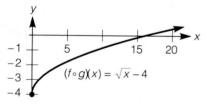

d. $g[f(x)] = \sqrt{x-1} - 3$; domain: $[1, \infty)$
e.

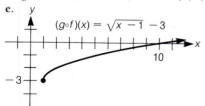

---

29. $(f \circ g)(x)$, where $f(x) = x^4$ and $g(x) = 3x - 1$ **31. a.** $(f \circ g)(x)$, where $f(x) = \sqrt[3]{x}$ and $g(x) = 3x + 4$
b. $(f \circ g)(x)$, where $f(x) = |x|$ and $g(x) = 2x - 3$ **c.** $(f \circ g)(x)$, where $f(x) = x^5$ and $g(x) = ax + b$ **d.** $(f \circ g)(x)$, where $f(x) = 1/x$ and $g(x) = \sqrt{x}$
33. a. $(b \circ c)(x)$ **b.** $(a \circ d)(x)$ **c.** $(c \circ d)(x)$
d. $(c \circ b)(x)$ **e.** $(c \circ a)(x)$ **f.** $(a \circ c)(x)$
g. $(b \circ d)(x)$ or $(d \circ b)(x)$
35. $(C \circ f)(t) = 2\pi/(t^2 + 1)$; $\pi/5$ ft
37. a. $100 + 450t - 25t^2$ **b.** \$1225
c. No, it is \$1900.
39. $f(x) = (x + 6)/4$ **41.** $a = -\frac{1}{2}$, $b = \frac{1}{2}$ **43. a.** $2x + 2a - 2$
b. $4x + 4a - 4$ **45. a.** $\frac{1}{2}x^2 + 1$
b. $\frac{1}{2}(x + 1)^2$ **c.** $\frac{1}{4}x^2 + 1$ **d.** $\frac{1}{4}(x + 1)^2$
e. $\frac{1}{2}(x^2 + 1)$
47. a. $p(x) = (g \circ f \circ h)(x)$
b. $q(x) = (h \circ g \circ f)(x)$
c. $r(x) = (f \circ g \circ h)(x)$
d. $s(x) = (h \circ f \circ g)(x)$
49. a. $(x + 1)/\sqrt{x^2 + 2x + 2}$
b. $10\sqrt{101}/101$
51. a.

$\circ$	i	a	b	c
i	i	a	b	c
a	a	i	c	b
b	b	c	i	a
c	c	b	a	i

b. yes **c.** i, i, i, c **d.** yes **e.** $\{i\}$
f. They both equal i. **g.** yes

Exercise Set 4.5

3. a. 4; **b.** -1; **c.** $\sqrt{2}$; **d.** $t + 1$;
e. $f(0) = 1$; $f^{-1}(1) = 0$;
f. $f(-1) = -2$; $f^{-1}(-2) = -1$
5. a. $f^{-1}(x) = (x + 1)/3$
c.

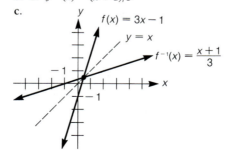

---

7. a. $f^{-1}(x) = x^2 + 1$ for $x \ge 0$
c.

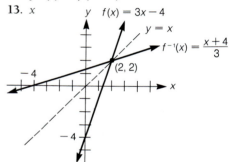

9. a. domain: all real numbers except 3; range: all real numbers except 1
b. $f^{-1}(x) = (3x + 2)/(x - 1)$
c. domain: all real numbers except 1; range: all real numbers except 3; domain f^{-1} = range f, and range f^{-1} = domain f
11. $f^{-1}(x) = \sqrt[3]{(x - 1)/2}$
13. x

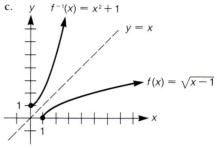

15. a. $f^{-1}(x) = \sqrt[3]{x + 1} + 3$
b.

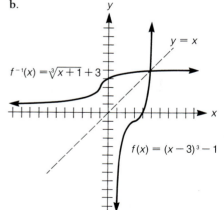

---

17. a.

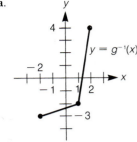

$y = g^{-1}(x)$

b.

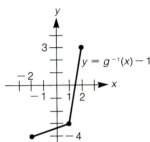

$y = g^{-1}(x) - 1$

c.

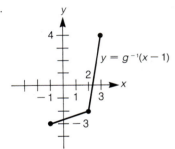

$y = g^{-1}(x - 1)$

d.

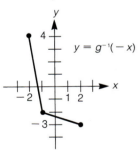

$y = g^{-1}(-x)$

e.

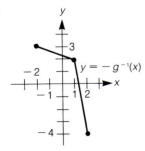

$y = -g^{-1}(x)$

f.

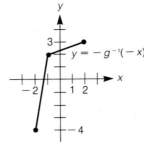

$y = -g^{-1}(-x)$

19. no   **21.** yes   **23.** yes
25. no   **27.** no   **29.** yes
33. a. $x = 7$ **b.** $x = -3$ **35.** $t = \frac{9}{4}$
37. a. $f^{-1}(x) = x^2$, domain: $[0, \infty)$
b. (i) f; (ii) f^{-1}; (iii) f; (iv) f^{-1}; (v) f;
(vi) f^{-1}; (vii) f^{-1}; (viii) f
43. $y = 2x - 6$

Exercise Set 4.6

1. a. $y = kx$ **b.** $A = k/B$
3. a. $x = kuv^2$ **b.** $z = kA^2B^3$
5. a. $F = k/r^2$ **b.** $V^2 = k(U^2 + T^2)$
7. a. -2 **b.** $A = -2/B$ **c.** $-\frac{8}{5}$
9. -24 **11.** six times the original
value   **13.** eight times the original
value   **15. a.** 4π; $S = 4\pi r^2$
b. 12π cm^2 **17.** 48 times the original
force   **19.** 122.5 m
21. a. $k = 2.05$, $V = 2.05/P$
b. 2.05 liters   **23. a.** 6 times the
original value **b.** $\sqrt{2}$ times the original
value   **25.** 110.6 lb   **27. a.** F;
b. C; **c.** D; **d.** A; **e.** B
29. shortened by $\frac{3}{4}L$ **31.** 4.24 ft
35. Mercury: 87.9693 days;
Venus: 224.7007 days;
Earth: 365.2564 days;
Mars: 686.9786 days;
Jupiter: 4336.6159 days;
Saturn: 10826.9994 days;
Uranus: 30873.7244 days;
Neptune: 60300.6863 days;
Pluto: 91814.3739 days   **37. a.** yes;
b. no; **c.** yes

Chapter Four Review Exercises

1. a. $(-\infty, 3]$ **b.** $\left(-\infty, \frac{1}{2}\right) \cup \left(\frac{1}{2}, \infty\right)$
3. yes   **5. a.** $-1/ax$ **b.** $1 - 4x - 2h$
7. a. $g^{-1}(x) = 1/(3x + 5)$

b.

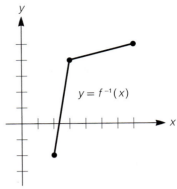

$y = f^{-1}(x)$

9. a. x-intercepts: -5, 1;
y-intercept: -1

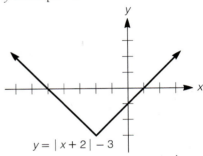

$y = |x + 2| - 3$

b. x-intercept: -1; y-intercept; $-\frac{1}{2}$

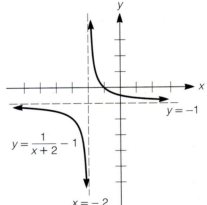

$y = \dfrac{1}{x + 2} - 1$

$y = -1$

$x = -2$

11. a. 5 **b.** $7 - 4\sqrt{2}$
13. $m(h) = 10 + h$

15.

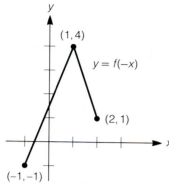

17. a. E; **b.** C; **c.** L; **d.** A; **e.** J; **f.** G;
g. B **h.** M; **i.** K; **j.** D; **k.** I; **l.** H;
m. N; **n.** F

19. x-intercepts: $1 \pm \sqrt{2}$; y-intercept: 1

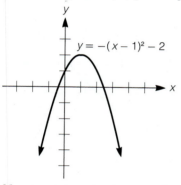

21. no x-intercept; y-intercept: 1

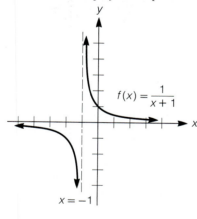

23. x-intercept: -3; y-intercept: 3

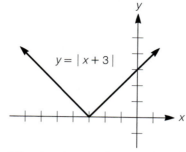

25. x-intercepts: ± 1; y-intercept: 1

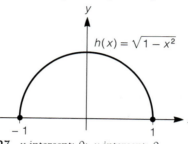

27. x-intercept: 0; y-intercept: 0

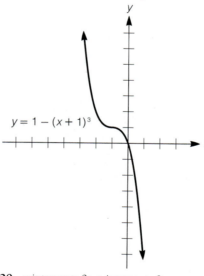

29. x-intercept: 0; y-intercept: 0

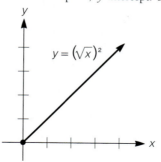

31. x-intercept: 1; no y-intercept

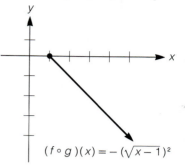

33. x-intercepts: -1, 0; y-intercept: 0

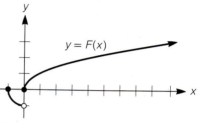

35. no x- or y-intercepts

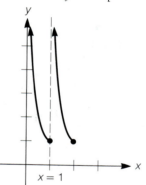

37. x-intercept: $\frac{1}{2}$; y-intercept: -1

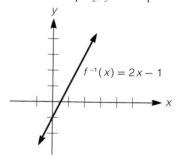

39. x-intercept: 0; y-intercept: 0

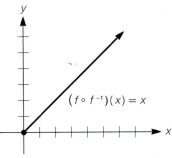

$(f \circ f^{-1})(x) = x$

41. $(-\infty, -3) \cup (-3, 3) \cup (3, \infty)$
43. $(-\infty, 4]$ **45.** $(-\infty, \infty)$
47. $(-\infty, -1] \cup [3, \infty)$
49. $(-\infty, 0) \cup (0, \infty)$
51. $\left(-\infty, \frac{1}{3}\right) \cup \left(\frac{1}{3}, \infty\right)$
53. $(-\infty, 0) \cup (0, \infty)$
55. $(-\infty, 2) \cup (2, \infty)$
57. $a(x) = (f \circ g)(x)$
59. $c(x) = (G \circ g)(x)$
61. $A(x) = (g \circ f \circ G)(x)$
63. $C(x) = (g \circ G \circ G)(x)$ **65.** 12
67. $-\frac{9}{19}$ **69.** $t^2 + t$
71. $x^2 - 5x + 6$ **73.** -4 **75.** 6
77. $x^4 - x^2$ **79.** $-2x^3 + 3x^2 - x$
81. $4x^2 - 2x$ **83.** $-2x^2 + 2x + 1$
85. $(2x + 2)/(2x - 5)$ **87.** $2x + h - 1$
89. $(4x + 3)/(1 - x)$ **91.** x **93.** x
95. $(1 + x)/2$ **97.** $\frac{22}{7}$ **99.** negative
101. -1 **103.** -1 **105.** $(0, -2)$, $(5, 1)$ **107.** $[-6, 0]$, $[5, 8]$
109. 0   **111.** no   **113.** 4
115. a. $x = 10$ **b.** $x = 0$
117. a. 5 **b.** -3 **c.** 4 **d.** $\frac{1}{4}$
119. $(f \circ f)(10)$ **121.** $[0, 4]$
123. 5   **125.** $(1, 3) \cup (6, 10)$
127. $(4, 7)$ **129.** 12 times original value   **131.** $2\sqrt{5} \approx 4.47$ ft
133. $\frac{4}{3}\sqrt{5}\sqrt[4]{27} \approx 6.80$ cm

Chapter Four Graphing-Calculator Exercises

1. a.

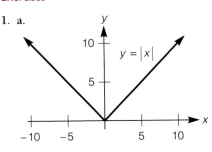

$y = |x|$

b.

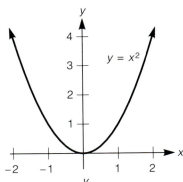

$y = x^2$

c.

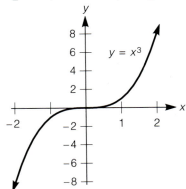

$y = x^3$

d. $(0, 0)$ and $(1, 1)$; x^2

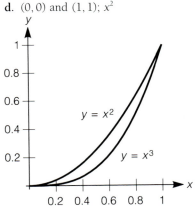

$y = x^2$
$y = x^3$

e. $x^3 > x^2$ on $(1, \infty)$

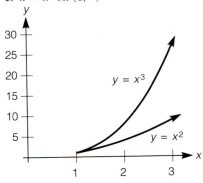

$y = x^3$
$y = x^2$

3. a.

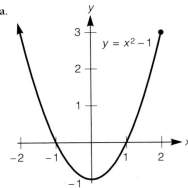

$y = x^2 - 1$

b.

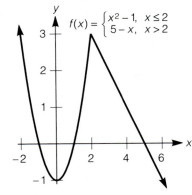

$f(x) = \begin{cases} x^2 - 1, & x \le 2 \\ 5 - x, & x > 2 \end{cases}$

5.

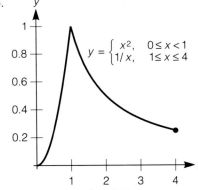

$y = \begin{cases} x^2, & 0 \le x < 1 \\ 1/x, & 1 \le x \le 4 \end{cases}$

7.

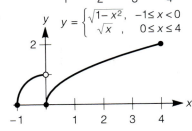

$y = \begin{cases} \sqrt{1 - x^2}, & -1 \le x < 0 \\ \sqrt{x}, & 0 \le x \le 4 \end{cases}$

9.

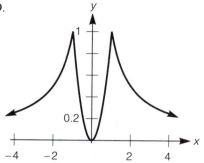

11. a.

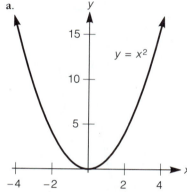

$y = x^2$

b. translation to the left 3 units

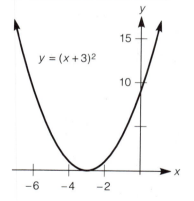

$y = (x + 3)^2$

c. translation to the left 3 units, reflection across the x-axis

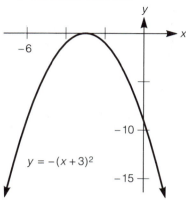

$y = -(x + 3)^2$

d. translation to the left 3 units, reflection across the y-axis

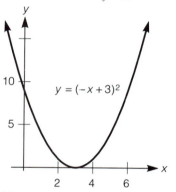

$y = (-x + 3)^2$

13. a.

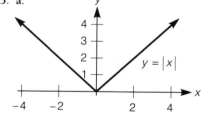

$y = |x|$

b. translation to the left 4 units

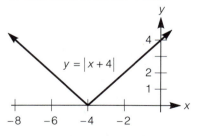

$y = |x + 4|$

c. translation to the left 4 units, reflection across the y-axis

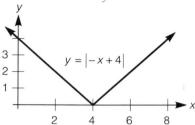

$y = |-x + 4|$

d. translation to the left 4 units, reflection across the y-axis, translation down 2 units

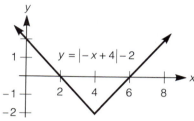

$y = |-x + 4| - 2$

15. a.

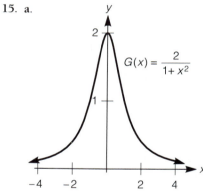

$G(x) = \dfrac{2}{1 + x^2}$

b. translation to the left 3 units

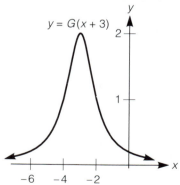

$y = G(x + 3)$

c. translation to the right 3 units

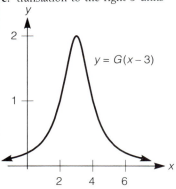

$y = G(x - 3)$

d. reflection across the x-axis, translation up 2 units

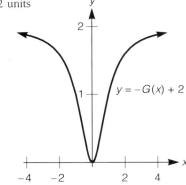

$y = -G(x) + 2$

17.

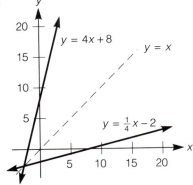

$y = 4x + 8$

$y = x$

$y = \frac{1}{4}x - 2$

19.

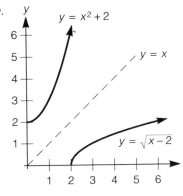

$y = x^2 + 2$

$y = x$

$y = \sqrt{x - 2}$

21.

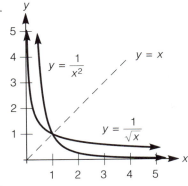

$y = \frac{1}{x^2}$

$y = x$

$y = \frac{1}{\sqrt{x}}$

23. not one-to-one

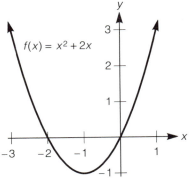

$f(x) = x^2 + 2x$

25. not one-to-one

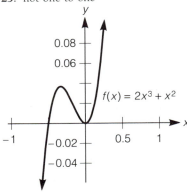

$f(x) = 2x^3 + x^2$

27. a. one-to-one

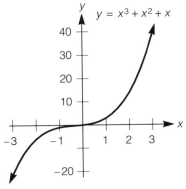

$y = x^3 + x^2 + x$

b. one-to-one

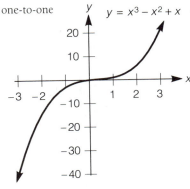

$y = x^3 - x^2 + x$

c. not one-to-one

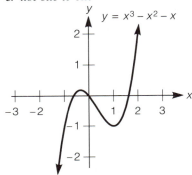

$y = x^3 - x^2 - x$

29. $F^{-1}(x) = -\sqrt{1 - x^2}, \; 0 \le x \le 1$

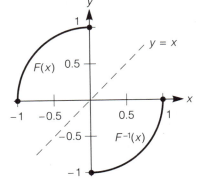

$F(x)$

$y = x$

$F^{-1}(x)$

31. a.

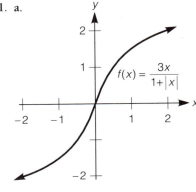

$f(x) = \dfrac{3x}{1 + |x|}$

b. $f^{-1}(x) = \begin{cases} x/(3-x) & \text{if} \quad 0 \le x < 3 \\ x/(3+x) & \text{if} \quad -3 < x < 0 \end{cases}$

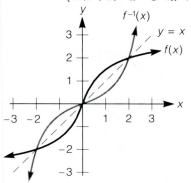

Chapter Four Test

1. $(-\infty, -1] \cup [6, \infty)$
2. $\left(-\infty, \frac{2}{3}\right) \cup \left(\frac{2}{3}, \infty\right)$
3. **a.** $2x^2 - 2x - 2$ **b.** $2x^2 - 5x + 2$
   c. 54   **4.** $-2/at$ **5.** $4x + 2h - 5$
6. $g^{-1}(x) = -x/(6x + 4)$ **7.** $-\frac{8}{5}$
8. **a.** x-intercepts: 4, 2; y-intercept: -2

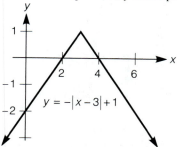

b. x-intercept: $-\frac{5}{2}$; y-intercept: $-\frac{5}{3}$

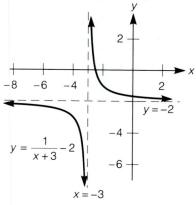

$y = \dfrac{1}{x+3} - 2$

9. **a.** $[-3, 1]$ **b.** $(-2, 1)$ **c.** -3
   d. The maximum is 1 at $x = -2$.
   e. $(-2, 2)$ **10. a.** $\frac{23}{4}$ **b.** $12 - 7\sqrt{3}$
11. domain: $(-\infty, -1) \cup (-1, \infty)$

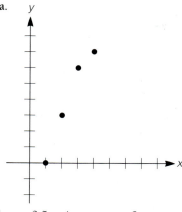

12.

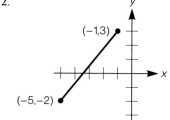

13. $t = 1$ **14.** $z = \frac{9}{4}$

CHAPTER FIVE

Exercise Set 5.1

1. $f(x) = \frac{2}{3}x + \frac{2}{3}$ **3.** $g(x) = \sqrt{2}x$
5. $f(x) = x - \frac{7}{2}$ **7.** $f(x) = \sqrt{3}$
9. $f(x) = \frac{1}{2}x - 2$ **11.** yes
13. $V(t) = -2375t + 20000$
15. **a.** $V(t) = -12000t + 60000$
b.

End of Year	Yearly Depreciation	Accumulated Depreciation	Value V
0	0	0	60,000
1	12,000	12,000	48,000
2	12,000	24,000	36,000
3	12,000	36,000	24,000
4	12,000	48,000	12,000
5	12,000	60,000	0

17. **a.** \$530 **b.** \$538 **c.** \$8/fan
19. **a.** \$50 **b.** \$50/player
   c. The answers are the same.
21. **a.** $\frac{4}{5}$ ft/sec **b.** 0 cm/sec **c.** 8 mph
23. **a.** B is traveling faster
   b. A is farther to the right **c.** $t = 8$ sec
25. **a.**

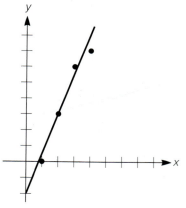

b. slope ≈ 2.5; y-intercept ≈ -2

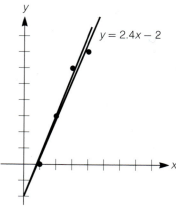

c.

$y = 2.4x - 2$

27. a. no **b.** 4,195,000 people
29. a. 3.6174 billion dollars **b.** no
31. $f(x) = 3x + 1$
35. a. $\Sigma x = 1 + 2 + 3 + 4 = 10$;
$\Sigma y = 0 + 3 + 6 + 7 = 16$
b. $\Sigma x^2 = 1 + 4 + 9 + 16 = 30$;
$\Sigma xy = 0 + 6 + 18 + 28 = 52$
c. $y = 2.4x - 2$ **37.** $y = 2.4x - 0.4$
39. $y = 0.084x + 37.241$

Exercise Set 5.2

1. vertex: $(-2, 0)$; axis: $x = -2$;
minimum: 0; x-intercept: -2;
y-intercept: 4

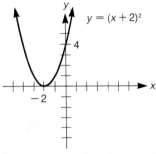

3. vertex: $(-2, 0)$; axis: $x = -2$;
minimum: 0; x-intercept: -2;
y-intercept: 8

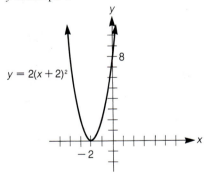

5. vertex: $(-2, 4)$; axis: $x = -2$;
maximum: 4; x-intercepts: $-2 \pm \sqrt{2}$;
y-intercept: -4

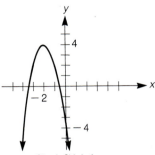

$y = -2(x + 2)^2 + 4$

7. vertex: $(2, -4)$; axis: $x = 2$;
minimum: -4; x-intercepts: 0, 4;
y-intercept: 0

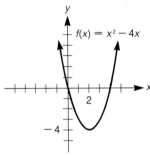

9. vertex: $(0, 1)$; axis: $x = 0$; maximum: 1;
x-intercepts: ± 1; y-intercept: 1

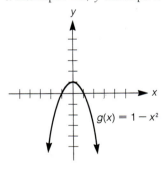

11. vertex: $(1, -4)$; axis: $x = 1$;
minimum: -4; x-intercepts: 3, -1;
y-intercept: -3

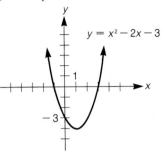

13. vertex: $(3, 11)$; axis: $x = 3$;
maximum: 11; x-intercepts: $3 \pm \sqrt{11}$;
y-intercept: 2

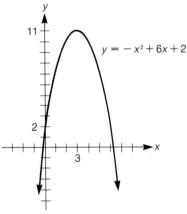

15. vertex: $(0, 0)$; axis: $t = 0$; minimum: 0;
t-intercept: 0; s-intercept: 0

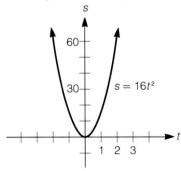

17. vertex: $\left(\frac{1}{6}, \frac{9}{4}\right)$; axis: $t = \frac{1}{6}$; maximum: $\frac{9}{4}$; t-intercepts: $-\frac{1}{3}, \frac{2}{3}$; s-intercept: 2

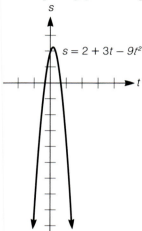

$s = 2 + 3t - 9t^2$

19. $x = 1$, smallest **21.** $x = \frac{3}{2}$, largest
23. $x = 0$, smallest
25. -13, minimum **27.** $\frac{25}{8}$,
29. 1000, maximum **31.** 5 units
33. quadratic **35.** neither
37. linear **39. a.** minimum at $(3, 8)$
b. mininum at $(3, 4)$
c. minimum at $\left(\pm\sqrt{3}, 64\right)$
41. a. 4 (at $x = 2$) **b.** $2\sqrt[3]{2}$ (at $x = 2$)
c. 16 $\left(\text{at } x = \pm\sqrt{2}\right)$
43. $y = -\frac{1}{2}(x - 2)^2 + 2$
45. $y = \frac{1}{4}(x - 3)^2 - 1$ **49.** $c = 1 + \sqrt{2}$
51. $x = (a + b)/2$

Exercise Set 5.3

1. a. $p(x) = 2x + 2\sqrt{144 - x^2}$
b. $A(x) = x\sqrt{144 - x^2}$
3. a. $D(x) = \sqrt{x^4 + 3x^2 + 1}$
b. $m(x) = (x^2 + 1)/x$
5. a. $A(y) = \frac{1}{4}\pi y^2$ **b.** $A(y) = \frac{1}{16}\pi^2 y^2$
7. a. $P(x) = 16x - x^2$
b. $S(x) = 2x^2 - 32x + 256$
c. $D(x) = (16 - x)^3 - x^3$
d. $A(x) = 8$ (it's constant)
9. $R(x) = -\frac{1}{4}x^2 + 8x$

11. a.

x	1	2	3	4
$P(x)$	17.88	19.49	20.83	21.86

x	5	6	7
$P(x)$	22.49	22.58	21.75

b. 22.58; $x = 6$ **c.** 22.63
13. a. $h(s) = \sqrt{3}\, s$ **b.** $A(s) = \sqrt{3}\, s^2$
c. $4\sqrt{3}$ cm **d.** $\frac{25}{4}\sqrt{3}$ in²
15. $V(r) = 2\pi r^3$ **17. a.** $h(r) = 12/r^2$
b. $S(r) = 2\pi r^2 + (24\pi/r)$
19. $V(S) = S\sqrt{\pi S}/6\pi$
21. $A(x) = \frac{1}{2}x\sqrt{400 - x^2}$
23. $d(x) = (x + 4)\sqrt{x^2 + 25}/x$
25. $x = 25$; $L = 25$

x	5	10	20	24	24.8	24.9
$A(x)$	225	400	600	624	624.96	624.99

x	25	25.1	25.2	45
$A(x)$	625	624.99	624.96	225

27. a. $x = 2$ yields the largest area

x	1	2	3	4
A	7.5	12	10.5	0

$x = 2.25$ yields the largest area

x	1.7500	2.0000	2.2500
A	11.3203	12.0000	12.3047

x	2.5000	2.7500
A	12.1875	11.6016

$x = 2.30$ yields the largest area

x	2.1500	2.2000	2.2500
A	12.2308	12.2760	12.3047

x	2.3000	2.3500
A	12.3165	12.3111

b. $x = 2.30$; $A = 12.316806$
29. $A(x) = \frac{17}{144}x^2 - \frac{1}{3}x + \frac{1}{2}$
31. a. $V(r) = \sqrt{3}\pi r^3/3$ **b.** $S(r) = 2\pi r^2$
33. a. $r(h) = 3h/\sqrt{h^2 - 9}$
b. $h(r) = 3r/\sqrt{r^2 - 9}$
35. $A(x) = [4x^2 + \pi(14 - x)^2]/16\pi$
37. $A(r) = \frac{1}{4}r(1 - 4\pi r)$
39. $A(x) = \frac{1}{3}\pi x^2$
41. $V(x) = 4x^3 - 28x^2 + 48x$
43. a. $A(r) = 32r - 2r^2 - \frac{1}{2}\pi r^2$
b. downward; yes
45. a. $y(s) = 3s/\sqrt{1 - s^2}$
b. $s(y) = y/\sqrt{y^2 + 9}$ **c.** $z(s) = 3/\sqrt{1 - s^2}$
d. $s(z) = \sqrt{z^2 - 9}/z$
47. a. $m(a) = (a^2 + 1)/a$
49. $A(x) = \frac{1}{4}(8x - x^2)$
51. $A(m) = (2m^2 - 8m + 8)/(m^2 - 4m)$
53. $A(m) = (ma - b)^2/(-2m)$

Exercise Set 5.4

1. $\frac{25}{4}$ **3.** $\frac{1}{2}$ **5.** $\frac{25}{4}$ m by $\frac{25}{4}$ m
7. 1250 in² **9. a.** 18 **b.** $\frac{23}{4}$ **c.** $\frac{47}{8}$
d. $\frac{95}{16}$ **11. a.** 16 ft; 12ft
b. 16 ft; 1 sec **c.** $\frac{1}{4}$ sec; $\frac{7}{4}$ sec
13. $\left(\frac{7}{2}, \frac{2 + \sqrt{6}}{2}\right)$; $d = \frac{\sqrt{7}}{2}$
15. a. $\frac{1}{2}$ **b.** $\frac{1}{4}$ **17.** 125 ft by 250 ft
19. 40
21. $x = 60$; $R = \$900$; $p = \$15$
23. a. $\frac{36}{13}$ **b.** $\sqrt{\frac{36}{13}} = \frac{6}{13}\sqrt{13}$
25. a. $\frac{225}{2}$ **27.** $x = \sqrt{2}/2$; area $= \frac{1}{2}$
29. $\frac{49}{12}$ **31.** $2R^2$ **33.** 100 yd by 150 yd **37. a.** $p(x) = -2x + 500$
b. $\$31,250$ when $p = \$250$
39. $\pm\sqrt{2}/2$ **41.** $\pm\sqrt{3}$
43. a. $A(x) = \dfrac{4 + \pi}{16\pi}x^2 - 2x + 16$
b. $x = 16\pi/(4 + \pi)$ **c.** $\pi/4$
45. a. $A(x) = \dfrac{4 + \pi}{16\pi}x^2 - \dfrac{L}{8}x + \dfrac{L^2}{16}$
b. $x = \pi L/(4 + \pi)$ **c.** $\pi/4$ **47.** 2

Exercise Set 5.5

1. This graph has four turning points, but a polynomial function of degree 3 can have at most two turning points.
3. As $|x|$ gets very large, our function should be similar to $f(x) = a_3 x^3$. But $f(x)$ does not have a parabolic shape like the

given graph. **5.** As $|x|$ gets very large with x negative, the graph should resemble $2x^5$. But the y-values of $2x^5$ are always negative when x is negative, contrary to the given graph. **7.** This graph has a corner, which cannot occur in the graph of a polynomial function. **9.** no x-intercepts; y-intercept: 5

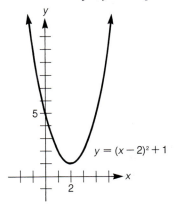

$y = (x - 2)^2 + 1$

11. x-intercept: 1; y-intercept: -1

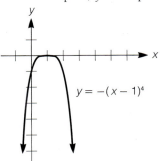

$y = -(x - 1)^4$

13. x-intercept: $4 + \sqrt[3]{2}$; y-intercept: -66

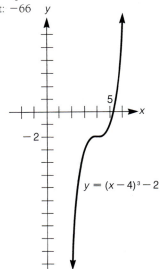

$y = (x - 4)^3 - 2$

15. x-intercept: -5; y-intercept: -1250

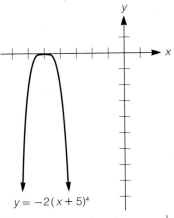

$y = -2(x + 5)^4$

17. x-intercept: -1; y-intercept: $\frac{1}{2}$

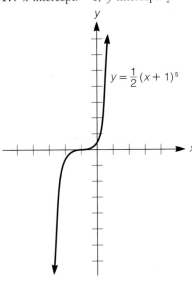

$y = \frac{1}{2}(x + 1)^5$

19. x-intercept: 0; y-intercept: 0

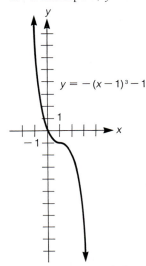

$y = -(x - 1)^3 - 1$

21. a. x-intercepts: 2, 1, -1; y-intercept: 2

b.

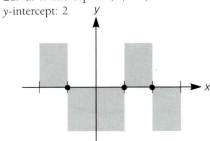

$y = (x - 2)(x - 1)(x + 1)$

23. a. x-intercepts: 0, 1, 2; y-intercept: 0

b.

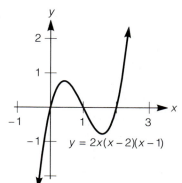

$y = 2x(x-2)(x-1)$

b.

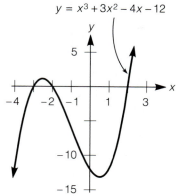

$y = x^3 + 3x^2 - 4x - 12$

31. a. *x*-intercepts: 1, 4; *y*-intercept: 128

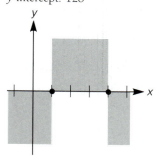

25. a. *x*-intercepts: 0, 5, −1; *y*-intercept: 0

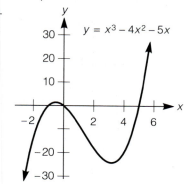

29. a. *x*-intercepts: 0, −2; *y*-intercept: 0

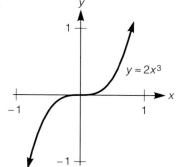

b.

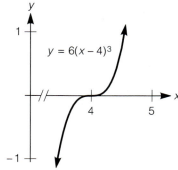

$y = 6(x-4)^3$

b.

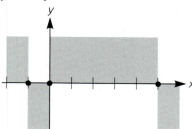

$y = x^3 - 4x^2 - 5x$

b.

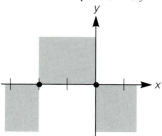

$y \approx 2x^3$

c.

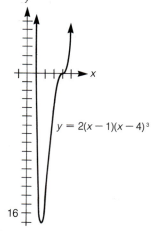

$y = 2(x-1)(x-4)^3$

27. a. *x*-intercepts: −3, −2, 2; *y*-intercept: −12

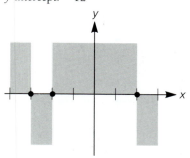

c.

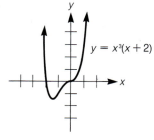

$y = x^3(x+2)$

33. a. *x*-intercepts: −1, 1, 3; *y*-intercept: 3

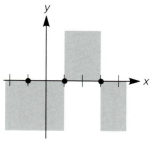

b.

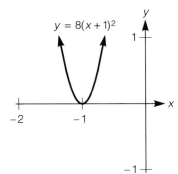

$y = 8(x + 1)^2$

c.

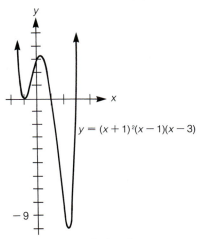

$y = (x + 1)^2(x - 1)(x - 3)$

35. a. x-intercepts: $0, 4, -2$;
y-intercept: 0

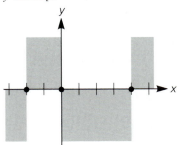

b.

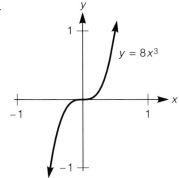

$y = 8x^3$

c.

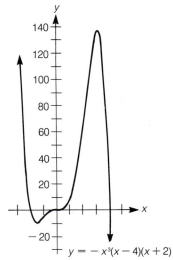

$y = -x^3(x - 4)(x + 2)$

37. a. x-intercepts: $0, 2, -2$;
y-intercept: 0

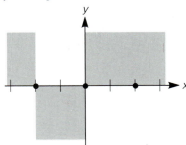

b.

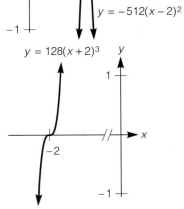

$y = -512(x - 2)^2$

$y = 128(x + 2)^3$

c.

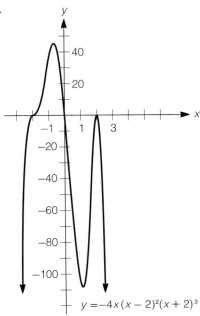

$y = -4x(x - 2)^2(x + 2)^3$

39. from left to right: $f(x)$, $g(x)$, $h(x)$,
$F(x)$, $G(x)$, $H(x)$ **41.** $[0, 0.68)$
43. no   **45.** yes, at $(100, 100)$
47. a.

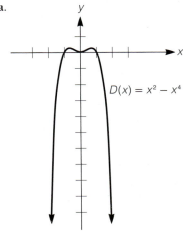

$D(x) = x^2 - x^4$

b. $\left(\pm\dfrac{\sqrt{2}}{2}, \dfrac{1}{4}\right)$, $(0,0)$ **c.** $\dfrac{1}{4}$

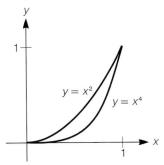

$y = x^2$

$y = x^4$

49. c. $(0, 6)$ **d.** maximum of $f(r)$: 273000; maximum volume: 522 cm^3

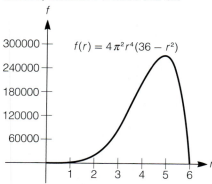

$f(r) = 4\pi^2 r^4 (36 - r^2)$

Exercise Set 5.6

1. domain: $(-\infty, 3) \cup (3, \infty)$; x-intercept: -5; y-intercept: $-\frac{5}{4}$
3. domain: $(-\infty, -2) \cup (-2, 3) \cup (3, \infty)$; x-intercepts: $9, -1$; y-intercept: $\frac{3}{2}$
5. domain: $(-\infty, 0) \cup (0, \infty)$; x-intercepts: ± 2, 1; no y-intercept
7. x-intercept: none; y-intercept: $\frac{1}{4}$; horizontal asymptote: $y = 0$; vertical asymptote: $x = -4$

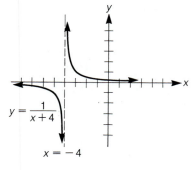

$y = \dfrac{1}{x+4}$

$x = -4$

9. x-intercept: none; y-intercept: $\frac{3}{2}$; horizontal asymptote: $y = 0$; vertical asymptote: $x = -2$

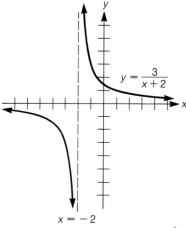

$y = \dfrac{3}{x+2}$

$x = -2$

11. x-intercept: none; y-intercept: $\frac{2}{3}$; horizontal asymptote: $y = 0$; vertical asymptote: $x = 3$

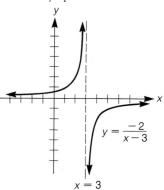

$y = \dfrac{-2}{x-3}$

$x = 3$

13. x-intercept: 3; y-intercept: 3; horizontal asymptote: $y = 1$; vertical asymptote: $x = 1$

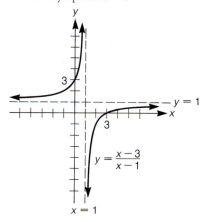

$y = 1$

$y = \dfrac{x-3}{x-1}$

$x = 1$

15. x-intercept: $\frac{1}{2}$; y-intercept: -2; horizontal asymptote: $y = 2$; vertical asymptote: $x = -\frac{1}{2}$

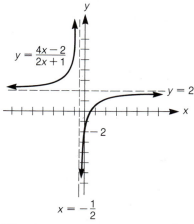

$y = \dfrac{4x-2}{2x+1}$

$y = 2$

$x = -\dfrac{1}{2}$

17. x-intercept: none; y-intercept: $\frac{1}{4}$; horizontal asymptote: $y = 0$; vertical asymptote: $x = 2$

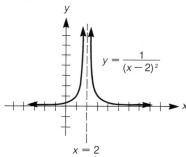

$y = \dfrac{1}{(x-2)^2}$

$x = 2$

19. x-intercept: none; y-intercept: 3; horizontal asymptote: $y = 0$; vertical asymptote: $x = -1$

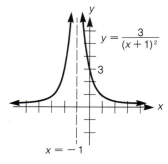

$y = \dfrac{3}{(x+1)^2}$

$x = -1$

21. x-intercept: none; y-intercept: $\frac{1}{8}$; horizontal asymptote: $y = 0$; vertical asymptote: $x = -2$

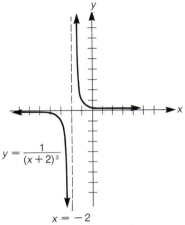

$$y = \frac{1}{(x+2)^3}$$

$x = -2$

23. x-intercept: none; y-intercept: $-\frac{4}{125}$; horizontal asymptote: $y = 0$; vertical asymptote: $x = -5$

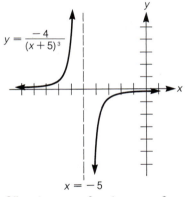

$$y = \frac{-4}{(x+5)^3}$$

$x = -5$

25. x-intercept: 0; y-intercept: 0; horizontal asymptote: $y = 0$; vertical asymptotes: $x = -2$, $x = 2$

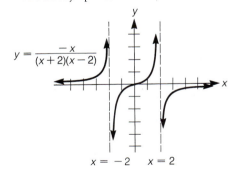

$$y = \frac{-x}{(x+2)(x-2)}$$

$x = -2 \qquad x = 2$

27. x-intercept: 0; y-intercept: 0; horizontal asymptote: $y = 0$; vertical asymptotes: $x = -3$, $x = 1$

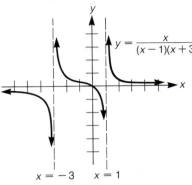

$$y = \frac{x}{(x-1)(x+3)}$$

$x = -3 \qquad x = 1$

29. a. x-intercepts: 2, 4; y-intercept: none; horizontal asymptote: $y = 1$; vertical asymptotes: $x = 0$, $x = 1$

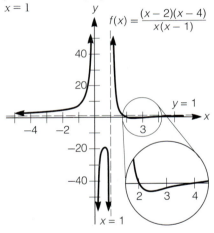

$$f(x) = \frac{(x-2)(x-4)}{x(x-1)}$$

$y = 1$

$x = 1$

b. x-intercepts: 2, 4; y-intercept: none; horizontal asymptote: $y = 1$; vertical asymptotes: $x = 0$, $x = 3$

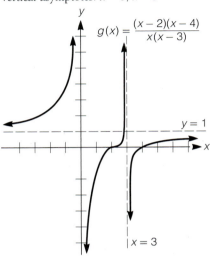

$$g(x) = \frac{(x-2)(x-4)}{x(x-3)}$$

$y = 1$

$x = 3$

31. x-intercepts: -2, 4; y-intercept: $-\frac{8}{3}$; horizontal asymptote: $y = 1$; vertical asymptotes: $x = 1$, $x = 3$; crosses the horizontal asymptote at $x = \frac{11}{2}$

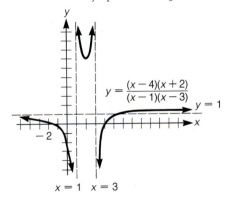

$$y = \frac{(x-4)(x+2)}{(x-1)(x-3)}$$

$y = 1$

$x = 1 \quad x = 3$

33. x-intercept: -1; y-intercept: $\frac{1}{3}$; horizontal asymptote: $y = 1$; vertical asymptotes: $x = 1$, $x = 3$; crosses the horizontal asymptote at $x = \frac{1}{3}$

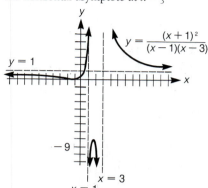

$$y = \frac{(x+1)^2}{(x-1)(x-3)}$$

$y = 1$

-9

$x = 3$
$x = 1$

37. a.

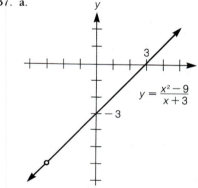

$$y = \frac{x^2 - 9}{x + 3}$$

b.

$$y = \frac{x^2 - 5x + 6}{x^2 - 2x - 3}$$

$y = 1$

$3 \quad 4$

-2

$x = -1$

c.

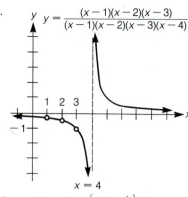

$$y = \frac{(x-1)(x-2)(x-3)}{(x-1)(x-2)(x-3)(x-4)}$$

1 2 3

-1

$x = 4$

39. turning point: $\left(-3, -\frac{1}{12}\right)$

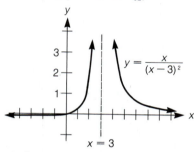

$$y = \frac{x}{(x-3)^2}$$

$x = 3$

41. b.

x	$x + 4$	$\dfrac{x^2 + x - 6}{x - 3}$
10	14	14.8571
100	104	104.0619
1000	1004	1004.0060

x	$x + 4$	$\dfrac{x^2 + x - 6}{x - 3}$
-10	-6	-6.4615
-100	-96	-96.0583
-1000	-996	-996.0600

c. vertical asymptote: $x = 3$;
x-intercepts: -3, 2; y-intercept: 2

d.

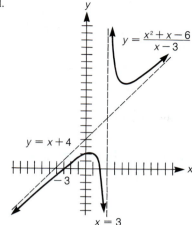

$$y = \frac{x^2 + x - 6}{x - 3}$$

$y = x + 4$

-3

$x = 3$

e. $\left(3 + \sqrt{6}, 7 + 2\sqrt{6}\right)$, $\left(3 - \sqrt{6}, 7 - 2\sqrt{6}\right)$

43.

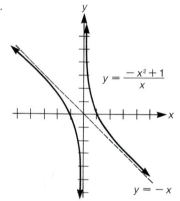

$$y = \frac{-x^2 + 1}{x}$$

$y = -x$

Chapter Five Review Exercises

1. -5 **2. a.** minimum value is -13 when $x = -1$ **b.** maximum value is 9 when $t = \pm\sqrt{3}$ **3.** \$20,000

5.

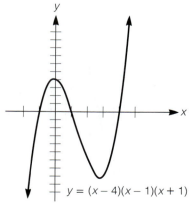

$$y = (x - 4)(x - 1)(x + 1)$$

7. $V(t) = -180t + 1000$

9. x-intercept: $-\frac{5}{3}$; y-intercept: $\frac{5}{2}$; horizontal asymptote: $y = 3$; vertical asymptote: $x = -2$

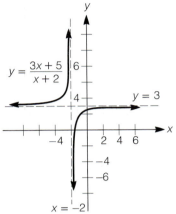

$$y = \frac{3x + 5}{x + 2}$$

11.

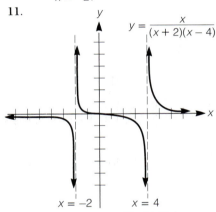

$$y = \frac{x}{(x + 2)(x - 4)}$$

13. $P(w) = 2w + 2\sqrt{144 - \pi^2 w^2}\big/\pi$

15. $f(x) = x + 2$ **17.** $f(x) = \frac{3}{8}x - \frac{5}{2}$

19. $f(x) = -\frac{3}{4}x + \frac{11}{4}$

21. vertex: $(-1, -4)$; x-intercepts: $1, -3$; y-intercept: -3

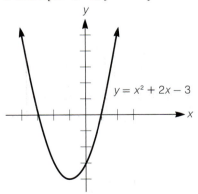

$$y = x^2 + 2x - 3$$

23. vertex: $\left(\sqrt{3}, 6\right)$; x-intercepts: $\sqrt{3} + \sqrt{6}, \sqrt{3} - \sqrt{6}$; y-intercept: 3

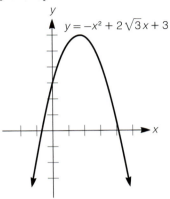

$$y = -x^2 + 2\sqrt{3}\,x + 3$$

25. vertex: $(2, 12)$; x-intercepts: $0, 4$; y-intercept: 0

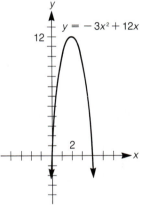

$$y = -3x^2 + 12x$$

27. 5   **29.** $\frac{3}{4}$ **31. a.** $\frac{v_0^2}{64}$ ft when

$t = \frac{v_0}{32}$ sec **b.** $\frac{v_0}{16}$ sec

33. a. $d(x) = \sqrt{x^4 - 3x^2 + 4}$

b. $\left(-\frac{\sqrt{6}}{2}, \frac{3}{2}\right)$ **35.** $b = \frac{17}{3}, -11$

37. 1   **39.** $\frac{225}{4}$ cm^2 **41.** $a = 1$

43. 400 units at $80 each

45. x-intercepts: $-4, 2$; y-intercept: -8

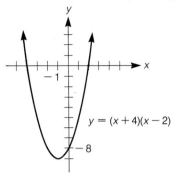

$$y = (x + 4)(x - 2)$$

47. x-intercept: -5; y-intercept: -125

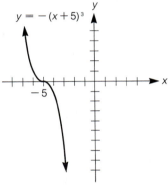

$$y = -(x + 5)^3$$

49. x-intercepts: $-1, 0$; y-intercept: 0

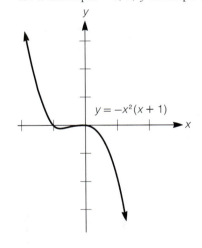

$$y = -x^2(x + 1)$$

51. x-intercepts: 0, 2, -2; y-intercept: 0

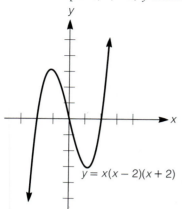

$$y = x(x - 2)(x + 2)$$

53. x-intercept: $-\frac{1}{3}$; y-intercept: none; horizontal asymptote: $y = 3$; vertical asymptote: $x = 0$

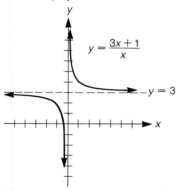

$$y = \frac{3x + 1}{x}$$

$y = 3$

55. x-intercept: none; y-intercept: -1; horizontal asymptote: $y = 0$; vertical asymptote: $x = 1$

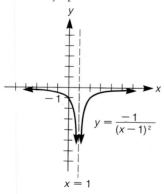

$$y = \frac{-1}{(x - 1)^2}$$

$x = 1$

57. x-intercept: 2; y-intercept: $\frac{2}{3}$; horizontal asymptote: $y = 1$; vertical asymptote: $x = 3$

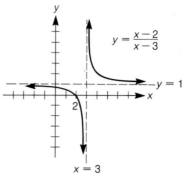

$$y = \frac{x - 2}{x - 3}$$

$y = 1$

$x = 3$

59. x-intercept: 1; y-intercept: $\frac{1}{4}$; horizontal asymptote: $y = 1$; vertical asymptote: $x = 2$

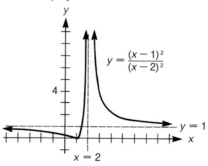

$$y = \frac{(x - 1)^2}{(x - 2)^2}$$

$y = 1$

$x = 2$

61. a. 1 **b.** $\sqrt{13}$ **c.** $\pm\sqrt{2}/2$

63. $k = 6$

65. $\left(-\infty, 4 - 2\sqrt{3}\right] \cup \left[4 + 2\sqrt{3}, \infty\right)$

67. $A(m) = m/2$

69. $A(x) = (1 - x)\sqrt{1 - x^2}$

71. a. y

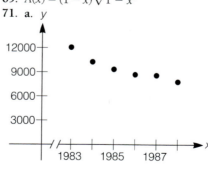

b. slope ≈ -770; y-intercept ≈ 1.5 million

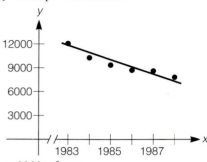

c. 2100 refugees

d. $f^{-1}(x) = (1541090 - x)/771.4$

e. 1996 **73. a.** reflections across the y-axis

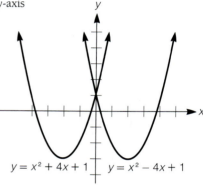

$$y = x^2 + 4x + 1 \qquad y = x^2 - 4x + 1$$

b. $f(-x) = ax^2 - bx + c$ **c.** Reversing the sign of b will reflect the graph of $f(x) = ax^2 + bx + c$ across the y-axis.

Chapter Five Graphing Calculator Exercises

1. $y = 4250$

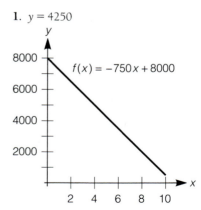

$$f(x) = -750x + 8000$$

3. a.

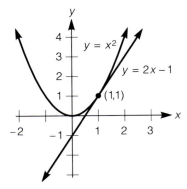

b.

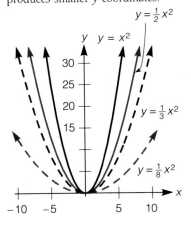

x	0.9	0.99	0.999
$2x - 1$	0.8	0.98	0.998
x^2	0.81	0.9801	0.998001

x	1.1	1.01	1.001
$2x - 1$	1.2	1.02	1.002
x^2	1.21	1.0201	1.002001

5. a. As the coefficient of x^2 decreases from 1 to $\frac{1}{8}$, the graph "flattens" and produces smaller y-coordinates.

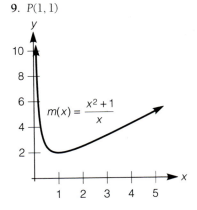

b. The graph is "flatter" than the others.

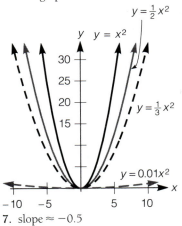

7. slope ≈ -0.5

$$A(m) = \frac{-4m^2 + 4m - 1}{2m}$$

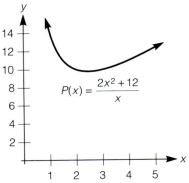

9. $P(1, 1)$

$$m(x) = \frac{x^2 + 1}{x}$$

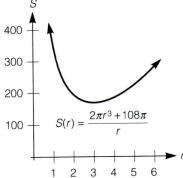

The line is tangent to the parabola at $(1, 1)$.

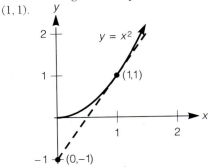

11. a. $P(x) = (2x^2 + 12)/x$
b. $x \approx 2.45$ m; length ≈ 2.45 m

$$P(x) = \frac{2x^2 + 12}{x}$$

13. a. $h(r) = 54/r^2$
b. $S(r) = (2\pi r^3 + 108\pi)/r$ **c.** $r \approx 3$ in.

$$S(r) = \frac{2\pi r^3 + 108\pi}{r}$$

d. $h = 6$ in.; h is double the value of r.

Chapter Five Test

1. $-\frac{18}{7}$ **2. a.** increasing on $(-\infty, 1)$; maximum $= 2$ **b.** $t = 0$

3. a. x-intercepts: $3, -4$;
y-intercept: -48

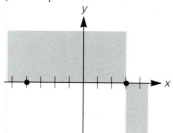

b.

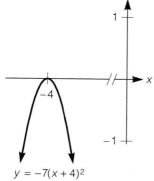

$$y = -7(x + 4)^2$$

c.

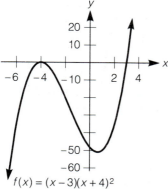

$$f(x) = (x - 3)(x + 4)^2$$

4. a.

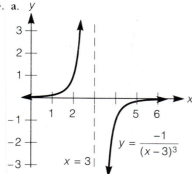

$$y = \frac{-1}{(x - 3)^3}$$

b.

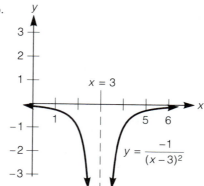

$$y = \frac{-1}{(x - 3)^2}$$

5. turning point: $\left(\frac{7}{2}, \frac{73}{4}\right)$;
x-intercepts: $\left(7 \pm \sqrt{73}\right)/2$; y-intercept: 6;
axis: $x = \frac{7}{2}$

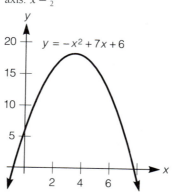

$$y = -x^2 + 7x + 6$$

6. revenue $= \$9600$; price $= \$40/$unit
7. $V(t) = -1325t + 14000$
8. x-intercept: 3; y-intercept: $-\frac{9}{2}$

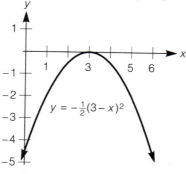

$$y = -\frac{1}{2}(3 - x)^2$$

9. x-intercept: $\frac{3}{2}$; y-intercept: -3;
horizontal asymptote: $y = 2$;
vertical asymptote: $x = -1$

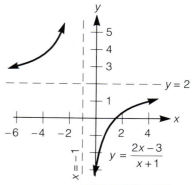

$$y = \frac{2x - 3}{x + 1}$$

10. a. $L(x) = \sqrt{10x^2 - 22x + 17}$
b. $x = 1.1$
11. a. vertical asymptotes: $x = 3$,
$x = -3$; horizontal asymptote: $y = 1$
b.

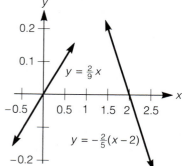

$$y = \frac{2}{9}x$$
$$y = -\frac{2}{5}(x - 2)$$

c.

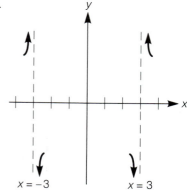

$x = -3$ $x = 3$

d.

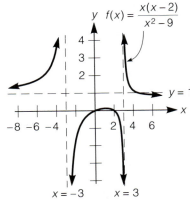

$$y \quad f(x) = \frac{x(x-2)}{x^2-9}$$

$y = 1$

$x = -3 \qquad x = 3$

12. $A(w) = \frac{1}{2}w\sqrt{64-w^2}$

13. a. As $|x|$ increases in size for positive values of x, the quantity $-x^3$ should be large negative values, which the function is not. **b.** This graph has four turning points, but a polynomial function with highest-degree term $-x^3$ can have at most two turning points.

14. a.

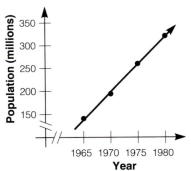

b. 381 million; too high

CHAPTER SIX

Exercise Set 6.1

1. a. 10^9; **b.** 10^{15} **3. a.** $x = 3$
b. $t = \frac{3}{2}$ **c.** $y = \frac{1}{4}$ **d.** $z = \frac{5}{2}$ **5.** $(-\infty, \infty)$
7. $(-\infty, \infty)$

9.

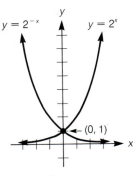

$y = 2^{-x}$ $y = 2^x$

$(0, 1)$

11.

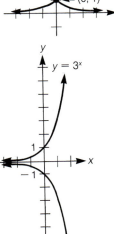

$y = 3^x$

$y = -3^x$

13.

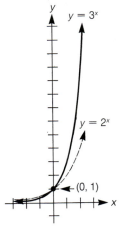

$y = 3^x$

$y = 2^x$

$(0, 1)$

15.

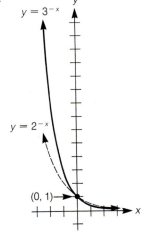

$y = 3^{-x}$

$y = 2^{-x}$

$(0, 1)$

17. domain: $(-\infty, \infty)$; range: $(-\infty, 1)$; x-intercept: 0; y-intercept: 0; asymptote: $y = 1$

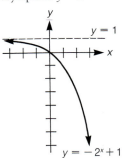

$y = 1$

$y = -2^x + 1$

19. domain: $(-\infty, \infty)$; range: $(1, \infty)$; x-intercept: none; y-intercept: 2; asymptote: $y = 1$

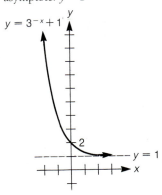

$y = 3^{-x} + 1$

$y = 1$

21. domain: $(-\infty, \infty)$; range: $(0, \infty)$; x-intercept: none; y-intercept: $\frac{1}{2}$; asymptote: $y = 0$

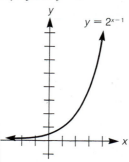

23. domain: $(-\infty, \infty)$; range: $(1, \infty)$; x-intercept: none; y-intercept: 4; asymptote: $y = 1$

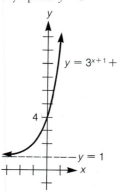

25. $-\frac{1}{3}$ **27.** $\frac{3}{2}$, 1
31. a.

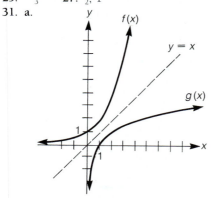

b. domain: $(0, \infty)$; range: $(-\infty, \infty)$; x-intercept: 1; y-intercept: none; asymptote: $x = 0$ **33. a.** 1.4 **b.** 1.41
35. a. 1.5 **b.** 1.52   **37. a.** 1.7
b. 1.73   **39. a.** 1.6 **b.** 1.62
41. 0.3   **43.** 0.7

45.

x	$\log_{10} x$
1	0.00
2	0.30
3	0.48
4	0.60
5	0.70
6	0.78
7	0.85
8	0.90
9	0.95
10	1.00

Exercise Set 6.2

1. domain: $(-\infty, \infty)$; range: $(0, \infty)$; x-intercept: none; y-intercept: 1; asymptote: $y = 0$

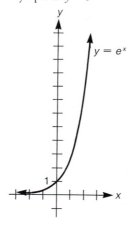

3. domain: $(-\infty, \infty)$; range: $(-\infty, 0)$; x-intercept: none; y-intercept: -1; asymptote: $y = 0$

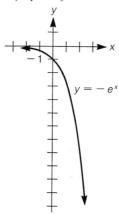

5. domain: $(-\infty, \infty)$; range: $(1, \infty)$; x-intercept: none; y-intercept: 2; asymptote: $y = 1$

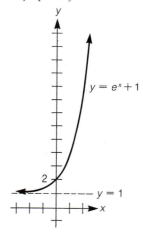

7. domain: $(-\infty, \infty)$; range: $(1, \infty)$; x-intercept: none; y-intercept: $e + 1$; asymptote: $y = 1$

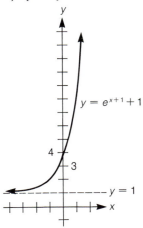

9. domain: $(-\infty, \infty)$; range: $(-\infty, 0)$; x-intercept: none; y-intercept: $-1/e^2$; asymptote: $y = 0$

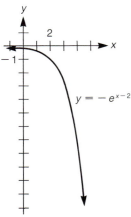

11. domain: $(-\infty, \infty)$; range: $(-\infty, e)$; x-intercept: 1; y-intercept: $e - 1$; asymptote: $y = e$

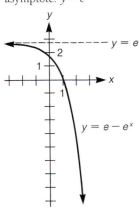

13.

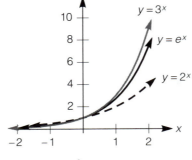

15. 2   **17.** $\sqrt[3]{4} \approx 1.587$ **19.** false
21. false   **23.** true   **25.** false
27. 1.2; 1.22   **29.** 2.2; 2.23
31. 1.6; 1.65   **33.** -0.2
35. a. $\sqrt{3}$ **b.** 486,000 bacteria
37. 6400 bacteria
39. a.

t (sec)	0	550	1100	1650	2200
$\mathcal{N}$ (g)	8	4	2	1	0.5

b.

t (yrs)	0	4.9×10^9	9.8×10^9
$\mathcal{N}$ (g)	10	5	2.5

t (yrs)	14.7×10^9	19.6×10^9
$\mathcal{N}$ (g)	1.25	0.625

41. 0.55 g   **43.** 97.15%
47. a. $(-\infty, \infty)$ **b.** 1
d.

$$C(x) = \frac{e^x + e^{-x}}{2}$$

$(0, 1)$

49.

n	$1 + \dfrac{1}{n}$	$\left(1 + \dfrac{1}{n}\right)^n$
1	2	2.0000
10	1.1	2.5937
100	1.01	2.7048
1000	1.001	2.7169
10000	1.0001	2.7181
100000	1.00001	2.7183

51. a. 3.36 billion **b.** too low
53. a.

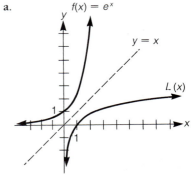

b. domain: $(0, \infty)$; range: $(-\infty, \infty)$; x-intercept: 1; asymptote: $x = 0$
c. (i) x-intercept: 1; asymptote: $x = 0$

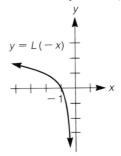

(ii) x-intercept: -1; asymptote: $x = 0$

(iii) x-intercept: 2; asymptote: $x = 1$

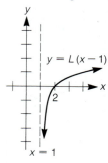

$y = L(x - 1)$

$x = 1$

Exercise Set 6.3

1. a. no **b.** yes **c.** yes **3. a.** $\dfrac{4x + 1}{2 - 3x}$

b. $\dfrac{3x + 4}{2x - 1}$ **c.** $\frac{1}{2}$ **d.** -4

5. $(-1, 3)$ and $(6, -1)$

7. a. $\log_3 9 = 2$ **b.** $\log_{10} 1000 = 3$
c. $\log_7 343 = 3$ **d.** $\log_2 \sqrt{2} = \frac{1}{2}$
9. a. $2^5 = 32$ **b.** $10^0 = 1$ **c.** $e^{1/2} = \sqrt{e}$
d. $3^{-4} = \frac{1}{81}$ **e.** $t^v = u$ **11.** $\log_5 30$
13. a. $\frac{3}{2}$ **b.** $-\frac{5}{2}$ **c.** $\frac{3}{2}$ **15. a.** $\frac{1}{16}$
b. $e^{-2} \approx 0.14$ **17. a.** $(0, \infty)$
b. $\left(-\infty, \frac{3}{4}\right)$ **c.** $(-\infty, 0) \cup (0, \infty)$ **d.** $(0, \infty)$
e. $(-\infty, -5) \cup (5, \infty)$
19. $A(0, 1)$, $B(1, 0)$, $C(4, 2)$, $D(2, 4)$
21. a. domain: $(0, \infty)$; range: $(-\infty, \infty)$;
x-intercept: 1; y-intercept: none;
asymptote: $x = 0$

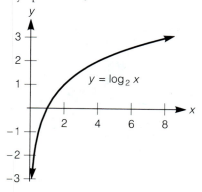

$y = \log_2 x$

b. domain: $(0, \infty)$; range: $(-\infty, \infty)$;
x-intercept: 1; y-intercept: none;
asymptote: $x = 0$

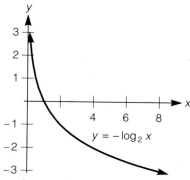

$y = -\log_2 x$

c. domain: $(-\infty, 0)$; range: $(-\infty, \infty)$;
x-intercept: -1; y-intercept: none;
asymptote: $x = 0$

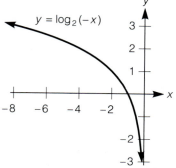

$y = \log_2 (-x)$

d. domain: $(-\infty, 0)$; range: $(-\infty, \infty)$;
x-intercept: -1; y-intercept: none;
asymptote: $x = 0$

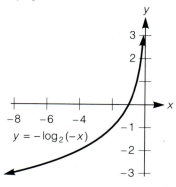

$y = -\log_2 (-x)$

23. domain: $(2, \infty)$; range: $(-\infty, \infty)$;
x-intercept: 5; y-intercept: none;
asymptote: $x = 2$

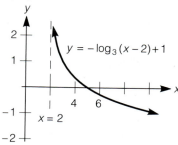

$y = -\log_3 (x - 2) + 1$

$x = 2$

25. domain: $(-e, \infty)$; range: $(-\infty, \infty)$;
x-intercept: $-e + 1$; y-intercept: 1;
asymptote: $x = -e$

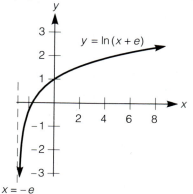

$y = \ln(x + e)$

$x = -e$

27. a. 4 **b.** -1 **c.** $\frac{1}{2}$
29. $\log_{10} 25 \approx 1.40$
31. $\pm\sqrt{\log_{10} 40} \approx \pm 1.27$
33. $\frac{1}{2}(-3 + \ln 10) \approx -0.35$
35. $\frac{1}{4}(1 - \ln 12.405) \approx -0.38$
37. a. $k \approx -1.54 \times 10^{-10}$
b. 99.9999846% **39. a.** $k = \ln \frac{1}{2}$
b. 3.32 yrs **41. a.** 0.51 hrs
b. 2.89 hrs **43.** $f^{-1}(x) = -1 + \ln x$
x-intercept: e; asymptote: $x = 0$

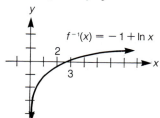

$f^{-1}(x) = -1 + \ln x$

45.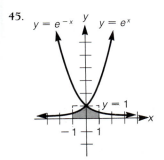
$y = e^{-x}$ y $y = e^x$
$y = 1$
-1 1

The area is less than that of the rectangle, which is two square units. **47.** $\ln 6$

49. 10^{30} **51.** $t = \dfrac{\ln(A_0/B_0)}{k_2 - k_1}$; this represents the time it takes for two radioactive substances to decay to the same mass. **53. a.** $P(10) = 4$; $P(18) = 7$; $P(19) = 8$

b.

x	$P(x)$	$\dfrac{x}{\ln x}$	$\dfrac{P(x)}{x/\ln x}$
10^2	25	22	1.151
10^4	1229	1086	1.132
10^6	78498	72382	1.084
10^8	5761455	5428681	1.061
10^9	50847534	48254942	1.054
10^{10}	455052512	434294482	1.048

c.

x	$P(x)$	$\dfrac{x}{\ln x - 1.08366}$	$\dfrac{P(x)}{x/(\ln x - 1.08366)}$
10^2	25	28	0.8804
10^4	1229	1231	0.9988
10^6	78498	78543	0.9994
10^8	5761455	5768004	0.9989
10^9	50847534	50917519	0.9986
10^{10}	455052512	455743004	0.9985

55.
y
8
7 $\ln y = x$
6
5
4
3
2
-2 -1 1 2 x

Exercise Set 6.4

1. 1 **3.** $\frac{1}{2}$ **5.** 4 **7.** 0 **9.** 4
11. $\log_{10} 60$ **13.** $\log_5 20$
15. a. $\ln 6$ **b.** $\ln \frac{3}{16384}$
17. $\log_b \dfrac{4(1+x)^3}{(1-x)^{3/2}}$
19. $\log_{10} \dfrac{27\sqrt{x+1}}{(x^2+1)^6}$
21. a. $2\log_{10} x - \log_{10}(1+x^2)$
b. $2\ln x - \frac{1}{2}\ln(1+x^2)$
23. a. $\frac{1}{2}\log_{10}(3+x) + \frac{1}{2}\log_{10}(3-x)$
b. $\frac{1}{2}\ln(2+x) + \frac{1}{2}\ln(2-x) - \ln(x-1) - \frac{3}{2}\ln(x+1)$ **25. a.** $\frac{1}{2}\log_b x - \frac{1}{2}$
b. $\ln(1+x^2) + \ln(1+x^4) + \ln(1+x^6)$
27. a. $a + 2b + 3c$ **b.** $1 + \frac{1}{2}a$
c. $\frac{1}{2}(1 + a + b + c)$ **d.** $1 + a - \frac{1}{2}b - \frac{1}{2}c$
29. a. $1 + t$ **b.** $u - t$ **c.** $\frac{3}{2}t + \frac{1}{2}u - 1$
d. $2 + t + \frac{1}{2}u$ **31.** $\frac{1}{2}(\ln 5 - \ln 2 + 1)$
33. $\ln 2 - \ln 3 - 1$ **35.** $\dfrac{\ln 9}{\ln 2}$
37. $\dfrac{\ln 5 + \ln 2}{\ln 5 - \ln 2}$ **39.** $\frac{1}{2}$ **41.** 3
43. $\frac{203}{99}$ **45.** 3 **47.** 7
49. a. $x = 10^y/(3 \cdot 10^y - 1)$
b. $x = (1 - y)/2$ **51.** $\dfrac{\log_{10} 5}{\log_{10} 2}$
53. $\dfrac{\log_{10} 3}{\log_{10} e}$ **55.** $\dfrac{\log_{10} 2}{\log_{10} b}$ **57.** $\dfrac{\ln 6}{\ln 10}$
59. $1/\ln 10$ **61.** $\dfrac{\ln(\ln x) - \ln(\ln 10)}{\ln 10}$
63. a. true **b.** true **c.** true **d.** false
e. true **f.** false **g.** true **h.** false **i.** true
j. false **k.** false **l.** true **m.** true
65. f. 2345.6 **g.** 0.123456
67. $k = -(\ln 243)/2.5$; $N_0 = 4$
73. x^3 **75.** $\frac{1}{2}$ **77.** $x = \beta e^{\alpha/3}$
85. $f^{-1}(x) = (e^{2x} - 1)/2e^x$

87.

x	10^2	10^3	10^6
y	0.4234	0.6589	0.9654

x	10^{20}	10^{50}	10^{99}
y	1.3428	1.5573	1.6918

Exercise Set 6.5

1. $1009.98 **3.** 8.45%
5. $767.27 **7. a.** $3869.68
b. $4006.39 **9.** 13 quarters
11. $3487.50 **13.** $2610.23
15. 5.83% **17.** the first investment
($13382.26) **19. a.** 14 years
b. 13.86 yrs **c.** 1.01%
21. $26.5 trillion **23. a.** 14 yrs
b.

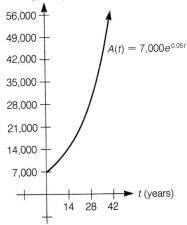
A (dollars)
56,000
49,000
42,000
35,000
28,000
21,000
14,000
7,000
$A(t) = 7{,}000 e^{0.05t}$
14 28 42 t (years)

25.

	1990 Population (billions)	% Population in 1990	Growth Rate (% per year)
World	5.321	100.0	1.8
More Dev.	1.214	22.8	0.5
Less Dev.	4.107	77.2	2.1

	Year 2000 Population (billions)	% of World Population in 2000
World	6.370	100.0
More Dev.	1.276	20.0
Less Dev.	5.067	80.0

27. a.

Country	1990 Population (millions)	Relative Growth Rate (percent per year)
Iraq	18.8	3.9
United Kingdom	57.4	0.2

Country	Year 2000 Population (millions)	Percent Increase in Population
Iraq	27.8	47.87
United Kingdom	58.6	2.09

b. 2020   **29. a.** $k \approx 0.0200$
b. $N \approx 170{,}853{,}155$ **c.** slower
31. a. 0.68% **b.** 0.96% **c.** 21,731,258
d. higher
33. a.

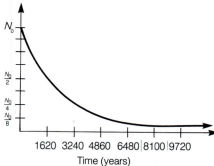

b.

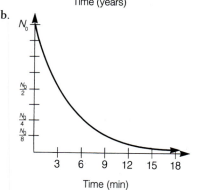

35. a. -0.0533 **b.** 58.7%; 0.5%
37. a. $k \approx -0.0248$ **b.** 279 yrs
c. 280 yrs   **39. a.** 2020 **b.** 2035
c. 2019   **41. a.** 201 yrs **b.** 132 yrs
43. a. $E = 10^{11.4} \cdot 10^{1.5M}$ **b.** 178
45. -1.4748×10^{-11}
47. 4.181 billion yrs   **49.** 15,505 yrs

1. $\log_5 126$ **3.** $\dfrac{4 \ln 1.5}{\ln 1.25}$ hrs
5. f is not one-to-one.

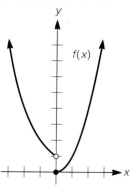

7. $x = (e+1)/(e-1)$ **9.** $y = e^x$;
domain: $(-\infty, \infty)$; range: $(0, \infty)$; $y = \ln x$;
domain: $(0, \infty)$; range: $(-\infty, \infty)$

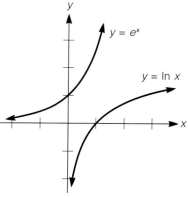

11. $-\frac{3}{2}$ **13.** -0.05 **15.** $2 - \ln \frac{12}{5}$
17. 2000
19. 7 yrs

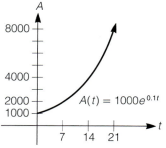

21. horizontal asymptote: $y = 0$;
vertical asymptote: none; x-intercept:
none; y-intercept: 1

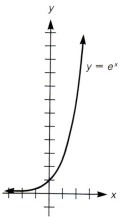

23. horizontal asymptote: none;
vertical asymptote: $x = 0$; x-intercept: 1;
y-intercept: none

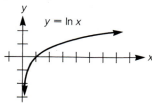

25. horizontal asymptote: $y = 1$;
vertical asymptote: none; x-intercept:
none; y-intercept: 3

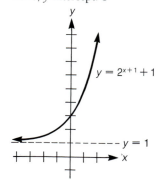

27. horizontal asymptote: $y = 0$; vertical asymptote: none; x-intercept: none; y-intercept: 1

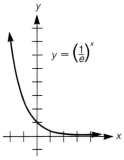

$y = \left(\frac{1}{e}\right)^x$

29. horizontal asymptote: $y = 1$; vertical asymptote: none; x-intercept: none; y-intercept: $e + 1$

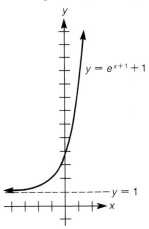

$y = e^{x+1} + 1$

$y = 1$

31. horizontal asymptote: none; vertical asymptote: none; x-intercept: 0; y-intercept: 0

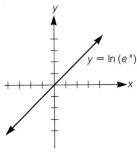

$y = \ln(e^x)$

33. 4 **35.** 3 **37.** $\frac{2}{3}$ **39.** $\sqrt[3]{3}$
41. $(1 - 2\ln 3)/10$ **43.** $\frac{200}{99}$ **45.** 2
47. $(0, \infty)$ **49.** 1 **51.** $\frac{1}{2}$ **53.** $\frac{1}{5}$
55. -1 **57.** 16 **59.** 4 **61.** 2
63. 2 **65.** $\frac{9}{14}$ **67.** $2a + 3b + \frac{1}{2}c$
69. $8a + 4b$ **71.** 2 and 3
73. 2 and 3 **75.** -2 and -3
77. a. The two curves intersect in the third quadrant.

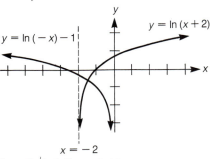

$y = \ln(-x) - 1$
$y = \ln(x + 2)$
$x = -2$

b. $-2e/(e + 1) \approx -1.46$
79. $k = \ln(1/2)/T$ **81.** 6.25%
83. $\dfrac{d \ln(c/b)}{\ln \frac{1}{2}}$ days **85.** $\log_{10} 2$
87. $\ln 10$ **89.** $\ln(x^a y^b)$
91. $\frac{1}{2}\ln(x - 3) + \frac{1}{2}\ln(x + 4)$
93. $3 \log_{10} x - \frac{1}{2}\log_{10}(1 + x)$
95. $\frac{1}{3}\log_{10} x - \frac{2}{3}$
97. $3 \ln(1 + 2e) - 3 \ln(1 - 2e)$
99. $\dfrac{\ln 2}{\ln[1 + (R/100)]}$ yrs **101.** 9.92%
103. $\dfrac{100 \ln 2}{R}$ yrs **105. a.** $7\frac{3}{4}$ yrs
b. $\dfrac{\ln n}{\ln[1 + (R/400)]}$ yrs **107. a.** $(0, \infty)$
b. $[1, \infty)$
109. a. $(-\infty, -3) \cup (-3, 5) \cup (5, \infty)$
b. $(-\infty, -3) \cup (5, \infty)$
111. $(-\infty, -1) \cup (1, \infty)$ **113.** -0.7
115. -2.2 **117.** 4.3 **119.** 2.7
121. 1.6
123. a.

Region	1990 Population (millions)	Growth Rate (%)	2025 Population
North America	275.2	0.7	351.6
Soviet Union	291.3	0.7	372.2
Europe	499.5	0.2	535.7
Nigeria	113.3	3.1	335.3

b. 222.0 million **c.** 193.5 million
d. They support the claim.
125. a. 1.0986 **b.** 0.00112%
c. $\ln\sqrt{3} \approx 0.5493$; $\ln 9 \approx 2.1972$; $\ln\frac{1}{3} \approx -1.0986$

Chapter Six Graphing Calculator Exercises

1. a.

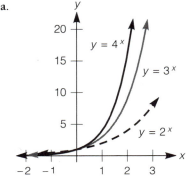

$y = 4^x$
$y = 3^x$
$y = 2^x$

b. It will be "narrower."

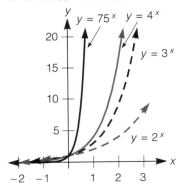

$y = 75^x$ $y = 4^x$
$y = 3^x$
$y = 2^x$

3. a. x-intercept: 2; y-intercept: $\frac{8}{9}$; asymptote: $y = 1$
b.

$y = 1$
$y = -3^{x-2} + 1$

5. a. *x*-intercept: none; *y*-intercept: 5; asymptote: $y = 4$

b.

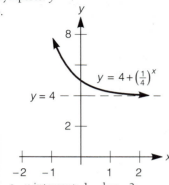

$$y = 4 + \left(\tfrac{1}{4}\right)^x$$

$$y = 4$$

7. a. *x*-intercept: $1 - \log_{10} 2$; *y*-intercept: 0.4; asymptote: $y = 0.5$

b.

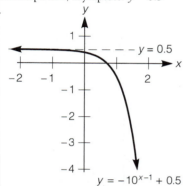

$$y = 0.5$$

$$y = -10^{x-1} + 0.5$$

9. a.

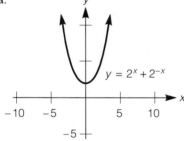

$$y = 2^x + 2^{-x}$$

b.

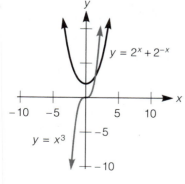

$$y = 2^x + 2^{-x}$$

$$y = x^3$$

11.

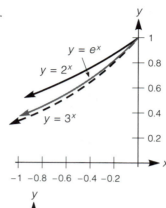

$$y = e^x$$
$$y = 2^x$$
$$y = 3^x$$

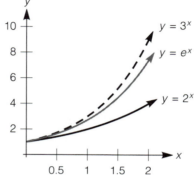

$$y = 3^x$$
$$y = e^x$$
$$y = 2^x$$

13. reflection across the *y*-axis

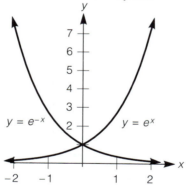

$$y = e^{-x}$$
$$y = e^x$$

15. reflection across the *x*-axis, reflection across the *y*-axis

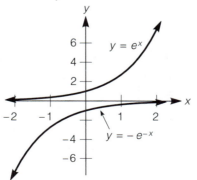

$$y = e^x$$
$$y = -e^{-x}$$

17. a. translation to the right one unit

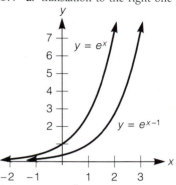

$$y = e^x$$
$$y = e^{x-1}$$

b. translation to the right one unit, reflection across the *y*-axis

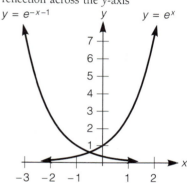

$$y = e^{-x-1}$$
$$y = e^x$$

19. a.

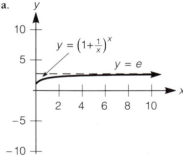

$$y = \left(1 + \tfrac{1}{x}\right)^x$$
$$y = e$$

b.

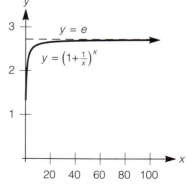

$$y = e$$
$$y = \left(1 + \tfrac{1}{x}\right)^x$$

c.

x	5	10	20
$\left(1+\dfrac{1}{x}\right)^x$	2.4883	2.5937	2.6533

x	50	100	10^3
$\left(1+\dfrac{1}{x}\right)^x$	2.6916	2.7048	2.7169

x	10^4	10^5	10^6
$\left(1+\dfrac{1}{x}\right)^x$	2.7181	2.7183	2.7183

21. a. Their graphs are reflections across $y = x$. **b.**

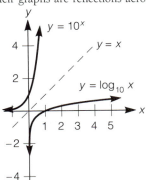

23. a. $(0, 1)$

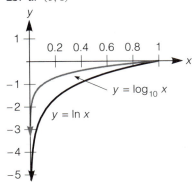

b. $(1, \infty)$

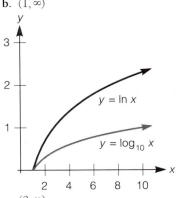

c. $(0, \infty)$

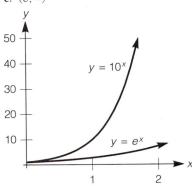

d. $(-\infty, 0)$

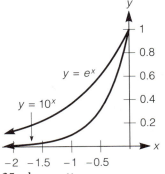

25. d.

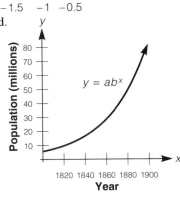

1. domain: $(-\infty, \infty)$; range: $(-3, \infty)$; x-intercept: $-(\ln 3)/(\ln 2)$; y-intercept: -2; asymptote: $y = -3$

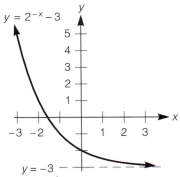

2. $\ln(5/3)/\ln 1.033 \approx 16$ hrs
3. $\log_2 17$; $\log_2 17 > 4 > \log_3 80$
4. $(\ln 15)/(\ln 2)$ **5.** 10^{12} **6.** 4 yrs
7. a. $(0, \infty)$ **b.**

b.

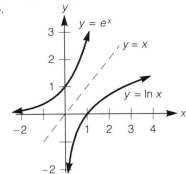

8. $3a - \frac{1}{2}b$ **9. a.** $-\frac{1}{2}$; **b.** -1
10. $0, -2, 2$ **11. a.** -0.1733
b. 0.35 g   **12.** $\ln\left[x^2/(x^2 + 1)^{1/3}\right]$
13. no solution   **14.** $g^{-1}(x) = 10^x + 1$; range: $(1, \infty)$ **15. a.** 12 yrs
b.

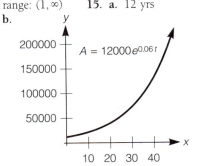

16. a. $(0, \infty)$ **b.** $1, e^2$

17. a.

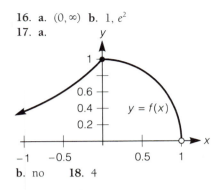

b. no **18.** 4

CHAPTER SEVEN

Exercise Set 7.1

1. a. $\sin \theta = \frac{15}{17}$, $\cos \theta = \frac{8}{17}$, $\tan \theta = \frac{15}{8}$, $\cot \theta = \frac{8}{15}$, $\sec \theta = \frac{17}{8}$, $\csc \theta = \frac{17}{15}$
b. $\sin \beta = \frac{8}{17}$, $\cos \beta = \frac{15}{17}$, $\tan \beta = \frac{8}{15}$, $\cot \beta = \frac{15}{8}$, $\sec \beta = \frac{17}{15}$, $\csc \beta = \frac{17}{8}$
3. a. $\sin \theta = \sqrt{5}/5$, $\cos \theta = 2\sqrt{5}/5$, $\tan \theta = \frac{1}{2}$, $\cot \theta = 2$, $\sec \theta = \sqrt{5}/2$, $\csc \theta = \sqrt{5}$ **b.** $\sin \beta = 2\sqrt{5}/2$, $\cos \beta = \sqrt{5}/5$, $\tan \beta = 2$, $\cot \beta = \frac{1}{2}$, $\sec \beta = \sqrt{5}$, $\csc \beta = \sqrt{5}/2$
5. a. $\cos A = 3\sqrt{13}/13$, $\sin A = 2\sqrt{13}/13$, $\tan A = \frac{2}{3}$ **b.** $\sec B = \sqrt{13}/2$ $\csc B = \sqrt{13}/3$, $\cot B = \frac{2}{3}$
7. $\sin B = \frac{12}{13}$, $\cos B = \frac{5}{13}$, $\tan B = \frac{12}{5}$, $\cot B = \frac{5}{12}$, $\sec B = \frac{13}{5}$, $\csc B = \frac{13}{12}$
9. a. $\sin B = \frac{4}{5}$, $\cos A = \frac{4}{5}$ **b.** $\sin A = \frac{3}{5}$, $\cos B = \frac{3}{5}$ **c.** 1 **11. a.** $\cos A = \frac{24}{25}$, $\sin A = \frac{7}{25}$, $\tan A = \frac{7}{24}$ **b.** $\cos B = \frac{7}{25}$, $\sin B = \frac{24}{25}$, $\tan B = \frac{24}{7}$ **c.** 1
13. $\sin \theta \approx 0.906$, $\cos \theta \approx 0.423$, $\tan \theta \approx 2.145$ **15.** $\sin \theta \approx 0.623$, $\cos \theta \approx 0.783$, $\tan \theta \approx 0.795$
17. $\sin \theta \approx 0.985$, $\cos \theta \approx 0.173$, $\tan \theta \approx 5.706$ **19.** $\sec \theta \approx 1.064$, $\csc \theta \approx 2.924$, $\cot \theta \approx 2.747$
21. $\sec \theta \approx 1.049$, $\csc \theta \approx 3.326$, $\cot \theta \approx 3.172$ **23.** $\sec \theta \approx 1.000$, $\csc \theta \approx 57.299$, $\cot \theta \approx 57.290$
25. Each side is $\frac{1}{2}$. **27.** Each side is $\frac{3}{2}$.
29. Each side is $\sqrt{3}/2$. **31.** Each side is $\frac{1}{2}$. **33.** Each side is $\sqrt{3}/3$.
35. Each side is 2.

37. a. $\cos 30° \approx 0.8660254038$; $\cos 45° \approx 0.7071067812$
b. $\sqrt{3}/2 \approx 0.8660254038$; $\sqrt{2}/2 \approx 0.7071067812$ **39. a.** $\cos \theta$
b. $\sec \beta$ **43.** $\sin 3° \approx 0.0523359562$; $\sin 6° \approx 0.1045284633$; $\sin 9° \approx 0.1564344650$; $\sin 12° \approx 0.2079116908$; $\sin 15° \approx 0.2588190451$; $\sin 18° \approx 0.3090169944$

Exercise Set 7.2

1. a. $11SC$ **b.** $11 \sin \theta \cos \theta$
3. a. $-8C^3S$ **b.** $-8 \cos^3 \theta \sin \theta$
5. a. $1 + 2T + T^2$
b. $1 + 2\tan \theta + \tan^2 \theta$
7. a. $T^2 + T - 6$ **b.** $\tan^2 \theta + \tan \theta - 6$
9. a. -1 **b.** -1 **11. a.** $(CS + 2)/S$
b. $\dfrac{\cos A \sin A + 2}{\sin A}$
13. a. $(T - 1)(T + 9)$
b. $(\tan \beta - 1)(\tan \beta + 9)$
15. a. $(2C + 1)(2C - 1)$
b. $(2 \cos B + 1)(2 \cos B - 1)$
17. a. $3ST^2(3ST + 2)$
b. $3 \sec B \tan^2 B (3 \sec B \tan B + 2)$
19. $\cos \theta = \sqrt{7}/4$, $\tan \theta = 3\sqrt{7}/7$, $\cot \theta = \sqrt{7}/3$, $\sec \theta = 4\sqrt{7}/7$, $\csc \theta = \frac{4}{3}$
21. $\sin \beta = \sqrt{22}/5$, $\tan \beta = \sqrt{66}/3$, $\cot \beta = \sqrt{66}/22$, $\sec \beta = 5\sqrt{3}/3$, $\csc \beta = 5\sqrt{22}/22$ **23.** $\cos A = \frac{12}{13}$, $\tan A = \frac{5}{12}$, $\cot A = \frac{12}{5}$, $\sec A = \frac{13}{12}$, $\csc A = \frac{13}{5}$ **25.** $\sin B = \frac{4}{5}$, $\cos B = \frac{3}{5}$, $\cot B = \frac{3}{4}$, $\sec B = \frac{5}{3}$, $\csc B = \frac{5}{4}$
27. $\sin C = \sqrt{5}/3$, $\cos C = \frac{2}{3}$, $\tan C = \sqrt{5}/2$, $\cot C = 2\sqrt{5}/5$, $\csc C = 3\sqrt{5}/5$ **29.** $\sin \alpha = \sqrt{3}/2$, $\cos \alpha = \frac{1}{2}$, $\tan \alpha = \sqrt{3}$, $\sec \alpha = 2$, $\csc \alpha = 2\sqrt{3}/3$ **31.** $\sin \theta \approx 0.887$, $\tan \theta \approx 1.916$, $\cot \theta \approx 0.522$, $\sec \theta \approx 2.162$, $\csc \theta \approx 1.128$
33. $\sin \theta \approx 0.768$, $\cos \theta \approx 0.640$, $\cot \theta \approx 0.833$, $\sec \theta \approx 1.562$, $\csc \theta \approx 1.302$ **35.** $\sin A + \cos A$
37. $\csc \theta$ **39.** $\cos^2 B$
41. $\cos A + 4$ **43.** $2 \csc \theta$ **45.** 0
47. 1 **49.** $\tan \theta$ **51.** -1
53. $\sin^2 \theta$ **55.** $-2 \cos \theta - 1$

Exercise Set 7.3

1. $BC = 30$ cm; $AC = 30\sqrt{3}$ cm
3. $AB = \frac{32}{3}\sqrt{3}$ cm; $BC = \frac{16}{3}\sqrt{3}$ cm
5. $AC \approx 11.5$ cm; $BC \approx 9.6$ cm
7. a. 15.59 ft **b.** 9 ft
9. 34 million miles **11.** 141.1 m
13. a. 27.3 ft **b.** 906.9 ft²
15. 1.50 in²
17. $\frac{7}{2} \sin (360°/7) \approx 2.736$ square units
19. $\pi - 2\sqrt{2} \approx 0.313$ square units
21. $\sqrt{3} \approx 1.732$ square units
25. 10,660 ft **27.** 136 m
29. $18(\sqrt{3} - 1)$ cm
31. a. $\angle BOA = 90° - \theta$; $\angle OAB = \theta$; $\angle BAP = 90° - \theta$; $\angle BPA = \theta$
b. $AO = \sin \theta$; $AP = \cos \theta$; $OB = \sin^2 \theta$; $BP = \cos^2 \theta$ **33. a.** $BC = 5/(\sin \theta)$
b. $AB = 4/(\cos \theta)$
c. $AC = 4 \sec \theta + 5 \csc \theta$
35. a. $DE = \sin \theta$ **b.** $OE = \cos \theta$
c. $CF = \tan \theta$ **d.** $OC = \sec \theta$
e. $AB = \cot \theta$ **f.** $OB = \csc \theta$
37. b. 1080 miles
39. b.

n	5	10	50	100
A_n	2.38	2.94	3.1333	3.1395

n	1,000	5,000	10,000
A_n	3.141572	3.1415918	3.1415924

c. As n gets larger, A_n becomes closer to the area of the circle, which is π.

Exercise Set 7.4

1. a. reference angle = 70°

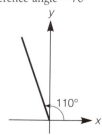

b. reference angle = 70°

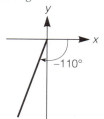

3. a. reference angle = 20°

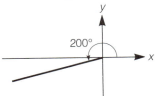

b. reference angle = 20°

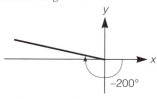

5. a. reference angle = 60°

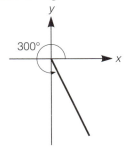

b. reference angle = 60°

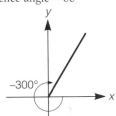

7. a. reference angle = 60°

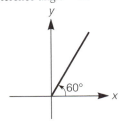

b. reference angle = 60°

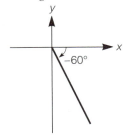

9. sin 270° = −1; cos 270° = 0; tan 270° is undefined; sec 270° is undefined; csc 270° = −1; cot 270° = 0 **11.** sin(−270°) = 1; cos(−270°) = 0; tan(−270°) is undefined; sec(−270°) is undefined; csc(−270°) = 1; cot(−270°) = 0 **13.** sin 810° = 1; cos 810° = 0; tan 810° is undefined; sec 810° is undefined; csc 810° = 1; cot 810° = 0 **15.** sin 10° ≈ 0.2 and sin(−10°) ≈ −0.2; sin 10° ≈ 0.17 and sin(−10°) ≈ −0.17 **17.** cos 80° ≈ 0.2 and cos(−80°) ≈ 0.2; cos 80° ≈ 0.17 and cos(−80°) ≈ 0.17 **19.** sin 120° ≈ 0.9 and sin(−120°) ≈ −0.9; sin 120° ≈ 0.87 and sin(−120°) ≈ −0.87 **21.** sin 150° = 0.5 and sin(−150°) = −0.5; sin 150° = 0.5 and sin(−150°) = −0.5 **23.** cos 220° ≈ −0.8 and cos(−220°) ≈ −0.8; cos 220° ≈ −0.77 and cos(−220°) ≈ −0.77 **25.** cos 310° ≈ 0.6 and cos(−310°) ≈ 0.6; cos 310° ≈ 0.64 and cos(−310°) ≈ 0.64 **27.** sin(40° + 360°) ≈ 0.6; sin(40° + 360°) ≈ 0.64 **29. a.** $\sqrt{2}/2$ **b.** $\sqrt{2}/2$ **c.** $-\sqrt{2}/2$ **d.** $\sqrt{2}/2$ **31. a.** $\frac{1}{2}$ **b.** $\frac{1}{2}$ **c.** $-\sqrt{3}/2$ **d.** $\sqrt{3}/2$ **33. a.** $-\sqrt{3}/2$ **b.** $-\sqrt{3}/2$ **c.** $-\frac{1}{2}$ **d.** $\frac{1}{2}$ **35. a.** $\sqrt{3}/2$ **b.** $\sqrt{3}/2$ **c.** $\frac{1}{2}$ **d.** $-\frac{1}{2}$ **37. a.** −2 **b.** $2\sqrt{3}/3$ **c.** $\sqrt{3}$ **d.** $-\sqrt{3}/3$

39.

θ	$\sin \theta$	$\cos \theta$	$\tan \theta$
0°	0	1	0
30°	1/2	$\sqrt{3}/2$	$\sqrt{3}/3$
45°	$\sqrt{2}/2$	$\sqrt{2}/2$	1
60°	$\sqrt{3}/2$	1/2	$\sqrt{3}$
90°	1	0	undefined
120°	$\sqrt{3}/2$	−1/2	$-\sqrt{3}$
135°	$\sqrt{2}/2$	$-\sqrt{2}/2$	−1
150°	1/2	$-\sqrt{3}/2$	$-\sqrt{3}/3$
180°	0	−1	0

41. $\frac{35}{4}\sqrt{3}$ cm² **43.** 302.7 cm² **45.** $108\sqrt{3} \approx 187.06$ cm²
47. a.

Terminal Side of Angle θ Lies In			
	Quadrant I	Quadrant II	Quadrant III	Quadrant IV
$\sin \theta$	positive	positive	negative	negative
$\cos \theta$	positive	negative	negative	positive
$\tan \theta$	positive	negative	positive	negative

49. a. sin 70° is larger. **b.** cos 10° is larger. **51. a.** cos 100° is larger. **b.** sin 100° is larger. **53. a.**

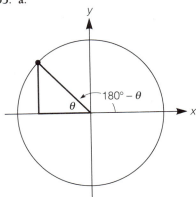

b. θ **c.** Since 180° − θ lies in the second quadrant where all y-coordinates are positive, sin(180° − θ) = sin θ.

55. d. $\sin 75° = (\sqrt{2} + \sqrt{6})/4$
f. $\sin 105° \approx 0.9659$;
$\sin 45° + \sin 60° \approx 1.5731 \neq \sin 105°$
57. a. $\sqrt{3}$ **b.** $-\sqrt{3}$ **c.** $\sqrt{3}/3$
59. a. 0.510 **b.** 0.510 **c.** 1

Exercise Set 7.5

1. a. $\cos \theta = -2\sqrt{6}/5$;
$\tan \theta = -\sqrt{6}/12$; $\cot \theta = -2\sqrt{6}$;
$\sec \theta = -5\sqrt{6}/12$; $\csc \theta = 5$
b. $\cos \theta = -2\sqrt{6}/5$; $\tan \theta = \sqrt{6}/12$;
$\cot \theta = 2\sqrt{6}$; $\sec \theta = -5\sqrt{6}/12$;
$\csc \theta = -5$ **3. a.** $\sin \theta = \frac{12}{13}$;
$\tan \theta = \frac{12}{5}$; $\cot \theta = \frac{5}{12}$; $\sec \theta = \frac{13}{5}$;
$\csc \theta = \frac{13}{12}$ **b.** $\sin \theta = -\frac{12}{13}$; $\tan \theta = \frac{12}{5}$;
$\cot \theta = \frac{5}{12}$; $\sec \theta = -\frac{13}{5}$; $\csc \theta = -\frac{13}{12}$
5. $\sin A = -\frac{1}{3}$; $\cos A = 2\sqrt{2}/3$;
$\tan A = -\sqrt{2}/4$; $\cot A = -2\sqrt{2}$;
$\sec A = 3\sqrt{2}/4$ **7.** $\cos B = -\frac{2}{3}$;
$\sin B = -\sqrt{5}/3$; $\tan B = \sqrt{5}/2$;
$\cot B = 2\sqrt{5}/5$; $\csc B = -3\sqrt{5}/5$
9. $\sin \theta = -\sqrt{9 - t^2}/3$;
$\tan \theta = -\sqrt{9 - t^2}/t$;
$\cot \theta = -t\sqrt{9 - t^2}/(9 - t^2)$; $\sec \theta = 3/t$;
$\csc \theta = -3\sqrt{9 - t^2}/(9 - t^2)$
11. $\cos \theta = -\sqrt{1 - 9u^2}$;
$\tan \theta = 3u\sqrt{1 - 9u^2}/(1 - 9u^2)$;
$\cot \theta = \sqrt{1 - 9u^2}/(3u)$;
$\sec \theta = -\sqrt{1 - 9u^2}/(1 - 9u^2)$;
$\csc \theta = -1/(3u)$
13. $\sin \theta = \sqrt{9 - 3u^2}/3$;
$\tan \theta = \sqrt{3 - u^2}/u$;
$\cot \theta = u\sqrt{3 - u^2}/(3 - u^2)$; $\sec \theta = \sqrt{3}/u$;
$\csc \theta = \sqrt{9 - 3u^2}/(3 - u^2)$
37. a. not an identity: let $\alpha = 30°$
b. is an identity
45. a. $(\cos \theta - \sin \theta)(1 + \cos \theta \sin \theta)$

Chapter Seven Review Exercises

1. $\sqrt{2}/2$ **3.** $-\sqrt{3}$ **5.** -2
7. -1 **9.** $\sqrt{2}/2$ **11.** 1
13. $-2\sqrt{3}/3$ **15.** 2
17. a. $\sin^2 (A + B) \neq \sin^2 A + \sin^2 B$
b. 1 **19.** $a = \frac{1}{2}$, $C = \sqrt{3}/2$
21. $b = \frac{35}{2}$ **23.** $\sin A = \sqrt{55}/8$;
$\cot A = 3\sqrt{55}/55$ **25.** 5 square units
27. $\frac{1}{5}$ **29.** $a = 16\sqrt{3}/3$

31. $c = 9$; area $= 9\sqrt{19}/4$
35. $\cos \theta = \dfrac{p^4 - q^4}{p^4 + q^4}$, $\tan \theta = \dfrac{2p^2q^2}{p^4 - q^4}$
37. $\sin A \cos A$ **39.** $\sin^2 A$
41. $\cos A + \sin A$ **43.** $\sin A \cos A$
45. $2 \sin A \cos A$
47. $\cos 10° \approx 0.9848$
49. $\sin 10° \approx 0.1736$
51. $\sin \theta = \frac{4}{5}$, $\tan \theta = \frac{4}{3}$
53. $\tan \theta = -\frac{24}{7}$ **55.** $\cot \theta = \frac{5}{12}$
57. $\tan(90° - \theta) = \dfrac{5t\sqrt{1 - 25t^2}}{1 - 25t^2}$
59. $\sin \theta = (2\sqrt{3} - \sqrt{6})/6$
63. 2521 cm^2 **79. a.** $PN = \sin \theta$
b. $ON = \cos \theta$ **c.** $PT = \tan \theta$
d. $OT = \sec \theta$ **e.** $NA = 1 - \cos \theta$
f. $NT = \sin \theta \tan \theta$
81. $A = (1 - \cos \theta)^2/(\sin \theta \cos \theta)$

Chapter Seven Test

1. a. $\sqrt{3}/3$ **b.** $\sqrt{2}$ **c.** 1 **d.** 0
2. a. negative **b.** negative **c.** positive
3. a. 1 **b.** -1 **c.** 0
4. $(2 \cot \theta + 3)(\cot \theta + 4)$ **5.** 18 cm^2
6. a. $\sqrt{2}/2$ **b.** $-\sqrt{3}/3$ **c.** -2
7. $\cos \theta = -2\sqrt{5}/5$, $\tan \theta = \frac{1}{2}$
8. $\cos \theta = \sqrt{9 - u^2}/3$,
$\cot \theta = -\sqrt{9 - u^2}/u$
9. $-2 \cos \theta - 1$
10. $\frac{81}{2} \sin 40° \approx 26.033 \text{ m}^2$ **11.** 1
12. $CD = 50(\tan 55° - \tan 25°)$
13. a. Since 85° has a much larger
y-coordinate than 5°, $\sin 85°$ is larger.
b. Since the x-coordinate is much larger
at 5° than the y-coordinate, $\cos 5°$ is
larger. **c.** Since the terminal side for 175°
lies in quadrant II, $\tan 175° < 0$. Since
the terminal side for 185° lies in quadrant
III, $\tan 185° > 0$. Thus $\tan 185°$ is larger.
14. a. $AD = 2.75 \cos 20°$;
$CD = 2.75 \sin 20°$;
$DB = 3.25 - 2.75 \cos 20°$
b. $CB \approx 1.15 \text{ cm}$

CHAPTER EIGHT

Exercise Set 8.1

1. a. $\pi/4 \approx 0.79$ radians
b. $\pi/2 \approx 1.57$ radians
c. $3\pi/4 \approx 2.36$ radians
3. a. 0 radians **b.** $2\pi \approx 6.28$ radians
c. $5\pi/2 \approx 7.85$ radians

5. a. 15° **b.** 30° **c.** 45°
7. a. 60° **b.** 300° **c.** 720°
9. a. 114.59° **b.** 171.89° **c.** 565.49°
11. smaller than a right angle
13. $30° = \pi/6$ radians, $45° = \pi/4$ radians,
$60° = \pi/3$ radians, $120° = 2\pi/3$ radians,
$135° = 3\pi/4$ radians, $150° = 5\pi/6$ radians
15. 2.5 radians **17.** 2 radians
19. $0.5 < \cos 1 < 0.6$; $0.8 < \sin 1 < 0.9$;
$\cos 1 \approx 0.54$; $\sin 1 \approx 0.84$
21. $0.5 < \cos(-1) < 0.6$;
$-0.9 < \sin(-1) < -0.8$; $\cos(-1) \approx 0.54$;
$\sin(-1) \approx -0.84$
23. $-0.7 < \cos 4 < -0.6$;
$-0.8 < \sin 4 < -0.7$; $\cos 4 \approx -0.65$;
$\sin 4 \approx -0.76$
25. $-0.7 < \cos(-4) < -0.6$;
$0.7 < \sin(-4) < 0.8$, $\cos(-4) \approx -0.65$;
$\sin(-4) \approx 0.76$
27. $0.8 < \sin(1 + 2\pi) < 0.9$;
$\sin(1 + 2\pi) \approx 0.84$ **29.** 4π ft
31. $\pi/2$ cm
33. a. $12\pi \text{ cm}^2 \approx 37.70 \text{ cm}^2$
b. $\frac{50}{9}\pi \text{ m}^2 \approx 17.45 \text{ m}^2$
c. $\frac{72}{5}\pi \text{ m}^2 \approx 45.24 \text{ m}^2$
d. $1.296\pi \text{ cm}^2 \approx 4.07 \text{ cm}^2$
35. $2\pi/5$ radians
37. a. 12π radians/sec **b.** 144π cm/sec
c. 72π cm/sec **39. a.** 6π radians/sec
b. 150π cm/sec **c.** 75π cm/sec
41. a. $50\pi/3$ radians/sec
b. 750π cm/sec **c.** 375π cm/sec
43. $180°/\pi \approx 57.30°$
45. a. 200π radians/min
b. 2000π cm/min **c.** $1000\pi/3$
radians/min **d.** $\frac{500}{3}$ rpm
47. 4930 miles **49.** 1470 miles
51. 2690 miles **53. a.** $r = 6/(1 + \theta)$
b. $A = 36\theta/(1 + \theta)^2$

Exercise Set 8.2

1.

θ	$\sin\theta$	$\cos\theta$	$\tan\theta$	$\csc\theta$	$\sec\theta$	$\cot\theta$
0	0	1	0	undef.	1	undef.
$\pi/6$	1/2	$\sqrt{3}/2$	$\sqrt{3}/3$	2	$2\sqrt{3}/3$	$\sqrt{3}$
$\pi/4$	$\sqrt{2}/2$	$\sqrt{2}/2$	1	$\sqrt{2}$	$\sqrt{2}$	1
$\pi/3$	$\sqrt{3}/2$	1/2	$\sqrt{3}$	$2\sqrt{3}/3$	2	$\sqrt{3}/3$
$\pi/2$	1	0	undef.	1	undef.	0
$2\pi/3$	$\sqrt{3}/2$	$-1/2$	$-\sqrt{3}$	$2\sqrt{3}/3$	-2	$-\sqrt{3}/3$
$3\pi/4$	$\sqrt{2}/2$	$-\sqrt{2}/2$	-1	$\sqrt{2}$	$-\sqrt{2}$	-1
$5\pi/6$	1/2	$-\sqrt{3}/2$	$-\sqrt{3}/3$	2	$-2\sqrt{3}/3$	$-\sqrt{3}$
π	0	-1	0	undef.	-1	undef.

13. $\cos t = -\frac{4}{5}$; $\tan t = \frac{3}{4}$
15. $\cos t = -\sqrt{13}/4$; $\tan t = -\sqrt{39}/13$
17. $\sec\alpha = \frac{13}{5}$; $\cos\alpha = \frac{5}{13}$; $\sin\alpha = \frac{12}{13}$
23. $\left(\sqrt{7}\cos\theta\right)/7$ **25. a.** $-\frac{2}{3}$ **b.** $\frac{1}{4}$
c. $\frac{1}{5}$ **d.** $-\frac{1}{5}$ **27. a.** $-\left(1 + 2\sqrt{2}\right)/3$
b. 1   **29. a.** $\sqrt{2}/2$ **b.** $\sqrt{3}/2$ **c.** 1
31. $\cos^2 t$ **33.** $\sin^2\theta$ **43.** $\frac{13}{31}$
47. a. 0 **b.** $\left(2 - \sqrt{2}\right)/2$ **c.** no
49.

θ	$1 - \dfrac{\theta^2}{2}$	$\cos\theta$
0.1	0.995	0.995004. . .
0.2	0.980	0.980066. . .
0.3	0.955	0.955336. . .

Exercise Set 8.3

1. period: 2; amplitude: 1
3. period: 4; amplitude: 6
5. period: 4; amplitude: 2
7. period: 6; amplitude: $\frac{3}{2}$
9. $(-7\pi/2, 1) \approx (-10.996, 1)$
11. $(5\pi/2, 1) \approx (7.854, 1)$
13. $(-4\pi, 0) \approx (-12.566, 0)$
15. $(-3\pi, 0) \approx (-9.425, 0)$
17. $(-\pi, 0) \approx (-3.142, 0)$
19. increasing   **21.** decreasing

23.

x	$\sin x$	$x - \sin x$
0.453	0.43766	0.01534
0.253	0.25031	0.00269
0.0253	0.02530	0.00000
0.00253	0.00253	0.00000

25. $(9\pi/2, 0) \approx (14.137, 0)$
27. $(-4\pi, 1) \approx (-12.566, 1)$
29. $(\pi/2, 0) \approx (1.571, 0)$
31. $(4\pi, 1) \approx (12.566, 1)$
33. $(-5\pi/2, 0) \approx (-7.854, 0)$
35. decreasing   **37.** increasing
39. a. $C(\cos\theta, \sin\theta)$

Exercise Set 8.4

1. a. amplitude: 2; period: 2π;
x-intercepts: 0, π, 2π;
increasing: $(0, \pi/2)$ and $(3\pi/2, 2\pi)$

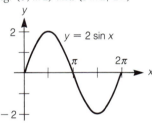

b. amplitude: 1; period: π;
x-intercepts: 0, $\pi/2$, π;
increasing: $(\pi/4, 3\pi/4)$

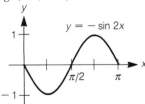

3. a. amplitude: 1; period: π;
x-intercepts: $\pi/4$, $3\pi/4$;
increasing: $(\pi/2, \pi)$

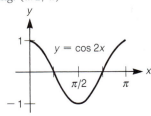

b. amplitude: 2; period: π;
x-intercepts: $\pi/4$, $3\pi/4$;
increasing: $(\pi/2, \pi)$

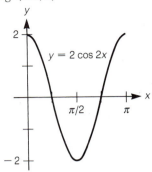

5. a. amplitude: 3; period: 4;
x-intercepts: 0, 2, 4;
increasing: $(0, 1)$ and $(3, 4)$

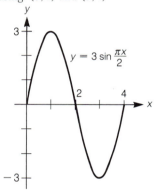

b. amplitude: 3; period: 4;
x-intercepts: 0, 2, 4; increasing: $(1, 3)$

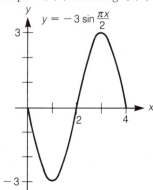

7. a. amplitude: 1; period: 1;
x-intercepts: $\frac{1}{4}$, $\frac{3}{4}$; increasing: $\left(\frac{1}{2}, 1\right)$

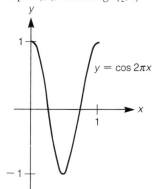

b. amplitude: 4; period: 1;
x-intercepts: $\frac{1}{4}$, $\frac{3}{4}$; increasing: $\left(0, \frac{1}{2}\right)$

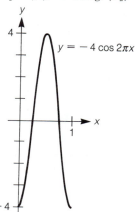

9. amplitude: 1; period: π;
x-intercept: $3\pi/4$;
increasing: $(0, \pi/4)$ and $(3\pi/4, \pi)$

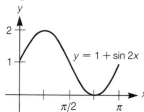

11. amplitude: 1; period: 6;
x-intercepts: 0, 6; increasing: $(0, 3)$

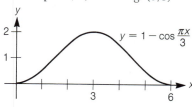

13. amplitude: 1; period: 2π;
phase shift: $\pi/6$;
x-intercepts: $\pi/6$, $7\pi/6$, $13\pi/6$;
high point: $(2\pi/3, 1)$;
low point: $(5\pi/3, -1)$

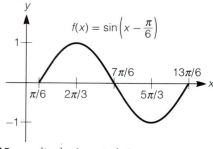

15. amplitude: 1; period: 2π;
phase shift: $-\pi/4$;
x-intercepts: $\pi/4$, $5\pi/4$;
high point: $(3\pi/4, 1)$;
low points: $(-\pi/4, -1)$ and $(7\pi/4, -1)$

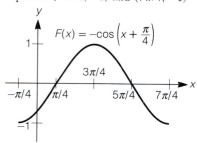

17. amplitude: 1; period: π;
phase shift: $\pi/4$;
x-intercepts: $\pi/4$, $3\pi/4$, $5\pi/4$;
high point: $(\pi/2, 1)$; low point: $(\pi, -1)$

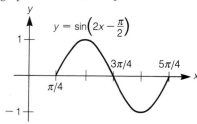

19. amplitude: 1; period: π;
phase shift: $\pi/2$; x-intercepts: $3\pi/4$, $5\pi/4$;
high points: $(\pi/2, 1)$ and $(3\pi/2, 1)$;
low point: $(\pi, -1)$

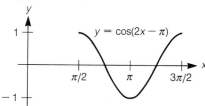

21. amplitude: 3; period: 4π;
phase shift: $-\pi/3$;
x-intercepts: $-\pi/3$, $5\pi/3$, $11\pi/3$;
high point: $(2\pi/3, 3)$;
low point: $(8\pi/3, -3)$

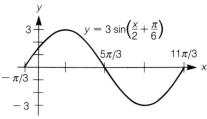

23. amplitude: 4; period: $2\pi/3$;
phase shift: $\pi/12$;
x-intercepts: $\pi/4$, $7\pi/12$;
high points: $(\pi/12, 4)$ and $(3\pi/4, 4)$;
low point: $(5\pi/12, -4)$

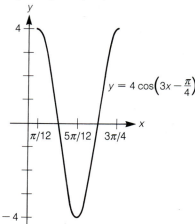

25. amplitude: $\frac{1}{2}$; period: 4;
phase shift: 2π;
x-intercepts: 2π, $2\pi + 2$, $2\pi + 4$;
high point: $\left(2\pi + 1, \frac{1}{2}\right)$;
low point: $\left(2\pi + 3, -\frac{1}{2}\right)$

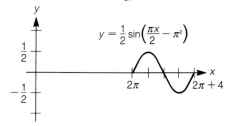

27. amplitude: 1; period: π;
phase shift: $\pi/6$;
x-intercepts: $\pi/6$ and $7\pi/6$;
high point: $(2\pi/3, 2)$;
low points: $(\pi/6, 0)$ and $(7\pi/6, 0)$

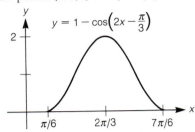

$$y = 1 - \cos\left(2x - \frac{\pi}{3}\right)$$

29. $A = 2, B = \frac{1}{2}$ **31.** $A = -3, B = \pi$
33. $A = -4, B = \frac{1}{5}$

35.

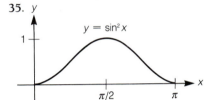

$$y = \sin^2 x$$

37.

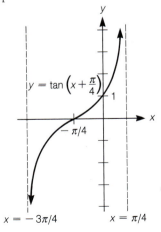

$$y = \sin x \cos x$$

Exercise Set 8.5

1. a. x-intercept: $-\pi/4$; y-intercept: 1;
asymptotes: $x = -3\pi/4$ and $x = \pi/4$

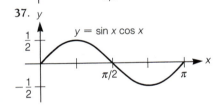

$$y = \tan\left(x + \frac{\pi}{4}\right)$$

b. x-intercept: $-\pi/4$; y-intercept: -1;
asymptotes: $x = -3\pi/4$ and $x = \pi/4$

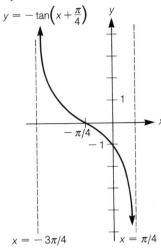

$$y = -\tan\left(x + \frac{\pi}{4}\right)$$

3. a. x- and y-intercepts: 0;
asymptotes: $x = -3\pi/2$ and $x = 3\pi/2$

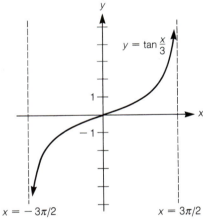

$$y = \tan\frac{x}{3}$$

b. x- and y-intercepts: 0;
asymptotes: $x = -3\pi/2$ and $x = 3\pi/2$

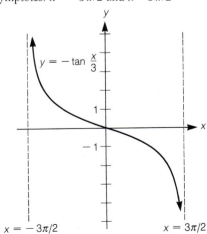

$$y = -\tan\frac{x}{3}$$

5. x- and y-intercepts: 0;
asymptotes: $x = -1$ and $x = 1$

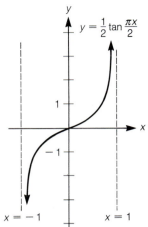

$$y = \frac{1}{2}\tan\frac{\pi x}{2}$$

7. x-intercept: 1; no y-intercept;
asymptotes: $x = 0$ and $x = 2$

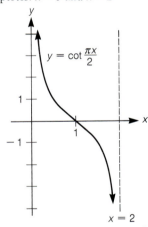

$$y = \cot\frac{\pi x}{2}$$

9. x-intercept: $3\pi/4$; y-intercept: 1;
asymptotes: $x = \pi/4$ and $x = 5\pi/4$

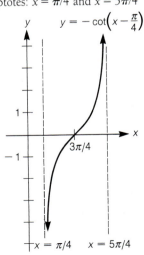

$$y = -\cot\left(x - \frac{\pi}{4}\right)$$

11. x-intercept: $\pi/4$; no y-intercept; asymptotes: $x = 0$ and $x = \pi/2$

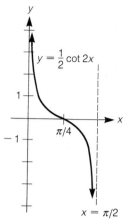

13. no x-intercept; y-intercept: $-\sqrt{2}$; asymptotes: $x = -3\pi/4$, $x = \pi/4$, and $x = 5\pi/4$

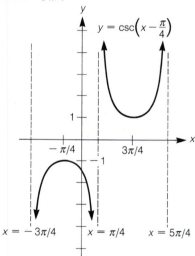

15. no x- or y-intercepts; asymptotes: $x = -2\pi$, $x = 0$, and $x = 2\pi$

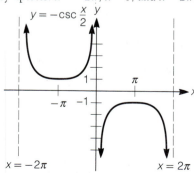

17. no x- or y-intercepts; asymptotes: $x = -1$, $x = 0$, and $x = 1$

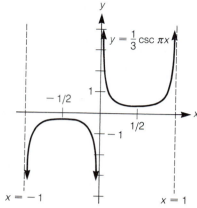

19. no x-intercept; y-intercept: -1; asymptotes: $x = -\pi/2$, $x = \pi/2$, and $x = 3\pi/2$

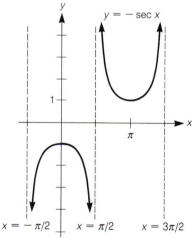

21. no x-intercept; y-intercept: -1; asymptotes: $x = \pi/2$, $x = 3\pi/2$, and $x = 5\pi/2$

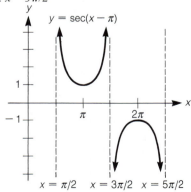

23. no x-intercept; y-intercept: 3; asymptotes: $x = -1$, $x = 1$, and $x = 3$

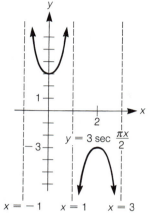

25. a. x-intercepts: $-\frac{11}{18}$, $-\frac{5}{18}$, $\frac{1}{18}$, $\frac{7}{18}$, and $\frac{13}{18}$; y-intercept: -1; no asymptotes

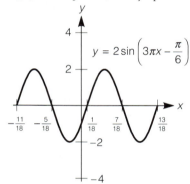

b. no x-intercepts; y-intercept: -4; asymptotes: $x = -\frac{11}{18}$, $x = -\frac{5}{18}$, $x = \frac{1}{18}$, $x = \frac{7}{18}$, and $x = \frac{13}{18}$

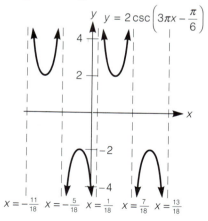

27. a. x-intercepts: $-\frac{5}{8}$, $-\frac{1}{8}$, $\frac{3}{8}$, and $\frac{7}{8}$; y-intercept: $-3\sqrt{2}/2$; no asymptotes

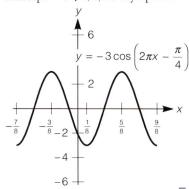

b. no x-intercepts; y-intercept: $-3\sqrt{2}$; asymptotes: $x = -\frac{5}{8}$, $x = -\frac{1}{8}$, $x = \frac{3}{8}$, and $x = \frac{7}{8}$

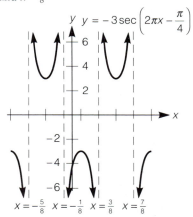

29.

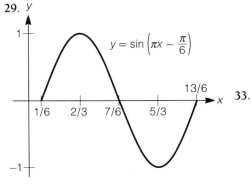

b.

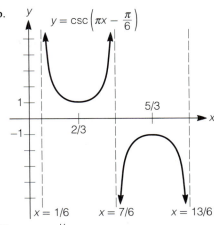

31. a.

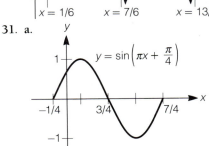

b.

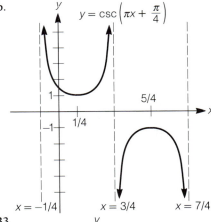

33.

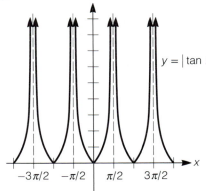

35.

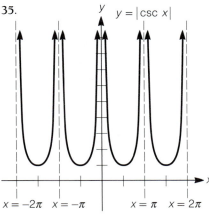

Exercise Set 8.6

1. $\sin 3\theta$ **3.** $\sin 2\theta$ **5.** $\cos 5u$
7. $\sqrt{3}/2$ **9.** $\sin B$ **11.** $\cos \theta$
13. $-\cos \theta$ **15.** $\sin t$
17. $(\sqrt{6} - \sqrt{2})/4$
19. $(\sqrt{6} + \sqrt{2})/4$ **21.** $\sqrt{2}\sin s$
23. $\sqrt{3}\sin \theta$ **25. a.** $-\frac{16}{65}$ **b.** $\frac{63}{65}$
27. a. $-\frac{44}{125}$ **b.** $-\frac{4}{5}$ **29. a.** $2\sqrt{6}/5$
b. $4\sqrt{6}/25$

31. $\sin(\theta + \beta) = \dfrac{2\sqrt{39} - 3\sqrt{13}}{26}$;

$\cos(\beta - \theta) = \dfrac{2\sqrt{13} - 3\sqrt{39}}{26}$

37. a. The graph of $y = \cos x$ is obtained by translating $y = \sin x$ by $\pi/2$ units to the left.
47. a.

t	1	2	3	4
$f(t)$	1.5	1.5	1.5	1.5

b. $f(t) = 1.5$ **53.** $\sqrt{3}/2$

Exercise Set 8.7

1. a. $\frac{4}{3}$ **b.** $\frac{44}{117}$ **3. a.** $3 - 2\sqrt{2}$
b. $3 - 2\sqrt{2}$ **5. a.** $\frac{336}{625}$ **b.** $-\frac{527}{625}$
c. $-\frac{336}{527}$ **7. a.** $-\frac{8}{17}$ **b.** $-\frac{15}{17}$ **c.** $\frac{8}{15}$
9. a. $\frac{1}{2}$ **b.** $\sqrt{3}/2$ **c.** $\sqrt{3}/3$
11. a. $2\sqrt{2}/3$ **b.** $\frac{1}{3}$ **c.** $2\sqrt{2}$
13. a. $-3\sqrt{7}/8$ **b.** $-\frac{1}{8}$
c. $\sqrt{8 + 2\sqrt{7}}/4$ **d.** $\sqrt{8 - 2\sqrt{7}}/4$

15. a. $4\sqrt{2}/9$ **b.** $-\frac{7}{9}$ **c.** $\sqrt{6}/3$
d. $-\sqrt{3}/3$ **17. a.** $\sqrt{2-\sqrt{3}}/2$
b. $\sqrt{2+\sqrt{3}}/2$ **c.** $2-\sqrt{3}$
19. a. $\sqrt{2+\sqrt{3}}/2$ **b.** $-\sqrt{2-\sqrt{3}}/2$
c. $-2-\sqrt{3}$ **21. a.** $\frac{24}{25}$ **b.** $\frac{7}{25}$ **c.** $\frac{24}{7}$
23. a. $\frac{24}{25}$ **b.** $-\frac{7}{25}$ **c.** $-\frac{24}{7}$
25. a. $\sqrt{10}/10$ **b.** $3\sqrt{10}/10$ **c.** $\frac{1}{3}$
27. a. $\sqrt{5}/5$ **b.** $2\sqrt{5}/5$ **c.** $\frac{1}{2}$
29. $\sin 2\theta = 2x\sqrt{25-x^2}/25$;
$\cos 2\theta = (25-2x^2)/25$
31. $\sin 2\theta = (x-1)\sqrt{3+2x-x^2}/2$;
$\cos 2\theta = (1+2x-x^2)/2$
33. $\frac{1}{8}(3-4\cos 2\theta + \cos 4\theta)$
35. $\frac{1}{8}(3-4\cos\theta + \cos 2\theta)$
63. a. $2\sin 8\theta + 2\sin 4\theta$
b. $\cos 2A - \cos 4A$ **c.** $\frac{1}{2}\cos 6\theta + \frac{1}{2}\cos 2\theta$

Exercise Set 8.8

1. yes **3.** no **5.** $(\pi/3)+2\pi k$ or
$(2\pi/3)+2\pi k$, where k is any integer
7. $(7\pi/6)+2\pi k$ or $(11\pi/6)+2\pi k$, where
k is any integer **9.** $\pi + 2\pi k$, where k
is any integer **11.** $(\pi/3)+\pi k$, where
k is any integer **13.** πk, where k is
any integer **15.** $(\pi/2)+\pi k$,
$(2\pi/3)+2\pi k$, or $(4\pi/3)+2\pi k$, where k
is any integer **17.** πk, where k is any
integer **19.** $(\pi/6)+2\pi k$,
$(5\pi/6)+2\pi k$, or $(3\pi/2)+2\pi k$, where k
is any integer **21.** $(\pi/4)+2\pi k$ or
$(3\pi/4)+2\pi k$, where k is any integer
23. $0°$, $120°$, or $240°$ **25.** $75°$, $105°$,
$195°$, $225°$, $315°$, or $345°$ **27.** $60°$,
$180°$, or $300°$ **29.** $30°$, $120°$, $210°$,
or $300°$ **31.** $14.5°$ or $165.5°$
33. $116.6°$ or $296.6°$ **35.** $128.2°$ or
$231.8°$ **37.** 1.39 or 4.90
39. 0.46 or 2.68 **41.** 1.41 or 4.55
43. 1.37 or 4.51 **45.** 1.11 or 5.18
47. 0 or π **49.** $2\pi/3$ or π
51. $(3\pi/16)+(k\pi/4)$, where k is any
integer **53.** $60.45°$
57. a. 1000.173 **b.** 1001.022

Exercise Set 8.9

1. $\frac{\pi}{3}$ **3.** $\frac{\pi}{3}$ **5.** $-\frac{\pi}{6}$ **7.** $\frac{\pi}{4}$
9. undefined **11.** $\frac{1}{4}$ **13.** $\frac{3}{4}$
15. $-\frac{\pi}{7}$ **17.** $\frac{\pi}{2}$ **19.** 0 **21.** $\frac{4}{3}$
23. $\frac{\sqrt{2}}{2}$ **25.** $\frac{12}{5}$ **27.** $\frac{1}{2}$

29. $\frac{2\sqrt{2}}{3}$ **31. a.** 0.84; **b.** 0.84;
c. 1.26; **d.** 0.90 **33.** $\sqrt{2}$
37. $\frac{1}{4}\sin^{-1}(3x/2) - \dfrac{3x\sqrt{4-9x^2}}{2}$
39. $\tan^{-1}\left(\dfrac{x-1}{2}\right) - \dfrac{2}{\sqrt{5-2x+x^2}}$
41. $\frac{8}{17}$ **43.** $\frac{31\sqrt{218}}{1090}$ **47.** $\frac{1}{2}$, 1
51. $\pm\frac{1}{3}$ **53.** $\frac{1}{2}$
55.

$x = \frac{3\pi}{2}$
π
$x = \frac{\pi}{2}$
$y = \sec x$

b.

$y = \frac{3\pi}{2}$
$y = \sec^{-1} x$
π
$y = \frac{\pi}{2}$

c. $\sec^{-1}\left(\frac{2}{\sqrt{3}}\right) = \frac{\pi}{6}$; $\sec^{-1}\left(-\frac{2}{\sqrt{3}}\right) = \frac{7\pi}{6}$
d. $\sec^{-1}\left(\sqrt{2}\right) = \frac{\pi}{4}$; $\sec^{-1}\left(-\sqrt{2}\right) = \frac{5\pi}{4}$
e. $\sec(\sec^{-1} 2) = 2$; $\sec^{-1}(\sec 0) = 0$

Chapter Eight Review Exercises

1. a. $-\sqrt{3}/2$ **b.** $-\sqrt{3}$ **3.** $2\sin t$
5.

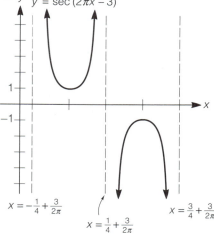

$y = \sec(2\pi x - 3)$
$x = -\frac{1}{4} + \frac{3}{2\pi}$
$x = \frac{1}{4} + \frac{3}{2\pi}$
$x = \frac{3}{4} + \frac{3}{2\pi}$

7. amplitude: 1; period: π;
phase shift: $\pi/2$

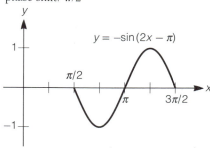

$y = -\sin(2x - \pi)$
$\pi/2$
π
$3\pi/2$

9. -1 **11.** $2\sqrt{3}/3$ **13.** $-\sqrt{3}/3$
15. $\frac{1}{2}$ **17.** 1 **19.** -2
21. 0.841 **23.** -1 **25.** 0.0123
31. $5\cos\theta$ **33.** $10\tan\theta$
35. $\left(\sqrt{5}/5\right)\cos\theta$ **37.** $-\frac{15}{8}$
39. $A = \pi - \theta - \sin\theta$
41. $(\pi - 2)/2$ cm^2 **43.** $y = 4\sin x$
45. $y = -2\cos 4x$
47. x-intercepts: $\pi/8$ and $3\pi/8$;
high point: $(\pi/4, 3)$;
low points: $(0, -3)$ and $(\pi/2, -3)$

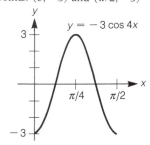

$y = -3\cos 4x$
$\pi/4$
$\pi/2$

49. x-intercepts: $\frac{1}{2}$, $\frac{5}{2}$, and $\frac{9}{2}$; high point: $\left(\frac{3}{2}, 2\right)$; low point: $\left(\frac{7}{2}, -2\right)$

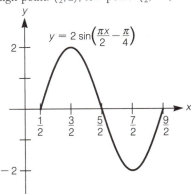

51. x-intercepts: $\frac{5}{2}$ and $\frac{11}{2}$; high points: $(1, 3)$ and $(7, 3)$; low point: $(4, -3)$

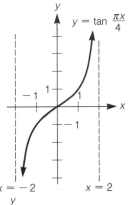

53. a.

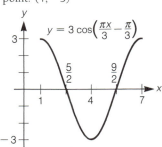

b.

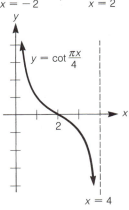

55. a.

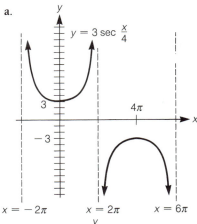

b.

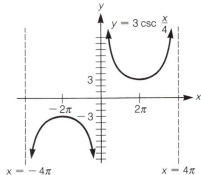

57. $-\cos x$ **59.** $-\cos x$
61. $-\sqrt{3}/2$ **63.** $\sqrt{3}\sin x$
65. -1 **67. a.** $2 - \sqrt{3}$ **b.** $2 + \sqrt{3}$
101. 1.34 or 4.48 **103.** 0.46 or 2.68
105. $\pi/3$, $2\pi/3$, $4\pi/3$, or $5\pi/3$
107. 0, π, $7\pi/6$, or $11\pi/6$
109. $\pi/3$, $2\pi/3$, $4\pi/3$, or $5\pi/3$
111. $\pi/4$, $\pi/2$, $3\pi/4$, $5\pi/4$, $3\pi/2$, or $7\pi/4$ **113.** $\frac{3\pi}{4}$ **115.** 0
117. $\frac{\sqrt{3}}{2}$ **119.** $\frac{\sqrt{3}}{3}$ **121.** $-\frac{3}{5}$
123. b. $x = 10$ yields the largest value for θ

x	0.1	1	2	3	10	100
θ	0.01	0.15	0.27	0.35	0.40	0.06

c. $2\sqrt{10} \approx 6.32$; $\theta \approx 0.44$; $(6.32, 0.44)$
129. a. 0.0625

1. a.

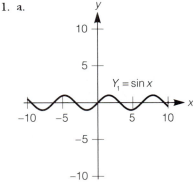

b.

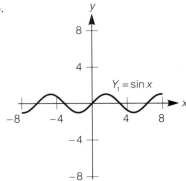

c. π, 2π **d.** $\pi/2$

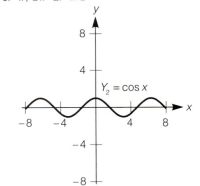

e.

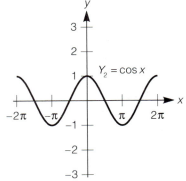

f. $\pi/2$ units

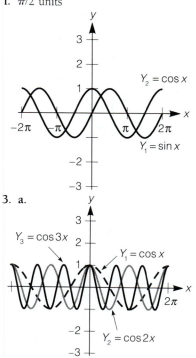

3. a.

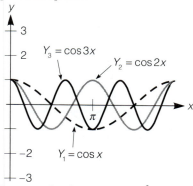

b. amplitude (all three): 1;
period of Y_1: 2π; period of Y_2: π;
period of Y_3: $2\pi/3$

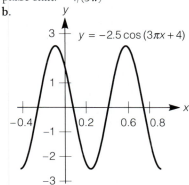

5. a. amplitude: 2.5; period: $\frac{2}{3}$;
phase shift: $-4/(3\pi)$

b.

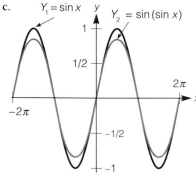

c. highest points:
$\left(\frac{1}{3} - \frac{4}{3\pi}, 2.5\right) \approx (-0.09, 2.5)$ and
$\left(1 - \frac{4}{3\pi}, 2.5\right) \approx (0.58, 2.5)$
lowest points:
$\left(-\frac{4}{3\pi}, -2.5\right) \approx (-0.42, -2.5)$,
$\left(\frac{2}{3} - \frac{4}{3\pi}, -2.5\right) \approx (0.24, -2.5)$,
$\left(\frac{4}{3} - \frac{4}{3\pi}, -2.5\right) \approx (0.91, -2.5)$

7. a. amplitude: 1; period: 4π;
phase shift: -1.5

b.

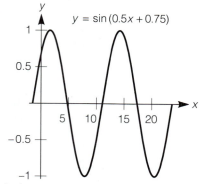

c. highest points:
$(-1.5 + \pi, 1) \approx (1.64, 1)$ and
$(-1.5 + 5\pi, 1) \approx (14.2, 1)$
lowest points:
$(-1.5 + 3\pi, -1) \approx (7.92, -1)$ and
$(-1.5 + 7\pi, -1) \approx (20.49, -1)$

9. a. amplitude: 0.02; period: 0.02;
phase shift: -0.04

b.

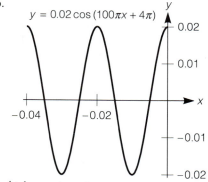

c. highest points: $(-0.04, 0.02)$,
$(-0.02, 0.02)$, and $(0, 0.02)$
lowest points: $(-0.03, -0.02)$ and
$(-0.01, -0.02)$

11. b. The two graphs are not identical.

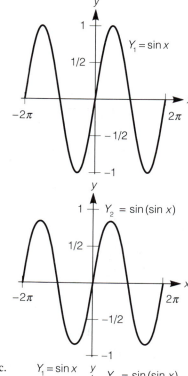

c.

d. $(1.58, 0.84)$. The y-coordinate gives
the amplitude. **e.** Since $\sin x$ is largest
when $x = \pi/2$ (then $\sin x = 1$), and $\sin x$
is increasing for $0 < x < \pi/2$, $\sin(\sin x)$
will have its largest value when $x = \pi/2$.
Its value then is $\sin[\sin(\pi/2)] = \sin 1$.
f. $\sin 1 \approx 0.84$

Section 8.5 Graphing Calculator Exercises

1.

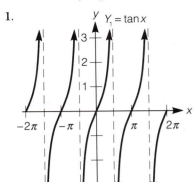
$Y_1 = \tan x$

3.

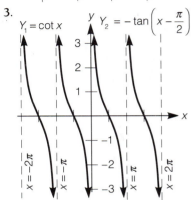
$Y_1 = \cot x$ $Y_2 = -\tan\left(x - \dfrac{\pi}{2}\right)$

5. a.

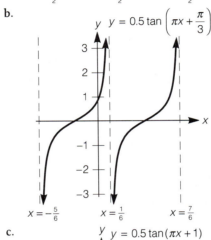
$y = \tan\dfrac{x}{4}$

b.

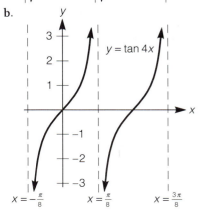
$y = \tan 4x$

7. a.

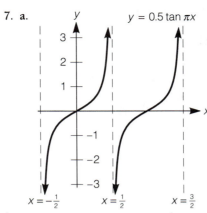
$y = 0.5\tan \pi x$

b.

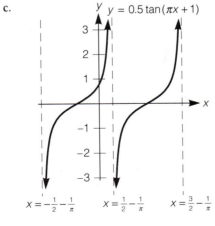
$y = 0.5\tan\left(\pi x + \dfrac{\pi}{3}\right)$

c.

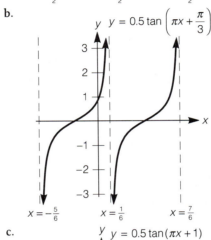
$y = 0.5\tan(\pi x + 1)$

9. a.

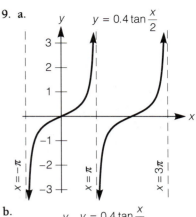
$y = 0.4\tan\dfrac{x}{2}$

b.

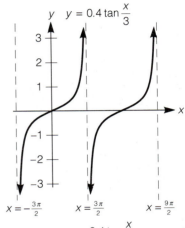
$y = 0.4\tan\dfrac{x}{3}$

c.

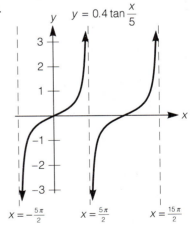
$y = 0.4\tan\dfrac{x}{5}$

11. $\sin x = \csc x$ at $-3\pi/2$, $-\pi/2$, $\pi/2$, and $3\pi/2$; there are no points at which $\sin x = -\csc x$.

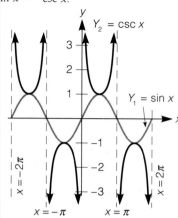

13.

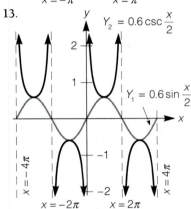

15.

$$Y_1 = 3\cos\left(2x - \frac{\pi}{6}\right)$$

$$Y_2 = 3\sec\left(2x - \frac{\pi}{6}\right)$$

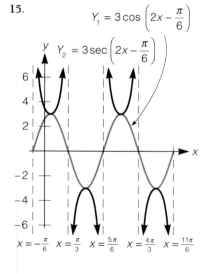

17. The two graphs are identical. This demonstrates that $\tan^2 x = \sec^2 x - 1$ is a trigonometric identity.

$$Y_1 = \tan^2 x$$
$$Y_2 = \sec^2 x - 1$$

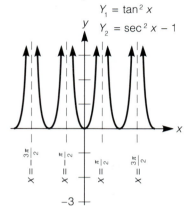

19. a.

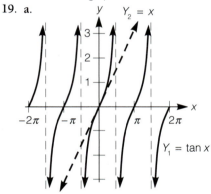

b.

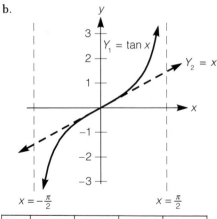

x	0.000123	0.01	0.05	0.1
$\tan x$	0.000123	0.010000	0.050042	0.100335

x	0.2	0.3	0.4	0.5
$\tan x$	0.202710	0.309336	0.422793	0.546302

c.

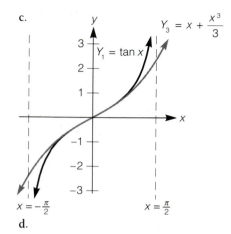

d.

x	0.000123	0.01	0.05	0.1
$\tan x$	0.000123	0.010000	0.050042	0.100335
$x + (x^3/3)$	0.000123	0.010000	0.050042	0.100333

x	0.2	0.3	0.4	0.5
$\tan x$	0.202710	0.309336	0.422793	0.546302
$x + (x^3/3)$	0.202667	0.309	0.421333	0.541667

21. a. The graphs intersect at $x \approx 4.5$.

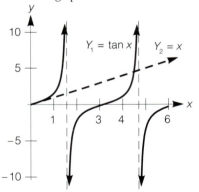

b. The graphs intersect at $x \approx 4.493$.

Section 8.8 Graphing Calculator Exercises

1. a.

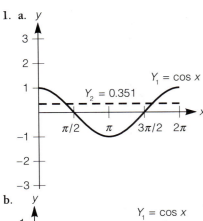

b.

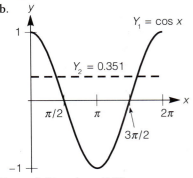

3. $x \approx 0.90$ and $x \approx 5.38$
5. $x \approx 0.43$ and $x \approx 2.71$
7. a. no solutions **b.** $x \approx 1.93$, $x \approx 2.28$, $x \approx 5.07$, and $x \approx 5.42$
9. $x \approx 0$, $x \approx 0.69$, and $x \approx 4.26$
11. $x \approx 67.5°$ and $x \approx 247.5°$
13. $x \approx 120.8°$ and $x \approx 329.2°$
15. $x = 0°$, $x \approx 1.0°$, and $x \approx 193.9°$

Section 8.9 Graphing Calculator Exercises

1. a. The graphs are the same.

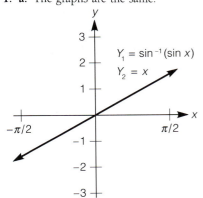

b. $\sin^{-1}(\sin x) = x$, and thus $\sin^{-1} x$ is the inverse function of $\sin x$ on the interval $-\pi/2 \le x \le \pi/2$.
3. a. The graph appears to be constant.

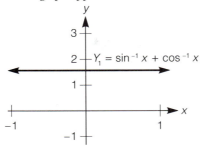

b. $\pi/2$; $\sin^{-1} x + \cos^{-1} x = \pi/2$
5. a. The graphs are identical. This demonstrates that $\tan(\tan^{-1} x) = x$, at least for the values of x that are graphed.

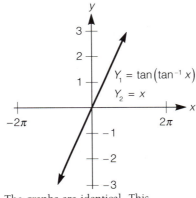

b. The graphs are identical. This demonstrates that $\tan^{-1}(\tan x) = x$ for $-\pi/2 < x < \pi/2$.

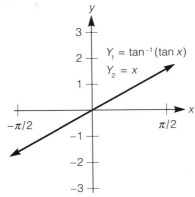

c. It was necessary to change the range settings since $\tan x$ is one-to-one on the interval $-\pi/2 < x < \pi/2$ **7.** $x \approx 0.79$
9. $x \approx 0.80$ **11.** $x = 0$, $x \approx 0.80$
13. a. $x = 0$, $x \approx 0.98$ **b.** $x = 0$
15. a. $x \approx 0.54$ **b.** $x \approx 0.50$

Chapter Eight Test

1. a. $-\frac{1}{2}$ **b.** -2 **c.** 1 **2.** $\frac{1}{4} \csc u$
3.

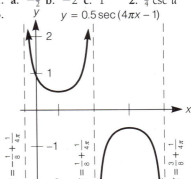

4. amplitude: 1; period: $2\pi/3$; phase shift: $\pi/12$

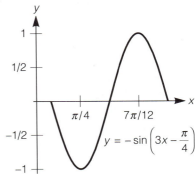

5.

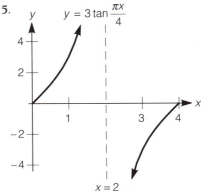

6. a. $\frac{35\pi}{36}$ radians **b.** $\frac{900°}{\pi}$
7. a. $\frac{25\pi}{12}$ cm **b.** $\frac{125\pi}{24}$ cm^2
8. a. 50π radians/sec **b.** 250π cm/sec
10. $-\cos\theta$ **11.** $-\frac{3}{5}$
12. 1.25 or 4.39 **13.** $7\pi/6$, $11\pi/6$
14. $\sqrt{5}/5$ **15.** $30°$
16. $\sqrt{18 + 12\sqrt{2}}/6$ **17. a.** $\pi/10$
b. 0 **18.** $\sqrt{7}/4$ **20.** x

Exercise Set 9.1

1. $4\sqrt{6}$ cm   **3.** 20 sin 50° cm
5. $a \approx 9.7$ cm; $c \approx 16.4$ cm
7. $A \approx 63.3°$; $C \approx 50.7°$; $c \approx 25.5$ cm
9. a. 45° or 135° **b.** 45° **c.** 14.5° or
165.5° **d.** 131.8°   **13. b.** $\angle C = 105°$;
$c \approx 1.93$ **c.** $\angle C = 15°$; $c \approx 0.52$
d. 0.68 and 0.18
15. $a = \dfrac{2 \sin 70°}{\sin 20°}$ cm; $b = \dfrac{2 \sin 50°}{\sin 20°}$ cm;

$c = \dfrac{2 \sin 50° \sin 70°}{\sin 20° \sin 85°}$ cm;

$d = \dfrac{2 \sin 50° \sin 15°}{\sin 20° \sin 85°}$ cm;   **17.** 160 ft
19. a. 7 cm **b.** $\sqrt{129}$ cm
21. a. 7.5 cm **b.** 17.7 cm
23. This is incorrect because x is not
the length of the side opposite the 130°
angle. The correct equation is
$6^2 = x^2 + 3^2 - 2(x)(3)\cos 130°$.
25. $\cos A = \frac{113}{140}$; $\cos B = \frac{29}{40}$; $\cos C = -\frac{5}{28}$
27. $A \approx 27.8°$; $B \approx 32.2°$; $C \approx 120°$
29. $A = 30°$; $B = 30°$; $C = 120°$
31. 5.9 units   **33. a.** $a \approx 4.2$ cm
b. $C \approx 29.3°$ **c.** $B \approx 110.7°$
35. approximately 31 miles
37. The distance from the plane to
lighthouse A is 1.26 miles and to
lighthouse B is 1.60 miles.
39. 860,000 miles   **41. b.** $a = 5$,
$b = 3$, and $c = 7$ (using $m = 2$ and $n = 1$)
43. 2:40 P.M.   **47.** 45° or 135°

Exercise Set 9.2

1. magnitude: $\sqrt{34}$

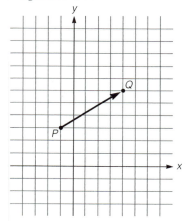

3. magnitude: $\sqrt{10}$

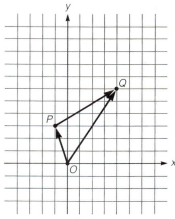

5. magnitude: $\sqrt{10}$

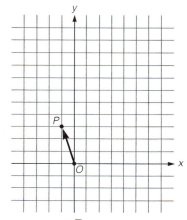

7. magnitude: $6\sqrt{2}$

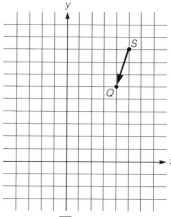

9. magnitude: $2\sqrt{13}$

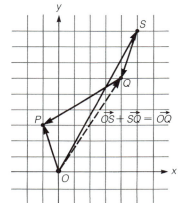

11. magnitude: $\sqrt{10}$

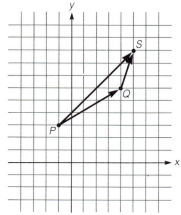

13. magnitude: 6

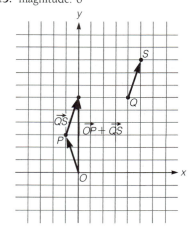

15. magnitude: 9

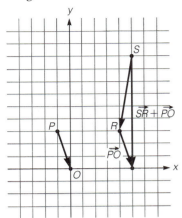

17. magnitude: $\sqrt{37}$

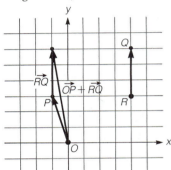

19. magnitude: $\sqrt{61}$

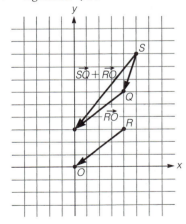

21. magnitude: $3\sqrt{5}$

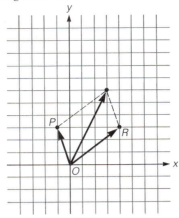

23. magnitude: $2\sqrt{13}$

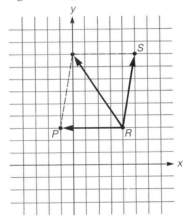

25. magnitude: $6\sqrt{5}$

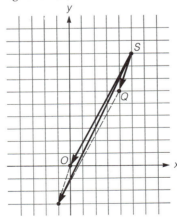

27. $|\mathbf{F} + \mathbf{G}| = \sqrt{41}$ N;
$\theta = \tan^{-1}\left(\frac{4}{5}\right) \approx 38.7°$
29. $|\mathbf{F} + \mathbf{G}| = 9\sqrt{2}$ N; $\theta = \tan^{-1}(1) = 45°$
31. $|\mathbf{F} + \mathbf{G}| = \sqrt{62.3525} \approx 7.90$ N;
$\theta = \tan^{-1}\left(\frac{3.22}{7.21}\right) \approx 24.1°$
33. magnitude ≈ 6.92 N; $\theta \approx 34.67°$
35. magnitude ≈ 39.20 N; $\theta \approx 21.46°$
37. magnitude ≈ 38.96 N; $\theta \approx 29.44°$
39. horizontal ≈ 13.86 cm/sec;
vertical $= 8$ cm/sec
41. horizontal ≈ 3.62 N;
vertical ≈ 13.52 N
43. horizontal ≈ -0.71 cm/sec;
vertical ≈ 0.71 cm/sec
45. horizontal ≈ -1.02 N;
vertical ≈ 0.72 N
47. ground speed ≈ 301.04 mph;
drift angle $\approx 4.76°$; bearing $\approx 25.24°$
49. ground speed ≈ 293.47 mph;
drift angle $\approx 8.82°$; bearing $\approx 91.18°$
51. perpendicular ≈ 9.83 lb;
parallel ≈ 6.88 lb
53. perpendicular ≈ 11.82 lb;
parallel ≈ 2.08 lb **55. a.** The initial
point of $(\mathbf{A} + \mathbf{B}) + \mathbf{C}$ is $(-1, 2)$ and the
terminal point is $(2, -3)$. **b.** The initial
point of $\mathbf{A} + (\mathbf{B} + \mathbf{C})$ is $(-1, 2)$ and the
terminal point is $(2, -3)$.

Exercise Set 9.3

1. length $= 5$

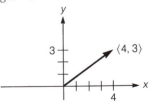

3. length $= 2\sqrt{5}$

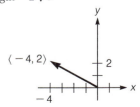

5. length = $\sqrt{13}/4$

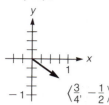

$\left\langle \frac{3}{4}, -\frac{1}{2} \right\rangle$

7. $\langle 1, 4 \rangle$ **9.** $\langle -1, 1 \rangle$ **11.** $\langle 8, -5 \rangle$
13. $\langle 7, 7 \rangle$ **15.** $\langle 24, 22 \rangle$ **17.** $\sqrt{130}$
19. $2\sqrt{17} - \sqrt{13} - \sqrt{37}$
21. $\langle 13, 6 \rangle$ **23.** $\langle 14, 21 \rangle$
25. $\langle -3, -1 \rangle$ **27.** $\langle 23, 12 \rangle$
29. $\langle -9, 0 \rangle$ **31.** -48 **33.** $3i + 8j$
35. $-8i - 6j$ **37.** $19i + 23j$
39. $\langle 1, 1 \rangle$ **41.** $\langle 5, -4 \rangle$
43. $\langle \sqrt{5}/5, 2\sqrt{5}/5 \rangle$
45. $\langle 2\sqrt{5}/5, -\sqrt{5}/5 \rangle$
47. $8\sqrt{145}/145\, i - 9\sqrt{145}/145\, j$
49. $u_1 = \sqrt{3}/2,\ u_2 = 1/2$
51. $u_1 = -1/2,\ u_2 = \sqrt{3}/2$
53. $u_1 = -\sqrt{3}/2,\ u_2 = 1/2$
59. a. Both are 8. **b.** Both are -14.
61. a. Both are 25. **b.** Both are 29.
63. $\cos \theta = 7/\sqrt{170}$;
$\theta \approx 57.53°$ or $\theta \approx 1.00$ radian
65. $\cos \theta = -57/\sqrt{3538}$;
$\theta \approx 163.39°$ or $\theta \approx 2.85$ radians
67. a. $\cos \theta = -7/\sqrt{170}$; $\theta \approx 122.47°$
or $\theta \approx 2.14$ radians **b.** $\cos \theta = 7/\sqrt{170}$;
$\theta \approx 57.53°$ or $\theta \approx 1.00$ radian
69. a. 0 **b.** The vectors are
perpendicular. **c.**

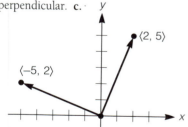

71. Since $\cos \theta = (\mathbf{A} \cdot \mathbf{B})/(|\mathbf{A}\|\mathbf{B}|)$, then
$\mathbf{A} \cdot \mathbf{B} = 0$ implies $\cos \theta = 0$. Thus $\theta = 90°$,
and the vectors are perpendicular.
73. $\langle \frac{5}{13}, \frac{12}{13} \rangle$ and $\langle -\frac{5}{13}, -\frac{12}{13} \rangle$

Exercise Set 9.4

1. $(2, 3)$ **3.** $\left(5\sqrt{3}/2, 1 \right)$
5. $\left(3\sqrt{2}/4, 3\sqrt{2}/4 \right)$

7.

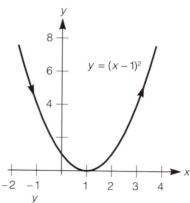

$y = (x - 1)^2$

9.

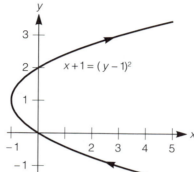

$x + 1 = (y - 1)^2$

11.

$\dfrac{x^2}{25} + \dfrac{y^2}{4} = 1$

13.

$\dfrac{x^2}{16} + \dfrac{y^2}{36} = 1$

15. a.

$x^2 + y^2 = 4$

b.

$\dfrac{x^2}{16} + \dfrac{y^2}{4} = 1$

17. a. $t = 1$: $x \approx 34.2$; $y \approx 83.0$;
$t = 2$: $x \approx 68.4$; $y \approx 128.9$;
$t = 3$: $x \approx 102.6$; $y \approx 142.9$
b. approximately 5.93 sec, approximately
203 ft   **21.** $x^{2/3} + y^{2/3} = 1$

Exercise Set 9.5

1. a. $\left(-\frac{3}{2}, 3\sqrt{3}/2 \right)$ **b.** $\left(2\sqrt{3}, -2 \right)$
c. $\left(2\sqrt{3}, -2 \right)$ **3. a.** $(0, 1)$ **b.** $(0, 1)$
c. $\left(-\dfrac{\sqrt{2 + \sqrt{2}}}{2}, -\dfrac{\sqrt{2 - \sqrt{2}}}{2} \right)$
5. $\left(\sqrt{2}, 5\pi/4 \right)$ **7.** $(x - 1)^2 + y^2 = 1$
9. $x^4 + x^2 y^2 - y^2 = 0$
11. $(x^2 + y^2)^3 = 9(x^2 - y^2)^2$
13. $(x^2/4) + (y^2/8) = 1$
15. $y = -\sqrt{3}x + 4$
17. $r = 2/(3 \cos \theta - 4 \sin \theta)$
19. $r = \tan^2 \theta \sec \theta$ **21.** $r^2 = \csc 2\theta$
23. $r^2 = 9/(9 \cos^2 \theta + \sin^2 \theta)$
25. a.

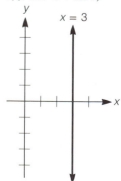

$x = 3$

b.

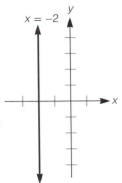

27.

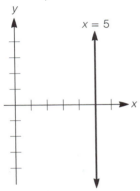

29.

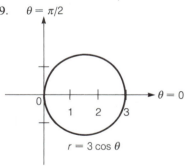

$r = 3 \cos \theta$

31.

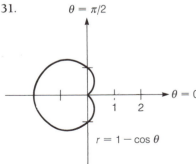

$r = 1 - \cos \theta$

33.

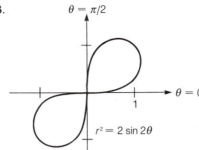

$r^2 = 2 \sin 2\theta$

35.

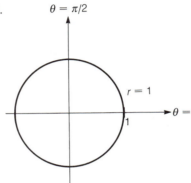

$r = 1$

37.

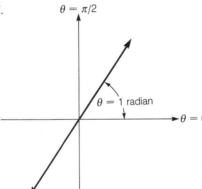

$\theta = 1$ radian

39.

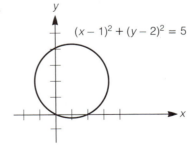

$(x - 1)^2 + (y - 2)^2 = 5$

41.

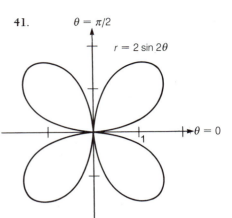

$r = 2 \sin 2\theta$

43.

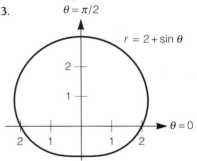

$r = 2 + \sin \theta$

45.

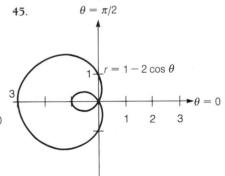

$r = 1 - 2 \cos \theta$

47.

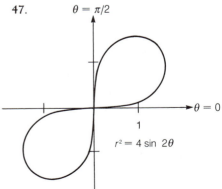

$r^2 = 4 \sin 2\theta$

49.

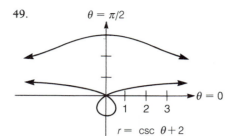

$\theta = \pi/2$

$r = \csc\theta + 2$

$\theta = 0$

51. a. C **b.** B **c.** D **d.** A

Exercise Set 9.6

1.

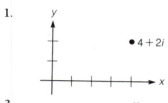

$\bullet\, 4 + 2i$

3.

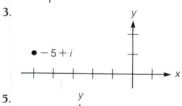

$\bullet -5 + i$

5.

7.

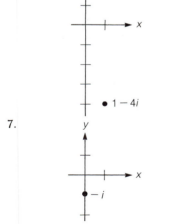

$\bullet\, 1 - 4i$

$\bullet -i$

9. $\sqrt{2} + \sqrt{2}i$ **11.** $-2\sqrt{3} + 2i$
13. $-1 - i$ **15.** $\sqrt{3}i$
17. $(\sqrt{6} - \sqrt{2}) + (\sqrt{6} + \sqrt{2})i$
19. $\cos(\pi/6) + i\sin(\pi/6)$
21. $2[\cos(2\pi/3) + i\sin(2\pi/3)]$
23. $4[\cos(7\pi/6) + i\sin(7\pi/6)]$
25. $6[\cos(3\pi/2) + i\sin(3\pi/2)]$
27. $\frac{1}{2}[\cos(11\pi/6) + i\sin(11\pi/6)]$
29. $3 + 3\sqrt{3}i$ **31.** $1 - \sqrt{3}i$
33. $3\sqrt{2}\cos(2\pi/7) + [3\sqrt{2}\sin(2\pi/7)]i$

35. $\frac{3\sqrt{2}}{2} + \frac{3\sqrt{2}}{2}i$ **37.** $\sqrt{3} + i$
39. 1 **41.** $\frac{243}{2} - \frac{243\sqrt{3}}{2}i$
43. $\frac{\sqrt{2}}{128} + \frac{\sqrt{2}}{128}i$ **45.** $-4i$
47. $-1 - i$ **49.** $\frac{1}{2} + \frac{\sqrt{3}}{2}i$
51. $128 + 128i$
53. $-128 + 128\sqrt{3}i$
55. $3i, -\frac{3}{2}\sqrt{3} - \frac{3}{2}i, \frac{3}{2}\sqrt{3} - \frac{3}{2}i$
57. $1, \frac{1}{2}\sqrt{2} + \frac{1}{2}\sqrt{2}i, i, -\frac{1}{2}\sqrt{2} + \frac{1}{2}\sqrt{2}i,$
$-1, -\frac{1}{2}\sqrt{2} - \frac{1}{2}\sqrt{2}i, -i, \frac{1}{2}\sqrt{2} - \frac{1}{2}\sqrt{2}i$
59. $4, -2 + 2\sqrt{3}i, -2 - 2\sqrt{3}i$
61. $3, \frac{3}{2} + \frac{3}{2}\sqrt{3}i, -\frac{3}{2} + \frac{3}{2}\sqrt{3}i, -3,$
$-\frac{3}{2} - \frac{3}{2}\sqrt{3}i, \frac{3}{2} - \frac{3}{2}\sqrt{3}i$
63. 92,236,816 **65.** $0.95 + 0.31i, i,$
$-0.95 + 0.31i, -0.59 - 0.81i,$
$0.59 - 0.81i$
67. $\frac{1}{2}(\sqrt{6} - \sqrt{2}) + \frac{1}{2}(\sqrt{6} + \sqrt{2})i,$
$-\frac{1}{2}(\sqrt{6} + \sqrt{2}) + \frac{1}{2}(\sqrt{6} - \sqrt{2})i,$
$\frac{1}{2}(\sqrt{2} - \sqrt{6}) - \frac{1}{2}(\sqrt{2} + \sqrt{6})i,$ and
$\frac{1}{2}(\sqrt{2} + \sqrt{6}) + \frac{1}{2}(\sqrt{2} - \sqrt{6})i$
69. a. $1, -\frac{1}{2} + \frac{1}{2}\sqrt{3}i, -\frac{1}{2} - \frac{1}{2}\sqrt{3}i$
71. -1 **73.** 1

Chapter Nine Review Exercises

1. $\angle C = 55°, a \approx 12.6$ cm, $b \approx 19.5$ cm
3. a. $\angle B \approx 62.4°, \angle C \approx 65.6°,$
$c \approx 9.2$ cm **b.** $\angle B \approx 117.6°, \angle LC \approx 10.4°,$
$c \approx 1.8$ cm **5.** $c \approx 7.7$ cm,
$\angle A \approx 108.5°, \angle B \approx 47.5°$
7. $\angle C \approx 106.6°, \angle B \approx 48.2°, \angle A \approx 25.2°$
9. 9.21 cm **11.** 32.48 cm^2
13. 55.23 cm^2 **15.** 7.89 cm
17. 15.43 cm **19.** 15 cm
21. 11 cm **23. a.** $\cos A = \frac{3}{4};$
$\cos C = \frac{1}{8}$ **25.** 36° **27.** 121.24°
29. 7.24 m **31. a.** 11.76 cm
b. 19.02 cm **35.** $a = 5$ and $b = 8$
37. magnitude = 25 N; direction $\approx$ 53.1°
39. horizontal $\approx$ 41.0 cm/sec;
vertical $\approx$ 28.7 cm/sec
41. weight $\approx$ 36.4 lb;
perpendicular $\approx$ 33.2 lb **43.** $\pm\sqrt{15}$
45. $\langle 10, 9 \rangle$ **47.** $\langle 12, 7 \rangle$ **49.** 48
51. $\langle 12, 8 \rangle$ **53.** $\langle -6, 2 \rangle$
55. $\langle -7, -6 \rangle$ **57.** $7i - 6j$
59. $\langle 3\sqrt{13}/13, 2\sqrt{13}/13 \rangle$

61. a.

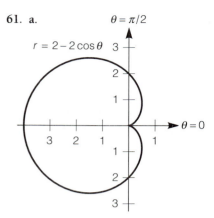

$\theta = \pi/2$

$r = 2 - 2\cos\theta$

$\theta = 0$

b.

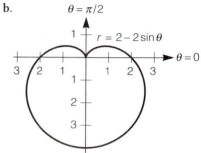

$\theta = \pi/2$

$r = 2 - 2\sin\theta$

$\theta = 0$

63. a. $\theta = \pi/2$

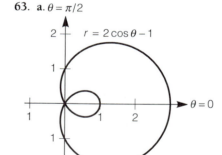

$r = 2\cos\theta - 1$

$\theta = 0$

b.

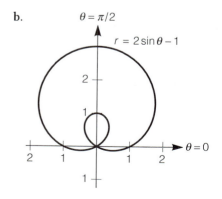

$\theta = \pi/2$

$r = 2\sin\theta - 1$

$\theta = 0$

65. a.

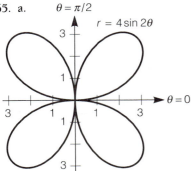

b.

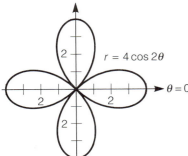

67. a.

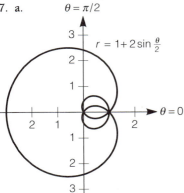

b.

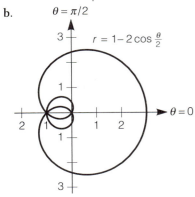

69.

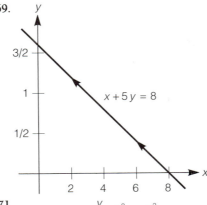

71.

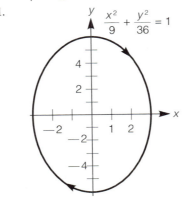

73.

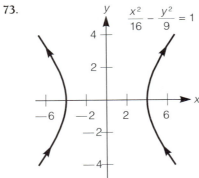

75. $\frac{3}{2} + \frac{3}{2}\sqrt{3}i$ **77.** $2^{-1/4} - 2^{-1/4}i$

79. $1[\cos(\pi/3) + i\sin(\pi/3)]$

81. $6[\cos(5\pi/4) + i\sin(5\pi/4)]$

83. $5\sqrt{2} + 5\sqrt{2}i$ **85.** $\sqrt{2} - \sqrt{2}i$

87. $\frac{27}{2} + \frac{27}{2}\sqrt{3}i$ **89.** $512 - 512\sqrt{3}i$

91. $1, \frac{1}{2} + \frac{1}{2}\sqrt{3}i, -\frac{1}{2} + \frac{1}{2}\sqrt{3}i, -1,$
$-\frac{1}{2} - \frac{1}{2}\sqrt{3}i,$ and $\frac{1}{2} - \frac{1}{2}\sqrt{3}i$

93. $-\frac{\sqrt{4 + 2\sqrt{2}}}{2} + \frac{\sqrt{4 - 2\sqrt{2}}}{2}i$ and
$\frac{\sqrt{4 + 2\sqrt{2}}}{2} - \frac{\sqrt{4 - 2\sqrt{2}}}{2}i$

95. $1.06 + 0.17i, 0.17 + 1.06i,$
$-0.95 + 0.49i, -0.76 - 0.76i,$ and
$0.49 - 0.95i$

Section 9.4 and 9.5 Graphing Calculator Exercises

1. a.

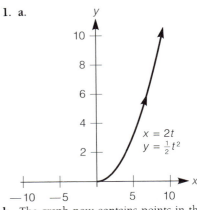

b. The graph now contains points in the second quadrant also.

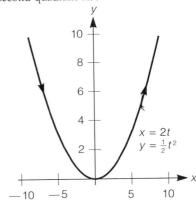

c.

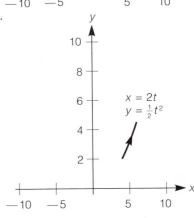

3.

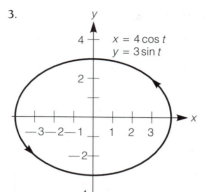

$x = 4\cos t$
$y = 3\sin t$

5.

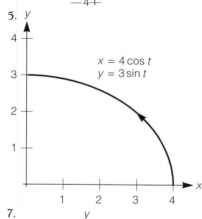

$x = 4\cos t$
$y = 3\sin t$

7.

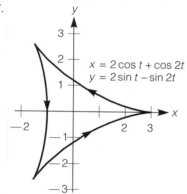

$x = 2\cos t + \cos 2t$
$y = 2\sin t - \sin 2t$

9.

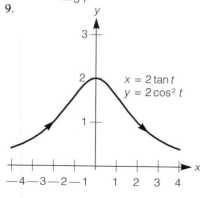

$x = 2\tan t$
$y = 2\cos^2 t$

11. a.

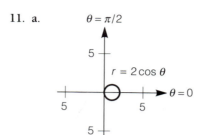

$r = 2\cos\theta$

b.

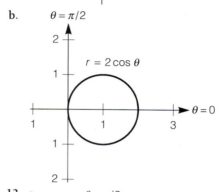

$r = 2\cos\theta$

13. a.

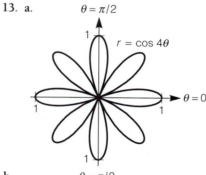

$r = \cos 4\theta$

b.

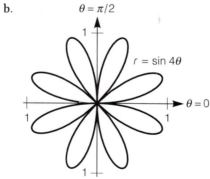

$r = \sin 4\theta$

15. a.

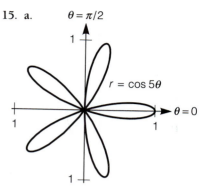

$r = \cos 5\theta$

b.

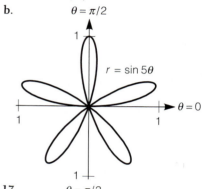

$r = \sin 5\theta$

17.

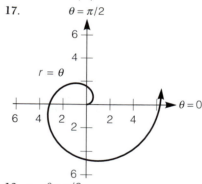

$r = \theta$

19. a.

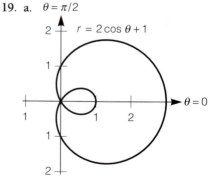

$r = 2\cos\theta + 1$

b. $\theta = \pi/2$

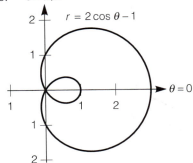

$r = 2\cos\theta - 1$

$\theta = 0$

c. $\theta = \pi/2$

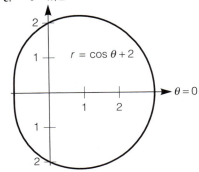

$r = \cos\theta + 2$

$\theta = 0$

d. $\theta = \pi/2$

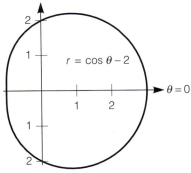

$r = \cos\theta - 2$

$\theta = 0$

e. $\theta = \pi/2$

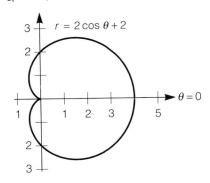

$r = 2\cos\theta + 2$

$\theta = 0$

f. $\theta = \pi/2$

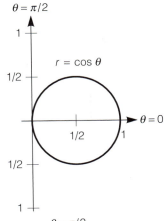

$r = \cos\theta$

$\theta = 0$

21. a. $\theta = \pi/2$

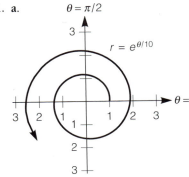

$r = e^{\theta/10}$

$\theta = 0$

b. $\theta = \pi/2$

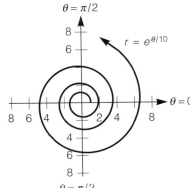

$r = e^{\theta/10}$

$\theta = 0$

c. $\theta = \pi/2$

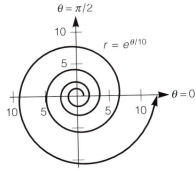

$r = e^{\theta/10}$

$\theta = 0$

1. 7 cm   **2.** $\cos\theta = -\frac{1}{4}$; The angle opposite the 4-cm side must be obtuse (not acute), since its cosine is negative.
3. 20 cm   **4.** $2\sqrt{17 - 4\sqrt{2}}$ cm
5. $a \approx 3.3$ cm; $C \approx 26.2°$; $B \approx 126.8°$
6. a. $2\sqrt{5}$ N **b.** $\tan\theta = 2$
7. a. $d = 4\sqrt{13 - 12\cos 110°}$ N

b. $\sin\theta = \dfrac{2\sin 110°}{\sqrt{13 - 12\cos 110°}}$

8. ground speed $= 50\sqrt{37}$ mph; $\tan\theta = \frac{1}{6}$ **9. a.** $\langle 13, 5\rangle$ **b.** $\sqrt{194}$
c. $\mathbf{i} - 3\mathbf{j}$
10. $\langle -11\sqrt{130}/130, -3\sqrt{130}/130\rangle$
11. $(x^2 + y^2)^2 = x^2 - y^2$
12. $\theta = \pi/2$

$r = 2(1 - \cos\theta)$

$\theta = 0$

13. y $\dfrac{x^2}{16} + \dfrac{y^2}{4} = 1$

x

14. $-1 + \sqrt{3}i$
15. $2[\cos(7\pi/4) + i\sin(7\pi/4)]$
16. $r^n(\cos n\theta + i\sin n\theta)$
17. $\frac{15}{2} + \frac{15}{2}\sqrt{3}i$ **18.** $2\sqrt{3} + 2i$, $-2\sqrt{3} + 2i$, and $-4i$

CHAPTER TEN

Exercise Set 10.1

1. a. yes **b.** no **c.** yes **d.** yes
3. yes **5.** no **7.** yes
9. $\left(-6, \frac{1}{2}\right)$ **11.** $\left(-\frac{5}{2}, \frac{13}{2}\right)$
13. $\left(\frac{1}{5}, \frac{7}{10}\right)$ **15.** $\left(-\frac{60}{13}, \frac{60}{13}\right)$
17. $\left(\sqrt{6}, 1\right)$ **19.** $\left(-\frac{2}{9}, \frac{23}{27}\right)$
21. $\left(-\frac{283}{242}, -\frac{3}{121}\right)$ **23.** $\left(-\frac{226}{25}, -\frac{939}{50}\right)$
25. $\left(\frac{5}{8}, 0\right)$ **27.** $(1, 2)$ **29.** $b = 3$;
$c = 4$ **31.** $A = -28$; $B = -22$
33. $\frac{49}{20}$ square units **35.** 80 cc of 10%
acid; 120 cc of 35% acid **37.** 8 lb of
$5.20 coffee; 8 lb of $5.80 coffee
39. $x = \dfrac{ab}{a + b}$; $y = \dfrac{ab}{a + b}$; $a \neq \pm b$
41. $x = \dfrac{a + b}{ab}$; $y = -\dfrac{1}{ab}$; $a \neq 0$; $b \neq 0$
43. $\left(-\frac{1}{11}, \frac{1}{9}\right)$ **45.** $(5, 2)$
47. $(1.32, -1.62)$ **49.** $(0.14, 0.05)$
51. $(1.61, 0)$ **53.** $(4, -8)$, $(4, 2)$,
$(-1, -8)$, $(-1, 2)$ **55.** 59
57. length 12 in.; width 5 in.
59. a. $(60, 40)$; **b.** $(60, 40)$
61. $(0, a)$ **63.** $(1/b, 1/a)$
65. $k = -\frac{28}{7}, 2$

Exercise Set 10.2

1. $(-3, -2, -1)$ **3.** $\left(-\frac{1}{60}, -\frac{2}{15}, \frac{3}{5}\right)$
5. $\left(\frac{25}{36}, \frac{5}{9}, -\frac{1}{3}\right)$ **7.** $(x, x/8, 0)$, for any
real number x **9.** $(-1, 0, 1, -5)$
11. $\left(\frac{11}{3}, \frac{8}{3}, \frac{17}{3}\right)$ **13.** $(1, 0, 1)$
15. $(2, 3, 1)$ **17.** inconsistent (no
solution) **19.** $\left(-\frac{1}{7}(z + 1), \frac{5}{7}(z + 1), z\right)$,
for any real number z
21. $\left(\dfrac{11 - 5z}{7}, \dfrac{-3z - 6}{7}, z\right)$ for any real
number z **23.** $(4, 1, -3, 2)$
25. $\left(\dfrac{17 - 17z}{5}, \dfrac{8z - 3}{5}, z\right)$, for any real
number z

27.
$\left(\dfrac{12 + 10w}{11}, \dfrac{146 + 19w}{55}, \dfrac{159 + 61w}{55}, w\right)$,
for any real number w
29. $\left(-\dfrac{5z}{12}, \dfrac{2z + 3}{3}, z\right)$, for any real
number z **31. a.** $A = \frac{1}{4}$, $B = -\frac{1}{4}$
b. $A = \frac{1}{2}$, $B = \frac{1}{2}$ **33.** $A = \frac{4}{3}$, $B = \frac{5}{3}$
35. a. $A = \frac{3}{8}$, $B = -\frac{3}{8}$ **b.** $A = \frac{3}{64}$, $B = -\frac{3}{64}$,
$C = -\frac{3}{8}$ **37.** $A = \frac{1}{3}$, $B = -\frac{1}{3}$,
$C = \frac{2}{3}$ **39.** $A = 4$, $B = 4$
41. $A = \frac{1}{2}$, $B = \frac{1}{2}$, $C = -\frac{1}{2}$, $D = \frac{1}{2}$
43. $17x^2 + 17y^2 - 49x + 65y - 166 = 0$
45. $x = \ln a$, $y = \ln 2a$, $z = \ln(a/2)$
49. a. $A = 1/(a - b)$, $B = 1/(b - a)$
b. $A = \dfrac{ap + q}{a - b}$, $B = \dfrac{bp + q}{b - a}$
51. $A = \dfrac{a^2 + pa + q}{(a - b)(a - c)}$,
$B = \dfrac{b^2 + pb + q}{(b - a)(b - c)}$, $C = \dfrac{c^2 + pc + q}{(c - a)(c - b)}$
53. 60 miles

Exercise Set 10.3

1. a. 2×3 **b.** 3×2 **3.** 5×4
5. coefficient matrix: $\begin{pmatrix} 2 & 3 & 4 \\ 5 & 6 & 7 \\ 8 & 9 & 10 \end{pmatrix}$
augmented matrix: $\begin{pmatrix} 2 & 3 & 4 & 10 \\ 5 & 6 & 7 & 9 \\ 8 & 9 & 10 & 8 \end{pmatrix}$
7. coefficient matrix:
$\begin{pmatrix} 1 & 0 & 1 & 1 \\ 1 & 1 & 0 & 2 \\ 0 & 1 & 1 & 1 \\ 2 & -1 & -1 & 0 \end{pmatrix}$
augmented matrix:
$\begin{pmatrix} 1 & 0 & 1 & 1 & -1 \\ 1 & 1 & 0 & 2 & 0 \\ 0 & 1 & 1 & 1 & 1 \\ 2 & -1 & -1 & 0 & 2 \end{pmatrix}$
9. $(-1, -2, 3)$ **11.** $(-5, 1, 3)$
13. $(3, 0, -7)$ **15.** $(8, 9, -1)$
17. $\left(\dfrac{9z + 5}{19}, \dfrac{31z - 6}{19}, z\right)$, for any real
number z **19.** $(2, -1, 0, 3)$
21. no solution **23.** $\begin{pmatrix} 3 & 2 \\ 2 & 4 \end{pmatrix}$

25. $\begin{pmatrix} 6 & 4 \\ 4 & 8 \end{pmatrix}$ **27.** $\begin{pmatrix} 11 & -2 \\ 11 & 1 \end{pmatrix}$
29. $\begin{pmatrix} 2 & 3 \\ -1 & 4 \end{pmatrix}$ **31.** undefined
33. $\begin{pmatrix} 10 & -2 \\ -8 & 0 \\ 4 & 6 \end{pmatrix}$
35. $\begin{pmatrix} 2 & 4 & 11 \\ -12 & 16 & 19 \\ 14 & 12 & 43 \end{pmatrix}$
37. $\begin{pmatrix} -9 & 10 & 10 \\ 4 & -8 & -12 \\ 10 & 4 & 21 \end{pmatrix}$
39. undefined **41.** $\begin{pmatrix} 0 & 0 & 0 \\ 0 & 0 & 0 \\ 0 & 0 & 0 \end{pmatrix}$
43. $\begin{pmatrix} 4 & 2 \\ 2 & 5 \end{pmatrix}$ **45.** undefined
47. $\begin{pmatrix} 1 & 18 \\ -6 & 13 \end{pmatrix}$ **49.** $\begin{pmatrix} -16 & 75 \\ -25 & 34 \end{pmatrix}$
51. a. $\begin{pmatrix} -13 & 1 & 40 \\ 43 & 17 & 0 \\ 89 & 61 & 60 \end{pmatrix}$
b. $\begin{pmatrix} -13 & 1 & 40 \\ 43 & 17 & 0 \\ 89 & 61 & 60 \end{pmatrix}$
c. $\begin{pmatrix} -52 & -82 & 61 \\ 87 & 141 & 0 \\ 216 & 318 & 165 \end{pmatrix}$
d. $\begin{pmatrix} -52 & -82 & 61 \\ 87 & 141 & 0 \\ 216 & 318 & 165 \end{pmatrix}$
53. a. $\begin{pmatrix} 16 & 20 \\ 24 & 28 \end{pmatrix}$ **b.** $\begin{pmatrix} 18 & 26 \\ 18 & 26 \end{pmatrix}$
c. $\begin{pmatrix} 14 & 14 \\ 30 & 30 \end{pmatrix}$ **d.** $\begin{pmatrix} 18 & 26 \\ 18 & 26 \end{pmatrix}$
55. a. $\begin{pmatrix} x \\ -y \end{pmatrix}$ **b.** $\begin{pmatrix} -x \\ y \end{pmatrix}$ **c.** $\begin{pmatrix} -x \\ -y \end{pmatrix}$,
reflection about the origin
57. a. $f(A) = -2$; $f(B) = 29$;
$f(AB) = -58$; yes

Exercise Set 10.4

1. $\begin{pmatrix} 4 & -1 \\ -5 & 2 \end{pmatrix} = A$

3. $\begin{pmatrix} 3 & 0 & -2 \\ 0 & 5 & 6 \\ 1 & 4 & -7 \end{pmatrix} = C$

5. $\begin{pmatrix} -5 & 9 \\ 4 & -7 \end{pmatrix}$ **7.** $\begin{pmatrix} -\frac{6}{23} & \frac{1}{23} \\ \frac{5}{23} & \frac{3}{23} \end{pmatrix}$

9. does not exist **11.** $\begin{pmatrix} 2 & -3 \\ 1 & 3 \end{pmatrix}$

13. $\begin{pmatrix} 2 & -1 \\ -3 & 2 \end{pmatrix}$ **15.** $\begin{pmatrix} \frac{6}{11} & 1 \\ -\frac{1}{11} & 0 \end{pmatrix}$

17. does not exist

19. $\begin{pmatrix} -1 & 2 & -3 \\ 2 & 1 & 0 \\ 4 & -2 & 5 \end{pmatrix}$

21. $\begin{pmatrix} 5 & -\frac{10}{3} & 1 \\ 0 & \frac{1}{3} & 0 \\ 4 & -\frac{8}{3} & 1 \end{pmatrix}$

23. $\begin{pmatrix} 2 & 1 & 4 \\ 3 & 2 & 5 \\ 0 & -1 & 1 \end{pmatrix}$

25. does not exist **27. a.** $x = -1$, $y = 1$ **b.** $x = -132$, $y = 48$
29. a. $x = 2$, $y = -1$, $z = 4$ **b.** $x = 1$, $y = 1$, $z = -2$

31. a. $I_3 = \begin{pmatrix} 1 & 0 & 0 \\ 0 & 1 & 0 \\ 0 & 0 & 1 \end{pmatrix}$ so $A^{-1} = A$.

b. $x = \frac{1}{2}$, $y = -1$, $z = 1$

35. a. $A^{-1} = \begin{pmatrix} -\frac{5}{2} & \frac{3}{2} \\ 2 & -1 \end{pmatrix}$,

$B^{-1} = \begin{pmatrix} 7 & -8 \\ -6 & 7 \end{pmatrix}$,

$B^{-1}A^{-1} = \begin{pmatrix} -\frac{67}{2} & \frac{37}{2} \\ 29 & -16 \end{pmatrix}$

b. $\begin{pmatrix} -\frac{67}{2} & \frac{37}{2} \\ 29 & -16 \end{pmatrix}$, $(AB)^{-1} = B^{-1}A^{-1}$

Exercise Set 10.5

1. a. 29 **b.** -29 **3. a.** 0 **b.** 0
5. -1 **7.** -60 **9.** 9
11. a. 314 **b.** 674 **c.** part (b)
13. a. 0 **b.** 0 **c.** 0 **d.** 0 **15.** 0
17. -3 **19.** 6848 **21.** 17120
23. $(y - x)(z - x)(z - y)$ **25.** xy
27. 20 **29.** 120 **33.** $(1, 1, 2)$
35. $(2, -3, 6)$ **37.** $(0, 0, 0)$

39. $\left(13 - \dfrac{11y}{3}, y, 13 - 4y\right)$, for any real number y **41.** $(1, 0, -10, 2)$
43. $4, -4, -1$ **55. a.** 1 **b.** 1
59. $\left(\dfrac{k(k - b)(k - c)}{a(a - b)(a - c)}, \dfrac{k(k - a)(k - c)}{b(b - a)(b - c)}, \right.$
$\left. \dfrac{k(k - a)(k - b)}{c(c - a)(c - b)}\right)$

61. $y = 2x + 5$ **65. b.** $\begin{pmatrix} -\frac{9}{61} & \frac{7}{61} \\ \frac{1}{61} & \frac{6}{61} \end{pmatrix}$

Exercise Set 10.6

1. $(0, 0), (3, 9)$ **3.** $\left(2\sqrt{6}, 1\right),$
$\left(-2\sqrt{6}, 1\right)$ **5.** $(-1, 1)$ **7.** $(2, 3),$
$(2, -3), (-2, 3), (-2, -3)$ **9.** $(1, 0),$
$(-1, 0)$

11. $\left(\dfrac{-1 + \sqrt{65}}{8}, \dfrac{1 + \sqrt{65}}{2}\right),$
$\left(\dfrac{-1 - \sqrt{65}}{8}, \dfrac{1 - \sqrt{65}}{2}\right),$

13. $\left(\sqrt{17}/17, \pm 1\right), \left(-\sqrt{17}/17, \pm 1\right)$
15. $(1, 0), \left(4, -\sqrt{3}\right)$ **17.** $(2, 4)$
19. $(100, 1000), \left(100, \frac{1}{1000}\right), \left(\frac{1}{100}, 1000\right),$
$\left(\frac{1}{100}, \frac{1}{1000}\right)$

21. $\left(\dfrac{2\ln 2 - 5\ln 3}{\ln 2 - \ln 3}, \dfrac{3\ln 2}{\ln 2 - \ln 3}\right)$

25. $(1/a, 1/b)$ **27.** $(9, 14), (14, 9)$

29. $\dfrac{p - \sqrt{2d^2 - p^2}}{2}$ by $\dfrac{p + \sqrt{2d^2 - p^2}}{2}$

31. $(5, 4), (5, -4)$ **33. a.** $k = \frac{1}{6}\ln 8$,
$N_0 = \frac{3}{2}$ **b.** $k = \frac{2}{7}\ln 10$, $N_0 = 10^{-1/7}$

35. $\left(\dfrac{p^2}{A}, \dfrac{q^2}{A}, \dfrac{r^2}{A}\right)$ and $\left(-\dfrac{p^2}{A}, -\dfrac{q^2}{A}, -\dfrac{r^2}{A}\right),$
where $A = \sqrt{p^2 + q^2 + r^2}$ **37.** 9 cm,
40 cm **39.** 3 cm by 20 cm
41. $(1, 2), (-1, -2), (2, 1), (-2, -1)$

43. $\left(\dfrac{1 + \sqrt{13}}{2}, \dfrac{-1 + \sqrt{13}}{2}\right),$
$\left(\dfrac{-1 + \sqrt{13}}{2}, \dfrac{1 + \sqrt{13}}{2}\right),$
$\left(\dfrac{1 - \sqrt{13}}{2}, \dfrac{-1 - \sqrt{13}}{2}\right),$
$\left(\dfrac{-1 - \sqrt{13}}{2}, \dfrac{1 - \sqrt{13}}{2}\right),$

45. $(2, 3), (-2, -3), (1, 2), (-1, -2)$

47. $(q, p), \left(\dfrac{-1 + \sqrt{3}}{2}q, \left(1 - \sqrt{3}\right)p\right),$
$\left(\dfrac{-1 - \sqrt{3}}{2}q, \left(1 + \sqrt{3}\right)p\right)$
49. $\left(e^{9/2}, e^3\right)$

Exercise Set 10.7

1. a. no **b.** yes
3.

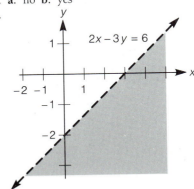

5.

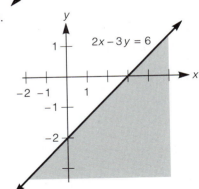

7.

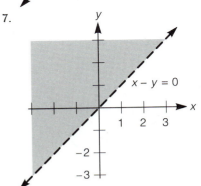

9.

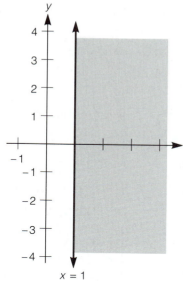

11.

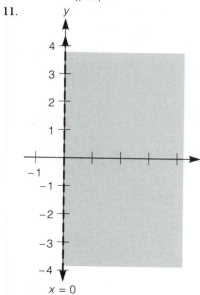

13.

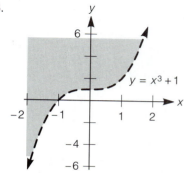

15.

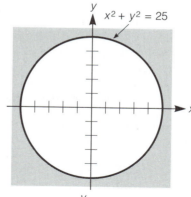

17.

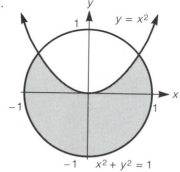

19.

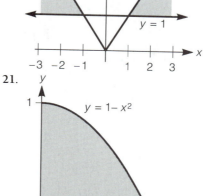

21.

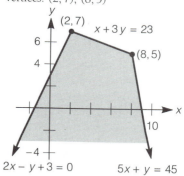

23. convex: yes; bounded: yes;
vertices: $(0, 0)$, $(7, 0)$, $(3, 8)$, $(0, 5)$

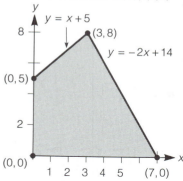

25. convex: yes; bounded: yes;
vertices: $(0, 0)$, $(0, 4)$, $(3, 5)$, $(8, 0)$

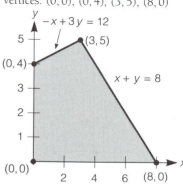

27. convex: yes; bounded: no;
vertices: $(2, 7)$, $(8, 5)$

29. convex: yes; bounded: no; vertex: $(6, 0)$

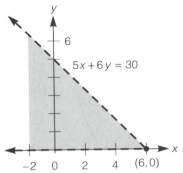

31. convex: yes; bounded: yes; vertices: $(0, 0)$, $(0, 5)$, $(6, 0)$

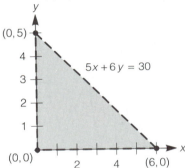

33. convex: yes; bounded: yes; vertices: $(5, 30)$, $(10, 30)$, $(20, 15)$, $(20, 20)$

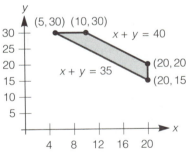

35. maximum $= 9$ at $(3, 4)$; minimum $= -5$ at $(5, 0)$
37. maximum $= 4$ at $(1, 2)$; minimum $= 0$ at $(0, 0)$
39. maximum $= 113$ at $(6, 6)$; minimum $= 27$ at $(2, 1)$
41. maximum $= 188$ at $(6, 7)$; minimum $= 100$ at $(0, 0)$
43. a. maximum $= 819$ at $(1, 8)$; minimum $= 0$ at $(0, 0)$
b. maximum $= 826$ at $(6, 7)$; minimum $= 0$ at $(0, 0)$

45. W_1 to D_1: 0; W_2 to D_1: 40; W_1 to D_2: 25; W_2 to D_2: 0; minimum cost $= \$10,150$
47. maximum $= \$766.20$ when $A = 833$ units, $B = 333$ units
49. $P = \$23,700$ when 150 acres of cherry tomatoes and 450 acres of regular tomatoes are planted.
51. maximum $= \$880$ with $\$4000$ in stock A, $\$8000$ in stock B
53. maximum $= \$540$ with $\$4000$ in stock A, $\$1000$ in stock B, $\$5000$ in stock C **55.** minimum $= \$7.60$ with 10 lb of food A, 4 lb of food B
57.

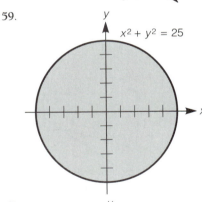

59.

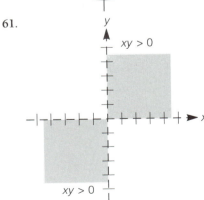

61.

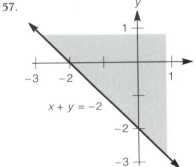

1. $(3, -5)$ **3.** $(-1, 4)$
5. $(-3, 15)$ **7.** $\left(-\frac{18}{5}, \frac{8}{5}\right)$
9. $\left(\frac{2}{3}, -\frac{1}{5}\right)$ **11.** $(-12, -8)$
13. $\left(\frac{1}{3}, -\frac{1}{4}\right)$
15. $\left(\dfrac{-1}{a^2 - 3a + 1}, \dfrac{1 - a}{a^2 - 3a + 1}\right)$, $a \neq \dfrac{3 \pm \sqrt{5}}{2}$ **17.** $(a^2, 1 - a^2)$
19. $(a^2 - b^2, a^2 + b^2)$
21. $\left(\dfrac{pq(p + q)}{p^2 + q^2}, \dfrac{p^3 - q^3}{p^2 + q^2}\right)$, where p and q are not both 0   **23.** $\left(\dfrac{a}{a - b}, \dfrac{b}{a + b}\right)$, $ab \neq 0$, $a \neq \pm b$ **25.** $(2, 3, 4)$
27. $(-1, -2, 0)$ **29.** no solution
31. $(x, 6 - 2x, -1)$, for any real number x
33. no solution   **35.** $(2b - z, a - b, z)$, for any real number z
37. $(4, 3, -1, 2)$ **39.** $A = \frac{1}{20}$, $B = -\frac{1}{20}$
41. $A = 2$, $B = -2$ **43.** $A = -\frac{5}{4}$, $B = \frac{5}{4}$
45. $A = \frac{1}{16}$, $B = -\frac{1}{16}$, $C = -\frac{1}{4}$
47. $A = 3$, $B = 1$, $C = 0$ **49.** $A = \frac{1}{48}$, $B = -\frac{1}{48}$, $C = \frac{1}{6}$ **51.** $A = 0$, $B = 1$, $C = a$ **53.** $A = b$, $B = -a$ **55.** 34
57. -56 **59.** 0   **61.** 24
67. $a = 2$, $b = 1$ **69. a.** $\left(\frac{2}{3}, 2\right)$ **b.** $\left(\frac{2}{3}, 2\right)$ **c.** $\left(\frac{2}{3}, 2\right)$ **d.** They all equal 2.
73. $\left(x + \frac{17}{6}\right)^2 + \left(y + \frac{8}{3}\right)^2 = \frac{245}{36}$
75. $\begin{pmatrix} 10 & -2 \\ 4 & 26 \end{pmatrix}$ **77.** $\begin{pmatrix} 8 & 4 \\ 4 & 32 \end{pmatrix}$
79. $\begin{pmatrix} 4 & -13 \\ 7 & 41 \end{pmatrix}$ **81.** $\begin{pmatrix} -3 & -14 \\ -4 & 3 \end{pmatrix}$
83. $\begin{pmatrix} 1 & 1 \\ 1 & 7 \end{pmatrix}$ **85.** $\begin{pmatrix} 1 & -11 \\ 6 & 36 \end{pmatrix}$
87. $\begin{pmatrix} 4 & 3 \\ 10 & 33 \end{pmatrix}$ **89.** $\begin{pmatrix} -42 & 58 \\ 5 & 20 \end{pmatrix}$
91. undefined   **93.** undefined
95. $\begin{pmatrix} 4 & -1 \\ 2 & 12 \end{pmatrix}$ **97.** $\begin{pmatrix} -4 & 13 \\ -7 & -41 \end{pmatrix}$
101. a. $\begin{pmatrix} -9 & 5 \\ 2 & -1 \end{pmatrix}$ **b.** $(-47, 10)$
103. a. $\begin{pmatrix} 10 & -2 & 5 \\ 6 & -1 & 4 \\ 1 & 0 & 1 \end{pmatrix}$ **b.** $x = 16$, $y = 15$, $z = 4$
105. $\begin{pmatrix} -65 & 20 & 9 & 133 \\ 3 & -1 & 0 & -6 \\ 26 & -8 & -4 & -53 \\ -23 & 7 & 3 & 47 \end{pmatrix}$

107. $D = -20$; $D_x = 12$; $D_y = 13$; $D_z = -31$; $(-\frac{3}{5}, -\frac{13}{20}, \frac{31}{20})$
109. $D = 0$; no solution   **111.** $D = 0$; $(-5 - 4y, y, -8 - 5y)$, for any real number y **113.** $D = 149$; $D_x = 596$; $D_y = 149$; $D_z = -149$; $D_w = 447$; $(4, 1, -1, 3)$ **115.** $(0, 0)$, $(6, 36)$
117. $(3, 0)$, $(-3, 0)$

119. $\left(\dfrac{5\sqrt{2}}{2}, \pm\dfrac{\sqrt{14}}{2}\right), \left(-\dfrac{5\sqrt{2}}{2}, \pm\dfrac{\sqrt{14}}{2}\right)$

121. $\left(\dfrac{-1 + \sqrt{5}}{2}, \dfrac{\sqrt{-2 + 2\sqrt{5}}}{2}\right)$

123. $\left(\pm\dfrac{\sqrt{-2 + 2\sqrt{17}}}{4}, \dfrac{-1 + \sqrt{17}}{4}\right)$

125. $(\sqrt{2}, 3\sqrt{2})$, $(-\sqrt{2}, -3\sqrt{2})$, $\left(\frac{7}{33}\sqrt{22}, \frac{31}{33}\sqrt{22}\right)$, $\left(-\frac{7}{33}\sqrt{22}, -\frac{31}{33}\sqrt{22}\right)$
127. $(5, 2)$, $(5, -4)$, $(1, 2)$, $(1, -4)$
129. $as/(a + b)$ and $bs/(a + b)$
131. $24, 60, 120$
133. $\frac{1}{2}\left(\sqrt{n + 2m} + \sqrt{n - 2m}\right)$ and $\frac{1}{2}\left(\sqrt{n + 2m} - \sqrt{n - 2m}\right)$, $\frac{1}{2}\left(\sqrt{n - 2m} - \sqrt{n + 2m}\right)$ and $-\frac{1}{2}\left(\sqrt{n - 2m} + \sqrt{n + 2m}\right)$
135. convex: no; bounded: no

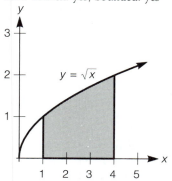

137. convex: yes; bounded: yes

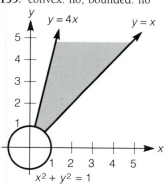

139. convex: no; bounded: no

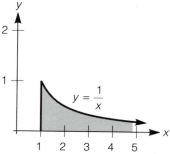

141. maximum $= \frac{1046}{7}$ at $\left(\frac{68}{7}, \frac{88}{7}\right)$; minimum $= 6$ at $(0, 0)$

Chapter Ten Graphing Calculator Exercises

1. a.

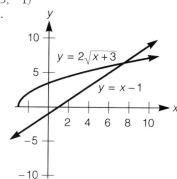

b. $(3, -1)$
3. a.

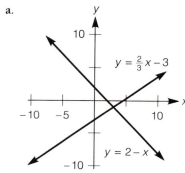

b. $(7.47, 6.47)$
5. a.

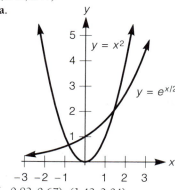

b. $(-0.82, 0.67)$, $(1.43, 2.04)$
7. a.

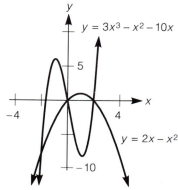

b. $(-2, -8)$, $(0, 0)$, $(2, 0)$
9. a.

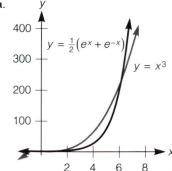

b. $(1.23, 1.85)$, $(6.14, 230.95)$

11. a. $\begin{pmatrix} -12 & 17 & -37 \\ -5 & 22 & -7 \\ 7 & -2 & 26 \end{pmatrix}$

b. $\begin{pmatrix} 19 & -7 & 6 \\ 5 & 17 & 4 \\ -5 & 16 & 0 \end{pmatrix}$

c. $\begin{pmatrix} -3 & 2 & -4 \\ -1 & 1 & -1 \\ 8 & -5 & 10 \end{pmatrix}$

d. $\begin{pmatrix} -.116 & .674 & -.767 \\ .035 & .198 & -.070 \\ -.081 & -.128 & .163 \end{pmatrix}$ **e.** 1

13. a. 1 **b.** -452 **c.** -452 **d.** -452
15. a. -0.0022123894
b. -0.0022123894

Chapter Ten Test

1. $(0, 3)$, $\left(-\frac{11}{4}, \frac{81}{16}\right)$ **2.** $(3, -5)$
3. a. $(1, -1, -3)$ **b.** $D = -19$;
$D_x = -19$; $D_y = 19$; $D_z = 57$; $(1, -1, -3)$

4. a. $\begin{pmatrix} 2 & -10 \\ 3 & -5 \end{pmatrix}$ **b.** $\begin{pmatrix} 8 & -4 \\ 7 & -6 \end{pmatrix}$

5. 12 square units **6.** $\left(\frac{1}{116}, -\frac{1}{144}\right)$

7. coefficient matrix: $\begin{pmatrix} 1 & 1 & -1 \\ 2 & -1 & 2 \\ 1 & -2 & 1 \end{pmatrix}$;

augmented matrix: $\begin{pmatrix} 1 & 1 & -1 & -1 \\ 2 & -1 & 2 & 11 \\ 1 & -2 & 1 & 10 \end{pmatrix}$

8. $(3, -3, 1)$ **9.** $y = -2x - 4$
10. $A = -\frac{3}{4}$, $B = \frac{3}{4}$, $C = -\frac{1}{2}$
11. a. 8 **b.** -8 **12.** -1120

13. $\left(\frac{5 + \sqrt{5}}{2}, \frac{5 - \sqrt{5}}{2}\right)$,

$\left(\frac{5 - \sqrt{5}}{2}, \frac{5 + \sqrt{5}}{2}\right)$,

$\left(\frac{-5 + \sqrt{5}}{2}, \frac{-5 - \sqrt{5}}{2}\right)$, and

$\left(\frac{-5 - \sqrt{5}}{2}, \frac{-5 + \sqrt{5}}{2}\right)$

14. $\left(1 - \frac{1}{5}C, -\frac{7}{5}C, C\right)$, where C is any real

number **15. a.** $\begin{pmatrix} 1 & -2 & 3 \\ 2 & -5 & 10 \\ -1 & 2 & -2 \end{pmatrix}$

b. $(12, 38, -9)$
16.

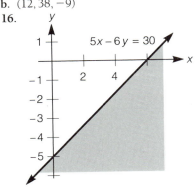

17. $p = -1$, $Q = -4$
18. bounded: no; convex: no

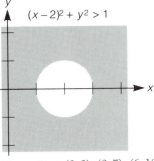

19. vertices: $(0, 0)$, $(0, 7)$, $(6, 10)$, $\left(\frac{261}{26}, \frac{225}{26}\right)$, $(11, 0)$

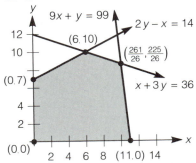

20. $1100 with 50 small and 25 large motors

CHAPTER ELEVEN

Exercise Set 11.1

1. quotient: $x - 1$; remainder: -7;
$x^2 - 6x - 2 = (x - 5)(x - 1) - 7$
3. quotient: $4x - 5$; remainder: 0;
$4x^2 - x - 5 = (x + 1)(4x - 5) + 0$
5. quotient: $6x^2 + 19x + 78$;
remainder: 313; $6x^3 - 5x^2 + 2x + 1 = (x - 4)(6x^2 + 19x + 78) + 313$
7. quotient: $x^2 + 2x + 4$; remainder: 7;
$x^3 - 1 = (x - 2)(x^2 + 2x + 4) + 7$
9. quotient: $x^4 - 2x^3 + 4x^2 - 8x + 16$;
remainder -33; $x^5 - 1 = (x + 2)(x^4 - 2x^3 + 4x^2 - 8x + 16) - 33$
11. quotient: $x^3 - 10x^2 + 40x - 160$;
remainder: 642; $x^4 - 6x^3 + 2 = (x + 4)(x^3 - 10x^2 + 40x - 160) + 642$
13. quotient: $x^2 + 6x + 57$;
remainder: 576; $x^3 - 4x^2 - 3x + 6 = (x - 10)(x^2 + 6x + 57) + 576$

15. quotient: $x^2 - 6x + 30$;
remainder: -150; $x^3 - x^2 = (x + 5)(x^2 - 6x + 30) - 150$
17. quotient: $54x^2 + 9x - 21$;
remainder: 0; $54x^3 - 27x^2 - 27x + 14 = \left(x - \frac{2}{3}\right)(54x^2 + 9x - 21) + 0$
19. quotient: $x^3 + 3x^2 + 12x + 36$;
remainder: 120; $x^4 + 3x^2 + 12 = (x - 3)(x^3 + 3x^2 + 12x + 36) + 120$
21. $(x - 2)(x^4 + 2x^3 + 4x^2 + 8x + 16)$
23. $(z - 3)(z^3 + 3z^2 + 9z + 27)$
25. quotient: $2x$; remainder: 1
27. quotient: $3x^2 - \frac{3}{2}x + \frac{3}{4}$; remainder: $\frac{1}{4}$
29. $k = 4$ **33.** $q(x) = x + (-4 + i)$;
$R(x) = -4i$ **35.** $q(x) = x + (-1 + i)$;
$R(x) = 0$ **37.** $-6\sqrt{3} + 57$ **39.** 0

Exercise Set 11.2

1. yes **3.** yes **5.** yes **7.** yes
9. yes **11.** no **13. a.** yes **b.** no
15. a. 1, 2 (multiplicity 3), 3
b. 1 (multiplicity 3)
c. 5 (multiplicity 6), -1 (multiplicity 4)
d. 0 (multiplicity 5), 1 **17.** -170
19. -9 **21.** $-3\sqrt{2} - 2$
23. -22 **25.** ± 3, 4
27. 1, $-1 \pm \sqrt{6}$ **29.** -2, $\frac{2}{3}$, 3
31. $-\frac{3}{2}$, $\frac{1}{2}(1 \pm \sqrt{5})$ **33.** 0, 5
35. -4, 3 **37.** -9, $1 \pm \sqrt{2}$
39. $\frac{1}{4}(3 \pm \sqrt{17})$ (multiplicity 2)
41. a. 1.125 **b.** -0.046875 **c.** $t - 1$
d. 1, $\frac{1}{2}(-1 \pm \sqrt{13})$
43. $x^3 - 4x^2 - 17x + 60 = 0$
45. $x^3 + 8x^2 + 13x + 6 = 0$
47. No such polynomial equation exists.
49. $ax^4 + (3a + b)x^3 + (-17a + 3b)x^2 + (6a - 17b)x + 6b = 0$, for any real
numbers a and b, where $a \neq 0$
51. 20.44 **53. a.** -0.05 **b.** 0.07
55. a. $x_2 = 2 - \sqrt{2}$, $x_3 = 2 + \sqrt{2}$
57. yes **59.** yes **61.** $a = -1$,
$b = -1$ **63.** $b = \pm\frac{3}{2}\sqrt{2}$
65. 2 (multiplicity 2), -4

Exercise Set 11.3

1. a, b, c **3.** $[x - (-1)](x - 3)$
5. $4\left(x - \frac{1}{4}\right)[x - (-6)]$
7. $\left[x - (-\sqrt{5})\right](x - \sqrt{5})$
9. $[x - (5 + i)][x - (5 - i)]$
11. $f(x) = x^3 + x^2 - 5x + 3$

13. $f(x) = x^4 - 16$
15. $f(x) = x^6 + 10x^4 - 87x^2 + 144$
17. $f(x) = -\frac{5}{42}x^2 + \frac{25}{42}x + \frac{30}{7}$
19. $f(x) = \frac{1}{30}x^3 - \frac{19}{30}x + 1$
21. $x^2 + (i + \sqrt{3})x + i\sqrt{3} = 0$
23. $x^2 - 3x - 54 = 0$
25. $x^2 - 2x - 4 = 0$
27. $x^2 - 2ax + a^2 - b = 0$
29. $(x + 2 - 2i)(x + 2 + 2i)(x - 2 - 2i)$
$(x - 2 + 2i)$ 33. $4, 3, -3$

Exercise Set 11.4

1. $\pm 1, \pm \frac{1}{2}, \pm \frac{1}{4}, \pm 3, \pm \frac{3}{2}, \pm \frac{3}{4}$
3. $\pm 1, \pm \frac{1}{2}, \pm \frac{1}{4}, \pm \frac{1}{8}, \pm 3, \pm \frac{3}{2}, \pm \frac{3}{4}, \pm \frac{3}{8},$
$\pm 9, \pm \frac{9}{2}, \pm \frac{9}{4}, \pm \frac{9}{8}$ 5. $\pm 1, \pm \frac{1}{2}, \pm 2, \pm 3,$
$\pm \frac{3}{2}, \pm 6$ 13. $1, -1, -3$
15. $-\frac{1}{4}, \pm \sqrt{5}$ 17. $-1, -\frac{2}{3}, -\frac{1}{3}$
19. $1, -1, (-1 \pm \sqrt{97})/2$
21. 1 (multiplicity 4)   23. $\frac{1}{2}, 6, -4$
25. $\frac{1}{4}, \frac{1}{5}, \pm \sqrt{6}/2$
27. **a.** upper bound: 2; lower bound: -1
b. upper bound: 2; lower bound: -1
c. upper bound: 6; lower bound: -2
29. between 0.68 and 0.69
31. between 2.88 and 2.89
33. between 4.31 and 4.32
35. between -2.15 and -2.14
37. between -5.27 and -5.26
41. $x \approx 0.32$

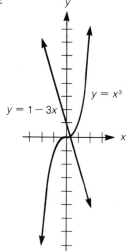

43. $x \approx 1.49$ 45. $p = 3, x = 1$
51. $b = 2$

Exercise Set 11.5

1. $7 + 2i$ 3. $5 - 2i, 3$
5. $-2 - i, -3$ (multiplicity 2)
7. $6 + 5i, -\frac{1}{4}$ 9. $4 - \sqrt{2}i, \pm \frac{3}{2}i$
11. $10 - 2i, 1 \pm \sqrt{5}$
13. $(1 - i\sqrt{2})/3, \frac{2}{5}$ 15. $3 + 2i,$
$-1 - i, -1 \pm \sqrt{2}$
17. $x^2 - 2x - 5 = 0$
19. $x^2 - \frac{4}{3}x - \frac{2}{3} = 0$
21. 2 complex roots, 1 negative real root
23. 4 complex roots, 1 negative real root
25. 2 complex roots, 1 positive real root,
1 negative real root   27. either 1
positive real root and 2 negative real
roots, or 1 positive real root and 2
complex roots   29. 1 positive real
root, 1 negative real root, 6 complex
roots   31. 1 positive real root, 8
complex roots   33. 1 positive real
root, 1 negative real root, 6 complex
roots   35. 1 positive real root, 1
negative real root, 4 complex roots
39. $f(x) = x^4 + 2x^2 + 49$

Chapter Eleven Review Exercises

1. $q(x) = x^3 + x^2 - 3x + 1; R(x) = -1$
3. quotient: $x^3 + 3x^2 + 7x + 21;$
remainder: 71
5. quotient: $2x^2 - 13x + 46;$
remainder: -187
7. quotient: $5x - 20$; remainder: 0
9. 99,904   11. $-\frac{999}{1000}$ 13. a^3
15. **a.** -0.24 **b.** -0.007 **c.** 0.00003
17. $a = -5$ 19. $a = -1$ or $a = -2$
23. $\pm 1, \pm 2, \pm 3, \pm 6, \pm 9, \pm 18$
25. $\pm 1, \pm \frac{1}{2}, \pm 2, \pm 4, \pm 8$
27. $\pm p, \pm 1$ 29. $2, -\frac{3}{2}, -1$
31. $\frac{5}{2}, -1 \pm \sqrt{3}$ 33. $\frac{2}{3}, \frac{1}{2}(-1 \pm i\sqrt{3})$
35. $-1, -\sqrt{7}$ (multiplicity 2),
$\sqrt{7}$ (multiplicity 2)
37. 2 (multiplicity 2), 5   39. **a.** Let
$p(x)$ and $d(x)$ be polynomials such that
$d(x) \neq 0$. Then there are unique
polynomials $q(x)$ and $R(x)$ such that
$p(x) = d(x) \cdot q(x) + R(x)$, where either
$R(x) = 0$ or the degree of $R(x)$ is less than
the degree of $d(x)$. **b.** When a
polynomial $f(x)$ is divided by $x - r$, the
remainder is $f(r)$. **c.** Let $f(x)$ be a
polynomial. If $f(r) = 0$, then $x - r$ is a

factor of $f(x)$. Conversely, if $x - r$ is a
factor of $f(x)$, then $f(r) = 0$. **d.** Every
polynomial equation of the form

$$a_n x^n + a_{n-1}x^{n-1} + \cdots + a_1 x + a_0 = 0$$
$$(n \geq 1, a_n \neq 0)$$

has at least one root among the complex
numbers. (This root may be a real
number.)   41. $6\left(x - \frac{4}{3}\right)\left[x - \left(-\frac{5}{2}\right)\right]$
43. $(x - 4)\left[x - \left(-\sqrt[3]{-5}\right)\right]$
$\left[x - \frac{1}{2}\left(\sqrt[3]{5} + i\sqrt{3\sqrt[3]{25}}\right)\right]$
$\left[x - \frac{1}{2}\left(\sqrt[3]{5} - i\sqrt{3\sqrt[3]{25}}\right)\right]$
45. $2 \pm 3i, 3$ 47. $1 \pm i\sqrt{2}, \pm\sqrt{7}$
49. 1 positive real root, 2 complex roots
51. 1 negative real root, 2 complex roots
53. 1 positive real root, 1 negative real
root, 2 complex roots   55. **c.** $-1, \pm i$
57. between 0.82 and 0.83
59. **c.** between 6.93 and 6.94
61. $x^2 - 8x + 11 = 0$
63. $x^4 - 12x^3 + 35x^2 + 60x - 200 = 0$
65. $x^4 - 4x^3 - 4x^2 + 16x - 8 = 0$
67. zeros: $0, 3, -1$

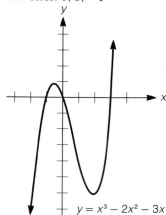

69. zeros: $0, -2, 2$

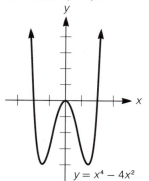

71. zeros: 1, 2

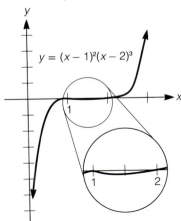

$$y = (x - 1)^2(x - 2)^3$$

73. a. yes **b.** yes **c.** yes **d.** yes

Chapter Eleven Graphing Calculator Exercises

1. a.

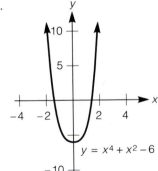

$$y = x^4 + x^2 - 6$$

b. ± 1.414

3. a.

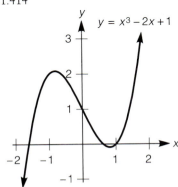

$$y = x^3 - 2x + 1$$

b. $-1.618, 0.618, 1$

5. a.

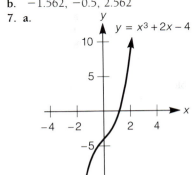

$$y = 2x^3 - x^2 - 9x - 4$$

b. $-1.562, -0.5, 2.562$

7. a.

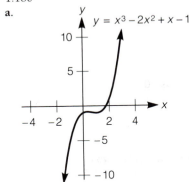

$$y = x^3 + 2x - 4$$

b. 1.180

9. a.

$$y = x^3 - 2x^2 + x - 1$$

b. 1.755

11.

$$f(x) = x^4 + 3x^2 - 7x - 5$$

Chapter Eleven Test

1. $f\left(\frac{1}{2}\right) = -\frac{3}{2}$ **2.** $-3, 1 + \sqrt{6},$
$1 - \sqrt{6}$ **3.** $\pm 1, \pm\frac{1}{2}, \pm 2, \pm 3,$
$\pm\frac{3}{2}, \pm 6$ **4.** $f(x) = 3x^2 + 21x - 24$
5. quotient: $4x^2 - 3x - 5$; remainder: 8
6. a. Let $f(x)$ be a polynomial. If
$f(r) = 0$, then $x - r$ is a factor of $f(x)$.
Conversely, if $x - r$ is a factor of $f(x)$,
then $f(r) = 0$. **b.** Every polynomial
equation of the form

$$a_n x^n + a_{n-1} x^{n-1} + \cdots + a_1 x + a_0 = 0$$
$$(n \geq 1, a_n \neq 0)$$

has at least one root among the complex
numbers. (This root may be a real
number.) **7. a.** 3 **b.** between 2.2
and 2.3 **8.** $1 \pm i, 3 \pm 2i, -2$
9. $q(x) = x^2 + 2x - 1$; $R(x) = -3x + 7$
10. $2\left[x - \frac{1}{2}(3 + i)\right]\left[x - \frac{1}{2}(3 - i)\right]$
11. a. $\pm 1, \pm 2, \pm 3, \pm 4, \pm 6, \pm 8,$
$\pm 12, \pm 24$ **c.** 1 **d.** none **12. a.** $\frac{3}{2}$
b. $\frac{3}{2}, \frac{1}{2}(-1 \pm i\sqrt{3})$ **13.** 1 positive real
root, 1 negative real root, and 2 complex
roots **14.** $x^3 + 6x + 20 = 0$
15. $f(x) = (x - 2)(x - 3i)^3\left[x - \left(1 + \sqrt{2}\right)\right]^2$

CHAPTER TWELVE

Exercise Set 12.1

1. focus: $(0, 1)$, directrix: $y = -1$,
focal width: 4

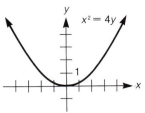

$$x^2 = 4y$$

3. focus: $(-2, 0)$, directrix: $x = 2$,
focal width: 8

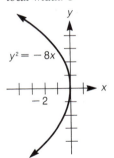

$$y^2 = -8x$$

5. focus: $(0, -5)$, directrix: $y = 5$, focal width: 20

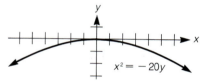

$x^2 = -20y$

7. focus: $(-7, 0)$, directrix: $x = 7$, focal width: 28

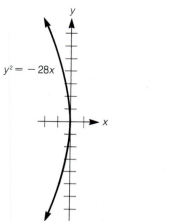

$y^2 = -28x$

9. focus: $\left(0, \frac{3}{2}\right)$, directrix: $y = -\frac{3}{2}$, focal width: 6

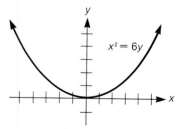

$x^2 = 6y$

11. focus: $\left(0, \frac{7}{16}\right)$, directrix: $y = -\frac{7}{16}$, focal width: $\frac{7}{4}$

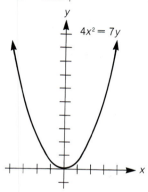

$4x^2 = 7y$

13. vertex: $(2, 3)$, focus: $(3, 3)$, directrix: $x = 1$, focal width: 4

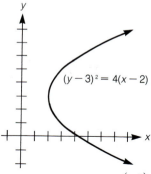

$(y - 3)^2 = 4(x - 2)$

15. vertex: $(4, 2)$, focus: $\left(4, \frac{9}{4}\right)$, directrix: $y = \frac{7}{4}$, focal width: 1

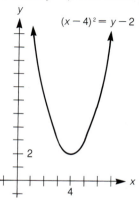

$(x - 4)^2 = y - 2$

17. vertex: $(0, -1)$, focus: $\left(\frac{1}{4}, -1\right)$, directrix: $x = -\frac{1}{4}$, focal width: 1

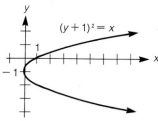

$(y + 1)^2 = x$

19. vertex: $(3, 0)$, focus: $\left(3, \frac{1}{8}\right)$, directrix: $y = -\frac{1}{8}$, focal width: $\frac{1}{2}$

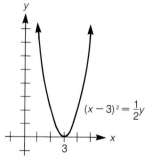

$(x - 3)^2 = \frac{1}{2}y$

21. vertex: $(4, 1)$, focus: $\left(4, \frac{9}{8}\right)$, directrix: $y = \frac{7}{8}$, focal width: $\frac{1}{2}$

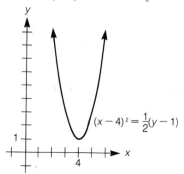

$(x - 4)^2 = \frac{1}{2}(y - 1)$

23. line of symmetry: $y = 1$

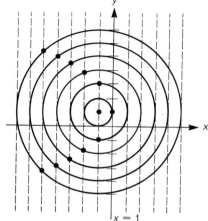

25. $x^2 = 12y$ **27.** $y^2 = 128x$

29. $y^2 = -9x$ **31. a.** $y = \frac{3}{4}x + 2$

b. $\left(-2, \frac{1}{2}\right)$ **c.** $\frac{25}{2}$

d. $(x - 3)^2 + \left(y - \frac{17}{4}\right)^2 = \frac{625}{16}$

e.

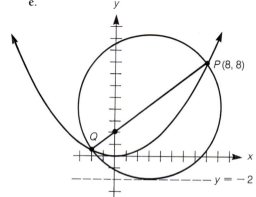

33. 11.25 ft   **35. a.** $y = \frac{9}{4}x - 1$
b. $y = -\frac{9}{8}x - \frac{1}{4}$ **37. a.** $Q\left(-\frac{1}{8}, \frac{1}{64}\right)$
b. $\left(\frac{15}{16}, \frac{257}{128}\right)$ **c.** $\frac{1}{2}$ **39.** length of side: $8\sqrt{3}\,p$ units, area: $48\sqrt{3}\,p^2$ sq. units

Exercise Set 12.2

1. length of major axis: 6, length of minor axis: 4, foci: $\left(\pm\sqrt{5}, 0\right)$, eccentricity: $\dfrac{\sqrt{5}}{3}$

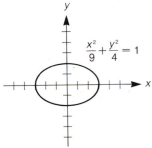

$$\frac{x^2}{9} + \frac{y^2}{4} = 1$$

3. length of major axis: 8, length of minor axis: 2, foci: $\left(\pm\sqrt{15}, 0\right)$, eccentricity: $\dfrac{\sqrt{15}}{4}$

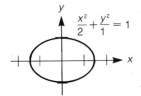

$$\frac{x^2}{16} + \frac{y^2}{1} = 1$$

5. length of major axis: $2\sqrt{2}$, length of minor axis: 2, foci: $(\pm 1, 0)$, eccentricity: $\dfrac{\sqrt{2}}{2}$

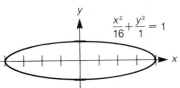

$$\frac{x^2}{2} + \frac{y^2}{1} = 1$$

7. length of major axis: 8, length of minor axis: 6, foci: $\left(0, \pm\sqrt{7}\right)$, eccentricity: $\dfrac{\sqrt{7}}{4}$

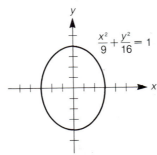

$$\frac{x^2}{9} + \frac{y^2}{16} = 1$$

9. length of major axis: $\dfrac{2\sqrt{15}}{3}$, length of minor axis: $\dfrac{2\sqrt{3}}{3}$, foci: $\left(0, \pm\dfrac{2\sqrt{3}}{3}\right)$, eccentricity: $\dfrac{2\sqrt{5}}{5}$

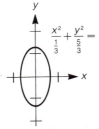

$$\frac{x^2}{\frac{1}{3}} + \frac{y^2}{\frac{5}{3}} = 1$$

11. length of major axis: 4, length of minor axis: $2\sqrt{2}$, foci: $\left(0, \pm\sqrt{2}\right)$, eccentricity: $\dfrac{\sqrt{2}}{2}$

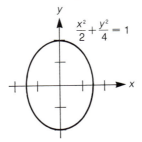

$$\frac{x^2}{2} + \frac{y^2}{4} = 1$$

13. center: $(5, -1)$, length of major axis: 10, length of minor axis: 6, foci: $(9, -1)$ and $(1, -1)$, eccentricity: $\frac{4}{5}$

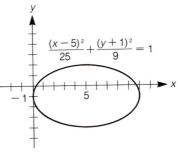

$$\frac{(x-5)^2}{25} + \frac{(y+1)^2}{9} = 1$$

15. center: $(1, 2)$, length of major axis: 4, length of minor axis: 2, foci: $\left(1, 2\pm\sqrt{3}\right)$, eccentricity: $\dfrac{\sqrt{3}}{2}$

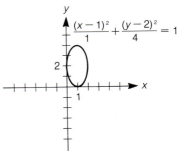

$$\frac{(x-1)^2}{1} + \frac{(y-2)^2}{4} = 1$$

17. center: $(-3, 0)$, length of major axis: 6, length of minor axis: 2, foci: $\left(-3\pm2\sqrt{2}, 0\right)$, eccentricity: $\dfrac{2\sqrt{2}}{3}$

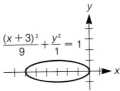

$$\frac{(x+3)^2}{9} + \frac{y^2}{1} = 1$$

19. center: $(1, -2)$, length of major axis: 4, length of minor axis: $2\sqrt{3}$, foci: $(2, -2)$ and $(0, -2)$, eccentricity: $\frac{1}{2}$

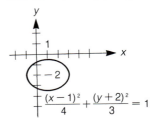

$$\frac{(x-1)^2}{4} + \frac{(y+2)^2}{3} = 1$$

21. center: $(4, 6)$

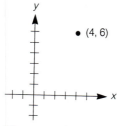

23. no graph

25. $\dfrac{x^2}{25} + \dfrac{y^2}{16} = 1$, or $16x^2 + 25y^2 = 400$

27. $\dfrac{x^2}{16} + \dfrac{y^2}{15} = 1$ or $15x^2 + 16y^2 = 240$

29. $\dfrac{x^2}{21} + \dfrac{y^2}{25} = 1$ or $25x^2 + 21y^2 = 525$

31. $\dfrac{x^2}{3^2} + \dfrac{y^2}{\left(\frac{3}{2}\right)^2} = 1$ or $x^2 + 4y^2 = 9$

33.

x	0	0.5	1.0	1.5
y	±2	±1.97	±1.89	±1.73

x	2.0	2.5	3.0
y	±1.49	±1.11	0

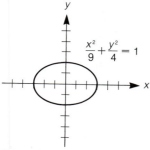

37. C_2 has the smallest percentage error

	Approximation Obtained	Percentage Error
C_1	25.531776	0.019
C_2	25.526986	0.000051
C_3	25.519489	0.029

Exercise Set 12.3

1. vertices: $(\pm2, 0)$, length of transverse axis: 4, length of conjugate axis: 2, asymptotes: $y = \pm\frac{1}{2}x$, foci: $(\pm\sqrt{5}, 0)$, eccentricity: $\dfrac{\sqrt{5}}{2}$

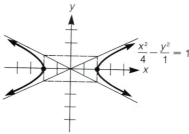

3. vertices: $(0, \pm2)$, length of transverse axis: 4, length of conjugate axis: 2, asymptotes: $y = \pm2x$, foci: $(0, \pm\sqrt{5})$, eccentricity: $\dfrac{\sqrt{5}}{2}$

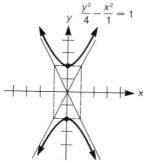

5. vertices: $(\pm5, 0)$, length of transverse axis: 10, length of conjugate axis: 8, asymptotes: $y = \pm\frac{4}{5}x$, foci: $(\pm\sqrt{41}, 0)$, eccentricity: $\dfrac{\sqrt{41}}{5}$

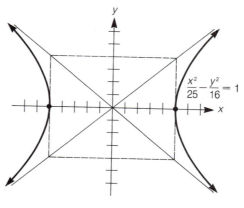

7. vertices: $\left(0, \pm\dfrac{\sqrt{2}}{2}\right)$, length of transverse axis: $\sqrt{2}$, length of conjugate axis: $\dfrac{2\sqrt{3}}{3}$, asymptotes: $y = \pm\dfrac{\sqrt{6}}{2}x$, foci: $\left(0, \pm\dfrac{\sqrt{30}}{6}\right)$, eccentricity: $\dfrac{\sqrt{15}}{3}$

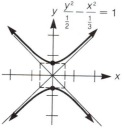

9. vertices: $(0, \pm5)$, length of transverse axis: 10, length of conjugate axis: 4, asymptotes: $y = \pm\frac{5}{2}x$, foci: $(0, \pm\sqrt{29})$, eccentricity: $\dfrac{\sqrt{29}}{5}$

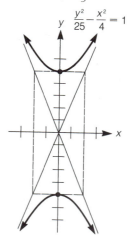

11. center: $(5, -1)$, vertices: $(10, -1)$ and $(0, -1)$, length of transverse axis: 10, length of conjugate axis: 6, asymptotes: $y = \frac{3}{5}x - 4$ and $y = -\frac{3}{5}x + 2$, foci: $\left(5 \pm \sqrt{34}, -1\right)$, eccentricity: $\frac{\sqrt{34}}{5}$

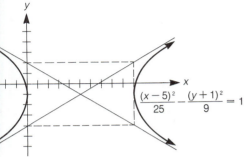

$$\frac{(x-5)^2}{25} - \frac{(y+1)^2}{9} = 1$$

13. center: $(1, 2)$, vertices: $(1, 4)$ and $(1, 0)$, length of transverse axis: 4, length of conjugate axis: 2, asymptotes: $y = 2x$ and $y = -2x + 4$, foci: $\left(1, 2 \pm \sqrt{5}\right)$, eccentricity: $\frac{\sqrt{5}}{2}$

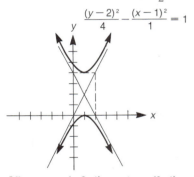

$$\frac{(y-2)^2}{4} - \frac{(x-1)^2}{1} = 1$$

15. center: $(-3, 4)$, vertices: $(1, 4)$ and $(-7, 4)$, length of transverse axis: 8, length of conjugate axis: 8, asymptotes: $y = x + 7$ and $y = -x + 1$, foci: $\left(-3 \pm 4\sqrt{2}, 4\right)$, eccentricity: $\sqrt{2}$

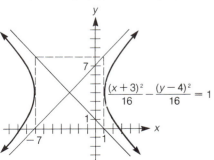

$$\frac{(x+3)^2}{16} - \frac{(y-4)^2}{16} = 1$$

17. center: $(0, 1)$, vertices: $(2, 1)$ and $(-2, 1)$, lengths of both transverse and conjugate axes: 4, asymptotes: $y = x + 1$ and $y = -x + 1$, foci: $\left(\pm 2\sqrt{2}, 1\right)$, eccentricity: $\sqrt{2}$

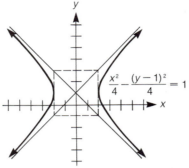

$$\frac{x^2}{4} - \frac{(y-1)^2}{4} = 1$$

19. center: $(2, 1)$, vertices: $(5, 1)$ and $(-1, 1)$, lengths of both transverse and conjugate axes: 6, asymptotes: $y = x - 1$ and $y = -x + 3$, foci: $\left(2 \pm 3\sqrt{2}, 1\right)$, eccentricity: $\sqrt{2}$

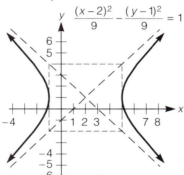

$$\frac{(x-2)^2}{9} - \frac{(y-1)^2}{9} = 1$$

21. center: $(0, -4)$, vertices: $(0, 1)$ and $(0, -9)$, length of transverse axis: 10, length of conjugate axis: 2, asymptotes: $y = 5x - 4$ and $y = -5x - 4$, foci: $\left(0, -4 \pm \sqrt{26}\right)$, eccentricity: $\frac{\sqrt{26}}{5}$

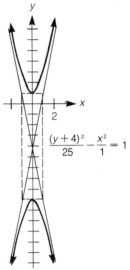

$$\frac{(y+4)^2}{25} - \frac{x^2}{1} = 1$$

23. two lines: $y = x + 3$ and $y = -x - 4$

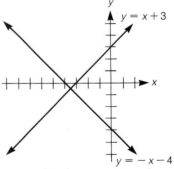

27. $15x^2 - y^2 = 15$ **29.** $x^2 - 4y^2 = 4$
31. $2x^2 - 5y^2 = 10$
33. $y^2 - 32x^2 = 49$ **35.** $y^2 - 9x^2 = 9$
39. **b.** $(0, \pm 6)$ **c.** $F_1P = 5, F_2P = 13$
41. **a.** The eccentricity is $\sqrt{2}$.
b. The eccentricity is $\sqrt{2}$.

Exercise Set 12.4

1. $F_1P = \dfrac{6\sqrt{19} - 8\sqrt{6}}{3}$,

$F_2P = \dfrac{6\sqrt{19} + 8\sqrt{6}}{3}$

3. $F_1P = 15 + 6\sqrt{2}, F_2P = 15 - 6\sqrt{2}$

5. a. foci: $\left(\pm\sqrt{7},0\right)$, eccentricity: $\dfrac{\sqrt{7}}{4}$,

directricies: $x=\pm\dfrac{16\sqrt{7}}{7}$ **b.** foci:

$(\pm5,0)$, eccentricity: $\frac{5}{4}$, directricies:
$x=\pm\frac{16}{5}$

7. a. foci: $(\pm1,0)$, eccentricity: $\dfrac{\sqrt{13}}{13}$,

directricies: $x=\pm13$ **b.** foci: $(\pm5,0)$,

eccentricity: $\dfrac{5\sqrt{13}}{13}$, directricies: $x=\pm\frac{13}{5}$

9. a. foci: $\left(\pm\sqrt{11},0\right)$,

eccentricity: $\dfrac{\sqrt{11}}{6}$,

directricies: $x=\pm\dfrac{36\sqrt{11}}{11}$

b. foci: $\left(\pm\sqrt{61},0\right)$, eccentricity: $\dfrac{\sqrt{61}}{6}$,

directricies: $x=\pm\dfrac{36\sqrt{61}}{61}$

11. $3x^2+4y^2=12$ **13.** $x^2-y^2=2$

Chapter Twelve Review Exercises

1. a. $y^2=16x$ **b.** $x^2=16y$
3. $x^2=12y$ **5.** $15x^2+16y^2=960$
7. $25x^2+9y^2=900$ **9.** $8x^2-y^2=32$
11. $240x^2-16y^2=135$
13. vertex: $(0,0)$, focus: $\left(0,\frac{5}{2}\right)$,
directrix: $y=-\frac{5}{2}$, focal width: 10

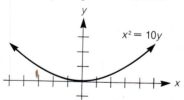

15. vertex: $(0,3)$, focus: $(0,0)$,
directrix: $y=6$, focal width: 12

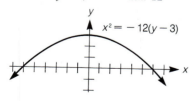

17. vertex: $(1,1)$, focus: $(0,1)$,
directrix: $x=2$, focal width: 4

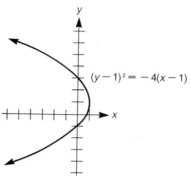

19. center: $(0,0)$, length of major axis: 4,
length of minor axis: $2\sqrt{2}$,

foci: $\left(\pm\sqrt{2},0\right)$, eccentricity: $\dfrac{\sqrt{2}}{2}$

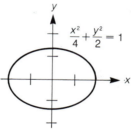

21. center: $(0,0)$, length of major axis: 14,
length of minor axis: 6,

foci: $\left(0,\pm2\sqrt{10}\right)$, eccentricity: $\dfrac{2\sqrt{10}}{7}$

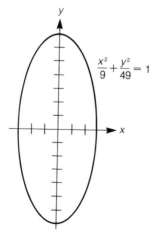

23. center: $(1,-2)$, length of major
axis: 10, length of minor axis: 6
foci: $(5,-2)$ and $(-3,-2)$, eccentricity: $\frac{4}{5}$

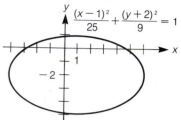

25. center: $(0,0)$, vertices: $(\pm2,0)$,

asymptotes: $y=\pm\dfrac{\sqrt{2}}{2}x$, foci: $(\pm6,0)$,

eccentricity: $\dfrac{\sqrt{6}}{2}$

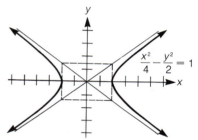

27. center: $(0,0)$, vertices: $(0,\pm3)$,
asymptotes: $y=\pm\frac{3}{7}x$, foci: $\left(0,\pm\sqrt{58}\right)$,

eccentricity: $\dfrac{\sqrt{58}}{3}$

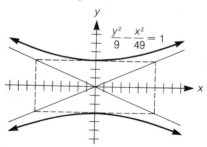

29. center: $(1,-2)$, vertices: $(6,-2)$ and
$(-4,-2)$, asymptotes: $y=\frac{3}{5}x-\frac{13}{5}$ and
$y=-\frac{3}{5}x-\frac{7}{5}$, foci: $\left(1\pm\sqrt{34},-2\right)$,

eccentricity: $\dfrac{\sqrt{34}}{5}$

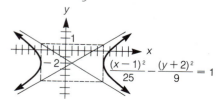

31. center: $(1, -2)$, length of major axis: 4, length of minor axis: $2\sqrt{3}$, foci: $(2, -2)$ and $(0, -2)$

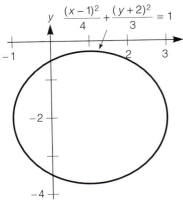

$$\frac{(x-1)^2}{4} + \frac{(y+2)^2}{3} = 1$$

33. vertex: $(4, -1)$, axis of symmetry: $y = -1$, focus: $(3, -1)$, directrix: $x = 5$

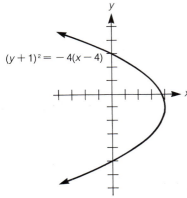

$(y+1)^2 = -4(x-4)$

35. center: $(1, 5)$, vertices: $(4, 5)$ and $(-2, 5)$, asymptotes: $y = \frac{4}{3}x + \frac{11}{3}$ and $y = -\frac{4}{3}x + \frac{19}{3}$, foci: $(6, 5)$ and $(-4, 5)$

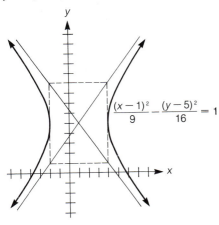

$$\frac{(x-1)^2}{9} - \frac{(y-5)^2}{16} = 1$$

37. point: $(4, 6)$

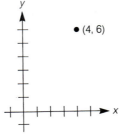

$\bullet\ (4, 6)$

39. two lines: $3x - 4y = 11$ and $3x + 4y = 19$

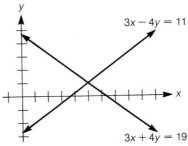

$3x - 4y = 11$

$3x + 4y = 19$

41. center: $(0, -4)$, vertices: $(0, 1)$ and $(0, -9)$, asymptotes: $y = 5x - 4$ and $y = -5x - 4$ foci: $\left(0, -4 \pm \sqrt{26}\right)$

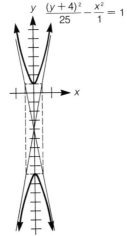

$$\frac{(y+4)^2}{25} - \frac{x^2}{1} = 1$$

43. no graph   **45. b.** 64

49. parabola: $y^2 = -8(x - 5)$,

ellipse: $\frac{x^2}{25} + \frac{y^2}{16} = 1$ **51. a.** $\frac{\sqrt{5}}{2}$

b. $\left(3 + \sqrt{5}, 4\right)$ and $x = 3 + \frac{4}{\sqrt{5}}$;

$\left(3 - \sqrt{5}, 4\right)$ and $x = 3 - \frac{4}{\sqrt{5}}$

c.

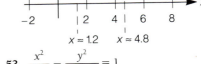

$$\frac{(x-3)^2}{4} - \frac{(y-4)^2}{1} = 1$$

$\left(3 - \sqrt{5}, 4\right)$ $\left(3 + \sqrt{5}, 4\right)$

$x \approx 1.2$ $x \approx 4.8$

53. $\dfrac{x^2}{81/4} - \dfrac{y^2}{405/16} = 1$

Chapter Twelve Graphing Calculator Exercises

1.

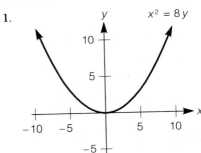

$x^2 = 8y$

3.

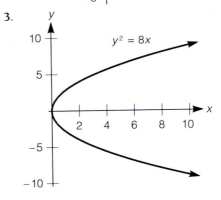

$y^2 = 8x$

5. b.

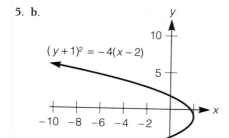

$(y+1)^2 = -4(x-2)$

7. $y = x - 2$

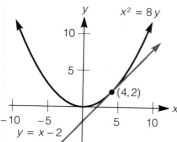

$x^2 = 8y$

$(4,2)$

$y = x - 2$

9.

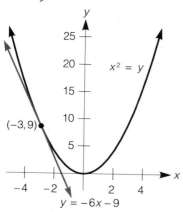

$x^2 = y$

$(-3,9)$

$y = -6x - 9$

11. a. symmetry with respect to the x-axis, the y-axis, and the origin

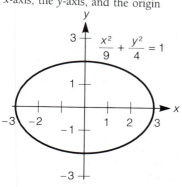

$\dfrac{x^2}{9} + \dfrac{y^2}{4} = 1$

b. symmetry with respect to the x-axis, the y-axis, and the origin

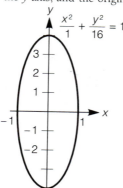

$\dfrac{x^2}{1} + \dfrac{y^2}{16} = 1$

13. c.

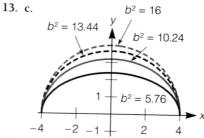

$b^2 = 16$

$b^2 = 13.44$

$b^2 = 10.24$

$b^2 = 5.76$

15. $y = x - 4$

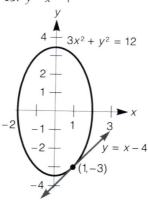

$3x^2 + y^2 = 12$

$y = x - 4$

$(1,-3)$

17. a.

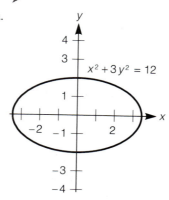

$x^2 + 3y^2 = 12$

b. $a = 2\sqrt{3}$, $b = 2$, $c = 2\sqrt{2}$

c. $x^2 + y^2 = 12$

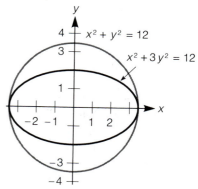

$x^2 + y^2 = 12$

$x^2 + 3y^2 = 12$

d. $x + y = 4$

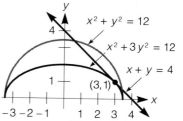

$x^2 + y^2 = 12$

$x^2 + 3y^2 = 12$

$x + y = 4$

$(3,1)$

e. $y = x + 2\sqrt{2}$

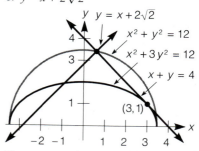

$y = x + 2\sqrt{2}$

$x^2 + y^2 = 12$

$x^2 + 3y^2 = 12$

$x + y = 4$

$(3,1)$

19. a. $\dfrac{(y-3)^2}{25} - \dfrac{(x-4)^2}{9} = 1$

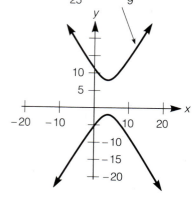

A-118

b.
$$\frac{(y-3)^2}{25} - \frac{(x-4)^2}{9} = 1$$

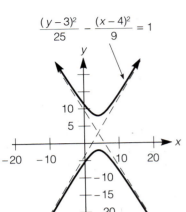

21. b.

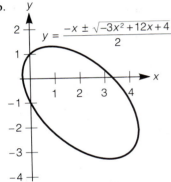

$$y = \frac{-x \pm \sqrt{-3x^2 + 12x + 4}}{2}$$

c. $(2, -1)$ **d.** $(2, -1)$, which agrees with our estimate from part (c)

Chapter Twelve Test

1. focus: $(-3, 0)$, directrix: $x = 3$

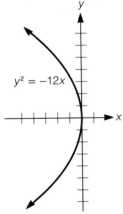

$y^2 = -12x$

2. foci: $(\pm\sqrt{5}, 0)$, asymptotes: $y = \pm\frac{1}{2}x$

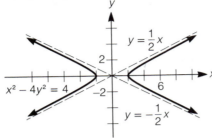

$y = \frac{1}{2}x$

$x^2 - 4y^2 = 4$

$y = -\frac{1}{2}x$

3. $\dfrac{x^2}{12} + \dfrac{y^2}{16} = 1$ **4.** $\dfrac{x^2}{3} - \dfrac{y^2}{1} = 1$

5. b. 64 **6.** length of major axis: 10, length of minor axis: 4, foci: $(\pm\sqrt{21}, 0)$

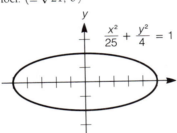

$$\frac{x^2}{25} + \frac{y^2}{4} = 1$$

7.

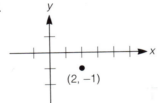

$(2, -1)$

8.

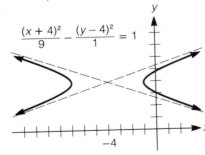

$$\frac{(x+4)^2}{9} - \frac{(y-4)^2}{1} = 1$$

9. focal width: 8, vertex: $(1, 2)$

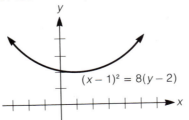

$(x - 1)^2 = 8(y - 2)$

---

10. $x = \frac{25}{4}$ **11. a.** $x = 3$ **b.** $\frac{13}{4}$
c. F_2: $(4, 0)$; F_2P: $\frac{13}{5}$
12. The ratio is $\frac{4}{5}$.

CHAPTER THIRTEEN

Exercise Set 13.1

27. a.

n	1	2	3	4	5
$f(n)$	1/2	2/3	3/4	4/5	5/6

b. $\frac{6}{7}$ **c.** $f(n) = \dfrac{n}{n+1}$

29. a.

n	1	2	3	4	5
$f(n)$	1	4	9	16	25

b. 36 **c.** $f(n) = n^2$

Exercise Set 13.2

1. a. $a^2 + 2ab + b^2$
b. $a^3 + 3a^2b + 3ab^2 + b^3$
3. $a^9 + 9a^8b + 36a^7b^2 + 84a^6b^3 + 126a^5b^4 + 126a^4b^5 + 84a^3b^6 + 36a^2b^7 + 9ab^8 + b^9$
5. $8A^3 + 12A^2B + 6AB^2 + B^3$
7. $1 - 12x + 60x^2 - 160x^3 + 240x^4 - 192x^5 + 64x^6$
9. $x^2 + 4x\sqrt{xy} + 6xy + 4y\sqrt{xy} + y^2$
11. $x^{10} + 5x^8y^2 + 10x^6y^4 + 10x^4y^6 + 5x^2y^8 + y^{10}$
13. $1 - \dfrac{6}{x} + \dfrac{15}{x^2} - \dfrac{20}{x^3} + \dfrac{15}{x^4} - \dfrac{6}{x^5} + \dfrac{1}{x^6}$
15. $\dfrac{x^3}{8} - \dfrac{x^2y}{4} + \dfrac{xy^2}{6} - \dfrac{y^3}{27}$
17. $a^7b^{14} + 7a^6b^{12}c + 21a^5b^{10}c^2 + 35a^4b^8c^3 + 35a^3b^6c^4 + 21a^2b^4c^5 + 7ab^2c^6 + c^7$
19. $x^8 + 8\sqrt{2}x^7 + 56x^6 + 112\sqrt{2}x^5 + 280x^4 + 224\sqrt{2}x^3 + 224x^2 + 64\sqrt{2}x + 16$
21. $5\sqrt{2} - 7$ **23.** $89\sqrt{3} + 109\sqrt{2}$
25. $12 - 24\sqrt[3]{2} + 12\sqrt[3]{4}$
27. $x^{10} - 10x^9 + 35x^8 - 40x^7 - 30x^6 + 68x^5 + 30x^4 - 40x^3 - 35x^2 - 10x - 1$
29. 120 **31.** 105 **33. a.** 10
b. 5 **35.** $n^2 + 3n + 2$ **37.** 0
39. $120a^2b^{14}$ **41.** $100x^{99}$ **43.** 45
45. 294,912 **47.** 28 **49.** 40,095

53. a.

k	0	1	2	3	4	5	6	7	8
$\binom{8}{k}$	1	8	28	56	70	56	28	8	1

Exercise Set 10.3

1. $\frac{1}{2}, \frac{2}{3}, \frac{3}{4}, \frac{4}{5}, \frac{5}{6}$ **3.** 0, 1, 4, 9, 16
5. $2, \frac{9}{4}, \frac{64}{27}, \frac{625}{256}, \frac{7776}{3125}$ **7.** $-1, 1, -1, 1, -1$ **9.** $1, -\frac{1}{2}, \frac{1}{6}, -\frac{1}{24}, \frac{1}{120}$ **11.** 3, 6, 9, 12, 15   **13.** 1, 4, 25, 676, 458329   **15.** 2, 2, 4, 8, 32   **17.** 1, 1, 2, 6, 24   **19.** 0, 1, 2, 4, 16
21.

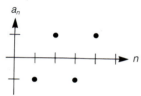

23.

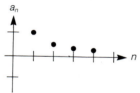

25.

27. 62   **29.** 40   **31.** $-\frac{19}{30}$
33. 903   **35.** 278   **37.** 3
39. 41   **41.** $x + x^2 + x^3$ **43.** $\frac{25}{12}$
45. -1 **47.** $\frac{6}{7}$ **49.** $\sum_{j=1}^{4} 5^j$
51. $\sum_{j=1}^{6} x^j$ **53.** $\sum_{k=1}^{12} \frac{1}{k}$
55. $\sum_{j=1}^{5} (-1)^{j+1} 2^j$ **57.** $\sum_{j=1}^{5} (-1)^{j+1} j$

59. b.

n	$F_1 + F_2 + \ldots + F_n$	$F_{n+2} - 1$
1	1	1
2	2	2
3	4	4
4	7	7
5	12	12

Exercise Set 10.4

1. a. 2 **b.** -4 **c.** $\frac{1}{3}$ **d.** $\sqrt{2}$ **3.** 131
5. 501   **7.** 998   **9.** $d = \frac{11}{6}; a = -\frac{23}{2}$
11. -190 **13.** $-\frac{1}{8}$ **15.** 500500
17. $91\pi/3$ **19.** $\frac{1}{2}$
21. $S_{16} = -768, d = -\frac{104}{15}$
23. $d = \frac{6}{5}, a = -\frac{17}{5}$ **25.** 900
27. 2,10,18 or 18,10,2   **29.** $-1,2,5$ or $5,2,-1$ **31. b.** $-6 - 9\sqrt{2}$
33. $\frac{n}{2(1-b)} \left[2 + (n-3)\sqrt{b}\right]$

Exercise Set 10.5

1. 6   **3.** 20, 100   **5.** 1   **7.** $\frac{256}{6561}$
9. ± 4 **11.** 7161   **13.** $63 + 31\sqrt{2}$
15. $\frac{1995}{64}$ **17.** 0.011111   **19.** $\frac{2}{5}$
21. 101   **23.** $\frac{5}{9}$ **25.** $\frac{61}{495}$ **27.** $\frac{16}{37}$
29. $r = -\frac{1}{2}, -2$ **35.** 12 ft

Exercise Set 10.6

1. 6 routes

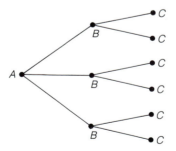

3. 12 routes

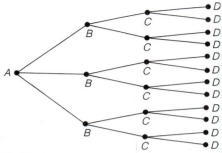

5. 12   **7.** 60   **9.** 108
11. 11,232,000   **13.** 3,628,800
15. 4   **17.** 8   **19.** 120
23. $P(101, 51)$ **25.** 240   **27.** 12
29. a. 900 **b.** 648   **31. a.** 729
b. 504 **c.** 56 **d.** 140   **33. a.** 12
b. 12   **35. a.** 56 **b.** 56
37. a. 792 **b.** 792
39. $C(n, n-2) = \frac{n(n-1)}{2}$ **41.** 56
43. a. 210 **b.** 80   **45.** 9   **47.** 56
49. 15 sums   **51. a.** 2352 **b.** 2649
53. In a polygon with 10 vertices, first choose a particular vertex (there are 9 other vertices to which a line could be drawn). If lines are drawn to the two adjacent vertices, the lines would become sides, not diagonals; thus, there are 7 diagonals from each vertex. Not wanting to count a diagonal from one vertex to another twice, we must divide by 2. So there are $\frac{7 \cdot 10}{2} = 35$ diagonals.

Exercise Set 10.7

1. a. $\{2, 4, 6\}$ **b.** $\{4, 5, 6\}$ **c.** $\{2, 4, 5, 6\}$
d. $\{4, 6\}$ **3. a.** $\{1, 3, 5\}$
b. $\{1, 2, 3, 4, 5, 6\}$ **c.** $\{1, 2, 3, 4, 5, 6\}$
d. $\varnothing$ **5.** no   **7.** yes   **9. a.** 3
b. 3 **c.** 4   **11. a.** 3 **b.** 6 **c.** 0
13. a. $\frac{1}{2}$ **b.** $\frac{1}{2}$ **15. a.** $\frac{1}{2}$ **b.** 1
17. a. $\frac{2}{3}$ **b.** 1   **19. a.** $\frac{5}{6}$ **b.** 1
21. $\frac{3}{8}$ **23.** $\frac{1}{2}$ **25.** $\frac{1}{4}$ **27.** $\frac{5}{8}$
29. $\frac{3}{4}$ **31.** 1   **33.** $\frac{5}{8}$ **35.** $\frac{1}{2}$
37. 0   **39.** $\frac{1}{4}$ **41.** $\frac{1}{3}$
43. a. $\frac{6}{7}$ **b.** $\frac{36}{91}$ **45.** $\frac{7}{12}$ **47.** $\frac{2}{9}$
49. a. $\frac{1}{8}$ **b.** $\frac{3}{8}$ **c.** $\frac{1}{2}$ **51.** $\frac{7}{10}$
53. $\frac{33}{66640}$

11. $81a^4 + 108a^3b^2 + 54a^2b^4 + 12ab^6 + b^8$

13. $x^4 + 4x^3\sqrt{x} + 6x^3 + 4x^2\sqrt{x} + x^2$

15. $x^{10} - 10x^8y^2 + 40x^6y^4 - 80x^4y^6 +$
$ 80x^2y^8 - 32y^{10}$

17. $1 + \dfrac{5}{x} + \dfrac{10}{x^2} + \dfrac{10}{x^3} + \dfrac{5}{x^4} + \dfrac{1}{x^5}$

19. $a^4b^2 - 4a^3b^2\sqrt{ab} + 6a^3b^3 -$
$ 4a^2b^3\sqrt{ab} + a^2b^4$

21. $15xy^8$ **23.** 84   **25.** 28

33. $1, \frac{4}{3}, \frac{3}{2}, \frac{8}{5}$

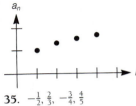

35. $-\frac{1}{2}, \frac{2}{3}, -\frac{3}{4}, \frac{4}{5}$

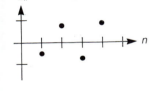

37. $-12, -48, -192, -768$

39. a. -5 **b.** $\frac{9}{10}$ **41.** $\displaystyle\sum_{k=1}^{5} \frac{5}{3^k}$

43. 73   **45.** $\frac{5}{1024}$ **47.** 177

49. 7,777,777,777   **51.** $-5\sqrt{10}$

53. $\frac{3}{4}$ **55.** $\frac{1}{10}$ **57.** $\frac{5}{11}$

63. a. 42,929 **b.** 42,925; 0.00932%
c. 6.48040×10^{10} **d.** 6.48027×10^{10};
$2 \times 10^{-3}\%$ **65.** 16

67. The permutations are $\{abc, acb, bac,$
$bca, cab, cba, abd, adb, bad, bda, dab, dba, acd,$
$adc, cad, cda, dac, dca, bcd, bdc, cbd, dbc, dcb\}$.
$P(4, 3) = 24$ **69.** 19,958,400

71. 480   **73.** 1620   **75.** 6/11

77. a. $\frac{5}{33}$ **b.** $\frac{7}{22}$ **c.** $\frac{31}{66}$ **79.** $\frac{1}{18}$

81. $\frac{1}{6}$ **83.** $\frac{2}{9}$ **85.** $\frac{1}{6}$ **87.** $\frac{25}{36}$

89. a. $\frac{7}{15}$ **b.** $\frac{8}{15}$ **c.** $\frac{2}{3}$ **d.** $\frac{2}{3}$

91. a. 252 **b.** $\frac{7}{25}$ **93.** 10,20,40 or
40,20,10   **97. a.** no value exists:

b. $x = \dfrac{b^2 - ac}{a + c - 2b}$

2. a. 27 **b.** -6

3. a. $S_n = \dfrac{a(1 - r^n)}{1 - r}$; **b.** $\frac{174075}{1024}$

4. a. 42240 **b.** $5280a^7b^{12}$

5. $243x^{10} + 405x^8y^3 + 270x^6y^6 + 90x^4y^9 +$
$ 15x^2y^{12} + y^{15}$

6. 177   **7.** $\frac{7}{9}$ **8.** $a_4 = 5, a_5 = 27$

9. $-5\sqrt{10}$ **10.** 224   **11.** 139

12. 121,080,960   **13.** $\frac{22}{35}$

14. a. $\frac{1}{15}$ **b.** $\frac{7}{15}$ **c.** $\frac{8}{15}$

INDEX

FACTORING TECHNIQUES

Technique	Example or Formula
Common factor	$3x^4 + 6x^3 - 12x^2 = 3x^2(x^2 + 2x - 4)$
	$4(x^2 + 1) - x(x^2 + 1) = (x^2 + 1)(4 - x)$
Difference of squares	$x^2 - a^2 = (x - a)(x + a)$
Trial and error	$x^2 + 2x - 3 = (x + 3)(x - 1)$
Difference of cubes	$x^3 - a^3 = (x - a)(x^2 + ax + a^2)$
Sum of cubes	$x^3 + a^3 = (x + a)(x^2 - ax + a^2)$
Grouping	$x^3 - x^2 + x - 1 = (x^3 - x^2) + (x - 1)$
	$= x^2(x - 1) + (x - 1) \cdot 1$
	$= (x - 1)(x^2 + 1)$

PROPERTIES OF LOGARITHMS

In the following properties, b, P, and Q are positive and $b \neq 1$.

1. (a) $\log_b b = 1$ (b) $\log_b 1 = 0$
2. $\log_b PQ = \log_b P + \log_b Q$
3. $\log_b (P/Q) = \log_b P - \log_b Q$
4. $\log_b P^n = n \log_b P$
5. $b^{\log_b P} = P$

FORMULAS FROM COORDINATE GEOMETRY

1. The distance formula:
$$d = \sqrt{(x_2 - x_1)^2 + (y_2 - y_1)^2}$$
2. The equation for a circle:
$$(x - h)^2 + (y - k)^2 = r^2$$
3. The definition of slope:
$$m = \frac{y_2 - y_1}{x_2 - x_1}$$
4. The point-slope formula:
$$y - y_1 = m(x - x_1)$$
5. The slope-intercept formula:
$$y = mx + b$$
6. The condition for two nonvertical lines to be parallel:
$$m_1 = m_2$$
7. The condition for two nonvertical lines to be perpendicular:
$$m_1 m_2 = -1$$
8. The midpoint formula:
$$(x_0, y_0) = \left(\frac{x_1 + x_2}{2}, \frac{y_1 + y_2}{2}\right)$$

THE GREEK ALPHABET

A	α	Alpha	N	ν	Nu
B	β	Beta	Ξ	ξ	Xi
Γ	γ	Gamma	O	o	Omicron
Δ	δ	Delta	Π	π	Pi
E	ϵ	Epsilon	P	ρ	Rho
Z	ζ	Zeta	Σ	σ	Sigma
H	η	Eta	T	τ	Tau
Θ	θ	Theta	Υ	υ	Upsilon
I	ι	Iota	Φ	ϕ	Phi
K	κ	Kappa	X	χ	Chi
Λ	λ	Lambda	Ψ	ψ	Psi
M	μ	Mu	Ω	ω	Omega